Quantitative Techniques in Management

Fourth Edition

About the Author

Dr. N. D. Vohra (b 1951) is currently Associate Professor in the Department of Commerce at Ramjas College, University of Delhi. He obtained his post graduate degree in commerce and a doctorate in finance from Delhi School of Economics. His areas of interest are quantitative analysis, including statistical analysis, and finance. He is teaching since 1974 and has a wide experience including that at the Department of Commerce, Delhi School of Economics. He has also taught at the Institute of Chartered Accountants of India and other institutions offering professional courses in management. Earlier, he served at the Foundation of Technical Institutes, Iraq. Besides, he has served as Director of Research with the BLB group of companies. He has written papers on derivatives and has co-authored a book titled Futures and Options which has been published by Tata McGraw Hill Education, New Delhi.

Quantitative Techniques in Management

Fourth Edition

N D Vohra
Associate Professor
Department of Commerce
Ramjas College
University of Delhi
Delhi

McGraw Hill Education (India) Private Limited
NEW DELHI

McGraw Hill Education Offices
New Delhi New York St Louis San Francisco Auckland Bogotá Caracas
Kuala Lumpur Lisbon London Madrid Mexico City Milan Montreal
San Juan Santiago Singapore Sydney Tokyo Toronto

McGraw Hill Education (India) Private Limited

Published by McGraw Hill Education (India) Private Limited
P-24, Green Park Extension, New Delhi 110 016

Quantitative Techniques in Management, 4/e

Ninth reprint 2014
RYAXCRDUDLQQD

ISBN (13 digit): 978-93-5134-015-7
ISBN (10 digit): 93-5134-015-5

Vice President and Managing Director: *Ajay Shukla*
Head—Higher Education (Publishing and Marketing): *Vibha Mahajan*
Publishing Manager— B&E/ HSSL: *Tapas K Maji*
Editorial Executive: *Hemant K Jha*
Development Editor: *Shalini Negi*
Assistant Manager (Editorial Services): *Anubha Srivastava*
Senior Production Manager: *Manohar Lal*
Production Executive: *Atul Gupta*

General Manager—Production: *Rajender P Ghansela*

Typeset at Script Makers, 19, AI-B, DDA Market, Paschim Vihar, New Delhi 110 063 and printed at
Rajkamal Electric Press, Kundli 131 028, Haryana

Cover Designer: K Anoop

Cover Printer: Rajkamal Electric Press

Dedicated
to the memory of
my parents

Preface to the Fourth Edition

For long, *Quantitative Techniques in Management* has been one of the most widely adopted books in the field of quantitative analysis. Therefore, it gives me pleasure and satisfaction to place this new edition in the hands of our esteemed readers.

This fourth edition of the book continues to retain the clear and crisp pedagogy of the previous three editions and there is no change in the overall format and approach. However, based on my class room experiences and suggestions from users of the previous editions, changes have been made to enhance the content and utility of the book. In particular, the following changes have been made in the book and the material included in accompanying website.

The chapter on Transportation and Transhipment problems has been rewritten for greater clarity and better comprehension of the concepts. It has also been extended to include a discussion on Sensitivity Analysis. Chapter 13 on Decision Analysis has also been modified.

Further, latest examination questions have been incorporated and also some new questions have been included in several chapters to provide greater variety.

Three hundred Power Point slides have been provided to assist teachers in teaching the subject. These may be adapted to facilitate classroom use. The solution manual has also been updated. Also included is the key to multiple choice questions.

Students' resources including chapter summaries, multiple choice questions, and a Mathematics and Statistics Refresher including Matrix Algebra, Differential Calculus, Probability and Probability Distributions, continue to be included.

In preparing this volume, I have benefitted from the comments and observations of many teachers. In particular, I thank Dr. S.K. Khandelwal, Associate Professor, Ram Lal Anand College, University of Delhi, for his continued interest in the project and for providing his valuable observations. My heartfelt thanks are due to Professor Madhu Vij, Faculty of Management Studies, University of Delhi for her help in bringing out this volume. Numerous mails from students all over the country have also contributed to make improvements in the book. I feel grateful to them all. My thanks are also due to the entire team at Tata McGraw Hill Education, New Delhi, who brought out this volume in a very professional manner.

I look forward to comments and observations from the readers of the book. These can be mailed to me at ndvohra@hotmail.com.

N D Vohra

Preface to the First Edition

Decision-making is a fundamental part of the management process and it pervades the activities of every business manager. In fact it is the manager's competence as a decision-maker that enables us to distinguish between a good manager and a bad one. Modern management is adopting and applying quantitative techniques to aid the process of decision-making in an ever increasing measure. This is due to the fact that an intelligent application of the appropriate tools can reduce an otherwise unwieldy and complex problem to one of manageable dimensions. Of course, the quantitative techniques are expected only to supplement—and not to supplant—the manager's sense of decision-making. He can use analytical tools in a wise manner only if he comprehends fully the underlying assumptions: what the analysis achieves, what compromises the model used makes with reality, and, above all, how the conclusions derived are to be adapted to the dynamic environmental conditions. Doubtless, however, a knowledge of quantitative analysis is a boon to the manager. Besides, the inclusion of operations research/quantitative techniques in management decision-making as an integral constituent of the management curricula of nearly all universities and professional institutions, such as the Institute of Chartered Accountants of India and the Institute of Cost and Works Accountants of India, is a pointer to the great significance that this field of study has assumed in the recent past.

In the present book, a modest attempt has been made to discuss some of the commonly used quantitative techniques in a wide spectrum of decision-making situations. It is addressed alike to the managers of today and tomorrow—the students pursuing post-graduate courses in commerce and management, the professional courses like CA and ICWA etc. The book is intended to provide a comprehensive presentation of the essential topics of quantitative decision-making. Towards this end, an attempt has been made to present the subject matter in a lucid style. It is aided by the inclusion of a large number of flow charts to display the schematics of the various algorithms. They have been included because they not only help to develop or extend skills that are fundamental to courses in computer science and management information systems, but also obviate the provision of algorithm summaries. An attempt has been made to develop various topics without resorting to rigorous mathematical sophistication but at the same time without an undue sacrifice in pedagogical foundation. Recondite notation and obscure manipulations have been scrupulously avoided. Wherever necessary, mathematical derivations have been relegated to the appendices. With a view to making the book self-contained, a review of the basic mathematics and statistics, required to understand the content of the book, has been provided. The book presents the application of various techniques through a large number of examples, review illustrations and exercises. A number of problems from various examinations have also been incorporated, Answers to the exercises with an asterisk (*) have been provided in Appendix A to the book.

The Instructors' Manual, containing detailed solutions to the exercises with asterisk mark and a few others, is also available.

Readers' views, observations, constructive criticism and suggestions are welcome.

N D Vohra

Contents

Quantitative Techniques in Management

CHAPTER 1

Decision-Making and Quantitative Techniques

CHAPTER OVERVIEW

Decision-making is an integral part of the process of management. It is the competence as a decision-maker that distinguishes a good manager from others. While traditionally, decision-making has been considered as an art, the growing complexity and competitive nature of environment in which managements have to operate now necessitate that decisions be made more systematically using as much quantitative information as possible, along with a consideration of qualitative factors. This book is about the use of quantitative techniques in managerial decision-making.

Broadly speaking, decision-making involves choosing a course of action from various available alternatives. The job of a manager, in the process of selecting from among available alternatives, is facilitated in a large measure by the application of appropriate quantitative techniques when, and to the extent, a problem is quantified. This approach to decision-making is known by several names like operations research, management science, quantitative analysis, etc.

The contents of this chapter will help a manager to understand questions like the following:

- What is operations research and how has it evolved? What are its characteristic features?
- What is the methodology used in operations research? In this context, what all is problem formulation, model building, acquisition of input data, solution and interpretation of the results obtained, model validation and implementation of the solution?
- What are different types of models, and what is the use of mathematical models in operations research?
- What are different classifications of solutions–feasible and infeasible; optimal and non-optimal; and unique and multiple optimal solutions?
- What is sensitivity analysis?
- How quantitative analysis is integral part of the modern computer-based information systems and how are quantitative tools used in each of the subsystems?

Thus, this introductory chapter gives some details about the decision-making process and an idea about the nature and methodology of the quantitative analysis. Finally, a plan of the book is presented which contains a brief account of the contents of each of the chapters to follow.

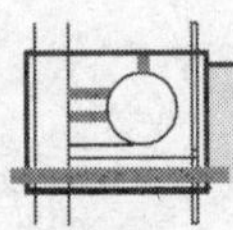

1.1 INTRODUCTION

Decision-making is an essential part of the management process. Although authorities differ in their definitions of the basic functions of management, everybody agrees that one is not a manager unless he or she has some authority to plan, organise and control the activities of an enterprise and behaviour of others. Thus, decision-making pervades the activities of every business manager. Further, since to carry out the key managerial functions of planning, organising, directing and controlling, the management is engaged in a continuous process of decision-making pertaining to each of them, we can go to the extent of saying that management may be regarded as equivalent to decision-making.

Traditionally, decision-making has been considered purely as an art, a talent that is acquired over a period of time through experience. It has been considered so because a variety of individual styles can be observed in the handling and successful solving of similar managerial problems by different people in actual business. However, the environment in which the management has to operate nowadays is complex and fast changing. There is a greater need for supplementing the art of decision-making by systematic and scientific methods. A systematic approach to decision-making is necessary because today's business and the environment in which it functions are far more complex than in the past, and the cost of making errors may be too high. Most of the business decisions cannot be made simply on the basis of rule of thumb, using common sense and/or 'snap' judgement. Common sense may be misleading and snap judgements may have painful implications. For large businesses, a single wrong decision may not only be ruinous but may also have ramifications in national economy. As such, the present-day management cannot rely solely on a trial-and-error approach and the managers have to be more sophisticated. They should employ scientific methods to help them make proper choices. Thus, the decision-makers in the business world today must understand the scientific methodology of making decisions. This calls for (1) defining the problem in a clear manner, (2) collecting pertinent facts, (3) analysing the facts thoroughly, and (4) deriving and implementing the solution.

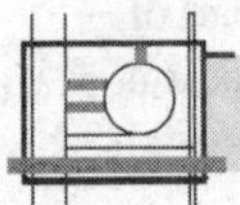

1.2 QUANTITATIVE APPROACH TO DECISION-MAKING: OR/MS

Managerial decision-making is a process by which the management, when faced with a problem, chooses a specific course of action from a set of possible options. In making a decision, a business manager attempts to choose the most effective course of action in the given circumstances in attaining the goals of the organisation. The various types of decision-making situations that a manager might encounter can be listed as follows:

1. Decisions under *certainty*, where all facts are known fully and for sure, or under *uncertainty* where the event that would actually occur is not known but probabilities can be assigned to various possible occurrences.
2. Decisions for one time period only, called *static* decisions, or a sequence of interrelated decisions made either simultaneously or over several time periods, called *dynamic* decisions.
3. Decisions where the opponent is *nature* (digging an oil well, for example) or a *rational opponent* (for instance, setting the advertising strategy when the actions of competitors have to be considered).

These classes of decision-making situations are not mutually exclusive and a given situation would exhibit characteristics from each class. Stocking of an item for sale in a certain trade fair, for instance, illustrates a static decision-making situation where uncertainty exists and nature is the opponent.

The elements of any decision are:

(a) a decision-maker, who could be an individual, group, organisation, or society;

(b) a set of possible actions that may be taken to solve the decision problem;

(c) a set of possible states that might occur;

(d) a set of consequences (pay-off) associated with various combinations of courses of action and the states that may occur; and

(e) the relationship between pay-off and the values of the decision-maker.

In an actual decision-making situation, the definition and identification of alternatives, the states and the consequences, are most difficult, albeit not the most crucial aspects of the decision problem.

In real life, some decision-making situations are simple while others are not. Complexities in decision situations arise due to several factors. These include the complicated manner of interaction of the economic, political, technological, environmental and competitive forces in society, the limited resources of an organisation, the values, risk attitudes and knowledge of the decision-makers and the like. For example, a company's decision to introduce a new product will be influenced by such considerations as market conditions, labour rates and availability, and investment requirements and availability of funds. The decision will be of multidimensional response, including the production methodology, cost and quality of the product, price, package design, and marketing and advertising strategy. The result of the decision would conceivably affect every segment of the organisation. The essential idea of the quantitative approach to decision-making is that if the factors that influence the decisions can be identified and quantified, it becomes easier to resolve the complexity of the tools of quantitative analysis. In fact, a large number of business problems have been given a quantitative representation with varying degrees of success and it has led to a general approach which is variedly designated as *operations research* (or *operational research), management science, systems analysis, decision analysis, decision science,* and so on. Quantitative analysis is now extended to several areas of business operations and represents probably the most effective approach to handling of some types of decision problems.

A significant benefit of attaining some degree of proficiency with quantitative methods is exhibited in the way problems are perceived and formulated. A problem has to be well defined before it can be formulated into a well-structured framework for solution. This requires an orderly and organised way of thinking.

Two observations may be made here. First, it should be understood clearly that, although quantitative analysis represents a scientific approach to decision-making, a decision by itself does not become a good and right decision for adoption merely because it is made within an orderly and mathematically precise framework. Quantification at best is an aid to business judgement and not its substitute. A certain degree of constructive skepticism is as desirable in considering a quantitative analysis of business decisions as it is in any other process of decision-making. Further, some allowances should be made for qualitative factors involving morale, motivation, leadership, and so on, which cannot be ignored. But they should not be allowed to dominate to such an extent that the quantitative analysis seems to be an interesting, but worthless, academic exercise. In fact, the manager should seek some balance between quantitative and qualitative factors.

Secondly, it may be noted that various names for quantitative analysis—operations research, management science, and so on—connote more or less the same general approach. We shall not attempt to discuss the differences among various labels, as it is prone to create more heat than light, but only state that the basic reason for so many titles is that the field is relatively new, and there is no consensus regarding which field of knowledge it includes. In this book, we use the terms *management science, operations research,* and *quantitative analysis* interchangeably.

We shall now briefly discuss operations research—its historical development, nature and characteristics, and methodology.

1.2.1 Historical Development of Operations Research (OR)

While it is difficult to mark the 'beginning' of the operations research/management science, the scientific approach to management can be traced back to the era of Industrial Revolution and even to periods before that. But operations research, as it exists today, was born during the second world war when the British military management called upon a group of scientists to examine the strategies and tactics of various military operations, with the intention of efficient allocation of scarce resources for the war effort. During this period, many new scientific and quantitative techniques were developed to assist military operations. The name *operational research* was derived directly from the context in which it was used—research activity on operational areas of the armed forces. British scientists spurred the American military management to similar research activities (where it came to be known as operations research). Among the investigations carried out by them were included the determination of (i) optimum convoy size to minimise losses from submarine attacks, (ii) the optimal way to deploy radar units in order to maximise potential coverage against possible enemy attacks, and (iii) the invention of new flight patterns, and the determination of the correct colour of the aircraft in order to minimise the chance of detection by the submarines.

After the war, operations research was adopted by the industry and some of the techniques that had been applied to the complex problems of war were successfully transferred and assimilated for use in the industrialised sector. The dramatic development and refinement of the techniques of operations research and the advent of digital computers are the two prime factors that have contributed to the growth and application of OR in the post-war period. In the 1950s, OR was mainly used to handle management problems that were clear-cut, well-structured and repetitive in nature. Typically, they were of a tactical and operational nature, such as inventory control, resource allocation, scheduling of construction projects, and so on. Since the 1960s, however, formal approaches have been increasingly adopted for the less well-structured planning problems as well. These problems are strategic in nature and are the ones that affect the future of the organisation. The development of corporate planning models and those relating to the financial aspects of the business, for example, are such type of problems. Thus, in the field of business and industry, operations research helps the management determine their tactical and strategic decisions more scientifically.

1.2.2 Nature and Characteristic Features of OR

In general terms, we can regard operations research as being the application of scientific methods to decision making. Thus, operations research attempts to provide a systematic and rational approach to the fundamental problems involved in the control of systems by making decisions which, in a sense, achieve the best results considering all the information that can be profitably used. A classical definition of OR is given by Churchman et al., "... Operations Research is the application of scientific methods, techniques and tools to problems involving the operations of systems so as to provide those in control of operations with optimum solutions to the problems."* Thus, it may be regarded as the scientific method employed for problem solving and decision-making by the management.

* Churchman, C.W., Ackoff, R.L. and Acnoff, *Introduction to Operations Research*, John Wiley & Sons, New York, 1977, p.9.

The significant features of operations research are given below.

1. Decision-making Primarily, OR is addressed to managerial decision-making or problem solving. A major premise of OR is that decision-making, irrespective of the situation involved, can be considered a general systematic process that consists of the following steps:

(a) Define the problem and establish the criterion, which will be used. The criterion may be the maximisation of profits, utility and minimisation of costs, and so on.
(b) Select the alternative courses of action for consideration.
(c) Determine the model to be used and the values of the parameters of the process.
(d) Evaluate the alternatives and choose the one that is optimal.

2. Scientific Approach OR employs scientific methods for the purpose of solving problems, and there is no place of whims and guesswork in it. It is a formalised process of reasoning and consists of the following steps:

(a) The problem to be analysed is defined clearly and the conditions for observations are determined.
(b) Observations are made under varying conditions to determine the behaviour of the system.
(c) On the basis of the observations, a hypothesis describing how the various factors involved are believed to interact and the best solution to the problem, is formulated.
(d) To test the hypothesis, an experiment is designed and executed. Observations are made and measurements are recorded.
(e) Finally, the results of the experiments are analysed and the hypothesis is either accepted or rejected. If the hypothesis is accepted, the best solution to the problem is obtained.

3. Objective OR attempts to locate the best or optimal solution to the problem under consideration. For this purpose, it is necessary that a measure of effectiveness be defined, which is based on the goals of the organisation. This measure is then used as the basis to compare the alternative courses of action.

4. Inter-disciplinary Team Approach OR is inter-disciplinary in nature and requires a team approach to a solution to the problem. No single individual can have a thorough knowledge of the myriad aspects of operations research and how the problems may be addressed. Managerial problems have economic, physical, psychological, biological, sociological and engineering aspects. This requires a blend of people with expertise in the areas of mathematics, statistics, engineering, economics, management, computer science, and so on. Of course, it is not always so. Some problem situations may be adequately handled even by one individual.

5. Digital Computer Use of a digital computer has become an integral part of the operations research approach to decision-making. The computer may be required due to the complexity of the model, volume of data required or the computations to be made. Many quantitative techniques are available in the form of 'canned' programmes.

1.2.3 Methodology of Operations Research

The basic and dominant characteristic feature of operations research is that it employs mathematical representations or models to analyse problems. This distinctive approach represents an adaptation of the scientific methodology used by other physical sciences. The scientific method translates a real given problem into a mathematical representation, which is solved and re-transformed into the original context. The OR approach to problem solving consists of the following steps.

1. Formulate the problem.
2. Determine the assumptions (model building) and formulate the problem in a mathematical framework.

3. Acquire the input data.
4. Solve the model formulated and interpret the results.
5. Validate the model.
6. Implement the solution obtained.

However, one step need not be completed fully before the next is started. In fact, one or more of the steps may be modified to some extent before final results are implemented. This would, of course, necessitate the subsequent steps being modified.

The steps are shown in Fig. 1.1 and we now discuss each of these one by one.

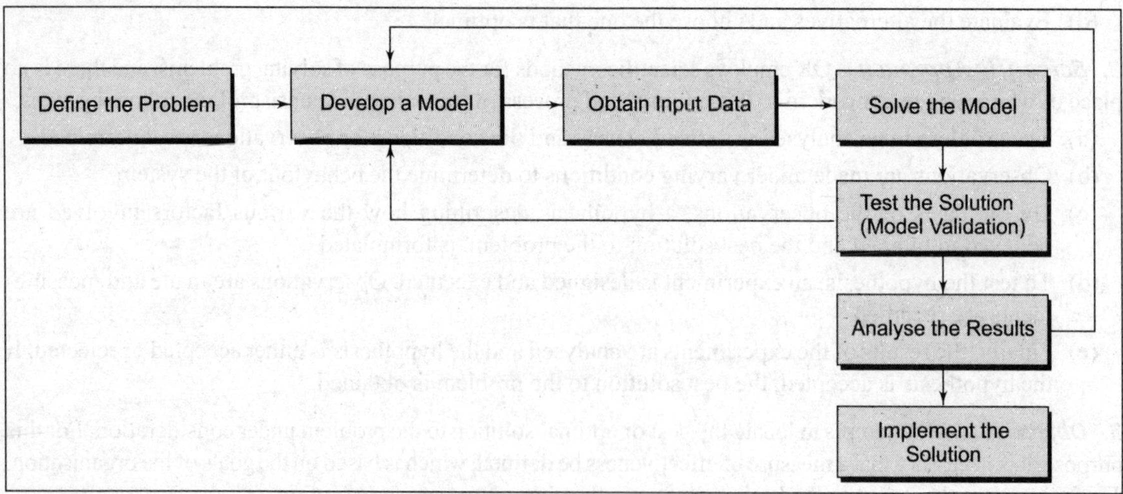

Figure 1.1 *Operations Research Methodology*

1. Problem Formulation The first step in quantitative analysis is to develop a clear and concise statement of the problem. In many cases, defining the problem proves to be the most important and difficult step. It is essential, therefore, that the root problem should be identified and understood in the first place. Logically speaking, we cannot expect to get the right answer if the problem is identified incorrectly. In that case, the solution derived from it is apt to be useless and all the efforts in that direction shall be a waste. The problem should be identified properly because often what is described as a problem may only be its symptom. For example, excessive costs *per se* do not constitute a problem. They are only an indication of some problem that may, for instance, be improper inventory levels, excessive wastage, and the like. Often the symptoms of a problem may extend beyond a single manager's control to other personnel and other departments in an organisation. It is essential, therefore, to go beyond symptoms of the problem and identify their actual causes. Also, one problem may be related to other problems and solving one problem without having regard to the others may make the matters worse. It is essential, therefore, that an analysis be made as to how the solution to one problem is likely to affect other problems or the situation in general. Further, it is likely for an organisation to have several problems. However, an analyst usually cannot deal with all the problems and has to focus only on a few of them. For most companies it translates in to selecting those problems whose solutions are likely to result in greatest profit increases or cost reductions. Thus, it is imperative that right problems be selected for solution. In sum, it is necessary for an analyst to understand that the formulation of a problem develops from a complicated interaction that involves the selection and interpretation of data between him and the management.

Once the problem has been identified, it is categorised as being standard or special. The *standard problems* are also known as *programmed problems*. As has already been mentioned, they are the well-structured problems characterised by routine, repetitive decisions that utilise specific decision-making techniques in their solution strategy. Standard solution procedures have been developed to handle such prototype problems. Examples of these problems include the assignment of workers to jobs, fixing the product-mix for a month and determination of the quantity of materials to be bought. On the other hand, there are *special* or *non-programmed* problems. They are unique and non-recurrent in nature and, therefore, ill-structured. Undertaking of a research and development project and the merger and consolidation decisions illustrate such type of decision situations.

2. Model Building Once the problem is defined, the next step is to build a suitable model. As already mentioned, the concepts of models and model building lie at the very heart of the operations research approach to problem solving. A model is a theoretical abstraction of a real-life problem. In fact, many real-life situations tend to be very complex because there are literally innumerable inherent factors in any given situation. Thus, the decision-maker has to abstract from the empirical situation those factors which are most relevant to the problem. Having selected the critical factors, he combines them in some logical manner so that they form a counterpart or a model of the actual problem.

Thus, a model is a simplified representation of a real-world situation that, ideally, strips a natural phenomenon of its bewildering complexity and replicates its essential behaviour. Models may be represented in a variety of ways. They can be classified as physical and symbolic models.

(a) Physical Models

A physical model is a physical or schematic representation of the real thing. There are two types of physical models: iconic and analogue.

(i) Iconic Models They are essentially the scaled-up/down versions of the particular thing they represent. A model aeroplane in a wind tunnel, a model of a proposed building provided by an architect, models of the sun and its planets housed in a planetarium, a model of a particular molecular structure of a chemical are examples of iconic models, because they look like what they represent (except size). Maps, pictures or drawings may also be categorised as iconic models since they represent essentially the images of certain things.

The chief merit of an iconic model is that it is concrete and specific. It resembles visually the thing it represents and, therefore, there are likely to be fewer problems in translating any 'findings' from the model into the real-life situation. However, the disadvantage of such models is that they often do not lend themselves to manipulation for experimental purposes.

(ii) Analogue Models The analogue models use one set of properties to represent another set. To illustrate, an electrical network model may be used as an analogue model to study the flows in a transportation system. Similarly, a barometer that indicates changes in atmospheric pressure through movements of a needle is an example of analogue model and the contour lines on a map are analogues of elevation. In general, the analogue models are less specific and concrete but they are easier to manipulate as compared to the iconic models.

(b) Symbolic Models

Many real-life problems can be described by symbolic models or mathematical forms. These are the most general and abstract types of models. They employ letters, numbers and other types of symbols to represent the variables and their interrelationships. As such, they are capable of experimental manipulation most easily. The symbolic models can be verbal or mathematical. Whereas the verbal models describe a situation in spoken language or written words, the mathematical models employ mathematical notation to represent, in a precise manner, the variables of the real situations. The mathematical models take the form of mathematical relationships that portray the structure of what they are intended to represent. The use of a verbal versus mathematical model could be shown by the formula for finding the perimeter of a rectangle. A verbal model would express this problem as follows. The perimeter (P) of a rectangle is equal to the sum total of two times the length (L) and two times the width (W) of the rectangle. In contrast, the advantage of the mathematical model is demonstrated by the following statement: $P = 2L + 2W$. If applied to the same rectangle, both models would yield identical results. However, a mathematical model is more precise.

Symbolic models are used in operations research because they are easier to manipulate and they yield more accurate results under manipulation compared to the iconic or analogue models.

Use of Mathematical Models Various types of mathematical models are used in modern operations research. Two broad categories of these are deterministic and probabilistic models. A *deterministic model* is the one in which all parameters in the mathematical formulation are fixed at predetermined values so that no uncertainty exists. In a *probabilistic model,* on the other hand, some or all the basic characteristics may be random variables (capable of assuming different values with given probabilities). In such models, uncertainty and errors are required to be given explicit consideration. Probabilistic models are also termed as stochastic or chance models.

The mathematical models comprise three basic components: decision variables, result variables and uncontrollable variables. The *decision variables* represent those factors where a choice could be made. These variables can be manipulated and, therefore, are controllable by the decision-maker. The *result variables* indicate the level of effectiveness of a system. They represent output of the system and are also termed as *dependent variables* but are beyond the control of the decision-maker. To illustrate, in the area of marketing, the decision variables may be the advertising budget, the number of regional salesmen employed, the number of products and so on; result variables may be the market share for the company, level of customer satisfaction and others; while the uncontrollable variables may be the competitors' strategies, consumer incomes, and so on.

As mentioned earlier, the different components of a mathematical model are tied together with the relationships in the form of equations, inequalities, and so on. Such a model consists of an objective function and describes how a dependent (result) variable is related to independent (decision) variables. For example, the profit function of a firm making two products can be stated as follows:

$$P = p_1x_1 + p_2x_2$$

in which P indicates the total profit of the firm, x_1 and x_2 are the number of units (independent) variables of the two products produced and sold, and p_1 and p_2 the profit per unit on the two products, respectively (the uncontrollable variables).

The objective function is called for to maximise (or minimise), subject to certain constraints (representing the uncontrollable variables). For example, in this case of production, the firm might be able to sell no more than a certain number of units, say 80. Then the marketing constraint (an uncontrollable variable) can be expressed as follows:

$$x_1 + x_2 \leq 80$$

Similarly, other constraints, if any, of the system can be expressed.

3. Obtaining Input Data Once an appropriate model has been formulated, the next step is to obtain the data to be used in the model as input. Since the quality of data determines the quality of output, the importance of obtaining accurate and complete data cannot be over-emphasised. Obviously, the finished product can be no better than the raw materials used. This situation may be described as GIGO: gold in, gold out or garbage in, garbage out.

Obtaining correct and relevant data may indeed be a difficult exercise when relatively large problems are involved. A number of sources, including company reports and documents, interviews with the company personnel and so on, may be used for collecting data.

4. Solution of Model Having formulated an appropriate model and collected the input data, the next stage in the analysis calls for the solution to the model and the interpretation of the solution in context of the given problem. A solution to a model implies determination of a specific set of decision variables that would yield a desired level of output. The desired level of output, in turn, is determined by the principle of choice adopted and represents the level which 'optimises'. Optimisation might mean maximising the level of goal attainment from a given set of resources or minimisation of cost as will satisfy the required level of goal attainment, or maximisation of ratio of the goal attainment to cost.

It may be noted that the solutions can be classified as being feasible or infeasible, optimal or non-optimal and unique or multiple.

(a) Feasible and Infeasible Solutions

A solution (a set of values of the decision variables, as already mentioned) which satisfies all the constraints of the problem is called a feasible solution, whereas an infeasible solution is the one which does not satisfy all the constraints. Since an infeasible solution fails to meet one or more of the system requirements, it is an unacceptable one. Only feasible solutions are of interest to the decision-maker.

(b) Optimal and Non-optimal Solutions

An optimal solution is one of the feasible solutions to a problem that optimises and is, therefore, the best among them. For example, for a multi-product firm working under some given constraints of capacity, marketing, finance and so on, the optimal solution would be that product-mix which would meet all the constraints and yield the maximum contribution margin towards profits. The feasible solutions other than the optimal solution are called non-optimal solutions. To continue with the example, several other product-mixes would satisfy the restrictions imposed and hence qualify for acceptance, but they would be ignored because lower contribution margins would be associated with them. They would be non-optimal.

(c) Unique and Multiple Solutions

If only one optimal solution to a given problem exists, it is called a unique solution. On the other hand, if two or more optimal solutions to a problem exist, which are equally efficient, then multiple optimal solutions are said to exist. Of course, these are preferable from the management's point of view as they provide a greater flexibility in implementation.

Once the principle of choice has been specified, the model is solved for optimal solution. For this, the feasible solutions are considered and of them the one (or more) that optimises is chosen. For this purpose, a complete enumeration may be made so that all the possible solutions are checked and evaluated. However, this approach is limited to those situations where the number of alternatives is small. Alternatively, and more commonly, methods involving algorithms may be used to get optimal solutions. It is significant to note that in contrast to complete enumeration, where all solutions are checked, an algorithm represents a trial-and-error process, where only a part of the feasible solutions are considered and the solutions are gradually improved until an optimal solution is obtained.

While algorithms exist for most of the standardised problems, there are also some numerical techniques, which yield solutions that are not necessarily optimal. Heuristics and simulation illustrate those methods. *Heuristics* are step-by-step logical *rules,* which, in a certain number of steps, yield some acceptable solution to a given problem. They are applied in those cases where no algorithms exist. Similarly, the technique of *simulation* is also applied where a given system is sought to be replicated and experimented with. Solutions using simulation need not be optimal because the technique is only descriptive in nature. Once a solution is obtained, it needs to be tested completely before it can be analysed and implemented. Because the solution depends on the input data and the model employed, both need testing. Such testing comprises of determining the accuracy and completeness of the data used by model and ensuring that the model used is logical and adequately represents the real situation.

Sensitivity Analysis In addition to the solution to the model formulated by any technique, sensitivity analysis, or post-optimality analysis, should also be performed. By sensitivity analysis we imply determination of the behaviour of the system to changes in the system inputs and specifications. Thus, it is *what if* analysis. This is done because the input data and the structural assumptions of the model may not be valid.

5. Model Validation The validation of a model requires determining whether the model can adequately and reliably predict the behaviour of the real system that it seeks to represent. Also, it involves testing the structural assumptions of the model to ascertain their validity. Usually, the validity of a model is tested by comparing its performance with the past data available in respect of the actual system. There is, of course, no assurance that the future performance of the system will continue to be in the same manner as in its past. Therefore, one must take cognisance of the changes in the system over time and adjust the model as required.

6. Implementation The final step is the implementation of the results. It is the process of incorporating the solution in the organisation. Implementation of solution is often more difficult than it may apparently seem. No standard prescription can be given, which would ensure that the solution obtained would automatically be adopted and implemented. This is because the techniques and models used in operations research may sound such and may be detailed in mathematical terms, but they generally do not consider the human aspects that are significant in implementation of solution. The impact of a decision may cut across various segments of the organisation, and the factors like resistance to change, desire to be consulted and informed, motivation, and so on may come in the way of implementation. Equally important as the skill and expertise needed in developing a model is the requirement of tackling issues related to the factors which may have a bearing on the implementation of a solution in a given situation. Thus, a model, that secures a moderate theoretical benefit and is implemented, is better than a model which ranks very high on obtaining theoretical advantage but cannot be implemented. In fact, the importance of having managers in the organisation who would act on the results of the study of a team that analyses the problem, can hardly be over-emphasised.

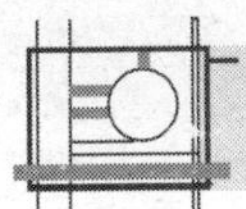

1.3 QUANTITATIVE ANALYSIS AND COMPUTER-BASED INFORMATION SYSTEMS

Quantitative analysis/operations research has become an integral part of the modern computer-based information systems. A computer-based information system comprises:

Hardware:	input, CPU, storage, and output,
Software:	general operating software, general and specialised application software,
Files:	tapes, disks, documents,
Procedures:	user, input and operating procedures,
People:	managers, analysts, technical support personnel and operating personnel, and
Database:	information about various facets of the organisation.

Such systems may include management information system (MIS), decision support systems (DSS) and the use of artificial intelligence. Quantitative analysis tools are used in each of these subsystems.

1. Management Information Systems A management information system represents an organised way of managing information and data, which are vital organisational resources that are essential to the efficient and effective operations. Thus, it comprises a body of organised procedures for identification, collection, processing, retrieval and dissemination of information. It aims at providing right information to the right people in the right place at the right time. Provision of right information can often involve help of quantitative analysis. To illustrate, if a manager needs help to take ordering and inventory stocking decisions, forecasting models for projecting demand, inventory models to determine order quantities, and so on, are going to be quite helpful. To be able to have the needed information, it is necessary that the manager is able to interact with the computers 'on-line'. In case of complexities, the quantitative analyst can aid the manager to retrieve the desired information from the system, by providing programmes to the manager for the same.

2. Decision Support Systems The advances in computer technology have witnessed development of decision support systems (DSS) and artificial intelligence. In a decision support system, a system is developed to aid management in improving its decision-making. It supports, rather than replaces, managerial judgement. The presence of such a system often implies the use of computers to assist the managers in decision-making for semi-structured problems. Decision support system is interactive and allows the manager the use of *what if* questions, so that she may try different decisions, use modified data and witness results of such changes quickly. Its emphasis is on effectiveness of decision-making rather than on efficiency. Break-even analysis and profitability decisions, decision tables and expected values, decision trees, relevance trees, etc., are all typical DSS models.

3. Artificial Intelligence and Expert Systems Although machine intelligence had been dreamt of for ages, the term *artificial* intelligence, to denote demonstration of intelligence by machines, evolved during 1950s. Artificial intelligence includes the ability of computer systems and technology to think, see, learn, understand and use common sense—in other words, to mimic the human behaviour. It has grown into several important subsystems that have broad implications for quantitative analysis. Expert systems, an offshoot of the research on artificial intelligence, are information systems that attempt to support or automate decision-making and act like intelligent and rational decision-makers. They do this by storing and using knowledge about a specific, limited topic and produce conclusions based on the data that they receive.

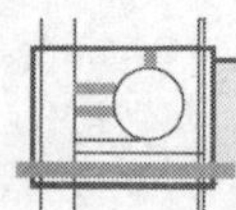

1.4 PLAN OF THE BOOK

As the title suggests, this book is an introduction to quantitative methods as an aid to managerial decision-making. Important quantitative techniques that may be employed by a manager in the evaluation of alternatives constitute the text of the book. The managerial problems and the tools that are used in handling them are discussed in the form of chapter scheme.

Chapters 2, 3 and 4 discuss ways and means of approaching the problems of resource allocation, under the general category of *linear programming.* The attempt here is to seek an optimal solution to allocations where (i) there are a number of activities to be performed, (ii) there exist at least two different ways to perform these activities, and (iii) resources are limited. The variables involved are assumed to be related linearly, and the boundaries or constraints of their interaction are also linear. Chapter 2 discusses the formulation of linear programming problems and the graphic approach to their solution, while Chapter 3 describes their solution using simplex method. Chapter 4 is devoted to post-optimality analysis and duality. *Transportation* and *transhipment* problems that deal with movement of a commodity from several origins to several destinations at a minimum cost are discussed in Chapter 5. While the transportation problem allows goods to be moved from various sources to different destinations, the transhipment problem permits goods being transhipped *en route* between different sources as well. The *assignment problem,* discussed in Chapter 6, is one where certain items, like people or activities, are to be assigned on a one-to-one basis to other items such as jobs or facilities, with the objective of minimising the cost or maximising the effectiveness of assignment.

Chapter 7 is addressed to *integer programming* problems, which restrict the decision variables to assume only integer values, and *goal programming* problems, which allow a manager to specify multiple goals instead of a single goal of, say, profit maximisation. In this chapter, the problem of determining the optimal way of scheduling the tour for a *travelling salesman* is also considered. Chapter 8 considers the problem of *sequencing*—the scheduling of different jobs to be performed on various machines. The next chapter deals with the *inventory* process and the question of how many units should be maintained in inventory, including how often and how many units should be ordered, and the holding of safety stocks.

Queuing theory is the subject of discussion in Chapter 10. Whenever persons or objects reach a service station, a waiting line is likely to form, particularly during the rush hours. Thus, there may be a queue at certain times while the service facility may be idle at others. In general, the larger the facility, the costlier the operation and smaller the waiting line. The problem is to determine such a service level that will minimise the relevant costs. The question of *replacement* of assets that wear out gradually and those that fail suddenly are considered in Chapter 11. Chapter 12 is addressed to the tools of *Programme Evaluation Review Technique* (PERT) and *Critical Path Method* (CPM), which are extremely useful for the purpose of planning and controlling complex projects.

Chapter 13 reviews the decision process in the structured framework of *decision theory*. Several principles of decision-making are discussed in the chapter. Also given are the posterior analysis, decision tree and utility approaches to decision-making. The next chapter contains a discussion of Markov chains, which deal basically with the problem of brand switching by the customers. It discusses how market shares are likely to be affected under conditions of customers shifting preferences from one brand to another. Chapter 15 is titled *theory of games* and deals with situations of conflict–where decisions are required to be made with the opponent being a rational decision-maker.

Dynamic programming, which is an important quantitative tool to deal with a large number of problems, is described in Chapter 16. In this method, a problem is solved by dissecting it into a number of sub-problems. The solution to the sub-problems one-by-one eventually leads to the solution to the given problem.

Chapter 17 introduces a powerful tool of quantitative analysis—the *simulation.* The Monte-Carlo analysis is introduced which is a technique of testing alternatives through trial and error approaches. In effect, a system is sought to be 'copied', and through the technique of selecting random sequences of numbers from a probability distribution, we are able to test or simulate the outcomes associated with the alternatives that do not lend themselves to direct analysis and comparison.

In Chapter 18 we consider the concept of the value of money, investment analysis and the break-even analysis. An attempt is made in this chapter to see how can the risk associated with an investment proposal be quantified and appropriate capital budgeting decisions made. Finally, the *forecasting techniques* are dealt with in Chapter 19. The forecasting techniques are used for making predictions of future demand or some other variable. They are helpful to the manager in as much as the manager is required to take decisions about future which is at best uncertain.

KEY POINTS TO REMEMBER

- Decision-making is an all-pervasive feature of management.
- A systematic approach to decision-making calls for application of scientific methods.
- Decision-making may involve decisions under certainty or uncertainty, under static or dynamic conditions, and against nature or some rational opponent.
- Qualitative factors should be given due consideration along with quantitative analysis of a problem.
- Quantitative analysis/operations research provides a systematic approach to decision-making.
- Features of operations research include decision-making, scientific approach, objectivity, inter-disciplinary approach and use of digital computers.
- The methodology of operations research comprises of (i) problem formulation, (ii) model building, (iii) data collection. (iv) solution of the model and interpretation of results, (v) model validation, and (vi) implementation of the solution obtained.
- Models may be physical (including iconic and analogue) or symbolic (mathematical).
- Mathematical models have three components: decision variables, result variables and uncontrollable variables.
- Quantitative analysis is now an integral part of the modern computer-based information systems, decision support systems and artificial intelligence and expert systems.

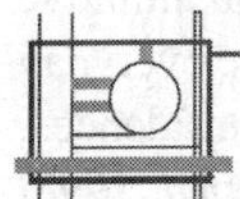

TEST YOUR UNDERSTANDING

Mark the following statements as T (*True*) or F (*False*)

1. Decision-making is purely an art even in the modern age.

2. Decisions taken only on the basis of quantitative analysis can be sound and correct.
3. For a real decision-making situation, definition and identification of alternatives, the states and consequences are the most difficult aspects.
4. An orderly and organised way of thinking is needed in order to formulate a problem in to a well-structured framework for solution.
5. Quantification of a problem situation represents sound business judgement and it enables the solution to be implemented without second thoughts.
6. Operations research is inter-disciplinary in nature and requires a team approach to the solution to a problem.
7. As real-life decision situations tend to be rather complex, the decision-maker has to abstract from a given empirical situation those factors which are most relevant to the problem and combine them in some logical manner as to form a model of it.
8. Non-programmed decision-making problems are ill structured.
9. Analogue models are essentially the scaled up/down versions of the particular thing they seek to represent.
10. A feasible solution is one which satisfies all the constraints of the given problem, while an infeasible solution is one which satisfies none of them.
11. Validation of a model implies determining if the model does adequately and reliably predict the behaviour of the real system that it seeks to represent.
12. Sensitivity analysis is *what if* analysis, and deals with determining how the output changes in response to changes in the inputs.
13. Implementation of solution determined from quantitative analysis is trivial because such a solution is the outcome of rigorous mathematical analysis.
14. A management information system (MIS) comprises a body of organised procedures for identification, collection, processing, retrieval and dissemination of information, aiming at providing right information to right people in right place at right time.
15. Decision support systems (DSS) are a good substitute for managerial judgement.

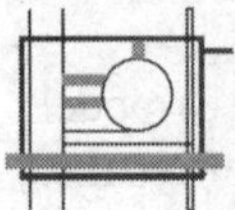

EXERCISES

1. It is said that management is equivalent to decision-making. Do you agree? Explain.
2. What are the essential characteristics of operations research? Mention different phases in an operations research study. Point out some limitations of operations research. Explain the role of computers in this field.
3. Discuss the role and scope of quantitative methods for scientific decision-making in business management.
4. "Executives at all levels in business and industry come across the problems of making decisions at every stage in their day-to-day activities. Operations Research provides them with various quantitative techniques for decision-making and enhances their ability to make long range plans and solve everyday problems of running a business and industry with greater efficiency, competence and confidence." Comment with examples. *(MBA, Delhi, April, 1998)*
5. Write an essay on the scope and methodology of operations research, explaining briefly the main phases of an OR study and techniques used in solving OR problems.

6. What is Operations Research? Account for the growing importance of Operations Research in business decisions. *(M Com, Delhi, 1997)*
7. Most operations research applications possess certain distinguishing characteristics. These could be identified as follows:
   (a) primary focus on decision-making;
   (b) an investigation based on some measurable criteria;
   (c) the use of a formal mathematical model; and
   (d) dependence upon computing facilities.
   Explain each of these.
8. (a) Define an OR model and give four examples. State their properties, advantages and limitations.
   (b) State the phases of an OR study and their importance in solving problems. *(CA Nov., 1986)*
9. Model building is a central element in operations research method. Give a description of the following basic types of models: (a) iconic, (b) analogue, and (c) mathematical (symbolic), with particular reference to the special insights provided by each.
   Give an account of the information requirements, assumptions and applications of any three of the following different type of mathematical models: (a) allocation, (b) queuing, (c) inventory, (d) replacement. *(ICMA, May, 1979)*
10. What is a 'model'? Distinguish between
    (a) analogue and iconic models, and
    (b) deterministic and stochastic models.
11. (a) "Statistics is the nerve centre for Operations Research." Discuss.
    (b) State any four areas for the application of OR techniques in Financial Management, and how it improves the performance of the organisation. *(CA, May, 1986)*
12. "All quantitative techniques have hardly any real-life applications." Do you agree with the statement? Discuss. *(MIB, Delhi, 1999)*
13. Comment on the following statements:
    (a) Operational Research is an art of winning war without actually fighting it.
    (b) Operational Research is no more than a quantitative analysis of the problem.
    (c) Operational Research is a war against *ad hocism*.
    (d) Operational Research is the art of finding bad answers where worst exist.
    (e) Operational Research advocates a systems-approach and is concerned with optimisation. It provides a quantitative analysis for decision-making.
    (f) Operational Research replaces management-by-personality.
14. It is common in business to insure against the occurrence of events, which are subject to varying degrees of uncertainty, for example, ill health of senior executives. At the same time, the use of formal analytical models to assist in the process of making decisions on business problems which are generally subject to uncertainty, does not appear to be very widespread.
    Describe the model building approach to the analysis of business problems, under conditions of uncertainty. Discuss the apparent inconsistency in company's willingness to insure when formal analytical of an Operations Research nature which allow for uncertainty are relatively rarely employed. *(ICMA, Nov., 1981)*

CHAPTER

2

Linear Programming I: Formulation and Graphic Solution

CHAPTER OVERVIEW

We live in a world of shortages since our resources which can be supplied are limited. This is all the more true at the micro level. Thus, there is always a problem as to how to allocate the given resources in the best possible manner. Linear programming is a technique which provides the answer in a wide variety of cases.

Some situations where a manager can use linear programming include the following:

- How to allocate the advertising budget among various alternate advertising media which have different degrees of effectiveness in reaching audiences and involve different costs?
- In case of a multi-product firm, what product-mix will yield the maximum profit, when different products are known to have different profitability coefficients and different resource requirements?
- How should the given funds be allocated between different investment opportunities that yield varying returns and involve different degrees of risk?
- How should a dietician decide about the foods that contain varying proportions of ingredients like carbohydrates, vitamins, proteins, etc. to be given to the patients so that their nutrition requirements are met with at the minimum cost?
- How should the land on an agricultural farm be allocated between different crops which involve different costs on account of labour, manure, seeds, etc. and have different yields, resulting in unequal profitability of the agricultural products produced?
- How should the HR manager of a hospital decide about the employment of nurses that involves lowest cost and yet meets the requirements at different times of the 24-hour day?

The next few chapters are devoted to a detailed account of linear programming. It must be kept in mind that the most important thing is to develop the skill and ability to translate a given real-life situation into a linear programming format, keeping in mind its assumptions and limitations. This chapter illustrates this with examples. Later in the chapter, the graphic solution to some such problems is provided.

The chapter heavily uses the inequalities of 'less than' and 'greater than' types and equations. You should be conversant with two-dimensional graphs and their use. Plotting of equations and inequalities on a graph and the ability to determine the space on the graph over which they are satisfied, called the feasible area, holds the key to successful graphic solution to the problems.

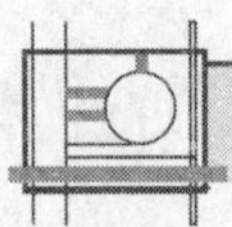

2.1 INTRODUCTION

A large number of decision problems faced by a business manager involve allocation of resources to various activities, with the objective of increasing profits or decreasing costs, or both. When resources are in excess, no difficulty is experienced. But such cases are very rare. Practically in all situations, the managements are confronted with the problem of scarce resources. Normally, there are several activities to perform but limitations of either of the resources or their use prevent each activity from being performed to the best level. Thus, the manager has to take a decision as to how best to allocate the resources among the various activities.

The decision problem becomes complicated when a number of resources are required to be allocated and there are several activities to perform. Rule of thumb, even of an experienced manager, in all likelihood, may not provide the right answer in such cases. The decision problems can be formulated, and solved, as mathematical programming problems.

Mathematical programming involves optimisation of a certain function, called the *objective function,* subject to certain *constraints.* For example, a manager may be faced with the problem of deciding the appropriate product mix of the four products. With the profitability of the products along with their requirements of raw materials, labour, etc. known, his problem can be formulated as a mathematical programming problem taking the objective function as the maximisation of profits obtainable from the mix, keeping in view the various constraints—the availability of raw materials, labour supply, market, and so on. The methods of mathematical programming can be divided into three groups: linear, integer, and non-linear programming.

This chapter and the next two are devoted to linear programming, while integer programming is discussed in Chapter 7. Non-linear programming is not considered in the book.

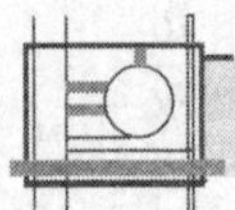

2.2 LINEAR PROGRAMMING

The linear programming method is a technique for choosing the best alternative from a set of feasible alternatives, in situations in which the objective function as well as the constraints can be expressed as linear mathematical functions. In order to apply linear programming, certain requirements have to be met. These are discussed here.

(a) There should be an objective which should be clearly identifiable and measurable in quantitative terms. It could be, for example, maximisation of sales, of profit, minimisation of cost, and so on.

(b) The activities to be included should be distinctly identifiable and measurable in quantitative terms, for instance, the products included in a production planning problem.

(c) The resources of the system, which are to be allocated for the attainment of the goal, should also be identifiable and measurable quantitatively. They must be in limited supply. The technique would involve allocation of these resources in a manner that would trade off the returns on the investment of the resources for the attainment of the objective.

(d) The relationships representing the objective as also the resource limitation considerations, represented by the objective function and the constraint equations or inequalities, respectively, must be *linear* in nature.

(e) There should be a series of feasible alternative courses of action available to the decision-maker which are determined by the resource constraints.

When these stated conditions are satisfied in a given situation, the problem can be expressed in algebraic form, called the *Linear Programming Problem* (LPP), and then solved for optimal decision. We shall first illustrate the formulation of linear programming problems and then consider the method of their solution.

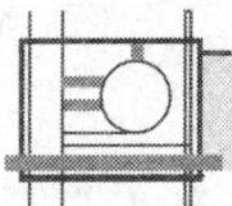

2.3 FORMULATION OF LINEAR PROGRAMMING PROBLEMS

1. The Maximisation Case Consider the following example.

Example 2.1 A firm is engaged in producing two products, *A* and *B*. Each unit of product *A* requires 2 kg of raw material and 4 labour hours for processing, whereas each unit of product *B* requires 3 kg of raw material and 3 hours of labour, of the same type. Every week, the firm has an availability of 60 kg of raw material and 96 labour hours. One unit of product *A* sold yields Rs 40 and one unit of product *B* sold gives Rs 35 as profit.

Formulate this problem as a linear programming problem to determine as to how many units of each of the products should be produced per week so that the firm can earn the maximum profit. Assume that there is no marketing constraint so that all that is produced can be sold.

The objective function The first major requirement of an LPP is that we should be able to identify the goal in terms of the objective function. This function relates mathematically the variables with which we are dealing in the problem. For our problem, the goal is the maximisation of profit, which would be obtained by producing (and selling) products *A* and *B*. If we let x_1 and x_2 represent the number of units produced per week of the products *A* and *B* respectively, the total profit, *Z*, would be equal to $40x_1 + 35x_2$, because the unit profit on the two products is, respectively, Rs 40 and Rs 35. Now $Z = 40x_1 + 35x_2$ is, then, the objective function, relating the profit and the output level of each of the two items. Notice that the function is a linear one. Further, since the problem calls for a decision about the optimal values of x_1 and x_2, these are known as the *decision variables.*

The constraints As has been laid, another requirement of linear programming is that the resources must be in limited supply. The mathematical relationship which is used to explain this limitation is inequality. The limitation itself is known as a *constraint.*

Each unit of product A requires 2 kg of raw material while each unit of product *B* needs 3 kg. The total consumption would be $2x_1 + 3x_2$, which cannot exceed the total availability of 60 kg every week. We can express this constraint as $2x_1 + 3x_2 \le 60$. Similarly, it is given that a unit of *A* requires 4 labour hours for its production and one unit of *B* requires 3 hours. With an availability of 96 hours a week, we have $4x_1 + 3x_2 \le 96$ as the labour hours constraint.

It is important to note that for each of the constraints, inequality* rather than equation has been used. This is because the profit maximising output might not use all the resources to the full—leaving some unused. Hence, the $\le$ sign. However, it may be noticed that all the constraints, like objective function, are also linear in nature.

Non-negativity condition Quite obviously, x_1 and x_2, being the number of units produced, cannot have negative values. Thus, both of them can assume values only greater-than-or-equal-to zero. This is the non-negativity condition, expressed symbolically as $x_1 \ge 0$ and $x_2 \ge 0$.

---

* The inequalities used are not strict ones. A strict inequality is expressed as, for example, $a < b$ (or $a > b$). Here the inequalities are loose—thus permitting *a* to be smaller (or greater) than, or equal to, *b*.

Now, we can write the problem in complete form as follows.

Maximise	$Z = 40x_1 + 35x_2$	Profit
Subject to		
	$2x_1 + 3x_2 \le 60$	Raw material constraint
	$4x_1 + 3x_2 \le 96$	Labour hours constraint
	$x_1, x_2 \ge 0$	Non-negativity restriction

2. The Minimisation Case Consider the following example.

Example 2.2 The Agricultural Research Institute suggested to a farmer to spread out at least 4800 kg of a special phosphate fertiliser and not less than 7200 kg of a special nitrogen fertiliser to raise productivity of crops in his fields. There are two sources for obtaining these—mixtures *A* and *B*. Both of these are available in bags weighing 100 kg each and they cost Rs 40 and Rs 24 respectively. Mixture *A* contains phosphate and nitrogen equivalent of 20 kg and 80 kg respectively, while mixture *B* contains these ingredients equivalent of 50 kg each.

Write this as a linear programming problem to determine how many bags of each type the farmer should buy in order to obtain the required fertiliser at minimum cost.

The objective function In the given problem, such a combination of mixtures *A* and *B* is sought to be purchased as would minimise the total cost. If x_1 and x_2 are taken to represent the number of bags of mixtures *A* and *B* respectively, the objective function can be expressed as follows:

Minimise $Z = 40x_1 + 24x_2$ Cost

The constraints In this problem, there are two constraints, namely, a minimum of 4,800 kg of phosphate and 7,200 kg of nitrogen ingredients are required. It is known that each bag of mixture *A* contains 20 kg and each bag of mixture *B* contains 50 kg of phosphate. The phosphate requirement can be expressed as $20x_1 + 50x_2 \ge 4{,}800$. Similarly, with the given information on the contents, the nitrogen requirement would be written as $80x_1 + 50x_2 \ge 7{,}200$.

Non-negativity condition As before, it lays that the decision variables, representing the number of bags of mixtures *A* and *B*, would be non-negative. Thus, $x_1 \ge 0$ and $x_2 \ge 0$.

The linear programming problem can now be expressed as follows:

Minimise	$Z = 40x_1 + 24x_2$	Cost
Subject to		
	$20x_1 + 50x_2 \ge 4800$	Phosphate requirement
	$80x_1 + 50x_2 \ge 7200$	Nitrogen requirement
	$x_1, x_2 \ge 0$	Non-negativity restriction

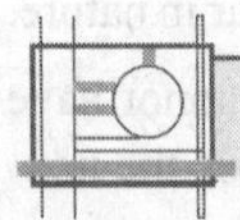

2.4 GENERAL STATEMENT OF LINEAR PROGRAMMING PROBLEMS

In general terms, a linear programming problem can be written as

Maximise $Z = c_1x_1 + c_2x_2 + \ldots + c_nx_n$ Objective function

Subject to

$$\left.\begin{array}{l} a_{11}x_1 + a_{12}x_2 + \ldots + a_{1n}x_n \le b_1 \\ a_{21}x_1 + a_{22}x_2 + \ldots + a_{2n}x_n \le b_2 \\ \vdots \\ a_{m1}x_1 + a_{m2}x_2 + \ldots + a_{mn}x_n \le b_m \end{array}\right\} \text{Constraints}$$

$$x_1, x_2, \ldots x_n \ge 0 \qquad \text{Non-negativity restriction}$$

where c_j, a_{ij}, b_i $(i = 1, 2 \ldots m; j = 1, 2, \ldots, n)$ are known as *constants* and x_j's are *decision variables*. The c_j's are termed as the *profit coefficients*, a_{ij}'s the *technological coefficients* and b_i's the *resource values*. In shorter term, the problem can be written as:

Maximise $\quad Z = \sum_{j=1}^{n} c_j x_j$

Subject to

$$\sum_{j=1}^{n} a_{ij}x_j \le b_i \qquad \text{for } i = 1, 2, \ldots, m$$

$$x_j \ge 0 \qquad \text{for } j = 1, 2, \ldots, n$$

Where the objective is to minimise a function, the problem is,

Minimise $\quad Z = \sum_{j=1}^{n} c_j x_j$

Subject to

$$\sum_{j=1}^{n} a_{ij}x_j \ge b_i \qquad \text{for } i = 1, 2, \ldots, m$$

$$x_j \ge 0 \qquad \text{for } j = 1, 2, \ldots, n$$

In matrix notation, an LPP can be expressed as follows:

	Maximisation Problem		*Minimisation Problem*
Maximise	$Z = cx$	Minimise	$Z = cx$
Subject to		Subject to	
	$ax \le b$		$ax \ge b$
	$x \ge 0$		$x \ge 0$

where

c = row matrix containing the coefficients in the objective function,

x = column matrix containing decision variables,

a = matrix containing the coefficients in the constraints,

b = column matrix containing the RHS values of the constraints.

Notes

1. Generally, the constraints in the maximisation problems are of the ≤ type, and of the ≥ type in the minimisation problems. But a given problem may contain a mix of the constraints, involving the signs ≤, ≥ and/or =.
2. Usually, the decision variables are non-negative. However, they need not always be so. To illustrate, in an investment problem, if we let x_1 represent the amount to be invested in the shares of a particular company,

then variable x_1 shall be non-negative since we may decide to invest ($x_1 > 0$) or not to invest ($x_1 = 0$). But, if we already hold such shares which may be sold, if the need be, then x_1 may take positive value (more investment), zero value (indicating no new investment in it) or negative value (implying disinvestment in this share). Hence, x_1 shall be *unrestricted in sign* or a *free variable.*

Formulation of some typical linear programming problems is given later in this chapter under Review Illustrations.

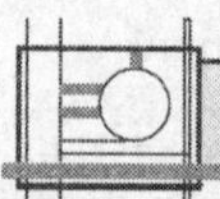

2.5 ASSUMPTIONS UNDERLYING LINEAR PROGRAMMING

A linear programming model is based on the assumptions of proportionality, additivity, continuity, certainty, and finite choices. These are explained here.

1. Proportionality A basic assumption of linear programming is that proportionality exists in the objective function and the constraint inequalities. For example, if one unit of a product is assumed to contribute Rs 10 toward profit, then the total contribution would be equal to $10x_1$, where x_1 is the number of units of the product. For 4 units, it would equal Rs 40 and for 8 units it would be Rs 80, thus if the output (and sales) is doubled, the profit would also be doubled. Similarly, if one unit takes 2 hours of labour of a certain type, 10 units would require 20 hours, 20 units would require 40 hours . . . . and so on. In effect, then, proportionality means that there are constant returns to scale—and there are no economies of scale.

2. Additivity Another assumption underlying the linear programming model is that in the objective function and constraint inequalities both, the total of all the activities is given by the sum total of each activity conducted separately. Thus, the total profit in the objective function is determined by the sum of the profit contributed by each of the products separately. Similarly, the total amount of a resource used is equal to the sum of the resource values used by various activities. This assumption implies that there is no interaction among the decision variables (interaction is possible when, for example, some product is a by-product of another one).

3. Continuity It is also an assumption of a linear programming model that the decision variables are continuous. As a consequence, combinations of output with fractional values, in the context of production problems, are possible and obtained frequently. For example, the best solution to a problem might be to produce $5\frac{2}{3}$ units of product *A* and $10\frac{1}{3}$ units of product *B* per week.

Although in many situations we can have only integer values, but we can deal with the fractional values, when they appear, in the following ways. *Firstly,* when the decision is a one-shot decision, that is to say, it is not repetitive in nature and has to be taken only once, we may round the fractional values to the nearest integer values. However, when we do so, we should evaluate the revised solution to determine whether the solution represented by the rounded values is a feasible solution and also whether the solution is the *best* integer solution. *Secondly,* if the problem relates to a continuum of time and it is designed to determine optimal solution for a given time period only, then the fractional values may not be rounded. For instance, in the context of a production problem, a solution like the one given earlier to make $5\frac{2}{3}$ units of *A* and $10\frac{1}{3}$ units of *B* per week, can be adopted without any difficulty. The fractional amount of production would be taken to be the work-in-progress and become a portion of the production of the following week. In this case, an output of 17 units of *A* and 31 units of *B* over a three-week period would imply $5\frac{2}{3}$ units of *A* and $10\frac{1}{3}$ units of *B* per week. *Lastly,* if we must insist on obtaining only integer values of the decision variables, we may restate the problem as an integer programming problem, forcing the solutions to be in integers only.

4. Certainty A further assumption underlying a linear programming model is that the various parameters, namely, the objective function coefficients, the coefficients of the inequality/equality constraints and the constraint (resource) values are known with certainty. Thus, the profit per unit of the product, requirements of materials and labour per unit, availability of materials and labour, etc. are given and known in a problem involving these. The linear programming is obviously *deterministic* in nature.

5. Finite Choices A linear programming model also assumes that a limited number of choices are available to a decision-maker and the decision variables do not assume negative values. Thus, only non-negative levels of activity are considered feasible. This assumption is indeed a realistic one. For instance, in the production problems, the output cannot obviously be negative, because a negative production implies that we should be able to reverse the production process and convert the finished output back into the raw materials!

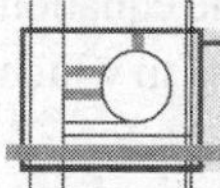

2.6 SOLUTION TO LINEAR PROGRAMMING PROBLEMS—GRAPHIC METHOD

Now we shall consider the solution to the linear programming problems. They can be solved by the graphic method or by applying the algebraic method, called the *Simplex Method.* The graphic approach is restricted in application—it can be used only when two variables are involved. Nevertheless, it provides an intuitive grasp of the concepts that are used in the simplex technique.

We shall discuss the graphic method here and the use of simplex algorithm shall be explained in the next chapter.

The Graphic Method To use the graphic method for solving linear programming problems, the following steps are required:

(a) Identify the problem—the decision variables, the objective function, and the constraint restrictions.

(b) Draw a graph that includes all the constraints/restrictions and identify the feasible region.

(c) Obtain the point on the feasible region that optimises the objective function—the optimal solution.

(d) Interpret the results.

We shall demonstrate the graphical approach first in respect of the maximisation and then for the minimisation problems.

1. The Maximisation Case

We shall consider Example 2.1 again. For this problem, the decision variables are x_1 and x_2, the number of units of the products *A* and *B* respectively. The objective function and the constraints are reproduced as,

Maximise	$Z = 40x_1 + 35x_2$	Profit
Subject to		
	$2x_1 + 3x_2 \leq 60$	Raw material constraint
	$4x_1 + 3x_2 \leq 96$	Labour hours constraint
	$x_1, x_2 \geq 0$	

Graphing the restrictions We shall first chart the given restrictions on the graph. The constraint of raw material availability, $2x_1 + 3x_2 \leq 60$, states that any combination of $2x_1 + 3x_2$ that does not exceed 60 is acceptable. With varying values of x_1 and x_2, the combination can assume a maximum value equal to 60. This constraint can be depicted graphically by plotting the straight line $2x_1 + 3x_2 = 60$. Since only two points are needed to obtain a straight line, we can get the two points by setting one of these variables in turn equal to zero, and calculating the value of the other. Thus, when $x_1 = 0$, x_2 would equal $60/3 = 20$, and when x_2 is set equal to 0, x_1 would be equal to $60/2 = 30$. Joining the points (0, 20) and (30, 0) we get the required straight line which is shown in Figure 2.1 as line PT. Any point (representing a combination of x_1 and x_2) that falls on this line or in the area below it, is acceptable insofar as this constraint is concerned. Since it is laid that both, x_1 and x_2, are non-negative, the triangle OPT formed by two axes and the line representing the equation $2x_1 + 3x_2 = 60$ is the region containing acceptable values of x_1 and x_2 in respect of this constraint.

Similarly, the other constraint $4x_1 + 3x_2 \leq 96$ can be plotted. The line JR in Figure 2.1 represents the equation $4x_1 + 3x_2 = 96$. The triangle formed by OJR, formed by the two axes and this line represents the area in which any point would satisfy this constraint of labour hours.

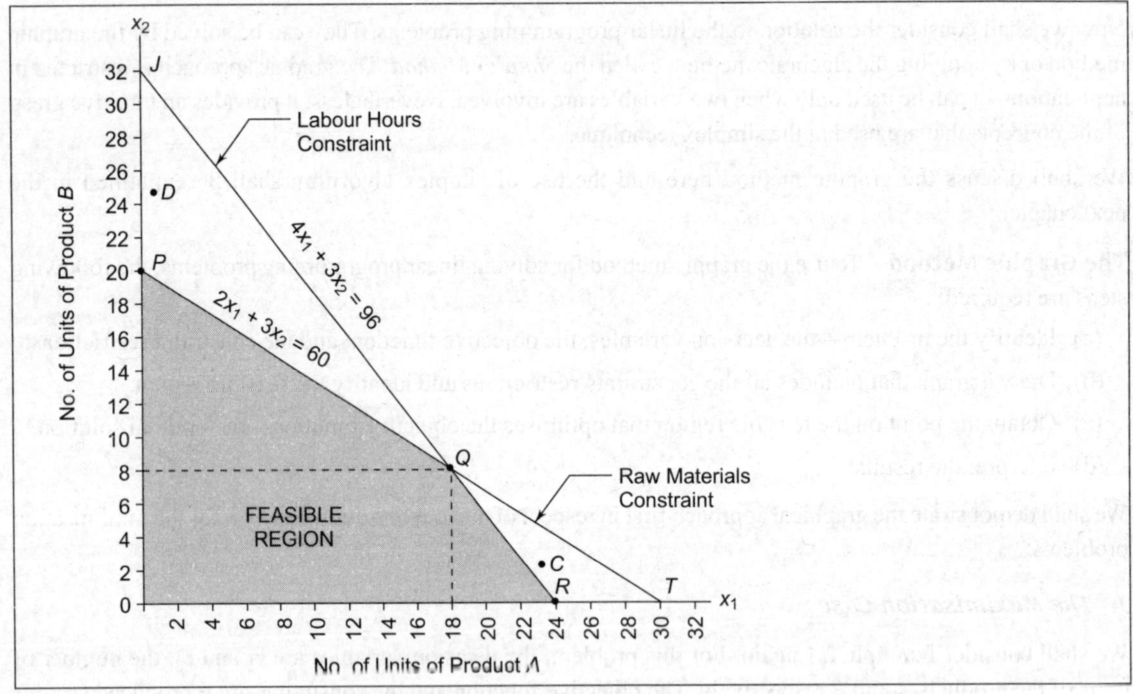

Figure 2.1 *Graphic Plot of Constraints*

Now, since both the constraints are to be satisfied, we have to consider the area on the graph that is bound by both of the constraints. A point like *C* on the graph would be unacceptable because only first of the constraints would be satisfied and not the other. Similarly, a point like *D* would also not be acceptable as it represents a combination of x_1 and x_2 which satisfies the second, and not the first, of the constraints. In fact, all points lying in the shaded area are the points of interest. They represent the combinations of x_1 and x_2 that would satisfy the given constraints. This region of acceptable values of the decision variables in relation to the given constraints (and the non-negativity restrictions) is called the *feasibility region.* In general terms, then, all inequalities of a linear programming problem, when plotted on the graph, define a space such that all points that lie within, or on the boundaries of that space simultaneously satisfy all the constraints, and the space is known as the *feasible region.* Thus, each point within this region would yield values of x_1 and x_2 as will satisfy *all* the constraints given in the problem. As such, corresponding to each point in the feasible region, we get a *feasible solution.* The feasible region of an LPP, therefore, gives a set of feasible solutions. Any point other than the one from this region will yield an *infeasible solution*—which would fail to satisfy one or more of the constraints.

Obtaining the optimal solution Now we shall see as to how the optimal solution to the problem can be obtained. Although all points in the feasible region represent feasible decision alternatives, they are not all equally attractive. Some provide a greater profit contribution than others. We have to select the best point from among the infinite number of points in the feasible region. In other words, we seek to obtain the feasible solution which optimises. It requires adding one more line to the graph, called an *iso-profit* or *constant-profit* line.

As its name implies, all the points on an iso-profit line yield the same profit. Suppose, for example, that we want to find an iso-profit line representing a profit of Rs 280. For this, we put $40x_1 + 35x_2 = 280$, and plot the line on the graph. We have $x_1 = 7$ when $x_2 = 0$, and $x_2 = 8$ when $x_1 = 0$. Joining the points (7, 0) and (0, 8), we get the iso-profit line *EF* in Figure 2.2, which is a reproduction of Figure 2.1. All points on this line, representing the various combinations of the variables x_1 and x_2 shall yield a profit of Rs 280.

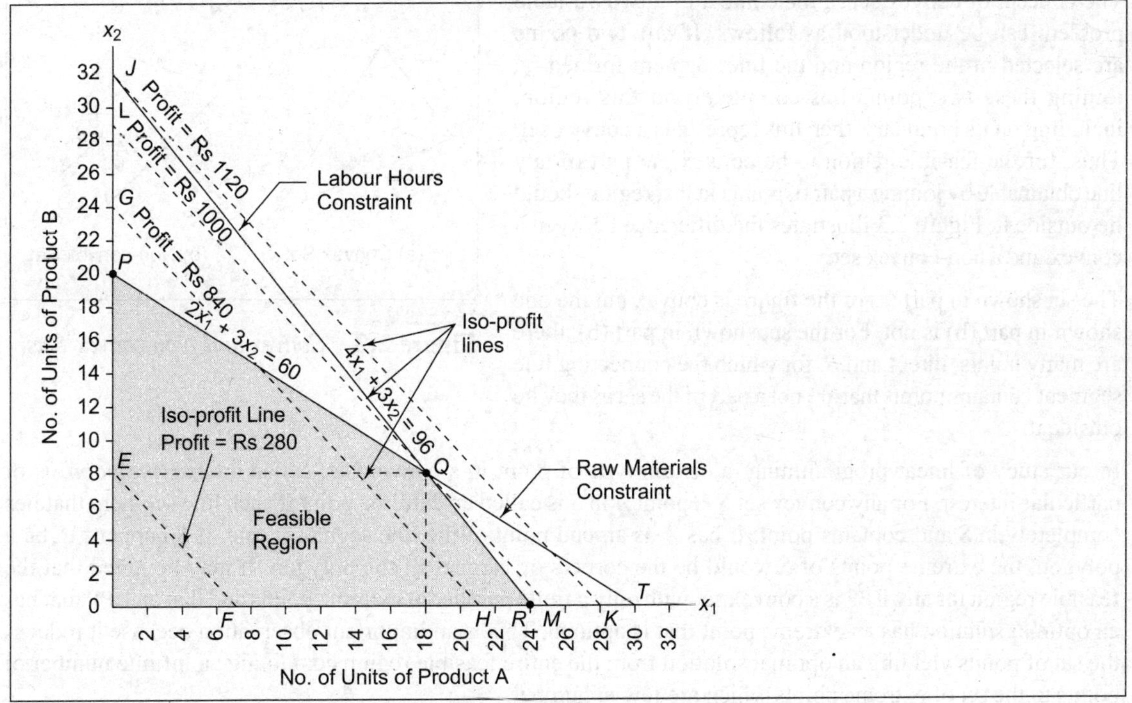

Figure 2.2 *Determination of Optimal Solution Using Iso-profit Line*

It may be observed that all the points that fall on this line (in the first quadrant of course) lie in the feasible region. So it is clearly possible for the firm to realise a profit of Rs 280. In fact, even greater profit can be realised by moving to other iso-profit lines, corresponding to higher profit values. Look at the Rs 840 iso-profit line, *GH.* Some of the points on this line fall outside the feasible region and, therefore, do not provide legitimate alternatives. However, other parts of this line fall in the feasible region. Therefore, a profit of Rs 840 is attainable. The iso-profit line, *JK* corresponding to the profit of Rs 1,120 lies beyond the feasible region, indicating that this profit cannot be attained.

Notice that the iso-profit lines are all parallel to each other and the farther are the lines removed from the origin of the graph, the greater is their contribution. Since all the lines have the same slope, the final step in the analysis is to continue constructing iso-profit lines that are successively farther away from the origin. This process would stop when a movement away from the origin would cause the line to lie beyond the feasible region. For example, the line that would yield the maximum profit is *LM* that passes through the point *Q.* This point gives a maximum contribution and thus represents the optimal solution. This decision is for the production of 18 units of product *A* and 8 units of product *B,* for a total profit of Rs 1000.

The production of 18 units of *A* and 8 units of *B* will consume $2 \times 18 + 3 \times 8 = 60$ kg of raw material and $4 \times 18 + 3 \times 8 = 96$ labour hours. Hence, both the resources shall be fully utilised by this production plan.

Convex Sets and LPPs Observe the feasible region from Figure 2.1 again. It is seen to be formed by a four-sided polygon, represented by *OPQR.* Besides the two sides provided by non-negativity restrictions, each of the other two sides of the polygon is provided by a constraint. It may be observed that the feasible region determined by the constraints of the given system is a *Convex Set.*

The concept of convex set in the context of a two-variable problem can be understood as follows. If *any* two points are selected in the region and the line segment formed by joining these two points lies completely in this region, including on its boundary, then this represents a convex set. Thus, for the feasible region to be convex, no part of any line obtainable by joining a pair of points in that region should lie outside it. Figure 2.3 illustrates the difference between a convex and a non-convex set.

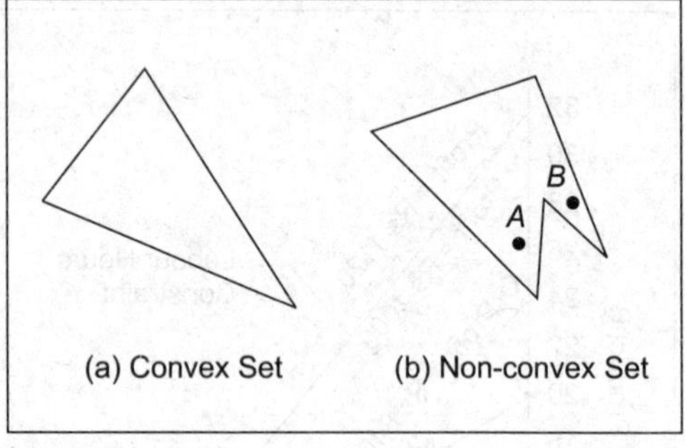

Figure 2.3 *Convex and Non-convex Sets*

The set shown in part (a) of the figure is convex but the one shown in part (b) is not. For the set shown in part (b), there are many points, like *A* and *B,* for which the connecting line segment contains points that are not a part of the set as they lie outside it.

In our study of linear programming, a certain type of point in a convex set, called an *extreme point,* is of particular interest. For any convex set *S*, a point *A* in *S* is called an extreme point if each line segment that lies completely in *S* and contains point *A,* has *A* as an end point of the line segment. And, if *S* happens to be a polygon, the extreme points of *S* would be the corners or vertices of the polygon. It may be noted that the feasible region for any LPP is a convex set with only a finite number of extreme points and that an LPP that has an optimal solution has an extreme point that is optimal. This is an important observation because it reduces the set of points yielding an optimal solution from the entire feasible region containing an infinite number of points to the set of extreme points which are few in number.

Thus, our task is simplified as the optimal solution to an LPP shall be given by some corner or extreme point of the feasible region. Accordingly, in lieu of drawing an iso-profit line, we may proceed to evaluate the extreme points only to obtain the optimal solution.

In our example, the feasible region is given by the polygon *OPQR*, so we shall obtain the ordinates of each of these points (giving the values of x_1 and x_2) and calculate the objective function value by substituting them into the objective function. The values of x_1 and x_2 at each point can be read directly from the graph. Also, in respect of those points which lie on the intersection of two (or more) lines, we can get the values by solving simultaneously the equations representing those lines. The ordinates at point *Q*, for example, can be determined by solving the following pair of equations simultaneously:

$$2x_1 + 3x_2 = 60$$

$$4x_1 + 3x_2 = 96$$

This gives $x_1 = 18$ and $x_2 = 8$.

The *Z*-values corresponding to the various points in respect of the given problem are shown here.

Point	x_1	x_2	$Z = 40x_1 + 35x_2$	
O	0	0	0	
P	0	20	700	
Q	18	8	1000	←Maximum
R	24	0	960	

Since *Z* is highest at point *Q*, the optimal solution is to produce 18 units of product *A* and 8 units of product *B* every week, to get the profit of Rs 1000. No other product mix under the given conditions could yield more profit than this.

2. The Minimisation Case

Now we shall consider the graphical solution to the linear programming problems of the minimisation nature. Here, Example 2.2 is reconsidered.

Minimise $Z = 40x_1 + 24x_2$ Total Cost

Subject to

$20x_1 + 50x_2 \geq 4,800$ Phosphate Requirement

$80x_1 + 50x_2 \geq 7,200$ Nitrogen Requirement

$$x_1, x_2 \geq 0$$

Here the decision variables x_1 and x_2 represent, respectively, the number of bags of mixture *A* and of mixture *B*, to be bought.

Graphing the restrictions The constraints are plotted in Figure 2.4. The first constraint of phosphate requirement, $20x_1 + 50x_2 \geq 4800$ can be represented as follows. We set $20x_1 + 50x_2 = 4800$. Putting $x_1 = 0$, we get $x_2 = 96$ and putting $x_2 = 0$, we have $x_1 = 240$. Joining the two points (0, 96) and (240, 0), we get the straight line corresponding to the above equation. The area beyond this line represents the feasible area in respect of this constraint—any point on the straight line or in the region above this line would satisfy the constraint.

Similarly, the second constraint can be depicted by plotting the straight line corresponding to the equation $80x_1 + 50x_2 = 7200$. Line *PT* in the figure represents the equation. The points falling on this line and in the area beyond it indicate the feasible values of x_1 and x_2 in respect of this constraint.

Since both the requirements are to be met, the feasible region in respect of the problem is as represented by the shaded area.

The feasible region here represents a convex set. However, it is not bounded from all the sides, as was in case of the maximisation problem. The region is unbounded on the upper side because none of the restrictions in the problem places an upper limit on the value of either of the decision variables. Obviously, if such limits are placed, the feasible region would be a bounded one.

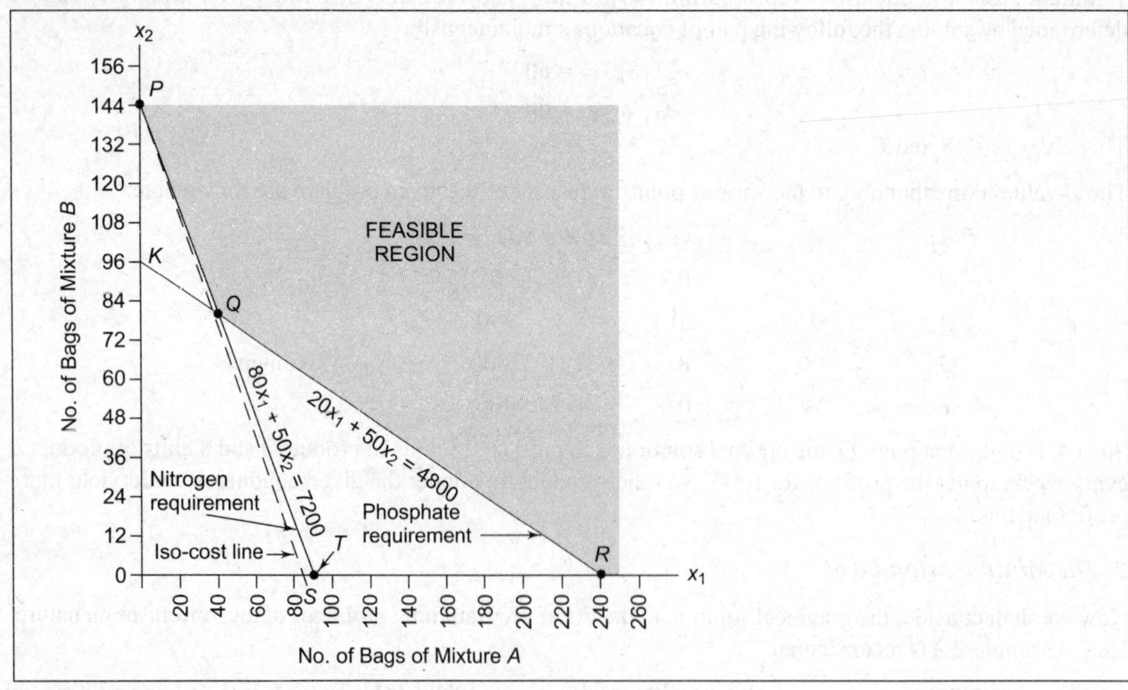

Figure 2.4 *Graphic Solution to the LPP*

Obtaining the optimal solution As in case of maximisation problems, the optimal solution can also be obtained by plotting the objective function on the graph in the form of *iso-cost lines.* For drawing an iso-cost line, we assume some cost value, equate the objective function with it, and plot the straight line corresponding to that equation on the graph. Obviously, each point on an iso-cost line would imply the same cost value.

Since the total cost is to be minimised, the iso-cost line touching the feasible region, which is closest to the origin shall be the line that would provide optimal solution. In our example, the iso-cost line *PS,* which entails a cost of Rs 3,456, is such line. Since it touches point *P*, the ordinates of this point (0, 144) provide the optimal solution.

With the optimal solution of 144 bags of mixture *B* only, the supply of phosphate would be $20 \times 0 + 50 \times 144 = 7{,}200$ kg, which is in excess of the minimum requirement of 4,800 kg. On the other hand, the nitrogen supply would equal $80 \times 0 + 50 \times 144 = 7{,}200$ kg, the required minimum.

Alternately, the optimal solution to the problem may be seen to be located at an extreme point—at point *P*, *Q*, or *R* in our example. We can evaluate the ordinates at each of these points as follows:

Point	x_1	x_2	$Z = 40x_1 + 24x_2$	
P	0	144	3,456	←Minimum
Q	40	80	3,520	
R	240	0	9,600	

Since the total cost is minimum at point *P*, the optimal solution to the problem is to buy 144 bags of mixture *B* only and none of mixture *A*. This would entail a total cost of Rs 3,456.

Binding and Non-binding Constraints

Once the optimal solution to an LPP is obtained, we may classify the constraints as being *binding* or *non-binding.* A constraint is termed as binding if the left-hand side and right-hand side of it are equal when optimal values of the decision variables are substituted into the constraint. On the other hand, if the substitution of the decision variables does not lead to an equality between the left- and the right-hand sides of the constraint, it is said to be non-binding. In Example 2.1, the optimal values of decision variables are: $x_1 = 18$ and $x_2 = 8$. Substituting these values in the two constraints, we get

$$2x_1 + 3x_2 \quad \text{or} \quad 2 \times 18 + 3 \times 8 = 60 = \text{RHS, and}$$

$$4x_1 + 3x_2 \quad \text{or} \quad 4 \times 18 + 3 \times 8 = 96 = \text{RHS.}$$

Thus, both the constraints are binding in nature. In Example 2.2, on the other hand, $x_1 = 0$ and $x_2 = 144$, the optimal values, may be substituted in the two constraints to get

$$20x_1 + 50x_2 \quad \text{or} \quad 20 \times 0 + 50 \times 144 = 7{,}200 \neq 4{,}800 \text{ (RHS), and}$$

$$80x_1 + 50x_2 \quad \text{or} \quad 80 \times 0 + 50 \times 144 = 7{,}200 = \text{RHS.}$$

Accordingly, the first of thc constraints is non-binding and the second one is a binding one.

Redundant Constraint(s)

As we have observed, plotting of each of the constraints on the graph serves to determine the feasible region of a given problem. If and when a constraint, when plotted, does not form part of the boundary marking the feasible region of the problem, it is said to be *redundant.* The inclusion or exclusion of a redundant constraint obviously does not affect the optimal solution to the problem.

Example 2.3 Continuing with Example 2.1, suppose that each of the products are required to be packed. Every unit of product *A* requires 4 hours while every unit of product *B* needs 3.5 hours for packaging. Suppose that in the packaging department, 105 hours are available every week.

Under these conditions, what product mix would maximise the profit?

The problem can be restated as follows:

Maximise	$Z = 40x_1 + 35x_2$	Profit
Subject		
	$2x_1 + 3x_2 \leq 60$	Raw material constraint
	$4x_1 + 3x_2 \leq 96$	Labour hours constraint
	$4x_1 + 3.5x_2 \leq 105$	Packaging hours constraint
	$x_1, x_2 \geq 0$	

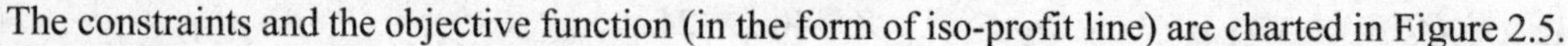

The constraints and the objective function (in the form of iso-profit line) are charted in Figure 2.5.

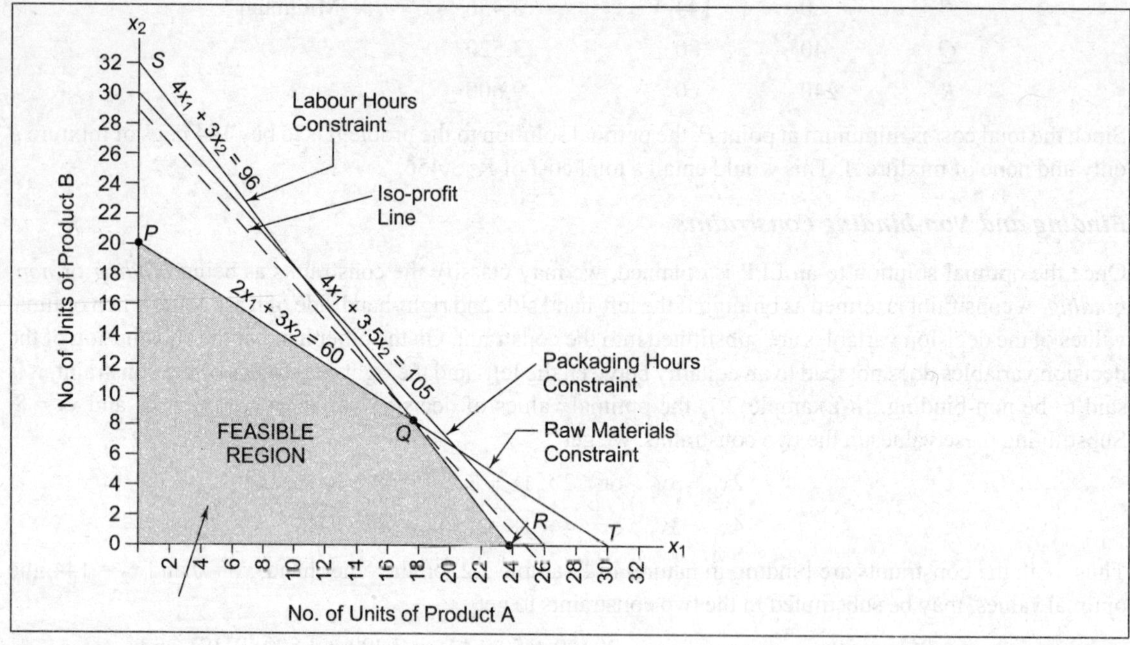

Figure 2.5 *Graphic Solution to LPP*

We observe that the inclusion of the packaging hours constraint does not provide a side of the polygon representing the feasible region. Thus, this constraint is of no consequence and hence is redundant. The optimal solution to the problem is the same as the optimal solution to the problem in Example 2.1. Thus, we have $x_1 = 18, x_2 = 8$ representing the optimal solution. The product mix given by this shall consume the entire 60 kg of raw material and 96 labour hours, but shall leave an unused capacity of $105 - (4 \times 18 + 3.5 \times 8) = 5$ hours in the packaging department.

2.7 SOME SPECIAL CASES

In each of the three examples discussed so far, we have obtained a unique optimal solution. We now consider three types of linear programming problems which do not have unique optimal solutions. These are problems having multiple optimal solutions, having no feasible solution, and having unbounded solutions.

1. Multiple Optimal Solutions As stated above, in each of the three examples that we have considered, we have observed that the optimal solution is given by an extreme point of the feasible region and the solution is unique. The uniqueness implies that no other solution to the given problem shall yield the same value of the objective function as given by that solution. It is, of course, possible that in a given problem there may be more than one optimal solution. Each of the multiple optima would naturally yield the same objective function value.

The solution (if it exists*) to a linear programming problem shall always be unique if the slope of the objective function (represented by the iso-profit lines) is different from the slopes of the constraints. In case the objective function has slope which is same as that of a constraint, then multiple optimal solutions might exist. There are two conditions that should be satisfied in order that multiple optimal solutions exist.

(a) The objective function should be parallel to a constraint that forms an edge or boundary on the feasible region; and

(b) The constraint should form a boundary on the feasible region in the direction of optimal movement of the objective function. In other words, the constraint must be a binding constraint.

To fully understand, let us consider the following examples.

Example 2.4 Solve graphically the following LPP:

Maximise $Z = 8x_1 + 16x_2$

Subject to

$$x_1 + x_2 \le 200$$
$$x_2 \le 125$$
$$3x_1 + 6x_2 \le 900$$
$$x_1, x_2 \ge 0$$

The constraints are shown plotted on the graph in Figure 2.6. Also, iso-profit lines have been graphed.

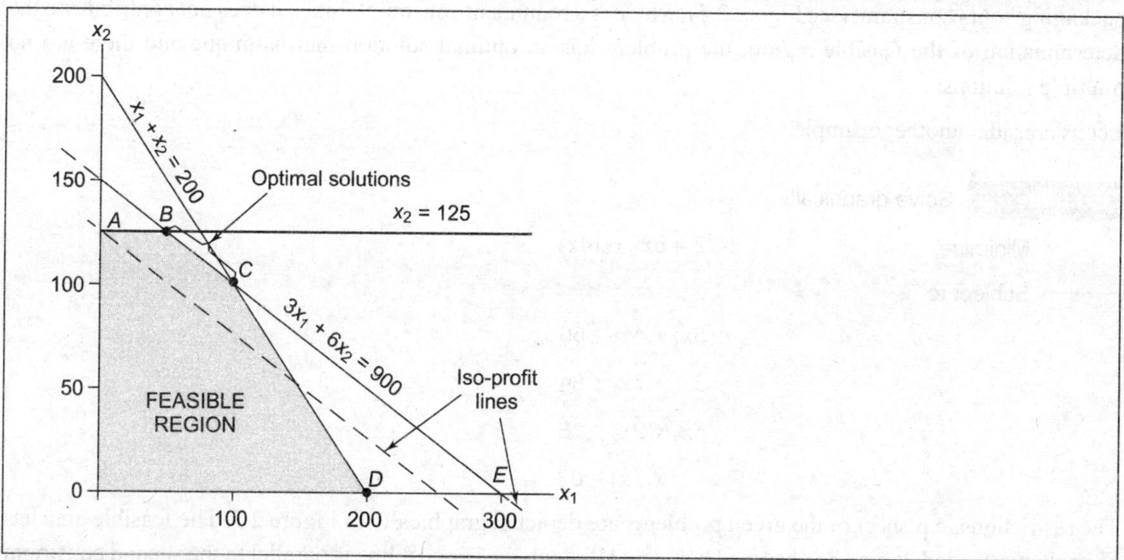

Figure 2.6 *Graphic Solution—Multiple Optima*

We observe that iso-profit lines are parallel to the equation for third constraint $3x_1 + 6x_2 = 900$. As we move the iso-profit line farther from the origin, it coincides with the portion *BC* of the constraint line that forms the

* See *Infeasibility*.

boundary of the feasible region. It implies that there are an infinite number of optimal solutions represented by all points lying on the line segment *BC*, including the extreme points represented by *B* (50, 125) and *C* (100, 100). Since the extreme points are also included in the solutions, we may disregard all other solutions and consider only these, to establish that the solution to a linear programming problem shall always lie at an extreme point of the feasible region.

The extreme points of the feasible region are given and evaluated here.

Point	x_1	x_2	$Z = 8x_1 + 16x_2$	
0	0	0	0	
A	0	125	2,000	
B	50	125	2,400	←Maximum
C	100	100	2,400	
D	200	0	1,600	

The points *B* and *C* clearly represent the optima.

In this example, the constraint to which the objective function was parallel, was the one which formed a side of the boundary of the space of the feasible region. As mentioned in condition (a), if such a constraint (to which the objective function is parallel) does not form an edge or boundary of the feasible region, then multiple solutions would not exist. In Example 2.3 for instance, the objective function has the same slope as that of the packaging hours constraint (see Figure 2.5), which is a redundant constraint. Since it does not contribute to the determination of the feasible region, the problem has an optimal solution that is unique and there are no multiple solutions.

Let us consider another example.

Example 2.5 Solve graphically:

Minimise $Z = 6x_1 + 14x_2$

Subject to

$$5x_1 + 4x_2 \geq 60$$

$$3x_1 + 7x_2 \leq 84$$

$$x_1 + 2x_2 \geq 18$$

$$x_1, x_2 \geq 0$$

The restrictions in respect of the given problems are depicted graphically in Figure 2.7. The feasible area has been shown shaded. It may be observed here that although the iso-cost line is parallel to the second constraint line represented by $3x_1 + 7x_2 = 84$, and this constraint does provide a side of the area of feasible solutions, yet the problem has a unique optimal solution, given by the point *D*. Here condition (*b*) mentioned earlier, is not satisfied. This is because, being a minimisation problem, the optimal movement of the objective function would be towards the origin and the constraint forms a boundary on the opposite side. Since the constraint is not a binding one, the problem does not have multiple optima.

We can show the uniqueness of the solution by evaluating various extreme points as done here.

Point	x_1	x_2	$Z = 6x_1 + 14x_2$	
A	8	5	118	
B	84/23	240/23	168	
C	28	0	148	
D	18	0	108	←Minimum

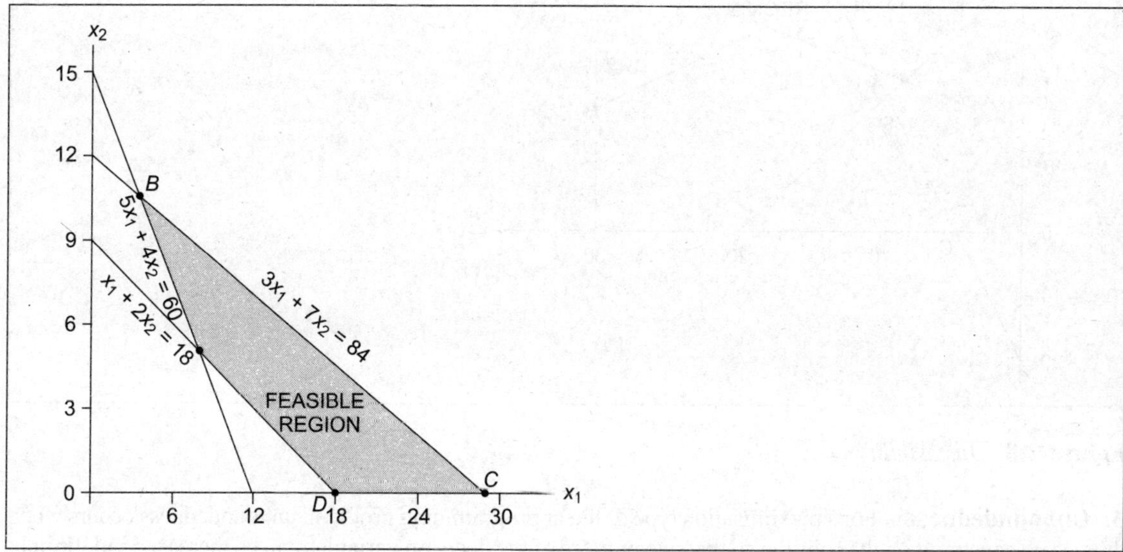

Figure 2.7 *Graphic Solution to LPP*

2. Infeasibility It has already been stated that a solution is called *feasible* if it satisfies all the constraints and the non-negativity conditions. Sometimes it is possible that the constraints may be inconsistent so that there is no feasible solution to the problem. Such a situation is called *infeasibility.*

In the graphic approach to the solution to an LPP, the infeasibility is evident if its feasible region is empty so that there is no feasible region in which all the constraints may be satisfied simultaneously. Consider the following example.

Example 2.6 Maximise $Z = 20x_1 + 30x_2$

Subject to

$$2x_1 + x_2 \leq 40$$
$$4x_1 - x_2 \leq 20$$
$$x_1 \geq 30$$
$$x_1, x_2 \geq 0$$

It is represented graphically in Figure 2.8. The feasible region corresponding to the first two constraints is bound by the convex set *OABC*, while the feasible region in respect of the third constraint is also shown shaded separately. We can easily observe that there is no common point in the areas shaded.

Therefore, all the constraints cannot be satisfied and, as such, there is no feasible solution to the given problem.

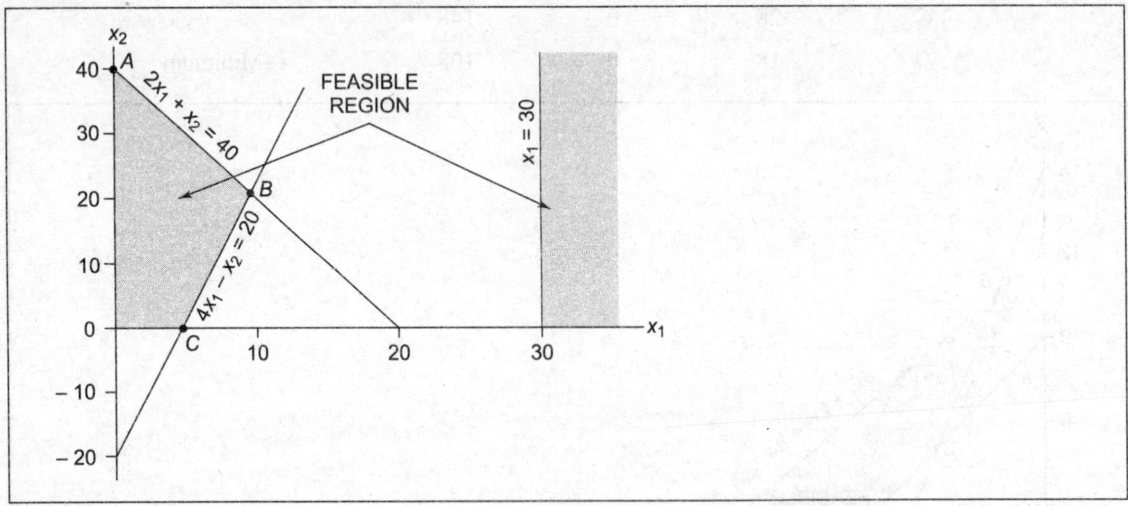

Figure 2.8 *Infeasibility*

3. Unboundedness For a maximisation type of linear programming problem, unboundedness occurs when there is no constraint on the solution so that one or more of the decision variables can be increased indefinitely without violating any of the restrictions (constraints). Thus, an unbounded LPP occurs if it is possible to find points in the feasible region with arbitrarily large *Z*-values (may be profit or revenue). This suggests that practically if we find the solution to be unbounded for a profit-maximising linear programming problem, it may be concluded that the problem has not been correctly formulated.

Consider the following example.

Example 2.7 Maximise $Z = 10x_1 + 20x_2$

Subject to

$$2x_1 + 4x_2 \geq 16$$
$$x_1 + 5x_2 \geq 15$$
$$x_1, x_2 \geq 0$$

This is represented graphically in Figure 2.9.

Clearly, here the objective function is not bound over the feasible region and we can move the iso-profit line upward without any limit. The problem has, therefore, unbounded solution.

For a minimisation LPP, unboundedness occurs when there are points in the feasible region with arbitrarily small values. To conclude, a maximisation LPP is unbounded if, moving parallel to the original iso-profit line in the direction of increasing *Z*, we never entirely leave the feasible region and, on the other hand, a minimisation problem is unbounded if we never leave the feasible region while moving in the direction of decreasing *Z*.

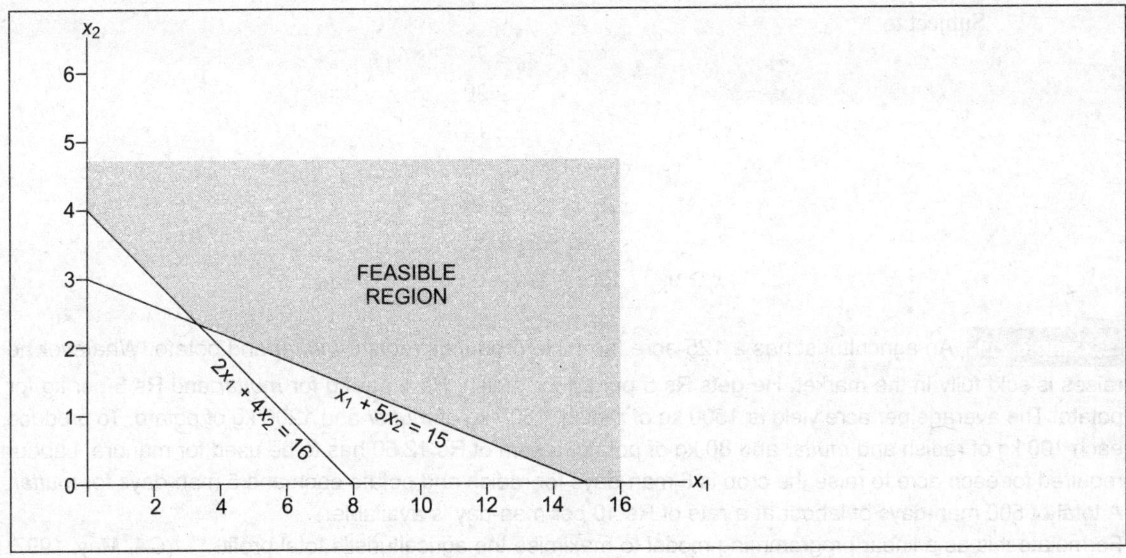

Figure 2.9 *Unboundedness*

2.7.1 Infeasibility vs Unboundedness

Both infeasibility and unboundedness have a similarity in that there is no optimal solution in either case. But there is a striking difference between the two: while in infeasibility there is not a single feasible solution, in unboundedness there are infinite feasible solutions but none of them can be termed as the optimal.

Review Illustrations

Example 2.8 A 24-hour supermarket has the following minimal requirements for cashiers:

Period	:	1	2	3	4	5	6
Time of day (24-hour clock)	:	3–7	7–11	11–15	15–19	19–23	23–03
Minimum number required	:	7	20	14	20	10	5

Period 1 follows immediately after period 6. A cashier works eight consecutive hours, starting at the beginning of one of the six time periods. Determine a daily employee worksheet which satisfies the requirements with the least number of personnel. Formulate the problem as a linear programming problem.

(M.B.A., Delhi, November, 1996)

Let x_1, x_2, x_3, x_4, x_5 and x_6 be the number of cashiers joining at the beginning of periods 1, 2, 3, 4, 5 and 6, respectively. Using the information given, the LPP may be expressed as follows:

Minimise $Z = x_1 + x_2 + x_3 + x_4 + x_5 + x_6$

Subject to

$$\begin{aligned}
x_1 \quad\quad\quad\quad\quad\quad + x_6 &\geq 7 \\
x_1 + x_2 \quad\quad\quad\quad &\geq 20 \\
x_2 + x_3 \quad\quad\quad &\geq 14 \\
x_3 + x_4 \quad\quad &\geq 20 \\
x_4 + x_5 \quad &\geq 10 \\
x_5 + x_6 &\geq 5 \\
x_i \geq 0, \quad \text{for } i = 1, 2, \ldots, 6
\end{aligned}$$

Example 2.9 An agriculturist has a 125-acre farm. He produces radish, *muttar* and potato. Whatever he raises is sold fully in the market. He gets Rs 5 per kg for radish, Rs 4 per kg for *muttar* and Rs 5 per kg for potato. The average per acre yield is 1500 kg of radish, 1800 kg of *muttar* and 1200 kg of potato. To produce each 100 kg of radish and *muttar* and 80 kg of potato, a sum of Rs 12.50 has to be used for manure. Labour required for each acre to raise the crop is 6 man-days for radish and potato each and 5 man-days for *muttar*. A total of 500 man-days of labour at a rate of Rs 40 per man-day is available.
Formulate this as a linear programming model to maximise the agriculturist's total profit. *(CA, May, 1997)*

Let x_1, x_2 and x_3 be the acrage for radish, *muttar* and potato, respectively. From the given information,

Output:
Output of radish = $1{,}500x_1$ kg
Output of *muttar* = $1{,}800x_2$ kg
Output of potato = $1{,}200x_3$ kg

Cost of Manure:
Radish : Rs 12.50 per 100 kg
Muttar : Rs 12.50 per 100 kg
Potato : Rs 12.50 per 80 kg

Accordingly,

$$\text{Total cost of manure} = \frac{12.50}{100} \times 1{,}500x_1 + \frac{12.50}{100} \times 1{,}800x_2 + \frac{12.50}{80} \times 1{,}200x_3$$

or $$= 187.5x_1 + 225x_2 + 187.5x_3$$

Labour cost:
Radish : $6x_1 \times 40 = 240x_1$
Muttar : $5x_2 \times 40 = 200x_2$
Potato : $6x_3 \times 40 = 240x_3$

Now,

$$\begin{aligned}
\text{Total Profit} &= \text{Revenue} - (\text{Cost of Manure} + \text{Labour Cost}) \\
&= 5 \times 1{,}500x_1 + 4 \times 1{,}800x_2 + 5 \times 1{,}200x_3 - (187.5x_1 + 225x_2 + 187.5x_3 + 240x_1 + 200x_2 + 240x_3) \\
&= 7072.5x_1 + 6775x_2 + 5572.5x_3
\end{aligned}$$

Now, with constraints on land availability and man-days availability, the LPP may be expressed as follows:

Maximise $$Z = 7{,}072.5x_1 + 6{,}775.0x_2 + 5{,}572.5x_3$$

Subject to

$$x_1 + x_2 + x_3 \leq 125 \quad \text{(Land availability)}$$
$$6x_1 + 5x_2 + 6x_3 \leq 500 \quad \text{(Man-days availability)}$$
$$x_1, x_2, x_3 \geq 0$$

Example 2.10 A manufacturing firm needs 5 component parts. Due to inadequate resources, the firm is unable to manufacture all its requirements. So the management is interested in determining as to how many, if any, units of each component should be purchased from outside and how many should be produced internally. The relevant data are given here.

Component	*M*	*A*	*T*	*TR*	*PP*	*PC*
C_1	4	1	1.5	20	48	30
C_2	3	3	2	50	80	52
C_3	1	1	0	45	24	18
C_4	3	1	0.5	70	42	31
C_5	2	0	0.5	40	28	16

M: Per unit milling time in hours

A: Per unit assembly time in hours

T: Per unit testing time in hours

TR: Total requirement in units

PP: Price per unit quoted in the market

PC: Per unit direct costs (including materials, labour, etc.)

Resources available are as follows:

Milling hours	:	300
Assembly hours	:	160
Testing hours	:	150

Formulate this as an LPP, taking the objective function as maximisation of saving by producing the components internally.

Let x_1, x_2, x_3, x_4, and x_5 represent the number of C_1, C_2, C_3, C_4, and C_5, respectively, produced internally.

Information Summary

Resources/ Constraints	*Component* C_1	C_2	C_3	C_4	C_5	*Total Availability*
Milling hrs (per unit)	4	3	1	3	2	300
Assembly hrs (per unit)	1	3	1	1	0	160
Testing hrs (per unit)	1.5	2	0	0.5	0.5	150
Requirement (units)	20	50	45	70	40	
Saving (per unit) (*PP* – *PC*)	18	28	6	11	12	

The problem would be:

Maximise $Z = 18x_1 + 28x_2 + 6x_3 + 11x_4 + 12x_5$

Subject to

$$\left.\begin{array}{r} 4x_1 + 3x_2 + x_3 + 3x_4 + 2x_5 \le 300 \\ x_1 + 3x_2 + x_3 + x_4 \qquad\quad \le 160 \\ 1.5x_1 + 2x_2 + 0.5x_4 + 0.5x_5 \le 150 \end{array}\right\} \text{Capacity constraints}$$

$$\left.\begin{array}{r} x_1 \le 20 \\ x_2 \le 50 \\ x_3 \le 45 \\ x_4 \le 70 \\ x_5 \le 40 \end{array}\right\} \text{'Total requirements' constraints}$$

$$x_1, x_2, x_3, x_4, x_5 \ge 0 \qquad \text{Non-negativity condition}$$

Example 2.11 A firm produces three products *A*, *B* and *C*. It uses two types of raw materials I and II of which 5,000 and 7,500 units, respectively, are available. The raw material requirements per unit of the products are given below:

Raw Material	*Requirement per Unit of Product*		
	A	B	C
I	3	4	5
II	5	3	5

The labour time for each unit of product *A* is twice as that of product *B* and three times that of product *C*. The entire labour force of the firm can produce the equivalent of 3,000 units. The minimum demand for the three products is 600, 650 and 500 units respectively. Also, the ratio of the number of units produced must be equal to 2 : 3 : 4. Assuming the profits per unit of *A*, *B* and *C* are 50, 50 and 80, respectively, formulate the problem as a linear programming problem in order to determine the number of units of each product which will maximise the profit. *(CA, November, 1997)*

Let x_1, x_2 and x_3 be the output of products *A*, *B* and *C*, respectively. The LPP, using the given information, may be expressed as follows:

Maximise $Z = 50x_1 + 50x_2 + 80x_3$ Total Profit

Subject to

$$\begin{array}{rl} 3x_1 + 4x_2 + 5x_3 \le 5{,}000 & \text{Raw Material I} \\ 5x_1 + 3x_2 + 5x_3 \le 7{,}500 & \text{Raw Material II} \end{array}$$

$$\left.\begin{array}{r} x_1 \ge 600 \\ x_2 > 650 \\ x_3 \ge 500 \end{array}\right\} \text{Minimum demand}$$

$$x_1 + \frac{1}{2}x_2 + \frac{1}{3}x_3 \leq 3{,}000 \qquad \text{Labour}$$

$$\frac{x_1}{2} = \frac{x_2}{3} = \frac{x_3}{4} \qquad \text{Output proportionality}$$

$$x_1, x_2, x_3 \geq 0$$

Notes

1. In the absence of a statement to the contrary, it is assumed that availability of labour is sufficient to produce equivalent to 3,000 units of product *A*.
2. In a given case, if the output of two products *A* and *B*, is desired to be in a given ratio, say 2 : 3, we may express it as $3A = 2B$ or $A/2 = B/3$. For example, if the outputs are 10 and 15 (in the ratio 2 : 3), then we have $3 \times 10 = 2 \times 15$ or $10/2 = 15/3$. Accordingly, when three products have to be in the ratio 2 : 3 : 4, we write $x_1/2 = x_2/3 = x_3/4$.

Example 2.12 The Marketing Department of Everest Company has collected information on the problem of advertising for its products. This relates to the advertising media available, the number of families expected to be reached with each alternative, cost per advertisement, the maximum availability of each medium and the expected exposure of each one (measured as the relative value of one advertisement in each of the media): The information is as given here:

Advertising Media	*No. of Families Expected to Cover*	*Cost per Ad (Rs)*	*Maximum Availability (No. of times)*	*Expected Exposure (Units)*
TV (30 sec)	3,000	8,000	8	80
Radio (15 sec)	7,000	3,000	30	20
Sunday edition of a daily (1/4 page)	5,000	4,000	4	50
Magazine (1 page)	2,000	3,000	2	60

Other information and requirements:

(a) The advertising budget is Rs 70,000.

(b) At least 40,000 families should be covered. (The families receiving messages could be common. But a family receiving three messages, for example, would be taken to be equivalent to three.)

(c) At least 2 insertions be given in Sunday edition of a daily but not more than 4 ads should be given on the TV.

Draft this as a linear programming problem. The company's objective is to maximise the expected exposure.

Let x_1 : the number of ads on TV,

x_2 : the number of ads on radio,

x_3 : the number of ads in Sunday edition of a daily, and

x_4 : the number of ads in a magazine.

Information Summary

Constraints/ Resources	*Advertising Media*				*Total*
	TV	*Radio*	*Daily*	*Magazine*	
No. of families (per ad)	3,000	7,000	5,000	2,000	40,000
Cost (per ad) and budget (Rs)	8,000	3,000	4,000	3,000	70,000
Max. availability (No.)	8	30	4	2	
Other information:					
	(i) at least two ads in daily				
	(ii) at most four ads on TV				
Expected exposure (Units per ad)	80	20	50	60	

The problem would be as under:

Maximise $Z = 80x_1 + 20x_2 + 50x_3 + 60x_4$

Subject to

$$3{,}000x_1 + 7{,}000x_2 + 5{,}000x_3 + 2{,}000x_4 \geq 40{,}000$$
$$8{,}000x_1 + 3{,}000x_2 + 4{,}000x_3 + 3{,}000x_4 \leq 70{,}000$$
$$x_1 \leq 8$$
$$x_2 \leq 30$$
$$x_3 \leq 4$$
$$x_4 \leq 2$$
$$x_3 \geq 2$$
$$x_1 \leq 4$$
$$x_1, x_2, x_3, x_4 \geq 0$$

Example 2.13 A multinational company has two factories that ship to three regional warehouses. The costs of transportation per unit are:

Warehouse	*Transportation Costs (Rs)*	
	Factory	
	F_1	F_2
W_1	2	4
W_2	2	2
W_3	5	3

Factory F_2 is old and has a variable manufacturing cost of Rs 20 per unit. Factory F_1 is modern and produces for Rs 10 per unit. Factory F_2 has a monthly capacity of 250 units, and Factory F_1 has a monthly capacity of 400 units. The requirements at the warehouses are:

Warehouse	*Requirement*
W_1	200
W_2	100
W_3	250

How should each factory ship to each warehouse in order to minimise the total cost? Formulate this problem as a linear programming model. Do not solve it. *(MBA, Delhi, April, 1996)*

The total cost (manufacturing plus transportation) matrix is given below:

Factory	*Warehouse*			*Availability*
	W_1	W_2	W_3	
F_1	12	12	15	400
F_2	24	22	23	250
Requirement	200	100	250	550/650

Let x_{ij} be the quantity shipped from ith factory to jth warehouse. The linear programming problem is;

Minimise $$Z = 12x_{11} + 12x_{12} + 15x_{13} + 24x_{21} + 22x_{22} + 23x_{23}$$

Subject to

$$x_{11} + x_{12} + x_{13} \leq 400$$
$$x_{21} + x_{22} + x_{23} \leq 250$$
$$x_{11} + x_{21} = 200$$
$$x_{12} + x_{22} = 100$$
$$x_{13} + x_{23} = 250$$
$$x_{ij} \geq 0 \quad \text{for } i = 1, 2 \quad \text{and } j = 1, 2, 3$$

Example 2.14 The Orient Manufacturing Company produces three types of typewriters: Tik-Tik, Mik-Mik, and Pik-Pik. All the three types are required first to be machined and then to be assembled. The time requirements for the various types are as follows:

Type	*Machine Time* (in hours)	*Assembly Time* (in hours)
Tik-Tik	15	4.4
Mik-Mik	13	3.5
Pik-Pik	12	4.0

The total available machine time and assembly times are, respectively, 4,000 and 1,240 hours per month. The data regarding selling price, costs, and the contribution margin for the three are:

	Tik-Tik	*Mik-Mik*	*Pik-Pik*
Selling prices	Rs 11,000	Rs 5,000	Rs 3,000
Labour, material & other variable expenses	8,000	2,400	1,500
Contribution margin	3,000	2,600	1,500

The company sells all the three on one month credit basis, but labour, material and other variable expenses must be paid in cash. Some further information follows. The company has taken a loan of Rs 40,000 from a co-operative bank and has just been informed by the bank that the loan is not likely to be renewed when it expires at the beginning of the next month, viz., 1st January. The Good Bank of India, from whom the company has borrowed Rs 60,000, has expressed its willingness to renew the loan provided that the company maintains a quick ratio (i.e. the ratio of cash to the short term liabilities) of at least 1.

The balance sheet of the company as on 31st December is as follows:

Balance Sheet as on 31st December

Liabilities	*Rs*	*Assets*	*Rs*
Loan from Coop. Bank	40,000	Cash	200,000
Loan from Good Bank of India	60,000	Receivables	50,000
		Plant & Machinery	250,000
Long-term Debt	250,000		
Net Worth	150,000		
	500,000		500,000

It is further given that the company has to pay a fixed sum of Rs 10,000 each for interest on debt and for top management salaries every month.

Now, upon being informed about the co-operative bank's decision, the senior manager of the company requires this problem to be put as a linear programming problem. Assuming that (a) he wants to maximise the profits, and (b) he wants to maximise the total revenue from sales, do it for him.

Let x_1, x_2, x_3 be the number of units of Tik-Tik, Mik-Mik, and Pik-Pik produced.

Information Summary

Constraints/Resources	*Typewriter*			*Total*
	Tik-Tik	*Mik-Mik*	*Pik-Pik*	
Machine Hours	15/unit	13/unit	12/unit	4000
Assembly Hours	4.4/unit	3.5/unit	4.0/unit	1240
Cash (Rs) (for labour, materials and other various expenses)	8000	2400	1500	130,000*
Selling Price per unit (Rs)	11000	5000	3000	
Contribution Margin (Rs) per unit (Selling Price – Variable Exp.)	3000	2600	1500	

*Cash available is calculated as:

Cash Balance + Amt. from Receivables (to be received in January) – Co-operative Bank loan to be paid off – Cash required for debt interest and top management salaries – Cash required for matching the loan from Good Bank of India (which wants the quick ratio to be equal to 1)

$$= 200{,}000 + 50{,}000 - 40{,}000 - 20{,}000 - 60{,}000 = 130{,}000$$

(a) When the objective is to maximise profits:

Maximise $Z = 3{,}000x_1 + 2{,}600x_2 + 1{,}500x_3$

Subject to

$$\left.\begin{aligned} 15x_1 + 13x_2 + 12x_3 &\leq 4{,}000 \\ 4.4x_1 + 3.5x_2 + 4.0x_3 &\leq 1{,}240 \end{aligned}\right\} \text{Capacity constraints}$$

$$8000x_1 + 2400x_2 + 1500x_3 \leq 130{,}000 \quad \text{Cash constraint}$$

$$x_1, x_2, x_3 \geq 0 \quad \text{Non-negativity condition}$$

(b) When the objective is to maximise total revenue, the objective function would be:

Maximise $Z = 11000x_1 + 5000x_2 + 3000x_3$

while all other constraints would be the same as shown in (a).

Example 2.15 A leading Chartered Accountant is attempting to determine a "best" investment portfolio and is considering six alternative investment proposals. The following table indicates point estimates for the price per share, the annual growth rate in the price per share, the annual dividend per share and a measure of the risk associated with each investment.

Portfolio Data

	Shares under Consideration					
	A	B	C	D	E	F
Current Price per Share (Rs)	80	100	160	120	150	200
Projected Annual Growth Rate	0.08	0.07	0.10	0.12	0.09	0.15
Projected Annual Dividend per Share (Rs)	4.00	4.50	7.50	5.50	5.75	0.00
Projected Risk in Return	0.05	0.03	0.10	0.20	0.06	0.08

The total amount available for investment is Rs 25 lakhs and the following conditions are required to be satisfied.

(i) The maximum rupee amount to be invested in alternative F is Rs 250,000.

(ii) No more than Rs 500,000 should be invested in alternatives A and B combined.

(iii) Total weighted risk should not be greater than 0.10 where

$$\text{Total Weighted Risk} = \frac{(\text{Amount Invested in Alternative } j)\ (\text{Risk of Alternative } j)}{(\text{Total Amount Invested in all the Alternatives})}$$

(iv) For the sake of diversity, at least 100 shares of each stock should be purchased.
(v) At least 10 percent of the total investment should be in alternatives A and B combined.
(vi) Dividends for the year should be at least Rs 10,000.

Rupee return per share of stock is defined as price per share one year hence *less* current price per share *plus* dividend per share. If the objective is to maximise total rupee return, formulate the linear programming model for determining the optimal number of shares to be purchased in each of the shares under consideration. You may assume that the time horizon for the investment is one year. The formulated LP problem is not required to be solved.

(CA, November, 1991)

Let x_1, x_2, x_3, x_4, x_5 and x_6 be the number of shares of companies *A, B, C, D, E* and *F*, respectively. The rupee-return for various shares is shown as follows.

	Share					
	A	*B*	*C*	*D*	*E*	*F*
Projected growth per share (Rs)	6.40	7.00	16.00	14.40	13.50	30.00
Projected dividend	4.00	4.50	7.50	5.50	5.75	0.00
Rupee-return per share	10.40	11.50	23.50	19.90	19.25	30.00

Accordingly, the objective function is:

Maximise $Z = 10.40x_1 + 11.50x_2 + 23.50x_3 + 19.90x_4 + 19.25x_5 + 30.00x_6$

The constraints are as follows:

$$80x_1 + 100x_2 + 160x_3 + 120x_4 + 150x_5 + 200x_6 \le 25{,}00{,}000 \quad \text{funds availability}$$

$$200x_6 \le 250{,}000 \quad \text{condition (i)}$$

Total Weighted Risk

$$= \frac{(80x_1)(0.05) + (100x_2)(0.03) + (160x_3)(0.10) + (120x_4)(0.20) + (150x_5)(0.06) + (200x_6)(0.08)}{80x_1 + 100x_2 + 160x_3 + 120x_4 + 150x_5 + 200x_6} \le 0.10$$

On simplification, it gives

$$-4x_1 - 7x_2 + 12x_4 - 6x_5 - 4x_6 \le 0$$

From condition (iv), $x_i \ge 100$ for $i = 1, 2, 3, 4, 5, 6$

Condition (v) requires $80x_1 + 100x_2 \ge 0.10(80x_1 + 100x_2 + 160x_3 + 120x_4 + 150x_5 + 200x_6)$

This becomes, $72x_1 + 90x_2 - 16x_3 - 12x_4 - 15x_5 - 20x_6 \ge 0$

The dividend requirement states that

$$4x_1 + 4.50x_2 + 7.50x_3 + 5.50x_4 + 5.75x_5 \ge 10{,}000$$

Accordingly, the LPP may be stated as follows:

Maximise $Z = 10.40x_1 + 11.50x_2 + 23.50x_3 + 19.90x_4 + 19.25x_5 + 30.00x_6$

Subject to

$$80x_1 + 100x_2 + 160x_3 + 120x_4 + 150x_5 + 200x_6 \le 25{,}00{,}000$$

$$200x_6 \le 2{,}50{,}000$$

$$80x_1 + 100x_2 \le 5{,}00{,}000$$

$$-4x_1 - 7x_2 \qquad + 12x_4 - 6x_5 - 4x_6 \leq 0$$
$$72x_1 + 90x_2 - 16x_3 - 12x_4 - 15x_5 - 20x_6 \geq 0$$
$$4x_1 + 4.5x_2 + 7.5x_3 + 5.50x_4 + 5.75x_5 \geq 10{,}000$$
$$x_1 \geq 100$$
$$x_2 \geq 100$$
$$x_3 \geq 100$$
$$x_4 \geq 100$$
$$x_5 \geq 100$$
$$x_6 \geq 100$$

Example 2.16 A company engaged in producing tinned food delicacies has 300 trained employees on its rolls each of whom can produce one can of food in a week. Due to the developing taste of the public for this kind of food, the company plans to add the existing labour force by employing 150 people in a phased manner, over the next five weeks. The newcomers would have to undergo a two-week training programme before being put to work. The training is to be given by employees from amongst the existing ones and it is known that one employee can train three trainees. Assume that there would be no production coming forth from the trainers and the trainees during training period as the training is off-the-job. However, the trainees would be remunerated at the rate of Rs 300 per week, the same rate as is for the trainers.

The company has booked the following number of cans to supply during the next five weeks.

Week	:	1	2	3	4	5
No. of cans	:	280	298	305	360	400

Assume that the production in any week would not be more than the number of cans ordered for so that every delivery of the delicacy would be 'fresh'.

Using this information, draft a LPP to develop a training schedule that will minimise the labour costs over the five week period.

Let x_1, x_2, x_3, x_4, and x_5 be the number of trainees appointed in beginning of week 1, 2, 3, 4, and 5 respectively.

Information Summary

	Week				
	1	*2*	*3*	*4*	*5*
Cans Required	280	298	305	360	400
Other Information:					
(a) every trainee to be trained for 2 weeks					
(b) one employee required to train 3 trainees					
(c) every trained worker producing 1 can per week but no production from trainers and trainees during training					
(d) number of people to be employed = 150					
(e) the production in any week not to exceed the cans required					
No. of weeks for which newcomer would be employed*	5	4	3	2	1

*Thus, workers employed at the beginning of the first week would get salary for all 5 weeks under consideration, those employed at the beginning of the second week would get it for 4 weeks, and so on. The problem would be:

Minimise $Z = 5x_1 + 4x_2 + 3x_3 + 2x_4 + x_5$

Subject to

$$\left.\begin{array}{ll} 300 - \frac{1}{3}x_1 & \geq 280 \\ 300 - \frac{1}{3}x_1 - \frac{1}{3}x_2 & \geq 298 \\ 300 + x_1 - \frac{1}{3}x_2 - \frac{1}{3}x_3 & \geq 305 \\ 300 + x_1 + x_2 - \frac{1}{3}x_3 - \frac{1}{3}x_4 & \geq 360 \\ 300 + x_1 + x_2 + x_3 - \frac{1}{3}x_4 - \frac{1}{3}x_5 & \geq 400 \end{array}\right\} \text{Capacity constraints}$$

$$x_1 + x_2 + x_3 + x_4 + x_5 = 150 \quad \text{New employment condition}$$

$$x_1, x_2, x_3, x_4, x_5 \geq 0 \quad \text{Non-negativity condition}$$

Notes

(a) All along, the 300 employees, capable of producing 300 cans a week remain employed. The production capacity in different weeks (as shown on LHS of the constraints) is affected (negatively) by the fact that some of them are away for training and is affected (positively) by the fact that newcomers join the team of experienced workers.

(b) Inequalities have been used in the constraints because some workers might remain idle in some week (s).

(c) The training schedule as obtained by solving this problem would involve minimum cost in respect of the trainees. The cost would be obtained by multiplying the objective function by 300 because each person would get a salary of Rs 300 per week.

Example 2.17 Solve graphically:

Maximise $Z = 10x_1 + 15x_2$

Subject to

$$2x_1 + x_2 \leq 26$$

$$2x_1 + 4x_2 \leq 56$$

$$x_1 - x_2 \geq -5$$

$$x_1, x_2 \geq 0$$

This problem is exhibited graphically in Fig. 2.10. The feasible region is shaded. This is significant to note here that we have not considered the area below the line corresponding to $x_1 - x_2 = -5$ for the negative values of x_1. This is because of the non negativity condition $x_1 \geq 0$, which implies that negative values of x_1 should not be considered.

The Z-values at different points are given here:

Point	x_1	x_2	$Z = 10x_1 + 15x_2$
A	0	5	75
B	6	11	225
C	8	10	230
D	13	0	130

Thus, Z is maximum (it is equal to 230) when $x_1 = 8$ and $x_2 = 10$.

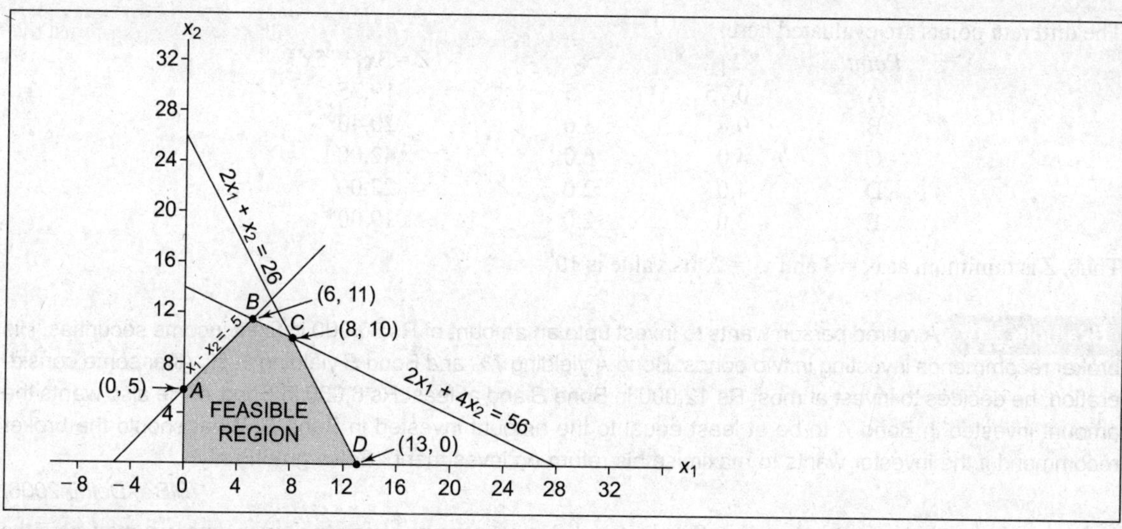

Figure 2.10 *Graphic Solution to the LPP*

Example 2.18 Solve graphically the following linear programming problem:

Minimise $Z = 3x_1 + 5x_2$

Subject to

$$-3x_1 + 4x_2 \leq 12; \qquad 2x_1 - x_2 \geq -2;$$

$$2x_1 + 3x_2 \geq 12; \qquad x_1 \leq 4;\ x_2 \geq 2, \text{ and}$$

$$x_1, x_2 \geq 0$$

(ICWA, June, 1985)

The given problem is depicted graphically in Figure 2.11. The feasible region is shaded and is bounded by *ABCDE.*

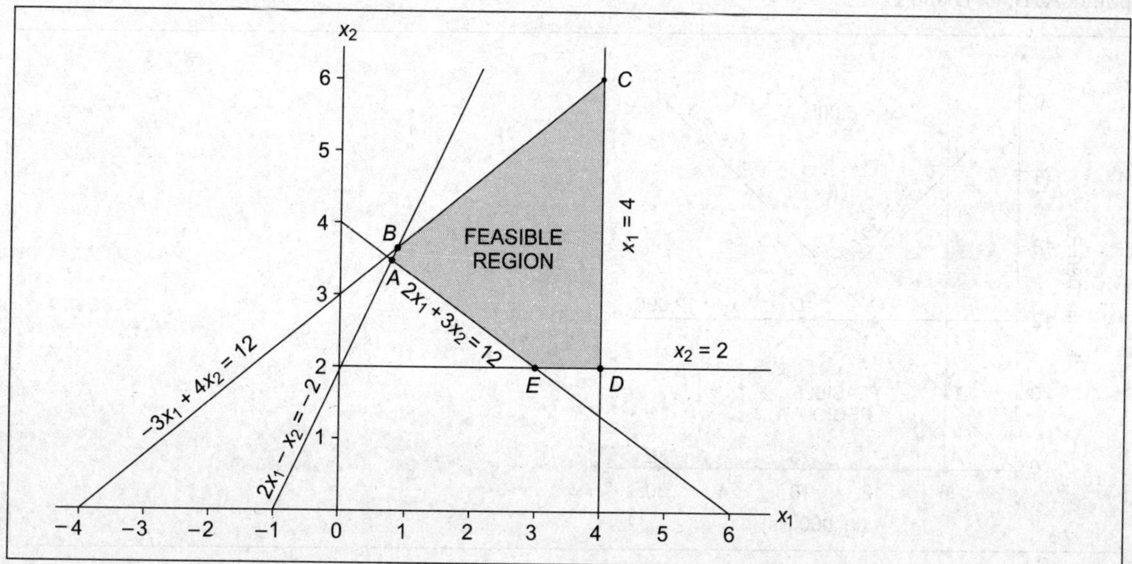

Figure 2.11 *Graphic Solution to the LPP*

The different points are evaluated here:

Point	x_1	x_2	$Z = 3x_1 + 5x_2$
A	0.75	3.5	19.75
B	0.8	3.6	20.40
C	4.0	6.0	42.00
D	4.0	2.0	22.00
E	3.0	2.0	19.00*

Thus, Z is minimum at $x_1 = 3$ and $x_2 = 2$. Its value is 19.

Example 2.19 A retired person wants to invest upto an amount of Rs 30,000 in fixed income securities. His broker recommends investing in two bonds: Bond *A* yielding 7% and Bond *B* yielding 10%. After some consideration, he decides to invest at most Rs 12,000 in Bond *B* and at least Rs 6,000 in Bond *A*. He also wants the amount invested in Bond *A* to be at least equal to the amount invested in Bond *B*. What should the broker recommend if the investor wants to maximise his return on investment? Solve graphically.

(MBA, Delhi, 2006)

Let x_1 and x_2 be the amount invested in Bonds *A* and *B*, respectively. Using the given data, we may state the problem as follows:

Maximise $Z = 0.07x_1 + 0.10x_2$

Subject to

$$x_1 + x_2 \leq 30{,}000$$
$$x_1 \geq 6{,}000$$
$$x_2 \leq 12{,}000$$
$$x_1 - x_2 \geq 0$$
$$x_1, x_2 \geq 0$$

The constraints are plotted graphically in Figure 2.12. The feasible region is shown shaded and is bound by points A, B, C, D and E.

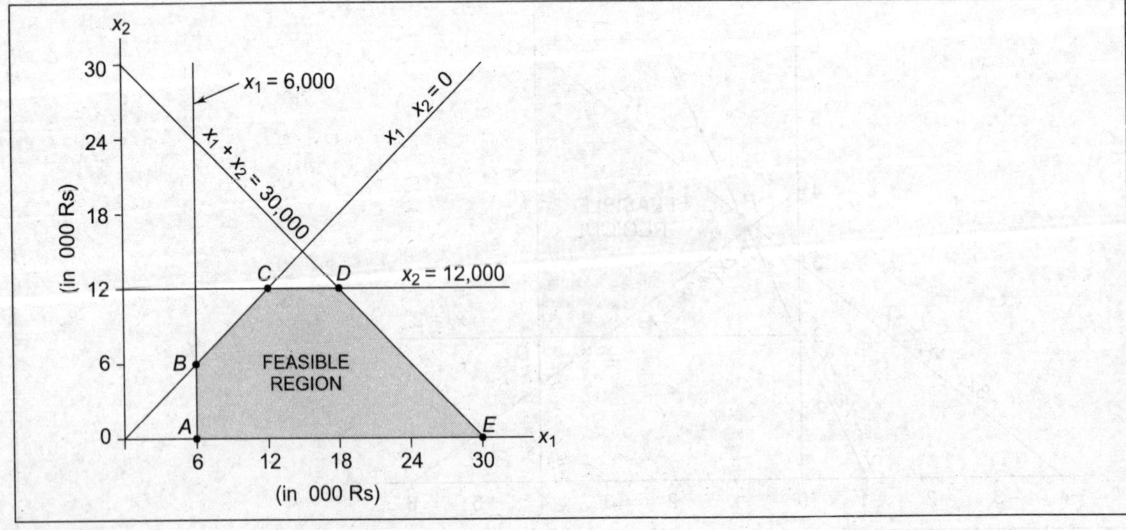

Figure 2.12 *Graphic Determination of Investment Mix*

The extreme points are evaluated here.

Point	x_1	x_2	$Z = 0.07x_1 + 0.10x_2$
A	6,000	0	420
B	6,000	6,000	1,020
C	12,000	12,000	2,040
D	18,000	12,000	2,460
E	30,000	0	2,100

The Z-value is maximum at point D. Accordingly, the optimal solution is: invest Rs 18,000 in Bond A and Rs 12,000 in Bond B. It would yield a return of Rs 2,460.

Example 2.20 A local travel agent is planning a charter trip to a major sea port. The eight day and seven night package includes the fare for round trip, surface transportation, board and lodging and selected tour options. The charter trip is restricted to 200 persons and past experience indicates that there will not be any problem for getting 200 clients. The problem for the travel agent is to determine the number of Deluxe, Standard and Economy tour packages to offer for this charter. These three plans differ according to seating and service for the flight, quality of accommodations, meal plans and tour options. The following table summarises the estimated prices for the three packages and the corresponding expenses for the travel agent. The travel agent has hired an aircraft for the flat fee of Rs 2,00,000 for the entire trip.

In planning the trip, the following considerations must be taken into account:

(i) At least 10% of the packages must be of the Deluxe type.

(ii) At least 35% but not more than 70% must be of the Standard type.

(iii) At least 30% must be of the Economy type.

(iv) The maximum number of Deluxe packages available in any aircraft is restricted to 60.

(v) The hotel desires that at least 120 of the tourists should be on the Deluxe and Standard packages taken together.

Prices and Costs for Tour Packages per Person

Tour Plan	*Price*	*Hotel Costs*	*Meal and Other Expenses*
Deluxe	10,000	3,000	4,750
Standard	7,000	2,200	2,500
Economy	6,500	1,900	2,200

The travel agent wishes to determine the number of packages to offer in each type so as to maximise the total profit.

(a) Formulate this as a linear programming problem.

(b) Restate the above LPP in terms of two decision variables, taking advantage of the fact that 200 packages will be sold.

(c) Find the optimal solution using graphical method for the restated problem and interpret your results.

(CA, May, 1991)

(a) Let x_1, x_2 and x_3 be the number of packages of Deluxe, Standard and Economy types, respectively. From the given information, unit profit for each of the three types can be obtained as under:

Deluxe Package : 10,000 – (3,000 + 4,750) = 2,250

Standard Package : 7,000 – (2,200 + 2,500) = 2,300

Economy Package : 6,500 – (1,900 + 2,200) = 2,400

With a hiring fee of Rs 2,00,000, the objective function can be expressed as:

Maximise $Z = 2{,}250x_1 + 2{,}300x_2 + 2{,}400x_3 - 2{,}00{,}000$

From the conditions given in the question, we have

$x_1 \geq 20$ condition (i)

$x_2 \geq 70$ and $x_2 \leq 140$ condition (ii)

$x_3 \geq 60$ condition (iii)

$x_1 \leq 60$ condition (iv) and

$x_1 + x_2 \geq 120$ condition (v)

Also, $x_1 + x_2 + x_3 = 200.$

Accordingly, the LPP can be stated as follows:

Maximise $Z = 2250x_1 + 2300x_2 + 2400x_3 - 200{,}000$

Subject to

$$20 \leq x_1 \leq 60$$
$$70 \leq x_2 \leq 140$$
$$x_3 \geq 60$$
$$x_1 + x_2 \geq 120$$
$$x_1 + x_2 + x_3 = 200$$
$$x_1, x_2, x_3 \geq 0$$

(b) Since $x_1 + x_2 + x_3 = 200$, we have $x_3 = 200 - x_1 - x_2$. Substituting it in the above relations, the LPP can be re-expressed as follows:

Maximise $Z = -150x_1 - 100x_2 + 280{,}000$

Subject to

$$20 \leq x_1 \leq 60$$
$$70 \leq x_2 \leq 140$$
$$120 \leq x_1 + x_2 \leq 140$$
$$x_1, x_2 \geq 0$$

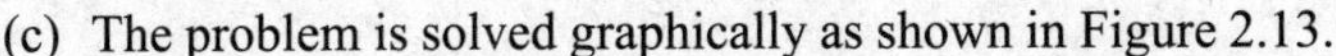

(c) The problem is solved graphically as shown in Figure 2.13.

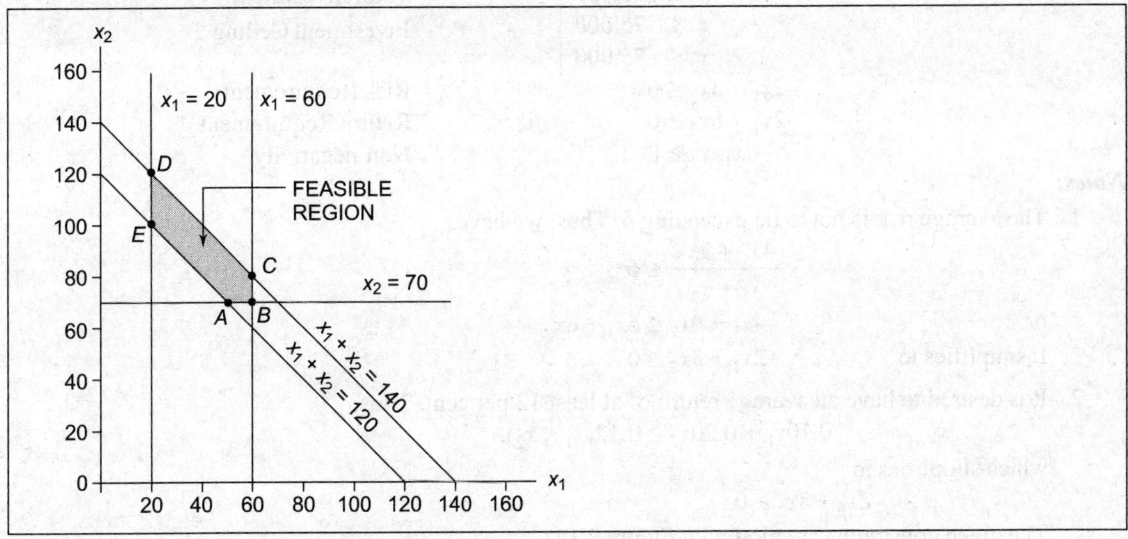

Figure 2.13 *Graphic Solution to the LPP*

Point	x_1	x_2	$Z(= -150x_1 - 100x_2 + 280{,}000)$
A	50	70	265,500
B	60	70	264,000
C	60	80	263,000
D	20	120	265,000
E	20	100	267,000

Thus, maximum profit of Rs 267,000 can be achieved when $x_1 = 20$, $x_2 = 100$, and $x_3 = 80$.

Example 2.21 Let us assume that you have inherited Rs 100,000 from your father-in-law that can be invested in a combination of only two stock portfolios, with the maximum investment allowed in either portfolio set at Rs 75,000. The first portfolio has an average return of 10%, whereas the second has 20%. In terms of risk factors associated with these portfolios, the first has a risk rating of 4 (on a scale from 0 to 10), and the second has 9. Since you want to maximise your return, you will not accept an average rate of return below 12% or a risk factor above 6. Hence, you then face the important question. How much should you invest in each portfolio?

Formulate this as a linear programming problem and solve it by graphic method. *(CA, May, 1999)*

Let x_1 : Investment in Portfolio 1

x_2 : Investment in Portfolio 2

The LPP may be expressed as under:

Maximise $Z = 0.10x_1 + 0.20x_2$ Total Return

Subject to

$$\left.\begin{aligned} x_1 + x_2 &\le 100{,}000 \\ x_1 &\le 75{,}000 \\ x_2 &\le 75{,}000 \end{aligned}\right\} \quad \begin{aligned} &\text{Total Investment} \\ &\text{Investment Ceiling} \end{aligned}$$

$$-2x_1 + 3x_2 \le 0 \qquad \text{Risk Requirement}$$

$$-2x_1 + 8x_2 \ge 0 \qquad \text{Return Requirement}$$

$$x_1, x_2 \ge 0 \qquad \text{Non-negativity}$$

Notes:

1. The average risk is not to be exceeding 6. Thus, we have,

$$\frac{4x_1 + 9x_2}{x_1 + x_2} \le 6$$

or $$4x_1 + 9x_2 \le 6x_1 + 6x_2$$

It simplifies to $$-2x_1 + 3x_2 \le 0.$$

2. It is desired to have an average return of at least 12 per cent. Thus,

$$0.10x_1 + 0.20x_2 \ge 0.12(x_1 + x_2),$$

which simplifies to

$$-2x_1 + 8x_2 \ge 0.$$

The given constraints are plotted in Figure 2.14.

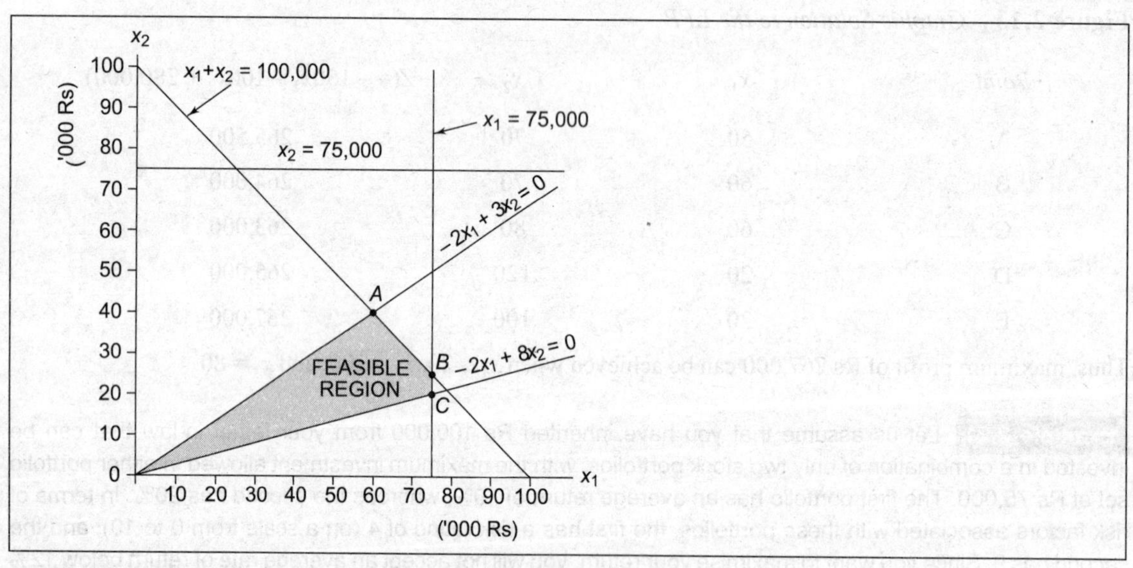

Figure 2.14 *Graphic Solution to Investment Problem*

The feasible region is given by the shaded region with vertices *O*, *A*, *B* and *C*. These points are evaluated below.

Point	x_1	x_2	*Z*
0	0	0	0
A	60,000	40,000	14,000
B	75,000	25,000	12,500
C	75,000	18,750	11,250

Evidently, Rs 60,000 should be invested in Portfolio 1 and Rs 40,000 in Portfolio 2. It would yield a maximum return of Rs 14,000, while meeting all the given requirements.

Example 2.22 A manufacturer employs three inputs: man hours, machine-hours and cloth material to manufacture two types of dresses. Type *A* dress fetches him a profit of Rs 160 per piece, while type *B*, that of Rs 180 per piece. The manufacturer has enough man-hours to manufacture 50 pieces of type *A* or 20 pieces of type *B* dresses per day while the machine-hours he possesses suffice only for 36 pieces of type *A* or for 24 pieces for type *B* dresses. Cloth material available per day is limited but sufficient enough for 30 pieces of either type of drsss. Formulate the linear programming model and solve it graphically. *(M Com, Delhi, 2005)*

Let x_1 and x_2 be the number of dresses of types A and B respectively. Using the given information, the linear programming problem may be stated as follows:

Maximise $Z = 160x_1 + 180x_2$ Total profit

Subject to $\frac{x_1}{50} + \frac{x_2}{20} \leq 1$ Man-hours

$\frac{x_1}{36} + \frac{x_2}{24} \leq 1$ Machine-hours

$\frac{x_1}{30} + \frac{x_2}{30} \leq 1$ Cloth material

$x_1, x_2 \geq 0$

The problem may be solved graphically now. The constraints are shown plotted in Fig. 2.15. The feasible solution is shown shaded and its vertices are evaluated below.

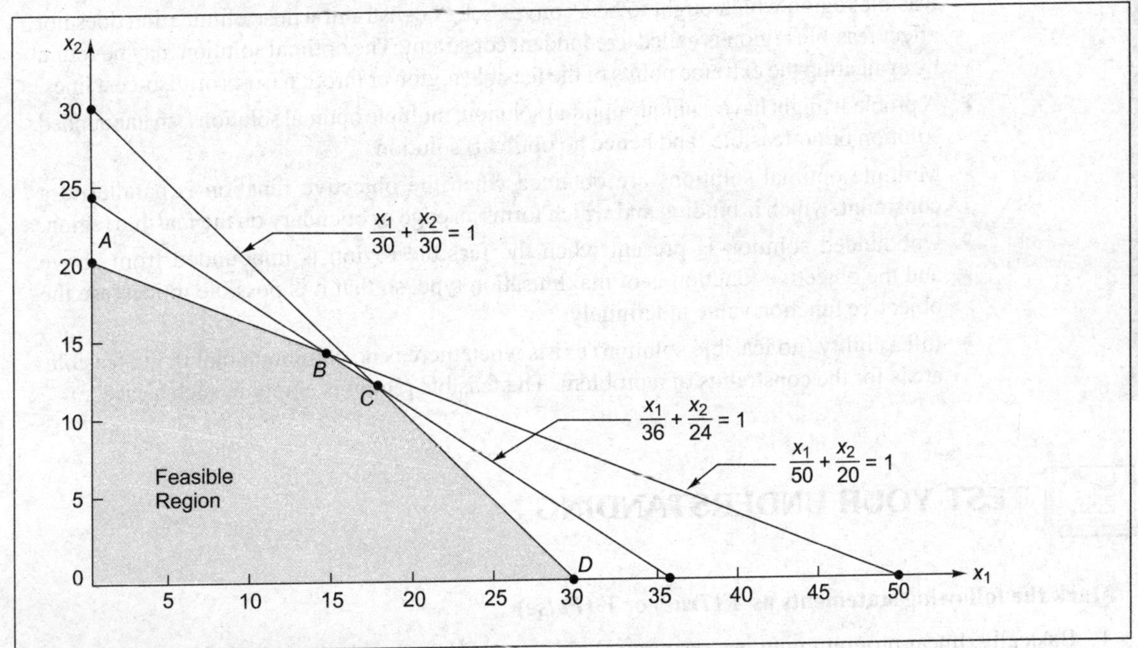

Figure 2.15 *Graphic Determination of Optimal Product Mix*

Point	x_1	x_2	Z
0	0	0	0
A	0	20	3,600
B	15	14	4,920
C	18	12	5,040
D	30	0	4,800

Thus, optimal solution calls for producing 18 units of type A dress and 12 units of type B dress. Total profit = Rs 5,040.

KEY POINTS TO REMEMBER

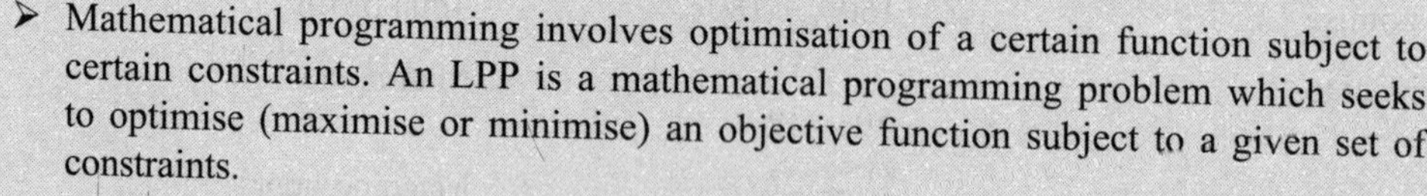

- Mathematical programming involves optimisation of a certain function subject to certain constraints. An LPP is a mathematical programming problem which seeks to optimise (maximise or minimise) an objective function subject to a given set of constraints.
- Formulating an LPP requires identification of decision variables, setting up the objective function and the constraints, and examining and stating if the variables are non-negative or not.
- An LPP model is based on the assumptions of (i) proportionality, (ii) additivity, (iii) continuity of decision variables, (iv) certainty and (v) finite choices.
- Solution of an LPP graphically requires plotting all the constraints and determining the feasible region which ought to be a convex set. A constraint whose elimination does not affect feasible region is called a redundant constraint. The optimal solution may be found by evaluating the extreme points of the feasible region or through iso-profit/iso-cost lines.
- A problem might have a unique optimal solution, multiple optimal solutions, an unbounded solution or no feasible (and hence no optimal) solution.
- Multiple optimal solutions are obtained when the objective function is parallel to a constraint, which is binding and which forms an edge or boundary on the feasible region.
- Unbounded solution is present when the feasible region is unbounded from above and the objective function is of maximisation type, so that it is possible to increase the objective function value indefinitely.
- Infeasibility (no feasible solution) exists when there is no common point in the feasible areas for the constraints of a problem. The feasible region is empty in such a case.

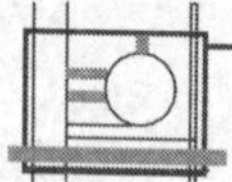

TEST YOUR UNDERSTANDING

Mark the following statements as T (*True*) or F (*False*)

1. Basically, linear programming is a resource allocation problem that deals with the best allocation of limited resources to a number of competing activities.

2. A prerequisite for applying linear programming is that there should be an objective which is clearly identifiable and which may, or may not, be measurable in quantitative terms.
3. Proportionality property in the linear programming context implies that per unit contribution of a variable in the objective function is independent of the size of the variable.
4. The solution to an LPP implicitly assumes that the variables are continuous, which may take fractional as well as integer values in the solution.
5. All constraints in an LPP as well as its objective function must be linear in nature.
6. A typical linear programming problem is characterised by an objective function to be maximised or minimised, and a set of constraints, and non-negativity condition.
7. For an *n* variables linear programming problem, there must be an equal number of constraints.
8. An LPP can have only two decision variables.
9. Linear programming is probabilistic in nature.
10. In real life, no variable can be unrestricted in sign.
11. An LPP must have all constraints of the "≤" or "≥" type.
12. A feasible solution is one which meets at least one of the constraints of the problem.
13. An optimal solution to a linear programming problem is a feasible solution which optimises.
14. The feasible region of a linear programming problem must be a convex set.
15. The graphic approach to the solution to LPPs cannot handle problems with more than three variables.
16. An iso-cost line cannot be parallel to the line of any constraint.
17. Iso-profit lines on a graph of an LPP would always be parallel to each other.
18. Replacing the '≤, sign of constraint by a '=' sign would improve the value of the objective function.
19. The feasible region for a constraint is restricted if its '≥' or '≤' sign is replaced by a '=' sign.
20. A constraint $4x_1 + 7x_2 \geq 57$ of an LPP is replaced by the constraint $4x_1 + 7x_2 \geq 40$. This would make the LPP more restrictive in nature.
21. Exclusion of a redundant constraint does not affect the optimal solution to an LPP. Thus, a redundant constraint represents an abundant resource.
22. A linear programming problem cannot have more than one redundant constraint.
23. For a linear programming model, the feasible region may change if non-binding constraints are deleted.
24. Every linear programming problem has a unique optimal solution.
25. It is possible for the objective function value of an LPP to be the same at two distinct extreme points.
26. When infeasibility does not exist, it is always possible to determine the optimal solution from a knowledge of all the extreme points of the polygon of the feasible region.
27. Changes in the objective function coefficients shall always result in changing the optimal values of the decision variables.
28. Infeasibility indicates that there are very few feasible solutions to an LPP and it is, therefore, difficult to say which of these is optimal.
29. An LPP with an unbounded feasible region would obviously have unbounded solution.
30. For a linear programming problem to be unbounded, its feasible region must be unbounded.

EXERCISES

1. What is a linear programming problem? Discuss the scope and role of linear programming in solving management problems.
2. Discuss and describe the role of linear programming in managerial decision-making bringing out limitations, if any. *(MBA, Delhi, 1999)*
3. "Linear Programming is one of the most frequently and successfully used Operations Research technique to managerial and business decisions." Elucidate this statement with some examples.
4. Give the mathematical and economic structure of the linear programming problem. What requirements should be met in order that the linear programming may be applied?
5. Briefly explain the major applications of linear programming in business.
6. What are the components of an LPP? What does the non-negativity restriction mean?
7. Give a general statement of a linear programming problem. Is it correct to say that the constraints should be of 'less than or equal to' form for the maximisation problems and of 'more than or equal to' form for the minimisation problems?
8. Discuss the assumptions of proportionality, additivity, continuity, certainty and finite choices in the context of LPPs.
9. In relation to linear programming, explain the implications of the following assumptions of the model:
   (a) linearity of objective function and constraints,
   (b) continuous variable,
   (c) certainty.
10. What steps are required in solving LPPs by graphic method? Discuss in brief.
11. What is feasibility region? Is it necessary that it should always be a convex set?
12. Define iso-profit line. How does it help to obtain solution to the linear programming problems?
13. What is a redundant constraint? What does it imply? Does it affect the optimal solution to an LPP?
14. How would you know whether the solution to a linear programming problem is unique or not? In this connection, state the conditions that should be satisfied for more than one optimal solution to a problem to exist.
15. With the help of suitable sketches, define convex, non-convex and infeasible regions in relation to the graphical solution of a linear programming problem. *(M.Com, Delhi, 2008)*

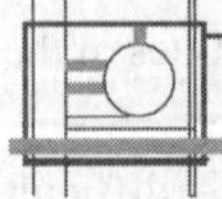

PRACTICAL PROBLEMS

1. Consider the production planning of The Super Fast Manufacturing Company which makes items P and V. The steel requirement for P is 400 gm per piece and that for V is 350 gm per piece. Both P and V, are machined on lathe which takes 85 and 50 minutes respectively, and are processed on a grinder which

requires 55 and 30 minutes respectively. Each unit of *P* consumes 20 minutes of polishing time. The resource availability is:

Total Machine Time : 1,450 hours
Total Steel : 250 kg

30 per cent of total machine time is that of lathe, 50 per cent of grinder and the remaining of polishing. Unit contribution to profits for *P* and *V* is Rs 40 and Rs 30, respectively.

Formulate this as a linear programming model for determining the number of units of *P* and *V* to be produced which would maximise the profits. Given also is the constraint that the company cannot sell more units of item *P* than of item *V*.

2. A company manufactures 3 types of parts which use precious metals platinum and gold. Due to shortage of these precious metals, the government regulates the amount that may be used per day. The relevant data with respect to supply, requirements, and profits are summarised in the table as follows:

Product	*Platinum required/unit* (gms)	*Gold required/unit* (gms)	*Profit/unit* (Rs)
A	2	3	500
B	4	2	600
C	6	4	1,200

Daily allotment of platinum and gold are 160 gm and 120 gm respectively. How should the company divide the supply of scarce precious metals?

Formulate it as a linear programming problem.

3. A manufacturer of purses makes four styles of purses: a three-compartment bag which takes 45 minutes to assemble; a shoulder-strap bag, taking one hour to assemble; a tote bag, needing 45 minutes for assembly, and pocket purse requiring 30 minutes to assemble. There are 32 hours of assembly time available per day. The profit contribution on the sale of a three-compartment bag is Rs 16, Rs 25 on a shoulder-strap bag, and Rs 12 each on tote bag and pocket purse.
Special kind of fancy pins are used in decorating pocket purses and they are available for only 30 pieces. Different type of pins are used in other three types of bags of which only 70 are in stock. Enough raw material is available for a total of 60 pocket purses and tote bags which need same quantity of raw material. The manufacturer estimates a minimum demand of 6 pocket purses and 10 shoulder strap bags every day. Formulate a lincar programming problem to optimise daily production.

4. An electronics company is engaged in the production of two components C_1 and C_2, used in radio sets. Each unit of C_1 costs the company Rs 5 in wages and Rs 5 in materials, while each unit of C_2 costs the company Rs 25 in wages and Rs 15 in materials. The company sells both products on one-period credit terms, but the company's labour and material expenses must be paid in cash. The selling price of C_1 is Rs 30 per unit and of C_2 it is Rs 70. Because of the strong monopoly of the company for these components, it is assumed that the company can sell at the prevailing prices as many units as it produces. The company's production capacity is, however, limited by two considerations. First, at the beginning of period 1, the company has initial balance of Rs 4,000 (cash *plus* bank credit *plus* collections from past credit sales). Second, the company has available, in each period, 2,000 hours of machine time and 1,400

hours of assembly time. The production of each C_1 requires 3 hours of machine time and 2 hours of assembly time, whereas the production of each C_2 requires 2 hours of machine time and three hours of assembly time.

Formulate the above problem as a linear programming problem.

5. At the beginning of a month, a lady has Rs 30,000 available in cash. She expects to receive certain revenues at the beginning of the months 1, 2, 3 and 4 and pay the bills after that, as detailed here:

Month	*Revenue*	*Bills*
1	Rs 28,000	Rs 36,000
2	Rs 52,000	Rs 31,000
3	Rs 24,000	Rs 40,000
4	Rs 22,000	Rs 20,000

It is given that any money left over may be invested for one month at the interest rate of 0.5%; for two months at 1.0% per month; for three months at 1.5% per month and for four months at 1.8% per month. Formulate her problem as a linear programming problem to determine an investment strategy that maximises cash in hand at the beginning of the month 5.

6. Formulate the following as a linear programming problem:

A publishing house publishes three weekly magazines—Daily life, Agriculture Today, and Surf's Up. Publication of one issue of each of these magazines requires the following amounts of production time and paper.

Magazine	*Production (hour)*	*Paper (kg)*
Daily Life	0.01	0.2
Agriculture Today	0.03	0.5
Surf's Up	0.02	0.3

Each week the publisher has available 120 hours of production time and 3,000 kg of paper. Total circulation for all three magazines must exceed 5,000 issue per week if the company is to keep its advertisers. The selling price per issue is Rs 22.50 for Daily Life, Rs 40.00 for Agriculture Today and Rs 15.00 for Surf's Up. Based on past sales, the publisher knows that the maximum weekly demand for Daily Life is 3,000 issues; for Agriculture Today it is 2,000 issues while for Surf's Up it is 6,000 issues. The production manager wants to know the number of issuses of each magazine to produce per week in order to maximize total sales revenue. *(MBA, Delhi, November, 2008)*

7. A Mutual Fund company has Rs 20 lakhs available for investment in Government bonds, blue chip stocks, speculative stocks and short-term bank deposits. The annual expccted return and risk factors are given as follows:

Type of investment	*Annual expected return*	*Risk factor (0 to 100)*
Government Bonds	14%	12
Blue Chip Stocks	19%	24
Speculative Stocks	23%	48
Short-term Deposits	12%	6

Mutual Fund is required to keep at least Rs 2 lakhs in short-term deposits and not to exceed an average risk factor of 42. Speculative stocks must be at most 20 percent of the total amount invested. How should Mutual Fund invest the funds so as to maximise its total expected annual return? Formulate this as a Linear Programming Problem. Do not solve it. *(CA, May, 1996)*

8. A company produces three types of parts for automatic washing machines. It purchases castings of the parts from a local foundry and then finishes the parts on drilling, shaping, and polishing machines. The selling prices of parts *A, B* and *C* respectively, are Rs 8, Rs 10 and Rs 14. All parts made can be sold. Castings for parts *A, B* and *C,* respectively cost Rs 5, 6 and 10. The company possesses only one of each type of machine. Costs per hour to run each of the three machines are Rs 20 for drilling, Rs 30 for shaping and Rs 30 for polishing. The capacities (parts per hour) for each part on each machine are shown in the following table:

Machine	*Capacity per hour*		
	Part A	*Part B*	*Part C*
Drilling	25	40	25
Shaping	25	20	20
Polishing	40	30	40

The manager of the company wants to know how many parts of each type to produce per hour in order to maximise profit for the hour's run. Formulate the above as a linear programming problem.

(MBA, Delhi, October, 1997)

9. Dr Shilpa Soni, the head administrator at XYZ hospital must determine a schedule for nurses to make sure there are enough nurses on duty throughout the day. During the day, the demand for nurses varies. Dr. Shilpa has broken the day in to 12 two-hour periods. The slowest time of the day encompasses the three periods from 12:00 AM to 6:00 AM, which, beginning at midnight, require a minimum of 30, 20 and 40 nurses respectively.

   The demand for nurses steadily increases during the next four daytime periods. Beginning with the 6:00 AM to 8:00 AM periods, a minimum of 50, 60, 80 and 80 nurses are required for these four periods respectively. For the five two-hour periods beginning at 2:00 PM and ending at midnight, 70, 70, 60, 50 and 50 nurses are required respectively. A nurse reports for duty at the beginning of one of the two hour periods and works eight consecutive hours (which is required in the nurses' contract). Dr Shilpa Soni wants to determine a nursing schedule that will meet hospital's minimum requirements throughout the day while using the minimum number of nurses.

   Formulate the above as an LPP. *(MBA, Delhi, November, 2008)*

10. Evening shift resident doctors in the B' Healthy Hospital work five consecutive days and have two consecutive days off. Their five days of work can start on any day of the week and the schedule rotates indefinitely. The hospital requires the following minimum number of doctors working:

S	*M*	*T*	*W*	*T*	*F*	*S*
35	55	60	50	60	50	45

No more than 40 doctors can start their five working days on the same day. Formulate a general linear programming model to minimise the number of doctors employed by the hospital.

11. Four products have to be processed through the plant, the quantities required for the next production period being:

Product 1	2,000 units
Product 2	3,000 units
Product 3	3,000 units
Product 4	6,000 units

There are three production lines on which the products could be processed. The rates of production in units per day and the total available capacity in days are given in the following table. The cost of using the lines is Rs 600, Rs 500 and Rs 400 per day respectively:

Production	*Product*				*Maximum line*
line	*1*	*2*	*3*	*4*	*Capacity (days)*
1	150	100	500	400	20
2	200	100	760	400	20
3	160	80	890	600	18
Total	2,000	3,000	3,000	6,000	

Formulate as a linear programming problem to minimise the cost of operation.

12. A certain firm has two plants. Orders from four customers have been received. The number of units ordered by each customer and the shipping cost from each plant are shown in the following table:

Customer	*Units ordered*	*Shipping cost (Rs)/unit*	
		From plant 1	*From plant 2*
A	500	15	40
B	300	20	30
C	1,000	30	25
D	200	35	20

Each unit of the product must be machined and assembled. These costs, together with the capacities at each plant, are shown below:

	Hours/Unit	*Cost(Rs)/Hour*	*Hours available*
Plant No. 1:			
Machining	0.10	40	120
Assembling	0.20	30	260
Plant No. 2:			
Machining	0.11	40	140
Assembling	0.22	30	250

Formulate a linear programming problem to minimise cost. *(MBA, Delhi, December, 1993)*

13. Formulate the following as a linear programming problem. Do not solve.

A trucking company with Rs 40,00,000 to spend on new equipment is contemplating three types of vehicles. Vehicle *A* has a 10-tonne pay-load and is expected to average 35 km per hour. It costs Rs 80,000. Vehicle *B* has a 20-tonne pay-load and is expected to average 30 km per hour. It costs Rs 1,30,000. Vehicle *C* is a modified form of *B;* it carries sleeping quarter for one driver, and this reduces its capacity to 18 tonnes and raises the cost to Rs 1,50,000. Vehicle *A* requires a crew of one man, and if driven on three shifts per day, could be run for an average of 18 hours per day. Vehicles *B* and *C* require a crew of two men each, but whereas *B* would be driven 18 hours per day with three shifts, *C* could average 21 hours per day. The company has 150 drivers available each day and would find it very difficult to obtain further crews. Maintenance facilities are such that the total number of vehicles must not exceed 30. How many vehicles of each type should be purchased if the company wishes to maximise its capacity in tonne-kms per day? *(MBA, Delhi, November, 1995)*

14. A refinery makes three grades of petrol *A, B, C* from three crude oils *d, e* and *f.* Crude oil *f* can be used in any grade but the others must satisfy the following specifications:

Grade	*Selling price per litre*	*Specification*
A	18.0	Not less than 50% crude *d*
		Not more than 25% crude *e*
B	16.5	Not less than 25% crude *d*
		Not more than 50% crude *e*
C	15.5	No specifications

There are capacity limitations on the amounts of the three crude elements that can be used:

Crude	*Capacity (kl)*	*Price per litre*
d	500	19.5
e	500	14.5
f	360	15.1

It is desired to obtain maximum profit. Formulate this as a Linear Programming Problem.

15. A company wants to plan production for the ensuing year so as to minimise the combined cost of production and inventory storage costs. In each quarter of the year, demand is anticipated to be 65, 80, 135 and 75 respectively. The product can be manufactured during regular time at a cost of Rs 16 per unit produced, or during overtime at a cost of Rs 20 per unit. The table given below gives data pertinent to production capacities. The cost of carrying one unit in inventory per quarter is Rs 2. The inventory level at the beginning of the first quarter is zero.

Quarter	*Capacities (units)*		*Quarterly demand*
	Regular time	*Overtime*	
1	80	10	65
2	90	10	80
3	95	20	135
4	70	10	75

Formulate the given problem to minimise the production plus storage costs for the year.

(MBA, Delhi, March, 2004)

16. Consider a company that must produce two products over a production period of three months of duration. The company can pay for the materials and labour from two sources: company funds and borrowed funds. The company faces three decisions:

 (1) How many units should it produce of Product 1?

 (2) How many units should it produce of Product 2?

 (3) How much money should it borrow to support the production of the two products?

 In making these decisions, the company wishes to maximise the profit contribution subject to the conditions given below:

 (i) Since the company's products are enjoying a seller's market, it can sell as many units as it can produce. The company would therefore like to produce as many units as possible subject to production capacity and financial constraints. The capacity constraints together with cost and price data are given in the table below.

Capacity, Price and Cost Data

	Selling price	*Cost of production*	*Required hours per unit in department*		
Product	*(Rs per Unit)*	*(Rs per Unit)*	*A*	*B*	*C*
1	14	10	0.5	0.3	0.2
2	11	8	0.3	0.4	0.1
Available hours per production period of three months			500	400	200

 (ii) The available company funds during the production period will be Rs 3 lakhs.

 (iii) A bank will give loans upto Rs 2 lakhs per production period at an interest rate of 20% per annum provided the company's acid (quick) test ratio is at least equal to 1 to 1 while the loan is outstanding. Take a simplified acid-test ratio given by

$$\frac{\text{Surplus cash on hand after production + Accounts receivable}}{\text{Bank borrowing + Interest accrued thereon}}$$

 (iv) Also make sure that the needed funds are made available for meeting the production costs.

 Formulate the above as a Linear Programming Problem. *(CA, November, 1992)*

17. A manufacturer currently produces four products. Recent recessionary trends cause a decline in demand and the company is laying off workers and discontinuing its third shift.

 The problem is that of rescheduling production during the first and second shifts for the remaining quarter of the year. The production involves various processes and one limiting resource in production is the availability of machine hours for a particular process *Z*. For this process, the four products require 4, 5, 5, and 7 hours respectively.

The sales manager has forecast the expected sales for each of the four products in the last quarter of the year. The estimates are shown in the following table:

Month	Forecast sales			
	Product #1	Product #2	Product #3	Product #4
October	8,000	19,000	4,000	7,000
November	7,000	19,000	15,000	7,000
December	6,000	18,000	17,000	7,000

The production capacity in terms of process *Z*, hours available, is expressed by month and shift.

Month	Process Z, hours available	
	Shift 1	Shift 2
October	110,000	100,000
November	130,000	120,000
December	115,000	116,000

The labour cost of operating the process *Z* machines is Rs 100 per hour during the first shift and Rs 120 for the second shift. The other relevant cost is storage. It costs Rs 40 per month to store one unit of any of the four products. It may be noted that it will be necessary to store some units of the four products as there is not enough labour available during the December demand.

Assuming that the company wishes to produce as many products as the sales manager has forecast, formulate an LP model to determine a production schedule that will meet the demand at minimum cost.

(Grad.Prg. in OR, ORSI, November 1983)

18. A manufacturer receives an order from a state transport corporation for six buses, to be delivered two at a time over the next three months. Production data for the manufacturer are shown in the following table:

	Months		
	1	2	3
Regular production capacity (in units)	1	2	3
Overtime production capacity (in units)	2	2	2
Regular production cost (Rs 10,000/unit)	35	43	40
Overtime production cost (Rs 10,000/unit)	39	47	45

Buses can be delivered to the city at the end of the same month in which they are assembled, or they can be stored by the manufacturer, at a cost of Rs 3,000 per bus per month for shipment during a later month. The manufacturer has no current inventory of these buses and desires none after the completion of this contract.

Formulate this problem as a linear programming problem to determine a production schedule that will meet the corporation demand at minimum cost to the manufacturer. *(MBA, Delhi, 1986)*

19. WELLTYPE Manufacturing Company produces three types of typewriters: Manual typewriters, Electronic typewriters, and Deluxe Electronic typewriters. All the three models are required to be machined first and then assembled. The times required for the various models are as follows:

Type	*Machine time (in hours)*	*Assembly time (in hours)*
Manual typewriter	15	4
Electronic typewriter	12	3
Deluxe Electronic typewriter	14	5

The total available machine time and assembly time are 3,000 hours and 1,200 hours respectively. The data regarding the selling price and variable costs for the three types are:

	Manual	*Electronic*	*Deluxe*
Selling Price (Rs)	4,100	7,500	14,600
Labour, material and other variable costs (Rs)	2,500	4,500	9,000

The company sells all the three on credit basis, but will collect the amounts on the first of next month. The labour, material and other variable expenses will have to be paid in cash. This company has a loan of Rs 40,000 from a co-operative bank and this company will have to repay it to the bank on 1st April, 1993. The TNC Bank from whom this company has borrowed Rs 60,000 has expressed its approval to renew the loan.

The Balance Sheet of this company as on 31.3.93 is as follows:

Liabilities	*Rs*	*Assets*	*Rs*
Equity Share Capital	150,000	Land	90,000
Capital Reserve	15,000	Building	70,000
General Reserve	110,000	Plant & Machinery	100,000
Profit & Loss a/c	25,000	Furniture & Fixtures	15,000
Long term loan	100,000	Vehicles	30,000
Loan from TNC Bank	60,000	Inventory	5,000
Loan from Co-op. Bank	40,000	Receivables	50,000
		Cash	140,000
Total	500,000	Total	500,000

The company will have to pay a sum of Rs 10,000 towards the salary for top management executives and other fixed overheads for the month. Interest on long-term loans is to be paid every month at 24% per annum. Interest on loans from TNC and Co-operative Banks may be taken to be Rs 1,200 for the month. Also, this company has promised to deliver 2 Manual typewriters and 8 Deluxe Electronic typewriters to one of its valued customers next month.

Also, make sure that the level of operations in this company is subject to the availability of cash next month. This company will also be able to sell all three types of typewriters in the market. The Senior Manager of this company desires to know as to how many units of each typewriter must be manufactured in the factory next month so as to maximise the profits of the company.

Formulate this as a linear programming problem. The formulated problem need not be solved.

(CA, May, 1993)

20. Obtain graphically the solution to the following LPP:

Maximise $Z = x_1 + 3x_2$

Subject to

$$x_1 + 2x_2 \le 9$$
$$x_1 + 4x_2 \le 11$$
$$x_1 - x_2 \ge 2$$
$$x_1, x_2 \ge 0$$

21. Find solution to the following LP problem:

Maximise $Z = 10x_1 + 8x_2$

Subject to

$$2x_1 + x_2 \le 20$$
$$x_1 + 3x_2 \le 30$$
$$x_1 - 2x_2 \ge -15$$
$$x_1, x_2 \ge 0$$

22. A firm makes two products X and Y, and has a total production capacity of 9 tonnes per day, X and Y requiring the same production capacity. The firm has a permanent contract to supply as least 2 tonnes of X and at least 3 tonnes of Y per day to another company. Each tonne of X requires 20 machine hours production time and each tonne of Y requires 50 machine hours production time. The daily maximum possible number of machine-hours is 360. All the firm's output can be sold, and the profit made is Rs 80 per tonne of X and 120 per tonne of Y. It is required to determine the production schedule for maximum profit and to calculate this profit. Use Graphical Method to get your solution.

(MBA, Delhi, April, 1996)

23. Cashewco has two grades of cashew nuts: Grade I—750 kg and Grade II—1,200 kg. These are to be mixed in two types of packages of one kilogram each—economy and special. The economy pack consists of grade I and grade II cashews in the proportion of 1 : 3, while the special pack combines the two in equal proportion. The profit margin on the economy and special packs is, respectively, Rs 5 and Rs 8 a pack.

(a) Formulate this as a linear programming problem.

(b) Ascertain graphically the number of packages of economy and special types to be made as will maximise the profits.

Would your answer be different if the profit margin on a special pack be Rs 10?

24. A company produces two types of pencils, say A and B. Pencil A is of superior quality and pencil B is of lower quality. Profits on the pencils A and B are Rs 5 and 3 per pencil respectively. Raw material required for each pencil A is twice as much as that of pencil B. The supply of raw material is sufficient only for 1000 pencils of type B per day. Pencil A requires a special clip and only 400 clips are available per day. For Pencil B only 700 clips are available per day. Use graphical method to find the product-mix that the company can make to make maximum profit. *(MBA, Delhi, March 2007)*

25. A company has two grades of inspectors 1 and 2, who are to be assigned for a quality control inspection. It is required that at least 2,000 pieces be inspected per 8- hour day. A grade 1 inspector can check pieces at the rate of 40 per hour, with an accuracy of 97 percent. A grade 2 inspector can check at the rate of 30 pieces per hour, with an accuracy of 95 percent.

    The wage rate of grade 1 inspector is Rs 5 per hour while that of a grade 2 inspector is Rs 4 per hour. An error made by inspector costs Rs 3 per piece. There are only 9 grade 1 inspectors and 11 grade 2 inspectors available in the company. The company wishes to assign work to the available inspectors so as to minimise the total cost of inspection. Formulate this problem as a linear programming model and solve it by using graphical method. *(MBA, Delhi, March 2004)*

26. Attempt graphically the following problem:

    Maximise $3x_1 + 2x_2$

    Subject to

    $$2x_1 + x_2 \leq 12;\ x_1 + x_2 \leq 10;\ -x_1 + 3x_2 \geq 6;$$

    and $x_1, x_2 \geq 0$

27. Minimise $Z = 3x_1 + 10x_2$

    Subject to

    $$15x_1 + 4x_2 \geq 60;\ 8x_1 + 8x_2 \geq 40;\ 4x_1 + 16x_2 \geq 32$$

    and further that both the variables are non-negative.

28. Solve graphically the following LPP:

    Minimise $Z = 4x_1 + 3x_2$

    Subject to

    $$x_1 + 3x_2 \geq 9$$
    $$2x_1 + 3x_2 \geq 12$$
    $$x_1 + x_2 \geq 5$$
    $$x_1, x_2 \geq 0$$

29. A company manufactures two kinds of machines, each requiring a different manufacturing technique. The deluxe machine requires 18 hours of labour, 9 hours of testing, and yields a profit of Rs 400. The standard machine requires 3 hours of labour, 4 hours of testing, and yields a profit of Rs 200. There are 800 hours of labour and 600 hours of testing available each month. A marketing forecast has shown the

monthly demand for the standard machine to be no more than 150. The management wants to know the numbers of each model to produce monthly that will maximise total profit. Formulate and solve this as a linear programming problem. *(MBA, Delhi, November, 2003)*

30. A farm is engaged in breeding pigs. The pigs are fed on various products grown on the farm. Because of the need to ensure certain nutrient constituents, it is necessary to buy one or two additional products, called A and B.

The nutrient constituent (vitamins and proteins) in each unit of the products are as follows:

Nutrient	*Nutrient contents in the product*		*Min. nutrient amount*
	A	*B*	
1	36	6	108
2	3	12	36
3	20	10	100

Product A costs Rs 20 and product B costs Rs 40 per unit. How much of products A and B should be purchased at the lowest possible cost so as to provide the pigs nutrients not less than the minimum required as given in the table?

31. Two kinds of food for children, F_1 and F_2, are being considered to be purchased. Food F_1 costs Rs 20 a unit while food F_2 is available at Rs 40 per unit. The nutrient contents of these foods are as follows:

Nutrients	*Nutrient content*	
	Food F_1	*Food* F_2
N_1	40	20
N_2	3	12
N_3	18	3

The minimum requirement of three nutrients is respectively, 200, 36 and 54 units.

Draft this as an LPP and find out graphically the quantities of food units which should be bought in order that the costs are minimised satisfying the given constraints.

32. (a) Is it necessary that the feasible region for a maximisation type of linear programming problem must always be a bounded one?

(b) Maximise $Z = 6x_1 - 2x_2,$

Subject to

$$8x_1 - 4x_2 \leq 8$$

$$0 \leq x_1 \leq 3$$

$$x_2 \geq 0$$

Solve graphically.

33. A medicine manufacturing company plans to market two syrups: Tonus-2,000, and Health-Wealth. There are sufficient ingredients available to make 20,000 bottles of Tonus-2,000 and 40,000 bottles of

Health-Wealth. It takes three hours to prepare enough material to fill 1,000 bottles of Tonus-2,000 and one hour to fill 1,000 bottles of Health-Wealth. Only 45,000 bottles and 66 hours are available for this operation next week. The profit expected is Rs 2.80 per bottle of Tonus-2000 and Rs 2.20 per bottle of Health-Wealth.

Formulate this problem as an LPP and determine graphically the optimum production programme.

34. A cold drinks company has two bottling plants, located at different places. Each plant produces three different drinks *A*, *B* and *C*. The capacities of the two plants in number of bottles per day are as follows:

	Product A	*Product B*	*Product C*
Plant I	3,000	1,000	2,000
Plant II	1,000	1,000	6,000

A market survey indicates that during any particular month there will be a demand of 24,000 bottles of *A*, 16,000 bottles of *B* and 48,000 bottles of *C*. The operating costs, per day, of running plants I and II are, respectively, Rs 6,000 and Rs 4,000. How many days should the company run each plant during the month so that the production cost is minimised while still meeting the market demand? Use graphical method to get the optimal solution.

35. A firm is engaged in producing two products: P_1 and P_2.
The relevant data are given here:

Per Unit:	Product P_1	Product P_2
(i) Selling price	Rs 200	Rs 240
(ii) Direct materials	Rs 45	Rs 50
(iii) Direct wages:		
Deptt A	8 hrs @ Rs 2/hr	10 hrs @ Rs 2/hr
Deptt B	10 hrs @ Rs 2.25/hr	6 hrs @ Rs 2.25/hr
Deptt C	4 hrs @ Rs 2.5/hr	12 hrs @ Rs 2.5/hr
(iv) Variable overheads	Rs 6.50	Rs 11.50

Fixed overhead = Rs 2,85,000 per annum.

No. of employees in the three departments: Deptt A = 20; Deptt B = 15; Deptt C = 18

No. of hours/employee/week = 40 in each department

No. of weeks per annum = 50.

(a) Formulate the given problem as a linear programming problem, and solve graphically to determine

(i) the product mix as will maximise the contribution margin of the firm.

(ii) the amount of contribution margin and profit obtainable per year.

(b) From the graph, do you observe any constraint that is redundant? Which one, if yes?

36. Solve graphically the following LPP:

Minimise $Z = -4x_1 + 3x_2$

Subject to

$$x_1 - 2x_2 \geq -4;\ 2x_1 + 3x_2 \geq 13;\ x_1 - x_2 \geq 4;\ \text{and } x_1, x_2 \geq 0$$

37. Subject to

$$2x_1 - 6x_2 \leq 0$$
$$-x_1 + 2x_2 \geq -2$$
$$-3x_1 - 3x_2 \geq -24$$
$$x_1 \geq 2$$
$$x_2 \geq 0$$

(a) Minimise $10x_1 - 4x_2$ (b) Maximise $10x_1 - 4x_2$ (c) Maximise $4x_1 + 10x_2$, (d) Minimise $4x_1 + 10x_2$

38. Solve graphically:

Minimise $Z = 12x_1 + 3x_2$

Subject to

$$4x_1 + 6x_2 \geq 24{,}000$$
$$x_1 + x_2 \geq 5{,}000$$
$$8x_1 + 2x_2 \geq 16{,}000$$
$$x_1, x_2 \geq 0$$

39. Maximise $3x_1 + 4x_2$ subject to the following constraints:

$$2x_1 + x_2 \leq 10;\ x_1 + 4x_2 \leq 36;\ x_1 + 2x_2 \leq 10;\ x_1 \geq 5;\ \text{and } x_2 \geq 7$$

40. Find the maximum and the minimum values of the function $Z = 8x_1 + 5x_2$, subject to the following:

$$3x_1 - 2x_2 \geq 6$$
$$-2x_1 + 7x_2 \geq 7$$
$$2x_1 - 3x_2 \leq 6$$
$$x_1, x_2 \geq 0$$

41. A local business firm is planning to advertise a special sale on radio and television during a particular week. A maximum budget of Rs 16,000 is approved for this purpose. It is found that radio commercials cost Rs 800 per 30-second spot with a minimum contract of five spots. Television commercials, on the other hand, cost Rs 4,000 per spot. Because of heavy demand only 4 television spots are still available in the week. Also it is believed that a TV spot is six times as effective as a radio spot in reaching consumers. How should the firm allocate its advertising budget to attract the largest number of consumers? How will the optimal solution be affected if the availability of TV spot is not constrained?

42. A firm employs man hours (M_1), a, machine hours (M_2), and raw materials (R_1 and R_2) manufacture two products. The maximum of M_1 and M_2 available are 120 and 270 hours respectively but raw materials R_1 and R_2 cannot be procured below 330 and 156 units on economic grounds. Man and machine hours as also the units of the two raw materials needed per unit of products A and B along with the profit margins allowed per unit of the two products are given in the following table.

Products	Requirements per unit of the product				Profit per unit
	M_1	M_2	R_1	R_2	
A	3	3	13	4	500
B	3	9	8	7	125
Availability	120	270	330	156	

Production of none of the two must exceed 25 units per day due to limitation of space. Given that the data pertaining to input requirements and availability refer to a typical day of production, formulate the problem and solve it graphically for optimal product-mix. (*M Com, Delhi*, 2008)

43. G Limited, manufacturer of superior garden ornaments, is preparing its production budget for the coming period. The company makes four types of ornaments, the data for which are as follows:

Product	Pixie	Elf	Queen	King
		(All values in Rs per unit)		
Direct Materials	25	35	22	25
Variable Overhead	17	18	15	16
Selling Price	111	98	122	326
Direct labour hours		(In hours per unit)		
Type 1	8	6	–	–
Type 2	–	–	10	10
Type 3	–	–	5	25

Fixed overhead amounts to Rs 15,000 per period. Each type of labour is paid Rs 5 per hour but because of the skills involved, an employee of one type cannot be used for work normally done by another type. The maximum hours available of each type are:

Type 1	8,000 Hours
Type 2	20,000 Hours
Type 3	25,000 Hours

The marketing department judges that, at the present selling prices, the demand for the products is likely to be:

Pixie	Unlimited Demand
Elf	Unlimited Demand
Queen	1,500 units
King	1,000 units

You are required

(a) to calculate the product mix that will maximise the profit, and the amount of the profit;

(b) to determine whether it would be worth while paying Type 1 Labour for overtime working at time and a half and, if so, to calculate the extra profit for each 1,000 hours of overtime;

(c) to comment on the principles used to find the optimum product mix in part (a), pointing out any possible limitations;

(d) to explain how a computer could assist in providing a solution for the data shown above.

(ICMA, November, 1993, Adapted)

44. The Hell Laboratories has a long history of production troubles. It produces two items, A and B, which are equally profitable. Recently, the company has entered into contract to supply 40 units of A and 20 units of B per week to another company. The technology of the chemical process implies that production of A must always be at least as large as of B. There are two raw material constraints to be satisfied:

$$5A + 8B \leq 400 \quad \text{and} \quad 55A + 50B \leq 2750.$$

Attempt the problem graphically and comment on the solution that you might obtain.

CHAPTER 3

Linear Programming II: Simplex Method

CHAPTER OVERVIEW

Once a linear programming problem is formulated, the next issue is to solve it for optimal values of the decision variables, which leads to maximisation (say, of profit) or minimisation (may be, of cost). The graphic method is indeed available but, as we know, it has its limitations. When only two items are produced by a firm, two investment opportunities are available to an investor, two advertising media are available to an advertising manager—in other words when only two decision variables are involved — then we can think of a graphic solution to an LPP. In case of more than two variables, however, the manager has to turn to the Simplex method. The Simplex method is a very powerful tool for solving linear programming problems. The Simplex and its variants can handle any complexities in the LPPs irrespective of the number and nature of the decision variables and constraints.

The application of Simplex method for solving a linear programming problem helps a manager to know answers to questions like the following:

- What mix of the products involved would yield the maximum profit?
- What maximum profit can be obtained from the optimal product-mix?
- Whether all resources would be fully utilised by the optimal mix, or are there any unutilised resources and to what extent?
- If the problem is of the minimisation type, then what mix of the products/items would minimise the cost and what is the cost involved?
- Does the problem have an alternate solution which is as good as the optimal solution obtained? If yes, what is that solution?
- Is any of the given restrictions of no consequence? To illustrate, a particular raw material may be available in plenty in a given situation.
- Does the problem have no solution that can meet all the requirements?

While the solution to relatively small or moderately large problem may be found while working manually with the method, large-scale problems can be solved with the help of computers where software are available. Real life problems obviously require a computer for solution.

Knowledge of simple algebraic manipulations and a good hand at arithmetic calculations are the prerequisites for applying Simplex algorithm. A command over arithmetic operations on fractional values is neccessary and desirable for Simplex calculations. Unfortunately, an ordinary electronic calculator is not of much help in this case.

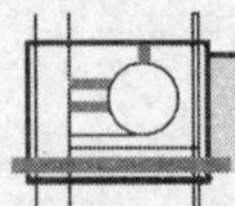

3.1 INTRODUCTION

In the previous chapter, we considered the formulation of linear programming problems and the graphic method of solving them. It was observed in the graphical approach to the solution of such problems that, in a given situation, the feasible region is determined by the set of constraints given in the problem while the objective function locates the optimal point—the one that *maximises* or *minimises,* as the case may be. Although this method is very efficient in developing the conceptual framework necessary for fully understanding the linear programming process, it suffers from the great limitation that it can handle problems involving only two decision variables. In the real-world situations, we frequently encounter cases where more than two variables are involved and, therefore, look for a method that can handle them. The *Simplex Method* provides an efficient technique which can be applied for solving LPPs of any magnitude—involving two or more decision variables. In this technique, the objective function is used to control the development and evaluation of each feasible solution to the problem.

The present chapter is devoted to a discussion of the simplex algorithm and demonstrates its application in solving linear programming problems.

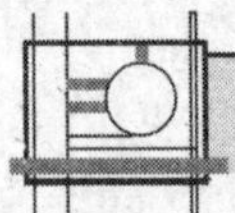

3.2 SIMPLEX METHOD

The *'simplex algorithm'* is an iterative procedure for finding, in a systematic manner, the optimal solution to a linear programming problem. We have seen earlier that if an optimal solution to the problem exists, it is located at a corner point of the feasible region determined by the constraints of the system. The simplex method, according to its iterative search, selects this optimal solution from among the set of feasible solutions to the problem. The algorithm is indeed very efficient, because it considers only those feasible solutions which are provided by the corner points, and that too *not* all of them. Thus, by using this technique, we have to consider a minimum number of feasible solutions to obtain an optimal one. Also, this technique has the merit to indicate whether a given solution is optimal or not.

For applying simplex method to the solution of an LPP, first of all, an appropriately selected set of variables is introduced into the problem. The iterative process begins by assigning values only to these variables and the primary (decision) variables of the problem are all set equal to zero. This assumption is analogous to starting the evaluation process in the graphic approach at the point of origin, where both x_1 and x_2 are equal to zero. The algorithm then replaces one of the initial variables by another variable—the variable which contributes most to the desired optimal value enters in, while the variable creating the bottleneck to the optimal solution goes out. This improves the value of the objective function. This procedure of substitution of variables is repeated until no further improvement in the objective function value is possible. The algorithm terminates there indicating that the optimal solution is reached, or that the given problem has no solution.

3.2.1 Conditions for Application of Simplex Method

In order that the simplex method may be applied to a linear programming problem, the following two conditions have to be satisfied.

(1) The R.H.S. of each of the constraints, b_i should be non-negative. If an LPP has a constraint for which a negative resource value is given, it should be, in the first step, converted into positive value by multiplying both sides of the constraint by –1. For example, if the given constraint is $8x_1 - 3x_2 \geq -6$, it shall change into $-8x_1 + 3x_2 \leq 6$. Notice that the direction of inequality changes in the process.

(2) Each of the decision variables of the problem should be non-negative. Sometimes, a problem might have a variable which is 'unrestricted in sign' or 'free', so that it can assume negative values as well as non-negative. This situation is handled by treating such a variable as the difference of two variables which are both non-negative, because such a difference may be positive, negative or zero. Consider the following example:

Maximise $Z = 5x_1 + 7x_2 - 2x_3$

Subject to

$$2x_1 + 5x_2 + 3x_3 \leq 80$$
$$5x_1 + 2x_2 - 2x_3 \leq 30$$
$$x_2 + 6x_3 \leq 42$$
$$x_1, x_2 \geq 0, x_3 \text{ unrestricted in sign}$$

Since x_3 is unrestricted in sign, we can represent it by a difference of two non-negative variable$, say x_4 and x_5. Substituting $x_3 = x_4 - x_5$ in the above and simplifying, we get

Maximise $Z = 5x_1 + 7x_2 - 2x_4 + 2x_5$

Subject to

$$2x_1 + 5x_2 + 3x_4 - 3x_5 \leq 80$$
$$5x_1 + 2x_2 - 2x_4 + 2x_5 \leq 30$$
$$x_2 + 6x_4 - 6x_5 \leq 42$$
$$x_1, x_2, x_4, x_5 \geq 0$$

After the solution is obtained, we shall substitute the difference of the values of x_4 and x_5 as the value of x_3.

The working of the *simplex method* proceeds by preparing a series of tables called *simplex tableaus.* We shall now discuss the application of the method—first for the *maximisation* and then for the *minimisation* case. Figure 3.2 gives the schematic of the simplex method.

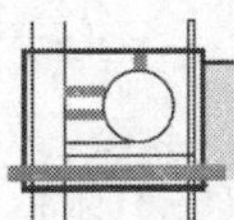

3.3 SOLUTION TO MAXIMISATION PROBLEMS

Consider the linear programming problem given in Example 2.1, reproduced below.

Example 3.1 Maximise $Z = 40x_1 + 35x_2$ Profit

Subject to

$2x_1 + 3x_2 \leq 60$ Raw Material Constraint

$4x_1 + 3x_2 \leq 96$ Labour Hours Constraint

$x_1, x_2 \geq 0$

The solution to this problem is illustrated below in a step-wise manner.

Standardisation of the Problem

The first step in applying simplex method is to standardise the problem. For this, inequalities of the constraints are converted into equations. The first inequality of the system is

$$2x_1 + 3x_2 \leq 60$$

To convert it into an equation, we add a variable S_1 on the left-hand side to get

$$2x_1 + 3x_2 + S_1 = 60$$

Now, if $2x_1 + 3x_2 = 60$, then $S_1 = 0$, and if $2x_1 + 3x_2 < 60$, then S_1 is equal to the difference. Therefore, S_1 can vary from 0 to 60 ($0 \leq S_1 \leq 60$) depending upon the value of $2x_1 + 3x_2$. The variable S_1 is referred to as a *slack variable* because it takes up any slack between the left- and the right-hand side of the inequality upon being converted into equation. In the context of the present problem, the value of S_1 shall indicate the amount of unused raw materials.

In a similar way, we introduce slack variable S_2 for converting the other inequality into equation, to get $4x_1 + 3x_2 + S_2 = 96$. Hencc, S_2 can vary between 0 and 96 ($0 \leq S_2 \leq 96$), depending upon the value of $4x_1 + 3x_2$, and it represents the labour hours not used.

We also need to modify the objective function here because the objective function for the simplex method should contain every variable in the system including slack or other variables (discussed later) added. We shall have to add S_1 and S_2 in the present case. But for this purpose, their coefficients also have to be determined. The coefficient may represent the cost involved in not using the raw materials or labour hours. We shall presume that the unused resources have no cost and, as such, they do not affect the profits. Thus, the coefficients assigned will be zeros.

The problem can now be expressed as follows:

Maximise $\quad Z = 40x_1 + 35x_2 + 0S_1 + 0S_2$

Subject to

$$2x_1 + 3x_2 + S_1 + 0S_2 = 60$$

$$4x_1 + 3x_2 + 0S_1 + S_2 = 96$$

$$x_1, x_2, S_1, S_2 \geq 0$$

This is the standardised form of the given problem.

Obtaining the Initial Tableau and Solution

Using the standardised form, the information in the problem is presented in a tableau. The initial simplex tableau is given in Table 3.1.

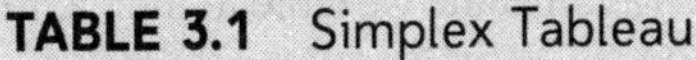

TABLE 3.1 Simplex Tableau

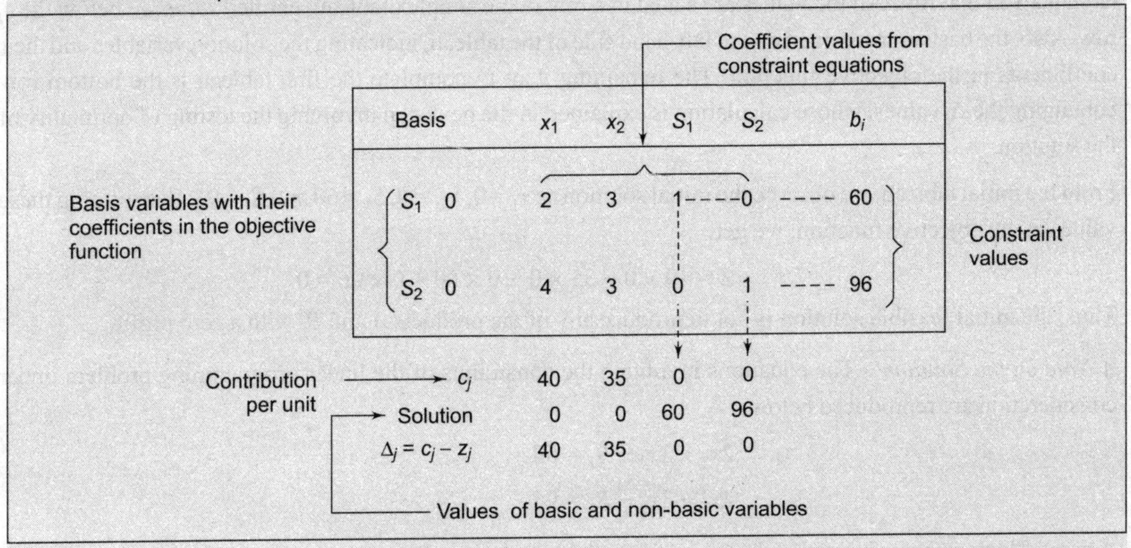

To set up the tableau, we first list horizontally all the variables contained in the problem. Here, there are four variables: x_1, x_2, S_1 and S_2. Next, the coefficients in the constraint equations are written listing vertically the coefficients under their respective variables. It may be noted that each of the slack variables appears only in one equation. Therefore, the coefficient of each of the slack variables is taken to be zero in all the equations except the one in which it appears. Thus, in the first constraint, $2x_1 + 3x_2 + S_1 + 0S_2$, the coefficients of x_1, x_2, S_1 and S_2 are 2, 3, 1 and 0 respectively. After putting the coefficients, the constraint values are mentioned on the right-hand side against the rows. Finally, the row titled c_j indicates the coefficients of the various variables in the objective function, mentioned respectively in the various columns representing the variables. Having set up the simplex tableau, the next step is to locate the *identity* (matrix) and the variables involved in it. The identity contains all zeros except a diagonal column of positive 1's. The identity must have this square form with all zeros and a diagonal of (plus) ones. The size of this square would be determined by the number of constraints in the system. Locating the identity is of prime importance as the solution is identified in reference to this. To determine the solution, representing the first feasible solution, we set all variables other than those in the identity, equal to zero and then assign the values of the constants (b_i's) to variables in the identity. The variables in the identity are called *basic variables* and the remaining ones are called *non-basic variables.* In general, if a linear programming model has n variables and m constraints, then m variables would be basic variables and $n - m$ variables would be non-basic*. The basic variables form the *basis* and are known as the *variables in the solution.*

For our problem, the identity is formed by variables S_1 and S_2 and, therefore, they constitute the basis. Here S_1 and S_2 are basic and x_1 and x_2 are non-basic. The variables x_1 and x_2 are assigned zero values while S_1 and S_2

---

* Under certain conditions some basic variables may also have zero values. This condition is called *degeneracy* and is discussed later.

equal 60 and 96 respectively. For each row, the constraint value, b_i, is written against the variable in the identity having a 1 in that row. All the values are stated in a row in the simplex tableau, entitled *solution,* below the c_j row. Also, the basis is mentioned on the left-hand side of the tableau, indicating the solution variables and their coefficients in the objective function. The remaining item to complete the first tableau is the bottom row containing the Δ_j values, whose calculation is explained in the next step involving the testing of optimality of the solution.

From the initial tableau, we observe the initial solution as $x_1 = 0, x_2, = 0, S_1 = 60$ and $S_2 = 96$. Substituting these values in the objective function, we get

$$Z = 40 \times 0 + 35 \times 0 + 0 \times 60 + 0 \times 96 = 0.$$

Thus, the initial feasible solution is not to produce any of the products *A* and *B,* with a zero profit.

A Note on the Solution The equations involving the constraints of the linear programming problem under consideration are reproduced below:

$$2x_1 + 3x_2 + S_1 = 60$$

$$4x_1 + 3x_2 + S_2 = 96$$

A careful look at these shows that there are four variables whose values are to be obtained while the equations are only two. It requires setting two variables equal to zero and then solving the equations for the other variables. In general, for a system $Ax = b$ of m linear equations in n variables, such that $n \geq m$, we need to set $n - m$ variables equal to zero and solve for the remaining m variables. Such a solution is called a *basic solution.* Obtaining a basic solution assumes that setting of the $n - m$ variables equal to zero each yields unique values for the remaining m variables, which implies that the columns for the remaining m variables are linearly independent. The set of $n - m$ variables is called *non-basic* variables and the m variables are termed the *basic variables.* Further, it may be noted that from n variables, a set of $n - m$ non-basic variables, or equivalently, m basic variables can be chosen in ${}^nC_m$ ways.

In the context of linear programming problems, when a problem has n variables and m constraints, we can have ${}^nC_m$ basic solutions. But since these problems require the variables to be non-negative, some of the basic solutions may not be feasible as they may involve negative values for some of the variables. Accordingly, a basic solution to $Ax = b$ in which all the variables are non-negative is termed as *basic feasible solution.* The intial solution for the problem under consideration is $x_1 = 0, x_2 = 0, S_1 = 60, S_2 = 96$, which indeed is a basic feasible solution.

The significance of basic feasible solution (bfs) stems from the fact that for any LPP, there is a unique extreme point of the feasible region corresponding to each *bfs* and also there is at least one *bfs* corresponding to each extreme point of the feasible region. To illustrate, for the example under consideration, the graphic representation of constraints is reproduced in Figure 3.1 and the constraints in the standardised form are presented below:

$$2x_1 + 3x_2 + S_1 + 0S_2 = 60$$

$$4x_1 + 3x_2 + 0S_1 + S_2 = 96$$

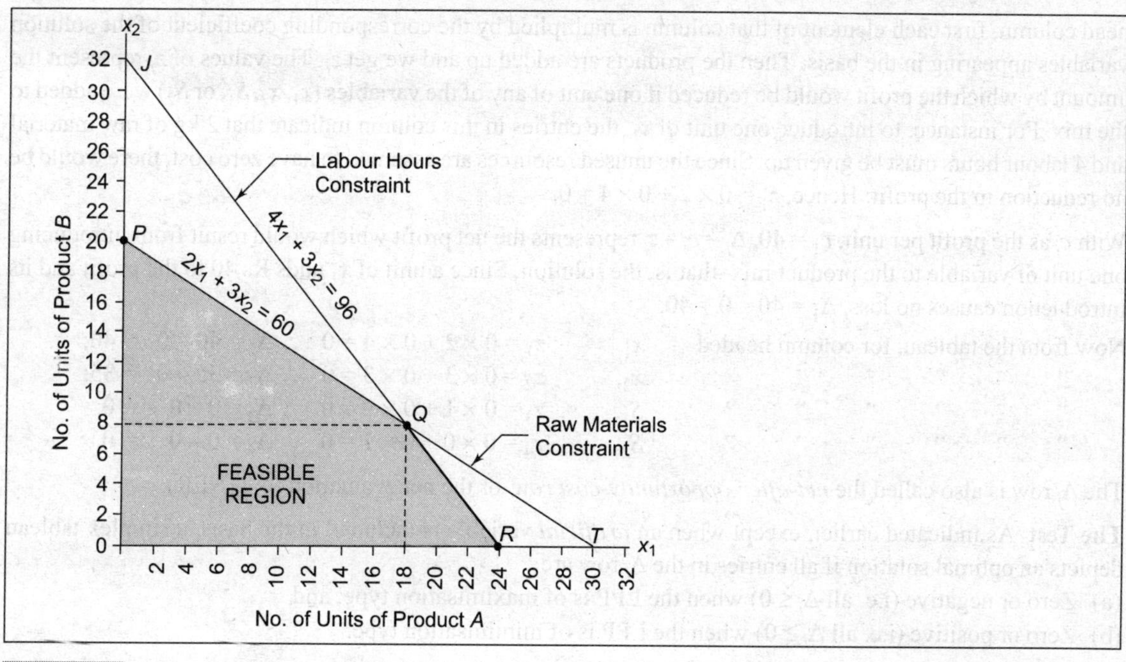

Figure 3.1 *Graphic Presentation of Constraints*

With 4 variables and 2 equations, there shall be ${}^4C_2 = 6$ basic solutions are detailed as follows:

Non-basic variables	*Basic variables*	*Solution*	*bfs?*	*Corner point*
S_1, S_2	x_1, x_2	$S_1 = 0,\ S_2 = 0,\ x_1 = 18,\ x_2 = 8$	Yes	Q
x_2, S_2	x_1, S_1	$x_2 = 0,\ S_2 = 0,\ x_1 = 24,\ S_1 = 12$	Yes	R
x_2, S_1	x_1, S_2	$x_2 = 0,\ S_1 = 0,\ x_1 = 30,\ S_2 = -24$	No	–
x_1, S_2	x_2, S_1	$x_1 = 0,\ S_2 = 0,\ x_2 = 32,\ S_1 = -36$	No	–
x_1, S_1	x_2, S_2	$x_1 = 0,\ S_1 = 0,\ x_2 = 20,\ S_2 = 36$	Yes	P
x_1, x_2	S_1, S_2	$x_1 = 0,\ x_2 = 0,\ S_1 = 60,\ S_2 = 96$	Yes	0

Observe here that two of the solutions are not basic feasible solutions since they involve negative value for a variable, and are not relevant, while each of the remaining are basic feasible solutions and they correspond to an extreme point on the graph of the feasible region.

Thus, extreme points of the feasible region of an LPP represent the basic feasible solutions to the problem and in looking for the optimal solution to an LPP, we need only locate the best basic feasible solution. Further, each simplex tableau represents a basic feasible solution except only when there is an 'artificial' variable in the basis (a situation discussed in Example 3.3).

Testing the Optimality

To test whether the solution obtained in simplex tableau 1 is optimal or not, we calculate Δ_j values as being equal to $c_j - z_j$, which are given at the bottom row of Table 3.1. To obtain the value of z_j under each variable head column, first each element of that column is multiplied by the corresponding coefficient of the solution variables appearing in the basis. Then the products are added up and we get z_j. The values of z_j represent the amount by which the profit would be reduced if one unit of any of the variables (x_1, x_2, S_1, or S_2) were added to the mix. For instance, to introduce one unit of x_1, the entries in this column indicate that 2 kg of raw material and 4 labour hours must be given up. Since the unused resources are assumed to have zero cost, there would be no reduction in the profit. Hence, $z_1 = 0 \times 2 + 0 \times 4 = 0$.

With c_j as the profit per unit, $c_1 = 40$, $\Delta_j = c_j - z_j$ represents the net profit which would result from introducing one unit of variable to the product mix–that is, the solution. Since a unit of x_1 adds Rs 40 to the profit and its introduction causes no loss, $\Delta_1 = 40 - 0 = 40$.

Now from the tableau, for column headed x_1, $z_1 = 0 \times 2 + 0 \times 4 = 0 \quad \therefore \Delta_1 = 40 - 0 = 40$;

" " " " " " x_2, $z_2 = 0 \times 3 + 0 \times 3 = 0 \quad \therefore \Delta_2 = 35 - 0 = 35$;

" " " " " " S_1, $z_3 = 0 \times 1 + 0 \times 0 = 0 \quad \therefore \Delta_3 = 0 - 0 = 0$;

" " " " " " S_2, $z_4 = 0 \times 0 + 0 \times 1 = 0 \quad \therefore \Delta_4 = 0 - 0 = 0$.

The Δ_j row is also called the *net-after-opportunity-cost row,* or the net evaluation row (NER).

The Test As indicated earlier, except when an *artificial* variable is included in the basis, a simplex tableau depicts an optimal solution if all entries in the Δ_j row are

(a) Zero or negative (i.e. all $\Delta_j \leq 0$) when the LPP is of maximisation type, and

(b) Zero or positive (i.e. all $\Delta_j \geq 0$) when the LPP is of minimisation type.

Now, looking at the Δ_j values in simplex tableau 1, we observe that the solution is not optimal because it contains positive values.

Deriving a Revised Tableau for Improved Solution

The presence of positive Δ_j values suggests that the solution can be improved upon by moving any one of the variables into the solution that are not there.

Each of the values in the Δ_j row signifies the amount of increase in the objective function that would occur if one unit of the variable (represented by the column head) were introduced into the solution. We select the variable that has the largest Δ_j value—the variable x_1. This is designated as the *incoming variable* and the selection of it is indicated by an arrow under the column headed x_1 in the Table 3.2 which reproduces the information in Table 3.1.

TABLE 3.2 Simplex Tableau 1: Non-optimal Solution

Basis		x_1	x_2	S_1	S_2	b_i	b_i/a_{ij}
S_1	0	2	3	1	0	60	30
S_2	0	4*	3	0	1	96	24 ← Outgoing variable (key row)
c_j		40	35	0	0		
Solution		0	0	60	96		
Δ_j		40	35	0	0		
		↑ Incoming variable (key column)					

The column corresponding to this variable is called the *key column*. Next, the b_i values are divided by the corresponding values in the key column, and we get the ratios b_i/a_{ij}, also called the *replacement ratios*. The row with the least non-negative quotient is then selected. It is called the *key row* and the variable corresponding to this represents the *outgoing variable*. The element which lies at the intersection of the key column and key row is termed as the *key element*. In the Table 3.2, we observe that the key row is represented by the arrow mark—the row with outgoing variable S_2. The key element is 4, with the asterisk mark (*).

Using the information on key column, key row and key element, another tableau is derived wherein the various elements are obtained as given here.

(a) Divide each element of the key row (including b_i) by the key element to get the corresponding values in the new tableau. The row of values so derived is called the *replacement row*.
For our example, the replacement row would be:
1 3/4 0 1/4 24

(b) For each row other than the key row,

$$\text{New row element} = \text{Old row element} - \begin{pmatrix}\text{Row element in the} \\ \text{key column}\end{pmatrix} \times \begin{pmatrix}\text{Corresponding replacement} \\ \text{row value}\end{pmatrix}$$

Accordingly, for the other row, the values shall be:

Old row element	*Minus*	*Row element in key column*	×	*Corresponding replacement row value*	=	*New element*
2	–	2	×	1	=	0
3	–	2	×	3/4	=	3/2
1	–	2	×	0	=	1
0	–	2	×	1/4	=	–1/2
60	–	2	×	24	=	12

The revised simplex tableau is given in Table 3.3.

TABLE 3.3 Simplex Tableau 2: Non-optimal Solution

Basis		x_1	x_2	S_1	S_2	b_i	b_i/a_{ij}
S_1	0	0	3/2*	1	–1/2	12	8 ← Outgoing variable
x_1	40	1	3/4	0	1/4	24	32
c_j		40	35	0	0		
Solution		24	0	12	0		
Δ_j		0	5	0	–10		
			↑ Incoming variable				

In this tableau, the c_j and the 'solution' values are written in the same way as in the previous tableau. According to this, the solution to the problem is: $x_1 = 24$, $x_2 = 0$, $S_1 = 12$ and $S_2 = 0$. Substituting these values in the objective function, we get

$$Z = 40 \times 24 + 35 \times 0 + 0 \times 12 + 0 \times 0 = 960.$$

The solution calls for producing 24 units of product *A* and none of the product *B*. A total of 12 kg of raw material would remain unutilised while the labour hours would be fully used by this decision which would bring in a profit of Rs 960. Notice that this decision corresponds to the corner point *R* in the graph in the Figure 3.1.

Now we shall test whether this solution is optimal or not. For this purpose, the Δ_j values are derived in a similar manner as before. Here they all are not less than, or equal to, zero. Therefore, the solution is not optimal.

Revised Tableau for Further Improved Solution

To obtain an improved solution to the problem, we proceed to derive a revised tableau as follows.

(a) Obtain the key column, key row and key element. The largest Δ_j, equal to 5, corresponds to the variable x_2. Thus, it is taken to be the incoming variable and the column is marked as *key column.* Using the a_{ij} values of the key column, the replacement ratios are found to be 8 and 32. Accordingly, the first row is the *key row* and the outgoing variable is S_1.

The intersection of the key column and key row yields the key element as 3/2, which is shown with an asterisk mark in Table 3.3.

(b) With information about key column, key row and key element, we can now obtain a revised simplex tableau in a manner similar to the one discussed earlier. This is contained in Table 3.4.

TABLE 3.4 Simplex Tableau 3: Optimal Solution

Basis		x_1	x_2	S_1	S_2	b_i
x_2	35	0	1	2/3	–1/3	8
x_1	40	1	0	–1/2	1/2	18
c_j		40	35	0	0	
Solution		18	8	0	0	
Δ_j		0	0	–10/3	–25/3	

The solution given by this corresponds to another extreme point *Q*, of the Figure 3.1. This solution, $x_1 = 18$, $x_2 = 8$, yields the objective function value $40 \times 18 + 35 \times 8 + 0 \times 0 + 0 \times 0 = 1{,}000$. By this, both the resources would be fully utilized. The values in the Δ_j row, being ≤ 0, indicate that the solution is optimal.

Besides indicating optimality, the Δ_j row values in the final tableau give significant information—the shadow prices of the resources. This is discussed in detail in the next chapter.

3.3.1 Justification and Significance of Elements in Simplex Tableau

We examine now the meaning and significance of the b_i values, the substitution rates (contained in the body of simplex tableau) and the Δ_j values in turn, first for Simplex Tableau 2 and then for Simplex Tableau 3.

For Simplex Tableau 2 The tableau is reproduced in Table 3.5.

TABLE 3.5 Simplex Tableau 2

Basis		x_1	x_2	S_1	S_2	b_i
S_1	0	0	3/2	1	–1/2	12
x_1	40	1	3/4	0	1/4	24
c_j		40	35	0	0	
Solution		24	0	12	0	
Δ_j		0	5	0	–10	

b_i Values In the simplex tableau 1 containing initial solution (see Table 3.2), we noted that the product A (x_1) yielded a profit of Rs 40 per unit while the product B (x_2) would contribute Rs 35 per unit. Hence, we decided to add x_1 to the next solution. With focus on the product A, the quantity to be added could be determined as:

$$\frac{\text{Number of units of raw material available}}{\text{Raw material required per unit of product } A} = \frac{60}{2} = 30,$$

$$\frac{\text{Number of labour hours available}}{\text{Labour hours required per unit of product } A} = \frac{96}{4} = 24$$

Thus, 24 units was determined as the largest quantity which could be produced without violating the resource availability restrictions.

The production of 24 units of product A would exhaust all the labour hours while this output would require $24 \times 2 = 48$ kg of raw material, leaving $60 - 48 = 12$ kg unused. The b_i values 12 for S_1 and 24 for x_1 in the simplex tableau 2 are indicative of this.

Substitution Rates Let us now understand about the substitution rates contained in the body of the tableau.

Column headed S_2: To begin with, the values under column headed S_2 may be interpreted. Since one unit of product A requires 4 hours of labour, the solution in the tableau uses up all the 96 labour hours available. Therefore, production of anything else using labour would require that some of the units of this product are given up. For example, if one unit of S_2, that is, one hour, is made available for other purposes, then 1/4 units of x_1 would have to be given up. In other words, every labour hour added to the solution reduces the output of product A, x_1, by 1/4 units. Now, the reduction in the output of product A would have an effect on the raw material, since both the products require each of the resources. Since product A requires 2 kg of raw material per unit and because adding 1 hour of S_2 reduces the output of A, x_1, by 1/4 units, $1/4 \times 2 = 1/2$ kg of raw material would be freed. The values –1/2 and 1/4 corresponding to S_1 and x_1, respectively, under S_2 indicate these facts.

Column headed S_1: A zero corresponding to x_1 in this column implies that reducing 1 kg of raw material has no effect on the output of A. This is because here labour hours is the limiting factor of production, all the hours having been used and making available 1 kg of raw material will cause no change in output of this product. With 12 kg of raw material still available, we can make 1 kg available without reducing the output of A, x_1.

However, *a* '1' corresponding to S_1 in this column implies that withdrawing 1 kg of raw material would remove 1 of the 12 kg of raw material as indicated by this solution.

Column headed x_1: A '1' corresponding to x_1 in this column implies that each unit of product *A* added to the current product mix would replace one unit of the same product in the current solution. Since a maximum of 24 units of the product can be produced in keeping with the availability of labour hours, it follows that to add a unit of *A*, while maintaining the restriction of labour hours, would require that we forgo 1 unit of the same product to make the necessary time available. Further, since adding a unit of *A* would necessarily involve reducing a unit of the same product, this would imply no change in requirement of raw material. This is indicated by the element zero against S_1.

Column headed x_2: The values 3/4 and 3/2 corresponding to x_1 and S_1, respectively in this column may be interpreted similarly. The addition of one unit of product *B*, x_2, would replace 3/4 unit of product *A*, x_1, because labour is the limiting resource (being fully used) and a unit of *B* requires 3 hours of labour while a unit of *A* requires 4. Thus, to produce a unit to *B*, the 3 labour hours needed would imply that we give up 3/4 unit of *A* to provide for the 3 hours. Also, adding a unit of *B* would replace 3/2 kg of raw material (S_1). It may be explained like this. Adding a unit of *B* replaces 3/4 units of *A*, and a unit of *A* requires 2 kg of raw material. Thus, giving up 3/4 units of *A* implies $2 \times 3/4 = 3/2$ kg would be released. Now, 3 kg of raw material required to produce a unit of *B minus* the 3/2 kg released equals $3 - 3/2 = 3/2$ kg net change.

Thus, the elements in the main body of the table are the marginal rates of substitution between the various variables involved, with positive rates indicating a decrease and the negative rates indicating a release/increase.

Δ_j Elements We may consider now the explanation of the elements in the Δ_j row outside the simplex tableau. Since Δ_j is defined as $c_j - z_j$, we first understand the meaning of the z_j values. A z_j represents the loss of profit that results from the addition of one unit of the variable heading a particular column. In simplex tableau 2, the z_j values for the columns headed x_1, x_2, S_1, and S_2 are 40, 30, 0 and 10, respectively.

For column headed x_1: As observed earlier, the addition of one unit of product $A(x_1)$ causes a reduction of one unit of the same product and a 0 change in S_1. The reduction of a unit *A* would result in a loss of profit of Rs 40. Hence, for x_1, the value equals 40.

For column headed x_2: The addition of one unit of product *B* (represented by x_2) results in two changes: (1) S_1 is reduced by 3/2 units and (2) x_1 is reduced by 3/4 units. Now, since profit per unit of unused material, S_1 is nil, the change in S_1 shall cause no change in the profit. However, since profit per unit of x_1 is Rs 40, a reduction of 3/4 units of x_1 would result in a loss of $40 \times 3/4 =$ Rs 30. Hence, the total loss of profit resulting from adding a unit of x_2, given by z_2, is $0 + 30 = 30$.

For column headed S_1: The addition of one unit of S_1 causes a substitution of 1 unit of the same and none of x_1. Therefore, the loss of profit is nil.

For column headed S_2: Adding one unit of S_2 produces two changes: (1) freeing of 1/2 unit of S_1, and (2) giving up of 1/4 units of x_1. The release of 1/2 units of S_1 has no effect on the profit because the profit rate for S_1 is zero. However, the reduction of 1/4 units of x_1 would cause a reduction of $40 \times 1/4 =$ Rs 10 (since Rs 40 is the unit profit of product *A*). Thus, an addition of S_2 involves a loss of Rs 10 of profit.

We may now turn to the Δ_j values. The c_j values for the various columns indicate the profits associated with the unit values of the variables involved. The values are 40, 35, 0 and 0, respectively, for x_1, x_2, S_1, and S_2. Thus, if one unit of x_1 is added to the solution, it would add Rs 40 to the profit whereas the addition of the unit also involves a loss of profit of Rs 40, as indicated by the z_j value for this variable. Therefore, the net change to profit on adding one unit of x_1, given by $\Delta_j = c_j - z_j = 40 - 40 = 0$. Similarly, for x_2 the unit profit is Rs 35, whereas the loss of profit is Rs 30, as noted earlier. Accordingly, the net change in profit resulting from the addition of one unit of x_2 (product *B*) is 35 – 30 = Rs 5. Now, for S_1, the net change is zero while for S_2, adding a unit in the solution would involve a net loss of Rs 10 (there being no profit on adding and only a loss of profit of Rs 10).

To conclude, then, a positive number in the Δ_j row indicates that the objective function value may be increased further while a negative element in the row reflects the loss (in profit) that would result if one unit of the variable heading the particular variable were added to the solution.

For Simplex Tableau 3 This is reproduced in Table 3.6 and an explanation of its elements follows:

TABLE 3.6 Simplex Tableau 3

Basis		x_1	x_2	S_1	S_2	b_i
x_2	35	0	1	2/3	–1/3	8
x_1	40	1	0	–1/2	1/2	18
c_j		40	35	0	0	
Solution		18	8	0	0	
Δ_j		0	0	–10/3	–25/3	

(i) The b_i values indicate the output of products *B* and *A* to be 8 and 18 units respectively.

(ii) The substitution rates and the Δ_j values under different variables indicate the following:

The values 0 and 1 under x_1 suggest that a unit of x_1 added to the current product-mix will call for replacing one unit of the same product in the solution with no change needed in the other product. Thus, addition to profit for adding a unit of product *A* (i.e. x_1) = Rs 40, and loss of profit for replacing a unit from the existing mix is also equal to Rs 40. Accordingly, the net change in profit, $\Delta_j = 40 - 40 = 0$.

A similar interpretation may be given for the values 1 and 0, and $\Delta_j = 0$ in respect of the variable x_2.

The substitution rates under S_1 are 2/3 against x_2 and –1/2 against x_1. These imply that releasing one kg of raw material will entail losing 2/3 units of product *B* and gaining 1/2 unit of product *A* (since one unit of product *B* requires 3 kg of raw material, reducing 2/3 unit of it will release 3 × 2/3 = 2 kg while adding 1/2 unit of x_1 would need one kg of raw material since one unit of product *A* requires 2 kg of raw material).

The loss of 2/3 unit of x_2 and the gain of 1/2 unit of x_1 would result in a net loss of (35 × 2/3) – (40 × 1/2) = Rs 10/3. This is indicated by $\Delta_j = -10/3$.

In a similar way, it may be observed that the values –1/3 and 1/2 under S_2 indicate the substitution rates with respect to x_2 and x_1 respectively. Withdrawing an hour of labour would cause a reduction of 1/2 unit of product *A* and addition of 1/3 unit of product *B*. The reduction of 1/2 unit of *A* would release 4 × 1/2 = 2 hours of labour while addition of 1/3 unit of *B* would consume 3 × 1/3 = 1 hour, resulting in a net release of one hour of labour as desired. These changes shall cause a net reduction of Rs 25/3 in profit – a loss of 40 × 1/2 = Rs 20 and gain of 35 × 1/3 = Rs 35/3 – as indicated by the Δ_j value.

Example 3.2 A firm produces three products *A*, *B*, and *C*, each of which passes through three departments: Fabrication, Finishing and Packaging. Each unit of product *A* requires 3, 4 and 2; a unit of product *B* requires 5, 4 and 4, while each unit of product *C* requires 2, 4 and 5 hours respectively in the three departments. Everyday, 60 hours are available in the fabrication department, 72 hours in the finishing department and 100 hours in the packaging department.

The unit contribution of product *A* is Rs 5, of product *B* is Rs 10, and of product *C* is Rs 8.

Required:

(a) Formulate the problem as an LPP and determine the number of units of each of the products, that should be made each day to maximise the total contribution. Also determine if any capacity would remain unutilised.

(b) If the optimal solution obtained does not require the production of some product, explain as to why such product would not be produced. In this context, indicate the quantity (quantities) of other product/s that would be foregone for producing such product.

(c) What would be the effect on the solution of each of the following:

(i) obtaining an order for 6 units of product *A*, which has to be met.

(ii) an increase of 20 percent capacity in the fabrication department.

(a) Let x_1, x_2 and x_3 represent the number of units of products *A, B* and *C* respectively. The given problem can be expressed as a LPP as follows:

Maximise $Z = 5x_1 + 10x_2 + 8x_3$ Contribution

Subject to

$3x_1 + 5x_2 + 2x_3 \leq 60$ Fabrication hours

$4x_1 + 4x_2 + 4x_3 \leq 72$ Finishing hours

$2x_1 + 4x_2 + 5x_3 \leq 100$ Packaging hours

$x_1, x_2, x_3 \geq 0$

Introducing slack variables, the augmented problem can be written as

Maximise $Z = 5x_1 + 10x_2 + 8x_3 + 0S_1 + 0S_2 + 0S_3$

Subject to

$$3x_1 + 5x_2 + 2x_3 + S_1 = 60$$
$$4x_1 + 4x_2 + 4x_3 + S_2 = 72$$
$$2x_1 + 4x_2 + 5x_3 + S_3 = 100$$
$$x_1, x_2, x_3, S_1, S_2, S_3 \geq 0$$

The solution to the problem using simplex algorithm is contained in Tables 3.7 to 3.9.

TABLE 3.7 Simplex Tableau 1: Non-optimal Solution

Basis		x_1	x_2	x_3	S_1	S_2	S_3	b_i	b_i/a_{ij}	
S_1	0	3	5*	2	1	0	0	60	12	← Outgoing variable
S_2	0	4	4	4	0	1	0	72	18	
S_3	0	2	4	5	0	0	1	100	25	
c_j		5	10	8	0	0	0			
Solution		0	0	0	60	72	100			
Δ_j		5	10	8	0	0	0			
			↑ Incoming variable							

TABLE 3.8 Simplex Tableau 2: Non-optimal Solution

Basis		x_1	x_2	x_3	S_1	S_2	S_3	b_i	b_i/a_{ij}	
x_2	10	3/5	1	2/5	1/5	0	0	12	30	
S_2	0	8/5	0	12/5*	– 4/5	1	0	24	10	← Outgoing
S_3	0	– 2/5	0	17/5	– 4/5	0	1	52	260/17	variable
c_j		5	10	8	0	0	0			
Solution		0	12	0	0	24	52			
Δ_j		– 1	0	4	– 2	0	0			
				↑						
				Incoming variable						

TABLE 3.9 Simplex Tableau 3: Optimal Solution

Basis		x_1	x_2	x_3	S_1	S_2	S_3	b_i
x_2	10	1/3	1	0	1/3	–1/6	0	8
x_3	8	2/3	0	1	–1/3	5/12	0	10
S_3	0	–8/3	0	0	1/3	–17/12	1	18
c_j		5	10	8	0	0	0	
Solution		0	8	10	0	0	18	
Δ_j		–11/3	0	0	–2/3	–5/3	0	

According to the Simplex Tableau 3, the optimal solution is: $x_1 = 0$, $x_2 = 8$, $x_3 = 10$. Thus, it calls for producing 8 and 10 units of products *B* and *C* respectively, each day. This mix would yield a contribution of $5 \times 0 + 10 \times 8 + 8 \times 10 =$ Rs 160. S_3 being equal to 18, an equal number of hours shall remain unutilised in the packaging department.

(b) Evidently, the optimal solution does not require the production of product *A*. It can be explained as follows. The production of one unit of *A* requires giving up 1/3 units of *B* (x_2) and 2/3 units of *C* (x_3). With the unit profit rates as 10 and 8, respectively, for these, the loss of profit would be $10 \times 1/3 + 8 \times 2/3 = 26/3$. Reduction of x_2 and x_3, together with production of one unit of *A*, would cause a net release of 8/3 hours in the packaging department as shown here:

Product	*Hours per Unit*	*Units Produced (reduced)*	*Capacity Used (released)*
A	2	1	2
B	4	(1/3)	(4/3)
C	5	(2/3)	(10/3)
		Net change	(8/3)

However, the released capacity would not affect the profit because the unit profit for it is nil. Hence, total loss of profit necessitated by substitution = 10/3 + 16/3 + 0 = 26/3. With a profit contribution of per unit of product being Rs 5, the net loss per unit equals 11/6, the difference of the two as given by the Δ_j value.

(c) To determine the effect on solution of each of the two changes, we proceed as follows:

(i) To meet the demand of 6 units of product *A*, the necessary changes may be known by reference to the substitution rates given in column headed x_1 in the following table.

Variable	Original Value		Change		New Value
x_2	8	–	(l/3 × 6)	=	6
x_3	10	–	(2/3 × 6)	=	6
S_3	18	–	(–8/3 × 6)	=	34
Z	160	–	(11/3 × 6)	=	138

Thus, forcing output of 6 units of product *A* would cause a reduction of profit by 11/3 × 6 = Rs 22 leading to a total profit equal to Rs 138. It would cause a reduction in the output of product *B* by 2 units and product *C* by 4 units. Of course, it will result in a lower utilisation of packaging hours by 16 hours.

(ii) Addition of capacity in fabrication would cause the following changes in the solution (obtained through substitution rates in column headed S_1).

Variable	Original Value		Change		New Value
x_2	8	+	(1/3 × 12)	=	12
x_3	10	+	(–1/3 × 12)	=	6
S_3	18	+	(1/3 × 12)	=	22
Z	160	+	(2/3 × 12)	=	168

Thus, adding 12 hours of fabrication would raise the output of product *B* by 4 units and reduce the output of product *C* by 4 units and release 4 hours in the packaging department. This would have a net effect of increasing profit by Rs 8.

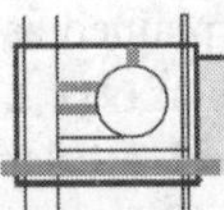

3.4 SOLUTION TO MINIMISATION PROBLEMS

The solution procedure for the linear programming problems that have the objective function of the minimisation type, is similar to the one for the maximisation problems, except for some differences. To illustrate, let us again consider the Example 2.2.

Example 3.3 Minimise $Z = 40x_1 + 24x_2$ Total cost

Subject to

$20x_1 + 50x_2 \geq 4{,}800$ Phosphate requirement

$80x_1 + 50x_2 \geq 7{,}200$ Nitrogen requirement

$x_1, x_2 \geq 0$

Following the approach already discussed, we first introduce some new variables to convert inequalities of the system into equations. The variable required for converting a 'greater than' type of inequality into an equation is called *surplus variable* and it represents the excess of what is generated (given by the LHS of the inequality) over the requirement (shown by the RHS value b_i). With surplus variables, S_1 and S_2 respectively for the first and the second constraints, the augmented problem shall be

Minimise $\quad Z = 40x_1 + 24x_2 + 0S_1 + 0S_2$

Subject to

$$20x_1 + 50x_2 - S_1 = 4{,}800$$
$$80x_1 + 50x_2 - S_2 = 7{,}200$$
$$x_1, x_2, S_1, S_2 \geq 0$$

Now, as soon as we proceed to the next step we experience a problem which is like this. We know that the simplex method needs an initial solution to get the process started. In this case, it is easy to visualize that an initial solution does not exist because, if we let x_1 and x_2 each equal to zero, we get $S_1 = -4{,}800$ and $S_2 = -7{,}200$, which is not feasible as it violates the *non-negativity* restriction. In terms of the simplex tableau, when we write all the information, we observe that we do not get identity because unlike in case of slack variables, the coefficient values of surplus variables S_1 and S_2 appear as minus one (–1).

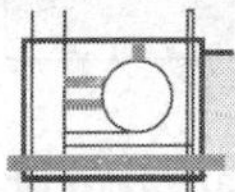

3.5 BIG-M METHOD

In a case where an identity is not obtained, as in the problem under consideration, a variant of the simplex method called the Big-M method is employed. In this method, we add *artificial variables* into the model to obtain an initial solution. Unlike *slack* or *surplus* variables, artificial variables have no tangible relationship with the decision problem. Their sole purpose is to provide an initial solution to the given problem.

When artificial variables are introduced in the problem under consideration, its constraints appear as

$$20x_1 + 50x_2 - S_1 + A_1 = 4{,}800$$
$$80x_1 + 50x_2 - S_2 + A_2 = 7{,}200$$

It is significant to understand that the artificial variables, which are not seen to disturb the equations already obtained since they are not 'real', are introduced for the limited purpose of obtaining an initial solution and are required for the constraints of '≥' type, or the constraints with '=' sign. It is not relevant whether the objective function is of the minimisation or the maximisation type. Obviously, since artificial variables do not represent any quantity relating to the decision problem, they must be driven out of the system and must not show in the final solution (and if at all they do, it represents a situation of infeasibility, which is discussed later in this chapter). This can be ensured by assigning an extremely high cost to them. Generally, a value M is assigned to each artificial variable, where M represents a number higher than any finite number. For this reason, the method of solving the problems where artificial variables are involved is termed as the *Big-M Method.* When the problem is of the minimisation nature, we assign in the objective function a coefficient of $+M$ to each of the artificial variables. On the other hand, for the problems with the objective function of maximisation type, each of the artificial variables introduced has a coefficient $-M$. Note that it is attempted to prohibit the appearance of artificial variables in the solution by assigning these coefficients: an extremely large value when the objective is to minimise and an extremely small (negative) value when it is desired to maximise the objective function.

For our present example, the objective function would appear as

$$\text{Minimise} \qquad Z = 40x_1 + 24x_2 + 0S_1 + 0S_2 + MA_1 + MA_2$$

It is to note that the initial solution obtained using artificial variables is *not* a feasible solution to the given problem. It only gives a starting point and the artificial variables are driven out in the normal course of applying the simplex algorithm. In fact, as long as an artificial variable is included in the basis, the solution is infeasible and only a solution to the problem which does not include an artificial variable in the basis, represents a feasible solution to the problem.

The initial simplex tableau giving the initial solution to our problem is given in Table 3.10.

TABLE 3.10 Simplex Tableau 1: Non-optimal Solution

Basis		x_1	x_2	S_1	S_2	A_1	A_2	b_i	b_i/a_{ij}
A_1	M	20	50*	–1	0	1	0	4,800	96 ←
A_2	M	80	50	0	–1	0	1	7,200	144
c_j		40	24	0	0	M	M		
Solution		0	0	0	0	4,800	7,200		
Δ_j		$40 - 100M$	$24 - 100M$ ↑	M	M	0	0		

For the minimisation problem, the optimal solution is indicated when the values in the Δ_j row are zero or positive. The presence of negative Δ_j values indicates that the solution can be improved.

Revised Solution To obtain a revised simplex tableau, the incoming variable is selected to be the one with the most negative Δ_j value and the column headed by this variable is called, as before, the key column. The selection of the key row (and the outgoing variable) is done exactly the same way as for maximisation problems—the row which has the least (non-negative) quotient is our row of interest. Finally, the revised simplex tableau is derived in the same way as discussed earlier. We proceed in this manner until optimal solution is obtained.

In respect of our problem, the initial solution is not optimal. Here the incoming variable is x_2 while A_1 is the outgoing variable. The revised tableau is given in Table 3.11. In a similar way, we proceed until the optimal solution is had. Table 3.13 contains the optimal solution.

TABLE 3.11 Simplex Tableau 2: Non-optimal Solution

Basis		x_1	x_2	S_1	S_2	A_1	A_2	b_i	b_i/a_{ij}
x_2	24	2/5	1	–1/50	0	1/50	0	96	240
A_2	M	60*	0	1	–1	–1	1	2,400	40 ←
c_j		40	24	0	0	M	M		
Solution		0	96	0	0	0	2,400		
Δ_j		$\frac{152}{5} - 60M$ ↑	0	$\frac{12}{25} - M$	M	$2M - \frac{12}{25}$	0		

TABLE 3.12 Simplex Tableau 3: Non-optimal Solution

Basis		x_1	x_2	S_1	S_2	A_1	A_2	b_i	b_i/a_{ij}
x_2	24	0	1	–2/75	1/150	2/75	–1/150	80	–3,000
x_1	40	1	0	1/60*	–1/60	–1/60	1/60	40	2,400 ←
c_j		40	24	0	0	M	M		
Solution		40	80	0	0	0	0		
Δ_j		0	0	–2/75	38/75	$M+\frac{2}{75}$	$M+\frac{38}{75}$		

TABLE 3.13 Simplex Tableau 4: Optimal Solution

Basis		x_1	x_2	S_1	S_2	A_1	A_2	b_i
x_2	24	8/5	1	0	–1/50	0	1/50	144
S_1	0	60	0	1	–1	–1	1	2,400
c_j		40	24	0	0	M	M	
Solution		0	144	2,400	0	0	0	
Δ_j		8/5	0	0	12/25	M	M – 12/25	

It may be noted that the solutions given in Tables 3.10 and 3.11 are both infeasible. The solution in Table 3.10 gives $x_1 = 0$ and $x_2 = 0$ which violates both the constraints while the solution given in Table 3.11 indicates $x_1 = 0$ and $x_2 = 96$, that does not satisfy the first constraint. The solution given in Table 3.12 is feasible since there is no artificial variable in the basis but is not optimal while the one in Table 3.13 is optimal (and, of course, feasible).

According to the optimal solution, the objective function value is

$$Z = 40 \times 0 + 24 \times 144 + 0 \times 2400 + 0 \times 0 + 0 \times M + 0 \times M = \text{Rs } 3{,}456$$

The value of $S_1 = 2{,}400$ indicates the surplus phosphate ingredient obtained by buying the least cost mix.

Problem with Mixed Constraints

We shall consider a problem with mixed constraints.

Example 3.4 Solve the following LPP.

Maximise $Z = 2x_1 + 4x_2$

Subject to

$$2x_1 + x_2 \le 18$$
$$3x_1 + 2x_2 \ge 30$$
$$x_1 + 2x_2 = 26$$
$$x_1, x_2 \ge 0$$

After introducing the necessary *slack, surplus,* and *artificial* variables, the augmented problem is given here:

Maximise $Z = 2x_1 + 4x_2 + 0S_1 + 0S_2 - MA_1 - MA_2$

Subject to

$$2x_1 + x_2 + S_1 = 18$$
$$3x_1 + 2x_2 - S_2 + A_1 = 30$$
$$x_1 + 2x_2 + A_2 = 26$$
$$x_1, x_2, S_1, S_2, A_1, A_2 \geq 0$$

The solution is contained in Tables 3.14 through 3.16.

TABLE 3.14 Simplex Tableau 1: Non-optimal Solution

Basis		x_1	x_2	S_1	S_2	A_1	A_2	b_i	b_i/a_{ij}
S_1	0	2	1	1	0	0	0	18	18
A_1	$-M$	3	2	0	−1	1	0	30	15
A_2	$-M$	1	2*	0	0	0	1	26	13 ←
c_j		2	4	0	0	$-M$	$-M$		
Solution		0	0	18	0	30	26		
Δ_j		$4M+2$	$4M+4$	0	$-M$	0	0		
			↑						

TABLE 3.15 Simplex Tableau 2: Non-optimal Solution

Basis		x_1	x_2	S_1	S_2	A_1	A_2	b_i	b_i/a_{ij}
S_1	0	3/2	0	1	0	0	−1/2	5	10/3
A_1	$-M$	2*	0	0	−1	1	−1	4	2 ←
x_2	4	1/2	1	0	0	0	1/2	13	26
c_j		2	4	0	0	$-M$	$-M$		
Solution		0	13	5	0	4	0		
Δ_j		$2M$	0	0	$-M$	0	$-2-2M$		
		↑							

TABLE 3.16 Simplex Tableau 3: Optimal Solution

Basis		x_1	x_2	S_1	S_2	A_1	A_2	b_i
S_1	0	0	0	1	3/4	−3/4	1/4	2
x_1	2	1	0	0	−1/2	1/2	−1/2	2
x_2	4	0	1	0	1/4	−1/4	3/4	12
c_j		2	4	0	0	$-M$	$-M$	
Solution		2	12	2	0	0	0	
Δ_j		0	0	0	0	$-M$	$-M-2$	

The optimal solution to the problem is: $x_1 = 2$ and $x_2 = 12$, $S_1 = 2$ and other variables = 0. The objective function value is $2 \times 2 + 4 \times 12 = 52$.

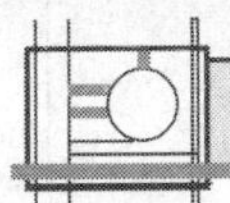

3.6 TWO-PHASE METHOD

As an alternative to the *Big-M* method, there is also available another method for dealing with linear programming problems involving artificial variables. This is called the *two-phase method* and, as its name implies, it separates the solution procedure into two phases. In phase-I, all the artificial variables are eliminated from the basis. If a feasible solution is obtained in this phase, which has no artificial variables in the basis in the final tableau, then we proceed to phase II. In this phase, we use the solution from phase I as the *initial* basic feasible solution and use the simplex method to determine the optimal solution. The method is discussed in the following paragraphs.

Suppose that we are given the following LPP:

Minimise $$Z = \sum_{j=1}^{n} c_j x_j$$

Subject to

$$\sum_{j=1}^{n} a_{ij}x_j \geq b_i \quad i = 1, 2, \ldots, m$$

$$x_j \geq 0 \quad j = 1, 2, \ldots, n$$

To solve it, we follow steps detailed here.

3.6.1 Phase I

Step 1 By subtracting a surplus variable, convert each of the constraints into equality relationships and then add an artificial variable.

Step 2 Assign zero coefficients to each of the primary (x_j) variables and to the surplus variables; and assign unit coefficients to each of the artificial variables (in a maximisation problem, the coefficients to each of the artificial variables shall be –1). This amounts to replacing the objective function of the original problem by the sum of the artificial variables.

The assignment of the stated coefficients yields the following auxiliary problem:

Minimise $$Y = \sum_{j=1}^{n} 0x_j + \sum_{i=1}^{m} 0S_i + \sum_{i=1}^{m} 1A_i \text{ or Min. } \Sigma A_i$$

Subject to

$$\sum_{j=1}^{n} a_{ij}x_j - S_i + A_i = b_i \quad i = 1, 2, \ldots, m$$

and $$x_j, S_i, A_i \geq 0 \text{ for all } i \text{ and } j$$

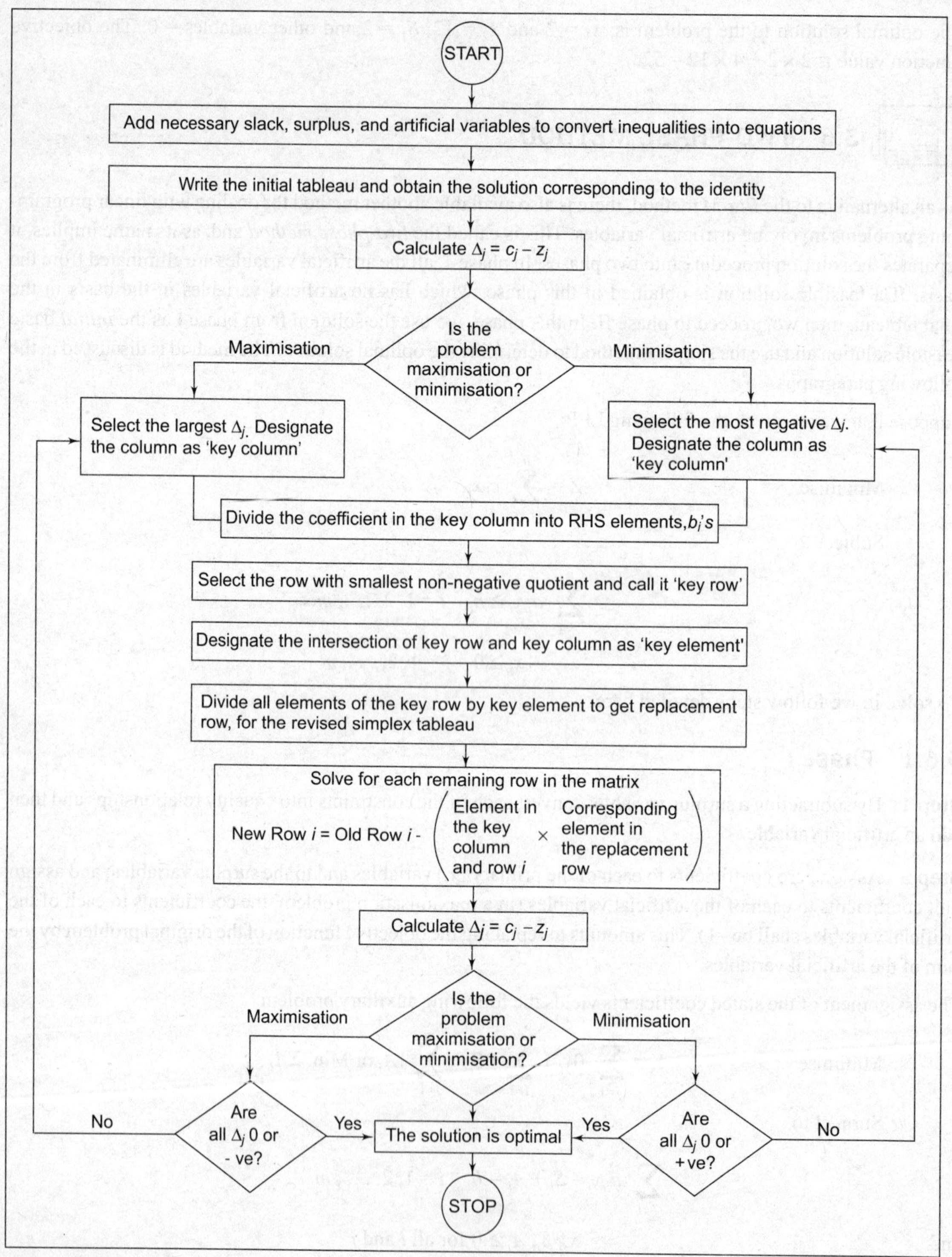

Figure 3.2 *Schematic Presentation of Simplex Method*

Step 3 Solve the auxiliary problem by applying the simplex method. If the original problem has a feasible solution, then this problem shall have an optimal solution with optimal value of the objective function equal to zero as each of the artificial variables $A_1, A_2, \ldots, A_m$ shall be equal to zero. The simplex method would remove all the artificial variables from the basis one by one.

It is significant to note that the value of objective function of the problem is bounded from below by zero, because the objective function represents the sum of artificial variables with positive unit coefficients. Thus, the auxiliarly problem cannot have unbounded solution and the simplex method should terminate after a finite number of iterations.

3.6.2 Phase II

In this phase, start with the (optimal) solution contained in the final simplex tableau of the phase I. Remove c_j row values at the bottom of the optimal tableau and replace them by the c_j values of the original problem. Further, eliminate the entries in the columns headed $A_1, A_2, \ldots, A_m$. Apply simplex algorithm to the problem contained in the new tableau to obtain the optimal solution. This would be the optimal solution to the given problem.

Example 3.5 Solve the problem given in Example 3.3 by using the two-phase method. The problem is:

Minimise $Z = 40x_1 + 24x_2$ Total cost

Subject to

$20x_1 + 50x_2 \geq 4{,}800$ Phosphate requirement

$80x_1 + 50x_2 \geq 7{,}200$ Nitrogen requirement

$x_1, x_2 \geq 0$

Phase I First we introduce surplus and artificial variables, and then rewrite the objective function by assigning a zero coefficient to the decision variables and a coefficient 1 to the artificial variables. The problem becomes

Minimise $Z = 0x_1 + 0x_2 + 0S_1 + 0S_2 + A_1 + A_2$

Subject to $20x_1 + 50x_2 - S_1 + A_1 = 4{,}800$

$80x_1 + 50x_2 - S_2 + A_2 = 7{,}200$

$x_1, x_2, S_1, S_2, A_1, A_2 \geq 0$

The solution to this problem shall be obtained using simplex method. It is given in Tables 3.17 through 3.19.

TABLE 3.17 Simplex Tableau 1: Non-optimal Solution

Basis		x_1	x_2	S_1	S_2	A_1	A_2	b_i	b_i/a_{ij}	
A_1	1	20	50	–1	0	1	0	4,800	240	
A_2	1	80*	50	0	–1	0	1	7,200	90	←
c_j		0	0	0	0	1	1			
Solution		0	0	0	0	4,800	7,200			
Δ_j		–100 ↑	–100	1	1	0	0			

TABLE 3.18 Simplex Tableau 2: Non-optimal Solution

Basis		x_1	x_2	S_1	S_2	A_1	A_2	b_i	b_i/a_{ij}
A_1	1	0	75/2*	–1	1/4	1	–1/4	3,000	80 ←
x_1	0	1	5/8	0	–1/80	0	1/80	90	144
c_j		0	0	0	0	1	1		
Solution		90	0	0	0	3,000	0		
Δ_j		0	–75/2	0	–1/4	0	5/4		

TABLE 3.19 Simplex Tableau 3: Optimal Solution

Basis		x_1	x_2	S_1	S_2	A_1	A_2	b_i
x_2	0	0	1	–2/75	1/150	2/75	–1/150	80
x_1	0	1	0	1/60	–1/60	–1/60	–1/60	40
c_j		0	0	0	0	0	1	
Solution		40	80	0	0	0	0	
Δ_j		0	0	0	0	1	1	

Phase II The Simplex Tableau 3 is reproduced here after replacing c_j row by the respective coefficients from the objective function of the original problem, and deleting the columns A_1 and A_2. The problem is then solved using simplex method as shown in Tables 3.20 and 3.21.

TABLE 3.20 Simplex Tableau 4: Non-optimal Solution

Basis		x_1	x_2	S_1	S_2	b_i	b_i/a_{ij}	
x_2	24	0	1	– 2/75	1/150	80	– 3,000	
x_1	40	1	0	1/60*	– 1/60	40	2,400	←
c_j		40	24	0	0			
Solution		40	80	0	0			
Δ_j		0	0	– 2/75	38/75			
				↑				

TABLE 3.21 Simplex Tableau 5: Optimal Solution

Basis		x_1	x_2	S_1	S_2	b_i
x_2	24	8/5	1	0	– 1/50	144
S_1	0	60	0	1	– 1	2400
c_j		40	24	0	0	
Solution		0	144	2,400	0	
Δ_j		8/5	0	0	12/25	

The optimal solution given by Simplex Tableau 5 is $x_1 = 0$ and $x_2 = 144$. It is the same as obtained earlier.

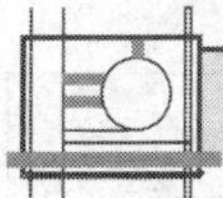

3.7 SOME SPECIAL TOPICS

3.7.1 Multiple Optimal Solutions

We have already seen in the previous chapter that the solution to a linear programming problem may or may not be unique. It was established in the graphic solution to the LPPs that when the iso-profit (or iso-cost) line has the same slope as that of a constraint, then the optimal solution to the given problem may not be unique and, instead, multiple optima may exist.

To demonstrate the existence of multiple optimal solutions using the simplex method, we consider again the data of Example 2.4 reproduced here.

Example 3.6 Maximise $Z = 8x_1 + 16x_2$

Subject to

$$x_1 + x_2 \le 200$$
$$x_2 \le 125$$
$$3x_1 + 6x_2 \le 900$$
$$x_1, x_2 \ge 0$$

Solution to this problem using simplex algorithm is contained in Tables 3.22 through 3.24.

TABLE 3.22 Simplex Tableau 1: Non-optimal Solution

Basis		x_1	x_2	S_1	S_2	S_3	b_i	b_i/a_{ij}	
S_1	0	1	1	1	0	0	200	200	
S_2	0	0	1*	0	1	0	125	125	←
S_3	0	3	6	0	0	1	900	150	
c_j		8	16	0	0	0			
Solution		0	0	200	125	900			
Δ_j		8	16	0	0	0			
			↑						

TABLE 3.23 Simplex Tableau 2: Non-optimal Solution

Basis		x_1	x_2	S_1	S_2	S_3	b_i	b_i/a_{ij}
S_1	0	1	0	1	−1	0	75	75
x_2	16	0	1	0	1	0	125	∞
S_3	0	3*	0	0	−6	1	150	50 ←
c_j		8	16	0	0	0		
Solution		0	125	75	0	150		
Δ_j		8 ↑	0	0	−16	0		

TABLE 3.24 Simplex Tableau 3: Optimal Solution

Basis		x_1	x_2	S_1	S_2	S_3	b_i
S_1	0	0	0	1	1	−1/3	25
x_2	16	0	1	0	1	0	125
x_1	8	1	0	0	−2	1/3	50
c_j		8	16	0	0	0	
Solution		50	125	25	0	0	
Δ_j		0	0	0	0	−8/3	

The values in the Δ_j row, being all zero or negative, suggest that the solution is an optimal one, with the objective function value equal to $8 \times 50 + 16 \times 125 = 2{,}400$.

It may be mentioned here that in every simplex tableau, the basic variables have all $\Delta_j = 0$, but not the non-basic ones. When a solution is indicated to be optimal and the Δ_j value for none of the non-basic variables is zero, the solution is *unique* in the sense that no other solution to the given problem exists, which would yield an identical value of the objective function. When a non-basic variable in an optimal solution has a zero value for Δ_j, then the solution is not unique, and *multiple* optimal solutions exist.

Notice that in Table 3.24, one of the non-basic variables, S_2, has a value zero in the Δ_j row. This variable has a net zero contribution and therefore it can be included in the basis without changing the value of the objective function. We can perform another iteration by taking the column headed by S_2 as the key column. The replacement ratios being 25, 125, and –25, the first row is the key row and S_1 the outgoing variable. The revised tableau would appear as follows.

TABLE 3.25 Simplex Tableau 4: Optimal Solution

Basis		x_1	x_2	S_1	S_2	S_3	b_i
S_2	0	0	0	1	1	−1/3	25
x_2	16	0	1	−10	0	1/3	100
x_1	8	1	0	2	0	−1/3	100
c_j		8	16	0	0	0	
Solution		100	100	0	25	0	
Δ_j		0	0	0	0	−8/3	

Here the solution is indicated to be optimal with objective function value equal to 8 × 100 + 16 × 100 = 2,400, the same as before. It may again be noticed that a non-basic variable, S_1, in this solution has a value zero in the Δ_j row, suggesting that the solution is one of the optima, and not unique.

To conclude, then, when a non-basic variable in the final tableau showing optimal solution to a problem has a net zero contribution then the optimal solution to given problem is not *one* but *multiple*.

3.7.2 Infeasibility

As already mentioned, infeasibility is said to exist when a given problem has no feasible solution, and is evident graphically when no common point is found in the feasible regions of all the constraints of a problem taken jointly. However, problems involving more than two variables cannot be graphed and it might not be readily known, when the model is constructed, that the problem has no feasible solution. The simplex method provides information as to where the infeasibility lies. In terms of the simplex algorithm, when in the final solution, an artificial variable is in the basis at a positive value then there is no feasible solution to the problem. We can demonstrate it by considering the data of Example 2.6 again.

Example 3.7 Maximise $Z = 20x_1 + 30x_2$

Subject to

$$2x_1 + x_2 \leq 40$$
$$4x_1 - x_2 \leq 20$$
$$x_1 \geq 30$$
$$x_1, x_2 \geq 0$$

Introducing the necessary slack, surplus and artificial variables, the augmented problem may be written as follows:

Maximise $Z = 20x_1 + 30x_2 + 0S_1 + 0S_2 + 0S_3 - MA_1$

Subject to

$$2x_1 + x_2 + S_1 = 40$$
$$4x_1 - x_2 + S_2 = 20$$
$$x_1 - S_3 + A_1 = 30$$
$$x_1, x_2, S_1, S_2, S_3, A_1 \geq 0$$

The solution is contained in Tables 3.26 through 3.28.

TABLE 3.26 Simplex Tableau 1: Non-optimal Solution

Basis		x_1	x_2	S_1	S_2	S_3	A_1	b_i	b_i/a_{ij}
S_1	0	2	1	1	0	0	0	40	20
S_2	0	4*	–1	0	1	0	0	20	5 ←
A_1	$-M$	1	0	0	0	–1	1	30	30
c_j		20	30	0	0	0	$-M$		
Solution		0	0	40	20	0	30		
Δ_j		$20 + M$ ↑	30	0	0	$-M$	0		

TABLE 3.27 Simplex Tableau 2: Non-optimal Solution

Basis		x_1	x_2	S_1	S_2	S_3	A_1	b_i	b_i/a_{ij}	
S_1	0	0	3/2*	1	– 1/2	0	0	30	20	← Outgoing
x_1	20	1	– 1/4	0	1/4	0	0	5	– 20	variable
A_1	– M	0	1/4	0	– 1/4	– 1	1	25	100	
c_j		20	30	0	0	0	– M			
Solution		5	0	30	0	0	25			
Δ_j		0	$35+\frac{M}{4}$ ↑	0	$-5-\frac{M}{4}$	– M	0			

TABLE 3.28 Simplex Tableau 3: Final, Non-optimal Solution

Basis		x_1	x_2	S_1	S_2	S_3	A_1	b_i
x_2	30	0	1	2/3	– 1/3	0	0	20
x_1	20	1	0	1/6	1/6	0	0	10
A_1	– M	0	0	– 1/6	– 1/6	– 1	1	20
c_j		20	30	0	0	0	– M	
Solution		10	20	0	0	0	20	
Δ_j		0	0	$-\frac{70}{3}-\frac{M}{6}$	$\frac{20}{3}-\frac{M}{6}$	– M	0	

Two-phase Method and Infeasibility

When the two-phase method is employed, infeasibility, if present, is detected in the phase I itself. In such a case, the objective function value is not equal to zero—it is positive for the minimisation and negative for the maximisation problems—and one artificial variable (or more) is found in the basis in the final tableau of phase I. In respect of the problem under consideration, Tables 3.29 through 3.31 contain the first phase solution. The phase I terminates in the third tableau. In this case, the objective function equals –20 and the artificial variable, A_1, remains in the basis at a positive level, at 20. Hence, we conclude that no feasible solution exists for the original problem and there is obviously no need to proceed to phase II.

Phase I Maximise $Y = 0x_1 + 0x_2 + 0S_1 + 0S_2 + 0S_3 - A_1$

Subject to

$$2x_1 + x_2 + S_1 = 40$$
$$4x_1 - x_2 + S_2 = 20$$
$$x_1 - S_3 + A_1 = 30$$
$$x_1, x_2, S_1, S_2, S_3, A_1 \geq 0$$

TABLE 3.29 Simplex Tableau 1: Non-optimal Solution

Basis		x_1	x_2	S_1	S_2	S_3	A_1	b_i	b_i/a_{ij}
S_1	0	2	1	1	0	0	0	40	20
S_2	0	4*	– 1	0	1	0	0	20	5 ←
A_1	– 1	1	0	0	0	– 1	1	30	30
c_j		1	0	0	0	0	– 1		
Solution		0	0	4	20	0	30		
Δ_j		1	0	0	0	– 1	0		
		↑							

TABLE 3.30 Simplex Tableau 2: Non-optimal Solution

Basis		x_1	x_2	S_1	S_2	S_3	A_1	b_i	b_i/a_{ij}
S_1	0	0	3/2*	1	– 1/2	0	0	30	20 ←
x_1	0	1	– 1/4	0	1/4	0	0	5	– 20
A_1	– 1	0	1/4	0	– 1/4	– 1	1	25	100
c_j		0	0	0	0	0	– 1		
Solution		5	0	30	0	0	25		
Δ_j		0	1/4	0	- 1/4	– 1	0		
			↑						

TABLE 3.31 Simplex Tableau 3: Final, Non-optimal Solution

Basis		x_1	x_2	S_1	S_2	S_3	A_1	b_i
x_2	0	0	1	2/3	– 1/3	0	0	20
x_1	0	1	0	1/6	1/6	0	0	10
A_1	–1	0	0	– 1/6	– 1/6	– 1	1	20
c_j		0	0	0	0	0	– 1	
Solution		10	20	0	0	0	20	
Δ_j		0	0	– 1/6	– 1/6	– 1	0	

3.7.3 Unboundedness

As discussed in Chapter 2, an LPP is said to have an unbounded solution if its objective function value can be increased (in case it is a maximisation problem) or decreased (if it is a minimisation problem) without limit. Now we consider as to how unboundedness may be discovered using the simplex method.

As we know, in applying simplex algorithm, to determine which variable would be the outgoing variable, when a non-optimal solution is sought to be improved, we select the row which has the smallest non-negative replacement ratio, b_i/a_{ij}. If, in a certain situation, there are no non-negative replacement ratios (so that all of them are negative) or they are equal to ∞ (for the reason of the denominator equal to zero), then the algorithm terminates. This indicates that the problem under consideration has unbounded solution. Consider the example given below.

Example 3.8 Maximise $Z = 10x_1 + 20x_2$

Subject to

$$2x_1 + 4x_2 \geq 16$$
$$x_1 + 5x_2 \geq 15$$
$$x_1, x_2 \geq 0$$

The solution to this problem is contained in Tables 3.32 through 3.35. It is first augmented using the necessary surplus and artificial variables as shown below.

Maximise $Z = 10x_1 + 20x_2 + 0S_1 + 0S_2 - MA_1 - MA_2$

Subject to

$$2x_1 + 4x_2 - S_1 + A_1 = 16$$
$$x_1 + 5x_2 - S_2 + A_2 = 15$$
$$x_1, x_2, S_1, S_2, A_1, A_2 \geq 0$$

TABLE 3.32 Simplex Tableau 1: Non-optimal Solution

Basis	x_1	x_2	S_1	S_2	A_1	A_2	b_i	b_i/a_{ij}
A_1 $-M$	2	4	-1	0	1	0	16	4
A_2 $-M$	1	5*	0	-1	0	1	15	3 ←
c_j	10	20	0	0	$-M$	$-M$		
Solution	0	0	0	0	16	15		
Δ_j	$10 + 3M$	$20 + 9M$ ↑	$-M$	$-M$	0	0		

TABLE 3.33 Simplex Tableau 2: Non-optimal Solution

Basis	x_1	x_2	S_1	S_2	A_1	A_2	b_i	b_i/a_{ij}
A_1 $-M$	6/5*	0	-1	4/5	1	$-4/5$	4	10/3 ←
x_2 20	1/5	1	0	$-1/5$	0	1/5	3	15
c_j	10	20	0	0	$-M$	$-M$		
Solution	0	3	0	0	4	0		
Δ_j	$6 + \frac{6}{5}M$ ↑	0	$-M$	$4 + \frac{4}{5}M$	0	$-4 - \frac{9}{5}M$		

TABLE 3.34 Simplex Tableau 3: Non-optimal Solution

Basis		x_1	x_2	S_1	S_2	A_1	A_2	b_i	b_i/a_{ij}
x_1	10	1	0	– 5/6	2/3	5/6	– 2/3	10/3	– 4
x_2	20	0	1	1/6*	– 1/3	– 1/6	1/3	7/3	14 ←
c_j		10	20	0	0	$-M$	$-M$		
Solution		10/3	7/3	0	0	0	0		
Δ_j		0	0	5	0	$-M-5$	$-M$		
				↑					

TABLE 3.35 Simplex Tableau 4: Unbounded Solution

Basis		x_1	x_2	S_1	S_2	A_1	A_2	b_i	b_i/a_{ij}
x_1	10	1	5	0	– 1	0	1	15	– 15
S_1	0	0	6	1	– 2	– 1	2	14	– 7
c_j		10	20	0	0	$-M$	$-M$		
Solution		15	0	14	0	0	0		
Δ_j		0	– 30	0	10	$-M$	$-M-10$		
					↑				

From the Simplex Tableau 4, we observe that the solution is not optimal because a positive value appears in the Δ_j row. According to this, S_2 is the incoming variable. Computing the replacement ratios, however, we notice that they both are negative. This implies that we can increase S_2 indefinitely without driving one of the basic variables to zero. The given solution is unbounded, therefore.

3.7.4 Degeneracy

Degeneracy in a LPP occurs when one or more of the basic variables assume a value of zero. As stated already, for a *n*-variable, *m*-constraint problem, there would be *m* basic and *n* – *m* non-basic variables, and the basic variables would assume positive values. However, in case a basic variable assumes a value equal to zero, then that variable and the solution are labelled *degenerate.* Thus, in condition of *degeneracy,* the solution would contain a smaller number of non-zero variables than the number of constraints, *m.*

Example 3.9 Maximise $Z = 28x_1 + 30x_2$

Subject to

$$6x_1 + 3x_2 \leq 18$$

$$3x_1 + x_2 \leq 8$$

$$4x_1 + 5x_2 \leq 30$$

$$x_1, x_2 \geq 0$$

This problem is first depicted graphically in Figure 3.3.

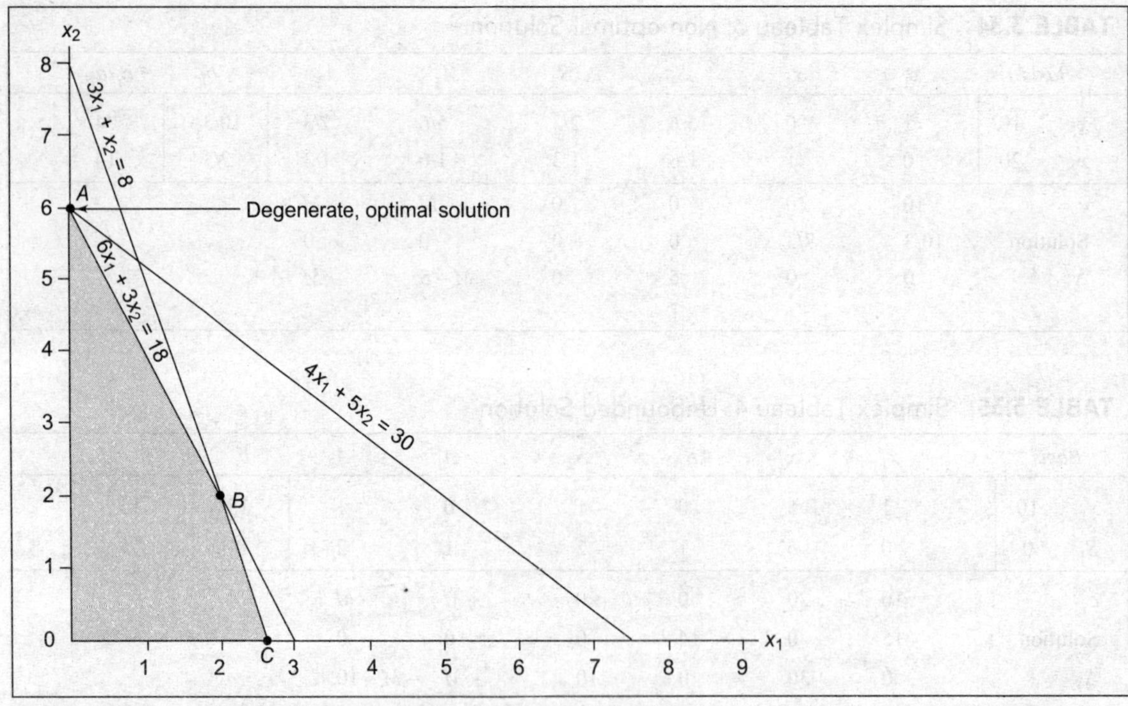

Figure 3.3 *Graphic Solution: Degenerate, Optimal*

The feasible region is bound by the polygon 0*ABC*. The extreme points are evaluated here.

Point	x_1	x_2	$Z = 28x_1 + 30x_2$
0	0	0	0
A	0	6	180*
B	2	2	116
C	8/3	0	$74\frac{2}{3}$

Thus, the solution is optimal at point *A*. A careful look would show that at this point two constraints are satisfied simultaneously and one of the decision variables, x_1 is equal to zero. This means that the slack variables in respect of the two constraints shall be equal to zero. Thus, in all, three variables would be equal to zero and the remaining two variables, the decision variable x_2 and the slack variable corresponding to the constraint $3x_1 + x_2 \leq 8$, shall be at positive level. Since the number of non-zero variables is less than the number of constraints in the problem, the solution is degenerate. Also, it is clear that the constraint $4x_1 + 5x_2 \leq 30$ is redundant in this case and is not required in determination of point *A*.

Now, let us consider the problem using simplex method for its solution. Introducing slack variables, the problem becomes:

Maximise $Z = 28x_1 + 30x_2 + 0S_1 + 0S_2 + 0S_3$

Subject to

$$6x_1 + 3x_2 + S_1 = 18$$
$$3x_1 + x_2 + S_2 = 8$$

$$4x_1 + 5x_2 + S_3 = 30$$
$$x_1, x_2, S_1, S_2, S_3 \geq 0$$

The initial solution is given in Table 3.36.

TABLE 3.36 Simplex Tableau 1: Non-optimal Solution

Basis		x_1	x_2	S_1	S_2	S_3	b_i	b_i/a_{ij}
S_1	0	6	3	1	0	0	18	6 ← Tie
S_2	0	3	1	0	1	0	8	8
S_3	0	4	5	0	0	1	30	6 ← Tie
c_j		28	30	0	0	0		
Solution		0	0	18	8	30		
Δ_j		28	30	0	0	0		
			↑					

In this tableau we observe that there is a tie between the first and the third rows and, therefore, either of the variables S_1 and S_3 could be taken to be the outgoing variable. We shall consider them one by one. By taking S_1 as the outgoing variable, the revised tableau may be drawn as shown in Table 3.37.

TABLE 3.37 Simplex Tableau 2: Optimal Solution

Basis		x_1	x_2	S_1	S_2	S_3	b_i
x_2	30	2	1	1/3	0	0	6
S_2	0	1	0	– 1/3	1	0	2
S_3	0	– 6	0	– 5/3	0	1	0
c_j		28	30	0	0	0	
Solution		0	6	0	2	0	
Δ_j		– 32	0	– 10	0	0	

This tableau gives the optimal solution $x_1 = 0$, and $x_2 = 6$, with objective function value equal to 180. Now, deleting S_3, the revised simplex tableau would appear as shown in Table 3.38.

TABLE 3.38 Simplex Tableau 3: Non-optimal Solution

Basis		x_1	x_2	S_1	S_2	S_3	b_i	b_i/a_{ij}
S_1	0	18/5*	0	1	0	– 3/5	0	0 ←
S_2	0	11/5	0	0	1	– 1/5	2	10/11
x_2	30	4/5	1	0	0	1/5	6	15/2
c_j		28	30	0	0	0		
Solution		0	6	0	2	0		
Δ_j		4	0	0	0	– 6		
		↑						

This solution is non-optimal because a value in the Δ_j row is positive. Here the objective function value is $28 \times 0 + 30 \times 6 = 180$. The revised tableau is given in Table 3.39.

TABLE 3.39 Simplex Tableau 4: Optimal Solution

Basis		x_1	x_2	S_1	S_2	S_3	b_i
x_1	28	1	0	5/18	0	– 1/6	0
S_2	0	0	0	– 11/18	1	1/6	2
x_2	30	0	1	– 2/9	0	1/3	6
c_j		28	30	0	0	0	
Solution		0	6	0	2	0	
Δ_j		0	0	– 10/9	0	– 16/3	

This gives an optimal solution because all entries in the row designated Δ_j are either zero or negative. The solution is: $x_1 = 0$, $x_2 = 6$, with objective function value $= 28 \times 0 + 30 \times 6 = 180$.

Some observations follow.

(a) In Table 3.37 the variable S_3 has a value zero although it is in the basis, while in Table 3.39 though x_1 is included in the solution it has a value equal to zero. The variable S_3 and the solution in the first case, and the variable x_1 and the solution in the second case are degenerate.

(b) Whenever there is a tie in the replacement ratios for determining outgoing variable, the next tableau would give a degenerate solution.

(c) We know that successive simplex tableaus represent improvements in the value of the objective function—increases are observed for the maximisation and decreases for the minimisation problems. However, when the outgoing variable happens to be a degenerate variable, the objective function value in the next tableau does not change. This is amply clear from Tables 3.38 and 3.39. The Z value for a 'non-optimal' solution is no different from the Z value obtained from the next, 'optimal' solution.

Here, a question arises naturally—if there is no improvement in the solution in terms of the objective function, then why should we not stop at the iteration where a degenerate solution appears? The answer to this is that one cannot be sure that the first appearance of a degenerate solution will coincide with the optimal solution. In other words, the problem may be temporarily degenerate. We may consider Example 3.10 where a degenerate solution is encountered but later on degeneracy disappears.

Example 3.10 Solve the problem given below, both graphically and using simplex method.

Maximise $Z = 5x_1 + 2x_2$

Subject to

$$4x_1 + 2x_2 \leq 16$$
$$3x_1 + x_2 \leq 9$$
$$3x_1 - x_2 \leq 9$$
$$x_1, x_2 \geq 0$$

The problem is shown graphically in Fig. 3.4.

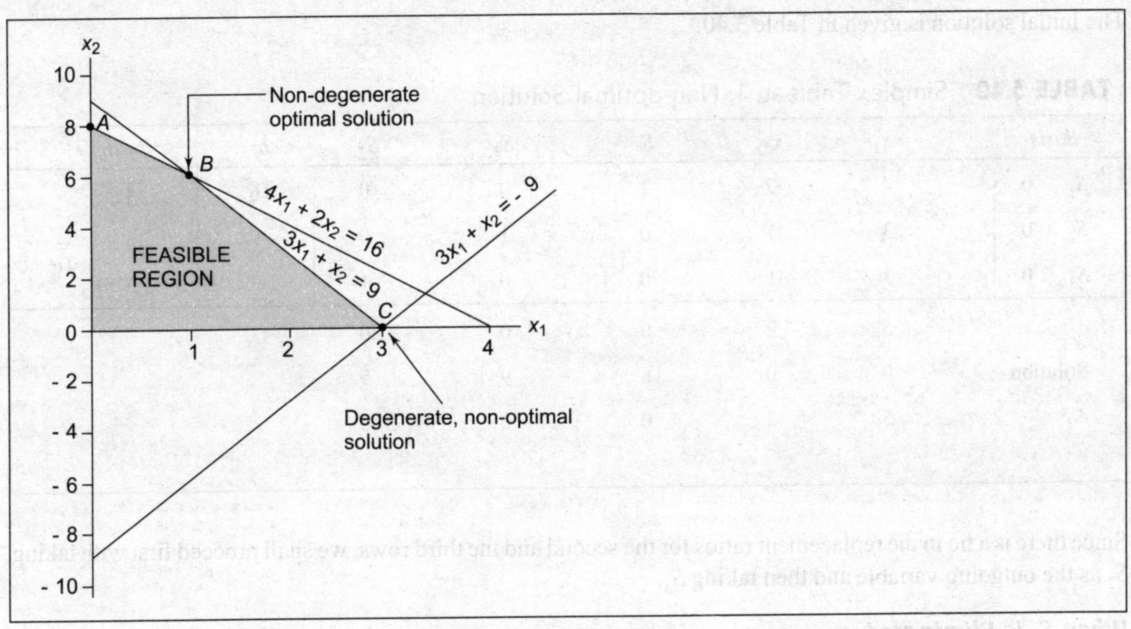

Figure 3.4 *Graphic Solution: Non-degenerate, Optimal*

The feasible solution is bounded by the polygon $0ABC$. The extreme points are evaluated as shown below.

Point	x_1	x_2	$Z = 5x_1 + 2x_2$
0	0	0	0
A	0	8	16
B	1	6	17*
C	3	0	15

The optimal solution lies at point *B*, lying at the intersection of the first two constraints. It is significant to note that in this case, the extreme point *C* is lying at the intersection of the second and the third constraints, but both of these are not required to determine this point. The third constraint here is redundant. Hence, if the simplex algorithm shows the solution corresponding to the extreme point *C* (where $x_1 = 3$ and $x_2 = 0$), then the solution would be indicated to be a degenerate one. The optimal solution at *B* is non-degenerate.

To consider the solution to the problem using simplex method for its solution, the problem is restated below in the augmented form.

Maximise $\quad Z = 5x_1 + 2x_2 + 0S_1 + 0S_2 + 0S_3$

Subject to

$$4x_1 + 2x_2 + S_1 = 16$$
$$3x_1 + x_2 + S_2 = 9$$
$$3x_1 - x_2 + S_3 = 9$$
$$x_1, x_2, S_1, S_2, S_3 \geq 0$$

The initial solution is given in Table 3.40.

TABLE 3.40 Simplex Tableau 1: Non-optimal Solution

Basis		x_1	x_2	S_1	S_2	S_3	b_i	b_i/a_{ij}
S_1	0	4	2	1	0	0	16	4
S_2	0	3	1	0	1	0	9	3 ← Tie
S_3	0	3	−1	0	0	1	9	3 ← Tie
c_j		5	2	0	0	0		
Solution		0	0	16	9	9		
Δ_j		5	2	0	0	0		
		↑						

Since there is a tie in the replacement ratios for the second and the third rows, we shall proceed first with taking S_2 as the outgoing variable and then taking S_3.

When S_2 is Eliminated

With S_2 as the outgoing variable, the solution follows as given in Tables 3.41 and 3.42.

TABLE 3.41 Simplex Tableau 2: Non-optimal Solution

Basis		x_1	x_2	S_1	S_2	S_3	b_i	b_i/a_{ij}
S_1	0	0	2/3*	1	−4/3	0	4	6 ←
x_1	5	1	1/3	0	1/3	0	3	9
S_3	0	0	−2	0	−1	1	0	–
c_j		5	2	0	0	0		
Solution		3	0	4	0	0		
Δ_j		0	1/3	0	−5/3	0		
			↑					

In this tableau, the solution has two basic variables, x_1 and S_1, with solution values greater than zero. Thus, this solution with $x_1 = 3$, $x_2 = 0$ and $Z = 15$ is degenerate, with the basic variable S_3 also being degenerate. Also, it is non-optimal. To improve it, the key column selected is the one headed x_2, while the replacment ratio chosen to determine the outgoing variable is 6, corresponding to S_1. It may be noted that while the numerator for the S_3 row is zero, it cannot be selected as the key row since the denominator is negative.

The improved solution is given in Table 3.42.

TABLE 3.42 Simplex Tableau 3: Optimal Solution

Basis		x_1	x_2	S_3	S_2	S_3	b_i
x_2	2	0	1	3/2	− 2	0	6
x_1	5	1	0	− 1/2	1	0	1
S_3	0	0	0	3	− 5	1	12
c_j		5	2	0	0	0	
Solution		1	6	0	0	12	
Δ_j		0	0	− 1/2	− 1	0	

This solution is optimal with $x_1 = 1$, $x_2 = 6$ and $Z = 17$. It is *non-degenerate.*

When S_3 is Eliminated

Beginning with the initial solution given in Table 3.40, if we select S_3 as the outgoing variable, the optimal solution may be reached as contained in Tables 3.43 through 3.45.

TABLE 3.43 Simplex Tableau 4: Non-optimal Solution

Basis		x_1	x_2	S_1	S_2	S_3	b_i	b_i/a_{ij}
S_1	0	0	10/3	1	0	− 4/3	4	6/5
S_2	0	0	2*	0	1	− 1	0	0 ←
x_1	5	1	− 1/3	0	0	1/3	3	− 1
c_j		5	2	0	0	0		
Solution		3	0	4	0	0		
Δ_j		0	11/3	0	0	− 5/3		
			↑					

For the solution in Table 3.43 with $x_1 = 3$ and $x_2 = 0$, we have $Z = 15$. The solution is degenerate and non-optimal. To improve the solution, we select the least non-negative replacement ratio equal to 0 for S_2, since the denominator here is positive. The improved solution is contained in Table 3.44.

TABLE 3.44 Simplex Tableau 5: Non-optimal Solution

Basis		x_1	x_2	S_1	S_2	S_3	b_i	b_i/a_{ij}
S_1	0	0	0	1	− 5/3	1/3*	4	12 ←
x_2	2	0	1	0	1/2	− 1/2	0	–
x_1	5	1	0	0	1/6	1/6	3	18
c_j		5	2	0	0	0		
Solution		3	0	4	0	0		
Δ_j		0	0	0	− 11/6	1/6		
						↑		

The solution in Table 3.44 involves $x_1 = 3$ and $x_2 = 0$ with $Z = 15$, same as in the previous solution. This is also non-optimal, as indicated by the Δ_j row. The incoming variable here is S_3 while the outgoing variable is S_1. Note that although the numerator for the row x_2 is zero, it cannot be taken as the key row because the a_{ij} value (in key column) in this row is negative ($= -1/2$).

TABLE 3.45 Simplex Tableau 6: Optimal Solution

Basis		x_1	x_2	S_1	S_2	S_3	b_i
S_3	0	0	0	3	– 5	1	12
x_2	2	0	1	3/2	– 2	0	6
x_1	5	1	0	– 1/2	1	0	1
c_j		5	2	0	0	0	
Solution		1	6	0	0	12	
Δ_j		0	0	– 1/2	– 1	0	

The solution in Table 3.45 is optimal, and is identical to the optimal solution obtained earlier in Table 3.42. It is non-degenerate, indeed.

Evidently, then, the degeneracy encountered in this problem has been only temporary.

Cycling As observed above, when the outgoing variable in a non-optimal solution is a degenerate variable, then the solution does not improve in terms of the objective function value. However, when the key rows do not involve degenerate variables, then successive solutions would be improved ones. Obviously, when solutions with higher values of Z are obtained in successive iterations, we can never have a simplex tableau repeated in the solution process. On the other hand, when a degenerate row is the key row, then, theoretically, there is a possibility that we obtain a sequence of tableaus (each of them having the same objective function value) that might bring us back to a tableau that has been considered earlier. When this happens, it implies that the simplex algorithm has cycled and, as such, the system would again move along the same route and the cycle would be repeated forever. In such cases, we cannot obtain optimal solution to the problem. Cycling is indeed very rare. However, the fact that it can occur points to the fact that there might be some problems which cannot be solved by the simplex method.

In general, then, degeneracy does not pose a problem in practice, except only when cycling occurs. Mostly, whenever there is a tie in the replacement ratios, selecting a row arbitrarily is sufficient. Besides, there are sophisticated rules to ensure that cycling would not occur. They are, however, beyond the scope of this book.

REVIEW ILLUSTRATIONS

Example 3.11 Consider the following linear programming model:

Maximise $\quad Z = 2X + 5Y$

Subject to

$$X + 4Y \leq 24$$
$$3X + Y \leq 21$$

$$X + Y \leq 9$$
$$X, Y \geq 0$$

Recast the model using the slack variables S_1, S_2, and S_3. Label the following solutions, giving reasons, as basic but not feasible, feasible but not basic, feasible as well as basic or as optimal basic feasible solution:

(i)	$X = 4$,	$Y = 5$,	$S_1 = 0$,	$S_2 = 4$,	$S_3 = 0$
(ii)	$X = 6$,	$Y = 0$,	$S_1 = 18$,	$S_2 = 3$,	$S_3 = 3$
(iii)	$X = 9$,	$Y = 0$,	$S_1 = 15$,	$S_2 = -6$,	$S_3 = 0$
(iv)	$X = 4$,	$Y = 5$,	$S_1 = 0$,	$S_2 = 4$,	$S_3 = 0$

The problem is restated below using slack variables:

Maximise $$Z = 2X + 5Y + 0S_1 + 0S_2 + 0S_3$$

Subject to

$$X + 4Y + S_1 = 24$$
$$3X + Y + S_2 = 21$$
$$X + Y + S_3 = 9$$
$$X, Y, S_1, S_2, S_3 \geq 0$$

With 5 variables and three equations, there shall be a total of ${}^5C_2 = 10$ basic solutions, each of which shall have two non-basic variables with solution value equal to zero. Recall that a solution is feasible if it meets all constraints and all variables are non-negative (if stipulated). The given solutions are considered here.

Solution (i): It is a basic feasible solution (bfs) as 2 variables are non-basic and 3 are basic, and all variables are non- negative in value. Whether or not it is optimal can be determined only after considering other bfs's.

Solution (ii): It is feasible but it is not basic as only one variable is non-basic.

Solution (iii): It is basic but not feasible since a variable has a negative solution value.

Solution (iv): It is same as solution (i)

Example 3.12 Solve the following LPP:

Maximise $$Z = 8x_1 - 4x_2$$

Subject to

$$4x_1 + 5x_2 \leq 20$$
$$-x_1 + 3x_2 \geq -23$$
$$x_1 \geq 0,\ x_2 \text{ unrestricted in sign}$$

As mentioned earlier, the twin necessary conditions for the application of the simplex algorithm are that the variables involved and the b_i's in the given problem should all be non-negative. In case of a negative b_i, the inequality/equation is multiplied by -1, while a variable unrestricted in sign is replaced by the difference of two non-negative variables. For the given problem, we multiply the second of the constraints by -1 and replace $x_2 = x_3 - x_4$. This yields the following problem:

Maximise $$Z = 8x_1 - 4x_3 + 4x_4$$

Subject to

$$4x_1 + 5x_3 - 5x_4 \leq 20$$

$$x_1 - 3x_3 + 3x_4 \leq 23$$
$$x_1, x_3, x_4 \geq 0$$

After introducing slack variables, the problem becomes:

Maximise $Z = 8x_1 - 4x_3 + 4x_4 + 0S_1 + 0S_2$

Subject to

$$4x_1 + 5x_3 - 5x_4 + S_1 = 20$$
$$x_1 - 3x_3 + 3x_4 + S_2 = 23$$
$$x_1, x_3, x_4, S_1, S_2 \geq 0$$

TABLE 3.46 Simplex Tableau 1: Non-optimal Solution

Basis		x_1	x_3	x_4	S_1	S_2	b_i	b_i/a_{ij}
S_1	0	4*	5	– 5	1	0	20	5 ←
S_2	0	1	– 3	3	0	1	23	23
c_j		8	– 4	4	0	0		
Solution		0	0	0	20	23		
Δ_j		8	– 4	4	0	0		
		↑						

TABLE 3.47 Simplex Tableau 2: Non-optimal Solution

Basis		x_1	x_3	x_4	S_1	S_2	b_i	b_i/a_{ij}
x_1	8	1	5/4	– 5/4	1/4	0	5	– 4
S_2	0	0	– 17/4	17/4*	– 1/4	1	18	72/17 ←
c_j		8	– 4	4	0	0		
Solution		5	0	0	0	18		
Δ_j		0	– 14	14	– 2	0		
				↑				

TABLE 3.48 Simplex Tableau 3: Optimal Solution

Basis		x_1	x_3	x_4	S_1	S_2	b_i
x_1	8	1	0	0	3/17	5/17	175/17
x_4	4	0	– 1	1	– 1/17	4/17	72/17
c_j		8	– 4	4	0	0	
Solution		175/17	0	72/17	0	0	
Δ_j		0	0	0	– 20/17	– 56/17	

Thus, optimal solution to the problem is: $x_1 = 175/17$, $x_2 = x_3 - x_4 = (0 - 72/17) = - 72/17$ and the objective function value is $(8 \times 175/17) - (4 \times (- 72/17)) = 1688/17$.

Example 3.13 Maximise $Z = 6x_1 + 20x_2$

Subject to

$$2x_1 + x_2 \leq 32$$
$$3x_1 + 4x_2 \leq 80$$
$$x_1 \geq 8$$
$$x_2 \geq 10$$

Sometimes, in a given problem it may be provided that one (or more) of the variables may not assume a value higher and/or lower than a particular value. The problem may be solved conveniently if the variables are restructured. In the given problem, the variable x_1 is sought to be at least equal to 8 and x_2 no less than 10. Although this problem can be solved by using slack variables in each of the first two constraints, and surplus and artificial variables in the other two, we can reframe the problem in the following way in order to save computational time and effort. Since $x_1 \geq 8$, we have $x_1 - x_3 = 8$ (x_3 being the surplus variable) with $x_1 = 8 + x_3$ and $x_2 = 10 + x_4$ (determined similarly), we can rewrite the problem as follows:

Maximise $Z = 6(8 + x_3) + 20(10 + x_4)$ or $Z = 6x_3 + 20x_4 + 248$

Subject to

$$2(8 + x_3) + (10 + x_4) \leq 32 \quad \text{or} \quad 2x_3 + x_4 \leq 6$$
$$3(8 + x_3) + 4(10 + x_4) \leq 80 \quad \text{or} \quad 3x_3 + 4x_4 \leq 16$$
$$x_3, x_4 \geq 0$$

Introducing the necessary slack variables, we get

Maximise $Z = 6x_3 + 20x_4 + 248 + 0S_1 + 0S_2$

Subject to

$$2x_3 + x_4 + S_1 = 6$$
$$3x_3 + 4x_4 + S_2 = 16$$
$$x_3, x_4, S_1, S_2 \geq 0$$

The solution is contained in Tables 3.49 and 3.50.

TABLE 3.49 Simplex Tableau 1: Non-optimal Solution

Basis		x_3	x_4	S_1	S_2	b_i	b_i/a_{ij}
S_1	0	2	1	1	0	6	6
S_2	0	3	4*	0	1	16	4 ←
c_j		6	20	0	0		
Solution		0	0	6	16		
Δ_j		6	20	0	0		
			↑				

TABLE 3.50 Simplex Tableau 2: Optimal Solution

Basis		x_3	x_4	S_1	S_2	b_i
S_1	0	5/4	0	1	– 1/4	2
x_4	20	3/4	1	0	1/4	4
c_j		6	20	0	0	
Solution		0	4	2	0	
Δ_j		– 9	0	0	– 5	

Corresponding to the optimal solution given in Table 3.50, we have $x_3 = 0$, and $x_4 = 4$, and the objective function value equal to $6 \times 0 + 20 \times 4 + 248 = 328$. Also $x_1 = 8 + x_3 = 8$, and $x_2 = 10 + x_4 = 10 + 4 = 14$.

Example 3.14 A finished product must weigh exactly 150 grams. The two raw materials used in manufacturing the product are *A*, with a cost of Rs 2 per unit and *B* with a cost of Rs 8 per unit. At least 14 units of *B* and not more than 20 units of *A* must be used. Each unit of *A* and *B* weighs 5 and 10 grams respectively.

How much of each type of raw material should be used for each unit of the final product in order to minimise the cost? Use Simplex method. *(M Com, Delhi, 1985)*

The given problem can be expressed as LPP as

Minimise $Z = 2x_1 + 8x_2$

Subject to

$$5x_1 + 10x_2 = 150$$
$$x_1 \le 20$$
$$x_2 \ge 14$$
$$x_1, x_2 \ge 0$$

Substituting $x_2 = 14 + x_3$ and introducing necessary slack and artificial variables, we have

Minimise $Z = 2x_1 + 8x_3 + 112 + MA_1 + 0S_1$

Subject to $5x_1 + 10x_3 + A_1 = 10$

$$x_1 + S_1 = 20$$
$$x_1, x_3, S_1, A_1 \ge 0$$

The solution is contained in Table 3.51 through 3.53.

TABLE 3.51 Simplex Tableau 1: Non-optimal Solution

Basis		x_1	x_3	A_1	S_1	b_i	b_i/a_{ij}
A_1	M	5	10*	1	0	10	1 ←
S_1	0	1	0	0	1	20	∞
c_j		2	8	M	0		
Solution		0	0	10	20		
Δ_j		$2 - 5M$	$8 - 10M$ ↑	0	0		

TABLE 3.52 Simplex Tableau 2: Non-optimal Solution

Basis		x_1	x_3	A_1	S_1	b_i	b_i/a_{ij}
x_3	8	1/2*	1	1/10	0	1	2 ←
x_4	0	1	0	0	1	20	20
c_j		2	8	M	0		
Solution		0	1	0	20		
Δ_j		−2 ↑	0	$M-\frac{8}{10}$	0		

TABLE 3.53 Simplex Tableau 3: Optimal Solution

Basis		x_1	x_3	A_1	S_1	b_i
x_1	2	1	2	1/5	0	2
x_4	0	0	−2	−1/5	1	18
c_j		2	8	M	0	
Solution		2	0	0	18	
Δ_j		0	4	$M-\frac{2}{5}$	0	

Thus, the optimal solution is: $x_1 = 2$ units, $x_2 = 14 + 0 = 14$ units, total cost $= 2 \times 2 + 8 \times 14 =$ Rs 116.

Example 3.15 Ashok Chemicals Company manufactures two chemicals *A* and *B* which are sold to the manufacturers of soaps and detergents. On the basis of the next month's demand, the management has decided that the total production for chemicals *A* and *B* should be at least 350 kilograms. Moreover, a major customer's order for 125 kilograms of product *A* must also be supplied. Product *A* requires 2 hours of processing time per kilogram and product *B* requires one hour of processing time per kilogram. For the coming month, 600 hours of processing time are available.

The company wants to meet the above requirements at minimum total production cost. The production costs are Rs 2 per kilogram for product *A* and Rs 3 per kilogram for product *B*.

Ashok Chemicals Company wants to determine its optimal product mix and the total minimum cost relevant thereto.

(i) Formulate the above as a linear programming problem.

(ii) Solve the problem with the simplex method.

(iii) Does the problem have multiple optimal solutions? Why? *(M Com, Delhi, 1998)*

Let x_1 and x_2 be the quantity in kg to be produced of products *A* and *B*, respectively. From the given information, the LPP can be written as follows:

Minimise	$Z = 2x_1 + 3x_2$	Total cost
Subject to		
	$x_1 + x_2 \geq 350$	Total output of *A* and *B*
	$x_1 \geq 125$	Customer order for *A*
	$2x_1 + x_2 \leq 600$	Time availability
	$x_2 \geq 0$	

For solving the problem by simplex method, we introduce the necessary slack, surplus and artificial variables, and restate the problem as:

Minimise $\quad Z = 2x_1 + 3x_2 + 0S_1 + 0S_2 + 0S_3 + MA_1 + MA_2$

Subject to

$$x_1 + x_2 - S_1 + A_1 = 350$$
$$x_1 - S_2 + A_2 = 125$$
$$2x_1 + x_2 + S_3 = 600$$
$$x_1, x_2, S_1, S_2, S_3, A_1, A_2 \geq 0$$

The solution is contained in Tables 3.54 through 3.57.

TABLE 3.54 Simplex Tableau 1: Non-optimal Solution

Basis		x_1	x_2	S_1	S_2	S_3	A_1	A_2	b_i	b_i/a_{ij}
A_1	M	1	1	−1	0	0	1	0	350	350
A_2	M	1*	0	0	−1	0	0	1	125	125 ←
S_3	0	2	1	0	0	1	0	0	600	300
c_j		2	3	0	0	0	M	M		
Solution		0	0	0	0	600	350	125		
Δ_j		$2 - 2M$ ↑	$3 - M$	M	M	0	0	0		

TABLE 3.55 Simplex Tableau 2: Non-optimal Solution

Basis		x_1	x_2	S_1	S_2	S_3	A_1	A_2	b_i	b_i/a_{ij}
A_1	M	0	1	−1	1	0	1	−1	225	225
x_1	2	1	0	0	−1	0	0	1	125	–
S_3	0	0	1	0	2*	1	0	−2	350	175 ←
c_j		2	3	0	0	0	M	M		
Solution		125	0	0	0	175	225	0		
Δ_j		0	$3 - M$	M	$2 - M$ ↑	0	0	$2M - 2$		

TABLE 3.56 Simplex Tableau 3: Non-optimal Solution

Basis		x_1	x_2	S_1	S_2	S_3	A_1	A_2	b_i	b_i/a_{ij}
A_1	M	0	1/2*	−1	0	−1/2	1	0	50	100 ←
x_1	2	1	1/2	0	0	1/2	0	0	300	600
S_2	0	0	1/2	0	1	1/3	0	−1	175	350
c_j		2	3	0	0	0	M	M		
Solution		300	0	0	175	0	50	0		
Δ_j		0	$2-\frac{M}{2}$ ↑	M	0	$\frac{M}{2}-1$	0	M		

TABLE 3.57 Simplex Tableau 4: Non-optimal Solution

Basis		x_1	x_2	S_1	S_2	S_3	A_1	A_2	b_i
x_2	3	0	1	−2	0	−1	2	0	100
x_1	2	1	0	1	0	1	−1	0	250
S_2	0	0	0	1	1	1	−1	−1	125
c_j		2	3	0	0	0	M	M	
Solution		250	100	0	125	0	0	0	
Δ_j		0	0	4	0	1	$M-4$	M	

Thus, the optimal solution is: $x_1 = 250$ and $x_2 = 100$, with $Z = 800$. The solution is unique optimal solution and no multiple solutions exist, because none of the non-basic variables has Δ_j equal to zero.

Example 3.16 A firm uses three machines in the manufacture of three products. Each unit of product *A* requires 3 hours on machine I, 2 hours on machine II and one hour on machine III. Each unit of product *B* requires 4 hours on machine I, one hour on machine II and 3 hours on machine III, while each unit of product *C* requires 2 hours on each of the three machines. The contribution margin of the three products is Rs 30, Rs 40 and Rs 35 per unit respectively. The machine hours available on three machines are 90, 54 and 93 respectively.

(i) Formulate the above problem as a linear programming problem.

(ii) Obtain optimal solution to the problem by using the simplex method. Which of the three products shall not by produced by the firm? Why?

(iii) Calculate the percentage of capacity utilisation in the optimal solution.

(iv) What are the shadow prices of the machine hours?

(v) Is the optimal solution degenerate?

(i) Let x_1, x_2 and x_3 represent the output of products *A*, *B* and *C*, respectively. With the given contribution margins, resource requirements and availability, the LPP can be expressed as follows:

Maximise $Z = 30x_1 + 40x_2 + 35x_3$ Contribution

Subject to

$$3x_1 + 4x_2 + 2x_3 \le 90 \quad \text{Machine I}$$
$$2x_1 + x_2 + 2x_3 \le 54 \quad \text{Machine II}$$
$$x_1 + 3x_2 + 2x_3 \le 93 \quad \text{Machine III}$$
$$x_1, x_2, x_3 \ge 0$$

(ii) To solve the problem using simplex method, we first introduce slack variables S_1, S_2 and S_3, and express the problem in standard form as follows:

Maximise $Z = 30x_1 + 40x_2 + 35x_3 + 0S_1 + 0S_2 + 0S_3$

Subject to

$$3x_1 + 4x_2 + 2x_3 + S_1 = 90$$
$$2x_1 + x_2 + 2x_3 + S_2 = 54$$
$$x_1 + 3x_2 + 2x_3 + S_3 = 93$$
$$x_1, x_2, x_3, S_1, S_2, S_3 \ge 0$$

The solution is given in Tables 3.58 to 3.60.

TABLE 3.58 Simplex Tableau 1: Non-optimal Solution

Basis		x_1	x_2	x_3	S_1	S_2	S_3	b_i	b_i/a_{ij}
S_1	0	3	4*	2	1	0	0	90	45/2 ←
S_2	0	2	1	2	0	1	0	54	54
S_3	0	1	3	2	0	0	1	93	31
c_j		30	40	35	0	0	0		
Solution		0	0	0	90	54	93		
Δ_j		30	40 ↑	35	0	0	0	$Z = 0$	

TABLE 3.59 Simplex Tableau 2: Non-optimal Solution

Basis		x_1	x_2	x_3	S_1	S_2	S_3	b_i	b_i/a_{ij}
x_2	40	3/4	1	1/2	1/4	0	0	45/2	45
S_2	0	5/4	0	3/2*	– 1/4	1	0	63/2	21 ←
S_3	0	– 5/4	0	1/2	– 3/4	0	1	51/2	51
c_j		30	40	35	0	0	0		
Solution		0	45/2	0	0	63/2	51/2		
Δ_j		0	0	15	– 10	0	0	$Z = 900$	

TABLE 3.60 Simplex Tableau 3: Optimal Solution

Basis		x_1	x_2	x_3	S_1	S_2	S_3	b_i
x_2	40	1/3	1	0	1/3	– 1/3	0	12
x_3	35	5/6	0	1	– 1/6	2/3	0	21
S_3	0	– 5/3	0	0	– 2/3	– 1/3	1	15
c_j		30	40	35	0	0	0	
Solution		0	12	21	0	0	15	
Δ_j		– 25/2	0	0	– 15/2	– 10	0	$Z = 1215$

From Table 3.60, it is evident that optimal solution to the problem is: $x_1 = 0$, $x_2 = 12$ and $x_3 = 21$, with $Z = 1{,}215$. This indicates that product *A* should not be produced by the firm. This is because, as indicated by the Δ_j value (equal to –25/2) under the variable x_1, each unit of it produced would reduce the profit by Rs 25/2 or Rs 12.50.

(iii) The capacity available and the capacity used as per optimal solution, is given below:

Machine	*Capacity*	*Capacity Unused**	*Capacity Utilised*	*Percentage Utilisation*
I	90	0	90	100%
II	54	0	54	100%
III	93	15	78	83.87%

*This is given by the solution values of the slack variables.

(iv) Shadow prices of the machine hours are given by the Δ_j values (ignoring signs) of the slack variables. Accordingly, they are:

Machine I	Rs 15/2 or Rs 7.50 per hour
Machine II	Rs 10 per hour
Machine III	Nil

(v) The optimal solution is not degenerate because none of the basic variables has a solution value equal to zero.

Example 3.17 Reconsider the problem given in Example 2.19. Solve it using the simplex method. Hence, show that each simplex tableau containing a feasible solution corresponds to a corner point of the feasible region obtained graphically.

The LPP is reproduced below:

Maximise $$Z = \frac{7}{100} x_1 + \frac{1}{10} x_2$$

Subject to

$$x_1 + x_2 \leq 30{,}000$$
$$x_1 \geq 6{,}000$$
$$x_2 \leq 12{,}000$$

$$x_1 - x_2 \geq 0$$
$$x_1, x_2 \geq 0$$

Introducing the necessary slack, surplus and artificial variables, the problem may be written as:

Maximise $$Z = \frac{7}{100}x_1 + \frac{1}{10}x_2 + 0S_1 + 0S_2 + 0S_3 + 0S_4 - MA_1 - MA_2$$

Subject to

$$x_1 + x_2 + S_1 = 30{,}000$$
$$x_1 - S_2 + A_1 = 6{,}000$$
$$x_2 + S_3 = 12{,}000$$
$$x_1 - x_2 - S_4 + A_2 = 0$$
$$x_1, x_2, S_1, S_2, S_3, S_4, A_1, A_2 \geq 0$$

Using simplex method, solution to the problem can be obtained as given in Table 3.61 through 3.65.

TABLE 3.61 Simplex Tableau 1: Infeasible Solution

Basis		x_1	x_2	S_1	S_2	S_3	S_4	A_1	A_2	b_i	b_i/a_{ij}
S_1	0	1	1	1	0	0	0	0	0	30,000	30,000
A_1	$-M$	1	0	0	-1	0	0	1	0	6,000	6,000
S_3	0	0	1	0	0	1	0	0	0	12,000	–
A_2	$-M$	1*	-1	0	0	0	-1	0	1	0	0 ←
c_j		7/100	1/10	0	0	0	0	$-M$	$-M$		
Solution		0	0	30,000	0	12,000	0	6,000	0		
Δ_j		$\frac{7}{100}+2M$ ↑	$\frac{1}{10}+M$	0	M	0	$-M$	0	0		

TABLE 3.62 Simplex Tableau 2: Infeasible Solution

Basis		x_1	x_2	S_1	S_2	S_3	S_4	A_1	A_2	b_i	b_i/a_{ij}
S_1	0	0	2	1	0	0	1	0	-1	30,000	15,000
A_1	$-M$	0	1*	0	-1	0	1	1	-1	6,000	6,000 ←
S_3	0	0	1	0	0	1	0	0	0	12,000	12,000
x_1	7/100	1	-1	0	0	0	-1	0	1	0	–
c_j		7/100	1/10	0	0	0	0	$-M$	$-M$		
Solution		0	0	30,000	0	12,000	0	6000	0		
Δ_j		0	$\frac{17}{100}+M$ ↑	0	$-M$	0	$\frac{7}{100}+M$	0	$-\frac{7}{100}-2M$		

TABLE 3.63 Simplex Tableau 3: Non-optimal Solution

Basis		x_1	x_2	S_1	S_2	S_3	S_4	A_1	A_2	b_i	b_i/a_{ij}
S_1	0	0	0	1	2	0	−1	−2	1	18,000	9,000
x_2	1/10	0	1	0	−1	0	1	1	−1	6,000	–
S_3	0	0	0	0	1*	1	−1	−1	1	6,000	6,000 ←
x_1	7/100	1	0	0	−1	0	0	1	0	6,000	–
c_j		7/100	1/10	0	0	0	0	$-M$	$-M$		
Solution		6,000	6,000	18,000	0	6,000	0	0	0	$Z = 1{,}020$	
Δ_j		0	0	0	$\frac{17}{100}$ ↑	0	−1/10	$-M-\frac{17}{100}$	$-M+\frac{1}{10}$		

TABLE 3.64 Simplex Tableau 4: Non-optimal Solution

Basis		x_1	x_2	S_1	S_2	S_3	S_4	A_1	A_2	b_i	b_i/a_{ij}
S_1	0	0	0	1	0	−2	1*	0	−1	6,000	6,000 ←
x_2	1/10	0	1	0	0	1	0	0	0	12,000	–
S_2	0	0	0	0	1	1	−1	−1	1	6,000	–
x_1	7/100	1	0	0	0	1	−1	0	1	12,000	–
c_j		7/100	1/10	0	0	0	0	$-M$	$-M$		
Solution		12,000	12,000	6,000	6,000	0	0	0	0	$Z = 2{,}040$	
Δ_j		0	0	0	0	$-\frac{17}{100}$	$\frac{7}{100}$ ↑	$-M$	$-M-\frac{7}{100}$		

TABLE 3.65 Simplex Tableau 5: Optimal Solution

Basis		x_1	x_2	S_1	S_2	S_3	S_4	A_1	A_2	b_i
S_4	0	0	0	1	0	−2	1	0	−1	6,000
x_2	1/10	0	1	0	0	1	0	0	0	12,000
S_2	0	0	0	1	1	−1	0	−1	0	12,000
x_1	7/100	1	0	1	0	−1	0	0	0	18,000
c_j		7/100	1/10	0	0	0	0	$-M$	$-M$	
Solution		18,000	12,000	0	12,000	0	6,000	0	0	$Z = 2{,}460$
Δ_j		0	0	−7/100	0	−3/100	0	$-M$	$-M$	

The solution given in Table 3.65 is evidently optimal since all Δ_j's are seen to be less than, or equal to, zero.

Now, Figure 2.13 is reproduced in Figure 3.5, with the feasible region shown shaded and having vertices *A*(6,000, 0), *B*(6,000, 6,000), *C*(12,000, 12,000), *D*(18,000, 12,000) and *E*(30,000, 0).

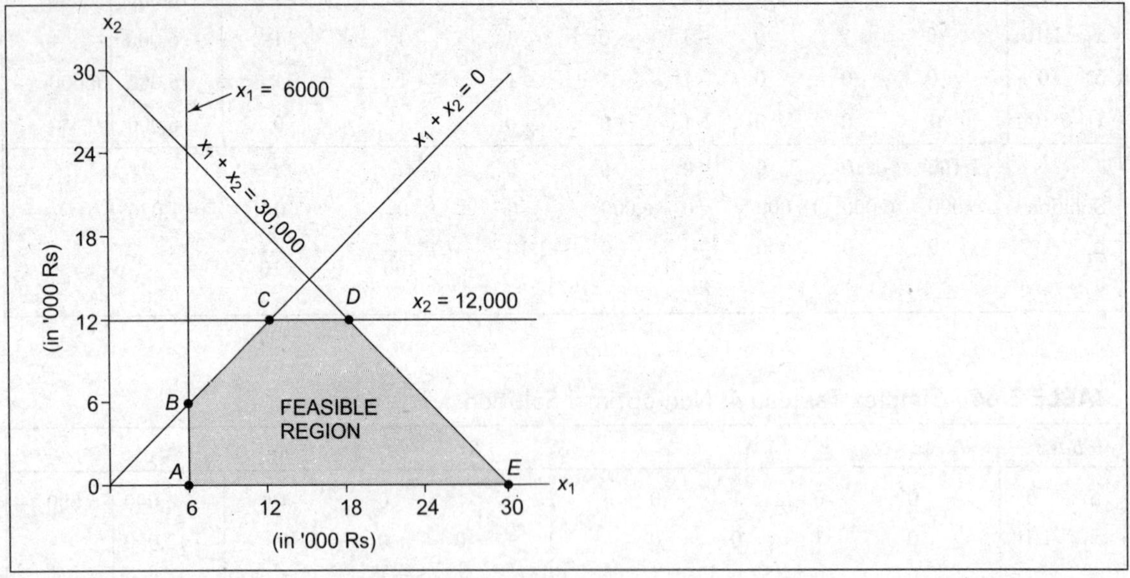

Figure 3.5 *Graphic Solution-Investment Mix*

A comparison of the solution obtained graphically and by simplex algorithm shows that solution in simplex tableau 3 corresponds to point *B* (with $x_1 = x_2 = 6{,}000$), simplex tableau 4 corresponds to point *C* (with $x_1 = x_2 = 12{,}000$) while simplex tableau 5 matches with the corner point *D* (with $x_1 = 18{,}000$ and $x_2 = 12{,}000$). It is evident, then, that the various solutions given by successive tableaus correspond to various corner points, although not all points need be represented.

A few additional points may be noted:

1. The solutions in Tables 3.61 and 3.62 are both degenerate since each one has a basic variable with a solution value equal to zero. The degeneracy, however, is temporary since the revised solutions do not have a basic variable equal to zero.
2. While selecting the outgoing variable for improving a given solution, the replacement ratios with denominator value as zero or negative are ignored.
3. Since $x_1 \geq 6{,}000$, we may set $x_1 = 6{,}000 + x_3$, and reformulate the problem as follows:

Maximise
$$Z = \frac{7}{100} x_3 + \frac{1}{10} x_2 + 420$$

Subject to

$$x_3 + x_2 \leq 24{,}000$$
$$x_2 \leq 12{,}000$$
$$-x_3 + x_2 \leq 6{,}000$$
$$x_3, x_2 \geq 0$$

This may be solved to get $x_2 = 12{,}000$ and $x_3 = 12000$. Accordingly, $x_1 = 6{,}000 + 12{,}000 = 18{,}000$ and $x_2 = 12{,}000$, the same as above.

Example 3.18 A company produces three products, P_1, P_2 and P_3 from two raw materials *A* and *B*, and labour *L*. One unit of product P_1 requires one unit of *A*, 3 units of *B* and 2 units of *L*. One unit of product P_2 requires 2 units of *A* and *B* each, and 3 units of *L*, while one unit of P_3 needs 2 units of *A*, 6 units of *B* and 4 units of *L*. The company has a daily availability of 8 units of *A*, 12 units of *B* and 12 units of *L*. It is further known that the unit contribution margin for the products is Rs 3, 2 and 5 respectively for P_1, P_2 and P_3.

Formulate this problem as a linear programming problem, and then solve it to determine the optimum product mix. Is the solution obtained by you unique? Identify an alternate optimum solution, if any.

If x_1, x_2 and x_3 be the output of the products P_1, P_2 and P_3, respectively, we may express the linear programming formulation as follows:

Maximise $\quad Z = 3x_1 + 2x_2 + 5x_3 \quad$ Contribution

Subject to

$$x_1 + 2x_2 + 2x_3 \leq 8 \qquad \text{Material } A$$

$$3x_1 + 2x_2 + 6x_3 \leq 12 \qquad \text{Material } B$$

$$2x_1 + 3x_2 + 4x_3 \leq 12 \qquad \text{Labour}$$

$$x_1, x_2, x_3, \geq 0$$

Introducing slack variables S_1, S_2 and S_3, we may write the problem as follows:

Maximise $\quad Z = 3x_1 + 2x_2 + 5x_3 + 0S_1 + 0S_2 + 0S_3$

Subject to

$$x_1 + 2x_2 + 2x_3 + S_1 = 8$$

$$3x_1 + 2x_2 + 6x_3 + S_2 = 12$$

$$2x_1 + 3x_2 + 4x_3 + S_3 = 12$$

$$x_1, x_2, x_3, S_1, S_2, S_3 \geq 0$$

The solution to the problem is contained in Tables 3.66 through 3.68.

TABLE 3.66 Simplex Tableau 1: Non-optimal Solution

Basis		x_1	x_2	x_3	S_1	S_2	S_3	b_i	b_i/a_{ij}
S_1	0	1	2	2	1	0	0	8	4
S_2	0	3	2	6*	0	1	0	12	2 ←
S_3	0	2	3	4	0	0	1	12	3
c_j		3	2	5	0	0	0		
Solution		0	0	0	8	12	12		
Δ_j		3	2	5	0	0	0		
				↑					

TABLE 3.67 Simplex Tableau 2: Non-optimal Solution

Basis		x_1	x_2	x_3	S_1	S_2	S_3	b_i	b_i/a_{ij}
S_1	0	0	4/3	0	1	– 1/3	0	4	–
x_3	5	1/2*	1/3	1	0	1/6	0	2	4 ←
S_3	0	0	5/3	0	0	– 2/3	1	4	–
c_j		3	2	5	0	0	0		
Solution		0	0	2	4	0	4	$Z = 10$	
Δ_j		1/2	1/3	0	0	– 5/6	0		
		↑							

TABLE 3.68 Simplex Tableau 3: Optimal Solution

Basis		x_1	x_2	x_3	S_1	S_2	S_3	b_i	b_i/a_{ij}
S_1	0	0	4/3	0	1	– 1/3	0	4	3
x_1	3	1	2/3	2	0	1/3	0	4	6
S_3	0	0	5/3*	0	0	– 2/3	1	4	12/5 ←
c_j		3	2	5	0	0	0		
Solution		4	0	0	4	0	4	$Z = 12$	
Δ_j		0	0	– 1	0	– 1	0		
			↑						

The solution contained in Table 3.68 is optimal with $x_1 = 4$, $x_2 = x_3 = 0$ and $Z = 12$. However, it is not unique since x_2, a non-basic variable, has Δ_j equal to zero. The problem, thus, has an alternate optimal solution. To obtain this, we revise the solution in Table 3.68 with x_2 as the entering variable. It is given in Table 3.69.

TABLE 3.69 Simplex Tableau 4: Alternate Optimal Solution

Basis		x_1	x_2	x_3	S_1	S_2	S_3	b_i
S_1	0	0	0	0	1	1/5	– 4/5	4/5
x_1	3	1	0	2	0	3/5	– 2/5	12/5
x_2	2	0	1	0	0	– 2/5	3/5	12/5
c_j		3	2	5	0	0	0	
Solution		12/5	12/5	0	4/5	0	0	$Z = 12$
Δ_j		0	0	– 1	0	– 1	0	

Example 3.19 From the following initial simplex tableau:

Basis	c_j	x_1	x_2	S_1	S_2	A_1	A_2	
		15	25	0	0	$-M$	$-M$	
A_1	$-M$	7	6	-1	0	1	0	20
S_2	0	8	5	0	1	0	0	30
A_2	$-M$	3	-2	0	0	0	1	18
z_j		$-10M$	$-4M$	M	0	$-M$	$-M$	$-38M$
$c_j - z_j$		$15 + 10M$	$25 + 4M$	$-M$	0	0	0	

Write down the original problem represented by the above tableau.

Find out the optimal solution of this problem. Is it a unique solution? Why? *(M Com, Delhi, 1996)*

The given information reveals that

(i) There are two decision variables x_1 and x_2, with objective function coefficients equal to 15 and 25, respectively.

(ii) There are three constraints with RHS values as 20, 30 and 18, and involving $\geq$, $\leq$ and $=$ signs, respectively. This is indicated by the slack, surplus and artificial variables.

(iii) The objective function is of maximisation type because the artificial variables bear negative coefficients ($=-M$) in the objective function.

The required LPP is, therefore,

Maximise $\quad Z = 15x_1 + 25x_2$

Subject to

$$7x_1 + 6x_2 \geq 20$$
$$8x_1 + 5x_2 \leq 30$$
$$3x_1 - 2x_2 = 18$$
$$x_1, x_2 \geq 0$$

The problem is solved as follows. Introducing the necessary slack, surplus and artificial variables, we have,

Maximise $\quad Z = 15x_1 + 25x_2 + 0S_1 + 0S_2 - MA_1 - MA_2$

Subject to

$$7x_1 + 6x_2 - S_1 + A_1 = 20$$
$$8x_1 + 5x_2 + S_2 = 30$$
$$3x_1 - 2x_2 + A_2 = 18$$
$$x_1, x_2, S_1, S_2, A_1, A_2 \geq 0$$

The solution is given in Tables 3.70 through 3.72.

TABLE 3.70 Simplex Tableau 1: Non-optimal Solution

Basis		x_1	x_2	S_1	S_2	A_1	A_2	b_i	b_i/a_{ij}
A_1	$-M$	7*	6	-1	0	1	0	20	20/7 ←
S_2	0	8	5	0	1	0	0	30	30/8
A_2	$-M$	3	-2	0	0	0	1	18	6
c_j		15	25	0	0	$-M$	$-M$		
Solution		0	0	0	30	20	18		
Δ_j		$15 + 10M$ ↑	$25 + 4M$	M	0	0	0		

TABLE 3.71 Simplex Tableau 2: Non-optimal Solution

Basis		x_1	x_2	S_1	S_2	A_1	A_2	b_i	b_i/a_{ij}
x_1	15	1	6/7	$-1/7$	0	1/7	0	20/7	–
S_2	0	0	$-13/7$	8/7*	1	$-8/7$	0	50/7	50/8 ←
A_2	$-M$	0	$-32/7$	3/7	0	$-3/7$	1	66/7	22
c_j		15	25	0	0	$-M$	$-M$		
Solution		20/7	0	0	50/7	0	66/7		
Δ_j		0	$\frac{85}{7} - \frac{32}{7}M$	$\frac{15}{7} + \frac{3}{7}M$ ↑	0	$-\frac{10M}{7} - \frac{15}{7}$	0		

TABLE 3.72 Simplex Tableau 3: Final, Non-optimal

Basis		x_1	x_2	S_1	S_2	A_1	A_2	b_i
x_1	15	1	5/8	0	1/8	0	0	15/4
S_1	0	0	$-13/8$	1	7/8	-1	0	50/8
A_2	$-M$	0	$-31/8$	0	$-3/8$	0	1	27/4
c_j		15	25	0	0	$-M$	$-M$	
Solution		15/4	0	50/8	0	0	27/4	
Δ_j		0	$\frac{125}{8} - \frac{31}{8}M$	0	$\frac{-15}{8} - \frac{3}{8}M$	$-M$	0	

Simplex tableau 3 contains final solution as shown by the Δ_j values which are all ≤ 0. However, since an artificial variable is one of the basic variables, the problem has no feasible solution.

KEY POINTS TO REMEMBER

- The simplex method is a step-by-step procedure for finding the optimal solution to a linear programming problem.
- To apply simplex method for solving a linear programming problem, it has to be ensured that all the variables in the problem are non-negative and all the values on the RHS of constraints are non-negative.
- For a constraint involving a negative b_i value, it is multiplied on each side by –1, and the direction of inequality is reversed. In case a variable is unrestricted in sign, it is replaced by the difference of two non-negative variables.
- The slack and surplus variables are used to convert ≤ and ≥ type of inequalities into equalities. Further, artificial variables are introduced if we fail to obtain an identity in the initial simplex tableau.
- Once an initial solution is obtained with the aid of identity matrix, it is tested for optimality in terms of $\Delta_j = c_j - z_j$.
- A non-optimal solution is improved by (i) selecting the variable with largest Δ_j (for a maximisation problem) and with most negative Δ_j (for a minimisation problem) as the incoming variable; (ii) selecting the variable with least non-negative replacement ratio as the outgoing variable, and (iii) obtaining a new simplex tableau. The solution is also tested for optimality. Similar improvements in the solutions are done until an optimal solution is reached.
- Where artificial variables are involved, the two-phase method can also be used. In Phase-I, all artificial variables are eliminated while in the other phase the optimal solution is obtained.
- Basic variables in any solution have Δ_j values equal to zero. In an optimal solution, if some non-basic variable(s) has $\Delta_j = 0$, the problem has multiple optimal solutions, while if none of the non-basic variables have $\Delta_j = 0$, the solution is unique optimal solution.
- Infeasibility is present when a problem has no feasible solution. In solving an LPP by simplex method, if a final solution is reached in terms of the Δ_j values and an artificial variable is present in the basis, then infeasibility is indicated.
- In improving a non-optimal solution, the outgoing variable is chosen as the one with the least non-negative replacement ratio. If all such ratios are negative, or a mix of negative and infinity type, then the problem has unbounded solution.
- A solution is said to be degenerate if some basic variable(s) has a solution value equal to zero. A degenerate solution may be optimal or non-optimal. When a degenerate solution is non-optimal and the outgoing variable happens to be a degenerate variable, there is no change in the objective function value for the solution obtained in the next tableau.

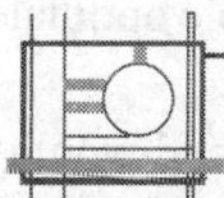

TEST YOUR UNDERSTANDING

Mark the following statements as T (*True*) or F (*False*).

1. The simplex method is an iterative process which involves the substitution of variables for obtaining successively better solutions.

2. To solve a linear programming problem by Simplex method, it is essential that *all* variables in it be non-negative.
3. Solution by Simplex method requires that an LPP should have no negative values in the right-hand-sides of the constraints.
4. For solving a linear programming problem by the Simplex method, it is necessary that any unrestricted variables are first replaced by non-negative variables.
5. For solving an LPP by Simplex method, all the constraints involving inequalities should be converted into equations by introducing slack/surplus variables.
6. In a maximisation problem, there is never a need to introduce artificial variables.
7. If a variable in an LPP is a free variable, it may be replaced by the difference of two non-negative variables.
8. In solving a linear programming problem, every constraint which involves an equality should be replaced equivalently by a pair of inequalities.
9. In improving a non-optimal solution, the key element may be positive, negative or zero.
10. An LPP may have a feasible solution although an artificial variable appears with a positive value in the final solution to the problem.
11. If the slack variable corresponding to a resource appears in the basis of the Simplex tableau containing optimal solution to an LPP, it is bound to have $\Delta_j = 0$.
12. In solution to an LPP by the Simplex method, once a variable is designated as an outgoing variable and it leaves the basis, it cannot re-enter the basis.
13. Successive solutions obtained in the Simplex algorithm *always* yield higher and higher values of Z.
14. In order to decide upon the outgoing variable in a Simplex tableau giving a non-optimal solution, the least non-negative replacement ratio is selected.
15. In the Simplex method, the optimal solution to a minimisation LPP is reached only when all Δ_j's $(= c_j - z_j)$ are found positive.
16. Surplus variables cannot appear in the basis of the optimal solution to an LPP.
17. The Big-M method and the Two-Phase method require the same number of iterations for solving a linear programming problem.
18. The artificial variables serve as *catalysts* in obtaining (optimal) solution to an LPP.
19. An artificial variable column can be dropped for further calculations once the artificial variable involved becomes non-basic.
20. For each of the basic variables in a given solution, whether optimal or not, the Δ_j equals zero.
21. The solution to a maximisation LPP is unique if $\Delta_j < 0$ for each of the non-basic variables.
22. The maximisation of a function Z is equivalent to the minimisation of $G = -Z$ with the same constraints except that min $G = \max - Z$.
23. If the outgoing variable does not correspond to the least non-negative replacement ratio, at least one basic variable would become negative in the next iteration.
24. A linear programming problem may have no optimal solution, one optimal solution, multiple optimal solutions or unbounded solution.
25. Multiple optimal solutions to an LPP would all have the same objective function values and would consume the same amount of each of the resources.
26. In the optimal solution, if a basic variable has a solution value equal to zero, then multiple optimal solutions are indicated.

27. If an LPP has an unbounded solution space, the objective function value will always be unbounded.
28. The optimal solution to a profit maximising LPP always implies maximum profit obtainable from the given resources and a full utilisation of them all.
29. If a certain solution to an LPP is degenerate and non-optimal, the next solution would necessarily be degenerate.
30. If the non-negative replacement ratios are tied, then multiple optimal solutions are indicated.

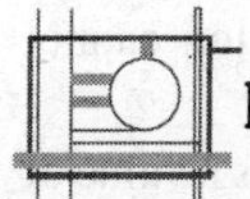

EXERCISES

1. Explain the concept and computational steps of the simplex method for solving linear programming problems. How would you identify whether an optimal solution to a problem obtained using simplex algorithm is unique or not?
2. (a) What is the difference between a feasible solution, a basic feasible solution, and an optimal solution of a linear programming problem?
   (b) What is the difference between simplex solution procedure for a 'maximisation' and a 'minimisation' problem?
   (c) Using the concept of net contribution, provide an intuitive explanation of why the criterion for optimality for maximisation problems is different from that of minimisation problems.
3. Outline the steps involved in the simplex algorithm for solving a linear programming maximisation problem. Also define the technical terms used therein.
4. What is the difference between slack, surplus, and artificial variables? How do they differ in their structure and use?
5. State the conditions required for applying simplex method to a linear programming problem. How do we proceed in a case when both of these are not met?
6. In solving a linear programming problem by simplex method, explain how you would move from a given basic feasible solution to another basic feasible solution with an improved value of the objective function.
7. How are key column and key row determined in a simplex tableau containing a non-optimal solution? How would you proceed if a tie is obtained in either of them?
8. Describe the two-phase method for solving the linear programming problems.
9. Consider each of the following statements and state whether it is true or false. In case it is false, write the correct statement.
   (i) Each inequality constraint in a linear programming problem adds only one variable, when it is solved by the simplex method.
   (ii) All the slack, surplus and artificial variables are included in the basis of the first simplex tableau.
   (iii) If optimal solution to an LPP is degenerate, it is of no consequence to the manager.
   (iv) Any solution which satisfies at least one of the constraints in an LPP is included in the feasible region of the problem.
   (v) For solving by simplex method, only an artificial variable is needed to be introduced if a constraint in an LPP involves an equation. Slack and surplus variables are not required in such a case.
10. Explain and graphically illustrate infeasibility and unboundedness. How can each of these be detected while applying simplex technique?
11. 'In a given problem, the value of objective function improves with successive iterations.' Is there any exception to this? Explain clearly.

12. Define the following and indicate their significance to decision-making with linear programming and the simplex method:
    (a) Key column, (b) Key row, (c) Degeneracy, (d) Cycling, and (e) Multiple optima.
13. Consider the following statements and state, in each case, whether it is true or false. In case the statement is false, state the correct position:
    (i) The addition of a new constraint in a linear programming problem can improve the value of the objective function.
    (ii) In the simplex method, any variable which is removed from the basis at an iteration cannot re-enter the basis at the next iteration.
    (iii) If, in the final simplex tableau of a linear programming problem, any basic variable has a value of zero in the quantity column, the problem has multiple optimal solutions.
    (iv) If an optimal solution is degenerate, it is meaningless to the manager. *(M Com, Delhi, 1997)*

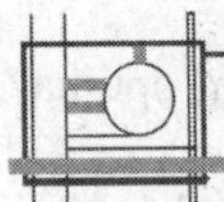

PRACTICAL PROBLEMS

1. Rewrite the following LPP in standardised form for application of simplex method:

   Maximise $Z = 8x_1 - 6x_2 + 7x_3 + 2x_4$

   Subject to

   $$4x_1 + 3x_2 + 6x_3 + x_4 \leq 40$$
   $$-x_1 + 2x_2 + 3x_3 + x_4 \leq 5$$
   $$9x_1 - 5x_2 + 7x_3 - x_4 \geq 60$$
   $$6x_2 + 2x_3 + 4x_4 = 47$$
   $$x_1, x_2, x_3, x_4 \geq 0$$

2. Solve the following problem using simplex method:

   Maximise $Z = 7x_1 + 14x_2$

   Subject to the constraints

   $$3x_1 + 2x_2 \leq 36$$
   $$x_1 + 4x_2 \leq 10$$
   $$x_1, x_2 \geq 0$$

3. Solve the following problem using simplex method:

   Maximise $Z = 21x_1 + 15x_2$

   Subject to the constraints

   $$-x_1 - 2x_2 \geq -6,$$
   $$4x_1 + 3x_2 \leq 12,$$
   $$x_1, x_2 \geq 0,$$

4. Maximise $Z = 20x_1 + 30x_2 + 5x_3$

   Subject to

   $$4x_1 + 3x_2 + x_3 \leq 40$$
   $$2x_1 + 5x_2 \leq 28$$

$$8x_1 + 2x_2 \leq 36$$
$$\text{and} \quad x_1, x_2, x_3 \geq 0$$

5. Solve the following linear programming problem using simplex method:

Maximise $Z = 8x_1 - 4x_2$

Subject to

$$4x_1 + 5x_2 \leq 20$$
$$-x_1 + 3x_2 \geq -23$$
$$x_1 \geq 0, x_2 \text{ unrestricted in sign.}$$

(MCA, RD University, 1997)

6. A pharmaceutical company produces two popular drugs *A* and *B* which are sold at the rate of Rs 9.60 and Rs 7.80, respectively. The main ingredients are *x*, *y* and *z* and they are required to the following proportions:

Drugs	*x*%	*y*%	*z*%
A	50	30	20
B	30	30	40

The total available quantities (gm) of different ingredients are 1,600 in *x*, 1,400 in *y* and 1,200 in *z*. The costs of *x*, *y* and *z* per gm are Rs 8, Rs 6 and Rs 4, respectively.
Estimate the most profitable quantities of *A* and *B* to produce, using simplex method.

(ICWA, June, 1994)

7. A company makes two kinds of leather belts. Belt *A* is a high quality belt, and belt *B* is of lower quality. The respective profits are Rs 20 and Rs 15 per belt. Each belt of type *A* requires twice as much time as belt of type *B*, and if all belts were of type *B*, the company could make 1,000 per day. The supply of leather is sufficient for only 800 belts per day (both *A* and *B* combined). Belt *A* requires a fancy buckle, and only 400 per day are available. There are only 700 buckles a day available for belt *B*. What should be the daily production of each type of belts to maximise profit. Use simplex method.
8. Show that the following linear programming problem is temporarily degenerate:

Maximise $Z = 3x_1 + 2x_2$

Subject to

$$4x_1 + 3x_2 \leq 12$$
$$4x_1 + x_2 \leq 8$$
$$4x_1 - 9x_2 \leq 8$$
$$x_1, x_2 \geq 0$$

(M Com, Kolkata, 1994)

9. Solve the following linear programming problem:

Maximise $Z = 3x_1 + 2x_2 + 3x_3$

Subject to constraints

$$2x_1 + x_2 + x_3 \leq 2$$
$$3x_1 + 4x_2 + 2x_3 \geq 8$$
$$x_1, x_2, x_3, \geq 0$$

Is there any alternate optimal solution to this problem? *(MIB, Delhi, 2005)*

10. A factory produces three different products viz. *A*, *B* and *C*, the profit (Rs) per unit of which are 3, 4 and 6 respectively. The products are processed in three operations viz. *X*, *Y* and *Z* and the time (hour) required in each operation for each unit is given below:

Operation	Products		
	A	*B*	*C*
X	4	1	6
Y	5	3	1
Z	1	2	3

The factory works 25 days in a month, at rate of 16 hours a day in two shifts. The effective working of all the processes is only 80% due to assignable causes like power cut and breakdown of machines. The factory has 3 machines in operation *X*, 2 machines in operation *Y* and one machine in operation *Z*. Find out the optimum product mix for the month. *(ICWA, December, 1993)*

11. A manufacturer of baby dolls makes three types of dolls: Doll *A*, Doll *B*, and Doll *C*. Processing of these dolls is done on three machines, M_1, M_2 and M_3. Doll *A* requires 2 hours on machine M_1 and 3 hours on machine M_3. Doll *B* requires 3 hours on machine M_1, 2 hours on machine M_2 and 2 hours on machine M_3, while doll *C* requires 5 hours on machine M_2 and 4 hours on machine M_3. There are 8 hours of time per day available on machine M_1, 10 hours of time available on machine M_2 and 15 hours of time per day available on machine M_3. The profit gained from doll *A* is Rs 3 per doll, from doll *B* is Rs 5 per doll and from doll *C* is Rs 4 per doll.

    What should be the daily production of each type of dolls? Which machine(s) would be idle according to this plan? By how much?

12. A factory engaged in the manufacturing of pistons, rings and valves for which the profits per unit are Rs 10, 6 and 4, respectively, wants to decide the most profitable mix. It takes one hour of preparatory work, ten hours of machining and two hours of packing and allied formalities for a piston. Corre sponding requirements for rings and valves are 1, 4 and 2, and 1, 5 and 6 hours, respectively. The total number of hours available for preparatory work, packing and allied formalities are 100, 600 and 300, respectively. Determine the most profitable mix, assuming that what all produced can be sold. *(CA, May 1995)*

13. Three products, *A*, *B* and *C* are produced at three machine centres, *X*, *Y* and *Z*. Each product involves operation at each of the machine centres. The time required for each operation on various products is indicated in the following table:

Product	Machine centres			Profit per unit
	X	*Y*	*Z*	*(Rs)*
A	10	7	2	12
B	2	3	4	3
C	1	2	1	1
Available Hours	100	77	80	

    (a) Formulate a linear programming problem on the basis of above information.
    (b) Find out the product mix so as to maximise profit.
    (c) Show that the total available hours of *X* and *Y* have been fully utilised and there is surplus hours of *Z*. How many hours are surplus in machine centre *Z*?

14. A pharmaceutical company has 100 kg of material *A*, 180 kg of material *B* and 120 kg of material *C* available per month. They can use these materials to make three basic pharmaceutical products, namely 5-10-5, 5-5-10 and 20-5-10, where the numbers in each case represent the percentage by weight of material *A*, material *B* and material *C* respectively, in each of the products and the balance represents inert ingredients. The cost of raw material is given below:

Ingredient	*Cost per kg (Rs)*
Material *A*	80
Material *B*	20
Material *C*	50
Inert ingredient	20

Selling price of these products is Rs 40.50, Rs 43 and Rs 45 per kg respectively. There is a capacity restriction of the company for the product 5-10-5, that is, they cannot produce more than 30 kg per month. Formulate a linear programming model for maximising the monthly profit.
Determine how much of each of the products should they produce in order to maximise their monthly profits. *(CS, June, 1992)*

15. Noah's Boats makes three different types of boats. All boats can be made profitably in this company, but the company's monthly production is constrained by the limited amount of labour, wood and screws available each month. The director will choose the combination of boats that maximises his revenue, in view of the information given in the following table:

Input	*Row boat*	*Canoe*	*Kayak*	*Monthly availability*
Labour (hours)	12	7	9	1,260 hours
Wood (Board feet)	22	18	16	19,008 board feet
Screws (kg)	2	4	3	396 kg
Selling price (in Rupees)	4,000	2,000	5,000	

(a) Formulate the above as a linear programming problem.
(b) Solve it by the simplex method.

From the optimal table of the solved linear programming problem, answer the following questions:

(c) How many boats of each type will be produced and what will be the resulting revenue?
(d) Which, if any, of the resources are not fully utilised? If so, how much of spare capacity is left?
(e) How much wood will be used to make all of the boats given in the optimal solution?

16. A trucking company with Rs 40,00,000 to spend on new equipment is contemplating three types of vehicles. Vehicle *A* has a l0-tonne pay-load and is expected to average 35 km per hour. It costs Rs 80,000. Vehicle *B* has a 20-tonne pay-load and is expected to average 30 km per hour. It costs Rs 130,000. Vehicle *C* is a modified form of *B;* it carries sleeping quarters for one driver and that reduces its capacity to 18 tonnes and raises the cost to Rs 150,000. Vehicle *A* requires a crew of one average-man, and if driven on three shifts per day, could be run for an average of 18 hours per day. Vehicles *B* and *C* require a crew of two men each, but whereas *B* would be driven 18 hours per day

with three shifts, *C* would average 21 hours per day. The company has 150 drivers available each day and would find it very difficult to obtain further crews. Maintenance facilities are such that the total number of vehicles must not exceed 30. How many vehicles of each type should be purchased if the company wishes to maximise its capacity in tonne-kms per day?

(MBA, Delhi, April, 1998)

17. Maximise $Z = 7x_1 + 2x_2 + 3x_3 + 4x_4,$

    Subject to
$$-4x_1 + 7x_2 + 6x_3 - 4x_4 \leq 20;$$
$$3x_1 - 3x_2 + 4x_3 + x_4 \leq 10;$$
$$8x_1 - 3x_2 + 4x_3 + 2x_4 \leq 25;$$
$$\text{and} \quad x_1, x_2, x_3, x_4 \geq 0.$$

18. A furniture company manufactures four models of desks. Each desk is first constructed in the carpentry shop and is next sent to the finishing shop, where it is varnished, waxed and polished. The number of man-hours of labour required in each shop is as follows:

	Desk I	Desk II	Desk III	Desk IV
Carpentry shop	4	9	7	10
Finishing shop	1	1	3	40

Because of limitations in capacity of the plant, no more than 6,000 man-hours can be expected in the carpentry shop and 4,000 in the finishing shop in the next six months.
The profit (revenue minus labour costs) from the sale of each desk is as follows:

Desk	I	II	III	IV
Profit	Rs 9	Rs 20	Rs 15	Rs 40

Assuming that raw materials and supplies are available in adequate quantities and all desks produced can be sold, set the linear programme of the problem of the company in terms of maximum profit and determine the optimal mix which will maximise profit.

19. Minimise $Z = 6x_1 + 4x_2$

    Subject to
$$3x_1 + 0.5x_2 \geq 12$$
$$2x_1 + x_2 \geq 16$$
$$x_1, x_2 \geq 0$$

20. Solve the LPP by two-phase method:

    Minimise $Z = x_1 + x_2$

    Subject to
$$2x_1 + x_2 \geq 4$$
$$x_1 + 7x_2 \geq 7$$
$$x_1, x_2 \geq 0$$

(M Sc, RD Univ., 1992)

21. Using two-phase method, solve the following problem:

    Minimise $150x_1 + 150x_2 + 100x_3$

    Subject to
$$2x_1 + 3x_2 + x_3 \geq 4$$

$$3x_1 + 2x_2 + x_3 \geq 3$$

and $x_1, x_2, x_3 \geq 0$

22. Minimise $Z = 60x_1 + 80x_2$

Subject to

$$20x_1 + 30x_2 \geq 900$$

$$40x_1 + 30x_2 \geq 1200$$

$$x_1,\ x_2 \geq 0.$$

23. A small metal press shop makes Ash Trays and Tea Trays. The process consists of buying sheet metal from the market, cutting out appropriate pieces for ash tray/tea tray on stamping machines, forming the stamped pieces into ash tray/tea tray and finally painting in the Spray Painting Booth. The finished products are sold through a distributor who is ready to take any amount that can be supplied. The stamping capacity is 30,000 sec/day and a tea tray takes 20 sec to stamp, while an ash tray takes 10 sec to stamp. Forming capacity is the same as stamping but forming a tea tray takes only 5 sec while an ash tray takes 15 sec to be formed. Painting capacity is 40,000 sec/day requiring 10 sec for an ash tray and 8 sec for a tea tray. The net realization for a tea tray is Rs 2.50 with a variable cost of Rs 2.20. An ash tray fetches Rs 1.50 against a variable cost of Rs 1.30. The fixed expenses are Rs 350 per day.
   (a) Formulate this product-mix problem as an LPP.
   (b) How many ash and tea trays should be made everyday?
   (c) If sheet metal is controlled with an allocation of 360 square metres per day and a tea tray requires 0.2 square metres against 0.16 square metres for an ash tray, what will be the new product-mix?

   (CA, November, 1976)

24. A company manufactures two products, *A* and *B*. Product *A* yields a contribution of Rs 30 per unit and product *B* Rs 40 per unit towards profits and fixed costs. It is estimated that the sales of product *A* for the coming month will not exceed 20. Sales of product *B* have not been estimated but the company does have a contract to supply at least 10 units to a regular customer.

   Machine hours available for the coming month are 100 and products *A* and *B* require 4 hours and 2 hours respectively, to produce one unit. Labour hours available are 180 and products *A* and *B* require 4 and 6 hours of labour, respectively. Materials available are restricted to 40 units and the two products each use one unit of material per unit. Determine the optimal product mix.

25. A company produces two products, *A* and *B*. The sales volume of product *A* is at least 60 percent of the total sales of the two products. Both the products use the same raw material of which the daily availability is limited to 100 tonnes. Products *A* and *B* use this material at the rate of 2 tonnes per unit and 4 tonnes per unit, respectively. The sales price for the two products are Rs 20 and Rs 40 per unit.
   (a) Construct a linear programming formulation of the problem.
   (b) Find the optimum solution by simplex method.
   (c) Find an alternative optimum, if any.

   (M Com, Kolkata, 1996)

26. A firm makes two products, *X* and *Y* and has total production capacity of 9 tonnes per day, *X* and *Y* requiring the same production capacity. The firm has a permanent contract to supply at least 2 tonnes of *X* and at least 3 tonnes of *Y* per day to another company. Each tonne of *X* requires 20 machine hours production time and each tonne of *Y* requires 50 machine hours production time; the daily maximum possible number of machine hours is 360. All the firm's output can be sold, and the profit made is Rs 80 per tonne of *X* and Rs 120 per tonne of *Y*. It is required to determine the production schedule for maximum profit and to calculate this profit. *(MBA, Delhi, April, 1994)*

27. A factory is engaged in producing two items, I_1 and I_2, each of which has to pass through three machines M_1, M_2 and M_3. One unit of I_1 requires one hour of machining on M_1, 2 hours on M_2 and 1 hour on M_3, while one unit of I_2 requires machining of an hour each on M_1 and M_2, and 3 hours on M_3. In a week, a maximum of 40 hours on machine M_1, 70 hours on M_2, and 90 hours on machine M_3 are available. It is given that the per unit profit on I_1 is Rs 40 and that on I_2 is Rs 60.

    The present production mix of the factory is: I_1 30 units and I_2 10 units, per week. This implies that the present profit of the factory is Rs 1,800 a week. Also, machines M_1 and M_2 are being fully utilised.

    In your opinion, is the present production mix optimal? If not, specify what should it be and find the increase in profit that would result if the new mix determined by you is adopted. Also determine the time for which some machine(s) may be idle in respect of the mix suggested by you.

28. An advertising company wants to cover two types of audiences: Type *A*, consisting of customers with annual income greater than Rs 150,000 and Type *B*, consisting of customers with annual income less than Rs 150,000. The total advertising budget of the company is Rs 2,10,000. One programme of TV advertising costs Rs 50,000 and one programme of radio advertising costs Rs 20,000. As per the terms of the contract, at least three programmes must be on TV and the number of radio programmes must be limited to five. On the basis of some surveys, it is found that a single TV programme reaches 4,50,000 customers of Type *A* and 50,000 customers of Type *B*, while one radio programme reaches 80,000 customers of Type *A* and 2,20,000 customers of type *B*. Use simplex method and determine the optimal media mix to maximise the total reach. What is the total reach in each type of audience?
    (M Com, Delhi, 2008)

29. A company is considering three magazines for releasing its advertisements. The relevant data are given here:

	Magazines		
	A	*B*	*C*
Readers	100,00,000	60,00,000	40,00,000
Prospective buyers the product	10%	15%	7%
Cost per advertisement (Rs)	10,000	5,000	6,000

    The company has allocated Rs 1,00,000 for magazine advertising. It has also decided that there should be at least two insertions each in magazines *A* and *C*, and that the number of insertions in magazine *B* should not exceed five. Formulate it as a LPP.
    Find out the optimal number of advertisements which the company should release for the three magazines so as to have the maximum exposures among the prospective buyers. How many such persons can be reached through the magazine advertisement?

30. Solve the following LPP:

    Minimise $Z = 120x_1 + 60x_2$

    Subject to

$$20x_1 + 30x_2 \geq 900$$
$$40x_1 + 30x_2 \geq 1200$$
$$x_1, x_2 \geq 0$$

    What would happen if the objective function of this problem was of the 'maximisation' type?

31. Two products *A* and *B* are processed on three machines M_1, M_2 and M_3. The processing times per unit, machine availability and profit per unit are:

Machine	Processing time (hours)		Availability (hours)
	A	B	
M_1	2	3	1,500
M_2	3	2	1,500
M_3	1	1	1,000
Profit	Rs 10	Rs 12	

Any unutilized time on machine M_3 can be given on rental basis to another firm at an hourly rate of Rs 1.50. Formulate the mathematical model, solve it using simplex technique to determine the maximum profit that can be made using the resources. What is the total profit earned?

32. A manufacturer produces three commodities, C_1, C_2 and C_3, which require, respectively, 2, 2 and 3 units of material and, respectively, 2, 2 and 1 hours of labour. The unit profit on C_1, C_2 and C_3 is Rs 6, Rs 3 and Rs 2 in that order.
    In a given week, if 300 units of material and 120 hours of labour are available, what product mix should he make in order to maximise the profit?
    Would the answer be different if the manufacturer decides to discontinue the production of the least profitable commodity, C_3? If yes, what would be the new product mix?

33. Given below are the objective function, the constraints, and the final simplex tableau for a linear programming product-mix problem:

    Objective function:

    Maximise $$Z = 2x_1 + 5x_2 + 8x_3$$

    Constraints:

    $$6x_1 + 8x_2 + 4x_3 \leq 96 \quad \text{(hours, department I)}$$
    $$2x_1 + x_2 + 2x_3 \leq 40 \quad \text{(hours, department II)}$$
    $$5x_1 + 3x_2 + 2x_3 \leq 60 \quad \text{(hours, department III)}$$
    $$x_1, x_2, x_3 \geq 0$$

Final Simplex Tableau

	C_j	2	5	8	0	0	0	
C_j	Product mix	x_1	x_2	x_3	S_1	S_2	S_3	Quantity
5	x_2	1/3	1	0	1/6	–1/3	0	8/3
8	x_3	5/6	0	1	–1/12	2/3	0	56/3
0	S_3	7/3	0	0	–1/3	–1/3	1	44/3
	Z_j	25/3	5	8	1/6	11/3	0	488/3
	$C_j - Z_j$	–19/3	0	0	–1/6	–11/3	0	

(a) Write the optimal product-mix and the profit contribution shown by the above solution.

(b) Is this solution feasible? Why? Give brief reason (s).

(c) Does the problem have any alternative optimal solution? If so, show one such solution.

(d) Indicate whether the solution given in the table is degenerate. If yes, which variable is degenerated?

(e) If the company produces six units of x_1, how many units of x_2 and x_3 shall have to be reduced, if any.

34. Maximise $Z = 10x_1 + 20x_2$,

Subject to

$$2x_1 + 5x_2 \geq 50$$
$$4x_1 + x_2 \leq 28$$
$$x_1, x_2 \geq 0$$

35. Solve the following problem by Simplex method:

Maximise $\pi = 22X + 30Y + 25Z$

Subject to

$$2X + 2Y \leq 100$$
$$2X + Y + Z \leq 100$$
$$X + 2Y + 2Z \geq 100$$
$$X, Y, Z \geq 0$$

(M Com, Delhi, 2008)

36. Solve the following LPP:

Maximise $Z = 40x_1 + 35x_2$,

Subject to

$$2x_1 + 3x_2 \geq 60$$
$$4x_1 + 3x_2 \geq 96$$
$$x_1, x_2 \geq 0$$

37. A company, during the festival season, combines two factors *A* and *B* to form a gift pack which must weigh 5 kg. At least 2 kg of *A* and no more than 4 kg of *B* should be used. The net profit contribution to the company is Rs 5 per kg for *A* and Rs 6 per kg for *B*. Determine the optimal product mix.

(MBA, Delhi, November, 1995)

38. Solve the following LPP:

Maximise $Z = 2x_1 + 4x_2$

Subject to

$$2x_1 + x_2 \leq 18,$$
$$3x_1 + 2x_2 \geq 30,$$
$$x_1 + 2x_2 = 25,$$
$$x_1, x_2 \geq 0$$

39. A firm manufactures two products *A* and *B*, which pass through manual and machine processes. The time required by each unit and the total available time is as given here:

	Per unit requirement (hours)		*Total hours available*
	A	*B*	
Labour	4	2	1,600
Machine	6	5	3,000

The enthusiastic production manager requires that at least 300 units of A ... nits of B should be produced. It is known that one unit of A yields Rs 10 of profit, while a ur ... ds Rs 8. In your opinion, what should be optimum production mix so that all the constraint ... d?

40. Maximise $Z = 50x_1 + 110x_2 + 120x_3$

Subject to

$$3x_1 + 3x_2 + 5x_3 \leq 100$$
$$x_1 + 3x_2 + 4x_3 \leq 80$$
$$2x_1 + 4x_2 + 3x_3 \leq 60$$
$$x_1, x_2, x_3 \geq 0$$

Linear Programming III: Duality and Sensitivity Analysis

CHAPTER OVERVIEW

The application of simplex method allows not only the solution to linear programming problems but also provides a fund of information which can be usefully employed by the manager in decision-making. The numbers generated in the course of applying simplex provide inputs to obtain answers to many 'what-if' questions.

Further, every linear programming problem has a mirror image problem which is also a linear programming problem. Using a set of rules, the mirror image problem, called the *dual* can be obtained for a given linear programming problem, known as *primal* problem in this context. This 'primal-dual' relationship is a very important one. Since the two problems are connected to each other, there is obviously a relationship to be expected between their solutions. There indeed is. This chapter explores and explains this, and does more. Apart from a mathematical connection between primal and dual, it explains the economic significance of the dual and helps a manager to answer to questions like the following:

- Is it advisable to produce all the items that can be produced in a given situation? If some item is left out in the optimal product-mix, then under what condition will it be advisable to produce it?
- What is the marginal profitability of each of the resources of the firm? In turn, this means by how much the profit will increase if more quantity of a particular resource is added or how much reduction in profit will result if its availability is reduced?
- To what extent will additional quantities of a resource (or its reduction) cause an increase (or decrease) at a uniform rate?
- Will the optimal product-mix need to change it the profitability of the various products changes? Obviously, when some small changes (increases or decreases) occur in the profitability of the products it would not cause changes in the quantities of the products being produced, but then what are the limits of these price changes?
- Is it advisable to introduce a new product given the amounts of resources required for its production and its profitability?
- How would the product-mix change, if at all, its technological changes cause the resource requirements of a product to change?

This chapter involves handling of inequalities; arithmetical operations on fractional values; plotting of equalities and inequalities on graphs, and their understanding. Further, a knowledge of the concepts of matrices and their transpose is needed.

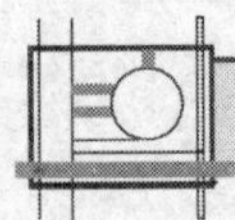

4.1 INTRODUCTION

In the previous two chapters, we have discussed the nature, formulation and solution to linear programming problems, using graphic approach and by application of simplex method. Besides yielding solution to a problem, the simplex method provides a host of additional information which is useful from the management viewpoint. The present chapter gives an account of this and is divided broadly into two distinct but related parts—duality and sensitivity analysis.

First we discuss about the *duality* which describes that linear programming problems exist in 'pairs' so that corresponding to every given linear programming problem, there is another LPP. The method of deriving the dual to a problem is followed by a discussion of the economic interpretation of the dual problem. Sensitivity analysis deals with the robustness of a solution to an LPP to changes in the inputs. It thus discusses how changes in the profit rates of the products, resource availability and resource requirements for the products would affect the solution to a linear programming problem.

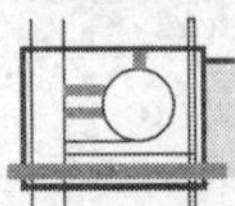

4.2 DUALITY IN LINEAR PROGRAMMING

As mentioned earlier, for every linear programming problem, there is another linear programming problem which is related to it and, therefore, may be derived from it. We now consider the method of deriving the dual to a given primal problem. Before writing the dual to an LPP, it is necessary to express it in standard form if it is not so. By standard form it is meant that (i) all the variables in the problem should be non-negative; and (ii) all the constraints should be of '$\leq$' type if the problem is of maximisation type, and of '$\geq$' type if it of the minimisation nature. The necessary changes may be made as follows:

When all variables in the LPP are not non-negative

If a variable in an LPP is unrestricted in sign so that it is a 'free' variable, then it may be replaced by the difference of two non-negative variables, each of which is non-negative. While each of the variables introduced is non-negative, their difference might be positive, negative or zero.

When all constraints are not in the 'right' direction

If a constraint involves an inequality in a direction opposite to the one desired, then it is multiplied by –1 throughout and the direction of the inequality is reversed. For example, if a constraint is given as $7x_1 - 5x_2 \geq 4$, and the desired inequality direction is $\leq$, then we replace this constraint as $-7x_1 + 5x_2 \leq -4$. It is easy to visualize as to why the direction of inequality reverses on multiplication by –1 by considering a simple example: as 10 is > 5 but – 10 is < – 5.

On the other hand, if a constraint involves an equation, then, to obtain inequalities, it is replaced by a pair of inequalities in opposite directions. To understand, if $A = 20$, then neither can we express it as $A \leq 20$ nor as $A \geq 20$, since both are incorrect individually. But, if we consider $A \leq 20$ and $A \geq 20$ together to satisfy, they can both be satisfied for only one value of A, equal to 20. Accordingly, if the linear programming problem has a constraint, say $7x_1 + 5x_2 = 42$, we would replace it by a pair of constraints as follows:

$$7x_1 + 5x_2 \leq 42, \text{ and}$$

$$7x_1 + 5x_2 \geq 42$$

One of these would be multiplied by –1 to bring the inequality in the desired direction.

4.2.1 Writing the Dual

Once the problem is known to be, or transformed into, standard form, we can write its dual. Suppose the primal problem is,

Maximise $Z = cx$

Subject to

$$ax \leq b$$

$$x \geq 0$$

where, c = row matrix containing the coefficients in the objective function,

x = column matrix containing the decision variables

a = matrix containing the coefficients in the constraints

b = column matrix containing the RHS values of the constraints

The dual corresponding to this problem is defined as

Minimise $G = b'y$

Subject to

$$a'y \geq c'$$

$$y \geq 0$$

where, b' = transpose of the b matrix of the primal problem,

a' = transpose of the coefficients matrix of the primal problem,

c' = transpose of the matrix of the objective function coefficients of the primal problem,

y = column matrix of the dual variables.

To see as to how the dual of a problem can be formulated, consider the following examples.

Example 4.1 For the LPP given in Example 3.1 reproduced below, write the dual.

Maximise $Z = 40x_1 + 35x_2$

Subject to

$$2x_1 + 3x_2 \leq 60$$

$$4x_1 + 3x_2 \leq 96$$

$$x_1, x_2 \geq 0$$

In accordance with above, its dual shall be

Minimise $G = 60y_1 + 96y_2$

Subject to

$$2y_1 + 4y_2 \geq 40$$

$$3y_1 + 3y_2 \geq 35$$

$$y_1, y_2 \geq 0$$

Observing a little closely, we find that

(a) the primal problem is of the maximisation type while the dual is of the minimisation type.

(b) the constraint values 60 and 96 of the primal have become the coefficients of the dual variables y_1 and y_2 in the objective function of the dual in that order; while the coefficients of the variables in the objective function of the primal have become the constraint values in the dual.

(c) the first column of the coefficients in the constraints in the primal has become the first row in the constraints in the dual, and the second column has similarly become the second row.

(d) the direction of inequalities in the dual is the reverse of that in the primal. Thus, while the inequalities in the primal are of the type ≤, they are of ≥ type in the dual.

We can represent them in matrix form as given in Chart 4.1.

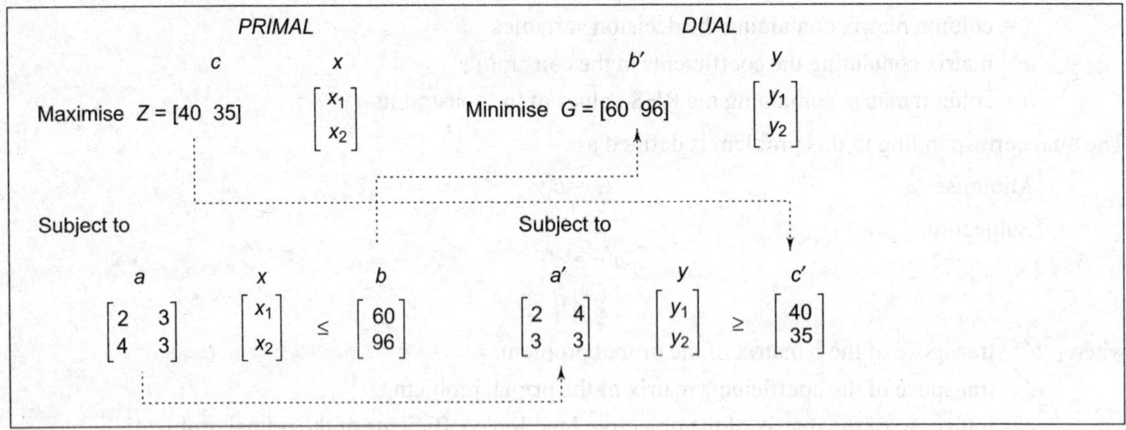

Chart 4.1 *The Primal and Dual Relationship*

4.2.2 Dual of LPP with Mixed Constraints

Sometimes a given LPP has mixed restrictions so that the inequalities given are not all in the right direction. In such a case, we should convert the inequalities in the wrong direction into those in the right direction. Similarly, if an equation is given in respect of a certain constraint, it should also be converted into inequality. To understand fully, consider the following examples.

Example 4.2 Write the dual of the following LPP:

Minimise $Z = 10x_1 + 20x_2$

Subject to

$$3x_1 + 2x_2 \geq 18$$
$$x_1 + 3x_2 \geq 8$$
$$2x_1 - x_2 \leq 6$$
$$x_1, x_2 \geq 0$$

Here, the first two inequalities are in the right direction (being ≥ type with a minimisation type of objective function) while the third one is not. Multiplying both sides by –1, this can be written as $-2x_1 + x_2 \geq -6$. Now, we can write the primal and dual as follows:

	Primal		*Dual*
Minimise	$Z = 10x_1 + 20x_2$	Maximise	$G = 18y_1 + 8y_2 - 6y_3$
Subject to		Subject to	
	$3x_1 + 2x_2 \geq 18$		$3y_1 + y_2 - 2y_3 \leq 10$
	$x_1 + 3x_2 \geq 8$		$2y_1 + 3y_2 + y_3 \leq 20$
	$-2x_1 + x_2 \geq -6$		
	$x_1, x_2 \geq 0$		$y_1, y_2, y_3 \geq 0$

Example 4.3 Obtain the dual of the LPP given here:

Maximise $Z = 8x_1 + 10x_2 + 5x_3$

Subject to

$$x_1 - x_3 \leq 4$$
$$2x_1 + 4x_2 \leq 12$$
$$x_1 + x_2 + x_3 \geq 2$$
$$3x_1 + 2x_2 - x_3 = 8$$
$$x_1, x_2, x_3 \geq 0$$

We shall first consider the constraints.

Constraints 1 and 2 : Since they are both of the type ≤, we do not need to modify them.

Constraint 3 : This is of type ≥. Therefore, we can convert it into ≤ type by multiplying both sides by –1 to become $-x_1 - x_2 - x_3 \leq -2$.

Constraint 4 : It is in the form of an equation. An equation, mathematically, can be represented by a pair of inequalities: one of ≤ type and the other of ≥ type. The given constraint can be expressed as

$$3x_1 + 2x_2 - x_3 \leq 8$$
$$3x_1 + 2x_2 - x_3 \geq 8$$

The second of these can again be converted into type ≤ by multiplying by –1 on both sides. Thus, it can be written as $-3x_1 - 2x_2 + x_3 \leq -8$.

Now we can write the primal and the dual as follows:

	Primal		*Dual*
Maximise	$Z = 8x_1 + 10x_2 + 5x_3$	Minimise	$G = 4y_1 + 12y_2 - 2y_3 + 8y_4 - 8y_5$
Subject to		Subject to	
	$x_1 - x_3 \leq 4$		$y_1 + 2y_2 - y_3 + 3y_4 - 3y_5 \geq 8$
	$2x_1 + 4x_2 \leq 12$		$4y_2 - y_3 + 2y_4 - 2y_5 \geq 10$
	$-x_1 - x_2 - x_3 \leq -2$		$-y_1 - y_3 - y_4 + y_5 \geq 5$
	$3x_1 + 2x_2 - x_3 \leq 8$		$y_1, y_2, y_3, y_4, y_5 \geq 0$
	$-3x_1 - 2x_2 + x_3 \leq -8$		
	$x_1, x_2, x_3 \geq 0$		

One point needs mention here. We know that corresponding to a n-variable, m-constraint primal problem, there would be m-variable, n-constraint dual problem. For this example involving three variables and four constraints, the dual should have four variables and three constraints. But we observe that the dual that we have obtained contains five variables. The seeming inconsistency can be resolved by expressing $(y_4 - y_5) = y_6$, a variable unrestricted in sign. Thus, although, y_4 and y_5 are both non-negative, their difference could be greater than, less than, or equal to zero. The dual can be rewritten as follows:

Minimise $\quad G = 4y_1 + 12y_2 - 2y_3 + 8y_6$

Subject to

$$\begin{aligned} y_1 + 2y_2 - y_3 + 3y_6 &\geq 8 \\ 4y_2 - y_3 + 2y_6 &\geq 10 \\ -y_1 \qquad - y_3 - y_6 &\geq 5 \\ y_1, y_2, y_3 &\geq 0, \ y_6 \text{ unrestricted in sign} \end{aligned}$$

Thus, whenever a constraint in the primal involves an equality sign, its corresponding dual variable shall be unrestricted in sign. Similarly, an unrestricted variable in the primal would imply that the corresponding constraint shall bear the = sign.

Example 4.4 Obtain the dual of the following LPP:

Maximise $\quad Z = 3x_1 + 5x_2 + 7x_3$

Subject to

$$\begin{aligned} x_1 + x_2 + 3x_3 &\leq 10 \\ 4x_1 - x_2 + 2x_3 &\geq 15 \\ x_1, x_2 \geq 0, \ x_3 &\text{ unrestricted in sign} \end{aligned}$$

First of all, we should convert the second restriction into the type ≤. This results in $-4x_1 + x_2 - 2x_3 \leq -15$.

Next, we replace the variable x_3 by the difference of two non-negative variables, say, x_4 and x_5. This yields the primal problem corresponding to which dual can be written, as shown against it.

Primal		*Dual*	
Maximise	$Z = 3x_1 + 5x_2 + 7x_4 - 7x_5$	Minimise	$G = 10y_1 - 15y_2$
Subject to		Subject to	
	$x_1 + x_2 + 3x_4 - 3x_5 \leq 10$		$y_1 - 4y_2 \geq 3$
	$-4x_1 + x_2 - 2x_4 + 2x_5 \leq -15$		$y_1 + y_2 \geq 5$
	$x_1, x_2, x_4, x_5 \geq 0$		$3y_1 - 2y_2 \geq 7$
			$-3y_1 + 2y_2 \geq -7$
			$y_1, y_2 \geq 0$

The fourth constraint of the dual can be expressed as $3y_1 - 2y_2 \leq 7$. Now, combining the third and the fourth constraints, we get $3y_1 - 2y_2 = 7$. The dual can be expressed as follows:

Minimise $\quad G = 10y_1 - 15y_2$

Subject to

$$y_1 - 4y_2 \geq 3$$
$$y_1 + y_2 \geq 5$$
$$3y_1 - 2y_2 = 7$$
$$y_1, y_2 \geq 0$$

The symmetrical relationship between the primal and dual problems, assuming the primal to be a 'maximisation' problem is depicted in Chart 4.2.

PRIMAL	*DUAL*
Maximisation	Minimisation
No. of variables	No. of constraints
No. of constraints	No. of variables
≤ type constraint	Non-negative variable
= type constraint	Unrestricted variable
Unrestricted variable	= type constraint
Objective function coefficient for *j*th variable	RHS constant for the *j*th constraint
RHS constant for *i*th constraint	Objective function coefficient for *i*th variable
Coefficient (a_{ij}) for *j*th variable in *i*th constraint	Coefficient (a_{ij}) for *i*th variable in *j*th constraint

Chart 4.2 *Symmetrical Relationship between Primal and Dual Problems*

4.2.3 Comparing the Optimal Solutions of Primal and Dual

Since the dual of a given primal problem is derived from and related to it, it is natural to expect that the (optimal) solutions to the two problems shall be related to each other in the same way. To understand this, let us consider the following primal and dual problems again and compare their optimal solutions.

	Primal		*Dual*
Maximise	$Z = 40x_1 + 35x_2$	Minimise	$G = 60y_1 + 96y_2$
Subject to		Subject to	
	$2x_1 + 3x_2 \leq 60$		$2y_1 + 4y_2 \geq 40$
	$4x_1 + 3x_2 \leq 96$		$3y_1 + 3y_2 \geq 35$
	$x_1, x_2 \geq 0$		$y_1, y_2 \geq 0$

The simplex tableau containing optimal solution to the primal problem is reproduced (from Table 3.4) in Table 4.1.

TABLE 4.1 Simplex Tableau: Optimal Solution

Basis		x_1	x_2	S_1	S_2	b_i
x_2	35	0	1	2/3	– 1/3	8
x_1	40	1	0	– 1/2	1/2	18
c_j		40	35	0	0	
Solution		18	8	0	0	
Δ_j		0	0	– 10/3	–25/3	

Now, let us consider the solution to the dual problem which is augmented, by introducing surplus and artificial variables, as follows:

Minimise $G = 60y_1 + 96y_2 + 0S_1 + 0S_2 + MA_1 + MA_2$

Subject to

$$2y_1 + 4y_2 - S_1 + A_1 = 40$$

$$3y_1 + 3y_2 - S_2 + A_2 = 35$$

$$y_1, y_2, S_1, S_2, A_1, A_2 \geq 0$$

The solution to it is contained in Tables 4.2 through 4.4.

Before comparing the solutions, it may be noted that there is a correspondence between variables of the primal and the dual problems. The structural variable x_1 in the primal, corresponds to the surplus variable S_1 in the dual, while the variable x_2 corresponds to S_2, the other surplus variable in the dual. In a similar way, the structural variables y_1 and y_2 in the dual correspond to the slack variables S_1 and S_2 respectively of the primal.

TABLE 4.2 Simplex Tableau 1: Non-optimal Solution

Basis		y_1	y_2	S_1	S_2	A_1	A_2	b_i	b_i/a_{ij}
A_1	M	2	4*	–1	0	1	0	40	10 ←
A_2	M	3	3	0	–1	0	1	35	35/3
c_j		60	96	0	0	M	M		
Solution		0	0	0	0	40	35		
Δ_j		$60 - 5M$	$96 - 7M$	M	M	0	0		
			↑						

TABLE 4.3 Simplex Tableau 2: Non-optimal Solution

Basis		y_1	y_2	S_1	S_2	A_1	A_2	b_i	b/a_{ij}
y_2	96	1/2	1	−1/4	0	1/4	0	10	20
A_2	M	3/2*	0	3/4	−1	−3/4	1	5	10/3 ←
c_j		60	96	0	0	M	M		
Solution		0	10	0	0	0	5		
Δ_j		$12-\dfrac{3M}{2}$ ↑	0	$24-\dfrac{3M}{4}$	M	$\dfrac{7M}{4}-24$	0		

TABLE 4.4 Simplex Tableau 3: Optimal Solution

Basis		y_1	y_2	S_1	S_2	A_1	A_2	b_i
y_2	96	0	1	−1/2	1/3	1/2	−1/3	25/3
y_1	60	1	0	1/2	−2/3	−1/2	2/3	10/3
c_j		60	96	0	0	M	M	
Solution		10/3	25/3	0	0	0	0	
Δ_j		0	0	18	8	$M-18$	$M-8$	

A comparison of the optimal solutions to the primal and the dual, and some observations follow.

(a) The objective function values of both the problems are the same. Thus, with $x_1 = 18$ and $x_2 = 8$, Z equals $40 \times 18 + 35 \times 8 = 1{,}000$. Similarly, with $y_1 = 10/3$ and $y_2 = 25/3$, the value of G would be $60 \times 10/3 + 96 \times 25/3 = 1{,}000$.

(b) The numerical value of each of the variables in the optimal solution to the primal is equal to the value of its corresponding variable in the dual contained in the Δ_j row. Thus, in the primal problem, $x_1 = 18$ and $x_2 = 8$, whereas in the dual $S_1 = 18$ and $S_2 = 8$ (in the Δ_j row).

Similarly, the numerical value of each of the variables in the optimal solution to the dual is equal to the value of its corresponding variable in the primal, as contained in the Δ_j row of it. Thus, $y_1 = 10/3$ and $y_2 = 25/3$ in the dual, and $S_1 = 10/3$ and $S_2 = 25/3$ (note that we consider only the absolute values) in the primal. Of course, we do not consider artificial variables because they do not correspond to any variables in the primal, and are introduced for a specific, limited purpose only.

Clearly then, if feasible solutions exist for both the primal and the dual problems then both problems have optimal solutions of which objective function values are equal. A peripheral relationship between them is that if one problem has an unbounded solution, its dual has no feasible solution.

Further, the optimal solution to the dual can be read from the optimal solution of the primal, and vice versa. The primal and dual need not both be solved, therefore, to obtain the solution. This offers a big computational advantage in some situations. For instance, if the primal problem is a minimisation one involving, say, 3 variables and 7 constraints, its solution would pose a big problem because a large number of surplus and artificial variables would have to be introduced. The number of iterations required for obtaining the answer would also be large. On the contrary, the dual, with 7 variables and 3 constraints can be solved comparatively much more easily.

Example 4.5 Given the following linear programming problem and the simplex tableau (Table 4.5) containing the solution to it, (i) formulate the dual to the given problem, and (ii) obtain the solution to the dual from the tableau. Also verify that the objective function values of both the problems are same.

Minimise $Z = x_1 + \frac{1}{2} x_2$

Subject to

$$6x_1 + 2x_2 \geq 24$$
$$3x_1 + 2x_2 \geq 18$$
$$x_1 + 3x_2 \geq 12$$
$$x_1, x_2 \geq 0$$

TABLE 4.5 Simplex Tableau: Optimal Solution

Basis		x_1	x_2	S_1	S_2	S_3	A_1	A_2	A_3	b_i
x_1	1	1	0	−1/3	1/3	0	1/3	−1/3	0	2
S_3	0	0	0	7/6	−8/3	1	−7/6	8/3	−1	8
x_2	1/2	0	1	1/2	−1	0	−1/2	1	0	6
c_j		1	1/2	0	0	0	M	M	M	
Solution		2	6	0	0	8	0	0	0	
Δ_j		0	0	1/12	1/6	0	$M - 1/12$	$M - 1/6$	M	

(i) The dual corresponding to the given problem:

Maximise $G = 24y_1 + 18y_2 + 12y_3$

Subject to

$$6y_1 + 3y_2 + y_3 \leq 1$$
$$2y_1 + 2y_2 + 3y_3 \leq \frac{1}{2}$$
$$y_1, y_2, y_3 \geq 0$$

(ii) The values of the dual variables y_1, y_2 and y_3 can be obtained from the Δ_j row of the simplex tableau given.

These are the respective values of S_1, S_2 and S_3. Thus, $y_1 = 1/12$, $y_2 = 1/6$ and $y_3 = 0$. Now, we can determine the objective function values of both the problems to get.

For primal, $Z = 2 + \frac{1}{2} \times 6 = 5$

For dual $G = 24 \times 1/12 + 18 \times 1/6 + 12 \times 0 = 5$

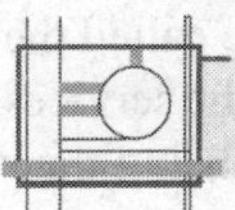

4.3 ECONOMIC INTERPRETATION OF DUAL

Not only are the primal and the dual related mathematically, they are also related in economic sense. We shall now consider the economic interpretation of the dual—first for a maximisation and then for a minimisation problem.

The Maximisation Problem Consider Example 4.1 again, whose pertinent data are reproduced here.

Maximise $Z = 40x_1 + 35x_2$ Profit

Subject to

$2x_1 + 3x_2 \leq 60$ Raw materials constraint

$4x_1 + 3x_2 \leq 96$ Labour hours constraint

$x_1, x_2 \geq 0$

The optimal solution to this problem indicates that producing 18 units of *A* and 8 units of *B* per week would yield the maximum profit equal to Rs 1,000.

Now, suppose that the firm is approached by an individual who would like to rent the facilities of the firm for one week. The firm has its assets in the form of 60 kg of raw material and 96 labour hours. If we let y_1 represent the rental rate per kg of raw material and y_2 the rental rate per labour hour, the firm would receive a total rental equal to $60y_1 + 96y_2$. We shall compute the minimum value of the rental so that the firm will know as to what minimum offer shall be economically acceptable to it. The constraints can be set up by keeping in mind that the alternative of renting must be at least as favourable as the other ones. The rental rates of the resources should be at least as attractive as producing products *A* and *B*. We know that production of one unit of *A* requires 2 kg of raw material and 4 labour hours. Thus, the total rental for these amounts of resources should be greater than, or equal to, the profit obtainable from one unit of the product, i.e. Rs 40. Thus, we should have

$$2y_1 + 4y_2 \geq 40$$

In a similar way, the resources consumed in producing one unit of product *B* are 3 kg of raw material and 3 labour hours. The total rental of these resources should equal at least Rs. 35, the unit profit of product *B*. Thus, we should have,

$$3y_1 + 3y_2 \geq 35$$

Besides, the rentals cannot be negative. Therefore, y_1 and y_2 should both be non-negative. In complete form, the problem can be expressed as:

Minimise $G = 60y_1 + 96y_2$

Subject to

$$2y_1 + 4y_2 \geq 40$$
$$3y_1 + 3y_2 \geq 35$$
$$y_1, y_2 \geq 0$$

This problem is absolutely the same as considered earlier—the dual to the given problem. These rental rates y_1 and y_2 are obtainable from the solution of the dual, as $y_1 = 10/3$ and $y_2 = 25/3$. Also, the values can be obtained from Δ_j row of the simplex tableau containing optimal solution to the primal problem. As observed previously, the objective function values of the primal and the dual would be identical. Thus, it is not surprising to observe that the minimum total rental acceptable by the firm is equal to the maximum profit that it can earn by producing the output itself using the given resources.

The individual rental rates, y_1 and y_2, the dual variables corresponding to the slack variables, are called the *shadow prices* or *imputed prices,* indicating the worth of the resources. These prices, assigned to the services of the two resources, materials and labour hours, are imputed from the profit obtained from utilising their services, and bear no relationship with the original cost of the two.

Of the two products *A* and *B,* we know that each unit of product *A* contributes Rs 40 to the profit. The imputed price of material and labour being, respectively, Rs 10/3 per kg and Rs 25/3 per hour, we observe that the total imputed cost of the resources used in making a unit of the product would be:

2 kg @ Rs 10/3 per kg	= Rs 20/3
4 hours @ Rs 25/3 per hour	= Rs 100/3
Total	= Rs 120/3 = Rs 40

Thus, the total imputed cost of producing one unit of product *A* equals the unit profit obtainable from it. Similarly, from each unit of product *B,* the total imputed cost of resources employed would be:

3 kg @ Rs 10/3 per kg	= Rs 10
3 hours @ Rs 25/3 per hour	= Rs 25
Total	= Rs 35

This, obviously equals the profit per unit of the product. It may again be emphasised that the services of the resources are valued in such a way that the total value of resources matches the profit realisable from the optimal utilisation of their services.

The shadow prices are also called the *marginal value products* or *marginal profitability* of the resources. Thus, if there were a market for renting materials and labour hours, the firm would be willing to take some materials if the price of the material were less than Rs 10/3 per kg, and labour hours, if the price payable is less than Rs 25/3 per hour.

This is because an additional kg of raw material would add Rs 10/3 to the profit while a reduction of one kg would reduce the profit by an equal amount. Similarly, an addition of one labour hour would cause the profit to increase by Rs 25/3 and a reduction of one hour would reduce profit by Rs 25/3.

If we denote marginal profitability of material as MP_M and the marginal profitability of labour as MP_L, respectively, the shadow prices of the two resources, we can write the dual as follows:

Minimise $G = 60MP_M + 96MP_L$

Subject to

$$2MP_M + 4MP_L \geq 40$$

$$3MP_M + 3MP_L \geq 35$$

$$MP_M, MP_L \geq 0$$

Now let us turn to the economic significance of the surplus variables S_1 and S_2 in the dual, whose numerical values can be obtained from Δ_j row in the optimal solution of the primal. The value of S_1 in the optimal solution represents the opportunity cost of the product *A* while the value of S_2 represents that of product *B*. For the product *A,* production of an additional unit will get the firm a profit of Rs 40 and, at the same time, the firm would use up resources worth $2 \times 10/3 + 4 \times 25/3 =$ Rs 40. Thus, the net effect of producing one unit of product would be $40 - 40 = 0$. Similarly, for product *B,* the opportunity cost equals zero.

Further, the Δ_j value equal to 18 under S_1 indicates that each rupee of additional profit in product *A* would increase the minimum rent acceptable by Rs 18 while in case of product *B*, the rent shall be up by Rs 8. Similarly, a reduction in unit profit in products *A* and *B* shall reduce the rent acceptable by Rs 18 and 8, respectively.

The Minimisation Problem A minimisation problem, Example 3.3, is reproduced here.

Minimise $Z = 40x_1 + 24x_2$ Total cost

Subject to

$20x_1 + 50x_2 \geq 4{,}800$ Phosphate requirement

$80x_1 + 50x_2 \geq 7{,}200$ Nitrogen requirement

$x_1, x_2 \geq 0$

From the optimal solution given in the Table 4.6, reproduced from Table 3.13, it may be noted that the fertiliser requirement of the farmer can be met with at a minimum cost of Rs 3,456, by buying 144 bags only of mixture *B*.

TABLE 4.6 Simplex Tableau: Optimal Solution

Basis		x_1	x_2	S_1	S_2	A_1	A_2	b_i
x_2	24	8/5	1	0	−1/50	0	1/50	144
S_1	0	60	0	1	−1	−1	1	2,400
c_j		40	24	0	0	M	M	
Solution		0	144	2,400	0	0	0	
Δ_j		8/5	0	0	12/25	M	$M - \frac{12}{25}$	

The dual corresponding to the problem can be written as:

Maximise $G = 4{,}800y_1 + 7{,}200y_2$

Subject to

$20y_1 + 80y_2 \leq 40$

$50y_1 + 50y_2 \leq 24$

$y_1, y_2 \geq 0$

The optimal solution to this problem is contained in the Table 4.7.

TABLE 4.7 Simplex Tableau: Optimal Solution

Basis		y_1	y_2	S_1	S_2	b_i
S_1	24	−60	0	1	−8/5	8/5
y_2	7,200	1	1	0	1/50	12/25
c_j		4,800	7,200	0	0	
Solution		0	12/25	8/5	0	
Δ_j		−2,400	0	0	−144	

From this table, we observe that $y_1 = 0$ and $y_2 = 12/25$ (as already mentioned, these values can also be obtained from the solution to the primal problem). Substituting these values in the objective function of the dual we get $4{,}800 \times 0 + 7{,}200 \times 12/25 =$ Rs 3,456.

Now, if we impute a price of 0 to each kg of phosphate and a price of Rs 12/25 to each kg of nitrogen, then the total value imputed to both of these would be Rs 3,456, which is equivalent to the minimum cost of obtaining fertiliser by the farmer. Thus, on the basis of their cost of production, we assign a value of zero and Rs 12/25 to each kg of phosphate and nitrogen in that order. These, again, are the shadow prices of the two fertilisers that are imputed from the (marginal) cost of producing them.

From the primal problem, we observe that the fertiliser requirements can be met with by using 144 bags of mixture *B* and none of the mixture *A*. The economic nature of the shadow prices of the two mixtures can be considered in more details as follows.

Mixture A Each bag of 100 kg contains 20 kg of phosphate and 80 kg of nitrogen. The contribution of each bag of the mixture can be valued as:

20 kg of Phosphate @ Re 0 per kg	= Re 0
80 kg of Nitrogen @ Rs 12/25 per kg	= Rs 192/5
Total	= Rs 192/5

Now, since the cost of each bag is Rs 40 and its contribution to the value is only Rs 192/5, obviously, each bag used means an increase of Rs 40 – Rs 192/5 = Rs 8/5 in cost. For this reason, this mixture has not been used.

Mixture B The cost of each bag of this mixture is Rs 24 and each bag contains 50 kg each of the two fertilisers. The total contribution of each bag of mixture *B* can be obtained as follows:

50 kg of Phosphate @ Re 0	= Re 0
50 kg of Nitrogen @ Rs 12/25	= Rs 24
Total	= Rs 24

For this mixture, the value matches the cost.

Finally, it may be observed that the fertiliser supply has been obtained by using mixture *B* only, whose 144 bags yield $144 \times 50 = 7{,}200$ kg of phosphate and an equal amount of nitrogen. Now, whereas the nitrogen supply exactly equals the required minimum, the phosphate is supplied in excess to the extent of $7{,}200 - 4{,}800 = 2{,}400$ kg. It should not be disturbing, therefore, that this has been assigned a shadow price equal to zero. This is obvious because this resource is in excess supply and economic value is attached to only those resources which are scarce.

4.3.1 A Problem with Unutilised Resources

Now we shall turn to the economic interpretation of the dual when some resources are not fully utilised in the optimal solution. For this, we reconsider Example 3.2.

Maximise $Z = 5x_1 + 10x_2 + 8x_3$

Subject to

$$
\begin{aligned}
3x_1 + 5x_2 + 2x_3 &\le 60 && \text{Fabrication hours}\\
4x_1 + 4x_2 + 4x_3 &\le 72 && \text{Finishing hours}\\
2x_1 + 4x_2 + 5x_3 &\le 100 && \text{Packaging hours}\\
x_1, x_2, x_3 &\ge 0
\end{aligned}
$$

The final tableau containing the optimal solution is shown in Table 4.8 here:

TABLE 4.8 Simplex Tableau: Optimal Solution

Basis		x_1	x_2	x_3	S_1	S_2	S_3	b_i
x_2	10	1/3	1	0	1/3	–1/6	0	8
x_3	8	2/3	0	1	–1/3	5/12	0	10
S_3	0	–8/3	0	0	1/3	–17/12	1	18
c_j		5	10	8	0	0	0	
Solution		0	8	10	0	0	18	
Δ_j		–11/3	0	0	–2/3	–5/3	0	

From this table, we see that the firm should produce none of product *A*, 8 units of product *B*, and 10 units of product *C* in order to get the maximum daily profit of Rs 160. The shadow prices of the hours in the three departments can be obtained as the numerical values associated with the slack variables in the Δ_j row of the tableau. These are: Rs 2/3 per hour for the fabrication department, Rs 5/3 per hour for the finishing department and nil for the packaging department. Alternatively, these can be obtained from the values of the dual variables y_1, y_2 and y_3 respectively. The dual corresponding to the problem and the final tableau depicting its solution are given here:

Minimise $G = 60y_1 + 72y_2 + 100y_3$

Subject to

$$3y_1 + 4y_2 + 2y_3 \geq 5$$
$$5y_1 + 4y_2 + 4y_3 \geq 10$$
$$2y_1 + 4y_2 + 5y_3 \geq 8$$
$$y_1, y_2, y_3 \geq 0$$

TABLE 4.9 Simplex Tableau: Optimal Solution

Basis		y_1	y_2	y_3	S_1	S_2	S_3	A_1	A_2	A_3	b_i
y_1	60	1	0	–1/3	0	–1/3	1/3	–1/6	1/3	–1/3	2/3
S_1	0	0	0	8/3	1	–1/3	–2/3	–7/6	1/3	2/3	11/3
y_2	72	0	1	17/12	0	1/6	–5/12	1/12	–1/6	5/12	5/3
c_j		60	72	100	0	0	0	*M*	*M*	*M*	
Solution		2/3	5/3	0	11/3	0	0	0	0	0	
Δ_j		0	0	18	0	8	0	$M + 4$	$M - 8$	$M - 10$	

As we can note from this tableau, the values of the dual variables y_1, y_2 and y_3 are the shadow prices of the hours in these three departments. With these imputed values, we can make the following observations:

(a) *For Product A* Here the decision reached is not to produce it. This is because it yields a profit of Rs 5 per unit, while the value of resources it consumes is:

3 hours @ Rs 2/3	= Rs 2
4 hours @ Rs 5/3	= Rs 20/3
2 hours @ Re 0	= Re 0
Total	= Rs 26/3

Since the total value of resources required by a unit of *A* is 26/3 while the profit associated is Rs 5, obviously it should not be produced. Any unit of product *A* produced would lead to a loss of 26/3 – 5 = Rs 11/3 (observe Δ_j value corresponding to x_1).

(b) *For Product B and C* The value of resources required for their production is as follows:

Produt B	Rs	*Product C*	Rs
5 hours @ Rs 2/3	= 10/3	2 hours @ Rs 2/3	= 4/3
4 hours @ Rs 5/3	= 20/3	4 hours @ Rs 5/3	= 20/3
4 hours @ Re 0	= 0	5 hours @ Re 0	= 0
Total	= Rs 10		= Rs 8
Profit per unit	= Rs 10		= Rs 8

For each of them, the opportunity cost is equal to zero.

(c) The hours in the packaging department have a zero opportunity cost because there is excess capacity in this department.

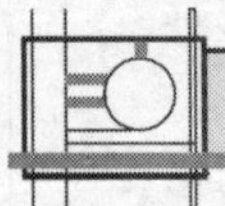

4.4 SENSITIVITY ANALYSIS

Equally important as obtaining the optimal solution to a linear programming problem is the question as to how the solution would be affected if the parameters c_j, b_i, or a_{ij} of the problem change. This question is answered by the *sensitivity analysis,* also called the *post-optimality analysis.* Of course, we may directly substitute the changed values in a given situation and re-solve the problem to determine the effect on the optimal solution. It will be appreciated, however, that it would be a lot advantageous if we could obtain this information graphically or from the tableau containing the final solution to the problem, without being required to carry out the whole exercise again. Let us consider as to how can we get the information of this type.

We shall first illustrate a graphic introduction to the sensitivity analysis and then consider the question through simplex tableau.

4.4.1 Graphic Approach to Sensitivity Analysis

Let us reconsider Example 3.1 reproduced below:

Maximise	$Z = 40x_1 + 35x_2$	Profit
Subject to		
	$2x_1 + 3x_2 \le 60$	Raw Material Constraint
	$4x_1 + 3x_2 \le 96$	Labour Hours Constraint
	$x_1, x_2 \ge 0$	

where x_1 is the number of units of product *A* and x_2 is the number of units of product *B*.

The graph depicting the constraints and iso-profit lines is shown in Figure 4.1.

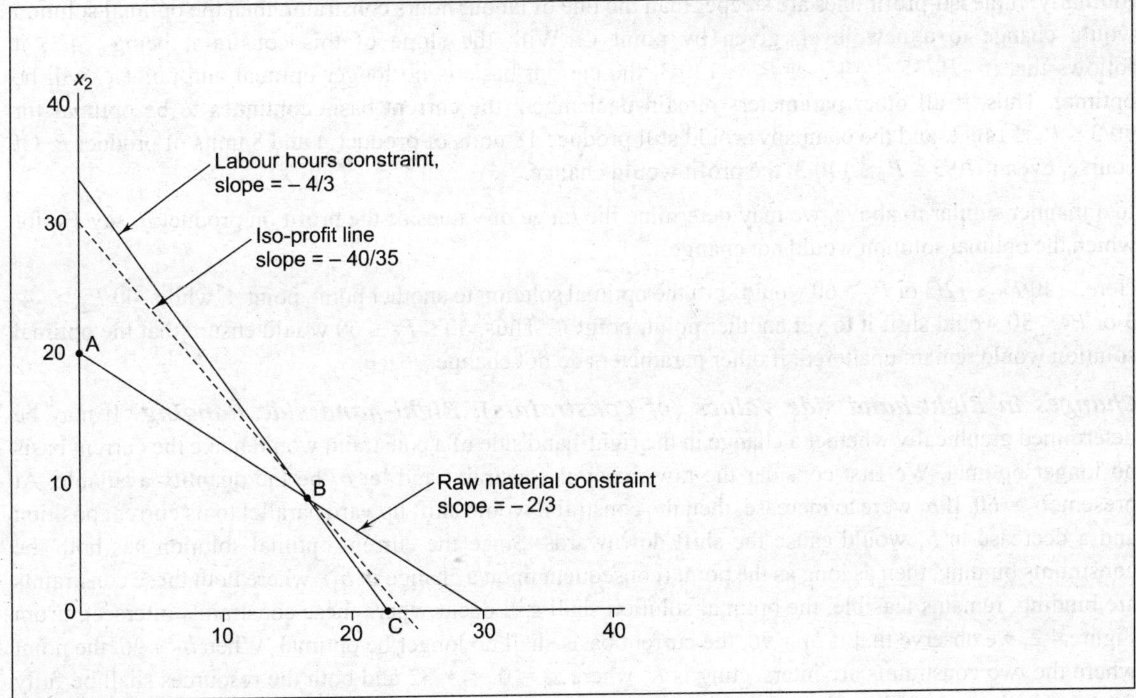

Figure 4.1 *Graphic Presentation—Sensitivity Analysis*

The optimal solution to the problem is $x_1 = 18$ and $x_2 = 8$ and $Z = 1{,}000$, given by point *B* in the figure and it has x_1 and x_2 as basic variables. Now, we shall examine in turn as to how would changes in the problem's objective function coefficients ($c_j's$) or right-hand sides ($b_i's$) change this solution.

Changes in Objective Function Coefficients In the current optimal solution, both the products are being produced. Now, if the profit on product *A* were to increase sufficiently, then it would be optimal to produce this product only, and if the profit on this were to decrease sufficiently, then it would be optimal for the company to not produce this product, and produce only product *B*. We may now determine the values of the profit for product *A* for which the current product mix will remain optimal. In other words, if P_1 be the profit per unit of product *A*, then for what values of P_1 the current basis shall remain optimal.

At present, $P_1 = 40$, and each iso-profit line is $40x_1 + 35x_2 = a$, constant, or $x_2 = (-40/35)x_1$ + constant, and thus has a slope of $-40/35$. [Remember, a line with an equation $y = bx + a$ has a slope equal to *b*. Following this, the slope of the line representing raw material constraints is $-2/3$, and the one representing labour hour constraints is $-4/3$]. It may be observed from the graph that if a change in P_1 causes the iso-profit lines to be flatter than the raw material constraint, then the optimal solution will change from the current level (point *B*) to a new level, point *A*. With a profit P_1, the slope of each iso-profit line will be $-P_1/35$, and with a slope of $-2/3$ for the raw material constraint, it follows that $-P_1/35$ should not exceed $-2/3$ for the solution to remain the same. Thus, iso-profit lines shall be flatter than the raw material constraint if $(-P_1/35) > -2/3$ or $P_1 < 70/3$, and the current basis will no longer be optimal.

Similarly, if the iso-profit lines are steeper than the line of labour hours constraint, then the optimal solution would change to a new level, given by point *C*. With the slope of this constraint being $-4/3$, it follows that if $-P_1/35 < -4/3$ or $P_1 > 140/3$, the current basis is no longer optimal and point *C* will be optimal. Thus, if all other parameters remain unchanged, the current basis continues to be optimal for $70/3 \le P_1 \le 140/3$, and the company would still produce 18 units of product *A* and 8 units of product *B*. Of course, even if $70/3 \le P_1 \le 140/3$, the profit would change.

In a manner similar to above, we may determine the range of values of the profit on product *B*, say P_2, for which the optimal solution would not change.

Here, $-40/P_2 > -2/3$ or $P_2 > 60$ would shift the optimal solution to another point, point *A*, while $-40/P_2, > -4/3$ or $P_2 < 30$ would shift it to yet another point, point *C*. Thus, $30 \le P_2 \le 60$ would ensure that the optimal solution would remain unaltered, if other parameters do not change.

Changes in Right-hand side Values (of Constraints): Right-hand side Ranging It may be determined graphically whether a change in the right-hand side of a constraint would make the current basis no longer optimal. We first consider the raw material constraint and let b_1 be the quantity available. At present, $b_1 = 60$. If b_1 were to increase, then the constraint would shift upward parallel to its current position and a decrease in b_1 would cause the shift downwards. Since the current optimal solution has both the constraints binding, then as long as the point (consequent upon a change in b_1), where both these constraints are binding, remains feasible, the optimal solution shall still occur where these constraints intersect. From Figure 4.2, we observe that if $b_1 > 96$, the current basis shall no longer be optimal. When $b_1 = 96$, the point where the two constraints are intersecting is *K*, where $x_1 = 0$, $x_2 = 32$ and both the resources shall be fully employed. Similarly, if $b_1 \le 48$, the current basis shall no longer hold to be optimal. At this value of b_1 (= 48), the two constraints are intersecting at point *C*. When b_1 is less than 48, these constraints will be binding only at an infeasible point with $x_2 < 0$ and the current basis shall no longer be optimal. Thus, it is evident that if all other parameters remain unchanged, the current basis remains optimal if $48 \le b_1 \le 96$. It may again be noted that although for $48 \le b_1 \le 96$ the current basis remains optimal, the values of the decision variables (x_1 and x_2) and the objective function do change. For instance, if $48 \le b_1 \le 60$, the optimal solution would change from point *B* to another point on the line segment *BC* while $60 \le b_1 \le 96$, then the optimal solution would shift from point *B* to some other point on the line segment *BK*.

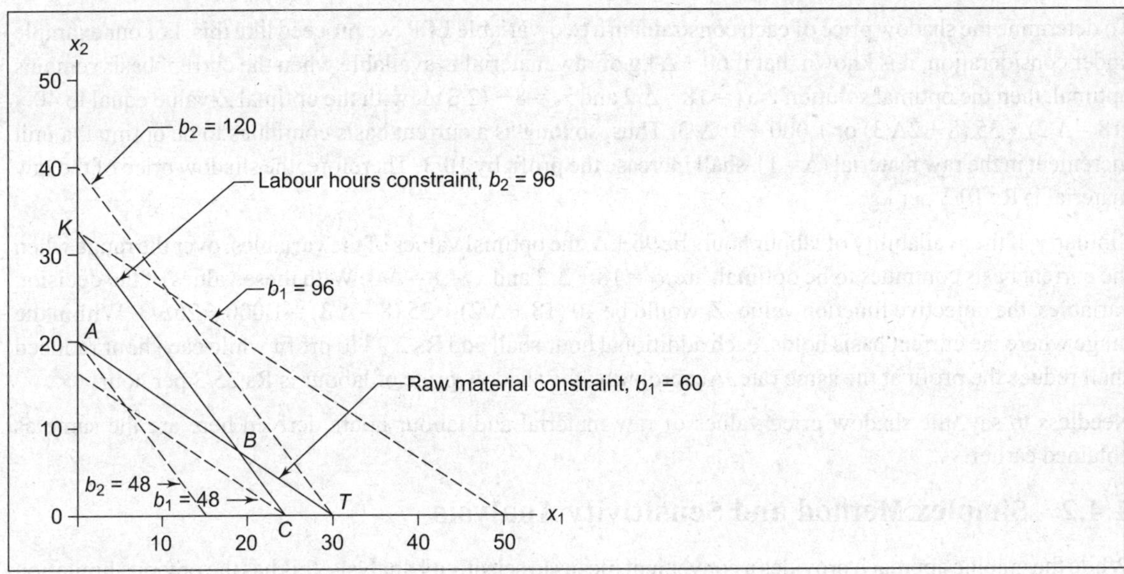

Figure 4.2 *Graphic Presentation: RHS Ranging*

We may also determine the range of values of the second constraint over which the current basis will remain unaffected. If we let RHS value be b_2, then, at present, $b_2 = 96$. An increase in b_2 can be made upto 120 so that both the constraints may remain binding, while the lower limit to which it may be revised is equal to 60 for this condition to hold. Accordingly, the optimality of the current basis shall be valid for $60 \leq b_2 \leq 120$.

For the range over which the current basis remains optimal, it may be easily determined as to how a change in the right-hand side of a constraint changes the value of the decision variables. Recall that on a change of b_1 to $60 + \Delta$, the current basis remains optimal for $-12 \leq \Delta \leq 36$. Since the current basis results in both the constraints to be binding, the new values of the decision variables can be found by solving the equations $2x_1 + 3x_2 = 60 + \Delta$ and $4x_1 + 3x_2 = 96$, which gives $x_1 = 18 - \Delta/2$, and $x_2 = 8 + (2/3)\Delta$. An increase in the raw material availability would, therefore, result in a decrease in the output of product *A* and increase in the output of product *B*. If the raw material supply increased to 72 kg, then $\Delta = 72 - 60 = 12$, and the optimal mix would be $x_1 = 18 - 12/2 = 12$ units and $x_2 = 8 + (2/3) \times 12 = 16$ units, for a total profit of Rs 1,040.

Similarly, if the availability of labour varies to $96 + \Delta$, then the current basis shall be optimal for $-36 \leq \Delta \leq 24$. Again, since the current basis involves both the constraints to be binding, the new optimal may be obtained by solving the equations $2x_1 + 3x_2 = 60$ and $4x_1 + 3x_2 = 96 + \Delta$ simultaneously. This yields $x_1 = 18 + \Delta/2$ and $x_2 = 8 - \Delta/3$. A total availability of 90 hours implies $\Delta = -6$ and, accordingly, optimal solution would be: $x_1 = 18 - 6/2 = 15$ and $x_2 = 8 - (-6/3) = 10$, and $Z = 950$.

Shadow Prices We have discussed about shadow prices in connection with the duality to observe that they indicate the 'worth' of the resources they represent, so that the shadow price of a resource means a change in profit caused by a unit change in the amount of that resource. Thus, shadow prices have to do with determining how changes in constraints' right-hand sides cause changes in the optimal *Z*-value. More formally, we define the shadow price of the *i*th constraint of an LPP to be the amount by which the optimal value of the objective function is improved (so that it increases the *Z*-value for a maximisation problem and decreases the *Z*-value for a minimisation problem) if the right-hand side of the constraint is increased by unity. This, however, is true only if the change in the right-hand side of the constraint *i* leaves the current basis optimal.

To determine the shadow price of each constraint in a two-variable LPP, we proceed like this. For our example under consideration, it is known that if 60 + Δ kg of raw material is available when the current basis remains optimal, then the optimal solution is $x_1 = 18 - \Delta/2$ and $x_2 = 8 + (2/3)\Delta$, with the optimal *Z*-value equal to 40 × (18 – Δ/2) + 35 (8 + 2Δ/3) or 1,000 + 10Δ/3. Thus, so long as a current basis continues to be optimal, a unit increment in the raw material (Δ = 1), shall increase the profit by 10/3. Therefore, the shadow price of the raw material is Rs 10/3 per kg.

Similarly, if the availability of labour hours be 96 + Δ, the optimal values of the variables, over the range when the current basis continues to be optimal, are $x_1 = 18 + \Delta/2$ and $x_2 = 8 - \Delta/3$. With these values of the decision variables, the objective function value, *Z*, would be 40 (18 + Δ/2) + 35 (8 – Δ/3) = 1,000 + 25Δ/3. Within the range where the current basis holds, each additional hour shall add Rs 25/3 to profit while each hour reduced shall reduce the profit at the same rate. Accordingly, the shadow price of labour is Rs 25/3 per hour.

Needless to say, the shadow price values of raw material and labour hours derived here are the same as obtained earlier.

4.4.2 Simplex Method and Sensitivity Analysis

While the graphic approach provides a convenient means to sensitivity analysis, this has the obvious limitation that its use is restricted to linear programming problems which involve two decision variables. We, therefore, consider a more generalised approach to the problem where relatively more information may be obtained with even greater convenience, once the optimal solution to the linear programming problem has been obtained.

We shall discuss the sensitivity analysis in the context of Example 3.2, which is reproduced here along with its optimal solution contained in Table 4.10. Once the concept is grasped, the reader may as well verify the values obtained graphically in respect of the Example 3.1.

Maximise $Z = 5x_1 + 10x_2 + 8x_3$ Profit

Subject to

$3x_1 + 5x_2 + 2x_3 \leq 60$ Fabrication hours

$4x_1 + 4x_2 + 4x_3 \leq 72$ Finishing hours

$2x_1 + 4x_2 + 5x_3 \leq 100$ Packaging hours

$x_1, x_2, x_3 \geq 0$

TABLE 4.10 Simplex Tableau: Optimal Solution

Basis		x_1	x_2	x_3	S_1	S_2	S_3	b_i
x_2	10	1/3	1	0	1/3	−1/6	0	8
x_3	8	2/3	0	1	−1/3	5/12	0	10
S_3	0	−8/3	0	0	1/3	−17/12	1	18
c_j		5	10	8	0	0	0	
Solution		0	8	10	0	0	18	
Δ_j		−11/3	0	0	−2/3	−5/3	0	

In context of this problem, we now consider the following:

(a) *Changes in Objective Function Coefficients, c_j's*

We first consider the question as to how the changes in the coefficients of the decision variables in the objective function, the profit rates in our case, shall influence the optimal solution. In the context of our problem, the variable x_1 is not in the solution—it is a non-basic variable, while the variables x_2 and x_3 are in the basis. We proceed in respect of each of these two types of variables, as follows.

(i) Variables Included in the Solution Consider whether the change of profit per unit of a product, that is currently being produced, causes a change in the optimal solution to the problem. Over a certain range, a change, positive or negative, in the unit profit would not cause a change in the optimal solution. To determine this, we divide the Δ_j row values (of the optimal solution) by the corresponding row values in the tableau in respect of the variable in question. For our example, in respect of x_2, we have,

Δ_j	:	−11/3	0	0	−2/3	−5/3	0
Coefficients in row x_2	:	1/3	1	0	1/3	−1/6	0
Quotient	:	−11	0	–	−2	10	–
					↑	↑	
					Least negative	Least positive	

The least positive quotient (= 10) provides the answer to how much the profit could increase without changing the solution. The least negative quotient (= −2), similarly, is the maximum decrease in the profit that would not cause a change in the solution. Accordingly, the range of profit is Rs 8 to Rs 20. In case the profit per unit falls below Rs 8 or becomes greater than Rs 20, the optimal solution would be different, otherwise not.

In a similar way, we can determine the profit range in respect of product *C*, represented by the variable x_3. For this, the range would be Rs 4 to Rs 10, since the least negative and positive quotients are −4 and +2 respectively as shown here:

Δ_j	:	−11/3	0	0	−2/3	−5/3	0
Coefficients in row x_3	:	2/3	0	1	−1/3	5/12	0
Quotient	:	−11/2	–	0	2	−4	–
					↑	↑	
					Least positive	Least negative	

(ii) Variables not Included in the Solution As stated previously, x_1 is not included in the basis of the optimal solution. The Δ_j value equal to − 11/3 in the tableau suggests that a unit of product *A* would cause a loss of Rs 11/3. This naturally implies that a profit margin of at least 11/3 must be increased in respect of this product before the firm would decide to include this item in its product mix in positive terms. This increased margin of 11/3 means that the profit margin of the product should be more than Rs 26/3(= 5 + 11/3) before its production could be undertaken. Thus, any decrease in the unit profit or an increase less than Rs 11/3 in the profit per unit of this product shall not cause it to be produced and would result in the present solution unchanged.

The information provided by the analysis is very useful to the marketing manager for he is aware of the pricing latitude he has, without interfering with the optimal solution.

Introducing a New Product Here we may consider a related issue of the introduction of a new product. Suppose that the firm in question is contemplating the introduction of a new product *D* with a likely profit margin of Rs 8. It is known that this product shall consume 3 hours in each of the fabrication and finishing departments, and 2 hours in the packaging department. Should the firm decide for the introduction of this product?

The analysis here would be similar to the one for the case where the product in question is not already in production. For the product *D*, the value of the hours required would be Rs 7, calculated as follows:

			Rs
3 hours in fabrication	@ Rs 2/3	=	2
3 hours in finishing	@ Rs 5/3	=	5
2 hours in packaging	@ Rs 0	=	0
Total		=	7

Since the profit per unit of *D* is greater than the value of resources consumed by it, the product should be included in the production list of the firm.

(b) Changes in the b_i Values: Right Hand Side Ranging

As mentioned previously, the net-of-contribution values, Δ_j, in respect of the variables S_1, S_2 and S_3 which are equal to 2/3, 5/3 and 0 respectively, indicate the shadow prices, or the marginal profitability, of these resources. Adding one hour in the fabrication department would increase the aggregate profit by Rs 2/3, while if one hour is reduced in this department, it would lower the profit by that amount. Similarly, adding an hour in the finishing department will raise the profit by Rs 5/3, while adding an hour in the packaging department will leave the profit unchanged. The question is, how many hours in a particular department may be increased, or decreased, that the shadow price would remain the same. In other words, our problem is to determine the range over which each of the shadow prices will remain valid. This analysis is sometimes called *RHS ranging.*

For the purpose of this analysis, first of all we divide the b_i values in the final tableau into the a_{ij} values in that tableau, in respect of the variable which represents the (slack) hours of the department for which we want to determine this range. Thus, for the fabrication department,

b_i	a_{ij}	b_i/a_{ij}	
8	1/3	24	← Least positive
10	– 1/3	– 30	← Least negative
18	1/3	54	

The least *positive* quotient here is the answer to the question as to how many hours can be *decreased* in the fabrication department, while the least *negative* quotient is the answer to the question as to how many hours in this department can be *increased.* Thus, we can conclude that with the present capacity of 60 hours, the shadow price of Rs 2/3 will remain valid over the range (60 – 24) = 36 to (60 + 30) 90 hours. Other things remaining the same, the shadow price of an hour of time in the fabrication department shall be equal to Rs 2/3 for a capacity between 36 and 90 hours, and shall be different when we cross either of these limits.

In a similar way, we can obtain the range for the finishing department as well. This is 48 hours to 85 hours (approximately).

It may be observed in respect of the packaging department that all hours in this department are not used—there are 18 hours unused. At least 18 hours should be reduced from the capacity of this department before a shortage would be experienced, and the number of hours can be increased without limit with no consequences on the solution. It may be concluded, therefore, that the right hand range for time in the packaging department is (100 – 18) = 82 hours to infinity.

(c) Changes in the Technological Coefficients, a_{ij}'s

In this section, we consider the question of changes in the coefficients a_{ij}'s of a linear programming problem and its effect on the optimal solution to the problem. In the context of the production problems, a_{ij} values are determined by the technological considerations. Due to technological improvements, the time required for processing a particular product may be reduced, which might, or might not, affect the optimal product mix determined previously.

In our discussion, we shall limit ourselves to the case of the variables that are not included in the solution—the non-basic ones, and shall not consider the case of variables that are in the solution, because of the very complicated nature of analysis it calls for. In respect of the production problem under consideration, x_1 is the non-basic variable, which requires 3, 4 and 2 hours respectively in the fabrication, finishing, and packaging departments. Suppose that because of technological innovations, the time required for processing one unit of the product in the finishing department is reduced by time t (hours). When we re-solve the problem, the final simplex tableau shall appear as given in Table 4.11.

TABLE 4.11 Simplex Tableau: Optimal Solution

Basis		x_1	x_2	x_3	S_1	S_2	S_3	b_i
x_2	10	$\frac{1}{3}+\frac{1}{6}t$	1	0	1/3	–1/6	0	8
x_3	8	$\frac{2}{3}-\frac{5}{12}t$	0	1	–1/3	5/12	0	10
S_3	0	$-\frac{8}{3}+\frac{17}{12}t$	0	0	1/3	–17/12	1	18
c_j		5	10	8	0	0	0	
Solution		0	8	10	0	0	18	
Δ_j		$-\frac{11}{3}+\frac{5}{3}t$	0	0	–2/3	–5/3	0	

When we compare this solution tableau with the optimal one of the original problem in Table 4.10, we observe that the two are identical except that for the column headed x_1. In this new tableau, the various values are equal to the old x_1 column values minus t times the old S_2 column values (remember that S_2 represents the slack hours in the finishing department). Thus,

New x_1 column = Old x_1 column minus t (Old S_2 column)

$$\begin{bmatrix} \frac{1}{3}+\frac{1}{6}t \\ \frac{2}{3}-\frac{5}{12}t \\ \frac{-8}{3}+\frac{17}{12}t \\ \frac{-11}{3}+\frac{5}{3}t \end{bmatrix} = \begin{bmatrix} \frac{1}{3} \\ \frac{2}{3} \\ \frac{-8}{3} \\ \frac{-11}{3} \end{bmatrix} - \begin{bmatrix} \frac{-1}{6}t \\ \frac{5}{12}t \\ \frac{-17}{12}t \\ \frac{-5}{3}t \end{bmatrix}$$

Now, the variable x_1 shall enter the basis if

$$-\frac{11}{3}+\frac{5}{3}t > 0$$

or

$$t > \frac{11}{3} \times \frac{3}{5} = \frac{11}{5} \text{ hours}$$

Thus, only if the technological innovation will reduce the processing time of the product *A* in the finishing department by more than 11/5 hours (to 4 – 11/5 = 9/5 hours) that it would be worthwhile to include it in the solution, and not otherwise.

Example 4.6 A firm has the choice of producing four similar products: P_1, P_2, P_3 and P_4 in any combination. These products have profit rates of Rs 70, 65, 80 and 75, respectively. They all require two types of raw materials R_1 and R_2, and two types of labour L_1 and L_2. The per unit requirements and availability of the resources every week is given in the following table:

Resource	*Per Unit Requirement*				*Availability*
	P_1	P_2	P_3	P_4	
R_1	4	4	3	7	90 kg
R_2	6	3	5	4	120 kg
L_1	5	2	3	3	60 hr
L_2	6	5	1	2	l00 hr

You are required to determine the optimal product mix for the firm and carry out the sensitivity analysis, for changes in the objective function coefficients and right-hand-side values of the constraints.

Let x_1, x_2, x_3 and x_4 be the number of units to be produced of the four products, respectively. The linear programming problem can be stated as follows:

Maximise $Z = 70x_1 + 65x_2 + 80x_3 + 75x_4$

Subject to

$$4x_1 + 4x_2 + 3x_3 + 7x_4 + \leq 90$$

$$6x_1 + 3x_2 + 5x_3 + 4x_4 + \leq 120$$
$$5x_1 + 2x_2 + 3x_3 + 3x_4 + \leq 60$$
$$6x_1 + 5x_2 + x_3 + 2x_4 + \leq 100$$
$$x_1, x_2, x_3, x_4 \geq 0$$

Introducing the necessary slack variables, the LPP can be written in the standard form as follows:

Maximise $Z = 70x_1 + 65x_2 + 80x_3 + 75x_4 + 0S_1 + 0S_2 + 0S_3 + 0S_4$

Subject to

$$4x_1 + 4x_2 + 3x_3 + 7x_4 + S_1 = 90$$
$$6x_1 + 3x_2 + 5x_3 + 4x_4 + S_2 = 120$$
$$5x_1 + 2x_2 + 3x_3 + 3x_4 + S_3 = 60$$
$$6x_1 + 5x_2 + x_3 + 2x_4 + S_4 = 100$$
$$x_1, x_2, x_3, x_4, S_1, S_2, S_3, S_4 \geq 0$$

The problem is solved as shown in Tables 4.12 to 4.14. From the Simplex Tableau 3, we observe that the optimal solution to the given LPP is: $x_1 = 0, x_2 = 15, x_3 = 10$ and $x_4 = 0$. With these values, $Z = 70 \times 0 + 65 \times 15 + 80 \times 10 + 75 \times 0 =$ Rs 1,775. Further, while R_1 and L_1 would be fully utilised, R_2 would remain unutilised to the extent of 25 kg and L_2 would not be used for 15 hours.

TABLE 4.12 Simplex Tableau 1: Non-optimal Solution

Basis		x_1	x_2	x_3	x_4	S_1	S_2	S_3	S_4	b_i	b_i/a_{ij}
S_1	0	4	4	3	7	1	0	0	0	90	30
S_2	0	6	3	5	4	0	1	0	0	120	24
S_3	0	5	2	3*	3	0	0	1	0	60	20 ←
S_4	0	6	5	1	2	0	0	0	1	100	100
c_j		70	65	89	75	0	0	0	0		
Solution0		0	0	0	90	120	60	100			
Δ_j		70	65	80 ↑	75	0	0	0	0		

TABLE 4.13 Simplex Tableau 2: Non-optimal Solution

Basis		x_1	x_2	x_3	x_4	S_1	S_2	S_3	S_4	b_i	b_i/a_{ij}
S_1	0	–1	2*	0	4	1	0	–1	0	30	15 ←
S_2	0	–7/3	–1/3	0	–1	0	1	–5/3	0	20	– 60
x_3	80	5/3	2/3	1	1	0	0	1/3	0	20	30
S_4	0	13/3	13/3	0	1	0	0	–1/3	1	80	240/13
c_j		70	65	80	75	0	0	0	0		
Solution		0	0	20	0	30	20	0	80		
Δ_j		–190/3	35/3 ↑	0	–5	0	0	–80/3	0		

TABLE 4.14 Simplex Tableau 3: Optimal Solution

Basis		x_1	x_2	x_3	x_4	S_1	S_2	S_3	S_4	b_i
x_2	65	–1/2	1	0	2	1/2	0	–1/2	0	15
S_2	0	–5/2	0	0	–1/3	1/6	1	–11/6	0	25
x_3	80	2	0	1	–1/3	–1/3	0	2/3	0	10
S_4	0	13/2	0	0	–23/3	–13/6	0	11/6	1	15
c_j		70	65	80	75	0	0	0	0	
Solution		0	15	10	0	0	25	0	15	
Δ_j		–115/2	0	0	–85/3	–35/6	0	–125/6	0	

Sensitivity Analysis

(i) Changes in Coefficients in the Objective Function

From the optimal solution it is clear that products P_1 and P_4 shall not be produced because every unit of P_1 produced will reduce the profit by Rs 115/2, while each unit of P_4 will result in a loss of Rs 85/3. Thus, unless the unit profit of P_1 rises by at least Rs 115/2, and, similarly, unless the profit for P_4 increases by Rs 85/3 at the minimum, it is not advisable to produce them, respectively. For products P_2 and P_3, we can make the calculations as follows:

Δ_j	:	–115/2	0	0	–85/3	–35/6	0	–125/6	0
a_{ij} Co-efficient	:	–1/2	1	0	2	1/2	0	–1/2	0
Ratio	:	115	0	–	–85/6	–70/6	–	125/3	–

Accordingly,

allowable increase (least positive ratio) = 125/3 = 41.67, and

allowable decrease (least negative ratio) = 70/6 = 11.67.

Similarly, for P_3, we have

Δ_j	:	–115/2	0	0	–85/3	–35/6	0	–125/6	0
a_{ij} Co-efficient	:	2	0	1	–1/3	–1/3	0	2/3	0
Ratio	:	–115/4	–	0	85	35/2	–	125/4	–

From these values, allowable increase = 35/2 = 17.50, and allowable decrease = 115/4 = 28.75.

In other words, the ranges over which the basis would remain unchanged are:

Product	*Variable*	*Current Co-efficient*	*Allowable Increase*	*Allowable Decrease*	*Range*
P_1	x_1	70	57.50	Infinity	$-\infty$ to 127.50
P_2	x_2	65	41.67	11.67	53.33 to 106.67
P_3	x_3	80	17.50	28.75	51.25 to 97.50
P_4	x_4	75	28.33	Infinity	$-\infty$ to 103.33

(ii) Changes in the Right-Hand-Side Values

The relevant ratios for each of the slack variables are shown calculated below:

Variable					*Ratio for*			
S_1	S_2	S_3	S_4	b_i	S_1	S_2	S_3	S_4
1/2	0	− 1/2	0	15	30	∞	− 30	∞
1/6	1	− 11/6	0	25	150	25	− 150/11	∞
− 1/3	0	2/3	0	10	− 30	∞	15	∞
− 13/6	0	11/6	1	15	− 90/13	∞	90/11	15

From the ratios shown in the last four columns, we can determine the allowable increase and allowable decrease by selecting, respectively, least negative and least positive quantities. Accordingly, the ranges over which the shadow prices would remain unchanged are shown calculated below.

Constraint	*Current RHS Value*	*Allowable Increase*	*Allowable Decrease*	*Range*
1	90	90/13	30	60 to 1,260/13
2	120	Infinity	25	95 to ∞
3	60	150/11	90/11	570/11 to 810/11
4	100	Infinity	15	85 to ∞

4.4.3 Sensitivity Analysis with Multiple Parameter Changes

In the sensitivity analysis discussed so far, it is assumed that all parameters of a problem remain constant except the one which is under consideration. For example, when we examine the range of a particular RHS value over which the current basis shall remain unchanged, it is implicitly assumed that all other RHS values shall remain unaltered. We now consider the case where more than one parameter is changed at a time.

(i) Changes in the Objective Function Coefficients

We first consider the case where objective function coefficients of more than one variable change. Such changes can be divided into two categories:

1. All the variables whose coefficients, c_j's, are changed are those with non-zero Δ_j values.
2. Of the variables whose coefficients, c_j's, are changed, at least one is such which has $\Delta_j = 0$.

(Of course, the Δ_j values in all cases are those which are obtained from the table of optimal solution.)

In the first case, the current basis remains optimal if and only if the coefficient of each variable remains within the allowable limits as determined in the sensitivity analysis discussed earlier. However, if the co-efficent of some variable (or variables) changes beyond the allowable limits, the current basis would no longer be optimal.

Example 4.7 Consider Example 4.6. Examine whether the current basis (*i.e.* the one of the optimal (solution) would remain unchanged under each of the two cases:

(a) The profit per unit of P_1 increases to Rs 90, and the profit per unit of P_4 decreases to Rs 62.

(b) The unit profit for P_1 increases to Rs 110, while for P_4, it increases to Rs 135.

In both the cases, the changes indicated are in P_1 and P_4, both of which have non-zero Δ_j values. Therefore, we shall consider this problem relating to category 1. From the analysis done already, we know that the ranges over which objective function coefficient changes would cause the current basis to continue to be optimal are:

$$P_1 : -\infty \text{ to } 127.50, \text{ and}$$
$$P_4 : -\infty \text{ to } 103.33$$

In (a), both the changes are within the allowable ranges. Thus, no change in the basis would take place.

In (b), however, while the change for P_1 is within the allowable limits, for P_4 it exceeds the limit. Accordingly, the problem needs to be re-solved as the current basis is no longer optimal.

In category 2, we may often show that in many cases when changes take place in the objective function coefficients, the current basis remains optimal, by what is known as the 100% *Rule*. Let us consider the rule.

Let c_j : the co-efficient of the variable x_j in the objective function,
Δc_j : the change in the co-efficient, and equal to the difference between the new and the old values
I_j : the maximum allowable increase in c_j for which the current basis remains unchanged
D_j : the maximum allowable decrease in c_j for which the current basis remains unchanged

For each variable, x_j, we define r_j, which measures the ratio of the actual change in c_j to the maximum allowable change in it that would keep the current basis optimal. Accordingly,

$$r_j = \Delta c_j / I_j, \text{ if } \Delta c_j > 0, \text{ and}$$
$$r_j = -\Delta c_j / D_j, \text{ if } \Delta c_j < 0$$

Of course, if the co-efficient of a certain variable does not change, so that $\Delta_j = 0$, then $r_j = 0$. Now, if only one c_j in the objective function changes while others remain the same, then the current basis would remain optimal if $r_j \leq 1$; that is to say, if r_j (expressed as a percentage) would not exceed 100 percent. The 100% Rule is a generalisation of this concept. According to this rule, in case of multiple changes in the objective function coefficients, if the summation of all the r_j's does not exceed 1 (i.e. 100 percent), then the current basis shall remain unaltered, while if it exceeded 1, then the current basis may, or may not, hold valid. Of course, although the basis would not change when $\Sigma r_j \leq 1$, the z – value will change due to changes in the coefficient values.

Example 4.8 For Example 4.6, determine whether the current basis would change for each of the following cases:

(a) The profit per unit of P_2 increases to Rs 85, while profit per unit of P_3 declines to Rs 70. What is the new optimal value of Z?

(b) The profit per unit of P_1, P_2 and P_3 change as follows:

P_1 : Rs 80 P_2 : Rs 77.5 P_3 : Rs 60

(a) Here, $\Delta c_2 = 85 - 65 = 20$
With $I_2 = 41.67$, we have $r_2 = 20/41.67 = 0.48$
Also, $\Delta c_3 = 70 - 80 = -10$
With $D_3 = 31.25$, we have, $r_3 = -(-10)/31.25 = 0.32$

Now, since r_1 and $r_4 = 0$, we get

$$\Sigma r_j = 0 + 0.48 + 0.32 + 0 = 0.80 \text{ or } 80\%.$$

Since the total is less than 100%, the current basis remains unchanged.

The new z – value is: $70 \times 0 + 85 \times 15 + 70 \times 10 + 75 \times 0 = \text{Rs } 1{,}975.$

(b) From the given changes:

$\Delta c_1 = 80 - 70 = 10$ $\quad I_1 = 57.50$ $\quad r_1 = 10/57.50 = 0.17$

$\Delta c_2 = 77.5 - 65 = 12.5$ $\quad I_2 = 41.67$ $\quad r_2 = 12.0/41.67 = 0.30$

$\Delta c_3 = 60 - 80 = -20$ $\quad D_3 = 31.25$ $\quad r_3 = 20/31.25 = 0.64$

$\Delta c_4 = 0$ $\quad r_4 = 0$

Here, $\Sigma\, r_j = 0.17 + 0.30 + 0.64 + 0.00 = 1.11$. Since the total exceeds unity, we cannot be sure whether the combined changes would affect optimal solution or not.

(ii) Changes in the RHS Values

The 100% Rule can be applied also when changes in the right-hand-side values are considered. In this case also, there are two cases to be investigated into. They are:

1. When *all* the constraints whose right-hand-side values are changed are non-binding constraints.
2. When at *least one* constraint whose right-hand-side values are changed is a binding constraint.

Recall that a binding constraint is one whose LHS exactly matches the RHS, so that it has a solution value equal to zero.

In the first case, the current basis remains optimal if each of the right-hand-side value remains within its allowable limits. In such a situation, the values of the decision variables and the optimal value of Z remain unchanged. However, if the right-hand-side value of any constraint is changed outside its allowable range, the current basis would no longer be optimal.

Example 4.9 Does the current basis remain optimal in each of the following cases, in context of the Example 4.6?

(a) Raw material R_2 reduced to 100 kg and labour L_2 increased to 150 hours.

(b) Supply of raw material R_2 increased to 160 kg and labour L_2 supply reduced to 80 hours.

In (a) above, both the right-hand-side values change within their respective allowable ranges. Hence, there is no change in the current basis, that is optimal.

(b) The raw material supply of R_2 is within allowable range while the labour supply has reduced by more than the allowable decrease. Hence, the current basis is no longer optimal.

For case 2, the 100% Rule can be used to seek the desired answer. Let,

b_i : the current RHS of the ith constraint,

Δb_i : the change in the RHS of the ith constraint, where

Δb_j = New b_i value – Given b_i value

I_i : the maximum allowable increase in the RHS of b_i for which the current basis remains unchanged

D_i : the maximum allowable decrease in the RHS of b_i for which the current basis remains unchanged

For each constraint, we compute the ratio of the change in b_i to the maximum allowable change in it that would keep the current basis optimal. Accordingly,

$$r_i = \Delta b_i / I_i, \text{ if } \Delta b_i > 0, \text{ and}$$

$$r_i = -\Delta b_i / D_i, \text{ if } \Delta b_i < 0$$

If $\Delta b_i = 0$ (*i.e.* if there is no change in b_i), then $r_i = 0$. Now, if one of the constraint values is changed then current basis remains optimal if $r_i \leq 1$. In generalization of this idea, the 100% Rule states that for multiple changes in the b_i values, the current basis would remain optimal if the sum of the r_i values does not exceed one. If, however, $\Sigma\, r_i > 1$, then we may not be sure as to whether the current basis would remain optimal.

Example 4.10 In the context of Example 4.6 data, state whether the current basis remains optimal in each of the following two cases:

(a) Availability of R_1 : 80 kg

Availability of L_1 : 67 hours

(b) Availability of various resources:

R_1 = 92 kg, R_2 = 100 kg, L_1 = 65 hours and L_2 = 120 hours.

Also, work out the new objective function value in each case, if possible.

(a) We have,

$\Delta b_1 = 80 - 90 = -10$ $\quad D_1 = 30$ $\quad r_1 = 10/30 = 0.33$

$\Delta b_3 = 67 - 60 = 7$ $\quad I_3 = 150/11$ $\quad r_3 = (7 \times 11)/150 = 0.51$

With $r_1 = 0.33$, $r_2 = 0.00$, $r_3 = 0.51$ and $r_4 = 0.00$, we get $\Sigma r_i = 0.84$. Since it does not exceed 1, the current optimal basis would not change. Remember, however, that although the basis does not change, the objective function value, Z, would change as the RHS values of the constraints change. With the Δ_j values for S_1 and S_3 being equal to –35/6 and –125/6, the net change in the Z-value can bc calculated as under:

	Change in Profit (Rs)
Decrease in R_1 10 kg @ 35/6	– 58.33
Increase in L_1 7 hours @ 125/6	145.83
Net Change	87.50

Accordingly, new profit, Z = Rs 1,775 + Rs 87.50 = Rs 1,862.50. (This may be confirmed by solving the problem again to find the optimal solution as $x_1 = 0$, $x_2 = 6.5$, $x_3 = 18$ and $x_4 = 0$.)

(b) Here,

$\Delta b_1 = 92 - 90 = 2$ $\quad I_1 = 90/13$ $\quad r_1 = 2 \times 13/90 = 0.29$

$\Delta b_2 = 100 - 120 = -20$ $\quad D_2 = 25$ $\quad r_2 = 20/25 = 0.80$

$\Delta b_3 = 65 - 60 = 5$ $\quad I_3 = 150/11$ $\quad r_3 = 5 \times 11/150 = 0.37$

$\Delta b_4 = 120 - 100 = 20$ $\quad I_4 = \infty$ $\quad r_4 = 25/\infty = 0.00$

$\Sigma r_i = 1.46$

Since $\Sigma r_i > 1$, it is not sure whether the current basis is optimal or not. Hence, the determination of new Z-value is not possible without re-solving the problem with revised values.

REVIEW ILLUSTRATIONS

Example 4.11 Find the dual problem for the following:

Minimise $\quad Z = 5x_1 - 6x_2 + 4x_3$

Subject to

$$3x_1 + 4x_2 + 6x_3 \geq 9$$

$$x_1 + 3x_2 + 2x_3 \geq 5$$

$$7x_1 - 2x_2 - x_3 \leq 10$$
$$x_1 - 2x_2 + 4x_3 \geq 4$$
$$2x_1 + 5x_2 - 3x_3 = 3$$
$$x_1, x_2, x_3 \geq 0$$

(ICWA, Dec. 1986)

Multiplying both sides of the third constraint by –1, we get $-7x_1 + 2x_2 + x_3 \geq -10$. Similarly replacing the equation of the last of the constraints by a pair of inequalities, we have $2x_1 + 5x_2 - 3x_3 \geq 3$ and $2x_1 + 5x_2 - 3x_3 \leq 3$. Now, we can rewrite the given problem as follows and write its dual, also given here:

Primal		*Dual*	
Minimise	$Z = 5x_1 - 6x_2 + 4x_3$	Maximise	$G = 9y_1 + 5y_2 - 10y_3 + 4y_4 + 3y_5 - 3y_6$
Subject to		Subject to	
	$3x_1 + 4x_2 + 6x_3 \geq 9$		$3y_1 + y_2 - 7y_3 + y_4 + 2y_5 - 2y_6 \leq 5$
	$x_1 + 3x_2 + 2x_3 \geq 5$		$4y_1 + 3y_2 + 2y_3 - 2y_4 + 5y_5 - 5y_6 \leq -6$
	$-7x_1 + 2x_2 + x_3 \geq -10$		$6y_1 + 2y_2 + y_3 + 4y_4 - 3y_5 + 3y_6 \leq 4$
	$x_1 - 2x_2 + 4x_3 \geq 4$		$y_i \geq 0, i = 1, 2, \cdots, 6$
	$2x_1 + 5x_2 - 3x_3 \geq 3$		
	$-2x_1 - 5x_2 + 3x_3 \geq -3$		
	$x_1, x_2, x_3 \geq 0$		

By substituting $y_5 - y_6 = y_7$, the dual can alternatively be expressed as:

Maximise $G = 9y_1 + 5y_2 - 10y_3 + 4y_4 + 3y_7$

Subject to

$$3y_1 + y_2 - 7y_3 + y_4 + 2y_7 \leq 5$$
$$-4y_1 - 3y_2 - 2y_3 + 2y_4 - 5y_7 \geq 6$$
$$6y_1 + 2y_2 + y_3 + 4y_4 - 3y_7 \leq 4$$
$$y_1, y_2, y_3, y_4 \geq 0, y_7 \text{ unrestricted in sign.}$$

Example 4.12 A timber merchant manufactures three types of plywood. The data below give the production hours per unit in each of three production operations, maximum time available, and profit per unit.

	Operation (Hours)			*Profit per Unit (Rs)*
Plywood	*I*	*II*	*III*	
Grade A	2	2	4	40
Grade B	5	5	2	30
Grade C	10	3	2	20
Max. Time Available	900	400	600	

How many units of each grade of plywood should be produced to maximise the total profit? Write the dual and use it to check the optimal solution. *(MBA, Delhi, 1999)*

Let x_1, x_2, and x_3 be the number of units of grade *A, B* and *C* of plywood, respectively. From the given information, the LPP may be stated as follows:

Maximise $Z = 40x_1 + 30x_2 + 20x_3$

Subject to

$$2x_1 + 5x_2 + 10x_3 \le 900$$
$$2x_1 + 5x_2 + 3x_3 \le 400$$
$$4x_1 + 2x_2 + 2x_3 \le 600$$
$$x_1, x_2, x_3 \ge 0$$

To solve the problem using simplex method, we introduce the slack variables to get the following:

Maximise $Z = 40x_1 + 30x_2 + 20x_3 + 0S_1 + 0S_2 + 0S_3$

Subject to

$$2x_1 + 5x_2 + 10x_3 + S_1 = 900$$
$$2x_1 + 5x_2 + 3x_3 + S_2 = 400$$
$$4x_1 + 2x_2 + 2x_3 + S_3 = 600$$
$$x_1, x_2, x_3, S_1, S_2, S_3 \ge 0$$

The solution is given in Tables 4.15 through 4.17.

TABLE 4.15 Simplex Tableau 1: Non-optimal Solution

Basis		x_1	x_2	x_3	S_1	S_2	S_3	b_i	b_i/a_{ij}
S_1	0	2	5	10	1	0	0	900	450
S_2	0	2	5	3	0	1	0	400	200
S_3	0	4*	2	2	0	0	1	600	150 ←
c_j		40	30	20	0	0	0		
Solution		0	0	0	900	400	600		$Z = 0$
Δ_j		40 ↑	30	20	0	0	0		

TABLE 4.16 Simplex Tableau 2: Non-optimal Solution

Basis		x_1	x_2	x_3	S_1	S_2	S_3	b_i	b_i/a_{ij}
S_1	0	0	4	9	1	0	–1/2	600	150
S_2	0	0	4*	2	0	1	–1/2	100	25 ←
x_1	40	1	1/2	1/2	0	0	1/4	150	300
c_j		40	30	20	0	0	0		
Solution		150	0	0	600	100	0		$Z = 6,000$
Δ_j		0	10 ↑	0	0	0	–10		

TABLE 4.17 Simplex Tableau 3: Optimal Solution

Basis		x_1	x_2	x_3	S_1	S_2	S_3	b_i
S_1	0	0	0	7	1	–1	0	500
x_2	30	0	1	1/2	0	1/4	–1/8	25
x_1	40	1	0	1/4	0	–1/8	5/16	137.5
c_j		40	30	20	0	0	0	
Solution		137.5	25	0	500	0	0	$Z = 6{,}250$
Δ_j		0	0	–5	0	–5/2	–35/4	

From the Simplex Tableau 3, the optimal solution is seen to be : $x_1 = 137.5, x_2 = 25$ and $x_3 = 0$, with $Z = 6{,}250$. Accordingly, the optimal product mix is:

Grade *A* : 137.5 units
Grade *B* : 25 units
Grade *C* : nil
Profit : Rs 6,250

Let y_1, y_2 and y_3 be the dual variables. With these, the dual LPP can be expressed as follows:

Minimise $G = 900y_1 + 400y_2 + 600y_3$

Subject to

$$\begin{aligned} 2y_1 + 2y_2 + 4y_3 &\geq 40 \\ 5y_1 + 5y_2 + 2y_3 &\geq 30 \\ 10y_1 + 3y_2 + 2y_3 &\geq 20 \\ y_1, y_2, y_3 &\geq 0 \end{aligned}$$

From Table 4.17, the optimal values of the dual variables (as given by the Δ_j values under the slack variables) are as follows:

$$y_1 = 0, y_2 = 5/2, \text{ and } y_3 = 35/4$$

Substituting these values in the objective function, we obtain the objective function value, $G = 900 \times 0 + 400 \times 5/2 + 600 \times 35/4 = 6{,}250$, which is the same as that of the maximisation, primal, problem.

Example 4.13 A company manufactures and sells three models of large sized pressure cookers for canteen use. While market demands pose no constraints, supplies of aluminum limited to 750 kgs per week and availability of machine time limited to 600 hours per week restrict the product-mix. The resource usage of the three models and their profitability are given below:

	Model		
	M_1	M_2	M_3
Aluminium/unit	6	3	5
Machine-time/unit	3	4	5
Contribution Rs/unit	60	20	80

(a) Formulate the problem as an LPP and solve for optimal solution.

(b) Using the information in optimal solution tableau, determine whether and how the current solution would be sensitive to the following changes. Treat each of the conditions given below independently.
   (i) An additional 150 kg of aluminum would become available.
   (ii) The machine hours available would reduce from current level of 600 hours to 450 hours.
   (iii) Following a reduction in selling price of M_3, its contribution margin decreases by Rs 15.
   (iv) What minimum contribution of M_2 would make it feature in the optimal solution?
   (v) A new model has been developed requiring 3 kg of aluminum and 3 hours of machine time per unit, with an estimated unit contribution of Rs 40. Would it be worthwhile manufacturing this new model?

(M Com, Delhi, 2004)

(a) Let x_1, x_2 and x_3 be the output of models M_1, M_2 and M_3 respectively. The problem is stated under:

Maximise $Z = 60x_1 + 20x_2 + 80x_3$

Subject to

$$6x_1 + 3x_2 + 5x_3 \leq 750$$

$$3x_1 + 4x_2 + 5x_3 \leq 600$$

$$x_1, x_2, x_3 \geq 0$$

To solve the problem by simplex method, slack variables S_1 and S_2 may be introduced in the given constraint inequalities as follows:

Maximise $Z = 60x_1 + 20x_2 + 80x_3 + 0S_1 + 0S_2$

Subject to

$$6x_1 + 3x_2 + 5x_3 + S_1 = 750$$

$$3x_1 + 4x_2 + 5x_3 + S_2 = 600$$

$$x_1, x_2, x_3\ S_1,\ S_2 \geq 0$$

The solution is given in Tables 4.18 through 4.20.

TABLE 4.18 Simplex Tableau 1: Non-optimal Solution

Basis		x_1	x_2	x_3	S_1	S_2	b_i	b_i/a_{ij}
S_1	0	6	3	5	1	0	750	150
S_2	0	3	4	5*	0	1	600	120 ←
c_j		60	20	80	0	0		
Solution		0	0	0	750	600		
Δ_j		60	20	80	0	0		
				↑				

TABLE 4.19 Simplex Tableau 2: Non-optimal Solution

Basis		x_1	x_2	x_3	S_1	S_2	b_i	b_i/a_{ij}
S_1	0	3*	–1	0	1	– 1	150	50 ←
x_3	80	3/5	4/5	1	0	1/5	120	200
c_j		60	20	80	0	0		
Solution		0	0	120	150	0		$Z = 9{,}600$
Δ_j		12 ↑	– 44	0	0	– 16		

TABLE 4.20 Simplex Tableau 3: Optimal Solution

Basis		x_1	x_2	x_3	S_1	S_2	b_i
x_1	60	1	– 1/3	0	1/3	– 1/5	50
x_3	80	0	1	1	– 1/5	2/5	90
c_j		60	20	80	0	0	
Solution		50	0	90	0	0	$Z = 10{,}200$
Δ_j		0	– 40	0	– 4	– 12	

Thus, optimal solution is : M_1 : 50 units and M_3: 90 units. The maximum profit obtainable is Rs 10,200.

(b) *RHS Ranging*

	For S_1		For S_2	
b_i	a_{ij}	b_i/a_{ij}	a_{ij}	b_i/a_{ij}
50	1/3	150	– 1/3	– 150
90	– 1/5	– 450	2/5	225

Range : (750 – 150) to (750 + 450)   (600 – 225) to (600 + 150)

i.e. 600 to 1,200   i.e. 375 to 750

(i) Since the change lies within the limits established, availability of additional 150 kg of aluminum would lead to additional profit of $150 \times 4 =$ Rs 600.

(ii) Here also the change is within the range obtained, reduction of machine hours by 150 will lead to a reduction of profit to the extent of $150 \times 12 =$ Rs 1,800.

(iii) For M_3 :

Δ_j	:	0	– 40	0	– 4	– 12
a_{ij}	:	0	1	1	– 1/5	2/5
Ratio	:	—	– 40	0	20	– 30

∴ Maximum allowable decrease = Rs 30.

Since the fall in contribution margin is within allowable limits, the basis will not change. The reduction in total profit will be 90 × 15 = Rs 1,350.

(iv) The minimum contribution of 20 + 40 = Rs 60 per unit will make M_2 feature in the optimal solution.

(v) Contribution to profit = Rs 40.

Loss of profit occurring from resource diversion = 3 × 4 + 3 × 12 = Rs 48.
Net loss on producing the new model = Rs 8/unit. Thus, it is not advisable to introduce it.

Example 4.14 The simplex tableau for a maximisation problem of linear programming is given as follows:

Product Mix						
		x_1	x_2	S_1	S_2	Quantity (b_i)
c_j	x_j					
5	x_2	1	1	1	0	10
0	S_2	1	0	–1	1	3
	c_j	4	5	0	0	
	z_j	5	5	5	0	50
	$c_j - z_j$	– 1	0	– 5	0	

Answer the following questions, giving reasons in brief:

(a) Is this solution optimal?

(b) Are there more than one optimal solutions?

(c) Is this solution degenerate?

(d) Is this solution feasible?

(e) If S_1 is slack in machine *A* (in hours/week) and S_2 is slack in machine *B* (in hours/week), which of these machines is being used to the full capacity when producing according to this solution?

(f) A customer would like to have one unit of product x_1 and is willing to pay in excess of the normal price in order to get it. How much should the price be increased in order to ensure no reduction of profits?

(g) How many units of the two products x_1 and x_2 are being produced according to this solution and what is the total profit?

(h) Machine *A* (associated with slack S_1, in hours/week) has to be shut down for repairs for 2 hours next week. What will be the effect on profits?

(i) How much would you be prepared to pay for another hour (per week) of capacity each on machine *A* and machine *B*?

(j) A new product is proposed to be introduced which would require processing time of 1/2 hour on machine *A* and 20 minutes on machine *B*. It would yield a profit of Rs 3 per unit. Do you think it is advisable to introduce this product?

The given simplex tableau is reproduced in Table 4.21.

TABLE 4.21 Simplex Tableau

Basis		x_1	x_2	S_1	S_2	b_i
x_2	5	1	1	1	0	10.
S_2	0	1	0	– 1	1	3
c_j		4	5	0	0	
Solution		0	10	0	3	
Δ_j		– 1	0	– 5	0	

(a) Yes, the given solution is optimal because all Δ_j's are less than, or equal to, zero.

(b) No, because for each of the non-basic variables x_1 and S_1, the Δ_j is strictly negative. Alternate optimal solution(s) exist when either of the non-basic variables has a zero Δ_j.

(c) This solution is not degenerate because none of the basic variables has a zero value.

(d) Yes, because the given solution has no artificial variables in the basis.

(e) Machine *A* is being used to the full capacity bacause, the slack variable S_1 corresponding to it has a zero value in the solution.

(f) Δ_j for x_1 being –1, production of each unit of x_1 would cause a reduction of 1 rupee (or any other given monetary unit). Thus, the price for x_1 should be increased by at least one rupee to ensure no reduction of profits.

(g) According to the given solution, none of x_1 and 10 units of x_2 are being produced. The total profit is $4 \times 0 + 5 \times 10 =$ Rs 50.

(h) The shadow price of hours on machine *A*, obtained from Δ_j, is Rs 5 per hour. Thus, reduction in profit of a 2-hour shut down $= 2 \times 5 =$ Rs 10.

(i) The shadow prices of hours on machines *A* and *B* being Rs 5 and Re 0, respectively, these are the maximum prices one would be prepared to pay for another hour of capacity for these two machines.

(j) Shadow prices of times on machines *A* and *B* are Rs 5 and Rs 0. Production of a unit of the proposed new product would, therefore, reduce the profit by $(1/2) \times 5 + (1/3) \times 0 =$ Rs 2.50. Since the product would yield a profit of Rs 3, it would result in a net increase in profit at a rate of 50 paise per unit. It is advisable, therefore to introduce it.

Example 4.15 (a) The following details are taken from the forecasts for 19xx of *XYZ* Limited.

Sales demand:	*Thousands of Units per Annum, Maximum*
Super Deluxe Model (x_1)	500
Deluxe Model (x_2)	750
Export Model (x_3)	400

Production: Two production facilities are required, machining and assembly, and these are common to each model. Capacity in each facility is limited by the number of direct labour hours available.

	Direct Labour, Total Hours Available in Millions	*Direct Labour Hours per Unit, for Each Model*		
		x_1	x_2	x_3
Machining (S_4)	1.4	0.5	0.5	1.0
Assembly (S_5)	1.2	0.5	0.5	2.0

Contribution, estimated to be

Model	x_1	x_2	x_3
Amount per Thousand Units (in Rs)	1,500	1,300	2,500

Set up the first tableau of a linear programme to determine the product mix which will maximise total contribution and then to complete the first iteration only.

(b) Interpret the following tableau, given that it is the final solution to the above problem. The *S* variables (S_1, S_2, S_3, S_4, S_5) relate to the constraints in the same sequence as presented in (a) earlier.

x_1	x_2	x_3	S_1	S_2	S_3	S_4	S_5	b_i
1	0	0	1	0	0	0	0	500
0	0	0	$\frac{1}{4}$	$\frac{1}{4}$	1	0	$-\frac{1}{2}$	112.5
0	0	1	$-\frac{1}{4}$	$-\frac{1}{4}$	0	0	$\frac{1}{2}$	287.5
0	0	0	$-\frac{1}{4}$	$-\frac{1}{4}$	0	1	$-\frac{1}{2}$	487.5
0	1	0	0	1	0	0	0	750
0	0	0	– 875	– 675	0	0	– 1,250	– 2,443,750

(ICMA, May 1979, Adapted)

(a) In terms of the production levels x_1, x_2 and x_3 (expressed in thousands of units) the linear programming problem is as follows:

Maximise $Z = 1{,}500x_1 + 1{,}300x_2 + 2{,}500x_3$

Subject to

$$\begin{aligned} x_1 &\le 500 \\ x_2 &\le 750 \\ x_3 &\le 400 \\ 0.5x_1 + 0.5x_2 + x_3 &\le 1{,}400 \\ 0.5x_1 + 0.5x_2 + 2x_3 &\le 1{,}200 \\ x_1, x_2, x_3 &\ge 0 \end{aligned}$$

The initial tableau is given in Table 4.22.

TABLE 4.22 Simplex Tableau 1 : Non-optimal

Basis		x_1	x_2	x_3	S_1	S_2	S_3	S_4	S_5	b_i	b_i/a_{ij}
S_1	0	1	0	0	1	0	0	0	0	500	∞
S_2	0	0	1	0	0	1	0	0	0	750	∞
S_3	0	0	0	1*	0	0	1	0	0	400	400 ←
S_4	0	0.5	0.5	1	0	0	0	1	0	1,400	1,400
S_5	0	0.5	0.5	2	0	0	0	0	1	1,200	600
c_j		1,500	1,300	2,500	0	0	0	0	0		
Solution		0	0	0	500	750	400	1,400	1,200		
Δ_j		1,500	1,300	2,500	0	0	0	0	0		
				↑							

In this table, S_1, S_2, S_3, S_4 and S_5 are the five slack variables introduced in the five constraints. We observe here that x_3 is the incoming and S_3 is the outgoing variable. The revised tableau is given in Table 4.23.

TABLE 4.23 Simplex Tableau 2 : Non-optimal Solution

Basis		x_1	x_2	x_3	S_1	S_2	S_3	S_4	S_5	b_i
S_1	0	1	0	0	1	0	0	0	0	500
S_2	0	0	1	0	0	1	0	0	0	750
x_3	2,500	0	0	1	0	0	1	0	0	400
S_4	0	0.5	0.5	0	0	0	−1	1	0	1,000
S_5	0	0.5	0.5	0	0	0	−2	0	1	400
c_j		1,500	1,300	2,500	0	0	0	0	0	
Solution	0	0	400	500	750	0	1,000	400		
Δ_j		1,500	1,300	0	0	0	− 2,500	0	0	

(b) According to the given tableau, the solution to the problem is x_1 = 500,000, x_2 = 750,000 and x_3 = 287,500 units, for a total contribution of Rs 2,443,750. In addition, the slack variables have the following values; $S_1 = 0$, $S_2 = 0$, $S_3 = 112.5$, $S_4 = 487.5$, and $S_5 = 0$. This implies that the production of the export model, x_3, is 112, 500 units short of the maximum demand for it, and also that a total of 487,500 hours of machining time is not being used. The shadow prices, as indicated by the final row of the table, associated with the first, second and fifth constraints are 875, 675, and 1,250 respectively. From these values, the following conclusions can be drawn:

(i) If the maximum demand for the super deluxe model could be enhanced by 1,000 units, it would result in an additional contribution of Rs 875.

(ii) If the maximum demand for the deluxe model could be increased by 1,000 units, this would yield an additional contribution of Rs 675.

(iii) If an additional 1,000 hours of assembly time were available, it would result in an additional contribution of Rs 1,250.

Example 4.16 A company has facilities for producing 5 products which require the same raw material and same type of production, finishing and packaging facilities. The unit contribution margin and the material and labour requirements for each of the products are given here:

Product	Contribution Margin	Raw Material (kg)	Labour Hours Production	Labour Hours Finishing & Packaging
P_1	150	10	10	30
P_2	120	10	20	20
P_3	160	20	10	20
P_4	160	30	10	20
P_5	100	20	20	10
Total availability ('000)		50	80	140

The manager of the company insists that the products P_3 and P_4 should be given top priority in the production, because they yield the maximum unit contribution.

Formulate this as a linear programming problem, obtain the optimal solution and state whether you agree with the manager's viewpoint. Further, attempt the following.

(a) Write the dual of this problem.

(b) Obtain the marginal profitability of (i) raw materials, (ii) the time in the production department, and (iii) the time in the finishing and packaging department.

(c) Over what range of values of the respective constraints the marginal profitabilities determined by you in (b) earlier would be valid?

(d) Obtain the optimal values of the dual variables.

(e) Verify that the objective function values of the primal and the dual problems are identical.

(f) The marketing manager informs that the selling price of the product P_2 has to be revised downward to Rs 116. What would be the new optimal product mix and the profit at it?

Let x_1, x_2, x_3, x_4 and x_5 be the number of units of P_1, P_2, P_3, P_4 and P_5, respectively, to be produced. The LPP corresponding to the given information can be expressed as:

Maximise $\quad Z = 150x_1 + 120x_2 + 160x_3 + 160x_4 + 100x_5$

Subject to

$$10x_1 + 10x_2 + 20x_3 + 30x_4 + 20x_5 \leq 50{,}000$$
$$10x_1 + 20x_2 + 10x_3 + 10x_4 + 20x_5 \leq 80{,}000$$
$$30x_1 + 20x_2 + 20x_3 + 20x_4 + 10x_5 \leq 140{,}000$$
$$x_i \geq 0,\ i = 1, 2, \ldots, 5$$

After introducing the slack variables, we get

Maximise $\quad Z = 150x_1 + 120x_2 + 160x_3 + 160x_4 + 100x_5 + 0S_1 + 0S_2 + 0S_3$

Subject to

$$10x_1 + 10x_2 + 20x_3 + 30x_4 + 20x_5 + S_1 = 50{,}000$$

$$10x_1 + 20x_2 + 10x_3 + 10x_4 + 20x_5 + S_2 = 80{,}000$$

$$30x_1 + 20x_2 + 20x_3 + 20x_4 + 10x_5 + S_3 = 140{,}000$$

$$x_i \geq 0,\ i = 1, 2, \ldots, 5,\ S_1, S_2, S_3 \geq 0$$

The solution is contained in Tables 4.24 through 4.27.

TABLE 4.24 Simplex Tableau 1: Non-optimal Solution

Basis		x_1	x_2	x_3	x_4	x_5	S_1	S_2	S_3	b_i	b_i/a_{ij}
S_1	0	10	10	20*	30	20	1	0	0	50,000	2,500 ←
S_2	0	10	20	10	10	20	0	1	0	80,000	8,000
S_3	0	30	20	20	20	10	0	0	1	140,000	7,000
c_j		150	120	160	160	100	0	0	1		
Solution		0	0	0	0	0	50,000	80,000	140,000		
Δ_j		150	120	160	160	100	0	0	0		
				↑							

TABLE 4.25 Simplex Tableau 2: Non-optimal Solution

Basis		x_1	x_2	x_3	x_4	x_5	S_1	S_2	S_3	b_i	b_i/a_{ij}
x_3	160	1/2	1/2	1	3/2	1	1/20	0	0	2,500	5,000
S_2	0	5	15	0	– 5	10	– 1/2	1	0	55,000	11,000
S_3	0	20*	10	0	– 10	– 10	– 1	0	1	90,000	4,500 ←
c_j		150	120	160	160	100	0	0	0		
Solution		0	0	2,500	0	0	0	55,000	90,000		
Δ_j		70	40	0	– 80	– 60	– 8	0	0		
		↑									

TABLE 4.26 Simplex Tableau 3: Non-optimal Solution

Basis		x_1	x_2	x_3	x_4	x_5	S_1	S_2	S_3	b_i	b_i/a_{ij}
x_3	160	0	1/4*	1	7/4	5/4	3/40	0	– 1/40	250	1,000 ←
S_2	0	0	25/2	0	– 5/2	25/2	– 1/4	1	– 1/4	32,500	2,600
x_1	150	1	1/2	0	– 1/2	– 1/2	– 1/20	0	1/20	4,500	9,000
c_j		150	120	160	160	100	0	0	0		
Solution		4,500	0	250	0	0	0	32,000	0		
Δ_j		0	5	0	– 45	– 25	– 120	0	– 7/2		
			↑								

TABLE 4.27 Simplex Tableau 4 : Optimal Solution

Basis		x_1	x_2	x_3	x_4	x_5	S_1	S_2	S_3	b_i
x_2	120	0	1	4	7	5	3/10	0	– 1/10	1,000
S_2	0	0	0	– 50	– 90	– 50	– 4	1	1	20,000
x_1	150	1	0	– 2	– 4	– 3	– 1/5	0	1/10	4,000
c_j		150	120	160	160	100	0	0	0	
Solution		4,000	1,000	0	0	0	0	20,000	0	
Δ_j		0	0	– 20	– 80	– 50	– 6	0	– 3	

The optimal solution, then, is to produce 4,000 units of P_1 and 1,000 units of P_2, and none of the other products, for a total profit of Rs 720,000. The manager's viewpoint is thus not acceptable.

(a) The dual of the given problem:

Minimise $G = 50{,}000y_1 + 80{,}000y_2 + 140{,}000y_3$

Subject to

$$10y_1 + 10y_2 + 30y_3 \geq 150$$
$$10y_1 + 20y_2 + 20y_3 \geq 120$$
$$20y_1 + 10y_2 + 20y_3 \geq 160$$
$$30y_1 + 10y_2 + 20y_3 \geq 160$$
$$20y_1 + 20y_2 + 10y_3 \geq 100$$
$$y_1, y_2, y_3 \geq 0$$

(b) The marginal profitability of the resources can be obtained from the Δ_j row of the optimal solution tableau. Thus, the marginal profitability values are: (i) materials—Rs 6 per kg, (ii) time in the production department—nil, and (iii) time in the finishing and packaging department—Rs 3 per hour.

(c) The range of values over which the marginal profitability values are valid are determined here.

For raw materials: In reference to Table 4.27, dividing b_i by a_{ij} values in column headed S_1, that is, b_i/a_{ij},

$$1{,}000 \div 3/10 = 10{,}000/3 \qquad \text{least positive}$$

$$20{,}000 \div (-4) = -5{,}000 \qquad \text{least negative}$$

$$4{,}000 \div (-1/5) = -20{,}000$$

Thus, the range required is 50,000 – 10,000/3, and 50,000 – (–5,000) or between 46,667 and 55,000 kg.

For production department: Since there are 20,000 hours unused in this department, the marginal profitability would remain valid over the range 80,000 – 20,000 = 60,000 upto ∞.

For finishing and packaging department: Dividing b_i by a_{ij} values corresponding to the column headed by S_3,

$$1{,}000 \div (-1/10) = -10{,}000 \qquad \text{least negative}$$

$$20{,}000 \div (1) = 20{,}000 \qquad \text{least positive}$$

$$4{,}000 \div (1/10) = 40{,}000$$

Thus, the range over which marginal profitability would be valid is 140,000 – 20,000, and 140,000 – (–10,000), or 120,000 and 150,000 hours.

(d) Optimal values of the dual variables:

$$y_1 = 6,\ y_2 = 0 \text{ and } y_3 = 3$$

(e) Objective function value of the dual is $6 \times 50{,}000 + 0 \times 80{,}000 + 3 \times 140{,}000 =$ Rs 720,000, which is the same as that of the primal.

(f) Other things remaining the same, the price range of P_2 over which the optimal solution would not change, is determined here:

Dividing the Δ_j values by the respective elements in the first row (except where both are zero), corresponding to x_2, we get –, 0, –5, –80/7, –10, –20, –30 and 30. The least negative of these values is –5 while the least positive is 30. Thus, the unit profit may be reduced upto Rs 5 and increased upto Rs 30, and the range of values is, therefore, Rs 115 and Rs 150.

Since the price is sought to be revised to Rs 116, there would be no change in the optimal solution. The profit would be:

$$150 \times 4{,}000 + 116 \times 1{,}000 = \text{Rs } 716{,}000$$

Example 4.17 The company, Portland p.l.c. has five products in its range and is currently running the following sales/production programme:

Details	Products				
	A	B	C	D	E
Sales in units	50,000	40,000	70,000	60,000	20,000
Per unit:					
Sales price (Rs)	3.50	3.00	4.50	5.00	2.00
Variable cost (Rs)	1.50	1.26	2.00	2.34	0.88
Labour hours	2.00	1.50	3.00	2.80	1.00
Machine hours	1.00	0.80	1.50	1.20	0.40

This programme fully utilises the availability of labour and machine time.

A linear programme reveals that the labour and machine hours have a shadow price of Re 1 and Re 0.30 per hour.

Determine the optimal production programme from this information and compare the contribution earned with that of the existing programme. *(ICMA, May 1982, Adapted)*

Since the current programme fully utilises both labour and machine time, the total availability of these resources can be calculated as follows:

$$\text{Labour} = 50 \times 2.0 + 40 \times 1.5 + 70 \times 3.0 + 60 \times 2.8 + 20 \times 1.0$$
$$= 558 \text{ i.e. } 558{,}000 \text{ hours}$$
$$\text{Machine} = 50 \times 1.0 + 40 \times 0.8 + 70 \times 1.5 + 60 \times 1.2 + 20 \times 0.4$$
$$= 267 \text{ i.e. } 267{,}000 \text{ hours}$$

Thus, the linear programme is

Maximise $Z=(3.50-1.50)A+(3.00-1.26)B+(4.50-2.00)C+(5.00-2.34)D+(2.00-0.88)E$

Subject to

$$2.0A + 1.5B + 3.0C + 2.8D + 1.0E \leq 558$$
$$1.0A + 0.8B + 1.5C + 1.2D + 0.4E \leq 267$$
$$A, B, C, D, E \geq 0$$

Its dual is,

Minimise $G = 558X + 267Y$

Subject to

$$2.0X + 1.0Y \geq 2.00$$
$$1.5X + 0.8Y \geq 1.74$$
$$3.0X + 1.5Y \geq 2.50$$
$$2.8X + 1.2Y \geq 2.66$$
$$1.0X + 0.4Y \geq 1.12$$
$$X, Y \geq 0$$

The shadow prices from the primal problem yield the solution to the dual problem so that $X = 1.0$ and $Y = 0.3$. Substitution of these values in the five constraints of the dual problem gives the following surplus variables:

$$S_1 = 0.3,\ S_2 = 0,\ S_3 = 0.95,\ S_4 = 0.5, \text{ and } S_5 = 0$$

The fact that there is no surplus in the second and the fifth constraints indicates that the non-zero variables in the primal problem are *B* and *E*. Therefore, from the primal problem,

$$1.5B + 1.0E = 558$$

$$0.8B + 0.4E = 267$$

Solving these equations simultaneously, we get $B = 219$ (thousand) and $E = 229.5$ (thousand).

Thus, the contribution under the optimal solution is

$$Z = 1.74 \times 219 + 1.12 \times 229.5 = 638.1 \text{ (thousand), or Rs } 638{,}100$$

Under the existing programme, the contribution is,

$$Z = 50 \times 2.00 + 40 \times 1.74 + 70 \times 2.50 + 60 \times 2.66 + 20 \times 1.12 = 526.6 \text{ (thousand)} = \text{Rs } 526{,}600.$$

Thus, contribution can be increased by Rs 111,500.

KEY POINTS TO REMEMBER

- For every LPP, there is another LPP, called as the primal and dual, respectively. When the primal is a maximisation problem, the dual is a minimisation problem and vice versa. For an *m*-variable and *n*-constraints primal, the dual has *n* variables and *m* constraints.
- To write the dual, it is first ensured that all its constraints are in the right direction (i.e. $\leq$ type when the objective function is of maximisation nature, and $\geq$ type when the objective function is of minimisation nature) and the variables are all non-negative. A constraint in a direction opposite to the one required is multiplied by –1 to reverse the direction, while a constraint with an "=" sign is replaced by a pair of inequalities in the opposite directions. An unrestricted variable is set equal to the difference of two non-negative variables.
- When both the conditions are satisfied, dual variables are introduced and the three matrices involved in the primal are transposed. For every unrestricted variable in the primal, the dual has an equation in one of the constraints, while for every constraint involving an equation, the dual has an unrestricted variable.
- If the primal problem has an optimal solution, then the dual also has an optimal solution with identical objective function value. If the primal problem has an unbounded solution, the dual has no feasible solution, and vice versa.
- The Δ_j values (ignoring minus signs, if any) corresponding to the slack/surplus variables of a problem represent optimal values of the dual variables.
- Also, the Δ_j values (ignoring the minus signs) corresponding to the slack/surplus variables indicate the marginal profitability or shadow prices of the resources/variables they are related to. Thus, the shadow price of a given constraint of an LPP is the amount by which the optimal value of the objective function is improved if the right-hand side of the constraint is increased by one unit.
- Sensitivity analysis answers questions like how would the solution of an LPP change when the coefficients in the objective function, the right-hand side values of the constraints or the coefficients in the left-hand side of the constraints change.
- Using graphic approach, to determine whether a change in the coefficient of a variable in the objective function would lead to a change in the basis or not, it should be noted

that a change in a coefficient causes a change in the iso-profit line (for a maximisation problem). The current basis continues to be optimal as long as the current optimal solution happens to be the last point in the feasible region to make contact with the iso-profit lines while we move in the direction of increasing values of the objective function. If the current basis remains optimal, the values of the decision variables remain what they have been, but the optimal Z-value may change.

- To determine whether a change in the right-hand side value of a constraint would cause a change in the optimality of the current basis, we begin by considering the constraints which are binding for the current optimal solution. For a change in the RHS of a constraint, the current basis remains optimal as long as the point where the constraints are binding remains feasible. However, even if the current basis remains optimal, the values of the decision variables and the optimal value of the objective function might change.
- Using simplex method, sensitivity analysis is done as follows:

  (a) *Change in c_j values.* For a variable included in the basis, divide the Δ_j row values by the corresponding row values in tableau for that variable. The range of c_j for which the current optimal solution would not change is obtained by adjusting the value by the least positive and negative quotients. For a variable not included in the basis, any decrease in the coefficient (profit) value or any increase by an amount less than the value shall cause no change in the solution.

  (b) *Change in b_i values.* Divide the b_i values in the optimal solution tableau by the a_{ij} values of the slack variable of interest. The least positive and the least negative values are then subtracted from the b_i value of the original problem. It gives the range of values for the given resource over which there shall be no change in its shadow price.

  (c) *Changes in a_{ij} values.* They can also be analysed to determine their effect on the optimal solution.
- To determine the advisability of the introduction of a new product in the existing set-up, we first obtain resource requirements and multiply them with the respective shadow prices. If the unit profit from the proposed product exceeds the sum of these multiplications, then it is desirable to add the new product.
- When multiple changes take place in the coefficients in the objective function or in the RHS values of the constraints, then the Cent-Percent Rule may be used to determine whether they would affect the current solution.

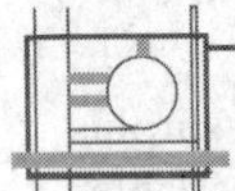

TEST YOUR UNDERSTANDING

Mark the following statements as T (*True*) or F (*False*).

1. It is necessary to convert the equality constraints into inequality constraints for writing the dual to a linear programming problem.
2. For a 4-variable and 5-constraint primal problem, the dual would be a 5-variable and 4-constraint problem.
3. The dual to given LPP would have as many variables as the number of constraints in the primal.
4. If the variables in a primal problem are all greater than, or equal to zero, then the variables in the dual problem would be all less than or equal to zero.

5. Corresponding to every unrestricted primal variable, an equality dual constraint is obtained.
6. The dual to the dual of an LPP is the primal LPP itself.
7. The optimal solution to the dual problem is readily available from the optimal solution to a primal problem.
8. If the number of primal variables is very small and the number of constraints is very large, then it is more efficient to solve the dual to obtain the optimal solution to the primal problem.
9. Duality plays a significant role in the development of sensitivity analysis.
10. At the optimal solution of a profit maximisation linear programming problem, the total (optimal) profit must be equal to the total worth of the resources.
11. For an unbounded primal problem, the dual would be infeasible.
12. The Δ_j values, in the Simplex tableau containing optimal solution, corresponding to the slack variables indicate the marginal profitability of the respective resources they represent.
13. A resource, whose worth is positive, must necessarily be a scarce one.
14. The optimal Simplex tableau provides information about the status and worth of the resources in addition to the optimal values of the decision variables.
15. If the slack variable corresponding to a given constraint has $\Delta_j = -5/2$ in the optimal solution, it implies that any number of units of this resource added will increase the profit by 5/2 and, similarly, any number of units of this resource reduced would reduce the profit by 5/2 per unit.
16. For any changes in the objective function coefficients, the optimal values of the decision variables would change.
17. Changes in the right-hand side values of the constraints within the allowable limits would neither change the basis nor the objective function value of an LPP.
18. The addition of a new constraint in an LPP can never improve the optimal value of the objective function.
19. The 100% Rule, as applied to changes in the coefficients in the objective function, states that if $\Sigma r_j \leq 1$, then there would be no change in the basis of the optimal solution.
20. The 100% Rule dictates that if $\Sigma r_i > 1$, then it cannot be stated for sure as to whether or not the set of changes in the b_i values would affect the current basis of the LPP.

EXERCISES

1. 'Every linear programming problem has a mirror image in the form of another linear programming problem, called its dual.' Do you agree? Explain the primal-dual relationship in detail. How is the knowledge of this relationship beneficial?
2. What is the significance of the duality theory of linear programming? Describe the general rules for writing the dual of a linear programming problem.
3. "Linear programming is one of the most frequently and successfully employed Operations Research techniques to managerial and business decisions." Elucidate this statement with some examples.

   (MBA, Delhi, March 1999)
4. (a) What is shadow price? How does the concept relate to the dual of an LP problem? How does it relate to the primal?

(b) 'The range of validity of a shadow price can have a finite lower bound and an infinite upper bound.' Do you agree? Explain.

5. Write a note on the economic interpretation of the dual.
6. (a) Explain the mechanism and managerial significance of post-optimality analysis of a simplex linear programming solution.
   (b) Explain the *proportionality* and *continuity* (of variables) assumptions of linear programming.
   (c) How does the simplex algorithm indicate that:
      (i) there is an alternate optimal solution?
      (ii) the problem has unbounded optimal solution?
      (iii) the problem has no feasible solution? *(M Com, Delhi, 2004)*
7. If the right-hand side constant of a constraint of an LP problem is replaced with another, within the range of validity of the shadow price for the relevant resource, and the problem is re-solved, the new optimal solution would have the same basic variables and the value of objective function. Do you agree? Explain.
8. Write a note on graphic approach to sensitivity analysis of LPPs. Take a two-variable linear programming problem and carry out the sensitivity analysis to determine the range within which (i) changes in the objective function coefficients, and (ii) changes in the right-hand side values of constraints would not cause a change in its basis.
9. A firm has a given product-mix, determined by application of simplex method. It is planning to introduce a new product with a certain profit rate and known requirement of resources already in use. How would you decide whether or not should this product be introduced?
10. How is sensitivity analysis carried out in case there are multiple changes in parameters? In this context, explain the 100% Rule.

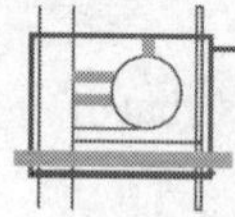

PRACTICAL PROBLEMS

1. Write dual of the following linear programming problem:

Maximise $Z = x_1 - x_2 + 3x_3$

Subject to

$$
\begin{aligned}
x_1 + x_2 + x_3 &\leq 10 \\
2x_1 \qquad - x_3 &\leq 2 \\
2x_1 - 2x_2 + 3x_3 &\leq 6 \\
\text{and } x_1, x_2, x_3 &\geq 0
\end{aligned}
$$

2. Write the dual:

(a) Maximise $Z = 10y_1 + 8y_2 - 6y_3$

Subject to

$$
\begin{aligned}
3y_1 + y_2 - 2y_3 &\leq 10 \\
-2y_1 + 3y_2 - y_3 &\geq 12 \\
y_1, y_2, y_3 &\geq 0
\end{aligned}
$$

(b) Minimise $Z = -4x_1 + 3x_2$

Subject to

$$x_1 - 2x_2 \leq -4$$
$$2x_1 + 3x_2 \geq 13$$
$$-x_1 + x_2 \leq -4$$
$$x_1, x_2 \geq 0$$

3. Write the dual corresponding to the following linear programming problem:

   Maximise $Z = 5x + 7y$

   Subject to

$$x + y \leq 4$$
$$3x + 8y \leq 26$$
$$10x + 7y \leq 35$$
and $x, y \geq 0$

   Next, find dual of the dual problem and show that it is the same as the given problem.

4. Give the dual in complete mathematical form for the following primal of a linear programming problem:

$$x_1 \leq 4; x_2 \leq 6; x_1 + x_2 \leq 5; -x_2 \leq -1 \text{ and } x_1, x_2 \geq 0$$

   Maximise $Z = 3x_1 - 2x_2$ *(ICWA, June, 1985)*

5. Using dual, convert the following problem into a maximisation problem:

   Minimise $Z = 2x_1 + 9x_2 + 3x_3$

   Subject to

$$x_1 + 4x_2 + 2x_3 \geq 5$$
$$3x_1 + x_2 + 2x_3 \geq 4$$
$$x_1, x_2 \geq 0, x_3 \text{ unrestricted in sign}$$

6. Write the dual of the following linear programming problem:

   Maximise $Z = 3x_1 + 4x_2 + 7x_3$

   Subject to

$$x_1 + x_2 + x_3 \leq 10$$
$$4x_1 - x_2 - x_3 \geq 15$$
$$x_1 + x_2 + x_3 = 7$$
$$x_1, x_2 \geq 0, x_3 \text{ unrestricted}$$

   (M Com., Delhi, 1982)

7. Given:

   Minimise $Z = 4x_1 + x_2$

   Subject to

$$3x_1 + x_2 = 2$$
$$4x_1 + 3x_2 \geq 6$$
$$x_1 + 2x_2 \leq 3$$
and $x_1, x_2 \geq 0$

(a) Write the dual problem for this linear programme.

(b) Solve it for the optimal values of x_1 and x_2.

8. Given the following problem:

Maximise $Z = 2x_1 + 3x_2$

Subject to

$$2x_1 + 3x_2 \le 30$$
$$x_1 + 2x_2 \le 10$$
$$x_1 - x_2 \le 0$$
$$x_1, x_2 \ge 0$$

(a) Construct the dual with y_1, y_2, y_3 as dual variables.

(b) The following is a list of solutions with respect to the primal and dual of the above problem:

	x_1	x_2	y_1	y_2	y_3
Solution 1	10	10/3	0	1	1
Solution 2	20	10	1	4	1
Solution 3	10/3	10/3	1	5/3	1/3

Determine whether the listed solutions are optimal.

(c) Construct the dual of the following problem:

Maximise $Z = 7x_1 + 5x_2 - 2x_3$

Subject to

$$x_1 + x_2 + 3x_3 = 10$$
$$2x_1 - x_2 + 3x_3 \le 16$$
$$3x_1 + x_2 - 2x_3 \ge 0$$
$$x_1 \ge 0,\ x_2 \ge 0,\ x_3 \text{ unrestricted}$$

(M Com., Kolkata, 1993)

9. A company manufactures three models of cars. There is a backlog of orders with the company. Model A requires 60, 100 and 80 worker-days in three production processes I, II, and III respectively. Model B requires 100, 240 and 100 worker-days, while Model C requires 200, 360, and 160 worker-days respectively in the three production processes. The number of workers employed in the three production processes are 15, 30 and 15 respectively, and an average worker is on the job for 200 working days in a year. The expected profit per car of the three models is Rs 7,500, Rs 15,000 and Rs 30,000 respectively.
   (i) Determine the optimal product-mix and the total maximum profit obtainable.
   (ii) What are the shadow prices of the resources? Determine the range over which each of these is valid.
   (iii) Write the dual to the above LPP and obtain the optimal values of the dual variables from the solution in (i).
   (iv) If there is any other optimal product-mix possible, identify that too. *(MFC, Delhi, 2004)*

10. A firm manufacturing office furniture provides the following information regarding resource consumption, availability, and profit contribution:

Resources	*Usage (per unit)*			*Daily*
	Tables	*Chairs*	*Book cases*	*Availability*
Timber (cubic feet)	8	4	3	640
Assembly Department (man-hours)	4	6	2	540
Finishing Department (man-hours)	1	1	1	100
Profit Contribution (per unit/Rs.)	30	20	12	

(a) The firm wants to determine its optimal product mix. Formulate the linear programming problem using the above data.

(b) Solve the problem with Simplex method and indicate the optimal product mix and the total maximun profit contribution.

(c) Indicate the shadow prices of the resources.

(d) Examine the sensitivity of the optimal solution to:

(i) Variation in profit contribution per unit for tables and chairs.

(ii) Variation in the availability of resources.

(e) What other information can be obtained from the optimal solution of the problem?

(M Com, Delhi, 2003)

11. A manufacturing company has three major departments for the manufacture of its two products A and B. The weekly capacities are given as follows:

Department	*Per unit time requirement (hours)*		*Hours available in a week*
	A	*B*	
I	2	3	42
II	7	7	70
III	7	5	70

The marginal profit per unit from models A and B are Rs 800 and Rs 500 respectively. Assuming that the company can sell any quantity that it produces, determine the optimum output for both the models. (Use simplex method.)

Write the dual of the linear programming problem formulated above. What is the economic interpretation of the dual variables that a manager can use for making better decisions?

(MBA, Delhi, November, 2008)

12. A company makes two products X and Y. Product X has a contribution of Rs 124 per unit and product Y Rs 80 per unit.

Both products pass through two departments for processing and the times in minutes per unit are:

	Product X	Product Y
Department 1	150	90
Department 2	100	120

Currently, there is a maximum of 225 hours per week available in department 1 and 200 hours in department 2. The company can sell all it can produce of *X* but EEC quotas restrict the sale of *Y* to a maximum of 75 units per week.

The company, which wishes to maximise contribution margin, currently makes and sells 30 units of *X* and 75 units of *Y* per week.

The company is considering several possibilities including

(i) altering the production plan if it could be proved that there is a better plan than the current one;

(ii) increasing the availability of either department 1 or department 2 hours. The extra cost involved in increasing capacity is Re 0.5 per hour for each department;

(iii) transferring some of their allowed sale quota for Product *Y* to another company. Because of commitments, the company would always retain a minimum sales level of 30 units.

You are required to

(a) calculate the optimum production plan using the existing capacities and state the extra contribution that would be achieved compared with the existing plan.

(b) advise the management whether they should increase the capacity of either department 1 or department 2 and, if so, by how many hours and what the resulting increase in contribution would be over that calculated in the improved production plan.

(c) calculate the minimum price per unit for which they could sell the rights to their own quota, down to the minimum level, given the plan in (a) as a starting point.

(ICMA, May 1988, Adapted)

13. A manufacturer makes three types of decorative lamps; model *A*, model *B*, and model *C*. The raw material requirement for all lamps is the same, but the cost of production differs due to different labour requirements. Each model *A* lamp requires 0.1 hr of assembly time, 0.2 hr of wiring time, and 0.1 hr of packaging time. Model *B* requires 0.2 hr of assembly time, 0.3 hr of wiring time, and 0.1 hr of packaging time. Model *C* requires 0.3 hr of assembly time, 0.4 hr of wiring time, and 0.1 hr of packaging time. The manufacturer makes a profit of Rs 120 on each model *A* lamp, Rs 190 on each model *B* lamp, and Rs 210 on each model *C* lamp. The manufacturer can schedule up to 80 hours of assembly labour, 120 hours of wiring labour, and 100 hours of packaging labour.

    Assuming that all lamps produced can be sold, determine the optimal quantity of each model and the shadow prices of each of the resources.

    Write the dual of the given problem and use it for checking the optimal solution.

14. A metal products company produces waste cans, filing cabinets, file boxes for correspondence, and lunch boxes. Its inputs are sheet metal of two different thickness, called *A* and *B*, and manual labour. Input-output relationships for the company are shown in the table given below:

	Waste cans	*Filing cabinets*	*Correspondence boxes*	*Lunch boxes*
Sheet Metal-*A*	6	0	2	3
Sheet Metal-*B*	0	10	0	0
Manual Labour	4	8	2	3

The sales revenue per unit of waste cans, filing cabinet, correspondence boxes and lunch boxes are Rs 20, Rs 400, Rs 90 and Rs 20, respectively. There are 225 units of sheet metal *A* available in the company's inventory, 300 of sheet metal B and 190 units of manual labour. What is the company's optimal sales revenue? Also write the dual of the given problem and use it to check the optimal solution.

(MBA, Delhi, April, 1998)

15. A chemical manufacturer is developing three fertiliser compounds for the agricultural industry. The product codes for these products are X_1, X_2 and X_3 and the relevant information is summarised below:

	Chemical constituents : percentage make-up per tonne			
	Nitrate	*Phosphate*	*Potash*	*Filler*
X_1	10	10	20	60
X_2	10	20	10	60
X_3	20	10	10	60

	Input prices per tonne	*Maximum available input in tonnes per month*
Nitrate	Rs 150	1,200
Phosphate	Rs 60	2,000
Potash	Rs 120	2,200
Filler	Rs 10	No limit

The fertilisers will be sold in bulk and managers have proposed the following prices per tonne:

$$X_1 : \text{Rs } 83;\ X_2 : \text{Rs } 81;\ \text{and } X_3 : \text{Rs } 81$$

The manufacturing costs of each type of fertiliser, excluding materials, are Rs 11 per tonne.

You are required to:

(a) formulate the above data into a linear programming model so that the company may maximise contribution;

(b) construct the initial tableau and state what is meant by 'slack variables' (define X_4, X_5, X_6 as the slack variables for X_1, X_2, and X_3, respectively);

(c) indicate, with explanations, which will be the 'entering variable' and 'leaving variable' in the first iteration; (you are not required to solve the model.)

(d) interpret the final matrix of the simplex solution given below:

Basic variable	X_1	X_2	X_3	X_4	X_5	X_6	*Solution*
X_1	1	0	3	20	−10	0	4,000
X_2	0	1	−1	−10	10	0	8,000
X_6	0	0	−0.4	−3	1	1	600
Z	0	0	22	170	40	0	284,000

(e) use the final matrix above to investigate:

(i) the effect of an increase in nitrate of 100 tonnes per month;

(ii) the effect of a minimum contract from an influential customer for 200 tonnes of X_3 per month to be supplied.

(ICMA, May 1989, Adapted)

16. Fill in the blanks:

Primal problem			*Dual problem*		
Variable	*Solution*	Δ_j	*Variable*	*Solution*	Δ_j
x_1	8/3	0	y_1	–	–
x_2	20/3	0	y_2	–	–
x_3*	80/3	0	y_3	–	–
x_4*	0	4/15	y_4	–	–
x_5*	0	1/15	y_5*	–	–
x_6*	160/3	0	y_6*	–	–

*surplus variable — * slack variable

Objective Function Value = 76/3. — Objective Function Value = –.

17. The Alloy Metal Company plans to purchase at least 200 quintals of scrap metal. The company decides that the scrap metal to be purchased must contain at least 100 quintals of a valuable metal M_1 and no more than 35 quintals of a base metal M_2. The company can purchase the scrap metal from two suppliers in unlimited quantities with thc following percentages, by weight, of M_1 and M_2.

Metal	Supplier X	Supplier Y
M_1	25%	75%
M_2	10%	20%

The price per quintal of X's scrap metal is Rs 2 and Y's scrap metal is Rs 4. The company wants to determine the quantities that it should buy from the two suppliers, so that cost is held to be minimum.

(a) Formulate the above as a linear programming problem to minimise the total cost, and

(b) Find its dual and solve it. Obtain also the optimal solution to the primal.

18. D Electronics produces three models of satellite dishes—Alpha, Beta and Gamma—which have contributions per unit of Rs 400, Rs 200 and Rs 100, respectively.

There is a two-stage production process and the number of hours per unit for each process are:

	Alpha	Beta	Gamma
Process 1	2	3	2.5
Process 2	3	2	2

There is an upper limit on process hours of 1,920 per period for Process 1 and 2,200 for Process 2. The Alpha dish was designed for a low-power satellite which is now fading and the sales manager thinks that sales will be no more than 200 per period. Fixed costs are Rs 40,000 per period.

You are required to

(a) formulate these data into a linear programming model using the following notation:

x_1 : number of alphas

x_2 : number of betas

x_3 : number of gammas

(b) formulate (but do not attempt to solve) the initial Simplex Tableau using

x_4 as slack for Process 1

x_5 as slack for Process 2

x_6 as slack for any sales limit

and describe the meaning of slack.

(c) interpret the final Simplex Tableau below:

	x_1	x_2	x_3	x_4	x_5	x_6	*Solution*
x_2	0	1	0.83	0.33	0	– 0.67	506.7
x_5	0	0	0.33	– 0.67	1	– 1.67	586.7
x_1	1	0	0	0	0	1	200.0
Z	0	0	66.67	66.67	0	266.7	181,333.3

Investigate the effect on the solution of each of the following:

(i) An increase of 20 hours per period in Process 1,

(ii) An increase of 10 units per period in the output of Alpha,

(iii) Receiving an order, which must be met, for 10 units of Gamma.

(ICMA, November 1989, Adapted)

19. A diet conscious housewife wishes to ensure certain minimum intake of vitamins A, B and C for the family. The minimum daily needs of vitamins A, B and C for the family are 60, 40 and 32 units, respectively. For the supply of these vitamins, the housewife relies on two fresh foods F_1 and F_2. One unit of F_1 contains 14, 10 and 4 units of vitamins A, B and C, respectively. One unit of F_2 contains 4, 8 and 16 units of the three vitamins, respectively. F_1 costs Rs 3 per unit and F_2 Rs 2 per unit. The problem is how many units of each food the housewife should buy every day to keep her food bill as low as possible.

    (a) Formulate it as a linear programming problem to minimise the total cost;

    (b) Write the dual of the problem as formulated in (a) and present the economic interpretation of the dual; and

    (c) Solve the dual by using simplex method and read therefrom the answer to the primal.

(M Com, Delhi, 1984)

20. A company produces three kinds of light bulbs: a 60-watt soft-lite bulb, a 60-watt regular bulb, and a 100-watt bulb. Each bulb takes one hour per case in production line 1. In production line 2, a case of the soft-lite takes 2 hours and a case of each of the others takes 1 hour. Production line 1 has 25 hours available per week and line 2 has 40 hours per week. The company has determined that the two 60-watt bulbs are considered substitute products by most people and that their combined demand will not be more than 25 cases per week. The demand for 100-watt bulbs will never be greater than 60 cases per week. If the soft-lite bulbs earn a profit of Rs 70 per case and other two bulbs earn a profit of Rs 50 per case, how many cases per week should the company produce of each type in order to maximise profits? Solve the above problem, write its dual and provide its economic interpretation. From the optimal solution tableau, obtain the optimal values of the dual variables.

21. An electronics firm is undecided as to the most profitable mix for its products. The products now manufactured are transistors, resistors, and electron tubes, with a profit (per 100 units) of Rs 100, Rs 60 and Rs 40, respectively.

    To produce a shipment of transistors containing 100 units requires 1 hour of engineering service, 10 hours of direct labour, and 2 hours of administration service. To produce 100 resistors requires 1 hour, 4 hours and 2 hours of engineering, direct labour and administration time respectively. To produce one shipment of electron tubes (100 units) requires 1 hour of engineering, 5 hours of direct labour and 6 hours of administration. There are 100 hours of engineering services available, 600 hours of direct labour and 300 hours of administration. What is the most profitable product mix? Write the dual of the given problem and give its economic interpretation. *(MBA, Delhi, November, 1996)*

22. A manufacturing company makes three products, each of which requires three operations as part of the manufacturing process. The company can sell all of the products it can manufacture but its production capability is limited by the capacity of its operations centres. Additional data concerning the company are as follows:

Product	*Manufacturing requirement (hours/unit)*			*Cost*	*Selling price*
	Centre 1	*Centre 2*	*Centre 3*	*(Rs)*	*(Rs)*
A	1	3	2	11	15
B	3	4	1	12	20
C	2	2	2	10	16
Hours available	160	120	80		

(a) What is the optimal product-mix? What is the maximum profit?
(b) What are the shadow prices of the resources? When the optimal solution has been reached, which resource has the highest marginal value?
(c) Over what range in each of the RHS values, are these shadow prices valid?
(d) What are the ranges over which the objective function co-efficients can vary for each of the two decision variables?
(e) State the dual to this problem and write its solution. *(M Com, Delhi, 2009)*

23. Given below is the simplex tableau for a maximisation type of linear programming problem:

Basis		x_1	x_2	S_1	S_2	b_i
x_2	4	1	1	1	0	6
S_2	0	1	0	− 1	1	2
	c_j	3	4	0	0	
	z_j	4	4	4	0	24
	$c_j - z_j$	− 1	0	− 4	0	

Giving reasons in brief, answer the following questions:
(i) Does the tableau represent an optimal solution?
(ii) Are there more than one optimal solution?
(iii) Is this solution degenerate?
(iv) Is this solution feasible?
(v) If S_1 is slack in machine *A* (in hours/week) and S_2 in machine *B* (in hours/week), which of these machines is used to the full capacity when producing according to this solution?
(vi) A customer would like to have one unit of product x_1 and is willing to pay in excess of the normal price in order to get it. How much should the price be increased to ensure no reduction in profit?
(vii) How many units of the two products x_1 and x_2 are being produced according to this solution and what is the total profit?
(viii) Machine *A* (associated with slack S_1, in hours/week) has to be shut down for repairs for 2 hours next week. How much will the reduction in profits be?
(ix) What is the maximum you would be prepared to pay for another hour (per week) of capacity each on machines *A* and *B*?
(x) What are the shadow prices of the machine hours?

24. A manufacturing firm has just discontinued production of an unprofitable product line which has resulted in excess capacity. The management is contemplating to use this capacity for the production of three products *A*, *B*, and *C*. The per unit contribution margin of the three products is, respectively, Rs 20, Rs 6 and Rs 8. Further, the available excess capacity and the capacity requirements of the three products are as given below:

Machine Type	*Capacity Requirement (machine hours per unit)*			*Available Excess Capacity (machine hours per week)*
	A	*B*	*C*	
Milling	8	2	3	250
Lathe	4	3	0	150
Grinder	2	0	1	50

Using the given information, determine the optimal product mix and obtain the maximum contribution that the company can earn. Also, determine the shadow prices of the three resources and the respective ranges over which they are valid.

25. When the primal is infeasible, its dual is unbounded. Prove, with the help of this illustration, that this rule is not always true:

Maximise $Z = 8x_1 + 6x_2$

Subject to

$$x_1 - x_2 \leq 3/5$$
$$x_1 - x_2 \geq 2$$
$$x_1, x_2 \geq 0$$

(M Com, Kolkata, 1993)

26. You are told that the objective row of the final tableau of a linear programming solution to a contribution maximisation problem showed shadow price values of +2 and 0, respectively, in the columns for products *A* and *B*. A third product which was included in the initial formulation is shown by separate analysis to have a dual value of –3 (i.e. a deficit value if produced). Product selling prices were Rs 20, Rs 30, Rs 15 per unit respectively.

You are required to provide:

(a) An interpretation of the values +2, 0, –3 which were based on the estimated demands for the products at the given prices;

(b) As assessment of the use the information in (a) may have in product planning.

(ICMA, May, 1983, Adapted)

27. Given below is a table obtained after a few iterations using simplex method to solve a linear programming problem to maximise total contribution margin from products *A* and *B*:

		8.5	10.5	0	0	0	C_j
Mix	*Total*	x_1	x_2	S_1	S_2	S_3	
x_2	300	0	1	3/5	–2/5	0	
x_1	300	1	0	–2/5	3/5	0	
S_3	400	0	0	–1/5	–1/5	1	

Give short answers to the following questions giving reasons as well:

(i) Is the above solution optimal?

(ii) Is the above solution feasible?

(iii) Does the problem have alternative optimal solution? If so, find another optimal solution.

(iv) Write the objective function of the problem.

(v) What are the shadow prices for the three resources?

(vi) If S represents the slack for the production capacity constraint, how much should the company be willing to pay for each additional unit of production capacity?

(vii) S_3 represents the slack for demand constraint. If the company is able to increase its total demand by 20 units, what will be the optimal mix and total contribution margin?

28. In a product-mix problem, x_1, x_2, x_3, and x_4 indicate the units of products A, B, C, and D respectively and we have:

Maximise profit $\quad Z = 4x_1 + 6x_2 + 3x_3 + x_4$

Subject to

$$1.5\,x_1 + 2x_2 + 4x_3 + 3x_4 \le 550 \quad \text{(Machine I hours)}$$
$$4x_1 + x_2 + 2x_3 + x_4 \le 700 \quad \text{(Machine II hours)}$$
$$2x_1 + 3x_2 + x_3 + 2x_4 \le 200 \quad \text{(Machine III hours)}$$
$$x_1, x_2, x_3, \text{ and } x_4 \ge 0$$

(a) Solve it with simplex method and find out the optimal product-mix as well as the total maximum profit contribution.

(b) Is there any other optimal solution to the problem?

(c) Write the dual to the given LP problem. Obtain the optimal values of the dual variables and verify that the primal and the dual problems have the same objective function values.

(d) What are the shadow prices of the machine hours on the three machines? If we wish to expand the production capacity, which of the three machines should be given priority?

(e) If a customer is prepared to pay a higher price for the product A, by how much should the price is increased so that the comapany's profit remains unchanged?

(f) If the profit contribution from product A increases by Rs 4 to 5 per unit, will it affect the product mix?

(M Com, Delhi, 2007)

29. A company is manufacturing two products, A and B. The manufacturing time required to make them, the profit, and capacity available at each work centre are as follows:

Product	*Work centre*			*Profit per unit*
	Machining	*Fabrication*	*Assembly*	*(Rs)*
A	1 hour	5 hours	3 hours	80
B	2 hours	4 hours	1 hour	100
Total capacity	720 hours	1,800 hours	900 hours	

If x_1 and x_2 represent the number of units of products A and B, respectively, while S_1, S_2 and S_3 represent the slack variables indicating the unused capacity in the three work centres, we can state the problem as follows:

Maximise $\quad$ Profit $= 80x_1 + 100x_2 + 0S_1 + 0S_2 + 0S_3$

Subject to

$$x_1 + 2x_2 + S_1 = 720;$$
$$5x_1 + 4x_2 + S_2 = 1{,}800;$$
$$3x_1 + x_2 + S_3 = 900$$
$$\text{and } x_1, x_2, S_1, S_2, S_3 \ge 0$$

The simplex algorithm, applied to this problem, yields the following final tableau:

Basis		x_1	x_2	S_1	S_2	S_3	b_i
x_2	100	0	1	5/6	−1/6	0	300
x_1	80	1	0	−2/3	1/3	0	120
S_3	0	0	0	7/6	−5/6	1	240
c_j		80	100	0	0	0	
Solution		120	300	0	0	240	
Δ_j		0	0	−30	−10	0	

Required:

(a) Identify the values of all the variables and the objective function for the optimal solution to the primal problem.

(b) Formulate the dual for this problem.

(c) Identify the values of all variables and the objective function for the solution to the dual problem.

(d) Provide an economic interpretation of the dual variables.

(e) Suppose that the cost of overtime in each of the departments is Rs 15 per hour. Would it be advisable to work either of the departments on an overtime basis? What would be the maximum amount of overtime authorised, if any?

(f) How would your answer in (e) change if the overtime costs were Rs 8 instead?

(g) Suppose that a price change is under consideration for the product *A*. This would raise the profit for this product to Rs 100 per unit. Would this change the optimal production plan? What is the maximum amount of change in profit for product *A* that would not cause a change in the optimum production plan?

(h) Suppose that there would be no change in the price of product *A*. How far the unit profit of *B* may vary without changing the optimal production plan?

(i) The company is planning to introduce a new product *C* with the following requirements per unit:

Machining : 2 hours
Fabrication : 3 hours
Assembly : 2 hours

What profit would be necessary before the company considers the production of this new product?

30. Solve the following LP problem and answer the questions regarding a firm that manufactures both lawn mowers (x_1) and snowblowers (x_2).

Maximise $Z = 30x_1 + 80x_2$

Subject to

$2x_1 + 4x_2 \leq 1{,}000$ labour hours available
$6x_1 + 2x_2 \leq 1{,}200$ lb. of steel available
$x_2 \leq 200$ snowblower engines available

(i) What is the best product mix? What is the optimal profit?

(ii) What are the shadow prices of the resources? When the optimal solution has been reached, which resource has the highest marginal value?

(iii) Over what range in each of the RHS values are these shadows valid?

(iv) What are the ranges over which the objective function coefficients can vary for each of the two decision variables?

(v) State the dual to this problem and write its solution. *(M Com, Delhi, 2004)*

CHAPTER

5

Specially Structured Linear Programmes I: Transportation and Transhipment Problems

CHAPTER OVERVIEW

The distribution of goods produced by a factory from various warehouses (or sources) to different markets (or other destinations) where they are required, causes problems to almost every business. The transportation method is developed to deal with the transportation of goods from different sources to different destinations, given the relevant data like available quantities at various sources, demand at each of the destinations, the cost of shipping along each route and non-availability of certain routes; if any. It may be interesting to note that such problems can also be handled as linear programming problems; the transportation method provides an efficient means to solve them.

The method allows the manager to seek answers to the questions like the following:

- What is the optimal way of shipping goods from various sources (warehouses) to different markets so as to minimise the total cost involved in the shipping?
- How to handle a situation when some routes are not available or when some units have to be necessarily transported from a particular source to a particular market?
- How would the optimal shipping schedule change if some routes become cheaper/costlier?
- If it were possible to increase supply, which of the sources should be preferred?
- Instead of allowing shipping of goods only from listed sources to different destinations, if it were possible to ship goods from a particular source to another source or destination and then from there to a further destination, how much cost can be saved? This is what is called a transhipment problem.
- If an item can be produced at different locations at varying costs and sold in different markets at different prices, then what production and shipping plan will yield maximum profit?
- How can the production be scheduled, given the cost of production of an item and the cost of carrying stock, so as to meet the requirements in different periods?

The use of transportation method is not limited to solution of the transportation problems alone. Problems of scheduling production, controlling inventory and management of funds over different time periods illustrate some other areas which lend themselves to handling by the transportation method.

This chapter basically requires elementary arithmetic calculations. However, familiarity with summation notation, a basic knowledge of inequalities, matrices and their transpose is also necessary. While working through this chapter, master the skill of drawing of a closed path.

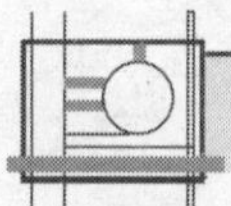

5.1 INTRODUCTION

We have discussed in the previous chapters solution to the general linear programming problem. Now we shall consider a special branch of linear programming that comprises problems of the transportation type, to be followed by a discussion of the problems of transhipment. Generally, transportation costs are involved in such problems but the scope of problems extends well beyond to cover situations which have nothing to do with these costs. The term *'transportation'* is associated with such problems principally because in studying efficient transportation routes, a special procedure was used which has come to be known as the *transportation method.*

A typical transportation problem is like this. Suppose that a manufacturer of refrigerators has three plants situated at places called *A, B* and *C*. Suppose further that his buyers are located in three regions *X, Y* and *Z* where he has to supply them the goods. Also, assume that the demands in the three regions and the production in different plants per unit time period are known and equal in aggregate, and further that the cost of transporting one refrigerator from each plant to each of the requirement centres is given and constant. The manufacturer's problem is to determine as to how he should route his product from his plants to the market places so that the total cost involved in the transportation is minimised. In other words, he has to decide as to how many refrigerators should be supplied from *A* to *X, Y*, and *Z;* how many from *B* to *X, Y*, and *Z*; and how many from *C* to *X, Y*, and *Z*, to achieve it at the least cost. This is illustrated in Fig. 5.1.

The places where the goods originate from (the plants in our example) are called the *sources* or the *origins* and places where they are to be supplied are the *destinations.* In this terminology, the problem of the manufacturer is to decide as to how many units be transported from different origins to different destinations so that the total transportation cost is the minimum.

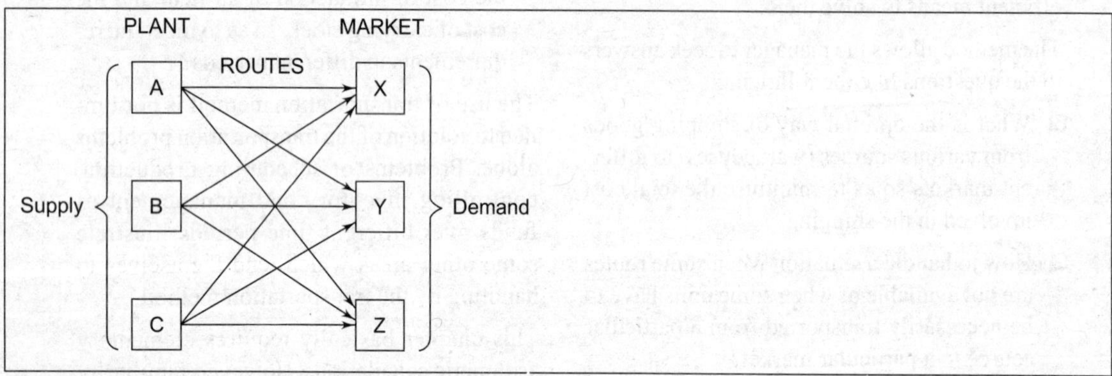

Figure 5.1 *Transportation Problem*

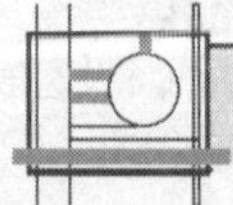

5.2 PROBLEM STATEMENT

The classical transportation problem can be stated mathematically as follows:

Let a_i = quantity of product available at origin i

b_j = quantity of product required at destination j

c_{ij} = the cost of transporting one unit of product from origin i to destination j

x_{ij} = the quantity transported from origin i to destination j

Assume that $\sum_{i=1}^{m} a_i = \sum_{j=1}^{n} b_j$ which means that the total quantity available at the origins is precisely equal to the total amount required at the destinations.

With these, the problem can be stated as a linear programming problem as:

Minimise Total cost $Z = \sum_{i=1}^{m} \sum_{j=1}^{n} c_{ij} x_{ij}$

Subject to

$$\sum_{j=1}^{n} x_{ij} = a_i \quad \text{for} \quad i = 1, 2, ..., m$$

$$\sum_{i=1}^{m} x_{ij} = b_j \quad \text{for} \quad j = 1, 2, ..., n$$

and $x_{ij} \geq 0$ for all $i = 1, 2, ..., m$, and $j = 1, 2, ..., n$

The transportation model can also be portrayed in a tabular form by means of a transportation tableau, shown in Table 5.1.

TABLE 5.1 Transportation Tableau

Origin (i)	*Destination (j)* 1	2	...	*n*	*Supply,* a_i
1	x_{11} c_{11}	x_{12} c_{12}	...	x_{1n} c_{1n}	a_1
2	x_{21} c_{21}	x_{22} c_{22}	...	x_{2n} c_{2n}	a_2
...	...	...	...	...	...
m	x_{m1} c_{m1}	x_{m2} c_{m2}	...	x_{mn} c_{mn}	a_m
Demand, b_j	b_1	b_2	...	b_n	$\Sigma a_i = \Sigma b_j$

This tableau can be thought of as a matrix within a matrix, of the dimension $m \times n$. The one is the per unit cost matrix which represents the unit transportation costs for each of the possible transportation routes. Its elements are given by c_{ij}, indicating the cost of shipping a unit from the ith origin to the jth destination. Superimposed on this matrix is the matrix in which each cell contains a transportation variable—that is, the number of units shipped from the row-designated *origin* to the column-designated *destination*. Each such variable is represented by x_{ij}, the amount shipped from ith source to jth destination. Right and bottom sides of the transportation tableau show, respectively, the amount of supplies a_i available at source i and the amount demanded b_j at each destination j. The a_i's and b_j's represent the supply and demand constraints.

The aggregate transportation cost is determined by multiplying the various x_{ij}'s with corresponding c_{ij}'s and then adding them up all. The solution to the transportation problem calls for determining values of x_{ij}'s as would yield the minimum aggregate transportation cost. When a problem is solved, some of the x_{ij}'s would assume positive values indicating utilised routes. The cells containing such values are called *occupied* or *filled* cells and each of them represents the presence of a *basic* variable. For the remaining cells, called the *empty* cells, x_{ij}'s would be zero. These are the routes that are not utilised by the transportation pattern and their corresponding variables (x_{ij}'s) are regarded to be *non-basic*.

Remember that the transportation costs are assumed to be linear in nature. Further, it is assumed here that the aggregate supply at sources (Σa_i) is equal to the aggregate demand at destinations (Σb_j). We shall consider later the situation when they do not match.

As observed earlier, the number of constraints in a transportation tableau is $m + n$ (the number of rows plus the number of columns). The number of variables required for forming a basis is one less, i.e. $m + n - 1$. This is so, because there are only $m + n - 1$ independent variables in the solution basis. In other words, with values of any $m + n - 1$ independent variables being given, the remaining would automatically be determined on the basis of those values. Also, considering the conditions of feasibility and non-negativity, the number of basic variables representing transportation routes that are utilised equals $m + n - 1$, and all other variables be non-basic, or zero, representing the unutilised routes. It means that a basic feasible solution of a transportation problem has exactly $m + n - 1$ positive components in comparison to the $m + n$ positive components required for a basic feasible solution in respect of a general linear programming problem in which there are $m + n$ structural constraints to satisfy.

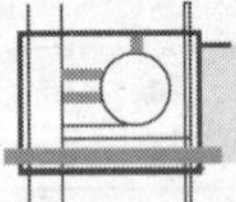

5.3 SOLUTION TO THE TRANSPORTATION PROBLEM

A transportation problem can be solved by two methods: (a) simplex method, and (b) transportation method. We shall illustrate these with the help of an example.

Example 5.1 A firm owns facilities at seven places. It has manufacturing plants at places *A*, *B* and *C* with daily output of 500, 300, and 200 units of an item respectively. It has warehouses at places *P*, *Q*, *R* and *S* with daily requirements of 180, 150, 350 and 320 units respectively. Per unit shipping charges on different routes are given below.

To:	*P*	*Q*	*R*	*S*
From *A*:	12	10	12	13
From *B*:	7	11	8	14
From *C*:	6	16	11	7

The firm wants to send the output from various plants to warehouses involving minimum transportation cost. How should it route the product so as to achieve its objective?

As a first step, the given information is presented in Table 5.2.

TABLE 5.2 Transportation Tableau

To → *From* ↓	*P*	*Q*	*R*	*S*	*Supply*
A	12	10	12	13	500
B	7	11	8	14	300
C	6	16	11	7	200
Demand	180	150	350	320	1000

5.3.1 The Simplex Method

The problem can be expressed as a linear programming problem as follows:

Let x_{ij} be the number of units shipped from plant *i* to warehouse *j*. With *Z* as the total cost, the problem can be stated as here:

Minimise

$$Z = 12x_{11} + 10x_{12} + 12x_{13} + 13x_{14} + 7x_{21} + 11x_{22} + 8x_{23} + 14x_{24} + 6x_{31} + 16x_{32} + 11x_{33} + 7x_{34}$$

Subject to

$$\left.\begin{aligned} x_{11} + x_{12} + x_{13} + x_{14} &= 500 \\ x_{21} + x_{22} + x_{23} + x_{24} &= 300 \\ x_{31} + x_{32} + x_{33} + x_{34} &= 200 \end{aligned}\right\} \text{Supply constraints}$$

$$\left.\begin{aligned} x_{11} + x_{21} + x_{31} &= 180 \\ x_{12} + x_{22} + x_{32} &= 150 \\ x_{13} + x_{23} + x_{33} &= 350 \\ x_{14} + x_{24} + x_{34} &= 320 \end{aligned}\right\} \text{Demand constraints}$$

$$x_{ij} \geq 0 \quad \text{for } i = 1, 2, 3 \text{ and } j = 1, 2, 3, 4$$

Note that this problem has a total of 12 decision variables (x_{ij}'s) and seven constraints: 3 of which are supply constraints and the rest are demand constraints. It may be mentioned that since the aggregate supply and aggregate demand are both equal (1,000), one of the constraints is redundant. This is because if any of the six constraints are satisfied, the seventh would automatically be. Accordingly, the problem may be modified by removing any one of the constraints. Thus, the problem can be restated by considering any of the six constraints. In general, a transportation problem with *m*-sources and *n*-destinations, with matching aggregate supply and demand, may be expressed as an LPP with $m \times n$ decision variables and $m + n - 1$ constraints.

Once a transportation problem is expressed as a linear programming problem, it can be solved by using simplex method. However, the solution process is likely to be very lengthy and cumbersome because of the involvement of a large number of decision variables and artificial variables. In solving the problem, as we progress from one tableau to another, the artificial variables are dropped one after the other as they become non-basic. A feasible solution is reached once they are all excluded and eventually the optimal solution is reached thereafter, with or without further iterations. What makes the solution procedure tedious is the requirement of at least as many iterations as the number of artificial variables on the large-sized simplex tableaus. To illustrate, the LPP at hand would involve 18 variables in all—12 of them the decision variables and 6 artificial variables. To solve it would need to carry out at least 6 iterations on a matrix of the order 6×18. It makes it virtually impossible (when working manually) to use simplex method for solving even this moderate-sized problem. Hence, we make for an alternate solution procedure, called the **transportation method** which is an efficient method that yields results faster and with much less computational effort.

5.3.2 The Transportation Method

It is a method specially developed to solve transportation problems. Unlike simplex algorithm, it does not require any artificial variables to introduce for obtaining an initial solution to the problem.

Broadly, the transportation method involves the following three steps:

(i) Obtaining an initial solution, that is to say making an initial assignment in such a way that a basic feasible solution is obtained;

(ii) Ascertaining whether it is optimal or not, by determining opportunity costs associated with the empty cells. If the solution is optimal then exit, and if it is not optimal, proceed to step (iii).

(iii) Revise the solution until an optimal solution is reached.

These steps are explained below. A detailed working of the method is shown schematically in Figure 5.4.

Step 1: Obtaining the Initial Feasible Solution The first step in using the transportation method is to obtain a feasible solution, namely, the one that satisfies the *rim* requirements (i.e. the requirements of demand and supply). The initial feasible solution can be obtained by several methods. The commonly used are the *North-West Corner* (NWC) *Rule, Least Cost Method* (LCM) and the *Vogel's Approximation Method* (VAM). We shall discuss these methods in turn.

North-West Corner Rule

The North-West Corner rule (N-W Corner rule) may be stated as follows.

Start with the north-west corner of the transportation tableau and consider the cell in the first column and the first row. Corresponding to this cell are the values a_1 and b_1 respectively in row 1 and column 1. Proceed as follows:

If $a_1 > b_1$, then assign quantity b_1 in this cell. This implies that we put $x_{11} = b_1$. If $a_1 < b_1$, then assign a_1 in the cell so that $x_{11} = a_1$. Simply speaking, put the lower of a_1 and b_1 as x_{11}. If $a_1 = b_1$, then $x_{11} = a_1 = b_1$. To illustrate, if supply $a_1 = 50$ and demand $b_1 = 30$, then assign 30 units as the quantity to be supplied from first source to the first destination. If supply, $a_1 = 30$ and demand, $b_1 = 50$, then $x_{11} = a_1 = 30$ and if $a_1 = 30$, $b_1 = 30$, then $x_{11} = 30$.

Now, if $a_1 > b_1$, then move horizontally to the next column in the first row; if $a_1 < b_1$, move vertically in the same column to the next row; and if $a_1 = b_1$, then move diagonally to the next column and next row. In operational terms, if $a_1 > b_1$, so that the supply is greater than demand, then having assigned quantity equal to demand (b_1), the remaining quantity is considered along with demand at the next destination (b_2), whereas if supply falls short of demand ($a_1 < b_1$), then having exhausted the available supply at source 1, consider obtaining from the next source (a_2). Obviously, if supply and demand match, then consider the next source and next destination (a_2 and b_2).

Once in the next cell, again compare the supply available at the source and demand at the destination, corresponding to the cell chosen, and assign lower of the two values. Move to the next cell appropriately as explained earlier and again assign the quantity considering demand and available supply.

Continue in the zigzag manner until the last source and last destination are covered, so that the south-east corner of the tableau is reached.

For Example 5.1, the initial basic solution using this method is obtained as follows. First, start with the cell on the intersection of *A* and *P*. The row total (a_1) corresponding to this is 500 and the column total (b_1) is 180. So, allocate 180 units to this cell. With the destination requirement having been satisfied, move horizontally to the cell *AQ*. With the supply available in the plant *A* (a_1) being 320 and the demand at warehouse *Q* (b_2) being 150 units, allocate 150 units to this cell and move further horizontally to cell *AR*. Here, the current row total (a_1) is 170 and the column total (b_3) is 350. So, assign 170 units and move vertically down to the cell *BR*. Now, the row and column totals are a_2 = 300 and b_3 = 180. Thus, assign 180 in the cell and shift to the cell *BS*. With current a_2 = 120 and b_4 = 320, allocate 120 units here. Finally, shift to the cell *CS* corresponding to which a_3 and b_4 are both equal to 200. So, assign this quantity to the cell. These assignments are shown in Table 5.3.

TABLE 5.3 Initial Feasible Solution: NWC Method

From ↓ / *To* →	*P*	*Q*	*R*	*S*	*Supply*
A	[180] 12	[150] 10	[170] 12	13	500
B	7	11	[180] 8	[120] 14	300
C	6	16	11	[200] 7	200
Demand	180	150	350	320	1000

Total cost = 12 × 180 + 10 × 150 + 12 × 170 + 8 × 180 + 14 × 120 + 7 × 200 = Rs 10,220

This routing of the units meets all the rim requirements and involves a total of 6 (= 3 + 4 – 1) shipments as there are six occupied cells. It involves a total cost of Rs 10,220. It may be mentioned that this method of obtaining initial solution is a relatively simple one but it is not considered very efficient in terms of cost minimising. This is because it takes into account only the available supply and demand requirements in making assignments and takes no account of the shipping costs involved. It thus ignores the very factor (i.e., the cost) which is sought to be minimised. The methods to be considered now do take costs into account.

Least Cost Method (LCM)

As the name suggests, this method entails making allocation at each step by selecting from the routes (that is, combinations of sources and destinations) available, the one with the minimum cost. To begin, we choose the cell which has the minimum cost, consider the supply available at the source (a_i) and the demand at the destination (b_j) corresponding to that, and allocate the lower of the two (a_i or b_j) to the cell. After this, we delete the row or column (or both if $a_i = b_j$) whichever is satisfied by this allocation. If the row is deleted, then the column value (b_j) is revised by subtracting the quantity assigned. Similarly, if the column is deleted, then the row total is revised. After this, we consider the routes remaining and again choose the one with the least cost, make assignment and adjust the row/column total. We continue in this manner until all the units are assigned.

It may be mentioned that at any stage, if there is a tie in the minimum cost, so that two or more routes have the same least cost of shipping, then, conceptually, either of them may be selected. However, a better initial solution is obtained if the route chosen is the one where largest quantity can be assigned. Thus, if there are three cells for which the (least) cost value is equal, then consider all of these one by one and determine the quantity (by reference to the demand and supply quantities given) which can be despatched, and choose the cell with the largest quantity. If there is still a tie, then either of them may be selected.

For Example 5.1, the initial solution by using least cost method is obtained as shown in Table 5.4. To begin with, the lowest of all cost elements is 6, for the cell *CP*, with corresponding supply and demand being 200 and 180 units respectively. Accordingly, assign 180 to this cell, delete the column headed *P* and adjust the supply in the row to 20. Of the remaining cost elements (excluding those of the deleted column), the least is 7, corresponding to the cell *CS*. The quantity assigned to this cell is 20, being lower of demand (= 320) and supply (= 20). Delete the row *C* and adjust the demand for *S* at 300. The least cost element in the remaining cells is 8, for *BR*. With the relevant supply and demand values equal to 300 and 350 units respectively, assign 300 to the cell *BR*, delete the row *B* and adjust the demand at *R* to 50. Now, with only one supply source remaining, the amount would be transferred to the current requiements at warehouses *Q*, *R*, and *S*: 150 to *Q*, 50 to *R*, and 300 to *S*.

TABLE 5.4 Obtaining Initial Feasible Solution: LC Method

From ↓ / To →	*P*	*Q*	*R*	*S*	*Supply*
A	12	[150] 10	[50] 12	[300] 13	500
B	7	11	[300] 8	14	300
C	[180] 6	16	11	[20] 7	~~200~~ 20
Demand	180	150	~~350~~ 50	~~320~~ 300	1000

The transportation schedule obtained is reproduced in Table 5.5. The solution has six occupied cells and it is seen to involve a total cost of Rs 9,620.

TABLE 5.5 Initial Feasible Solution: Least Cost Method

From ↓ / To →	*P*	*Q*	*R*	*S*	*Supply*
A	12	10 [150]	12 [50]	13 [300]	500
B	7	11	8 [300]	14	300
C	6 [180]	16	11	7 [20]	200
Demand	180	150	350	320	1000

Total cost = 10 × 150 + 12 × 50 + 13 × 300 + 8 × 300 + 6 × 180 + 7 × 20 = Rs 9,620

Vogel's Approximation Method (VAM)

The Vogel's Approximation Method, commonly known as VAM, is another cost-based method used for obtaining initial solution to a transportation problem. Actually, it uses cost differentials rather than absolute costs, to select the appropriate cell to make allocations. The method is shown schematically in Figure 5.2 and is explained here.

First, consider every row of the cost matrix individually and find the difference between two least cost cells in it. Then, repeat the exercise for every column. Consider all cost differences and identify the row or column with the largest value. Now, focus on the cell with the smallest cost in the row or column so chosen and make allocation in that. The number of units allocated would be lower of the corresponding supply and demand values. In case of a tie in the largest cost difference, although either of them may be chosen, it is preferable to choose the cost difference corresponding to which the largest number of units may be assigned or corresponding to which the cell chosen has minimum cost. To illustrate, suppose the largest cost difference is found to be tied for a row and a column. In applying the first of the rules, determine the quantity (considering the available supply and demand) which can be assigned in each case and select the one where larger quantity can be assigned. In the other case, unit cost values of the two least cost cells, in the row and column, and choose the one which has a lower value.

Delete the row or column which has been satisfied by the allocation and adjust the quantity of demand/supply. Recalculate the cost differences for the reduced matrix and proceed in the same manner as discussed earlier. Continue with the process until all units have been assigned. The initial solution to Example 5.1 using VAM is obtained in Table 5.6.

TABLE 5.6 Obtaining Initial Feasible Solution: VAM

To → / From ↓	P	Q	R	S	Supply	Iteration I	Iteration II	Iteration III
A	12	[150] 10	[230] 12	[120] 13	500	2	2	2
B	[180] 7	11	[120] 8	14	~~300~~ 120	1	1	3
C	6	16	11	[200] 7	200	1	–	–
Demand	180	150	~~350~~ 230	~~320~~ 120	1000			
I	1	1	3	(6)				
II	(5)	1	4	1				
III	–	1	(4)	1				

To begin with, the cost differences between the pairs of least cost cells are taken for each row and column (iteration I). The largest of these being 6 (= 13 – 7), the column designated *S* is selected. In the lowest-cost cell of this column, *CS*, a value 200 is assigned and the row C is deleted. The requirement at warehouse *S* is revised at 120. In iteration II, the cost differences are recalculated. The largest value is for the column headed *P* and the least cost cell is *BP*. Considering supply and demand values, 180 is allocated to this cell, column headed *P* is deleted and the supply at *B* revised down to 120. In the next iteration, the highest cost difference is 4 which is in respect of the column headed *R*. Here the least cost cell is *BR*, wherein a quantity 120 in the cell *BR* and the row designated *B* is deleted. The column total is put at 230. Finally, only one row is left and, therefore, no cost differences need to be recalculated. The allocations for each of the cells remaining are made having reference to the requirements: 150 to *AQ*, 230 to *AR* and 120 to *AS*.

It may be noted in operation of this method that when a row is deleted, the other row differences do not change and the column cost differences need to be recalculated, and if a column is deleted then the row differences require recalculation. In the event of both row and column being deleted (when $a_i = b_j$) then all cost differences need to be calculated again.

The solution obtained by VAM to the Example 5.1 is reproduced in Table 5.7. This also is seen to have a total of six occupied cells. It involves a total cost of Rs 9,440.

TABLE 5.7 Initial Feasible Solution: VAM

From ↓ / *To* →	*P*	*Q*	*R*	*S*	*Supply*
A	12	150 10	230 12	120 13	500
B	180 7	11	120 8	14	300
C	6	16	11	200 7	200
Demand	180	150	350	320	1000

Total cost = 10 × 150 + 12 × 230 + 13 × 120 + 7 × 180 + 8 × 120 + 7 × 200 = Rs 9,440

The following points may be noted.

(i) The Vogel's approximation method is also called the *Penalty Method* because the cost differences that it uses are nothing but the penalties of not using the least-cost routes. Since the objective function is the minimisation of the transportation cost, in each iteration that route is selected which involves the maximum penalty of *not* being used.

(ii) The initial solution (to Example 5.1) obtained by VAM is found to be the best of all as it involves the lowest total cost among all the three initial solutions. Of course, it does not come as a rule, but usually VAM provides the best initial solution. Therefore, this method is generally preferred in solving the problems.

Step 2: Testing the Optimality After obtaining the initial basic feasible solution, the next step is to test whether it is optimal or not. There are two methods of testing the optimality of a basic feasible solution. One of these is called the *stepping-stone method* in which optimality test is applied by calculating the opportunity cost of each empty cell. The other method is called as the *modified distribution method* (MODI). This one is easier to apply and more efficient than the stepping-stone method. It is based on the concept of dual variables that are used to evaluate the empty cells. Using these dual variables, the opportunity cost of each of the empty cells is determined. Thus, while both the methods involve determining opportunity costs of empty cells, the methodology employed by them differs. The opportunity cost values indicate whether the given solution is optimal or not.

Both the methods can be used only when the solution is a basic feasible solution, so that it has $m + n - 1$ basic variables. Refer to section 5.4.4 if a solution involves smaller than $m + n - 1$ number of occupied cells (basic variables). Incidently, none of the methods used to find initial solution would yield a solution with greater than $m + n - 1$ number of occupied cells.

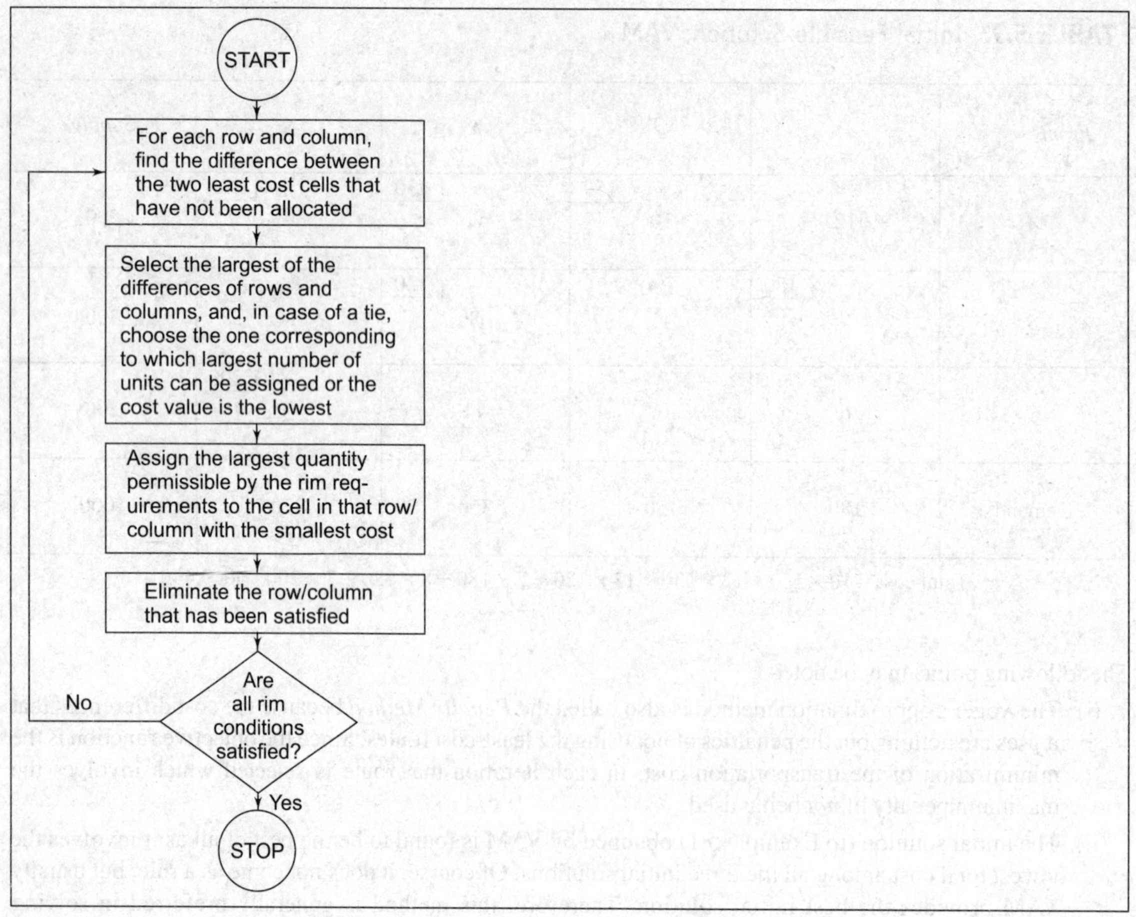

Figure 5.2 *Schematic Presentation of VAM*

Step 3: Improving the Solution By applying either of the methods, if the solution is found to be optimal, then the process terminates as the problem is solved. If the solution is not seen to be optimal, then a revised and improved basic feasible solution is obtained. This is done by exchanging a non-basic variable for a basic variable. In simple terms, re-arrangement is made by transferring units from an occupied cell to an empty cell that has the largest opportunity cost, and then adjusting the units in other related cells in a way that all the rim requirements are satisfied. This is done by first tracing a *closed loop* (discussed in detail later).

The solution so obtained is again tested for optimality (step 2) and revised, if necessary. We continue in this manner until an optimal solution is finally obtained.

We now discuss the stepping-stone and the MODI methods in turn.

Stepping-stone Method

This is a procedure for determining the potential, if any, of improving upon each of the non-basic variables in terms of the objective function. To determine this potential, each of the non-basic variables (empty cells) is considered one-by-one. For each such cell, we find what effect on the total cost would be if one unit is assigned to this cell. With this information, then, we come to know whether the solution is optimal or not. If not, we improve that solution. To understand it, we refer to the Table 5.8 which is a reproduction of the Table 5.3 containing the initial basic feasible solution obtained by NW Corner Rule.

TABLE 5.8 Initial Feasible Solution: Testing for Optimality

From ↓ / To →	P	Q	R	S	Supply
A	180 12	150 10	170 12 −	13 +	500
B	7	11	180 8 +	120 14 −	300
C	6	16	11	200 7	200
Demand	180	150	350	320	1000

Total cost = 12 × 180 + 10 × 150 + 12 × 170 + 8 × 180 + 14 × 120 + 7 × 200 = Rs 10,220

In this table, there are six empty cells: *AS, BP, BQ, CP, CQ* and *CR,* with corresponding non-basic variables as $x_{14}, x_{21}, x_{22}, x_{31}, x_{32}$ and x_{33}. To start with, let us assess the potential for improvement of the non-basic variable x_{14}. Unfortunately, the effect of sending one unit on this route cannot be directly identified because the supplies of the source *A* are already assigned to *P*, *Q* and *R* while the demand at *S* is to be met from *B* and *C*. To assign a unit in this cell, therefore, needs adjustments among some of the already occupied cells. To illustrate, a unit in *AS* can be assigned by diverting (reducing) from *AR*. This would necessitate increase of one unit in the cell *BR* (to maintain the column total) as also reduction from the cell *BS* to bring balance in the row *B* and column *S* totals. These adjustments are shown marked by arrows in the table, forming a closed loop.

Having traced the path of adjustments, we can now analyse the cost effect. A shipment of one item on the route *AS* would cause an increase of Rs 13, but it will also mean a reduction of one unit from x_{13} and, thereby, a reduction of Rs 12. Further, an increase of a unit in x_{23} would raise the cost by Rs 8 while a unit lesser in x_{24} would save Rs 14. The net effect of the operation would be a saving of Rs 5 (= 13 – 12 + 8 – 14). This implies that a reduction of Rs 5 can be effected by adopting the route *AS*. The opportunity cost of this route is, therefore, Rs 5. Since the opportunity cost here is positive, it means that it is worth considering making variable x_{14} a basic variable.

Before we digress on the drawing of a closed loop, we evaluate each of the remaining empty cells in Table 5.8 in the manner we did for the cell *AS*. This is done here:

Cell	Closed Loop	Net Cost Change			Opportunity Cost
BP	BP– BR – AR – AP	7 – 8 + 12 – 12	=	–1	1
BQ	BQ – BR – AR – AQ	11 – 8 + 12 – 10	=	5	–5
CP	CP – CS – BS – BR – AR – AP	6 – 7 + 14 – 8 + 12 – 12	=	5	–5
CQ	CQ – CS – BS – BR – AR – AQ	16 – 7 + 14 – 8 + 12 – 10	=	17	– 17
CR	CR – CS – BS – BR	11 – 7 + 14 – 8	=	10	– 10

Tracing a Closed Loop

When we intend to assign some units to an unoccupied cell, it calls for making adjustments in some other occupied cells because such an assignment affects the row and column totals of supply and demand respectively that need to be maintained for a feasible solution. In effect, the assignment of one or more units in a (unoccupied) cell produces a chain effect and a closed loop, or a *closed path* as it is called, attempts to capture this.

When a closed loop is to be drawn, start with the empty cell to be evaluated (or intended to be included in the solution). Then, moving clockwise, draw an arrow from this cell to an occupied cell in the same row or column, as the case may be. After that, move vertically (or horizontally) to another occupied cell and draw an arrow. Again, move horizontally (or vertically) to another occupied cell and continue the procedure until reaching the original empty cell. Essentially then, for drawing a closed loop, begin with an empty cell and move alternately horizontally and vertically, through occupied cells only, until reaching back to the starting point. In the process of moving from one occupied cell to another, (a) move only horizontally or vertically, but never diagonally; and (b) step over empty and, if the need be, over occupied cells without changing them. Thus, a closed loop would always have right-angled turns with corners only on the occupied cells.

Having traced the path, place *plus* and *minus* signs alternately in the cells on each turn of the loop, beginning with a plus (+) sign in the empty cell. An important restriction is that there must be exactly one cell with a plus sign and one cell with a minus sign in any row or column in which the loop takes a turn. This restriction ensures that the rim requirements would not be violated when units are shifted among cells.

The following points may also be noted in connection with the closed loops:

(a) An even number of at least four cells must participate in a closed loop and an occupied cell can be considered only once and not more.

(b) If there exists a basic feasible solution with $m + n - 1$ positive variables, then there would be one and only one closed loop for each cell. This is irrespective of the size of the matrix given.

(c) All cells that receive a plus or a minus sign, except the starting empty cell, must be the occupied cells.

(d) Closed loops may or may not be square or rectangular in shape. In larger transportation tables, the closed loops may have peculiar configurations and a loop may cross over itself (see Example 5.7).

(e) Although, as mentioned earlier, movement on the path set by the loop is generally clockwise, even if the progression on the path is anticlockwise, it would not affect the result.

Continuing with testing optimality, the rule is: if none of the empty cells has a positive opportunity cost, the solution is optimal. The solution in question, therefore, is not optimal. Now, we proceed to revise this solution and see how an improved solution may be obtained. For the purpose of revising a given solution, the cells for which the opportunity cost is negative are not considered because their inclusion would increase the transportation cost. If a cell has a zero opportunity cost, it has the implication that the inclusion of that particular route would leave the objective function value unaffected. We focus, therefore, only on the cells that have positive opportunity cost and would choose the one with the highest opportunity cost as it would lead to largest per unit cost reduction. Remember that in one iteration only one of the cells is selected although multiple number of cells might have cost-reduction potential.

In our illustration, the most favourable cell is *AS* for it has the largest opportunity cost, equal to 5. We decide, therefore, to include x_{14} as a basic variable in the solution. Further, since each unit shipped on this route has the effect of saving the cost by Rs 5, we should transfer the largest number of units possible through it. To determine this number, understand that in a closed loop, the cells with a '+' sign are the ones where units are to be shifted while the cells with a '–' sign are those wherefrom the units are to be shifted. Among the cells with a '–', we look for the one with *minimum* quantity. This is the maximum number of units which can be shifted over the cells on the closed loop. Accordingly, in case of the closed loop in question, this number is 120 (being lower of 120 and 170). Shifting these units results in the revised solution contained in Table 5.9. Notice the variable x_{14} has replaced x_{24}.

TABLE 5.9 Improved Solution: Non-Optimal

From ↓ / *To* →	*P*	*Q*	*R*	*S*	*Supply*
A	180 12	150 10	50 12	120 13	500
B	7	11	300 8	14	300
C	6	16	11	200 7	200
Demand	180	150	350	320	1000

Total cost = 12 × 180 + 10 × 150 + 12 × 50 + 13 × 120 + 8 × 300 + 7 × 200 = Rs 9,620

This solution involves a total cost of Rs 9, 620 which is lower by Rs 600 (= 5 × 120) in comparison to the initial solution. However, it is not sure if this solution is optimal. We again apply step 2 to determine optimality. Application of this yields the following results:

Cell	Closed Loop	Net Cost Change	Opportunity Cost
BP	*BP– BR – AR – AP*	7 – 8 + 12 – 12 = – 1	+ 1
BQ	*BQ – BR – AR – AQ*	11 – 8 + 12 – 10 = + 5	– 5
BS	*BS – AS – AR – BR*	14 – 13 + 12 – 8 = + 5	– 5
CP	*CP – CS – AS – AP*	6 – 7 + 13 – 12 = 0	0
CQ	*CQ – CS – AS – AQ*	16 – 7 + 13 – 10 = + 12	– 12
CR	*CR – CS – AS – AR*	11 – 7 + 13 – 12 = + 5	– 5

Since a cell has a positive opportunity cost, the solution is not optimal. Further, since the cell *BP* has it, we will include this cell in the transportation schedule. With the cells *BR* (300) and *AP* (180) with a '–' sign, the quantity to be shifted along the closed loop is 180. The revised solution with this change is given in Table 5.10.

TABLE 5.10 Improved Solution: Optimal

From ↓ / To →	P	Q	R	S	Supply
A	12	150 10	230 12	120 13	500
B	180 7	11	120 8	14	300
C	6	16	11	200 7	200
Demand	180	150	350	320	1000

Total cost = 10 × 150 + 12 × 230 + 13 × 120 + 7 × 180 + 8 × 120 + 7 × 200 = Rs 9,440

This solution also involves six occupied cells and involves a total cost of Rs 9,440. The solution would now be tested for optimality. This is done here:

Cell	Closed Loop	Net Cost Change	Opportunity Cost
AP	*AP– AR – BR – BP*	12 – 12 + 8 – 7 = + 1	– 1
BQ	*BQ – BR – AR – AQ*	11 – 8 + 12 – 10 = + 5	– 5
BS	*BS – AS – AR – BR*	14 – 13 + 12 – 8 = + 5	– 5
CP	*CP – CS – AS – AR – BR – BP*	6 – 7 + 13 – 12 + 8 – 7 = + 1	– 1
CQ	*CQ – CS – AS – AQ*	16 – 7 + 13 – 10 = + 12	– 12
CR	*CR – CS – AS – AR*	11 – 7 + 13 – 12 = + 5	– 5

Since the opportunity costs of all the empty cells are negative, the solution obtained is optimal. A careful look at the solution reveals that it is identical to the initial solution obtained by VAM as given in Table 5.7. Thus, the initial solution by applying VAM is optimal and does not require any improvement.

Before discussing MODI method of testing optimality, we consider the initial solution by least cost method and test its optimality using stepping stone method. The solution is reproduced in Table 5.11.

TABLE 5.11 Initial Feasible Solution (LCM): Testing for Optimality

From ↓ \ To →	P	Q	R	S	Supply
A	12	150 10	50 12	300 13	500
B	7	11	300 8	14	300
C	180 6	16	11	20 7	200
Demand	180	150	350	320	1000

Total cost = 10 × 150 + 12 × 50 + 13 × 300 + 8 × 300 + 6 × 180 + 7 × 20 = Rs 9,620

The empty cells are considered one by one and the results are given here:

Cell	*Closed Loop*	*Net Cost Change*		*Opportunity Cost*
AP	*AP– AS – CS – CP*	12 – 13 + 7 – 6	= 0	0
BP	*BP – CP – CS – AS – AR – BR*	7 – 6 + 7 – 13 + 12 – 8	= – 1	+ 1
BQ	*BQ – BR – AR – AQ*	11 – 8 + 12 – 10	= + 5	– 5
BS	*BS – AS – AR – BR*	14 – 13 + 12 – 8	= + 5	–5
CQ	*CQ – CS – AS – AQ*	16 – 7 + 13 – 10	= + 12	– 12
CR	*CR – CS – AS – AR*	11 – 7 + 13 – 12	= +5	–5

The solution is seen to be non-optimal. Considering the cell *BP*, the only one with a positive opportunity cost, we shift 180 units along the closed loop *BP – CP – CS – AS – AR – BR*. The revised solution is presented in Table 5.12.

TABLE 5.12 Improved Solution: Optimal

From ↓ / To →	P	Q	R	S	Supply
A	12	150 10	230 12	120 13	500
B	180 7	11	120 8	14	300
C	6	16	11	200 7	200
Demand	180	150	350	320	1000

Total cost = 10 × 150 + 12 × 230 + 13 × 120 + 7 × 180 + 8 × 120 + 7 × 200 = Rs 9,440

It may be noted that this solution matches the one given in Table 5.10 and hence is optimal.

Modified Distribution Method (MODI)

As mentioned previously, the MODI method is an efficient method of testing the optimality of a transportation solution. It may be recalled that in the application of the stepping stone method, each of the empty cells is evaluated for the opportunity cost by drawing a closed loop. In situations where a large number of sources and destinations are involved, this would be a very time-consuming and intricate exercise. The MODI method avoids this kind of extensive scanning and reduces the number of steps required in the evaluation of the empty cells. This method gives a straightforward computational scheme whereby we can determine the opportunity cost of each of the empty cells. The method is shown schematically in Fig. 5.3 and its operation is discussed and illustrated here.

Step 1 Add to the transportation tableau a column on the RHS titled u_i and a row in the bottom of it labelled v_j. The objective is to obtain a value for each row and column.

Step 2 Calculate the u_i and v_j values with the help of occupied cells only. For this, assign a value arbitrarily to a row or column variable u_i or v_j. The value may be positive, negative or zero. Generally, a zero is assigned to the first row i.e., $u_1 = 0$.

After assigning initial value, consider the row (column) to which this is assigned, select an occupied cell in it and place in the corresponding column (row) a value such that the row and column values add up to the cost element of the cell. Repeat the exercise for any other occupied cells in the row (column) in consideration. Next, using the u_i and v_j values obtained, determine the remaining values of these variables with the help of other occupied cells. At the end, make sure that for each of the occupied cells in the table, the row value (u_i) and column value (v_j) equals the cost element c_{ij}. Thus, $u_i + v_j = c_{ij}$ for every occupied cell. Remember, however, that all the u_i and v_j values can be obtained only when the solution is a basic feasible solution, with $m + n - 1$ occupied cells in a $m \times n$ problem. If the occupied cells are smaller than $m + n - 1$ in number, the solution is called degenerate. This is discussed in Section 5.4.4.

The assignment of u_i and v_j values is illustrated using initial solution to Example 5.1 by North West Corner rule, which is reproduced in Table 5.13. To begin with, set $u_1 = 0$. Now, since the cells *AP*, *AQ* and *AR* in this row are all occupied, consider them one-by-one and obtain the corresponding column values, v_j. Accordingly, $v_1 = 12$ since $c_{11} = 12$; $v_2 = 10$ since $c_{12} = 10$ and $v_3 = 12$ because $c_{13} = 12$. Next, v_3 can be used to find u_2. With $c_{23} = 8$, $u_2 = 8 - 12 = -4$. Similarly, with $c_{24} = 14$, $v_4 = 14 - (-4) = 18$, and $u_3 = c_{34} - v_4 = 7 - 18 = -11$.

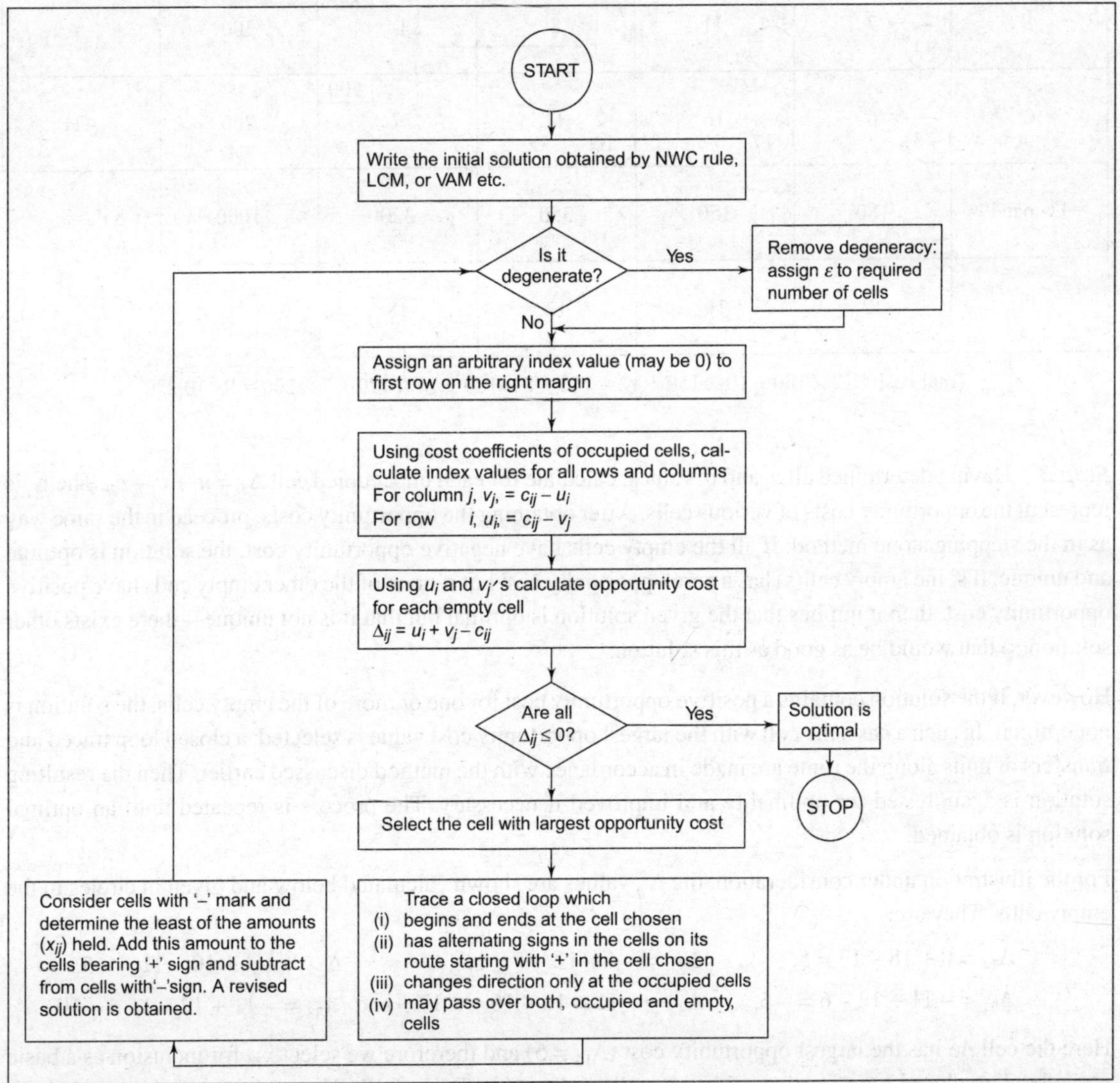

Figure 5.3 *Schematic Presentation of MODI*

TABLE 5.13 Initial Solution: Testing for Optimality

To → / From ↓	P	Q	R	S	Supply	u_i
A	[180] 12	[150] 10	[170] 12 –	13 + (+5)	500	0
B	7 (+1)	11 (–5)	[180] 8 +	[120] 14 –	300	–4
C	6 (–5)	16 (–17)	11 (–10)	[200] 7	200	–11
Demand	180	150	350	320	1000	
v_j	12	10	12	18		

Total cost = 12 × 180 + 10 × 150 + 12 × 170 + 8 × 180 + 14 × 120 + 7 × 200 = Rs 10,220

Step 3 Having determined all u_i and v_j values, calculate for each unoccupied cell $\Delta_{ij} = u_i + v_j - c_{ij}$. The Δ_{ij}'s represent the opportunity costs of various cells. After obtaining the opportunity costs, proceed in the same way as in the stepping stone method. If all the empty cells have negative opportunity cost, the solution is optimal and unique. If some empty cell(s) has a zero opportunity cost but if none of the other empty cells have positive opportunity cost, then it implies that the given solution is optimal but that it is not unique—there exists other solution(s) that would be as good as this solution.

However, if the solution contains a positive opportunity cost for one or more of the empty cells, the solution is not optimal. In such a case, the cell with the largest opportunity cost value is selected, a closed loop traced and transfers of units along the route are made in accordance with the method discussed earlier. Then the resulting solution is again tested for optimality and improved if necessary. The process is repeated until an optimal solution is obtained.

For the illustration under consideration, the Δ_{ij} values are shown calculated below and given in circles in the empty cells. They are:

$\Delta_{14} = 0 + 18 - 13 = 5$, $\Delta_{21} = -4 + 12 - 7 = 1$, $\Delta_{22} = -4 + 10 - 11 = -5$,
$\Delta_{31} = -11 + 12 - 6 = -5$, $\Delta_{32} = -11 + 10 - 16 = -17$ and $x_{33} = -11 + 12 - 11 = -10$.

Here the cell *AS* has the largest opportunity cost ($\Delta_{14} = 5$) and therefore we select x_{14} for inclusion as a basic variable. The closed loop starting with this cell is shown in Table 5.13. The revised solution is given in Table 5.14.

TABLE 5.14 Improved Solution: Non-optimal

To → / From ↓	P	Q	R	S	Supply	u_i
A	180 12	150 10	50 12	120 13	500	0
B	7 (+1)	11 (−5)	300 8	14 (−5)	300	−4
C	6 (0)	16 (−12)	11 (−5)	200 7	200	−6
Demand	180	150	350	320	1000	
j	12	10	12	13		

Total cost = 12 × 180 + 10 × 150 + 12 × 50 + 13 × 120 + 8 × 300 + 7 × 200 = Rs 10,220

The solution in this table is tested for optimality and is found to be non-optimal. Here, the cell *BP* has a positive opportunity cost and, therefore, a closed loop is traced starting with it. This enables 180 units to be transferred to this route. The resulting solution is given in Table 5.15.

TABLE 5.15 Improved Solution: Optimal

To → / From ↓	P	Q	R	S	Supply	u_i
A	12 (−1)	150 10	230 12	120 13	500	0
B	180 7	11 (−5)	120 8	14 (−5)	300	−4
C	6 (−1)	16 (−12)	11 (−5)	200 7	200	−6
Demand	180	150	350	320	1000	
v_j	11	10	12	13		

Total cost = 10 × 150 + 12 × 230 + 13 × 120 + 7 × 180 + 8 × 120 + 7 × 200 = Rs 9,440

This solution is found to be optimal. It involves a total cost of Rs 9,440. Further, since all Δ_{ij} values are negative, the solution is unique.

The following points may be noted:

1. The solution given in Table 5.15 is the same as the initial solution obtained using VAM. Hence, the solution obtained by VAM is also optimal and needs no revision.
2. The opportunity costs values obtained both by stepping stone and MODI methods are the same. The difference lies in the way they are calculated. In stepping stone method, they are obtained by drawing closed paths (loops) whereas in MODI method, they are derived with the help of u_i and v_j values.
3. The u_i and v_j values, also called the *index values*, in a given solution are not independent. On the contrary, they are interdependent and a change in the initial index value, or the location of its insertion, causes all of them to change systematically. To illustrate, if the initial row value is increased by k, then each of the row values will be up by k, while each of the column values will be down by this amount. It can be verified in respect of Table 5.15 where it can be seen that if the initial value u_1 is set to equal to, say 8, then the u_i and v_j values get revised as: $u_2 = 4$, $u_3 = 2$, $v_1 = 3$, $v_2 = 2$, $v_3 = 4$ and $v_4 = 5$. With the revised values, there is no change in the Δ_{ij} values.
4. Both the methods involve drawing closed loops. The stepping stone method requires them for determining opportunity cost values and choosing one of them for improving upon the solution if it is not optimal. On the other hand, MODI method requires tracing a closed loop only for improving a non-optimal solution.
5. We have seen earlier that a transportation problem can be expressed as a linear programming problem. Further, like any other problem, this LPP also would have a dual (see Section 5.5). The u_i and v_j values in the optimal solution tableau are in fact optimal values of the dual variables. This implies that the summation of the products of the various availability (or supply) values and the corresponding u_i values, together with the demand values and the corresponding v_j values, would match the total cost value in respect of the optimal solution. Accordingly, from Table 5.15, we have: $500 \times 0 + 300\,(-4) + 200\,(-6) + 180 \times 11 + 150 \times 10 + 350 \times 12 + 320 \times 13 = 9440$, which is equal to the total cost.

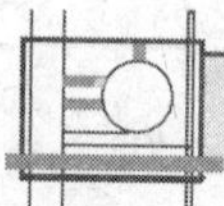

5.4 SOME SPECIAL TOPICS

5.4.1 Unbalanced Transportation Problems

The methods we have discussed for solving a transportation problem require that the aggregate supply (Σa_i) is equal to the aggregate demand (Σb_j). In practice, however, situations may arise when the two are unequal. The two such possibilities are, first, when the aggregate supply exceeds the aggregate demand (i.e. $\Sigma a_i > \Sigma b_j$) and, second, when aggregate supply falls short of the aggregate demand (i.e. $\Sigma a_i < \Sigma b_j$). Such problems are called unbalanced transportation problems. Balancing must be done before they can be solved.

When the aggregate supply exceeds the aggregate demand, the excess supply is assumed to go to inventory and costs nothing for shipping. A column of slack variables is added to the transportation tableau which represents a dummy destination with a requirement equal to the amount of excess supply and the transportation costs equal to zero. On the other hand, when the aggregate demand exceeds the aggregate supply in a transportation problem, balance is restored by adding a dummy origin. The row representing it is added with an assumed total availability equal to the difference between the total demand and supply, and with each of the cells having a zero unit cost. In some cases, however, when the penalty of *not* satisfying the demand at a particular destination(s) is given, then such penalty value should be considered and not zero.

Once the transportation problem is balanced, its solution proceeds in exactly the same manner as discussed earlier.

5.4.2 Prohibited Routes

Sometimes in a given transportation problem some route(s) may not be available. This could be due to a variety of reasons like the unfavourable weather conditions on a particular route, strike on a particular route, etc. In such situations, there is a restriction on the routes available for transportation. To handle a situation of this type, we assign a very large cost represented by M to each of such routes which are not available. Then the problem is solved in the usual way. The effect of adding a large cost element would be that such routes would automatically be eliminated in the final solution.

5.4.3 Unique vs Multiple Optimal Solutions

The optimal solution to a given problem may, or may not, be unique. Recall that the solution to a transportation problem is optimal if all the Δ_{ij} values (in terms of the MODI method) are less than, or equal to zero. For a given solution, if all the Δ_{ij} values are negative, then it is unique. If, however, some cell (or cells) has $\Delta_{ij} = 0$, then multiple optimal solutions are indicated so that there exist transportation pattern(s) other than the one obtained which can satisfy all the rim requirements for the same cost.

To obtain an alternate optimal solution, trace a closed loop beginning with a cell having $\Delta_{ij} = 0$, and get the revised solution in the same way as a solution is improved. It may be observed that this revised solution would also entail the same total cost as before. It goes without saying that for every 'zero' value of Δ_{ij} value in the optimal solution tableau, a revised solution can be obtained.

Example 5.2 Given the following transportation problem:

Warehouse	*Market*			*Supply*
	A	*B*	*C*	
1	10	12	7	180
2	14	11	6	100
3	9	5	13	160
4	11	7	9	120
Demand	240	200	220	

It is known that currently nothing can be sent from warehouse 1 to market *A* and from warehouse 3 to market *C*. Solve the problem and determine the least cost transportation schedule. Is the optimal solution obtained by you unique? If not, what is/are the other optimal solution/s?

The given transportation problem is unbalanced since aggregate demand = 240 + 200 + 220 = 660 units while the aggregate supply = 180 + 100 + 160 + 120 = 560 units. Thus, a dummy warehouse (5) is introduced with supply of 660 – 560 = 100 units, and zero transportation cost since no penalties are provided for not satisfying the demand at each market. Further, since routes 1-*A* and 3-*C* are given as prohibited, the cost element for each of these is replaced by *M*. The information is presented in Table 5.16.

TABLE 5.16 Transportation Problem: Unbalanced, with Prohibited Routes

Warehouse	*Market* A	B	C	*Supply*
1	M	12	7	180
2	14	11	6	100
3	9	5	M	160
4	11	7	9	120
5	0	0	0	100
Demand	240	200	220	660

The initial solution is contained in Table 5.17. It is obtained using VAM.

TABLE 5.17 Initial Basic Feasible Solution by VAM

Warehouse	*Market* A	B	C	*Supply*	*Iteration* I	II	III	IV	V
1	*M*	12	7 [180]	180	5	(5)	–	–	–
2	14 [60]	11	6 [40]	100 / 60	5	5	(5)	3	3
3	9	5 [160]	*M*	160	4	4	4	(4)	–
4	11 [80]	7 [40]	9	120 / 80	2	2	2	4	4
5	0 [100]	0	0	100	0	–	–	–	–
Demand	240 / 140	200 / 40	220 / 40	660					
I	(9)	5	6						
II	2	2	1						
III	2	2	3						
IV	2	2	-						
V	3	(4)	-						

In the first step, cost differences are calculated and the largest one, 9, is selected and 100 units are placed in the cell 5-*A*. The row 5 is deleted and differences are recalculated. Now, a tie is observed in the largest value of the differences, that is 5. If the row 1 is selected then the appropriate cell would be 1-*C*, where, considering demand and supply, 180 units can be sent, while if row 2 is chosen, then in the least cost cell 2-*C*, 100 units can be placed. Thus, we may preferably choose 1-*C*. After this, we proceed in the usual manner-cost differences are recalculated and quantities assigned to the least cost cells of the rows/columns selected. The solution is reproduced in Table 5.18 and tested for optimality. Since all $\Delta_{ij} \leq 0$, the solution is seen to be optimal. Note that Δ_{ij} is not calculated for the prohibited cells *1-A* and 3-*C*. Even if Δ_{ij} is calculated for a prohibited route, it is bound to be extremely negative since it would involve '–*M*'.

The solution given in Table 5.18 is not unique because in cell 3-*A*, $\Delta_{ij} = 0$. To obtain an alternative optimal solution, we draw a closed loop as shown in this table. The revised solution is given in Table 5.19. It also involves a total cost of Rs 4,300.

TABLE 5.18 Initial Feasible Solution: Optimal

Warehouse	*Market*			*Supply*	u_i
	A	*B*	*C*		
1	*M*	12 (−1)	7 [180]	180	15
2	14 [60]	11 (−1)	6 [40]	100	14
3	9 (0) +	5 [160] −	*M*	160	9
4	11 [80] −	7 [40] +	9 (−6)	120	11
5	0 [100]	0 (−4)	0 (−8)	100	0
Demand	240	200	220	660	
v_j	0	− 4	− 8		

5.4.4 Degeneracy

We have already seen that a basic feasible solution of a transportation problem has $m + n - 1$ basic variables, which means that the number of occupied cells in such a solution is *one* less than the number of rows plus the number of columns. It may happen sometimes that the number of occupied cells is smaller than $m + n - 1$. Such a solution is called a *degenerate solution.*

Degeneracy in a transportation problem can figure in two ways. Degeneracy can arise in the first instance when an initial feasible solution is obtained (see Example 5.3). Secondly, the problem may become degenerate when two or more cells are vacated simultaneously in the process of transferring units along the closed path.

TABLE 5.19 Alternate Optimal Solution

Warehouse	Market A	Market B	Market C	Supply	u_i
1	M	(−1) 12	[180] 7	180	15
2	[60] 14	(−1) 11	[40] 6	100	14
3	[80] 9	[80] 5	M	160	9
4	(0) 11	[120] 7	(−6) 9	120	11
5	[100] 0	(−4) 0	(−8) 0	100	0
Demand	240	200	220	660	
v_j	0	−4	−8		

The difficulty, when a solution is degenerate, is that it cannot be tested for optimality. Both, the stepping stone method and the modified distribution method (MODI) are inoperative in such a case. The former cannot be applied because for some of the empty cells the closed loops cannot be traced, while the latter fails because all u_i and v_j values cannot be determined.

To overcome degeneracy, we proceed by assigning an infinitesimally small amount, close to zero, to one (or more, if the need be) empty cell and treat the cell as an occupied cell. This amount is represented by a Greek letter e (epsilon) and is taken to be such an insignificant value that would not affected the total cost. Thus, it is big enough to cause the particular route to which it is assigned to be considered as a basic variable but not large enough to cause a change in the toal cost and the other non-zero amounts. Although e is, theoreticaly, non-zero, the operations with it in the context of problem at hand are given here:

$$k + \varepsilon = k; \qquad \varepsilon + \varepsilon = \varepsilon;$$
$$k - \varepsilon = k; \qquad \varepsilon - \varepsilon = 0; \text{ and}$$
$$0 + \varepsilon = \varepsilon; \qquad k \times \varepsilon = 0.$$

It is important to remember that an epsilon cannot be placed in any randomly selected unoccupied cell. Instead, the cell chosen for inserting epsilon must be an independent cell, originating from which a closed loop cannot be traced. Placing an epsilon in a cell that is not independent will fail to yield unique u_i and v_j values and, consequently, would not resolve the problem of degeneracy. It may further be noted that if a given problem requires two (or more) epilons, then a cell in which an epsilon has already been placed will be treated as occupied while determining independence of cells for inserting an epsilon subsequently.

While a number of independent cells may be available in a given situation, it may be a good idea to choose the one with the smallest unit cost value. From a practical standpoint, it is not necessary to evaluate each and every empty cell for independence to insert epsilon. Instead, a cell with minimum cost may be examined whether it is independent or not. If it is found to be independent, then place epsilon in it and if not then move to another cell.

After placing epsilon(s) in the required number of cells, we proceed to solve the problem in the usual manner.

Example 5.3 Determine optimal solution to the problem given below. Obtain the initial solution by VAM.

		To Market				
		M_1	M_2	M_3	M_4	Supply
	P_1	6	4	9	1	40
From	P_2	20	6	11	3	40
Plant	P_3	7	1	0	14	50
	P_4	7	1	12	6	90
Demand		90	30	50	30	

Since the aggregate supply is 220 units and the aggregate demand is 200 units, we introduce a dummy market, M_5, for an amount equal to 20 (the difference between the aggregate supply and demand), with all cost elements equal to zero. The initial solution is given in Table 5.20.

It is obtained using VAM and its degenerate because it contains only seven basic variables, since there are only 7 occupied cells and not 8 (= 4 + 5 – 1), required for a non-degenerate solution in this case. Here empty cells P_1M_3, P_2M_3, P_3M_1, P_3M_2, P_3M_4, P_3M_5 and P_4M_3 are independent, while others are not. For removing degeneracy, let us place ε in the cell P_3M_5 and then test it for optimality. This is done in Table 5.20. (It may be interesting for the reader to try by inserting ε in any of the cells which are not independent and see the result.) The solution is found to be non-optimal. The improved solution is given in Table 5.21.

TABLE 5.20 Initial Feasible Solution: Non-optimal

From ↓ / To →	M_1	M_2	M_3	M_4	M_5	Supply	u_i
P_1	[30] 6 (–)	4 (–4)	9 (–11)	[10] 1 (+)	0 (–2)	40	0
P_2	20 (–12)	6 (–4)	11 (–11)	[20] 3 (–)	[20] 0 (+)	40	2
P_3	7 (1) (+)	1 (1)	[50] 0	14 (–11)	[ε] 0 (–)	50	2
P_4	[60] 7	[30] 1	12 (–13)	6 (–4)	0 (–1)	90	1
Demand	90	30	50	30	20	220	
v_j	6	0	– 2	1	– 2		

Total cost: 6 × 30 + 1 × 10 + 3 × 20 + 0 × 20 + 0 × 50 + 7 × 60 + 1 × 30 = Rs 700

TABLE 5.21 Revised Solution: Optimal

From ↓ / To →	M_1	M_2	M_3	M_4	M_5	Supply	u_i
P_1	30 / 6	(−4) 4	(−10) 9	10 / 1	(−2) 0	40	0
P_2	(−12) 20	(−4) 6	(−10) 11	20 / 3	20 / 0	40	2
P_3	ε / 7	(0) 1	50 / 0	(−12) 14	(−1) 0	50	1
P_4	60 / 7	30 / 1	(−12) 12	(−4) 6	(−1) 0	90	1
Demand	90	30	50	30	20	220	
v_j	6	0	−1	1	−2		

Total cost = 6 × 30 + 1 × 10 + 3 × 20 + 0 × 50 + 7 × 60 + 1 × 30 = Rs 700

Notice that in improving the solution, only ε moves, as this is the minimum quantity in the cells bearing '–' sign on the closed loop. The optimality test, MODI, suggests that the solution given in Table 5.21 is an optimal one. The solution can be stated as:

$$x_{11} = 30, x_{14} = 10, x_{24} = 20, x_{33} = 50, x_{41} = 60, x_{42} = 30, \text{ all other } x_{ij} = 0; \text{ Total cost} = \text{Rs } 700.$$

When the problem becomes degenerate at the solution-revision stage, epsilon (ε) is placed in one (or more, if necessary) of the recently vacated cells, with the minimum cost. And then we proceed with the problem in the usual manner.

Example 5.4 Solve the following transportation problem. Obtain the initial solution by NW corner rule.

		To 1	2	3	4	Supply
From	A	7	3	8	6	60
	B	4	2	5	10	100
	C	2	6	5	1	40
Demand		20	50	50	80	200

The initial basic feasible solution is given in Table 5.22. It is *not* optimal.

TABLE 5.22 Initial Basic Feasible Solution: NW Corner Rule

From ↓ / To →	1	2	3	4	Supply	u_i
A	[20] 7	[40] 3 −	8 (−2)	6 + (5)	60	0
B	4 (2)	[10] 2 +	[50] 5	[40] 10 −	100	−1
C	2 (−5)	6 (−13)	5 (−9)	[40] 1	40	−10
Demand	20	50	50	80	200	
v_j	7	3	6	11		

Total cost = 7 × 20 + 3 × 40 + 2 × 10 + 5 × 50 + 10 × 40 + 1 × 40 = Rs 970

This revised solution is degenerate because there are only 5 occupied cells and not 6. The two recently vacated cells are A2 and B4. The per unit cost of these being Rs 3 and Rs 10, respectively, we may assign to the cell A2 to cope with degeneracy. Thus, solution is also found to be non-optimal. Hence, we move to the next iteration and further revise the solution as shown in Table 5.24.

TABLE 5.23 Revised Solution: Non-optimal

From ↓ / To →	1	2	3	4	Supply	u_i
A	[20] 7 −	[ε] 3 +	8 (−2)	[40] 6	60	0
B	4 + (2)	[50] 2 −	[50] 5	10 (−5)	100	−1
C	2 (0)	6 (−8)	5 (−4)	[40] 1	40	−5
Demand	20	50	50	80	200	
v_j	7	3	6	6		

Total cost = 7 × 20 + 6 × 40 + 2 × 50 + 5 × 50 + 1 × 40 = Rs 770

TABLE 5.24 Revised Solution: Optimal

From ↓ / To →	1	2	3	4	Supply	u_i
A	7 (−2)	3 [20]	8 (−2)	6 [40]	60	0
B	4 [20]	2 [30]	5 [50]	10 (−5)	100	− 1
C	2 (−2)	6 (−8)	5 (−4)	1 [40]	40	− 5
Demand	20	50	50	80	200	
v_j	5	3	6	6		

Total cost = 3 × 20 + 6 × 40 + 4 × 20 + 2 × 30 + 5 × 50 + 1 × 40 = Rs 730

The optimality test indicates that this solution is optimal. The optimal transportation plan, then, is:

$x_{12} = 20, x_{14} = 40, x_{21} = 20, x_{22} = 30, x_{23} = 50, x_{34} = 40$, all other $x_{ij} = 0$, for a total cost of Rs 730.

5.4.5 Maximisation Problem

As previously mentioned, the classical transportation problem is one of the minimisation type. However, a transportation tableau may contain unit profits instead of unit costs and the objective function be maximisation of total profits. To solve a maximisation type of problem, the transportation method is not applied directly. The given problem is first converted into an equivalent minimisation problem. This is done by subtracting each element of the given (profit) matrix from a constant value *k*, which can be any number. Usually, to avoid working with negative values, the profit values are subtracted from the largest one among them. The procedure yields what is termed as the *opportunity loss matrix* to which the transportation method is applied. The entries in the opportunity loss matrix indicate how much each of the values is away from that constant (the highest profit value). The minimisation of the opportunity loss ('distance' from the maximum value) automatically leads to maximisation of total profit. However, it should be remembered that while the transportation method is applied to the opportunity loss matrix, the total profit value is obtained by using the transportation pattern of the solution on the unit profit matrix, and not through the opportunity loss matrix.

It may be noted, however, that if a maximisation type of transportation problem is unbalanced, then it should be balanced by introducing the necessary dummy row/column for a source/destination, before converting it into a minimisation problem. Similarly, if such a problem has a prohibited route, then the payoff element for such a route should be substituted by –*M* before proceeding to convert to minimisation type.

Example 5.5 Solve the following transportation problem for maximum profit.

Per Unit Profit (Rs)

		Market			
		A	*B*	*C*	*D*
	X	12	18	6	25
Warehouse	*Y*	8	7	10	18
	Z	14	3	11	20

Availability at warehouses:

X : 200 units
Y : 500 units
Z : 300 units

Demand in the markets:

A : 180 units
B : 320 units
C : 100 units
D : 400 units

To solve this problem, we first convert the unit profit matrix into opportunity loss matrix by subtracting each of the values in it from the largest value equal to 25. The information is presented in Table 5.25.

TABLE 5.25 Opportunity Loss Matrix: Initial Solution

Warehouse	*Market* *A*	*B*	*C*	*D*	*Availability*	u_i
X	13 (−17)	[200] 7	19 (−20)	0 (−4)	200	0
Y	17 (−10)	[100] + 18	15 (−5)	[400] − 7	500	11
Z	[180] 11	[20] − 22	[100] 14	+ 5 (+6)	300	15
Demand	180	320	100	400	1000	
v_j	− 4	7	− 1	− 4		

Total profit = 18 × 200 + 7 × 100 + 18 × 400 + 14 × 180 + 3 × 20 + 11 × 100 = Rs 15,180

Also given in the table is the initial feasible solution obtained using VAM. The solution is tested for optimality using MODI method and is found to be non-optimal. A closed loop is traced beginning with the cell ZD which has the largest positive opportunity cost. The revised solution is given in Table 5.26.

TABLE 5.26 Revised Solution: Non-optimal

Warehouse	*Market*				*Availability*	u_i
	A	*B*	*C*	*D*		
X	13 (−11)	[200] 7	19 (−14)	0 (−4)	200	0
Y	17 (−4)	[120] 18	15 (+1) +	[380] 7 −	500	11
Z	[180] 11	22 (−6)	[100] 14 −	[20] 5 +	300	9
Demand	180	320	100	400	1000	
v_j	2	7	5	−4		

Total profit = 18 × 200 + 7 × 120 + 18 × 380 + 14 × 180 + 11 × 100 + 20 × 20 = Rs 15,300

This solution is tested for optimality and is also found to be non-optimal. It is further revised by making x_{23} a basic variable. Given in Table 5.27, this is tested to be optimal.

TABLE 5.27 Revised Solution: Optimal

Warehouse	*Market*				*Availability*	u_i
	A	*B*	*C*	*D*		
X	13 (−11)	[200] 7	19 (−15)	0 (−4)	200	0
Y	17 (−4)	[120] 18	[100] 15	[280] 7	500	11
Z	[180] 11	22 (−6)	14 (−1)	[120] 5	300	9
Demand	180	320	100	400	1000	
v_j	2	7	4	−4		

Total profit = 18 × 200 + 7 × 120 + 10 × 100 + 18 × 280 + 14 × 180 + 20 × 120 = Rs 15,400

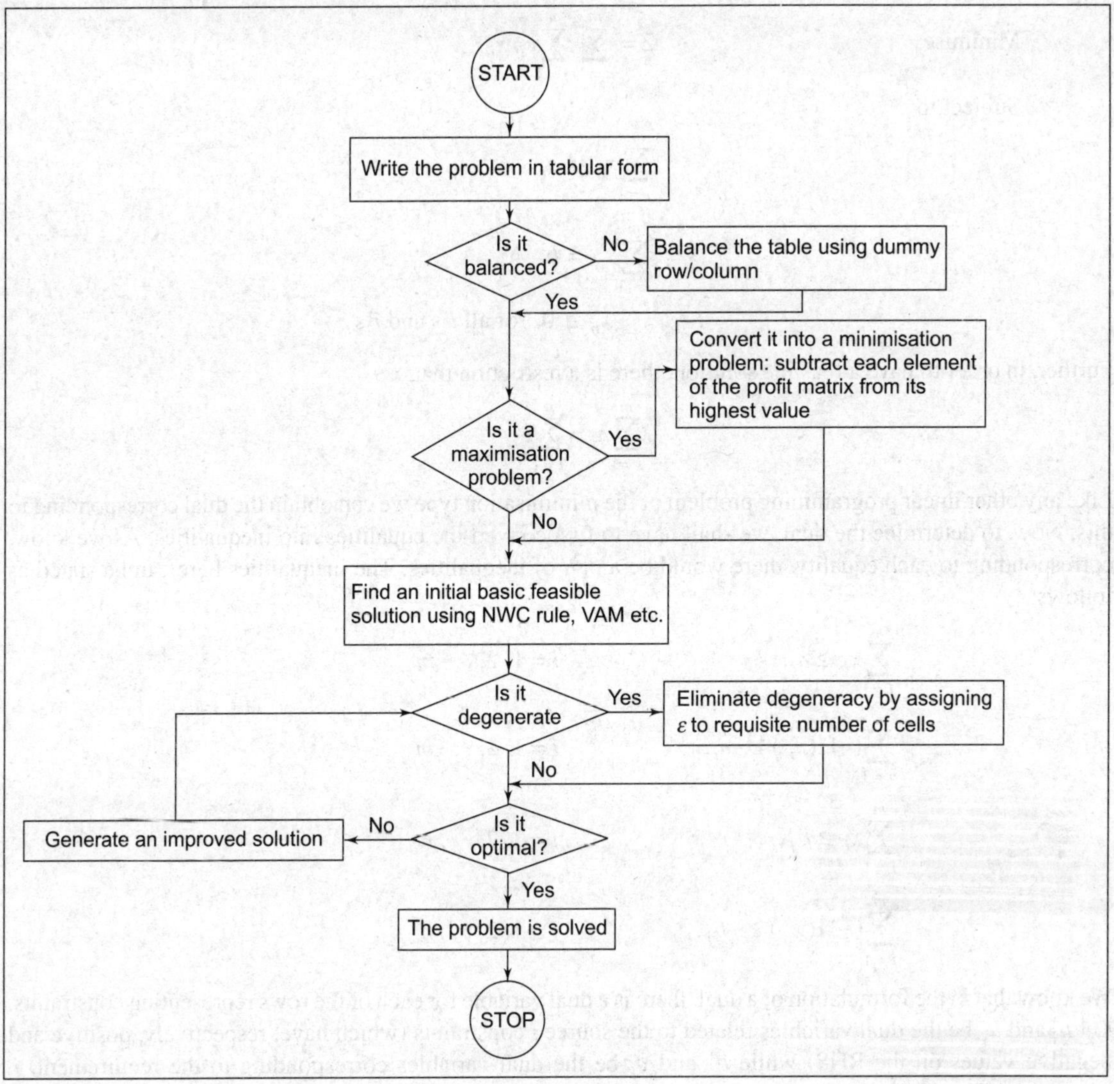

START
Write the problem in tabular form
Is it balanced?
No
Balance the table using dummy row/column
Yes
Convert it into a minimisation problem: subtract each element of the profit matrix from its highest value
Is it a maximisation problem?
Yes
No
Find an initial basic feasible solution using NWC rule, VAM etc.
Is it degenerate
Yes
Eliminate degeneracy by assigning ε to requisite number of cells
No
Is it optimal?
No
Generate an improved solution
Yes
The problem is solved
STOP

Figure 5.4 *Schematic Presentation of Transportation Method*

Accordingly, the maximum profit obtainable is Rs 15,400. Observe that the total profit at each stage is found with reference to the profit matrix given in the problem.

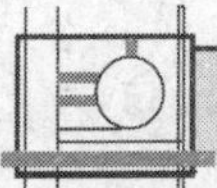

5.5 DUAL OF THE TRANSPORTATION MODEL

As already observed, the transportation problem is only a special case of linear programming problems. It has, therefore, a dual with the usual pricing interpretations, as discussed in the previous chapter. As we have seen, a transportation problem can be written as an LPP as follows:

Minimise $$Z = \sum_{i=1}^{m} \sum_{j=1}^{n} c_{ij}\, x_{ij}$$

Subject to

$$\sum_{j=1}^{n} x_{ij} = a_i$$

$$\sum_{i=1}^{m} x_{ij} = b_j$$

$$x_{ij} \geq 0, \text{ for all } i\text{'s and } j\text{'s}$$

Further, in order to have a feasible solution, there is a restriction that

$$\sum_{i=1}^{m} a_i = \sum_{j=1}^{n} b_j$$

Like any other linear programming problem of the minimisation type we can obtain the dual corresponding to this. Now, to determine the dual, we shall have to first convert the equalities into inequalities. As we know, corresponding to each equality, there would be a pair of inequalities. The inequalities here can be stated as follows:

$$\sum_{j=1}^{n} x_{ij} \geq a_i, \qquad i = 1, 2, \ldots m$$

$$\sum_{j=1}^{n} (-1)(x_{ij}) \geq -a_i, \qquad i = 1, 2, \ldots m$$

$$\sum_{i=1}^{m} x_{ij} \geq b_j, \qquad j = 1, 2, \ldots n$$

$$\sum_{i=1}^{m} (-1)(x_{ij}) \geq -b_j, \qquad j = 1, 2, \ldots n$$

We know that in the formulation of a dual, there is a dual variable for each of the rows representing constraints. Let u_i^+ and u_i^- be the dual variables related to the source i constraints (which have, respectively, positive and negative values on the RHS) while v_j^+ and v_j^- be the dual variables corresponding to the requirement, j, constraints. The dual can be written as follows:

Maximise $$Z = \sum_{i=1}^{m} \left(u_i^+ - u_i^-\right) a_i + \sum_{j=1}^{n} \left(v_j^+ - v_j^-\right) b_j$$

Subject to

$$(u_i^+ + u_i^-) + (v_j^+ - v_j^-) \leq c_{ij} \qquad \text{for all } i, j\text{'s}$$

$$u_i^+, u_i^-, v_j^+, v_j^- \geq 0 \qquad \text{for all } i, j\text{'s}$$

Putting $(u_i^+ - u_i^-) = u_i$ and $(v_j^+ - v_j^-) = v_j$, the dual can alternatively be expressed as follows

Maximise $$Z = \sum_{i=1}^{m} u_i a_i + \sum_{j=1}^{n} v_j b_j$$

Subject to

$$u_i + v_j \leq c_{ij} \quad \text{for all } i, j\text{'s}$$
$$u_i, v_j \quad \text{unrestricted in sign}$$

The u_i and v_j's here are the same as in MODI method (in fact, MODI method is based on the dual). In the dual model, the dual variables are the shadow prices for the resources (i.e. the capacities or supplies) and the requirements. Here u_i can be interpreted as the value of the product, free on board, at the *i*th origin and v_j as its value (delivered) at the *j*th destination. We can also say that u_i is the relative implicit contribution (or value) of an additional unit of capacity at source *i* and v_j is the relative implicit contribution of an additional unit sent to requirement destination *j*.

The dual corresponding to Example 5.1 can be formulated as follows:

Maximise $Z = 500u_1 + 300u_2 + 200u_3 + 180v_1 + 150v_2 + 350v_3 + 320v_4$

Subject to

$$u_1 + v_1 \leq 12 \qquad u_2 + v_3 \leq 8$$
$$u_1 + v_2 \leq 10 \qquad u_2 + v_4 \leq 14$$
$$u_1 + v_3 \leq 12 \qquad u_3 + v_1 \leq 6$$
$$u_1 + v_4 \leq 13 \qquad u_3 + v_2 \leq 16$$
$$u_2 + v_1 \leq 7 \qquad u_3 + v_3 \leq 11$$
$$u_2 + v_2 \leq 11 \qquad u_3 + v_4 \leq 7$$
$$u_1, u_2, u_3, v_1, v_2, v_3, v_4 \text{ unrestricted in sign.}$$

For this dual, if we consider the first constraint, we have

$$u_1 + v_1 \leq \text{Rs } 12 \text{ or } v_1 = 12 - u_1.$$

Thus, the delivered value of the product at destination *P* cannot be greater than the cost of transporting one unit from *A* to *P* minus the f.o.b. (per unit) value at *A*.

Now, the values of u_i and v_j's can be obtained by solving this LPP. These can also be taken from the table containing optimal solution to the example using MODI, as demonstrated earlier. To repeat, from Table 5.15, we have $u_1 = 0$, $u_2 = -4$, $u_3 = -6$, $v_1 = 11$, $v_2 = 10$, $v_3 = 12$ and $v_4 = 13$. According to these, the objective function value, Z, equals $500 \times 0 + 300(-4) + 200(-6) + 180 \times 11 + 150 \times 10 + 350 \times 12 + 320 \times 13$ or Rs 9,440.

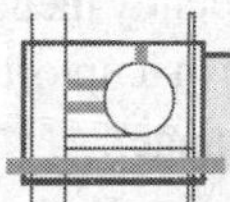

5.6 SENSITIVITY ANALYSIS

Having considered obtaining optimal solution to a transportation problem, we now proceed to investigate the effects on the solution of events like changes in the cost elements, increasing or decreasing availability and demand on a given route, etc. Here we discuss the following aspects of this sensitivity analysis in respect of a transportation problem:

(a) Change in the objective function coefficient of a non-basic variable—which means a change in the cost element of an unoccupied cell.

(b) Change in the objective function coefficient of a basic variable—a change in the cost element of an occupied cell.

(c) Changing a particular source supply and a particular destination demand by a given amount.

(d) Setting upper or lower limits on the values of x_{ij} variables. In other words, the effect of a decision to dispatch a certain maximum or minimum number of units on a given route.

We illustrate these changes in the context of Example 5.1. The optimal solution to the problem is reproduced in Table 5.28.

TABLE 5.28 Optimal Solution

From ↓ / To →	P	Q	R	S	Supply	u_i
A	12 (−1)	150 / 10	230 / 12	120 / 13	500	0
B	180 / 7	11 (−5)	120 / 8	14 (−5)	300	−4
C	6 (−1)	16 (−12)	11 (−5)	200 / 7	200	−6
Demand	180	150	350	320	1000	
v_j	11	10	12	13		

Change in the Objective Function coefficient of a Non-basic Variable Recall that the u_i and v_j values are determined by reference to occupied cells only. Now, if there is a change in the cost element of an unoccupied cell, it will have no effect on the u_i and v_j values and the solution will remain unchanged as long as the Δ_{ij} values remain non-positive. To illustrate, the cost element c_{11} is changed from 12 to 12 + k. To see for what values of k will there be no change in the solution, we set $\Delta_{11} = u_1 + v_1 - c_{11} = 0 + 11 - (12 + k) = -1 - k$. Thus, the current basis remains unchanged for $-1 - k \leq 0$ or $k \geq -1$ and $c_{11} \geq 12 - 1$ or 11. Evidently then, an increase in the cost element c_{11} will in any case not affect the current solution and so will a decline in it upto 1 (also indicated by the Δ_{11} value). If the cost were to decline below Rs 11, then the solution would change.

Change in the Objective Function Coefficient of a Basic Variable A change in the cost coefficient of a basic variable, that is to say an occupied cell, would cause a change in the u_i and v_j values. This will in turn affect the Δ_{ij} values of the unoccupied cells. However, so long as one or more of the Δ_{ij} values do not turn to be positive, the solution would remain unchanged. To understand this, suppose we change c_{13} from 12 to 12 + k. This means that $u_1 + v_3 = 12$ will change to $u_1 + v_3 = 12 + k$. Now, to find the value of k, we solve the following equations:

$$u_1 = 0, \quad u_1 + v_2 = 10, \quad u_1 + v_3 = 12 + k, \quad u_1 + v_4 = 13,$$
$$u_2 + v_1 = 7, \quad u_2 + v_3 = 8 \quad \text{and} \quad u_3 + v_4 = 7.$$

Solving these equations, we get $u_1 = 0$, $u_2 = -4 - k$, $u_3 = -6$, $v_1 = 11 + k$, $v_2 = 10$, $v_3 = 12 + k$ and $v_4 = 13$. With these values, we recalculate $\Delta_{ij} = u_i + v_j - c_{ij}$ values of the unoccupied cells and set them all ≤ 0 as follows:

$$\Delta_{11} = 0 + 11 + k - 12 = -1 + k \leq 0 \quad \text{or} \quad k \leq 1$$

$$\Delta_{22} = -4 - k + 10 - 11 = -5 - k \leq 0 \quad \text{or} \quad k \geq -5$$

$$\Delta_{24} = -4 - k + 13 - 14 = -5 - k \leq 0 \quad \text{or} \quad k \geq -5$$

$$\Delta_{31} = -6 + 11 + k - 6 = -1 + k \leq 0 \quad \text{or} \quad k \leq 1$$

$$\Delta_{32} = -6 + 10 - 16 = -12$$

$$\Delta_{33} = -6 + 12 + k - 11 = -5 + k \leq 0 \quad \text{or} \quad k \leq 5$$

From the calculations, it is evident that the least positive value for k is 1 while the least negative value is – 5. Accordingly, $-5 \leq k \leq 1$ will cause no change in the current optimal solution. Thus, $\bar{C}_{13} = 12 + k$ or $12 - 5 = 7$ and $12 + 1 = 13$. In simple words, if the cost element, 12, of the route AR changes between 7 and 13, no change in the current solution would take place.

Change in a Source Supply by k and a Destination Demand by k If an equal change, k, occurs in the ith source of supply and the jth destination of demand, then it keeps the transportation problem balanced. Now, since u_i's and v_j's may be thought of as negatives of shadow prices of the supply and demand constraints, then: New Z-value = old z-value + $ku_i + kv_j$, if the current basis remains optimal. To illustrate, in our example, if we increase the supply at plant A by 2 units and demand at warehouse Q by 2 units, then revised total cost = 9,440 + 2 × 0 + 2 × 5 = Rs 9,450.

Let us consider how the new values of the decision variables (the quantities, x_{ij}'s sent on various routes) would change.

(a) If x_{ij} is a basic variable, so that the route represented by source and destination where change takes place is an occupied cell, then the x_{ij} is revised to $x_{ij} + k$. To illustrate, if the supply at plant A is increased by 3 units and the demand at warehouse Q by 3 units, then we revise the quantity sent on this route to 153. This is shown in Table 5.29.

The new total cost is equal to 9,440 + 0 × 3 + 10 × 3 = Rs 9,470.

(b) If the route where change takes place is an unoccupied cell, then we draw a closed loop starting with that cell. Once this is done, we select the cell (on the closed loop) in the same row in which supply is increased and add k to the x_{ij} value in that cell. Then, we go round the loop, alternately subtracting and adding k to the x_{ij} values in those cells.

To illustrate, suppose we add 10 units to the supply of plant C and demand at warehouse P. Now, since the route CP is empty, we draw a closed loop as: $CP – CS – AS – AR – BR – BP – CP$. The occupied cell in the plant C row on the closed loop is CS. So we add 10 units to this cell, subtract 10 units from AS, and further add to AR, subtract from BR and, finally, add to the cell BP. The revised solution is given in Table 5.30.

TABLE 5.29 Transportation Tableau: Sensitivity Analysis

From ↓ / To →	P	Q	R	S	Supply	u_i
A	12 (−1)	[153] 10	[230] 12	[120] 13	503	0
B	[180] 7	11 (−5)	[120] 8	14 (−5)	300	−4
C	6 (−1)	16 (−12)	11 (−5)	[200] 7	200	−6
Demand	180	153	350	320	1003	
v_j	11	10	12	13		

TABLE 5.30 Transportation Tableau: Sensitivity Analysis

From ↓ / To →	P	Q	R	S	Supply	u_i
A	12 (−1)	[150] 10	[240] 12 +	[110] 13 −	500	0
B	[190] 7 +	11 (−5)	[110] 8 −	14 (−5)	300	−4
C	6 − (−1)	16 (−12)	11 (−5)	[210] 7 +	210	−6
Demand	190	150	350	320	1010	
v_j	11	10	12	13		

This change has brought about an increase of 10(– 6) + 10 × 11 = Rs 50 in the total cost which works out to be Rs 9,490.

It may be mentioned here that the maximum number of 120 units can be increased in the supply source plant *C* and warehouse *P* demand. Can you answer why?

Setting Limits on the x_{ij} Variables We now examine how the solution to a transportation problem is affected when it is designed to assign a certain number of units on a particular route. We examine this in the context of the solution given in Table 5.28. To begin with, we consider an unoccupied cell, say, *BS*. If it is desired to send 50 units on this route, then it will be done by drawing a closed loop beginning with this cell, namely, *BS* – *AS* – *AR* – *BR* – *BS*. After adjusting, the x_{ij} values for these cells would be: *BS* – 50, *AS* – 70, *AR* – 280 and *BR* – 70. It will have the effect of increasing the cost by Rs 250 Also, it may be noted that a maximum of 120 units can be dispatched on the route *BS* in context of the present solution. However, if more than 120 units, say 200 units, are planned to be sent on this route, then the problem would need to be re-solved after reducing the supply at *B* and the demand at *S* by 200. To the total cost obtained from such a solution, we shall add the cost of sending 200 units on the route *BS*.

Now, suppose that it is desired to send not more than 200 units on the route *AR*. This can be done by reducing the supply on *A* and the demand on *R* by 200 units (which means setting $k = -200$) and proceeding in the manner discussed earlier. This can alternatively be handled by reducing the supply at *A* and demand at *R* by 200 each and re-solving the problem after replacing the cost element of the route *AR* by *M*, thus making it a prohibited route.

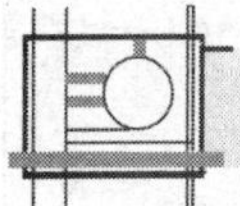

5.7 PRODUCTION SCHEDULING AND INVENTORY CONTROL

The transportation model can usefully be applied to the problem of scheduling production and controlling inventory over different time periods. For illustration, let us consider the case of a manufacturer whose product has a seasonal demand. To meet the demand, the production plan of the manufacturer could either be (a) completely uniform production (and absorbing the seasonal fluctuations by inventory build ups or depletions); (b) variable production, completely matching the demand (and, therefore, no inventories); or (c) something between the two represented by (a) and (b).

The drawing up of the optimal production plan for him would involve and call for a trade-off between overtime costs and inventory holding costs. In the formulation of the problem, we define for each time period the regular and overtime production capacities in each period as 'origins' and demands for the periods as the 'destinations'. The cost associated with each cell is taken to be the sum of the production cost (which will depend upon the period, normal or overtime, in which the item is produced) and the cost of holding a unit in inventory (if held). Here the decision variables would be in terms of the production quantities in each given period, manufactured by a particular method, namely regular or overtime, for meeting the current or future demand. The following example demonstrates the formulation.

Example 5.6 A manufacturer of a certain component has the following estimates of the demand for its product:

Period	:	1	2	3	4
Demand	:	50	70	85	75

The regular production capacity for each period is 60 units while with overtime working, and additional of upto 20 units can be produced in each period. The unit cost of production in regular time is worked out to be Rs 6 whereas it would cost Rs 10 per unit for the items manufactured during overtime. You may assume that the costs would be identical in all the periods. Any production in excess of the requirement in the current period may be held at a cost of Rs 3 per unit per period. No holding cost is involved if a unit is sold in the period in which it is produced.

(a) Formulate the problem as a transportation problem; and

(b) Draw the optimal production plan for the manufacturer.

The given information is presented as a transportation problem in Table 5.31. The cost elements in each cell are determined by adding the production cost (Rs 6 or Rs 10 depending upon whether the production is in regular period or in overtime) and the holding cost (equal to Rs 3 per unit each period). Thus, in the first row, the cost of Rs 15 is obtained as Rs 6 plus Rs 9, being the production and holding costs. And, since the output of any period cannot be used in a period preceding it, the cost element *M* is written in the appropriate cells. Further, a column (dummy destination) has been introduced as the supply exceeds the demand.

TABLE 5.31 Production Planning Problem

Origin	Destination					Supply
	1	2	3	4	D	
R_1	6	9	12	15	0	60
O_1	10	13	16	19	0	20
R_2	M	6	9	12	0	60
O_2	M	10	13	16	0	20
R_3	M	M	6	9	0	60
O_3	M	M	10	13	0	20
R_4	M	M	M	6	0	60
O_4	M	M	M	10	0	20
Demand	50	70	85	75	40	320

R: Regular Time, O: Overtime D: Dummy

Now, the problem can be solved as a transportation problem.

The solution is left as an exercise for the reader. Table 5.32 contains final solution to the problem, that is an optimal one. Accordingly, the production schedule, involving a total cost of Rs 1,885 is as given in Table 5.33.

TABLE 5.32 Final, Improved Solution: Optimal

Origin	*Destination* 1	2	3	4	*D*	*Supply*
R_1	(50) 6	(10) 9	12	15	0	60
O_1	10	13	16	19	(20) 0	20
R_2	*M*	(60) 6	9	12	0	60
O_2	*M*	10	(5) 13	16	(15) 0	20
R_3	*M*	*M*	(60) 6	9	0	60
O_3	*M*	*M*	(20) 10	13	0	20
R_4	*M*	*M*	*M*	(60) 6	0	60
O_4	*M*	*M*	*M*	(15) 10	(5) 0	20
Demand	50	70	85	75	40	320

TABLE 5.33 Production Schedule

Produce in Month		*Units*	*For use in Month* *1*	*2*	*3*	*4*
1	R	60	50	10		
	O.T.	nil				
2	R	60		60		
	O.T.	5			5	
3	R	60			60	
	O.T.	20			20	
4	R	60				60
	O.T.	15				15
Demand			50	70	85	75

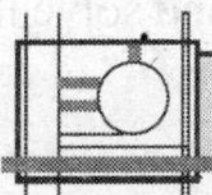

5.8 TRANSHIPMENT PROBLEM

In the transportation problems discussed thus far, it was assumed that a source (factory, plant, etc.) acts only as a shipper of the goods and a destination (market, etc.) acts only as receiver of the goods. We shall now consider the broader class of transportation problem, called as *transhipment problem,* which allows for the shipment of goods both from one source to another, and from one destination point to another. Thus, we allow for the possibility of transhipment—the goods produced/available at one source and destined for some destination point may reach there via other sources and/or destinations and transhipped at these points. This is obviously a more realistic statement of the distribution problem faced by a business/industrial house. For example, a multi-plant firm may find it necessary to send some goods from one plant to another in order to meet the substantial increase in the demand in the second market. The second plant here would act both as a source and a destination and there is no real distinction between source and destination.

A transportation problem can be converted into a transhipment problem by relaxing the restrictions on the receiving and sending the units on the origins and destinations respectively. An *m*-origin, *n*-destination, transportation problem, when expressed as a transhipment problem, shall become an enlarged problem: with $m + n$ origins and an equal number of destinations. With minor modifications, this problem can be solved using the transportation method.

If we let the sources and destinations in a transhipment problem as *T*, then x_{ij} would represent the amount of goods shipped from the *i*th terminal (T_i) to the *j*th terminal (T_j), and c_{ij} would represent the unit cost of such shipment. Naturally, x_{ii} would equal to zero because no units would be shipped from a terminal to itself. Now, assume that at *m* terminals ($T_1, T_2, \ldots, T_m$), the total outshipment exceeds the total inshipment by amounts equal to $a_1, a_2, \ldots, a_m$ respectively and at the remaining *n* terminals ($T_{m+1}, T_{m+2}, \ldots, T_{m+n}$) the total inshipment exceeds the total outshipment by amounts $b_{m+1}, b_{m+2}, \ldots, b_{m+n}$ respectively. If the total inshipment at terminals $T_1, T_2, \ldots, T_m$ be $t_1, t_2, \ldots, t_m$ respectively and the total outshipment at the terminals $T_{m+1}, T_{m+2}, \ldots, T_{m+n}$ be $t_{m+1}, t_{m+2}, \ldots, t_{m+n}$ respectively, we can write the transhipment problem as:

Minimise $$Z = \sum_{i=1}^{m+n} \sum_{j=1}^{m+n} c_{ij} x_{ij}$$

Subject to

$$x_{i1} + x_{i2} + \ldots + x_{i,\,i-1} + x_{i,\,i+1} + \ldots + x_{i,\,m+n} - t_i = a_i \qquad (i = 1, 2, \ldots, m) \qquad (1)$$

$$x_{1i} + x_{2i} + \ldots + x_{i-1,\,i} + x_{i+1,\,i} + \ldots + x_{m+n,\,i} = t_i \qquad (i = 1, 2, \ldots, m) \qquad (2)$$

$$x_{1j} + x_{2j} + \ldots + x_{j-1,\,j} + x_{j+1,\,j} + \ldots + x_{m+n,\,j} - t_j = b_j \qquad (j = m+1, m+2, \ldots, m+n) \qquad (3)$$

$$x_{j1} + x_{j2} + \ldots + x_{j,\,j-1} + x_{j,\,j+1} + \ldots + x_{j,\,m+n} = t_j \qquad (j = m+1, m+2, \ldots, m+n) \qquad (4)$$

$$\sum_{i=1}^{m} a_i = \sum_{j=m+1}^{m+n} b_j \qquad (5)$$

As can be easily observed, these constraints are similar to the constraints of a transportation problem with $m + n$ sources and $m + n$ destinations, with the differences that here there are no x_{ii} and x_{jj} terms, and that $b_j = 0$ for $j = 1, 2, \ldots, m$ and $a_i = 0$ for $i = m + 1, m + 2, \ldots, m + n$. The terms t_i and t_j in these constraints may be seen as the algebraic equivalents of x_{ii} and x_{jj}. Now, we can view this problem as an enlarged problem and solve it by using the transportation method.

The transhipment problem can be depicted in tabular form as shown in Table 5.34.

TABLE 5.34 Transhipment Problem

Terminal $T_i \downarrow T_j \rightarrow$	*1*	*2*	...	*m*	*m + 1*	...	*m + n*	*Supply* a_i
1	$-t_1$	x_{12}		x_{1m}	$x_{1,\,m+1}$		$x_{1,\,m+n}$	a_1
2	x_{21}	$-t_2$		x_{2m}	$x_{2,\,m+1}$		$x_{2,\,m+n}$	a_2
.								
.								
.								
m	x_{m1}	x_{m2}		$-t_m$	$x_{m,\,m+1}$		$x_{m,\,m+n}$	a_m
$m+1$	$x_{m+1,\,1}$	$x_{m+1,\,2}$		$x_{m+1,\,m}$	$-t_{m+1}$		$x_{m+1,\,m+n}$	0
.								
.								
.								
$m+n$	$x_{m+n,\,1}$	$x_{m+n,\,2}$		$x_{m+n,\,m}$	$x_{m+n,\,m+1}$		$-t_{m+n}$	0
Demand b_j	0	0		0	b_{m+1}		b_{m+n}	$\Sigma a_i = \Sigma b_j$

The first *m* rows represent the *m* constraints given in (1) while the remaining *n* rows show the constraints given in (4). The constraints in (2) and (3) are represented by the first *m* columns and the remaining *n* columns respectively. All the *t* values are placed on the diagonal from left top to right bottom. Each of them bears negative sign which must be considered carefully when a *t* is involved in the readjustment (during the solution process).

We shall now consider Example 5.1 as a transhipment problem with the following additional cost data.

From Plant	*To Plant* A	B	C	*From Warehouse*	*To Warehouse* P	Q	R	S
A	0	2	8	*P*	0	6	5	8
B	5	0	7	*Q*	8	0	3	2
C	10	9	0	*R*	5	4	0	10
				S	9	4	8	0

The given problem is written as a transhipment problem in Table 5.35. We start with the optimal solution obtained previously. Application of the MODI method for determining the optimality of the solution suggests that the route *AB* be brought into the solution. Since the *t* values bear the negative sign in this (transhipment problem), the maximum amount is limited here by the cell *AR*. The next tableau is derived as shown in Table

5.36, and it has the total transhipment cost equal to Rs 8,980, which is smaller than the total cost of Rs 9,440 obtained earlier. Another improvement leads to optimal solution to the problem. This is given in Table 5.37.

TABLE 5.35 Initial Feasible Solution: Non-optimal

	A	*B*	*C*	*P*	*Q*	*R*	*S*	*Supply*	u_i
A	0 [−0]	+ 2 (+2)	8 (−2)	12 (−1)	10 [150]	− 12 [230]	13 [120]	500	0
B	5 (−9)	− 0 [−0]	7 (−5)	7 [180]	11 (−5)	+ 8 [120]	14 (−5)	300	− 4
C	10 (−16)	9 (−11)	0 [−0]	6 (−1)	16 (−12)	11 (−5)	7 [200]	200	− 6
P	12 (−23)	7 (−14)	6 (−11)	0 [−0]	6 (−7)	5 (−6)	8 (−6)	0	− 11
Q	10 (−20)	11 (−17)	16 (−20)	8 (−7)	0 [−0]	3 (−1)	2 (+1)	0	− 10
R	12 (−24)	8 (−16)	11 (−17)	5 (−6)	4 (−6)	0 [−0]	10 (−9)	0	− 12
S	13 (−26)	14 (−23)	7 (−14)	9 (−11)	4 (−7)	8 (−9)	0 [−0]	0	− 13
Demand	0	0	0	180	150	350	320	1000	
v_j	0	4	6	11	10	12	13		

Total cost = 10 × 150 + 12 × 230 + 13 × 120 + 7 × 180 + 8 × 120 + 7 × 200 = Rs 9,440

From Table 5.37, the optimal solution is:

Send from plant *A*	:	230 units to plant *B* and 270 units to warehouse *Q*,
Send from plant *B*	:	180 units to warehouse *P* and 350 units to warehouse *R*,
Send from plant *C*	:	200 units to warehouse *S*, and
Send from warehouse *Q*	:	150 units to warehouse *S*.

This transportation pattern would involve a total cost of Rs 8,860, resulting in a saving of Rs 9,440 – Rs 8,860 = Rs 580 on account of the possibility of transhipment.

TABLE 5.36 Revised Solution: Non-optimal

(Values in square brackets are boxed allocations; values in parentheses are circled; + / − mark the loop.)

	A	*B*	*C*	*P*	*Q*	*R*	*S*	*Supply*	u_i
A	0 [– 0]	2 [230]	8 (– 2)	12 (– 3)	+ 10 [150]	12 (– 2)	– 13 [120]	500	0
B	5 (– 7)	0 [– 230]	7 (– 3)	7 [180]	11 (– 3)	8 [350]	14 (– 3)	300	– 2
C	10 (– 16)	9 (– 13)	0 [– 0]	6 (– 3)	16 (– 12)	11 (– 7)	7 [200]	200	– 6
P	12 (– 21)	7 (– 14)	6 (– 9)	0 [– 0]	6 (– 7)	5 (– 4)	8 (– 4)	0	– 9
Q	10 (– 20)	11 (– 19)	16 (– 20)	8 (– 9)	– 0 [– 0]	3 (– 3)	+ 2 (+ 1)	0	– 10
R	12 (– 22)	8 (– 16)	11 (– 15)	5 (– 6)	4 (– 4)	0 [– 0]	10 (– 7)	0	– 10
S	13 (– 26)	14 (– 25)	7 (– 14)	9 (– 13)	4 (– 7)	8 (– 11)	0 [– 0]	0	– 13
Demand	0	0	0	180	150	350	320	1000	
v_j	0	2	6	9	10	10	13		

Total cost = 2 × 230 + 10 × 150 + 13 × 120 + 7 × 180 + 8 × 350 + 7 × 200 = Rs 8,980

TABLE 5.37 Revised Solution: Optimal

	A	*B*	*C*	*P*	*Q*	*R*	*S*	*Supply*	u_i
A	0 [−0]	2 [230]	8 (−3)	12 (−3)	10 [270]	12 (−2)	13 (−1)	500	0
B	5 (−7)	0 [−230]	7 (−4)	7 [180]	11 (−3)	8 [350]	14 (−4)	300	−2
C	10 (−15)	9 (−12)	0 [−0]	6 (−2)	16 (−11)	11 (−6)	7 [200]	200	−5
P	12 (−21)	7 (−14)	6 (−10)	0 [−0]	6 (−7)	5 (−4)	8 (−5)	0	−9
Q	10 (−20)	11 (−19)	16 (−21)	8 (−9)	0 [−120]	3 (−3)	2 [120]	0	−10
R	12 (−22)	8 (−16)	11 (−16)	5 (−6)	4 (−4)	0 [−0]	10 (−8)	0	−10
S	13 (−25)	14 (−24)	7 (−14)	9 (−12)	4 (−6)	8 (−10)	0 [−0]	0	−12
Demand	0	0	0	180	150	350	320	1000	
v_j	0	2	5	9	10	10	12		

Total cost = 2 × 230 + 10 × 270 + 7 × 180 + 8 × 350 + 7 × 200 + 2 × 120 = Rs 8,860

REVIEW ILLUSTRATIONS

Example 5.7 Solve the following problem using transportation method, obtaining the initial feasible solution by VAM. The cell entries in the table are unit costs (in rupees).

From	*To* 1	2	3	4	5	*Supply*
1	80	69	103	64	61	12
2	47	100	72	65	40	16
3	16	103	87	36	94	20
4	86	15	57	19	25	8
5	27	20	72	94	19	8
Demand	16	14	18	6	10	

This initial feasible solution, using VAM, is displayed in Table 5.38. Since the number of occupied cells (8) is less than $m + n - 1 = 9$, an ε is introduced in a cell to test whether the solution is optimal. This solution is found non-optimal. It is improved and presented in Table 5.39. This one is also found non-optimal. The next revised solution presented in Table 5.40 is found to be optimal.

TABLE 5.38 Initial Feasible Solution: Non-optimal

From ↓ / *To* →	1	2	3	4	5	*Supply*	u_i
1	80 (−24)	69 (3)	103 [12] −	64 (12) +	61 (10)	12	0
2	47 (−22)	100 (−59)	72 [6] +	65 (−20)	40 [10] −	16	− 31
3	16 [16]	103 (−71)	87 (−24)	36 [4]	94 (−63)	20	− 40
4	86 (−87)	15 [6] +	57 (−11)	19 [2] −	25 (−11)	8	− 57
5	27 (−23)	20 [8] −	72 (−21)	94 (−70)	19 [ε] +	8	− 52
Demand	16	14	18	6	10	64	
v_j	56	72	103	76	71		

Total cost = 103 × 12 + 72 × 6 + 40 × 10 + 16 × 16 + 36 × 4 + 15 × 6 + 19 × 2 + 20 × 8 = Rs 2,756

TABLE 5.39 Improved Solution: Non-optimal

From ↓ / To →	1	2	3	4	5	Supply	u_i
1	(−36) 80	(3) 69	− 103 [10]	64 [2]	(10) + 61	12	0
2	(−34) 47	(−59) 100	+ 72 [8]	(−32) 65	− 40 [8]	16	− 31
3	16 [16]	(−59) 103	(−12) 87	36 [4]	(−51) 94	20	− 28
4	(−99) 86	15 [8]	(−11) 57	(−12) 19	(−11) 25	8	− 57
5	(−35) 27	20 [6]	(−21) 72	(−82) 94	19 [2]	8	− 52
Demand	16	14	18	6	10	64	
v_j	44	72	103	64	71		

Total cost = 103 × 10 + 64 × 2 + 72 × 8 + 40 × 8 + 16 × 16 + 36 × 4 + 15 × 8 + 20 × 6 + 19 × 2 = Rs 2,732

TABLE 5.40 Improved Solution: Optimal

From ↓ / To →	1	2	3	4	5	Supply	u_i
1	(−36) 80	(−7) 69	103 [2]	64 [2]	61 [8]	12	0
2	(−34) 47	(−69) 100	72 [16]	(−32) 65	(−10) 40	16	− 31
3	16 [16]	(−69) 103	(−12) 87	36 [4]	(−61) 94	20	− 28
4	(−89) 86	15 [8]	(−1) 57	(−2) 19	(−11) 25	8	− 47
5	(−25) 27	20 [6]	(−11) 72	(−72) 94	19 [2]	8	− 42
Demand	16	14	18	6	10	64	
v_j	44	62	103	64	61		

Total cost = 103 × 2 + 64 × 2 + 61 × 8 + 72 × 16 + 16 × 16 + 36 × 4 + 15 × 8 + 20 × 6 + 19 × 2 = Rs 2,652

Example 5.8 A company has three plants at locations *A*, *B* and *C* which produce the same product. It has to supply this to buyers located at *D*, *E* and *F*. The weekly plant capacities for *A*, *B* and *C* are 100, 800 and 150 units respectively, while the buyer requirements are 750, 200 and 500 units respectively for *D*, *E* and *F*. The unit shipping costs (in Rs) are given here.

	Buyer		
Plant	*D*	*E*	*F*
A	8	4	10
B	9	7	9
C	6	5	8

Assume that the penalty for failing to supply buyer requirement is Rs 4, Rs 3 and Rs 3 per unit in respect of *D*, *E* and *F* respectively.

Determine the optimal distribution for the company so as to minimise the cost of transportation and penalty payable.

This problem is an unbalanced transportation problem wherein aggregate requirement is Rs 1,450 units while aggregate supply available is 1,050 units. To solve, the problem is first balanced by adding a dummy plant *T* with supply value of 400 units and cost elements equal to the given penalty values 4, 3 and 3 in columns headed *D*, *E* and *F* respectively. The information is shown in Table 5.41. The initial feasible solution to the problem using VAM is also given in the table. Upon testing, the solution is found to be optimal. It involves a total cost of transportation equal to Rs 8,300 and a penalty cost of Rs 1,200.

TABLE 5.41 Initial Solution–VAM

Plant	Buyer *D*	*E*	*F*	*Supply*	u_i
A	8 (−2)	[100] 4	10 (−4)	100	0
B	[600] 9	[100] 7	[100] 9	800	3
C	[150] 6	5 (−1)	8 (−2)	150	0
T	4 (−1)	3 (−2)	[400] 3	400	−3
Demand	750	200	500	1,450	
v_j	6	4	6		

Total cost = 4 × 100 + 9 × 600 + 7 × 100 + 9 × 100 + 6 × 150 = Rs 8,300 and Penalty = 3 × 400 = Rs 1,200

Example 5.9 Consider the following transportation problem:

Source	P	Q	R	S	Supply
	Destination				
A	6	11	9	8	200
B	13	10	8	15	300
C	9	9	14	12	500
D	12	10	12	10	100
Demand	350	250	300	200	

(a) Find optimal solution to this problem and determine the total cost of transportation.
(b) Is the optimal solution unique? If there is an alternate optimal solution, identify it.
(c) If the transporter agrees to reduce the transportation charges on the route *B-S* by 40 per cent, how will it affect the optimal solution?
(d) It is decided to send a total of 150 units on the route *A-P*. How will the optimal solution and total cost be affected?
(e) Trace the effect of sending 100 units on the route *C-S*.
(f) Within what limits a change in the cost element of the route *C-P* will have no effect on the current optimal solution?

(a) The initial solution using VAM is given in Table 5.42. The solution is seen to be degenerate. Accordingly, an ε is placed in the cell *A-R* as it is an independent cell. The solution is tested for optimality and is found not to be optimal. A closed path is drawn as shown in the table and a revised solution is obtained as given in Table 5.43. It is tested and found to be optimal. The total cost of transportation involved is Rs 9,300.

TABLE 5.42 Initial Solution-VAM

Source	*P*	*Q*	*R*	*S*	*Supply*	u_i
	Destination					
A	6 [200] {−}	11 (−5)	9 [ε]	8 (+1) {+}	200	0
B	13 (−8)	10 (−5)	8 [300]	15 (−7)	300	−1
C	9 [150] {+}	9 [250]	14 (−2)	12 [100] {−}	500	3
D	12 (−5)	10 (−3)	12 (−2)	10 [100]	100	1
Demand	350	250	300	200	1100	
v_j	6	6	9	9		

Total cost = 6 × 200 + 8 × 300 + 9 × 150 + 9 × 250 + 12 × 100 + 10 × 100 = Rs 9,400

TABLE 5.43 Revised Solution–Optimal

Source	*Destination*				*Supply*	u_i
	P	*Q*	*R*	*S*		
A	6 [100]	11 (−5)	9 [ε]	8 [100]	200	0
B	13 (−8)	10 (−5)	8 [300]	15 (−8)	300	−1
C	9 [250]	9 [250]	14 (−2)	12 (−1)	500	3
D	12 (−4)	10 (−2)	12 (−1)	10 [100]	100	2
Demand	350	250	300	200	1100	
v_j	6	6	9	8		

Total cost = 6 × 100 + 8 × 100 + 8 × 300 + 9 × 250 + 9 × 250 + 10 × 100 = Rs 9,300

(b) The optimal solution obtained in Table 5.43 is a unique one since no Δ_{ij} is equal to zero.

(c) A 40 per cent reduction on the route *BS* implies that the unit cost on this route will be revised to Rs 9. In the optimal solution shown in Table 5.43, the Δ_{ij} = – 8 with a cost value of Rs 15 for this route would be revised to – 2 with the unit cost of Rs 9. However, since Δ_{ij} continues to be negative, no cost saving is possible by supplying goods on this route. Hence, the optimal solution will not be affected by this change.

(d) The optimal solution calls for transferring 100 units on the route *AP*. Now, if it is desired to send 150 units on this route, we shall reduce the supply at source *A* and the demand at destination *P* by 150 units and resolve the problem. For this solution, we make the route *AP* prohibited since no more units can be sent on it.
The solution is given in Table 5.44. It is tested for optimality and found to be optimal. Accordingly, the total cost of transportation works out to be: 6 × 150 + 8 × 50 + 8 × 300 + 9 × 200 + 9 × 250 + 12 × 50 + 10 × 100 = Rs 9,350.

(e) To transfer 100 units on the route *CS*, we draw a closed loop beginning with this cell in Table 5.43, to see if it can indeed be done. The closed path is $CS \rightarrow AS \rightarrow AP \rightarrow CP \rightarrow CS$. With given quantities, it is evident that it is possible to do so. The quantities of these cells would be revised to – *AS*: 0, *AP*: 200, *CP*: 150 and *CS*: 100. Further, since Δ_{ij} = – 1 for the cell *CS*, this transfer of 100 units will cause the total cost to increase by 100 × 1 = Rs 100.

TABLE 5.44 Revised Solution–2

Source	*Destination*				*Supply*	u_i
	P	*Q*	*R*	*S*		
A	*M*	(–6) 11	[ε] 9	[50] 8	50	0
B	(–9) 13	(–6) 10	[300] 8	(–8) 15	300	–1
C	[200] 9	[250] 9	(–1) 14	[50] 12	500	4
D	(–5) 12	(–3) 10	(–1) 12	[100] 10	100	2
Demand	200	250	300	200	950	
v_j	5	5	9	8		

Total cost = 6 × 150 + 8 × 50 + 8 × 300 + 9 × 200 + 9 × 250 + 12 × 50 + 10 × 100 = Rs 9,350

(f) Since *CP* is an occupied cell, the required answer can be obtained as follows. Let the cost element of this cell change to $9 + k$. To find k, we solve the following equations: $u_1 = 0$, $u_1 + v_1 = 6$, $u_1 + v_3 = 9$, $u_1 + v_4 = 8$, $u_2 + v_3 = 8$, $u_3 + v_1 = 9 + k$, $u_3 + v_2 = 9$, and $u_4 + v_4 = 10$. This gives $u_1 = 0$, $u_2 = -1$, $u_3 = 3 + k$, $u_4 = 2$, $v_1 = 6$, $v_2 = 6 - k$, $v_3 = 9$, and $v_4 = 8$. With these values,

$$\Delta_{12} = 0 + 6 - k - 11 = -5 - k \le 0 \quad \text{or} \quad k \ge -5$$

$$\Delta_{21} = -1 + 6 - 13 = -8$$

$$\Delta_{22} = -1 + 6 - k - 9 = -4 - k \le 0 \quad \text{or} \quad k \ge -4$$

$$\Delta_{24} = -1 + 8 - 15 = -8$$

$$\Delta_{33} = 3 + k + 9 - 14 = k - 2 \le 0 \quad \text{or} \quad k \le 2$$

$$\Delta_{34} = 3 + k + 8 - 12 = k - 1 \le 0 \quad \text{or} \quad k \le 1$$

$$\Delta_{41} = 2 + 6 - 12 = -4$$

$$\Delta_{42} = 2 + 6 - k - 10 = -k - 2 \le 0 \quad \text{or} \quad k \ge -2$$

$$\Delta_{43} = 2 + 9 - 12 = -1$$

The least positive value of k is 1 while least negative value is – 2. Accordingly, $-2 \le k \le 1$ will cause no change the current optimal solution. Thus, $c_{31} = 9 + k$ or $9 - 2 = 7$ and $9 + 1 = 10$. It means that no change in the current optimal solution will occur if the unit cost of the route *CP* varies between Rs 7 and Rs 10.

Example 5.10 A company has seven manufacturing units situated in different parts of the country. Due to recession it is proposing to close four of these and to concentrate production in the remaining three units.

Production in these units will actually be increased from present levels and will require an increase in the personnel employed in them. Personnel at the units to be closed have expressed willingness to move to any one of the three remaining units and the company is willing to provide them with removal costs.

The technology of production is different to some degree at each unit and retraining expenses will be incurred on transfer. Not all existing personnel can be absorbed by transfer and a number of redundancies will arise. Cost of redundancy is given as a general figure at each unit to be closed.

	Units to be Closed			
	A	*B*	*C*	*D*
Number Employed	200	400	300	200
Retraining Costs: Rs '000 per person				
Transfer to:	*A*	*B*	*C*	*D*
Unit *E*	0.5	0.4	0.6	0.3
Unit *F*	0.6	0.4	0.6	0.3
Unit *G*	0.5	0.3	0.7	0.3
Removal Costs: Rs '000 per person				
Transfer to:	*A*	*B*	*C*	*D*
Unit *E*	2.5	3.6	3.4	4.7
Unit *F*	2.4	4.6	3.4	1.7
Unit *G*	2.5	2.7	3.3	2.7
Redundancy Payments: Rs '000 per person	*A*	*B*	*C*	*D*
	6.0	5.0	6.0	7.0
Additional personnel required at units remaining open:				
	E	F	G	
	350	450	200	

(a) Obtain a solution to the problem of the cheapest (least cost) means to transfer personnel from the units to be closed to those which will be expanded. (You should show all workings clearly indicating how transfers from one unit to another have been evaluated.)

(b) State with reasons whether or not the solution you obtained is optimal and unique.

(c) State the costs of the initial and final solution. *(ICMA, November, 1981, Adapted)*

(a) Since the number of people currently employed in the units to be closed exceed the additional requirement, the surplus (= 100) shall be laid off. Hence, a dummy unit *R* shall be introduced. The cost of transferring an individual between any two points is obtained by adding the retraining and removal costs when he/she is transferred to units *E, F* or *G*. The redundancy costs are similarly to be considered when an employee is to be laid off. The appropriate cost matrix (with figures in Rs '000) along with the requirement and availability are shown in Table 5.45. Also, the initial basic feasible solution is obtained using VAM. It is tested for optimality and found to be optimal.

TABLE 5.45 Initial Feasible Solution: Optimal

From ↓ \ To →	*E*	*F*	*G*	*R*	*Availability*	u_i
A	(0) 3	[200] 3	(−1) 3	(−2) 6	200	0
B	[100] 4	(−1) 5	[200] 3	[100] 5	400	1
C	[250] 4	[50] 4	(−1) 4	(−1) 6	300	1
D	(−3) 5	[200] 2	(−2) 3	(−4) 7	200	−1
Requirement	350	450	200	100	1100	
v_j	3	3	2	4		

Total cost = 3 × 200 + 4 × 100 + 3 × 200 + 5 × 100 + 4 × 250 + 4 × 50 + 2 × 200 = 3,700 (thousand) rupees

According to the solution obtained,

For Unit	*Transfer*
A	200 to *F*
B	100 to *E,* 200 to *G,* and 100 lay off
C	250 to *E,* 50 to *F*
D	200 to *F*

(b) The solution obtained is indeed optimal. However, it is not unique for the reason that a 0 (zero) appears in one of the Δ_{ij} values calculated, in the cell *AE*. This implies that another solution exists which is as efficient as this one. The alternative solution can be obtained by changing numbers along the path *AE-AF-CF-CE*. This would be:

For Unit	*Transfer*
A	200 to *E*
B	100 to *E,* 200 to *G,* and 100 lay off
C	50 to *E,* 250 to *F*
D	200 to *F*

(c) Since the initial solution is optimal, the cost of each is Rs 37,00,000.

Example 5.11 A manufacturer of jeans is interested in developing an advertising campaign that will reach four different age groups. Advertising campaigns can be conducted through TV, radio and magazines. The following table gives the estimated cost in paise per exposure for each age group according to the medium employed. In addition, maximum exposure levels possible in each of the media, namely, TV, radio and Magazines are 40, 30 and 20 millions, respectively. Also, the minimum desired exposures within each age group, namely 13–18,19–25, 26–35, 36 and older, are 30, 25, 15, and 10 millions. The objective is to minimise the cost of obtaining the desired minimum exposure level in each age group.

Media	*Age Groups*			
	13-18	*19-25*	*26-35*	*36 and older*
TV	12	7	10	10
Radio	10	9	12	10
Magazines	14	12	9	12

(i) Formulate the above as a transportation problem, and find the optimal solution.

(ii) Solve this problem if the policy is to provide at least 4 million exposures through TV in the 13–18 age group and at least 8 million exposures through TV in the age group 19–25. *(CA, May, 1991)*

(i) The given problem is formulated as a transportation problem and shown in Table 5.46. With the aggregate of 'maximum exposures available' equal to 90 million and 'minimum exposures required' totalling as 80 million, the problem is unbalanced. Hence, a dummy category to age-groups is added to make the problem balanced.

TABLE 5.46 Advertising Campaign Schedule

Media	*Age Groups*					*Max. Exposures available (in millions)*
	13–18	*19–25*	*26–35*	*36 & older*	*Dummy*	
TV	12	7	10	10	0	40
Radio	10	9	12	10	0	30
Magazines	14	12	9	12	0	20
Min. Exposures (in millions)	30	25	15	10	10	90

The initial solution using VAM is presented in Table 5.47. The solution is degenerate because there are 6 occupied cells against the required number of 7(= 5 + 3 – 1). Degeneracy is resolved by placing an ε in the cell represented by last column of the second row.

The solution is tested for optimality and is found to be non-optimal. Thus, a closed path is drawn as shown.

TABLE 5.47 Initial Solution: Non-optimal

Media	*Age-groups*					*Max. Exp.*	u_i
	13–18	19–25	26–35	36 & over	Dummy		
TV	12 (−1)	7 [25]	10 [5]	10 [10]	+ 0 (1)	40	0
Radio	10 [30]	9 (−3)	12 (−3)	10 (−1)	0 [ε]	30	−1
Magazines	14 (−4)	12 (−6)	+ 9 [10]	12 (−3)	− 0 [10]	20	−1
Min. Exp	30	25	15	10	10	90	
v_j	11	7	10	10	1		

The improved solution is given in Table 5.48. It is tested and found to be optimal. It may be noted, however, it is not unique since the cell (2, 4) $\Delta_{ij} = 0$.

The solution is:

Through TV : 25 m people to reach in the age group 19–25, 10 m people to reach in the age group 36 and over

Through Radio : 30 m people to reach in the age group 13–18

Through Magazines : 15 m people to reach in the age group 26–35

Total minimum cost = Rs 71 lakh

TABLE 5.48 Improved Solution: Optimal

Media	*Age-groups*					*Max. Exp.*	u_i
	13–18	19–25	26–35	36 & over	Dummy		
TV	12 (−2)	7 [25]	10 (−1)	10 [10]	0 [5]	40	0
Radio	10 [30]	9 (−2)	12 (−3)	10 (0)	0 [ε]	30	0
Magazines	14 (−4)	12 (−5)	9 [15]	12 (−2)	0 [5]	20	0
Min. Exp	30	25	15	10	10	90	
v_j	10	7	9	10	0		

(ii) Since the optimal solution requires reaching 25 m people in the age group 19–25 through TV, it entirely meets the requirement of at least 8 m such exposures. Now, to ensure 4 m exposures through TV in the age group 13–18, we place a 4 in the cell (1, 1) and draw a closed path and make adjustments in the cells which lie on it. The cells include (1, 5), (2, 5) and (2, 1). The resulting solution is provided in Table 5.49.

TABLE 5.49 Revised Solution

Media	*Age-groups*					*Max. Exp.*
	13–18	19–25	26–35	*36 & over*	*Dummy*	
TV	12 [4]	7 [25]	10	10 [10]	0 [1]	40
Radio	10 [26]	9	12	10	0 [4]	30
Magazines	14	12	9 [15]	12	0 [5]	20
Min. Exp	30	25	15	10	10	90

The revised plan is:

Through TV	:	4 m people to reach in the age group 13–18, 25 m people to reach in the age group 19–25, 10 m people to reach in the age group 36 and above
Through Radio	:	26 m people to reach in the age group 13–18
Through Magazines	:	15 m people to reach in the age group 26–35

This would involve a total cost of Rs 71.8 lakh.

Example 5.12 A cement company has three factories which manufacture cement which is then transported to four distribution centres. The quantity of monthly production of each factory, the demand of each distribution centre and the associated transportation cost per quintal are given as follows:

		Distribution Centres				*Monthly Production (in quintals)*
		W	*X*	*Y*	*Z*	
	A	10	8	5	4	7,000
Factories	*B*	7	9	15	8	8,000
	C	6	10	14	8	10,000
Monthly Demand (in quintals)		6,000	6,000	8,000	5,000	

(i) Suggest the optimal transportation schedule.

(ii) Is there any other transportation schedule which is equally attractive? If so, write that.

(iii) If the company wants at least 5,000 quintals of cement are transported from factory *C* to distribution centre *Y*, will the transportation schedule be any different? If so, what will be the new optimal schedule and the effect on cost?

(iv) Suppose the company desires to send at most 500 quintals of cement from factory *C* to distribution centre *Y*, what will the optimal schedule be? Also, obtain the total transportation cost in such a case.

(v) The monthly production at factory *C* and the monthly demand at distribution centre *X* increase by 1,000 quintals each. What effect will it have on the optimal solution and on the total cost?

(i) The initial solution to the given problem, using VAM, is shown in Table 5.50.

TABLE 5.50 Initial Solution: VAM

Factory	*Distribution Centres*				*Supply*	u_i
	W	*X*	*Y*	*Z*		
A	10 (−14)	8 (−9)	[7,000] 5	4 (−6)	7,000	0
B	7 (−1)	[6,000] 9	[1,000] 15 −	[1,000] 8 +	8,000	10
C	[6,000] 6	10 (−1)	14 (1) +	[4,000] 8 −	10,000	10
Demand	6,000	6,000	8,000	5,000	25,000	
v_j	−4	−1	5	−2		

This solution is tested for optimality and found to be non-optimal. An improved solution is given in Table 5.51. This solution is found to be optimal.

The optimal solution is indicated below:

From ↓ To →	W	X	Y	Z
A			7,000	
B		6,000		2,000
C	6,000		1,000	3,000

Total cost = 5 × 7,000 + 9 × 6,000 + 8 × 2,000 + 6 × 6,000 + 14 × 1,000 + 8 × 3,000 = Rs 179,000

TABLE 5.51 Improved Solution: Optimal

Factory	Distribution Centres				Supply	u_i
	W	X	Y	Z		
A	10 (−13)	8 (−8)	[7,000] 5	4 (−5)	7,000	0
B	7 (−1)	[6,000] 9	15 (−1)	[2,000] 8	8,000	9
C	[6,000] 6	10 (−1)	[1,000] 14	[3,000] 8	10,000	9
Demand	6,000	6,000	8,000	5,000	25,000	
v_j	−3	0	5	−1		

(ii) No. There is no other transportation schedule which is equally attractive. The above solution is unique since all Δ_{ij} values are negative, and none equals zero.

(iii) The optimal solution calls for transporting 1,000 units from factory *C* to centre *Y*. If the company wants to send at least 5,000 quintals on this route, it would affect the present schedule and increase cost. The problem can be solved by reducing this quantity from supply at *C* and from demand at *Y* because it is intended to send this quantity on this route.

The solution to the revised problem is given in Table 5.52. It is obtained using VAM.

TABLE 5.52 Revised Problem I: Initial Solution

Factory	Distribution Centres				Supply	u_i
	W	X	Y	Z		
A	10 (−7)	8 (−3)	[3,000] 5	[4,000] 4	7,000	0
B	[1,000] 7	[6,000] 9	15 (−6)	[1,000] 8	8,000	4
C	[5,000] 6	10 (−2)	14 (−6)	8 (−1)	5,000	3
Demand	6,000	6,000	3,000	5,000	20,000	
v_j	3	5	5	4		

The solution is seen to be optimal. The total transportation cost is: 5 × 3,000 + 4 × 4,000 + 7 × 1,000 + 9 × 6,000 + 8 × 1,000 + 6 × 5,000 = Rs 130,000 plus 14 × 5,000 (route *C–Y*) = Rs 200,000. The schedule is:

From ↓ To →	*W*	*X*	*Y*	*Z*
A			3,000	4,000
B	1,000	6,000		1,000
C	5,000		5,000	

(iv) The optimal solution obtained in (i) involves despatching 1,000 quintals from factory *C* to centre *Y*. However, if not more than 500 quintals is to be sent on this route, then we would replace the original problem by one where supply at *C* and demand at *Y* would be reduced by 500 (sending 500 units on route *C–Y*) and replacing its cost element by *M* (so that no more units may be sent on this route). The revised problem is shown in Table 5.53, wherein initial solution is also provided. The solution, obtained through VAM, is found to be optimal.

The total cost involved is Rs 179,500.

TABLE 5.53 Revised Problem II: Initial Solution

Factory	*Distribution Centres*				*Supply*	u_i
	W	*X*	*Y*	*Z*		
A	10 (–14)	8 (–9)	[7,000] 5	4 (–6)	7,000	0
B	7 (–1)	[6,000] 9	[500] 15	[1,500] 8	8,000	10
C	[6,000] 6	10 (–1)	*M*	[3,500] 8	9,500	10
Demand	6,000	6,000	7,500	5,000	24,500	
v_j	–4	–1	5	–2		

(v) Reconsider the solution given in Table 5.51. The cell *CX* is an empty cell here. So, we draw a closed loop beginning with this cell as: *CX* → *CZ* → *BZ* → *BX* → *CX*. Thus, we add 1,000 to the quantities of the cells *CZ* and *BX* each and subtract it from the quantity of the cell *BZ*. The revised solution is:

	W	*X*	*Y*	*Z*	*Total*
A			7,000		7,000
B		7,000		1,000	8,000
C	6,000		1,000	4,000	11,000
Total	6,000	7,000	8,000	5,000	26,000

The total cost would increase by 1,000 × 9 + 1,000 × 0 = Rs 9,000 to become Rs 188,000. This can be verified by direct calculation as well.

Example 5.13 **ABC Enterprises has three plants manufacturing dry cells, located at different locations. Production cost differs from plant to plant. There are five offices of the company located in different regions of the country. The selling prices can differ from region to region. The shipping cost from each plant to each sales office and other information is given ahead.**

Production Information

Plant Number	*Maximum Capacity (in number of units)*	*Production Cost per unit*
1	150	20
2	200	22
3	125	18

Shipping Cost (Rs)

Plant Number	*Sales Office*				
	A	*B*	*C*	*D*	*E*
1	1	1	5	9	4
2	9	7	8	3	6
3	4	5	3	2	7

Demand and Selling Prices

	A	*B*	*C*	*D*	*E*
Demand (units):	80	100	75	45	125
Selling Price (Rs):	30	32	31	34	29

Find out the most profitable distribution schedule to the company.

(MIB, Delhi, 2008)

Using the given information, we first derive the profit matrix indicating the profit per unit obtainable when produced and sold in various combinations of plants and sales offices. To illustrate, a unit produced in plant 1 costs Rs 20 and its shipping to sales office *A* costs Re 1, where it sells for Rs 30. Accordingly, a unit produced in plant 1 and sold in sales office *A* would yield a profit of Rs 30 – (Rs 20 + Re1) = Rs 9. The profits for other combinations are similarly derived and are given in Table 5.54.

Further, since *AD* = 425 and *AS* = 475, the given problem is unbalanced. A dummy sales office *F* is therefore shown created, with demand equal to 475 – 425 = 50 units and zero profit values.

TABLE 5.54 Profit Matrix

Plant	*Sales Office*						*Supply*
	A	*B*	*C*	*D*	*E*	*F*	
1	9	11	6	5	5	0	150
2	– 1	3	1	9	1	0	200
3	8	9	10	14	4	0	125
Demand	80	100	75	45	125	50	475

To solve the problem, we first convert the profit matrix into an opportunity loss matrix, by subtracting each element from the largest value, equal to 14. The opportunity loss matrix is shown in Table 5.55. The table also contains initial solution using VAM, which is tested and found to be not optimal.

TABLE 5.55 Initial Solution—Non-optimal

Plant	*Sales Office*						*Supply*	u_i
	A	*B*	*C*	*D*	*E*	*F*		
1	[50] 5	[100] 3	8 (–5)	9 (–8)	9 (0)	14 (–4)	150	0
2	15 (–6)	11 (–4)	13 (–6)	[45] 5 –	[105] 13 +	[50] 14	200	4
3	[30] 6	5 (–1)	[75] 4	0 (2) +	[20] 10 –	14 (–3)	125	1
Demand	80	100	75	45	125	50	475	
v_j	5	3	3	1	9	10		

The revised solution is shown in Table 5.56. It is also tested for optimality and found to be optimal.

TABLE 5.56 Revised Solution—Optimal

Plant	*Sales Office*						*Supply*	u_i
	A	*B*	*C*	*D*	*E*	*F*		
1	[50] 5	[100] 3	8 (−5)	9 (−10)	9 (−2)	14 (−6)	150	0
2	15 (−4)	11 (−2)	13 (−4)	[25] 5	[125] 13	[50] 14	200	6
3	[30] 6	5 (−1)	[75] 4	[20] 0	10 (−2)	14 (−5)	125	1
Demand	80	100	75	45	125	50	475	
v_j	5	3	3	−1	7	8		

For the solution obtained in Table 5.56, the total profit can be calculated by summation of the products of the quantities in various cells and the unit profit values given in Table 5.54. The total profit works out to be equal to Rs 3,170.

Example 5.14 The *XYZ* Tobacco Company purchases tobacco and stores in warehouses located in the following four cities:

Warehouse Location	*Capacity (Tonnes)*
City *A*	90
City *B*	50
City *C*	80
City *D*	60

The warehouses supply tobacco to cigarette companies in three cities that have the following demand:

Cigarette Company	*Demand (Tonnes)*
Bharat	120
Janata	100
Red Lamp	110

The following railroad shipping costs per tonne (in hundred rupees) have been determined:

From	To →	Bharat	Janata	Red Lamp
A		7	10	5
B		12	9	4
C		7	3	11
D		9	5	7

Because of railroad construction, shipments are temporarily prohibited from warehouse at city *A* to Bharat Cigarette Company.

(a) Find the optimal distribution for *XYZ* Tobacco Company.
(b) Are there multiple optimal solutions? If there are alternative optimal solutions, identify them.
(c) Write the dual of the given transportation problem and use it for checking the optimal solution.

(MBA, Delhi, November, 2001)

(a) The given information is portrayed in Table 5.57. The problem is balanced by adding a dummy row and, further, a cost element of 7 in the given table, corresponding the cell *A*—Bharat, is replaced by *M* because this route is prohibited. The table also contains the initial solution given by Vogel's Approximation Method, which, when tested for optimality, is found to be optimal. Thus, the optimal solution is: $x_{13} = 90$, $x_{21} = 30$, $x_{23} = 20$, $x_{32} = 80$, $x_{41} = 40$, and $x_{42} = 20$, for a total cost of Rs 1,59,000.

TABLE 5.57 Initial Feasible Solution: Optimal

From ↓ / To →	Bharat	Janata	Red Lamp	Supply	u_i
A	M	(−1) 10	5 [90]	90	0
B	12 [30]	(−1) 9	4 [20]	50	−1
C	(0) 7	3 [80]	(−12) 11	80	−6
D	9 [40]	5 [20]	(−6) 7	60	−4
Dummy	0 [50]	(−4) 0	(−8) 0	50	−13
Demand	120	100	110	330	
v_j	13	9	5		

(b) Since $\Delta_{31} = 0$, there exists an alternate optimal solution. The alternate solution is: $x_{13} = 90$, $x_{21} = 30$, $x_{23} = 20$, $x_{31} = 40$, $x_{32} = 40$, and $x_{42} = 60$, for a total cost of Rs 1,59,000.

(c) The dual of the problem is:

Maximise $\quad Z = 90u_1 + 50u_2 + 80u_3 + 60u_4 + 50u_5 + 120v_1 + 100v_2 + 110v_3$

Subject to

$$\begin{array}{lll} u_1 + v_1 \le M & u_2 + v_3 \le 4 & u_4 + v_2 \le 5 \\ u_1 + v_2 \le 10 & u_3 + v_1 \le 7 & u_4 + v_3 \le 7 \\ u_1 + v_3 \le 5 & u_3 + v_2 \le 3 & u_5 + v_1 \le 0 \\ u_2 + v_1 \le 12 & u_3 + v_3 \le 11 & u_5 + v_2 \le 0 \\ u_2 + v_2 \le 9 & u_4 + v_1 \le 9 & u_5 + v_3 \le 0 \end{array}$$

u_i, v_j unrestricted in sign

$i = 1, 2, 3, 4, 5$ and $j = 1, 2, 3$

Using optimal values of u_i's and v_j's in the objective function, we get 90(0) + 50(–1) + 80(–6) + 60(–4) + 50(–13) + 120(13) + 100(9) + 110(5) = 1590 (hundred). This is the same as seen earlier.

KEY POINTS TO REMEMBER

- A transportation problem is a special type of linear programming problem, and hence can be formulated and solved as such.
- A transportation problem is said to be *balanced* if the aggregate supply is equal to the aggregate demand, and *unbalanced* if they are unequal.
- An unbalanced problem is balanced first by introducing a dummy source or destination, with cost elements equal to zero. If penalties are provided for not fulfilling the demand at destinations, then such penalties should be taken as the cost elements.
- If certain routes are prohibited then their cost elements are replaced by M—an extremely large value. If the problem is of the maximisation type, then the profit elements are replaced by $-M$.
- A maximisation type of problem is first converted into a minimisation one by subtracting each value of the given matrix from the largest value.
- To solve a problem by transportation method, first obtain an initial solution by either of the following methods.

  North-West Corner Rule, wherein starting with north-west corner of the table and moving in a zigzag manner ending at the south-east corner, quantities are assigned to various cells by considering demand and supply.

  Least Cost Method, in which least cost cell is detected and quantity is assigned. After this, the row/column satisfied is deleted and the least cost cell from the remaining ones is chosen. Move in the same manner until all requirements are met.

  Vogel's Approximation Method, in which difference between two least cost values for each row and column is obtained and quantity assigned to the least cost cell of the row/column with the largest difference. After deleting the row/column satisfied, move in the same manner until all requirements are met.
- The solution is then tested to determine whether it is optimal or not. A solution may be tested for optimality if the number of occupied cells equals $m + n - 1$. For this, the row

and column values are calculated from which Δ_{ij} are obtained. If they are all seen to be negative or zero, the solution is optimal.

- If some Δ_{ij} is equal to zero in the optimal solution, then the problem has multiple optimal solutions.
- If a solution is non-optimal, an improved solution may be obtained. For this, a closed path, beginning with the cell with the largest Δ_{ij} value, is traced with cells marked alternately as + and – signs. The least quantity in the cells with – sign is transferred between the cells and a new solution is obtained. It is again tested for optimality and improved if necessary.
- The solution to a problem is said be *degenerate* if the number of occupied cells is less than $m + n - 1$. The degeneracy, if present, is first removed by placing an infinitesimally small value in each of the required number of independent unoccupied cells.
- Once optimal solution to a transportation problem is obtained, it is possible to investigate the effects of changes in the cost elements of various routes, changes in available supply at a source or demand at a destination, and dispatching a certain maximum or minimum number of units on a given route.
- A transportation problem is regarded as a transhipment problem when shipments of goods is also allowed from one source to another and from one destination to another. Thus, a transportation problem with m-origins and n-destinations becomes a transhipment problem with $m + n$ sources and an equal number of destinations. With some minor modifications, it is solved as a transportation problem.

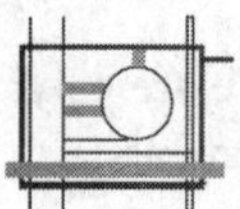

TEST YOUR UNDERSTANDING

Mark the following statements as T (*True*) or F (*False*).

1. A transportation problem is a special type of linear programming problem.
2. It is not necessary for the aggregate demand to be equal to the aggregate supply in a transportation problem.
3. A transportation problem is said to be unbalanced when the number of origins (sources) does not match with the number of destinations (markets).
4. An unbalanced transportation problem must be converted into a balanced problem before solving it.
5. It is possible that in some cases both, the dummy source and dummy destination, be required to be introduced to convert an unbalanced transportation problem into a balanced one.
6. The cost elements in the dummy row/column shall *always* be taken equal to zero.
7. The cost differences in the Vogel's Approximation Method indicate the penalties for not using the respective least cost routes.
8. VAM cannot be used to find an initial solution to a transportation problem if some routes are given to be prohibited.
9. The initial solution obtained by the least-cost-method would invariably be optimal.

10. The transportation method essentially uses the same steps as of the Simplex method.
11. To determine u_i and v_j values, an initial value has to be supplied. Different initial values would lead to different u_i and v_j values, and consequently, to different incoming variables.
12. In a non-optimal solution to a transportation problem, a certain cell with $i = 2$ and $j = 3$, has $\Delta_{23} = 4$. This implies that sending a unit from source 2 to destination 3 shall save a cost of 4.
13. A closed loop would always involve an even number of cells, subject to a minimum of 4.
14. The maximum number of units which can be transferred along the closed path is equal to the *minimum* quantity chosen from among the cells bearing a negative sign on the closed path.
15. The u_i and v_j values may be determined by initially inserting any finite number which may be positive, negative or zero, to a row/column.
16. Units sent from a dummy source to various markets represent the shortfall in supply to those markets.
17. The summation of the products of the u_i values with the corresponding supply column values and of the v_j values with the corresponding demand row values would be equal to the total cost of transportation.
18. A transportation problem solution is said to be degenerate if the number of occupied cells is smaller than the number of rows *plus* the number of columns *minus* 1 (one).
19. Once non-optimal degenerate solution is obtained, the next solution is bound, also, to be degenerate.
20. In improving a non-optimal solution to a transportation problem, it is possible that more than one cell may get vacated, although normally, only one cell gets vacated and gets filled.
21. A degenerate solution may or may not be optimal.
22. To remove degeneracy, an infinitesimally small quantity is placed in each of the required number of independent cells.
23. Multiple optimal solutions are indicated if there are multiple zeros for u_i and v_j values.
24. If each cost element in a transportation problem is increased by a constant amount, it will not affect the optimal solution to the problem.
25. For an optimal solution to a transportation problem, the u_i and v_j values represent the optimal values of the dual problem.
26. If all the cost elements, c_{ij}, are multiplied by a constant, the total cost of transportation in the optimal solution shall be multiplied by the same constant.
27. A cost reduction by an amount greater than the absolute value of Δ_{ij} for a given cell would make that route a preferable one.
28. If a constant is subtracted from each value of the matrix of a profit-maximising transportation problem, it is converted into a "minimisation problem."
29. A transhipment problem allows for the shipment of goods from one source to another, and from one destination to another.
30. An m-source, n-destination transportation problem, when written as a transhipment problem would have $m + n$ sources and n destinations.
31. In the context of transportation problems, sensitivity analysis deals only with investigating the effect of changes in the cost elements of some routes.
32. Once the optimal solution is obtained, any change in the cost element of an unoccupied cell does not alter the u_i and v_j values, as also the optimal solution.
33. If case of a change in the cost element of an occupied cell, the optimal solution does not change as long as changes in the u_i and v_j values do not cause any Δ_{ij} values to turn positive.

34. If the supply at ith source and the demand at jth destination are both increased by k units and the current basis remains optimal, then new Z value = old Z value + $ku_i + kv_j$.
35. In a transportation problem, suppose supply at source 3 and demand at destination 2 are increased by 300 units each and x_{32} is an unoccupied cell in the optimal solution, then the new optimal solution would have $x_{32} = 300$, and total cost increase by $300\,u_3 + 300\,v_2$.

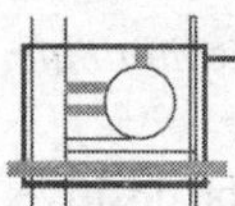

EXERCISES

1. Describe the transportation problem and give its mathematical model.
2. Explain, by taking an illustration, the North-West Corner Rule, the Least Cost Method and the Vogel's Approximation Method to obtain the initial feasible solution to a transportation problem.
3. Discuss the various methods of finding initial feasible solution of a transportation problem and state the advantages, disadvantages, and two areas of application for them.
4. Explain the transportation method of solving a transportation problem. Also give its schematic.
5. (a) What is meant by an optimality test? How would you determine whether a given transportation solution is optimal or not?

   (b) Compare the Stepping Stone and the MODI methods of testing the optimality of a solution to a transportation problem. Give suitable illustrations.
6. Write a note on tracing a closed loop. What are the characteristic features of a closed loop?
7. (a) What do you understand by unbalanced transportation problem? How would you convert it into balanced transportation problem?

   (b) What is the indication that a given transportation problem has multiple optimal solutions?

   (c) What would be the number of occupied routes in the optimal solution of an 'm-origins, n-destinations' transportation problem'?

   (d) What is the indication that a given transportation solution is degenerate?

   (e) How would you apply North-West Corner rule to find initial feasible solution if the first-source, first-destination route is prohibited?

   (f) How would you explain the interpretation of the optimum solution of an unbalanced transportation problem?
8. What is degeneracy? How does the problem of degeneracy arise in a transportation problem? How can we deal with this problem?
9. Why can degeneracy arise in the solution of

   (i) a transportation problem?

   (ii) a linear programming problem?

   In the case of a transportation problem, why is it necessary to resolve degeneracy before testing any basic feasible solution for optimality? *(M Com, Delhi, 1994)*
10. How can the transportation method be applied to a transportation problem where the objective function is called to be maximised?

11. Write the dual of a standard transportation problem and give its economic interpretation.
12. Write a detailed note on sensitivity analysis in the context of transportation problems.
13. Explain as to how we can use the transportation algorithm for scheduling of production in a manufacturing organisation whose product is subject to seasonal variation.
14. In what way is a transhipment problem different from a transportation problem? How can we use the transportation method for solving the transhipment problem?

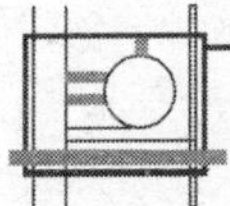

PRACTICAL PROBLEMS

1. The table below records transportation costs per unit of a product from origins O_1, O_2, O_3 and O_4 to destinations D_1, D_2, D_3, D_4 and D_5. The capacities of the four origins are respectively 55, 45, 30 and 50 while the requirements of the five destinations are respectively 40, 20, 50, 30 and 40. Employing Vogel's Approximation method, or otherwise, make the initial allocations to the origins to satisfy the requirements of the destinations and test the optimality of these allocations. Use Modified Distribution method (MODI) for obtaining an optimal solution that would minimise the total cost of transportation.

Origin	*Destination*					*Capacity*
	D_1	D_2	D_3	D_4	D_5	
O_1	12	4	9	5	9	55
O_2	8	1	6	6	7	45
O_3	1	12	4	7	7	30
O_4	10	15	6	9	1	50
Requirement	40	20	50	30	40	

Find the total cost implied by the solution. Is the solution unique? If not, find an alternative solution as well.

(M Com, Delhi, 2008)

2. A transportation problem has the supplies at four sources and requirements at five destinations. The following table shows the cost of shipping one unit from a particular source to a particular destination:

Source	*Destination*				
	1	2	3	4	5
1	15	7	12	8	12
2	11	4	9	9	10
3	4	15	7	10	10
4	13	18	9	12	4

The following feasible transportation pattern is proposed: $x_{11} = 25$, $x_{14} = 30$, $x_{22} = 20$, $x_{23} = 25$, $x_{31} = 15$, $x_{33} = 15$, $x_{43} = 10$ and $x_{45} = 40$, and all other $x_{ij} = 0$.

You are required to test whether this pattern has the least possible transportation cost. If so, how? If not, determine the optimal transportation pattern.

3. A Steel Company has three open-hearth furnaces and five rolling mills. Transportation cost (Rs per quintal) for shipping steel from furnaces to rolling mills are shown in the following table.

	M_1	M_2	M_3	M_4	M_5	*Supply*
F_1	4	2	3	2	6	8
F_2	5	4	5	2	1	12
F_3	6	5	4	7	7	14
Demand	4	4	6	8	8	

What is the optimum shipping schedule? *(CA, May, 1992)*

4. Solve the following transportation problem:

From	*To*				*Availability*
	1	*2*	*3*	*4*	
1	8	8	5	12	7
2	6	9	11	9	7
3	10	15	6	13	10
4	6	8	7	8	6
5	11	10	11	13	5
6	8	14	5	12	6
Demand	9	10	8	14	

5. The data relating to production capacities of plants, orders from warehouses and freight costs for a company manufacturing consumer products are given below:

Production capacities of plants

Plant	*Unit cost (Rs)*	*Capacities ('000 units)*
X	12	20
Y	10	16
Z	15	25

Warehouse order sizes

Warehouse	*Order size ('000 units)*
P	15
Q	18
R	12
S	14

Freight costs/unit

	P	*Q*	*R*	*S*
X	3	1	2	4
Y	6	4	3	2
Z	4	5	7	3

From the above information, determine the optimum distribution for minimising the total costs to make and ship the product to warehouses. *(MBA, Delhi, November, 1995)*

6. Solve the following transportation problem:

Source	*Destination*				*Supply*
	1	*2*	*3*	*4*	
1	15	18	22	16	30
2	15	19	20	14	40
3	13	16	23	17	30
Demand	20	20	25	35	100

Is the optimal solution obtained by you a unique one? If not, why? What are the alternate optima then?

7. Consider the following data for the transportation problem:

Factory	*Destination*			*Supply to be exhausted*
	1	*2*	*3*	
A	8	4	10	10
B	9	7	9	80
C	6	5	8	15
Demand	75	20	50	

Since there is not enough supply, some of the demands at the three destinations may not be satisfied. For the unsatisfied demands let the penalty costs be rupees 4, 5 and 6 for destinations (1), (2) and (3), respectively. Find the optimal allocation that minimises the transportation and penalty costs.

8. A company is spending Rs 1,200 on transportation of its units from three plants to four distribution centres. The supply and demand of units with unit cost of transportation are given as under:

Plants	Distribution centres				Supply
	1	2	3	4	
P_1	20	30	50	17	7
P_2	70	35	40	60	10
P_3	40	12	60	25	18
Demand	5	8	7	15	

What can be the maximum saving by optimal scheduling?

9. A company has three cement plants from which cement has to be transported to four distribution centres. With identical costs of production at the three plants, the only variable costs involved are transportation costs. The monthly demand at the four distribution centres and the distance from the plants to the distribution centres (in km) are given below:

Plants	Distribution centres				Monthly production (tonnes)
	W	X	Y	Z	
A	500	1,000	150	800	10,000
B	200	700	500	100	12,000
C	600	400	100	900	8,000
Monthly demand (tonnes)	9,000	9,000	10,000	4,000	

The transportation charges are Rs 10 per tonne per km. Suggest optimal transportation schedule and indicate the total minimum transportation cost.

If, for certain reasons, route from plant *C* to distribution centre *X* is closed down, will the transportation schedule change? If so, suggest the new schedule and effect on total cost.

(M Com, Delhi, 1998)

10. A company has four factories manufacturing the same commodity, which are required to be transported to meet the demands in four warehouses. The supplies and demands as also the cost per transportation from factory to warehouse in rupees per unit of product are given in the following table:

Factory	Warehouses				Supply (units)
	X	Y	Z	W	
A	25	55	40	60	60
B	35	30	50	40	140
C	36	45	26	66	150
D	35	30	41	50	50
Demand (units)	90	100	120	140	

(a) Derive an optimal strategy of transportation of goods from factories to warehouses and assess the optimal cost.

(b) If a new transporter agrees to transport goods from factory *C* to warehouse *W* at a unit cost of Rs 50, analyse the impact of this on your current optimal solution. *(DMS, Mumbai, 1998)*

11. The table given below has been taken from the solution procedure of a transportation problem, involving minimisation of cost (in rupees):

Factories	*Stockists*			*Monthly capacity (units)*
	X	*Y*	*Z*	
A	31 4	25 8	8	56
B	41 16	24	41 16	82
C	8	77 16	24	77
Monthly demand (Units)	72	102	41	

(i) Show that the above solution in not optimal.

(ii) Find an optimal solution.

(iii) Does the problem have multiple optimal solutions? Give reasons. If so, find one more optimal solution.

(iv) Comment upon the managerial significance of multiple optimal solutions.

(v) If it is considered necessary to transport 20 units from factory *A* to stockist *Z*, what will be the least-cost distribution schedule and the effect on cost?

(vi) If the transport cost from factory *A* to stockist *Z* is increased by Rs 5 per unit, will the solution change? If so, find the new solution. *(MBA, Ruhailkhand, 2003)*

12. The following table gives one of the possible solutions to a transportation probelm involving three sources and four destinations:

	D_1	D_2	D_3	D_4
S_1	10	200 6	100 18	23
S_2	150 4	9	50 13	10
S_3	7	13	50 15	350 5

(i) Is the above solution degenerate?

(ii) Is the above solution optimal? Is it a unique solution? If yes, why?

(iii) What is the opportunity cost of transporting one unit from source S_1 to destination D_4?

(iv) How would the cost be affected if it is decided to transport one unit from S_2 to D_2?

(MIB, Delhi, 2009)

13. ABC Limited has three production shops supplying a product to five warehouses. The cost of production varies from shop to shop and cost of transportation from one shop to a warehouse also varies. Each shop has a specific production capacity and each warehouse has certain amount of requirement. The costs of transportation are given as follows:

Shop	*Warehouses*					*Supply*
	I	*II*	*III*	*IV*	*V*	
A	6	4	4	7	5	100
B	5	6	7	4	8	125
C	3	4	6	3	4	175
Demand	60	80	85	105	70	

The cost of manufacturing the product at different production shops is:

Shop	:	*A*	*B*	*C*
Variable cost (Rs)	:	14	16	15

(a) Find the optimum distribution pattern so as to minimise cost.

(b) Identify alternate solution(s), if any.

(c) Write dual of the given transportation problem and use it for checking the optimal solution.

(MBA, Delhi, November 1997)

14. The purchase manager, Mr Shah of the State Road Transport Corporation must decide on amounts of fuel to buy from three possible vendors. The corporation refuels its buses regularly at four depots within the area of its operations.

The three oil companies have said that they can furnish up to the following amounts of fuel during the coming month: 2,75,000 litres by oil company 1; 5,50,000 litres by oil company 2; and 6,60,000 litres by oil company 3. The required amount of fuel is 1,10,000 litres at depot 1; 2,20,000 litres at depot 2: 3,30,000 litres at depot 3; and 4,40,000 litres at depot 4.

When the transportation costs are added to the bid price per litre supplied, the combined cost per litre of fuel from each vendor servicing a specific depot is shown here:

	Company 1	*Company 2*	*Company 3*
Depot 1	5.00	4.75	4.25
Depot 2	5.00	5.50	6.75
Depot 3	4.50	6.00	5.00
Depot 4	5.50	6.00	4.50

Determine the optimal schedule. Write the dual of the given problem and check optimal solution by using the dual. *(MBA, Delhi, 1984)*

15. Given:

Minimise $Z = 12x_{11} + 8x_{12} + 2x_{13} + 9x_{21} + 10x_{22} + 9x_{23} + 7x_{31} + 15x_{32} + 6x_{33}$

Subject to

$$\left.\begin{aligned} x_{11} + x_{12} + x_{13} &= 40 \\ x_{21} + x_{22} + x_{23} &= 30 \\ x_{31} + x_{32} + x_{33} &= 30 \end{aligned}\right\} \text{Supply constraints}$$

$$\left.\begin{aligned} x_{11} + x_{21} + x_{31} &= 20 \\ x_{12} + x_{22} + x_{32} &= 50 \\ x_{13} + x_{23} + x_{33} &= 30 \end{aligned}\right\} \text{Demand constraints}$$

$$x_{ij} \geq 0,\ i = 1, 2, 3;\quad j = 1, 2, 3$$

(a) Formulate and solve it as a transportation problem for optimal transportation pattern.

(b) Write the dual of the transportation problem and, from this, verify the objective function value obtained in (a) above.

16. The following table shows all necessary information on the availability of supply to each warehouse, the requirement of each market and the unit transportation cost from each warehouse to each market:

Warehouse	*Market*				*Supply*
	P	*Q*	*R*	*S*	
A	6	3	5	4	22
B	5	9	2	7	15
C	5	7	8	6	8
Requirement	7	12	17	9	

The shipping clerk has worked out the following schedule from experience: 12 units from *A* to *Q*, 1 unit from *A* to *R*, 9 units from *A* to *S*, 15 units from *B* to *R*, 7 units from *C* to *P* and 1 unit from *C* to *R*.

(a) Check and see if the clerk has the optimal schedule.

(b) Find the optimal schedule and minimum total transport cost.

(c) If the clerk is approached by a courier to route *C* to *Q*, who offers to reduce his rate in the hope of getting some business, by how much the rate should be reduced that the clerk will offer him the business?

17. A company has four warehouses and five stores. The warehouses have total surplus of 430 units of a given commodity that is divided among them as follows:

Warehouse	:	W_1	W_2	W_3	W_4
Surplus	:	150	30	120	130

The five stores have, in all, a requirement of 450 units of the commodity. Individual requirements are:

Store	:	S_1	S_2	S_3	S_4	S_5
Requirement	:	80	60	20	210	80

Cost of shipping one unit from the *i*th warehouse to the *j*th store is as displayed in the following table:

Warehouse	*Store* S_1	S_2	S_3	S_4	S_5
W_1	9	12	10	10	6
W_2	5	18	12	11	2
W_3	10	M	7	3	20
W_4	5	6	2	M	8

M in this table indicates that the route is not available.

How should the company arrange to transport the units so that the transportation cost is minimised?

18. A firm is faced with a short-term cash flow problem, which would necessitate obtaining a loan from its bank. The bank loan will be used to balance the cash inflows from accounts receivable and cash outflows from accounts payable, which are estimated as follows:

	Accounts receivable (in Rs lakh)	*Accounts payable (in Rs lakh)*
October	18	20
November	27	32
December	35	43

It may be assumed that both accounts receivable and accounts payable are due at the end of the relevant month and that all accounts payable are to be settled by the end of the year (December). While in any month funds from accounts receivable would flow at a time such that sufficient time would be available to finance firm's own payments of the months, in October and November, the payments to be made to the suppliers could be delayed by at the most one month. However, the delay in the payments would make the firm lose the 2 per cent discount receivable for making payment within one month.

Information about the bank loan is as follows:

(i) Any amount of loan may be taken by the firm but a loan be agreed at the start of a given month.

(ii) The loan would carry an interest rate at the rate of 2.5 per cent per month and attract a minimum of one month's interest.

Further, any surplus cash may be invested with the bank, which would earn an interest 1 per cent per month.

Using the above information, formulate the problem as a transportation problem by appropriately defining the costs and flows. Also determine the optimal solution to the problem.

19. A transportation problem involving three sources and four distribution centres is presented in the following table.

    You are required to answer the following questions:

    (a) Is the solution feasible?

    (b) Is the solution degenerate?

    (c) Is the solution optimal? Is it unique? Why?

    (d) What is the opportunity cost of transporting one unit from source S_1 to D_4?

    (e) What are the optimal values of the dual variables to the given problem? Show that the primal and dual both have same objective function values.

    (f) What do u_i and v_j values indicate?

    (g) How would the cost be affected if it is decided to transport one unit from S_2 to D_2?

    (h) The transporter is willing to provide a discount of 25 per cent on the freight on the route S_1 D_4, provided that at least 50 units are sent on this route. Should the offer be accepted?

    (i) For some reason, the output at source S_1, is reduced by two units and, to compensate for the loss of output, the supply at source S_2 in increased by as many units. Would it lead to a change in the cost? By how much?

    (j) The management is contemplating to embark on an advertisement campaign in one of the distribution centres. Where sould it concentrate?

Source	*Distribution Centre* D_1	D_2	D_3	D_4	*Supply*	u_i
S_1	12 (−1)	[250] 8	[150] 20	25 (−15)	400	10
S_2	[200] 6	11 (−8)	[100] 15	12 (−7)	300	5
S_3	9 (−1)	15 (−10)	[100] 17	[400] 7	500	7
Demand	200	250	350	400	1,200	
v_j	1	−2	10	0		

20. A company has three plants in which it produces a standard product. It has four agencies in different parts of the country where this product is sold. The production cost varies from factory to factory and the selling price from market to market. The shipping cost per unit of the product from each market to each of the agencies is known and stable. The relevant data are given in the following table:

(a) Plant	*Weekly production capacity (Units)*	*Unit production cost (Rs)*
1	400	18
2	300	24
3	800	20

(b) Shipping cost (in Rs) per unit:

		To agency			
		1	*2*	*3*	*4*
From plant	1	2	5	7	3
	2	8	4	6	2
	3	3	4	4	5

(c) Agency	*Demand (Units)*	*Selling price (Rs)*
1	300	32
2	400	35
3	300	31
4	500	36

Determine the optimal plan so as to maximise the profits.

21. A company wishes to determine an investment strategy for each of the next four years. Five investment types have been selected, investment capital has been allocated for each of the coming four years, and maximum investment levels have been established for each investment type. An assumption is that amounts invested in any year will remain invested until the end of the planning horizon of four years. The following table summarises the data for this problem. The values in the body of the table represent net return on investment of one rupee to the end of the planning horizon. For example, a rupee invested in investment type *B* at the beginning of year 1 will grow to Rs 1.90 by the end of the fourth year, yielding a net return of Rs 0.90.

Investment made at the beginning of year	*Investment type* *A*	*B*	*C*	*D*	*E*	*Rupees available (in '000)*
	(Net return data)					
1	0.80	0.90	0.60	0.75	1.00	500
2	0.55	0.65	0.40	0.60	0.50	600
3	0.30	0.25	0.30	0.50	0.20	750
4	0.15	0.12	0.25	0.35	0.10	800
Maximum rupee investment ('000)	750	600	500	800	1,000	

The objective in this problem is to determine the amount to be invested in the beginning of each year in an investment type so as to maximise the net rupee return for the four-year period.

Solve the above transportation problem and get an optimal solution. Also calculate the net return on investment for the planning horizon of four-year period. *(CA, May, 1993)*

22. A company is producing three products P_1, P_2 and P_3 at two of its plants situated in cities *A* and *B*. The company plans to start a new plant either in city *C* or in city *D*. The unit profits from the various plants are listed in the table, along with the demand for various products and capacity available in each of the plants.

Plant	*Product*			*Capacity*
	P_1	P_2	P_3	
A	35	24	20	600
B	30	28	25	1,000
C	20	25	37	800
D	24	32	28	800
Demand	500	800	600	

The company would set up the new plant on the basis of maximising aggregate profits from the three cities' plants. Using the transportation method determine in which city would the plant be set up and what would the corresponding profit be.

23. Stronghold Construction Company is interested in taking loans from banks for some of its projects, *P*, *Q*, *R*, *S*, *T*. The rates of interest and the lending capacity differ from bank to bank. All these projects are to be completed. The relevant details are provided in the following table. Assuming the role of a consultant, advise this company as to how it should take the loans so that the total interest payable will be the least. Are there alternate optimum solutions? If so, indicate one such solution.

Bank	*Interest rate in percentage for project*					*Max. credit (in thousands)*
	P	*Q*	*R*	*S*	*T*	
Pvt. bank	20	18	18	17	17	Any amount
Nationalised	16	16	16	15	16	400
Co-operative Bank	15	15	15	13	14	250
Amount required (in thousands)	200	150	200	125	75	

24. XYZ and Co has provided the following data seeking your advice on optimum investment strategy:

Investment made at the beginning of year	*Net return data (in paise) of selected investments*				*Amount available (Lac)*
	P	*Q*	*R*	*S*	
1	95	80	70	60	70
2	75	65	60	50	40
3	70	45	50	40	90
4	60	40	40	30	30
Maximum Investment (Lac)	40	50	60	60	

The following additional information is also provided:

(i) *P*, *Q*, *R* and *S* represent the selected investments.

(ii) The company has decided to have four years' investment plan.

(iii) The policy of the company is that amount invested in any year will remain so until the end of the fourth year.

(iv) The values (paise) in the table represent net return on investment of one Rupee till the end of the planning horizon (for example, a Rupee invested in Investment *P* at the beginning of year 1 will grow to Rs 1.95 by the end of the fourth year, yielding a return of 95 paise).

Using the above, determine the optimum investment strategy. *(CA, November, 1996)*

25. The personnel manager of a manufacturing company is in the process of filling 175 jobs in six different entry level skills due to the establishment of a third shift by the company. Union wage scales and requirements for the skills are shown in the following table:

		Pay scales and skill requirements					
Entry level skill	:	*A*	*B*	*C*	*D*	*E*	*F*
Wage scale (Rs PM)	:	1,000	1,100	1,200	1,300	1,400	1,500
No. required	:	25	29	31	40	33	17

Two hundred and thirty applicants for jobs have been tested and their aptitudes and skills for the jobs in question have been matched against company standard and evaluated. The applicants have been grouped into four categories by their abilities, the grouping and values of each category to the company are shown in the table that follows:

Applicant category	*Category value (Rs PM)*						*No. of applicants*
	A	*B*	*C*	*D*	*E*	*F*	
I	1,000	1,100	1,500	1,400	1,400	1,450	54
II	1,200	1,250	1,200	1,350	1,400	1,400	57
III	1,000	1,100	1,200	1,400	1,500	1,600	45
IV	1,500	1,500	1,600	1,400	1,400	1,500	74

How many applicants of each category should the personnel manager hire and for which jobs?

(MBA, Delhi, December, 1986)

26. The demand and production costs vary from month to month in an industry. The following table contains budgeted information of a firm in this industry on the quantity demanded, the production cost per unit and the production capacity in each of the coming five months.

Month	:	*Jan*	*Feb*	*Mar*	*Apr*	*May*
Demand	:	200	250	150	80	120
Production Cost	:	24	27	32	30	34
Capacity	:	250	225	250	200	225

It is known that the production in any month can meet demand in that month or can be held for the future. The holding cost is Rs 5 per unit per month.

Using this information (a) formulate a transportation model to assist in the production planning, and (b) determine what would the optimal plan be.

27. A manufacturer must produce a product in sufficient quantity to meet contractual sales in the next four months. The production capacity and unit cost of production vary from month to month. The product produced in one month may be held for sale in later months, but at an estimated storage cost of Re 1 per unit per month. No storage cost is incurred for goods sold in the same month in which they are produced. There is no opening inventory and none is desired at the end of the four months. The necessary details are given in the following table:

Month	*Contracted sales*	*Maximum production*	*Unit cost of production*
1	20	40	14
2	30	50	16
3	50	30	15
4	40	50	17

The management wishes to know how much to produce each month in order to minimise total cost.

(a) Formulate the problem as a transportation problem by constructing the appropriate cost and requirements table.

(b) Solve the problem as formulated in (a).

28. A company produces two models of a product: Standard and Deluxe. The production cost of the standard model during normal working hours is Rs 400, while the deluxe model costs Rs 750. It is possible to produce both these models through overtime working at a unit cost of Rs 450 and Rs 820, respectively. Inventory carrying cost of a unit of standard model is Rs 6 per month while that for a deluxe unit is Rs 10. According to the present capacity each month, during the normal working time, a maximum of 300 units (both standard and deluxe combined) can be produced, while the overtime capacity is 200 units per month. Given the following demand projections, what production schedule should the management adopt so that the total demand may be met at the lowest cost?

Model	*Month*		
	Jan	*Feb*	*Mar*
Standard	250	300	320
Deluxe	100	180	175

29. (a) Find optimal solution to the transportation problem given in the following table:

Origin	*Destination*					*Availability*
	B_1	B_2	B_3	B_4	B_5	
A_1	71	70	57	21	50	400
A_2	55	68	97	50	53	800
A_3	58	50	42	58	27	400
A_4	66	51	93	35	33	400
Requirement	280	360	460	680	220	2,000

(b) For the problem given in (a), assume that transhipment will be allowed from any origin to another origin and from any destination to another destination or origin. The unit transportation costs between various points are given below:

Origin	A_2	A_3	A_4	*Destination*	B_2	B_3	B_4	B_5
A_1	20	23	19	B_1	48	28	24	38
A_2		34	17	B_2		39	54	45
A_3			55	B_3			36	25
				B_4				52

Reformulate the problem and solve it for optimal solution.

30. Using the information given here, determine the optimal solution under the assumption (a) that inter-plant and inter-warehouse transfers are not allowed, and (b) that the units may be transhipped.

Manufacturing Plants	:	P_1		P_2
Capacity	:	240 units		160 units
Warehouses	:	W_1	W_2	W_3
Requirements	:	80	120	200
Transportation Costs	:	Exactly proportional to the distances (given in miles in the following figure)		

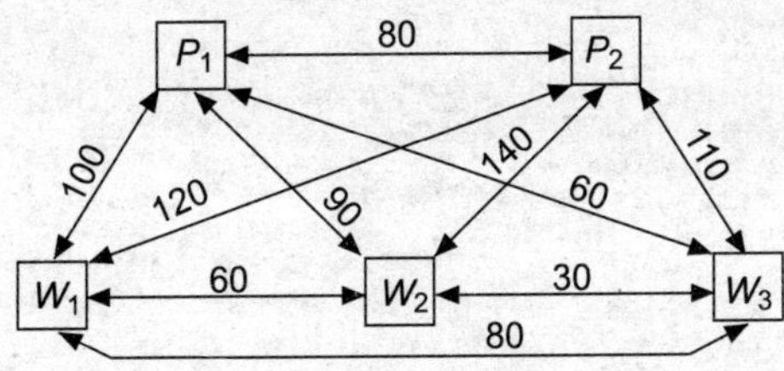

Distance Profile

31. Solve the following transhipment problem for minimum cost:

Origin	*Destination*			*Availability*
	D	*E*	*F*	
A	6	4	1	50
B	3	8	7	40
C	4	4	2	60
Demand	20	95	35	150

Origin	*A*	*B*	*C*	*Destination*	*D*	*E*	*F*
A	0	3	2	*D*	0	2	5
B	3	0	4	*E*	2	0	1
C	2	4	*0*	*F*	5	1	0

CHAPTER

6

Specially Structured Linear Programmes II: Assignment Problem

CHAPTER OVERVIEW

In every workplace, there are jobs to be done and there are people available to do them. But everyone is not equally efficient at every job. Someone may be more efficient on one and less efficient on the other job, while it might be otherwise for someone else. The relative efficiency is reflected in terms of the time taken for, or the cost associated with, performance of different jobs by different people. An obvious problem for a manager to handle is to assign jobs to various workers in a manner that they can be done in the most efficient way. Interestingly, such problems can be formulated as linear programming problems or as transportation problems and solved as such, but a method, called Hungarian Assignment Method provides an easy route. It allows a manager to obtain answers to the questions like the following:

- How to assign the given jobs to some workers on a one-to-one basis when completion time of performance is given for each combination and it is desired that the jobs are completed in the least time or at the least cost. Further, in the assignment pattern so obtained, is only one option available or there are other equally attractive ones also?
- How to deal with situations when the number of jobs do not match with the number of job performers, when some job(s) cannot be performed by, or is not be given to, a particular performer, or when a certain job has to be given to a particular individual?
- How should the salesmen of a company be assigned to different sales zones so that the total expected sales are maximised?
- Given the order of preferences that different managers have for the different rooms in a hotel on one of its floors, what pattern of assignment of rooms to the managers will satisfy their requirements the most?
- How to schedule the flights or the bus routes between two cities so that the layover times for the crew can be minimised?

In fact, the assignment method works for any problems in which one-to-one matching is called for in the light of the given payoffs, where the total payoff is sought to be minimised or maximised.

Only ordinary arithmetical skills are required for working through this chapter. Also needed for some part of it is familiarity with the summation notation, algebraic inequalities, matrices and their transpose.

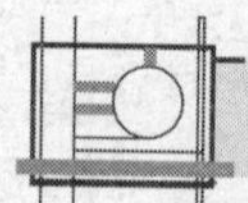

6.1 INTRODUCTION

In the previous chapter, we discussed about solution to the transportation problem. Now we consider another type of special linear programming problem, called the *assignment problem.*

There are many situations where the assignment of people, machines, and so on may be called for. Assignment of workers to machines, clerks to various checkout counters, salesmen to different sales areas, service crews to different districts, are typical examples of these. Assignment is a problem because people possess varying abilities for performing different jobs and, therefore, the costs of performing those jobs by different people are different. Obviously, if all persons could do a job in the same time or at the same cost then it would not matter who among them is assigned the job. Thus, in an assignment problem, the question is how the assignments should be made in order that the total cost involved is minimised (or the total value is maximised when pay-offs are given in terms of, say, profits). A typical assignment problem follows:

Example 6.1 A production supervisor is considering how he should assign the four jobs that are to be performed, to four of the workers. He wants to assign the jobs to the workers such that the aggregate time to perform the jobs is the least. Based on previous experience, he has the information on the time taken by the four workers in performing these jobs, as given in Table 6.1.

TABLE 6.1 Time Taken by Workers on Various Jobs (in minutes)

Worker	*Job*			
	A	*B*	*C*	*D*
1	45	40	51	67
2	57	42	63	55
3	49	52	48	64
4	41	45	60	55

We shall consider analytically the solution to this problem of the supervisor in the later sections.

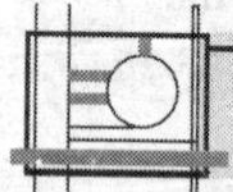

6.2 ASSIGNMENT PROBLEM: A VARIANT OF THE TRANSPORTATION PROBLEM

The assignment problem can be expressed as a transportation problem also. The various cost (or time) elements are taken as given, and the availability at each of the sources and the requirement at each of the destinations is taken to be 1. Further, as assignments are made on a one-to-one basis, the numbers of origins and destinations should be made exactly equal. The x_{ij} values should equal 0 or 1, so that for a job assigned to a worker x_{ij} would be 1 and other values in that (*i*th) row and the (*j*th) column should equal to 0.

Thus, an assignment problem is a variation of the transportation problem with two characteristics. First, the pay-off matrix for the problem would be a square matrix, and, second, the optimal solution to the problem would always be such that there would be only one assignment in a given row or column of the pay-off matrix.

The information given in Example 6.1 is depicted in Table 6.2 as a transportation model. This is a 4 × 4 assignment problem. Here the c_{ij}'s are the cost coefficients and a particular value of c_{ij} indicates the cost (although times taken by different workers are given, they are taken to be 'cost' since it is sought to minimise the total time and hence the cost) of assigning *j*th job to the *i*th worker. The solution to this problem consists of assigning x_{ij}'s one of the two values—0 or 1—in such a manner that the total assignment cost is the minimum.

TABLE 6.2 The Assignment Problem as a Transportation Problem

Worker	*Job*				*Supply*
	A	*B*	*C*	*D*	
1	45	40	51	67	1
2	57	42	63	55	1
3	49	52	48	64	1
4	41	45	60	55	1
Demand	1	1	1	1	4

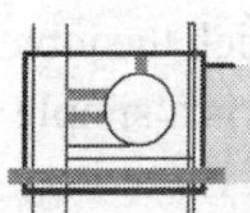

6.3 SOLUTION TO THE ASSIGNMENT PROBLEM

There are four methods of solving an assignment problem: (a) complete enumeration method; (b) transportation method; (c) simplex method; and (d) Hungarian assignment method. We shall discuss these methods now.

6.3.1 Complete Enumeration Method

In this method, all possible assignments are listed out and the assignment involving the minimum cost (or maximum profit if the problem is of the maximisation type) is selected. This represents the optimal solution. In case there are more than one assignment patterns involving the same least cost, they all shall represent optimal solutions—the problem has multiple optima then.

For the problem given in Example 6.1, there are a total of 24 assignments possible. These are listed in Table 6.3 wherein the time involved with each assignment pattern is also given.

It is clear that assignment number 20 listed in the table that involves worker-job assignments 4*A*, 1*B*, 3*C* and 2*D* is the optimal one as it involves the least total time, equal to 184 minutes.

TABLE 6.3 Worker-Job Assignments

S.No.	Assignment	Time	S.No.	Assignment	Time
1.	1A, 2B, 3C, 4D	190	13.	3A, 1B, 2C, 4D	207
2.	1A, 2B, 4C, 3D	211	14.	3A, 1B, 4C, 2D	204
3.	1A, 3B, 2C, 4D	215	15.	3A, 2B, 4C, 1D	218
4.	1A, 3B, 4C, 2D	212	16.	3A, 2B, 1C, 4D	197
5.	1A, 4B, 2C, 3D	217	17.	3A, 4B, 1C, 2D	200
6.	1A, 4B, 3C, 2D	193	18.	3A, 4B, 2C, 1D	224
7.	2A, 1B, 3C, 4D	200	19.	4A, 1B, 2C, 3D	208
8.	2A, 1B, 4C, 3D	221	20.	4A, 1B, 3C, 2D	184*
9.	2A, 3B, 1C, 4D	215	21.	4A, 2B, 1C, 3D	198
10.	2A, 3B, 4C, 1D	236	22.	4A, 2B, 3C, 1D	198
11.	2A, 4B, 3C, 1D	207	23.	4A, 3B, 1C, 2D	199
12.	2A, 4B, 1C, 3D	217	24.	4A, 3B, 2C, 1D	223

It may be noted that, in general, for a n-worker n-job problem, there are $n!$ number of ways in which the jobs can be assigned to the workers. Thus, the listing and evaluation of all the possible assignments is a simple matter when n is a small number. But, when n is an even moderately large number, this method is not very practical. For example, for a 5-person and 5-job problem, a total of $5! = 120$ assignments will need to be evaluated, while a problem involving 8 persons and 8 jobs will require enumeration and evaluation of as many as $8! = 40,320$ assignments, and a problem with 10 workers and as many jobs, the number of possible assignments works out to be 36,28,800. Therefore, the method is not suitable for real world situations.

6.3.2 Transportation Method

We have observed earlier that an assignment problem can be represented as a classical transportation problem. As such, the problem is amenable to solution by applying the transportation method discussed in the previous chapter. But there is a significant point to note here. The solution obtained by applying this method would be severely degenerate. This is because the optimality test in the transportation method requires that there must be $m + n - 1$ ($= 2n - 1$, when $m = n$) basic variables. For an assignment problem of the order $n \times n$, however, there would be only n basic variables in the solution because here n assignments are required to be made. Thus, a large number of epsilons may be required to be introduced in the solution in order to proceed with this method.

Example 6.2 Solve the assignment problem given in Example 6.1 using transportation method.

The problem is reproduced in Table 6.4, wherein the initial solution using VAM is contained. There are only 4 occupied cells against the required number of 4 + 4 – 1 = 7, and hence the solution is degenerate.

TABLE 6.4 Initial Solution: VAM

Worker	*Job*				*Supply*
	A	*B*	*C*	*D*	
1	[ε] 45	[ε] 40	51	[1] 67	1
2	57	[1] 42	63	55	1
3	[ε] 49	52	[1] 48	64	1
4	[1] 41	45	60	55	1
Demand	1	1	1	1	4

To remove degeneracy, an ε is placed in three independent cells. The solution is reproduced in Table 6.5 and tested for optimality. With a total cost equal to 67 + 42 + 48 + 41 = 198, this solution is found to be non-optimal. Accordingly, a closed path is drawn, starting with the cell 2-*D*, which has the largest Δ_{ij} value equal to 14.

The revised solution is given in Table 6.6. It is tested for optimality. The Δ_{ij} values indicate that it is indeed optimal.

The solution is:

Worker	*Job*	*Time*
1	*B*	40
2	*D*	55
3	*C*	48
4	*A*	41
		Total 184

It is identical to the one obtained earlier.

TABLE 6.5 Initial Solution: Non-optimal

Worker	Job A	Job B	Job C	Job D	Supply	u_i
1	[ε] 45	[ε] + 40	51 (−7)	[1] − 67	1	0
2	57 (−10)	[1] − 42	63 (−17)	+ 55 (14)	1	2
3	[ε] 49	52 (−8)	[1] 48	64 (7)	1	4
4	[1] 41	45 (−9)	60 (−20)	55 (8)	1	−4
Demand	1	1	1	1	4	
v_j	45	40	44	67		

TABLE 6.6 Revised Solution: Optimal

Worker	Job A	Job B	Job C	Job D	Supply	u_i
1	[ε] 45	[1] 40	51 (−7)	67 (−14)	1	0
2	57 (−10)	[ε] 42	63 (−17)	[1] 55	1	2
3	[ε] 49	52 (−8)	[1] 48	64 (−7)	1	4
4	[1] 41	45 (−9)	60 (−20)	55 (−6)	1	−4
Demand	1	1	1	1	4	
v_j	45	40	44	53		

6.3.3 Simplex Method

We have seen earlier that an assignment problem can be formulated as a transportation problem which, in turn, is a special type of linear programming problem. Accordingly, an assignment problem can be formulated as a linear programming problem (with integer variables) and solved using a modified simplex method or otherwise. This is discussed in detail in Chapter 7. It may be noted that its formulation requires that the decision variables take only one of the two values 1 or 0, accordingly as an assignment (of a worker to a job) is made or not.
In general, let

$$x_{ij} = \begin{cases} 1 \text{ if } i\text{th person is assigned the } j\text{th job.} \\ 0 \text{ if } i\text{th person is not assigned the } j\text{th job} \end{cases}$$

The problem would be:

Minimise $$Z = \sum_{i=1}^{n} \sum_{j=1}^{n} c_{ij} x_{ij}$$

Subject to

$$\sum_{j=1}^{n} x_{ij} = 1 \qquad \text{for } i = 1, 2, \ldots, n$$

$$\sum_{i=1}^{n} x_{ij} = 1 \qquad \text{for } j = 1, 2, \ldots, n$$

$$x_{ij} = 0 \text{ or } 1 \qquad \text{for all } i \text{ and } j.$$

Example 6.1 data can be formulated as a LPP as follows:

Minimise $$Z = 45x_{11} + 40x_{12} + 51x_{13} + 67x_{14} + 57x_{21} + 42x_{22} + 63x_{23} + 55x_{24} + 49x_{31} + 52x_{32} + 48x_{33} + 64x_{34} + 41x_{41} + 45x_{42} + 60x_{43} + 55x_{44}$$

Subject to

$$x_{11}+x_{12}+x_{13}+x_{14} = 1, \qquad x_{11}+x_{21}+x_{31}+x_{41} = 1,$$

$$x_{21}+x_{22}+x_{23}+x_{24} = 1, \qquad x_{12}+x_{22}+x_{32}+x_{42} = 1,$$

$$x_{31}+x_{32}+x_{33}+x_{34} = 1, \qquad x_{13}+x_{23}+x_{33}+x_{43} = 1,$$

$$x_{41}+x_{42}+x_{43}+x_{44} = 1, \qquad x_{14}+x_{24}+x_{34}+x_{44} = 1,$$

$$\text{all } x_{ij}\text{'s} = 0 \text{ or } 1$$

It may be easily visualised that it is relatively tedious to solve the problem in this format. Thus, this approach to the solution is not considered.

6.3.4 Hungarian Assignment Method (HAM)

It may be observed that none of the three working methods discussed earlier to solve an assignment is efficient. A method, designed specially to handle the assignment problems in an efficient way, called the *Hungarian*

Assignment Method, is available, which is based on the concept of opportunity cost. It is shown schematically in Figure 6.1, and is discussed here. For a typical balanced assignment problem involving a certain number of persons and an equal number of jobs, and with an objective function of the minimisation type, the method is applied as listed in the following steps:

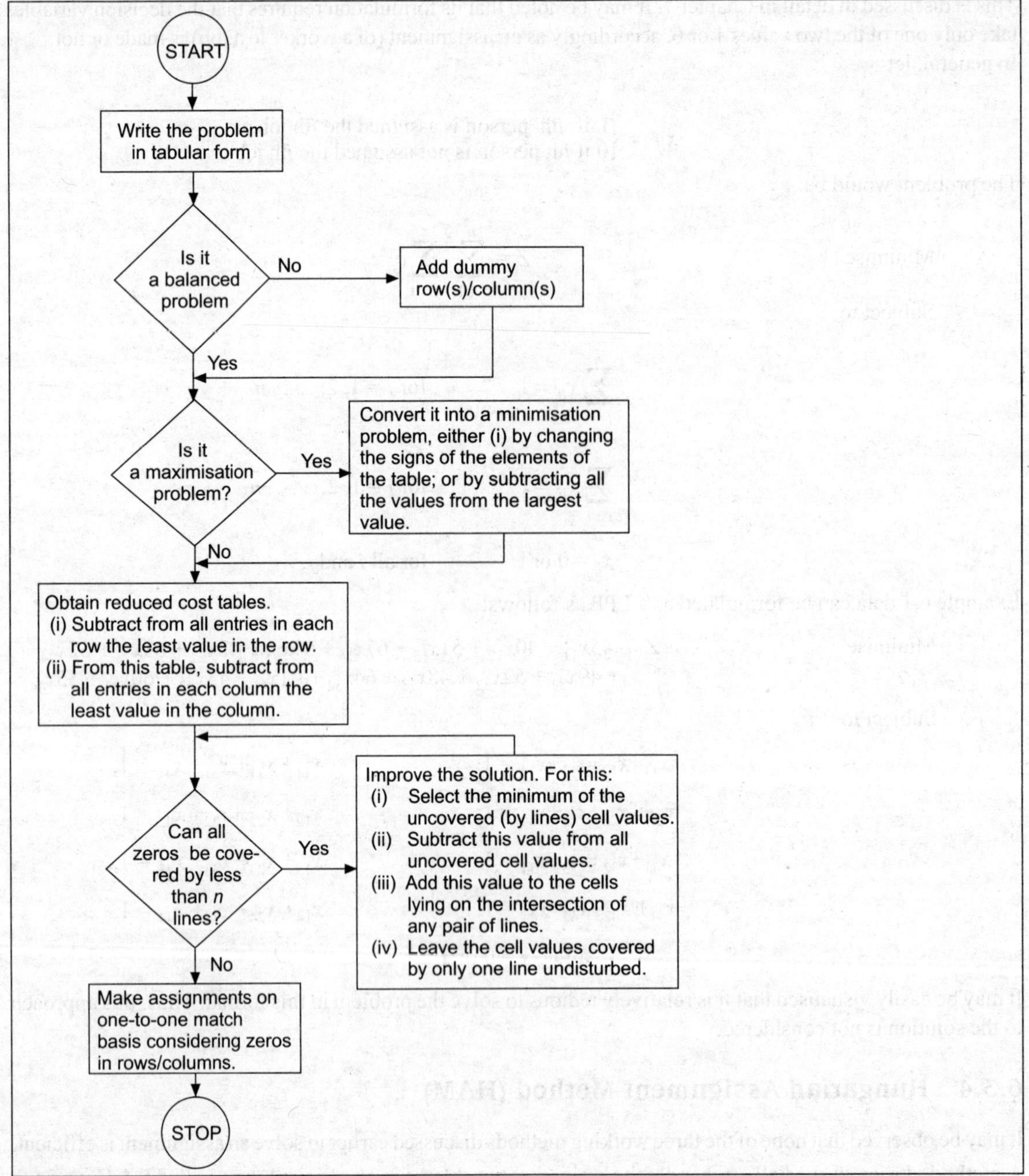

Figure 6.1 *Schematic Presentation of Hungarian Assignment Method*

Step 1 Locate the smallest cost element in each row of the cost table. Now subtract this smallest element from each element in that row. As a result, there shall be at least one zero in each row of this new table, called the *Reduced Cost Table.*

Step 2 In the reduced cost table obtained, consider each column and locate the smallest element in it. Subtract the smallest value from every other entry in the column. As a consequence of this action, there would be at least one zero in each of the rows and columns of the second reduced cost table.

Step 3 Draw the *minimum* number of horizontal and vertical lines (not the diagonal ones) that are required to cover all the 'zero' elements. If the number of lines drawn is equal to n (the number of rows/columns) the solution is optimal, and proceed to step 6. If the number of lines drawn is smaller than n, go to step 4.

Step 4 Select the *smallest uncovered* (by the lines) cost element. *Subtract* this element from all uncovered elements including itself and *add* this element to each value located at the intersection of any two lines. The cost elements through which only one line passes remain unaltered.

Step 5 Repeat steps 3 and 4 until an optimal solution is obtained.

Step 6 Given the optimal solution, make the job assignments as indicated by the 'zero' elements. This is done as follows:

(a) Locate a row which contains only one 'zero' element. Assign the job corresponding to this element to its corresponding person. Cross out the zeros, if any, in the column corresponding to the element, which is indicative of the fact that the particular job and person are no more available.
(b) Repeat (a) for each of such rows which contain only one zero. Similarly, perform the same operation in respect of each column containing only one 'zero' element, crossing out the zero(s), if any, in the row in which the element lies.
(c) If there is no row or column with only a single 'zero' element left, then select a row/column arbitrarily and choose one of the jobs (or persons) and make the assignment. Now cross the remaining zeros in the column and row in respect of which the assignment is made.
(d) Repeat steps (a) through (c) until all assignments are made.
(e) Determine the total cost with reference to the original cost table.

Example 6.3 Solve the assignment problem given in Example 6.1 for optimal solution using HAM. The information is reproduced in Table 6.7.

TABLE 6.7 Time Taken (in minutes) by 4 Workers

Worker	*Job*			
	A	*B*	*C*	*D*
1	45	40	51	67
2	57	42	63	55
3	49	52	48	64
4	41	45	60	55

The solution to this problem is given here in a step-wise manner.

Step 1 The minimum value of each row is subtracted from all elements in the row. It is shown in the reduced cost table, also called *Opportunity Cost Table,* given in Table 6.8.

TABLE 6.8 Reduced Cost Table 1

Worker	*Job*			
	A	*B*	*C*	*D*
1	5	0	11	27
2	15	0	21	13
3	1	4	0	16
4	0	4	19	14

Step 2 For each column of this table, the minimum value is subtracted from all the other values.

Obviously, the columns that contain a zero would remain unaffected by this operation. Here only the fourth column values would change. Table 6.9 shows this.

TABLE 6.9 Reduced Cost Table 2

Worker	*Job*			
	A	*B*	*C*	*D*
1	5	0	11	14
2	15	0	21	0
3	1	4	0	3
4	0	4	19	1

Step 3 Draw the minimum number of lines covering all zeros. As a general rule, we should first cover those rows/columns which contain larger number of zeros. Table 6.9 is reproduced in Table 6.10 and the lines are drawn.

TABLE 6.10 Reduced Cost Table 3

Worker	*Job*			
	A	*B*	*C*	*D*
1	5	0	11	14
2	15	0	21	0
3	1	4	0	3
4	0	4	19	1

Step 4 Since the number of lines drawn is equal to 4(= *n*), the optimal solution is obtained. The assignments are made after scanning the rows and columns for unit zeros. Assignments made are shown with squares, as shown in Table 6.11.

TABLE 6.11 Assignment of Jobs

Worker	*Job*			
	A	*B*	*C*	*D*
1	5	[0]	11	14
2	15	~~0~~	21	[0]
3	1	4	[0]	3
4	[0]	4	19	1

Assignments are made in the following order. Rows 1, 3, and 4 contain only one zero each. So assign 1-*B*, 3-*C*, and 4-*A*. Since worker 1 has been assigned job *B*, we cross the zero in the second column of the second row. After making these assignments, only worker 2 and job *D* arc left for assignment. The final pattern of assignments is 1-*B*, 2-*D*, 3-*C*, and 4-*A*, involving a total time of 40 + 55 + 48 + 41 = 184 minutes. This is the optimal solution to the problem—the samc as obtained by enumeration and transportation methods.

Example 6.4 Using the following cost matrix, determine (a) optimal job assignment, and (b) the cost of assignments.

Machinist	*Job*				
	1	*2*	*3*	*4*	*5*
A	10	3	3	2	8
B	9	7	8	2	7
C	7	5	6	2	4
D	3	5	8	2	4
E	9	10	9	6	10

Iteration 1 Obtain row reductions.

TABLE 6.12 Reduced Cost Table 1

Machinist	*Job*				
	1	*2*	*3*	*4*	*5*
A	8	1	1	0	6
B	7	5	6	0	5
C	5	3	4	0	2
D	1	3	6	0	2
E	3	4	3	0	4

Iteration 2 Obtain column reductions and draw the minimum number of lines to cover all zeros.

TABLE 6.13 Reduced Cost Table 2

Machinist	*Job*				
	1	*2*	*3*	*4*	*5*
A	7	0	0	0	4
B	6	4	5	0	3
C	4	2	3	0	0
D	0	2	5	0	0
E	2	3	2	0	2

Since the number of lines covering all zeros is less than the number of columns/rows, we modify the Table 6.13. The least of the uncovered cell values is 2. This value would be subtracted from each of the uncovered values and added to each value lying at the intersection of lines (corresponding to cells *A*-4, *D*-4, *A*-5 and *D*-5). Accordingly, the new table would appear as shown in Table 6.14.

Iteration 3

The optimal assignments can be made as the least number of lines covering all zeros in Table 6.14 equals 5. Considering rows and columns, the assignments can be made in the following order:

(i) Select the second row. Assign machinist *B* to job 4. Cross out zeros at cells *C*-4 and *E*-4.

(ii) Consider row 4. Assign machinist *D* to job 1. Cancel the zero at cell *E*-1.

(iii) Since there is a single zero in the fifth row, put machinist *E* to job 3 and cross out the zero at *A*-3.

(iv) There being only a single zero left in each of the first and third rows, we assign job 2 to machinist *A* and job 5 to *C*.

The total cost associated with the optimal machinist-job assignment pattern *A*-2, *B*-4, *C*-5, *D*-1 and *E*-3 is 3 + 2 + 4 + 3 + 9 = 21.

TABLE 6.14 Reduced Cost Table 3

Machinist	*Job*				
	1	*2*	*3*	*4*	*5*
A	7	[0]	~~0~~	2	6
B	4	2	3	[0]	3
C	2	~~0~~	1	~~0~~	[0]
D	[0]	2	5	2	2
E	~~0~~	1	[0]	~~0~~	2

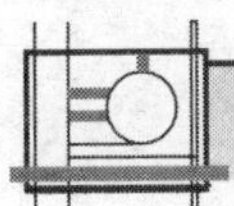

6.4 SOME SPECIAL CASES

6.4.1 Unbalanced Assignment Problems

The Hungarian method of solving an assignment problem requires that the number of columns should be equal to the number of rows. When they are equal, the problem is a *balanced* problem, and when not, it is called an *unbalanced* problem. Thus, when there are 5 workers and 4 machines, or when there are 4 workers and 6 machines, for instance, we have unbalanced situations in which one-to-one match is not possible. In case the machines are in excess, the excess machine(s) would remain idle and so is the case when men are in excess—the number of excess people would not get an assignment.

In such situations, dummy column(s)/row(s), whichever is smaller in number, are inserted with zeros as the cost elements. For example, when the given cost matrix is of the dimension 4 × 5, a dummy row would be included. In each column in respect of this row, a 'zero' would be placed. After this operation of introducing dummy columns/rows, the problem is solved in the usual manner.

6.4.2 Constrained Assignment Problems

Sometimes a worker cannot perform a certain job or is not to be assigned a particular job. To cope with this situation, the cost of performing that job by such person is taken to be extremely large (which is written as *M*). Then the solution to the assignment problem proceeds in the manner discussed earlier. The effect of assigning prohibitive cost to such person-job combinations is that they do not figure in the final solution.

Example 6.5 You are given the information about the cost of performing different jobs by different persons. The job-person marking × indicates that the individual involved cannot perform the particular job. Using this information, state (i) the optimal assignment of jobs, and (ii) the cost of such assignment.

Person	Job				
	J_1	J_2	J_3	J_4	J_5
P_1	27	18	×	20	21
P_2	31	24	21	12	17
P_3	20	17	20	×	16
P_4	22	28	20	16	27

Balancing the problem and assigning a high cost to the pairings $P_1 - J_3$ and $P_3 - J_4$, we have the cost table given in Table 6.15.

Now we can derive the reduced cost table as shown in Table 6.16. Note that the cells with prohibited assignments continue to be shown with the cost element *M,* since *M* is defined to be extremely large so that subtraction or addition of a value does not practically affect it. To test optimality, lines are drawn to cover all zeros. Since the number of lines covering all zeros is less than *n,* we select the lowest uncovered cell, which equals 4. With this value, we can obtain the revised reduced cost table, shown in Table 6.17.

TABLE 6.15 Cost Table

Person	Job				
	J_1	J_2	J_3	J_4	J_5
P_1	27	18	M	20	21
P_2	31	24	21	12	17
P_3	20	17	20	M	16
P_4	22	28	20	16	27
P_5 (Dummy)	0	0	0	0	0

TABLE 6.16 Reduced Cost Table 1

Person	Job				
	J_1	J_2	J_3	J_4	J_5
P_1	9	0	M	2	3
P_2	19	12	9	0	5
P_3	4	1	4	M	0
P_4	6	12	4	0	11
P_5 (Dummy)	0	0	0	0	0

TABLE 6.17 Reduced Cost Table 2

Person	Job				
	J_1	J_2	J_3	J_4	J_5
P_1	9	[0]	M	6	3
P_2	15	8	5	[0]	1
P_3	4	1	4	M	[0]
P_4	2	8	[0]	~~0~~	7
P_5 (Dummy)	[0]	~~0~~	~~0~~	4	~~0~~

The number of lines covering zeros is equal to $5(= n)$, hence the optimal assignment can be made. The assignment is: $P_1 - J_2$, $P_2 - J_4$, $P_3 - J_5$, $P_4 - J_3$, while job J_1 would remain unassigned. This assignment pattern would cost $18 + 12 + 16 + 20 = 66$ in the aggregate.

6.4.3 Unique vs Multiple Optimal Solutions

In the process of making assignments, it was stated earlier that we select a row/column with only a single zero to make an assignment. However, a situation may arise wherein the various rows and columns, where assignments are yet to be done, have all multiple zeros. In such cases, we get multiple optimal solutions to the given problem. In any of the problems discussed so far, we have not experienced such a situation. Hence, each one of them has had a unique optimal solution. When a problem has a unique optimal solution, it means that no other solution to the problem exists which yields the same objective function value (cost, time, profit, etc.) as the one obtained from the optimal solution derived. On the other hand, in a problem with multiple optimal solutions, there exist more than one solution which are all optimal and equally attractive. Consider the following example.

Example 6.6 Solve the following assignment problem and obtain the minimum cost at which all the jobs can be performed.

Worker	*Job (Cost in '00 Rs)*				
	1	*2*	*3*	*4*	*5*
A	25	18	32	20	21
B	34	25	21	12	17
C	20	17	20	32	16
D	20	28	20	16	27

This problem is unbalanced since number of jobs is 5 while the number of workers is 4. We first balance it by introducing a dummy worker *E*, as shown in Table 6.18.

TABLE 6.18 Balancing the Assignment Problem

Worker	*Job*				
	1	*2*	*3*	*4*	*5*
A	25	18	32	20	21
B	34	25	21	12	17
C	20	17	20	32	16
D	20	28	20	16	27
E	0	0	0	0	0

Step 1 Obtain reduced cost values by subtracting the minimum value in each row from every cell in the row. This is given in Table 6.19.

TABLE 6.19 Reduced Cost Table 1

Worker	Job 1	2	3	4	5
A	7	0	14	2	3
B	22	13	9	0	5
C	4	1	4	16	0
D	4	12	4	0	11
E	0	0	0	0	0

Since there is at least one zero in each row and column, we test it for optimality. Accordingly, lines are drawn. All zeros are covered by 4 lines, which is less than 5 (the order of the given matrix). Hence, we proceed to improve the solution. The least uncovered value is 4. Subtracting from every uncovered value and adding it to every value lying at the intersection of lines, we get the revised values as shown in Table 6.20.

The solution given in Table 6.20 is optimal since the number of lines covering zeros matches with the order of the matrix. We can, therefore, proceed to make assignments. To begin with, since each of the columns has multiple zeros, we cannot start making assignments considering columns and have, therefore, to look through rows. The first row has a single zero. Thus, we make assignment *A*-2 and cross out zero at *E*-2. Further, the second and the third rows have one zero each. We make assignments *B*-4 and *C*-5, and cross out zeros at *D*-4 and *E*-5. Now, both the rows left have two zeros each and so have both the columns. This indicates existence of multiple optimal solutions. To obtain the solutions, we select zeros arbitrarily and proceed as discussed below.

TABLE 6.20 Reduced Cost Table 2

Worker	Job 1	2	3	4	5
A	7	[0]	14	6	3
B	18	9	5	[0]	1
C	4	1	4	20	[0]
D	[0]	8	~~0~~	~~0~~	7
E	~~0~~	~~0~~	[0]	4	~~0~~

(i) Select the zero at *D*-1, make assignment and cross out zeros at *D*-3 and *E*-1 (as both, worker *D* and job 1, are not available any more). Next, assign worker *E* to job 3, corresponding to the only zero left. Evidently, selecting the zero at *E*-3 initially would have the effect of making same assignments.

(ii) Select the zero at *D*-3, make assignment and cross out zeros at *D*-1 and *E*-3. Next, make assignment at the only zero left at *E*-1. Obviously, selecting the zero at *E*-1 making assignment in the first place would lead to the same assignments.

To conclude, the problem has two optimal solutions as given below.

Solution 1			*Solution 2*		
		('00 Rs)			*('00 Rs)*
Worker	*Job*	*Cost*	*Worker*	*Job*	*Cost*
A	2	18	*A*	2	18
B	4	12	*B*	4	12
C	5	16	*C*	5	16
D	1	20	*D*	3	20
Job left	3		Job left	1	
	Total	66		Total	66

6.4.4 Maximisation Case

In some situations, the assignment problem may call for maximisation of profit, revenue, etc. as the objective. For example, we may be faced with the problem of assignment of salesmen in different regions in which they can display different qualities in making sales (reflected in amounts of sales executed by them). Obviously, assignment would be made in such a way that the total expected revenue is maximised.

For dealing with a maximisation problem, we first change it into an equivalent minimisation problem. This is achieved by subtracting each of the elements of the given pay-off matrix from a constant (value) *K*. Thus, we may simply put a negative sign before each of the pay-off values (which is equivalent to subtracting each value from zero). Usually, the largest of all values in the given matrix is located and then each one of the values is subtract from it (the largest value is taken so as to avoid the appearance of negative signs). Then the problem is solved the same way as a minimisation problem is.

Example 6.7 A company plans to assign 5 salesmen to 5 districts in which it operates. Estimates of sales revenue in thousands of rupees for each salesman in different districts are given in the following table. In your opinion, what should be the placement of the salesmen if the objective is to maximise the expected sales revenue?

Expected Sales Data

Salesman	*District*				
	D_1	D_2	D_3	D_4	D_5
S_1	40	46	48	36	48
S_2	48	32	36	29	44
S_3	49	35	41	38	45
S_4	30	46	49	44	44
S_5	37	41	48	43	47

Since it is a maximisation problem, we would first subtract each of the entries in the table from the largest one, which equals 49 here. The resultant data are given in Table 6.21.

TABLE 6.21 Opportunity Loss Matrix

Salesman	*District*				
	D_1	D_2	D_3	D_4	D_5
S_1	9	3	1	13	1
S_2	1	17	13	20	5
S_3	0	14	8	11	4
S_4	19	3	0	5	5
S_5	12	8	1	6	2

Now we shall proceed as usual.

Step 1 Subtract minimum value in each row from every value in the row. The resulting values are given in Table 6.22.

TABLE 6.22 Reduced Cost Table 1

Salesman	*District*				
	D_1	D_2	D_3	D_4	D_5
S_1	8	2	0	12	0
S_2	0	16	12	19	4
S_3	0	14	8	11	4
S_4	19	3	0	5	5
S_5	11	7	0	5	1

Step 2, 3 Subtract minimum value in each column in reduced cost table 1 from each value in the column. Test for optimality by drawing lines to cover zeros. These are shown in Table 6.23.

TABLE 6.23 Reduced Cost Table 2

Salesman	*District*				
	D_1	D_2	D_3	D_4	D_5
S_1	8	0	0	7	0
S_2	0	14	12	14	4
S_3	0	12	8	6	4
S_4	19	1	0	0	5
S_5	11	5	0	0	1

Since the number of lines covering all zeros is fewer than n, we select the least uncovered cell value, which equals 4. With this, we can modify the table as given in Table 6.24.

Step 4, 5, 6 Find improved solution. Test for optimality and make assignments.

TABLE 6.24 Reduced Cost Table 3

Salesman	District				
	D_1	D_2	D_3	D_4	D_5
S_1	12	[0]	~~0~~	7	~~0~~
S_2	[0]	10	8	10	~~0~~
S_3	~~0~~	8	4	2	[0]
S_4	23	1	[0]	~~0~~	5
S_5	15	5	~~0~~	[0]	1

There are more than one optimal assignments possible in this case because of the existence of multiple zeros in different rows and columns. The assignments possible are:

$$S_1 - D_2, S_2 - D_1, S_3 - D_5, S_4 - D_3, S_5 - D_4; \text{ or}$$
$$S_1 - D_2, S_2 - D_5, S_3 - D_1, S_4 - D_3, S_5 - D_4; \text{ or}$$
$$S_1 - D_2, S_2 - D_5, S_3 - D_1, S_4 - D_4, S_5 - D_3; \text{ or}$$
$$S_1 - D_2, S_2 - D_1, S_3 - D_5, S_4 - D_4, S_5 - D_3.$$

Each of these assignment patterns would lead to an expected aggregated sales equal to 231 thousand rupees.

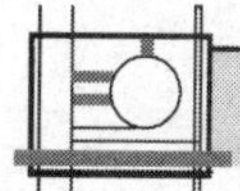

6.5 DUAL OF THE ASSIGNMENT PROBLEM

We know that in solving assignment problems, we determine, using the given cost matrix $C = \| c_{ij} \|$ of the order $n \times n$ with all its elements $c_{ij} \geq 0$ for i and $j = 1, 2 \ldots, n$; the assignment matrix, $X = \| x_{ij} \|$ such that

$$Z = \sum_{ij} c_{ij} x_{ij} = \text{a minimum}$$

$$x_{ij} = x_{ij}^2 \text{ for } i, j = 1, 2 \ldots, n \qquad \text{(i)}$$

$$\sum_{i=1}^{n} x_{ij} = \sum_{j=1}^{n} x_{ij} = 1 \qquad \text{(ii)}$$

These two equations are the conditions that jointly specify that

$$x_{ij} = \begin{cases} 1 \text{ if worker } i \text{ is assigned to job } j \\ 0 \text{ if worker } i \text{ is not assigned to job } j \end{cases}$$

and further that each row and each column of the matrix X has a single unity element and all other elements are equal to zero.

Now, to determine the dual of the given assignment problem, we proceed as in case of a transportation problem and determine a set of u_i and v_j's such that the following conditions are satisfied:

$$c_{ij} \geq u_i + v_j \quad \text{for } x_{ij} = 0$$

$$c_{ij} = u_i + v_j \quad \text{for } x_{ij} > 0$$

or

$$Z \geq \sum_i (u_i \times \sum_j x_{ij}) + \sum_j (v_j \times \sum_i x_{ij})$$

Since

$$\sum_j x_{ij} = \sum_i x_{ij} = 1,$$

we have

$$Z \geq \sum_i u_i + \sum_j v_j$$

If we denote RHS of this expression by G, the problem associated with minimising $\sum_{ij} c_{ij}x_{ij}$ is identical to the following dual problem:

Maximise $$G = \sum_i u_i + \sum_j v_j$$

Subject to $$c_{ij} \geq u_i + v_j \; i, j = 1, 2, \ldots, n$$

u_i and v_j being unrestricted in sign.

Example 6.8 Write the dual of the problem given in Example 6.1 and obtain the optimal values of the dual variables.

The dual problem corresponding to the given problem is as follows:

Maximise $G = u_1 + u_2 + u_3 + u_4 + v_1 + v_2 + v_3 + v_4$

Subject to

$$u_1 + v_1 \leq 45 \qquad u_3 + v_1 \leq 49$$

$$u_1 + v_2 \leq 40 \qquad u_3 + v_2 \leq 52$$

$$u_1 + v_3 \leq 51 \qquad u_3 + v_3 \leq 48$$

$$u_1 + v_4 \leq 67 \qquad u_3 + v_4 \leq 64$$

$$u_2 + v_1 \leq 57 \qquad u_4 + v_1 \leq 41$$

$$u_2 + v_2 \leq 42 \qquad u_4 + v_2 \leq 45$$

$$u_2 + v_3 \leq 63 \qquad u_4 + v_3 \leq 60$$

$$u_2 + v_4 \leq 55 \qquad u_4 + v_4 \leq 55$$

$$u_i\; v_j \geq 0,\; i = 1, 2, 3, 4;\; j = 1, 2, 3, 4$$

We can obtain the optimal values of the dual variables, u_i and v_j's, from the optimal solution to the problem given in Table 6.6 and reproduced in Table 6.25.

TABLE 6.25 Obtaining Optimal Values of Dual Variables

Worker	*Job*				u_i
	A	*B*	*C*	*D*	
1	45 [ε]	40 [1]	51	67	0
2	57	42 [ε]	63	55 [1]	2
3	49 [ε]	52	48 [1]	64	4
4	41 [1]	45	60	55	– 4
v_j	45	40	44	53	

The optimal values are: $u_1 = 0$, $u_2 = 2$, $u_3 = 4$, $u_4 = -4$, $v_1 = 45$, $v_2 = 40$, $v_3 = 44$, and $v_4 = 53$. These sum upto 184, which is obviously the same as the optimal value of Z. It may further be observed that they satisfy all the constraints and are not restricted in sign.

Review Illustrations

Example 6.9 To stimulate interest and provide an atmosphere for intellectual discussion, the finance faculty in a management school decides to hold special seminars on four contemporary topics—leasing, portfolio management, private mutual funds, swaps and options. Such seminars would be held once per week in the afternoons. However, scheduling these seminars (one for each topic, and not more than one seminar per afternoon) has to be done carefully so that the number of students unable to attend is kept to a minimum. A careful study indicates that the number of students who cannot attend a particular seminar on a specific day is as follows:

	Leasing	*Portfolio Management*	*Private Mutual Funds*	*Swaps and Options*
Monday	50	40	60	20
Tuesday	40	30	40	30
Wednesday	60	20	30	20
Thursday	30	30	20	30
Friday	10	20	10	30

Find an optimal schedule of the seminars. Also find out the total number of students who will be missing at least one seminar.

(CA, November, 1992)

Since this is an unbalanced problem, we first balance it by adding a dummy topic, as shown below:

	Leasing A	*Portfolio Management* B	*Private Mutual Funds* C	*Swaps and Options* D	*Dummy* E
Monday	50	40	60	20	0
Tuesday	40	30	40	30	0
Wednesday	60	20	30	20	0
Thursday	30	30	20	30	0
Friday	10	20	10	30	0

Subtracting the minimum value in each column from each of the values of the column (note that each row has a least value equal to zero already), we obtain RCT-l as shown in Table 6.26.

TABLE 6.26 Reduced Cost Table 1

A	*B*	*C*	*D*	*E*
40	20	50	0	0
30	10	30	10	0
50	0	20	0	0
20	10	10	10	0
0	0	0	10	0

The minimum number of lines covering all zeros is 4, which is less than the order of the given matrix (= 5). Thus, we cannot obtain optimal solution from here. The revised table is given in Table 6.27, which is obtained by subtracting the least uncovered cell value 10 from each of the uncovered values and adding to each of the values lying at the intersection of lines.

TABLE 6.27 Reduced Cost Table 2

A	*B*	*C*	*D*	*E*
30	20	40	[0]	~~0~~
20	10	20	10	[0]
40	[0]	10	~~0~~	~~0~~
10	10	[0]	10	~~0~~
[0]	10	~~0~~	20	10

Since the minimum number of lines to cover all zeros equals 5, we may make assignments. The assignments have been made in the following order: Row 5, Column 1; Row 3, Column 2; Row 4, Column 3; Row 1, Column 4; and Row 2, Column 5. Accordingly, the optimal schedule is:

Monday	: Swaps and Options	Thursday	: Private Mutual Funds
Tuesday	: No Seminar	Friday	: Leasing
Wednesday	: Portfolio Management	No. of Students Missing = 20 + 0 + 20 + 20 + 10 = 70	

Example 6.10 A solicitor's firm employs typists on hourly piece-rate basis for their daily work. There are five typists and their charges and speed are different. According to an earlier understanding, only one job is given to one typist and the typist is paid for a full hour even when he works for a fraction of an hour. Find the least cost allocation for the following data:

Typist	*Rate/hour (Rs)*	*Number of Pages Typed/hour*	*Job*	*No. of Pages*
A	5	12	*P*	199
B	6	14	*Q*	175
C	3	8	*R*	145
D	4	10	*S*	298
E	4	11	*T*	178

(MBA, Delhi, April, 1998)

Using the given information, we first obtain the cost matrix, when different jobs are performed by different typists. This is shown in Table 6.28. The elements of the matrix are obtained as follows. To illustrate, if typist *A* is given job *P*, he would require $199/12 = 16\frac{7}{12}$ hours and, hence, be paid for 17 hours @ Rs 5 per hour. This results in a cost of Rs 85 for this combination.

TABLE 6.28 Total Cost Matrix

Typist	*Job*				
	P	*Q*	*R*	*S*	*T*
A	85	75	65	125	75
B	90	78	66	132	78
C	75	66	57	114	69
D	80	72	60	120	72
E	76	64	56	112	68

Subtracting the minimum element of each row from all its elements, we obtain RCT-1, as shown in Table 6.29.

TABLE 6.29 Reduced Cost Table 1

Typist	*Job*				
	P	*Q*	*R*	*S*	*T*
A	20	10	0	60	10
B	24	12	0	66	12
C	18	9	0	57	12
D	20	12	0	60	12
E	20	8	0	56	12

Now, subtracting the minimum element of each column from all the elements, we get RCT-2, given in Table 6.30.

TABLE 6.30 Reduced Cost Table 2

Typist	*Job*				
	P	*Q*	*R*	*S*	*T*
A	2	2	0	4	0
B	6	4	0	10	2
C	0	1	0	1	2
D	2	4	0	4	2
E	2	0	0	0	2

Here, the minimum number of lines to cover all zeros is equal to 4, which is smaller than the order, 5, of the given matrix. Accordingly, the revised table is prepared by considering the least uncovered value, equal to 1, and adjusting it with uncovered cell values and those lying at the intersection of lines. Table 6.31 contains RCT-3.

TABLE 6.31 Reduced Cost Table 3

Typist	*Job*				
	P	*Q*	*R*	*S*	*T*
A	2	1	0	3	0
B	6	3	0	9	2
C	0	0	0	0	2
D	2	3	0	3	2
E	3	0	1	0	3

In Table 6.31 as well, 4 lines can cover all zeros. Accordingly, RCT-4 is obtained drawn up as the revised table. This is given in Table 6.32.

TABLE 6.32 Reduced Cost Table 4

Typist	Job				
	P	*Q*	*R*	*S*	*T*
A	2	1	2	3	[0]
B	4	1	[0]	7	~~0~~
C	~~0~~	[0]	2	~~0~~	2
D	[0]	1	~~0~~	1	~~0~~
E	3	~~0~~	3	[0]	3

In this case, the minimum number of lines to cover all zeros equals 5, which matches with the order of the matrix. Accordingly, assignments have been made as described below:

Typist	*Job*	*Cost*
A	*T*	75
B	*R*	66
C	*Q*	66
D	*P*	80
E	*S*	112
	Total	399

This optimal solution, however, is not unique.

Example 6.11 Welldone Company has taken the third floor of a multi-storeyed building for rent with a view to locate one of their zonal offices. There are five main rooms in this to be assigned to five managers. Each room has its own advantages and disadvantages.

Some have windows, some are closer to the washrooms or to the canteen or secretarial pool. The rooms are of all different sizes and shapes. Each of the five managers were asked to rank their room preferences amongst the rooms 301, 302, 303, 304 and 305. Their preferences were recorded in a table as indicated below:

MANAGER				
M_1	M_2	M_3	M_4	M_5
302	302	303	302	301
303	304	301	305	302
304	305	304	304	304
	301	305	303	
		302		

Most of the managers did not list all the five rooms since they were not satisfied with some of these rooms and they have left these from the list. Assuming that their preferences can be quantified by numbers, find out as to which manager should be assigned to which room so that their total preference ranking is a minimum.

(CA, November, 1990)

In the first step, we formulate the assignment problem using preference ranks. This is shown in Table 6.33. Notice that the rooms not ranked by a manager are represented by *M*—as prohibited assignments.

TABLE 6.33 Assignment Problem

Room	Manager				
	M_1	M_2	M_3	M_4	M_5
301	*M*	4	2	*M*	1
302	1	1	5	1	2
303	2	*M*	1	4	*M*
304	3	2	3	3	3
305	*M*	3	4	2	*M*

Since greater preferences are shown by lower numbers, the optimal solution calls for minimising the total preference ranking. To solve, we obtain RCT-l by subtracting the least value from each value of the row, for each of the rows. This is given in Table 6.34.

TABLE 6.34 Reduced Cost Table 1

Room	Manager				
	M_1	M_2	M_3	M_4	M_5
301	*M*	3	1	*M*	0
302	0	0	4	0	1
303	1	*M*	0	3	*M*
304	1	0	1	1	1
305	*M*	1	2	0	*M*

Since each of the columns as well as rows has a zero, lines are drawn to cover all zeros. Further, the number of lines being equal to 5, the order of the given matrix, assignments can be made, as shown in the Table 6.35.

TABLE 6.35 Assignment of Rooms to Managers

Room	Manager				
	M_1	M_2	M_3	M_4	M_5
301	*M*	3	1	*M*	[0]
302	[0]	~~0~~	4	~~0~~	1
303	1	*M*	[0]	3	*M*
304	1	[0]	1	1	1
305	*M*	1	2	[0]	*M*

Thus, the optimal assignment pattern is:

Room	*Manager*	*Ranking*
301	M_5	1
302	M_1	1
303	M_3	1
304	M_2	2
305	M_4	2

Example 6.12 A company has four sales representatives who are to be assigned to four different sales territories. The monthly sales increase estimated for each sales representative for different sales territories (in lakhs of rupees), are shown in the following table:

Sales Representatives	*Sales Territories*			
	I	*II*	*III*	*IV*
A	200	150	170	220
B	160	120	150	140
C	190	195	190	200
D	180	175	160	190

Suggest optimal assignment and the total maximum sales increase per month.

If for certain reasons, sales representative *B* cannot be assigned to sales territory III, will the optimal assignment schedule be different? If so, find that schedule and the effect on total sales.

(M Com, Delhi, 1997)

The given problem, being of the maximisation type, needs to be converted first into minimisation type. This is shown in Table 6.36, wherein the various elements are obtained by subtracting each of the elements in the given table from 220.

TABLE 6.36 Opportunity Loss Matrix

	I	*II*	*III*	*IV*
A	20	70	50	0
B	60	100	70	80
C	30	25	30	20
D	40	45	60	30

Subtracting minimum value in each row from each of the row elements, we derive RCT-1, shown in Table 6.37.

TABLE 6.37 Reduced Cost Table 1

	I	*II*	*III*	*IV*
A	20	70	50	0
B	0	40	10	20
C	10	5	10	0
D	10	15	30	0

Now, subtracting minimum value in each column from each of the column elements of RCT-1, we get RCT-2, as shown in Table 6.38.

TABLE 6.38 Reduced Cost Table 2

	I	*II*	*III*	*IV*
A	20	65	40	0
B	0	35	0	20
C	10	0	0	0
D	10	10	20	0

Since the minimum number of lines to cover all zeros (3) is smaller than the matrix order (4), RCT-3 is derived as a revised table. This is given in Table 6.39.

TABLE 6.39 Reduced Cost Table 3

	I	*II*	*III*	*IV*
A	10	55	30	[0]
B	[0]	35	~~0~~	30
C	10	~~0~~	[0]	10
D	~~0~~	[0]	10	0

With the number of lines to cover all zeros being equal to the order of the given matrix, assignments can be made as shown in Table 6.39. However, the problem has an alternative optimal solution as well. Both of these are given below:

Alternative 1		
Salesman	*Territory*	*Sales*
A	IV	220
B	I	160
C	III	190
D	II	175
	Total	745

Alternative 2		
Salesman	*Territory*	*Sales*
A	IV	220
B	III	150
C	II	195
D	I	180
	Total	745

If salesman B cannot be assigned to territory III (alternative 2), then alternative 1 above may be adopted without adverse effect on sale.

Example 6.13 A firm produces four products. There are four operators who are capable of producing any of these four products. The firm records 8 hours a day and allows 30 minutes for lunch. The processing time in minutes and the profit for each of the products are given below:

Operator	*Products*			
	A	*B*	*C*	*D*
1	15	9	10	6
2	10	6	9	6
3	25	15	15	9
4	15	9	10	10
Profit (Rs) per unit	8	6	5	4

Find the optimal assignment of products to operators. *(CA, November, 1997)*

An 8-hour working day, with a 30-minutes lunch time allowed, implies that net working time available per day is 7 hours and 30 minutes, that is 450 minutes. The number of units of different products which could be produced by the four operators can be calculated by dividing 450 by the given processing times. With the profit per unit of each product being given, we may calculate the profit resulting from each possible assignment. The profit matrix is given in Table 6.40. The values in this matrix are derived as follows. For example, operator 1 can produce 450/15 = 30 units of product A which, at a profit rate of Rs 8 per unit, implies a total profit of Rs 240.

TABLE 6.40 Profit Matrix

Operator	*Product*			
	A	*B*	*C*	*D*
1	240	300	225	300
2	360	450	250	300
3	144	180	150	200
4	240	300	225	180

To solve the problem, it is first converted into a minimisation problem by obtaining opportunity loss matrix by subtracting each value from 450 – the highest profit value in the table. It is given in Table 6.41.

TABLE 6.41 Opportunity Loss Matrix

Operator	*Product*			
	A	*B*	*C*	*D*
1	210	150	225	150
2	90	0	200	150
3	306	270	300	250
4	210	150	225	270

Using Hungarian method, we first obtain zeros in every row and column as given in Tables 6.42 and 6.43.

TABLE 6.42 Reduced Cost Table 1

Operator	Product			
	A	*B*	*C*	*D*
1	60	0	75	0
2	90	0	200	150
3	56	20	50	0
4	60	0	75	120

TABLE 6.43 Reduced Cost Table 2

Operator	Product			
	A	*B*	*C*	*D*
1	4	0	25	0
2	34	0	150	150
3	0	20	0	0
4	4	0	25	120

It is observed that all zeros are covered by three lines, which is one less than the order of the matrix. The improved matrix is obtained as shown in Table 6.44, with adjustment with the least uncovered value of 4.

TABLE 6.44 Reduced Cost Table 3

Operator	Product			
	A	*B*	*C*	*D*
1	0	0	21	[0]
2	30	[0]	146	150
3	0	24	[0]	4
4	[0]	0	21	120

Since the minimum number of lines to cover all zeros is 4, which matches with the order of the matrix, assignments can be made as shown in Table 6.44. Thus, the optimal assignment is:

Operator	*Product*	*Profit*
1	*D*	300
2	*B*	450
3	*C*	150
4	*A*	240
	Total	1,140

Example 6.14 An airline, operating seven days a week, has given the following schedule of its flights between New Delhi and Mumbai. The crews should have a minimum of five hours between the flights. Obtain the pairing of flights that minimises layover time away from home. For any given pairing, the crew will be based at the city that results in the smallest layover.

New Delhi—Mumbai			*Mumbai—New Delhi*		
Flight No.	*Departure*	*Arrival*	*Flight No.*	*Departure*	*Arrival*
101	5 A.M.	7 A.M.	201	7 A.M.	9 A.M.
102	7 A.M.	9 A.M.	202	8 A.M.	10 A.M.
103	1 P.M.	3 P.M.	203	1 P.M.	3 P.M.
104	7 P.M.	9 P.M.	204	6 P.M.	8 P.M.

(M Com, Delhi, 2008)

To begin with, we first assume that all the crew is based at New Delhi. Using this assumption, we can obtain the layover times of various combinations of flights, as shown in Table 6.45. To illustrate, the flight 101 which starts from New Delhi at 5 A.M. reaches Mumbai at 7 A.M. If it is to return as 201, the scheduled time for which is 7 A.M., then it can do so only after 24 hours since a minimum layover time of 5 hours is required. Similarly, layover times for other flight combinations can be obtained as shown in the table.

TABLE 6.45 Layover Time—Crew at New Delhi

Flight	*201*	*202*	*203*	*204*
101	24	25	6	11
102	22	23	28	9
103	16	17	22	27
104	10	11	16	21

The layover times for various flight combinations, when crew is assumed to be based at Mumbai, are similarly calculated and shown in Table 6.46.

TABLE 6.46 Layover Time—Crew at Mumbai

Flight	*201*	*202*	*203*	*204*
101	20	19	14	9
102	22	21	16	11
103	28	27	22	17
104	10	9	28	23

Now, since the crew can be based at either of the stations, minimum layover times can be obtained for different flight combinations by selecting the corresponding lower value out of the above two tables. For instance, in combining 101–102, we select 20, which is lower of the two values 24 and 20 (Tables 6.45 and 6.46). The values are shown in Table 6.47.

TABLE 6.47 Layover Time—Crew at New Delhi/Mumbai

Flight	*201*	*202*	*203*	*204*
101	20*	19*	6	9*
102	22	21*	16*	9
103	16	17	22	17*
104	10	9*	16	21

*when crew is based at Mumbai.

To continue, we can obtain the optimal pairing of flights so as to minimise the total layover time, using the Hungarian Assignment Method. Table 6.48 gives the RCT-1.

TABLE 6.48 Reduced Cost Table 1

Flight	*201*	*202*	*203*	*204*
101	14	13	[0]	3
102	13	12	7	[0]
103	[0]	1	6	1
104	1	[0]	7	12

With a zero in each column as well, there is no need to perform column reductions. Further, since the number of lines covering all zeros is equal to the order of the given matrix, we can obtain the optimal assignment as shown in the table.

The optimal pairing of flights is given below:

Flight No.	*Flight No.*	*Layover Time*	*Crew Based At*
101	203	6	New Delhi
102	204	9	New Delhi
103	201	16	New Delhi
104	202	9	Mumbai
		Total 40	

Example 6.15 A company has just developed a new item for which it proposes to undertake a national television promotional campaign. It has decided to schedule a series of one-minute commercials during peak audience viewing hours of 1 to 5 P.M. To reach the widest possible audience, the company wants to schedule one commercial on each of the networks and to have only one commercial appear during each of the four one-hour time blocks. The exposure ratings for each hour, which represent the number of viewers per Rs 10,000 spent, are given below.

Viewing Hours	Network			
	A	B	C	D
1–2 P.M.	27.1	18.1	11.3	9.5
2–3 P.M.	18.9	15.5	17.1	10.6
3–4 P.M.	19.2	18.5	9.9	7.7
4–5 P.M.	11.5	21.4	16.8	12.8

(a) Which network should be scheduled each hour to provide the maximum audience exposure?
(b) How would the schedule change if it is decided not to use network A between 1 and 3 P.M.?

(M Com, Delhi, 2004)

(a) To solve this problem, we first multiply each value in the matrix by 10 to express exposure ratings per Rs 1 lakh. It simplifies the calculation work somewhat. Further, being a maximisation problem, we subtract each value from the largest value to get the opportunity loss matrix. The result of these steps is given in Table 6.49.

TABLE 6.49 Opportunity Loss Matrix

Viewing Hours	*Network*			
	A	*B*	*C*	*D*
1–2 P.M.	0	90	158	176
2–3 P.M.	82	116	100	165
3–4 P.M.	79	86	172	194
4–5 P.M.	156	57	103	143

Now we apply HAM to solve the problem. Row reductions and column reductions are shown in Tables 6.50 and 6.51 respectively.

TABLE 6.50 Reduced Cost Table 1

Viewing Hours	*Network*			
	A	*B*	*C*	*D*
1–2 P.M.	0	90	158	176
2–3 P.M.	0	34	18	83
3–4 P.M.	0	7	93	115
4–5 P.M.	99	0	46	86

TABLE 6.51 Reduced Cost Table 2

Viewing Hours	*Network*			
	A	*B*	*C*	*D*
1–2 P.M.	0	90	140	93
2–3 P.M.	0	34	0	0
3–4 P.M.	0	7	75	32
4–5 P.M.	99	0	28	3

Since number of lines covering all zeros is 3, we improve the solution, as shown in Tables 6.52 and 6.53.

TABLE 6.52 Reduced Cost Table 3

Viewing Hours	*Network*			
	A	*B*	*C*	*D*
1–2 P.M.	0	90	137	90
2–3 P.M.	3	37	0	0
3–4 P.M.	0	7	72	29
4–5 P.M.	99	0	25	0

TABLE 6.53 Reduced Cost Table 4

Viewing Hours	*Network*			
	A	*B*	*C*	*D*
1–2 P.M.	[0]	83	130	83
2–3 P.M.	10	37	[0]	~~0~~
3–4 P.M.	~~0~~	[0]	65	22
4–5 P.M.	106	~~0~~	25	[0]

Assigments made are shown in Table 6.53. The optimal solution is:

Viewing Hour	Network
1–2 P.M.	A
2–3 P.M.	C
3–4 P.M.	B
4–5 P.M.	D

For the given restriction, the prohibited time-slots are replaced by M, in the Table 6.49 and the problem is solved. Table 6.54 shows the effect of row and column reductions both. Since number of lines covering all zeros is 3, further calculations lead to values given in Table 6.55.

TABLE 6.54 Reduced Cost Table 5

Viewing Hours	*Network*			
	A	*B*	*C*	*D*
1–2 P.M.	M	0	68	21
2–3 P.M.	M	16	0	0
3–4 P.M.	0	7	93	50
4–5 P.M.	99	0	46	21

TABLE 6.55 Reduced Cost Table 6

Viewing Hours	*Network*			
	A	*B*	*C*	*D*
1–2 P.M.	M	[0]	47	~~0~~
2–3 P.M.	M	37	[0]	~~0~~
3–4 P.M.	[0]	28	93	50
4–5 P.M.	78	~~0~~	25	[0]

Since the number of lines covering zeros is equal to n, the assignments can be made. There are two optimal solutions as given here:

Viewing Hour	*Network*	*Network*
1–2 P.M.	B	D
2–3 P.M.	C	C
3–4 P.M.	A	A
4–5 P.M.	D	B

KEY POINTS TO REMEMBER

- Assignment problems deal with assignment of persons to jobs that they can perform with varying efficiency.
- The solution to an assignment problem can be obtained by complete enumeration and evaluation of all possible assignments. However, the method becomes impractical for even moderately large values of n.
- An assignment problem may also be formulated as a transportation problem or an LPP with the variables restricted to assuming one of the two values—0 and 1 (technically known as zero-one integer programming problem).
- An assignment problem can be solved using transportation method, but it is not very convenient as it produces degenerate solutions. For example, a 10-person, 10-job problem would have 10 occupied cells, against the requirement of 10 + 10 – 1 = 19.
- An assignment problem may be a balanced one when the number of jobs matches with the number of workers, or unbalanced when they do not match.
- An unbalanced problem has to be balanced first by introducing the required number of dummy workers/jobs.
- For every prohibited assignment, the given cost element is replaced by M, which is a very large value.
- If a job is already decided to be given to a particular worker, then the job and the worker are dropped from the problem and the remaining problem is solved.
- The Hungarian assignment method is mainly used to solve assignment problems. It involves the three steps of (i) obtaining zeros in every row and column; (ii) covering all zeros with minimum number of horizontal and vertical lines to check whether assignments can be made; and (iii) making assignments at zeros.
- Modifications are needed if the minimum number of lines covering all zeros is smaller than n.
- Multiple zeros in all columns and rows are indicative of multiple optimal solutions.
- If a problem is of maximisation type, (i) it is, first of all, balanced if it is unbalanced; (ii) the pay-off value for a job-person combination which is prohibited, is replaced by $-M$, (iii) it is transformed into a minimisation problem by subtracting each of the values from the largest value in the pay-off matrix; and (iv) then it is solved as a regular problem.
- The dual to an assignment problem is given by the u_i and v_j values of the optimal solution to the transformed transportation problem.

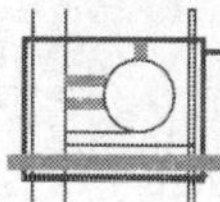

TEST YOUR UNDERSTANDING

Mark the following statements as T (*True*) or F (*False*).

1. An assignment problem involving 6 workers and 5 jobs has 720 assignments possible, while an assignment problem with 6 workers and 4 jobs has 360 assignments possible.
2. The cost elements in the matrix of a typical assignment problem indicate the cost of performing the particular jobs by particular personnel.
3. An assignment problem is said to be *balanced* when the number of rows in the given matrix matches with the number of columns.
4. An assignment problem is always solved on the assumption that the assignments are to be done on a one-to-one basis.
5. A job assignment problem is *unbalanced* when each of the given workers cannot do each of the given jobs.
6. An assignment problem can be designed and solved as a transportation problem.
7. The transportation method can be used to solve an assignment problem.
8. *Any* unbalanced assignment problem can be balanced by introducing a dummy row or a dummy column.
9. All dummy rows/columns *must* have cost elements equal to zero.
10. Balancing of an unbalanced assignment problem might involve the introduction of dummy row as well as columns.
11. The solution to an assignment problem by transportation method would always be degenerate.
12. The relevant cost element is replaced by a zero in case a certain worker is not to be assigned a particular job.
13. The minimum number of lines required to cover all zeros cannot be more than n, the number of rows/columns.
14. If the minimum number of lines covering all the zeros is smaller than n, then the least uncovered value in the table is added to each value lying at the intersection of lines and subtracted from every other value in the table.
15. A maximising assignment problem can be converted into an equivalent minimisation problem by subtracting each element of the given matrix from a constant, K.
16. In an assignment problem involving assignment of salesmen to different zones to exploit their sales potential fully, if a salesman cannot be assigned to a particular zone, then the relevant cell (involving that salesman and the sales zone) value would be replaced by M for solving the problem.
17. If a constant is subtracted from each element of the cost matrix of an assignment problem, it would not affect the optimal assignment schedule.
18. If all entries in a row of the cost matrix are increased by a constant, it will not affect the optimal solution to the problem.
19. In case multiple zeros are obtained in all rows and columns, multiple optimal solutions are indicated.
20. The variables of the dual problem of a given assignment problem shall be unrestricted in sign.

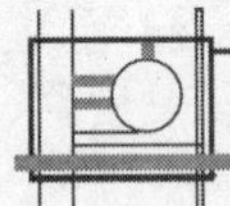

EXERCISES

1. What is an assignment problem? Is it true to say that it is a special case of the transportation problem? Explain.
2. How can you formulate an assignment problem as a standard linear programming problem? Illustrate.
3. What do you understand by an assignment problem? Give a brief outline for solving it.
4. State and discuss the methods employed for solving an assignment problem.
5. Explain the Hungarian Assignment Method. Is it better than other methods of solving assignment problem? How?
6. How would you deal with the assignment problems where
   (a) some assignments are prohibited?
   (b) the objective function is of maximisation type?
7. What is an unbalanced assignment problem? How is the Hungarian Assignment Method applied in respect of such a problem?
8. Can there be multiple optimal solutions to an assignment problem? How would you identify, if possible, the existence of multiple solutions in the (a) complete enumeration method, and (b) Hungarian assignment method?
9. Illustrate, by taking a small example, how you would write the dual of an assignment problem.
10. Discuss the similarities in the solution procedures for transportation and assignment models.

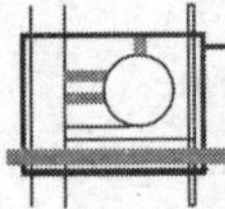

PRACTICAL PROBLEMS

1. Write the following as (a) a transportation problem, and (b) a linear programming problem.

	Time taken (in Hrs) in performing		
	Job 1	*Job 2*	*Job 3*
Worker 1	3	4.5	7
Worker 2	6	4	6
Worker 3	5	–	4

(–) indicates that worker 3 cannot perform job 2.

2. Convert the following LPP into an assignment problem and then solve it by the complete enumeration method.

   Minimise $Z = 8x_{11} + 5x_{12} + 6x_{13} + 6x_{21} + 9x_{22} + 4x_{23} + 5x_{31} + 7x_{32} + 7x_{33}$

   Subject to

   $x_{11} + x_{12} + x_{13} = 1, \qquad x_{11} + x_{21} + x_{31} = 1,$

$$x_{21}+x_{22}+x_{23}=1, \qquad x_{12}+x_{22}+x_{32}=1,$$

$$x_{31}+x_{32}+x_{33}=1, \qquad x_{13}+x_{23}+x_{33}=1,$$

all $x_{ij}=0, 1$

Note: $x_{ij}=0$ if ith worker is not assigned jth job

$=1$ if ith worker is assigned jth job

3. Solve the following assignment problem by (a) enumeration method, and (b) Hungarian assignment method.

Time (in minutes)

Worker	*Job 1*	*Job 2*	*Job 3*
A	4	2	7
B	8	5	3
C	4	5	6

4. *ABC* company is engaged in manufacturing 5 brands of packed snacks. It has five manufacturing setups, each capable of manufacturing any of its brands one at a time. The costs to make a brand on these setups vary according to the following table.

	S_1	S_2	S_3	S_4	S_5
B_1	4	6	7	5	11
B_2	7	3	6	9	5
B_3	8	5	4	6	9
B_4	9	12	7	11	10
B_5	7	5	9	8	11

Assuming five setups are S_1, S_2, S_3, S_4 and S_5 and five brands are B_1, B_2, B_3, B_4 and B_5. Find the optimum assignment of products on these setups resulting in the minimum cost.

(CA, November, 1998)

5. Five employees of a company are to be assigned to five jobs which can be done by any of them. Because of different number of years with the firm, the workers get different wages per hour. These are: Rs 15 per hour for *A, B* and *C* each, and Rs 13 per hour for *D* and *E* each. The amount of time taken (in hours) by each employee to do a given job is given in the following table. Determine the assignment pattern that (a) minimises the total time taken, and (b) minimises the total cost of getting five units of work done.

Job	*Employee*				
	A	*B*	*C*	*D*	*E*
1	7	9	3	3	2
2	6	1	6	6	5
3	3	4	9	10	7
4	1	5	2	2	4
5	6	6	9	4	2

6. There are five jobs to be assigned to five machines. Costs of completion of the jobs on the respective machines are as given in the table below:

Job	Machine				
	M_1	M_2	M_3	M_4	M_5
J_1	65	40	90	80	90
J_2	60	35	100	85	85
J_3	60	38	105	90	95
J_4	70	45	120	90	100
J_5	65	40	105	87	90

Employ the Hungarian method of optimal assignment of jobs to the machines so as to minimise the total cost of all the jobs. Is the solution obtained by you unique? If not, work out an alternative solution and calculate the total cost implied by it. *(M Com, Delhi, 2005)*

7. A dispatcher of the police department has received four requests for police assistance. Currently, six patrol-cars are available for assignment and the estimated response time (in minutes) are shown in the table that follows:

Incident	Patrol unit					
	1	2	3	4	5	6
I	6	5	3	4	5	6
II	8	6	2	3	7	6
III	4	4	7	6	5	5
IV	3	7	9	8	4	7

(a) Which patrol units should respond?
(b) What will be the average response time?

8. A fast-food chain wants to build four stores. In the past, the chain has used six different construction companies, and satisfied with each, has invited them to bid on each job. The final bids (in lakhs of rupees) were as shown in the following table:

	Construction companies					
	1	2	3	4	5	6
Store 1	85.3	88.0	87.5	82.4	89.1	86.7
Store 2	78.9	77.4	77.4	76.5	79.3	78.3
Store 3	82.0	81.3	82.4	80.6	83.5	81.7
Store 4	84.3	84.6	86.2	83.3	84.4	85.5

Since the fast-food chain wants to have each of the new stores ready as quickly as possible, it will award at most one job to one company. What assignment results in minimum total cost to the fast-food chain? *(MBA, Delhi, November, 2008)*

9. In the following situation calling for assignment of a-truck-to-a-territory, what assignment would involve the lowest aggregate cost?

Truck	*Territory*				
	1	*2*	*3*	*4*	*5*
T_1	200	190	180	190	110
T_2	150	210	190	200	100
T_3	80	90	200	90	70
T_4	220	80	90	140	60
T_5	170	110	90	150	40
T_6	90	100	120	90	60
T_7	120	200	130	80	60
T_8	160	90	160	110	80

10. A company solicits bids on each of four projects from five contractors. Only one project may be assigned to any contractor. The bids received (in thousands of rupees) are given in the accompanying table. Contractor *D* feels unable to carry out project 3 and, therefore, submits no bid.

Project	*Contractor*				
	A	*B*	*C*	*D*	*E*
1	18	25	22	26	25
2	26	29	26	27	24
3	28	31	30	–	31
4	26	28	27	26	29

(i) Use the Hungarian method to find the set of assignments with the smallest possible total cost.

(ii) What is the minimum total achievable cost? *(MBA, Delhi, October, 1996)*

11. The following information is available regarding four different jobs to be performed and about the clerks capable of performing the jobs:

Clerks	*Jobs (Time taken in hours)*			
	A	*B*	*C*	*D*
I	4	7	5	6
II	–	8	7	4
III	3	–	5	3
IV	6	6	4	2

Clerk II cannot be assigned to job A and clerk III cannot be assigned to job B. You are required to find out the optimal assignment schedule and the total time taken to perform the jobs. Also find whether the given problem has more than one optimal assignment schedule. *(MIB, Delhi, 2005)*

12. In the modification of a plant layout of a factory, four new machines M_1, M_2, M_3, and M_4 are to be installed in a machine shop. There are five vacant places *A*, *B*, *C*, *D*, and *E* available. Because of limited space, machine M_2 cannot be placed at *C* and M_3 cannot be placed at *A*. The cost of locating of machine *i* to place *j*, in rupees, is shown here. Find the optimal assignment schedule.

	A	*B*	*C*	*D*	*E*
M_1	89	91	95	90	91
M_2	92	89	–	90	89
M_3	–	91	94	91	87
M_4	94	88	92	87	88

13. Five swimmers are eligible to compete in a relay team which is to consist of four swimmers swimming four different swimming styles; back stroke, breast stroke, free style and butterfly. The time taken for the five swimmers—Anand, Bhaskar, Chandru, Dorai and Eswar—to cover a distance of 100 metres in various swimming styles are given below in minutes and seconds. Anand swims the back stroke in 1:09, the breast stroke in 1:15 and has never competed in the free style or butterfly. Bhaskar is a free style specialist averaging 1:01 for the 100 metres but can also swim the breast stroke in 1:16 and butterfly in 1:20. Chandru swims all styles—back 1:10, butterfly 1:12, free style 1:05 and breast stroke 1:20. Dorai swims only the butterfly 1:11, while Eswar swims the back stroke 1:20, the breast stroke 1:16, the free style 1:06 and the butterfly 1:10. Which swimmers should be assigned to which swimming style? Who will not be in the relay? *(CA, November, 1991)*

14. The purchase manager of a government department is analysing a set of bids that he has received for five projects from six bidders. The guidelines established for selecting the successful bidders require the manager to (a) minimise the total cost to complete the projects; (b) not award more than one contract to each bidder; and (c) disregard the lowest bid whenever it is more than 25 per cent of and below the next lowest bid (the assumption being that the quality of the work will not be up to standards). The bids on the projects (in thousands of rupees) are shown in the table that follows:

Bidder	*Contract*				
	A	*B*	*C*	*D*	*E*
1	7	8	8	12	7
2	9	13	10	14	5
3	3	7	6	13	11
4	17	17	7	8	8
5	8	12	7	15	16
6	10	10	10	16	8

Now,

(i) identify which particular bidder(s) shall not be eligible for which particular contract(s) in pursuance of the requirement in (c).

(ii) identify the successful bidder for each of the five contracts.

(iii) state how much this five contract assignment will cost the department.

(iv) work out the optimal assignment pattern if the requirement (c) is waived.

15. A manufacturer of a complex electronic equipment has just received a sizable contract and plans to subcontract part of the job. He has solicited bids for six subcontracts from four firms. Each job is sufficiently large so that any one firm can take on only one job. The below table shows the bids and the cost estimates (in Rs 10,000's) for doing the jobs internally. Note that no more than two jobs can be performed internally.

		Job					
		1	*2*	*3*	*4*	*5*	*6*
	1	48	72	36	52	50	65
Firm	2	44	67	41	53	48	64
	3	46	69	40	55	45	68
	4	43	73	37	51	44	62
Internal		50	65	35	50	46	63

(a) How do you complete this table so that the problem can be solved by Hungarian method?

(b) Solve the problem by Hungarian method, minimising total costs. *(MBA, Delhi, 1987)*

16. Solve the following unbalanced assignment problem of minimising total time for doing all the jobs:

Operator	*Job*				
	1	*2*	*3*	*4*	*5*
1	6	2	5	2	6
2	2	5	8	7	7
3	7	8	6	9	8
4	6	2	3	4	5
5	9	3	8	9	7
6	4	7	4	6	8

Hence, determine the optimal values of the dual variables. *(CA, May, 1985)*

17. MCS Inc is a software company, which has three projects of Y2K with the departments of health, education and housing of Maharashtra Government. Based on the background and experiences of the project leaders, they differ in terms of their performance at various projects. The performance score matrix is given below:

Project leaders	*Projects*		
	Health	*Education*	*Housing*
P_1	20	26	42
P_2	24	32	50
P_3	32	34	44

Help the management by determining the optimal assignment which maximises the total performance score. *(DMS, Mumbai, 1998)*

18. Suggest the optimal assignment schedule for the following assignment problem:

Sale (Rs in lakh)

Salesmen	*Markets*			
	I	*II*	*III*	*IV*
A	80	70	75	72
B	75	75	80	85
C	78	78	82	78

What will be the total maximum sale? *(M Com, Delhi, 2001)*

19. The owner of a small machine shop has four machinists available to assign to jobs for the day. Five jobs are offered with an expected profit in rupees for each machinist on each job as follows.

	A	*B*	*C*	*D*	*E*
1	62	78	50	101	82
2	71	84	61	73	59
3	87	92	111	71	81
4	48	64	87	77	80

Determine the assignment of machinists to jobs that will result in a maximum profit. Which job should be declined? *(MBA, Delhi, 1984)*

20. A sales manager wishes to assign four sales territories to four sales persons. The sales persons differ in their sales acumen and, consequently, the sales expected to be effected in each territory are different for each sales person. The estimates of sales per month for each sales person in each territory are given below:

Sales person	*Estimated monthly sales (in lacs of rupees) in territory*			
	W	*X*	*Y*	*Z*
A	120	140	145	130
B	150	140	155	140
C	145	140	142	150
D	148	150	142	145

Suggest optimal assignment of the four sales persons to various territories and the total maximum monthly sales.

21. A company has four salesmen who are to be assigned to four different sales territories. The monthly increases, estimated for each salesman in different territories (in lakh rupees), are shown in the table below:

Salesmen	*Sales territories*				
	I	*II*	*III*	*IV*	*V*
A	75	80	85	70	90
B	91	71	82	75	85
C	78	90	85	80	80
D	65	75	88	85	90

Suggest optimal assignment for the salesmen. Which sales territory will remain unassigned? What will be the maximum sales increase every month?

If for certain reasons, salesman *D* cannot be assigned to territory III, will the optimal assignment schedule be different? If so, show the new assignment schedule/schedules.

(M Com, Delhi, 1992)

22. A college department chairman has the problem of providing teachers for all courses offered by his department at the highest possible level of educational quality. He has got three professors, and one teaching assistant (TA). Four courses must be offered and after appropriate introspection and evaluation he has arrived at the following relative ratings (100 = best rating) regarding the ability of each instructor to teach each of the four courses, respectively.

	Course 1	*Course 2*	*Course 3*	*Course 4*
Prof 1	70	50	70	80
Prof 2	30	70	60	80
Prof 3	30	40	50	70
TA	40	20	40	50

How should he assign his staff to the courses to maximise educational quality in his department?

23. The Captain of a cricket team has to allot five middle batting positions to five batsmen. The average runs scored by each batsman at these positions are as follows:

Batsmen	*Batting positions*				
	I	*II*	*III*	*IV*	*V*
P	40	40	35	25	50
Q	42	30	16	25	27
R	50	48	40	60	50
S	20	19	20	18	25
T	58	60	59	55	53

(i) Find the assignment of batsmen to positions, which would give the maximum number of runs.

(ii) If another batsman *'U'* with the following average runs in batting positions as given below:

Batting position :	I	II	III	IV	V
Average runs :	45	52	38	50	49

is added to the team, should he be included to play in team? If so, who will be replaced by him?

(CA, May, 1992)

24. An agency is booking four groups, each of which would give one performance on a particular day. Because of the local preferences of the people, the agency is expecting different audience sizes for the different groups. Five major cities are under consideration of the agency. Considering the estimates of sales of tickets likely to be made, what group-city combination would you advise?

(Estimated sale of tickets ('000 Rs))

Group	*City*				
	C_1	C_2	C_3	C_4	C_5
G_1	58	56	21	50	45
G_2	16	34	18	25	15
G_3	39	44	30	64	36
G_4	82	102	71	110	73

25. A company has five plants P_1, P_2, P_3, P_4 and P_5. It now wants to introduce three products *A, B* and *C,* but not more than one in a plant. The unit production and distribution costs are shown in tables that follow:

Unit production cost:

Product	*In plant*				
	P_1	P_2	P_3	P_4	P_5
A	20	24	21	35	28
B	12	28	40	25	20
C	21	17	30	22	16

Unit distribution cost:

Product	*From plant*				
	P_1	P_2	P_3	P_4	P_5
A	12	6	13	3	6
B	14	6	8	10	12
C	17	8	12	18	16

The estimated sales and prices of the three products are as follows:

Product	*Sale (No. of units)*	*Selling price*
A	800	Rs 50
B	1,000	Rs 50
C	400	Rs 70

Using this information, state which product should be produced in which plant so that the aggregate profits are maximised. What is the amount of this profit?

26. A firm marketing a product has four salesmen S_1, S_2, S_3 and S_4. There are three customers to which a sale of a unit each is to be made. The chances of making a sale to a customer depend on the salesman-customer rapport. The following table depicts the probability with which each of the salesmen can make a sale to each of the customers.

Customer	*Salesman*			
	S_1	S_2	S_3	S_4
C_1	0.7	0.4	0.5	0.8
C_2	0.5	0.8	0.6	0.7
C_3	0.3	0.9	0.6	0.2

If only one salesman is assigned to each of the customers, what combination of salesmen and customers shall be optimal, given further that the profit obtainable by selling one unit to C_1 is Rs 500, whereas it is, respectively, Rs 450 and Rs 540 for sale to C_2 and C_3? What is the expected profit?

27. The market research team of the Look Forward Company requires some household data from four different cities. The team has to perform this job in two days—the next Saturday and Sunday. It plans to spend half a day in each of the cities. The relevant data are given as follows:

Day and time	*Probability of a household contact*			
	City 1	*City 2*	*City 3*	*City 4*
Saturday Morning	0.32	0.85	0.16	0.64
Saturday Evening	0.60	0.56	0.95	0.80
Sunday Morning	0.70	0.35	0.40	0.62
Sunday Evening	0.10	0.72	0.64	0.90
Number of households expected to interview	150	100	200	200

How should the team plan its visit to the four cities so that the expected response may be maximised? State this expected response.

28. A trip from Chennai to Bangalore takes six hours by bus. A typical timetable of the bus service in both the directions is given below:

Departure from Chennai	*Route number*	*Arrival at Bangalore*	*Departure from Bangalore*	*Route number*	*Arrival at Chennai*
0600 hrs	*A*	1200 hrs	0530 hrs	1	1130 hrs
0730 hrs	*B*	1330 hrs	0900 hrs	2	1500 hrs
1130 hrs	*C*	1730 hrs	1500 hrs	3	2100 hrs
1900 hrs	*D*	0100 hrs	1830 hrs	4	0030 hrs
0030 hrs	*E*	0630 hrs	0000 hrs	5	0600 hrs

The cost of providing this service by the transport company depends upon the time spent by the bus crew (driver and conductor) away from their places in addition to service times. There are five crews. There is a constraint that every crew should be provided with more than 4 hours of rest before the return trip again and should not wait for more than 24 hours for the return trip. The company has residential facilities for the crew at Chennai as well as at Bangalore. Find which crew be assigned which line of service or which line of service be connected with which other line, so as to reduce the waiting time to be the minimum?

CHAPTER

7

Extensions of Linear Programming: Integer Programming and Goal Programming

CHAPTER OVERVIEW

In continuation with the linear programming problems that seek to maximise or minimise the objective function, we now move on to the situations that are closer to reality. Formulation and solution of such problems as linear programming problems are based on two simplifying assumptions that there is only a single objective to achieve (such as maximisation of profit) and that the variables involved are continuous so that, for example, solution to a profit maximising problem of a multi-product firm may call for producing in numbers involving fractional values as well, like 18.7, 32.6, etc. It obviously sounds impractical, even nonsense, to produce fractional unit, give fractional number of advertisements in a media-mix problem, engage employees in fractional numbers... and so forth; and the firm may have multiple objectives to achieve. The present chapter addresses both these issues.

To ensure that the solution involves only integer values, we use *integer programming*, and the other extension of linear programming covered in this chapter is *goal programming*, which postulates setting up of multiple goals (may be conflicting in nature) instead of a single objective that calls for maximisation or minimisation. The contents of this chapter help a manager to obtain answers to the questions like the following:

- What is the optimal mix of the products in terms of exact number of completed units to be produced for each one?
- What combination of the investment proposals is best to undertake?
- How to obtain optimal assignment of workers to jobs by using a different approach and how to determine the optimal route for a travelling salesman who wants to cover a number of cities in a tour?
- How best can the multiple objectives be achieved by assigning penalties for not satisfying each of the objectives and weaving these penalties in the objective function?

This chapter requires higher skills in formulating the problem—both integer and goal programming—in the first instance. It will be evident from the variety of IPPs given and discussed. Further, in the context of goal programming, difference between goals and constraints should be understood clearly for a correct formulation of the problems. For solution to both, the integer as well as goal programming problems, while a knowledge of the Simplex method is essential, much more needs to be understood. This is in keeping with the requirements for each of the two types of problems.

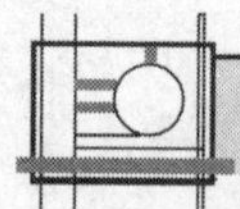

7.1 INTRODUCTION

Linear programming problems involve optimisation (minimisation or maximisation) of a given objective function subject to a set of constraints. Further, the variables involved in a problem are non-negative and may assume values which are fractional as well as integer. We now relax the conditions present in linear programming problems and consider situations where the variables may be allowed to take only integer values, and those where, instead of a single objective, multiple goals are sought to be achieved. The present chapter deals with both of these. The first of them is discussed under *integer programming* while the other one is given in *goal programming.*

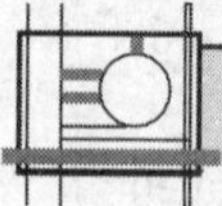

7.2 INTEGER PROGRAMMING

Integer Programming Problems (IPP) are a special class of linear programming problems where all or some of the variables in the optimal solution are restricted to non-negative integer values. It may be recalled that linear programming problems are based on the assumption that the decision variables are continuous, with the result that they may take fractional values in the optimal solution: 15¾ units of product *A* and $25\frac{1}{2}$ units of product *B* per day, for example. However, many situations in real life may not allow for this and warrant for the decision variables to take only integer values. For example, in production, manufacturing is frequently scheduled in terms of lots, batches or production runs. In allocation of goods, a shipment of goods involves a discrete number of trucks, wagons or aircraft, and fractional values are meaningless. Similarly, situations like sequencing, scheduling and routing decisions necessitate the use of integer programming models. An example is the travelling salesman problem, where it is aimed to determine a least distance route so that the salesman must visit each of a given set of cities, starting and ending his journey from a particular city (may be his home town). Capital budgeting provides another situation calling for the application of integer programming models. Hence, the need for restricting variables to assume only integer values and integer programming. The linear programming problems with variables restricted to non-negative integers are called *pure, all-integer,* or *mixed-integer* problems depending, respectively, on whether all or some of the variables are restricted to integer values. Also, there are '0–1' integer programming problems, which constitute an important class of such problems.

7.2.1 Pure and Mixed Integer Programming Problems

An integer programming problem in which *all* variables are required to be integers is called a *pure* or *all-integer programming* problem. For example,

Maximise $Z = 20x_1 + 8x_2$

Subject to

$$5x_1 + 7x_2 \le 63$$
$$3x_1 + 5x_2 \le 42$$
$$x_1, x_2 \ge 0 \; x_1, x_2 \text{ integer}$$

is a pure integer programming problem.

On the other hand, an IPP in which only some of the variables are required to be integers, is known as a *mixed* integer programming problem. For example,

Maximise $Z = 20x_1 + 8x_2$

Subject to

$$5x_1 + 7x_2 \leq 63$$
$$3x_1 + 5x_2 \leq 42$$
$$x_1, x_2 \geq 0 \; x_1 \text{ integer}$$

is a mixed integer programming problem since x_2 is not required to be an integer.

Example 7.1 The owner of a ready-made garments store sells two types of premium shirts, known as Zee-shirts and Star-shirts. He makes a profit of Rs 200 and Rs 300 per shirt on Zee- and Star-shirts, respectively. He has two tailors, *A* and *B*, at his disposal to stitch the shirts. Tailor *A* can devote a total of 17 hours per day, while tailor *B* can give at the most 15 hours a day. Both type of shirts are stitched by both the tailors. The time needed for stitching a Zee-shirt is 2 hours by tailor *A* and 3 hours by tailor *B*. Similarly, a Star-shirt requires 4 hours by tailor *A* and 3 hours by tailor *B*. How many shirts of each type should be stitched in order to maximise daily profit? Formulate it as an integer programming problem.

Let the daily output of Zee-shirts and Star-shirts be x_1 and x_2 units, respectively. Considering the profitability of shirts and the availability of the tailor time, the problem may be set up as follows:

Maximise $Z = 200x_1 + 300x_2$

Subject to

$$2x_1 + 4x_2 \leq 17$$
$$3x_1 + 3x_2 \leq 15$$
$$x_1, x_2 \geq 0 \text{ and integer}$$

7.2.2 Zero–One Model

A special type of integer programming problem is a case where the values of the decision variables are limited to two logical variables, like yes or no, match or no match and so on, which are symbolised by the values zero and one. An IPP where all the variables must equal zero or one is called a 0–1 integer programming problem. Surprisingly, there is a large number of situations giving rise to 0 – 1 IPPs. A few are given below.

1. The Assignment Problem An example of 0 – 1 IPP is provided by the LP formulation of an assignment problem. The formulation of assignment problem as an integer programming problem is shown below:

Minimise $Z = \sum_{i=1}^{n} \sum_{j=1}^{n} c_{ij} x_{ij}$

Subject to

$$\sum_{j=1}^{n} x_{ij} = 1 \quad \text{for } i = 1, 2, ..., n$$
$$\sum_{i=1}^{n} x_{ij} = 1 \quad \text{for } j = 1, 2, ..., n$$
$$x_{ij} = 0 \text{ or } 1 \quad \text{for all } i \text{ and } j$$

Consider the following example.

Example 7.2 Suppose we have an assignment problem where there are three workers and an equal number of jobs, and the cost matrix as given below:

Worker	Job J_1	J_2	J_3
W_1	4	7	8
W_2	6	5	10
W_3	7	7	9

Formulate this as an IPP.

The problem is expressible as a 0–1 IPP as follows, in which x_{ij}'s represent assignment of ith worker to the jth job.

Minimise $Z = 4x_{11} + 7x_{12} + 8x_{13} + 6x_{21} + 5x_{22} + 10x_{23} + 7x_{31} + 7x_{32} + 9x_{33}$

Subject to

$$x_{11} + x_{12} + x_{13} = 1 \qquad x_{11} + x_{21} + x_{31} = 1$$
$$x_{21} + x_{22} + x_{23} = 1 \qquad x_{12} + x_{22} + x_{32} = 1$$
$$x_{31} + x_{32} + x_{33} = 1 \qquad x_{13} + x_{23} + x_{33} = 1$$
$$x_{ij} \geq 0 \text{ for } i = 1, 2, 3; j = 1, 2, 3$$
$$x_{ij} = 0 \text{ (if assignment is not made) or}$$
$$1 \text{ (if assignment is made)}$$

2. The Knapsack Problem An IPP which has only one constraint is referred to as knapsack problem.

Example 7.3 Four items are considered for loading on an aeroplane, which has the capacity to load up to 15 tonnes. The weights and values of the items are indicated below:

Item:	A	B	C	D
Weight (tonnes):	3	5	6	4
Per unit Value:	20	25	32	22

Which items and what quantities should be loaded on the plane so as to maximise the value of the cargo transported?

Let x_1, x_2, x_3 and x_4 be the number of items A, B, C and D, respectively, that are loaded on the plane. Accordingly, the problem is

Maximise $Z = 20x_1 + 25x_2 + 32x_3 + 22x_4$

Subject to

$$3x_1 + 5x_2 + 6x_3 + 4x_4 \leq 15$$
$$x_i = 0, 1 \ (i = 1, 2, 3, 4)$$

Here the decision variables are 0 (do not load) or 1 (load).

3. Capital Budgeting/Investment Problem Frequently, companies face the situation of selecting one or more research and development projects or other investment opportunities among several competing projects. The ones accepted would get the value 1 and those not accepted would be assigned zero.

Example 7.4 A company is considering five investments. The net present value (NPV) and the initial cash outflow required for each of them is given below:

Investment	*Net Present Value (NPV)*	*Cash Outflow*
I	Rs 20,000	Rs 12,000
II	28,000	14,000
III	18,500	7,000
IV	27,500	13,000
V	31,000	16,000

The company has a total of Rs 48,000 to invest. Further, it is required that

(i) the company should not make more than 3 investments.

(ii) the company should invest in investment II if it invests in investment III.

(iii) if the company invests in investment IV, then it should not invest in investment V.

Formulate this as an integer programming problem, taking the objective function as maximisation of NPV.

Let x_1, x_2, x_3, x_4 and x_5 be the variables indicating investment in investments I, II, III, IV and V, respectively. Each of the variables can take a value of 0 (if an investment is not made) or 1 (if the investment is made). Thus, we define

$$x_i \ (i = 1, 2, 3, 4, 5) = 0 \text{ if investment is not made}$$
$$= 1 \text{ otherwise}$$

The total NPV of all investments = $20{,}000x_1 + 28{,}000x_2 + 18{,}500x_3 + 27{,}500x_4 + 31{,}000x_5$.

Constraints Since the cash availability is Rs 48,000, the constraint may be expressed as

$$12{,}000x_1 + 14{,}000x_2 + 7{,}000x_3 + 13{,}000x_4 + 16{,}000x_5 \le 48{,}000$$

Requirement (i) may be expressed as $x_1 + x_2 + x_3 + x_4 + x_5 \le 3$.

In terms of variables x_2 and x_3, the requirement (ii) states that if $x_3 = 1$, then x_2 must also equal 1. Now, if we add the constraint $x_3 \le x_2$ or $x_3 - x_2 \le 0$, it will take care of this requirement. It may be shown why it is so. For this, we consider the two possibilities: either $x_3 = 1$ or $x_3 = 0$. In the first case when $x_3 = 1$, it implies that x_2 must be greater than or equal to 1 for $x_3 - x_2 \le 0$ to hold. Since x_2 must equal only 0 or 1, it implies that $x_2 = 1$. Similarly, if $x_3 = 0$, then the condition $x_3 - x_2 \le 0$ implies that $x_2 \ge 0$ which allows $x_2 =$ 0 or 1. In short, if $x_3 = 0$, the given condition does not restrict the value of x_2 which is consistent with the given requirement.

Finally, the requirement that investment in both investments IV and V cannot be made, may be taken care of by adding the condition $x_4 + x_5 \le 1$. Thus, if $x_4 = 1$, x_5 would have to be zero, and if $x_4 = 0$, then x_5 may be 0 or 1.

Accordingly, the IPP may be stated as follows:

Maximise $\quad Z = 20{,}000x_1 + 28{,}000x_2 + 18{,}500x_3 + 27{,}500x_4 + 31{,}000x_5$

Subject to

$$12{,}000x_1 + 14{,}000x_2 + 7{,}000x_3 + 13{,}000x_4 + 16{,}000x_5 \le 48{,}000$$
$$x_1 + x_2 + x_3 + x_4 + x_5 \le 3$$
$$-x_2 + x_3 \le 0$$
$$x_4 + x_5 \le 1$$
$$x_1, x_2, x_3, x_4 \text{ and } x_5 = 0, 1$$

4. The Fixed Charge Problem Many problems in real life involve a combination of fixed and variable costs. The fixed costs are incurred only if certain projects are undertaken or a certain capacity level is exceeded. A common example is machine scheduling.

Example 7.5 Three machines are available to make 5000 units which are required for the internal use of a company as components to another product. If the production is to be done by any one of these machines, a set-up cost is incurred apart from the cost of making each unit on different machines. The cost data are as follows:

Machine	*Set-up Cost (Rs)*	*Cost per Unit (Rs)*	*Maximum Production*
1	8,000	5	4,000
2	5,000	4	3,000
3	4,000	8	1,000

What is the best production strategy? Only formulate. *(MBA, Delhi, 1994)*

Let x_1, x_2 and x_3 represent the quantities to be produced on machines 1, 2 and 3, respectively, and d_1, d_2 and d_3 indicate whether a machine is to be used (1) or not (0). Accordingly, the fixed cost would be $8{,}000d_1 + 5{,}000d_2 + 4{,}000d_3$, while the variable cost would be $5x_1 + 4x_2 + 8x_3$. The integer programming problem is:

Minimise $\quad Z = 8{,}000d_1 + 5{,}000d_2 + 4{,}000d_3 + 5x_1 + 4x_2 + 8x_3$

Subject to

$$x_1 + x_2 + x_3 \geq 5{,}000$$

$$x_1 \leq 4{,}000d_1$$

$$x_2 \leq 3{,}000d_2$$

$$x_3 \leq 1{,}000d_3$$

$$x_1, x_2 \text{ and } x_3 \geq 0;\ d_1, d_2 \text{ and } d_3 = 0, 1$$

5. The Facility Location Problem Another application of integer programming is facility location problems. Such a problem involves deciding where to locate a limited number of facilities such as plants and warehouses, given a larger number of potential locations. Consider a company engaged in the production of two-wheeler scooter and having a wide range geographical area to serve through its sales outlets. In designing a production and distribution system, the company faces a fundamental problem of determining where to locate its plants in order to strike a balance between fixed costs of operating the plants and transportation costs in the distribution of its scooters to various outlets. On the one extreme, it may decide to have one plant adjacent to each outlet—which does not appear to be worthwhile because while transportation costs would be very low, the fixed operating costs would be very high. The other extreme calls for having only one plant to serve all the sales outlets, which is equally unattractive because it would involve low fixed operating costs but very large transportation costs. The optimal solution to the problem would be somewhere between the two extremes and could be obtained by using integer programming. Consider the following example.

Example 7.6 The Standard Cars Limited currently has four plants for manufacturing and assembling cars and six warehouses where these are transported for subsequent delivery to its dealers/showrooms. The company management has collected the following information:

Annual production capacity of each plant,

Annual demand of cars at various warehouses,

Costs involved in transporting cars from various plants to different warehouses, and

Annual fixed operating costs for each of the plants.

The information is summarised in the following table:

Transportation Costs (in '000 Rs) per '000 Cars

Plant	*Warehouse*						*Annual Production Capacity ('000 Cars)*	*Annual Fixed Operating Costs ('000 Rs)*
	1	*2*	*3*	*4*	*5*	*6*		
1	88	15	29	67	38	120	42	418
2	59	82	37	7	19	54	30	302
3	21	80	21	15	32	30	48	362
4	38	30	18	27	21	94	40	168
Annual Demand ('000 Cars)	21	12	17	6	8	20		

To illustrate, plant 3 has a total annual capacity of 48 thousand cars; the demand at warehouse 2 is 12,000 cars per annum; the cost of transporting 1,000 cars from plant 3 to warehouse 5 is Rs 32,000; and plant 4 has an annual fixed operating cost equal to Rs 168,000. Currently, the company is sending the cars as per the following schedule:

From: Plant	*To: Warehouse*					
	1	*2*	*3*	*4*	*5*	*6*
1	12	3				
2				6	8	
3	9	9	17			
4						20

This schedule involves a total cost of Rs 4,441,000 by way of transportation and Rs 1,250,000 as annual fixed operating costs.

The management is contemplating to change the present schedule so as to attempt to reduce the total cost to the minimum, even considering the closure of some plant/s. You are required to formulate the problem as an IPP.

Let x_{ij} be the number of cars (in thousands) to be transported from ith plant to jth warehouse. With the objective to minimise the total cost including transportation cost and operating cost, the problem may be stated as follows:

$$\text{Minimise } Z = 88x_{11} + 15x_{12} + 29x_{13} + 67x_{14} + 38x_{15} + 120x_{16} + 59x_{21} + 82x_{22} + 37x_{23} + 7x_{24} + 19x_{25} + 54x_{26} + 21x_{31} + 80x_{32} + 21x_{33} + 15x_{34} + 32x_{35} + 30x_{36} + 38x_{41} + 30x_{42} + 18x_{43} + 27x_{44} + 21x_{45} + 94x_{46} + 418y_1 + 302y_2 + 362y_3 + 168y_4$$

(Transportation and Operating Cost)

Subject to

$$x_{11} + x_{12} + x_{13} + x_{14} + x_{15} + x_{16} \le 42y_1$$
$$x_{21} + x_{22} + x_{23} + x_{24} + x_{25} + x_{26} \le 30y_2$$
$$x_{31} + x_{32} + x_{33} + x_{34} + x_{35} + x_{36} \le 48y_3$$
$$x_{41} + x_{42} + x_{43} + x_{44} + x_{45} + x_{46} \le 40y_4$$
$$x_{11} + x_{21} + x_{31} + x_{41} = 21$$
$$x_{12} + x_{22} + x_{32} + x_{42} = 12$$
$$x_{13} + x_{23} + x_{33} + x_{43} = 17$$
$$x_{14} + x_{24} + x_{34} + x_{44} = 6$$
$$x_{15} + x_{25} + x_{35} + x_{45} = 8$$
$$x_{16} + x_{26} + x_{36} + x_{46} = 20$$
$$x_{ij} \ge 0, \quad i = 1, 2, 3, 4; \quad j = 1, 2, 3, 4, 5, 6$$
$$y_i = 0, 1 \quad i = 1, 2, 3, 4$$

In this problem, y_i would be equal to zero, if a particular plant is not run, and equal to one if it is run. If a plant is not operated, its capacity would be reduced to zero, as would be its operating cost.

6. Integer Programming and Piece-wise Linear Functions The 0 – 1 variables may also be used to model optimisation problems that involve piece-wise linear functions. Such a function consists of several straight line segments. An example of such function is provided where quantity discounts are allowed. To illustrate, suppose the price of an item is quoted like this: Rs 25 per unit for the first 600 units, the next 600 units at Rs 22 per unit and the next 800 units at a rate of Rs 20 per unit, where a maximum of 2000 can be purchased. Now, if x is the number of units bought and $C(x)$ be the cost of buying x units, we can determine $C(x)$ as follows:

For $x < 0$, $C(x) = 0$, for $0 \le x \le 600$, $C(x) = 25x$,

while for $600 \le x \le 1200$, $C(x) = 15000 + 22(x - 600)$, where 15,000 represents the cost of first 600 units @ Rs 25 per unit and $22\ (x - 600)$ represents the cost of units in excess of 600. This simplifies to $1800 + 22x$.

Similarly, for $1200 \le x \le 2000$, $C(x) = 28200 + 20\ (x - 1200)$, which works out to be $4200 + 20x$. The cost of buying the first 1200 units equals Rs 28,200, which is included in the function and along with it the cost of units in excess of 1200 @ Rs 20 per unit is included.

It is clear that the cost here is not a linear function of the number of units purchased. The function has breaks at 0, 600, 1200 and 2000.

Now, since a piece-wise linear function is not linear, it may appear that we may not be able to use linear programming to solve optimisation problems using such functions. However, it is not so because by 0–1 variables, such functions can be represented in linear form. This may be done as follows. Suppose that a piece-wise linear function has break points $b_1, b_2, \ldots b_n$. For some $k (k = 1, 2, \ldots, n - 1)$, $b_k \le x \le b_k + 1$. Then, for some number z_k (a fraction value, between zero and one), x may be written as $x = z_k b_k + (1 - z_k) b_k + 1$, and since the function $f(x)$ is linear for $b_k \le x \le b_{k+1}$, we may write $f(x) = z_k f(b_k) + (1 - z_k) f(b_k + 1)$.

For instance, if $x = 1000$, then we have $b_1 = 600$, $b_2 = 1200$, and $b_1 \le x \le b_2$ or $600 \le x \le 1200$, and it is possible to write $x = 1/3 \times (600) + 2/3 \times (1200)$. [In fact, we set $x = b_1 z_k + b_2(1 - z_k)$ and solve for z_k. Thus, here $1000 = 600z_k + 1200(1 - z_k)$ and solve it to get $z_k = 1/3$].

Accordingly,
$$C(x) = C(1{,}000) = 1/3C(600) + 2/3C(1{,}200)$$
$$= 1/3 \times 15{,}000 + 2/3 \times 28{,}200 = 23{,}800$$

Now we can discuss the expression of a piece-wise linear function through linear constraints and 0 – 1 variables. As a first step, replace $f(x)$ by $z_1 f(b_1) + z_2 f(b_2) + \cdots + z_n f(b_n)$. After this, add the following constraints to the problem:

$$z_1 \le y_1;\ z_2 \le y_1 + y_2;\ z_3 \le y_2 + y_3 \ldots z_{n-1} \le y_{n-2} + y_{n-1};\ z_n \le y_{n-1}$$
$$y_1 + y_2 + \cdots + y_n = 1$$
$$z_1 + z_2 + \cdots + z_n = 1$$
$$x = z_1 b_1 + z_2 b_2 + \cdots + z_n b_n$$
$$y_i = 0 \text{ or } 1 \text{ (for } i = 1, 2, \ldots, n - 1);\ z_i \ge 0 \text{ (for } i = 1, 2, \ldots, n)$$

Example 7.7 A company produces two products *A* and *B*. Each unit of product *A* requires 1 hour of engineering service and 5 hours of machine time. To produce 1 unit of product B requires 2 hours of engineering and 8 hours of machine time. There are 100 hours of engineering and 400 hours of machine time available. The cost of production is a non-linear function of the quality produced as given in the following table:

Product A		*Product B*	
Production (Units)	Unit Cost (Rs)	Production (Units)	Unit Cost (Rs)
0–50	10	0–40	7
51–100	8	41–100	3

The unit selling price of product *A* is Rs 12 and the unit selling price of product *B* is Rs 14. The company would like a production plan which gives the number of units of *A* and the number of units of *B* to be produced that will maximise profit. Formulate an integer programming model for this problem.

(MBA, Delhi, 1996)

Let x_1 and x_2 represent the number of units of products *A* and *B*, respectively. The total revenue would be $12x_1 + 14x_2$, while the total cost would be

$$C(x_1) = \begin{cases} 10x_1 & (0 \le x_1 \le 50) \\ 8x_1 + 100 & (50 \le x_1 \le 100) \end{cases}$$

$$C(x_2) = \begin{cases} 7x_2 & (0 \le x_1 \le 40) \\ 3x_2 + 160 & (40 \le x_1 \le 100) \end{cases}$$

The objective function and constraints can be stated as follows:

Maximise $Z = 12x_1 + 14x_2 - C(x_1) - C(x_2)$

Subject to

$$x_1 + 2x_2 \leq 100 \quad \text{Engineering Hours}$$
$$5x_1 + 8x_2 \leq 400 \quad \text{Machine Time}$$
$$x_1, x_2 \geq 0$$

Now, since $C(x_1)$ and $C(x_2)$ are piece-wise linear functions, with break points at 0, 50 and 100; and 0, 40 and 100 respectively, we proceed as follows:

Replace $C(x_1)$ by $C_1(x_1) = z_1C_1(0) + z_2C_1(50) + z_3C_1(100)$ and

$C(x_2)$ by $C_2(x_1) = z_4C_2(0) + z_5C_2(40) + z_6C_2(100)$.

Further, we add the following constraints:

$$x_1 = 0z_1 + 50z_2 + 100z_3$$
$$z_1 \leq y_1;\ z_2 \leq y_1 + y_2;\ z_3 \leq y_2$$
$$z_1 + z_2 + z_3 = 1$$
$$y_1 + y_2 = 1$$
$$y_i = 0 \text{ or } 1 \text{ (for } i = 1, 2);\ z_i \geq 0 \text{ (for } i = 1, 2, 3).$$
$$x_2 = 0z_4 + 40z_5 + 100z_6$$
$$z_4 \leq y_3;\ z_5 \leq y_3 + y_4;\ z_6 \leq y_4$$
$$z_4 + z_5 + z_6 = 1$$
$$y_3 + y_4 = 1$$
$$y_i = 0 \text{ or } 1 \text{ (for } i = 3, 4);\ z_i \geq 0 \text{ (for } i = 4, 5, 6)$$

Accordingly, the new formulation gives the following integer programming problem:

Maximise $Z = 12x_1 + 14x_2 - z_1C_1(0) - z_2C_1(50) - z_3C_1(100) - z_4C_2(0) - z_5C_2(40) - z_6C_2(100)$

Subject to

$$x_1 + 2x_2 \leq 100$$
$$5x_1 + 8x_2 \leq 400$$
$$z_1 \leq y_1$$
$$z_2 \leq y_1 + y_2$$
$$z_3 \leq y_2$$
$$z_4 \leq y_3$$
$$z_5 \leq y_3 + y_4$$
$$z_6 \leq y_4$$
$$z_1 + z_2 + z_3 = 1$$
$$z_4 + z_5 + z_6 = 1$$
$$y_1 + y_2 = 1$$
$$y_3 + y_4 = 1$$

$y_i = 0$ or 1 (for $i = 1, 2, 3, 4$); $z_i \geq 0$ (for $i = 1, 2, 3, 4, 5, 6$) $x_1, x_2 \geq 0$, integer.

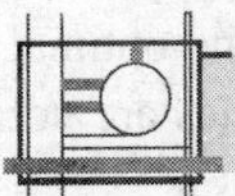

7.3 SOLUTION TO IPPs

It may appear that it would suffice to obtain a non-integer solution to the integer programming problems using the regular graphic or simplex methods and then rounding off the fractional values occurring in the optimal solution. Unfortunately, however, this rarely works. Usually, the deviations from actual optimal integer values (due to rounding off) may become large enough to cause the solution infeasible. Hence, there is a need to develop a systematic procedure for obtaining optimal solution to such problems, meeting the integer requirements.

The first of the systematic approaches to solve integer programming problems was suggested by RE Gomory in 1956. While initially only the pure IPPs could be handled, the method was extended later to deal with the mixed IPPs. The method uses *cutting plane algorithm* and ensures the optimal integer solution in a finite number of iterations. Another significant approach to solving such problems is that of *branch-and-bound* method. This is also used to solve the 0–1 IPPs.

7.3.1 Cutting Plane Algorithm

To solve an integer programming problem (IPP) using the cutting plane algorithm, we first drop the integer requirements and obtain the linear programming (LP) relaxation. The LP relaxation is solved using the simplex method in the usual way. If all the variables in the optimal solution assume integer values, we have found an optimal solution to the IPP. Otherwise, we introduce a cut using a constraint in the optimal solution which has a fractional right-hand side (b_i) value. The cut is then introduced into the optimal solution to LP relaxation and the revised problem is solved using simplex method, with some modification. Cuts are introduced until all the variables assume integer values.

The use of cutting plane algorithm is illustrated below.

Example 7.8 Solve the problem given in Example 7.1 using cutting plane algorithm.

The problem is restated below:

Maximise $Z = 200x_1 + 300x_2$

Subject to

$$2x_1 + 4x_2 \leq 17$$

$$3x_1 + 3x_2 \leq 15$$

$$x_1, x_2 \geq 0 \text{ and integer}$$

The LP relaxation of the problem is:

Maximise $Z = 200x_1 + 300x_2$

Subject to

$$2x_1 + 4x_2 \leq 17$$

$$3x_1 + 3x_2 \leq 15$$

$$x_1, x_2 \geq 0$$

The solution to the LP relaxation using simplex method is given in Tables 7.1 through 7.3. The optimal solution to the problem is: $x_1 = 3/2$ and $x_2 = 7/2$ and $Z = 1350$. Since the values of the variables are not integers, it does not represent the solution to the IPP.

TABLE 7.1 Simplex Tableau 1

Basis		x_1	x_2	S_1	S_2	b_i	b_i/a_{ij}
S_1	0	2	4*	1	0	17	17/4 ←
S_2	0	3	3	0	1	15	5
c_j		200	300	0	0		
Solution		0	0	17	15	Z = 0	
Δ_j		200	300 ↑	0	0		

TABLE 7.2 Simplex Tableau 2

Basis		x_1	x_2	S_1	S_2	b_i	b_i/a_{ij}
x_2	300	1/2	1	1/4	0	17/4	17/2
S_2	0	3/2*	0	−3/4	1	9/4	3/2 ←
c_j		200	300	0	0		
Solution		0	17/4	0	9/4	Z = 1,275	
Δ_j		50 ↑	0	−75	0		

TABLE 7.3 Simplex Tableau 3

Basis		x_1	x_2	S_1	S_2	b_i
x_2	300	0	1	1/2	−1/3	7/2
x_1	200	1	0	−1/2	2/3	3/2
c_j		200	300	0	0	
Solution		3/2	7/2	0	0	Z = 1,350
Δ_j		0	0	−50	−100/3	

To apply the cutting plane method, we begin by choosing any constraint in the LP relaxation's tableau containing the optimal solution in which a basic variable is fractional. In the Simplex Tableau 3, let us (arbitrarily) select the first constraint which is

$$0x_1 + x_2 + 1/2S_1 - 1/3S_2 = 7/2$$

We now define (*y*) to be the largest integer less than or equal to *y*. For example, (2.80) = 2 and (– 2.30) = – 3. Any number *y* may be expressed in the form of (*y*) + *f*, where *f* is the non-negative fractional part, and hence, $0 \le f \le 1$. Thus, 2.80 = 2 + 0.80 and – 2.30 = – 3 + 0.70. Now we may write the chosen constraint as follows:

$$0x_1 + (1 + 0)x_2 + (0 + 1/2)S_1 + (-1 + 2/3)S_2 = 3 + 1/2$$

Putting all integer coefficients on the right-hand side and all terms with fractional coefficients on the left-hand side gives

$$1/2S_1 + 2/3S_2 = 1/2 + (3 - 1x_2 + 1S_2)$$

We may think of this equation as

$$1/2S_1 + 2/3S_2 = 1/2 + \text{some integer}$$

If the LHS is equal to 1/2 + some integer, then it is obvious that the LHS is greater than or equal to 1/2 since S_1 and S_2 are both non-negative. Thus, $1/2S_1 + 2/3S_2 \ge 1/2$.

The cutting plane algorithm now suggests adding this constraint to the LP relaxation's optimal solution tableau as $-1/2S_1 - 2/3S_2 \le -1/2$ or $-1/2S_1 - 2/3S_2 + S_3 = -1/2$. Such a constraint is known as a *Gomory constraint* or *cut*.

A cut derived in this way has two properties. First of these is that the current optimal solution to the LP relaxation shall not satisfy the cut. For our example, the optimal solution contained in Table 7.3 has $S_1 = 0$ and $S_2 = 0$. These values obviously do not satisfy the cut which requires $1/2S_1 + 2/3S_2 \ge 1/2$. A second property of the cut is that any feasible point for the IPP will satisfy the cut. In effect, a cut truncates the current optimal solution to the LP relaxation but not any feasible solutions to the IPP. When a cut is added to the LP relaxation, we expect to obtain a solution where all variables assume integer values. If it is so, we have found the optimal solution to the original IPP. If the new optimal solution (to the LP relaxation and the cut) has some fractional values, another cut is generated and the process is continued. Gomory showed that this process yields an optimal solution to the IPP after a finite number of iterations in the form of cuts.

For the example under consideration, the effect of the cut $1/2S_1 + 2/3S_2 \ge 1/2$ may be observed in Figure 7.1.

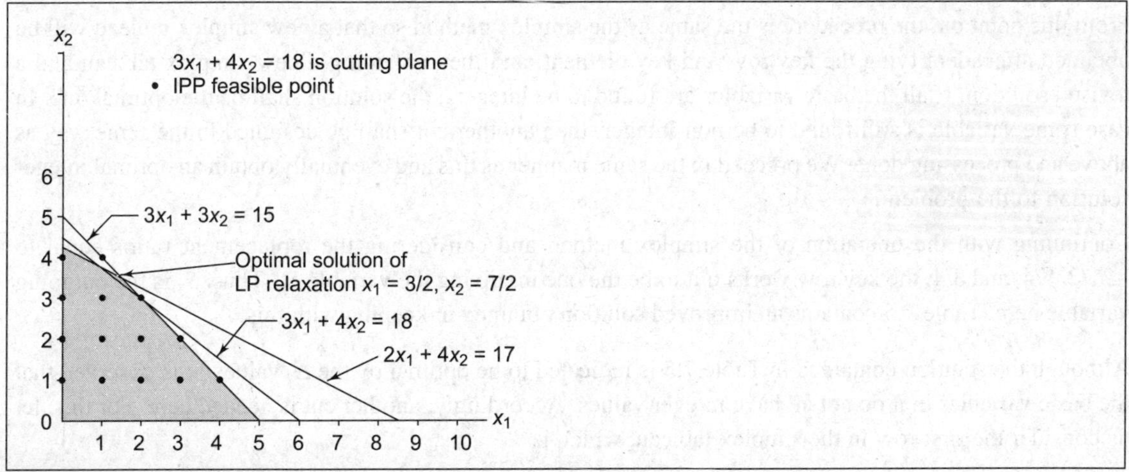

Figure 7.1 *Graphic Representation of IPP—Cutting Plane*

It may be seen that all points feasible for the IPP satisfy the cut but the current optimal solution to the LP relaxation, $x_1 = 3/2$ and $x_2 = 7/2$, does not satisfy it. The graph of the cut is obtained as follows. The first constraint of the IPP states $2x_1 + 4x_2 + S_1 = 17$, while the second one is $3x_1 + 3x_2 + S_2 = 15$. Thus, $S_1 = 17 - 2x_1 - 4x_2$ and $S_2 = 15 - 3x_1 - 3x_2$. Substituting these values in the relation $1/2S_1 + 2/3S_2 \geq 1/2$ and simplifying we get $3x_1 + 4x_2 \geq 18$.

To continue with the solution process, this cut is added to the simplex tableau in Table 7.3 to get the new simplex tableau shown in Table 7.4 and the solution is updated.

TABLE 7.4 Revised Simplex Tableau 1

Basis		x_1	x_2	S_1	S_2	S_3	b_i	b_i/a_{ij}
x_2	300	0	1	1/2	–1/3	0	7/2	–21/2
x_1	200	1	0	–1/2	2/3	0	3/2	9/4
S_3	0	0	0	–1/2	–2/3*	1	–1/2	3/4
c_j		200	300	0	0	0		
Solution		3/2	7/2	0	0	–1/2		
Δ_j		0	0	–50	–100/3	0		
					↑			

In terms of the simplex algorithm, here we have a problem in trying to determine as to which variable shall be the incoming variable since all the values are less than, or equal to zero. To get the solution started, we divide the Δ_j values by the corresponding negative values in the S_3 row (the third row in the simplex tableau) and consider the variable with the smallest quotient. We do it to get [– 50 ÷ (– 1/2) = 100] and [– 100/3 ÷ (– 2/3) = 50]. Accordingly, S_2 is the incoming variable (the key column).

From this point on, the procedure is the same as the simplex method so that a new simplex tableau will be obtained after identifying the key row and key element, and then obtaining a new simplex tableau and a revised solution. If all the basic variables are found to be integers, the solution shall be the optimal one. In case some variable is still found to be non-integer, then another cut shall be designed in the same way as above and processing done. We proceed in the same manner as this and eventually obtain an optimal integer solution to the problem.

Continuing with the operation of the simplex method and considering the replacement ratios equal to –21/2, 9/4, and 3/4, the key row works out to be the one involving the variable S_3. Thus, S_3 is the outgoing variable here. Table 7.5 contains an improved solution obtained in keeping with this.

Although the solution contained in Table 7.5 is indicated to be optimal by the Δ_j values, it is observed that the basic variables in it do not all have integer values. Accordingly, another cut is needed here. For this, let us consider the first row in the simplex tableau, which is

$$0x_1 + 1x_2 + 3/4S_1 + 0S_2 - 1/2S_3 = 15/4$$

TABLE 7.5 Revised Simplex Tableau 2

Basis		x_1	x_2	S_1	S_2	S_3	b_i
x_2	300	0	1	3/4	0	−1/2	15/4
x_1	200	1	0	−1	0	1	1
S_2	0	0	0	3/4	1	−3/2	3/4
c_j		200	300	0	0	0	
Solution		1	15/4	0	3/4	0	
Δ_j		0	0	−25	0	−50	

It may be rewritten as

$$0x_1 + 1x_2 + 3/4S_1 + 0S_2 - S_3 + 1/2S_3 = 3 + 3/4$$

Thus, $$3/4S_1 + 1/2S_3 = 3/4 + (3 - x_2 + S_3)$$

Accordingly,

$$3/4S_1 + 1/2S_3 \geq 3/4$$

or $$-3/4S_1 - 1/2S_3 \leq -3/4$$

or $$-3/4S_1 - 1/2S_3 + S_4 = -3/4$$

where S_4 is a slack variable.

Simplex tableau is updated by introduction of this constraint as shown in Table 7.6.

TABLE 7.6 Revised Simplex Tableau 3

Basis		x_1	x_2	S_1	S_2	S_3	S_4	b_i	b_i/a_{ij}
x_2	300	0	1	3/4	0	−1/2	0	15/4	5
x_1	200	1	0	−1	0	1	0	1	−1
S_2	0	0	0	3/4	1	−3/2	0	3/4	1
S_4	0	0	0	−3/4	0	−1/2	1	−3/4	1 ←
c_j		200	300	0	0	0	0		
Solution		1	15/4	0	3/4	0	−3/4		
Δ_j		0	0	−25 ↑	0	−50	0		

From this table, we first determine the incoming variable, in accordance with the rule described earlier, as follows:

for S_1: $-25 \div (-3/4) = 100/3$, and

for S_3: $-50 \div (-1/2) = 100$

Accordingly, S_1 is the incoming variable whereas S_4 may be taken as the outgoing variable as seen from the replacement ratio values. The solution is contained in Table 7.7.

TABLE 7.7 Revised Simplex Tableau 4: Optimal Solution

Basis		x_1	x_2	S_1	S_2	S_3	S_4	b_i
x_2	300	0	1	0	0	−1	1	3
x_1	200	1	0	0	0	5/3	−4/3	2
S_2	0	0	0	0	1	−1	−2	0
S_1	0	0	0	1	0	2/3	−4/3	1
c_j		200	300	0	0	0	0	
Solution		2	3	1	0	0	0	
Δ_j		0	0	0	0	−100/3	−100/3	

From the solution, it may be observed that each of the basic variables has an integer solution value. Accordingly, the solution is the desired optimal solution to the IPP. In terms of this solution, $x_1 = 2, x_2 = 3$ and $Z = 1300$.

It may be recalled that a cut does not eliminate any points that are feasible for the integer programming problem. It implies that when we solve LP relaxation to an IPP, with any cut/s and find an optimal solution wherein all the variables are integers, we have solved original IPP. Since x_1 and x_2 are integers in the current optimal solution, this point will also be optimal for the original IPP.

Two points are noteworthy here:

(i) The algorithm requires that all the coefficients in the constraints and all the right-hand side values of the constraints be integers—it would in fact ensure that when decision variables are integers, the slack and surplus variables should also be integers. A constraint like $3.2x_1 + x_2 \leq 15.6$ would have to be replaced as $32x_1 + 10x_2 \leq 156$.

(ii) At any stage of working of the algorithm, when two or more constraints have fractional values on the RHS, it is desirable to generate the cut by using the constraint whose RHS has a fractional part closest to 1/2.

7.3.2 Branch and Bound Method

The branch and bound method is another method used to solve integer programming problems. It may be mentioned at the outset that the branch and bound method is not a particular method for solving a particular problem. It refers to a certain procedure for finding an optimal solution and is applied differently in different kinds of problems. This is generally used in what may be called as *combinatorial problems*—problems where there are a finite number of solutions. By applying some rules, these solutions are divided into two parts—one that most probably contains the optimal solution and therefore, should be examined further; and the second part that would not contain the optimal solution and, thus, be left out of further consideration.

Thus, branching and bounding essentially keeps dividing up the feasible region in to smaller and smaller parts until the solution that optimises (maximises or minimises, as the case may be) is determined. Here, we shall demonstrate the use of the method for solving assignment problems (0–1 integer programming problems) and travelling salesman problems.

7.3.3 Solution to Assignment Problems

We have considered in the previous chapter, in solution to assignment problems that since such problems have a finite number of possible solutions, it is possible to solve them by enumeration method. As indicated, in this method all the possible solutions to a problem are listed and then the one which optimises is chosen. However, since the number of possible solutions rises very rapidly as people and jobs are added to the problem, it becomes very difficult to handle the problems. Even when computers may be employed, it becomes excessively costly and time-consuming. For example, a problem involving ten workers and ten jobs requires a consideration of 36,28,800 feasible solutions! The branch and bound method also employs enumeration but it does this so efficiently that only a small number of the total number of possible solutions need to be examined individually. The use of this method for solving assignment problems is illustrated with the following example.

Example 7.9 The cost to perform different jobs by different workers is given as follows:

Worker	*Job*			
	1	*2*	*3*	*4*
A	90	18	48	50
B	72	28	85	80
C	53	92	12	78
D	20	70	70	25

Obtain the optimal assignment of jobs to workers.

Here we have four jobs, each of which can be done by four workers *A*, *B*, *C* and *D*. Our objective is to use branch and bound method for solving the problem to obtain least cost assignment. This is achieved in a step-wise manner as follows.

Step I The given problem has $4! = 24$ possible solutions. For these, we first obtain a lower bound on the total cost of the assignment, that is, a cost below which the total cost would not fall. This need not be feasible solution so that there may be more than one job assigned to a worker. The initial lower bound calculation attempts to provide a floor below which the cost to do all jobs cannot fall. This may be achieved in the quickest manner by adding up the minimum cost value in each of the four columns. It gives $20 + 18 + 12 + 25 = 75$, and implies assignment of job 1 to *D*, job 2 to *A*, job 3 to *C* and job 4 to *D*. It is obviously not a feasible solution because *D* is scheduled to get two of the jobs (a feasible solution requires a one-to-one job assignment pattern). In the steps to follow, we proceed to find the least cost feasible solution.

Step II In this step, we may divide the process of looking for solutions. We begin by assigning job 1 in turn to each of the workers *A*, *B*, *C* and *D* and then considering the smallest value in each of the remaining three columns, without considering whether such an assignment is feasible or not. To begin with, let us first assign job 1 to *A*, for a cost of Rs 90. Now ignoring values in the first column (for job 1) and the first row(for worker *A*), we consider the minimum value in second, third and fourth columns, equal to 28, 12 and

25, respectively. Similarly, when we assign job 1 to *B*, we ignore the first column and second row values and then select the minimum values from each column. The results of this operation are given in the Table 7.8.

TABLE 7.8 Assignment of Job 1 to each worker

Assignment	*Lower Bound on the Total*	*Feasibility*
1 – *A* 2 – *B* 3 – *C* 4 – *D*	90 + 28 + 12 + 25 = 155	Feasible
1 – *B* 2 – *A* 3 – *C* 4 – *D*	72 + 18 + 12 + 25 = 127	Feasible
1 – *C* 2 – *A* 3 – *A* 4 – *D*	53 + 18 + 48 + 25 = 144	Infeasible
1 – *D* 2 – *A* 3 – *C* 4 – *A*	20 + 18 + 12 + 50 = 100	Infeasible

It may be observed that the first two of these are feasible assignments while the other two are not. The outcomes of the two steps are shown in Figure. 7.2 in the form of a tree. The value 75 is shown marked, being the lowest cost bound obtained in step I, followed by four branches indicating four solutions obtained in step II, of which the least value of 100 is shown marked.

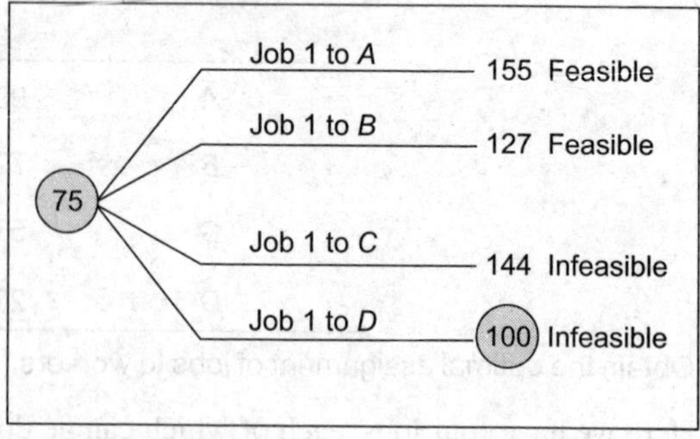

Figure 7.2 *Tree Diagram: Assignment of Job 1 to Each worker*

Since 155, representing the least cost when job 1 is assigned to worker *A*, is greater than 127 (when job 1 is assigned to worker *B*) and this latter value represents a feasible solution, all other solutions which may be obtained by assigning job 1 to worker *A* can be logically ignored as none of them would entail a cost lower than 155. By the same logic, since no solution would involve a cost lower than 144, when job 1 is assigned to worker *C*, we can discard all those solutions which may emanate from the branch following assignning job 1 to *C*. Thus, evidently the branch and bound method involves evaluation of a small proportion of total number of alternatives without discarding the one which is optimal.

Step III In the third step, we consider the least cost branch of the tree obtained in step II when job 1 is assigned to worker *D*. Although with a total cost of 100, this branch does not involve a feasible solution, we may nevertheless attempt to search other low cost feasible solution/s. For this, we let job 2 be assigned to *A*, *B* and *C* in turn and consider the minimum value in each of the columns, leave values contained in the first column and the fourth row (having already assigned job 1 to *D*), and without having regard to whether the solution so obtained is feasible or otherwise. The result is contained in Table 7.9.

TABLE 7.9 Assignment of Job 1 to Worker D and Job 2 to Each Worker

Assignment	*Lower-bound on the Total*	*Feasibility*
1 – *D* 2 – *A* 3 – *C* 4 – *A*	20 + 18 + 12 + 50 = 100	Infeasible
1 – *D* 2 – *B* 3 – *C* 4 – *A*	20 + 28 + 12 + 50 = 110	Feasible
1 – *D* 2 – *C* 3 – *C* 4 – *A*	20 + 92 + 12 + 50 = 174	Infeasible

This operation yields one feasible solution, viz. job 1 – worker *D*, job 2 – worker *B*, job 3 – worker *C* and job 4 – worker *A*, with a total cost of 110. The tree after this is updated as shown in Figure 7.3. Here the shaded value 100 represents the least-cost solution obtained in this step.

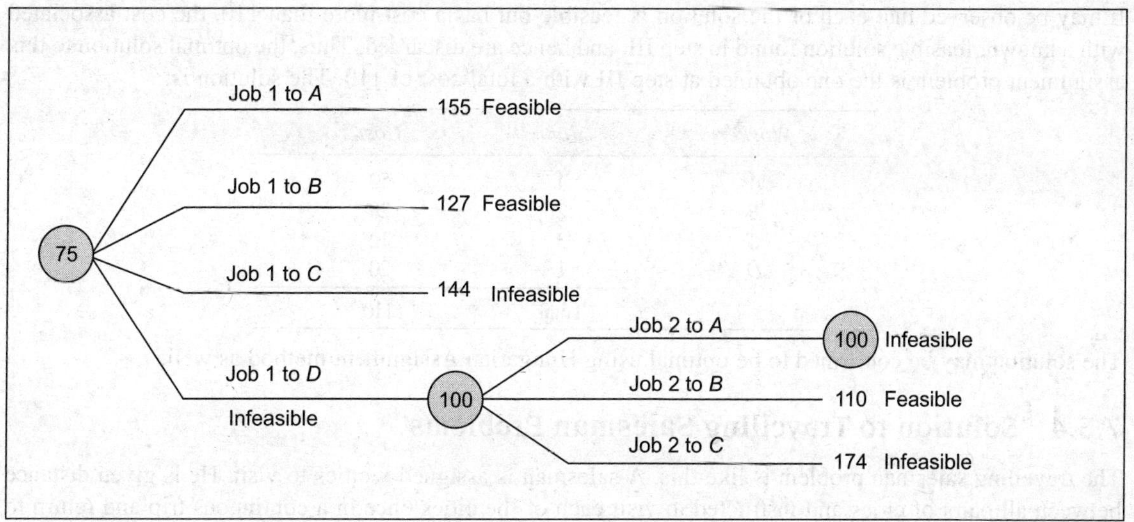

Figure 7.3 *Tree Diagram: Assignment of Job 1 to D, Job 2 to Each Worker*

The solutions possible with job 1 to worker *D* and job 2 to worker *C* have a minimum cost of 174 and need not be further investigated into since a lower cost solution emanating from the branch showing assignment of job 1 to *D* and job 2 to *B* involving a total cost of 110, which is available that is feasible.

Step IV To continue with the problem, we consider the branch indicating an infeasible solution with job 1 to *D* and job 2 to *A* with a total cost of 100, and investigate if it is possible to find some solution that is feasible with a total cost lower than the one obtained so far. With jobs 1 and 2 assigned to workers *D* and *A*, respectively, there are two other possibilities: job 3 is assigned to *B* and job 4 to *C*, or job 3 is assigned to *C* and job 4 is given to *B*. Each of these is feasible with the cost as 20 + 18 + 85 + 78 = 201 and 20 + 18 + 12 + 80 = 130 respectively. The complete branch and bound tree is shown in Figure 7.4.

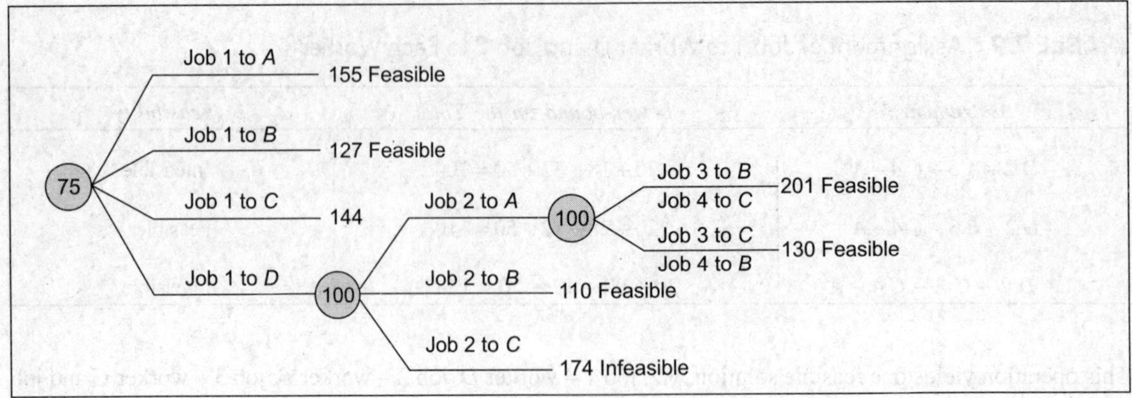

Figure 7.4 *Branch and Bound Tree: Optimal Assignment*

It may be observed that each of the solution is feasible but has a cost more than 110, the cost associated with a known, feasible solution found in step III, and hence are discarded. Thus, the optimal solution to this assignment problem is the one obtained at step III with a total cost of 110. The solution is:

Worker	*Job*	*Cost*
A	4	50
B	2	28
C	3	12
D	1	20
	Total	110

The solution may be confirmed to be optimal using Hungarian Assignment method as well.

7.3.4 Solution to Travelling Salesman Problems

The travelling salesman problem is like this. A salesman is assigned *n* cities to visit. He is given distance between all pairs of cities and instructed to visit each of the cities once in a continuous trip and return to the original, using the shortest route. In this context, we define the visit to the cities in a sequence, ending in the same city where it begins, including a visit to each of the cities only once, as a *tour*. Since a complete cycle is involved, it makes no difference which city is the starting city. While the problem seems trivial, there is no general algorithm available for its solution. There are several methods of solving such a problem, but all are enumerative to some degree or other and hence become impractical for problems of large size. Here we shall illustrate the application of the *branch and bound* procedure for solving the problem. This is done with the following example.

Example 7.10 Mr Iyer is a salesman with Delite Manufacturing Company. He wants to visit six cities, say, 1, 2, 3, 4, 5 and 6, starting with city 1 where he is stationed. The distances between various cities are given in the Table 7.10.

Mr Iyer wants to develop a tour through the five other cities and return to his home city in such a way that he has to travel the minimum distance.

It may be noted that the matrix of the distances between different cities is not a symmetrical one. For instance, the distance from city 1 to 2 is not given to be the same as the distance from city 2 to 1. Such situations may develop because of several reasons. In the one-way traffic system, for example, the distance to be covered to reach a point might be different from the distance to be covered while returning.

TABLE 7.10 Inter-city Distances (in km)

From City	*To City*					
	1	*2*	*3*	*4*	*5*	*6*
1	–	25	18	35	50	39
2	21	–	28	16	30	13
3	22	28	–	14	16	20
4	35	12	14	–	12	12
5	50	30	16	12	–	8
6	39	15	20	12	7	–

To consider the solution to the given problem, to begin with, it may be easily visualised that there would be a finite number of tours that are possible. We can enumerate and evaluate them all for finding the answer to his problem. But since there would be a large number of tours (120 to be precise) possible, it is impractical to look at them all. Let us see how we can adopt the branch and bound procedure to solve his problem.

To begin with, suppose Mr Iyer travels to the cities in the numerical order listed and makes his tour as 1–2–3–4–5–6–1. This would entail a total travelling of 25 + 28 + 14 + 12 + 8 + 39 = 126 km. Now, since this tour is a feasible solution to the problem, the optimal solution should not require him to travel any more than 126 km in all on his tour. The tours which would require travelling greater than this distance are automatically ignored. This value of 126 km is the *upper bound* on the value of the solution to the problem.

We also set a *lower bound* on the solution value. This is done by considering the matrix containing inter-city distances and solving it as a usual assignment problem of the minimisation type. The distances are taken to be the costs of making assignments, and since a particular city cannot be 'assigned' to itself, we place an M in each of the cells in the distance matrix lying on the diagonal, proceeding from 1 through 6. The value of the solution obtained would be the lower bound because no other assignment will yield a value smaller than this.

If the pattern of assignments is such that it constitutes a tour, then the solution would be the optimal one. To look to what we do in case the optimal solution does not yield a tour, let us consider the solution to the given problem as an assignment problem. This is given in Tables 7.11 and 7.12.

TABLE 7.11 Reduced Cost Table 1

City	*City*					
	1	*2*	*3*	*4*	*5*	*6*
1	*M*	7	0	17	32	21
2	8	*M*	15	3	17	0
3	8	14	*M*	0	2	6
4	23	0	2	*M*	0	0
5	42	22	8	4	*M*	0
6	32	8	13	5	0	*M*

TABLE 7.12 Reduced Cost Table 2

City	City					
	1	2	3	4	5	6
1	M	7	[0]	17	32	21
2	[0]	M	15	3	17	⊠
3	⊠	14	M	[0]	2	6
4	15	[0]	2	M	⊠	⊠
5	34	22	8	4	M	[0]
6	24	8	13	5	[0]	M

Assignments : 1–3, 2–1, 3–4, 4–2, 5–6, 6–5
Sub-tours : 1–3–4–2–1 and 5–6–5
Length : 65 km 15 km

In Table 7.12, the minimum number of lines covering all zeros equals 6, the number of rows. Hence, optimal solution is obtained. From the assignments 1–3, 2–1, 3–4, 4–2, 5–6, and 6–5, we obtain a total distance of 80 km. (This is calculated in reference to the inter-city distance matrix). This value is a lower bound on the solution because no assignment pattern connecting one city to another in the present case would involve a distance of less than 80 km. But the solution is not a feasible one for the salesman as it does not provide for a tour. In fact, it involves two *sub-tours* 1–3–4–2–1 and 5–6–5.

Now, the upper bound is 126 kms and the lower bound is 80 km. What we do next is to reduce the gap between the upper and the lower bounds by formulating and solving a series of assignment problems that would aim at reducing the upper bound and raising the lower bound. And we end where no more gap reduction is possible. Since the upper bound always involves a feasible solution, the solution given by the upper bound in that event would be the optimal one.

For the purpose of shortening the gap between the upper and lower bounds, we resort to *branching* in such a way that the sub-tours are broken and joined to form a tour. In a particular situation, we break the smallest of the sub-tours. For example, the two sub-tours given by the solution are 1–3–4–2–1 and 5–6–5 which involve distances of 65 and 15 km respectively. Since 5–6–5 is smaller of the two, we break this sub-tour. Here there would be two branches—one making the route 5–6 and the other making 6–5 as unacceptable. (There would be as many branches in a given situation as the number of routes involved in the sub-tour being broken).

This is shown in Figure 7.5.

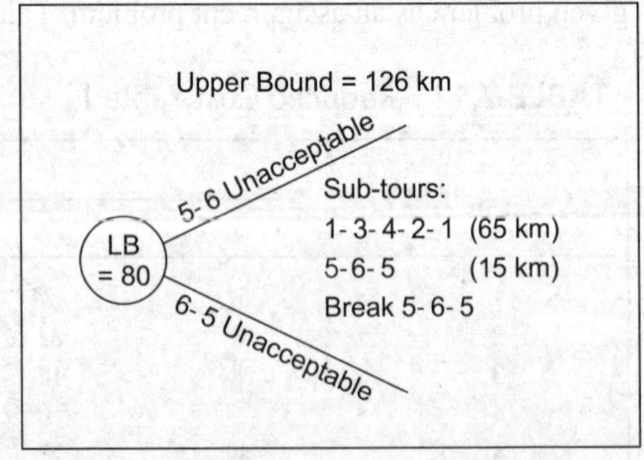

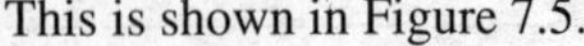

Figure 7.5 *Initial Branch and Bound Tree*

The working of the method is like this. First, start with the optimal solution obtained earlier, insert M in the cell corresponding to the route 5–6 in the table, making the route impossible. Now, solve it as an assignment problem. It is done in Tables 7.13 and 7.14.

TABLE 7.13 Reduced Cost Table 1

City	*City*					
	1	*2*	*3*	*4*	*5*	*6*
1	M	7	0	17	32	21
2	0	M	15	3	17	0
3	0	14	M	0	2	6
4	15	0	2	M	0	0
5	34	22	8	4	M	M
6	24	8	13	5	0	M

TABLE 7.14 Reduced Cost Table 2

City	*City*					
	1	*2*	*3*	*4*	*5*	*6*
1	M	7	[0]	17	32	21
2	~~0~~	M	15	3	17	[0]
3	[0]	14	M	~~0~~	2	6
4	15	[0]	2	M	~~0~~	~~0~~
5	30	22	4	[0]	M	M
6	24	8	13	5	[0]	M

Assignments : 1–3, 2–6, 3–1, 4–2, 5–4, 6–5

Sub-tours : 1–3–1, and 2–6–5–4–2

Length : 40 km 44 km

As we observe, the optimal solution to this problem leads to two sub-tours 1–3–1, and 2–6–5–4–2, with a total distance value of 84 km.

Next again start with the optimal solution of the problem obtained earlier (Table 7.12) and put an M in the cell corresponding to the route 6–5. This makes this route unacceptable. Now solve this problem also as an assignment problem. This is shown in Tables 7.15 through 7.17.

TABLE 7.15 Reduced Cost Table 1

City	City 1	2	3	4	5	6
1	M	7	0	17	32	21
2	0	M	15	3	17	0
3	0	14	M	0	2	6
4	15	0	2	M	0	0
5	34	22	8	4	M	0
6	24	8	13	5	M	M

TABLE 7.16 Reduced Cost Table 2

City	City 1	2	3	4	5	6
1	M	7	0	17	32	21
2	0	M	15	3	17	0
3	0	14	M	0	2	6
4	15	0	2	M	0	0
5	34	22	8	4	M	0
6	19	3	8	0	M	M

TABLE 7.17 Reduced Cost Table 3

City	City 1	2	3	4	5	6
1	M	7	[0]	19	32	23
2	[0]	M	13	3	15	~~0~~
3	~~0~~	12	M	~~0~~	[0]	6
4	17	[0]	2	M	~~0~~	~~0~~
5	34	20	6	4	M	[0]
6	19	1	6	[0]	M	M

Assignments : 1–3, 2–1, 3–5, 4–2, 5–6, 6–4
Tour : 1–3–5–6–4–2–1
Length : 87 km

The optimal pattern of assignments in Table 7.17 clearly involves a tour 1 – 3 – 5 – 6 – 4 – 2 – 1 with a total length equal to 87 km.

Considering the results of both the branches, we reduce the upper bound to 87 (because movement along the branch making 6–5 an unacceptable route results in a feasible solution to the problem). Similarly, the lower bound would be revised to 84 km. We can show the modified branch and bound tree as in Figure. 7.6.

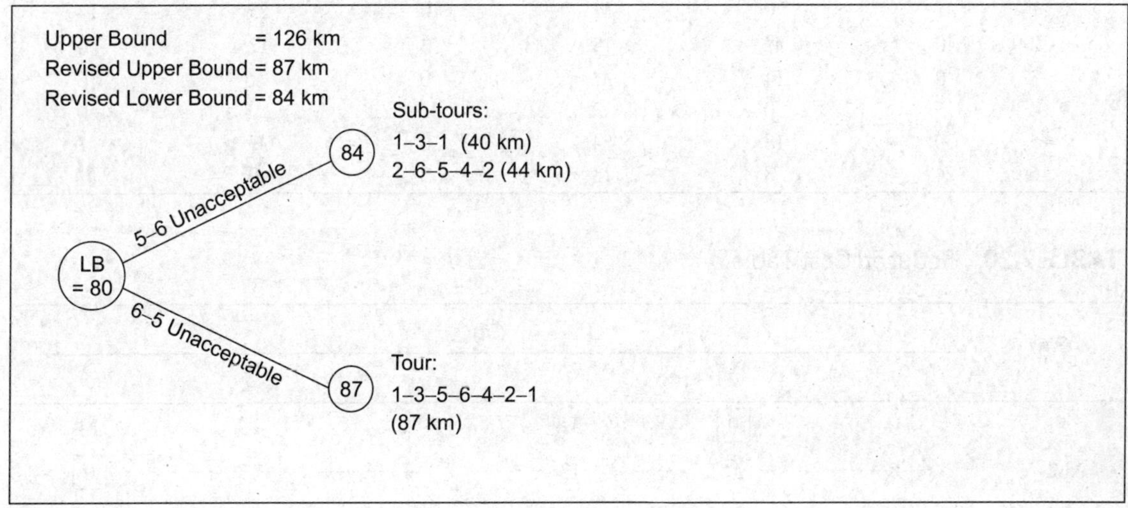

Figure 7.6 *Modified Branch and Bound Tree*

Now, since a gap exists between the lower and the upper bounds, we shall attempt to find the possibility of narrowing it. For this again, we break the smallest of the sub-tours—in the present case the sub-tour 1–3–1. There would again be two branches: one that makes the route 1–3 unacceptable and the other that makes the route 3–1 not acceptable. For this purpose, we start with the optimal solution given in Table 7.14 and put *M* in the cell indicating the route 1–3. This makes this route impossible. The problem is then solved in the manner of an assignment problem, as shown in Tables 7.18–7.20.

TABLE 7.18 Reduced Cost Table 1

City	*City*					
	1	*2*	*3*	*4*	*5*	*6*
1	*M*	7	*M*	17	32	21
2	0	*M*	15	3	17	0
3	0	14	*M*	0	2	6
4	15	0	2	*M*	0	0
5	30	18	4	0	*M*	*M*
6	24	8	13	5	0	*M*

TABLE 7.19 Reduced Cost Table 2

City	City 1	2	3	4	5	6
1	*M*	0	*M*	10	25	14
2	0	*M*	15	3	17	0
3	0	14	*M*	0	2	6
4	15	0	2	*M*	0	0
5	30	18	4	0	*M*	*M*
6	24	8	13	5	0	*M*

TABLE 7.20 Reduced Cost Table 3

City	City 1	2	3	4	5	6
1	*M*	0	*M*	10	25	14
2	0	*M*	13	3	17	0
3	0	14	*M*	0	2	6
4	15	0	0	*M*	0	0
5	30	18	2	0	*M*	*M*
6	24	8	1	5	0	*M*

Assignments : 1–2, 2–6, 3–1, 4–3, 5–4, 6–5
Tour : 1–2–6–5–4–3–1
Length : 93 km

Making the route 1–3 unacceptable results in a tour 1–2–6–5–4–3–1 with a total distance of 93 km. Tables 7.21–7.23 give the solution when the route 3–1 is taken not to be an acceptable one.

TABLE 7.21 Reduced Cost Table 1

City	City 1	2	3	4	5	6
1	*M*	7	0	17	32	21
2	0	*M*	15	3	17	0
3	*M*	14	*M*	0	2	6
4	15	0	2	*M*	0	0
5	30	18	4	0	*M*	*M*
6	24	8	13	5	0	*M*

TABLE 7.22 Reduced Cost Table 2

City	*City*					
	1	*2*	*3*	*4*	*5*	*6*
1	*M*	7	0	21	36	21
2	0	*M*	15	7	21	0
3	*M*	10	*M*	0	2	2
4	15	0	2	*M*	4	0
5	26	14	0	0	*M*	*M*
6	20	4	0	5	0	*M*

TABLE 7.23 Reduced Cost Table 3

City	*City*					
	1	*2*	*3*	*4*	*5*	*6*
1	*M*	5	0	21	34	19
2	0	*M*	17	9	21	0
3	*M*	8	*M*	0	0	0
4	15	0	4	*M*	4	0
5	24	12	0	0	*M*	*M*
6	20	4	11	7	0	*M*

Assignments : 1–3, 2–1, 3–6, 4–2, 5–4, 6–5

Tour : 1–3–6–5–4–2–1

Length : 90 km

The result of making the route 3–1 unacceptable is also a tour 1–3–6–5–4–2–1 involving a total distance of 90 km.

Clearly, then, the movement along both the branches of this node results in feasible solutions, but each one of them involves a plan that would make the salesman travel longer than how much he would if he accepts the plan already at hand. To conclude, therefore, the upper bound cannot be reduced any further and the optimal solution to the problem is the tour in the order 1–3–5–6–4–2–1.

The final branch and bound tree to this problem is shown in Figure 7.7.

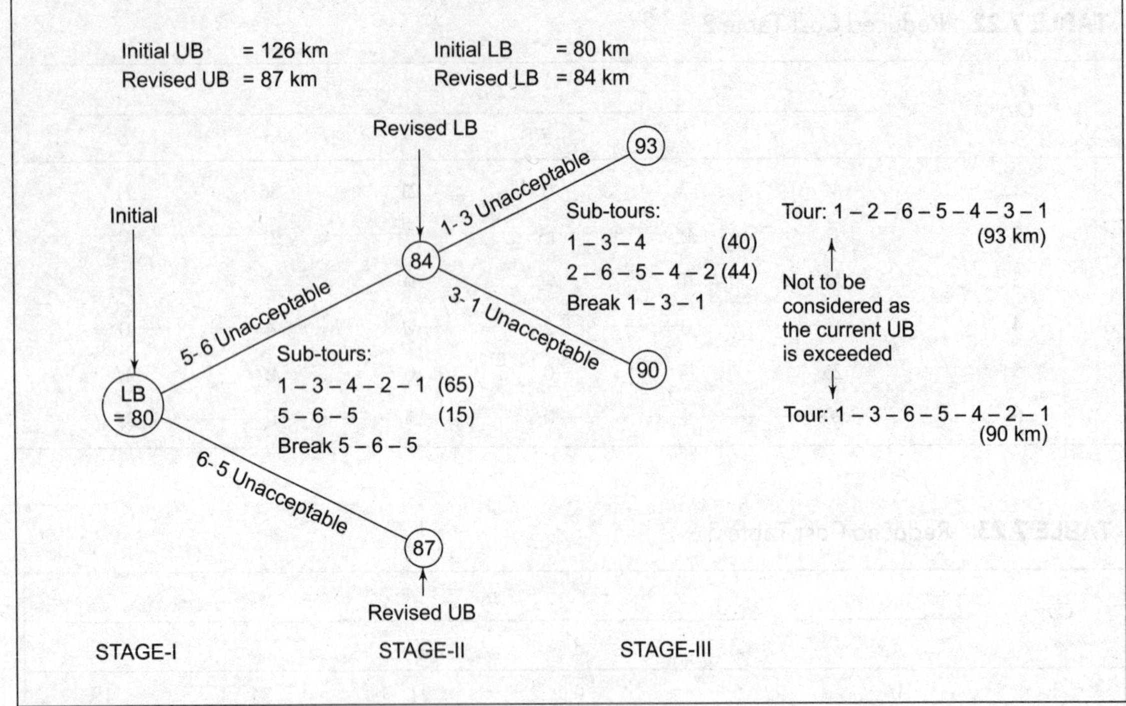

Figure 7.7 *Final Branch and Bound Tree of the Travelling Salesman's Problem*

The branch and bound algorithm for travelling salesman problem can be shown schematically as in Figure 7.8.

7.4 GOAL PROGRAMMING

As observed earlier, traditional mathematical programming models, including linear programming and integer programming, are based on the assumption that the decision maker has a single, quantifiable, objective such as maximisation of profit or minimisation of inefficiency or cost. However, often there are situations, where instead of posing a single objective, managers use multiple criteria in decision-making. Thus, instead of setting only one objective, multiple goals may be set. Specification of multiple goals creates difficulties in the solution to a given problem because the objectives are usually conflicting and incommensurate. For example, increasing a portfolio's annual return calls for making more risky investments. Thus, a portfolio manager's objective to maximise returns would conflict with minimising risk. Further, objectives may be incommensurate if they involve different units. Thus, maximisation of profits, expressed in monetary terms is an economic objective; maintaining a certain workforce level is an employment objective which is input in terms of the number of workers employed, while an environmental objective such as pollution may be measured in terms of damage to surroundings.

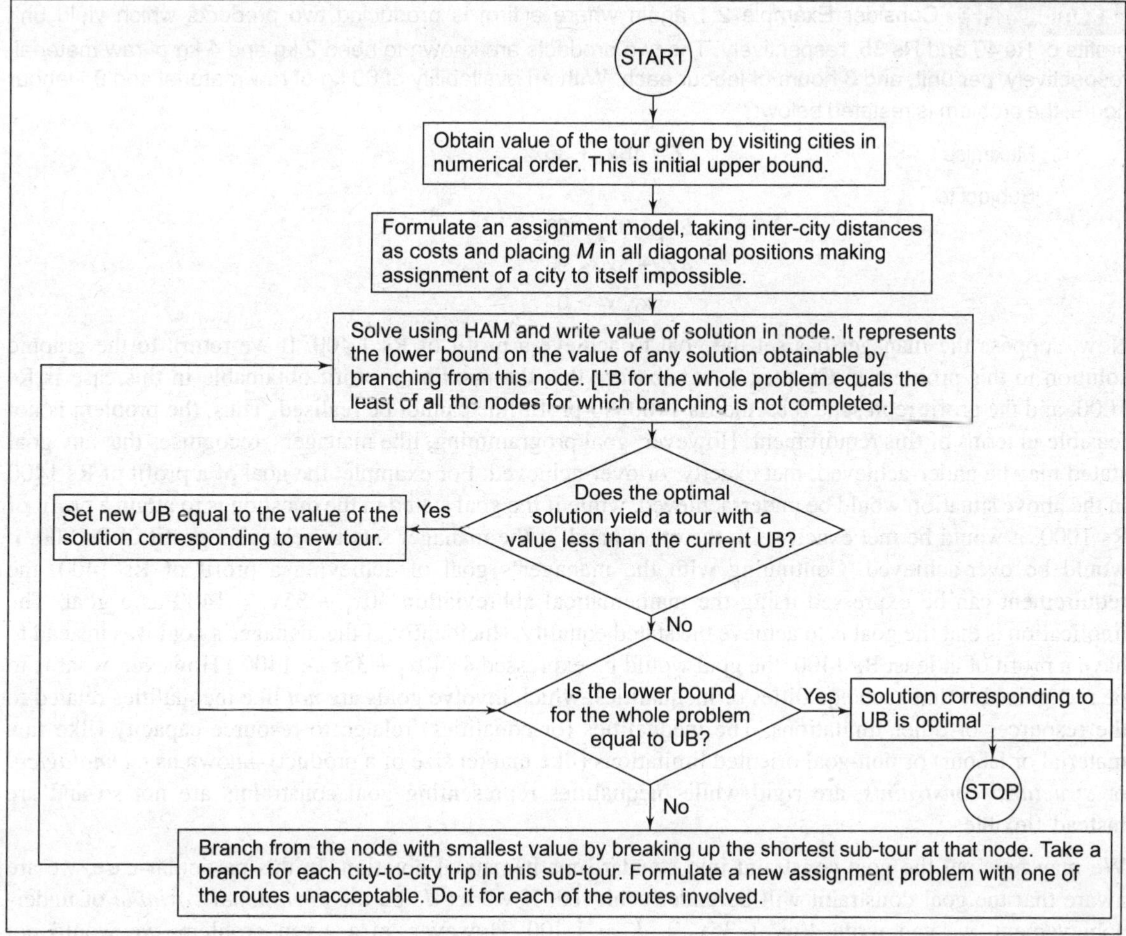

Figure 7.8 *Branch and Bound Algorithm for Travelling Salesman's Problem*

Goal programming allows handling such multi-objective situations. It uses the concept of penalties in the format of linear programming. To apply goal programming, first a target value, a goal, is set for each of the objectives/goals. Since all the goals are unlikely to be satisfied simultaneously, a penalty is assigned for each unit of deviation from the target value in each direction. Thus, a revised linear programming problem, using 'deviational' variables is formulated whose optimal solution comes as close as possible to achieving the stated goals in the sense that it minimises the sum of penalties incurred for under-achieving or over-achieving the various goals. Thus, the central concept of goal programming is to determine the individual preferences of the decision-makers in terms of the goals stated and to establish a single (an overall) function, which is to be minimised or a series of goal functions which are to be minimised in order of their relative importance.

7.4.1 Goal Programming with a Single Goal

Although goal programming is meant to deal with problems with multiple goals, we may begin with a problem with a single goal to develop some basic ideas.

Example 7.11 Consider Example 2.1 again where a firm is producing two products which yield unit profits of Rs 40 and Rs 35, respectively. The two products are known to need 2 kg and 4 kg of raw material, respectively, per unit, and 3 hours of labour each. With an availability of 60 kg of raw material and 96 labour hours, the problem is restated below:

Maximise $Z = 40x_1 + 35x_2$

Subject to

$$2x_1 + 3x_2 \leq 60$$
$$4x_1 + 3x_2 \leq 96$$
$$x_1, x_2 \geq 0$$

Now, suppose the manager has set the goal to achieve a profit of Rs 1,400. If we return to the graphic solution to this problem in Chapter 2, we observe that the maximum profit obtainable in this case is Rs 1000, and the profit represented by the Rs 1400 iso-profit line cannot be realised. Thus, the problem is not feasible in terms of this requirement. However, goal programming, like managers, recognises that any goal stated may be under-achieved, met exactly, or over-achieved. For example, the goal of a profit of Rs 1400 in the above situation would be under-achieved, while if the goal stated in the question is to attain a profit of Rs 1000, it would be met exactly. On the other hand, if the manager sets to achieve a profit of Rs 800, it would be over-achieved. Continuing with the manager's goal of achieving a profit of Rs 1400, the requirement can be expressed using the mathematical abbreviation $40x_1 + 35x_2 = 1400$ as a goal. The implication is that the goal is to achieve the stated equality. (Incidently, if the manager's goal was instead to have a profit of at least Rs 1400, the goal would be expressed as $40x_1 + 35x_2 \geq 1400$.) However, what is to be recognised is that such equalities or inequalities, which involve goals are *not* like inequalities related to the resources or other limitations. The inequalities (or equalities) related to resource capacity (like raw material or labour) or non-goal oriented limitations (like market size of a product), known as *technological* or *structural constraints*, are rigid while inequalities representing goal constraints are not so and are instead flexible.

We may now put the goal constraint into standard mathematical notation. In this particular case, we are aware that the goal constraint will be under-achieved. If we let d^- be the *deviational variable* of under-achievement, we can write $40x_1 + 35x_2 + d^- = 1{,}400$. However, in a given problem, we would not know whether a goal will be under-achieved or be over-achieved or exactly met, we introduce two deviational variables—one of under-achievement and other of over-achievement. Thus, if we designate the deviational variable of over-achievement by d^+, we can express the above goal constraint as

$$40x_1 + 35x_2 + d^- - d^+ = 1{,}400$$

where deviational variables d^- and d^+ are (i) either positive or zero, and (ii) either d^- or d^+ is zero, or (iii) both d^- and d^+ are zero. This relationship can always be met. For example, $x_1 = 0$, $x_2 = 0$, $d^- = 1{,}400$ and $d^+ = 0$ provides a solution.

It may be noted here that when a goal is met exactly, both the deviational variables d^- and d^+ would be equal to zero. Further, we are assured that we never get a solution in which d^- and d^+ are both non-zero, since the objective is to minimise their sum—the combination that minimises will always have one of them at a zero level.

Thus, flexible goal constraints are always expressed in the way the given constraint has been shown above, by introducing deviational variables.

Now, reconsidering the example, we observe that we have two resource constraints and one converted goal constraint. To find the optimal solution, we first convert this problem to a standard minimisation or maximisation problem. This is effected by the concept of goal function. It is stipulated that we want to be as close to the profit of Rs 1,400 as possible and, therefore, minimise d^-, the under-achievement. Accordingly, the problem may be set as follows:

Minimise $Z = d^-$

Subject to

$$2x_1 + 3x_2 \leq 60$$
$$4x_1 + 3x_2 \leq 96$$
$$40x_1 + 35x_2 + d^- - d^+ = 1400$$
$$x_1, x_2, d^-, d^+ \geq 0$$

We may solve this LPP to get $x_1 = 18$, $x_2 = 8$, $d^- = 400$ and $d^+ = 0$, and profit = Rs 1,000, implying thereby that the profit goal is under-achieved by Rs 400 and that the maximum profit is Rs 1,000. The solution is the same as that of the original problem. The solution to the problem is contained in Tables 7.24 through 7.26.

TABLE 7.24 Simplex Tableau 1: Non-optimal Solution

Basis		x_1	x_2	S_1	S_2	d^-	d^+	b_i	b_i/a_{ij}
S_1	0	2	3	1	0	0	0	60	30
S_2	0	4*	3	0	1	0	0	96	24 ←
d^-	1	40	35	0	0	1	−1	1,400	35
c_j		0	0	0	0	0	0		
Solution			0	0	60	96	1,400	0	
Δ_j		−40	−35	0	0	0	1		
		↑							

TABLE 7.25 Simplex Tableau 2: Non-optimal Solution

Basis		x_1	x_2	S_1	S_2	d^-	d^+	b_i	b_i/a_{ij}
S_1	0	0	3/2*	1	−1/2	0	0	12	8 ←
x_1	0	1	3/4	0	1/4	0	0	24	32
d^-	1	0	5	0	−10	1	−1	440	88
c_j		0	0	0	0	1	0		
Solution			24	0	12	0	440	0	
Δ_j		0	−5	0	10	0	1		
			↑						

TABLE 7.26 Simplex Tableau 3: Optimal Solution

Basis		x_1	x_2	S_1	S_2	d^-	d^+	b_i
x_2	0	0	1	2/3	−1/3	0	0	8
x_1	0	1	0	−1/2	1/2	0	0	18
d^-	1	0	0	−10/3	−25/3	1	−1	400
c_j		0	0	0	0	1	0	
Solution		18	8	0	0	400	0	
Δ_j		0	0	10/3	25/3	0	1	

7.4.2 Goal Programming with Multiple Goals

We now consider the usual goal programming situations which involve multiple goals. As indicated earlier, the basic approach of goal programming is to establish a specific goal or target (in quantitative terms) for each objective, formulate an objective function for each objective and then look for a solution that minimises the deviations of these objective functions from their respective goals.

The goals that one may set may be either *one-sided* or *two-sided*. Further, one-sided goal can be a *lower*, which sets a lower limit that we do not want to fall under (but exceeding the limit is all right) or an *upper*, which sets an upper limit that we do not want to exceed (and falling short of which is just fine). Similarly, a two-sided goal sets a specific target, missing which from either side is not desired.

The deviational variables for any goal constraint measures the amount of discrepancy between the value of the objective and the goal specified in respect thereof.

Depending upon how the goals compare, goal programming problems may be categorised as being non-preemptive and preemptive. Non-preemptive goal programming involves situations where all the goals are of roughly comparable importance, whereas preemptive goal programming deals with cases where there is a hierarchy of priority levels for various goals so that there are goals of primary importance which receive attention before others which are of secondary importance. Thus, different goals are considered one by one according to their relative importance. We shall discuss the two types in turn.

7.4.3 Non-preemptive Goal Programming

In non-preemptive goal programming, we first establish a goal for each objective and then seek a solution that minimises the sum of the deviations of these objectives from their respective goals. The goal programming assumes that the decision-maker has a linear utility function with respect to the objectives, that is to say, the marginal rate of substitution between the objectives is linear, regardless of the extent of deviation from the goal. Further, the deviations may be given different weights, called *penalty weights*, in accordance with the relative significance of the objectives, and the solution sought may be the one which minimises the weighted sum of the deviations. The weights used in a goal programming model are indicative of the decision-maker's utility for the various objectives. Specifically, they measure the marginal rate of substitution between objectives and the degree of importance in attaining the goal of each objective relative to the others.

We shall illustrate the non-preemptive goal programming with the help of some examples. Consider the following:

Example 7.12 The production manager of a company wants to schedule a week's production run for two products P_1 and P_2, each of which requires the labour and materials as shown below:

	Product	
	P_1	P_2
Labour Hours	2	4
Material M_1 (kg)	4	5
Material M_2 (kg)	5	4

The weekly availability of resources is limited to 600 labour hours, 1000 kg of material M_1 and 1,200 kg of M_2. The unit profit for P_1 and P_2 is Rs 20 and Rs 32, respectively.

Products P_1 and P_2 are, in fact, new models and are replacements of the older ones which have been discontinued very recently. The manager would like to maximise profit but he is equally concerned with maintaining workforce of the division at nearly constant level in the interest of employee morale. The workforce, which consists of people engaged in production, sales, distribution, peons and other general staff, consisted of 108 persons in all. From a detailed study, it is known that production of one unit of P_1 would maintain 0.3 person in the workforce, while one unit of P_2 would maintain 0.75 person.

Had the production manager been considering only maximising profit, without regard to maintaining the workforce, he would do so by producing 167.67 units of P_1 and 66.67 units of P_2 (this can be checked by solving the problem as an LPP). On the basis of the available capacity, this would yield a profit equal to 167.67 × 20 + 66.67 × 32 or Rs 5,486.67. However, this would maintain 100.3 people in the workforce. The manager feels that probably he could increase the workforce requirement to desired level by accepting somewhat lower profit. In keeping with this, the following two goals are established: a profit of Rs 5,400 per work and a workforce of 108 persons. Formulate and solve this as a goal programming problem.

If we let x_1 and x_2 represent the number of units of P_1 and P_2, respectively, to be produced every week, the goals and constraints of the problem can be stated as follows:

$$\begin{aligned}
20x_1 + 32x_2 &= 5{,}400 && \text{Goal 1}\\
0.3x_1 + 0.75x_2 &= 108 && \text{Goal 2}\\
2x_1 + 4x_2 &\le 600 && \text{Labour}\\
4x_1 + 5x_2 &\le 1{,}000 && \text{Material } M_1\\
5x_1 + 4x_2 &\le 1{,}200 && \text{Material } M_2
\end{aligned}$$

This problem may not have a feasible solution which satisfies both the goals. Further, to be able to solve this problem using simplex method, we need to introduce deviational variables in each of the constraints involving goals. They may be defined as follows:

Let d_1^- = number of rupees below the profit goal of Rs 5,400

d_1^+ = number of rupees above the profit goal of Rs 5,400

d_2^- = number of people below the workforce goal of 108

d_2^+ = number of people above the workforce goal of 108

Now, considering the nature of goals, it may be presumed that overachievement of either of them would attract no penalty, and we may seek a solution which would minimise the underachievement of both of them. Accordingly, the problem may be stated as follows:

Minimise $Z = d_1^- + d_2^-$

Subject to

$$20x_1 + 32x_2 + d_1^- - d_1^+ = 5{,}400$$

$$0.3x_1 + 0.75x_2 + d_2^- - d_2^+ = 108$$

$$2x_1 + 4x_2 + S_1 = 600$$

$$4x_1 + 5x_2 + S_2 = 1{,}000$$

$$5x_1 + 4x_2 + S_3 = 1{,}200$$

$$x_1, x_2, S_1, S_2, S_3, d_1^-, d_1^+, d_2^- \text{ and } d_2^+ \geq 0$$

It may be observed here that, as discussed earlier, the goals are written as constraints. Further, in solving a problem by simplex method, while artificial variables are ordinarily introduced when equations are involved, it is not so in the context of the constraints involving goals. Here, expressing the deviation from goal as the difference between two non-negative variables ($d_1^- - d_2^-$, for example) provides the model sufficient flexibility to permit the underachievement or overachievement of the goal. To solve the problem using simplex method, the initial solution will be provided by the two deviational variables d_1^- and d_2^-, and the three slack variables S_1, S_2 and S_3 introduced in the three constraints involving resources: labour and materials. The detailed solution to the problem is contained in Tables 7.27 through 7.31.

The optimal solution to the problem as obtained from Table 7.31 is: $x_1 = 150$, $x_2 = 75$, $d_2^- = 27/4 = 6.75$, $S_2 = 25$ and $S_3 = 150$, with the objective function value equal to 6.75. From this, it is evident that workforce shall be 101.25, with the employment goal being underachieved to the extent of 6.75 people, while 25 kg of material M_1 and 150 kg of material M_2 would remain unutilised. The other variables are non-basic so that all the available labour hours shall be used and the profit goal be met exactly.

TABLE 7.27 Simplex Tableau 1

Basis		x_1	x_2	d_1^-	d_1^+	d_2^-	d_2^+	S_1	S_2	S_3	b_i	b_i/a_{ij}
d_1^-	1	20	32	1	−1	0	0	0	0	0	5,400	168.75
d_2^-	1	3/10	3/4*	0	0	1	−1	0	0	0	108	144←
S_1	0	2	4	0	0	0	0	1	0	0	600	150
S_2	0	4	5	0	0	0	0	0	1	0	1,000	200
S_3	0	5	4	0	0	0	0	0	0	1	1,200	300
c_j		0	0	1	0	1	0	0	0	0		
Solution		0	0	5,400	0	108	0	600	1,000	1,200		
Δ_j		−213/10	−131/4	0	0	1	0	0	0	0		

TABLE 7.28 Simplex Tableau 2

Basis		x_1	x_2	d_1^-	d_1^+	d_2^-	d_2^+	S_1	S_2	S_3	b_i	b_i/a_{ij}
d_1^-	1	36/5	0	1	–1	–128/3	128/3	0	0	0	792	299/16
x_2	0	2/5	1	0	0	4/3	–4/3	0	0	0	144	–
S_1	0	2/5	0	0	0	–16/3	16/3*	1	0	0	24	9/2 ←
S_2	0	2	0	0	0	–20/3	20/3	0	1	0	280	42
S_3	0	17/5	0	0	0	–16/3	16/3	0	0	1	624	117
c_j		0	0	1	0	1	0	0	0	0		
Solution		0	144	792	0	0	0	24	280	624		
Δ_j		–36/5	0	0	1	131/3	–128/3	0	0	0		
							↑					

TABLE 7.29 Simplex Tableau 3

Basis		x_1	x_2	d_1^-	d_1^+	d_2^-	d_2^+	S_1	S_2	S_3	b_i	b_i/a_{ij}
d_1^-	1	4	0	1	–1	0	0	–8	0	0	600	150
x_2	0	1/2	1	0	0	0	0	1/4	0	0	150	300
d_2^+	0	3/40*	0	0	0	–1	1	3/16	0	0	9/2	60 ←
S_2	0	3/2	0	0	0	0	0	–5/4	1	0	250	500/3
S_3	0	3	0	0	0	0	0	–1	0	1	600	200
c_j		0	0	1	0	1	0	0	0	0		
Solution		0	150	600	0	0	9/2	0	250	600		
Δ_j		–4	0	0	1	0	0	8	0	0		
		↑										

TABLE 7.30 Simplex Tableau 4

Basis		x_1	x_2	d_1^-	d_1^+	d_2^-	d_2^+	S_1	S_2	S_3	b_i	b_i/a_{ij}
d_1^-	1	0	0	1	–1	160/3*	– 160/3	–18	0	0	360	27/4 ←
x_2	0	0	1	0	0	20/3	– 20/3	–1	0	0	120	18
x_1	0	1	0	0	0	–40/3	40/3	5/2	0	0	60	–
S_2	0	0	0	0	0	20	– 20	–5	1	0	160	8
S_3	0	0	0	0	0	40	– 40	–17/2	0	1	420	21/2
c_j		0	0	1	0	1	0	0	0	0		
Solution		60	120	360	0	0	0	0	160	420		
Δ_j		0	0	0	1	–157/3	160/3	18	0	0		
						↑						

TABLE 7.31 Simplex Tableau 5: Optimal Solution

Basis		x_1	x_2	d_1^-	d_1^+	d_2^-	d_2^+	S_1	S_2	S_3	b_i
d_2^-	1	0	0	3/160	–3/160	1	– 1	–27/80	0	0	27/4
x_2	0	0	1	– 1/8	1/8	0	0	5/4	0	0	75
x_1	0	1	0	1/4	– 1/4	0	0	–2	0	0	150
S_2	0	0	0	–3/8	3/8	0	0	7/4	1	0	25
S_3	0	0	0	–3/4	3/4	0	0	5	0	1	150
c_j		0	0	1	0	1	0	0	0	0	
Solution		150	75	0	0	27/4	0	0	25	150	
Δ_j		0	0	157/160	3/160	0	1	27/80	0	0	

Graphic Presentation

The problem has been depicted graphically in Figure 7.9 in which solution obtained above is shown. The feasible area corresponding to the three resource constraints, bounded by *OPQRS*, is shown shaded, while the goal constraints are shown by broken lines. It may be noted that for plotting the goal constraints we set each of the deviational variables equal to zero.

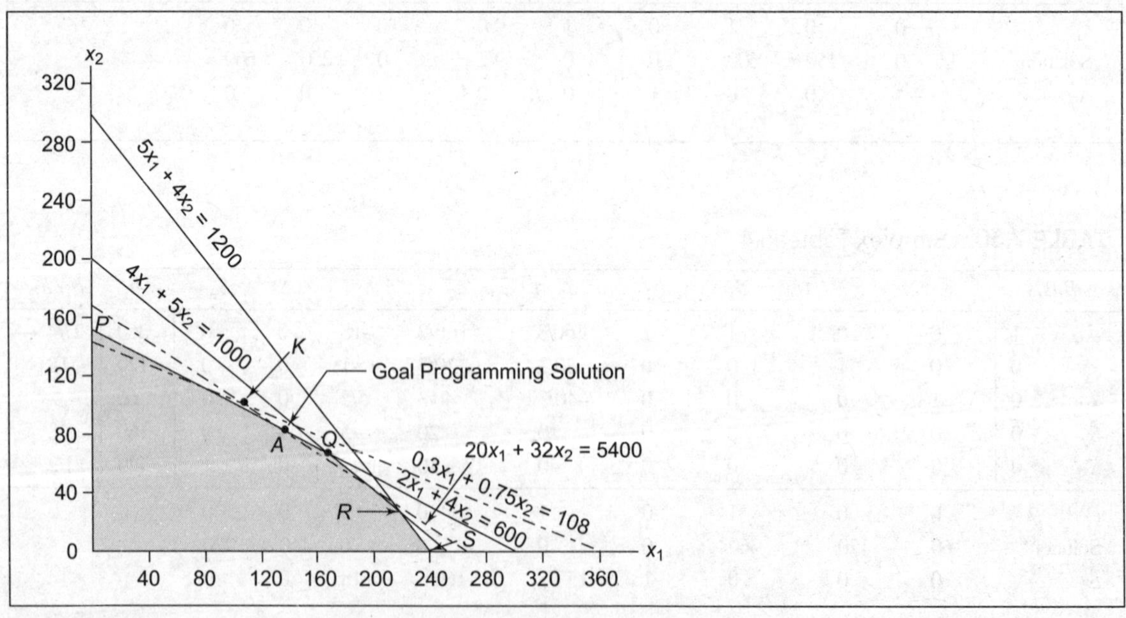

Figure 7.9 *Graphic Presentation of Goal Programming Problem*

The solution obtained by goal programming is indicated by point A which is *not* a corner point of the feasible region given by the two-dimensional space. (It would indeed be a corner point of a six-dimensional space of x_1, x_2, d_1^-, d_1^+, d_2^- and d_2^+.) Notice that the solution represented by intersection of the dashed lines, point K, where both the goals are exactly met, is not in the feasible region. This implies that both the goals cannot be met. However, there do exist points where it is possible to overachieve the profit goal but we do not consider such points because overachievement of goals has been precluded. Of course, it may be observed that some increased labour hours could result in achievement of both the goals. Since the two goal constraint lines intersect at the point where $x_1 = 110$, $x_2 = 100$, total labour requirement would be $2 \times 110 + 4 \times 100 = 620$ hours. With the existing availability being 600 hours, another 20 hours would ensure this, while other resources, materials M_1 and M_2, are sufficient for producing this product mix.

Differential Weighting

It may be noted that the objective function in the above example postulates that no weightage (penalty) is assigned to the over achievement of any goal and that equal weights are placed on the underachievement of each of the goals. It implies that a one-rupee underachievement of profit goal and a one person underachievement of the employment goal are weighted the same. Evidently, it does not seem to be a fair one. An underachievement of Rs 5 for profit would mean an underachievement by 5/5,400 or 0.09% while an underachievement of 5 persons workforce would imply a 5/108 or 4.63% underachievement of the employment goal. Thus, to avoid the problems associated with having large magnitudes difference, we may standardise the weights so as to seek minimising the percentage deviation from the goals in each objective. If we suppose for our problem that percentage deviations are of interest to the production manager, then the weights that would give equal weighting to each objective would be:

for $\quad d_1^- : 1/5{,}400 \quad$ and $\quad$ for $d_2^- : 1/108$.

Further, if the decision-maker feels that achieving the profit goal is twice as important as employment goal, he would give change the weighted as follows:

for $\quad d_1^- : 2 \times (1/5400) = 1/2{,}700 \quad$ and for $d_2^- : 1/108$.

However, weights to be assigned to different goals may be arrived at in other ways. To continue with the problem, suppose that in terms of the managerial assessment of the trade-off between different goals, it is held that a one-person underachievement of the employment goal is deemed equal to Rs 100 in profit. This means that we may be willing to give up Rs 100 for getting one more person into workforce (or that we may be ready to give up one person in exchange of an additional Rs 100 profit per week). In addition, we continue to assume that overachievement of any of the two goals has no weightage. We may rewrite the objective function as:

Minimise $\quad Z = d_1^- + 100d_2^-$

With the changed objective function, and the set of constraints as given earlier, the solution to the revised problem may be obtained similarly as before. It is given in Tables 7.32 through 7.35. In terms of the optimal solution contained in Table 7.35, we have $x_1 = 60$ and $x_2 = 120$, for a total profit of $20 \times 60 + 32 \times 120$ = Rs 5,040, and $d_1^- = 360$, implying an underachievement of the profit goal of Rs 5,400 by Rs 360. However, since d_2^- is equal to zero, it has the implication that the employment goal of 108 would be met. In terms of the solution, the labour hours will be fully utilised while the raw materials would remain unutilised: 160 kg of material M_1 and 420 kg of material M_2.

TABLE 7.32 Simplex Tableau 1

Basis		x_1	x_2	d_1^-	d_1^+	d_2^-	d_2^+	S_1	S_2	S_3	b_i	b_i/a_{ij}
d_1^-	1	20	32	1	−1	0	0	0	0	0	5,400	168.75
d_2^-	100	3/10	3/4*	0	0	1	−1	0	0	0	108	144 ←
S_1	0	2	4	0	0	0	0	1	0	0	600	150
S_2	0	4	5	0	0	0	0	0	1	0	1,000	200
S_3	0	5	4	0	0	0	0	0	0	1	1,200	300
c_j		0	0	1	0	100	0	0	0	0		
Solution		0	0	5,400	0	108	0	600	1000	1200		
Δ_j		−50	−107	0	1	0	100	0	0	0		
			↑									

TABLE 7.33 Simplex Tableau 2

Basis		x_1	x_2	d_1^-	d_1^+	d_2^-	d_2^+	S_1	S_2	S_3	b_i	b_i/a_{ij}
d_1^-	1	36/5	0	1	−1	−128/3	128/3	0	0	0	792	299/16
x_2	0	2/5	1	0	0	4/3	−4/3	0	0	0	144	–
S_1	0	2/5	0	0	0	−16/3	16/3*	1	0	0	24	9/2 ←
S_2	0	2	0	0	0	−20/3	20/3	0	1	0	280	42
S_3	0	17/5	0	0	0	−16/3	16/3	0	0	1	624	117
c_j		0	0	1	0	100	0	0	0	0		
Solution		0	144	792	0	0	0	24	280	624		
Δ_j		−36/5	0	0	1	428/3	−128/3	0	0	0		
							↑					

TABLE 7.34 Simplex Tableau 3

Basis		x_1	x_2	d_1^-	d_1^+	d_2^-	d_2^+	S_1	S_2	S_3	b_i	b_i/a_{ij}
d_1^-	1	4	0	1	−1	0	0	− 8	0	0	600	150
x_2	0	1/2	1	0	0	0	0	1/4	0	0	500	500
d_2^+	0	3/40*	0	0	0	−1	1	3/16	0	0	9/2	60 ←
S_2	0	3/2	0	0	0	0	0	−5/4	1	0	250	500/3
S_3	0	3	0	0	0	0	0	−1	0	1	600	200
c_j		0	0	1	0	100	0	0	0	0		
Solution		0	150	600	0	0	9/2	0	250	600		
Δ_j		− 4	0	0	1	100	0	8	0	0		
		↑										

TABLE 7.35 Simplex Tableau 4: Optimal Solution

Basis		x_1	x_2	d_1^-	d_1^+	d_2^-	d_2^+	S_1	S_2	S_3	b_i
d_1^-	1	0	0	1	−1	160/3	−160/3	−18	0	0	360
x_2	0	0	1	0	0	20/3	−20/3	−1	0	0	120
x_1	0	1	0	0	0	−40/3	40/3	5/2	0	0	60
S_2	0	0	0	0	0	20	−20	−5	1	0	160
S_3	0	0	0	0	0	40	−40	−17/2	0	1	420
c_j		0	0	1	0	100	0	0	0	0	
Solution		60	120	360	0	0	0	0	160	420	
Δ_j		0	0	0	1	140/3	160/3	18	0	0	

7.4.4 Preemptive Goal Programming

As the name implies, in case of preemptive goal programming, there is a hierarchy of priority levels for the various goals. Such a situation arises when one or more of the goals are specified to be more significant than others, in contrast to non-preemptive goal programming where all goals are approximately of comparable importance. In preemptive goal programming, the initial focus is on achieving, as closely as possible, the top priority goal and then considering the lower priority ones, if necessary. Thus, after an optimal solution with respect to the first-priority goals, we can break any ties for the optimal solution by considering the second-priority goals. Similarly, any ties remaining after this revised optimisation are broken by considering third-priority goals, and so on.

To apply this, first the goals are ranked from the most important (goal 1) to the least important (goal n) and the objective function coefficient assigned for the (deviational) variables representing goal i is P_i. It is assumed that $P_1 >>> P_2 >>> P_3 >>> \ldots >>> P_n$. In fact, P's are not assigned any actual values. Rather, they are a convenient way of indicating that one goal is more important than the other. These coefficients indicate that the weight of goal 1 is much larger than the weight of goal 2, which in turn is much larger than that of goal 3, and so on. Thus, a lower priority goal is never sought to be achieved at the expense of higher-priority goal. The definition of $P_1, P_2, \ldots, P_n$ ensures that a decision-maker first tries to satisfy the most important goal. After this, among all the points that satisfy goal 1, he attempts to come as close as possible to satisfying goal 2. Then he proceeds to break any ties remaining if the revised solution is obtained after satisfying goal 2, by considering third-priority goal. The decision-maker continues in this manner until the only way he can come closer to satisfying a goal is to increase the deviation from a higher-priority goal.

Graphic Solution to GPP

A goal programming problem may be solved graphically as follows. First, we set each of the deviational variables equal to zero and use the constraint equations to graph the goal and technological restrictions. The decision variable combinations (that is, points on the graph) are identified that satisfy all the technological constraints and the portion on the graph is shaded. Of course, if there is no such region, the problem does not have a feasible region and we cannot proceed further. Otherwise, we go to the next step to eliminate from the (shaded) region all points that involve any undesirable deviation from the top-priority goal. Next, within this shaded region points that provide a zero undesirable deviation from the next-highest priority goal are determined. If there are no such points, then the corresponding goal constraint line is moved in a parallel fashion until it touches the feasible region boundary. This is repeated in order of priority for each goal in the given problem.

Next, the point (or points) on the boundary of the shaded area that best meets the hierarchy of goals is identified. This point, respresenting decision variable combination, represents solution to the GPP. Finally, the solution values are substituted into each constraint and the resulting equations are solved for the values of the deviational variables.

Example 7.13 A firm is engaged in producing two products *A* and *B*, which yield a unit profit of Rs 400 and Rs 300, respectively. Both of these require a certain raw material whose supply is limited to 300 kg per month. The raw material requirement is one kg per unit of each product. Further, it is known that one unit of *A* requires 2 hours of processing while each unit of *B* needs one hour. The firm has a normal processing capacity of 400 hours per month. The marketing department has stated that according to the prevailing conditions, and likely to continue in the foreseeable future, the maximum number of units of *A* and *B* which can be sold each month is 150 and 350, respectively. The manager of the firm has set the following goals in order of their importance—the priorities being preemptive.

Goal 1: Avoid under-utilisation of normal productive capacity.

Goal 2: Sell the maximum number of units of products *A* and *B*. However, since the two products yield profit in the ratio 400 : 300, the manager would give the weightage to the sale of the products in the same ratio.

Goal 3: Minimise overtime of processing capacity.

You are required to formulate this as goal programming problem and solve it graphically.

Let x_1 and x_2 represent the number of units of products *A* and *B* respectively to be produced every month.

Constraints and Goals

Raw Material: With a one kg requirement per unit of *A* and *B*, and availability of 300 kg, the raw material constraint is $x_1 + x_2 \leq 300$.

Processing Capacity: Since overtime operation is possible in terms of the given information, the processing capacity constraint can be stated as follows:

$$2x_1 + x_2 + d_1^- - d_1^+ = 400$$

where d_1^- is the number of hours unutilised capacity and d_1^+ is the number of hours overtime operation of the normal processing capacity.

Market: Since the sales goals given are in terms of the maximum possible sales volumes, they can be expressed as

$$x_1 + d_2^- = 150; \text{ and } x_2 + d_3^- = 350.$$

Notice that since the given sales levels are the maximum ones, deviational variables are provided only for their underachievement and not for overachievement.

With P_1, P_2 and P_3 being the preemptive priority factors, the goal programming problem may be stated as follows:

Minimise $$Z = P_1 d_1^- + P_2(4d_2^- + 3d_3^-) + P_3 d_1^+$$

Subject to

$$x_1 + x_2 \leq 300$$

$$2x_1 + x_2 + d_1^- - d_1^+ = 400$$

$$x_1 + d_2^- = 150$$
$$x_2 + d_3^- = 350$$
$$x_1, x_2, d_1^-, d_1^+, d_2^- \text{ and } d_3^- \geq 0$$

To solve the problem graphically, the raw material constraint is first plotted, as shown in Figure 7.10, given by line *AT* and the feasible region is shown shaded. Then the constraints involving goals are plotted (by setting deviational variables equal to zero). Since the production capacity constraint may be under- or over-achieved, the feasible area lies on both sides of the line, as indicated by d_1^- and d_1^+. Thus, it does not alter the feasible region. The first goal would, thus, be met and $d_1^- = 0$. Therefore, the focus shifts to the second goal.

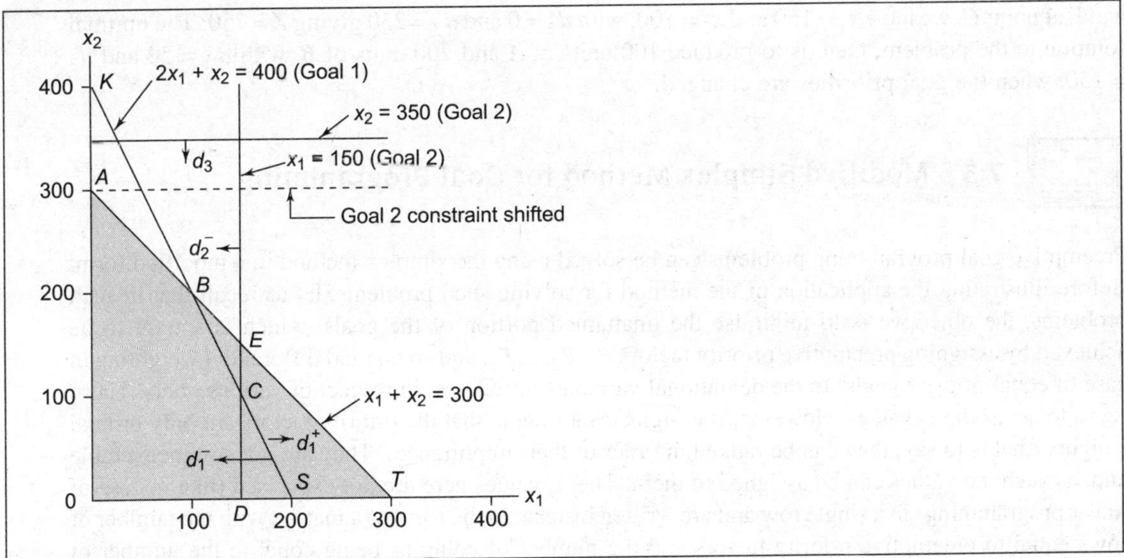

Figure 7.10 *Graphic Solution to GPP: Preemptive*

Observing the lines corresponding to the goal constraints, we note that the area DET represents the undesirable deviation and is, therefore, not to be considered. On the other hand, the line depicting the maximum output of the product *B* (x_2 = 350) lies beyond the feasible region. Accordingly, the goal constraint line is shown shifted until it touches the boundary of the feasible region at point *A*. Since the product *A* yields a greater profit of Rs 400 over product *B* which yields Rs 300 per unit, a higher penalty is stipulated for deviating from the output target of *A*. Between points *A*, *B*, and *E*, we may obtain the optimal solution by computing the values of d_2^- and d_3^-, and the value of *Z* therefrom.

Point	x_1	x_2	d_2^-	d_3^-	$z = 4d_2^- + 3d_3^-$
A	0	300	150	50	750
B	100	200	50	150	650
E	150	150	0	200	600 ← Minimum

Thus, the optimal solution to the problem is: x_1 = 150 and x_2 = 150. Substituting these values in the goal 1 constraint equation yields $2 \times 150 + 150 - d_1^+ = 400$ or $d_1^+ = 50$, implying that 50 hours of processing capacity shall be worked overtime. Evidently, since the second-priority goal has not been met fully, there has been no consideration of the next-priority goal.

To continue with the problem, suppose the ranking of the goals is changed so that goals 2 and 3 are interchanged, it would have the effect that while d_1^- would continue to be equal to zero since goal 1 is met, the variable d_1^+ needs to be put equal to zero since a positive value of d_1^+ is an undesirable deviation, in the new objective function: Minimise $Z = P_1d_1^- + P_2d_1^+ + P_3(4d_2^- + 3d_3^-)$. Variables d_1^- and d_1^+ would both be equal to zero on line segment BC (segment KB of this line is not considered because it falls beyond the feasible region given by the technological constraint of material availability while CS is not considered because of the market limitation for product A). In view of the multiple points where the second goal is equally met, consideration is given to the third goal. Accordingly, at B we have $x_1 = 100$ and $x_2 = 200$, leaving $d_2^- = 50$, $d_3^- = 150$ and $Z = 650$. On the other hand, at point C, we have $x_1 = 150$ and $x_2 = 100$, with $d_2^- = 0$ and $d_3^- = 250$ giving $Z = 750$. The optimal solution to the problem, then, is to produce 100 units of A and 200 units of B, with $d_2^- = 50$ and $d_3^- = 150$, when the goal priorities are changed.

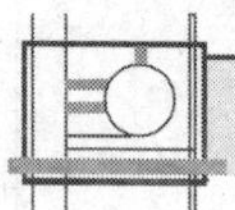

7.5 Modified Simplex Method for Goal Programming

Preemptive goal programming problems can be solved using the simplex method in a modified form. Before illustrating the application of the method for solving such problems, let us recall that in such problems, the objective is to minimise the unattained portion of the goals, which is sought to be achieved by assigning preemptive priority factors $P_1, P_2, \ldots P_n$, and so on (and differential weighting in case of equal-priority goals) to the deviational variables introduced in respect of various goals. These serve to act as the c_j values. However, it is significant to note that the priority factors are only ordinal weights (that is to say, they can be ranked in order of their importance). They are not commensurable and, as such, no values can be assigned to them. The Δ_j values here are not expressed (like in case of linear programming) in a single row and are written instead, in the form of a matrix, with the number of rows equal to preemptive priority factors and the number of columns being equal to the number of variables involved. The expression of values as a matrix calls for a modified way of selecting the key column to identify the incoming variable. Since $P_j >>> P_{j+1}$, the selection procedure involves initiation from P_j and moving to lower priority goals.

Let us consider the following example.

Example 7.14 Solve the goal programming problem given in Example 7.13, which is reproduced below:

Minimise $$Z = P_1d_1^- + P_2(4d_2^- + 3d_3^-) + P_3d_1^+$$

Subject to

$$x_1 + x_2 \leq 300$$
$$2x_1 + x_2 + d_1^- - d_1^+ = 400$$
$$x_1 + d_2^- = 150$$
$$x_2 + d_3^- = 350$$
$$x_1, x_2, d_1^-, d_1^+, d_2^- \text{ and } d_3^- \geq 0$$

To solve this problem, we first express the given information as shown in Table 7.36. Observe that the basic variables in Simplex Tableau 1 (Table 7.36) are S_1, d_1^-, d_2^-, and d_3^- with solution values equal to 300, 400, 150 and 350 respectively. Evidently, since both x_1 and x_2 are equal to zero, the zero output implies various negative deviational variables d_1^-, d_2^-, and d_3^- are assuming positive values. The rows under the c_j row are listed according to the priority levels of the goals, with the highest priority goal P_1 at the bottom and the lower ranked goals P_2 and P_3 placed above it in the ascending order. The values in these indicate the Δ_j values. Let us understand the calculation of Δ_j $(= c_j - z_j)$ values in the table. For calculating z_j values, consider the weights (coefficients) mentioned with P_1, P_2, P_3, etc. First, look to the column headed x_1. Since the value in this column corresponding to the row involving P_1 (basic variable d_1^+) is 2, we get $z_j = 2 \times 1 = 2$ in respect of P_1. With $c_j = 0$ for x_1, the $\Delta_j = 0 - 2 = -2$ is written in the bottom row marked P_1. Now, for P_2, the values in the column are 1 and 0, with the corresponding basic values being 4 and 3, we get $z_j = 1 \times 4 + 0 \times 3 = 4$. With $c_j = 0$ for x_1, the $\Delta_j = 0 - 4 = -4$ is shown in the row marked P_2. There being no P_3 in the basis the $\Delta_j = 0$ in respect of the row marked P_3 under x_1. As another example, consider the column headed d_1^+. Since the value in this column opposite the basic variable d_1^- is -1, we get $z_j = 1 \times (-1) = -1$, where 1 is the weightage associated with P_1. Accordingly, Δ_j in the row marked P_1 is found to be $0 - (-1) = 1$. Since both the values in the column headed d_1^+ are zero, $z_j = 4 \times 0 + 3 \times 0 = 0$ and $\Delta_j = 0 - 0 - 0$ against the row marked P_2. Finally, since no variable in the basis has the priority level P_3, we have $z_j = 0$ and $\Delta_j = 1 - 0 = 1$.

The values against the rows marked P_1, P_2, and P_3 under the column headed b_i are the z_j values and they are derived as follows. The solution value of d_1^- (with priority level 1) is 400. With the weightage of P_1 being 1, the z-value is $400 \times 1 = 400$. Similarly, with $d_2^- = 150$ and $d_3^- = 350$ in the table and the weightage being 4 and 3 respectively, the z_j value for this priority goal works out to be $150 \times 4 + 350 \times 3 = 1{,}650$. Finally, there being no basic variable with priority P_3, we have $d_1^+ = 0$ and z-value for this priority equal to $1 \times 0 = 0$.

Test of Optimality and Derivation of Revised Tableau

To test the optimality, we first look to the Δ_j values in the bottom row, headed P_1. The objective function being of minimisation type, the solution is not optimal since all Δ_j's are not greater than, or equal to zero. To obtain a revised simplex tableau, we consider the most negative of the Δ_j values in the P_1 row in the Simplex Tableau 1. It being equal to -2 for the column headed x_1, the variable x_1 is the incoming variable. Using a_{ij} values in this column, we calculate replacement ratios b_i/a_{ij} and select the least non-negative one, which in this case is 150, corresponding to the variable d_2^-. Thus, d_2^- is taken to be the outgoing variable.

TABLE 7.36 Simplex Tableau 1: Non-optimal Solution

Basis		x_1	x_2	S_1	d_1^-	d_1^+	d_2^-	d_3^-	b_i	b_i/a_{ij}
S_1	0	1	1	1	0	0	0	0	300	300
d_1^-	P_1	2	1	0	1	−1	0	0	400	200
d_2^-	$4P_2$	1*	0	0	0	0	1	0	150	150 ←
d_3^-	$3P_2$	0	1	0	0	0	0	1	350	–
c_j		0	0	0	P_1	P_3	$4P_2$	$3P_2$		
Δ_j	P_3	0	0	0	0	1	0	0	0	
	P_2	− 4	−3	0	0	0	0	0	1650	
	P_1	− 2	−1	0	0	1	0	0	400	
		↑								

TABLE 7.37 Simplex Tableau 2: Non-optimal Solution

Basis		x_1	x_2	S_1	d_1^-	d_1^+	d_2^-	d_3^-	b_i	b_i/a_{ij}
S_1	0	0	1	1	0	0	−1	0	150	150
d_1^-	P_1	0	1*	0	1	−1	−2	0	100	100 ←
x_1	0	1	0	0	0	0	1	0	150	–
d_3^-	$3P_2$	0	1	0	0	0	0	1	350	350
c_j		0	0	0	P_1	P_3	$4P_2$	$3P_2$		
Δ_j	P_3	0	0	0	0	1	0	0	0	
	P_2	0	−3	0	0	0	4	0	1050	
	P_1	0	−1	0	0	1	2	0	100	
			↑							

TABLE 7.38 Simplex Tableau 3: Non-optimal Solution

Basis		x_1	x_2	S_1	d_1^-	d_1^+	d_2^-	d_3^-	b_i	b_i/a_{ij}
S_1	0	0	0	1	−1	1*	1	0	50	50 ←
x_2	0	0	1	0	1	−1	−2	0	100	–
x_1	0	1	0	0	0	0	1	0	150	–
d_3^-	$3P_2$	0	0	0	−1	1	2	1	250	250
c_j		0	0	0	P_1	P_3	$4P_2$	$3P_2$		
Δ_j	P_3	0	0	0	0	1	0	0	0	
	P_2	0	0	0	3	−3	−2	0	750	
	P_1	0	0	0	1	0	0	0	0	
						↑				

It may be mentioned here that if there is a tie in the most negative Δ_j values of a given priority level, we consider the corresponding Δ_j values in the next (lower) priority level and select the variable which has a greater numerical value (ignoring sign) of Δ_j.

Once the incoming and outgoing variables (key column and key row) are determined, the revised simplex tableau can be obtained in the same way as in case of linear programming problems. The resulting tableau is contained in Table 7.37. We have, output of product A, $x_1 = 150$; unused materials $S_1 = 150$ kg; underachievement of goal 2, $d_3^- = 350$, being the output of product B short by as many units of the goal. For this solution as well, the Δ_j values are calculated in the same way as discussed in the context of solution given in Table 7.36. As there is a negative Δ_j value (= – 1) for the variable x_2, it is taken to be the incoming variable. With replacement ratios calculated with the help of a_{ij} values of the key column, it is evident that d_1^- is the outgoing variable. The next solution is contained in Table 7.38.

TABLE 7.39 Simplex Tableau 4: Optimal Solution

Basis		x_1	x_2	S_1	d_1^-	d_1^+	d_2^-	d_3^-	b_i
d_1^+	P_3	0	0	1	–1	1	1	0	050
x_2	0	0	1	1	0	0	–1	0	150
x_1	0	1	0	0	0	0	1	0	150
d_3^-	$3P_2$	0	0	–1	0	0	1	1	200
c_j		0	0	0	P_1	P_3	$4P_2$	$3P_2$	
Δ_j	P_3	0	0	–1	0	0	0	0	50
	P_2	0	0	3	0	0	1	0	600
	P_1	0	0	0	1	0	0	0	0

For the solution given in Table 7.38, it is evident that there are no negative Δ_j values in the row marked P_1, implying that optimality has been achieved in relation to the top priority goal. Hence, we consider the Δ_j values in the row marked P_2. Here, the two negative values seen are –3 and –2 under d_1^+ and d_2^-, respectively and we select the more negative value of the two, that is, –3. Thus, the variable d_1^+ enters and S_1 leaves as the least non-negative replacement ratio corresponds to this variable.

Finally, the revised simplex tableau is given in Table 7.39 which gives the optimal solution. It is optimal in the sense that it enables the manager to attain the set goals as closely as possible within the given decision scenario and the priority structure. It may be noted that all Δ_j's are greater than or equal to zero in the rows marked P_1 and P_2. However, there is a negative Δ_j (= – 1) under the variable S_1 in the P_3 row. Apparently, we can improve the solution by considering this negative Δ_j value by introducing S_1 in the solution. But, it may be observed that this can be done only at the expense of achieving the higher priority goal, as a positive Δ_j value, equal to 3, for this variable is present in the row P_2. The rule, then, is that if there is a negative Δ_j value at a lower priority level, then the variable in that column cannot be selected as the incoming variable if there is a positive Δ_j value at a higher priority level.

From the simplex tableau 4, the optimal solution may be stated as follows: $x_1 = 150$, $x_2 = 150$, $d_1^+ = 50$, and $d_3^- = 200$, which implies that 150 units of each of the products *A* and *B* be produced so that the target of selling 350 units of product *B* shall be underachieved by 200 units and an overtime work of 50 hours shall be involved.

REVIEW ILLUSTRATIONS

Example 7.15 Write constraints to satisfy each of the following conditions in a project selection model. The projects are numbered 1, 2, 3, 4, 5, 6, 7, 8, 9 and 10.

(i) Exactly one project from the set (1, 2, 3) must be selected.

(ii) Project 2 can be selected only if number 10 is selected. However, 10 can be selected without 2 being selected.

(iii) No more than one project from the set (1, 3, 5, 7, 9) can be selected.

(iv) If number 4 is selected, then number 8 cannot be selected.

(v) Projects 4 and 10 must both be selected or both be rejected. *(MBA, Delhi, November, 2001)*

Let x_i $(i = 1, 2 \ldots, 10)$ be the variable indicating the project *i*. Each of these can take the value 1 or 0 accordingly as the project is selected or not. Thus,

$$x_i\ (i = 1, 2, \ldots, 10) = 1, \text{ if the project is selected}$$
$$= 0 \text{ otherwise}$$

Now, the given conditions can be expressed as constraints as shown below:

(i) Since exactly one of the three given projects is to be selected, we have

$$x_1 + x_2 + x_3 = 1$$

(ii) This requirement can be stated as follows:

$$x_{10} - x_2 \geq 0$$

It permits selecting project 10 without selecting project 2; selecting none of them and selecting both of them. It does not allow project 2 when project 10 is not selected (since 0 – 1 = – 1)

(iii) To select no more than one project from the set (1, 3, 5, 7, 9), we write

$$x_1 + x_3 + x_5 + x_7 + x_9 \leq 1$$

(iv) $x_4 + x_8 \leq 1$ would ensure that only one of the projects 4 and 8 can be selected.

(v) If both the projects are selected or both are rejected, their variable values have to be identical and, thus, their difference equal to zero. Thus,

$$x_4 - x_{10} = 0$$

Example 7.16 A manufacturer of toys makes two types of toys, A and B. Processing of these two toys is done on two machines X and Y. Toy A requires 4 hours on machine X and six hours on machine Y, while Toy B requires eight hours on machine X and five hours on machine Y. There are thirty-two hours of time per day available on machine X and thirty hours on machine Y. The profit obtained on each of the toys is Rs 30 per unit. What should be the daily production of each of the toys for maximum profit? A non-integer solution to the problem is not acceptable.

Let x_1 and x_2 be the number of toys of type A and B respectively. From the given information, the problem is:

Maximise $\quad Z = 30x_1 + 30x_2 \quad$ Profit

Subject to

$$\left.\begin{aligned} 4x_1 + 8x_2 &\leq 32 \\ 6x_1 + 5x_2 &\leq 30 \end{aligned}\right\} \text{Capacity constraints}$$

$$x_1, x_2 \geq 0 \quad \text{and integer}$$

We first consider its LP relaxation and solve by simplex method. This is given below:

Maximise $\quad Z = 30x_1 + 30x_2 + 0S_1 + 0S_2$

Subject to

$$4x_1 + 8x_2 + S_1 = 32$$
$$6x_1 + 5x_2 + S_2 = 30$$
$$x_1, x_2, S_1, S_2 \geq 0$$

The solution is contained in Tables 7.40 through 7.42.

TABLE 7.40 Simplex Tableau 1: Non-optimal Solution

Basis		x_1	x_2	S_1	S_2	b_i	b_i/a_{ij}
S_1	0	4	8*	1	0	32	4 ←
S_2	0	6	5	0	1	30	6
c_j		30	30	0	0		
Solution		0	0	32	30	Z = 0	
Δ_j		30	30 ↑	0	0		

TABLE 7.41 Simplex Tableau 2: Non-optimal Solution

Basis		x_1	x_2	S_1	S_2	b_i	b_i/a_{ij}
x_2	30	1/2	1	1/8	0	4	8
S_2	0	7/2*	0	–5/8	1	10	20/7 ←
c_j		30	30	0	0		
Solution		0	8	0	10	Z = 240	
Δ_j		15 ↑	0	–15/2	0		

TABLE 7.42 Simplex Tableau 3: Optimal Solution

Basis		x_1	x_2	S_1	S_2	b_i
x_2	30	0	1	3/14	–1/7	18/7
x_1	30	1	0	–5/28	2/7	20/7
c_j		30	30	0	0	
Solution		20/7	18/7	0	0	Z = 1140/7
Δ_j		0	0	–15/14	–30/7	

The optimal solution to the LPP obtained in Simplex Tableau 3 evidently does not involve integer values of the variables. We therefore attempt to obtain the solution in terms of Gomory constraints. To begin with, we add a Gomory constraint by considering the first row in the Simplex Tableau 3, as shown in Table 7.43. The constraint is:

$$-\frac{3}{14}S_1 - \frac{6}{7}S_2 + S_3 = -\frac{4}{7}$$

TABLE 7.43 Simplex Tableau 4: Non-optimal Solution

Basis		x_1	x_2	S_1	S_2	S_3	b_i	
x_2	30	0	1	3/14	–1/7	0	18/7	
x_1	30	1	0	–5/28	2/7	0	20/7	
S_3	0	0	0	–3/14*	–6/7	1	–4/7	←
c_j		30	30	0	0	0		
Solution		20/7	18/7	0	0	– 4/7		
Δ_j		0	0	–15/14	–30/7	0		
				↑				

With the key column and key row marked, the improved solution is shown in Table 7.44. However, since the variables are not seen to involve integer values, another Gomory constraint is introduced in reference to the second row of this tableau, which is shown in Table 7.45. Finally, the all-integer solution is obtained in Table 7.46.

TABLE 7.44 Simplex Tableau 5: Non-optimal Solution

Basis		x_1	x_2	S_1	S_2	S_3	b_i
x_2	30	0	1	0	–1	1	2
x_1	30	1	0	0	1	–5/6	10/3
S_1	0	0	0	1	4	–14/3	8/3
c_j		30	30	0	0	0	
Solution		10/3	2	8/3	0	0	$Z = 160$
Δ_j		0	0	0	0	–5	

TABLE 7.45 Simplex Tableau 6: Non-optimal Solution

Basis		x_1	x_2	S_1	S_2	S_3	S_4	b_i	
x_2	30	0	1	0	–1	1	0	2	
x_1	30	1	0	0	1	–5/6	0	10/3	
S_1	0	0	0	1	4	–14/3	0	8/3	
S_4	0	0	0	0	0	–1/6	1	–1/3	←
c_j		30	30	0	0	0	0		
Solution		10/3	2	8/3	0	0	–1/3		
Δ_j		0	0	0	0	–5	0		
						↑			

TABLE 7.46 Simplex Tableau 7: Optimal Solution

Basis		x_1	x_2	S_1	S_2	S_3	S_4	b_i
x_2	30	0	1	0	–1	0	6	0
x_1	30	1	0	0	1	0	–5	5
S_1	0	0	0	1	4	0	–28	12
S_3	0	0	0	0	0	1	–6	2
c_j		30	30	0	0	0	0	
Solution		5	0	12	0	20	$Z=150$	
Δ_j		0	0	0	0	0	–30	

Thus, optimal solution to the problem is : $x_1 = 5$, $x_2 = 0$ and $Z = 150$.

Example 7.17 A company produces two products, each of which requires stamping, assembly and painting operations. Total productive capacity by operation if it were solely devoted to one product or the other is shown below:

	Productive Capacity (units/week)	
Operation	*Product A*	*Product B*
Stamping	50	75
Assembly	40	80
Painting	90	45

Pro rata allocation of productive capacity is permissible so also combinations of production of the two products. Demand for the two products is unlimited and the profit on the product A and product B is Rs 150 and Rs 120 respectively. Determine the optimal product-mix. (Non-integer solution for this problem will not be accepted.) *(MBA, Delhi, November 2004)*

Let x_1 and x_2 be the number of units of products A and B respectively, produced every week. The IPP is:

Maximise $Z = 150x_1 + 120x_2$ Total profit

Subject to

$$\frac{x_1}{50} + \frac{x_2}{75} \le 1 \quad \text{Stamping}$$

$$\frac{x_1}{40} + \frac{x_2}{80} \le 1 \quad \text{Assembly}$$

$$\frac{x_1}{90} + \frac{x_2}{45} \le 1 \quad \text{Painting}$$

$$x_1, x_2 \ge 0 \text{ and integer}$$

The problem is restated below with constraints expressed in a simplified manner.

Maximise $Z = 150x_1 + 120x_2$

Subject to

$$3x_1 + 2x_2 \leq 150$$
$$2x_1 + x_2 \leq 80$$
$$x_1 + 2x_2 \leq 90$$
$$x_1, x_2 \geq 0 \text{ and integer}$$

We consider the LP relaxation of this problem and solve it as such. Accordingly, introducing slack variables S_1, S_2 and S_3, the problem is restated below.

Maximise $Z = 150x_1 + 120x_2 + 0S_1 + 0S_2 + 0S_3$

Subject to

$$3x_1 + 2x_2 + S_1 = 150$$
$$2x_1 + x_2 + S_2 = 80$$
$$x_1 + 2x_2 + S_3 = 90$$
$$x_1, x_2, S_1, S_2, S_3 \geq 0$$

The solution to this problem is given in Tables 7.47 through 7.49.

TABLE 7.47 Simplex Tableau 1: Non-optimal Solution

Basis		x_1	x_2	S_1	S_2	S_3	b_i	b_i/a_{ij}	
S_1	0	3	2	1	0	0	150	50	
S_2	0	2*	1	0	1	0	80	40	←
S_3	0	1	2	0	0	1	90	90	
c_j		150	120	0	0	0			
Solution		0	0	150	80	90			
Δ_j		150	120	0	0	0			
		↑							

TABLE 7.48 Simplex Tableau 2: Non-optimal Solution

Basis		x_1	x_2	S_1	S_2	S_3	b_i	b_i/a_{ij}	
S_1	0	0	1/2	1	–3/2	0	30	60	
x_1	150	1	1/2	0	1/2	0	40	80	
S_3	0	0	3/2*	0	–1/2	1	50	100/3	←
c_j		150	120	0	0	0			
Solution		40	0	30	0	50	Z=6,000		
Δ_j		0	45	0	–75	0			
			↑						

TABLE 7.49 Simplex Tableau 3: Optimal Solution

Basis		x_1	x_2	S_1	S_2	S_3	b_i
S_1	0	0	0	1	–4/3	–1/3	40/3
x_1	150	1	0	0	2/3	–1/3	70/3
x_2	120	0	1	0	–1/3	2/3	100/3
c_j		150	120	0	0	0	
Solution		70/3	100/3	40/3	0	0	Z=7,500
Δ_j		0	0	0	–60	–30	

From Table 7.49, we observe that the optimal solution to the LP relaxation is: $x_1 = 70/3$, and $x_2 = 100/3$, which is not the solution to IPP as the values of x_1 and x_2 are not integers. To apply the cutting plane algorithm, we select the first constraint in Table 7.49. The constraint is:

$$0x_1 + 0x_2 + 1S_1 - \frac{4}{3}S_2 - \frac{1}{3}S_3 = \frac{40}{3}$$

This may be rewritten as

$$0x_1 + 0x_2 + (1 + 0S_1) + \left(-2 + \frac{2}{3}\right)S_2 + \left(-1 + \frac{2}{3}\right)S_3 = 13\frac{1}{3}$$

Re-arranging by putting all fractional values on the LHS and all integers on RHS, we get

$$\frac{2}{3}S_2 + \frac{2}{3}S_3 = \frac{1}{3} + (13 - 1S_1 + 2S_2 + S_3)$$

Accordingly,

$$\frac{2}{3}S_2 + \frac{2}{3}S_3 \geq \frac{1}{3} \quad \text{or}$$

$$-\frac{2}{3}S_2 - \frac{2}{3}S_3 \leq -\frac{1}{3}$$

And,

$$-\frac{2}{3}S_2 - \frac{2}{3}S_3 + S_4 = -\frac{1}{3}$$

With this constraint added, the Simplex Tableau 3 of Table 7.49 is shown in Table 7.50.

TABLE 7.50 Revised Simplex Tableau 1

Basis		x_1	x_2	S_1	S_2	S_3	S_4	b_i	b_i/a_{ij}
S_1	0	0	0	1	–4/3	–1/3	0	40/3	–
x_1	150	1	0	0	2/3	–1/3	0	70/3	–
x_2	120	0	1	0	–1/3	2/3	0	100/3	50
S_4	0	0	0	0	–2/3	–2/3*	1	–1/3	1/2 ←
c_j		150	120	0	0	0	0		
Solution		70/3	100/3	40/3	0	0	–1/3		
Δ_j		0	0	0	– 60	– 30 ↑	0		

TABLE 7.51 Revised Simplex Tableau 2

Basis		x_1	x_2	S_1	S_2	S_3	S_4	b_i
S_1	0	0	0	1	–1	0	–1/2	27/2
x_1	150	1	0	0	1	0	–1/2	47/2
x_2	120	0	1	0	–1	0	1	33
S_3	0	0	0	0	1	1	–3/2	1/2
c_j		150	120	0	0	0	0	
Solution		47/2	33	27/2	0	1/2	0	Z=7,485
Δ_j		0	0	0	–30	0	– 45	

The solution given in Table 7.51 is optimal in terms of Δ_j values but not acceptable because of the fractional values involved. For introducing another cut, we consider the first of the constraints given in Table 7.51,

$$0x_1 + 0x_2 + 1S_1 - 1S_2 + 0S_3 - \frac{1}{2}S_4 = \frac{27}{2}$$

which simplifies to

$$\frac{1}{2}S_4 = \frac{1}{2} + [13 - S_1 + S_2 + S_4]$$

or

$$\frac{1}{2}S_4 \leq -\frac{1}{2}$$

or

$$-\frac{1}{2}S_4 + S_5 = -\frac{1}{2}$$

With this cut introduced, the revised Simplex Tableau is shown in Table 7.52.

TABLE 7.52 Revised Simplex Tableau 3

Basis		x_1	x_2	S_1	S_2	S_3	S_4	S_5	b_i	b_i/a_{ij}	
S_1	0	0	0	1	–1	0	–1/2	0	27/2	–	
x_1	150	1	0	0	1	0	–1/2	0	47/2	–	
x_2	120	0	1	0	–1	0	1	0	33	33	
S_3	0	0	0	0	1	1	–3/2	0	1/2	–	
S_5	0	0	0	0	0	0	–1/2*	1	–1/2	1	←
c_j		150	120	0	0	0	0	0			
Solution		47/2	33	27/2	0	1/2	0	–1/2			
Δ_j		0	0	0	–30	0	–45	0			
							↑				

The revised solution is given in Table 7.53.

TABLE 7.53 Revised Simplex Tableau 4

Basis		x_1	x_2	S_1	S_2	S_3	S_4	S_5	b_i
S_1	0	0	0	1	–1	0	0	–1	14
x_1	150	1	0	0	1	0	0	–1	24
x_2	120	0	1	0	–1	0	0	2	32
S_3	0	0	0	0	1	1	0	–3	2
S_4	0	0	0	0	0	0	1	–2	1
c_j		150	120	0	0	0	0	0	
Solution		24	32	14	0	2	1	0	Z=7,440
Δ_j		0	0	0	–30	0	0	– 90	

This solution is found to be optimal for the IPP at hand. Accordingly, $x_1 = 24$, $x_2 = 32$ and Z corresponding to these values being 7,440.

Example 7.18 A salesman must travel from city to city to maintain his accounts. This week he has to leave his home base and visit four other cities and return home. The table shows the distances (in km) between the various cities. The home city is city *A*. Use the assignment method to determine the tour that will minimise the total distances of visiting all cities and returning home.

From City	To City				
	A	B	C	D	E
A	–	375	600	150	190
B	375	–	300	350	175
C	600	300	–	350	500
D	160	350	350	–	300
E	190	175	500	300	–

(*MBA, Delhi, 2004*)

The given information is reproduced in Table 7.54 for solution as assignment problem. For the purpose, each of the elements on the diagonal is represented by *M*, to prohibit travelling from a particular city to itself. The row and column reductions are then carried out in Tables 7.55 and 7.56.

TABLE 7.54 Distance Profile

City	*A*	*B*	*C*	*D*	*E*
A	*M*	375	600	150	190
B	375	*M*	300	350	175
C	600	300	*M*	350	500
D	160	350	350	*M*	300
E	190	175	500	300	*M*

TABLE 7.55 Reduced Cost Table 1

City	*A*	*B*	*C*	*D*	*E*
A	*M*	225	450	0	40
B	200	*M*	125	175	0
C	300	0	*M*	50	200
D	0	190	190	*M*	140
E	15	0	325	125	*M*

TABLE 7.56 Reduced Cost Table 2

City	*A*	*B*	*C*	*D*	*E*
A	*M*	225	325	0	40
B	200	*M*	0	175	0
C	300	0	*M*	50	200
D	0	190	65	*M*	140
E	15	0	200	125	*M*

In RCT-2, the number of lines covering all zeros is smaller than *n*. Thus, improvement is made as shown in Table 7.57. Since the number of lines is smaller than *n* here as well, further improvements are made as shown in the subsequent tables. Finally, assignments are made as depicted in Table 7.60.

TABLE 7.57 Reduced Cost Table 3

City	*A*	*B*	*C*	*D*	*E*
A	*M*	240	325	0	40
B	200	*M*	0	175	0
C	285	0	*M*	35	185
D	0	205	65	*M*	140
E	0	0	185	110	*M*

TABLE 7.58 Reduced Cost Table 4

City	*A*	*B*	*C*	*D*	*E*
A	*M*	275	325	0	40
B	235	*M*	0	175	0
C	285	0	*M*	0	150
D	0	205	30	*M*	105
E	0	0	150	75	*M*

TABLE 7.59 Reduced Cost Table 5

City	*A*	*B*	*C*	*D*	*E*
A	*M*	275	295	0	10
B	265	*M*	0	205	0
C	285	0	*M*	0	120
D	0	205	0	*M*	75
E	0	0	120	75	*M*

TABLE 7.60 Reduced Cost Table 6

City	A	B	C	D	E
A	M	275	295	[0]	10
B	265	M	~~0~~	205	[0]
C	285	[0]	M	~~0~~	120
D	~~0~~	205	[0]	M	75
E	[0]	~~0~~	120	75	M

The assignments made, as shown in Table 7.60, are: *A–D*, *B–E*, *C–B*, *D–C* and *E–A*. These assignments, when put together, are *A–D–C–B–E–A*, which represent a tour. This is optimal solution to the given problem. Accordingly, the salesman should tour the cities in the order *A* to *D* to *C* to *B* to *E* to *A*, covering a total distance 150 + 350 + 300 + 175 + 190 = 1,165 km.

Example 7.19 A salesman is planning to tour cities *B*, *C*, *D* and *E*, starting from his home city *A*. The inter-city distances are shown in the following table:

City	A	B	C	D	E
A	0	103	188	136	38
B	103	0	262	176	52
C	187	262	0	85	275
D	136	176	85	0	162
E	38	52	275	162	0

How should he plan his tour so that (i) he visits each of the cities only once, and (ii) and travels the minimum distance?

As a first step, we place an *M* in each of the elements on the principal diagonal, to prevent travelling from a city to itself, and solve the resulting problem as an assignment problem. This is shown in Tables 7.61 through 7.64.

TABLE 7.61 Distance Profile

City	A	B	C	D	E
A	M	103	188	136	38
B	103	M	262	176	52
C	187	262	M	85	275
D	136	176	85	M	162
E	38	52	275	162	M

TABLE 7.62 Reduced Cost Table 1

City	*A*	*B*	*C*	*D*	*E*
A	*M*	65	149	98	0
B	51	*M*	210	124	0
C	102	177	*M*	0	190
D	51	91	0	*M*	77
E	0	14	237	124	*M*

TABLE 7.63 Reduced Cost Table 2

City	*A*	*B*	*C*	*D*	*E*
A	*M*	51	149	98	0
B	51	*M*	210	124	0
C	102	163	*M*	0	190
D	51	77	0	*M*	77
E	0	0	237	124	*M*

TABLE 7.64 Reduced Cost Table 3

City	*A*	*B*	*C*	*D*	*E*
A	*M*	0	149	98	[0]
B	[0]	*M*	210	124	0
C	51	112	*M*	[0]	190
D	0	26	[0]	*M*	77
E	0	[0]	288	175	*M*

Assignments	:	*A–E, B–A, C–D, D–C, E–B*
Sub-tours	:	*A–E–B–A, C–D–C*
Length	:	193 + 170 = 363

Evidently, the solution to the problem does not lead to a tour and we proceed to set the total length of the sub-tours *A–E–B–A* and *C–D–C*, equal to 363, as the lower bound. Now, suppose the salesman travels cities in the order *A–B–C–D–E–A*, which constitutes a tour, covering a total distance of 650. This may be set as the upper bound. Thus, to begin with, we have the following:

Lower bound	:	363
Upper bound	:	650

Now, we break the sub-tour *C–D–C* given by the problem (which we call as sub-problem 1) above. For this, RCT-3 given in Table 7.64 is considered and two sub-problems are formulated: sub-problem 2 where the route *C–D* is made unacceptable and sub-problem 3, where route *D–C* is made unacceptable, by placing an *M* in the appropriate cell. The cost table for sub-problem 2 is shown in Table 7.65 and for sub-problem 3 in Table 7.66. Along with, the solutions obtained from both of them are given.

TABLE 7.65 Reduced Cost Table 4: Sub-problem 2

City	*A*	*B*	*C*	*D*	*E*
A	*M*	0	149	98	0
B	0	*M*	210	124	0
C	51	112	*M*	*M*	190
D	0	26	0	*M*	77
E	0	0	288	175	*M*

Assignments : *A–D, B–E, C–A, D–C, E–B*

Sub-tours : *A–D–C–A, B–E–B*

Total Distance : 512

TABLE 7.66 Reduced Cost Table 5: Sub-problem 3

City	*A*	*B*	*C*	*D*	*E*
A	*M*	0	149	98	0
B	0	*M*	210	124	0
C	51	112	*M*	0	190
D	0	26	*M*	*M*	77
E	0	0	288	175	*M*

Assignments : *A–C, B–E, C–D, D–A, E–B*

Sub-tours : *A–C–D–A, B–E–B*

Total Distance : 513

Since none of the two sub-problems yields a tour, we need to revise the lower bound upward to 512 and further break the sub-tours. At this stage, the branch and bound tree appears as shown in Fig. 7.11.

We first consider the solution to the sub-problem 2 and generate two sub-problems: one with route *B–E* unacceptable (sub-problem 4) and the other with route *E–B* unacceptable (sub-problem 5). The cost tables and the solutions to these are contained in Tables 7.67 and 7.68, respectively.

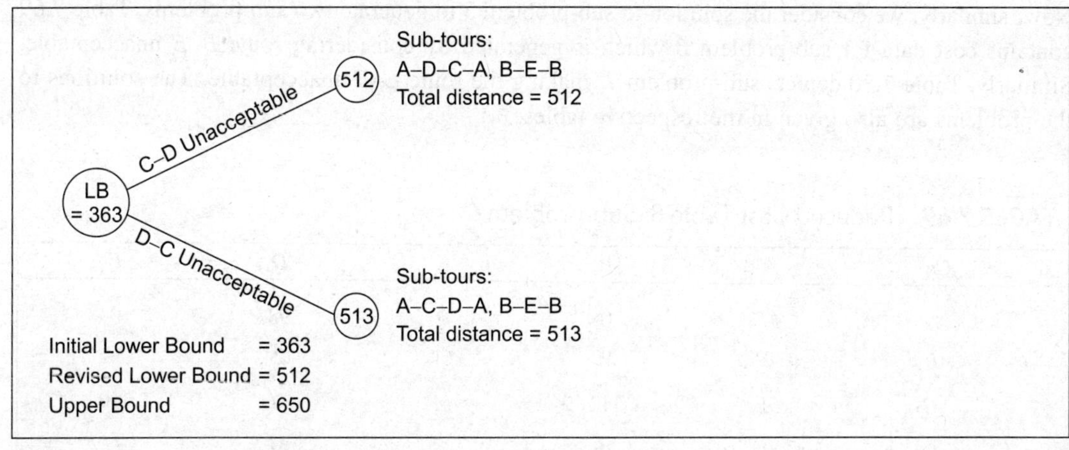

Figure 7.11 *Branch and Bound Tree*

TABLE 7.67 Reduced Cost Table 6: Sub-problem 4

City	*A*	*B*	*C*	*D*	*E*
A	*M*	0	149	0	0
B	0	*M*	210	26	*M*
C	0	61	*M*	*M*	190
D	0	26	0	*M*	77
E	0	0	288	77	*M*

Assignments	:	*A–E, B–D, C–A, D–C, E–B*
Tour	:	*A–E–B–D–C–A*
Total Distance	:	538

TABLE 7.68 Reduced Cost Table 7: Sub-problem 5

City	*A*	*B*	*C*	*D*	*E*
A	*M*	0	149	0	0
B	0	*M*	210	26	0
C	0	61	*M*	*M*	190
D	0	26	0	*M*	77
E	0	*M*	288	77	*M*

Assignments	:	*A–D, B–E, C–B, D–C, E–A*
Tour	:	*A–D–C–B–E–A*
Total Distance	:	573

Now, similarly, we consider the solution to sub-problem 3 to generate two sub-problems. Table 7.69 contains cost data for sub-problem 6 which is generated by considering route *B–E* unacceptable. Similarly, Table 7.70 depicts sub-problem 7, making the route *E–B* unacceptable. The solutions to the problems are also given in the respective tables.

TABLE 7.69 Reduced Cost Table 8: Sub-problem 6

City	*A*	*B*	*C*	*D*	*E*
A	*M*	0	0	98	0
B	0	*M*	61	124	*M*
C	51	112	*M*	0	190
D	0	26	*M*	*M*	77
E	0	0	139	175	*M*

Assignments : *A–E, B–C, C–D, D–A, E–B*

Tour : *A–E–B–C–D–A*

Total Distance : 573

TABLE 7.70 Reduced Cost Table 9: Sub-problem 7

City	*A*	*B*	*C*	*D*	*E*
A	*M*	0	0	98	0
B	0	*M*	61	124	0
C	51	112	*M*	0	190
D	0	26	*M*	*M*	77
E	0	*M*	139	175	*M*

Assignments : *A–C, B–E, C–D, D–B, E–A*

Tour : *A–C–D–B–E–A*

Total Distance : 539

It may be noted that each of the four sub-problems yield a tour. Thus, no further branching is needed. Of all the tours, the shortest one involves a distance of 538. Thus, the lower and upper bounds would both be revised to 538. The complete branch-and-bound tree is shown in Figure 7.12.

From the above, it is clear that the optimal solution to the problem is the route *A–E–B–D–C–A* and the total distance covered is 538.

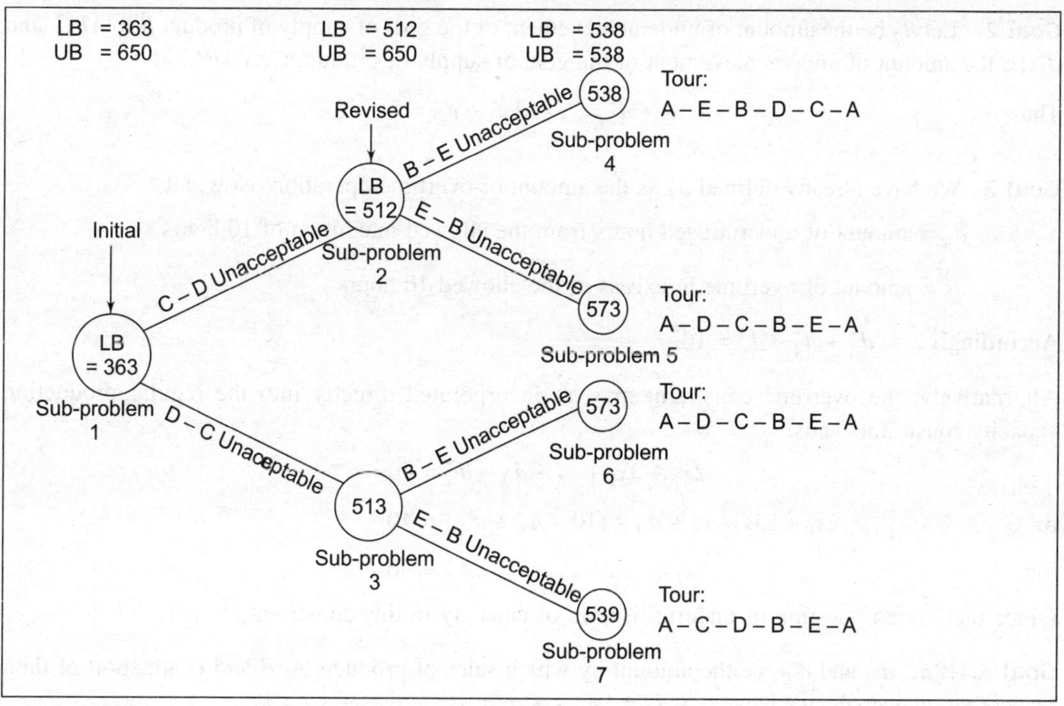

Figure 7.12 *Final Branch and Bound Tree*

Example 7.20 A company produces three items *A*, *B* and *C*. All these are processed in a central plant. Production of one unit of A needs 2 hours, while production of a unit of *B* and *C* requires, respectively, 3 and 1 hours. The regular plant capacity is 40 hours per week. The marketing department has informed that maximum sales per week of *A*, *B* and *C* is 10, 10 and 12 units, respectively. The chief executive of the company has established the following goals according to their importance:

(i) Avoid any underutilisation of production capacity.
(ii) Meet the order of *ABC* stores for 7 units of *B* and 5 units of *C* per week.
(iii) Avoid the overtime operation of the plant beyond 10 hours.
(iv) Achieve the sales goal of 10 units of *A*, 10 units of *B* and 12 units of *C*.
(v) Minimise the overtime operation as much as possible.

Formulate this as a goal programming problem.

Let x_1, x_2 and x_3 be the number of units of products *A*, *B* and *C*, respectively. Using the given information, we may obtain the set of goals and constraints in terms of deviational variables as follows.

Goal 1 Let d_1^- be the amount of underutilisation of normal plant capacity, and d_1^+ be the amount of overutilisation (overtime) of normal plant capacity.

We have, thus
$$2x_1 + 3x_2 + x_3 + d_1^- - d_1^+ = 40$$

Goal 2 Let d_2^- be the amount of underachievement of the goal of supply of product B to ABC, and d_3^- be the amount of underachievement of the goal of supply of product C to ABC.

Thus, $x_2 + d_2^- = 7$, and $x_3 + d_3^- = 5$

Goal 3 We have already defined d_1^+ as the amount of overtime operation. Now, let

d_4^- = amount of underutilised hours from the allowed maximum of 10 hours

d_4^+ = amount of overtime in excess of the allowed 10 hours

Accordingly, $d_1^+ + d_4^- - d_4^+ = 10.$

Alternatively, the overtime constraint may be incorporated directly into the regular production capacity constraint. Thus,

$$2x_1 + 3x_2 + x_3 + d_1^- - d_1^+ = 40$$

or
$$2x_1 + 3x_2 + x_3 + d_1^- - (10 - d_4^- + d_4^+) = 40$$

or
$$2x_1 + 3x_2 + x_3 + d_4^- - d_4^+ = 50$$

(since there is no question of underutilisation of capacity in this constraint).

Goal 4 If d_5^-, d_6^- and d_7^- be the amount by which sales of products A, B and C are short of their respective sales goals, we have $x_1 + d_5^- = 10$; $x_2 + d_6^- = 10$; and $x_3 + d_7^- = 12$.

Now, the goal programming problem may be stated as under:

Minimise
$$Z = P_1 d_1^- + P_2 d_2^- + P_2 d_3^- + P_3 d_4^- + P_4 d_5^- + P_4 d_6^- + P_4 d_7^- + P_5 d_1^+$$

Subject to

$$2x_1 + 3x_2 + x_3 + d_1^- - d_1^+ = 40$$
$$x_2 + d_2^- = 7$$
$$x_3 + d_3^- = 5$$
$$2x_1 + 3x_2 + x_3 + d_4^- - d_4^+ = 50$$
$$x_1 + d_5^- = 10$$
$$x_2 + d_6^- = 10$$
$$x_3 + d_7^- = 12$$
$$x_1, x_2, x_3, d_1^-, d_1^+, d_2^-, d_3^-, d_4^-, d_4^+, d_5^-, d_6^-, d_7^- \geq 0$$

Example 7.21 Using the data given below, develop a linear programming (goal programming) model to evaluate the jobs. The model should include constraints covering:

(i) The total score for the Job 1 (highest) should be close to 2000 but not more.

(ii) The score assigned to Job 16 should be as near as possible to 1200.

(iii) The score for the Job 30 (lowest) should be as close to 400 but not less.

(iv) The differential in scores between the jobs listed above should be at least 300.
(v) The scores should be in conformity to the stipulated rankings of the jobs.

Points by Each Factor

Benchmark Jobs	*A*	*B*	*C*	*D*	*E*
1	70	65	70	70	70
9	45	40	50	60	55
16	40	45	40	50	40
24	30	35	35	40	30
30	20	20	15	25	15

(MMS, Mumbai, 1998)

Let the weights of the factors A, B, C, D and E be x_1, x_2, x_3, x_4 and x_5 respectively. With usual deviational variables introduced, the goal programming model is shown formulated here:

Minimise $Z = d_1^+ + d_2^- + d_2^+ + d_3^- + d_4^- + d_5^- + d_6^- + d_7^-$

Subject to

$$70x_1 + 65x_2 + 70x_3 + 70x_4 + 70x_5 + d_1^- - d_1^+ = 2000$$

$$40x_1 + 45x_2 + 40x_3 + 50x_4 + 40x_5 + d_2^- - d_2^+ = 1200$$

$$20x_1 + 20x_2 + 15x_3 + 25x_4 + 15x_5 + d_3^- - d_3^+ = 400$$

$$30x_1 + 20x_2 + 30x_3 + 20x_4 + 30x_5 + d_4^- - d_4^+ = 300$$

$$20x_1 + 25x_2 + 25x_3 + 25x_4 + 25x_5 + d_5^- - d_5^+ = 300$$

$$5x_1 - 5x_2 + 10x_3 + 10x_4 + 15x_5 + d_6^- - d_6^+ = 0$$

$$10x_1 + 15x_2 + 20x_3 + 15x_4 + 15x_5 + d_7^- - d_7^+ = 0$$

$$x_i, d_i^-, d_i^+ \geq 0 \quad i = 1, 2, \ldots 5$$

Notes:

(1) The constraint (iv) given in the problem requires that the total score of job 1 should be greater than that of job 16 by at least 300, and that the total score of job 16 should be greater than the total score of job 30 by as much amount. Thus, the two requirements are:

Total score (Job 1) – Total score (Job 16) ≥ 300, and
Total score (Job 16) – Total score (Job 30) ≥ 300.

(2) In terms of the given constraint (v), the jobs 1, 9, 16, 24 and 30 should have scores in the descending order of magnitude. Since differences between jobs 1, 16 and 30 are already considered in order in constraint (iv), we need to put the following two constraints as well:

Total score (Job 9) – Total score (Job 16) ≥ 0, and
Total score (Job 24) – Total score (Job 30) ≥ 0.

Example 7.22 A firm produces two products *P* and *Q*, which yield a contribution margin of Rs 200 and Rs 300 per unit, respectively. The firm has a limited capacity in the two departments where these products need processing. The availability and requirements are given below:

Department	Processing Time (hours)		Daily Availability (hours)
	Product P	Product Q	
I	4	2	45
II	4	4	70

The management of the firm has specified the following goals:

Goal	Priority
1. Produce a product-mix to make a daily profit of at least Rs 4800	P_1
2. Achieve a daily sales of at least 15 units of product Q	P_2
3. Achieve a daily sales of at least 5 units of product P	P_3

Formulate and solve it as a goal programming problem.

Let x_1 and x_2 be the daily output of the products *P* and *Q* respectively. Further, the following deviational variables are defined:

d_1^- = underachievement of the daily profit target
d_1^+ = overachievement of the daily profit target
d_2^- = underachievement of the target daily sales of product *Q*
d_2^+ = overachievement of the target daily sales of product *Q*
d_3^- = underachievement of the target daily sales of product *P*
d_3^+ = overachievement of the target daily sales of product *P*

Also, let S_1 and S_2 be the number of hours unutilised in the departments I and II, respectively. The goal programming problem is stated below:

Maximise $Z = P_1 d_1^- + P_2 d_2^- + P_3 d_3^-$

Subject to

$$200x_1 + 300x_2 + d_1^- - d_1^+ = 4800 \quad \text{Profit goal}$$
$$x_2 + d_2^- - d_2^+ = 15 \quad \text{Product } Q$$
$$x_1 + d_3^- - d_3^+ = 5 \quad \text{Product } P$$
$$4x_1 + 2x_2 + S_1 = 45 \quad \text{Deptt. I}$$
$$4x_1 + 4x_2 + S_2 = 70 \quad \text{Deptt. II}$$
$$x_1, x_2, d_1^-, d_1^+, d_2^-, d_2^+, d_3^-, d_3^+, S_1, S_2 \geq 0$$

The solution to this problem, using modified simplex method, is contained in Tables 7.71 through 7.75.

TABLE 7.71 Simplex Tableau 1: Non-optimal Solution

Basis		x_1	x_2	d_1^-	d_1^+	d_2^-	d_2^+	d_3^-	d_3^+	S_1	S_2	b_i	b_i/a_{ij}
d_1^-	P_1	200	300	1	−1	0	0	0	0	0	0	4800	16
d_2^-	P_2	0	1*	0	0	1	−1	0	0	0	0	15	15 ←
d_3^-	P_3	1	0	0	0	0	0	1	−1	0	0	5	–
S_1	0	4	2	0	0	0	0	0	0	1	0	45	45/2
S_2	0	4	4	0	0	0	0	0	0	0	1	70	35/2
c_j		0	0	P_1	0	P_2	0	P_3	0	0	0		
Δ_j	P_3	−1	0	0	0	0	0	0	1	0	0		
	P_2	0	−1	0	0	0	1	0	0	0	0		
	P_1	−200	−300	0	1	0	0	0	0	0	0		
			↑										

TABLE 7.72 Simplex Tableau 2: Non-optimal Solution

Basis		x_1	x_2	d_1^-	d_1^+	d_2^-	d_2^+	d_3^-	d_3^+	S_1	S_2	b_i	b_i/a_{ij}
d_1^-	P_1	200	0	1	−1	−300	300*	0	0	0	0	300	1 ←
x_2	0	0	1	0	0	1	−1	0	0	0	0	15	–
d_3^-	P_3	1	0	0	0	0	0	1	−1	0	0	5	–
S_1	0	4	0	0	0	−2	2	0	0	1	0	15	15/2
S_2	0	4	0	0	0	−4	4	0	0	0	1	10	5/2
c_j		0	0	P_1	0	P_2	0	P_3	0	0	0		
Δ_j	P_3	−1	0	0	0	0	0	0	1	0	0		
	P_2	0	0	0	0	1	0	0	0	0	0		
	P_1	−200	0	0	1	300	−300	0	0	0	0		
							↑						

TABLE 7.73 Simple x Tableau 3: Non-optimal Solution

Basis		x_1	x_2	d_1^-	d_1^+	d_2^-	d_2^+	d_3^-	d_3^+	S_1	S_2	b_i	b_i/a_{ij}
d_2^+	0	2/3 *	0	1/300	– 1/300	–1	1	0	0	0	0	1	3/2 ←
x_2	0	2/3	1	1/300	–1/300	0	0	0	0	0	0	16	24
d_3^-	P_3	1	0	0	0	0	0	1	– 1	0	0	5	5
S_1	0	8/3	0	–1/150	1/150	0	0	0	0	1	0	13	39/8
S_2	0	4/3	0	–1/75	1/75	0	0	0	0	0	1	6	9/2
c_j		0	0	P_1	0	P_2	0	P_3	0	0	0		
Δ_j	P_3	– 1	0	0	0	0	0	0	1	0	0		
	P_2	0	0	0	0	1	0	0	0	0	0		
	P_1	0	0	1	0	0	0	0	0	0	0		
		↑											

TABLE 7.74 Simplex Tableau 4: Non-optimal Solution

Basis		x_1	x_2	d_1^-	d_1^+	d_2^-	d_2^+	d_3^-	d_3^+	S_1	S_2	b_i	b_i/a_{ij}
x_1	0	1	0	1/200	–1/200	–3/2	3/2	0	0	0	0	3/2	–
x_2	0	0	1	0	0	1	– 1	0	0	0	0	15	–
d_3^-	P_3	0	0	– 1/200	1/200	3/2	– 3/2	+ 1	– 1	0	0	7/2	700
S_1	0	0	0	– 1/50	1/50	4	– 4	0	0	1	0	9	450
S_2	0	0	0	– 1/50	1/50*	2	– 2	0	0	0	1	4	200 ←
c_j		0	0	P_1	0	P_2	0	P_3	0	0	0		
Δ_j	P_3	0	0	1/200	– 1/200	– 3/2	3/2	0	1	0	0		
	P_2	0	0	0	0	1	0	0	0	0	0		
	P_1	0	0	1	0	0	0	0	0	0	0		
					↑								

TABLE 7.75 Simplex Tableau 5: Optimal Solution

Basis		x_1	x_2	d_1^-	d_1^+	d_2^-	d_2^+	d_3^-	d_3^+	S_1	S_2	b_i
x_1	0	1	0	0	0	– 1	1	0	0	0	1/4	5/2
x_2	0	0	1	0	0	1	– 1	0	0	0	0	15
d_3^-	P_3	0	0	0	0	1	– 1	1	– 1	0	– 1/4	5/2
S_1	0	0	0	0	0	2	– 2	0	0	1	– 1	5
d_1^+	0	0	0	– 1	1	100	–100	0	0	0	50	200
c_j		0	0	P_1	0	P_2	0	P_3	0	0	0	
Δ_j	P_3	0	0	0	0	– 1	1	0	1	0	1/4	
	P_2	0	0	0	0	1	0	0	0	0	0	
	P_1	0	0	1	0	0	0	0	0	0	0	

From the optimal solution in Simplex Tableau 5, it is evident that it calls for producing 5/2 units of product *P* and 15 units of *Q* daily. This product-mix satisfies the first two goals and results in minimum deviation (non-achievement) of goal 3.

KEY POINTS TO REMEMBER

- Integer programming problems are special type of linear programming problems where all or some of the variables are restricted to be integer variables.
- Integer programming problems are termed as *pure* or *mixed-integer* problems accordingly as all or some of the variables are sought to be integer variables.
- In the 0–1 type of integer programming problems, the variables can take only either of these two values.
- Obtaining optimal solution to a problem using graphic approach or simplex method and then rounding off the fractional values to get integer solution appears to be an easy way but it generally does not work.
- The cutting plane algorithm promises optimal integer solution in a finite number of iterations.
- In using cutting plane algorithm, the problem is first solved as an LPP relaxation by dropping the integer requirements. Then a cutting plane is introduced by using a constraint in the optimal solution which bears a fractional right-hand side value. Cuts are introduced until all the variables assume integer values.
- A cut does not eliminate any points which are feasible for the IPP.
- The cutting plane algorithm requires all right-hand side values as well as all the coefficients in the constraints to be integers.
- Branch and bound method is very usefully employed in problems where there are a finite number of solutions.
- In applying branch and bound method, essentially what is done is to divide the feasible region into smaller and smaller parts until the optimal solution is located.

- Goal programming is used where multiple goals, related to profit, employment etc., are specified instead of only maximising or minimising of a single objective function done in linear/integer programming.
- A value for each goal is specified and deviational variables are introduced. The goals are put as constraints along with technological constraints and the objective function is expressed as an overall function to be minimised. This is done when all the goals are considered to be of equal importance. However, differential weights may be given to the various goals in line with their relative significance.
- In case of preemptive goal programming, we proceed in the same way but here goals are assigned priority levels. First of all, the first-priority goal is considered and an attempt is made to achieve it as closely as possible. Only when an optimal solution with respect to the first-priority goal is achieved that we consider the next-priority goal. Similarly, the next-in-line goals are considered only when the optimality with respect to the higher priority goals is obtained. We continue in this manner until the only way we can come closer to satisfying a goal is to increase the deviation from a higher-priority goal.
- For solving a preemptive goal programming problem, the simplex method is modified and instead of writing a single row for showing the Δ_j values, a matrix is provided consisting of various goals expressed according to their priority level. The Δ_j values are obtained in respect of each of the goals.
- Only when optimality is achieved with respect to a high priority goal that the next-in-line is considered due to different priority levels involved—a high priority goal must be considered first and then only a movement to the next priority goal can be made.

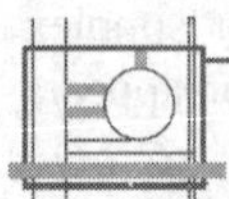

TEST YOUR UNDERSTANDING

Mark the following statements as T (*True*) or F (*False*).

1. An integer programming problem is a linear programming problem with all decision variables restricted to integer values.
2. A linear programming problem with only one decision variable restricted to integer value is not an integer programming problem.
3. A mixed IPP is one where mixed constraints are involved.
4. An IPP where all the variables must be equal to zero or one is called a "0-1 IPP."
5. Capital budgeting problems may be handled as 0-1 type of integer programming problems.
6. The solution to the integer programming problems can always be obtained by rounding off the fractional values obtained in solution to the problems by using graphic/simplex method.
7. If the solution obtained by simplex method to a given problem yields only integer values, it may or may not represent optimal solution to the corresponding integer programming problem.

8. In a pure IPP, solution to all variables, decision and slack or surplus, would always be in integer values, provided that all the coefficients in the constraints and all the right hand side values of the constraints be integers.
9. In using cutting-plane algorithm, a cut is introduced by using a constraint in the optimal solution, obtained by simplex method, which has a fractional b_i value.
10. When two or more constraints have fractional values in the RHS, it is desirable to generate the cut by using the constraint whose RHS has a fractional part closest to 1.
11. A cut may or may not eliminate any points that are feasible for the IPP.
12. Branch and bound method is not a standardised method. It is applied differently for different problems.
13. Basically, branching and bounding keeps dividing up the feasible region into smaller and smaller parts until the solution that optimises is determined.
14. An assignment problem solved by branch and bound method may or may not have the same optimal solution as obtained by Hungarian assignment method.
15. In *branch and bound* algorithm for solving a travelling salesman problem, we may initially consider a tour by considering travelling in the order 1–2–3–...–*n*. The distance covered in this tour would provide a lower bound on the total distance of the optimal tour.
16. Goal programming is used where multiple criteria, instead of a single objective, are used in decision-making.
17. The multiple goals specified in a goal programming problem are usually conflicting and incommensurate.
18. The main idea of goal programming is to determine the individual preferences of the decision-makers in terms of goals stated and achieve them all.
19. Each of the goals in a goal programming problem is either underachieved or overachieved.
20. The deviational variables for any goal constraint measure the amount of discrepancy between the value of the objective and the goal specified in respect thereof.
21. The inequalities related to structural constraints are rigid while those representing goals are flexible.
22. Both the deviational variables related to a goal—one of underachievement and the other of overachievement—can together assume positive values.
23. One of the deviational variables for a goal constraint must equal zero.
24. One-sided goals set a lower or an upper limit which may not fall short of or exceeded, respectively. In two-sided goals, both lower and upper limits are set.
25. Non-preemptive goal programming deals with situations where all goals are of comparable importance and it assumes that the marginal rate of substitution between the objectives is linear.
26. Penalty weights may be assigned to deviational variables related to different goals, in accordance with their relative importance to the decision-maker.
27. The weights used in non-preemptive goal programming are indicative of decision-maker's utility for the various objectives.
28. Preemptive goal programming deals with situations where goals have different priority levels, so that some of them are considered more important than others.
29. Various goals in preemptive goal programming are given equal or differential weightage.

30. In solving a preemptive goal programming problem, once the Δ_j values are found to be all greater than or equal to zero, with respect to the first-priority goal, the optimal solution to the problem is reached.

EXERCISES

1. Discuss integer linear programming. Give an example each of a *pure* and a *mixed* integer linear-programming problem.
2. "It may not be possible to obtain an optimal integer solution by rounding off the continuous optimum of a linear programming problem." Comment.
3. What is integer linear programming? Explain the merits and demerits of 'rounding off' a continuous optimal solution to an LP problem to obtain an integer solution.
   (MBA, Delhi, 1998)
4. State the Travelling Salesman problem. Discuss the branch and bound method of solving it. State the linear programming formulation of an assignment problem.
5. Discuss the *cutting-plane algorithm* for solving integer programming problems.
6. Give a detailed account of the solution of 0–1 integer programming problems by the branch-and-bound method.
7. Differentiate between non-preemptive and preemptive goal programming problems. How is the solution method different in the two cases?
8. "In goal programming, we attempt to 'satisfy' or come as close as possible to satisfying, the various goals." Discuss.
9. "Goal programming appears to be the most appropriate, flexible and powerful quantitative technique for complex decision problems involving conflicting objectives." Discuss this statement.

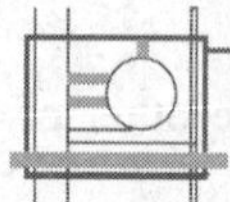

PRACTICAL PROBLEMS

1. ABC Construction Company is faced with the problem of determining which projects it should undertake over the next 4 years. The following table gives information about each project:

Project	Estimated present value	Capital requirements			
		Year 1	Year 2	Year 3	Year 4
A	18,00,000	3,00,000	4,00,000	4,00,000	3,00,000
B	2,00,000	1,20,000	80,000	0	40,000
C	7,20,000	3,00,000	2,00,000	2,00,000	2,00,000
D	8,00,000	2,00,000	4,00,000	4,00,000	1,00,000
Funds available		6,50,000	8,00,000	8,00,000	5,00,000

Formulate a zero-one programming model to maximise the estimated present value.
(MBA, Delhi, October, 2002)

2. Three machines are available to make 5,000 units which are required for the internal use of a company as components to another product. If the production is to be made by any one of these machines, a set-up cost is incurred apart from the cost of making each unit on different machines. The cost data are given below:

Machine	*Set-up cost (Rs)*	*Cost per unit (Rs)*	*Maximum production*
1	9,000	11	4,000
2	6,000	10	3,000
3	4,500	16	1,000

What is the best production strategy? Formulate only.

3. An air-conditioning and refrigeration company has been awarded a contract for the air-conditioning of a new computer installation. The company has to make a choice between two alternatives: (a) hire one or more refrigeration technicians for 6 hours a day, or (b) hire one or more part time refrigeration apprentice technicians for 4 hours a day.
   The rate of wages of the refrigeration technicians is Rs 20 per hour, while the corresponding rate for apprentice technicians is Rs 8 per hour. The company wants to engage the technicians on work for not more than 25 man-hours per day and also limit the charges to technicians to Rs 240. The company estimates that the productivity of a refrigeration technician is 8 units and that of a part-time apprentice technician is 3 units. Formulate the integer programming problem to enable the company to select the optimum number of technicians and apprentices.
4. In a project selection model, eight projects, numbered 1, 2, ... 8, are in consideration. The following conditions are given. You are required to write constraints to satisfy each of them.
   (a) Only one of the projects 3 and 8 has to be selected.
   (b) If only project 4 is selected that the project 7 can be selected, but if project 4 is selected then project 7 may or may not be selected.
   (c) Not more than two of the projects 1, 3, 5, and 8 may be selected.
   (d) Identical decision is required in respect of projects 3 and 6.
   (e) Not more than 4 and no less than 2 of the projects 1, 2, 4, 5, 6, and 8 may be selected.
   (f) Exactly five of the projects are to be selected.
5. A manufacturing company faces the following problem. Should they make or buy each of several products? The company policies specify that they will either make or buy the whole lot of each product in a complete lot. The company has four products to make or buy with six machines involved in making these products if they are made in the shop. The times (in hours) per unit required are as follows:

Product	*Machine*					
	A	*B*	*C*	*D*	*E*	*F*
1	0.04	0.02	0.02	0.00	0.03	0.06
2	0.00	0.01	0.05	0.15	0.09	0.06
3	0.02	0.06	0.00	0.06	0.20	0.02
4	0.06	0.04	0.15	0.00	0.00	0.05

Forty hours are available on each machine. One houndred ten units of each product are needed. The costs of making the products are:

Product:	1	2	3	4
Cost/unit(Rs):	2.25	2.22	4.50	1.90

The costs to buy the product are:

Product:	1	2	3	4
Cost/unit(Rs):	3.10	2.60	4.75	2.25

Formulate the given problem as a zero-one programming problem. Do not solve it.

(MBA, Delhi, 2006)

6. A company is considering four investment opportunities. The following table presents information about the investments (in Rs thousands):

Project	*Present value of expected return*	*Capital required by the projects*		
		Year 1	*Year 2*	*Year 3*
1	1250	720	560	400
2	1320	880	550	360
3	620	300	170	110
4	740	350	210	-
Capital available for investment		1400	800	440

In addition, it is given that projects 1 and 2 are mutually exclusive, while project 4 is contingent upon the prior acceptance of project 3. Formulate an integer programming model to determine which projects to accept and which ones to reject so as to maximize the present value from accepted projects.

7. An exporter of ready-made garments makes two types of shirts: *X* and *Y*. He makes a profit of Rs 10 and Rs 40 per shirt on *X* and *Y*, respectively. He has two tailors, *A* and *B*, at his disposal to stitch these shirts. Tailors *A* and *B* can devote at the most 7 hours and 15 hours per day, respectively. Both these shirts are to be stitched by both the tailors. Tailor *A* and Tailor *B* spend 2 hours and 5 hours, respectively, in stitching an *X* shirt and 4 hours and 3 hours, respectively, in stitching a *Y* shirt. How many shirts of both types should be stitched in order to maximise daily profits? (A non-integer solution for this problem will not be accepted.)

(MBA, Delhi, November, 1995)

8. Find the optimal solution to the following integer programming problem, using Gomory's cutting plane method.

Maximise $Z = 7x_1 + 9x_2$

Subject to

$$-x_1 + 3x_2 \text{ LE } 18$$

$$7x_1 + x_2 \text{ LE } 105$$

x_1 and x_2 to be non-negative integers

LE means "less than or equal to"

Given below is the optimal tableau of the associated Linear Programme, with no restriction on the integrality of x's.

C_B			7	9	0	0
	Basis	b_i	x_1	x_2	S_1	S_2
9	x_2	10.5	0	1	7/22	1/22
7	x_1	13.5	1	0	–1/22	3/22
$C_j - Z_j$		–189	0	0	–28/11	–15/11

(MMS, Mumbai, 2001)

9. A manufacturer of baby dolls makes two types of dolls. One is sold under the brand name Molina and the other Suzie. Processing of these two dolls is done on two machines A and B. The processing time for each Molina is 3 hours and 4 hours on machines A and B respectively, and that for each Suzie is 1 hour and 4 hours on machines A and B respectively. There is 50 hours of time per week available on machine A and 90 hours on machine B. The profit contribution for a Molina is Rs 6 and that for Suzie is Rs 18. Formulate and solve this problem to determine the optimal weekly production schedule of the two dolls. (Provide an all-integer solution.)

   (MBA, Delhi, March 2004)

10. Maximise $Z = 2x_1 + 4x_2 + 3x_3$

    Subject to

$$3x_1 + 4x_2 + 2x_3 \leq 60$$
$$2x_1 + x_2 + 2x_3 \leq 40$$
$$x_1 + 3x_2 + 2x_3 \leq 80$$
$$x_1, x_2, x_3 \geq 0 \text{ and integer}$$

11. A manufacturer of toys makes two types of toys, *A* and *B*. Processing of these toys is done on two machines *X* and *Y*. Toy *A* requires 2 hours on machine *X* and 6 hours on machine *Y*. Toy *B* requires 4 hours on machine *X* and 5 hours on machine *Y*. There are 18 hours of time per day available on machine *X* and 30 hours on machine *Y*. The profit obtained on both the toys is same, that is, Rs 25 per toy. What should be the daily production of each of the two toys? (A non-integer solution for this problem will not be accepted.) *(MBA, Delhi, November 2008)*

12. A company manufactures two products P_1 and P_2, which require time on three different machines M_1, M_2 and M_3. Product P_1 requires 2 hours on machine M_1 and 1 hour on M_1. Product P_2 requires 4 hours on machine M_2 and 2 hours on machine M_3. There are 50 hours available on machine M_1, 16 hours on machine M_2 and 20 hours on machine M_3 per week. The per unit profit on products P_1 and P_2 is Rs 30 and Rs 50, respectively. Determine the output mix that will maximise the total profit per week, when only integer solution is acceptable.

13. The dietician at the local hospital is planning the breakfast menu for the maternity ward patients. She is planning a special nonfattening diet, and has chosen cottage cheese and scrambled eggs for breakfast. She is primarily concerned with Vitamin E and iron requirements in planning the breakfast.

    According to the State Medical Association (SMA) new mothers must get at least 12 milligrams of Vitamin E and 24 milligrams of iron from breakfast. The (SMA) handbook reports that a scoop of cottage cheese contains 3 milligrams of Vitamin E and 3 milligrams of iron. An average scoop of scrambled egg contains 2 milligrams of Vitamin E and 8 milligrams of iron. The SMA handbook recommends that new mothers should eat at least two scoops of cottage cheese for their breakfast. The dietician considers this as one of the model constraints.

The hospital accounting department estimates that a scoop of cottage cheese costs Rs 2 and scoop of scrambled egg also costs Rs 2. The dietician is attempting to determine the optimum breakfast menu that satisfies all the requirements and minimises total cost. The cook insists that he can serve foods by only full scoop, thus necessitating an integer solution. Determine the optimum-integer solution to the problem. *(MBA, Delhi, April, 1994)*

14. The following table is the cost matrix for travel among a particular set of locations. Determine a minimum cost, travelling salesman itinerary:

States	*1*	*2*	*3*	*4*	*5*
1	–	15	22	17	18
2	15	–	22	16	17
3	15	17	–	19	15
4	16	19	20	–	19
5	19	17	21	20	–

15. A salesman has to visit five cities *A*, *B*, *C*, *D*, and *E*. The distances (in hundred km) between the five cities are as follows:

	A	*B*	*C*	*D*	*E*
A	–	17	16	18	14
B	17	–	18	15	16
C	16	18	–	19	17
D	18	15	19	–	18
E	14	16	17	18	–

If the salesman starts from city *A* and has to come back to city *A*, which route should he select so that the total distance travelled by him is minimum.

16. A machine operator processes four types of items on his machine and he must choose a sequence for them. The set-up cost per change depends on the items currently on machine and the set-up made according to the following table:

From item	*To*			
	A	*B*	*C*	*D*
A	–	4	7	3
B	4	–	6	3
C	7	6	–	7
D	3	3	7	–

If he processes each of the items once each week, then how should he sequence the items on his machine? Use the method for the problem of travelling salesman.

(MBA, Delhi, November, 1999)

17. The expected times required to be taken by a salesman in travelling from one city to another are as follows:

	Time (in hours)				
From/To	C_1	C_2	C_3	C_4	C_5
C_1	–	10	13	11	Not Possible
C_2	10	–	12	10	12
C_3	14	13	–	13	11
C_4	11	10	14	–	10
C_5	12	11	12	10	–

How should the salesman plan his trip so that he covers each of these cities no more than once, and completes his trip in minimum possible time required for travelling?

18. The *ABC* Ice Cream Company has a distribution depot in Greater Kailash Part I for distributing ice cream in South Delhi. There are four vendors located in different parts of South Delhi (call them *A*, *B*, *C* and *D*) who have to be supplied ice cream everyday. The following matrix displays the distance (in kms) between the depot and the four vendors:

From	*To*				
	Depot	*Vendor A*	*Vendor B*	*Vendor C*	*Vendor D*
Depot	–	3.5	3.0	4.0	2.0
Vendor A	3.5	–	4.0	2.5	3.0
Vendor B	3.0	4.0	–	4.5	3.5
Vendor C	4.0	2.5	4.5	–	4.0
Vendor D	2.0	3.0	3.5	4.0	–

What route should the company van follow that the total distance covered is minimised?

19. You have been retained as a consultant to formulate a model for job evaluation in an organisation. The data collected for 5 representative benchmark jobs across the hierarchy of all blue collar jobs evaluated on a 5-point rating scale for four factors viz. Skill, Effort, Working Condition, and Responsibility.

Job Grade	*Skill*	*Effort*	*Working Condition*	*Responsibility*
A	4	5	5	5
G	4	4	4	4
M	3	4	3	3
S	2	3	2	2
Z	1	2	2	1

Find weights to be assigned to the four factors satisfying the following broad guidelines:

(i) Grade A: Total points should be as near as possible to 600 but not in excess.

(ii) Grade M: Total points should be as near as possible to 360 but may be higher or lower.

(iii) Grade Z: Total points should be at least as near as possible to 120 but not lower.

(iv) The difference between two benchmark grades should be at least 120.

Formulate a goal programming model. *(MMS, Mumbai, 2000)*

20. A rural clinic hires its staff from nearby cities and towns on a part-time basis. The clinic attempts to have a general practitioner (GP), a nurse and an internist on duty during at least a portion of each week. The clinic has a weekly budget of Rs 1,200. A GP charges the clinic Rs 40 per hour, a nurse charges Rs 20 per hour, and an internist charges Rs 150. The clinic has established the following goals in order of priority.
    (i) A nurse should be available for at least 30 hours per week.
    (ii) The weekly budget of Rs 1,200 should not be exceeded.
    (iii) A GP or internist should be available at least 20 hours per week.
    (iv) An internist should be available at least 6 hours per week.

    Formulate a goal programming model for determining the number of hours to hire each staff member in order to satisfy the various goals. *(MBA, Delhi, March 2004)*
21. A farmer has 1,000 acres of land on which he intends to plant corn, wheat and soyabeans. Each acre of corn costs Rs 10,000 for preparation, requires 7 worker days of labour, and yields a profit of Rs 3,000. An acre of wheat costs Rs 12,000 to prepare, requires 10 worker days of labour, and yields Rs 4,000 profit. An acre of soyabeans costs Rs 7,000 to prepare, requires 8 worker days of labour, and yields Rs 2,000 profit. The farmer has taken out a loan of Rs 80,00,000 for crop preparation and has contracted with a union for 6,000 worker days of labour. A granary has agreed to purchase 200 acres of corn, 500 acres of wheat and 300 acres of soyabeans. The farmer has established the following goals, in order of their importance:
    (i) In order to maintain good relations with the union, the labour contract must be honoured; that is, the full 6,000 must be used.
    (ii) Preparation costs should not exceed the loan amount so that additional loans will not have to be secured.
    (iii) The farmer desires a profit of at least Rs 1,05,00,000 in order to remain in good financial condition.
    (iv) Contracting for excess labour should be avoided.
    (v) The farmer would like to use as much of the available acreage as possible.
    (vi) The farmer would like to meet the sales with the granary. However, the goal should be weighted according to the profit returned by each crop.

    Formulate a goal programming model for determining the number of acres of each crop the farmer should plant in order to satisfy the goals in the best possible way.

    (MBA, Delhi, March 2003)
22. Amro Mutual Fund has Rs 20 lakh available for investment in Government bonds, blue chips, speculative stocks, and gold. The expected annual return on these investments is given below:

Investment	*Expected Return*
Government bonds	8% interest
Blue chip stocks	7% dividend plus 5% capital gains
Speculative stocks	4% dividend plus 12% capital gains
Gold	14% capital gains

The investment policy of Amro is:

(a) The amount of investment can in no case be more than Rs 20 lakh.
(b) Never invest more than 15 per cent of the funds in gold.
(c) Achieve a total return of at least Rs 2,50,000.

(d) Do not invest more than Rs 4 lakh in speculative stocks.
(e) Invest at least Rs 3 lakh in Government bonds.
(f) Derive at least Rs 60,000 in capital gains.

Further, the current investment guidelines stipulate that

(i) the total returns be considered twice as important as the investment ceiling in speculative stocks,
(ii) investment limitation in speculative stocks and Government bonds is important as their respective returns, and
(iii) the target of capital gains is one-fourth as important as the investment limitation in Government bonds.

Formulate the above problem as a goal programming problem.

23. The PQR manufacturing Company produces four different airplane parts from fabricated sheet metal for several major aircraft companies. The manufacturing process consists of four operations—stamping, assembly, finishing and packaging. The proccssing times per unit for each operation and total available hours per year to produce these parts are as follows:

	Part (hour/unit)				*Total Hours*
Operation	*1*	*2*	*3*	*4*	*per year*
Stamping	0.06	0.17	0.10	0.14	600
Assembly	0.18	0.20	--	0.14	500
Finishing	0.07	0.20	0.08	0.12	550
Packaging	0.09	0.12	0.07	0.15	450

The sheet metal required for each part, the estimated annual demand, and the profit per part are as follows:

Part	*Sheet Metal (ft²)*	*Estimated Annual Demand*	*Profit (Rs)*
1	2.6	2,600	9,000
2	1.4	1,800	10,000
3	2.5	4,100	8,000
4	3.2	1,200	12,000

The company has 15,000 square feet of fabricated metal delivered each month. The company has the following prioritized production goals:

(i) Avoid overtime, which would erode profit levels.
(ii) Meet parts demand.
(iii) Achieve an annual profit of Rs 7,00,00,000.
(iv) Avoid ordering more material since a surcharge is required by the supplier for changing the standard monthly order.

Formulate a goal programming model to determine the amount of each part to produce in order to achieve the company's objectives. *(MBA, Delhi, March 2007)*

24. Consider the following Goal Programming model:

Minimise $Z = P_1 + d_1^+ + P_1 d_2^+ + 30P_2 d_3^- + 40P_2 d_4^- + P_3 d_5^-$

Subject to

$$2x_1 + 4x_2 + d_1^- - d_1^+ = 80$$

$$3x_1 + 3x_2 + d_2^- - d_2^+ = 80$$

$$x_1 + \quad d_3^- - d_3^+ = 10$$

$$x_2 + d_4^- - d_4^+ = 10$$

$$30x_1 + 40x_2 + d_5^- - d_5^+ = 1,200$$

All variables to be non-negative.

The modified simplex tableau giving IBFS is shown below. Carry out the simplex iterations until an optimal solution is reached. State clearly the optimal solution and to what extent the goals have been achieved.

			0	0	0	P_1	0	P_1	$30P_2$	0	$40P_2$	0	P_3	0
C_B	X_B	b	x_1	x_2	d_1^-	d_1^+	d_2^-	d_2^+	d_3^-	d_3^+	d_4^-	d_4^+	d_5^-	d_5^+
0	d_1^-	80	2	4	1	–1	0	0	0	0	0	0	0	0
0	d_2^-	80	3	3	0	0	1	–1	0	0	0	0	0	0
$30P_2$	d_3^-	10	1	0	0	0	0	0	1	–1	0	0	0	0
$40P_2$	d_4^-	10	0	1	0	0	0	0	0	0	1	–1	0	0
P_3	d_5^-	1,200	30	40	0	0	0	0	0	0	0	0	1	–1
	P_3	1,200	– 30	– 40	0	0	0	0	0	0	0	0	0	1
C_j–Z_j	P_2	700	– 30	– 40	0	0	0	0	0	30	0	40	0	0
	P_1	0	0	0	0	1	0	1	0	0	0	0	0	0

(MMS, Mumbai 2001)

25. The Super-Star Company produces three models of microcomputers—SS100, SS150 and SS250. Most of the components are imported from the Far East, and the company only assembles the microcomputers. The operation hours required to produce one unit of the microcomputers are 2 hours for SS100, 3 hours for SS150 and 4 hours for SS250. The normal capacity of the assembly line is 400 hours per month. The profits per unit are Rs 10,000 for SS100, Rs 15,000 for SS150 and Rs 25,000 for SS250. The president of the company has set the following goals, according to their order of importance:

    (i) Avoid the underutilisation of production capacity.

    (ii) Meet the outstanding orders—30 units for SS100, 20 units for SS150 and 50 units for SS250.

    (iii) Avoid the overutilisation of the production capacity.

    (iv) Maximise total profit as much as possible.

Formulate a goal programming model for the problem and solve it (up to three iterations only).

(MBA, Delhi, December, 1994)

26. The Soft-Ad Company is planning for the next month an advertising campaign for one of its clients, Amco. Soft-Ad is considering advertising during sports programmes and during soap operas on TV. Amco has specified the following goals:

    Goal 1 : Reach at least 20 lakh high-income-males (HIM)

    Goal 2 : Reach at least 30 lakh medium-income-families (MIF)

    Goal 3 : Reach at least 15 lakh high-income-females (HIF)

    The information about cost and viewers for each one-minute ad slots is given below:

Ad-slot	*Number of viewers per Ad (in lakhs)*			*Cost per slot (Rs)*
	HIM	*MIF*	*HIF*	
Sports	4	5	2	2,00,000
Soap opera	3	8	4	3,00,000

Further, the company has budgeted Rs 12 lakhs for these advertisements, which should not be exceeded. It has also estimated that each lakh of viewers by which it falls shorts of goal 1 costs it Rs 2 lakhs on account of lost sales. Similarly, this penalty for goals 2 and 3 is Rs 1 and 0.8 lakh respectively. How should Soft-Ad plan for Amco? Formulate and solve this as a goal linear programming problem.

27. A production manager faces the problem of job allocation between two assembly lines. The production rate of assembly line 1 is 10 units per hour, and the production rate of assembly line 2 is 12 units per hour. The normal working period for both lines is 8 hours per day. The production manager has set the following goals for the next day, listed in order of their importance:

    (i) Avoid any underachievement of the production level, which is set at 200 units of product.

    (ii) Avoid any overtime operation of line 2 beyond 4 hours.

    (iii) Avoid any underutilisation of regular working hours (assign differential weights according to the relative productivity of the two lines).

    (iv) Minimise overtime in both assembly lines.

    Formulate a goal programming model and solve it. *(MBA, Delhi, April, 1994)*

CHAPTER

8

Sequencing

CHAPTER OVERVIEW

When a number of jobs are given to be done and they require processing on two or more machines (in the same order, of course) the question arises as to what order, or sequence, they should be processed in. It may seem to be no problem or at best a trivial problem, considering the fact that the time needed for processing all jobs is known with certainty. However, interestingly, the order of processing the jobs does matter and it affects the total amount of time in which all of these can be completed. A manager, therefore, has to decide upon the optimal sequence of the job performance.

This chapter helps the manager to know the answers to questions like the following:

- What is a sequencing problem and why is it a problem?
- How can relatively simple problems be handled graphically?
- What is Johnson's rule and how can it be used to determine the optimal sequence of performing a given set of jobs when they all need to pass through two machines in the same order?
- How to obtain the total time of performing all the jobs, called the total elapsed time, and what is the idle time on the second of the machines on which the jobs are done?
- What are the conditions under which we can determine optimal sequence of a number of jobs to be processed on three machines in a certain given order? What is the methodology for solving such problems?
- How should two maintenance teams be scheduled to work on a number of projects/jobs, on which they are required to operate one after the other?
- How should work be scheduled when two jobs are required to be processed on a given set of machines, but the order of processing of each job is different?

This is probably one of the simplest topics to consider as the nature and scope of problems which can be solved is rather limited. Perhaps the only relatively tricky thing to understand is the calculation of the total time which would be needed in completing all jobs.

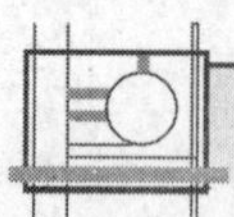

8.1 INTRODUCTION

Sequencing problems arise when we are concerned with situations where there is a choice as to the order in which a number of tasks can be performed. In such cases, the effectiveness is a function of the order or sequence in which the tasks are performed. A sequencing problem could involve jobs in a manufacturing plant, aircraft waiting for landing and clearance, maintenance scheduling in a factory, programmes to be run on a computer centre, customers in a bank, and so forth.

The sequencing problems do get solved, since most of the tasks are performed, the aircraft do land and the bank customers transact business. However, most of these problems are solved casually or automatically without an explicit recognition that a problem even existed, much less than that a solution was obtained. Sometimes an ordering is determined by chance, more often the tasks are performed in the order in which they arise. A common basis evolved out of a sense of fair play is the 'first come first served' basis of solution of sequencing problems. This basis may be appropriate for customers in a bank, but it may not necessarily be applied to inanimate jobs on a factory floor.

We shall consider the sequencing problems in respect of the jobs to be performed in a factory and study the method of their solution. Such sequencing problems can be broadly divided in two groups. In the first one, there are *n* jobs to be done, each of which requires processing on some or all of the *k* different machines. We can determine the effectiveness of each of the sequences that are technologically feasible (that is to say, those satisfying the restrictions on the order in which each job must be processed through the machines) and choose a sequence which optimises the effectiveness. To illustrate, the timings of processing of each of the *n* jobs on each of the *k* machines, in a certain given order, may be given and the time for performing the jobs may be the measure of effectiveness. We shall select the sequence(s) for which the total time taken in processing all the jobs on the machines would be the minimum.

The second type of sequencing problems deal with the situations where we have a system with a certain number of machines and a list of jobs to be done, and every time that a machine completes a job on which it is engaged, we have to decide on the next job to be started.

Our concern in the present chapter is with the *first type* of problems.

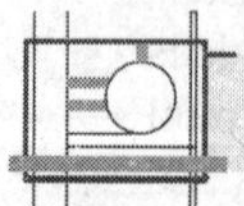

8.2 THE SEQUENCING PROBLEM

The following is a typical sequencing problem. A company has to complete ten jobs in one week. The jobs must be processed on three different machines M_1, M_2 and M_3. They are all processed first on machine M_1, then on M_2, and finally on M_3. The production hours, including processing, set-up and movement time, required for each job on each machine are shown in the table given on the next page.

The objective is to determine the sequence of jobs which will allow all jobs to be performed so that the time, from the beginning of the first job till the completion of the last job, is the minimum. Of course, there may be more than one such sequence in a given case. The answer to the sequencing problem would also indicate to the management information like which of the machine(s) would, if at all, be idle and for how long; whether the work on all the jobs can be done in the given week without overtime, and if not, which jobs will need overtime work.

Job	Production Time (in Hrs)		
	M_1	M_2	M_3
1	2	5	4
2	6	8	2
3	7	6	4
4	3	7	2
5	5	4	4
6	2	1	8
7	6	3	2
8	5	5	5
9	4	3	3
10	2	8	1

In particular, we shall consider the following types of sequencing problems:

(a) Processing *n* jobs on two machines
(b) Processing *n* jobs on three machines
(c) Processing *n* jobs on *k* machines
(d) Processing 2 jobs on *k* machines

8.2.1 The Assumptions

There are some general assumptions made to solve the sequencing problems. These are given here:

(a) The processing times on various machines are independent of the order in which different jobs are processed on them.
(b) The time taken by different jobs in going from one machine to another is negligible.
(c) A job once started on a machine would be performed to the point of completion uninterrupted.
(d) A machine cannot process more than one job at a given point of time.
(e) A job would start on a machine as soon as the job and the machine on which it is to be processed are both free.

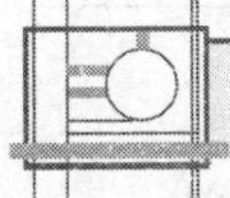

8.3 SOLUTION TO SEQUENCING PROBLEMS

Sequencing problems can be solved using the *Gantt Chart* and by applying an algorithm. First, we shall consider the Gantt Charts which are used for handling relatively simpler problems involving two machines.

8.3.1 Gantt Charts

We shall illustrate the use of the Gantt Charts with the help of the following examples.

Example 8.1 Suppose that there are two jobs J_1 and J_2, each requiring work on two machines M_1 and M_2, in this order, with the required processing times given as follows:

Processing Time (Hours)

Job	Machine	
	M_1	M_2
J_1	4	10
J_2	8	7

What order of performance of the jobs will involve the least time?

For processing the two jobs, two orders are possible J_1–J_2 and J_2–J_1. Both the alternatives are evaluated in Figure. 8.1. The first part of the figure depicts the job sequences J_1–J_2. According to this sequence, the two jobs can be completed in 21 hours. The other sequence J_2–J_1, is shown in the second part wherein it is clear that the two jobs would take 25 hours to finish.

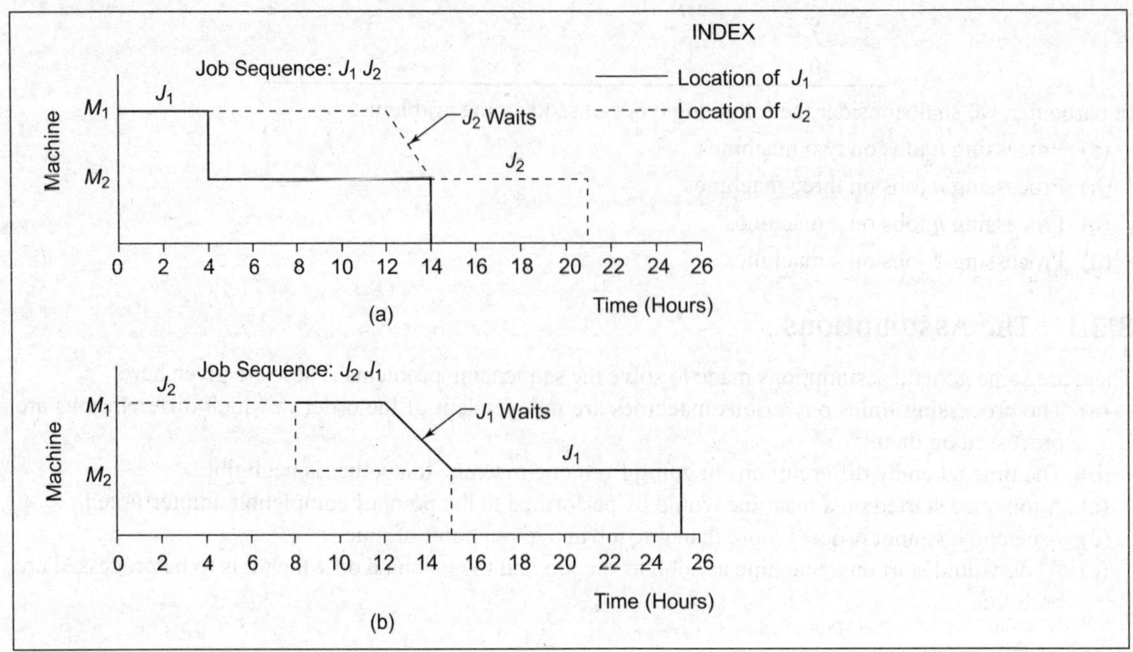

Figure 8.1 *Gantt Chart*

Clearly then, the job sequence J_1–J_2 is optimal, since the total elapsed time, T, is the lower of the two.

Example 8.2 Three jobs, 1, 2 and 3, are to be processed on two machines M_1 and M_2 in the order $\overrightarrow{M_1, M_2}$ for which the processing times are given below.

Processing Time (Hours)

Job	Machine	
	M_1	M_2
1	3	5
2	6	4
3	5	8

In what sequence should the jobs be performed so that the elapsed time, T, is the least?

With three jobs to be processed on two machines, we have six possible orders. These are 123, 132, 213, 231, 312 and 321. Each of these sequences is evaluated in the chart shown in Figure 8.2. Clearly, the sequence 132 takes the minimum processing time in aggregate, that is 20 hours, and hence is optimal.

The use of Gantt Chart becomes disproportionately difficult as the number of jobs increases. In fact, when there are n jobs, each of which is to be processed over two machines in the same order, there are $n!$ possible sequences. When there are 6 jobs to be processed, we shall have to evaluate $6! = 720$ possible orders in an attempt to find the optimal one and for 8 such jobs, the number of sequences would be $8! = 40{,}320$. Hence, we need a more efficient method by which the sequence desired can be determined.

8.3.2 Algorithm for Solving Sequencing Problems: Johnson's Rule

The method shall be discussed in respect of processing of n jobs through (a) two machines, (b) three machines, and (c) k machines.

Processing n jobs through 2 machines As mentioned earlier, we have the following information:

(a) Only two machines are involved.

(b) Each job is to be processed in the order AB so that first the work would be performed on machine A and then on machine B.

(c) The processing times for different jobs on first machine $A_1, A_2, ..., A_n$ are given and so are the processing times on the second machine $B_1, B_2, ..., B_n$.

With the objective to determine the sequence in which the jobs may be performed so that the total time taken (i.e. the total elapsed time) is the minimum, we proceed as follows. Let

A_i = time required by job i on machine A;

B_i = time required by job i on machine B;

T = total elapsed time for jobs 1, 2, ..., n; and

t_i = idle time on machine B, from end of job $i - 1$ to start of job i.

The objective is to determine the sequence S in which the n jobs may be performed so that T is minimised. We know that the total elapsed time, T, is determined by the point of time at which the first job goes on machine A and the point of time on which the last job comes off machine B. At any moment of time, the machine B is either working or idle. The total time for which the machine B has to work is $\sum_{i=1}^{n} B_i$ (which is determined by the given job times and has nothing to do with the sequence in which they are performed). Thus,

$$T = \sum_{i=1}^{n} B_i + \sum_{i=1}^{n} t_i$$

Because the first of the terms of the RHS of this equation is a constant, the problem is to minimise the second of the terms, the total idle time on machine B. If $D_n(S)$ represents the total idle time on machine B for sequence S, the problem is that of finding a sequence S^* of jobs (1, 2, ..., n) such that $D_n(S^*) \leq D_n(S_0)$ for any S_0. Without proof, it may be stated that an optimal sequence is obtained by using the following rule.

Job J precedes job $J + 1$ if $\text{Min}(A_j, B_{J+1}) < \text{Min}(A_{J+1}, B_J)$.

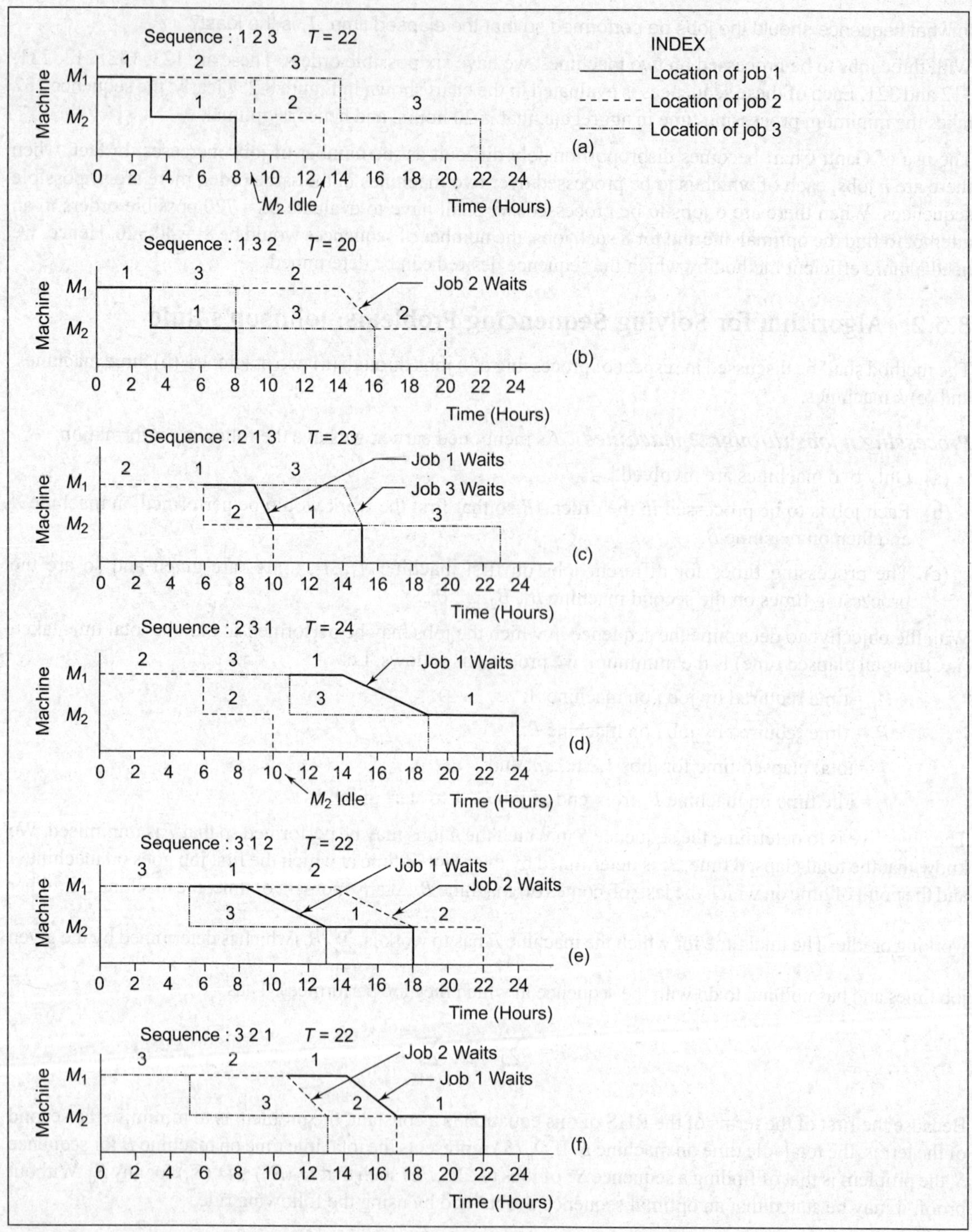

Figure 8.2 *Gantt Chart*

For example, suppose there are two jobs whose times are: *Job 1*: Machine *A* 4 hours, Machine *B* 10 hours; *Job 2*: Machine *A* 8 hours, and Machine *B* 7 hours (same as in Example 8.1). Thus, $A_1 = 4$, $A_2 = 10$, $B_1 = 8$, and $B_2 = 7$. Here, minimum of $A_1, B_2 (= 4, 7)$ is 4 and minimum of $A_2, B_1 (= 10, 8)$ is 8. Now, since Min (A_1, B_2) is less than Min (A_2, B_1) in this case, it is clear that job 1 should precede job 2. If this condition were not satisfied, job 2 would have preceded job 1.

Further, if Min (A_j, B_{j+1}) = Min (A_{j+1}, B_j), then job j would be indifferent to job $j + 1$; any of them could precede the other.

This rule can be extended to find the optimal sequence of a set of given jobs. Starting with any sequence S_0, the optimal sequence S^* can be obtained by successive interchange of consecutive jobs by applying this rule.

In operational terms, the algorithm can be stated in a stepwise manner as follows:

Step 1 Select the smallest processing time, considering $A_1, A_2, ..., A_n$ and $B_1, B_2, ..., B_n$ together.

Step 2

(a) If the minimum is for A_r, that is to say, for the *r*th job on machine *A*, do the *r*th job the first.
(b) If the minimum is for B_s, that is, for the *s*th job on machine *B*, do the *s*th job in the end.
(c) In case of a tie between A_r and B_s, perform the *r*th job first and the *s*th job the last.
(d) If there is a tie between 2 or more timings in either of the series, select either of the jobs involved and perform the first or last, accordingly as the tie is in $A_1, A_2, ..., A_n$, or $B_1, B_2, ..., B_n$.

Step 3 After the job(s) has/have been assigned, apply steps 1 and 2 to the reduced set of processing times obtained by deleting the machine times corresponding to the job(s) already assigned.

Step 4 Continue in this manner until all jobs are assigned.

The sequence of jobs to be performed obtained in this way shall be optimal, involving the least aggregate time for completion of the jobs.

Example 8.3 In a factory, there are six jobs to perform, each of which should go through two machines *A* and *B*, in the order *AB*. The processing timings (in hours) for the jobs are given here. You are required to determine the sequence for performing the jobs that would minimise the total elapsed time, *T*. What is the value of *T*?

Job	*Machine A*	*Machine B*
1	7	3
2	4	8
3	2	6
4	5	6
5	9	4
6	8	1

(a) The least of all the times given in the table is for job 6 on machine *B*. So, perform job 6 in the end. It is last in the sequence. Now delete this job from the given data.
(b) Of all timings now, the minimum is for job 3 on machine *A*. So, do the job 3 first.
(c) After deleting job 3 also, the smallest time of 3 hours is for job 1 on machine *B*. Thus, perform job 1 in the end (before job 6).

(d) Having assigned job 1, we observe that the smallest value of 4 hours is shared by job 2 on machine *A* and job 5 on machine *B*. So, perform job 2 first and job 5 in the end.

(e) Now, the only job remaining is job 4, it shall be assigned the only place left in the sequence. The resultant sequence of jobs is, therefore, as follows:

3 2 4 5 1 6

This sequence is the optimal one. The total elapsed time, *T*, is obtained in Table 8.1 as equal to 36 hours.

TABLE 8.1 Calculation of Total Elapsed Time (T)

Job	*Machine A*		*Machine B*	
	In	*Out*	*In*	*Out*
3	0	2	2	8
2	2	6	8	16
4	6	11	16	22
5	11	20	22	26
1	20	27	27	30
6	27	35	35	36

As shown in this table, the first job, job 3, starts at time 0 on the machine *A* and is over by time 2, when it passes to machine *B* to be worked on till time 8. The job 2 starts on the machine *A* at time 2 as the machine is free at that time. It is completed at time 6 and has to wait for 2 hours before it is processed on machine *B*, starting at time 8 when this machine is free. Similarly, the various jobs are assigned to the two machines and the *in* and *out* times are obtained.

Example 8.4 Ten jobs are required to be processed on two machines M_1 and M_2 in the order 'M_1 followed by M_2'. Processing times are given here. Determine the optimal sequence(s) and evaluate for total elapsed time.

Processing Time (hours)

Job	:	J_1	J_2	J_3	J_4	J_5	J_6	J_7	J_8	J_9	J_{10}
Machine M_1	:	7	8	10	3	7	4	5	8	5	6
Machine M_2	:	4	2	6	6	5	7	2	6	7	6

(a) Minimum time is 2 hours each for jobs J_2 and J_7 on machine M_2. So rank J_7 in the end, preceded by J_2. Else, put J_2 in the end, preceded by J_7. Delete J_2 and J_7.

(b) From the remaining, minimum time is 3 hours for J_4 on machine M_1. So do J_4 first.

(c) Now, the minimum time is 4 hours for job J_1, on machine M_2 and for job J_6 on machine M_1. Therefore, order J_6 at the first and J_1 at the last available places.

(d) Having deleted jobs J_4, J_6, J_1, J_2 and J_7, the minimum time is 5 hours on machine M_1 in respect of the job J_9. So order J_9 immediately after J_6.

(e) In the jobs remaining, the least time is 6 hours for the job J_{10} on machines M_1 and M_2 each, for job J_3 on machine M_2 and the job J_8 on machine M_2. Thus, rank J_{10} and J_8 is the end preceded or, alternately, succeeded by job J_3.

We have the following alternate sequences, referred to as S_1, S_2, S_3 and S_4.

S_1	J_4	J_6	J_9	J_{10}	J_3	J_8	J_5	J_1	J_2	J_7
S_2	J_4	J_6	J_9	J_{10}	J_3	J_8	J_5	J_1	J_7	J_2
S_3	J_4	J_6	J_9	J_{10}	J_8	J_3	J_5	J_1	J_2	J_7
S_4	J_4	J_6	J_9	J_{10}	J_8	J_3	J_5	J_1	J_7	J_2

We shall now evaluate first of the sequences for determining the value of T. This is done in Table 8.2.

TABLE 8.2 Determination of T

Job	*Machine A*		*Machine B*	
	In	*Out*	*In*	*Out*
J_4	0	3	3	9
J_6	3	7	9	16
J_9	7	12	16	23
J_{10}	12	18	23	29
J_3	18	28	29	35
J_8	28	36	36	42
J_5	36	43	43	48
J_1	43	50	50	54
J_2	50	58	58	60
J_7	58	63	63	65

The total elapsed time (T) is 65 hours for this sequence. Other sequences can also be evaluated similarly.

Processing n jobs through 3 machines No general solution is available (except enumeration, of course) for the sequencing problems of processing n jobs through three machines. However, we do have a method applicable under the condition that no passing of jobs is permissible (that is to say, the same order over each machine is maintained) and if *either* or *both* the following conditions is/are satisfied.

For three machines A, B, and C on which some jobs are to be performed in the order ABC.

Condition 1

The minimum of the times for different jobs on machine *A* is at least equal to the maximum of the times of different jobs on machine *B*.

Condition 2

The minimum of the times of various jobs on machine *C* is at least equal to the maximum of the times of the different jobs on machine *B*.

If Min A_i represents the least of the job times on machine *A*, Min C_i represents the least of the job times on machine *C*, and Max B_i indicates the largest of the job times on machine *B*, then if *either* Min $A_i \geq$ Max B_i or Min $C_i \geq$ Max B_i, or *both* are satisfied then the following method can be applied.

The Method

Step 1 Replace the given problem with an equivalent problem involving *n* jobs and 2 fictitious machines *G* and *H*, and define the corresponding processing times G_i and H_i as follows:

$$G_i = A_i + B_i \qquad (i = 1, 2, ..., n), \text{ and}$$
$$H_i = B_i + C_i \qquad (i = 1, 2, ..., n)$$

Thus, for the job 1, $G_1 = A_1 + B_1$, $H_1 = B_1 + C_1$; for the job 2, $G_2 = A_2 + B_2$, $H_2 = B_2 + C_2$ ... and so on.

Step 2 To the problem obtained in step (1), apply the method discussed for processing *n* jobs through 2 machines. The optimal sequence resulting from this shall also be optimal for the given (original) problem.

The schematic of the method is given in Fig. 8.3.

Example 8.5 You are given the following data regarding the processing times of some jobs on three machines, *I*, *II* and *III*. The order of processing is *I*-*II*-*III*. Determine the sequence that minimises the total elapsed time (*T*) required to complete the jobs. Also evaluate *T* and the idle time of *II* and *III*.

Job	Processing Time (Hours)		
	Machine		
	I	*II*	*III*
A	3	4	6
B	8	3	7
C	7	2	5
D	4	5	11
E	9	1	5
F	8	4	6
G	7	3	12

According to the given information,

$$\text{Min } I_i = 3$$
$$\text{Max } II_i = 5; \qquad \text{and}$$
$$\text{Min } III_i = 5$$

Clearly, since Min III_i = Max II_i, the second of the conditions specified is met. Now, we can solve the problem as follows.

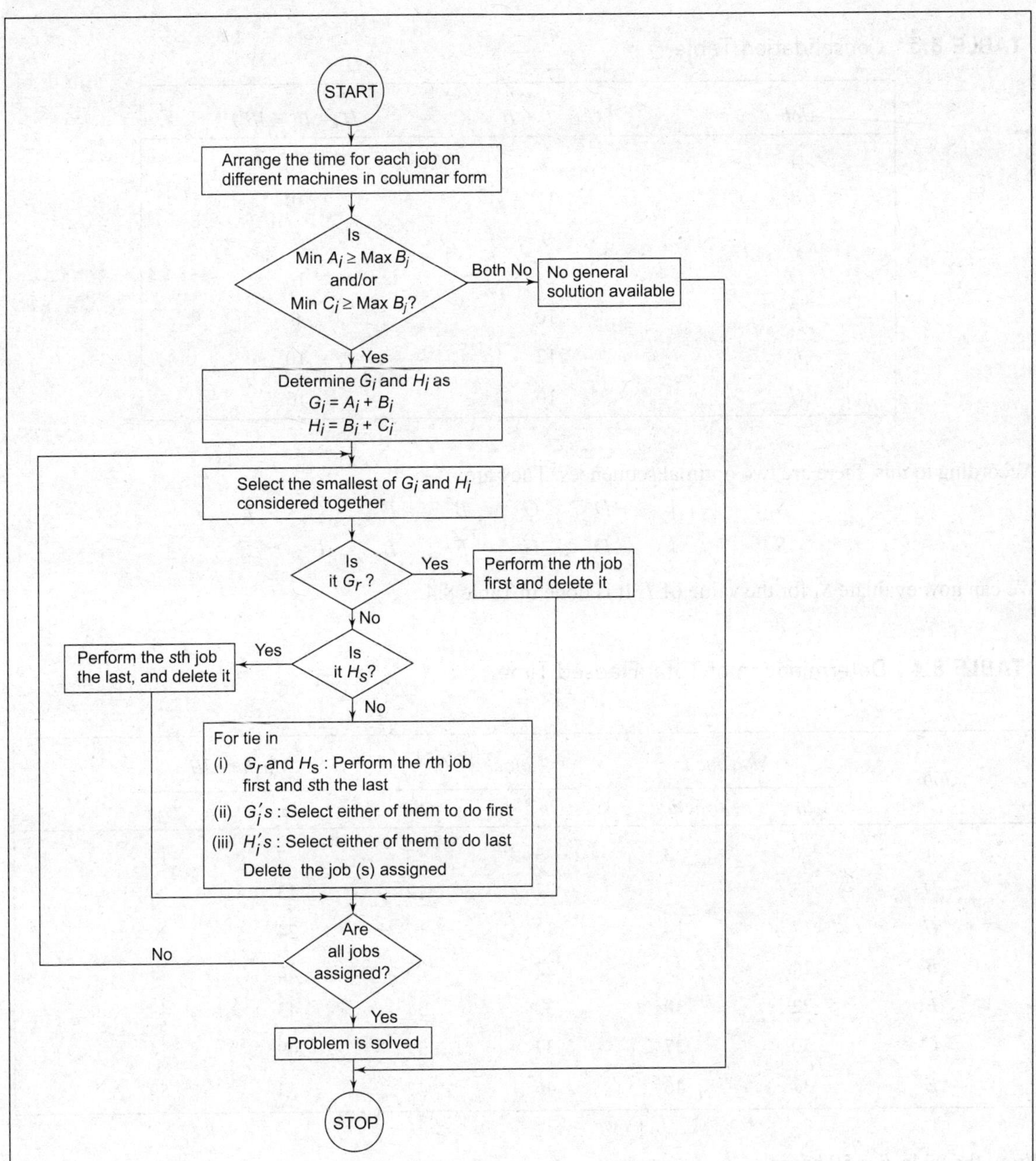

Figure 8.3 *Schematic Presentation of the Algorithm for Processing n Jobs through 3 Machines A, B and C in the Order ABC*

First the consolidation table is drawn as shown in Table 8.3.

TABLE 8.3 Consolidation Table

Job	$G_i (= I_i + II_i)$	$H_i (= II_i + III_i)$
A	7	10
B	11	10
C	9	7
D	9	16
E	10	6
F	12	10
G	10	15

According to this, there are two optimal sequences. They are:

S_1 : *A* *D* *G* *B* *F* *C* *E*

S_2 : *A* *D* *G* *F* *B* *C* *E*

We can now evaluate S_1 for the value of *T*. It is done in Table 8.4.

TABLE 8.4 Determination of Total Elapsed Time

Job	*Machine I*		*Machine II*		*Machine III*	
	In	*Out*	*In*	*Out*	*In*	*Out*
A	0	3	3	7	7	13
D	3	7	7	12	13	24
G	7	14	14	17	24	36
B	14	22	22	25	36	43
F	22	30	30	34	43	49
C	30	37	37	39	49	54
E	37	46	46	47	54	59

From the table, $T = 59$ hours.

The idle time for various machines is given as follows:

Machine II : 3 + 2 + 5 + 5 + 3 + 7 + 12 = 37 hours

(0–3) (12–14) (17–22) (25–30) (34–37) (39–46) (47–59)

Machine III : 7 hours (0–7)

Processing *n* jobs on *k* machines There is no general method available by which we can obtain optimal sequence(s) in problems involving processing of *n* jobs on *k* machines. They can be handled only by enumeration, which is a very lengthy and time consuming exercise because a total of $(n!)^k$ different sequences would require consideration in such a case. However, like in case of processing *n* jobs over three machines, if *either* or *both* of the conditions stipulated below are met with, we can apply the algorithm.

Let there be *n* jobs to be processed on *k* machines, *A*, *B*, ..., *J*, *K* in the order *ABC* ... *JK*, then we have

Condition 1

Min $A_i \geq$ Max B_i, Max C_i, ..., and Max J_i

That is, the minimum of the job times on machine *A* should be at least equal to the maximum of the job times on machine *B*, machine *C*, ..., and machine *J*. Similarly,

Condition 2

Min $K_i \geq$ Max B_i, Max C_i, ... and Max J_i

The algorithm works as follows:

Step 1 Obtain timings on two fictitious machines *X* and *Y* such that

$$X_i = A_i + B_i + ... + J_i; \text{ and}$$
$$Y_i = B_i + C_i + ... + J_i + K_i$$

Step 2 Determine the optimal sequence of performance of the jobs on *X* and *Y*, in accordance with the rules given earlier in respect of processing *n* jobs on two machines. The resulting sequence shall be optimal for the given problem.

Example 8.6 Determine the optimal sequence of performing 5 jobs on 4 machines. The machining of each job is required in the order *ABCD* and the process timings are as follows:

Job	*Machine*			
	A	*B*	*C*	*D*
1	8	3	4	7
2	9	2	5	5
3	6	4	5	8
4	12	5	1	9
5	7	1	2	3

Here Min $A_i = 6$, Max $B_i = 5$, Max $C_i = 5$ and Min $D_i = 3$.

Also since Min $A_i >$ Max B_i, Max C_i the first of the conditions is satisfied. Thus, we can apply the algorithm. First, we prepare the consolidation table as given in Table 8.5.

TABLE 8.5 Consolidation Table

Job	$X_i = A_i + B_i + C_i$	$Y_i = B_i + C_i + D_i$
1	15	14
2	16	12
3	15	17
4	18	15
5	10	6

According to the information in the consolidation table, the optimal sequence of jobs would be as follows:

3 4 1 2 5

This sequence would involve a total completion time of 50 hours, as shown in the Table 8.6.

TABLE 8.6 Calculation of Total Elapsed Time (T)

Job	*Machine A*		*Machine B*		*Machine C*		*Machine D*	
	In	*Out*	*In*	*Out*	*In*	*Out*	*In*	*Out*
3	0	6	6	10	10	15	15	23
4	6	18	18	23	23	24	24	33
1	18	26	26	29	29	33	33	40
2	26	35	35	37	37	42	42	47
5	35	42	42	43	43	45	47	50

Processing 2 jobs through *k* machines Now we shall consider a case where two jobs are to be processed on *k* machines and the two require different order in which they could be processed on these machines.

Let the machines be $A, B, ..., K$ and the jobs to be processed as J_1 and J_2. The processing times are known to be $A_1, A_2, B_1, B_2, ..., K_1$ and K_2. The objective here is to find the order of performing the jobs on the different machines that will minimise the total elapsed time, assuming that no alternate sequencing of machining is permissible for either job, and each machine can work only on a single job at a time.

The optimal sequence can be determined using the graphic method. On a two-axis graph, one job is represented horizontally and the other vertically. The sequence (of machines) in which job J_1 is to be performed is taken on *X*-axis while the sequence in which job J_2 is required to be done is depicted on *Y*-axis. The time is scaled on the two axes. Then areas are identified and marked in respect of each machine considering the time at which it will be used for jobs J_1 and J_2. In a way, these areas would represent overlaps which should be avoided because of our assumption that only one job can be handled by a machine at a given point in time. After shading these areas, we mark the working on the two jobs. A horizontal movement on the graph, shown by horizontal line, shall represent work on job J_1 while job J_2 remains idle, whereas a vertical movement shall mean work on job J_2 while job J_1 waits to be worked on. A 45° line would represent simultaneous work on both the jobs. The shortest line, composed of combinations of the 45° line, the horizontal and the vertical lines, from the origin (i.e. the start) to the goal (i.e. the end of the jobs) would represent the optimal sequence.

Example 8.7 Using graphical method, determine the minimum time needed to process the two jobs on six machines. The information about the machine sequence and the time required by each job on each machine is given here.

Job 1		Job 2	
Machine Sequence	*Time (Hours)*	*Machine Sequence*	*Time (Hours)*
A	4	*B*	6
B	5	*A*	3
C	1	*C*	2
D	3	*F*	4
E	6	*D*	3
F	5	*E*	5

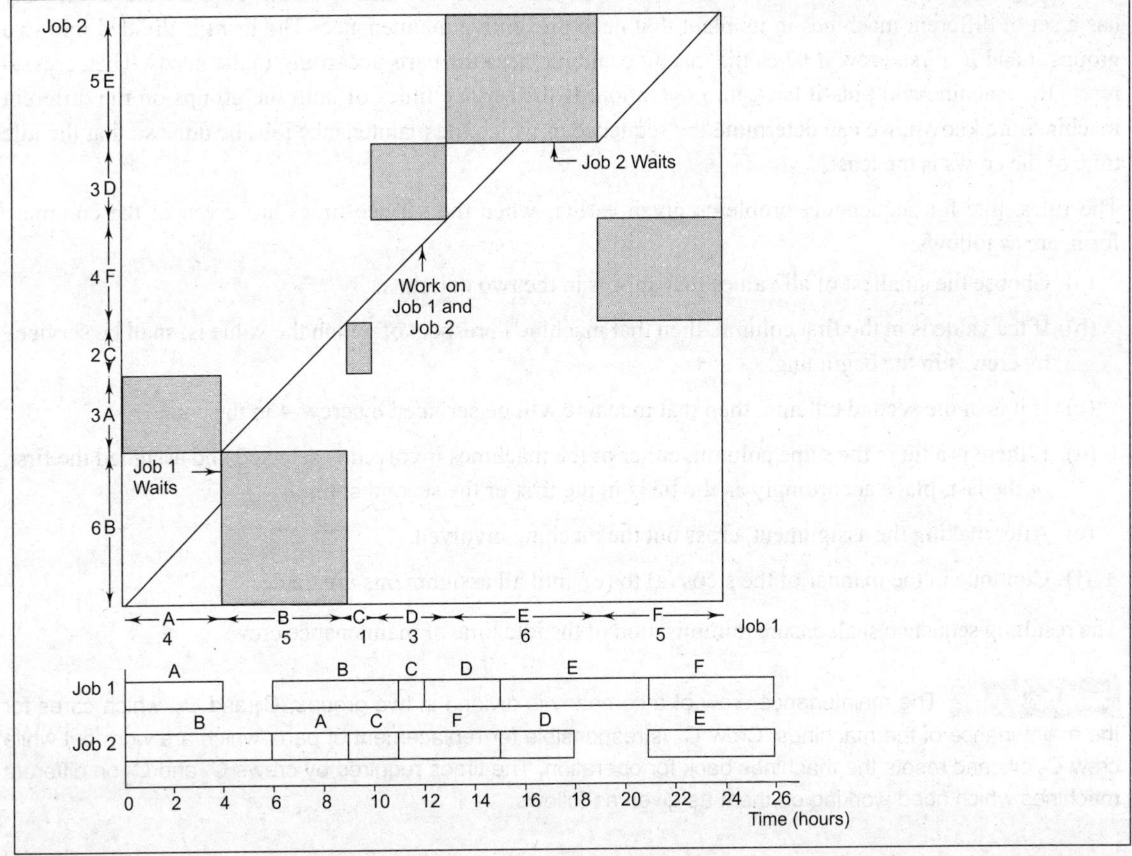

Figure 8.4 *Graphic Solution of the 2-jobs K-machines Problem*

The given information is depicted in Fig. 8.4 in which crosshatched rectangular blocks represent the overlaps (or clashes) which are to be avoided. The thick line in the figure depicts the work on the jobs. Insofar as possible, we move diagonally, which represents work on both the jobs simultaneously.

In the figure, it is clear that the work can go on both the jobs together for the first four hours after which job 1 has to wait for two hours since the machine required for its processing further is not available. After this, both jobs can be simultaneously processed for the next 12 hours, after which job 2 has to wait for 3 hours to get the machine needed for processing it. After this, both the jobs can be processed simultaneously till each one finishes.

Work on the two jobs is shown by Gantt chart in the lower (b) part of the figure. The total time required to complete is 26 hours: the minimum.

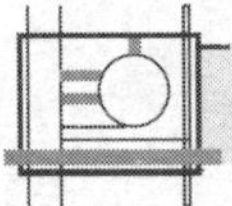

8.4 MAINTENANCE CREW SCHEDULING—AN APPLICATION

The method of solving sequencing problems relating to the scheduling of jobs in a factory can be extended to scheduling the factory maintenance crews in such a way as to minimise their idle time. Suppose that a company has a set of different machines in its plant that need preventive maintenance. The team is divided into two groups *A* and *B*. First, crew *A* takes the machine and replaces the parts according to the needs. Then crew *B* resets the machine and puts it back into operation. If the service times of both the groups on the different machines are known, we can determine the sequence in which the maintenance jobs be done so that the idle time of the crews is the least.

The rules, like for sequencing problems given earlier, when the service times are given in the columnar form, are as follows:

(a) Choose the smallest of all values that appear in the two columns.

(b) If the value is in the first column, then that machine in respect of which the value is, shall be serviced by crew *A* in the beginning.

(c) If it is in the second column, then that machine will be serviced by crew *A* in the end.

(d) If there is a tie in the same column, either of the machines involved is selected and assigned the first, or the last, place accordingly as the tie is in the first or the second column.

(e) After making the assignment, cross out the machine involved.

(f) Continue in the manner of the steps (a) to (e) until all assignments are made.

The resulting sequence shall ensure minimisation of the idle time of maintenance crew.

Example 8.8 The maintenance crew of a company is divided in two groups, C_1 and C_2, which cares for the maintenance of the machines. Crew C_1 is responsible for replacement of parts which are worn out while crew C_2 oils and resets the machines back for operation. The times required by crews C_1 and C_2 on different machines which need working on them as given as follows:

Machine	Maintenance Time (Hrs) Crew C_1	Crew C_2
M_1	8	5
M_2	6	3
M_3	10	7
M_4	11	12
M_5	10	8
M_6	14	6
M_7	4	7

In what order should the machines be handled by crews C_1 and C_2 so that the total time taken is minimised?

In accordance with the rules given earlier, the order in which machines should be handled is as follows:

$$M_7 \quad M_4 \quad M_5 \quad M_3 \quad M_6 \quad M_1 \quad M_2$$

The total elapsed time is obtained in Table 8.7.

TABLE 8.7 Determination of Total Elapsed Time

Machine	Crew C_1		Crew C_2	
	Start	Finish	Start	Finish
M_7	0	4	4	11
M_4	4	15	15	27
M_5	15	25	27	35
M_3	25	35	35	42
M_6	35	49	49	55
M_1	49	57	57	62
M_2	57	63	63	66

The total time required by the maintenance crew to handle the seven machines is 66 hours. The working and the idle hours of the crews C_1 and C_2 are shown in Fig. 8.5.

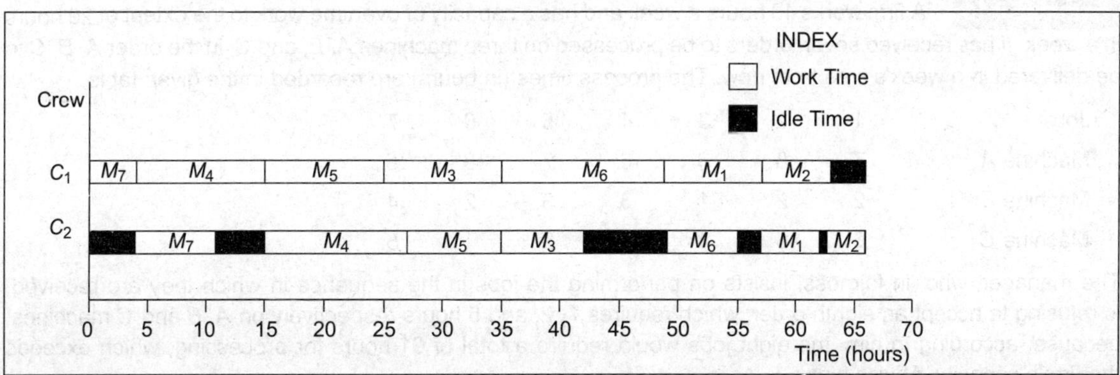

Figure 8.5 *Gantt Chart Showing Working of Crew*

REVIEW ILLUSTRATIONS

Example 8.9 A company plans to fill six positions. Since the positions are known to vary considerably with respect to skill and responsibility, different types of aptitude tests and interviews are required for each. While the aptitude tests are conducted by people from the clerical positions, the job interviews are held by the personnel from the management cadre. The job interviews immediately follow the aptitude test. The time required (in minutes) by each of the positions is given here:

Position	:	P_1	P_2	P_3	P_4	P_5	P_6
Aptitude Test	:	140	180	150	200	170	100
Job Interview	:	70	120	110	80	100	90

If it is desired to minimise the waiting time of the management personnel, in what order the position filling be handled?

From the given information, the optimal sequence can be determined using the algorithm. This would be: P_2, P_3, P_5, P_6, P_4 and P_1. The total elapsed time T is equal to 1010 minutes, as calculated in Table 8.8, while the idle time for the management personnel would be: $180 + 30 + 60 + 110 + 60 = 440$ minutes.

TABLE 8.8 Calculation of Total Elapsed Time T

Position	*Aptitude Test*		*Job Interview*	
	Start	*Finish*	*Start*	*Finish*
P_2	0	180	180	300
P_3	180	330	330	440
P_5	330	500	500	600
P_6	500	600	600	690
P_4	600	800	800	880
P_1	800	940	940	1,010

Example 8.10 A firm works 40 hours a week and has a capacity of overtime work to the extent of 20 hours in a week. It has received seven orders to be processed on three machines *A*, *B*, and *C*, in the order *A*, *B*, *C* to be delivered in a week's time from now. The process times (in hours) are recorded in the given table.

Job	:	1	2	3	4	5	6	7
Machine *A*	:	7	8	6	6	7	8	5
Machine *B*	:	2	2	1	3	3	2	4
Machine *C*	:	6	5	4	4	2	1	5

The manager, who, in fairness, insists on performing the jobs in the sequence in which they are received, is refusing to accept an eighth order, which requires 7, 2, and 5 hours respectively on *A*, *B* and *C* machines, because, according to him, the eight jobs would require a total of 61 hours for processing, which exceeds the firm's capacity. Advise him.

The processing of 8 jobs according to the manager's plan will indeed take 61 hours. It needs to be examined, however, if this plan is optimal.

To obtain the optimal sequence, first the timings of the processing of all the eight jobs would be tabulated as given in Table 8.9.

TABLE 8.9 Processing Time (Hours)

Job	*Machine A*	*Machine B*	*Machine C*
1	7	2	6
2	8	2	5
3	6	1	4
4	6	3	4
5	7	3	2
6	8	2	1
7	5	4	5
8	7	2	5

From the table,

$$\text{Min } A_i = 5, \text{ Max } B_i = 4, \text{ and Min } C_i = 1.$$

Since Min $A_i >$ Max B_i, the first of the conditions laid previously is satisfied. We proceed now to make the consolidation table, as shown in Table 8.10.

TABLE 8.10 Consolidation Table

Job	$G_i = A_i + B_i$	$H_i = B_i + C_i$
1	9	8
2	10	7
3	7	5
4	9	7
5	10	5
6	10	3
7	9	9
8	9	7

According to this, several optimal sequences are possible, of which one is as follows:

7, 1, 8, 4, 2, 5, 3, 6.

The total elapsed time T can be obtained as shown in the Table 8.11. It equals 57 hours. Thus, the jobs can be processed within the given capacity level.

TABLE 8.11 Determination of Total Elapsed Time, T

Job	Machine A		Machine B		Machine C	
	In	Out	In	Out	In	Out
7	0	5	5	9	9	14
1	5	12	12	14	14	20
8	12	19	19	21	21	26
4	19	25	25	28	28	32
2	25	33	33	35	35	40
5	33	40	40	43	43	45
3	40	46	46	47	47	51
6	46	54	54	56	56	57

Example 8.11 Find the sequence that minimises the total elapsed time required (T) in completing the following jobs. Each job is processed in the order ABC. Also, calculate T.

Job	:	1	2	3	4	5	6	7
Machine A	:	10	8	12	6	9	11	9
Machine B	:	6	4	6	5	3	4	2
Machine C	:	8	7	5	9	10	6	5

(M Com, Delhi, 2008)

From the given information,

$$\text{Min } A_i = 6, \text{ Max } B_i = 6, \text{ and Min } C_i = 5.$$

Since Min $A_i \geq$ Max B_i, we can solve the problem as a 2-machine problem using Johnson's rule. Setting $G_i = A_i + B_i$ and $H_i = B_i + C_i$, we get

Job	G_i	H_i
1	16	14
2	12	11
3	18	11
4	11	14
5	12	13
6	15	10
7	11	7

Considering times on G and H, the optimal sequence is: 4-5-1-2-3-6-7 or 4-5-1-3-2-6-7. The total elapsed time is shown calculated in Table 8.12. It equals 72.

TABLE 8.12 Determination of Total Elapsed Time, T

Job	Machine A		Machine B		Machine C	
	In	Out	In	Out	In	Out
4	0	6	6	11	11	20
5	6	15	15	18	20	30
1	15	25	25	31	31	39
2	25	33	33	37	39	46
3	33	45	45	51	51	56
6	45	56	56	60	60	66
7	56	65	65	67	67	72

KEY POINTS TO REMEMBER

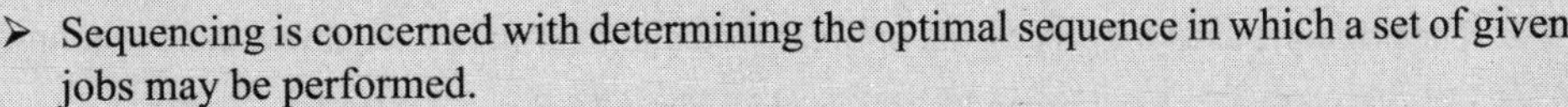

- Sequencing is concerned with determining the optimal sequence in which a set of given jobs may be performed.
- The problems dealt with are basically those where *n* jobs are to be processed on two machines—first on the first machine and then on the second machine. The optimal sequence is the one which results in minimum total time in which the jobs may be done and minimum idle time on the second machine.
- Gantt charts can be used to determine optimal sequence in relatively small-sized problems.
- In terms of the algorithm used, we locate the minimum processing time from all the jobs, do the job first if the minimum time happens to be on the first machine (*A*) and do it at the end if it is on second machine (*B*). Select either in case of a tie. Once the optimal sequence is obtained, total elapsed time, *T*, is obtained by considering the performance of jobs in that order using the given times.
- Processing of *n* jobs on three machines is also considered in some limited cases where Min $A_i \geq$ Max B_i and/or Min $C_i \geq$ Max B_i. In such cases, the problem is converted into a 2-machine problem by creating hypothetical machines *G* and *H*, where $G_i = A_i + B_i$ and $H_i = B_i + C_i$. The optimal sequence is then found as for a 2-machine problem and *T* is then determined by using times on three machines considering the performance of jobs in that order using the given times.
- In case of two jobs needing processing on some machines in varying order, the problem may be solved graphically, by first plotting overlappings on different machines on the graph and then determining work on the jobs through horizontal, vertical and diagonal lines drawn to connect the lower left corner and the upper right corner of the graph.

TEST YOUR UNDERSTANDING

Mark the following statements as T (*True*) or F (*False*).

1. In sequencing problems, the effectiveness is a function of the order in which the given tasks are performed.
2. First-come-first-served basis of job performance is the ideal one since it involves a sense of fair play.
3. In sequencing problems involving n jobs and 3 machines, the algorithm is based on the assumption that each of the jobs requires processing on three machines in the same order.
4. The optimal sequence for an n-job, 2-machine problem implies that the sequence of performance of jobs which would ensure their completion in the minimum time as well as minimisation of idle time on the second machine.
5. Gantt Chart provides a very effective technique for solving large-sized sequencing problems.
6. Ten jobs required to be performed on three machines A, B and C, in that order, can be performed in 3,628,800 ways.
7. Total elapsed time is determined by the point of time at which the first of the n jobs goes to machine A until the point on which the last job comes off machine B.
8. Since the times that the different jobs take on different machines are known and constant, it makes no difference as to in what order are they performed.
9. The principle used for processing optimally certain jobs on two machines can be summarised as follows:
   Job j precedes job $j + 1$ if Min $(A_j, B_{j+1}) <$ Min (A_{j+1}, B_j).
10. In a two-machine sequencing problem, if the lowest times are both on the second machine and they are tied for jobs, say 2 and 7, then the job 2 should be done in the end while job 7 in the first place.
11. Every sequencing problem must have a unique optimal solution.
12. There is no general solution available for solving sequencing problems involving processing of n jobs through 3 machines.
13. In an n-job, 3-machine sequencing problem, if the minimum times on the first and the third machines are both equal to or greater than the maximum time on the second machine, then the problem can be solved by a standard method.
14. If Min $A_i \geq$ Max B_i or Min $C_i \geq$ Max B_i or both, then the problem is replaced by an equivalent problem involving n jobs and 2 fictitious machines G and H, where $G_i = A_i + B_i$ and $H_i = B_i + C_i$.
15. If an n-job, 3-machine sequencing problem is solved as a two-machine problem, then the total elapsed time, T, whether obtained from the 3 machines or the two fictitious machines data would be the same.
16. If some jobs are processed on K machines in the order $A, B, C, \ldots, K$, then a standard method can be employed when Min $A_i \geq$ Max $B_i \geq$ Max $C_i \geq \ldots \geq$ Max J_i and/or Min $K_i \geq$ Max $B_i \geq$ Max $C_i \geq \ldots \geq$ Max J_i.
17. Sequencing problems involving processing of 2 jobs on K machines can be solved graphically.
18. In case of sequencing problems involving 2 jobs on K machines, it is necessary that the processing of the two jobs must be in the same order.
19. A 45-degree line on a graph, where two jobs are shown on the two axes, indicates simultaneous work on the jobs.

20. Maintenance crew scheduling problems can also be programmed as sequencing problems.

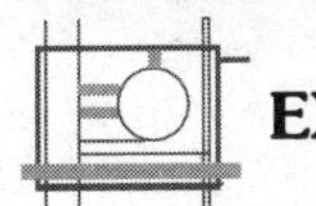

EXERCISES

1. (a) What is a sequencing problem? Explain and illustrate.
   (b) What assumptions are generally made in solving sequencing problems?
2. Give the various sequencing models that are available for solving sequencing problems. Give suitable examples.
3. How would you use Gantt chart for solving sequencing problems? Why is it not employed for solving larger problems?
4. How does the sequencing technique help the manager? Draw a flow chart to show the method of solution of sequencing problems.
5. Explain the general method for solving sequencing problem where *n* jobs are required to be processed on a pair of machines, with the machining to be done on all the jobs in the same order.
6. State the condition(s) to be satisfied in order to apply the algorithm used in solving sequencing problems involving processing of *n* jobs on 2 machines, to the problems involving processing of *n* jobs on 3 machines. Also explain the method of solution in such cases.

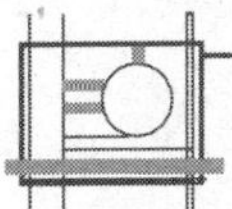

PRACTICAL PROBLEMS

1. The following are the timings in regard to two jobs J_1 and J_2, each of which requires to be processed on two machines *A* and *B* in the order '*A* following *B*'. In what sequence should they be performed so that the total processing time involved is the least? Also obtain the time involved.

Job	*Time (hours)*	
	Machine A	*Machine B*
J_1	6	8
J_2	7	3

Use graphical method.

2. Use Gantt chart technique to obtain the sequence in which the three jobs J_1, J_2, and J_3 should be performed on two machines M_1 and M_2 for completion in the minimum time. The machining is to be done in the order M_1–M_2 and the timings are given in the following table.

Job	*Processing time (hours)*	
	Machine M_1	*Machine* M_2
J_1	5	7
J_2	6	3
J_3	4	6

3. Determine the optimal sequencing of jobs from the following data and obtain the value of T, the total elapsed time involved:

Job	:	1	2	3	4	5	6	7	8
Machine M_1	:	7	3	6	8	9	5	4	3
Machine M_2	:	8	8	2	4	7	5	6	8

Processing order for all jobs is M_1–M_2.

4. Find the sequence that minimises the total elapsed time required to complete the following tasks on machines M_1 and M_2, in the order M_1, M_2. Also, find the minimum total elapsed time.

Task	:	*A*	*B*	*C*	*D*	*E*	*F*	*G*	*H*	*I*
M_1	:	2	5	4	9	6	8	7	5	4
M_2	:	6	8	7	4	3	9	3	8	11

(MBA, Chennai, April, 1997)

5. Seven jobs are required to be processed through two machines *A* and *B*. The processing time (in hours) of each jobs on the two machines is given below:

Job	*Processing time*	
	Machine A	*Machine B*
1	10	5
2	20	21
3	5	4
4	25	15
5	15	14
6	12	12
7	6	9

Suggest optimal sequence of processing the jobs and total minimum elapsed time.

6. A company has categorised its investment proposals into seven types. The financial analysts are needed to analyse the risk and return characteristics of these types. Then the proposals are examined by a committee for approval. The time that the analysts and the committee take is based on the size of the project. If it is required to minimise the time taken in evaluation of the seven proposals, how should they be scheduled? The requirements of time are:

Investment	:	*A*	*B*	*C*	*D*	*E*	*F*	*G*
Analysis (by analysts)	:	8	5	10	8	14	10	7
Evaluation (by committee)	:	5	3	7	4	8	6	5

What is the total time needed for evaluation?

7. There are six jobs which must go through two machines *A* and *B* in the order *A*–*B*. Processing time (in hours) is given here:

Job	:	1	2	3	4	5	6
Machine *A*	:	8	10	11	12	16	20
Machine *B*	:	7	15	10	14	13	9

Determine the optimum sequence and the total elapsed time. *(M Com, Delhi, 1987)*

8. A refrigeration company has six plants located in different parts of a city. Every year, it is necessary for each plant to be completely overhauled. The overhauling is carried out in two stages *A* and *B*, and each stage requires a crew of workmen with completely different skills. The work on stage *B* can start only when stage *A* has been completed. A plant has to be closed for the entire period of its overhauling.

The present schedule of the overhaul of the six plants is as follows:

Time required by crew (Days)

Plant	:	P_1	P_2	P_3	P_4	P_5	P_6
Crew A	:	6	6	4	6	5	8
Crew B	:	4	2	10	5	3	6

(a) Determine the optimal sequence.

(b) If down-time of any of the six plants costs Rs 800 per day for a plant, idle time for crew A costs Rs 250 per day, and idle time for crew B costs Rs 430 per day, which of the two schedules, the present one and the one determined in (a), will be more economical? What are their respective costs?

9. A book-binder has one printing press, one binding machine and the manuscripts of a number of different books. The time required to perform the printing and binding operations for each book are shown as follows:

Book	:	1	2	3	4	5	6
Printing time (hrs)	:	30	120	50	20	90	100
Binding time (hrs)	:	80	100	90	60	30	10

Determine the order in which the books should be processed in order to minimise the total time required to turn out all the books.

10. A lathe operator has to perform two operations, turning and knurling, on a number of different jobs. The time required to perform these operations (in minutes) for each job is known. Determine the sequence in which the jobs should be processed in order to minimise the total time required to turn out all the jobs.

Job	:	J_1	J_2	J_3	J_4	J_5	J_6	J_7
Time for turning	:	3	9	12	16	6	11	10
Time for knurling	:	8	3	10	10	6	1	12

11. Find the sequence that minimises the total elapsed time required to complete the following jobs. Each job is processed in the order *ACB*.

Jobs	:	J_1	J_2	J_3	J_4	J_5	J_6
Time on Machine A	:	12	8	7	11	10	5
Time on Machine B	:	7	10	9	6	10	4
Time on Machine C	:	3	4	2	5	2	4

12. With the help of the following example, describe the algorithm for solving an '*n*-job, 3-machines' sequencing problem:

Job	*Processing time (in hours)*		
	Machine A	*Machine B*	*Machine C*
1	3	3	5
2	8	4	8
3	7	2	10
4	5	1	7
5	2	5	6

(M Com, Delhi, 1996)

13. There are five jobs, each of which must go through machines *A*, *B* and *C* in the order *ABC*. Processing times (in hours) are given below:

Job	:	1	2	3	4	5
Machine *A* (A_i)	:	10	11	8	7	6
Machine *B* (B_i)	:	6	4	5	3	2
Machine *C* (C_i)	:	9	5	4	6	8

Determine the optimal processing sequence.

14. A firm has three groups of workers, performing exclusively the jobs of either cutting and planing, or chiselling and fitting, or finishing and polishing. It has received six orders which are to be completed and delivered at the earliest. The orders are marked as 1, 2, ..., 6, in the sequence in which they have been received. The manager insists that the jobs must be done in the order in which they were received. As a consultant, what would be your suggestion?

Job	*Time (in hours) required*		
	Cutting & planing	*Chiselling & fitting*	*Finishing & polishing*
1	12	6	7
2	10	5	6
3	9	6	6
4	14	4	5
5	7	2	4
6	9	4	4

15. A ready-made garments manufacturer has to process 7 items through two stages of production, viz. cutting and sewing. The time taken by each of these items at the different stages are given in appropriate units.

Item	*Processing time*	
	Cutting	*Sewing*
1	5	2
2	7	6
3	3	7
4	4	5
5	6	9
6	7	5
7	12	8

(a) Find an order in which these items are to be processed so as to minimise the total processing time.

(b) Suppose a third stage of production is added, viz. ironing and packing, with processing time for these items as follows:

Item	:	1	2	3	4	5	6	7
Processing Time (ironing and packing)	:	10	12	11	13	12	10	11

Find an order in which these seven items are to be processed so as to minimise the time taken to process all the items through all the three stages.

16. Six jobs, $J_1, J_2, ..., J_6$, are required to be processed on four machines, M_1, M_2, M_3 and M_4, in that order. Using the processing times given below, determine the optimal sequence(s) of job performance.

Job	*Machine*			
	M_1	M_2	M_3	M_4
J_1	12	6	6	8
J_2	8	3	6	12
J_3	13	4	6	9
J_4	9	2	4	14
J_5	10	6	7	12
J_6	12	7	2	10

17. Solve the following sequencing problem, giving an optimal solution, if passing is not allowed.

Job	*Machines*			
	M_1	M_2	M_3	M_4
A	13	8	7	14
B	12	6	8	19
C	9	7	5	15
D	8	5	6	15

(MBA, Chennai, November, 1996)

18. Using graphical method, find the minimum elapsed time to perform jobs J_1 and J_2 on machines *A* through *D*, the sequence and process timing for which are given as follows:

Job 1	Sequence	:	*A*	*B*	*C*	*D*
	Time (Hrs)	:	2	4	5	1
Job 2	Sequence	:	*B*	*C*	*D*	*A*
	Time (Hrs)	:	6	5	2	3

19. Determine the least time in which the two jobs can be performed on the eight machines. The processing times and the machine sequence for the jobs are given here:

Job 1	Machine sequence	:	*A*	*B*	*C*	*D*	*E*	*F*	*G*	*H*
	Time (Hours)	:	6	4	5	5	7	2	1	5
Job 2	Machine sequence	:	*C*	*B*	*A*	*D*	*E*	*G*	*F*	*H*
	Time (Hours)	:	5	6	3	4	4	6	1	4

20. Two jobs, *A* and *B*, are to be processed on 6 machines. The sequence of machines and the processing times are given in the following table.

Job A	Machine sequence	:	M_1	M_2	M_3	M_4	M_5	M_6
	Time (Hours)	:	6	4	5	3	4	2
Job B	Machine sequence	:	M_1	M_3	M_5	M_4	M_2	M_6
	Time (Hours)	:	4	8	4	3	6	4

What is the minimum time in which both the jobs can be completed?

CHAPTER

9

Inventory Management

CHAPTER OVERVIEW

The efficient management of inventories is a very important factor in the successful functioning of manufacturing and retailing organisations. Understandably, since inventories require investment of funds, there are painful implications of over-stocking as also of under-stocking. There are many facets of inventory management. Among the questions that will be answered in this chapter are the following:

- What does inventory mean and what are the different types of inventories?
- What inventory decisions are to be taken and what are the costs involved?
- What are the inventory management systems?
- When the purchases should be made—what should be the timing of purchases? In other words, should the purchases be made periodically at fixed intervals or as soon as the stock level reaches a pre-determined level?
- What inventory models are available to handle different situations? In particular, what quantity should be ordered every time that a purchase is made when:
    - Only one price rules for all order sizes and no discounts are available on large-sized orders?
    - One or more price cuts are available on big orders?
    - The item is not purchased from outside suppliers and is instead produced internally by a sister department?
    - Back-ordering is possible so that orders are recorded when out-of-stock conditions exist and are supplied when the purchased lot arrives?
- What is safetly stock and how is the level of safety stock to be kept determined taking due consideration of uncertainty of supply time, the rate of demand and the costs involved in being out-of-stock?
- What are the different bases, like usage, cost, seasonality, etc., on which inventories may be classified, so as to exercise a better control over them?

This chapter involves a good amount of calculation work and numerous formulae. Understanding the derivation of cost functions in respect of different models is important for this chapter. Besides, knowledge of normal distribution and calculation of areas under the normal curve is required. The derivations given in the appendix require an understanding of one-variable and more-than-one-variable differential calculus.

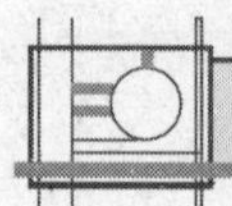

9.1 INTRODUCTION

Inventories are vital to the successful functioning of manufacturing and retailing organisations. They may consist of raw materials, work-in-progress, spare parts/consumables, and finished goods. It is not necessary that an organisation has all these inventory classes. But, whatever may be the inventory items, they need efficient management as, generally, a substantial share of its funds is invested in them.

Different departments within the same organisation adopt different attitudes towards inventory. This is mainly because the particular functions performed by a department influence the department's motivation. For example, the sales department might desire large stock in reserve to meet virtually every demand that comes. The production department similarly would ask for large stocks of materials so that the production system runs uninterrupted. On the other hand, the finance department would always argue for a minimum investment in stocks so that the funds could be used elsewhere for other better purposes.

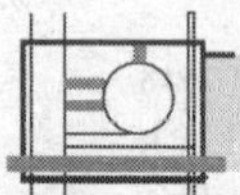

9.2 TYPES OF INVENTORY

Inventories may be held for a variety of purposes, but, in general, there are five types of inventories that an organisation can use for serving these purposes. These are:

(a) Movement inventories;
(b) Buffer inventories;
(c) Anticipation inventories;
(d) Decoupling inventories; and
(e) Cycle inventories.

We shall consider these now.

(a) Movement Inventories Movement inventories are also called *transit* or *pipeline inventories*. Their existence owes to the fact that transportation time is involved in transferring substantial amounts of resources. For example, when coal is transported from the coal fields to an industrial town by trains, then the coal, while in the transit, cannot provide any service to the customers for power generation or for burning in furnaces.

(b) Buffer Inventories Buffer inventories are held to protect against the uncertainties of demand and supply. An organisation generally knows the average demand for various items that it needs. However, the actual demand may not exactly match the average and could well exceed it. To meet this kind of a situation, inventories may be held in excess of the average or expected demand. Similarly, the average delivery time (that is, the time elapsing between placing an order and having the goods in stock ready for use, and technically called as the *lead time*) may be known. But unpredictable events could cause the actual delivery time to be more than the average. Thus, excess stocks might be kept in order to meet the demand during the time for which the delivery is delayed. These inventories which are in excess of those necessary just to meet the average demand (during the average lead time period), held for protecting against the fluctuations in demand and lead time are known also by the term *safety stocks*.

The idea of keeping buffer stocks is to render a higher level of customer service and consequently reduce the number of stockouts and back-orders. A stockout occurs when a customer is denied fulfilment of an order because the inventory of the item(s) has runout. In some situations back-ordering is possible (i.e. the order for goods demanded may be fulfilled as soon as the next shipment of the item(s) is received) while in others it is not and the demand might be lost forever leading to temporary/permanent loss of customer goodwill.

It is almost inevitable to keep buffer stocks for the simple reason that perfect prediction of demand and lead time is an exception rather than the rule. We shall discuss later how the optimum level of the buffer stocks to be held can be determined.

(c) Anticipation Inventories Anticipation inventories are held for the reason that a future demand for the product is anticipated. Production of specialised items like crackers well before Diwali, umbrellas and rain-coats before rains set in, fans while summers are approaching, or the piling up of inventory stocks when a strike is on the anvil, are all examples of anticipation inventories. The underlying idea is to smoothen the production process for a longer duration on a continuous scale rather than operating with excessive overtime in one period and then letting the system be idle or closing down for the reason of inadequate/no demand for another period.

(d) Decoupling Inventories The idea of the decoupling inventories is to decouple, or disengage, different parts of the production system. As we can observe easily, different machines/equipment and people normally work at different rates—some slower and some faster. A machine, for example, might be producing half the output of the machine on which the item being handled is to be processed the next. Inventories in between the various machines are held in order to disengage the processing on those machines. In the absence of such inventories, different machines and people cannot work simultaneously on a continuous basis. When such inventories are held, then, even if a machine breaks down, the work on others would not stop.

Indeed, if all the machines and people work at the same rate (a rare case, of course) then there would be no need for such inventories. But then, in that case if a machine breaks down, very soon the entire work might come to a standstill.

Clearly, therefore, the decoupling inventories act as shock absorbers and have a cushioning effect in the face of varying work-rates, and machine breakdowns and failures, and so on.

(e) Cycle Inventories Cycle inventories are held for the reason that purchases are usually made in lots rather than for the exact amounts which may be needed at a point of time. Of course, if all purchases are made exactly as and when the item is required, there would be no cycle inventories. But, practically, then the cost involved in obtaining the items would be very large.

As the cycle inventories are related to the purchases in lots, they are also called lot-size inventories and, obviously, the larger the lot-size the greater would be the level of cycle inventories.

9.2.1 Inventory Decisions

In an inventory control situation, there are three basic questions to be answered. They are:

(a) How much to order? That is to say, what is the optimal quantity of an item that should be ordered whenever an order is placed?

(b) When should the order be placed?

(c) How much safety stock should be kept? Thus, what quantity of an item in excess of the expected requirements should be held as buffer stock in anticipation of the variations in its demand and/or the time involved in acquiring fresh supplies?

9.2.2 Inventory Costs

In determining an optimal inventory policy, the criterion most often is the cost function. The classical inventory analysis identifies four major cost components. Depending on the structure of an inventory situation, some, or all, of these are included in the objective function.

1. Purchase Cost This refers to the *nominal* cost of inventory. It is the *purchase price* for the items that are bought from outside sources, and the *production cost* if the items are produced within the organisation. This may be constant per unit, or it may vary as the quantity purchased/produced increases or decreases. Quite often, situations are found when it may be stipulated that, for example, the unit price is Rs 20 for an order upto 100 units and Rs 19.50 if the order is for more than 100 units.

If the unit cost is constant, it does not affect the inventory control decisions because then whether all the requirements are bought (produced) just once, or whether they are obtained in instalments, the total amount of money involved would be the same. However, we do consider the quantity discounts when they occur, because they affect these decisions.

2. Ordering Cost/Set-up Cost Ordering cost is incurred whenever the inventory is replenished. It includes costs associated with the processing and chasing of the purchase order, transportation, inspection for quality, expediting overdue orders, and so on. It is also known as the *procurement cost.*

The parallel of the ordering cost when units are produced within the organisation is the *set-up cost.* It refers to the cost incurred in relation to developing the production schedules, the resources employed in making the production system ready, and so on.

The ordering (and set up) cost is likely, and taken, to be independent of the order size. Therefore, the unit ordering/set-up cost declines as the purchase order/production run increases in size.

3. Carrying Cost Also known as the *holding cost* or the *storage cost*, carrying cost represents the cost that is associated with storing an item in inventory. It is proportional to the amount of inventory and the time over which it is held. The elements of carrying cost include the opportunity cost of capital invested in the stock; the costs directly associated with storing goods (like storemen's salary, rates, heating and lighting, racking and pelletisation, protective clothing, store's transport, etc.); the obsolescence cost (including scrapping and possible rework); deterioration costs and costs incurred in preventing deteriorations; and fire and general insurance, etc.

The carrying cost is usually expressed as a rate per unit or as a percentage of the inventory value. It is taken to be fixed for each unit of a certain item of inventory held for a unit time.

4. Stockout Cost Stockout cost means the cost associated with not serving the customers. Stockouts imply shortages. If the stockout is internal (that is, in the production system) it would imply that some production is lost, resulting in idle time for men and machines, or that the work is delayed which might attract some penalty. While if the stockout is external, it would result in a loss of potential sales and/or loss of customer goodwill. A shortage can evoke different reactions from customers. It would result in a backorder or a lost sales. In case of backorder the sales are not lost, they are only delayed. When the new shipment arrives, a customer who was denied earlier would be immediately supplied the goods. But it would involve costs like those of expediting and, maybe, also packaging and shipment costs. On the other hand, when the sales are lost forever, it is difficult to assess the costs involved in terms of the profit on potential sales lost, profit on whatever the customer would have bought in all future periods in case he decides not to turn to the organisation for anything in future, and so on and so forth. Such costs can be exorbitant indeed.

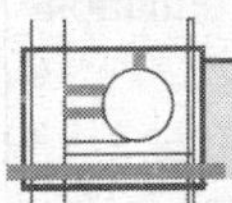

9.3 INVENTORY MANAGEMENT SYSTEMS

Basically, there are two inventory management systems. They are:

(a) *Fixed order quantity system*, or the *re-order point system*, in which a level called the *re-order level* is determined. As soon as the stock level reaches this point, an order for a predetermined number of units is placed.

(b) *Periodic review system*, where the inventories are replenished at fixed intervals of time. Whereas the size of order is fixed in the previous system and the time is not, in this system the time after which the supplies are ordered is fixed but not the quantity to be ordered.

There is another system, called the Ss system, which combines the features of both of these. We shall discuss the three systems in turn.

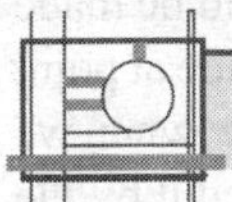

9.4 FIXED-ORDER QUANTITY SYSTEM

The fixed-order quantity system of inventory management is also called the *Q-system*. As pointed out earlier, under this system a re-order point is established and as soon as the stock level reaches this point, a fresh order for supplies is triggered. We shall consider this system under conditions of certainty (i.e. when the demand and the lead time are known with certainty) and then extend it to the probabilistic conditions when demand and/or lead time are uncertain. Four models, based on varying underlying conditions shall be developed to study the operations of the system under deterministic conditions.

Model 1: The Classical EOQ Model

To begin with, we shall develop the classical *economic order quantity model* (EOQ model) also known as the *Wilson formulation*, which is the most elementary of all the inventory models. This model is an analytical one. For this, first a cost model is developed and then it is manipulated to form an inventory model.

Assumptions The model is based on the following basic assumptions:

(a) The demand for the item is certain, continuous and constant over time.

(b) The lead time, that is the time between placing an order and its delivery, is known and fixed. Thus, when the lead time is zero, the delivery of item is instantaneous.

(c) Within the range of the quantities to be ordered, the per unit holding cost and the ordering cost (per order) are constant and thus independent of the quantity ordered.

(d) The purchase price of the item is constant, that is to say, no discount is available on purchases of large lots.

(e) The inventory is replenished immediately as the stock level reaches exactly equal to zero. Consequently, there are no stock overages or shortages.

With this set of assumptions, the inventory level would vary over time as depicted in Figure 9.1.

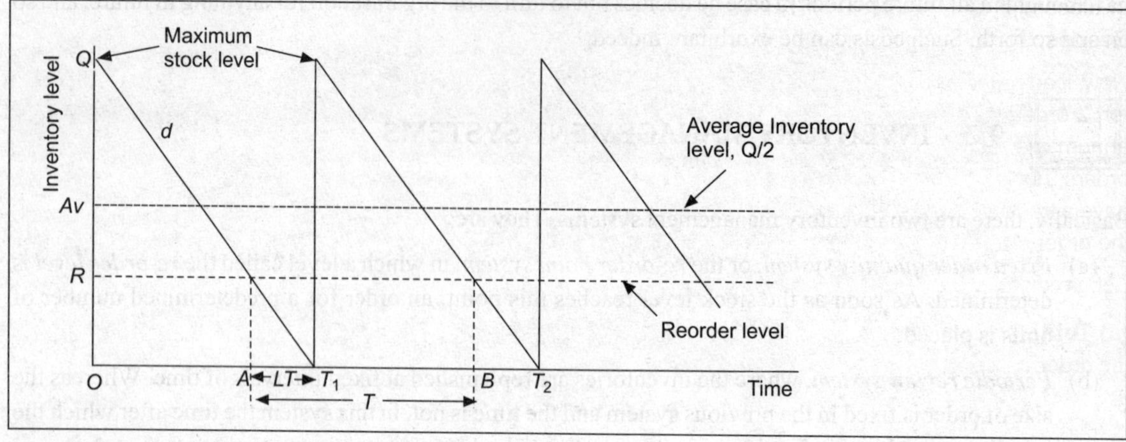

Figure 9.1 *Inventory Profile of Classical EOQ Model*

Suppose that we begin with a stock of Q on the timc 0. This will be consumed at the rate of d units per day, say. If the stock can be replenished instantaneously, that is, the lead time is zero, then a fresh order would be made and the inventory obtained at point T_1. Similarly, when this stock is consumed, an order would be made at point T_2. On the other hand, if the lead time is positive, equal to LT, we would place an order at point A, i.e. when we have in stock an amount equal to the demand during the lead time so that we would have consumed it by the time the fresh delivery is due to arrive. This is called the *re-order level*, or the *reorder point*, R.

The interval between two successive points when orders are placed (for example A and B), or the time elapsed in consuming the entire lot of items received, Q, is called the *inventory cycle*. It is represented in the Figure 9.1 by T. The maximum inventory held would be Q while the minimum be zero, and hence the average inventory level would equal $Q/2$.

Because of the first and the second assumptions, there is no need for maintaining a safety stock. Also, because of the first assumption, the inventory decision, as to when to order, is completely specified when the quantity to order is known. Therefore, the basic issue to settle is the determination of the order quantity, Q.

For determining the optimum order quantity, we shall consider two types of costs, viz. the ordering and the holding cost. Since the purchase price of the units is uniform (assumption d), it does not affect the decision as to the quantity of the item to be ordered for purchase and, hence, is irrelevant for the purpose.

According to this, our cost model would, assuming a period of one year, be

$$T(Q) = O(Q) + H(Q)$$

where,

Q = the ordering quantity

$T(Q)$ = total (variable) annual inventory cost

$O(Q)$ = total annual ordering cost

$H(Q)$ = total annual holding cost

Let us develop the model with the following example.

Example 9.1 Easton Electronics Co produces 2,000 TV sets in a year for which it needs an equal number of tubes of a certain type. Each tube costs Rs 10 and the cost to hold a tube in stock for a year is Rs 2.40. Besides, the cost of placing an order is Rs 150, which is not related to its size.

Now, if an order for 2,000 tubes is placed, only one order per annum is required. When 1,000 units are ordered for, 2 orders in a year are needed, while if 500 units are ordered to be supplied, then a total of 4 orders per annum are required. Naturally, as the number of orders placed increases the ordering cost goes up. More orders, however, would also imply smaller order quantity and, therefore, decreasing holding costs. Thus, we have a trade-off between the ordering and the holding cost. What we attempt in our EOQ model is, then, to find the order size that minimises the cost function $T(Q)$.

Let us consider for our illustration the computation in a step-wise function.

(a) Total Annual Ordering Cost This is given by the number of times an order is placed, N, multiplied by the ordering (or acquisition) cost per order, A. Thus,

$$O(Q) = N \times A$$

The value of N itself is dependent on the order quantity Q, and the annual demand, D. Here N would be equal to D/Q. Accordingly,

$$O(Q) = \frac{D}{Q} A$$

For our example, when

$$N = 1, Q = 2{,}000 \quad \text{and} \quad O(Q) = 1 \times 150 = \text{Rs } 150,$$
$$N = 2, Q = 1{,}000 \quad \text{and} \quad O(Q) = 2 \times 150 = \text{Rs } 300,$$
$$N = 4, Q = 500 \quad \text{and} \quad O(Q) = 4 \times 150 = \text{Rs } 600,$$
$$N = 5, Q = 400 \quad \text{and} \quad O(Q) = 5 \times 150 = \text{Rs } 750.$$

Figure 9.2 depicts the total ordering cost. The larger the order quantity, the smaller the total ordering cost, and vice versa.

(b) Total Annual Holding Cost The total annual holding cost is obtained by multiplying the unit holding cost (measured as rupees per unit per annum), h, by the average number of units held in the inventory. As has been pointed out earlier, the average inventory held equals $Q/2$. Consequently, the total cost of holding inventory, per annum would be:

$$H(Q) = \frac{Q}{2} h$$

For our example, when

$$Q = 2{,}000, H(Q) = \frac{2{,}000}{2} \times 2.40 = \text{Rs } 2{,}400,$$
$$Q = 1{,}000, H(Q) = \frac{1{,}000}{2} \times 2.40 = \text{Rs } 1{,}200,$$
$$Q = 500, H(Q) = \frac{500}{2} \times 2.40 = \text{Rs } 600,$$
$$Q = 400, H(Q) = \frac{400}{2} \times 2.40 = \text{Rs } 480.$$

Obviously, the total holding cost varies in direct proportion to the order quantity. This is displayed in Figure 9.2.

It may be pointed out that although the cost of holding a unit in an inventory is given in this question, more often the holding cost is expressed as a proportion, or percentage of the value of inventory. It may be stated for example, that the inventory holding costs are 15% per annum of the value of an item. It implies that if an item costs Rs 40, then the holding cost would be 15% of Rs 40 = Rs 6 per unit per year. Thus, if i is the proportion (or percentage), which is also known as the holding rate, and c is the cost of the item, then unit holding cost, $h = ic$.

(c) Total (Variable) Annual Inventory Cost Both the cost components can now be added up and we can obtain the total cost of inventory. The calculations for some order quantities are given in Table 9.1.

TABLE 9.1 Calculation of Total Cost

Order Quantity	*Annual Ordering Cost*	*Annual Holding Cost*	*Total Annual Cost*
100	3,000	120	3,120
200	1,500	240	1,740
400	750	480	1,230
500	600	600	1,200
1,000	300	1,200	1,500
2,000	150	2,400	2,550

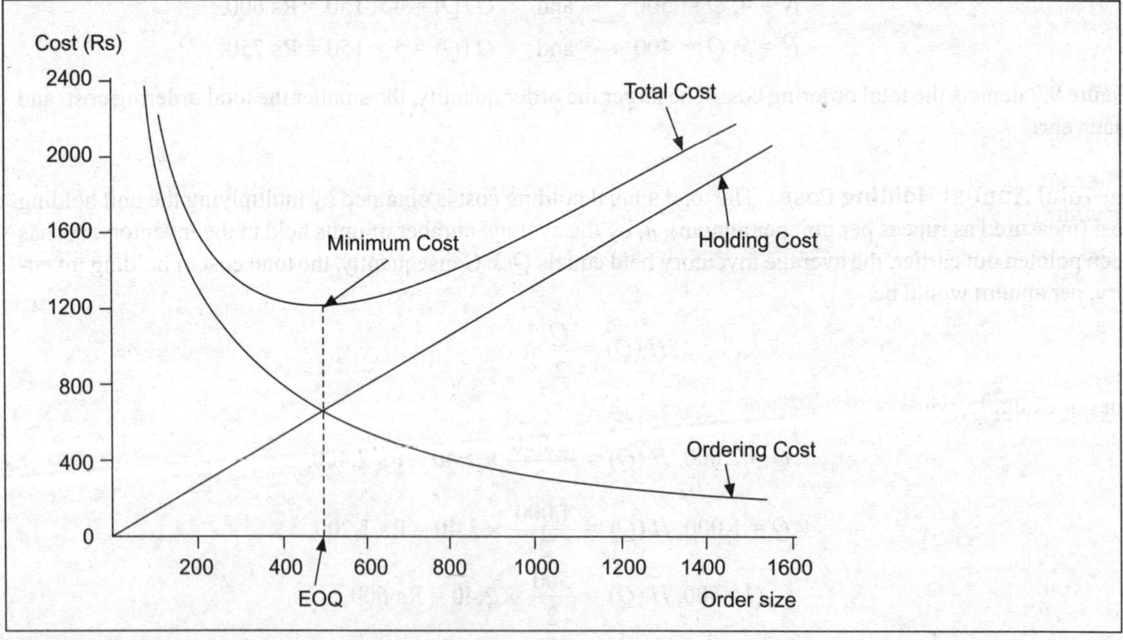

Figure 9.2 *Graphic Determination of EOQ*

In Figure 9.2, the total cost curve is obtained by adding the two components $O(Q)$ and $H(Q)$. The minimum point on this curve would determine the optimal quantity, the Economic Order Quantity (Q^*) for which an order be placed each time. This would ensure minimisation of the total cost.

From the figure, it is clear that the total cost is minimum when $Q = 500$ units, as also is clear from the Table 9.1. Therefore, EOQ, also represented as Q^*, is equal to 500 units. It may further be noted that at this order level, the holding and ordering costs are also equal. From the table, we notice that each of these equals Rs 600.

The graphic and tabulation methods of determining the EOQ are quite cumbersome. We can obtain this value using the following formulation (see appendix to this chapter, 9*A*, for its derivation):

$$Q^* = \sqrt{\frac{2AD}{h}} \quad \text{or} \quad Q^* = \sqrt{\frac{2AD}{ic}} \quad \text{(where the holding cost is given is given as a proportion)}$$

For Example 9.1, we have A = Rs 150 per order, h = Rs 2.40 per unit per annum, D = 2,000 units. Thus,

$$Q^* = \sqrt{\frac{2 \times 150 \times 2{,}000}{2.40}}$$

$$= \sqrt{2{,}50{,}000} = 500 \text{ units}$$

It is the same as obtained earlier.

Determination of the Re-order Level The re-order level would be established at a point such that the stock in hand would be just sufficient to meet the demand during the lead time. Suppose that for the TV example, the following additional data are given:

No. of working days in the year	= 250
Lead time	= 15 working days
With this information, the daily demand	= 2,000/250 = 8 tubes
Demand during lead time	= 15 × 8 = 120 tubes
∴ Re-order level	= 120 tubes

Other Information In addition to the optimal order quantity and the re-order level, we can have the following information:

(a) ***Annual Total Variable Inventory Cost*** The minimum annual inventory cost can be determined by substituting Q^* for Q in the cost equation. Thus,

$$T(Q^*) = \frac{D}{Q^*}A + \frac{Q^*}{2}h$$

$$= \frac{DA}{\sqrt{\frac{2AD}{h}}} + \frac{\left(\sqrt{\frac{2AD}{h}}\right)h}{2}$$

On simplification,

$$T(Q^*) = \sqrt{2ADh}$$

For our example

$$T(Q^*) = \sqrt{2 \times 150 \times 2{,}000 \times 2.40}$$

$$= \sqrt{14{,}40{,}000} = \text{Rs } 1{,}200$$

It may be noted that when the holding cost is expressed in the 'proportion' form, we have,

$$T(Q^*) = \sqrt{2ADic}$$

(b) *Inventory Cycle* With a uniform, constant demand D, and the economic order quantity Q^*, the problem of *optimal interval between the successive orders* can be answered easily. If T^* represents the optimal interval between any two consecutive orders, we have,

$$T^* = Q^*/D$$

T^* is also called the *inventory cycle time.*

For our example, $T^* = 500/2{,}000 = 0.25$ year. And, since there are 250 working days in the year, this equals $0.25 \times 250 = 62.5$ days.

It follows, therefore, that an order would be placed every 62.5 working days, implying further that the quantity ordered is sufficient to meet 62.5 days' demand.

(c) *Number of Orders* The optimal number of orders placed per year, N^*, can also be obtained. It equals the reciprocal of T^*. Thus, $N^* = 1/T^*$. Thus, when $T^* = 0.25 = 1/4$, $N^* = 4$ orders per year.

(d) *Rupee Values* The monetary value of the optimal order quantity and of the average inventory held can also be determined, if needed. We have,

$$\text{Rupee value of the EOQ} = Q^* \times c \text{ (where } c \text{ is the unit price)}$$

$$= 500 \times 10 = \text{Rs } 5{,}000$$

$$\text{Rupee value of the average inventory} = \frac{Q^* c}{2}$$

$$= \frac{500}{2} \times 10 = \text{Rs } 2{,}500.$$

Many a time, the demand is expressed in monetary terms rather than in units. In such cases, if the unit price is known, then the demand may be converted in units by dividing the rupee demand by this unit cost price. Where, however, the unit cost is not given, this cannot be done and we can determine the economic order quantity in rupee terms. When the demand is given in monetary terms, the holding cost must be expressed as a proportion (of the value of the inventory).

Thus, if

cD = the annual demand in rupee terms,

A = the acquisition cost,

i = the holding rate,

we have,

$$\text{EOQ (in rupees)} = \sqrt{\frac{2AcD}{i}}$$

$$\text{Total cost, } T(Q^*) = \sqrt{2AcDi}$$

Obviously, the total cost, $T(Q^*)$ is the same as given in the earlier formulation.

Example 9.2 Using the following data, obtain the EOQ and the total variable cost associated with the policy of ordering quantities of that size.

Annual demand = Rs 20,000
Ordering cost = Rs 150 per order
Inventory carrying cost = 24% of average inventory value.

Here

$$cD = \text{Rs } 20{,}000$$

$$A = \text{Rs } 150/\text{order}$$

$$i = 24\% = 0.24$$

$$\text{EOQ (in rupees)} = \sqrt{\frac{2 \times 150 \times 20{,}000}{0.24}}$$

$$= \text{Rs } 5{,}000$$

$$\text{Total cost, } T(Q^*) = \sqrt{2 \times 150 \times 20{,}000 \times 0.24}$$

$$= \text{Rs } 1{,}200$$

A closer look at the example would reveal that it is based on Example 9.1, and, therefore, has the same solution.

Robustness of the EOQ Model An important characteristic of the EOQ model is the robustness which draws from the fact that the total cost curve is relatively flat at the bottom. The model, therefore, tends to give reasonably good results even when parameter values are in error or they vary. We can investigate the sensitivity of the model by considering changes in the ordering quantity and corresponding changes in the total cost. This is useful in some cases as it gives the manager some flexibility in deciding on an order quantity that might better meet the constraints, if any. For example, the EOQ might turn out to be 10.33 crates while the supplies can be in 10 or 11 crates. Here the decision-maker would do well to determine the costs when 10 or 11 crates are ordered and see which of the alternatives suits him better. Not only this, generally as the total cost curve is comparatively flat, as mentioned earlier, even significant departures from EOQ would not result in appreciable rise in the costs.

Suppose we wish to determine the change in the cost associated with an order quantity which is k times the EOQ (thus if we wish to determine the effect of a 20% increase in the order size, $Q = 1.2Q^*$).

Since $Q = kQ^*$, we have

$$K = \frac{Q}{Q^*}$$

The cost associated with quantity Q would be

$$T(Q) = \frac{AD}{kQ^*} + \frac{khQ^*}{2}$$

The new cost (at Q) in relation to the cost at Q^* is

$$\frac{T(Q)}{T(Q^*)} = \frac{\frac{AD}{kQ^*} + \frac{khQ}{2}}{\frac{AD}{Q^*} + \frac{hQ^*}{2}}$$

Since $\dfrac{AD}{Q^*} = \dfrac{hQ^*}{2}$,

we have
$$\frac{T(Q)}{T(Q^*)} = \frac{\frac{hQ^*}{2k} + \frac{khQ^*}{2}}{\frac{hQ^*}{2} + \frac{hQ^*}{2}}$$

$$= \frac{hQ^* + k^2hQ^*}{2khQ^*}$$

$$= \frac{1}{2k} + \frac{k}{2}$$

$$= \frac{1}{2}\left(\frac{1}{k} + k\right)$$

For our TV picture tubes example, suppose we wish to determine the effect of a 25% increase in the order quantity on the cost. Here $Q^* = 500$ units, $Q = 625$ units (the desired level), and

$$k = \frac{625}{500} = 1.25$$

The cost ratio would be

$$\frac{T(Q)}{TQ^*} = \frac{1}{2}\left(\frac{1}{1.25} + 1.25\right) = \frac{1}{2}(0.8 + 1.25) = 1.025$$

which implies that the cost would go up by 2.5% when the order quantity is increased by 25%. The cost would similarly go up when we consider $Q < Q^*$. From the format of the relationship $T(Q)/T(Q^*)$ it is clear that the same result would be obtained when $k = 0.8$ (as when $k = 1.25$) i.e. when $Q = 0.8 \times 500 = 400$ units. So if the order quantity is decreased by 20% (from EOQ), then also the cost would go up by 2.5% (this may be verified from the information given in Table 9.1) For the Example 9.1, then, when the order quantities vary between 400 and 625 units, the total cost would increase by a margin of equal to or less than 2.5%.

Review of the Assumptions of EOQ Model

1. In the EOQ model we assumed that the demand for the item under consideration is certain, continuous and constant. In reality, however, the demand is more likely to be uncertain, discontinuous and variable. We shall consider such situations later.
2. Another assumption in the EOQ model is that all demand is supplied immediately and there is no question of a shortage. However, even when the demand and lead time are known and constant, stockouts may be deliberately permitted. Model 4 considers this situation.
   When demand and lead time are not certain, the model needs modification and safety stocks are required to be kept. We shall discuss in detail the question of safety stocks later in the chapter.
3. The EOQ model is based on the assumption that the unit price is the same. The analysis can be extended to cover situations when quantity discounts are available. The next model deals with this problem.
4. Uniformity of the unit holding cost and of the cost of placing an order is the other underlying assumption. If either or both of these assumptions are not satisfied, we can consider the EOQ model as modified in (3) given earlier.
5. The implicit assumption in the EOQ model that the entire quantity ordered for would be received in a single lot, may not hold true sometimes. If the supply of the goods is gradual (specially when the source of supply is internal production), the model needs adjustment—see Model 3.

Model 2: EOQ with Price Breaks

As mentioned earlier, the classical EOQ model is based on the assumption that the unit cost of the item under consideration is uniform. In real life, however, it is not uncommon to find price discounts that are based on the quantity for which order is placed. In general, lower rates are quoted if the orders are of a big size. There may be one or more than one price break that may be offered. Clearly, in cases like these, the quantity ordered has to be determined carefully taking into consideration the price levels for the different quantity ranges. When the unit cost price is uniform, the purchasing cost is irrelevant in determining the order size. But under the conditions of price break(s), the item cost, being a function of the order quantity, is the incremental cost and must be included in the cost model. As such, the cost model would include the ordering, the holding and the purchasing cost of the items. Thus,

$$T(Q) = \frac{D}{Q}A + \frac{Q}{2}h + c_i D$$

in which

$$c_i = \begin{cases} c_0 & \text{if } Q < q_1 \\ c_1 & \text{if } q_1 \le Q < q_2 \\ c_2 & \text{if } q_2 \le Q < q_3 \\ \cdot & \\ \cdot & \\ \cdot & \\ c_{n-1} & \text{if } Q \ge q_n \end{cases}$$

Thus, when the order quantity (Q) is less than a certain size q_1, the per unit price is c_0; when the order quantity is at least equal to q_1, but less than q_2, the unit price is c_1, ... and so forth. This cost model is clearly a step function, and not continuous like the one given earlier. To understand how the optimal order quantity can be determined in such a case, we shall continue with our TV tubes examples where $D = 2{,}000$ units per annum, $A =$ Rs 150 per order, and $h =$ Rs 2.40 per unit per annum. Suppose now that the supplier informs that if the order size is at least 800 units, he is prepared to supply the tubes at a discounted price of Rs 9.80 per tube. We shall now examine whether the offer of the supplier is attractive or not.

With the EOQ = 500 units already established, the total cost involved would be

$$T(Q) = \frac{2{,}000}{500} \times 150 + \frac{500}{2} \times 2.40 + 10 \times 2{,}000$$

$$= \text{Rs } 21{,}200$$

The price of the tubes has been taken to be Rs 10 per tube as the lower price is available only for the order sizes of at least 800 units. Thus, we shall determine the total cost involved at the minimum quantity at which the discount is available. At $Q = 800$, we have,

$$T(Q) = \frac{2{,}000}{800} \times 150 + \frac{800}{2} \times 2.40 + 9.80 \times 2{,}000$$

$$= \text{Rs } 20{,}935$$

Since this cost is lower than the earlier one, we conclude that it is prudent to order for 800 units. The total cost curve can be shown as in Figure 9.3.

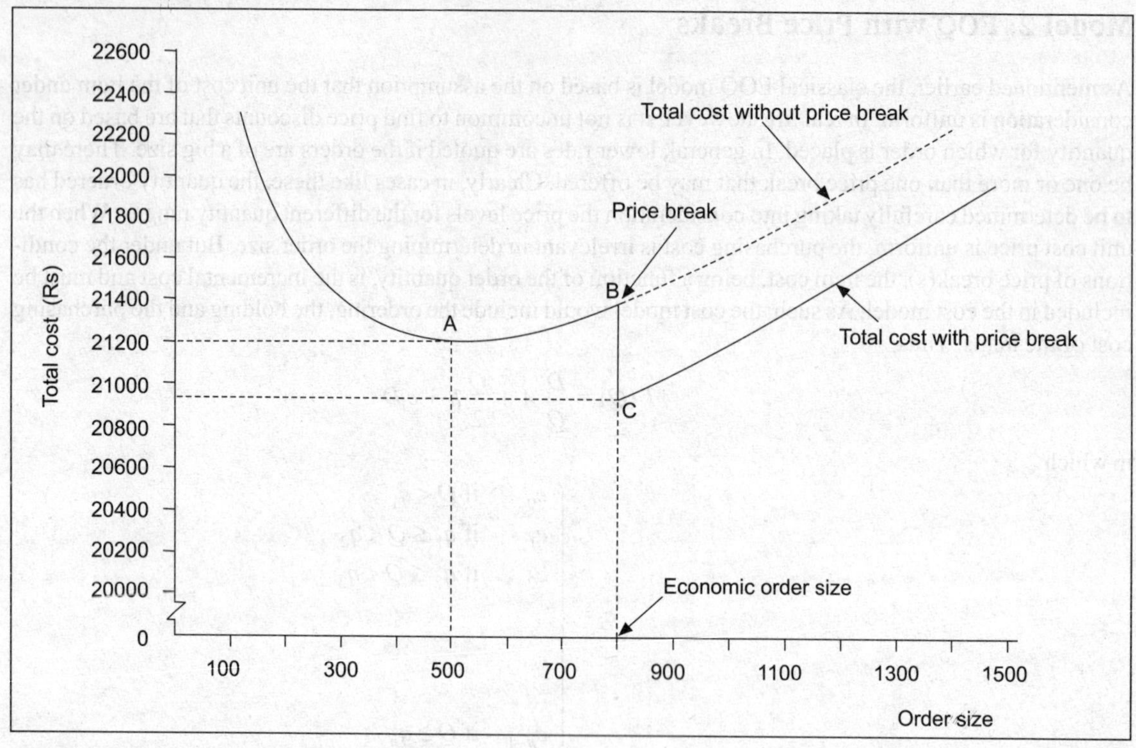

Figure 9.3 *Cost Curve for the Price Break Model*

Clearly, the curve shows a sizeable drop in the cost due to the price discount at a quantity of 800 units. At this level, the total cost is lower than the total cost corresponding to 500 units.

Multiple Price Breaks Sometimes more than one price break may be available. The mechanism in that case is simply an extension of the one in the single price break situation. Suppose the price schedule given by the supplier is as follows:

Quantity (Q) in Units	*Price per Unit (Rs)*
less than 300	10.00
$300 \leq Q < 600$	9.90
$600 \leq Q < 1{,}000$	9.70
$1{,}000 \leq Q < 1{,}500$	9.50
$Q \geq 1{,}500$	9.40

The first break is at a quantity equal to 300 units, and we shall refer it to as q_1. Similarly, q_2, q_3, and q_4 would be 600, 1,000, and 1,500 units.

The EOQ, as calculated earlier, is 500 units and the unit price relevant to this order size is Rs 9.90. The first price break which occurs at $Q = 300$ is, therefore, not important for our purpose. We shall determine the total cost at each of the successive quantity levels at which the price break occurs. At whatever level it is the least, we would call that as the optimum order size. Clearly, the various order quantities are tested to determine if the savings in the purchase cost of the item and the reduced order cost will offset the increases in the holding cost.

The total cost at different levels is calculated and shown in Table 9.2.

TABLE 9.2 Calculation of Total Annual Cost: Price Break Model

Category	*Unit Cost* c_i	*Order Quantity* Q^* *or* q_i	*Annual Holding Cost*	*Annual Ordering Cost*	*Annual Purchase Cost*	*Annual Total Cost*
1	10.00	Not relevant	–	–	–	–
2	9.90	500	600	600	19,800	21,000
3	9.70	600	720	500	19,400	20,620
4	9.50	1,000	1,200	300	19,000	20,500
5	9.40	1,500	1,800	200	18,800	20,800

Clearly, the total cost is minimum at a quantity of 1,000 units. The inventory decision, then, is that to minimise the total cost, we should order 1,000 units every time.

Sometimes the holding cost may be given as a certain proportion or percentage of the inventory value. In such situations, the unit holding cost would vary with the cost of the item. The mode of determination of the optimal ordering quantity in such cases is slightly different. It is as follows.

We first determine the EOQ with the lowest cost price available. If it is feasible, that is to say, if the EOQ falls in the slab of the lot size of which the price is used, then it represents the optimal solution. But if it is infeasible, the next lowest price is taken into consideration for determining EOQ. If it is infeasible too, then we try the next lowest. Similarly, we proceed and whenever a feasible EOQ is obtained, the total cost corresponding to this quantity level is determined. Also, the total cost of each of the order quantity values corresponding to the cut-off points of the subsequent lot size categories is determined. Among these total cost values, the minimum is selected. The quantity on which the least cost total is based is the *optimal ordering quantity*.

Example 9.3 A hardware store procures and sells hardware items. Information on an item is given here:

Expected annual sales	= 8,000 units
Ordering cost	= Rs 180 per order
Holding cost	= 10% of the average inventory value

The item can be purchased according to the following schedule:

Lot Size	Unit Price (Rs)
1– 999	Rs 22.00
1,000 – 1,499	Rs 20.00
1,500 – 1,999	Rs 19.00
2,000 and above	Rs 18.50

You are required to determine the best order size.

We have, $$\text{EOQ} = \sqrt{\frac{2AD}{h}}$$

in which h = 10% of the unit price. Using the lowest unit price of Rs 18.50, we get

$$\text{EOQ} = \sqrt{\frac{2 \times 180 \times 8{,}000}{0.10 \times 18.50}} = 1{,}248 \text{ units}$$

But it is not feasible because the unit price of Rs 18.50 is not available for an order size of 1,248 units. Now, with the unit price equal to Rs 19, we get

$$\text{EOQ} = \sqrt{\frac{2 \times 180 \times 8{,}000}{0.10 \times 19}} = 1{,}231 \text{ units}$$

This again is not feasible.

With the unit price of Rs 20, we have

$$\text{EOQ} = \sqrt{\frac{2 \times 180 \times 8{,}000}{0.10 \times 20}} = 1{,}200 \text{ units}$$

This is a feasible order quantity.

Now, the total cost corresponding to 1,200 units,

$$TC(1{,}200) = 8{,}000 \times 20 + \frac{8{,}000}{1{,}200} \times 180 + \frac{10}{100} \times \frac{1{,}200}{2} \times 20$$

$$= 1{,}60{,}000 + 1{,}200 + 1{,}200 = \text{Rs } 1{,}62{,}400$$

We shall determine the total cost at cut-off points 1,500 and 2,000.

$$TC(1{,}500) = 8{,}000 \times 19 + \frac{8{,}000}{1{,}500} \times 180 + \frac{10}{100} \times \frac{1{,}500}{2} \times 19$$

$$= 1{,}52{,}000 + 960 + 1425 = \text{Rs } 1{,}54{,}385$$

$$TC(2{,}000) = 8{,}000 \times 18.50 + \frac{8{,}000}{2{,}000} \times 180 + \frac{10}{100} \times \frac{2{,}000}{2} \times 18.50$$

$$= 1{,}48{,}000 + 720 + 1{,}850 = \text{Rs } 1{,}50{,}570$$

Since the total cost is minimum at Q = 2,000, it represents the optimal order quantity.

Model 3: Build-up Model/EOQ Model for Production Runs

It is assumed in the classical EOQ model that the entire quantity of the item ordered for is received in a single lot. We will now relax this assumption and consider situations in which the goods are received for inventory at a constant rate over time and they are also being consumed at a constant rate. This is particularly relevant for situations when the item in question is being produced internally rather than being procured from external suppliers. When the production begins, a constant number of units are supposed to be added to the inventory each day till the time the production run is completed. Simultaneously, the items would be demanded at a constant rate, as stipulated earlier. Obviously, the rate at which they are produced has to be higher than the rate of their depletion, for only then can we think of the inventory build-up.

The inventory profile in this model is depicted in Figure 9.4.

Here the total (order) quantity Q is produced over a period, t_p, which is defined by the production rate p. Since the inventory does not pile up in one shot but rather continuously over a time period and is also consumed simultaneously, the average inventory level would be determined not only by the lot size Q, but also be affected by the production rate p and depletion (demand) rate d.

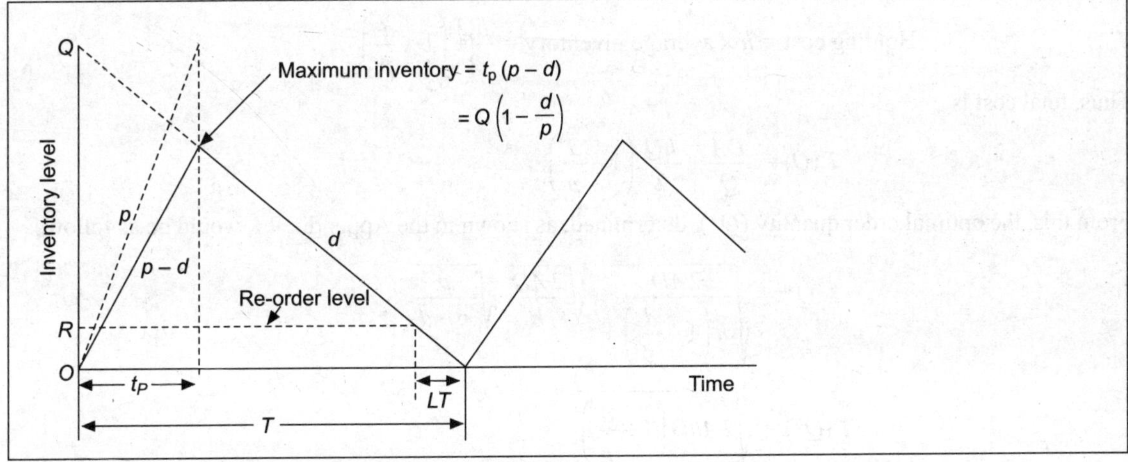

Figure 9.4 *Inventory Profile—Build-up Model*

To determine the average inventory, we proceed as follows:

we have, $t_p = \dfrac{Q}{p}$

where Q is the order quantity, or quantity produced, p is the rate of production and t_p is the length of the production run (i.e. the time required to produce the entire lot).

The inventory shall build up at a rate of $(p - d)$ per unit of time (in which p and d are expressed). Accordingly, the maximum inventory level shall be equal to $t_p(p - d)$. With this,

$$\text{average inventory} = \frac{t_p\,(p-d)}{2}$$

Since $t_p = Q/p$, we have

$$\text{average inventory} = \frac{Q\,(p-d)}{2p}$$

$$= \frac{Q}{2}\left(1-\frac{d}{p}\right)$$

The Cost Function For this model, as with the classical model, we have Total (Variable) Cost = Ordering Cost + Holding Cost.

Ordering Cost As before, the ordering cost equals DA/Q. However, as mentioned earlier, in the context of the production situation, it is referred to as production set-up cost including the cost of developing production plans, performing the essential paper work, time and resources employed in preparing production system for operation. It is a fixed charge which does not depend on the production lot of size Q. Thus, with a total demand (per annum) D, an order size Q and the set up cost equal to A, total annual 'ordering' cost shall be equal to, as stated above, DA/Q.

Holding Cost As before, with h as the holding cost per unit per year, the total annual holding cost would be as follows:

$$\text{Holding cost} = h \times \text{average inventory} = \frac{hQ}{2}\left(1 - \frac{d}{p}\right)$$

Thus, total cost is

$$T(Q) = \frac{DA}{Q} + \frac{hQ}{2}\left(1 - \frac{d}{p}\right)$$

From this, the optimal order quantity (Q^*), determined, as shown in the Appendix 9*A*, would be as follows:

$$Q^* = \sqrt{\frac{2AD}{h\left(1 - \frac{d}{p}\right)}} = \sqrt{\frac{2AD}{h}}\sqrt{\frac{p}{p-d}}$$

$$T(Q^*) = \sqrt{2AhD\left(1 - \frac{d}{p}\right)}$$

In Example 9.1, suppose that the delivery of the tubes is not made in a single lot and, instead, they are supplied at a rate of 50 tubes per working day. Under the conditions postulated, we have

$$\text{Economic Order Quantity, } Q^* = \sqrt{\frac{2 \times 150 \times 2{,}000}{2.40}}\sqrt{\frac{50}{50-8}}$$

$$= 545 \text{ units}$$

Total cost (of holding and ordering) corresponding to $Q^* = 545$ units,

$$T(Q^*) = \sqrt{2 \times 150 \times 2.40 \times 2{,}000\left(\frac{50-8}{50}\right)}$$

$$= \text{Rs } 1{,}100$$

Note that this cost is lower than the total cost of Rs 1,200 under the regular EOQ model.

Maximum Stock Level In the context of the present problem, t_p would be equal to the number of days required to deliver the entire lot. At the supply rate of 50 units a day, it would take 545/50 = 10.9 days to supply the lot. Thus,

$$\text{Maximum Stock Level} = t_p(p - d) = 10.9(50 - 8) = 458 \text{ units}$$

$$\begin{aligned}\text{Re-order Level} &= \text{consumption during the lead time}\\ &= \text{consumption rate} \times \text{lead time}\\ &= 15 \times 8 = 120 \text{ units.}\end{aligned}$$

Model 4: Inventory Model with Planned Shortages

In general inventory situations, a shortage (or stockout) is considered undesirable and is avoided, if possible. This is because shortages may, and are likely to, mean loss of customer goodwill, reduction in future orders, unfavourable change in the market share, and so on. While in some situations the customers shift to other sources for their requirements, and so may be lost forever, in some others the customers may not withdraw the orders and wait until the next shipment arrives. This latter situation is called, as mentioned earlier, the *back-ordering situation*. We shall now develop an inventory model under the assumption that back-ordering is possible. The EOQ model assumes that the inventory is replenished precisely when the inventory level falls off to zero. There is no question of shortages and, therefore, the cost of shortage is not considered in that model. With the assumption of back-ordering, shortages may in fact be deliberately planned to occur. It may be advisable on economic considerations, specially when the value of the item in question is very high with consequent high holding cost. Basically, then, it is a question of setting off the cost of shortages against the saving in the holding cost.

The inventory profile of this model is shown in Figure 9.5.

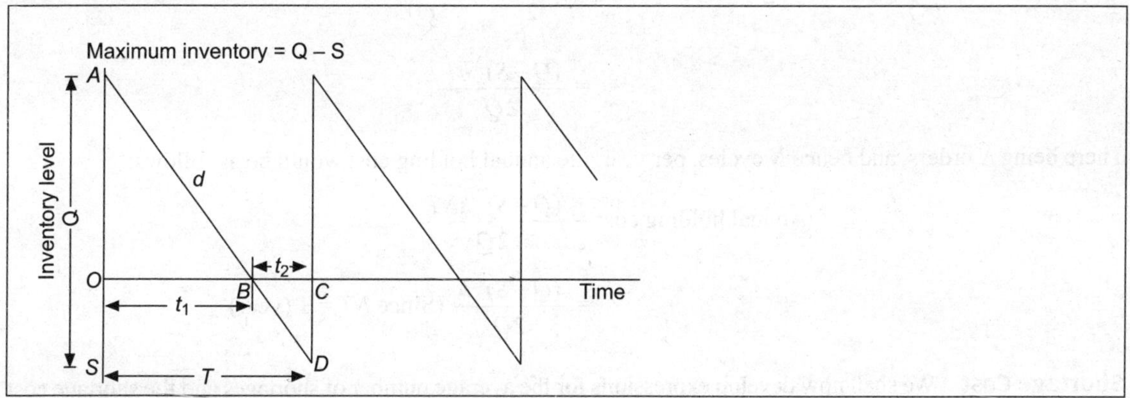

Figure 9.5 *Inventory Profile—Planned Shortages Model*

S below the zero level indicates negative inventory i.e. the number of units backordered. As soon as the lot of Q items is received, the customers whose orders are pending would be supplied their requirements immediately and as such the maximum inventory level would be $Q - S$. The inventory cycle T would be divided in two phases: t_1—the time when inventory is on hand and orders filled as and when they occur, and t_2—when there is a stockout and all the orders are placed on backorder.

In developing the cost function, we would consider the cost of shortages in addition to the holding and the ordering costs. Cost of shortages or the backordering cost is incurred in terms of the labour and special delivery expenses and the loss of customer goodwill (which may be taken to be a function of the time a customer has to wait). Thus,

$$\text{Total (variable) cost} = \text{ordering cost} + \text{holding cost} + \text{shortage cost}$$

Ordering Cost As seen before, if the cost of placing an order be A, and the total demand be D, we have,

$$\text{Annual ordering cost} = \frac{D}{Q} A$$

Holding Cost As noted earlier, t_1 is the period in a given inventory cycle when positive inventory is held. Since the maximum inventory, M, is $Q - S$, the average inventory level equals $(Q - S)/2$. Thus,

$$\text{Holding cost during a given cycle } T = \frac{Q-S}{2} ht_1$$

From the Figure 9.5, we observe that the quantity $Q - S$ is sufficient to last a period t_1. Thus, $Q - S = t_1 d$, where d is the usage rate. Similarly, a quantity Q is adequate to last a full cycle T, and, therefore, $Q = Td$. Dividing the first of these equations by the second, we get

$$\frac{Q-S}{Q} = \frac{t_1 d}{Td}$$

or

$$t_1 = \frac{T(Q-S)}{Q}$$

Substituting the value of t_1 in this expression for holding cost, we get

$$\text{Holding cost during a given cycle } T = \frac{Q-S}{2} h \frac{T(Q-S)}{Q}$$

$$= \frac{(Q-S)^2 hT}{2Q}$$

There being N orders, and hence N cycles, per year, the annual holding cost would be as follows:

$$\text{Annual holding cost} = \frac{(Q-S)^2 hNT}{2Q}$$

$$= \frac{(Q-S)^2 h}{2Q} \quad \text{(Since } NT = 1 \text{ (year))}$$

Shortage Cost We shall now develop expressions for the average number of shortages and the shortage cost with the help of which we shall determine the annual shortage cost. Since S represents the maximum level of shortages, the average level of shortages, during the period when there is a shortage (t_2) shall be $S/2$. If we let b represent the backorder cost, that is to say, the shortage cost per unit of shortage per year, we have,

$$\text{Shortage cost in a given cycle } T = \frac{S}{2} bt_2$$

$$\text{As before, we note that } d = \frac{S}{t_2} = \frac{Q}{T}$$

or

$$t_2 = \frac{TS}{Q}$$

Substituting this value in the earlier expression,

$$\text{Shortage cost per cycle} = \frac{S}{2} b \frac{TS}{Q}$$

$$= \frac{bTS^2}{2Q}$$

$$\text{Annual shortage cost} = \frac{bTNS^2}{2Q} \quad \text{(when there are } N \text{ cycles)}$$

$$= \frac{bS^2}{2Q} \quad (TN \text{ being equal to one})$$

From this analysis, then, the total cost expression would be as follows:

$$T(Q) = \frac{D}{Q}A + \frac{(Q-S)^2 h}{2Q} + \frac{bS^2}{2Q}$$

Derived from this expression, as shown in the Appendix 9*A* to this chapter, the EOQ, Q^*, would be

$$Q^* = \sqrt{\frac{2AD}{h}\left(\frac{h+b}{b}\right)} = \sqrt{\frac{2AD}{h}}\sqrt{\frac{h+b}{b}}$$

Also,

$$S^* = Q^*\left(\frac{h}{h+b}\right) \text{ or } S^* = \sqrt{\frac{2ADh}{hb+b^2}} \text{ ; and}$$

$$\text{Maximum Inventory Level, } M = Q^* - S^* = Q^*\left(\frac{b}{b+h}\right)$$

$$T(Q^*) = \sqrt{2AhD}\sqrt{\frac{b}{h+b}}$$

It may be noted that the Q^* in this model is different from Q^* in the classical EOQ model by the added square root term $\sqrt{(h+b)/b}$. Obviously, as the backordering cost, b, becomes larger relative to the holding cost h, the expressions $(h + b)/b$ and $b/(h + b)$ in the earlier stated equations tend to approach unity. Therefore, in such situations, the results shall be close to the EOQ model results. Similarly, $h/(h + b)$ would, in such a case, be very small and as such S^* would also be quite small. As, however, the holding cost becomes larger in relation to the back-ordering cost, the number of back-orders would be large. That explains why many an item that are costly are handled on a back-order basis.

To continue with the TV tubes example, let us suppose that the shortage cost is known to be Rs 1.60 per unit per annum. We reproduce here the entire information. We have,

D = 2,000 units per annum,
A = Rs 150 per order,
h = Rs 2.40 per unit per annum,
b = Rs 1.60 per unit per annum,
No. of working days = 250
Lead Time = 15 days

Using this information, we have

$$\text{Economic Order Quantity, } Q^* = \sqrt{\frac{2 \times 150 \times 2{,}000}{2.40}}\sqrt{\frac{2.40 + 1.60}{1.60}}$$

$$= 500 \times 1.58 = 790 \text{ units}$$

$$\text{Optimal level of shortages, } S^* = 790\left(\frac{2.40}{2.40+1.60}\right)$$

$$= 474 \text{ units}$$

$$\text{Maximum inventory level, } M = 790 - 474 = 316 \text{ units}$$

$$\text{Total (variable) cost, } T(Q^*) = \sqrt{2 \times 150 \times 2.40 \times 2{,}000}\ \sqrt{\frac{1.60}{2.40 + 1.60}}$$

$$= 1{,}200 \times 0.632$$

$$= \text{Rs } 759.$$

Determination of the Re-order level The optimal shortage being 474 units and the consumption during the lead time being equal to 8 × 15 = 120 units, the re-order level would be established at a point where the shortage reaches 474 – 120 = 354 units.

∴ Re-order level = – 354 units (i.e. at a shortage level of 354 units)

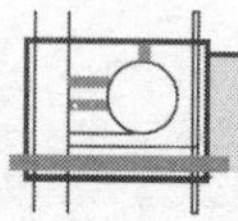

9.5 THE QUESTION OF SAFETY STOCK

The four inventory models discussed so far are all based on the twin common assumptions of constant and known demand for the item and the lead time. Because of these assumptions, these models are called *deterministic*. We shall now consider the situations in which the demand and/or lead time are not known with certainty and they need not be constant. They shall be taken as random variables, capable of assuming varying values whose probability distributions may be known.

In the deterministic models, the stock is replenished as soon as the stock reaches the point of exhaustion, due to the assumptions underlying them. Under such idealistic situations, there is no need to maintain any extra stock because the supplies would reach the moment the stock level reduces to zero and there would be no stockouts (unless they are intentionally allowed to occur). However, when the demand is variable and so is the lead time, there is a need to provide for the *safety* or *buffer stocks* in order to meet either or both the contingencies, viz. that the demand rate during the lead time is in excess of what was expected/forecasted; and that the delivery of goods is delayed. Figure 9.6 illustrates this point. The safety stock, then, acts as a cushion against stockouts caused by random perturbations of nature.

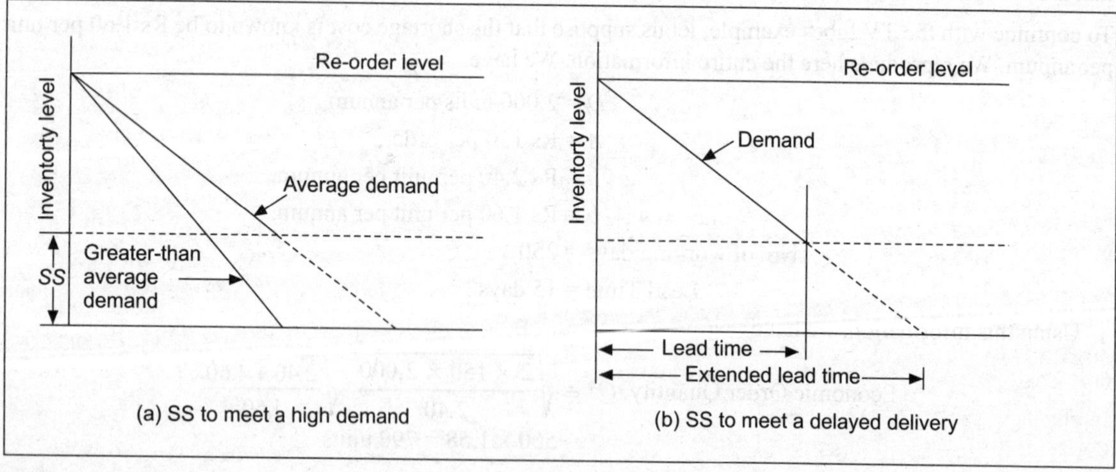

Figure 9.6 *Safety Stock*

The safety stock is an important constituent of the re-order point. In the fixed order quantity inventory system under consideration, the re-order level would be determined as the expected demand of the item during lead time* plus the safety stock. If the demand varies about the mean daily demand equal to d with the expected lead time equal to L days, and we set the re-order level R at $L\ \bar{d}$ units, then we should expect a shortage to occur in about half the lead time periods. To reduce this 50% probability of being out of stock, the safety stock SS would be required to be kept. Thus,

$$\text{Re-order level, } R = L\ \bar{d} + SS$$

An ideal inventory system where the actual demand is the same as expected demand and shows no variation from the forecasts, would be as depicted in Figure 9.7.

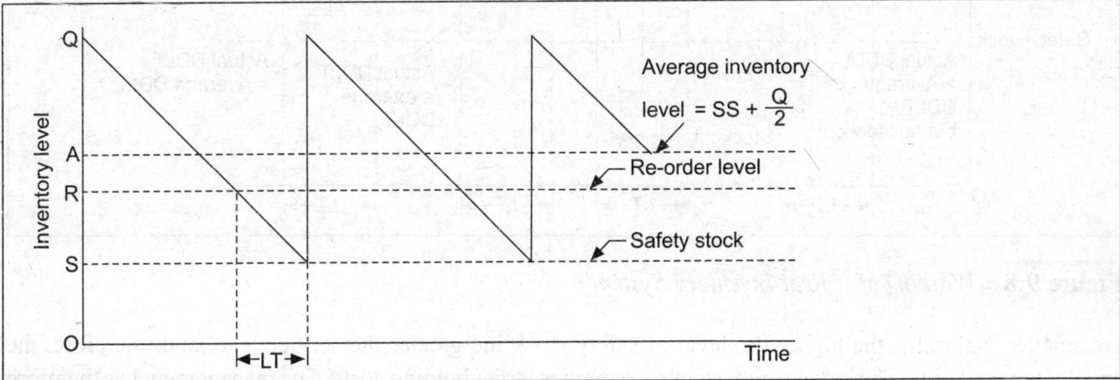

Figure 9.7 *An Ideal Inventory System*

Here, fresh supplies are received as soon as the stock level reaches the safety stock level. In this kind of a situation, the average stock held would be exactly equal to $SS + Q/2$. However, a realistic inventory situation would look like the one shown in Figure 9.8. In this, the demand is not continuous and uniform. Instead, it is discrete and irregular. Displayed in the figure are four inventory cycles. In the first cycle, the demand during lead time (DDLT) is more than expected but there is sufficient safety stock to meet the demand. In the second cycle, the demand is so great that the amount of safety stock cannot meet it. The result is a stockout. The third cycle portrays a condition where the DDLT is less than the expected demand. Lastly, the fourth cycle illustrates a situation wherein the DDLT just matches the expected demand.

Investment in safety stock is a fairly permanent investment in stock. Even when the demand is irregular and unequal, as illustrated earlier, the amount of (average) stock held would be nearly $Q/2$ plus the safety stock. Notice that the DDLT would be sometimes more, and sometimes less than the expected demand. Therefore, an amount equal to the safety stock would nearly always be carried. The idea of keeping the safety stock is clearly to prevent stockouts and it is the amount of stock that the organisation would always like to preserve for meeting extraordinary situations. As such, investment in safety stocks is made on considerations other than the ones on which the working stock is invested in. In general, higher safety stocks would be called for in situations where costs of stockouts are large; higher level of service (i.e. meeting greater proportion of demand) is sought; significant variations are observed in the lead time and/or lead time demand; and where holding costs

* We know that in this system, an order is placed as soon as the stock reaches the re-order level. Therefore, how high or low is the rate of demand before the re-order level reaches is of little consequence. What is significant is the level of demand during the lead time.

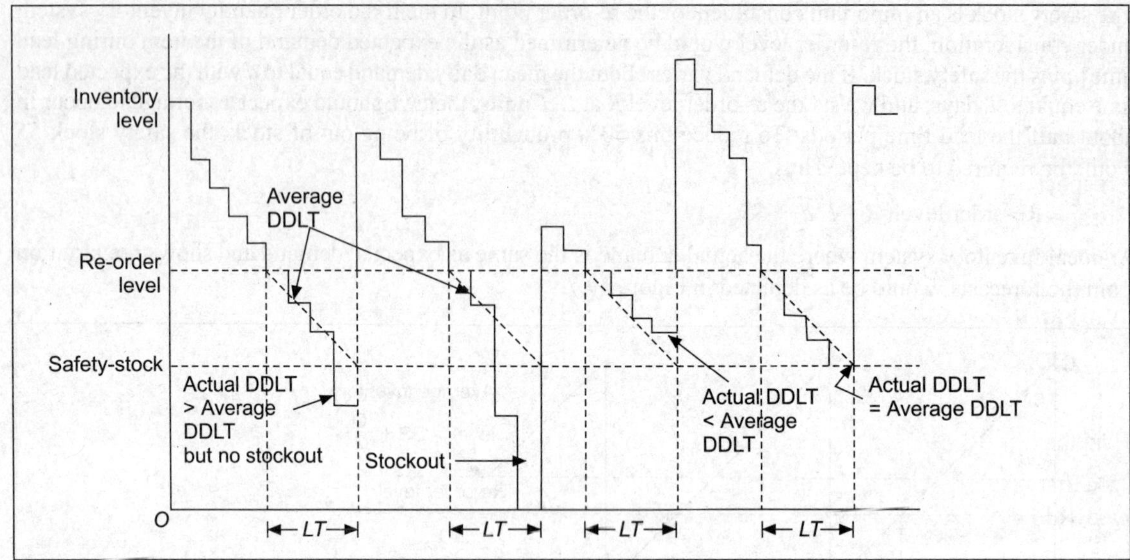

Figure 9.8 *Working of a Real Inventory System*

are smaller. Naturally, the higher the level of safety stock the greater the service level and, therefore, the smaller the stockouts. The larger safety stock also implies larger holding costs. The management has therefore to strike a balance between the two. The optimal safety stock level is determined where successively declining stockout costs and successively rising holding costs, caused by the successive units added to the safety stock, would balance.

There is no rigid formulation for determining the optimum level of safety stock. The different approaches available for the purpose are based on the demand, the lead time and the stockout costs. The complexity of the situation is determined by the extent and nature of information available about these factors.

9.5.1 Determination of Safety Stock

We shall now consider the question of determination of the safety stock and the re-order point. The analysis shall be done under each of the two conditions, first, when the stockout costs are known and, second, when they are not.

(a) When the Stockout Costs are Known We begin with the determination of an appropriate level of safety stock under the assumptions that the stockout costs are known, the lead time is constant and the probability distribution of demand during lead time is known. For the distribution of DDLT, the management must keep a record of the actual DDLT in terms of the quantities and the frequencies with which they are demanded. The frequencies are then converted and expressed as relative frequencies which indicate the probabilities of different DDLT values. Evidently, if the lead time is constant, a stockout can result only as a result of increase in demand after the order to replenish inventory has been placed (that is to say, during the lead time period). Obviously, if the increase had occurred before the re-order level was reached; an order would have been placed at the time when the stock level fell to re-order level.

Once the inputs are ready, the various safety stock levels are considered and the optimal level is chosen by using the given probabilities, cost of stock outs and carrying cost of the safety stock. This is illustrated with the following example.

Example 9.4 Using the EOQ model, an inventory manager has established the economic order quantity in respect of an item with annual usage of 24,000 units on the average, to be equal to 4800 units, with a mean usage of 80 units per day. The lead time is known to be 5 working days. It is also known that the cost of carrying one unit of the item in safety stock is Rs 15 and the cost of being out-of-stock for this particular item is Rs 50 for each unit of the item. From the past records, the manager finds that the demand during lead time over the last 50 re-order periods has been as follows:

DDLT (units):	250	300	350	400	450	500	550	600
Frequency:	1	3	6	18	10	7	3	2

Help the manager to establish an appropriate level of safety stock and re-order level.

As a first step, we derive the probability distribution of the DDLT as given below. The probabilities are derived by dividing each frequency by 50, the total frequency.

DDLT	*Frequency*	*Probability*
250	1	0.02
300	3	0.06
350	6	0.12
400	18	0.36
450	10	0.20
500	7	0.14
550	3	0.06
600	2	0.04

With the average demand at 80 units per day and the lead time equal to 5 days, the expected DDLT is equal to 80 $\times 5 = 400$ units. If the manager carries no safety stock and fixes the re-order level at 400 units, he would run the risk of being out-of-stock (when DDLT is 450, 500, 550 or 600 units) equal to $0.20 + 0.14 + 0.06 + 0.04 = 0.44$ or 44 per cent. In keeping with the given demand levels, the manager has the following options on the amount of safety stock, in addition to the option of having no safety stock kept.

(a) 50 units: This would cover a usage of 450 units during the lead time. Out-of-stock situation would result when the usage is 500, 550 or 600 units. The chances are $0.14 + 0.06 + 0.04 = 0.24$.

(b) 100 units: This would cover a usage of 450 or 500 units during the lead time durations. Out-of-stock situation would result in case of a usage of 550 or 600 units. The chances for this are 0.10.

(c) 150 units: This level of safety stock would cover a lead time usage of upto 550 units. Only possibility of stockout is when the usage is 600 units, which has a chance of 0.04.

(d) 200 units: This level of safety stock would ensure no out-of-stock conditions as it can meet the usage of upto 600 units, which is the maximum demand.

Further, since the annual usage is 24,000 units and the EOQ is 4,800 units, it follows that there would be $24{,}000/4{,}800 = 5$ inventory cycles (orders) in the year and the danger of running out-of-stock would also be as many times.

The cost of being out-of-stock for various options are shown calculated in Table 9.3.

TABLE 9.3 Determination of Stockout Costs

Safety Stock	*No. of Units Short*	*Prob.*	*Expected Annual Cost (Rs)**	*Total Annual Stock-out Costs (Rs)*
0	50	0.20	50 × 0.20 × 50 × 5 = 2,500	
	100	0.14	100 × 0.14 × 50 × 5 = 3,500	
	150	0.06	150 × 0.06 × 50 × 5 = 2,250	
	200	0.04	200 × 0.04 × 50 × 5 = 2,000	10,250
50	50	0.14	50 × 0.14 × 50 × 5 = 1,750	
	100	0.06	100 × 0.06 × 50 × 5 = 1,500	
	150	0.04	150 × 0.04 × 50 × 5 = 1,500	4,750
100	50	0.06	50 × 0.06 × 50 × 5 = 750	
	100	0.04	100 × 0.04 × 50 × 50 = 1,000	1,750
150	50	0.04	50 × 0.04 × 50 × 5 = 500	500
200	0	0	0	0

*Expected Annual Cost = No. of units short × Prob. × Stockout cost per unit × No. of orders per year

Further, the total cost in respect of various safety stock options is calculated and shown in Table 9.4.

TABLE 9.4 Determination of Optimal Level of Safety Stock

Safety Stock	*Stockout Costs*	*Carrying Cost*	*Total Annual Cost (Rs)*
0	10,250	0	10,250
50	4,750	50 × 15 = 750	5,500
100	1,750	100 × 15 = 1,500	3,250
150	500	150 × 15 = 2,250	2,750
200	0	200 × 15 = 3,000	3,000

It is evident that the total cost is lowest when safety stock of 150 units is kept. This is the optimal level of safety stock to carry. Accordingly,

$$\text{Re-order level} = \text{Expected DDLT} + \text{Safety stock}$$
$$= 80 \times 5 + 150 = 550 \text{ units.}$$

The analysis can be extended to cover situations where lead time is also variable. In cases where discrete probability distributions of the lead time and demand rate are given and they are known be independent, we can determine the joint probability distribution to obtain the required information. The joint probability distribution describes the probability distribution of DDLT by considering various combinations of lead time and demand rate, and obtaining the products of their probabilities. Consider the following examples.

Example 9.5 An examination of past records gives the following distributions of lead time and daily demand. Determine the distribution of demand during the lead time. Hence, compute the mean and variance of demand during lead time.

Lead Time Distribution

Lead Time (Days)	:	1	2	
Frequency	:	4	6	

Demand Distribution

Demand (units per day)	:	6	9	12
Frequency	:	5	3	2

(*CA, May, 1985*)

The joint probability distribution of DDLT is derived and given in Table 9.5. The values of DDLT in column (v) are obtained by considering various combinations of lead time and the daily demand.

TABLE 9.5 Probability Distribution of DDLT, and its Mean and Variance

Lead Time	*Prob.*	*Daily Demand*	*Prob.*	*DDLT* X_i	*Probability* p_i	$p_i X_i$	$p_i X_i^2$
1	0.4	6	0.5	6	0.200	1.200	7.200
1	0.4	9	0.3	9	0.120	1.080	9.720
1	0.4	12	0.2	12	0.230	2.760	33.120
2	0.6	6	0.5	15	0.180	2.700	40.500
2	0.6	9	0.3	18	0.174	3.132	56.376
2	0.6	12	0.2	21	0.072	1.512	31.752
				24	0.024	0.576	13.824
					Total	12.960	192.492

$$\text{Mean, } \overline{X} = \Sigma p_i X_i = 12.96$$

$$\text{Variance, } \sigma^2 = \Sigma p_i X_i^2 - (\Sigma p_i X_i)^2$$

$$= 192.492 - (12.960)^2 = 24.53$$

Thus, the distribution of demand during the lead time is contained in columns (v) and (vi) of the table and the mean and variance values of this distribution are 12.960 units and 24.53 (units)^2, respectively.

Example 9.6 DB plc operates a conventional stock control system based on re-order levels and economic order quantities. The various control levels were set originally based on estimates which did not allow for any uncertainty and this has caused difficulties because, in practice, lead times, demands and other factors do vary.

As part of a review of the system, a typical stock item, part No. X206, has been studied in detail as follows:

Data for part No. X206

Lead Times	*Probability*
15 working days	0.2
20 working days	0.5
25 working days	0.3
Demand per working day	*Probability*
5,000 units	0.5
7,000 units	0.5

Note: It can be assumed that the demands would apply for the whole of the appropriate lead time.

DB plc works for 240 days per year and it costs Re 0.15 p.a. to carry a unit of X 206 in stock. The re-order level for this part is currently 150,000 units and the re-order cost is Rs 1,000.

You are required to

(a) calculate the level of buffer stock implicit in a re-order level of 150,000 units;

(b) calculate the probability of a stockout;

(c) calculate the expected annual stockouts in units;

(d) calculate the stockout cost per unit at which it would be worthwhile raising the re-order level to 175,000 units. *(ICMA, May, 1990, Adapted)*

(a) From the given data, we may obtain the distribution of demand during lead time (DDLT) by multiplying various combinations of the lead time and demand per day with their corresponding probabilities. Finally, the expected DDLT may be calculated. This is done below:

Demand		*Probability*	*Expected Value*
15 × 5,000 =	75,000	0.2 × 0.5 = 0.10	7,500
15 × 7,000 =	105,000	0.2 × 0.5 = 0.10	10,500
20 × 5,000 =	100,000	0.5 × 0.5 = 0.25	25,000
20 × 7,000 =	140,000	0.5 × 0.5 = 0.25	35,000
25 × 5,000 =	125,000	0.3 × 0.5 = 0.15	18,750
25 × 7,000 =	175,000	0.3 × 0.5 = 0.15	26,250
		Total 1.00	123,000

Thus, expected demand during lead time, Exp. DDLT = 123,000 units.

Safety stock (Buffer stock) = Re-order Level – Exp. DDLT

= 150,000 – 123,000 = 27,000 units

(b) Probability of a stockout can be obtained from the distribution of DDLT in (a) above. Evidently, since probability of the demand exceeding 150,000 units is 0.15 (the probability of demand level of 175,000 units), it is the required probability.

(c) From the given data, we have

$$\text{EOQ} = \sqrt{\frac{2 \times (6{,}000 \times 240) \times 1000}{0.15}} = 138{,}564 \text{ units}$$

(**Note:** Expected daily demand = $5{,}000 \times 0.5 + 7{,}000 \times 0.5 = 6{,}000$ units, and annual demand = $6{,}000 \times 240$)

$$\text{Thus, number of orders per annum} = \frac{6{,}000 \times 240}{138{,}564} = 10.39$$

Accordingly, expected stockouts per annum = 0.15 (175,000 – 150,000) × 10.39 = 38,962 units

(d) At ROL = 150,000, stockout is 38,962 units, and at ROL = 175,000, stockout is nil (as obtained from distribution of DDLT in (a) above).

$$\text{Additional cost} = 25{,}000 \times 0.15 = \text{Rs } 3{,}750$$

∴ Unit cost at which increased stock is justified = Rs 3,750/38,962 = Re 0.096 or 9.6 paise per unit.

(b) When Stockout Costs are not Known: Service Level Policies Another approach to establish the safety stock level is based on the premise that the stockout costs cannot be easily and objectively determined. In this approach, the amount of safety stock is to be kept in accordance with the level of service the management aims to achieve.

It may be mentioned that service level can be measured severally but the most commonly used measure is as follows*:

$$\text{Service Level, } S = \frac{\text{Number of units supplied without delay}}{\text{Number of units demanded}}$$

The quantity may be measured over any interval of time. S would vary between 0 and 1, the former indicating complete dis-service and the latter a cent per cent service level.

A minimum service level in respect of DDLT may be established in advance on the basis of which the safety stock level would be fixed. This service level would mean the minimum proportion of the customers' orders during the lead time that would be filled out of the finished goods inventory, or the minimum proportion of the orders of the production department which shall be filled out of raw materials inventory, as the case may be. A 90% service level to the customers, for example, would imply that 90% of thc customers' orders, on the average, are sought to be filled out of inventory and 10% of the units demanded shall be refused because of stockouts during lead times.

To consider as to how the safety stock level may be determined, we shall assume the lead time to be constant and consider the distribution of DDLT.

DDLT Distributions

Sometimes sufficient historical data may be available to obtain the distribution of DDLT. In such a situation, the determination of the safety stock is fairly straightforward. Consider the following example.

Example 9.7 Given here is the distribution of the actual demand of an inventory item of the Cutex Manufacturing Company, over a long period of time.

---

* There are other measures like percentage of stockouts, stockout days, shortages, etc. For details *see* Buchan J and Koenigsberg E., *Scientific Inventory Management*, Prentice-Hall Inc.

Actual DDLT (in units)	Relative Frequency
81–100	0.06
101–120	0.19
121–140	0.28
141–160	0.40
161–180	0.05
181–200	0.02

The average lead time (with negligible variation) is 10 days.

Assuming that the distribution adequately represents the system,

(a) determine the safety stock required if it is desired to have no stockouts.

(b) determine the safety stock if it is desired to supply 95% of the DDLT (i.e. the service level is equal to 95%). Also obtain the re-order point corresponding to this.

The safety stock at any given service level is equal to the difference between the maximum demand during the lead time at that service level and the expected demand during lead time.

(a) When no stockouts are desired,

Safety Stock = maximum demand during lead time – average demand during lead time

From the given distribution, we observe that the maximum DDLT is 200 units while the average DDLT is equal to 135.5 units (as shown in the Table 9.6).

∴ Safety Stock = 200 – 135.5 = 64.5 units

(b) When a 95% service level is desired,

Safety Stock = maximum DDLT at 95% service level – average DDLT

Maximum demand at 95% service level shall be given by the 95th percentile of the given distribution.

TABLE 9.6 Calculation of Average and 95th Percentile

DDLT (Units)	*X*	*Frequency (f)*	*fX*	*Cumulative Frequency*
81–100	90.5	0.06	5.430	0.06
101–120	110.5	0.19	20.995	0.25
121–140	130.5	0.28	36.540	0.53
141–160	150.5	0.40	60.200	0.93
161–180	170.5	0.05	8.525	0.98
181–200	190.5	0.02	3.810	1.00
Total		1.00	135.500	

Average DDLT = 135.5 units.

By interpolation,

$$\text{95th percentile} = 160.5 + \frac{0.95 - 0.93}{0.05} \times 20$$

$$= 168.5 \text{ units.}$$

Thus, safety stock = 168.5 – 135.5 = 33 units.

The process of continuously compiling the distribution of historical demand/sales data for every item in the inventory in order to set the safety stock level may be very tedious. It can be done away with if the demand of the item in question be characterised by one of the certain well-defined theoretical distributions, whose properties can be used as a short-cut to obtain the desired information. Usually, the normal distribution is seen to fit well to the data about demand. Accordingly, we shall assume that the DDLT follows a normal distribution and see how the safety stock level may be determined.

Setting safety stock level at a desired level for constant lead time and normally distributed DDLT

When the DDLT may be approximated by a normal distribution with a certain mean (μ) and standard deviation (σ), the re-order point is obtained by locating the value of DDLT that has a probability of 5% of being exceeded—the probability of being less than this DDLT level is 95%, when the desired service level is 95%. The safety stock level would then be obtained by subtracting the average DDLT from this valuc. This idea is represented in Fig. 9.9 in which the extent of dis-service, is shown equal to α.

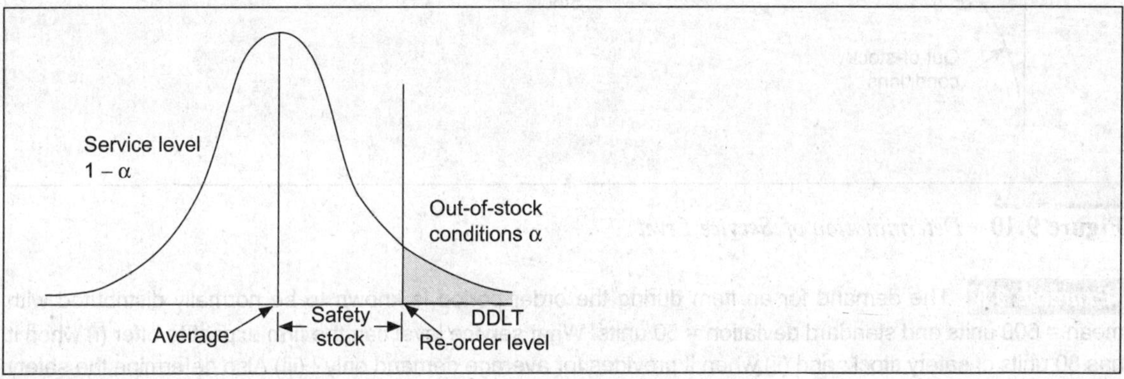

Figure 9.9 *Normally Distributed DDLT and Safety Stock*

The safety stock can also be directly obtained as:

$$SS = z\sigma$$

where z is the standard normal variate whose value depends upon the service level. Obtained from the area table (the Table B1 containing area under the normal curve, given at the end of the book in appendix B) are the z-values corresponding to some selected service levels as shown here:

Service Level	z
80 %	0.84
90 "	1.28
95 "	1.645
99 "	2.33

Similarly, for a particular level of demand rate (which is higher than the average rate) during the lead time, the safety stock and the service level corresponding thereto can be determined. The cutoff point on the normal curve established by the high demand rate where the stock would be reduced to nil shall suggest the level of service. This is depicted in Figure 9.10.

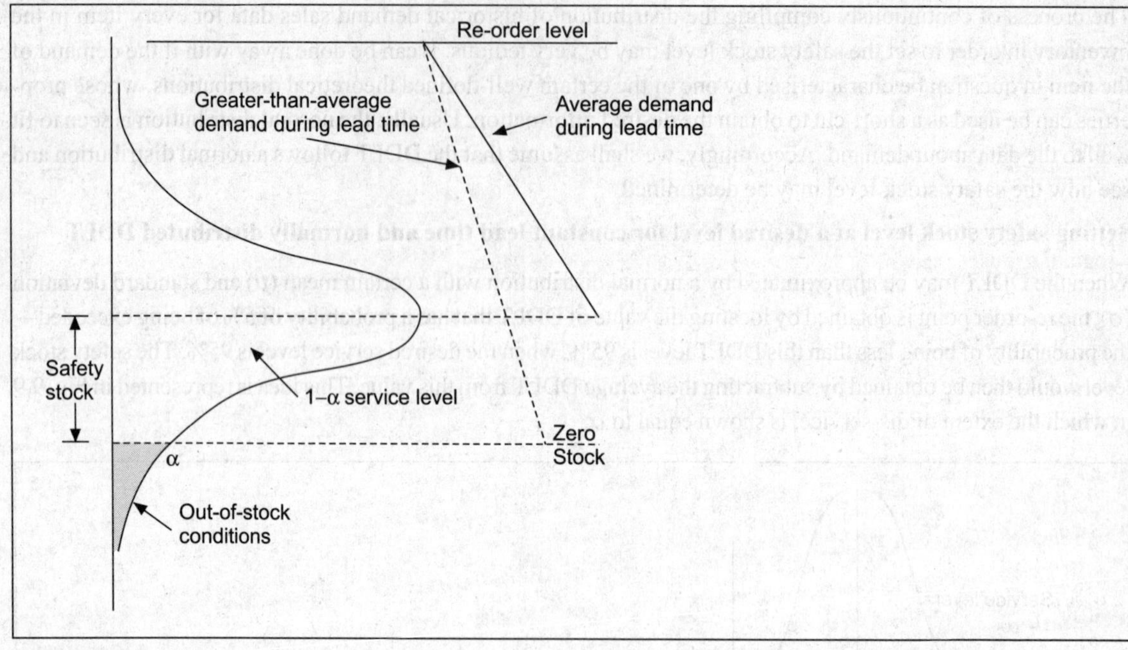

Figure 9.10 *Determination of Service Level*

Example 9.8 The demand for an item during the order period is known to be normally distributed with mean = 600 units and standard deviation = 50 units. What service level can the firm expect to offer (i) when it has 80 units of safety stock, and (ii) when it provides for average demand only? (iii) Also determine the safety stock if it is sought to provide a 99% service level to the customers.

(i) The given information is depicted in Figure 9.11.

To determine the required probability, we shall find the area to the left of $X = 680$. We have,

$$z = \frac{X - \mu}{\sigma} = \frac{680 - 600}{50} = 1.6$$

From the normal area tables, the area between μ and $z = 1.6$ equal 0.4452 and, thus, the area to the left of $X = 680$ is equal to $0.5 + 0.4452 = 0.9452$.

$\therefore$ Service level at 80 units $SS = 94.52\%$.

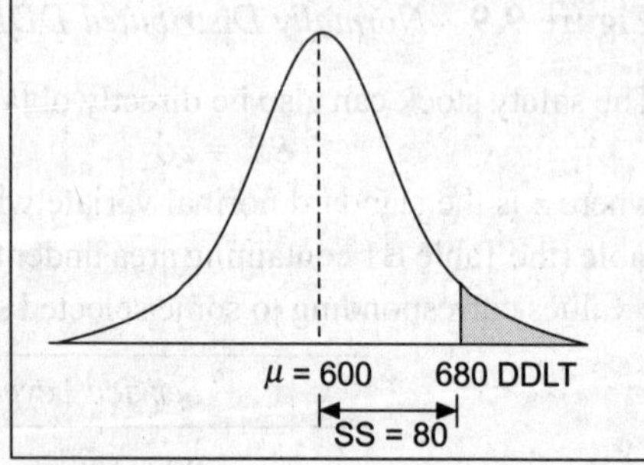

Figure 9.11 *Distribution of Demand During Lead Time*

(ii) When provision is made for average demand only, the service level is naturally 50%.

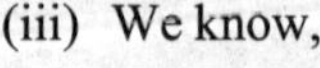

(iii) We know,

Safety Stock $= z\sigma$

The z-value corresponding to 99% service level is 2.33. Thus, $SS = 2.33 \times 50 = 116.5$ or 117 units (approx).

In the situation when both, the daily demand and the lead time, show significant variations (about their respective averages) the determination of the safety stock becomes very complicated. Even if their distributions be known, yet the relation of safety stock to stockouts cannot be directly established. A technique known as

Monte Carlo Simulation can be used in such cases to simulate the effects of the interactions between fluctuating demand and fluctuating lead times (see Chapter 17).

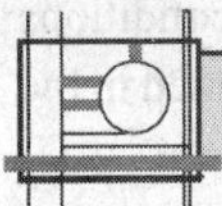

9.6 PERIODIC REVIEW SYSTEM

As already mentioned, in the periodic review system, the stock is replenished after reviewing the stock level at regular time intervals. It is unlike the fixed order quantity system where replenishment is triggered by the stock level reaching the re-order point. In this system, the stock is reviewed after a fixed interval of time and an order is placed for a quantity sufficient to replenish inventory to a predetermined level, referred to as *target inventory* (TI) level. After the re-order lead time, the shipment arrives and goes into inventory. While the review period is the same, the quantities ordered may be, and generally are, different in each period, being calculated as follows:

$$Q = \text{TI} - \text{inventory on hand} - \text{previous orders not yet received (if any)}$$

The target inventory (TI) level is obtained by adding the average demand during the review period and the average DDLT, as also the safety stock, if any.

The key issue in this system is the determination of the optimal length of review period *t*.

The procedure in this system has the advantage of simplicity in the sense that the person(s) reviewing would check the stock level, say, every three weeks and just note the quantity that the stock level is short of the level determined previously and orders would be sent for such amount. Besides, if the same supplier furnishes goods to the firm, the firm may save on the transportation charges as joint shipment of several items of stocks may be possible.

Under deterministic conditions, when the assumptions of the classical EOQ model are met, we can determine the optimal review period length as follows. We have, D = annual demand, A = ordering cost per order, h = holding cost per unit per annum, T = the interval between successive orders.

Since the annual demand is uniformally distributed, the demand during an inventory cycle of duration T (expressed as a fraction of a year) would equal DT, while the average inventory held during a cycle would be $DT/2$. With this, we can derive the cost equation as follows:

$$\text{Annual ordering cost} = \frac{D}{DT} \times A = \frac{A}{T};$$

$$\text{Annual holding cost} = \frac{DT}{2} \times h = \frac{hDT}{2}; \text{ and}$$

$$\text{Total cost} = \frac{A}{T} + \frac{hDT}{2}$$

The value of T that would minimise the total cost can be derived as:

$$T^* = \sqrt{\frac{2A}{DH}}$$

With the data given in Example 9.1, namely, D = 2,000 units per year, A = Rs 150 per order, h = Rs 2.40 per unit per year, and number of working days during the year = 250, we have,

$$T^* = \sqrt{\frac{2 \times 150}{2{,}000 \times 2.40}} = 0.25 \text{ year (as before)}$$

For a 250-working-days year, the optimal review period = 0.25 × 250 = 62.5 working days. In simple words, every 63rd working day should an order be placed.

A careful reader will note that this model is not different from the classical EOQ model when the conditions underlying the latter are satisfied. The difference is only of the emphasis—order quantity is emphasized in the EOQ model while the emphasis is on time in the review model.

For situations in which the demand and/or the lead time are *not certain*, safety stock is required to be kept. Under such conditions,

Target Inventory = average review period demand + average lead time demand + safety stock

The safety stock in such a case is determined in the following way. Suppose the demand during review period is normally distributed with mean μ_1 and standard deviation σ_1, and the demand during lead time, DDLT, is normally distributed with mean μ_2 and standard deviation, σ_2; then we first obtain the variance of the demand over the full cycle by combining the two variances as follows:

$$\sigma^2 = \sigma_1^2 + \sigma_2^2$$

From this, σ can be obtained by extracting its square root. Once this is calculated, the safety stock, *SS*, value can be determined as: $SS = z\sigma$. The value of *z*, as before, depends upon the service level chosen.

Evidently, the system of periodic reviews generally would require larger provisions for safety stock, than do the fixed order quantity models. This is because, in the fixed order quantity models, there is a continuous monitoring of the stocks and orders are placed as soon as the stock falls to the point of re-order. On the contrary, in the periodic review system, safety stocks must be provided as a protection against the stockouts which may occur during the interval prior to the time point when stocks are to be reviewed, as well as the stockouts that may occur during the lead time. It is amply brought out by the fact that, in a fixed order quantity model, the safety stock is based on the standard deviation of the DDLT alone, while in the periodic review method, the safety stock is based on the standard deviation value which is large in magnitude as it is derived from the combination of the two standard deviation values.

Example 9.9 The Chi-Square Manufacturing Company has adopted the periodic review system for its inventory management. For an item PS-700, the review period is 25 days during which the demand for this item is known to be normally distributed with an average of 1,450 units and standard deviation equal to 150 units, while the lead time demand is also normally distributed with parameters mean = 430 units and standard deviation = 80 units. How much safety stock should be kept in respect of this item when a 90% service level is desired?

From the given information,

$$\mu_1 = 1{,}450,\ \sigma_1 = 150;$$

$$\mu_2 = 430,\ \sigma_2 = 80$$

$$\sigma = \sqrt{\sigma_1^2 + \sigma_2^2} = \sqrt{150^2 + 80^2} = 170 \text{ units}$$

z (corresponding to 90% service level) = 1.28

$$\therefore \quad \text{Safety Stock} = z\sigma = 1.28 \times 170 = 217.6 \text{ or } 218 \text{ units (app.)}$$

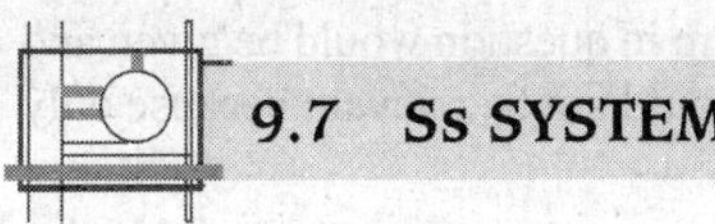

9.7 Ss SYSTEM

A third inventory management system is called the Ss system, also termed sometimes as *optional replenishment policy*. This system represents a modified periodic review system where a lower limit is placed on the size of the variable re-order quantity. It uses the formulation of the periodic review system as also the re-order point of the fixed order quantity system. In this system, the rule of operation is:

At the time of each periodic review, order a quantity q such that

$$q = TI - (I + q_0) \text{ whenever } I + q_0 \leq R,$$

where

TI = target inventory, in units,

I = inventory on hand, in units,

q_0 = quantity on order, in units, if any, and

R = re-order level:

and do not order otherwise.

This policy incorporates the re-order feature of the fixed order quantity system with variable order quantity characteristic of the periodic review system, and is defined by the following three parameters:

- the length of the review period,
- the maximum inventory level, and
- the re-order level.

The inventory profile of this system is displayed in Figure. 9.12.

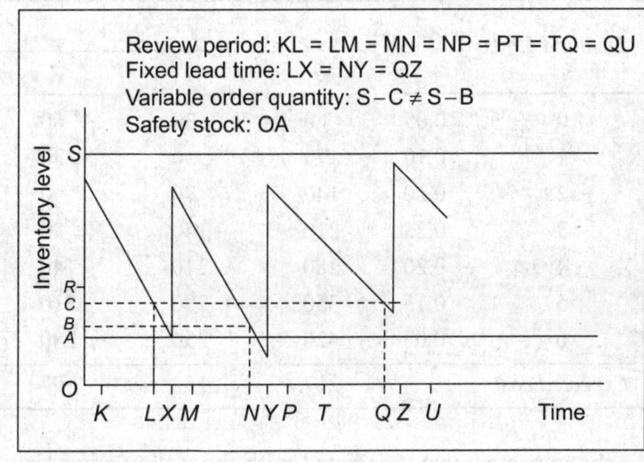

Figure 9.12 *Inventory Profile of Ss System*

In the figure, S represents the maximum stock level—the target inventory, while the re-order level is given by R. The review period is fixed, which equals KL (or LM, MN, etc.) while the lead time is also fixed and equals LX (or NY, QZ, etc.). In case the stock level is at R or below, an order is placed for the item. When, however, at the review time, the stock level is greater than R, no order is placed. The safety stock carried is indicated by OA.

This system permits the orders to be placed in efficient quantities, and avoids the placement of frequent orders and costs resulting therefrom. It may be noted that this system may require substantial safety stocks. Thus, in case the inventory level at the time of review is slightly above the re-order point, the coverage would be required for two order intervals plus the lead time.

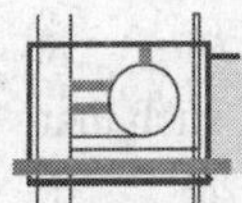

9.8 ONE-PERIOD MODEL OF INVENTORY MANAGEMENT

The inventory management systems that we have considered are all for controlling *over time*. Now we shall consider situations where the ordering quantity is to be decided upon for *single* periods only, under conditions of uncertain demand. Here the probability distribution of demand for the item in question would be given and also the information about holding and shortage costs. The ordering cost would not be relevant because only one order is to be placed and the ordering size would have no bearing on it.

Under these conditions, there are two approaches to the determination of the right ordering size—one in which a pay-off matrix is developed and the decision is taken in light of the expected cost (pay-off) associated with each strategy (see Chapter 13); and in the second approach, the *marginal analysis*, or *incremental analysis*, as it is called, is employed to get the required answer. We shall illustrate these with the following example.

Example 9.10 A TV dealer finds that the cost of holding a TV in stock for a week is Rs 30 and the cost of unit shortage is Rs 70. For a particular model of TV, the probability distribution is as follows:

Weekly Sales	:	0	1	2	3	4	5	6
Probability	:	0.05	0.10	0.20	0.25	0.20	0.15	0.05

How many units per week should the dealer order?

Since the weekly demand varies between 0 and 6 sets, the cost matrix for each of the strategies of holding 0 through 6 sets is given in Table 9.7. Next, the costs associated with various demand levels in respect of each of the strategies are multiplied by the respective probabilities and added up to get the expected cost.

TABLE 9.7 Determination of Optimal Order Quantity

Demand Units	*Prob.*	*Order Strategy*						
		0	*1*	*2*	*3*	*4*	*5*	*6*
0	0.05	0	30	60	90	120	150	180
1	0.10	70	0	30	60	90	120	150
2	0.20	140	70	0	30	60	90	120
3	0.25	210	140	70	0	30	60	90
4	0.20	280	210	140	70	0	30	60
5	0.15	350	280	210	140	70	0	30
6	0.05	420	350	280	210	140	70	0
Expected cost		217	152	97	62	52*	62	87

As the expected cost associated with the policy of ordering 4 sets is the least, this quantity is the optimal quantity to order every week.

Marginal Analysis Using this, the optimal quantity can be determined as follows:

Step 1 Determine the cumulative probabilities for the demand for the item, r, to be greater than or equal to Q. Thus, find $p(r \geq Q)$.

Step 2

Let h = over ordering or holding cost per unit for the period

b = under-ordering or shortage cost per unit for the period and calculate the ratio, R, such that

$$R = \frac{h}{h+b}$$

Call the ratio as *critical probability*.

Step 3 Compare the cumulative probabilities with the critical probability R. The largest value of Q for which the cumulative probability is equal to, or greater than, the critical probability value, would be the required ordering quantity.

In general terms, the optimal ordering quantity, O, is determined as

$$O = \text{Max. } P(r \geq Q) \geq R, \text{ where}$$

$$R = \frac{h}{h + b}$$

For Example 9.10 we have the cumulative probabilities given in Table 9.8.

TABLE 9.8 Determination of Optimal Ordering Quantity

Demand Q	*Prob. p*	$p(r \geq Q)$
0	0.05	1.00
1	0.10	0.95
2	0.20	0.85
3	0.25	0.65
4	0.20	0.40*
5	0.15	0.20
6	0.05	0.05

$$R = \frac{30}{30 + 70} = 0.30$$

Comparing the cumulative probabilities with R, we find that the maximum value of Q where $P(r \geq Q) \geq R$ is 4. This is the optimal quantity to order.

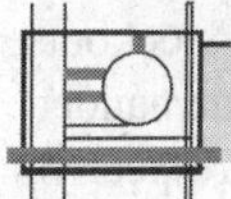

9.9 SELECTIVE APPROACHES TO INVENTORY CONTROL

In practice, all items of inventory cannot, and need not, be controlled with equal attention. An effective inventory control calls for an understanding and knowledge of the nature of inventories. Some inventory items may be very important while some may to be too small or may be too unimportant to warrant a tight and intensive control. Managements often lay down priorities for different items for the extent and nature of control required in respect of each of them. For this purpose, items may be severally classified on different bases. Here we shall consider the following types:

ABC; VED; HML; SDE; S-OS; FNS; and XYZ.

9.9.1 ABC Analysis

In this analysis, the inventory items in an organisation are classified on the basis of their usage in monetary terms. It is very common to observe that usually a small number of items account for a large share of the total cost of materials and a comparatively large number of items involve an insignificant share. Based on this criterion, the items are divided into three categories as follows:

A : high consumption value items;
B : moderate consumption value items; and
C : low consumption value items.

The division of the items into various categories is accomplished by plotting the usage value of the items to obtain the *ABC distribution curve*, which is also called the *Pareto Curve* or *the curve of maldistribution*. The mechanics of the *ABC* analysis is given in a step-wise manner.

Step 1 Obtain a list of items along with information on their unit cost and the periodic (usually annual) consumption.

Step 2 Determine the annual usage value for each of the items by multiplying unit cost with number of units and rank them in descending order on the basis of their respective usage values.

Step 3 Express the value for each item as a percentage of the aggregate usage value. Now, cumulate the per cent of annual usage values.

Step 4 Obtain the percentage value for each of the items. For n items, each item shall represent $100/n$ per cent. Thus, if there are 20 items involved in classification, then each item would represent $100/20 = 5\%$ of the materials.

Next, cumulate these percentage values as well.

Step 5 Using the data on cumulated values of items and the cumulated percentage usage values, plot the curve by showing these, respectively, on X and Y axes.

Step 6 Determine appropriate divisions for the *A*, *B* and *C* categories. The curve would rise steeply upto a point. This point is marked and the items upto that point constitute the *A*-type items. Similarly, the curve would only be moderately sloped towards upright. The point beyond which the slope is negligible is marked and the items covered beyond that point are classed as *C*-type items because they cause only a negligible increase in the cost. The other items are the *B*-type items for which the curve depicts a gradual upward rise.

After the items are so classified, the inventory decisions are made on the basis of this classification. '*A*' items would call for a strict control and should be delivered near the time of use. The protection against their stockouts may be set as high as 98%. The '*C*' items might be kept in open storage available and as they are demanded they might be issued without formalities. For these, the re-order quantities might be larger than their respective EOQs. Periodic review system may be invoked for the procurement of such items. In regard to the *B*-class items, the policy would be of a fairly tight control—not as tight as for *A*-type items of course.

Example 9.11 Perform ABC analysis using the following data:

Item	*Units*	*Unit price (Rs)*	*Item*	*Units*	*Unit price (Rs)*
1	700	5.00	7	6,000	0.20
2	2,400	3.00	8	300	3.50
3	150	10.00	9	30	8.00
4	60	22.00	10	2,900	0.40
5	3,800	1.50	11	1,150	7.10
6	4,000	0.50	12	410	6.20

As a first step, we determine the usage values for the various items. These are shown in the fourth column of the Table 9.9. The percentage of usage value for each item to the aggregate usage value is contained in the next column of the table. The item rankings on the basis of usage values are given in the last column.

TABLE 9.9 Determination of Usage Values

Item	*Units*	*Unit Cost*	*Usage (Rs)*	*Percentage*	*Rank*
1	700	5.00	3,500	9.8	4
2	2.400	3.00	7,200	20.2	2
3	150	10.00	1,500	4.2	7
4	60	22.00	1,320	3.7	8
5	3,800	1.50	5,700	16.0	3
6	4,000	0.50	2,000	5.6	6
7	6,000	0.20	1,200	3.4	9
8	300	3.50	1,050	3.0	11
9	30	8.00	240	0.7	12
10	2,900	0.40	1,160	3.3	10
11	1,150	7.10	8,165	23.0	1
12	410	6.20	2,542	7.1	5
		Total	35,577	100.0	

In the Table 9.10, the items are rearranged on the basis of usage value percentages, as shown in the first column. The cumulative percentage values are obtained and also the percentages of items along with their cumulated values are determined, as shown in the next columns.

TABLE 9.10 Determination of Cumulative Percentages

Item	*% Value*	*Cum. Value*	*% Number*	*Cum. Value*
11	23.0	23.0	8.3	8.3
2	20.2	43.2	8.3	16.6
5	16.0	59.2	8.4	25.0
1	9.8	69.0	8.3	33.3
12	7.1	76.1	8.3	41.6
6	5.6	81.7	8.4	50.0
3	4.2	85.9	8.3	58.3
4	3.7	89.6	8.3	66.6
7	3.4	93.0	8.4	75.0
10	3.3	96.3	8.3	83.3
8	3.0	99.3	8.3	91.6
9	0.7	100.0	8.4	100.0

The *ABC* distribution curve is drawn in Figure 9.13 and *A*, *B*, and *C* classification is marked thereon. Items 11, 2 and 5 are type *A*; 1, 12, and 6 are type *B* and the remaining are type *C*.

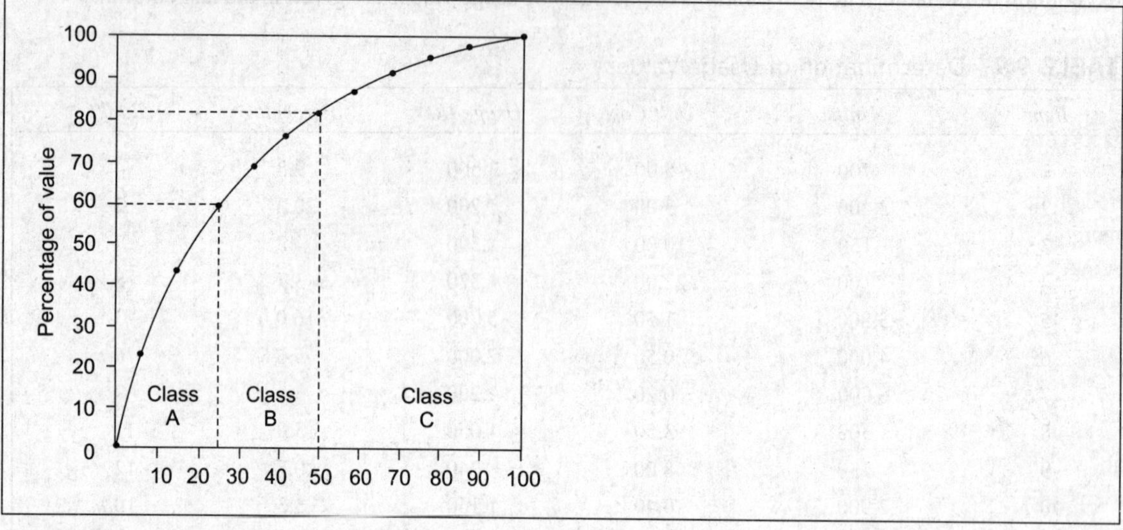

Figure 9.13 *ABC Classification of Items*

9.9.2 VED Analysis

In *VED* analysis, the items are classified on the basis of their *criticality* to the production process or other services. In the *VED* classification of materials, *V* stands for *Vital* items without which the production process would come to a standstill. *E* in the system denotes *Essential* items whose stockout would adversely affect the efficiency of the production system. Although the system would not altogether stop for want of these items, yet their non-availability might cause temporary losses in, or dislocation of, production. The *D* items are the *Desirable* items which are required but do not immediately cause a loss of production.

The *VED* analysis is done mainly in respect of spare parts.

9.9.3 HML Analysis

This is similar to the *ABC* analysis except that, in this analysis, the items are classified on the basis of unit cost rather than their usage value. The items are classified accordingly as their cost per unit is *H*-high, *M*-medium, or *L*-low. This type of analysis is useful for keeping control over materials consumption at the departmental level.

9.9.4 SDE Analysis

This uses the criterion of the availability of items. In this analysis, *S* stands for *Scarce* items which are in short supply, *D* refers to the *Difficult* items—meaning the items that might be available in the indigenous market but cannot be procured easily; while *E* represents *Easily available* items, from the local markets may be.

9.9.5 S-OS Analysis

S-OS analysis is based on the nature of supplies, wherein *S* represents the *Seasonal* items and *OS* represents the *Off-Seasonal* items. This classification of items is done with the aim of determining proper procurement strategies.

9.9.6 FSN Analysis

Based on the consumption pattern of the items, the *FSN* classification calls for classification of items, as *Fast-moving, Slow-moving,* and *Non-moving*. Some analysts classify the items as *FNSD*: Fast-moving, Normal-moving, Slow-moving, and Dead (or non-moving). This 'speed' classification helps in the arrangement of stocks in the stores and in determining the distribution and handling patterns.

9.9.7 XYZ Analysis

XYZ analysis is based on the closing inventory value of different items. Items whose inventory values are high, are classed as *X* items while those with low investment in them are termed as *Z* items. Other items are the *Y* items whose inventory value is neither too high nor too low.

It can be easily visualised that the various types of analysis discussed are not mutually exclusive. They can be, and often are, used jointly to ensure better control over materials. For example, *ABC* and *XYZ* analyses may be combined to classify and control depending on whether the items are *AX, BY, CZ, AY*. . . and so on. Similarly, XYZ-FSN combine classification exercise will help in timely prevention of obsolescence.

REVIEW ILLUSTRATIONS

Example 9.12 A manufacturing company has determined from an analysis of its accounting and production data for a certain part that (i) its demand is 9,000 units per annum and is uniformly distributed over the year, (ii) its cost price is Rs 2 per unit, (iii) its ordering cost is Rs 40 per order, (iv) the inventory carrying charge is 9% of the inventory value.
Further, it is known that the lead time is uniform and equals 8 working days, and that the total working days in a year are 300.

Determine:

(a) The economic order quantity, EOQ;
(b) The optimum number of orders per annum;
(c) The total ordering and holding cost associated with the policy of ordering an amount equal to EOQ;
(d) The re-order level;
(e) The number of days' stock at re-order level;
(f) The length of the inventory cycle;
(g) The amount of savings that would be possible by switching to the policy of ordering EOQ determined in (a) from the present policy of ordering the requirements of this part thrice a year; and
(h) The increase in the total cost associated with ordering (i) 20% more, and (ii) 40% less than the EOQ.

We are given that D = 9,000 units/year, A = Rs 40/order, i = 0.09, c = Rs 2/unit, and, therefore, $h = i \times c = 0.09 \times 2 = 0.18$. Also, lead time = 8 working days, and total working days in the year = 300.

(a) EOQ, $Q^* = \sqrt{\dfrac{2AD}{h}} = \sqrt{\dfrac{2 \times 40 \times 9{,}000}{0.18}} = 2{,}000$ units

(b) Optimum number of orders per year, $N^* = D/Q^*$
$= 9{,}000/2{,}000 = 4.5$

(c) Total variable cost, $T(Q^*) = \sqrt{2ADh} = \sqrt{2 \times 40 \times 9{,}000 \times 0.18} =$ Rs 360

(d) Re-order level = lead time in days × demand per day
$= 8 \times 9{,}000/300 = 240$ units

(e) No. of days' stock at the re-order level = 8 (equal to lead time)

(f) Length of inventory cycle, $T^* = Q^*/D = 2{,}000/9{,}000$
$= 0.222$ year or $0.222 \times 300 = 66.7$ days

Alternatively, T^* (in days) = Q^*/demand per day
$= 2{,}000/30 = 66.7$ days

(g) For the present policy of an order quantity = 3,000 units,
Ordering cost = 40×3 = Rs 120
Holding cost = $(3{,}000/2) \times 0.18$ = Rs 270
∴ $T(3{,}000) = 120 + 270$ = Rs 390

Thus, saving in cost = Rs 390 – Rs 360 = Rs 30 per year.

(h) (i) *Ordering 20% higher than EOQ:*

$$\text{Ordering quantity} = \frac{120}{100} \times 2{,}000 = 2{,}400 \text{ units.}$$

With $Q^* = 2{,}000$ and $Q = 2{,}400$, $k = 2{,}400/2{,}000 = 1.2$

We have $$\frac{T(Q)}{T(Q^*)} = \frac{1}{2}\left(\frac{1}{k} + k\right) = \frac{1}{2}\left(\frac{1}{1.2} + 1.2\right) = \frac{61}{60}$$

Thus, the cost would increase by 1/60th or $360 \times 1/60$ = Rs 6.

(ii) *Ordering 40% lower than EOQ:*

In such a situation, $k = 0.60$, and

$$\frac{T(Q)}{T(Q^*)} = \frac{1}{2}\left(\frac{1}{0.60} + 0.60\right) = \frac{17}{15}$$

Thus, the increase in cost would be 2/15th over the cost for EOQ, and would equal $360 \times 2/15$ = Rs 48.

Example 9.13 Yogesh keeps his inventory in special containers. Each container occupies 10 sq ft of store space. Only 5,000 sq ft of the storage space is available. The annual demand for the inventory item is 9,000 containers, priced at Rs 8 per container. The ordering cost is estimated at Rs 40 per order, and the annual carrying costs amount to 25% of the inventory value.

Would you recommend to Yogesh to increase his storage space? If so, how much should be the increase?

(M Com, Delhi, 1990)

Given, $D = 9{,}000$ units, $A = $ Rs 40/order, $c = $ Rs 8/unit and $h = 25\%$ of Rs 8 = Rs 2/unit/year, we have

$$\text{EOQ} = \sqrt{\frac{2DA}{h}}$$

$$= \sqrt{\frac{2 \times 9{,}000 \times 40}{2}} = 600 \text{ units}$$

With EOQ = 600 units, the maximum level of inventory would be 600 units and the space required to hold it equals 600 × 10 = 6,000 square feet. The present capacity being 5,000 square feet, he should increase the storage capacity by 1,000 square feet.

Example 9.14 A wholesaler supplies 30 stuffed dolls each day to various shops. Dolls are purchased from the manufacturer in lots of 120 each at Rs 1,200 per lot. Every order incurs a handling charge of Rs 60 plus a freight charge of Rs 250 per lot. Multiple and fractional lots can also be ordered, and all orders are met the next day. The incremental cost is Re 0.60 per year to store a doll in inventory. The wholesaler finances inventory investments by paying its holding company 2% monthly for borrowed funds.

How many dolls should be ordered at a time in order to minimise the total annual inventory cost? Assume that there are 250 week-days in a year. How frequently should he order? *(ICWA, December, 1984)*

From the given information, annual demand, $D = 30 \times 250 = 7{,}500$ units, $A = 60 + 250 = $ Rs 310/order, $h = 0.60 + 24\%\ (= 2\% \times 12)$ of Rs 10 = Rs 3/unit/year. Thus,

$$\text{Economic order quantity, } Q^* = \sqrt{\frac{2 \times 310 \times 7{,}500}{3}} = 1{,}245 \text{ units}$$

The frequency of ordering is given by the duration of the inventory cycle. We have,

$$T^* = Q^*/D = 1{,}245/7{,}500 = 0.166 \text{ year or } 0.166 \times 250 = 41.5 \text{ working days.}$$

Example 9.15 The purchasing manager of a distillery company is considering three sources of supply for oak barrels. The first supplier offers any quantity of barrels at Rs 150 each. The second supplier offers barrels in lots of 150 or more at Rs 125 per barrel. The third supplier offers barrels in lots of 250 or more at Rs 100 each. The distillery uses 1,500 barrels a year at constant rate. Carrying costs are 40 per cent, and it costs the purchasing agent Rs 400 to place an order. Calculate the total annual cost for the orders placed to the probable suppliers, and find out the supplier to whom orders should be placed. *(CA, May, 1991)*

The given inputs are:

Annual demand, $D = 1{,}500$ barrels,

Ordering cost, $A = $ Rs 400/order

Carrying cost, $h = 40\%$ of unit price

Unit price:

Supplier	*Lot size*	*Price per barrel*
1	Any quantity	Rs 150
2	150 or more	Rs 125
3	250 or more	Rs 100

We first calculate the EOQ for each of the suppliers.

$$\textit{Supplier 1: } \text{EOQ} = \sqrt{\frac{2 \times 1{,}500 \times 400}{0.40 \times 150}} = 141.42 \simeq 141 \text{ barrels}$$

$$\textit{Supplier 2: } \text{EOQ} = \sqrt{\frac{2 \times 1{,}500 \times 400}{0.40 \times 125}} = 154.9 \simeq 155 \text{ barrels}$$

$$\textit{Supplier 3: } \text{EOQ} = \sqrt{\frac{2 \times 1{,}500 \times 400}{0.40 \times 100}} = 173.2 \simeq 173 \text{ barrels}$$

Now, since the EOQ for supplier 2 falls within the given range, we calculate total cost for order size of 155 barrels. For supplier 3, since the EOQ is below the minimum limit of the range (250 or more barrels), we may obtain total cost at the price break, that is, 250. Note that since a feasible EOQ is available at a lower price (at Rs 125), there is no point in considering supplier 1 whose supply price is Rs 150 per barrel.

Now,

$$\text{Total cost} = \text{purchasing cost} + \text{ordering cost} + \text{carrying cost}$$

Supplier 2:

$$\text{Total Cost} = 1{,}500 \times 125 + \frac{1{,}500}{155} \times 400 + \frac{155}{2} \times 125 \times 0.40$$

$$= 1{,}87{,}500 + 3{,}871 + 3{,}875 = \text{Rs } 1{,}95{,}246$$

(The difference between ordering and carrying costs at EOQ is due to rounding.)

Supplier 3:

$$\text{Total Cost (250)} = 1{,}500 \times 100 + \frac{1{,}500}{250} \times 400 + \frac{250}{2} \times 100 \times 0.40$$

$$= 1{,}50{,}000 + 2{,}400 + 5{,}000 = \text{Rs } 1{,}57{,}400$$

Since total cost is minimum for supplier 3, the order should be placed with supplier 3. The order quantity is equal to 250.

Example 9.16 A company uses 8,000 units of a product as raw material, costing Rs 10 per unit. The administrative cost per purchase is Rs 40. The holding costs are 28% of the average inventory. The company is following an optimal purchase policy and places orders according to the EOQ. It has been offered a quantity discount of one per cent if it purchases its entire requirement only four times a year.

Should the company accept the offer of quantity discount of one per cent? If not, what minimum discount should the company demand? *(M Com, Delhi, 1993)*

With D = 8,000 units, A = Rs 40/order, and h = 28% of Rs 10 = Rs 2.80/unit/year, we have

$$\text{EOQ, } Q^* = \sqrt{\frac{2AD}{h}}$$

$$= \sqrt{\frac{2 \times 40 \times 8{,}000}{2.80}} \simeq 478 \text{ units}$$

$$\text{Total Cost} = DC + \frac{D}{EOQ} \times A + \frac{EOQ}{2} \times h$$

$$= 8{,}000 \times 10 + \frac{8{,}000}{478} \times 40 + \frac{478}{2} \times 2.80$$

$$= \text{Rs } 81{,}340$$

Discount offer

With a 1% discount offered, the unit price = 99% of Rs 10 = Rs 9.90

$$\text{Total Cost} = 8{,}000 \times 9.90 + \frac{8{,}000}{2{,}000} \times 40 + \frac{2{,}000}{2} \times 0.28 \times 9.90 = \text{Rs } 82{,}132$$

Since the total cost is higher when 1% discount is offered on a condition, the company should not accept the offer.

Determination of the minimum discount acceptable: Let x be the percentage minimum discount acceptable to the company. It may be determined by setting the total cost equal to the total cost associated with the policy of ordering EOQ. Accordingly,

$$8{,}000 \times \left(\frac{100 - x}{100}\right) \times 10 + 4 \times 40 + \frac{2{,}000}{2} \times 0.28 \times \left(\frac{100 - x}{100}\right) \times 10 = 81{,}340$$

On simplification, $$\left(\frac{100 - x}{100}\right) \times 82{,}800 = 81{,}340 - 160 = 81{,}180$$

or $$\frac{100 - x}{100} = \frac{81{,}180}{82{,}800} = 0.98$$

Thus, $$x = 100 - 98 = 2$$

Thus, minimum discount acceptable to the company is 2%.

Example 9.17 An aircraft uses high tensile bolts at an approximately constant rate of 50,000 numbers per year. The bolts cost Rs 20 each and the purchase department estimated the cost at Rs 200 to place an order. The opportunity cost on working capital is 20% per year. No shortages are allowed.

(i) How frequently should orders be placed and what is economic order quantity?

(ii) If orders could be executed only once in two months, the ordering quantity would be higher than the optimal quantity. By this, what would be the percentage change in the total relevant cost?

(iii) The company finds, to its error, that the cost of placing an order was Rs 5,000 and carrying cost was 15% per year and not the earlier data, how much was the company losing per year on inventory, because of imperfect information?

(iv) Working on the new ordering cost and carrying cost, the company receives the following offer from the supplier:

Upto 20,000 pieces per order, the price is	Rs 20.00 per unit
Above 20,000 pieces upto 30,000	Rs 19.50 per unit
Above 30,000 pieces upto 45,000	Rs 19.25 per unit
Above 45,000	Rs 19.00 per unit

Should they make use of this offer?

(v) If the entire requirement has to be bought in a single order, what should be the justifiable unit price offer to the company? *(DMS, Mumbai, 1993)*

(i) With $D = 50{,}000$ units, c = Rs 20 per unit, A = Rs 200 per order, and h = 20% of Rs 20 = Rs 4/unit/year, we have

$$\text{EOQ} = \sqrt{\frac{2AD}{h}}$$

$$= \sqrt{\frac{2 \times 200 \times 50{,}000}{4}} \simeq 2{,}236 \text{ units}$$

$$\text{Length of order cycle} = \frac{EOQ}{D} = \frac{2{,}236}{50{,}000} = 0.045 \text{ year} = 16 \text{ days approx}$$

(ii) When orders are placed once in two months:
Number of orders = 6
Ordering quantity = 50,000/6 = 8,333 units

$$\text{Total relevant cost} = \text{Ordering cost} + \text{Holding cost} = 6 \times 200 + \frac{8{,}333}{2} \times 4 = \text{Rs } 17{,}866$$

Total relevant cost under policy of ordering EOQ

$$= \sqrt{2ADh}$$

$$= \sqrt{2 \times 200 \times 50{,}000 \times 4} = \text{Rs } 8{,}944$$

$$\% \text{ change in the total relevant cost} = \frac{17{,}866 - 8{,}944}{8{,}944} \times 100 = 99.75\%$$

(iii) With revised values, $A = 5{,}000$ and h = 15% of 20 = 3,

$$\text{EOQ} = \sqrt{\frac{2 \times 5{,}000 \times 50{,}000}{3}} = 12{,}910 \text{ units}$$

$$\text{Total relevant cost} = \sqrt{2 \times 5{,}000 \times 50{,}000 \times 3} = \text{Rs } 38{,}730$$

$$\text{Actual total relevant cost under policy in (i) above} = \frac{50{,}000}{2{,}236} \times 5{,}000 + \frac{2{,}236}{2} \times 0.15 \times 20$$

$$= \text{Rs } 1{,}15{,}161$$

$\therefore$ Loss on account of imperfect information = Rs 1,15,161 – Rs 38,730 = Rs 76,431

(iv) From (iii) it is clear that EOQ = 12,910 units is feasible with a cost price of Rs 20 per unit. Since the price break is available at a quantity of 20,000 prices only, it is evident that lower prices would not yield feasible results. Accordingly, we may determine total cost of EOQ = 12,910, and at each price break. This is done below.

$$TC(12{,}910) = 50{,}000 \times 20 + \frac{50{,}000}{12{,}910} \times 5{,}000 + \frac{12{,}910}{2} \times 0.15 \times 20$$

$$= \text{Rs } 10{,}38{,}730$$

$$TC(20{,}000) = 50{,}000 \times 19.50 + \frac{50{,}000}{20{,}000} \times 5{,}000 + \frac{20{,}000}{2} \times 0.15 \times 19.50$$

$$= \text{Rs } 10{,}16{,}750$$

$$TC(30,000) = 50,000 \times 19.25 + \frac{50,000}{30,000} \times 5,000 + \frac{30,000}{2} \times 0.15 \times 19.25$$

$$= \text{Rs } 10,14,146$$

$$TC(45,000) = 50,000 \times 19.00 + \frac{50,000}{45,000} \times 5,000 + \frac{45,000}{2} \times 0.15 \times 19.00$$

$$= \text{Rs } 10,19,681$$

Since the total cost is lowest at quantity of 30,000 units, the order should be placed for a quantity just above 30,000 (note that the unit price of Rs 19.25 is available for a quantity *exceeding* 30,000).

Thus, the company should make use of the offer.

(v) If the requirement has to be bought in a single lot, the price may be obtained by setting the total cost with that price equal to the total cost which is available with the offer in (iv) above. Thus, with unit price set equal to x, we have

$$50,000x + \frac{50,000}{50,000} \times 5,000 + \frac{50,000}{2} \times 0.15x = 10,14,146$$

or $$53,750x = 10,09,146$$

or $$x = 10,09,146/53,750 = \text{Rs } 18.77$$

Thus, a unit price offer of Rs 18.77 would be justifiable as an alternative to the proposal given in (iv).

Example 9.18 A manufacturing company needs 2,500 units of a particular component every year. The company buys it at the rate of Rs 30 per unit. The order-processing cost for this part is estimated at Rs 15 and the cost of carrying a part in stock comes to about Rs 4 per year.

The company can manufacture this part internally. In that case, it saves 20% of the price of the product. However, it estimates a set-up cost of Rs 250 per production run. The annual production rate would be 4,800 units. However, the inventory holding costs remain unchanged.

(i) Determine the EOQ and the optimal number of orders placed in a year.

(ii) Determine the optimum production lot size and the average duration of the production run.

(iii) Should the company manufacture the component internally or continue to purchase it from the supplier?

(M Com, Delhi, 1993)

(i) With $D = 2,500$ units, $A = \text{Rs } 15/\text{order}$ and $h = \text{Rs } 4/\text{unit/year}$, we have,

$$\text{EOQ} = \sqrt{\frac{2AD}{h}} = \sqrt{\frac{2 \times 15 \times 2,500}{4}} \simeq 137 \text{ units}$$

$$\text{Optimal number of orders} = \frac{D}{EOQ} = \frac{2,500}{137} \simeq 18$$

(ii) Given, $D = 2,500$ units, $S = \text{Rs } 250/\text{set up}$, $h = \text{Rs } 4/\text{unit/year}$, $p = 4,800$ units and $d = 2,500$ units, we get

$$\text{Economic Lot Size (ELS)} = \sqrt{\frac{2DS}{h}}\sqrt{\frac{p}{p-d}}$$

$$= \sqrt{\frac{2 \times 2{,}500 \times 250}{4}}\sqrt{\frac{4{,}800}{4{,}800 - 2{,}500}} \simeq 808 \text{ units}$$

$$\text{Average duration of the production run} = \frac{808}{2{,}500} = 0.32 \text{ year}$$

(iii) *When item is purchased from outside:*

$$\text{Total cost} = Dc + \frac{D}{EOQ} \times A + \frac{EOQ}{2} \times h$$

$$= 2{,}500 \times 30 + \frac{2{,}500}{137} \times 15 + \frac{137}{2} \times 4$$

$$= \text{Rs } 75{,}548$$

When item is produced internally:

$$\text{Cost per unit} = 80\% \text{ of Rs } 30 = \text{Rs } 24$$

$$\text{Set-up cost, } S = \text{Rs } 250 \text{ per set-up}$$

$$\text{Total cost} = Dc + \frac{D}{ELS} \times S + \frac{ELS}{2} \times \frac{p-d}{p} \times h$$

$$= 2{,}500 \times 24 + \frac{2{,}500}{808} \times 250 + \frac{808}{2} \times \frac{2{,}300}{4{,}800} \times 4$$

$$= \text{Rs } 61{,}548$$

Evidently, the company should manufacture the product internally.

Example 9.19 The annual demand for a component *Z* is 2,08,000 units at a steady weekly rate of 4,000 units. An appropriate formula for calculating the economic batch quantity for production of a component which is being used (at a rate of *s*) and produced (at a rate of *r* per week) at the same time is

$$\text{EBQ} = \sqrt{\frac{2AC_0}{(1 - s/r)ip}}$$

The initial cost of installing the line for producing *Z* was Rs 6,000 for a maximum production capacity of 8,000 per week. The operating costs at full capacity are Rs 100 per week for labour, Rs 600 per week for material, Rs 300 per week for variable overhead and Rs 250 per week for fixed overhead. The cost of preparing the production order, producing drawings and so on is Rs 40 each time production is required. Storage costs including interest have been calculated as Rs 2 per unit per annum.

Now,

(a) Calculate the most economic quantity that should be produced each time the line is set up.

(b) Advise the management if it now thinks that there is an opportunity to produce a special one-off order for 50,000 *Z*'s for delivery in six months' time. Your answer should consider quantitative and qualitative factors. *(ICMA, May 1983, Adapted)*

(a) The relevant costs are:

Set-up cost C_0 = Rs 40 per set-up

Storage cost ip = Rs 2 per unit per year

The other costs are not relevant to determine the batch size as they are not related to, and do not affect it. Also, we have:

Annual Demand A = 2,08,000 units

Production rate r = 8,000 units/week

Usage rate s = 4,000 units/week

$$\text{Thus, EBQ} = \sqrt{\frac{2 \times 2{,}08{,}000 \times 40}{\left(1 - \frac{4{,}000}{8{,}000}\right) \times 2}} = 4{,}079 \text{ units}$$

In reality, this would be rounded down to 4,000 units, which is the weekly consumption. Thus, the production pattern would be to have a production run each week to produce the weekly requirement.

(b) At present, the variable costs associated with this component are:

Labour	:	Rs 100
Material	:	Rs 600
Variable Overhead	:	Rs 300
		Rs 1,000 per week

The maximum capacity being 8,000 units per week, the cost per unit at this level = 1,000/8,000 = Rs 0.125. However, the storage cost is Rs 2 per unit per year which is excessively high in relation to the unit production cost. As such, the production should be deferred till it could be. With the extra capacity of 4,000 units per week over and above the current level of production, the extra order would take 50,000/4,000 = 12.5 weeks to produce so that the production of this special order should be delayed for 26 (or 6 months) – 12.5 = 13.5 weeks.

Example 9.20 A dealer supplies you the following information with regard to a product dealt with by him:

Annual demand	:	10,000 units
Ordering cost	:	Rs 10 per order
Inventory carrying cost	:	20% of the value of inventory per year
Price	:	Rs 20 per unit

The dealer is considering the possibility of allowing some back-order (stockout) to occur. He has estimated that the annual cost of back-ordering will be 25% of the value of inventory.

(a) What should be the optimum number of units of the product he should buy in one lot?

(b) What quantity of the product should be allowed to be back-ordered, if any?

(c) What should be the maximum quantity of inventory at any time of the year?

(d) Would you recommend to allow back-ordering? If so what would be the annual cost saving by adopting the policy of back-ordering? *(MIB, Delhi, 1999)*

According to the given information, D = 10,000 units, A = Rs 10/order, h = 20% of Rs 20 = Rs 4/unit/year and b = 25% of Rs 20 = Rs 5/unit/year.

(a) Economic order quantity, (i) when stockouts are not permitted,

$$Q^* = \sqrt{\frac{2AD}{h}} = \sqrt{\frac{2 \times 10 \times 10{,}000}{4}} = 223.6 \text{ units}$$

(ii) when back-ordering is permitted,

$$Q^* = \sqrt{\frac{2AD}{h}}\sqrt{\frac{h+b}{b}} = \sqrt{\frac{2\times 10\times 10{,}000}{4}\left(\frac{4+5}{5}\right)} = 300 \text{ units}$$

(b) Optimal quantity of the product to be back-ordered,

$$S^* = \sqrt{\frac{2ADh}{hb+b^2}} = \sqrt{\frac{2\times 10\times 10{,}000\times 4}{4\times 5+5^2}} = 133 \text{ units}$$

(c) Maximum inventory level,

$$M = Q^* \frac{b}{b+h} = 300\times\frac{5}{5+4} = 167 \text{ units}$$

Alternatively, $M = Q^* - S^* = 300 - 133 = 167$ units

(d) We have,

$$TC(223.6) = \sqrt{2ADh} = \sqrt{2\times 10\times 10{,}000\times 4} = \text{Rs } 894.43$$

$$\text{TC}(300) = \sqrt{2ADh}\sqrt{b/(b+h)} = \sqrt{2\times 10\times 10{,}000\times 4}\sqrt{5/(5+4)}$$

$$= \text{Rs } 666.67$$

Since the total cost with back-ordering permitted is lower than when it is not permitted, the dealer should accept the proposal for back-ordering. This will result in a saving of 894.43 – 666.67 = Rs 227.76 per annum.

Example 9.21 A dealer supplies you the following information pertaining to an item of inventory:

Annual demand	:	800 units
Buying cost	:	Rs 150 per order
Inventory carrying cost	:	Rs 3 per unit per year
Back-ordering cost	:	Rs 20 per unit per year

(i) What will be the optimal number of units of the inventory item he should buy in one lot?
(ii) What quantity he should allow to be back-ordered?
(iii) What will be the cost-savings, if any, resulting from back-ordering?
(iv) What would be the maximum inventory of the item at any time of the year?
(v) If the dealer wants that no more than 25% of the units can be back-ordered, should the policy of back-ordering be adopted? *(M Com, Delhi, 1996)*

Here, $D = 800$ units, $A =$ Rs 150/order, $h =$ Rs 3/unit/year and $b =$ Rs 20/unit/year. With these inputs,

(i)
$$Q^* = \sqrt{\frac{2AD}{h}\left(\frac{h+b}{b}\right)}$$

$$= \sqrt{\frac{2\times 150\times 800}{3}\left(\frac{20+3}{20}\right)} \simeq 303 \text{ units}$$

Thus, optimal number of units to buy in one lot = 303.

(ii) Optimal number of units to be back-ordered,

$$S^* = \sqrt{\frac{2ADh}{hb+b^2}}$$

$$= \sqrt{\frac{2\times 150\times 800\times 3}{20\times 3+20^2}} \simeq 40 \text{ units}$$

(iii) Cost saving from back-ordering can be calculated as follows:

$$\text{Total relevant cost (with back-orders)} = \sqrt{2ADh}\ \sqrt{b/(b+h)}$$
$$= \sqrt{2 \times 150 \times 800 \times 3}\ \sqrt{20/(20+3)} = \text{Rs } 791$$

$$\text{Total relevant cost (without back-orders)} = \sqrt{2ADh}$$
$$= \sqrt{2 \times 150 \times 800 \times 3} = \text{Rs } 848$$

Thus, cost saving by back-ordering = Rs 848 – Rs 791 = Rs 57

(iv) Maximum inventory level $= Q^* - S^*$
$= 303 - 40 = 263$ units

(v) Here, the optimal number of units which can be back-ordered by the dealer is 40, which is less than 25% (of the ordering quantity). Since the percentage of back-orders is within the limit, specified, the given statement has no effect on the results obtained. The policy of back-ordering should obviously be adopted.

Example 9.22 The demand for an item is deterministic and constant over time at 600 units per year. The item costs Rs 50 per unit and the cost of placing an order is estimated to be Rs 5. The inventory carrying cost is 20% and the shortage cost is Re 1 per unit per month. Find the optimal ordering quantity if stockouts are permitted and the units can be back-ordered at the shortage cost indicated. What will the company lose if no stockouts are permitted? *(ICWA, December, 1985)*

Here, $D = 600$ units, $A =$ Rs 5/order, $i = 0.20$, $c =$ Rs 50/unit, $b =$ Re 1/unit/month or Rs 12/unit/year. Thus, $h = 0.20 \times 50 =$ Rs 10/unit/year. We have, when back-ordering is permitted,

$$\text{EOQ, } Q^* = \sqrt{\frac{2AD}{h}}\ \sqrt{\frac{h+b}{b}}$$
$$= \sqrt{\frac{2 \times 5 \times 600}{10}}\ \sqrt{\frac{10+12}{12}} = 33 \text{ units (app.)}$$

Total variable cost, $T(Q^*) = \sqrt{2 \times 5 \times 600 \times 10}\ \sqrt{12/(10+12)} = \text{Rs } 181 \text{ (app.)}$

When back-ordering is not permitted, we have

$$\text{EOQ, } Q^* = \sqrt{\frac{2 \times 5 \times 600}{10}} = 24 \text{ units (app.)}$$
$$T(Q^*) = \sqrt{2 \times 5 \times 600 \times 10} = \text{Rs } 245 \text{ (app.)}$$

Thus, additional cost when back-ordering is not allowed = Rs 245 – Rs 181 = Rs 64.

Example 9.23 You are given the following information in regard to an item:

Annual usage	=	20,000 units
Ordering cost	=	Rs 160 per order
Carrying cost	=	20% of the average inventory investment
Unit cost	=	Rs 2
Lead time	=	10 working days
Total working days	=	250 per annum

It is observed from the past data that the demand during lead time has been upto a maximum level of 150 units per day. Keeping this level in mind, what safety stock would you recommend? Also determine (a) the re-order level when the safety stock level suggested by you is kept in stock, (b) average level of inventory stock held, and (c) the ordering and carrying costs associated with this fixed order inventory policy.

We know, safety stock, SS = max DDLT – average DDLT

With maximum demand per day = 150 units,

average demand per day = 20,000/250 = 80 units,

lead time = 10 days,

we have, Maximum DDLT = 150 × 10 = 1,500 units, and

Average DDLT = 80 × 10 = 800 units

∴ SS = 1500 – 800 = 700 units

(a) Re-order level = SS + Average DDLT = 700 + 800 = 1,500 units

(b) Average stock level = $SS + Q^*/2$

From the given information, $Q^* = \sqrt{\dfrac{2 \times 160 \times 20{,}000}{0.20 \times 2}} = 4{,}000$ units

Thus, average stock level = 700 + 4,000/2 = 2,700 units

(c) Ordering cost $= \dfrac{20{,}000}{4{,}000} \times 160$ = Rs 800

Carrying cost = 2,700 × 0.40 = Rs 1,080

∴ Total cost = 800 + 1,080 = Rs 1,880

Example 9.24 For a Fixed Order Quantity System, find out the various parameters for an item with the following data:

Annual consumption – 10,000 units, cost of one unit – Re 1, set up cost – Rs 12 per production run, the inventory carrying cost – Re 0.24 per unit. Past lead times: 15 days, 25 days, 12 days, 14 days, 30 days, 17 days.

(ICWA, June, 1985)

From the given information,

(a) EOQ, $Q^* = \sqrt{\dfrac{2AD}{h}} = \sqrt{\dfrac{2 \times 12 \times 10{,}000}{0.24}}$ = 1,000 units

(b) Determination of safety stock (SS):

We know, SS = (max lead time – avg lead time) demand rate

Here max lead time = 30 days

Avg. led time = (15 + 25 + 12 + 14 + 30 + 17)/6 = 18.8 days

(Some people prefer to calculate the average lead time as 'modified average lead time' by excluding the extreme values. Without considering 25 and 30, and considering only 15, 12, 14 and 17 days, the average here may be taken to be 58/4 = 14.5 days. If this is done, the calculations would change accordingly).

Demand rate = 10,000/360 = 27.8 units/day

(This is by assuming 360 working days in a year)

$\therefore \quad SS = (30 - 18.8)\, 27.8 = 312$ units

Now, expected DDLT = average lead time × demand per day

$= 18.8 \times 27.8 = 523$ units

Thus, re-order level, ROL $= SS +$ expected DDLT

$= 312 + 523 = 835$ units

(c) Average stock $= SS + \dfrac{1}{2}$ EOQ

$= 312 + \dfrac{1}{2} \times 1{,}000 = 812$ units.

Example 9.25 Daily demand for a product *AX*-303 is normally distributed with mean = 60 units and a standard deviation of 6 units. The lead time is constant at 9 days (working). The cost of placing an order is Rs 20 and the annual holding costs are 20% of the unit price of Rs 10. A 95% service level is desired for the customers, who place orders during the re-order period. You are required to determine the order quantity and the re-order level for the item in question, assuming that there are 300 working days during a year.

According to the given information:

A = Rs 20/order, $D = 60 \times 300 = 18{,}000$ units, $h = 0.20 \times 10$ = Rs 2 per unit per annum.

Average demand during lead time, $\mu = 60 \times 9 = 540$ units

Variance of DDLT, $\quad \sigma^2 = \sum_{i=1}^{9} \sigma_i^2 = 9 \times 6^2 = 324$ units^2

Standard deviation of DDLT, $\quad \sigma = \sqrt{324} = 18$ units

Now,

$$\text{EOQ}, \quad Q^* = \sqrt{\frac{2AD}{h}} = \sqrt{\frac{2 \times 20 \times 18{,}000}{2}} = 600 \text{ units}$$

Re-order level, X, corresponding to a service level of 95% can be determined as follows:

$$z = \frac{X - \mu}{\sigma} \qquad (z = 1.645 \text{ for 95\% service level})$$

$$\therefore \quad 1.645 = \frac{X - 540}{18} \quad \text{or} \quad X = 540 + 29.6 \simeq 570 \text{ units}$$

Here average DDLT = 540 units, and SS = 30 units.

Example 9.26 A company uses a component for which the forecast of usage is known to be normally distributed with mean absolute deviation (MAD) = 40 units. The company management adopts a policy which limits the stockouts to be no more than one order cycle in a year. Required: (a) How much safety stock should be kept if the order quantity is generally a 2 weeks' supply? (b) How much safety stock should be kept if the order quantity is 4 weeks' supply?

For a normal distribution, MAD = $\left(\sqrt{2/\pi}\right)\sigma$. Thus,

$$\sigma = \text{MAD}/\sqrt{2/\pi} = 1.25 \text{ MAD}.$$

Since here MAD = 40, we have

$$\sigma = 1.25 \times 40 = 50.$$

(a) Since the order quantity is a 2 weeks' supply, the number of orders during the year, and therefore, the number of order cycles = 52/2 = 26. Thus, the probability of being out of stock = 1/26.

$\therefore$ Service level = 1 – dis-service (stockout) = 1 – 1/26 = 25/26 = 0.96

Here we are required to determine safety stock amount, X, so that a service level of 0.96 may be obtained. We have

$$z = \frac{X - 0}{50}$$

Since z corresponding to area 0.46 (= 0.96 – 0.50) is 1.75, we have

$$1.75 = \frac{X-0}{50} \quad \text{or } X = 50 \times 1.75 = 87.5 \text{ or } 88 \text{ units.}$$

(b) In case of a 4-week supply order size, the number of order cycles = 52/4 = 13. Thus, the probability of being out-of-stock = 1/13 = 0.0769.

The z value corresponding to the service level 1 – 0.0769 = 0.9231 is 1.43. Thus, safety stock X, required for this service level can be obtained as follows:

$$1.43 = \frac{X-0}{50} \quad \text{or} \quad X = 50 \times 1.43 = 71.5 \text{ or } 72 \text{ units}$$

Example 9.27 CD Ltd sells an electric motor but finds that it runs out of stock on occasions and thus loses the contribution on missed sales. The following information is available:

Estimated demand	12,000 per year
Purchase price	Rs 100 each
Selling price	Rs 155 each
Lead time	5 days guaranteed
Cost of holding a motor	Rs 20 per year
EOQ	1,200 motors

The company works a five-day week for 48 weeks a year, that is 240 days. The demand figures have been analysed for the last 27 weeks.

Motors sold	:	30	40	50	60	70	80	90	Total
Number of days level of sales occurred	:	10	20	50	30	15	5	5	135

At present, CD Ltd uses a re-order level of 250 motors and does not carry any safety stock because of the guaranteed delivery time. Ideally, it wishes to satisfy customers on average at least 95% of the time whilst minimising the associated costs.

You are required

(a) to estimate the annual stockout costs of using the present re-order level;

(b) to calculate what re-order level would enable the company to meet its 95% requirement.

(ICMA, November, 1992 Adapted)

To calculate the required values, we first obtain various inputs as follows:

Units Sold	*No. of Days*	*Probability*	*Units × Probability*
30	10	0.07	2.1
40	20	0.15	6.0
50	50	0.37	18.5
60	30	0.22	13.2
70	15	0.11	7.7
80	5	0.04	3.2
90	5	0.04	3.6
	135	1.00	54.3

Expected demand = 54.3 units per day

Expected demand during lead time = 54.3 × 5 = 271.5 ≃ 272 units

No. of orders per annum = 12,000/1,200 = 10

∴ Stockout cost per unit = Rs 155 – Rs 100 = Rs 55

(a) The calculation of cost of stockouts is shown in Table 9.11.

TABLE 9.11 Calculation of Stockout Costs

Re-order Level	*Safety Stock*	*Holding Cost (Rs)*	*Possible Shortage*	*Prob.*	*No. of Orders*	*Shortage Cost (Rs)*
250	0	0	50	0.22	10	6,050
			100	0.11	10	6,050
			150	0.04	10	3,300
			200	0.04	10	4,400
			Total Cost = Rs 19,800			
300	30	600	50	0.11	10	3,025
			100	0.04	10	2,200
			150	0.04	10	3,300
			Total Cost = Rs 9,125			
350	80	1,600	50	0.04	10	1,100
			100	0.04	10	2,200
			Total Cost = Rs 4,900			
400	130	2,600	50	0.04	10	1,100
			Total Cost = Rs 3,700			
450	180	3,600	0	0.0	10	0
			Total Cost = Rs 3,600			

From the table, it is evident that the annual stockout cost of using present re-order level of 250 is equal to Rs 19,800.

(b) It may be observed from the table that the re-order level of 450 units has the minimum cost and it enables the company to meet its requirement of 95% level of customer satisfaction.

Example 9.28 Your company's experience of being out-of-stock in respect of a key item is as follows:

Stockout (No. of Units)	*Number of Times*	
2,000	4	(1)
1,600	8	(2)
1,000	12	(3)
400	16	(4)
200	40	(10)
0	320	(80)
	400	(100)

Figures in brackets represent percentage of times the item was out of stock. Assume that the stockout costs are Rs 100 per unit. The carrying cost of inventory is Rs 50 per unit. Determine the optimal level of stockout inventory (safety stock). *(ICWA, June, 1983)*

The pay-off matrix when different levels of safety stock are kept and there are stockouts of various levels, is contained in Table 9.12. To illustrate, when a stock of 1,000 units is carried and the stockout of 2,000 units is experienced (so that the demand is for 2,000 units), then the total cost of inventory holding would be 1,000 × 50 = 50,000 and the cost of being out-of-stock would be (2,000 – 1,000) × 100 = Rs 1,00,000. Thus, for this combination, the total cost would be Rs 1,50,000.

The expected costs are obtained by the summation of the products of cost values and their respective probabilities, for each of the strategies. We observe that the optimal policy is not to keep any safety stock.

TABLE 9.12 Determination of Optimal Safety Stock Level

Stockout	*Prob.*	*Safety Stock (No. of units)*					
(No. of units)		*2,000*	*1,600*	*1,000*	*400*	*200*	*0*
2,000	0.01	1,00,000	120,000	1,50,000	1,80,000	1,90,000	2,00,000
1,600	0.02	1,00,000	80,000	1,10,000	1,40,000	1,50,000	1,60,000
1,000	0.03	1,00,000	80,000	50,000	80,000	90,000	1,00,000
400	0.04	1,00,000	80,000	50,000	20,000	30,000	40,000
200	0.10	1,00,000	80,000	50,000	20,000	10,000	20,000
0	0.80	1,00,000	80,000	50,000	20,000	10,000	0
Expected cost		1,00,000	80,400	52,200	25,800	17,800	11,800*

KEY POINTS TO REMEMBER

- Inventories are held for a variety of reasons. Usually, large amounts are invested in them and, therefore, a proper control over them is necessary.
- In context of inventory management, the three questions needed to be addressed to are: (i) How much to order? (ii) When to order? and (iii) How much safety stock be kept?
- The decisions are taken in the context of two inventory management systems, Fixed Order Quantity system and Periodic Review system, having reference to the total cost comprising purchase cost, ordering/set up cost, carrying cost and stockout cost.
- In situations where the demand is given and uniform and there is no variation in the lead time, no safety stock is required to be kept.
- Under the Fixed Order Quantity system, with uniform demand, known lead time and no shortages permitted, the order quantity is determined so as to minimise the cost. For the classical EOQ model, the ordering and carrying costs are the relevant costs and the least-cost order quantity (the economic order quantity, EOQ) is such where they are equal. An order quantity greater than the EOQ saves the ordering cost and increases carrying cost, while an order quantity smaller than EOQ has the opposite effect on these cost elements. In either case, however, the total cost increases. The order level is set where the stock left is equal to demand during lead time.
- A critical assumption of the classical EOQ model is that unit cost is the same whether the order quantity is small or large. The price break model deals with situations where discount is available for large order quantities. In such cases, the order quantity is established where the total cost including purchase cost, ordering cost and carrying cost is the minimum.
- In cases where the source of supply is internal, the build-up model provides solution to the problem of determining the optimal lot size and the frequency of production runs.
- In situations where back-ordering is possible at a cost, shortages of inventory are in fact planned to occur. The planned shortages model analyses such situations to determine the optimal quantity to order and the optimal number of units to be back-ordered.
- Safety stock is needed to be kept when there is variation in the demand and/or lead time. In general, greater the degree of variation, greater the amount of safety stock required. Its level also depends on the service level to be provided. If the distribution of demand during lead time may be approximated by a distribution such as normal, then safety stock and service level calculations can be done with reference to the given information.
- The Periodic Review system of inventory management, where ordering intervals are fixed but the order size may vary, usually necessitates greater level of safety stock to be kept.
- For better and effective control over inventory, items are usually classified into different categories. The most commonly used classification is the one based on their usage value, labelled *ABC*. Here, *A*: high, *B*: moderate, and *C*: low consumption value. The *A* class items are subjected to the highest degree of control while control is not stringent for *B* class items and there is virtually no control over *C* class items.
- Other classifications include VED, HML, SDE, S-OS, FNS and XYZ.

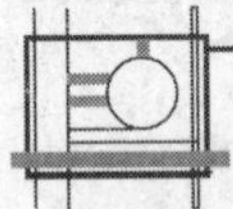

TEST YOUR UNDERSTANDING

Mark the following statements as T (*True*) or F (*False*).

1. *Lead time* is the time elapsing between placing an order and having the goods ready in stock for use.
2. The stock which is kept to meet the total demand during lead time is called *safety stock*.
3. The amount of safety stock required depends on the variations in the demand rate and the lead time.
4. Ordering cost is assumed to be independent of the order quantity.
5. The ordering cost per unit is variable while the holding cost per unit per time period is fixed.
6. Stockout cost is the easiest to estimate.
7. In periodic review system of inventory management, the stock is usually replenished at equal time intervals.
8. The Wilson formulation assumes, among other things, that the demand for the item is fixed and continuous over time.
9. When it is assumed that there are no variations in the demand rate and the lead time, the re-order level would be equal to the demand during lead time, and no safety stock is needed to be kept.
10. *Inventory cycle* is the time interval between the successive points of time when orders are placed.
11. In the classical inventory model, the ordering and the holding costs are equal at EOQ.
12. The relevant cost in the classical EOQ model is the sum of the ordering and holding costs.
13. With increases in the carrying cost, the optimal order quantity also increases.
14. The total cost curve in the price break model is discontinuous.
15. In case of multiple price breaks, once a feasible EOQ is determined, the total cost function is evaluated at the EOQ so determined and each of the points at which the breaks are available. The point at which the total cost is the minimum, gives the optimal order quantity.
16. With same inputs of annual demand, ordering/set up cost, and holding cost, the economic lot size in the build-up model is greater than economic order quantity in the classical EOQ model.
17. In the inventory model applicable to production runs, the production rate, p, must be greater than the demand rate, d.
18. An increase in set-up cost per production run leads to a decrease in the economic lot size.
19. For economic lot size, the total set-up cost need not be equal to the total holding cost.
20. It is permissible to incur both holding and shortage costs in the same inventory model.
21. The inventory model with planned shortages is applicable only when back-ordering is possible.
22. If back-ordering is allowed, the shortage cost can be an important cost component in the inventory model.
23. The total variable cost in the planned shortages model is split evenly between ordering cost on the one hand and holding cost and shortage cost on the other.
24. The economic order quantity increases with an increase in the back-ordering cost.
25. The total variable cost for the planned shortages model would be greater than that for the classical EOQ model because of the inclusion of an additional component of shortage cost.

26. The economic order quantity for the planned shortages model would be greater than that for the classical EOQ model with the same input data.
27. As the holding cost becomes large in relation to the back-ordering cost, the number of units back-ordered would be large.
28. The four inventory models are deterministic in nature.
29. Under the assumptions made in the inventory models, the stock is always replenished at the designated time and no safety stock is needed.
30. A 100% customer service level requires an amount of safety stock determined as follows:

    Safety Stock = maximum DDLT – expected DDLT
31. Where demand during lead time is assumed to be normally distributed with a certain mean and standard deviation, the service level corresponding to a given safety stock level shall be equal to the area under the normal curve to the left of the re-order point.
32. Both mean and standard deviation are necessary to be known for determining the service level corresponding to a given amount of safety stock, when the DDLT is given to be normally distributed.
33. The system of periodic reviews generally requires larger provisions for safety stock than does the fixed order quantity system.
34. The Ss system embodies the re-order feature of the fixed order quantity system and variable order quantity characteristic of the periodic review system.
35. The Ss system requires high safety stock volumes.
36. In ABC analysis, the inventory items are classified into three categories on the basis of their usage quantities.
37. The *A*-category items are those which fall in steep-rise segment of the curve of *mal-distribution.*
38. The VED analysis deals with classification of items on the basis of their availability.
39. HML classifies items accordingly as they are high-usage, medium-usage or low-usage.
40. FNSD is the *speed* classification of items, accordingly as they are fast-moving, normal-moving, slow-moving and dead items.

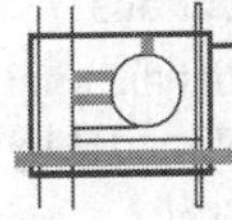

EXERCISES

1. What are different types of inventories? Explain
2. What functions does inventory perform? State the two basic inventory decisions management must make as they attempt to accomplish the functions of inventory just described by you. *(CA, May, 1995)*
3. Distinguish between the fixed order quantity system and the periodic review system. Also explain the Ss system.
4. With the help of quantity-cost curve, explain the significance of EOQ. What are the limitations of using the formula for an EOQ?
5. Discuss the assumptions underlying the basic EOQ formula. Also, state the economic order quantity model, discuss its sensitivity, and explain its major extensions.
6. (a) Give the different motives to keep inventory in an organisation. Do you consider hoarding as inventory?

   (b) Give the role of inflation and credit system in inventory management. *(CA, May, 1984)*

7. "Although the classical inventory decision model, known as the EOQ model, is too oversimplified to represent many of the real-world situations, it is an excellent starting point to develop more realistic and complex inventory decision models." In the light of this statement, explain the major limitations of the EOQ model. Also explain the major techniques developed by management scientists to overcome these. *(M Com, Delhi, 2009)*
8. (a) Explain, in detail, what constitutes the ordering cost and carrying cost. Draw a rough sketch to show the movement of these two cost curves in opposite directions with increase in the order quantity.
   (b) What is a re-order point? Draw a rough sketch of a simple inventory model showing the re-order point, order quantity and procurement time. *(M Com, Meerut, 1983)*
9. Discuss, in brief, the costs associated with inventory management. Derive the formula for EOQ, Q^*.

$$Q^* = \sqrt{\frac{2kb}{h(1 - b/a)}}$$

where, k = ordering cost per set-up
h = holding cost of unit per time
b = usage rate per unit of time
a = production rate per unit of time ($a \geq b$). *(CA, May, 1983)*

10. In a certain manufacturing situation, the production is instantaneous and the demand per day is R. Show that the optimal order quantity q per run which minimises that total cost is,

$$q = \sqrt{\frac{2RC_3(C_1 + C_2)}{C_1C_2}}$$

where, C_1 is the cost of holding one unit of inventory per day
C_2 is the unit cost of shortage per day
C_3 is the set-up cost per run. *(M Com, Bombay, 1982)*

11. What are the relevant costs for inventory decisions? How are they obtained in real life situations?
12. "An important characteristic of the EOQ model is the robustness which draws from the fact that the total cost curve is relatively flat at the bottom." In light of this, investigate the sensitivity of the model by considering the effect of a 30 per cent change (on either side) in order quantity on the total cost.
13. 'EOQ models, however complex, are restricted by so many assumptions that they have very limited practical value.' Do you agree with this statement? Illustrate your answer.
14. (i) The annual demand for an item of inventory is D units, ordering costs are B rupees per order, transportation costs per order (irrespective of the number of units transported) is T rupees, annual interest charges are x % of the value of inventory, insurance and warehousing charges amount to y % of the value, and the price of the product is P rupees per unit. What will be the EOQ?
   (ii) Would the EOQ in the case of gradual receipt of goods ordered be smaller or larger than the EOQ in the case of instantaneous receipt, all other relevant data remaining the same? Why?
   (iii) List the major weaknesses of the classical EOQ model of inventory.
   (iv) List three factors affecting the re-order level of inventory.
15. What is 'safety stock'? Why should it be kept by an organisation?

16. State whether the following statements are correct. Give reasons:
    (i) Safety stock increases as demand increases.
    (ii) In ABC analysis high cost items are most likely to fall in category *A*, and least cost items are likely to fall in Category *C*.
    (iii) To protect against stockouts, a large batch size is a must.
    (iv) EOQ is based on a balancing between inventory carrying costs and shortage costs.
    (v) Lead time is the time interval elapsing between the placement of a replenishment order and the receipt of last instalment of goods against the order.
17. Discuss the marginal analysis to the determination of the optimal ordering size. In what conditions is it employed?
18. Explain the basics of selective inventory control and state different selection techniques adopted in Inventory Control System. Give a brief note on each. *(ICWA, December, 1983)*
19. (a) "ABC analysis is a very useful approach for selective inventory control but has some major limitations." Do you agree with this statement? Explain how these limitations, if any, can be overcome.
    (b) Explain the concept of the Q-system, the P-system and the Two-Bin system for management of inventories, by giving appropriate examples. *(MBA, Delhi, November, 1999)*
20. Define selective inventory control. Explain the ABC, VED, HML, SDE, S-OS, and FSN bases of inventory classification. Are they mutually exclusive?

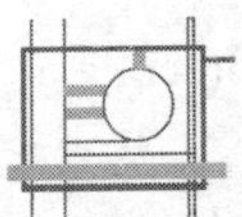

PRACTICAL PROBLEMS

1. A company has determined from its analysis of production and accounting data that, for a part number KC-438, the annual demand is equal to 10,000 units, the cost to purchase the item is Rs 36 per order, and the holding cost is Rs 2/unit/year.
   Determine
   (a) What should the Economic Order Quantity be?
   (b) What is the number of days' supply per optimum order?
2. A factory follows an economic order quantity system for maintaining stocks of one of its component requirements. The annual demand is for 24,000 units, the cost of placing an order is Rs 300, the component cost is Rs 60 per unit. The factory has imputed 24% as the inventory carrying rate.
   (i) Find the optimal interval for placing orders, assuming a year is equivalent to 360 days.
   (ii) If it is decided to place only one order per month, how much extra cost would the factory incur per year as a consequence of this decision? *(DMS, Mumbai, 1998)*
3. (a) For an item, the annual demand is known to be 3,000 units which is uniformly distributed over the year. The unit cost of the item is Rs 300 and the holding cost is 10% of the value. It costs Rs 450 to place an order for this product.
   Determine
   (i) the economic ordering quantity, EOQ,
   (ii) the change in EOQ when the ordering cost changes to Rs 600,
   (iii) the change in EOQ when the holding cost becomes 7.5% of the item value,
   (iv) both, the ordering and the holding costs change as in (ii) and (iii),
   (v) the per cent rise in cost associated with ordering 600 units rather than the EOQ obtained in (i) above (using the cost figures given originally).

(b) In a certain inventory system, the EOQ is 1,200 units. If the demand for the next year increases by 50%, what will be the new EOQ? If the inventory carrying costs increase (in addition to the increase in the demand) from 25% to 40% of the value of the inventory, what will be the new EOQ?

(c) The annual demand, buying cost per order, and carrying cost per year as a percentage of the value of inventory in the case of two products *A* and *B* are same. However, the price of *A* is 5 times as much as that of *B*. If the EOQ for *A* is 100 units, what will be the EOQ for *B*?

4. (a) Determine EOQ from the following data:

Ordering cost	= Rs 200 per order
Carrying cost	= 20% of the unit cost per year
Unit cost	= Rs 25
Annual requirement	= Rs 50,000

(b) Now, if the annual requirement becomes as large as Rs 200,000 (i.e. 4 times the demand), would the EOQ become four times as large as obtained in (a)? Why?

5. The following is the general information and information pertaining to a product component TT-4 used by Kay Engineering Works:

Purchase price	= Rs 60 per 100
Annual usage	= 1,20,000 units
Purchase office cost	= Fixed Rs 16,320 per annum, Variable Rs 12 per order
Rent and heating of component store	= Rs 1,850 per annum
Interest	= 20%; and Insurance = 0.1% per annum based on total purchases
Deterioration	= 2% per annum of all items purchased

Required:

(a) calculate the EOQ for the component TT-4,

(b) calculate the percentage change in total annual variable costs (of ordering and holding) relating to the component TT-4 if the annual usage was (i) 25% more, and (ii) 25% less than the current usage of 120,000 units.

6. A company uses a certain component at the rate of 8,000 units per year. Each component part is valued at Rs 18. The company estimates that the cost of holding inventory is 20% of the value of the item, per year. The company has the option to produce the part on either of the two machines, *A* or *B*. Machine *A* has a set-up cost of Rs 200, machine *B* has a set-up cost of only Rs 100. However, it costs 10 paise more per unit to produce using the machine *B* than it does using machine *A*. That is to say, the cost of producing the part on machine *B* would equal Rs 18.10 per unit.

Which machine should the company use? What is the optimum lot size?

7. The daily demand for a product is approximately 100 units. The ordering cost per order is Rs 100 and the average daily holding cost per unit is 2 paise. If the lead time is 14 days, determine the economic lot size for placing an order and also the re-order level.

If the lead time is reduced to 7 days, what will be the re-order level? How will the re-order level change if the lead time increases to 21 days? *(M Com, Delhi, 1985)*

8. Fill in the blanks:

Item	Annual demand (units)	Ordering cost (Rs)	Unit holding cost (Rs)	Lead time	EOQ	Re-order level
A	8,000	15	0.06	10 days	?	?
B	?	40	0.18	6 days	?	216
C	7,500	?	30.00	?	300	210

For your calculations, assume the number of working days in the year equal to 250.

9. The annual demand of a particular item used by a company is 10,000 units. This item may be obtained from either an outside supplier or subsidiary company. The relevant data for the procurement of the item are given below:

Costs	From outside supplier Rs	From subsidiary company Rs
Cost per unit	12	13
Cost of placing an order	10	10
Cost of receiving an order	20	15
Storage and all carrying cost, including capital cost per unit per annum	2	2

(i) What purchase quantity and from which source should you recommend to procure?
(ii) What would be the minimum total cost in that case? *(CA, May, 1992)*

10. The Heavy Nitro Company is considering the optimal batch size for re-order of concentrated sulphuric acid. The management has supplied the following information:
The purchase price of H_2SO_4 is Rs 2 per gallon. The clerical and data processing costs are Rs 10.625 per order. All the transport is done by rail. A charge of Rs 40 is made each time the special line to the factory is opened. A charge of 20 paise per gallon is also made. The company use 40,000 gallons per annum. Maintenance costs of stock are Rs 4 per gallon per year. Each gallon requires 1/2 square foot of storage space. If warehouse space is not used, it can be rented out to Manganese Ltd. at Rs 2 per square foot per annum. Available warehouse space is 1,000 square feet, the overhead costs being Rs 2000 per annum.
Assume that all the free warehouse space can be rented out.
(a) Calculate the economic re-order size.
(b) Calculate the minimum total annual cost of holding and re-ordering stock.
(ICMA, May, 1976, Adapted)

11. (a) Finance managers regularly deal with the problem of determining the optimal level of cash to keep in hand. Consider a company that has regular cash needs during a period. The problem for the finance manager of this company is to determine the optimal amount of new funds to obtain from borrowing. The objective is to minimise the cost of going to the market for funds and the opportunity cost associated with "carrying" the funds.

Let Q = amount of cash obtained from each bond issue (Rs per issue).
C_o = fixed cost associated with going to the market for funds by floating a bond issue (Rs per issue).
C_h = annual opportunity (carrying) cost associated with having cash in hand.
D = annual cash needs (Rs per year).

Assume that cash needs are uniform during the year, that funds can be obtained in a deterministic period of time, and that interest rates on bonds are constant and unaffected by the size of the bond issue. Shortage of funds is not permitted.

(i) Write a general expression for the total annual cost associated with borrowing and holding cash.

(ii) If D = Rs 100 million, C_o = Rs 100,000, C_h = 8% per annum, determine the optimal size of each bond issue. How many times a year should bond issues be floated?

For the data in (ii), determine:

(iii) What is the annual cost of floating bonds? The annual opportunity cost associated with holding cash?

(iv) At what level of cash should a bond issue be initiated if it takes five days to float an issue? Assume 255 'trading' days in a year (i.e. days when the bond market and the firm actively engage in their affairs).

(b) Because of the dynamics of the money market, analysts concede that the percentage of the opportunity cost associated with holding cash may fluctuate between 8% and 12% per annum. Conduct sensitivity analysis to determine the effects on your answers to parts (ii), (iii) and (iv) if C_h assumes the extreme value of 12% per annum. *(CA, November, 1992)*

12. The Purchase Manager of Sigma Company is contacted by a new supplier who offers a quantity discount on an item KR-100 being used by the company. The ordering cost is Rs 80 per order and the holding cost 25 per cent of the average inventory value, on a yearly basis. The annual demand for this item is 40 thousand units at a constant daily rate. The supplier states a price of Rs 80 per unit (which is being currently paid by the company) and a discount of Rs 4 per unit if the company would order for a least two thousand units of the item at a time. The manager is in a fix as to whether to take advantage of this discount offer. What would you suggest him? Show a comparative analysis of the cost involved in both the alternatives.

13. A purchase manager has decided to place an order for a minimum quantity of 500 units of a particular item of inventory in order to get a discount of 10%. Past records reveal that last year 8 orders (each of 200 units) were placed. Given the ordering cost = Rs 500 per order, inventory carrying cost = 40% of the inventory value, and price of the item = Rs 400 per unit, is the manager justified in his decision? What is the effect of his decision to the company? *(M Com, Delhi, 1995)*

14. Obtain the optimal order quantity using the information given here:
Annual requirement: 1 million units
Ordering and processing cost: Rs 28.80 per order
Holding cost: 20% of the unit cost
Price Schedule:

Order Size	*Unit Price*
0-9,999	Rs 2.00
10,000-19,999	Rs 1.60
20,000 and above	Rs 1.40

15. The annual demand for an item of inventory is 2400 units. The order processing costs amount to Rs 350 per order. Inventory holding costs are estimated to be 2 per cent per month of the value of inventory. The normal price of the product is Rs 10 per unit. However, the supplier offers a quantity discount of 7.5 per cent on an order of at least 400 units, and 12.5 per cent on an order of 600 units or more. Determine the EOQ. *(M Com, Delhi, 2002)*

16. A shopkeeper has a uniform demand of an item at the rate of 50 units per month. He buys the item from the supplier at the cost of Rs 6 per unit and the cost of placing an order is Rs 10 every time. If the stock holding cost is 20% p.a. of the stock value, how frequently should he replenish his stock? Further, suppose the supplier offers a 5% discount on orders between 200 units and 999 units of the item and a discount of 10% on the orders exceeding or equal to 1,000 units. Is it advisable for the shopkeeper to avail either of these discount offers? *(MIB, Delhi, 2008)*
17. The Honolulu Company has a contract to supply 5,000 units of an item per year to a dealer. For this item, the company estimates that the ordering cost is Rs 150 every time that an order is made while the carrying cost (p.a.) is reckoned to be 20 per cent of the unit price.
The company is negotiating with a dealer who offers to give the following quantity discount.

Order size	*Price per unit (Rs)*
Less than 1,000	500
1,000-2,999	450
3,000-4,999	400
5,000 or more	350

Recommend to the company the best inventory policy with regard to this item.
18. A manufacturing company of microwave ovens uses Rs 75,000 worth of LED readout circuits annually in its production process. Cost per order is Rs 45, and the carrying charge assessed against this classification of inventory is 25% of the average balance per year. This company follows an EOQ purchasing system and to date has not been offered by discounts on these circuits. Now the supplier has indicated that if the company would buy its circuits four times a year in equal quantities, a discount of 1.5% off list price would be given in return. Would you advise this company to accept this offer? In order to maintain the present total cost, what should be the minimum discount acceptable to the company if four orders of equal sizes are placed in a year? *(CA, November, 1991)*
19. For one of the A-class items, the following data are available:

Annual demand = 1,000 units

Ordering cost per order = Rs 400

Inventory carrying cost = 45%

Cost per item = Rs 15

The purchase manager has placed an order for 500 items in the beginning of the year availing 5% discount. At the beginning of the seventh month he procured 250 items without any discount. At the beginning of the eighth month he procured another order for 250 items with a discount of 8%. Had he followed economic order quantity policy what would have been the gain or loss for the organisation? *(ICWA, December, 1995)*

20. (a) In respect of a component costing Rs 10 each, the annual demand is known to be 5,000 units. The cost of placing an order is Rs 200 and the total holding cost is 20% of the average inventory investment.

Determine,

(i) the most economical order size,

(ii) the optimal number of orders during a year,

(iii) the cost of this policy exclusive of the component cost.

(b) A supplier is willing to supply this component and prepared to give discount as follows: For ordering at least 1,800, 2,000 and 2,400 units a discount of 0.5, 1 and 1.5%, respectively. Assuming that the carrying cost per unit would not change because of the discount offered, what is the most economical quantity that should be ordered?

21. An item sells for Rs 4 per unit but 10% discount is offered on lots of 150 units or more. A manufacturing unit that consumes this item at the rate of 20 items per day wants to decide whether or not to take advantage of the discount. The setup cost for ordering a lot is Rs 50 and the holding cost per unit per day is Rs 0.30. Will it be economical for the manufacturing unit to take advantage of the discount? If yes, then determine the range on the percentage discount in the price of the item that when offered for lots of size 150 or more will not result in any financial advantage to the manufacturing unit. (Assume that the unit works for all the 365 days of the year.) *(CA, November, 1994)*

22. The annual demand for a product is 1,00,000 units. The rate of production is 2,00,000 units per year. The set-up cost per production run is Rs 5,000 and the variable production cost of each item is Rs 10. The annual holding cost per unit is 20% of its value. Find the optimum production lot size and the length of the production run.

23. A company is presently having a production run of 500 units every 3 months. He is considering a review of its production-lot size decision. The relevant information is given below:

Annual demand of the item	:	2,000 units
Rate of production	:	8,000 units per year
Set-up cost	:	Rs 300 per run
Inventory holding cost	:	Rs 1.60 per unit per year

Would you recommend a change in the current production-lot size? Why?

What will be the cost saving, if any, as a result of change? *(M Com, Delhi, 2003)*

24. Kobo Bearing Limited is committed to supply 24,000 bearings per annum to M/s Deluxe Fans on a daily basis. It is estimated that it costs 10 paise as inventory holding cost per bearing per month and that the set-up cost per run of bearing manufacture is Rs 324.
    (a) What should be the optimal run size for bearing manufacture?
    (b) What would be the interval between two consecutive optimum runs?
    (c) Find out the minimum inventory cost.

25. You are given the following information regarding the production-lot size of a particular product:

Annual demand	:	5,000 units
Set-up costs	:	Rs 100 per set-up
Daily demand	:	17 units
Production rate	:	50 units per day
Optimal production-lot size	:	275 units

Rising interest rates and other costs have caused a 10% increase in the annual holding costs. Determine the new optimum production lot size for the product. *(M Com, Delhi, 1995)*

26. The Clean-Well Company of India has introduced a new detergent KIN of which it can produce 20,000 packs per month of 20 working days. The annual demand for this detergent is forecasted to be 1,92,000 packs, at a uniform rate. The set-up cost for the production of KIN is Rs 1,080 while the monthly cost for holding a pack in inventory is Re 0.30.

Determine,

(a) the optimum lot size,

(b) the optimum number of production runs,

(c) the time interval between the start of successive production runs, and

(d) the total variable cost, per year, associated with the lot size as determined in (a).

27. Company A wants to know that production cost its major competitor, Company B, has assigned to product item P_7. After a bit of investigation, Company A has collected the following data about company B's production of item P_7:

Production-lot size	:	2,600 units
Set-up cost	:	Rs 135
Annual demand	:	30,000 units
Daily demand	:	100 units
Production rate	:	200 units per day
Inventory holding cost	:	28% of the average value per year

Company A has further learnt that company B produces according to 'economic lot size' model. What is company B's cost of producing product item P_7? *(M Com, Delhi, 1999)*

28. A manufacturer of food processors has annual sales of 20,000 units. The manufacturer operates 250 days a year and produces 200 units per day. The setup cost is Rs 600 for a production run. The food processors cost Rs 400 to produce. The inventory holding cost fraction is 0.025 per year.

(a) What is the most economical batch size (EBQ)?

(b) What is the minimum annual total of inventory holding costs and setup costs?

(c) What is the number of production runs per year?

(d) How many days are required for the production of each batch?

(e) What is the maximum inventory on hand for the EBQ?

(M Com, Delhi, 2008)

29. An item used by a company may be purchased from a supplier for Rs 32 per unit or manufactured in its plant at a rate of 10,000 units per annum, for Rs 30 per unit. If the item is purchased, the ordering cost is Rs 12 per order while if it is produced internally, the set-up cost per production run is Rs 60. The annual demand for the production is 2,500 units—a uniform 10 units per day. The holding cost is 12% of the inventory value. Should the item be purchased from outside or produced internally?

30. A manufacturing company needs 2,500 units of a particular part every year. The company buys it at the rate of Rs 30 per unit. The cost of placing an order for the part is estimated at Rs 15 and the cost of carrying a part in stock comes to about Rs 4 per year.

The company can manufacture the part internally. In that case, it saves 20% of the price of the product. However, it estimates a set-up cost of Rs 250 per production run. The annual production rate would be 4,800 units. The carrying costs remain unchanged.

(i) What would be the EOQ and the optimal number of orders placed in a year?

(ii) What would be the EPLS and the average length of the production run?

(iii) Should the company manufacture the part internally or continue to purchase it from the supplier?

31. The demand of an item is uniform at a rate of 20 units per month. The fixed cost is Rs 10 each time a production run is made. The production cost is one rupee per item and the inventory carrying cost is Re 0.25 per item per month. If the shortage cost is Rs 15 per item per year, determine how often to make a production run and of what size it should be? *(MBA, Chennai, 1997)*

32. (a) A retailer gets discount for large orders. The discount is 4 per cent if the quantity ordered is 500 or more, and an additional 1 per cent discount is received if the quantity ordered is 1,000 or more. The product costs Rs 50 and sells uniformly at the rate of 2,500 per year. The fixed order cost is Rs 50, while the inventory carrying cost is 20% of the value of the average inventory held.

    (i) What is the best order quantity?

    (ii) What is the minimum total annual cost, including the purchasing cost?

    (b) A manufacturer of a line of microwave ovens must decide the manufacturing batch sizes. The setup cost for manufacturing a particular microwave oven is Rs 15,000. The annual rate of sales is 10,000 and the rate of sales is constant during the year. The annual manufacturing capacity is 20,000. The cost of a completed microwave oven is Rs 15,000. The inventory holding cost fraction is 0.01 per year.

    (i) What is the most economical batch size (EBQ)?

    (ii) What is the minimum annual total of inventory holding and set-up costs?

    (iii) What is the number of production runs per year?

    (iv) What is the amount of time required for the production of each batch?

    (v) What is the maximum inventory on hand for the EBQ?

    (c) Assuming that the total annual demand, carrying cost of inventory and ordering/set-up cost are the same, in which of the two conditions will the EOQ be larger and why:

    (i) Instantaneous receipt of inventory

    (ii) Gradual receipt of inventory

    (d) In a certain inventory system, the EOQ is 1000 units. If the demand for the next year increases by 50 per cent, how will be EOQ change? If the inventory carrying cost increases, in addition to the increase in demand, from 25% to 40% of the value of the inventory, what will be the revised EOQ?

    (e) The lead time demand for a continuous review inventory system has a normal distribution, with a mean of 50 units and a standard deviation of 10 units. The order quantity is 300 units and the annual demand is 3600 units.

    (i) If it is desired that the probability of shortage in a re-order cycle be 0.05, what should be the reorder level?

    (ii) If it is desired that the probability of shortage in a re-order cycle be 0.01, what should be the safety stock?

    (iii) What service level can be expected with a safety stock of 25 units? *(M Com, Delhi, 2004)*

33. A furniture dealer sells special typist chairs. Each purchase order costs Rs 50 to the dealer and the holding costs amount to Rs 80 per chair per year. The dealer sells 90 chairs per month. He had estimated a back-ordering cost of Rs 20 per chair per year.

    (i) What is the EOQ for the chair?

    (ii) When the supply of the chairs comes, how many chairs on an average are expected to be delivered to customers immediately?

    (iii) What is the total annual cost of ordering, holding and back-ordering? *(M Com, Delhi, 1991)*

34. The manager of a company is facing an inventory problem with regard to an item whose demand is known to be evenly distributed with an annual value of 8,000 units. The cost of placing an order is Rs 50 while the annual carrying cost of one unit in inventory is Rs 5. Because of the high carrying cost, the manager is planning to have stockouts but he does not want to lose customer goodwill. He has determined that unit shortage cost is Rs 10. Using the present information, determine,
    (a) *When stockouts are not permitted*
        (i) the optimum ordering quantity,
        (ii) the total inventory cost associated with the policy of ordering this quantity.
    (b) *When stockouts are permitted*
        (i) the optimum order quantity,
        (ii) the maximum level of inventory,
        (iii) the optimum number of shortage units,
        (iv) the total inventory cost involved with the policy, and
        (v) represent graphically the inventory profile for the first three inventory cycles.
35. Using the relevant data from the following information, calculate economic order quantity and re-order level when (i) stockouts are not permitted, and (ii) when stockouts are permitted.

Uniformly distributed annual demand for the item	= 24,000 units
Cost of placing an order	= Rs 20
Cost of receiving, inspecting and checking a delivery	= Rs 70
Cost of carrying stock	= Rs 3 per unit per annum
Shortage cost per unit per year	= Rs 2
Lead time	= 1/24th of a year
Interest paid by the company on funds employed	= 12% per annum
Cost of extra storage facilities	= Rs 600 per annum.

36. A dealer supplies you the following information with regard to a product dealt with by him:

Annual demand	= 5,000 units
Buying cost	= Rs 250 per order
Inventory carrying cost	= 30% per year
Price	= Rs 100 per unit

The dealer is considering the possibility of allowing some back-orders to occur for the product. He has estimated that the annual cost of back-ordering (allowing shortage of) the product will be Rs 10 per unit.
    (i) What is the optimum number of units of the product that he should buy in one lot?
    (ii) What quantity of the product should he allow to be back-ordered?
    (iii) How much additional cost will he have to incur on inventory if he does not permit back-ordering?
37. Higley Radio Components Company has a product for which the assumptions of the inventory model with backorders are valid. information obtained by the company is as follows:

Inventory carrying charge = 20 percent per annum
Units cost = Rs 50
Ordering cost = Rs 250 per order
Backorder cost = Rs 30 per unit per year

Calculate (i) economic order quantity; (ii) maximum stock level; (iii) maximum number of units back-ordered; and (iv) total relevant cost per annum.

If the company decides to prohibit backorders and adopts the regular EOQ model, what effect will it have on order quantity; maximum sotck level and total relevant cost?

(M Com, Delhi, 2009)

38. The annual requirement of a commodity is 3,000 units, the cost of placing an order is Rs 300 and the cost of carrying an item in inventory for one year is Rs 20.
    (a) Determine the EOQ.
    (b) Determine the re-order level if the number of working days in a year is 300 and the lead time is 15 days.
    (c) If the maximum lead time is 20 days and the maximum daily demand is 15 units, determine the safety stock required to prevent stockouts.
39. Obtain (i) Economic Order Quantity, (ii) Number of orders, (iii) Re-order level, (iv) Safety stock, for the following inventory problem:

    Annual demand = 36,000 units
    Cost per unit = Re 1
    Ordering cost = Rs 25
    Cost of capital = 15%
    Store charge = 5%
    Lead time = 1/2 month
    Safety stock = 1 month consumption.
40. The daily demand for an electronic machine is approximately 25 items. Everytime an order is placed, a fixed cost of Rs 25 is incurred. The daily holding cost per item inventory is Re 0.40. If the lead time is 16 days, determine the economic lot size and re-order point. *(ICWA, December, 1986)*
41. A company purchases a component from another factory. Based on the past records, the EOQ for the component is 3,600 units, with an average usage of 50 units a day. The normal lead time is 6 days and the carrying cost per unit is Rs 10 per year. The cost of being out-of-stock for this component is estimated to be Rs 50 per unit and the company places 5 orders a year.
    An analysis of the past inventory records for the component has revealed the following usage pattern:

Usage during the re-order period (units)	*No. of times this quantity was used*
150	3
200	4
250	6
300	68
350	9
400	7
450	3
Total	100

Determine the optimal safety stock and the optimal re-order level for the component.

(M Com, Delhi, 1998)

42. A company is using fixed-order quantity system of inventory management. For one of the items, the re-order quantity is 400 units and the re-order level is 350 units. The weekly demand of the product is known to be normally distributed with a mean of 100 units and a standard deviation equal to 40 units. Given that the lead time is equal to 3 weeks, determine the service level offered. If a service level of 97.7% is required, what re-order level be selected?

43. The daily demand for a component assembly items is normally distributed with a mean of 90 and a standard deviation of 10 units. Further, the source of supply is dependable and maintains a constant lead time of five days. If the cost of placing the order is Rs 30 and the annual holding costs are Re 0.8 per unit, find the re-order quantity and re-order point to satisfy 80% demand of the customers who place orders during the re-order period.

44. (a) The following information relates to the Beta company:

Units required per annum = 30,000
Cost of placing an order = Rs 400
Carrying cost = Rs 600 per unit per annum.

Assuming that the demand for the product is uniformly distributed throughout the year, what is the EOQ?

(i) 200, (ii) 300, (iii) 400, (iv) 500, (v) 600.

(b) A company requires 10,000 units of an item, the demand for which is evenly distributed over the year. The set-up cost every time that a production run is made is Rs 80 while the cost of carrying a unit in inventory for a one-year period amounts to 40 paise. Assuming that the item is to be produced at the lowest possible cost and that each run would be for the same number of units, how many production runs would be made?

(i) 3, (ii) 4, (iii) 5, (iv) 8, (v) 10.

(c) A company desires to know the amount of safety stock that it should maintain for a certain product that would result in the lowest cost. Each stockout would cost Rs 80, and the carrying cost of each unit of safety stock will be Rs 1.20. If the product is to be ordered 5 times a year, which of the following will produce the lowest cost?

(i) a safety stock of 20 units which is associated with a 50% probability of running out-of-stock during lead time.

(ii) a safety stock of 30 units which is associated with a 30% chance of running out-of-stock during an order period.

(iii) a safety stock of 40 units that is associated with a 10% probability of running out-of-stock during an order period.

(iv) a safety stock of 60 units that is associated with a 5% probability of running out-of-stock during an order period.

(d) For a product *X* what safety stock amount will result in the lowest cost? You are provided with the following information:

Stockout cost	:	Rs 80 per occurrence
Carrying cost of safety stock	:	Rs 2 per unit
Number of orders per annum	:	5

Units of safety stock	*Probability of being out of safety stock*
10	0.50
20	0.40
30	0.30
40	0.20
50	0.10
60	0.05

The number of units of the safety stock that would result in the lowest total cost would be:
(i) 10, (ii) 20, (iii) 30, (iv) 40, (v) 50, (vi) 60.

45. Auto Care Limited provides several auto parts to local garages in a small city. The company purchases parts from manufacturers in accordance with the EOQ model and sends the parts from its regional warehouse direct to its customers. For a particular part, the analysis recommends orders with EOQ = 50 boxes to satisfy an annual demand equal to 800 boxes. The company has 250 working days per year and the lead time averages 15 days.
   (a) What is the re-order level if Auto Care assumes a constant demand rate?
   (b) A further analysis of Auto Care demand for the product in question shows that lead time demand follows normal probability distribution with $\mu = 25$ and $\sigma = 4$ boxes. If the company management can tolerate one stock-out per year, what is the revised re-order level? If holding cost per box per year is Rs 5, what is the extra cost due to uncertainty of demand?
   (c) What is Auto Care's re- order level if the company is willing to tolerate two stock-outs per year?
   (d) What is Auto Care's re-order level if the company wants to restrict the probability of stock-out on any one cycle to at most one percent?

46. A firm with an annual demand of approximately 3,000 units has an ordering cost of Rs 30 per order and a holding cost of Rs 8 per unit per year. The demand exhibits some variability such that the lead time demand follows normal distribution with $\mu = 120$ units and $\sigma = 20$ units.
   (a) What is the recommended order quantity?
   (b) What is the re-order point and the safety stock if the firm desires a 2% probability of stockout in a given order cycle?
   (c) If the re-order point is set at 140 units, what is the probability of a stockout in any given order cycle? How many times would you expect to stockout during the year if this re-order point were used?

47. An item is ordered by New Star Manufacturing Company in quantities of five thousand units at a time. The cost price of the item is Rs 4 per unit. The average weekly demand for this item is 400 units with a mean absolute deviation (MAD) of 250 units. The lead time is known to be equal to 2 weeks while the holding cost per unit per week is Re 0.01. The company has established a 95% service level in respect of this item.

Using the information given, determine

(a) level of safety stock,
(b) re-order point,
(c) annual cost of maintaining the safety stock.

48. The distribution of lead time in respect of a certain component is given in the following table:

Lead time (weeks)	*Frequency*
0–1	4
1–2	8
2–3	20
3–4	24
4–5	16
5–6	4
6–7	4

The management of the company wishes to set safety stock level that would ensure that the risk of being out of stock for reason of delayed delivery should be no more than 10%. Assuming that the demand is known and uniform at 200 units per week, calculate the amount of safety stock which should be provided so as to meet the desired service level.

49. The daily demand of an item VP-100 is known to be normally distributed with a mean of 20 units and MAD of 5 units. For the item, the lead time is 9 days. You are required to determine:

(a) the re-order point corresponding to a 50% service level,
(b) the service level corresponding to a zero safety stock level,
(c) the re-order point corresponding to a service level of 90%.

50. A company uses a component for which its forecast of usage is normally distributed with MAD = 40 units. The company management adopts a policy which limits the stockouts to be no more than one order cycle during a year.

(i) How much safety stock should be kept if the order quantity is generally a two weeks' supply?
(ii) How much safety stock should be kept if the order quantity is generally a four weeks' supply?

51. A ship building company has launched a programme for the construction of a new class of ships. Certain spare units like prime mover, each costing Rs 2,00,000 have to be purchased. If these units are not available when needed, a very serious loss is incurred, which is of the order of Rs 10,00,000 in each instance.

Requirements of the spares with corresponding probabilities are given here:

No. of spares	:	0	1	2	3	4	5
Probability of requirement	:	0.876	0.062	0.041	0.015	0.005	0.001

How many spares should the company buy in order to optimise the inventory decision?

52. The following information is known about a group of items. Classify the materials in ABC classification:

Model No.	*Annual consumption (in pieces)*	*Unit price (in paise)*
501	30,000	10
502	2,80,000	15
503	3,000	10
504	1,10,000	5
505	4,000	5
506	2,20,000	10
507	15,000	5
508	80,000	5
509	60,000	15
510	8,000	10

53. What is selective inventory control? From the following details, draw a plan of ABC selective control.

Item	:	1	2	3	4	5	6	7	8	9	10	11	12
Units ('000)	:	7	24	1.5	0.6	38	40	60	3	0.3	29	11.5	4.1
Unit cost	:	5	3	10	22	1.5	0.5	0.2	3.5	8	0.4	7.1	6.2

(CA, May, 1983)

54. Ten items kept in inventory by the School of Management Studies at State University are listed below. Which items should be classified as '*A*' items, '*B*' items and '*C*' items? What percentage of items is in each class? What percentage of total annual value is in each case?

Item	*Annual usage*	*Value per unit (Rs)*
1	200	40.00
2	100	360.00
3	2,000	0.20
4	400	20.00
5	6,000	0.04
6	1,200	0.80
7	120	100.00
8	2,000	0.70
9	1,000	1.00
10	80	400.00

(CA, May, 1994)

Appendix 9A

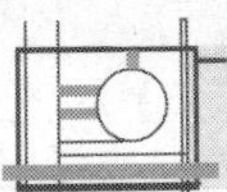

9A.1 DETERMINATION OF EOQ(Q*) FOR THE CLASSICAL EOQ MODEL

The total cost equation for this model is

$$T(Q) = \frac{D}{Q}A + \frac{Q}{2}h$$

To determine the value of Q corresponding to lowest $T(Q)$ value, we shall take the first derivative of this w.r.t. Q and set it equal to zero. Thus,

$$\frac{dT(Q)}{d(Q)} = -\left(\frac{AD}{Q^2}\right) + \frac{h}{2} = 0$$

or

$$\frac{AD}{Q^2} = \frac{h}{2}$$

or

$$Q^2 = \frac{2AD}{h}$$

$\therefore$

$$Q = Q^* = \sqrt{\frac{2AD}{h}}$$

To verify that the point is the minimum point, we check for a positive second derivative, i.e.

$$\frac{d^2T(Q)}{dQ^2} = \frac{2AD}{Q^3} > 0$$

A, D and Q all being positive, the second derivative $2AD/Q^3$ is positive and the point is the minimum point.

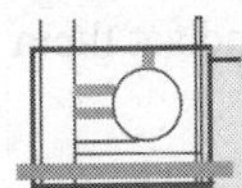

9A.2 DETERMINATION OF THE OPTIMAL LOT SIZE Q* FOR THE PRODUCTION-LOT SIZE MODEL

The total annual cost function for the production-lot size model is

$$T(Q) = \frac{1}{2}\left(1 - \frac{d}{p}\right)Qh + \frac{D}{Q}A$$

To find the quantity Q that minimises the total cost, we set the derivative $dT(Q)/dQ$ equal to zero and solve.

$$\frac{dT(Q)}{d(Q)} = \frac{1}{2}\left(1 - \frac{d}{p}\right)h - \frac{D}{Q^2}A = 0$$

or
$$\frac{1}{2}\left(1-\frac{d}{p}\right)h = \frac{D}{Q^2}A$$

or
$$\left(1-\frac{d}{p}\right)h = \frac{2AD}{Q^2}$$

or
$$Q^2 = \frac{2AD}{\left(1-\frac{d}{p}\right)h}$$

or
$$Q^2 = \frac{2AD}{h}\left(\frac{p}{p-d}\right)$$

$$Q^* = \sqrt{\frac{2AD}{h}\left(\frac{p}{p-d}\right)} = \sqrt{\frac{2AD}{h}}\sqrt{\frac{p}{p-d}}$$

The second derivative is

$$\frac{d^2T(Q)}{dQ^2} = \frac{2AD}{Q^3}$$

D, A, and *Q* being all greater than zero, the value of the second derivative would be positive, the solution is the least cost solution.

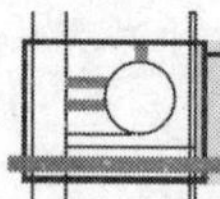

9A.3 DETERMINATION OF Q^* AND S^* FOR THE PLANNED SHORTAGES MODEL

The total cost expression for the inventory model with planned shortages is

$$T(Q) = \frac{D}{Q}A + \frac{(Q-S)^2h}{2Q} + \frac{bS^2}{2Q}$$

Here there are two inventory decision variables, viz. *Q* and *S*. To determine *Q* and *S* values that would minimise *T*(*Q*) in the above equation we shall obtain the two partial derivatives $\partial T(Q)/\partial Q$ and $\partial T(Q)/\partial S$ and set them equal to zero. The total cost equation above can also be written as:

$$T(Q) = \frac{D}{Q}A + \frac{(Q^2-2QS+S^2)h}{2Q} + \frac{bS^2}{2Q}$$

$$= \frac{D}{Q}A + \frac{Q^2h-2QSh+S^2h}{2Q} + \frac{bS^2}{2Q}$$

$$= \frac{D}{Q}A + \frac{Qh}{2} - Sh + \left(\frac{h+b}{2Q}\right)S^2$$

Setting $\partial T(Q)/\partial S = 0$, we get

$$\frac{\partial T(Q)}{\partial S} = \left(\frac{h+b}{Q}\right)S - h = 0$$

$$\therefore \qquad S^* = Q\left(\frac{h}{h+b}\right)$$

Now, setting $\partial T(Q)/\partial Q = 0$, we get

$$\frac{\partial T(Q)}{\partial (Q)} = \frac{-(h+b)\,S^2}{2Q^2} + \frac{h}{2} - \frac{DA}{Q^2} = 0$$

Substituting the value of S^* from above in this equation, we get

$$\frac{\partial T(Q)}{\partial (Q)} = \frac{-(h+b)\,Q^2\,(h^2)/(h+b)^2}{2Q^2} + \frac{h}{2} - \frac{DA}{Q^2} = 0$$

or $$\frac{-h^2}{2(h+b)} + \frac{h}{2} = \frac{DA}{Q^2}$$

or $$\frac{-h^2 + h(h+b)}{2(h+b)} = \frac{DA}{Q^2}$$

or $$Q^2 = \frac{2(h+b)DA}{hb}$$

$$= \frac{2hDA}{hb} + \frac{2bDA}{hb}$$

$$= \frac{2DA}{h}\left(\frac{h}{b} + \frac{b}{b}\right)$$

Thus, $$Q^* = \sqrt{\frac{2DA}{h}\left(\frac{h+b}{b}\right)}$$

$$= \sqrt{\frac{2DA}{h}}\sqrt{\frac{h+b}{b}}$$

Second order conditions can be used to show that equations of S^* and Q^* provide the minimum cost solution.

CHAPTER 10

Queuing Theory

CHAPTER OVERVIEW

Queuing is a common occurrence—in banks, public transportation, hospitals, hair saloons, theatres, airways, workshops ... everywhere. Understandably, queues are formed when there are more arrivals than what the service facility can handle, and no queues are formed when the arrivals are less than that. The uncertainty of arrivals and the time they require for getting service accounts for the fact that at times there are long queues while at some other times there are no queues at all. Further, it can be easily understood that the greater the average rate of arrivals for a given rate of service which can be provided, the lengthier would the queues formed and more would be the time required by a typical arrival to get service. Similarly, for a given rate of arrivals, the greater the rate at which service can be provided, the smaller would the queues be and the time needed to get the service. However, one cannot quantify that for a given set of arrival and service rates, how long the queues will generally be and how much time a customer will take to get the service. A study of queuing theory does that.

Specifically, the theory helps the manager to seek answers to the questions like the following when actual/expected arrival and service rates are known and a set of assumptions is made.

- ❑ What fraction of time is the system providing service expected to be busy (or idle)?
- ❑ What are the chances of a certain number of customers being present in the system at a random point in time?
- ❑ On the average, how many customers are expected to be present in the system?
- ❑ What is the average length of queues formed from time to time?
- ❑ How much time is a customer expected to spend in the system—in waiting for and getting the service?
- ❑ How much time does a customer typically spend in the queue?
- ❑ What is the expected total cost of providing service for various alternative service levels and, accordingly, what is the optimal level of service to provide?

An understanding of this chapter requires knowledge of the concepts of probability, expected value, as also the exponential and Poisson distributions. However, the derivations of the various formulae are not given, as the focus is on applications, which simplifies learning.

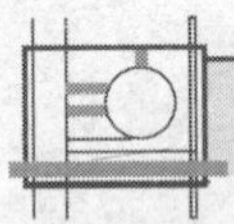

10.1 INTRODUCTION

The queuing theory, also called the *waiting line theory*, owes its development to A K Erlang's efforts to analyse telephone traffic congestion with a view to satisfying the randomly arising demand for the services of the Copenhagen automatic telephone system, in the year 1909. The theory is applicable to situations where the 'customers' arrive at some 'service station (s)' for some service; wait (occasionally not); and then leave the system after getting the service. In such 'arrival and departure' problems, the *customers* might be people waiting to deposit their electricity bills at a cash counter, machines waiting to be repaired in a factory's repair shop, aeroplanes waiting to land at an airport, patients in a hospital who need treatment ... and so on. The *service stations* in such problems are the cash counters in the electricity office, repairmen in the shop, runways at the airport and doctors attending the patients, respectively. Some more examples of queuing situations are given in Table 10.1.

TABLE 10.1 Queuing Examples

Situation	*Arriving Customers*	*Service Facility*
(a) Passage of customers through a supermarket checkout	Shoppers	Checkout counters
(b) Flow of automobile traffic through a road network	Automobiles	Road network
(c) Transfer of electronic messages	Electronic messages	Transmission lines
(d) Banking transactions	Bank patrons	Bank tellers
(e) Flow of computer programmes through a computer system	Computer programmes	Central processing unit
(f) Sale of theatre tickets	Theatre-goers	Ticket booking windows
(g) Arrival of trucks to carry fruits and vegetables from a central market	Trucks	Loading crews and facilities
(h) Registration of unemployed at employment exchange	Unemployed personnel	Registration assistants
(i) Occurrences of fires	Fires	Firemen and equipment
(j) Flow of ships to the seashore	Ships	Harbour and docking facilities
(k) Calls at police control room	Service calls	Policemen

The waiting lines develop because the service to a customer may not be rendered immediately as the customer reaches the service facility. Thus, lack of adequate service facility would cause waiting lines of customers to be formed. The only way that the service demand can be met with ease is to increase the service capacity (and raising the efficiency of the existing capacity if possible) to a higher level. The capacity might be built to such a high level as can always meet the peak demand with no queues. But adding to capacity may be costly affair and uneconomic after a stage because then it shall remain idle to varying degrees when there are no or few customers. A manager, therefore, has to decide on an appropriate level of service which is neither too low nor too high. Inefficient or poor service would cause excessive waiting which has a cost in terms of customer frustration, loss of goodwill in the long run, direct cost of idle employees (where, for example, the employees have to wait near the store to obtain the supplies of materials, parts or tools needed for their work), or loss associated with poor employee morale resulting from being idle. On the other hand, too high a service level would result in very high set-up cost and idle time for the service station(s). Thus, the goal of queuing modelling is the achievement of an economic balance between the cost of providing service and the cost associated with the wait required for that service.

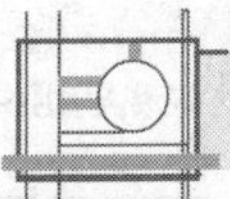

10.2 GENERAL STRUCTURE OF QUEUING SYSTEM

The general structure of a queuing system is depicted in Figure. 10.1.

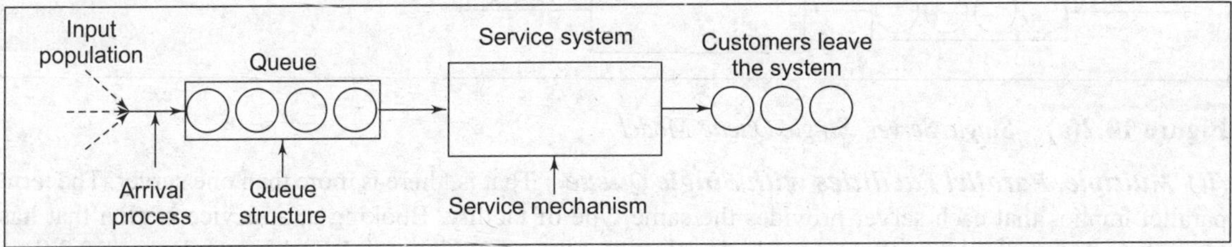

Figure 10.1 *General Structure of the Queuing System*

We shall discuss in more details the various elements of a queuing system and then present mathematical results for some specific systems. The elements of a system are:

1. Arrival Process The arrivals from the input population may be classified on different bases as follows:

(a) *According to source* The source of customers for a queuing system can be infinite or finite. For example, all people of a city or state (and others) could be the potential customers at a superbazar. The number of people being very large, it can be taken to be infinite. On the other hand, there are many situations in business and industrial conditions where we cannot consider the population to be infinite—it is *finite*. Thus, the ten machines in a factory requiring repairs and maintenance by the maintenance crew would examplify finite population. Removing one machine from a small, finite, population like this will have a noticeable effect on the calls expected to be made (for repairing) by the remaining machines than if there were a large number of machines, say 500.

(b) *According to numbers* The customers may arrive for service individually or in groups. Single arrivals are illustrated by customers visiting a beautician, students reaching at a library counter, and so on. On the other hand, families visiting restaurants, ship discharging cargo at a dock are examples of bulk, or batch, arrivals.

(c) *According to time* Customers may arrive in the system at known (regular or otherwise) times, or they might arrive in a random way. The queuing models wherein customers' arrival times are known with certainty are categorised as *deterministic models* (insofar as this characteristic is concerned) and are easier to handle. On the other hand, a substantial majority of the queuing models are based on the premise that the customers enter the system stochastically, at random points in time.

With random arrivals, the number of customers reaching the system per unit time might be described by a probability distribution. Although the arrivals might follow any pattern, the frequently employed assumption, which adequately supports many real world situations, is that the arrivals are *Poisson* distributed.

2. Service System There are two aspects of a service system—(a) structure of the service system, and (b) the speed of service.

(a) Structure of the Service System

By structure of the service system we mean how the service facilities exist. There are several possibilities. For example, there may be

(i) A Single Service Facility A library counter is an example of this. The models that involve a single service facility are called *single server models*. Figure 10.2 (a) illustrates such a model.

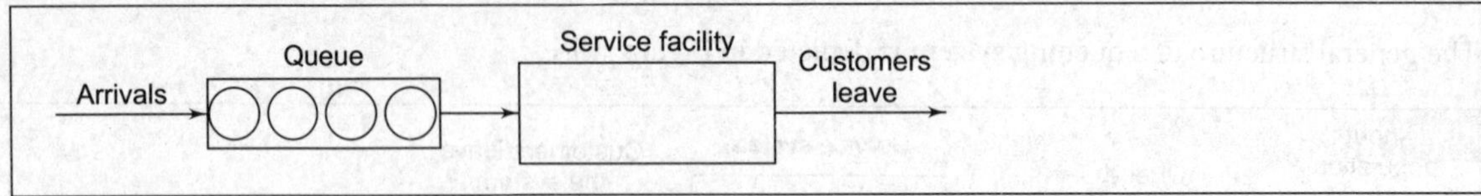

Figure 10.2(a) *Single Server, Single Queue Model*

(ii) Multiple, Parallel Facilities with Single Queue That is, there is more than one server. The term parallel implies that each server provides the same type of facility. Booking at a service station that has several mechanics, each handling one vehicle, illustrates this type of model. It is shown in Figure 10.2 (b).

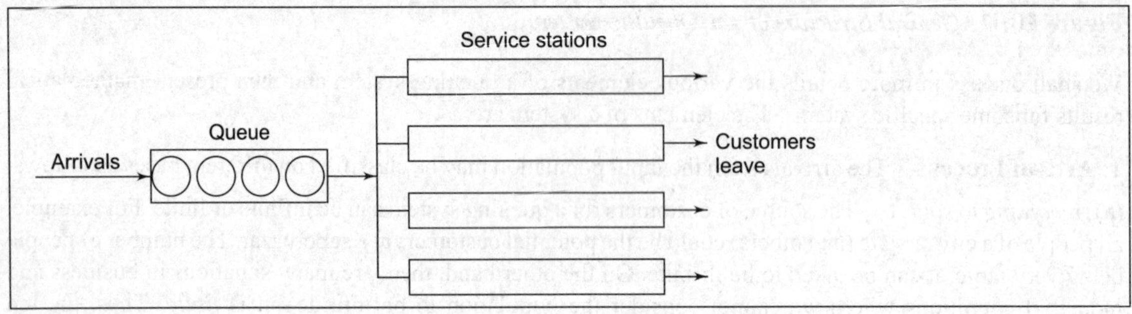

Figure 10.2(b) *Multiple, Parallel Servers, Single Queue Model*

(iii) Multiple, Parallel Facilities with Multiple Queues This type of model is different from the earlier one only in that each of the servers has a different queue. Different cash counters in an electricity office where the customers can make payment in respect of their electricity bills provide an example of this type of model. Figure 10.2(c) portrays such a model.

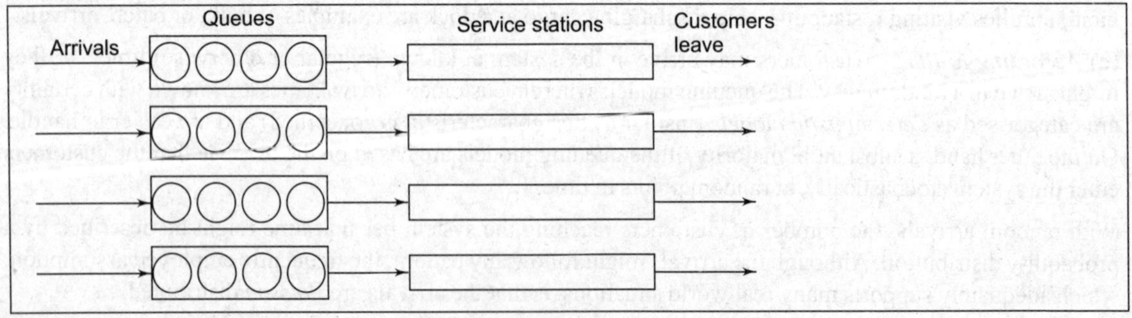

Figure 10.2(c) *Multiple, Parallel Servers, Multiple Queues Model*

(iv) Service Facilities in a Series In this, a customer enters the first station and gets a portion of service and then moves on to the next station, gets some service and then again moves on to the next station ... and so on, and finally leaves the system, having received the complete service. For example, machining of a certain steel item may consist of cutting, turning, knurling, drilling, grinding, and packaging operations, each of which is performed by a single server in a series. Figure 10.2(d) shows such a situation.

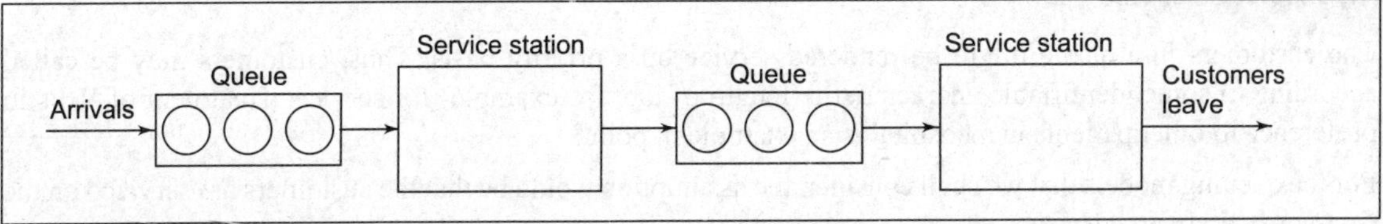

Figure 10.2(d) *Multiple Servers in Series*

Besides these, there may be other possibilities as well.

(b) Speed of Service

In a queuing system, the speed with which service is provided can be expressed in either of two ways—as *service* rate and as *service time*. The service rate describes the number of customers serviced during a particular time period. The service time indicates the amount of time needed to service a customer. Service rates and times are reciprocals of each other and either of them is sufficient to indicate the capacity of the facility. Thus, if a cashier can attend, on the average, to 10 customers in an hour, the service rate would be expressed as 10 customers/hour and service time would be equal to 6 minutes/customer. Generally, however, we consider the service time only.

If these service times are known exactly, the problem can be handled easily. But, as generally happens, if these are different and not known with certainty, we have to consider the distribution of the service times in order to analyse the queuing system. Generally, the queuing models are based on the assumption that service times are *exponentially* distributed about some average service time.

3. Queue Structure Another element of a queuing system is the queue structure. In the queue structure, the important thing to know is the queue discipline which means the order by which customers are picked up from the waiting line for service. There are a number of possibilities. They are:

(a) First-come-first-served

When the order of service of customers is in the order of their arrival, the queue discipline is of the first-come-first-served type. For example, with a queue at the bus stop, the people who came first will board the bus first.

(b) Last-come-first-served

Sometimes, the customers are serviced in an order reverse of the order in which they enter so that the ones who join the last are served first. For example, assume that letters to be typed, or order forms to be processed accumulate in a pile, each new addition being put on the top of them. The typist or the clerk might process these letters or orders by taking each new task from the top of the pile. Thus, a just arriving task would be the next to be serviced provided that no fresh task arrives before it is picked up. Similarly, the people who join an elevator last are the first ones to leave it.

(c) Service-in-random-order (SIRO)

Random order of service is defined as: whenever a customer is chosen for service, the selection is made in a way that every customer in the queue is equally likely to be selected. The time of arrival of the customers is, therefore, of no consequence in such a case.

(d) Priority Service

The customers in a queue might be rendered service on a priority basis. Thus, customers may be called according to some identifiable characteristic (length of job, for example) for service. Treatment of VIPs in preference to other patients in a hospital is an example in point.

For the queuing models that we shall consider, the assumption would be that the customers are serviced on the first-come-first-served basis.

Another thing to consider in the queuing structure is the behaviour or attitude of the customers entering the queuing system. On this basis, the customers may be classified as being (a) patient, or (b) impatient. If the customers join a queue, when it exists, and wait till they enter the service station for getting service, they are called *patient* customers. On the other hand, the queuing systems may enjoy customer behaviour in the form of defections from the queue. The customers may not select queues randomly (if there are multiple queues) and look for the shortest queue. There may be *jockeying* among the many queues, that is the customers may switch to other queues which are moving 'fast', and also *reneging* is possible—when a customer stands in the queue for sometime and then leaves the system because it is working 'too slowly'. There may also be *bribing* or *cheating* by some customers for queue positions. Besides, some customers may, upon their arrival, not join the queue for some reason and decide to return for service at a later time, or may even abandon the input population altogether. In terms of the queuing theory, this is known as *balking*, and occurs particularly when there are limits on the time and the extent of storage capacity available to hold waiting customers. Unless otherwise specified, the storage capacity is taken to be infinite. In the queuing models that we consider, we shall assume that there is no balking or jockeying and that the customers leave the system only after receiving service, and not before. Mathematical models give way to simulation when this assumption breaks.

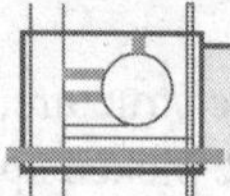

10.3 OPERATING CHARACTERISTICS OF QUEUING SYSTEM

An analysis of a given queuing system involves a study of its different operating characteristics. This is done using queuing models. Some of the more commonly considered characteristics are discussed below.

1. *Queue length*—the average number of customers in the queue waiting to get service. Large queues may indicate poor server performance while small queues may imply too much server capacity.

2. *System length*—the average number of customers in the system, those waiting to be and those being serviced. Large values of this statistic imply congestion and possible customer dissatisfaction and a potential need for greater service capacity.

3. *Waiting time in the queue*—the average time that a customer has to wait in the queue to get service. Long waiting times are directly related to customer dissatisfaction and potential loss of future revenues, while very small waiting times may indicate too much service capacity.

4. *Total time in the system*—the average time that a customer spends in the system, from entry in the queue to completion of service. Large values of this statistic are indicative of the need to make adjustment in the capacity.

5. *Server idle time*—the relative frequency with which the service system is idle. Idle time is directly related to cost. However, reducing idle time may have adverse effects on the other characteristics mentioned above.

We now proceed to discuss some of the queuing models. It may be mentioned here that the results obtained from various models are based on the assumption that the service system is operating under equilibrium or *steady state* conditions. For many systems, the operating day begins in *transient state* with no customers in the system. It takes some initial time interval for enough customers to arrive such that a steady state balance is reached. It should be clearly understood that a steady state does not mean that the system will reach a point where the number of customers in the system never changes. Even when the system reaches equilibrium, fluctuations will occur. A steady state condition really implies that various system performance measures (the operating characteristics) would reach stable values.

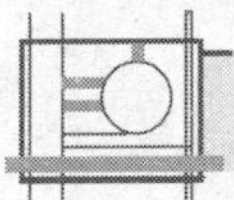

10.4 QUEUING MODELS

The queuing models can be categorised as being *deterministic* or *probabilistic*. If each customer arrives at known intervals and the service time is known with certainty, the queuing model shall be deterministic in nature. The vast majority of the queuing models are, however, based on the assumption that one or more elements of the queuing system can be expressed only in probabilistic terms. Hence, nearly all of the queuing models are of probabilistic type.

10.4.1 Deterministic Queuing Model

Let us first consider the case where the customers arrive in the queuing system at regular intervals and the service time for each customer is known and constant.

Suppose that customers come to a bank's teller counter every 5 minutes. Thus, the interval between the arrival of any two successive customers is exactly 5 minutes. Suppose further that the banker takes exactly 5 minutes to serve a customer. Here the arrival and the service rates are each equal to 12 customers per hour. In this situation there shall never be a queue and the banker shall always be busy with work.

Now, suppose that the banker can serve 15 customers per hour. The consequence of this higher service rate would be that the banker would be busy 4/5th of the time and idle in 1/5th of his time. He shall take 4 minutes to serve a customer and wait for 1 minute for the next customer to come. There would be, as before, no queue.

If, on the other hand, the banker can serve only 10 customers per hour, then the result would be that he would be always busy and the queue length will increase continuously without limit with the passage of time. It is easy to visualise that when the service rate is less than the arrival rate, the service facility cannot cope with all the arrivals and eventually the system leads to an explosive situation. The problem in such situations can be resolved by providing additional service station(s). Symbolically, let the arrival rate be λ customers per unit time and the service rate is μ customers per unit time.

Then,

if $\lambda > \mu$ the waiting line shall be formed which will increase indefinitely; the service facility would always be busy; and the service system will eventually fail; and

if $\lambda \leq \mu$ there shall be no waiting time; the proportion of time the service facility would be idle is $1 - \lambda/\mu$.

The ratio $\lambda/\mu = \rho$ is called the *average utilisation*, or the *traffic intensity*, or the *clearing ratio*.

For our present model,

if $\rho > 1$, the system would ultimately fail, and

if $\rho \leq 1$, the system works and ρ is the proportion of time it is busy.

We can easily visualise that the condition of uniform arrival and uniform service rates has a very limited practicability. Such conditions may exist when we are dealing, for example, with movements of items for processing in highly automated plants. However, generally, and more particularly when human beings are involved, the arrivals and servicing time are variable and uncertain. Thus, variable arrival rates and servicing times are the more realistic assumptions. The probabilistic queuing models, as mentioned previously, are based on these assumptions.

10.4.2 Probabilistic Queuing Models

Of the numerous queuing models available, we shall consider the following models:

(a) Poisson-exponential, single server model-infinite population;
(b) Poisson-exponential, single server model-finite population; and
(c) Poisson-exponential, multiple server model-infinite population.

In each of these, the words 'Poisson-exponential' indicate that the customer arrivals follow Poisson distribution while the service times are distributed exponentially. To recapitulate, if the arrivals are independent, with the average arrival rate equal to λ per period of time, then, according to the Poisson probability distribution, the probability that n customers will arrive in the system during a given interval T, is given by the following:

$$P\,(n \text{ customers during period } T) = e^{-m}\frac{m^n}{n!}$$

where $m = \lambda T$, and $e = 2.7183$

Example 10.1 On an average, 5 customers reach a barber's shop every hour. Determine the probability that exactly 2 customers will reach in a 30-minute period, assuming that the arrivals follow Poisson distribution.

Here $\lambda = 5$ customers/hour

$T = 30$ minutes $= 0.5$ hour

Thus, $m = \lambda T = 5 \times 0.5 = 2.5$ customers

$$P\,(2 \text{ customers}) = 2.7183^{-2.5} \times \frac{(2.5)^2}{2!}$$
$$= 0.257$$

Similarly, when the times taken to serve different customers are independent, the probability that no more than T periods would be required to serve a customer is given by exponential distribution as follows:

$$P\,(\text{no more than } T \text{ time period needed to service a customer}) = 1 - e^{-\mu T},$$

where μ refers to the average service rate.

Example 10.2 The manager of a bank observes that, on the average, 18 customers are served by a cashier in an hour. Assuming that the service time has an exponential distribution, what is the probability that (a) a customer shall be free within 3 minutes, (b) a customer shall be serviced in more than 12 minutes?

(a) Here $\mu = 18$ customers/hour

$T = 3$ minutes $= 0.05$ hour

$\therefore$ P (less than 3 minutes) $= 1 - 2.7183^{-(18)(0.05)}$

$= 0.593$

(b) With $\mu = 18$ customers/hour,

$T = 12$ minutes $= 0.20$ hour,

P (more than 12 minutes) $= e^{-\mu T}$

$= 2.7183^{-(18)(0.20)}$

$= 0.027$

Further, in each of the models, the customer service is assumed to be in first-come-first-served order.

Now we shall describe and analyse these three models in turn.

A. Poisson-exponential Single Server Model—Infinite Population This model is based on the following assumptions:

(a) The arrivals follow Poisson distribution, with a mean arrival rate of, say, λ.
(b) The service time has exponential distribution. Let the average service rate be μ.
(c) Arrivals are from infinite population.
(d) The customers are served on a first-come-first-served basis.
(e) There is only a single service station.

To evaluate the model, we begin with the first question whether the service station can handle the customer demand for service. The answer to this question lies in the values of μ and λ. If $\lambda \geq \mu$, the waiting line would increase without limit, leading to the breakdown of the system ultimately. For a workable system, it is necessary that $\lambda < \mu$. It is interesting to observe that even if $\lambda = \mu$, then system shall not work. This can be shown mathematically.

As pointed out earlier, the ratio $\rho = \lambda/\mu$ is defined as the traffic intensity. This indicates the proportion of time, or the probability, that the service station is busy. From this, the probability that the system is idle, that is, there are no customers in the system, equals $P_0 = 1 - \rho$.

We can show that the probability of having exactly one customer in the system is $P_1 = \rho P_0$. Similarly, the probability of having exactly 2 customers in the system would be $P_2 = \rho P_1 = \rho^2 P_0$.

To generalise, the probability of having exactly n customers in the system shall be:

$$P_n = \rho^n P_0 = \rho^n(1 - \rho)$$

We can use this result to determine the expected number of customers in the system.

$$L_s = \sum_{n=0}^{\infty} nP_n$$

This can be solved to obtain

$$L_s = \frac{\lambda}{\mu - \lambda} \quad \text{or} \quad \frac{\rho}{1 - \rho}$$

Expected number of customers in the queue shall be equal to the difference between the expected number of customers in the system and the expected number of customers being served. Now, since the server is busy, or is serving one unit, λ/μ of the time, and is serving 0 unit (i.e., it is idle) $1 - \lambda/\mu$ of the time, the expected number being served equals $1\left(\frac{\lambda}{\mu}\right) + 0\left(1 - \frac{\lambda}{\mu}\right) = \frac{\lambda}{\mu}$.

Consequently, expected number of customers in the queue,

$$L_q = L_s - \frac{\lambda}{\mu} = \frac{\lambda}{\mu - \lambda} - \frac{\lambda}{\mu}$$

$$= \frac{\lambda^2}{\mu(\mu - \lambda)} \quad \text{or} \quad \frac{\rho^2}{1 - \rho}$$

It is significant to note that L_q is the average length of *all* queues including empty queues. The average length of non-empty queues, that is, those which contain at least one customer, $L_{q'}$ is given by the following expression.

$$L_{q'} = \frac{1}{1 - \rho} \quad \text{or} \quad \frac{\mu}{\mu - \lambda}$$

With an average arrival rate of λ, the average time between the arrivals is $1/\lambda$. Thus, the mean waiting time in queue, W_q, is the product of the average time between the arrivals and the average queue length. Symbolically,

$$W_q = \frac{1}{\lambda} \cdot L_q$$

Substituting $\lambda^2/\mu(\mu - \lambda)$ for L_q and simplifying, we get

$$W_q = \frac{\lambda}{\mu(\mu - \lambda)} = \frac{\rho}{\mu - \lambda}$$

Similarly, the mean time in the system, W_s, is equal to the product of the average time between arrivals and the average number of customers in the system. Thus,

$$W_s = \frac{1}{\lambda} \times L_s$$

Putting $L_s = \lambda/(\mu - \lambda)$ in this equation and simplifying, we get

$$W_s = \frac{1}{\mu - \lambda}$$

Stated another way, since the mean service rate is μ, the average (expected) time for completing the service is $1/\mu$. Therefore, the expected time a customer would spend in the system shall be equal to the expected waiting time in the queue plus the average processing/servicing time. Thus,

$$W_s = W_q + \frac{1}{\mu}$$

$$= \frac{\lambda}{\mu(\mu - \lambda)} + \frac{1}{\mu} = \frac{1}{\mu - \lambda}$$

which is the same as shown earlier.

Finally, the probability that a customer spends more than t units of time in the system, $W_s(t) = e^{-t/W_s}$. And, the probability that a customer spends more than t units of time in the queue, $W_q(t) = \rho e^{-t/W_s}$.

It is significant to note that although the time spent in the system is distributed exponentially, and so is the service time, but the difference of the two, that is to say, the time spent in the queue, $W_q(t)$, is not exponentially distributed.

Example 10.3 A tailor specialises in ladies' dresses. The number of customers approaching the tailor appear to be Poisson distributed with a mean of 6 customers per hour. The tailor attends the customers on a first-come-first-served basis and the customers wait if the need be. The tailor can attend the customers at an average rate of 10 customers per hour with the service time exponentially distributed.

Required,

(a) Find the probability of the number of arrivals (0 through 5) during (i) a 15-minute interval, and (ii) a 30-minute interval.
(b) The utilisation parameter.
(c) The probability that the queuing system is idle.
(d) The average time that the tailor is free on a 10-hour working day.
(e) The probability associated with the number of customers (0 through 5) in the queuing system.
(f) What is the expected number of customers in the tailor shop?
(g) What is the expected number of customers waiting for tailor's services?
(h) What is the average length of queues that have at least one customer?
(i) How much time should a customer expect to spend in the queue?
(j) What is the expected time a customer would spend in the tailor's shop?
(k) Assuming that $n > 0$ (i.e. customers are in the system) what is the probability that the waiting time (excluding service time) of a customer in the queue shall be more than 10 minutes?
(l) Assuming that the customers are in the system, what is the probability that a customer shall be in the shop for more than 15 minutes?

(a) The probabilities in two cases (i) when $T = 1/4$ hour and (ii) when $T = 1/2$ hour, of 0 through 5 customers are given in the Table 10.2. These are calculated by using the formula

$$P(n) = e^{-m} \times \frac{m^n}{n!}$$

where $m = \lambda T$ and $e = 2.7183$

with $\lambda = 6$, we have $m = 6 \times 1/4 = 1.5$ in (i), and $m = 6 \times 1/2 = 3$ for (ii).

TABLE 10.2 Calculation of Probabilities of Arrivals

n	*When T = 1/4 hr.* $P(n)$	*When T = 1/2 hr.* $P(n)$
0	0.2231	0.0498
1	0.3347	0.1497
2	0.2510	0.2240
3	0.1255	0.2240
4	0.0471	0.1680
5	0.0141	0.1008

(b) Utilisation parameter $\rho = \frac{\lambda}{\mu} = \frac{6}{10} = 0.6$

(c) Probability that system is idle, $P_0 = 1 - \frac{\lambda}{\mu} = 1 - 0.6 = 0.4$

(d) Average time the tailor is free on a 10-hour working day = $P_0 \times$ No. of hours = $0.4 \times 10 = 4$ hours

(e) Probability of n customers in the system, $P_n = \rho^n(1 - \rho)$.

Using this rule, probabilities are given in Table 10.3.

TABLE 10.3 Calculation of Probabilities of Customers in System

n	*Probability P(n)*
0	0.40
1	0.24
2	0.144
3	0.0864
4	0.05184
5	0.031104

(f) Expected number of customers in the shop,

$$L_s = \frac{\rho}{1-\rho} = \frac{0.6}{1-0.6} = 1.5$$

(g) Expected number of customers in the queue,

$$L_q = \frac{\rho^2}{1-\rho} = \frac{0.6^2}{1-0.6} = 0.36/0.4 = 0.9$$

(h) Expected length of non-empty queues,

$$L_{q'} = \frac{1}{1-\rho} = \frac{1}{1-0.6} = 2.5$$

(i) Expected waiting time in the queue,

$$W_q = \frac{\lambda}{\mu(\mu-\lambda)} = \frac{6}{10(10-6)} = \frac{6}{40} \text{ hr.} = 9 \text{ minutes}$$

(j) Expected time a customer spends in the system,

$$W_s = \frac{1}{\mu-\lambda} = \frac{1}{10-6} = \frac{1}{4} \text{ hr.} = 15 \text{ minutes}$$

(k) Probability that a customer shall wait for more than 10 minutes in the queue,

$$W_q(10) = \rho e^{-t/W_s} = 0.6 \times 2.7183^{-10/15} = 0.31$$

(l) Probability that a customer shall be in the shop for more than 15 minutes,

$$W_s(t) = e^{-t/W_s} = 2.7183^{-15/15} = 0.368$$

Example 10.4 Arrivals at a telephone booth are considered to be Poisson, with an average time of 10 minutes between one arrival and the next. The length of a phone call is assumed to be distributed exponentially, with mean 3 minutes. Find:

(i) the probability that an arrival finds that four persons are waiting for their turn;
(ii) the average number of persons waiting and making telephone calls; and
(iii) the average length of the queue that is formed from time to time. *(CS, June, 1998)*

The average inter-arrival time being 10 minutes, the average arrival rate = 60/10 = 6 customers per hour. Similarly, average length of the phone call being 3 minutes, the average service rate = 60/3 = 20 customers per hour. With $\lambda = 6$ customers/hour and $\mu = 20$ customers/hour, we have $\rho = \lambda/u = 6/20 = 0.3$.

(i) With four people waiting for their turn in the queue implies a total of 5 people in the system. Thus, the probability that an arrival finds four persons in the queue is,

$$P(n = 5) = \rho^n(1 - \rho)$$
$$= 0.3^5 (1 - 0.3)$$
$$= 0.0017$$

(ii) The average number of people waiting and making calls is given by the expected length of the system. Thus,

$$L_s = \frac{\rho}{1 \rho}$$
$$= \frac{0.3}{1 - 0.3} = 0.43$$

(iii) The average length of queue that is formed from time to time,

$$L_{q'} = \frac{1}{1 - \rho}$$
$$= \frac{1}{1 - 0.3} = 1.43$$

Example 10.5 The Taj Service Station has a central store where service mechanics arrive to take spare parts for the jobs they work upon. The mechanics wait in queue if necessary and are served on a first-come-first-served basis. The store is manned by one attendant who can attend 8 mechanics in an hour on an average. The arrival rate of the mechanics averages 6 per hour. Assuming that the pattern of mechanics' arrivals is Poisson distributed and the servicing time is exponentially distributed, determine W_s, W_q and L_q where the symbols carry their usual meaning.

According to the given information,

$$\lambda = 6 \text{ mechanics/hour, and } \mu = 8 \text{ mechanics/hour}$$

∴ Average utilisation, $\rho = 6/8$, which is less than unity.

With these values,

(i) Expected time spent by a mechanic in the system,

$$W_s = \frac{1}{\mu - \lambda} = \frac{1}{8 - 6} = 1/2 \text{ hour} = 30 \text{ minutes}$$

(ii) Expected time spent by a mechanic in the queue,

$$W_q = \frac{\lambda}{\mu(\mu - \lambda)} = \frac{6}{8(8 - 6)} = \frac{6}{16} \text{ hour} = 22.5 \text{ minutes}$$

(iii) Expected number of mechanics in the queue,

$$L_q = \frac{\lambda^2}{\mu(\mu - \lambda)} = \frac{6^2}{8(8 - 6)} = \frac{36}{16} = 2.25 \text{ mechanics.}$$

10.4.3 Cost Analysis

The information provided by a queuing model can be usefully employed in determining the appropriate level of service. For instance, in Example 10.5, the manager of the station might be concerned about the time that the mechanics have to wait in the queue in order to obtain the spare parts or tools that they need, because it costs to the service station (to pay for their idle time). The manager may consider employing additional attendants in the storeroom so that service may be provided at a faster rate and, consequently, the mechanics don't have to wait long for obtaining supplies. This decision would have the twin effects—first, it reduces the idle time and therefore the cost (associated with it), and, second, it increases the cost of providing service through payment to the additional hand employed. The manager would do well to employ additional attendant(s) if and so long as the saving in cost more than offsets the increase in cost. The optimum level of service would naturally be determined where the total of the waiting time cost and cost of providing service is the minimum. These ideas are portrayed in Figure 10.3 where it is clear that increasing the service level would result in increasing the cost of service and reducing the cost of waiting time.

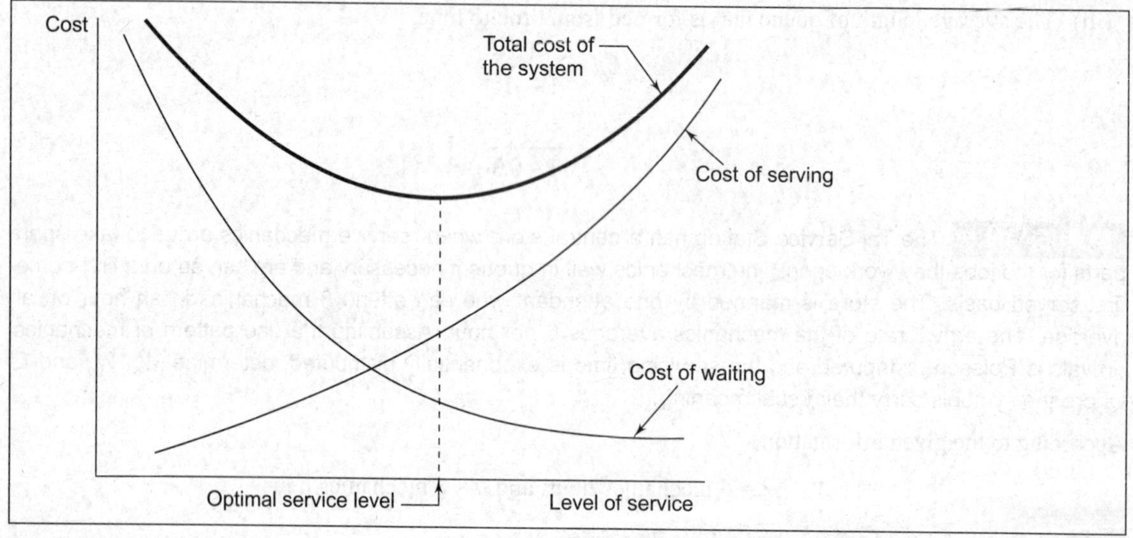

Figure 10.3 *Determination of Optimal Service Level*

The thick curve shows that the total cost declines upto a point and then rises. The service level corresponding to the minimum point on it is the optimal service level.

Let us analytically consider the manager's problem. Suppose that the service mechanics are paid at Rs 8 per hour and the storeroom attendant's wages are Rs 5 per hour. In a typical 8-hour day, the total arrivals would be $8 \times 6 = 48$. We know that an arriver has to wait 1/2 hour before he obtains parts of his requirement. Thus, the total waiting time in the system for a day equals $48 \times 1/2 = 24$ hours.

Now, the daily cost of service provided by one attendant = cost of service + cost of waiting.

Cost of service = No. of attendants × Hourly rate × No. of hours

$$= 1 \times 5 \times 8 = \text{Rs } 40$$

$$\text{Cost of waiting} = \text{No. of hours waiting} \times \text{Hourly rate (for mechanics)}$$

$$= 24 \times 8 = 192$$

$$\therefore \quad \text{Total cost} = 40 + 192 = \text{Rs } 232 \text{ per day}$$

Suppose now that the manager believes that if one more attendant is hired for the storeroom, the service rate can be increased from 8 to 12 mechanics an hour. The expected time spent by a mechanic in the system would equal

$$W_s = \frac{1}{\mu - \lambda} = \frac{1}{12-6} = \frac{1}{6} \text{ hour}$$

Total waiting time in the system for a day = 48 × 1/6 = 8 hours.

Accordingly,

Service cost $= 2 \times 5 \times 8 =$ Rs 80

Cost of waiting $= 8 \times 8 =$ Rs 64

∴ Total cost $= 80 + 64 =$ Rs 144 per day.

Thus, the total cost of providing service reduces from Rs 232 to Rs 144 per day by hiring an additional attendant.

Further, suppose that the manager believes that by employing yet another person, the service rate can be stepped up to 16 mechanics per hour. With this,

$$W_s = \frac{1}{\mu - \lambda} = \frac{1}{16-6} = \frac{1}{10} \text{ hour}$$

Total waiting time in the system $= 48 \times \frac{1}{10} = 4.8$ hours per day.

Service cost	$= 3 \times 5 \times 8$	= Rs 120.0
Waiting cost	$= 4.8 \times 8$	= Rs 38.4
Total cost		= Rs 158.4

Clearly, it is not worthwhile employing the third person. To minimise the total cost, therefore, the manager should employ two attendants.

Example 10.6 A repairman is to be hired by a company to repair machines that break down following a Poisson process, with an average rate of four per hour. The cost of non-productive machine time is Rs 90 per hour. The company has the option of choosing either a fast or a slow repairman. The fast repairman charges Rs 70 per hour and will repair machines at an average rate of 7 per hour, while the slow repairman charges Rs 50 per hour and will repair machines at an average rate of 6 per hour. Which repairman should be hired?

For solving this problem, we compare the total expected daily cost for both the repairmen. This would equal the total wages paid plus the cost of non-productive machine hours.

We have, Total wages = Hourly rate × No. of hours

For fast repairman,

Total wages = 70 × 8 (assuming 8-hour shift) = Rs 560

For slow repairman,

Total wages = 50 × 8 = Rs 400

Cost of non-productive time can be calculated as under:

Expected No. of machines in the system × Cost of idle machine hour × No. of hours

Expected number of machines in the system, $L_s = \dfrac{\lambda}{\mu - \lambda}$

Cost of idle machine = Rs 90/hour

No. of hours (assuming 8-hour shift) = 8

For fast repairman,

$\lambda = 4$ machines/hour and $\mu = 7$ machines/hour

$$\therefore \quad L_s = \frac{4}{7-4} = \frac{4}{3} \text{ machines}$$

Thus, cost of non-productive machine time = $\dfrac{4}{3} \times 90 \times 8$ = Rs 960

For slow repairman,

$\lambda = 4$ machines/hour and $\mu = 6$ machines/hour

$$\therefore \quad L_s = \frac{4}{6-4} = 2 \text{ machines}$$

Accordingly, cost of non-productive machine time = 2 × 90 × 8 = Rs 1,440

The cost of non-productive machine time can be alternatively calculated as follows:

Expected time a machine spends in the system × No. of arrivals per day × Cost of idle time per hour

The product of the first two elements gives the total machine hours lost per day which, when multiplied by the hourly rate, yields the cost of idle (non-productive) time per day.

Now,

$$W_s \text{ (for fast)} = \frac{1}{7-4} = \frac{1}{3} \text{ hour}$$

$$W_s \text{ (for slow)} = \frac{1}{6-4} = \frac{1}{2} \text{ hour}$$

No. of arrivals per day of 8-hours = 8 × 4 = 32

Cost of idle time per hour = Rs 90

Accordingly, the cost of idle time is as follows:

for fast repairman : $\dfrac{1}{3} \times 32 \times 90$ = Rs 960

for slow repairman : $\dfrac{1}{2} \times 32 \times 90$ = Rs 1,440

Now, Total cost for each of the repairmen is as follows:

Fast repairman : Rs 560 + Rs 960 = Rs 1,520

Slow repairman : Rs 400 + Rs 1,440 = Rs 1,840

It is evident that the fast repairman should be employed by the manager.

B. Poisson-exponential, Single-server Model—Finite Population This queuing model is based on similar assumptions as the earlier discussed model except that the input population is finite. For this model, the system structure is such that we have a total of M customers; a customer is either in the system (consisting of a queue and a single service station), or outside the system and, in a sense, arriving. In particular, when a customer is in the arriving condition, then the time it takes him to arrive is a random variable having an exponential distribution with mean = $1/\lambda$. As a result, when there are n customers in the system, then there are $M - n$ customers in the arriving state and, therefore, the total average rate of arrivals in the system is $\lambda(M - n)$. It may be observed that this system is self-regulating. By this, we mean that when the system gets busy, with many of these customers in the queue, then the rate at which additional customers arrive is, in fact, reduced, thus, lowering the further congestion in the system.

To illustrate, suppose that a factory has five machines which are required to be handled by the maintenance staff for repairing and working. The number of customers (the machines here), or the population, is finite, a small countable set. It has the implication that the removal of one machine from production can drastically change the probability of another machine being removed. Thus, there exists a strong dependency relationship between the arrivals and they are no longer independent. Of course, if there were a large number of machines in the factory, say three hundred, removal of one machine would have but an insignificant impact on the chances of another machine being removed for repairs. In such a situation, the assumption of independence of arrivals nearly sustains and, therefore, we can use the first model for studying various operating characteristics of the queuing system.

For the reason of dependency relationships between arrivals, the Poisson probability law cannot be strictly applied when the input population is finite. Instead of the arrivals' statement as an average for the population, we classify them as an average of a unit time. Hence, as mentioned, the inter-arrival time of customers is stated to be following exponential distribution with mean = $1/\lambda$.

The operating characteristics of such a system are given here:

With M as the number of customers in the source population, $1/\lambda$ as the average inter-arrival time between successive arrivals, and μ as the service rate, we have

(a) Probability that the system shall be idle,

$$P_0 = \left[\sum_{i=0}^{M}\left(\frac{M!}{(M-i)!}\right)\left(\frac{\lambda}{\mu}\right)^i\right]^{-1}$$

(b) Probability that there shall be n customers in the system,

$$P_n = \begin{cases} P_0\left(\dfrac{\lambda}{\mu}\right)^n \dfrac{M!}{(M-n)!} & 0 < n \le M \\ 0 & n > M \end{cases}$$

(c) Expected length of the queue,

$$L_q = M - \frac{\lambda+\mu}{\lambda}(1-P_0)$$

(d) Expected number of customers in the system,

$$L_s = L_q + (1-P_0)$$

$$= M - \frac{\mu}{\lambda}(1-P_0)$$

(e) Expected waiting time of a customer in the queue,

$$W_q = \frac{L_q}{\mu(1 - P_0)}$$

$$= \frac{1}{\mu}\left(\frac{M}{1 - P_0} - \frac{\lambda + \mu}{\lambda}\right)$$

(f) Expected time a customer spends in the system,

$$W_s = W_q + \frac{1}{\mu}$$

$$= \frac{1}{\mu}\left[\frac{M}{1 - P_0} - \frac{\lambda + \mu}{\lambda} + 1\right]$$

Example 10.7 Past records indicate that of the five machines that a factory owns, breakdowns occur at random and the average time between the breakdowns is 2 days. Assuming that the repairing capacity of the workman is one machine a day and the repairing time is distributed exponentially, determine the following:

(a) the probability that the service facility will be idle,
(b) the probability of various number of machines (0 through 5) to be, and being repaired,
(c) the expected length of the queue,
(d) the expected number of machines waiting to be, and being repaired,
(e) the expected time that a machine shall wait in the queue to be repaired, and
(f) the expected time a machine shall be idle for reason of waiting to be repaired and being repaired.

According to the given information,

$$M = 5 \text{ machines}$$
$$\mu = 1 \text{ machine/day}$$
$$\lambda = 1/2 \text{ machine/day}$$
$$\therefore \quad \rho = \lambda/\mu = 0.5$$

Since P_0 is basic to all the formulae for this model, we shall first determine its value. It is shown in Table 10.4.

TABLE 10.4 Calculation of P_0

i	$M!/(M-i)!$	$(\lambda/\mu)^i$	$\frac{M!}{(M-i)!}\left(\frac{\lambda}{\mu}\right)^i$
0	1	1.00000	1.00
1	5	0.50000	2.50
2	20	0.25000	5.00
3	60	0.12500	7.50
4	120	0.06250	7.50
5	120	0.03125	3.75
		Total	27.25

We have,

$$P_0 = \left[\sum_{i=0}^{M}\left(\frac{M!}{(M-i)!}\left(\frac{\lambda}{\mu}\right)^i\right)\right]^{-1}$$

$$= 27.25^{-1}$$

$$= 0.0367$$

Thus,

(a) The probability that the service facility will be idle,

$$P_0 = 0.0367$$

(b) The probability of various number of customers, n, being in the system is shown in the Table 10.5. It is obtained by using the following rule:

$$P_n = P_0\left(\frac{\lambda}{\mu}\right)^n\left(\frac{M!}{(M-n)!}\right)$$

TABLE 10.5 Probabilities of Various Number of Machines in the System

n	$\left(\frac{\lambda}{\mu}\right)^n\left(\frac{M!}{(M-n)!}\right)$	*Probability**
0	1.00	0.036700
1	2.50	0.091750
2	5.00	0.183500
3	7.50	0.275250
4	7.50	0.275250
5	3.75	0.137620

* The sum total of all these probabilities should naturally equal 1. Here it is slightly different because of the approximation error.

(c) Expected length of the queue,

$$L_q = M - \frac{\lambda+\mu}{\lambda}(1-P_0)$$

$$= 5 - \frac{0.5+1}{0.5}(1-0.0367)$$

$$= 5 - 2.89 = 2.11 \text{ machines}$$

(d) Expected number of machines waiting to be, and being repaired,

$$L_s = M - \frac{\mu}{\lambda}(1-P_0)$$

$$= 5 - \frac{1}{0.5}(1-0.0376)$$

$$= 5 - 1.93 = 3.07 \text{ machines}$$

(e) Expected time a machine will wait in queue to be repaired,

$$W_q = \frac{1}{\mu}\left[\frac{M}{1-P_0} - \frac{\lambda+\mu}{\lambda}\right]$$

$$= \frac{1}{1}\left[\frac{5}{1-0.0367} - \frac{0.5+1}{0.5}\right]$$

$$= \frac{5}{0.9633} - 3$$

$$= 5.1905 - 3 = 2.19 \text{ days}$$

(f) Expected time a machine will be in the system,

$$W_s = W_q + \frac{1}{\mu} = 2.19 + \frac{1}{1} = 3.19 \text{ days}$$

Cost Analysis To continue with Example 10.7, suppose that the machine idle time cost is Rs 160 per machine per day; the operator's cost is Rs 40 per day; while the workman's charges are Rs 70 per day. Using these data, we can obtain the expected cost per breakdown as follows:

Total Cost = Machine idle time cost + Lost productivity of machine operator + Repair cost

$$= (\text{Idle time}) \times (\text{Idle time cost}) + \frac{\text{Idle time of a machine} \times \text{Operator cost}}{\text{No. of machines}}$$

$$+ (\text{Repair time})\,(\text{Repair rate})$$

$$= W_s \times 160 + \frac{W_s \times 40}{5} + (W_s - W_q) \times 70$$

$$= 3.19 \times 160 + \frac{3.19 \times 40}{5} + (3.19 - 2.19) \times 70$$

$$= 510.4 + 25.52 + 70 = \text{Rs } 605.92$$

Thus, the average cost per machine breakdown would be equal to Rs 605.92.

Sensitivity Analysis We can now examine the sensitivity of our previous example to the number of machines which can be repaired by a repairman everyday (of course, sensitivity can be tested by changing any of the parameters of the system, but here we confine to the changes in the service rate). To do this, we will allow the service rate to be $\mu = 0.8$, 1 (already discussed) and 1.2 machines per day, with the number of machine breakdowns per day being held constant at $\lambda = 0.5$ machines per day. Using the relationships discussed earlier for this finite population model, we can make computations for various operating characteristics of the system. The results are given here:

μ	P_0	L_s	L_q	W_s	W_q
0.8	0.0177	3.43	2.45	4.36	3.11
1.0	0.0367	3.07	2.11	3.19	2.19
1.2	0.0625	2.75	1.81	2.44	1.61

The system is moderately sensitive to the changes in the average customer service rate. For example, when the service rate decreases from $\mu = 1$ machine per day to $\mu = 0.8$ machine per day, a decrease of 20 per cent, the expected number of customers in the system increases from 3.07 to 3.43 machines, an increase of 12 per cent. The expected number of customers in the system queue increases from 2.11 to 2.45 machines, indicating a rise of 16 per cent. The expected waiting time in the queuing system increases from 3.19 to 4.36 days — an increase of 37 per cent. Finally, the expected waiting time in the queue rises from 2.19 to 3.11 days, suggesting a rise of 42 per cent. On the other hand, when the service rate increases by the same amount, the percentage decreases in various parameters are: $L_s = 10$, $L_q = 14$, $W_s = 24$ and $W_q = 26$. Clearly, an equal degree of change in the service rate on either side does not produce the same effect. The system is more volatile when there is an unfavourable change in the service rate.

Further, using the same cost data, as considered earlier, the average cost per breakdown for $\mu = 0.8$ works out to Rs 819.98 while for $\mu = 1.2$, it equals Rs 468.02. Thus, for a 20 per cent fall in the working efficiency of the repairman per day, the cost per breakdown would rise by about 35%, while an equal amount of increase in the efficiency would produce a cost saving of 23%.

C. Poisson-Exponential, Multiple Server Model—Infinite Population Each of the previous two models were based on the assumption that there is only one service facility in the system. Now we shall consider a model in which there are multiple service stations, which are giving parallel, identical service. It covers situations where, for example, there may be more than one doctor in a hospital whom the patients can visit; there may be more than one runway at an airport for takeoff and landing, ... and so on.

In cases where multiple servers are involved, an important point to consider is whether there is a separate, independent queue for each server or the customers form a single queue form which they are picked up for service. Indeed if there are as many queues as the service stations, equal to, say, K, the system can be thought of, and analysed as one composed of K different single server queuing systems. For our purposes here, we shall presume that the customers form a single queue.

Assumptions The model is based on the following assumptions:

(a) The arrival of customers follows Poisson probability law, the average arrival rate being λ.
(b) The service time has an exponential distribution.
(c) There are K service stations, each of which provides identical service.
(d) A single waiting line is formed.
(e) The input population is infinite.
(f) The service is on a first-come-first-served basis.
(g) The arrival rate is smaller than the combined service rate of all K service facilities.

Operating Characteristics We shall use the following expressions:

λ = average rate of arrivals

μ = mean service rate of each of the servers

K = number of servers

$K\mu$ = mean combined service rate of all the servers

$$\rho = \frac{\lambda}{K\mu} = \text{utilisation factor of the entire system}$$

The operating characteristics are given here:

(a) Probability that the system shall be idle,

$$P_0 = \left[\sum_{i=0}^{k-1} \frac{(\lambda/\mu)^i}{i!} + \frac{(\lambda/\mu)^K}{K!(1-\rho)}\right]^{-1}$$

(b) Probability that there shall exactly be n customers in the system,

$$P_n = (P_0)\,\frac{(\lambda/\mu)^n}{n!}, \text{ when } n \leq K,$$

and

$$P_n = (P_0)\,\frac{(\lambda/\mu)^n}{K!\,K^{n-K}}, \text{ when } n > K$$

(c) The expected number of customers in the waiting line,

$$L_q = \frac{(\lambda/\mu)^K \rho}{K!(1-\rho)^2}\,(P_0)$$

(d) The expected number of customers in the system,

$$L_s = L_q + \frac{\lambda}{\mu}$$

(e) The expected waiting time in the queue,

$$W_q = \frac{L_q}{\lambda}$$

(f) The expected time a customer spends in the system,

$$W_s = W_q + \frac{1}{\mu}$$

Table 10.6 provides the probabilities P_0 associated with different values of ρ and K, which ease the computations.

TABLE 10.6 P_0 Values for Various Combinations of ρ and K

$\frac{\lambda}{K\mu} = \rho$	*Number of Channels, K*							
	2	3	4	5	6	7	8	10
.02	.9608	.9418	.9231	.9048	.8869	.8694	.85214	.81873
.04	.9231	.8869	.8521	.8187	.7866	.7558	.72615	.67032
.06	.8868	.8353	.7866	.7408	.6977	.6570	.61878	.54881
.08	.8519	.7866	.7261	.6703	.6188	.5712	.52729	.44933
.10	.8182	.7407	.6703	.6065	.5488	.4966	.44933	.36788
.12	.7857	.6975	.6188	.5488	.4868	.4317	.38289	.30119
.14	.7544	.6568	.5712	.4966	.4317	.3753	.32628	.24660
.16	.7241	.6184	.5272	.4493	.3829	.3263	.27804	.20190
.18	.6949	.5821	.4866	.4065	.3396	.2837	.23693	.16530
.20	.6667	.5479	.4491	.3678	.3012	.2466	.20189	.13534

(*Contd*)

TABLE 10.6 *(Contd)*

$\frac{\lambda}{K\mu} = \rho$	Number of Channels, K							
	2	3	4	5	6	7	8	10
.22	.6393	.5157	.4145	.3328	.2671	.2144	.17204	.11080
.24	.6129	.4852	.3824	.3011	.2369	.1864	.14660	.09072
.26	.5873	.4564	.3528	.2723	.2101	.1620	.12492	.07427
.28	.5625	.4292	.3255	.2463	.1863	.1408	.10645	.06081
.30	.5385	.4035	.3002	.2228	.1652	.1224	.09070	.04978
.32	.5152	.3791	.2768	.2014	.1464	.1064	.07728	.04076
.34	.4925	.3561	.2551	.1821	.1298	.0925	.06584	.03337
.36	.4706	.3343	.2351	.1646	.1151	.0804	.05609	.02732
.38	.4493	.3137	.2165	.1487	.1020	.0698	.04778	.02236
.40	.4286	.2941	.1993	.1343	.0903	.0606	.04069	.01830
.42	.4085	.2756	.1834	.1213	.0800	.0527	.03465	.01498
.44	.3889	.2580	.1686	.1094	.0708	.0457	.02950	.01226
.46	.3699	.2414	.1549	.0987	.0626	.0397	.02511	.01003
.48	.3514	.2255	.1422	.0889	.0554	.0344	.02136	.00820
.50	.3333	.2105	.1304	.0801	.0490	.0298	.01816	.00671
.52	.3158	.1963	.1195	.0721	.0432	.0259	.01544	.00548
.54	.2987	.1827	.1094	.0648	.0381	.0224	.01311	.00448
.56	.2821	.1699	.0999	.0581	.0336	.0194	.01113	.00366
.58	.2658	.1576	.0912	.0521	.0296	.0167	.00943	.00298
.60	.2500	.1460	.0831	.0466	.0260	.0144	.00799	.00243
.62	.2346	.1349	.0755	.0417	.0228	.0124	.00675	.00198
.64	.2195	.1244	.0685	.0372	.0200	.0107	.00570	.00161
.66	.2048	.1143	.0619	.0330	.0175	.0092	.00480	.00131
.68	.1905	.1048	.0559	.0293	.0152	.0079	.00404	.00106
.70	.1765	.0957	.0502	.0259	.0132	.0067	.00338	.00085
.72	.1628	.0870	.0450	.0228	.0114	.0057	.00283	.00069
.74	.1494	.0788	.0401	.0200	.0099	.0048	.00235	.00055
.76	.1364	.0709	.0355	.0174	.0085	.0041	.00195	.00044
.78	.1236	.0634	.0313	.0151	.0072	.0034	.00160	.00035
.80	.1111	.0562	.0273	.0130	.0061	.0028	.00131	.00028
.82	.0989	.0493	.0236	.0111	.0051	.0023	.00106	.00022
.84	.0870	.0428	.0202	.0093	.0042	.0019	.00085	.00017
.86	.0753	0.366	.0170	.0077	.0035	.0015	.00067	.00013
.88	.0638	.0306	.0140	.0063	.0028	.0012	.00052	.00010
.90	.0526	.0249	.0113	.0050	.0021	.0009	.00039	.00007
.92	.0417	.0195	.0087	.0038	.0016	.0007	.00028	.00005
.94	.0309	.0143	.0063	.0027	.0011	.0005	.00019	.00003
.96	.0204	.0093	.0040	.0017	.0007	.0003	.00012	.00002
.98	.0101	.0045	.0019	.0008	.0003	.0001	.00005	.00001

Example 10.8 The Taj Service Station has five mechanics each of whom can service a scooter in 2 hours on an average. The scooters are registered at a single counter and then sent for servicing to different mechanics. Scooters arrive at the service station at an average rate of 2 scooters per hour.

Assuming that the scooter arrivals are Poisson distributed and the servicing times are distributed exponentially, determine,

(a) utilisation factor,
(b) the probability that the system shall be idle,
(c) the probability that there shall be 3 scooters in the service centre,
(d) the probability that there shall be eight scooters in the centre,
(e) the expected number of scooters waiting in the queue,
(f) the expected number of scooters in the service centre,
(g) the average waiting time in the queue, and
(h) the average time spent by a scooter in waiting and getting serviced.

Here

$$\lambda = 2 \text{ scooters/hour}$$
$$\mu = 1/2 \text{ scooter/hour}$$
$$K = 5$$

With these values,

(a) Utilisation factor, $\rho = \dfrac{\lambda}{K\mu} = \dfrac{2}{5 \times 1/2} = 0.8$

(b) The probability that the system is idle,

$$P_0 = \left[\sum_{i=0}^{k-1} \frac{(\lambda/\mu)^i}{i!} + \frac{(\lambda/\mu)^K}{K!(1-\rho)}\right]^{-1}$$

$$= \left[\sum_{i=0}^{5-1} \frac{\left(\frac{2}{0.5}\right)^i}{i!} + \frac{\left(\frac{2}{0.5}\right)^5}{5!(1-0.8)}\right]^{-1}$$

Now,

$$\sum_{i=0}^{5-1} \frac{\left(\frac{2}{0.5}\right)^i}{i!} = \frac{4^0}{0!} + \frac{4^1}{1!} + \frac{4^2}{2!} + \frac{4^3}{3!} + \frac{4^4}{4!} = \frac{103}{3},$$

and

$$\frac{\left(\frac{2}{0.5}\right)^5}{5!\,(1-0.8)} = \frac{4^5}{120 \times 0.2} = \frac{128}{3}$$

$\therefore P_0 = \left(\dfrac{103}{3} + \dfrac{128}{3}\right)^{-1} = 77^{-1} = 0.0130$ (this value may be read directly from the Table 10.6)

(c) The probability that there shall be 3 scooters in the service centre;

We have, $$P_n = \frac{(\lambda/\mu)^n}{n!} \times (P_0) \quad \text{when } n \le K$$

$\therefore$ $$P_3 = \frac{\left(\frac{2}{0.5}\right)^3}{3!} \times 0.0130 = 0.1387$$

(d) The probability that there shall be 8 scooters in the system,

$$P_n = \frac{(\lambda/\mu)^n}{K!\,K^{n-K}} \times (P_0), \qquad \text{when } n > K$$

$$\therefore \qquad P_8 = \frac{\left(\dfrac{2}{0.5}\right)^8}{5!\,5^{8-5}} \times 0.0130$$

$$= \frac{4^8}{120 \times 5^3} \times 0.0130 = 0.0568$$

(e) Expected number of scooters waiting in the queue,

$$L_q = \frac{(\lambda/\mu)^K \rho}{K!(1-\rho)^2} \times (P_0)$$

$$= \frac{\left(\dfrac{2}{0.5}\right)^5 (0.8)}{5!\,(1-0.8)^2} \times 0.0130 = 2.22 \text{ scooters}$$

(f) Expected number of scooters in the system,

$$L_s = L_q + \frac{\lambda}{\mu} = 2.22 + \frac{2}{0.5} = 6.22 \text{ scooters}$$

(g) Expected waiting time in the queue,

$$W_q = L_q/\lambda = 2.22/2 = 1.11 \text{ hours}$$

(h) Expected time a scooter is in the system,

$$W_s = W_q + \frac{1}{\mu} = 1.11 + \frac{1}{0.5} = 3.11 \text{ hours}$$

Cost Analysis Suppose now that the manager of the service station is considering to engage a sixth mechanic in the system. It is given that cost of ill-will created by customer waiting is valued at Rs 12 per customer per hour, while each mechanic is paid Rs 5 per hour. To decide as to whether it would be economical for the manager to add a new mechanic, we shall first work out the operating characteristics of the system with $K = 6$, for the same arrival rate $\lambda = 2$ customers/hour. Then we shall determine the cost implications.

For $K = 6$, $\mu = 1/2$ customer/hour, and $\lambda = 2$ customers/hour, we have $\rho = 2/3 = 0.67$. From Table 10.6, we can obtain the value of P_0 by interpolation. This works out to be 0.001635. The values for other operating characteristics are given here, along with values for $K = 5$ (already worked out):

$K = 5 : L_s = 6.22 \quad L_q = 2.22 \qquad W_s = 3.11 \qquad W_q = 1.11$

$K = 6 : L_s = 4.558 \quad L_q = 0.558 \qquad W_s = 2.279 \qquad W_q = 0.279$

Now we can obtain the cost of the two systems as follows:

System 1: When $K = 5$

Hourly cost of ill-will = Expected number of arrivals per hour × Waiting time is system × Hourly rate

$= 2 \times 3.11 \times 12 = \text{Rs } 74.64$

Hourly cost of providing service = Number of mechanics × Hourly rate

$= 5 \times 5 = \text{Rs } 25.00$

∴ Total cost = 74.64 + 25.00 = Rs 99.64 per hour

System 2: When $K = 6$

Hourly cost of ill-will = $2 \times 2.279 \times 12 = \text{Rs } 54.70$

Hourly cost of providing service = $6 \times 5 = \text{Rs } 30.00$

∴ Total cost = 54.70 + 30.00 = Rs 84.70 per hour.

From the results, it is evident that it would be economical to add a sixth mechanic. Similar computations can be done to find the advisability of engaging more mechanics. Calculations will show that adding a seventh mechanic with same efficiency, with arrival rate remaining the same at 2 customers per hour, would raise the average hourly cost to Rs 85.20. Hence, it is not economical to have more than 6 servers in the present case.

REVIEW ILLUSTRATIONS

Example 10.9 Customers arrive at the First Class Ticket counter of a Theatre at a rate of 12 per hour. There is one clerk serving the customers at a rate of 30 per hour.

(i) What is the probability that there is no customer in counter (i.e. that the system is idle)?

(ii) What is the probability that there are more than 2 customers in the counter?

(iii) What is the probability that there is no customer waiting to be served?

(iv) What is the probability that a customer is being served and nobody is waiting?

(CA, November, 1990)

Here, $\lambda = 12$ customers/hour

$\mu = 30$ customers/hour

∴ $\rho = \lambda/\mu = 12/30 = 0.4$

(i) P (system is idle), $P(0) = 1 - \rho = 1 - 0.4 = 0.6$

(ii) $P(n > 2) = 1 - P(n \le 2)$

$= 1 - [P(0) + P(1) + P(2)]$

$= 1 - [0.6 + 0.4 \times 0.6 + 0.4^2 \times 0.6]$

$= 1 - (0.6 + 0.24 + 0.096)$

$= 1 - 0.936 = 0.064$

Also, $P(> n) = \rho^{n+1}$

∴ $P(> 2) = 0.4^{2+1} = 0.4^3 = 0.064$

(iii) P (no customer waiting to be served) $= P(0) + P(1)$

$$= 0.6 + 0.24$$

$$= 0.84$$

(iv) P (a customer is being served and none is waiting) $= P(1) = 0.24$

Example 10.10 A TV repairman finds that the time spent on his job has an exponential distribution with mean 30 minutes. If he repairs sets in the order in which they come and if the arrival of sets is approximately Poisson with an average rate of 10 per 8-hour day, what is his expected idle time each day? How many jobs are ahead of the set just brought in?

We are given,

The expected arrival rate, $\lambda = 10$ sets/day.

The expected service rate, $\mu = 16$ sets/day.

$\therefore$ $\rho = \lambda/\mu = 10/16 = 5/8$

The probability for the repairman to be idle,

$$P(0) = 1 - \rho = 1 - \frac{5}{8} = \frac{3}{8}$$

$\therefore$ Expected idle time per day $= 8 \times 3/8 = 3$ hours.

To determine the expected number of sets ahead of the set just brought in, we shall find the expected number of sets in the system. Thus,

$$L_s = \frac{\rho}{1-\rho} = \frac{\frac{5}{8}}{1-\frac{5}{8}} = \frac{5}{3} = 1\frac{2}{3} \text{ sets}$$

Example 10.11 Assume that at a bank teller window the customers arrive in their cars at the average rate of twenty per hour according to a Poisson distribution. Assume also that the bank teller spends an average of two minutes per customer to complete a service, and the service time is exponentially distributed. Customers, who arrive from an infinite population, are served on a first-come-first-served basis, and there is no limit to possible queue length.

(i) What is the expected waiting time in the system per customer?

(ii) What is the mean number of customers waiting in the system?

(iii) What is the probability of zero customers in the system?

(iv) What value is the utilisation factor? *(DMS, Mumbai, May, 1998)*

Here, arrival rate, $\lambda = 20$ customers/hour,

service rate, $\mu = 30$ customers/hour.

Thus, $\rho = \lambda/\mu = 20/30 = 2/3$.

(i) Expected waiting time in the system per customer,

$$W_s = \frac{1}{\mu - \lambda} = \frac{1}{30-20} = \frac{1}{10} \text{ hour or 6 minutes}$$

(ii) Mean number of customers waiting in the system,

$$L_q = \frac{\rho^2}{1-\rho} = \frac{(2/3)^2}{1-2/3} = \frac{4}{3}$$

(iii) Probability of zero customers in the system,

$$P(0) = 1 - \rho = 1 - \frac{2}{3} = \frac{1}{3}$$

(iv) Utilisation factor, $\rho = 2/3$.

Example 10.12 The rate of arrival of customers at a public telephone follows Poisson distribution, with an average time of ten minutes between one customer and the next. The duration of a phone call is assumed to follow exponential distribution with a mean time of three minutes.

(i) What is the probability that a person arriving at the booth will have to wait?

(ii) What is the average length of the queue?

(iii) The Mahanagar Telephone Nigam Ltd. will install another booth when it is convinced that the customers would have to wait for at least three minutes for their turn to make a call. How much should be the flow of customers in order to justify a second booth? *(M Com, Delhi, 1999)*

From the given information,

arrival rate, $\lambda = 6$ customers/hour

service rate, $\mu = 20$ customers/hour

Accordingly, $\rho = \lambda/\mu = 6/20 = 0.3$

(i) A customer arriving at the booth will have to wait if the booth is busy. Thus,

$$P\text{(a customer has to wait)} = \rho = 0.3$$

(ii) Average length of the queue,

$$L_q = \frac{\rho^2}{1-\rho}$$

$$= \frac{0.3^2}{1-0.3}$$

$$= \frac{0.09}{0.70} = \frac{9}{70} \text{ customers}$$

(iii) $$W_q = \frac{\lambda}{\mu(\mu-\lambda)}$$

At present, $$W_q = \frac{6}{20(20-6)} = \frac{3}{140} \text{ hour or } \frac{9}{7} \text{ minutes}$$

Let the new arrival rate be λ'. Setting $W_q = 3$ minutes or 3/60 hour, we get

$$\frac{3}{60} = \frac{\lambda'}{20(20-\lambda')}$$

or $20 - \lambda' = \lambda'$ or $\lambda' = 10$ customers/hour

Thus, as soon as the arrival rate increases to 10 customers/hour, another booth may be installed.

Example 10.13 The average time between successive arrivals to a repair shop, which works 8 hours a day, is 30 minutes. The shop has one mechanic who can repair the incoming vehicles at an average rate of 3 per hour. The mechanic is paid at Rs 14 per hour while the cost of waiting time, in terms of customer dissatisfaction and lost goodwill, is Rs 20 per hour of the time spent waiting in the queue. The owner is contemplating to replace the mechanic by another one who demands Rs 18 per hour and who can repair 4 vehicles on the average. Under conditions of the single server model, calculate the total cost per day at present and the total cost per day if the present mechanic is replaced. Is it advisable to replace the existing mechanic? *(M Com, Delhi, 2005)*

According to the given information,

arrival rate, $\lambda = 2$ customers/hour

Thus, expected number of arrivals per day, $N = 8 \times 2 = 16$

Cost of waiting time in queue, $c =$ Rs 20 per hour.

With existing mechanic:

With $\mu = 3$ customers/hour, expected waiting time in queue, $W_q = \dfrac{\lambda}{\mu(\mu - \lambda)}$

$$= \frac{2}{3(3-2)} = \frac{2}{3} \text{ hour}$$

$$\text{Cost of waiting per day} = N \times W_q \times c$$

$$= 16 \times \frac{2}{3} \times 20 = \text{Rs } 213.33$$

$$\text{Total cost per day} = \text{Mechanic's charges} + \text{Cost of waiting}$$

$$= 14 \times 8 + 213.33 = \text{Rs } 325.33$$

With new mechanic:

Here, $\mu = 4$ customers/hour.

Thus, expected waiting time in queue, $W_q = \dfrac{2}{4(4-2)} = \dfrac{1}{4}$ hour

$$\text{Cost of waiting per day} = N \times W_q \times c$$

$$= 16 \times \frac{1}{4} \times 20 = \text{Rs } 80$$

$$\text{Total cost per day} = 18 \times 8 + 80 = \text{Rs } 224$$

Since the total cost is lower with the new mechanic, it is advisable to replace the existing mechanic.

Example 10.14 A warehouse has only one loading dock manned by a three person crew. Trucks arrive at the loading dock at an average rate of 4 trucks per hour and the arrivals are Poisson distributed. The loading of a truck takes 10 minutes on an average and the loading time can be assumed to be exponentially distributed about this average. The operating cost of a truck is Rs 100 per hour and the members of the loading crew are paid at a rate of Rs 25 per hour. Assuming that the addition of new crew members would reduce the loading time to 7.5 minutes, would you advise the truck owner to add another crew of three persons?

According to given information,

$\lambda = 4$ trucks/hour,

$\mu = 6$ trucks/hour.

We have,

Total hourly cost = Loading crew cost + Cost of waiting time

At present

Loading crew cost = No. of loaders × Hourly wage rate

= 3 × 25 = Rs 75 per hour

Cost of waiting time = Expected waiting time per truck (W_s) × Expected arrivals per hour (λ) × Hourly waiting cost

$$= \frac{1}{6-4} \times 4 \times 100 = \text{Rs } 200 \text{ per hour}$$

Alternatively, cost of waiting time = Expected number of trucks in the system (L_s) × Hourly waiting cost

$$= \frac{4}{6-4} \times 100 = \text{Rs } 200 \text{ per hour}$$

∴ Total cost = Rs 75 + Rs 200 = Rs 275 per hour

With proposed crew addition

Loading crew cost = 6 × 25 = Rs 150 per hour

$$\text{Cost of waiting time} = \frac{4}{8-4} \times 100 = \text{Rs } 100 \text{ per hour}$$

∴ Total cost = Rs 150 + Rs 100 = Rs 250 per hour

Conclusion: It is advisable to add a crew of three loaders.

Example 10.15 (a) A typist at an office receives, on an average, 22 letters per day of typing. The typist works 8 hours a day and it takes on an average 20 minutes to type a letter. The company has determined that the cost of a letter waiting to be mailed (opportunity cost) is 80 paise per hour and the equipment operating cost plus the salary of the typist will be Rs 40 per day.

(i) What is the typist's utilisation rate?
(ii) What is the average number of letters waiting to be typed?
(iii) What is the average waiting time needed to have a letter typed?
(iv) What is the total cost of waiting letters to be mailed?

(b) Forced to improve the letter typing service, the above company is planning to take lease of one of the two models of an automated typewriter available in the market. The daily costs and the resulting increase in the typist's efficiency are displayed in the table given below:

Model	*Additional Cost per Day*	*Increase in Typist's Efficiency*
I	Rs 37	50%
II	Rs 39	75%

What action should the company take to minimise the total daily costs of waiting letters to be mailed?

(CA, May, 1993)

(a) From the given information,

arrival rate, $\lambda = 22$ letters/day

service rate, $\mu = \dfrac{8 \times 60}{20} = 24$ letters/day

(i) The typist's utilisation rate, $\rho = \lambda/\mu = 22/24 = 0.917$

(ii) Expected number of letters waiting to be typed,

$$L_q = \frac{\lambda^2}{\mu(\mu - \lambda)}$$

$$= \frac{22^2}{24(24-22)} = 10.08 \text{ letters}$$

(iii) Expected time needed to have a letter typed,

$$W_s = \frac{1}{\mu - \lambda}$$

$$= \frac{1}{24-22} = \frac{1}{2} \text{ day or 4 hours}$$

(iv) Total daily cost of waiting letters to be mailed = Expected number of letters in the system $(L_s) \times$ Opportunity cost + Salary of the typist

Now,
$$L_s = \frac{\lambda}{\mu - \lambda}$$

$$= \frac{22}{24-22} = 11 \text{ letters}$$

Opportunity cost to mail a letter is 80 p per hour or 8×80 p = Rs 6.40 per day. Thus, total opportunity cost of waiting letters = Rs 6.40×11 = Rs 70.40.

Accordingly

Total daily cost = Rs 70.40 + Rs 40.00 = Rs 110.40

(b) *Model I* Increase in efficiency is given to be 50% with this model. Therefore, service rate = 24×1.5 = 36 letters per day. With $\lambda = 22$ letters per day,

$$L_s = \frac{\lambda}{\mu - \lambda}$$

$$= \frac{22}{36-22} = \frac{11}{7} \text{ letters}$$

Now,

Total daily cost = $L_s \times$ Opportunity cost + Equipment operating cost + Salary of the typist

$$= \frac{11}{7} \times 6.40 + 37 + 40$$

= Rs 87.06

Model II With a 75% increase in the service rate with this model, we have $\mu = 24 \times 7/4 = 42$ letters per day. The given arrival rate being 22 letters per day, we get

$$L_s = \frac{22}{42-22} = \frac{11}{10} \text{ letters}$$

Accordingly,

$$\text{Total daily cost} = \frac{11}{10} \times 6.40 + 39 + 40 = \text{Rs } 86.04$$

Conclusion: It is advisable to use model II typewriter in order to minimise the total daily cost of waiting letters to be mailed.

Example 10.16 A factory operates for 8 hours everyday and has 240 working days in the year. It buys a large number of small machines which can be serviced by its maintenance engineer at a cost of Rs 4 per hour for the labour and spare parts. The machines can, alternatively, be serviced by the supplier at an annual contract price of Rs 20,000 including the labour and spare parts needed. The supplier undertakes to send a repairman as soon as a call is made but in no case more than one repairman is sent. The service times of the maintenance engineer and the supplier's repairman are both exponentially distributed with respective means of 1.7 and 1.5 days. The machine breakdowns occur randomly and follow Poisson distribution, with an average of 2 in 5 days. Each hour that a machine is out of order, it costs the company Rs 8. Which servicing alternative would you advise it to opt for?

We shall calculate the total cost involved for each of the alternatives for taking the decision.

Total Annual Cost = Cost of idle machinery time + Cost of labour and spare parts

Cost of Idle Machinery Time = Expected no. of machines in the system × No. of working hours per year × Hourly idle time cost

Alternative 1: Maintenance by company engineer

We have, from the given information,

$$\lambda = 2/5 \text{ machines per day}$$

$$\mu = 1/1.7 \text{ machines per day}$$

Expected number of machines in the system,

$$L_s = \frac{\lambda}{\mu - \lambda} = \frac{2/5}{\frac{1}{1.7} - \frac{2}{5}} = \frac{17}{8} \text{ machines}$$

$$\text{Cost of Idle Machinery Time} = \frac{17}{8} \times 1920 \times 8 = \text{Rs } 32{,}640$$

Cost of Labour and Spare Parts = No. of machines in the system × No.of working hours per year × Hourly cost

$$= \frac{17}{8} \times 1920 \times 4 = \text{Rs } 16{,}320$$

$$\therefore \qquad \text{Total cost} = 32{,}640 + 16{,}320 = \text{Rs } 48{,}960 \text{ p.a.}$$

Alternative 2: Maintenance by supplier

We have,

$$\lambda = 2/5 \text{ machines per day}$$

$$\mu = 1/1.5 \text{ machines per day}$$

Expected number of machines in the system,

$$L_s = \frac{\lambda}{\mu - \lambda} = \frac{2/5}{\frac{1}{1.5} - \frac{2}{5}} = \frac{3}{2} \text{ machines}$$

$$\text{Cost of Idle Machine Time} = \frac{3}{2} \times 1920 \times 8 = \text{Rs } 23{,}040$$

Cost of Labour and Spare Parts = Rs 20,000

$\therefore$ Total cost = 23,040 + 20,000 = Rs 43,040 p.a.

Clearly, then, the maintenance of the machines by the supplier is the better alternative.

Example 10.17 On an average, 4 patients per hour require the service of an emergency clinic. Also, on an average, a patient requires 10 minutes of active attention. The clinic can handle only one emergency at a time. Suppose that it costs the clinic Rs 300 per patient treated to obtain an average servicing time of 10 minutes and that each minute of decrease in this average time would cost Rs 50 per patient treated. How much would have to be budgeted by the clinic to decrease the average size of the queue from 4/3 patients to 1/2 patients?

According to the given information,

Mean arrival rate, $\lambda = 4$ patients/hour

Mean service rate, $\mu = 6$ patients/hour

$$\therefore \rho = \frac{4}{6} = \frac{2}{3}$$

At present, the expected length of queue,

$$L_q = \frac{\rho^2}{1-\rho}$$

$$= \frac{(2/3)^2}{1-\frac{2}{3}} = \frac{4/9}{1/3} = \frac{4}{3} = 1\frac{1}{3} \text{ patients}$$

To obtain the planned average queue length, let the service rate be μ'. With this,

$$L_q = \frac{\lambda^2}{\mu'(\mu' - \lambda)}$$

Substituting the known values,

$$\frac{1}{2} = \frac{4^2}{\mu'(\mu'-4)}$$

or $\mu'^2 - 4\mu' = 32$

or $\mu'^2 - 4\mu' - 32 = 0$

By factorisation,

$\mu'^2 - 8\mu' + 4\mu' - 32 = 0$

or $(\mu' - 8)(\mu' + 4) = 0$

Since negative value of μ' is not possible, we have $\mu' = 8$ (customers/hour).

The service rate of 8 customers per hour implies a service time equal to 7.5 minutes per customer.

Thus, reduction in service time = 10 – 7.5 = 2.5 minutes. Cost of reduction being Rs 50 per minute, the budget requirement equals 300 + 50 × 2.5 = 425 rupees per patient.

$\therefore$ Total budget = 4 × 24 × 425 = Rs 40,800 per day

Example 10.18 The Tooth-Care Hospital provides free dental service to the patients on every Saturday morning. There are three dentists on duty, who are equally qualified and experienced. It takes, on an average, 20 minutes for a patient to get treatment, and the actual time taken is known to vary approximately exponentially around this average. The patients arrive according to the Poisson distribution with an average of 6 per hour. The administrative officer of the hospital wants to investigate the following:

(a) The expected number of patients waiting in the queue.
(b) The average time that a patient spends at the clinic.
(c) The average percentage idle time for each of the dentists.
(d) The fraction of time at least one dentist is idle.

Help him.

From the given information,

$K = 3$, $\lambda = 6$ patients per hour, $\mu = 1/20$ patients per minute or 3 patients per hour.

Thus, $\rho = 6/(3 \times 3) = 2/3$

Further,

$$P_0 = \left[1 + 2 + \frac{2^2}{2!} + \frac{3(2)^3}{3!(3-2)}\right]^{-1} = \frac{1}{9}$$

Now,

(a) Expected number of patients waiting in the queue,

$$L_q = \frac{(\lambda/\mu)^K \rho}{K!(1-\rho)^2} \times P_0$$

$$= \left[\frac{2^3 \times 2/3}{3!\left(1 - \frac{2}{3}\right)^2}\right]\left[\frac{1}{9}\right] = \frac{8}{9} = 0.889$$

(b) Expected time a patient spends in the system, $W_s = W_q + \dfrac{1}{\mu}$

Here, $$W_q = L_q/\lambda = \frac{8}{9} \times \frac{1}{6} = \frac{4}{27}$$

$\therefore$ $$W_s = \frac{4}{27} + \frac{1}{3} = \frac{13}{27} \text{ hours or 28.9 minutes}$$

(c) To determine the percentage of idle time of the dentists, we realise that when the system is empty, all the three dentists would be idle; when only one person is in the clinic, then two of them are idle; when two patients are in the clinic, only one doctor is idle; and, finally, when three or more patients are in the clinic, then all the dentists are busy. Assuming random selection of dentists, when more than one dentist is idle,

$$P\text{ (idle dentist)} = 1 \times P_0 + \frac{2}{3}P_1 + \frac{1}{3}P_2$$

$$= \frac{1}{9} + \left(\frac{2}{3}\right)\left(\frac{2}{1!}\right)\left(\frac{1}{9}\right) + \left(\frac{1}{3}\right)\left(\frac{2^2}{2!}\right)\left(\frac{1}{9}\right) = \frac{1}{3}$$

Thus, one-third of his time, a dentist has no patients to examine. Of course, this probability can be directly calculated as equal to $1 - \rho = 1 - 2/3 = 1/3$.

(d) Probability of at least one idle dentist $= \sum_{n=0}^{K-1} P_n$

$$= P_0 + P_1 + P_2 = 5/9.$$

Example 10.19 Customers arrive at the rate of twenty per hour and the present serving arrangements can cope with thirty per hour for an eight-hour day.

Using the queuing formulae provided, you are required to calculate and state:

(a) the average time in the queue;

(b) the implied value of customers' time if the owner of the service has considered but rejected a faster service arrangement which would cost an extra Rs 20 for an eight-hour day and would raise the service rate to forty per hour.

The following formulae are given:

Average time in system: $\dfrac{1}{1-r} \cdot \dfrac{1}{m}$

Average time in queue: $\dfrac{r}{1-r} \cdot \dfrac{1}{m}$. *(ICMA, May, 1983, Adapted)*

(a) Here $\lambda = 20/\text{hour}, \mu = 30/\text{hour}$

$$\therefore \rho = 20/30 = 2/3.$$

$$\text{Average time in queue} = \frac{2/3}{1-\frac{2}{3}} \times \frac{1}{30} \text{ hours} = 4 \text{ minutes}$$

(b) With a service rate of 40 per hour, we have = 20/40 = 0.5

$$\text{Previous average time in system} = \frac{1}{1-0.67} \times \frac{1}{30} \times 60 = 6 \text{ minutes}$$

$$\text{New average time in system} = \frac{1}{1-0.5} \times \frac{1}{40} \times 60 = 3 \text{ minutes.}$$

If the break-even value of customers' waiting time is Rs K per hour, then

$$8 \times \left(\frac{6}{60} - \frac{3}{60}\right) \times \lambda \times K < 20$$

or

$$8 \times \frac{3}{60} \times 20 \times K < 20$$

Therefore, $K < \dfrac{60}{24}$ or Rs 2.50 per hour.

Example 10.20 A bank has two tellers working on savings accounts. The first teller handles withdrawals only while the second teller handles depositors only. It has been found that the service time distributions of both depositors and withdrawals are exponential with a mean service time of 3 minutes per customer. Depositors and withdrawers are found to arrive in a Poisson process throughout the day with mean arrival rate of 16 and 8 per hour respectively. What would be the effect on the average waiting time for depositors and withdrawers if each teller could handle both withdrawals and deposits? What would be the effect if this could only be accomplished by increasing the service time to 4 minutes?

The existing situation is as follows:

	For Withdrawers	*For Depositors*
Mean arrival rate	$\lambda = 16$ customers/hr.	$\lambda = 8$ customers/hr.
Mean service rate	$\mu = 20$ customers/hr.	$\mu = 20$ customers/hr.
Utilisation parameter	$\rho = 16/20 = 0.8$	$\rho = 8/20 = 0.4$
Waiting time, $W_q = \dfrac{\rho}{\mu - \lambda}$	$W_q = \dfrac{0.8}{20 - 16} = \dfrac{1}{5}$ hour	$W_q = \dfrac{0.4}{20 - 8} = \dfrac{1}{30}$ hour
	$= 12$ minutes	$= 2$ minutes

When there are two tellers rendering the same service,

$$\lambda = 16 + 8 = 24 \text{ customers/hr.}$$

$$\mu = 20 \text{ customers/hr.}$$

$$K = 2$$

Accordingly,

$$\rho = \frac{\lambda}{K\mu} = \frac{24}{2 \times 20} = 0.60$$

From the Table 10.6, P_0 for $\rho = 0.60$ and $K = 2$ is 0.2500. We have,

$$W_q = L_q/\lambda, \text{ where}$$

$$L_q = \frac{(\lambda/\mu)^K \rho}{K!(1 - \rho)^2} P_0$$

Thus,

$$L_q = \frac{(24/20)^2 \times 0.60}{2!(1 - 0.60)^2} \times 0.2500 = 0.675$$

and

$$W_q = 0.675/24 = 0.0281 \text{ hour or } 1.69 \text{ minutes}$$

When service time averages 4 minutes,

$$\mu = 60/4 = 15 \text{ customers/hr.}$$

$$\rho = \frac{24}{2 \times 15} = 0.80$$

For $\rho = 0.80$ and $K = 2$, we have $P_0 = 0.1111$

Thus,

$$L_q = \frac{(24/15)^2 \times 0.80}{2!(1 - 0.80)^2} \times 0.1111 = 2.8442$$

Hence,

$$W_q = 2.8442/24 = 0.1185 \text{ hour or } 7.11 \text{ minutes}$$

Example 10.21 A company currently has two tool cribs, each having a single clerk, in its manufacturing area. One tool crib handles only the tools for the heavy machinery, while the second one handles all other tools. It is observed that for each tool crib the arrivals follow a Poisson distribution with a mean of 20 per hour, and the service time distribution is negative exponential with a mean of 2 minutes.

The tool manager feels that if the tool cribs are combined in such a way that either clerk can handle any kind of tool as demand arises, it would be more efficient and the waiting problem could be reduced to some extent. It is believed that the mean arrival rate at the two tool cribs will be 40 per hour, while the service rate will remain unchanged.

Compare in status of queue and the proposal with respect to the total expected number of mechanics at the tool crib(s), the expected waiting including service time for each mechanic and probability that he has to wait for service. *(MBA, Delhi, November, 1999)*

The existing system is as follows:

Inputs	*Tool Crib 1*	*Tool Crib 2*
Mean arrival rate, λ	20 mechanics/hour	20 mechanics/hour
Mean service rate, μ	30 mechanics/hour	30 mechanics/hour

Inputs	*Tool Crib 1*	*Tool Crib 2*
Operating characteristics:		
Utilisation parameter, $\rho = \lambda/\mu$	20/30 = 2/3	20/30 = 2/3
Probability that a mechanic has to wait, P (busy) = ρ	2/3 = 0.67	2/3 = 0.67
Expected No. of mechanics at the crib, $L_s = \dfrac{\rho}{1-\rho}$	$\dfrac{2/3}{1-2/3} = 2$	$\dfrac{2/3}{1-2/3} = 2$
Expected No. of mechanics in the queue, $L_q = \dfrac{\rho^2}{1-\rho}$	$\dfrac{(2/3)^2}{1-2/3} = 1.33$	$\dfrac{(2/3)^2}{1-2/3} = 1.33$
Expected waiting time in the system, $W_s = \dfrac{1}{\mu - \lambda}$	$\dfrac{1}{30-20} = \dfrac{1}{10}$ hour = 6 minutes	$\dfrac{1}{30-20} = \dfrac{1}{10}$ hour = 6 minutes

For the proposed system:

Inputs:

Mean arrival rate,	$\lambda = 40$ mechanics/hour
Service rate,	$\mu = 30$ mechanics/hour
No. of servers,	$K = 2$

Operating characteristics:

Utilisation parameter, $\rho = \dfrac{\lambda}{K\mu} = \dfrac{40}{2 \times 30} = \dfrac{2}{3}$

It may be observed from Table 10.6 that, for $\rho = 0.66$ and $K = 2$, $P_0 = 0.2048$, while for $\rho = 0.68$ and $K = 2$, it equals 0.1905. Thus, for $\rho = 0.67$, the estimated $P_0 = (0.2048 + 0.1905)/2 = 0.19765$. With this, Expected number of mechanics in the queue,

$$L_q = \frac{(\lambda/\mu)^K \rho}{K!(1-\rho)^2}(P_0)$$

Substituting the known values,

$$L_q = \frac{(40/30)^2 \times 0.67}{2!(1-0.67)^2} \times 0.19765 = 1.067$$

Further,

$$L_s = L_q + \frac{\lambda}{\mu} = 1.067 + 4/3 = 2.4$$

$$W_q = L_q/\lambda = 1.067/40 = 0.0267 \text{ hour or } 1.6 \text{ minutes}$$

$$W_s = W_q + 1/\mu = 0.0267 + 1/30 = 0.060 \text{ hour or } 3.6 \text{ minutes}$$

Thus, it is evident that the system would be more efficient if both the tool cribs provide similar service.

KEY POINTS TO REMEMBER

- Queuing theory deals with situations where customers arrive, wait for the service, get the service and leave the system. Customers, human or non-human, may come individually or in groups from large/small population, at known/variable times, form one or more queues and move in a certain order then to the service station/s providing service whose speed may be fixed or variable.
- Queuing structures are analysed for determining the optimal service level, where the total cost of providing service and waiting, is minimised. An increase in the service level increases the cost of providing service but reduces the cost of waiting, while a decrease in the service level induces opposite changes.
- Operating characteristics of a queuing system include queue length, system length, waiting time in the queue, total time in the system and the server idle time.
- While queuing models can be deterministic as well, the probabilistic ones are most commonly occurring and analysed. Three such models include:
  (a) Poisson-exponential, single server model—infinite population
  (b) Poisson-exponential, single server model—finite population
  (c) Poisson-exponential, multiple server model—infinite population
- For a workable queuing system, it is necessary that the arrival rate of customers per unit of time should be less than the service rate.
- For the three queuing models, operating characteristics are summarised below:

Characteristic	*Model A*	*Model B*	*Model C*
L_q	$\dfrac{\rho^2}{1-\rho}$	$M - \dfrac{\lambda+\mu}{\lambda}(1-P_0)$	$\dfrac{(\lambda/\mu)^K \rho}{K!(1-\rho)^2} P_0$
L_s	$\dfrac{\rho}{1-\rho}$	$M - \dfrac{\mu}{\lambda}(1-P_0)$	$L_q + \dfrac{\lambda}{\mu}$
W_q	$\dfrac{\rho}{\mu-\lambda}$	$\dfrac{1}{\mu}\left[\dfrac{M}{1-P_0} - \dfrac{\lambda+\mu}{\mu}\right]$	L_q/λ
W_s	$\dfrac{1}{\mu-\lambda}$	$\dfrac{1}{\mu}\left[\dfrac{M}{1-P_0} - \dfrac{\lambda+\mu}{\mu} + 1\right]$	$W_q + \dfrac{1}{\mu}$
	λ: arrival rate	M: total customers	λ: arrival rate
	μ: service rate	λ: arrival rate	μ: service rate
	$\rho = \lambda/\mu$	μ: service rate	K: number of servers
			$\rho = \lambda/K\mu$

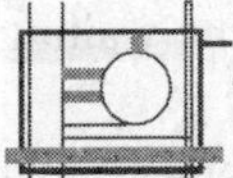

TEST YOUR UNDERSTANDING

Mark the following statements as T (*True*) or F (*False*).

1. In queuing theory, *customers* might include humans, machines, automobiles, and so on.
2. A study of queuing theory helps the manager to establish an optimal level of service.
3. Arrival of passengers from aeroplanes at an airport is an example of individual arrivals of customers.
4. Arrival of patients with appointment to a dentist can be described as arrival by Poisson process.
5. The service rate and service time are reciprocals of each other.
6. Multiple servers may be in *series* or in *parallel.*
7. The only way the customers are serviced in queuing situations is the *first-come-first-served* basis.
8. If a waiting customer becomes impatient, it may decide to *renege*.
9. An expectation of a long waiting time, particularly when there are limits on the time and extent of storage capacity available, an arriving customer may *balk*.
10. *Jockeying* is exercised by the customers in shifting to the "fast" moving queues in an attempt to save the waiting time.
11. A queuing model where customer arrivals are at known intervals and the service time is also certain, is a deterministic model.
12. In a deterministic queuing model, the arrival rate must not exceed the service rate, but in a probabilistic model it can.
13. The term *Poisson-exponential* in the context of a queuing model indicates that the arrival rate of customers is Poisson distributed while the service rate is distributed exponentially.
14. If the arrivals occur according to a Poisson distribution, the inter-arrival times would be exponentially distributed.
15. An arrival rate of 10 customers per hour according to Poisson process implies an average inter-arrival time of six minutes.
16. The arrival rate in the Poisson distribution is equal to the mean of the exponential inter-arrival time.
17. In the Poisson-exponential single server model, the system is not workable if the arrival and service rates are equal.
18. In a single server queuing situation, steady state is reached after a sufficiently long period of time if the service rate is greater than the arrival rate.
19. In a Poisson-exponential single server queuing model, the probability of having at least *n* customers in the system at a random point of time is equal to $\rho^n (1 - \rho)$.
20. For a single server model, $L_{q'} > L_s > L_q$ and $W_s > W_q$.
21. The expected length of system, L_s, should be equal to the expected length of queue *plus* one.
22. It is necessary that at the optimal level of service, the cost of waiting equals cost of servicing.
23. In the Poisson-exponential single server model with finite source population, there is a strong dependency relationship between the arrivals.
24. Equal degree of change in the service rate on higher and lower sides produces skewed effect in terms of changes in various parameters.
25. In both single server and multi-server models, *rho* is defined as the ratio of *lambda* to *mu*.

26. The distribution of waiting time is not related to queue discipline used in selecting the waiting customers for service.
27. The Poisson-exponential multiple server model is based on the assumption, among others, that multiple queues are formed.
28. In a multiple server model, the arrival rate of customers may be larger than the service rate of an individual server but it should be smaller than the combined service rate of all K service facilities.
29. *Parallel service stations* implies multiple servers providing same service.
30. The probability of n customers arriving during a time interval T, according to the Poisson law, is given by

$$P(n) = \frac{e^{-m} \, m^n}{n!}, \text{ where } m = \lambda T.$$

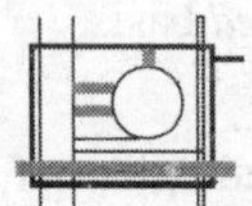

EXERCISES

1. What is queuing theory? What type of questions are sought to bc answered in analysing a queuing system?
2. Give a general structure of the queuing system and explain. Illustrate some queuing situations.
3. What is queuing theory? In what types of problem situations can it be applied successfully? Discuss giving examples.
4. Analyse the following queuing systems by describing their various system properties:
   (a) Hospital Emergency Room,
   (b) Traffic Light,
   (c) Computer System at a University.
5. What is a queuing problem? What are the basic characteristics of a queuing system?
6. Describe a single-server waiting line model. Give an example from real life, for each of the following queuing models:
   (i) First-come-first-served,
   (ii) Last-come-first-served,
   (iii) Random pick service,
   (iv) Customers stay only if served instantly.
7. What do you understand by queuing structure? Explain (a) first-come-first-served, (b) last-come-first-served, and (c) service-in-random-order basis of customer handling.
8. Explain the basic queuing process. State the meaning of queue discipline and give its role in queuing problem.
9. Distinguish between deterministic and probabilistic queuing models. Discuss some probabilistic queuing models with examples.
10. What is traffic intensity? If traffic intensity of a system is given to be 0.76, what per cent of time the system would be idle? Can the traffic intensity be more than 1?
11. Write the expressions for (a) expected length of the queue, and (b) expected waiting time of customers in the queue, in the single server model (i) with infinite input population, and (ii) with finite input population.
12. Give the role of queuing theory in decision-making and discuss its application.

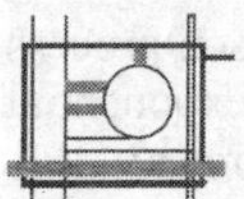

PRACTICAL PROBLEMS

1. At a certain petrol pump, customers arrive in a Poisson process with an average time of five minutes between successive arrivals. The time taken at the petrol pump to serve customers follows exponential distribution with an average of two minutes. You are required to obtain the following:
   (a) Arrival and service rates.
   (b) The utilisation parameter.
   (c) Probability that there shall be four customers in the system.
   (d) Probability that there are more than four customers in the system.
   (e) Expected queue length.
   (f) Expected number of customers in the system.
   (g) Expected time that a customer has to wait in the queue.
   (h) Expected time that a customer has to spend in the system.
   (i) It is decided to open a new service point when the customers' expected waiting time rises to three times the present level. What increased flow of customers will justify this?

2. The secretary of the Excellent Corporation receives letters for typing, the arrival of which follows the Poisson law with an average arrival rate $\lambda = 3$ letters/hour.
   (a) Find the probability that she will receive no more than 2 letters in an hour, at least one letter in an hour.
   (b) Determine the number of letters expected to be received in a two-hour interval.

3. On a national highway, automobiles arrive for toll tax payment at an average rate of 3 in 5 minutes, according to the Poisson probability law. The attendant receives the tax in an average time of one minute per customer. The service (collecting the tax) time is exponentially distributed. You are required to determine:
   (a) the probability of arrivals of 0 through 5 customers in a 10-minute interval.
   (b) the percentage of time the attendant at the toll gate shall be idle.
   (c) the average time that the attendant is free in his 8-hour duty time.
   (d) the probability of 0 through 5 customers in the system.
   (e) the expected number of customers in the system.
   (f) the expected number of customers waiting in the queue to pay tax.

4. (a) For a single server queuing model, the average service time is 3 minutes and service is idle for 25% of the time. The inter-arrival time of customers for such a system will be:
   (i) 12 minutes (ii) 4.5 minutes (iii) 4 minutes (iv) indeterminate
   (b) For a single server queuing model, the average service time is 15 minutes and the inter-arrival time of customers is 20 minutes. What is the average waiting time of a customer in the queue?
   (c) If on an average, 10 customers join a queue in one hour and the average service time per customer is 6 minutes, then the average waiting time of a new arrival in a *M*/*M*/1 (single server) queuing system is one hour.
   (i) True, (ii) False, (iii) Cannot say.
   (d) In a single server queuing situation, the service system is occupied 75% of the time and the inter-arrival time of customers is 4 minutes. For this, the average service time will be:
   (i) 4.5 minutes (ii) 4 minutes (iii) 3 minutes (iv) 2 minutes

(e) Customers arrive at a sales counter manned by a single person according to a Poisson process with a mean rate of 20 per hour. The time required to serve a customer has an exponential distribution with a mean of 100 seconds. What is the average waiting time of a customer?

5. A TV repairman finds that the time spent on his jobs has an exponential distribution with a mean 30 minutes. If he repairs sets on the first-come-first-served basis and if the arrival of sets is with an average rate of 10 per 8-hour day, what is repairman's expected idle time each day? Also obtain average number of units in the system. *(M Com, Delhi, 1984)*

6. A foreign bank is considering opening a drive-in window for customer service. Management estimates that customers will arrive for service at the rate of 12 per hour. The teller, whom it is considering to staff the window, can serve customers at the rate of one every three minutes. Assuming Poisson arrivals and exponential service, find (i) utilisation of teller, (ii) average number in the system, (iii) average waiting time in the line, and (iv) average waiting time in the system. *(CA, November, 1999)*

7. The mechanic at CarPoint is able to install new mufflers at an average of three per hour while customers needing this service arrive at an average rate of 2 per hour. Assuming that the conditions for a single-server infinite population model are all satisfied, calculate the following:
   (a) The utilisation parameter.
   (b) The average number of customers in the system.
   (c) The average time a customer spends in the system.
   (d) The average time a customer spends in the queue.
   (e) The probability that there are more than three customers in the system.

   (M Com, Delhi, 2008)

8. A company distributes its products by trucks loaded at its only loading station. Both, company's trucks and contractor's trucks are used for this purpose. It was found that on an average, every five minutes, one truck arrived and the average loading time was three minutes. 50% of the trucks belong to the contractor. Find out:
   (i) the probability that a truck has to wait;
   (ii) the waiting time of a truck that waits; and
   (iii) the expected waiting time of contractor's trucks per day, assuming a 24 hours shift.

   (CA, November, 1996)

9. Customers for a local bakery arrive randomly following a Poisson process. The single salesman can attend customers at an average rate of 12 customers per hour, the service time being distributed exponentially. The mean arrival rate of the customers is 20 per hour.
   Is the above queuing system workable? If not, why? If yes, determine the following:
   (a) the mean number of customers in the bakery.
   (b) the mean time spent by a customer in the bakery.
   (c) the expected number of customers waiting to be served.
   (d) the mean waiting time of a typical customer in the queue.

10. Customers arrive at a one-window drive according to a Poisson distribution with mean of 10 minutes and service time per customer is exponential with mean of 6 minutes. The space in front of the window can accommodate only three vehicles including the serviced one. Other vehicles have to wait outside this space. Calculate:
    (i) Probability that an arriving customer can drive directly to the space in front of the window.
    (ii) Probability that an arriving customer will have to wait outside the directed space.
    (iii) How long an arriving customer is expected to wait before getting the service?

    (CA, May, 1998)

11. Customers arrive to a typist, who is known for quality typing, according to Poisson probability law with an average inter-arrival time of 20 minutes. The customers wait if the typist is not free. The typist

completes the typing jobs in an average time of 15 minutes, the times taken being distributed exponentially. From this information, determine:

(a) What fraction of time is the typist busy?
(b) What is the probability of having less than 3 customers with the typist at any time?
(c) What is the expected number of customers with the typist?
(d) What is the expected number of customers waiting in the queue?
(e) How much time a customer is expected to spend in the queue?
(f) What time would, on an average, elapse between a customer reaching and leaving the typist?
(g) The expected time a customer would spend in the system.
(h) The probability that a customer shall wait for more than 10 minutes in the queue.
(i) The probability that a customer shall be in the system for more than 10 minutes.

12. In the central railway station 15 computerised reservation counters are available. A customer can book his/her ticket in any train on any day in any one of these computerised reservation counters. The average time spent on each customer by each clerk is five minutes. Average arrivals per hour during three types of activity periods have been calculated and customers have been surveyed to determine how long they are willing to wait during each type of period.

Type of period	*Arrivals per hour*	*Customers' acceptable waiting time*
Peak	110	15 minutes
Normal	60	10 minutes
Low	30	5 minutes

Making suitable assumptions on this queuing process, determine how many counters should be kept open during each type of period. *(CA, Nov. 1991)*

13. A repair shop attended to by a single mechanic has an average of four customers an hour who bring small appliances for repair. The mechanic inspects them for defects and quite often can fix them right away or otherwise render a diagnosis. This takes him six minutes, on the average. Arrivals are Poisson and service time has exponential distribution. You are required to:
(a) Find the proportion of time during which the shop is empty.
(b) Find the probability of finding at least one customer in the shop.
(c) What is the average number of customers in the system?
(d) Find the average time spent, including service. *(M Com, Delhi, 2008)*

14. Arrivals at the enquiry counter of Roshanara Transport Company are Poisson distributed with an average of six per hour. The time that the customers spend in seeking information from the clerk stationed at the counter is known to be exponentially distributed with an average of three minutes. Using this information determine
(a) the probability that a customer reaching the counter shall have to wait for getting the needed information.
(b) the probability that a queue shall be formed.
(c) the expected time that a customer shall wait in the queue to obtain information.

Suppose that the company manager employs another clerk if he is convinced that a customer has to wait for at least 4 minutes for receiving information. What arrival flow rate would justify employing the second clerk?

15. At a tool counter, the supply clerk is paid daily wage at Rs 100 per day for an 8-hour working day. The employees who come to him to take and return the tools, their daily charges are Rs 120 per day. On an average, 25 workers come to the counter every hour for replacing the tools. The supply clerk takes, on an average, 2 minutes either for issuing or for replacing the tools. The company is considering a semi-

automatic device at the tool counter so that the service rate becomes faster. It will involve an additional expenditure of Rs 100 per day besides the clerk. But, at the same time, it will increase the service rate by 33.33 per cent (one-third). Should the company install the semi-automatic device?

(MIB, Delhi, 2008)

16. In an engineering workshop, motors breakdown at an average rate of 5 per day, the number of breakdowns being Poisson distributed. The present unqualified mechanic can repair motors at an average rate of 6 per day and is paid a daily wage of Rs 100. A qualified mechanic offers his services at a daily wage of Rs 200 and is capable of repairing, on the average, 8 motors per day. Whenever a motor is idle, there is a downtime cost incurrence at the rate of Rs 100 per day. Would it be worthwhile to employ the qualified mechanic in lieu of the present mechanic? Justify on cost/benefit analysis.

(DMS, Mumbai, 1993)

17. Unique Tyre Company decided to hire a new mechanic to handle all tyre changes for customers ordering a new set of tyres. Two mechanics applied for the job. One mechanic has limited experience, can be hired for Rs 14 per hour, and can service an average of three customers per hour. The other mechanic has several years of experience, can service an average of four customers per hour, but needs to be paid Rs 20 per hour. Assume that the customers arrive at the company garage at the rate of two customers per hour.

    Calculate the following characteristics of the queuing system using each mechanic, assuming Poisson arrivals and exponential service times:

    (i) Expected length of queue,

    (ii) Expected length of system,

    (iii) Expected waiting time in queue, and

    (iv) Expected waiting time in the system.

    If the company assigns a customer waiting time cost of Rs 30 per hour, which mechanic provides the lower operating cost? *(M Com, Delhi, 2009)*

18. PQR Company has been experiencing machine breakdown for sometime and now wants to hire a mechanic on a permanent basis as against the present practice of engaging a mechanic as and when a breakdown occurs. The company has a large number of machines and, on an average, two machines breakdown per day. The machine down-time is estimated to cost the company Rs 800 per day.

    Two mechanics *A* and *B*, are available. *A* will cost the company Rs 40 per day and can on average repair three machines per day, whereas *B* will cost Rs 50 per day and his average service rate is four machines per day. Describing any assumptions you might make, determine which repairman should be hired so that total expected cost per day is minimised.

19. A large firm has on an average, 30 pieces of an equipment requiring repairs per week. The probability of an equipment failure is approximately constant so that the arrivals are distributed according to the Poisson law. The repair time is exponentially distributed. The down-time cost of the equipment is estimated at Rs 100 per work day. There are 6 working days and the firm works 50 weeks per annum. The firm has two repair facilities to choose from. The details are:

	Facility	
	F_1	F_2
Installation Charges	Rs 120,000	Rs 200,000
Operating Expenses p.a.	Rs 40,000	Rs 50,000
Economic Life	5 Years	5 Years
Scrap Value	nil	nil
Service Rate	40 per week	80 per week

Assuming that there is no time value of money, which facility would you suggest the firm to choose?

20. A mechanic is to be recruited to repair scooters that breakdown at an average rate of 6 per hour. The breakdowns follow a Poisson distribution. Non-productive time of a scooter costs Rs 40 per hour. Mechanics A and B have been interviewed for the purpose. Mechanic A charges Rs 20 per hour, and services scooters at the rate of 8 per hour. Mechanic B charges Rs 28 per hour, and services 12 scooters per hour, on an average. Which one of the two should be recruited and why? Assume an 8 hours of work each day. *(M Com, Delhi, 2007)*
21. The Hi-Power Company provides electric power, supplied by four big generators, to an industrial area. The generators are subject to an occasional breakdown and, from the past data, it is known that a generator develops problem, on an average, once in every 12 months. The company has a repair crew which is entrusted with the task of repairing the faulty generators. From the past records, it is known that it takes 2 months, on an average, for a generator to be repaired. Calculate
   (a) the proportion of time there shall be no generators to be repaired,
   (b) the proportion of time there shall be one, two, three, and four defective generators,
   (c) the average number of generators waiting repairs,
   (d) the average number of generators out of service,
   (e) the average waiting time for repair of a defective generator to begin, and
   (f) the average down-time per generator.
22. A laundry has six washing machines which breakdown randomly and the average time interval between successive breakdowns is 4 hours. The machines breaking down require, on an average, forty-five minutes to repair, the service time being exponentially distributed. Using this information, you are required to determine:
   (a) the probabilities of 0, 1, 2, 3, 4, 5 and 6 machines in the system,
   (b) the expected waiting time of a machine in the queue,
   (c) the expected number of machines in the system, and
   (d) the expected number of machines in the queue.
23. A telephone exchange has two long distance operators. It is observed that, during the peak load, long distance calls arrive in a Poisson fashion at an average rate of 15 per hour. The length of service on these calls is approximately exponentially distributed with mean length of 5 minutes.
   (i) What is the probability that a subscriber will have to wait for this long distance call during the peak hours of the day?
   (ii) If the subscribers will wait and are serviced in turn, what is the expected waiting time? *(MBA, Delhi, 1998)*
24. Patients arriving in the OPD of the Delhi Institute of Medical Sciences are registered at the counter after which they can be attended to by any one of the panel of 5 equally efficient doctors. The patients arrive at an average interval of 1.5 minutes and the arrival rate follows the Poisson probability law. A doctor attends to a patient for an average time of six minutes, the time taken being distributed exponentially. Assuming the steady state conditions, determine the following:
   (a) The traffic intensity.
   (b) The probability that none of the doctors is busy.
   (c) The probability of there being exactly 3 patients in the hospital.
   (d) The expected length of the queue of patients waiting for checkup.
   (e) The expected number of patients in the hospital.
   (f) The probability of there being exactly 8 patients in the hospital.
   (g) The average waiting time in the queue.
   (h) The average time a patient spends in the hospital, waiting for and getting doctor's service.
25. A company has two manufacturing shops and two tool cribs, one for each shop. Both the tool cribs handle almost identical jobs, gauges, and measuring instruments. Analysis of service time shows a

negative exponential distribution with a mean of 2.5 minutes per workman. Arrivals of workmen follows Poisson distribution with mean of 18 per hour.

The production manager feels that if tool cribs are combined for both shops, efficiency will improve and waiting time in the queue will reduce. Do you agree with his opinion?

26. Two typists have identical jobs. Each types letters dictated by a manager. Suppose that letters to be typed arrive at random (following Poisson distribution) at a rate of three per hour for each typist. Suppose that each typist can type four letters on the average per hour (following exponential distribution).
    (a) Assuming that each typist does her own work, what is the expected waiting time for a letter (time before work is started on a letter)?
    (b) Suppose that the two typists are 'pooled'. That is, letters are sent to the two together and are done by whoever is free, in order of arrival. What is the expected waiting time for a letter under this arrangement? *(MBA, Delhi, April, 1996)*
27. An insurance company has three claim adjusters in its Delhi branch office. Claimants against the company are found to arrive in a Poisson fashion, at an average rate of 20 per 10-hour day. The amount of time a server (the adjuster) spends with a customer is found to be exponentially distributed with mean service time of 40 minutes. Claims are processed in the order in which they appear.
    (a) How many hours a week can an adjuster expect to spend with claimants? Assume a 5-day week.
    (b) How much time, on the average, does a claimant spend in the office?
28. Customers arrive at a barber's shop according to Poisson distribution. The average time between successive arrivals is 6 minutes. There are three barbers, all of same efficiency. The service time of the customers is exponentially distributed with a mean equal to 10 minutes per customer. Find (a) the expected number of customers in the shop, (b) the expected time a customer spends in the shop, (c) the average time a customer has to wait in the queue, and (d) the expected number of barbers idle.
29. A post office has two counters, the first one handles money-orders and registration letters, while the second handles all other business. It has been found that service time distributions for both the counters are exponential with a mean service time of 4 minutes per customer. The customers are found to arrive at the first counter in a Poisson fashion with mean arrival rate of 10 per hour while the customers arrive at the second counter in a Poisson fashion with mean arrival rate of 12 per hour. What would be the effect on the average waiting time for the two types of customers if each counter can handle all types of business?
30. The customers in the Raja Ji Departmental Store, which is being renovated, order their requirements in the different sections, pay cash at the cash counters and then, finally, obtain delivery from the 'delivery counter'. After taking delivery, the customers queue up to leave through a checkout lane.

    The manager is in a fix whether to provide 2 checkout lanes or 3. He needs your help and provides you with the following information.

    "It is estimated that the mean arrival of the customers at the checkout would be 7 per one-minute period and the service rate would be 4 customers per one-minute period per lane. Further, due to goodwill, customers' waiting time costs the store 10 p per minute. The cost to operate a lane is Rs 2.65 per hour."

    For your analysis, assume that (i) customers form a single queue, and (ii) the arrival and service rates are Poisson distributed.
    (a) Show to the manager the comparative values of $P(0)$, L_s, W_q and W_s for 2 and 3 lanes, respectively.
    (b) Suggest to the manager whether to provide 2 lanes or 3.
31. A steel fabrication plant was considering the installation of a second tool crib in the plant to save walking time of the skilled craftsmen who check out equipment at the tool cribs. The Poisson/exponential assumptions about arrivals are justified in this case. The time of the craftsman is valued at

Rs 20 per hour. The current facility receives an average of ten calls per hour, with two cribs, each with average five calls per hour. Currently there are two attendants, each of whom services one craftsman at a time, each has a service rate of eight craftsman per hour. Each could do just as well in a separate tool crib. There would be added average inventory costs over the year of Rs 2/hour with the separate tool cribs. However, each craftsman would require six minutes less walking time per call. Evaluate the proposal to set up a new crib so that each attendant would run one crib. *(MBA, Delhi, 2003)*

32. The customers arrive into a large departmental store and wander around the place selecting items of household and other use, and then proceed to checkout stands with a basket of items. There are 4 checkout counters each of which can serve a customer in six minutes, on the average. Assume that the customers arrive at a rate of one every three minutes.
    Calculate the following:
    (a) the probabilities of 0, 1, 2, 3, 4, and 5 customers in the system.
    (b) the average number of customers in the store.
    (c) the average number of customers waiting for a checkout.
    (d) the average number of customers being serviced.
    (e) the average waiting time in the system per customer.
    (f) the average waiting time in the queue.
33. (a) The arrivals at a desk are Poisson distributed with a mean of one customer every two minutes. The service times are exponentially distributed with an average of 15 seconds per customer. You are required to calculate:
    (i) Overall system utilisation.
    (ii) Expected number of customers in the system.
    (iii) Expected length of queues.
    (iv) Expected time a customer spends in the system.
    (v) Expected time a customer spends in the queue.
    (vi) Probability that the system is completely idle.
    (vii) Probability that an arriving customer waits.
    (b) Recalculate the parameters given in (a) assuming, first, that the server works twice as fast and, second, that two servers each work at the original rate, and compare them.
34. A post office has two clerks, either of whom averages 1.5 minutes per customer transaction (the service time being distributed exponentially). The arrival rate of the customers to the post office is one customer per minute. Compute
    (a) the probability that both the clerks would be idle.
    (b) the probability that there shall be one customer in the post office.
    (c) the probability that there shall be five customers in the post office.
    (d) the average number of customers waiting in the queue.
    (e) the average number of customers being serviced.
    (f) the average time a customer spends waiting for service.
    (g) the average time a customer spends in the post office.
35. If the mean arrival rate is 24 per hour, consider from the customers's point of view, the time spent in the system, whether three channels in parallel with a mean service rate of 10 per hour is better or worse than a single channel with mean service rate of 30 per hour.
36. A certain queuing system has a Poisson input with a mean arrival rate of two calling units per hour. The service time distribution is exponential with a mean of 0.4 hour. The marginal cost of providing each server is Rs 4 per hour, where it is estimated that the cost which is incurred by having each calling unit idle (i.e. in the queuing system) is Rs 100 per hour. Determine the number of servers that should be assigned to the system in order to minimise the expected total cost per hour. *(MBA, Delhi, 1974)*

CHAPTER 11

Replacement Theory

CHAPTER OVERVIEW

Machines, equipment, and vehicles all do not work forever and they need to be replaced sooner or later. With passage of time, the working efficiency of such assets goes down and increasingly greater amounts of money are required to keep them running effciently. But, at the same time, the longer the period an asset of this type is used, the better it is because the original cost gets spread over a longer time and we can postpone the cost of replacing it. The question facing a manager is as to when such an asset should be replaced: should it be used until it breaks down and cannot be used any more or should it be replaced before this stage arrives? Further, while the assets like machines and vehicles wear and tear gradually over fairly predictable range of time, there are others like bulbs and electronic components which fail suddenly and need to be replaced immediately as they fail. The present chapter considers this problem of replacement. It helps the manager to find answers to questions like the following:

- For a gradually depreciating asset, given the cost of running and maintenance along with its cost price and the reasale value at different intervals of time, after what intervals of time should it be replaced?
- During the currency of use of an asset, if a better model or a new substitute becomes available, should the old asset be replaced with the new one? In case we decide for this, then at what time should the replacement be done—immediately or after some specified time?
- If the time value of money is taken into consideration, then after what period would the replacement of the asset be due?
- How to plan for the replacement of the items like bulbs and electronic components which are used in bulk and may fail randomly? At what regular intervals of time it may be advisable to replace them all?
- How is replacement of staff of an organisation, which may be necessitated by reasons of death, departure, and retirement, etc., of the empolyees, to be planned?

The answers to such questions are given by the theory of replacement on the cost considerations. Computation of average costs, discounting of future costs/revenues, understanding application of probability, probability distribution and expected value are the essentials for this chapter.

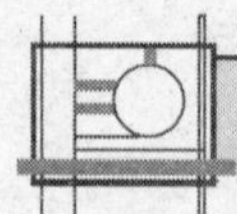

11.1 INTRODUCTION

In any establishment, sooner or later all equipment need to be replaced. A replacement is called for whenever new equipment offers more efficient or economical service than the old, existing one. For example, the old equipment might fail and work no more, or is worn out and needs substantial expenditure on its maintenance. The problem, in such situations, is to determine the best policy to be adopted with respect to replacement of the equipment. The replacement theory provides answer to this question in terms of *optimal replacement period*.

In our analysis, we shall consider the question of replacement in the following situations:

(a) In respect of equipment/assets that deteriorate in performance over a time and may be restored in whole or in part by expending maintenance costs. Thus, the existing assets might be good technically, yet on economic considerations, it may not be worthwhile continuing with them and hence replacement may be called for. Machinery, equipment, buildings fall in this category.

(b) In respect of units that perform adequately until sudden complete failure. The length of their service until failure varies randomly over some predictable range. Car bulbs, tubes, some electronic components, etc. are the items that are considered in this category.

(c) In respect of replacement of staff of an organisation which diminishes gradually due to death, retirement, retrenchment and other reasons.

We shall consider the three situations one by one.

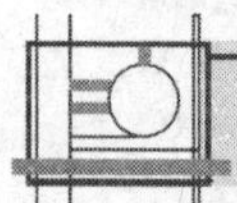

11.2 REPLACEMENT POLICY FOR EQUIPMENT WHICH DETERIORATES GRADUALLY

To determine the optimal replacement period, the period after which an equipment/machine should be appropriately replaced, we proceed in the following manner. Initially, we shall assume that there is no time value of money. That is to say, a rupee received at the end of a period, say 5 years, is regarded as good as a rupee received now.

We know that the cost of a piece of equipment over a given time period, say n years, has three elements:

Purchase price – Value remaining after n years + Maintenance cost for n years

Let C = the purchase price of the equipment,

S = the scrap value of the equipment at the end of n years,

and M_t = the maintenance cost of the equipment in year t.

The total cost, $T(n)$, of owning and maintaining the equipment for n years shall be

$$T(n) = C - S + \sum_{t=1}^{n} M_t$$

Correspondingly, the average cost, $A(n)$, would be defined as,

$$A(n) = \frac{1}{n}\left[C - S + \sum_{t=1}^{n} M_t\right]$$

In such a case, the optimal replacement period would be the one corresponding to which $A(n)$ would be the minimum. To find $A(n)$, we shall, first of all, determine the maintenance costs accumulated to the nth year. To this, the cost of equipment, net of its salvage value, would be added. The two would then be aggregated to get the total cost, $T(n)$, and then the total cost divided by n, the number of years, would give the average cost.

Example 11.1 A firm is using a machine whose purchase price is Rs 13,000. The installation charges amount to Rs 3,600 and the machine has a scrap value of only Rs 1,600 because the firm has a monopoly of this type of work. The maintenance cost in various years is given in the following table.

Year :	1	2	3	4	5	6	7	8	9
Cost (Rs) :	250	750	1000	1500	2100	2900	4000	4800	6000

The firm wants to determine after how many years should the machine be replaced on economic considerations, assuming that the machine replacement can be done only at the year ends.

Cost of machine, C = Rs 13000 + Rs 3600 = Rs 16,600

Scrap value, S = Rs 1600

We can determine the optimal replacement period as in the Table 11.1.

TABLE 11.1 Determination of Optimal Replacement Period

Year	*Maintenance Cost, M_t*	*Cum. Main. Cost, $\sum M_t$*	$C-S$	$T(n)$	$A(n)$
(i)	(ii)	(iii)	(iv)	(v) = (iii) + (iv)	(vi) = (v)/n
1	250	250	15,000	15,250	15,250
2	750	1,000	15,000	16,000	8,000
3	1,000	2,000	15,000	17,000	5,667
4	1,500	3,500	15,000	18,500	4,625
5	2,100	5,600	15,000	20,600	4,120
6	2,900	8,500	15,000	23,500	3,917*
7	4,000	12,500	15,000	27,500	3,929
8	4,800	17,300	15,000	32,300	4,038
9	6,000	23,300	15,000	38,300	4,256

Here the lowest average cost, $A(n)$, is Rs 3,917 approximately, which corresponds to $n = 6$. In Table 11.1, it is indicated by an asterisk mark, (*). Therefore, the machine may best be replaced every 6 years.

In this and other examples, we implicitly assume that a machine is to be used to do a job that has to be done forever and we are concerned with doing the job as cheaply as possible (and not with whether the job should be done or not). In other words, in replacement analysis, we do not view machine or equipment as a capital investment decision[1]. Because when we consider it so, we have to look to the profit arising from the product/s the equipment produces and not merely the costs involved.

Example 11.2 The Simple Engineering Company has a machine whose purchase price is Rs 80,000. The expected maintenance costs and resale price in different years are as given here:

[1]This problem is discussed in Chapter 18.

Year	:	1	2	3	4	5	6	7
Maintenance Cost Rs	:	1,000	1,200	1,600	2,400	3,000	3,900	5,000
Resale Value ('000 Rs)	:	75	72	70	65	58	50	45

After what time interval, in your opinion, should the machine be replaced?

The optimal replacement period determination is shown in Table 11.2.

TABLE 11.2 Determination of Optimal Replacement Period

Year	Maintenance Cost, M_t	Cum. Main. Cost, $\sum M_t$	$C - S$	$T(n)$	$A(n)$
(i)	(ii)	(iii)	(iv)	(v) = (iii) + (iv)	(vi) = (v)/n
1	1,000	1,000	5,000	6,000	6,000
2	1,200	2,200	8,000	10,200	5,100
3	1,600	3,800	10,000	13,800	4,600*
4	2,400	6,200	15,000	21,200	5,300
5	3,000	9,200	22,000	31,200	6,240
6	3,900	13,100	30,000	43,100	7,183
7	5,000	18,100	35,000	53,100	7,586

Here minimum $A(n)$ = Rs 4600, for $n = 3$ (shown by * mark). The machine should, therefore, be replaced every three years.

Example 11.3 (a) A machine M_1, costing Rs 9,000 has a maintenance cost of Rs 200 in the first year of its operation which rises by Rs 2,000 in each of the successive years. Assuming that the machine replacement can be done only at the end of a year, determine the best age at which the machine be replaced.

(b) There is an offer to replace the machine M_1 which is a year old, by another machine M_2 which costs Rs 8,000. The machine M_2 needs Rs 2,000 to be spent on installation, has no salvage value, and requires Rs 400 on maintenance in the first year followed by an increase of Rs 800 per annum in the yearly expenditure on maintenance. Should the machine M_1 be replaced by machine M_2? If so, when?

(a) The calculation of optimal replacement period is shown in Table 11.3.

TABLE 11.3 Determination of Optimal Replacement Period for M_1

Year	Maintenance Cost, M_t	Cum. Maint. Cost, $\sum M_t$	$C - S$	$T(n)$	$A(n)$
1	200	200	9,000	9,200	9,200
2	2,200	2,400	9,000	11,400	5,700
3	4,200	6,600	9,000	15,600	5,200*
4	6,200	12,800	9,000	21,800	5,450
5	8,200	21,000	9,000	30,000	6,000
6	10,200	31,200	9,000	40,200	6,700
7	12,200	43,400	9,000	52,400	7,485
8	14,200	57,600	9,000	66,600	8,325

Thus, optimal replacement period = 3 years.

(b) To determine as to whether M_1 should be replaced by M_2, we shall obtain the least average cost, $A(n)$, value. If it is found to be smaller than Rs 5,200 (for M_1) we shall replace, otherwise not. Its calculation is shown in Table 11.4.

TABLE 11.4 Determination of Minimum Average Cost for M_2

Year	Maintenance Cost, M_t	Cum. Maint. Cost, $\sum M_t$	$C - S$	$T(n)$	$A(n)$
1	400	400	10,000	10,400	10,400
2	1,200	1,600	10,000	11,600	5,800
3	2,000	3,600	10,000	13,600	4,533
4	2,800	6,400	10,000	16,400	4,100
5	3,600	10,000	10,000	20,000	4,000*
6	4,400	14,400	10,000	24,400	4,067
7	5,200	19,600	10,000	29,600	4,229

Since the minimum average cost for M_2 is lower than that for M_1, the machine should be replaced.

To decide the time of replacement, we should compare the minimum average cost for M_2 (Rs 4,000) with yearly cost of maintaining and using the machine M_1. Since there is no salvage value of the machine, we shall consider only the maintenance cost. We would keep the machine M_1 so long as the yearly maintenance cost is lower than Rs 4,000 and replace when it exceeds the 4,000 mark.

On the one-year old machine M_1, Rs 2,200 would be required to be spent in the next year; while Rs 4,200 would be needed in the year following. Thus, we should keep the machine M_1 for one more year and replace it thereafter.

When time value of money is considered

In the problems discussed so far, we have assumed that there is no time value of money. When we drop this assumption, we have to discount the future payments on maintenance costs (and receipts possible due to salvage value) so as to express them all in terms of present value.

Now, if the rate of interest be r, then the present value of a rupee paid at the end of one year would be $1/(1 + r)$, paid at the end of two years would be $1/(1 + r)^2$ ... and so on.* The present value of one rupee paid at the end of 1, 2, ..., n years is called the *present value factor* (PV factor, in short) or discount factor in respect of 1, 2, ..., n years. The table of present value factors is given in Appendix *B* (Table B3) from which values for

*We have the compound interest formula to determine the amount A, which a present amount P would become after n periods when the rate of interest is r, as $A = P(1 + r)^n$. With $A = 1$, the present value P equal $1/(1 + r)^n$. For details, refer Chapter 18.

different combinations of time periods and interest rates may be read off directly. Any payments made at different times be converted into their present worth by multiplying them by appropriate PV factors.

To consider the problem of replacement when time value of money is taken into consideration, we shall assume that (i) the equipment in question has no salvage value, and (ii) the maintenance costs are incurred in the beginning of the different time periods (which we shall take to be years). Working under these assumptions, we first determine the *annualised cost*, or the weighted average of costs, for different years of the life of the equipment. This is done as follows:

Step 1 Find the present value of the maintenance cost for each of the years, by multiplying the cost value by an appropriate present value factor (based on the given rate of discount and the time).

Step 2 Accumulate present values obtained in *Step 1* upto each of the years 1, 2, 3, ... Add the cost of the equipment to each of these values.

Step 3 Accumulate present value factors upto each of the years 1, 2, 3, ...

Step 4 Divide the cost plus cumulative maintenance costs for each year, obtained in *Step 2*, by the corresponding cumulative present value factors in *Step 3*. This gives the annualised costs for the various years. It may be noted that the annualised cost is also the weighted average of costs—the PV factors serve as weights here and the average is calculated, as in the usual way, by dividing the summation of the products of the costs and their respective weights by the summation of the weights.

Once the weighted average of costs for different years is obtained, the following rules (the proof of which is given in Appendix 11A) are followed to decide on the replacement.

(a) Do not replace if the next period's cost is less than the weighted average of the previous costs, and

(b) Replace if the next period's cost is greater than the weighted average of the previous costs.

In operational terms, the criterion for optimal replacement is: replace every n years where A_n, the annualised cost of replacing every n years (that is, the weighted average of costs for n years) is the minimum.

Example 11.4 A machine *X* costs Rs 5,000. Its maintenance cost is Rs 1,000 in each of the first four years and then it increases by Rs 200 every year. Assuming that the machine has no salvage value and the maintenance cost is incurred in the beginning of each year, determine the optimal replacement time for the machine assuming that the time value of money is 10% p.a.

The given information about the costs is contained in the second column of the Table 11.5. The present value factors corresponding to the 10% interest rate, for various years are given in the third column. The fourth column represents the present value of each year's maintenance cost, which is computed by applying the discount factor to the maintenance cost. In the fifth column of the table is presented the present value of the Rs 5,000 purchase price plus the cumulative maintenance cost (at present value) to the end of each year. The cumulative sums of present value factors of the third column are given in the sixth column while the seventh column represents the annualised costs obtained by dividing fifth column values by the corresponding column values.

TABLE 11.5 Determination of Optimal Replacement Period

Year	*Maint. Cost, M_t*	*P V Factor*	*P Value of M_t*	*Cost Plus Cum. M_t*	*Cum. P V Factor*	*Annualised Cost*
1	1,000	1.0000	1,000.0	6,000.0	1.0000	6,000
2	1,000	0.9091	909.1	6,909.1	1.9091	3,619
3	1,000	0.8264	826.4	7,735.5	2.7355	2,828
4	1,000	0.7513	751.3	8,486.8	3.4868	2,434
5	1,200	0.6830	819.6	9,306.4	4.1698	2,232
6	1,400	0.6209	869.2	10,175.6	4.7907	2,124
7	1,600	0.5645	903.2	11,078.8	5.3552	2,069
8	1,800	0.5132	923.8	12,002.6	5.8684	2,045
9	2,000	0.4665	933.0	12,935.6	6.3349	2,042*
10	2,200	0.4241	933.0	13,868.6	6.7590	2,052
11	2,400	0.3855	925.2	14,793.8	7.1445	2,071
12	2,600	0.3505	911.3	15,705.1	7.4950	2,095

In the table, we compare the second column, wherein values are constantly increasing, with the last column, where values first decrease and then increase. Looking to this, we should decide to replace the machine in the latest year that the seventh column value exceeds the corresponding second column value. Here, the year 9 is the last because in the year next to it, the maintenance cost of Rs 2,200 exceeds the value of Rs 2,052 in the seventh column. It is the year in which the annualised cost is the least. Thus, optimal replacement interval for the machine is 9 years.

Notes

1. In the analysis, we have assumed that the equipment in question has no salvage value. We can drop this assumption to incorporate into analysis the salvage value that an equipment might have. For this purpose, the given salvage value at the end of a particular year is translated into present value terms by applying an appropriate discount factor (keeping in mind that for a given year, the maintenance cost is assumed to be incurred in the beginning of the year while the salvage value represents the asset value at the end of the year, so that the same discount rate should not be applied to both of them). This value is then subtracted from the cost of the equipment and the difference added to the cumulative maintenance costs upto the year in question.
2. A second assumption made in this analysis is that the costs are incurred in the beginning of each year. If the costs are incurred during the year instead, they may be discounted to the start of the year. If we could assume that the costs are incurred continuously over the year, then discounting can be done using integral calculus. A more reasonable assumption for most problems of the type considered, sufficient accuracy may be achieved if we assume that the expenditure takes place half way through the year. If j be the rate of interest per half year, and r is the rate per annum, then one rupee invested for six months would become $(1 + j)$ rupees and if this amount is invested for a second six-month period, the total at the end of the year would be $(1 + j)^2$ rupees. Thus, $1 + r = (1 + j)^2$ or $1 + j = \sqrt{1 + r}$. Thus, one rupee due in six months would have a present value of $1/\sqrt{1 + r}$.

3. As mentioned earlier also, an implicit assumption in our analysis is that the equipment/machine concerned is to be used for ever. If the equipment is required for a certain fixed time period, then the formulation needs to be modified.

   Example 11.9 illustrates a case where the machine is required to be used for a given period of time.

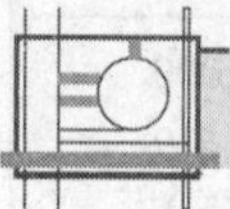

11.3 REPLACEMENT OF ITEMS THAT FAIL SUDDENLY

Now we shall consider the question of determining optimal policy for replacing items used in large numbers which do not deteriorate gradually but fail all of a sudden. They may not involve any maintenance cost, but fail without warning. Light bulbs and fluorescent tubes are examples of this type. It may, however, be observed that although the failure of a single unit, a bulb for instance, is random, the behaviour of the group as a whole is expected to be fairly stable.

Generally, in such situations, the replacement of individual units, upon their failure, costs relatively high, whereas if the entire group is replaced, the unit cost works out to be comparatively a much lower value. Thus, we might be inclined to replace the whole lot at certain intervals together with the individual replacements as and when needed. If the group replacement is done at short intervals, the cost per period shall be high. As the time interval between group replacements increases, the cost of group replacement per period declines. But, the cost of individual replacements increases because of larger number of replacements needed. Thus, the total cost starts declining upto a certain point when the time interval increases, and then it starts rising where the individual replacement component of the cost exerts a greater influence on the cost. Our problem is to determine this optimal interval of time. Figure 11.1 portrays these ideas.

GR represents the sum of the cost of group replacement per unit of time (G/T) and the cost of individual replacement of items which fail during the replacement interval (also per unit of time) and, hence, equal to

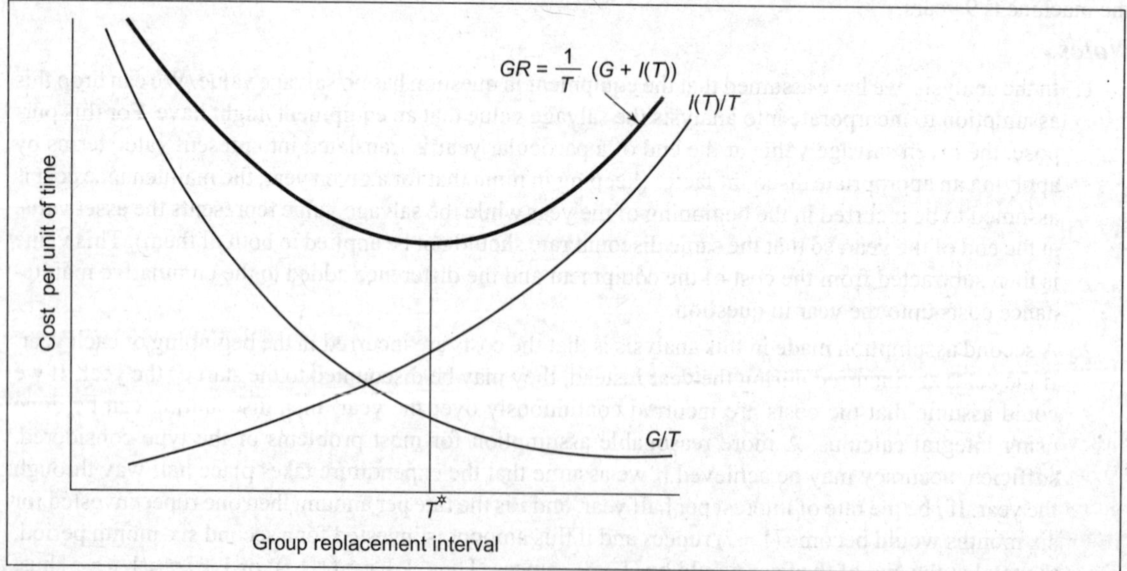

Figure 11.1 *Determination of Optimal Replacement Interval*

I(T/T). In the figure, the optimal group replacement interval is determined as T^*, corresponding to the lowest point on the GR curve.

We shall illustrate the determination of the optimal replacement period by means of the following example.

Example 11.5 A large establishment has an installation with 1,000 bulbs of a certain type. From the past data, it has observed the failure rates of these bulbs as detailed here:

End of week	:	1	2	3	4	5
Probability of failure to date	:	0.10	0.25	0.50	0.70	1.00

It is given that the cost of replacing an individual bulb is Rs 3 while if the entire group of bulbs is replaced, the cost would be Re 1 per bulb. It is decided to replace all the bulbs simultaneously at fixed intervals of time and also to replace the individual bulbs that fail in between.

Assuming that all the bulbs failing during a week might fail at any time of the week and that the group replacements can be made only at the end of a week, you are required to determine the optimal interval between the group replacements.

Also establish if the policy, as determined by you, adopted is superior to the policy of replacing bulbs as and when they fail, there being nothing like 'group replacement'.

First of all, we obtain the probability distribution of the lives of the bulbs—indicating the probability of failure of a bulb during a certain week of its life. It is given as follows:

Week	:	1	2	3	4	5
Probability of failure	:	0.10	0.15	0.25	0.20	0.30

From this distribution we observe that 10 per cent of the bulbs are expected to be burnt during the first week of their life. In the lot of 1,000 bulbs, therefore, 100 bulbs are expected to fail in the first week. Similarly, of this lot, 150 (i.e. 15%) bulbs are expected to fail in the second week, 250 bulbs in the third week, 200 in the fourth week, and the remaining 300 bulbs are likely to fail in the fifth week.

Of the 100 bulbs that would be replaced in the first week, 10 per cent, i.e. 10 bulbs, would fail in the first week of their life, that is to say, week 2; 15 per cent (= 15 bulbs) would fail in week 3, and so on. Similarly, of the total replacements numbering 160 (equal to 150 of the original lot plus 10 out of the 100 replacements in the first week) during the second week, 10 per cent would fail during the third week, 15 per cent during the week 4, ... and so on. The total replacements are shown diagrammatically in the form of *failure tree* in Figure 11.2.

Starting with the lot of 1,000 bulbs in the beginning, at $t = 0$, 900 bulbs are expected to survive and 100 to fail in week 1. The horizontal line depicts the former and the diagonal line shows the latter. Starting at $t = 1$, 750 of the original bulbs would survive and 150 would fail during the period from $t = 1$ to $t = 2$. Again, the number of surviving bulbs is shown horizontally and the number of those failing is shown diagonally. Of the 100 replacements at $t = 1$, 90 would be surviving at $t = 2$ (shown horizontally) and 10 would fail by $t = 2$ (shown diagonally). In a similar way, horizontal and diagonal lines are drawn for each week.

Next, the sum total of the numbers shown on the diagonals in respect of each week is obtained as written at the bottom. These totals represent the number of replacements in each time period.

Before we see how this information about the number of replacements required every week could be used to take a decision about the optimal interval for replacement, it may be mentioned that it is not necessary to draw a failure tree for obtaining this information. The number of failures expected in different weeks can also be determined algebraically.

In general, if K is the maximum number of periods of life that a unit can have, the number of failures, N_i, in any given period i, can be expressed as follows:

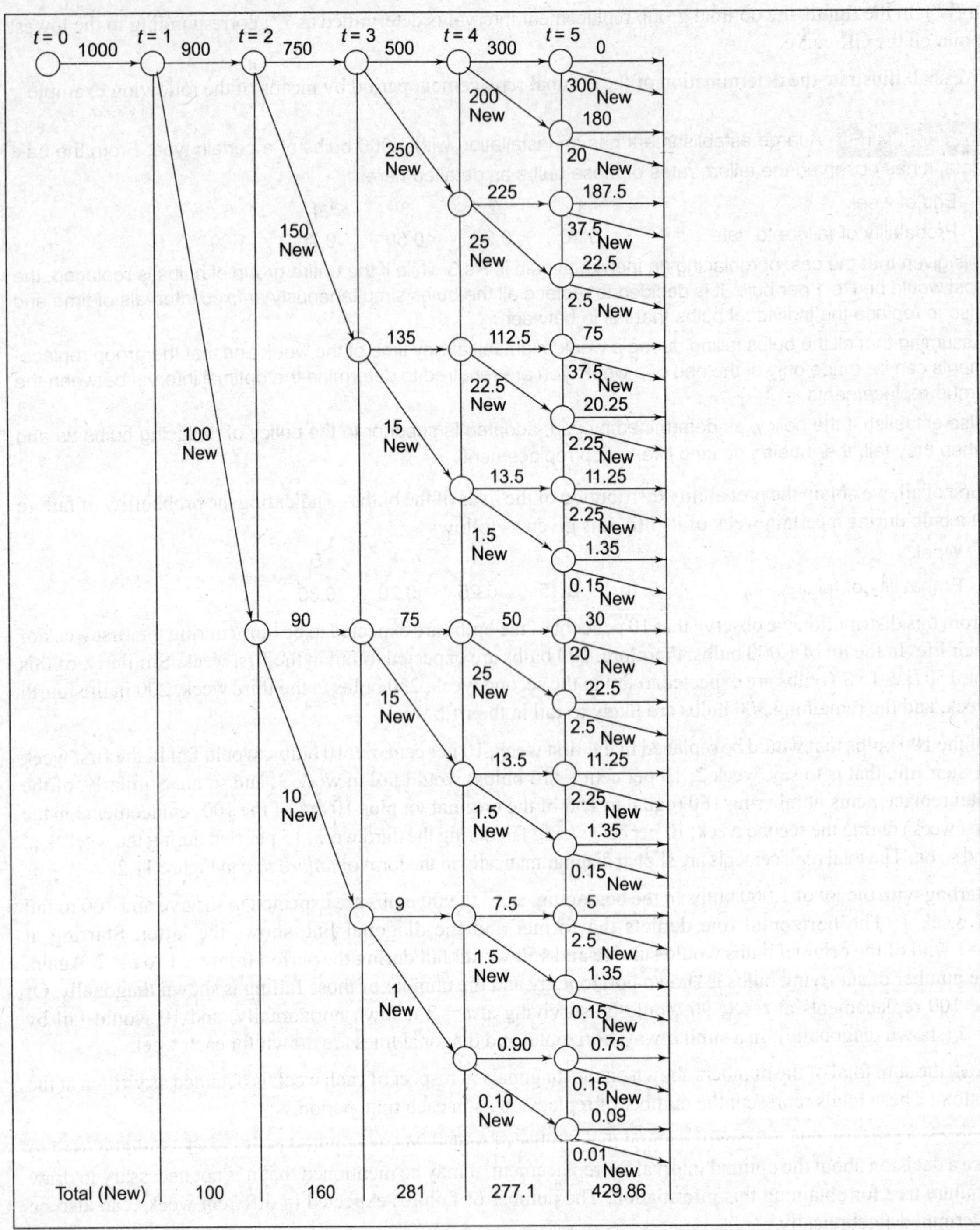

Figure 11.2 *Failure Tree*

$$N_i = \begin{cases} N_0 \times P_i + \sum_{t=1}^{i-1} N_{i-t}\, p_t & \text{for } i \leq K \\ \sum_{t=1}^{K} N_{i-t}\, p_t & \text{for } i > K \end{cases}$$

in which p_t's are the failure probabilities and N_0 is the total number of items. Using the formula, we can easily determine the number of failures in different weeks for our example as given in the Table 11.6.

TABLE 11.6 Determination of Number of Failures in Different Weeks

Week		*Expected No. of Failures*	
1	$N_1 = N_0 p_1$	$= 1000 \times 0.10$	$= 100$
2	$N_2 = N_0 p_2 + N_1 p_1$	$= 1000 \times 0.15 + 100 \times 0.10$	$= 160$
3	$N_3 = N_0 p_3 + N_1 p_2 + N_2 p_1$	$= 1000 \times 0.25 + 100 \times 0.15 + 160 \times 0.10$	$= 281$
4	$N_4 = N_0 p_4 + N_1 p_3 + N_2 p_2 + N_3 p_1$	$= 1000 \times 0.20 + 100 \times 0.25$ $+ 160 \times 0.15 + 281 \times 0.10$	$= 277.1$
5	$N_5 = N_0 p_5 + N_1 p_4 + N_2 p_3 + N_3 p_2 + N_4 p_1$	$= 1000 \times 0.30 + 100 \times 0.20 + 160 \times 0.25$ $+ 281 \times 0.15 + 277.1 \times 0.10$	$= 429.8$
6	$N_6 = N_5 p_1 + N_4 p_2 + N_3 p_3 + N_2 p_4 + N_1 p_5$	$= 429.8 \times 0.10 + 277.1 \times 0.15 + 281 \times 0.25$ $+ 160 \times 0.20 + 100 \times 0.30$	$= 216.8$
7	$N_7 = N_6 p_1 + N_5 p_2 + N_4 p_3 + N_3 p_4 + N_2 p_5$	$= 216.8 \times 0.10 + 429.8 \times 0.15 + 277.1 \times 0.25$ $+ 281 \times 0.20 + 160 \times 0.30$	$= 259.6$

It may be noticed that the number of bulbs failing each week increases until week 5 and then it decreases. Further calculations on the same basis would show that the failures would increase. In fact, the number of failures would continue to oscillate until finally it would settle to a steady state in which the proportion of bulbs failing each week would be equal to the reciprocal of the average life of the bulbs.

Now we can determine the total and the average cost associated with the policy of replacing bulbs every week, every two weeks, ... and so on. This is shown in Table 11.7.

TABLE 11.7 Determination of Average Cost of Different Policies

Upto Week	*No. of Individual Replacements*	*Cost of Replacements*			
		Individual	*Group*	*Total*	*Average Cost*
1	100	300	1,000	1,300	1,300.0
2	260	780	1,000	1,780	890.0
3	541	1,623	1,000	2,623	874.3
4	818.1	2,454.3	1,000	3,454.3	863.6*
5	1,247.9	3,743.7	1,000	4,743.7	948.7
6	1,464.7	4,394.1	1,000	5,394.1	899.0
7	1,724.3	5,172.9	1,000	6,172.9	881.8

From the table, it is clear that the minimum cost per week is obtained by replacing all bulbs (whether failed or not) after every 4 weeks. Thus, optimal replacement time interval = 4 weeks.

When the policy of 'replacing the bulbs as and when they fail' is adopted, there are no group replacements. This policy, however, is but a special case of the policy of group replacements in which the interval between the group replacements is infinite. In such a case, the expected number of replacements per week is determined. In the steady state, this would be equal to the product of the number of bulbs and the reciprocal of the average life of a bulb (or as number of bulbs/average life of a bulb). The cost of such a policy shall be determined by the expected number of replacements (i.e. failures) times the cost of an individual replacement.

For our example, since it is assumed that a bulb can fail at any time during the week, the mean life of a bulb can be obtained as follows:

TABLE 11.8 Determination of Average Life

Life in Weeks X_i	*Probability* p_i	$p_i X_i$
0.5	0.10	0.050
1.5	0.15	0.225
2.5	0.25	0.625
3.5	0.20	0.700
4.5	0.30	1.350
	Total	2.950

Mean life of a bulb = 2.95 weeks

$$\text{Expected number of failures during a week} = \frac{\text{No. of bulbs}}{\text{Mean Life}}$$

$$= \frac{1{,}000}{2.95} = 338.98$$

Weekly replacement cost = Expected number of failures (replacements) × Cost of a replacement

= 338.98 × 3 = Rs 1,016.94

Obviously, since the cost of the policy of individual replacements is higher than that of the group replacements, it is advisable to adopt the policy of group replacements.

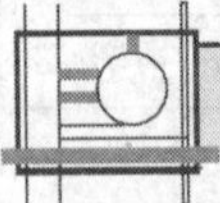

11.4 STAFF REPLACEMENT

As mentioned earlier, the staff of an organisation calls for replacements because people leave the organisation for several reasons. For planning a suitable recruitment policy, historical data are collected to estimate the likely stay of individuals with the organisation through time. Here also, the stay of an individual employee may be a random variable but the characteristics of the group of employees are likely to be fairly stable. We use this fact for the purpose at hand.

Consider the following example.

Example 11.6 A company plans to have a strength of 80 sales persons over time. The people employed by the company leave their job, the distribution of which is assumed to be as follows:

Year	:	1	2	3	4	5	6	7	8	9	10	11	12
% Expected to Leave by End of Period	:	5	18	35	45	62	75	88	90	95	96	99	100

Using this information, answer the following questions:

(a) What recruitment level per annum would be necessary to maintain this strength?

(b) If 15 of the sales persons are to be promoted to the next higher post, for which length of service in the organisation is the primary criterion, at what average length of service can a new entrant expect to be promoted?

(a) With an intake of 100 sales people every year, the distribution of the length of service of the sales-force, when equilibrium has been reached, shall be as follows:

Year	:	0	1	2	3	4	5	6	7	8	9	10	11	12	Total
No. of People Continuing	:	100	95	82	65	55	38	25	12	10	5	4	1	0	492

Recruitment of 100 sales people every year would thus give a total strength of 492. To maintain a strength of 80 sales people, we need to recruit $80 \times 100/492 = 16.26$ persons per year.

This assumes, however, that those who complete x years of service leave immediately before completing $x + 1$ years of service. If we assume that the people leave as soon as they complete x years of service, the total of the sales force would be 392. According to this, the recruitment necessary to have a sales force strength of 80 shall be equal to $80 \times 100/392 = 20.41$ people per year. Since, in practice, the sales persons may leave at any time of the year, the number of recruitments should be $(16.26 + 20.41)/2$ or 18 (app).

(b) With 80 persons in the sales team, the distribution of the completed length of service would be as follows:

Year :	0	1	2	3	4	5	6	7	8	9	10	11	12
No. of Persons:	16	15	13	11	9	6	4	2	2	1	1	0	0

Since 15 of the sales people are to be promoted and the promotion would be on the basis of the length of their service, the cumulation of the frequencies in this distribution reveals that a sales person can expect to be promoted after 5 years of service.

Limitations The assumption in the models relating to the replacement of depreciable assets that the future costs and resale prices of an equipment are predictable is quite unrealistic. In practice, they are likely to be random variables calling for probabilistic approach to the analysis. Similarly, the assumption that the historical costs in respect of an equipment requiring replacement are appropriate estimates of the costs for the equipment by which it is sought to be replaced may not hold. Besides, the maintenance costs do not always follow a smooth pattern over time. Very often, they are incurred in discrete lumps caused by such things as, for example, installing a new gearbox in a van, or overhauling and reconditioning the body of a truck. In such cases, the best policy might be to decide whether to scrap the vehicles (or other equipment) or undertake the major repair work and use it further.

In respect of the group replacement decisions in case of items that fail suddenly, a practical objection is that all the items are replaced (when group replacement takes place) whether they are working (or not)—even if they are the ones that were installed only recently. Further, a common limitation of both the replacement situations discussed is another assumption that the replacements could be carried out only at certain fixed time—at the year-ends or weekends, for example. In reality, the replacement can be, and is, done at any point during a given time.

REVIEW ILLUSTRATIONS

Example 11.7 The data on the operating costs per year and resale prices of equipment *A* whose purchase price is Rs 10,000 are given here:

Year	:	1	2	3	4	5	6	7
Operating cost (Rs)	:	1,500	1,900	2,300	2,900	3,600	4,500	5,500
Resale value (Rs)	:	5,000	2,500	1,250	600	400	400	400

(a) What is the optimum period for replacement?

(b) When equipment *A* is 2 years old, equipment *B*, which is a new model for the same usage, is available. The optimum period for replacement is 4 years with an average cost of Rs 3,600. Should we change equipment *A* with that of *B*? If so, when? *(CA, May, 1986)*

(a) The determination of the optimal period of replacement of equipment *A* is given in Table 11.9.

TABLE 11.9 Determination of Optimal Replacement Period

Year	M_t	Cum M_t	$C - S$	$T(n)$	$A(n)$
1	1,500	1,500	5,000	6,500	6,500.0
2	1,900	3,400	7,500	10,900	5,450.0
3	2,300	5,700	8,750	14,450	4,816.7
4	2,900	8,600	9,400	18,000	4,500.0
5	3,600	12,200	9,600	21,800	4,360.0*
6	4,500	16,700	9,600	26,300	4,383.3
7	5,500	22,200	9,600	31,800	4,542.9

Since the average cost corresponding to the 5-yearly period is the least, the optimal period for replacement = 5 years.

(b) As the minimum average cost for equipment *B* is smaller than that for equipment *A*, it is prudent to change the equipment. To decide the time of change, we would determine the cost of keeping the equipment in its 3rd, 4th and 5th year of life and compare each of these values with Rs 3,600 (the average cost for equipment *B*). The equipment *A* shall be held as long as the marginal cost of holding it would be smaller than the minimum average cost for equipment *B*. The calculations are given here:

Year	*Operating Cost*	*Depreciation*	*Total Cost*
3	2,300	1,250 (= 2,500 – 1,250)	3,550
4	2,900	650 (= 1,250 – 600)	3,550
5	3,600	200 (= 600 – 400)	3,800

Since the cost incurred in keeping the equipment *A* in the third and the fourth years is less than the average cost for equipment *B*, the replacement should be done after 2 years.

Example 11.8 The Mini Transport Company owns three mini buses, two of which are two years old while the third one is only a year old. Each of these buses was purchased for Rs 80,000. The company contemplates replacing the three buses by two full-sized buses, each such bus containing 50% more seating capacity than a mini bus. Cost of each is Rs 1,20,000. Using the following data on the running costs and the resale value of both the types of buses, state whether the mini buses be replaced by the full-sized buses. If not, state why? If yes, state when?

Year	For a Mini Bus		For a Full-sized Bus	
	Running Cost	*Resale Value*	*Running Cost*	*Resale Value*
1	3,000	70,000	3,400	100,000
2	3,600	61,000	3,900	92,000
3	4,800	55,000	4,700	86,000
4	5,000	49,000	5,800	81,000
5	8,000	32,000	7,200	76,000
6	11,200	20,000	9,000	66,000
7	15,000	10,000	12,000	54,000
8	20,000	5,000	16,000	40,000

We shall first calculate the minimum average cost for each type of the buses. This is given in Tables 11.10 and 11.11.

For mini bus

TABLE 11.10 Determination of Average Cost

Year	M_t	Cum. M_t	$C-S$	$T(n)$	$A(n)$
1	3,000	3,000	10,000	13,000	13,000
2	3,600	6,600	19,000	25,600	12,800
3	4,800	11,400	25,000	36,400	12,133
4	5,000	16,400	31,000	47,400	11,850*
5	8,000	24,400	48,000	72,400	14,480
6	11,200	35,600	60,000	95,600	15,933
7	15,000	50,600	70,000	120,600	17,229
8	20,000	70,600	75,000	145,600	18,200

For full-sized bus

TABLE 11.11 Determination of Average Cost

Year	M_t	*Cum.* M_t	$C - S$	*T(n)*	*A(n)*
1	3,400	3,400	20,000	23,400	23,400
2	3,900	7,300	28,000	35,300	17,650
3	4,700	12,000	34,000	46,000	15,333
4	5,800	17,800	39,000	56,800	14,200
5	7,200	25,000	44,000	69,000	13,800*
6	9,000	34,000	54,000	88,000	14,667
7	12,000	46,000	66,000	112,000	16,000
8	16,000	62,000	80,000	142,000	17,750

Thus, the minimum average cost for a mini bus is Rs 11,850 p.a. and Rs 13,800 p.a. for a full-sized bus. However, these two should not be compared directly because three mini buses are equivalent to two full-sized buses. Thus,

$$\text{average cost for all 3 mini buses} = 11,850 \times 3 = 35,550$$
$$\text{average cost for 2 large buses} = 13,800 \times 2 = 27,600.$$

Clearly, then, it is prudent to replace the mini buses by the full-sized buses.

To decide the timing at which the replacement be done, we shall first find the total yearly costs for the new buses. The year in which the average cost of the new buses shall be lower than the total cost of maintaining and running the old ones shall be the year when the replacement should be done. The calculations are given in Table 11.12.

TABLE 11.12 Determination of Yearly Cost of a Mini Bus

Year	*Running Cost*	*Depreciation*	*Total Cost*
1	3,000	10,000	13,000
2	3,600	9,000	12,600
3	4,800	6,000	10,800
4	5,000	6,000	11,000
5	8,000	17,000	25,000
6	11,200	12,000	23,200
7	15,000	10,000	25,000
8	20,000	5,000	25,000

Total cost for next year would be: $2 \times 10,800 + 12,600 = \text{Rs } 34,200$ (since two of the buses would be running in the third year and the third one in the second year).

Total cost for the subsequent years shall be:

$$2 \times 11{,}000 + 10{,}800 = \text{Rs } 32{,}800$$
$$2 \times 25{,}000 + 11{,}000 = \text{Rs } 61{,}000 \text{ ... etc.}$$

Since the total average cost of running the two buses is Rs 27,600 whereas in the years to come the cost of owning and running the old buses would be greater than this, the conclusion is that the buses should be replaced immediately.

Example 11.9 A firm has a machine whose purchase price is Rs 20,000. Its maintenance cost and resale price at the end of different years are as given here:

Year	:	1	2	3	4	5	6
Maintenance Cost	:	1,500	1,700	2,000	2,500	3,500	5,500
Resale Price	:	17,000	15,300	14,000	12,000	8,000	3,000

(a) Obtain the economic life of the machine and the minimum average cost.

(b) The firm has obtained a contract to supply the goods produced by the machine, for a period of 5 years from now. After this time period, the firm does not intend to use the machine. If the firm has a machine of this type that is one year old, what replacement policy should it adopt if it intends to replace the machine not more than once?

(a) The calculation of average cost is shown in Table 11.13.

TABLE 11.13 Determination of Optimal Replacement Interval

Year	M_t	*Cum.* M_t	$C - S$	*T(n)*	*A(n)*
1	1,500	1,500	3,000	4,500	4,500
2	1,700	3,200	4,700	7,900	3,950
3	2,000	5,200	6,000	11,200	3,733*
4	2,500	7,700	8,000	15,700	3,925
5	3,500	11,200	12,000	23,200	4,640
6	5,500	16,700	17,000	33,700	5,617

Since the minimum average cost, Rs 3,733, corresponds to year 3, the economic life of the machine is 3 years.

(b) To consider the situation regarding the one-year old machine, we shall first obtain the yearly cost of keeping this machine in second, third, ... year of its life. This is given in Table 11.14.

TABLE 11.14 Determination of Yearly Costs for Existing Machine

Year	*Maint. Cost*	*Depreciation*	*Total Cost*	*Cum. Total Cost*
1	1,700	1,700	3,400	3,400
2	2,000	1,300	3,300	6,700
3	2,500	2,000	4,500	11,200
4	3,500	4,000	7,500	18,700
5	5,500	5,000	10,500	29,200

Alternate policies and the costs associated with each one are as follows:

Policy 1 : Keep old machine for the zero year and buy a new one for full 5 years.

Total cost = 0 + 23,200 = Rs 23,200

Policy 2 : Keep old machine for 1 year and buy a new one for 4 years

Total cost = 3,400 + 15,700 = Rs 19,100

Policy 3 : Keep old machine for 2 years and buy a new one for 3 years

Total cost = 6,700 + 11,200 = Rs 17,900

Policy 4 : Keep old machine for 3 years and buy a new one for 2 years

Total cost = 11,200 + 7,900 = Rs 19,100

Policy 5 : Keep old machine for 4 years and buy a new one for 1 year

Total cost = 18,700 + 4,500 = Rs 23,200

Policy 6 : Keep old machine for 5 years and don't buy a new one

Total cost = 29,200 + 0 = Rs 29,200

Therefore, policy 3 is the optimal policy.

Example 11.10 Consider Example 11.4 again. Suppose that there is another machine *Y* which costs Rs 8,000. It requires maintenance cost equal to Rs 200 in the first year, which increases by Rs 400 in each succeeding year. Assuming that other conditions are same, determine the optimal replacement time for machine *Y*. If a choice is to be made between *X* and *Y*, which one would be preferred?

The determination of the replacement period is shown in Table 11.15.

TABLE 11.15 Determination of Optimal Replacement Period

Year	M_t	*P V Factor*	*P V of* M_t	*Cum. P V of* M_t *+ Cost*	*Cum. P V Factor*	*Annualised cost*
1	200	1.0000	200.0	8,200	1.0000	8,200
2	600	0.9091	545.5	8,745.5	1.9091	4,581
3	1,000	0.8264	826.4	9,571.9	2.7355	3,499
4	1,400	0.7513	1,051.8	10,623.7	3.4868	3,047
5	1,800	0.6830	1,229.4	11,853.1	4.1698	2,843
6	2,200	0.6209	1,366.0	13,219.1	4.7907	2,759
7	2,600	0.5645	1,467.7	14,686.8	5.3552	2,743*
8	3,000	0.5132	1,539.6	16,226.4	5.8684	2,765
9	3,400	0.4665	1,586.1	17,812.5	6.3349	2,812

Optimal replacement interval for the machine, therefore, is 7 years.
Since minimum average cost for X (= 2,042) is less than the corresponding cost for Y (= 2,743), machine X would be preferable to machine Y.

Example 11.11 The cost of a new car is Rs 10,000. Compare the optimum moment of replacement assuming the following cost informations.

Age of Car (years)	*Repair cost per Year*	*Salvage Value at the End of the Year*
1	Rs 5,000	Rs 8,000
2	Rs 10,000	Rs 6,400
3	Rs 10,000	Rs 5,120

Assume that the repairs are made at the end of each year only if the car is to be retained and are not necessary if the car is to be sold for its salvage value. Also assume that the rate of discount is 10%.

(CA, November, 1982)

Here there are three replacement policies to be compared viz. replacement on a yearly, two-yearly and three-yearly basis. For this purpose, we shall consider a six-year period (LCM of 1, 2 and 3) so that each one of the policies would have completed a certain number of cycles. The calculations are shown in Table 11.16.

TABLE 11.16 Determination of Optimal Replacement Policy

Year	*Purchase Cost*	*Maintenance Cost*	*Salvage Value*	*Net Cost*	*P V Factor*	*P V of Cost*
Yearly Replacement Policy						
0	10,000	–	–	10,000	1.0000	10,000.0
1	10,000	–	(8,000)	2,000	0.9091	1,818.2
2	10,000	–	(8,000)	2,000	0.8264	1,652.8
3	10,000	–	(8,000)	2,000	0.7513	1,502.6
4	10,000	–	(8,000)	2,000	0.6830	1,366.0
5	10,000	–	(8,000)	2,000	0.6209	1,241.8
6	–	–	(8,000)	(8,000)	0.5645	(4,516.0)
				Total Cost		13,065.4
Two-Yearly Replacement Policy						
0	10,000			10,000	1.0000	10,000.0
1		5,000		5,000	0.9091	4,545.5
2	10,000		(6,400)	3,600	0.8264	2,975.0
3		5,000		5,000	0.7513	3,756.5
4	10,000		(6,400)	3,600	0.6830	2,458.8
5		5,000		5,000	0.6209	3,104.5
6	–		(6,400)	(6,400)	0.5645	(3,612.8)
				Total Cost		23,227.5

(Contd.)

TABLE 11.16 Contd.

Year	Purchase Cost	Maintenance Cost	Salvage Value	Net Cost	P V Factor	P V of Cost
Three-Yearly Replacement Policy						
0	10,000			10,000	1.0000	10,000.0
1		5,000		5,000	0.9091	4,545.5
2		10,000		10,000	0.8264	8,264.0
3	10,000		(5,120)	4,880	0.7513	3,666.3
4		5,000		5,000	0.6830	3,415.0
5		10,000		10,000	0.6209	6,209.0
6			(5,120)	(5,120)	0.5645	(2,890.2)
				Total Cost		33,209.6

Since the total cost in respect of yearly replacement policy is the minimum, this is the optimal policy.

Example 11.12 A group of process plants in an oil refinery are fitted with valves. Over a period of time, the failure pattern of these 400 valves has been observed and it is as follows:

Month	:	1	2	3	4	5	6	7	8	Total
Number of valves failing	:	8	20	48	104	120	56	32	12	400

It costs Rs 100 to replace each valve individually. If all the valves are replaced at a time, it costs Rs 50 per valve. The maintenance department is considering following replacement policies:

(a) to replace all valves simultaneously at fixed intervals, in addition to replacing valves as and when they fail.
(b) to replace valves as and when they fail.

Suggest an optimal replacement policy. *(MBA, Bangalore, June, 1998)*

First the probability distribution of the lives of valves is obtained and the expected life is calculated. This is done below:

Life (months)	X_i	*No. of Valves failing*	p_i	p_iX_i
0–1	0.5	8	0.02	0.010
1–2	1.5	20	0.05	0.075
2–3	2.5	48	0.12	0.300
3–4	3.5	104	0.26	0.910
4–5	4.5	120	0.30	1.350
5–6	5.5	56	0.14	0.770
6–7	6.5	32	0.08	0.520
7–8	7.5	12	0.03	0.225
		400	Exp. Life	4.160

Using the probability values given in the fourth column, we calculate the number of replacements required to be done in different months. This is shown in Table 11.17.

TABLE 11.17 Determination of Expected Number of Replacements

Month	*Expected Number of Replacements*		N_i
1	N_0p_1	$= 400 \times 0.02$	$= 8.00$
2	$N_0p_2 + N_1p_1$	$= 400 \times 0.05 + 8 \times 0.02$	$= 20.16$
3	$N_0p_3 + N_1p_2 + N_2p_1$	$= 400 \times 0.12 + 8 \times 0.05 + 20.16 \times 0.02$	$= 48.80$
4	$N_0p_4 + N_1p_3 + N_2p_2 + N_3p_1$	$= 400 \times 0.26 + 8 \times 0.12 + 20.16 \times 0.05$ $+ 48.80 \times 0.02$	$= 106.94$
5	$N_0p_5 + N_1p_4 + N_2p_3 + N_3p_2 + N_4p_1$	$= 400 \times 0.30 + 8 \times 0.26 + 20.16 \times 0.12$ $+ 48.80 \times 0.05 + 106.94 \times 0.02$	$= 129.08$
6	$N_0p_6 + N_1p_5 + N_2p_4 + N_3p_3 + N_4p_2 + N_5p_1$	$= 400 \times 0.14 + 8 \times 0.30 + 20.16 \times 0.26$ $+ 48.80 \times 0.12 + 106.94 \times 0.05$ $+ 129.08 \times 0.02$	$= 77.43$
7	$N_0p_7 + N_1p_6 + N_2p_5 + N_3p_4 + N_4p_3 + N_5p_2 + N_6p_1$	$= 400 \times 0.08 + 8 \times 0.14 + 20.16 \times 0.30$ $+ 48.80 \times 0.26 + 106.94 \times 0.12$ $+ 129.08 \times 0.05 + 77.43 \times 0.02$	$= 72.69$
8	$N_0p_8 + N_1p_7 + N_2p_6 + N_3p_5 + N_4p_4 + N_5p_3 + N_6p_2 + N_7p_1$	$= 400 \times 0.03 + 8 \times 0.08 + 20.16 \times 0.14$ $+ 48.80 \times 0.30 + 106.94 \times 0.26$ $+ 129.08 \times 0.12 + 77.43 \times 0.05$ $+ 72.69 \times 0.02$	$= 78.72$

Now we can determine the average cost for different group replacement policies. This is shown in Table 11.18.

TABLE 11.18 Determination of Average Cost

Month	*No. of Replacements*		*Cost of Replacements*			
	Group	*Individual*	*Group*	*Individual*	*Total*	*Av. Cost*
1	400	8.00	20,000	800	20,800	20,800
2	400	28.16	20,000	2,816	22,816	11,408
3	400	76.96	20,000	7,696	27,696	9,232
4	400	183.90	20,000	18,390	38,390	9,598
5	400	312.98	20,000	31,298	51,298	10,260
6	400	390.41	20,000	39,041	59,041	9,840
7	400	463.10	20,000	46,310	66,310	9,473
8	400	541.82	20,000	54,182	74,182	9,273

The average cost is lowest against month 3. Thus, optimal replacement policy calls for group replacement of valves every 3 months, together with individual replacements as and when required. The cost involved is Rs 9,232 per month.

If only individual replacements are done:

$$\text{Average cost of replacements per month} = \frac{\text{No. of valves}}{\text{Average life of valves}} \times \text{Cost per replacement}$$

$$= \frac{400}{4.160} \times 100 = \text{Rs } 9{,}615$$

Hence, the policy of group replacements is better.

Example 11.13 A large computer installation contains 2,000 components of identical nature which are subject to failure as per probability distribution that follows:

Month End	:	1	2	3	4	5
% Failure to Date	:	10	25	50	80	100

Components which fail have to be replaced for efficient functioning of the system. If they are replaced as and when failures occur, the cost of replacement per unit is Rs 45. Alternatively, if all components are replaced in one lot at periodic intervals and individually replace only such failures as occur between group replacements, the cost of component replaced is Rs 15.

(a) Assess which policy of replacement would be economical.

(b) If group replacement is economical at current costs, then assess at what cost of individual replacement would group replacement be uneconomical.

(c) How high can the cost per unit in group replacement be to make a preference for individual replacement policy?

(a) From the given probability distribution, the distribution indicating the failures of components in each month is derived and provided and the average life of component obtained.

Month (X_i)	*Prob. of Failure* (p_i)	p_iX_i*
1	0.10	0.10
2	0.15	0.30
3	0.25	0.75
4	0.30	1.20
5	0.20	1.00
	Total	3.35

*This is assuming that failures are towards month-ends.

$$\text{Average Number of Replacements per month} = \frac{\text{No. of components}}{\text{Mean life of components}}$$

$$= \frac{2{,}000}{3.35} = 597$$

With the policy of individual replacements followed, the average cost per month would work out to be equal to $597 \times 45 = \text{Rs } 26{,}865$.

When the policy of group replacements, along with individual replacements, is followed, the cost and the optimal policy can be determined as follows. First we determine the expected number of failures in each month, as shown in Table 11.19.

TABLE 11.19 Determination of Expected Number of Replacements

Month (i)		*Expected No. of Replacements*	N_i
1	$N_0 p_1$	$= 2{,}000 \times 0.10$	= 200
2	$N_0 p_2 + N_1 p_1$	$= 2{,}000 \times 0.15 + 200 \times 0.10$	= 320
3	$N_0 p_3 + N_1 p_2 + N_2 p_1$	$= 2{,}000 \times 0.25 + 200 \times 0.15 + 320 \times 0.10$	= 562
4	$N_0 p_4 + N_1 p_3 + N_2 p_2 + N_3 p_1$	$= 2{,}000 \times 0.30 + 200 \times 0.25 + 320 \times 0.15 + 562 \times 0.10$	= 754.2
5	$N_0 p_5 + N_1 p_4 + N_2 p_3 + N_3 p_2 + N_4 p_1$	$= 2{,}000 \times 0.20 + 200 \times 0.30 + 320 \times 0.25 + 562 \times 0.15$ $+ 754.2 \times 0.10$	= 699.72

Now we can determine the average cost for different group replacement policies. This is shown in Table 11.20.

TABLE 11.20 Determination of Average Cost

Upto Month	*Ind. Replacements*	*Cost*		*Total Cost*	*Average Cost*
		Ind. Rep.	*Group Rep.*		
1	200	9,000	30,000	39,000	39,000
2	520	23,400	30,000	53,400	26,700
3	1,082	48,690	30,000	78,690	26,230
4	1,836.2	82,629	30,000	112,629	28,157
5	2,535.92	114,116	30,000	144,116	28,823

Since the average cost is lowest against month 3, the optimal interval between group replacement = 3 months. And, since the average cost is less than Rs 26,865 for individual replacement policy, the policy of group replacement is better.

(b) For a group replacement policy (GR policy) to be uneconomical, the average cost per month of this policy should be greater than the average cost per month of the policy of individual replacements (IR policy). If K be the cost of an individual replacement, we can determine its value in respect of each of the GR policies and then decide on the basis of the least of all the K values obtained.

Policy Group Rep. Every	*Condition*				*K Value*
	Av. Cost of GR Policy	>	*Av. Cost of IR Policy*		
1 month	(30,000 + 200 *K*)/1	>	597 *K*	→	$K < 75.57$
2 months	(30,000 + 520 *K*)/2	>	597 *K*	→	$K < 44.51$
3 months	(30,000 + 1,082 *K*)/3	>	597 *K*	→	$K < 42.31$
4 months	(30,000 + 1,836.2 *K*)/4	>	597 *K*	→	$K < 54.37$
5 months	(30,000 + 2,535.92 *K*)/5	>	597 *K*	→	$K < 66.80$

Since the least of the K values is 42.31, it follows that if the cost of individual replacement is anything smaller than Rs 42.31, the GR policy would be uneconomical.

(c) As in (b) a preference for IR policy implies that the average cost per month of a GR policy should be greater than the average monthly cost under this policy. With P as the per unit cost of item replacement under GR policy, we have,

Policy GR Every	Condition				P Value
1 month	$(2{,}000P + 9{,}000)/1$	>	26,865	→	$P > 8.93$
2 months	$(2{,}000P + 23{,}400)/2$	>	26,865	→	$P > 15.17$
3 months	$(2{,}000P + 48{,}690)/3$	>	26,865	→	$P > 15.95$
4 months	$(2{,}000P + 82{,}629)/4$	>	26,865	→	$P > 12.42$
5 months	$(2{,}000P + 114{,}116.4)/5$	>	26,865	→	$P > 10.10$

Since the largest of the P values is 15.95, it follows that per unit cost (of replacement on a group basis) of any value greater than Rs 15.95 would imply that an IR policy would be preferable to a GR policy.

KEY POINTS TO REMEMBER

- Generally, there are two types of replacement problems:
  (i) for items that deteriorate gradually, and
  (ii) for items that fail suddenly.
- For the items that deteriorate gradually, operating and maintenance costs steadily increase with passage of time, whereas depreciation per year decreases with time. In view of this behaviour of the two elements of cost, the problem in such cases is to determine at what point in time should an asset be replaced so that the total cost may be minimised.
- Assuming (i) the asset can be replaced at the same cost, (ii) there is no time-value of money, and (iii) replacements are done only at the year-ends (not during a year), the problem of obtaining the optimal replacement interval is solved as follows:
  (a) Find the cumulative operating/maintenance cost.
  (b) Find the total depreciation (i.e. Purchase Cost *minus* Resale Price).
  (c) Add (a) and (b) to get Total Cost.
  (d) Divide Total Cost by n, the number of years, to get average cost. The optimal period of replacement is the one corresponding to which average cost is the minimum.
- In case the time value of money is considered, the present value of the maintenance cost and resale values need to be obtained first. After this, the weighted average cost (using the present value factors as weights) is determined. The replacement in such a case is done at a point when it is found that the next period's cost is greater than the weighted average of the previous costs.
- For items that fail suddenly, we first obtain the expected number of failures per period of time. This is done by applying the appropriate probabilities. Once the individual replacements are ascertained, we consider the various alternate policies of group replacement and determine the cost implications of each one of them. The policy with the least average cost is chosen to be the appropriate one.

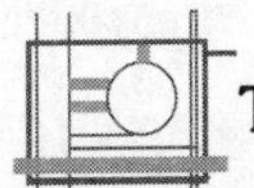

TEST YOUR UNDERSTANDING

Mark the following statements as T (*True*) or F (*False*).

1. The replacement theory is concerned with replacement of assets when they stop working.
2. For the assets that deteriorate in performance over time, the maintenance costs usually rise significantly with time.
3. While total depreciation of depreciable assets increases with passage of time, the successive years' depreciation usually declines.
4. The total cost of a piece of equipment over a given period of n years would equal Purchase price + maintenance cost for n years + value of the asset after n years.
5. Optimal replacement interval is the minimum time elapsing between the successive replacements.
6. The replacement interval calculations are based on the implicit assumption that the replacement of the asset is repetitive in nature.
7. An optimal replacement interval of six years indicates that the asset in question should be replaced at the beginning of every six years.
8. When the time value of money is considered, costs need to be discounted.
9. Timing of incurrence of costs is important when the time value of money is considered.
10. Even if it is decided in principle to replace an existing machine with a new one, the replacement should be postponed up to the point where the old machine can be run at a relatively lower cost.
11. Replacement models considered in the text are based on the assumption that replacements, individual or group, can be done only at fixed intervals of time.
12. The present value factors serve as the weights when time value of money is considered.
13. When time value of money is considered, weighted average costs are obtained and the optimal policy is "Do not replace if the next period's cost is less than the weighted average of the previous costs, and replace if the next period's cost is greater than the weighted average of the previous costs."
14. The models relating to replacement of depreciable assets are based on rather unrealistic assumption that future costs and resale prices of an equipment are predictable accurately.
15. No replacement policy can be laid for items that fail suddenly, because the failure time of each unit is uncertain.
16. The optimal group replacement interval is determined by the point where the sum of group replacement per unit of time and the cost of individual replacements is the minimum.
17. From the following probability distribution, it can be said that for a group of 1,000 bulbs, the number of bulbs expected to fail in the first week are $1{,}000 \times 0.10 = 100$ and $(1{,}000 - 100 =)\ 900 \times 0.15 = 135$ in the second week.

Week	:	1	2	3	4	5
Probability of failure	:	0.10	0.15	0.30	0.25	0.20

18. Expected cost of a replacement policy of individual replacements can be obtained as follows: Expected cost per time period = (No. of units/Expected life of asset) × Replacement cost per unit
19. For a hotel with 2,000 bulbs, the replacement policy of "group replacement at fixed intervals together with individual replacements during the intervening period" would always be inferior to the policy of "replacement of individual failures as and when needed".
20. Staff replacement policies can be easily formulated because people retire at known times.

EXERCISES

1. What is a replacement problem? Describe some important replacement situations and policies.
2. Briefly explain the costs which are relevant to decisions for replacement of depreciable assets. Illustrate their behaviour and explain how the optimal time for replacement of an asset can be determined. *(M Com, Delhi, 1993)*
3. In the replacement of items which deteriorate with time, prove that "Do not replace the item if the next period's cost is less than the weighted average of the previous costs."
4. How would you deal with replacement of items that fail completely and suddenly?
5. Explain briefly the difference in replacement policies of items which deteriorate gradually and items which fail completely.

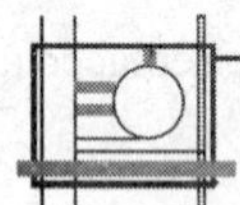

PRACTICAL PROBLEMS

1. An electro-mechanical equipment has a purchase price of Rs 7,000. Its running costs per year and resale values are given here:

Year	1	2	3	4	5	6	7	8
Running cost (Rs)	2,000	2,100	2,300	2,600	3,000	3,500	4,100	4,600
Resale value (Rs)	4,000	3,000	2,200	1,600	1,400	700	700	700

At which year is the replacement due?

2. Determine the optimum replacement interval of an equipment which costs Rs 5,200 and whose resale values and running costs are as follows:

Year	1	2	3	4	5	6	7
Resale value (Rs)	3,500	2,700	1,800	1,000	850	600	425
Running cost (Rs)	600	850	1,000	1,250	1,400	1,475	2,000

3. Determine the age at which the following type of machine be replaced:

Cost price = Rs 8,000
Operating costs = Rs 1,000 for the first year, increasing by Rs 500 every year
Resale value = Rs 4,000 for the first year, decreasing by Rs 500 every year

4. A firm is considering replacement of a machine, whose cost price is Rs 12,200; and the scrap value is only Rs 200. The maintenance costs are found as follows:

Year	1	2	3	4	5	6	7	8
Maintenance cost (Rs)	200	500	800	1,200	1,800	2,500	3,200	4,000

When should the machine be replaced?

5. The cost of a machine is Rs 6,500 and its scrap value is only Rs 500 whenever disposed. The maintenance costs are found to be as under on the basis of experience:

Year	1	2	3	4	5	6	7	8
Maintenance cost (Rs)	100	250	400	600	900	1,250	1,800	2,500

When should the machine be replaced to get maximum advantage? *(CA, November, 1994)*

6. Determine the optimal replacement period and the average yearly cost associated with it, in respect of a machine whose details are as follows:

   Cost of the machine including installation charges = Rs 4,000

   Maintenance charges expected to be paid, $M_t = 200 + 100t + 10t^2$ (for $t = 1, 2, ..., n$)

   Resale value at the year end, $S_t = 4000 - 500t$ (for $t = 1, 2, ..., 7$)

7. A truck owner finds from the past records that the maintenance cost of a truck (whose purchase price is Rs 3,00,000) during the first 8 years of its life and the resale prices at the end of each year, are as follows:

Year	*Maintenance cost (Rs)*	*Resale price (Rs)*
1	36,000	200,000
2	48,000	150,000
3	60,000	100,000
4	72,000	80,000
5	84,000	70,000
6	96,000	60,000
7	108,000	50,000
8	120,000	40,000

Suggest optimal policy for the replacement of the truck. *(M Com, Delhi, 1990)*

8. A truck owner finds from his past experience that the maintenance costs are Rs 200 for the first year and then increase by Rs 2,000 every year. The cost of the truck type *A* is Rs 9,000. Determine the best age at which to replace the truck. If the optimum replacement is followed, what will be the average yearly cost of owning and operating the truck? Truck type *B* costs Rs 10,000. Annual operating costs are Rs 400 for the first year and then increase by Rs 800 every year. The truck owner has now the truck type *A* which is one year old. Should it be replaced with *B* type, and if so, when? *(ICWA, June, 1985)*

9. The maintenance costs and the resale price of machine *M* whose purchase price is Rs 12,000 are given as:

Year	1	2	3	4	5	6	7
Maintenance cost (Rs)	2,600	3,000	3,400	4,000	4,700	5,600	6,600
Resale price (Rs)	7,000	4,500	3,250	2,600	2,400	2,400	2,400

   (i) Suggest optimal period for the replacement of the machine.

   (ii) When this machine is 2 years old, another machine *N*, which is a new model of machine *M*, is available. The optimal period for replacement of this machine, *N*, is 4 years, with an average cost of Rs 4,700. Should we change machine *M* with *N*? If so, when?

10. A company is intending to purchase a small machine in the near future. It is not necessary that a new machine be purchased and for economic considerations estimates of selling and purchasing prices from new to 5 years' old have been obtained. These are displayed in the following table along with information on the estimates of maintenance costs.

Year end	:	1	2	3	4	5
Maintenance cost	:	192	244	308	384	496
Selling price	:	3,360	2,880	2,440	2,080	1,800
Purchase price	:	3,720	3,160	2,680	2,280	1,880

The cost of a new machine is Rs 4,000. What would be the best replacement policy?

(M Com, Delhi, 1986)

11. A machine shop has a press which is to be replaced as it wears out. A new press is to be installed now. Further, an optimum replacement plan is to be found for the next 7 years after which the press is no longer required. The following data is given:

Year	*Cost of installing a new press at the beginning of a year (Rs)*	*Salvage value at the end of year (Rs)*	*Operating cost during the years (Rs)*
1	200	100	60
2	210	50	80
3	220	30	100
4	240	20	120
5	260	15	150
6	290	10	180
7	320	0	230

Find an optimum replacement policy and the corresponding minimum cost. *(ICWA, June, 1986)*

12. A fleet owner finds from the past records that the costs per year of running a truck whose purchase price is Rs 60,000 are given as follows:

Year	*Running costs (Rs)*	*Resale price (Rs)*
1	10,000	30,000
2	12,000	15,000
3	14,000	7,500
4	18,000	3,750
5	23,000	2,000
6	28,000	2,000
7	34,000	2,000
8	40,000	2,000

At what age is a replacement due?

Now, suppose that the truck owner in this case has three trucks, two of which are two years old and the third one is one year old. He is considering a new type of truck with 50% more capacity than one of the old ones at a unit price of Rs 80,000. He estimates the running costs and resale price for the new truck will be as follows:

Year	*Running cost (Rs)*	*Resale price (Rs)*
1	12,000	40,000
2	15,000	20,000
3	18,000	10,000
4	24,000	5,000
5	31,000	3,000
6	40,000	3,000
7	50,000	3,000
8	60,000	3,000

Assuming that the loss of flexibility due to fewer trucks is of no importance, and that he will continue to have sufficient work for three of the old trucks, what should his policy be? *(MBA, Delhi, 1981)*

13. For a machine, the following data are available:

Year	*0*	*1*	*2*	*3*	*4*	*5*	*6*
Cost of spares (Rs)	–	200	400	700	1,000	1,400	1,600
Salary of maintenance staff (Rs)	–	1,200	1,200	1,400	1,600	2,000	2,600
Losses due to breakdown (Rs)	–	600	800	700	1,000	1,200	1,600
Resale value (Rs)	12,000	6,000	3,000	1,500	800	400	400

Determine the optimum period for replacement of the above machine. *(CA, November, 1987)*

14. A piece of equipment costs Rs 7,500 initially and requires Rs 400 to be spent on its maintenance in the first year. The maintenance cost would increase by Rs 500 per year in each of the subsequent years. Determine the optimal replacement for the machine when (i) future costs are not discounted, and (ii) future costs are discounted at the rate of 10% p.a.

15. A manufacturer has to decide between two alternative machines M_1 and M_2, about which pertinent information is given below:

	M_1	M_2
(a) Cost	Rs 5,000	Rs 2,500
(b) Maintenance cost	Rs 800 p.a. for years 1, 2, ..., 5, increasing by Rs 200 every year thereafter.	Rs 1,200 p.a. for years 1, 2, ..., 6, increasing by Rs 200 every year thereafter.
(c) Scrap value	Nil	Nil
(d) Cost of Capital	(to be used as discount rate) 10% p.a.	

Determine,

(i) Optimal replacement period of M_1 and of M_2.

(ii) Which of the two is a better alternative?

16. Goodlite Company has installed 2,000 electric bulbs of a certain brand. The company follows the policy of replacing the bulbs as and when they fail. Each replacement costs Rs 2. The probability distribution of the life of the bulbs is as given here:

Life of bulb (weeks)	:	1	2	3	4	5
% of bulbs	:	0.10	0.30	0.45	0.10	0.05

Determine the cost/week of the replacement policy in the long run.

17. The management of a large hotel is considering the periodic replacement of light bulbs fitted in its rooms. There are 500 rooms in the hotel and each room has 6 bulbs. The management is now following the policy of replacing the bulbs as they fail at a total cost of Rs 3 per bulb. The management feels that this cost can be reduced to Re 1 by adopting the periodic replacement method. On the basis of information given here, evaluate the alternatives and make a recommendation to the management.

Month of use	:	1	2	3	4	5
% of bulbs failing by that month	:	10	25	50	80	100

(CA, November, 1984)

18. An electronic machine has a component that is subject to failure and needs replacement immediately when this occurs. A total of 200 such machines are used in a company. The cost of each replacement is Rs 80 per item and for group replacement it is Rs 5,000 for all the 200 machines. Records of previous failure rates indicate the following pattern:

10 per cent would fail during the first month
20 per cent would fail during the second month
20 per cent would fail during the third month
30 per cent would fail during the fourth month
20 per cent would fail during the fifth month.

Assuming that group replacement can be carried out only at the month end, what group replacement policy would you suggest? Consider an 8-month period for cost evaluation.

19. A large computer system has 100 identical components which are subject to random failure over time. The failure pattern is given as follows:

Month end	:	1	2	3	4	5	6
Failure to date (percentage)	:	12	28	50	75	90	100

Failed components have to be replaced immediately for the effective functioning of the computer. If they are replaced individually, as and when they fail, then the cost per replacement is estimated as Rs 4. Alternatively, if all the components are replaced in one lot at periodical intervals, then the cost per unit is estimated as Rs 1.50.

(i) Determine the more economical of the two policies of complete individual replacement or periodic group replacement.

(ii) Would the above decision remain valid if the unit cost in group replacement is given to be Rs 2? Justify your answer. *(DMS, Mumbai, 1992)*

20. The distribution of life of a special type of bulb used by a company is as follows:

Life in months	:	1	2	3	4	5	6
Proportion of bulbs failing	:	0.08	0.12	0.20	0.30	0.20	0.10

The cost of replacing an individual bulb when it fails is Rs 5 while the cost when replaced on a group basis is Rs 2 per bulb. The company has one thousand bulbs of this type installed.

(a) Calculate average life of a bulb.

(b) Determine if the policy of group replacement is superior to the policy of individual replacement. Consider the periods of 1 month, 2 months, and 3 months, for replacement.

21. An electronic assembly has 200 units of a particular chip. The failure of any of these causes the assembly to stop functioning. Accordingly, any failed component need to be immediately replaced. Based on the past data, the mortality pattern of this chip is found to be as follows:

Month	:	1	2	3	4	5
% failures by the month end	:	10	30	50	80	100

The cost of chip replacement is Rs 7 per failure. If all of the chips are replaced simultaneously, the cost works out to be Rs 3 per unit. Currently, the company is following the policy of replacing chips as and when they fail. It is contemplating to switch over to group replacement policy (involving group replacements at regular invervals and individual replacements as and when necessary). Find the optimal replacement interval and cost, and suggest whether it is advisable to change the policy.

(M Com, Delhi, 2004)

22. Perfect Machine Tools Ltd. has 40 lathes in service. The cost of repairing a breakdown is Rs 1,000 per lathe. A preventive maintenance programme costing Rs 300 per lathe per month can prevent the breakdowns. Records indicate that the possibility of a breakdown is as follows:

Months until breakdown	:	1	2	3	4	5	6	7	8
Prob. of breakdown	:	0.10	0.05	0.10	0.15	0.20	0.25	0.10	0.05

(a) What is the expected breakdown cost/month?

(b) If preventive maintenance is followed, how often should the lathes be serviced?

23. An electronic assembly contains 1,600 units of a particular component, failure of any of them stops the functioning of that assembly and hence the failed components need immediate replacement when the failure occurs.

The mortality pattern of these components based on the past data is as follows:

Month end	:	1	2	3	4	5	6
% failures by month end	:	5	25	45	70	85	100

The cost of the component replaced under the presently followed individual replacement policy is Re 1 each, and the cost of disruption is Rs 9 per failure. If all the components are replaced concurrently, the cost per replacement including disruption cost works out to be Rs 7.

Decide whether it would be worthwhile to switch over group replacement policy from the existing individual replacement policy and, if so, find the optimal replacement period and cost.

(DMS, Mumbai, 1995)

24. The administrator for a large mid-western city continually seeks ways to reduce costs without cutting service. The administrator has asked all department heads to review their operations to determine if cost saving procedures can be implemented.

The Department of Streets is responsible for the proper functioning of the city's computerised traffic control system including the replacement of 50,000 bulb units in the traffic lights. The department has kept detailed records regarding the failure rate of the bulb units over the past 18 months. The pattern of bulb failures that has been experienced is as follows:

Failure occurs within quarter of replacement	*Probability*
First	0.1
Second	0.3
Third	0.6

The Department of Streets has been replacing the bulb units as they have failed. The estimated cost to replace the bulb units, using this procedure, exclusive of the cost of the bulb unit, is Rs 5.40 per unit. The manager of the Department of Streets is considering replacing all of the bulb units at once, for example, at the beginning of every quarter, plus replacing bulbs as they fail. The manager estimates that the cost to replace all bulb units at once would be Rs 1.40 per unit, exclusive of the cost of the bulb. The cost to replace each unit as it failed would still be Rs 5.40 per unit.

Each bulb costs Re 1 regardless of the replacement procedure used.

(a) Calculate the annual bulb and replacement cost to the Department, if the present policy of replacing the bulb units as they fail is continued.

(b) Calculate the estimated annual bulb and replacement cost if all bulb units are replaced when they fail *and all bulb units are replaced* on a regular basis at the beginning of (i) every quarter, (ii) every second quarter, (iii) every third quarter. Show your calculations for all the three alternatives.

(c) Why would there be such a large difference between the two estimated replacement costs per unit (replace as failure occurs, Rs 5.40; replace all at once, Rs 1.40)? Explain your answer.

(CMA, Adapted)

25. A computer has a large number of components all of which have to be fully working and failures have to be replaced to maintain operational efficiency. Past observations have shown the following mortality pattern. If the component is replaced individually, then cost of a single replacement is Rs 20, whereas if all components are replaced concurrently, the cost of a single component is Rs 7. Decide whether it would be worthwhile to have a group replacement policy and, if so, find the optimal replacement frequency.

Mortality Statistics: Observed Number-500 Components

End of week	1	2	3	4	5	6
No. of components servicing	455	375	250	75	15	0

(DMS, Mumbai, 1993)

26. An organisation is considering periodic replacement of 2,000 units of an item which follows sudden failure mechanism. It is now following a policy of replacing items as they fail at a cost of Rs 800 per item. The organisation feels that it can cut its per item replacement cost by Rs 150 using the periodic replacement method. By making use of the information given in the table below, evaluate these alternatives and make a recommendation to this organisation:

End of the month	1	2	3	4	5
Per cent of original items which fail	10	20	40	70	100
During the month	10	10	20	30	30

(CA, November, 1991)

27. Given that cost of failure replacement is 3 times compared to the cost of common preventive replacement of Rs 100 per item in a system of 1,000 items and the following data, suggest the strategy between failure replacement and common preventive replacement that should be followed for cost reduction.

Week	:	0	1	2	3	4	5
No. of items surviving at the beginning of the week	:	100	90	75	40	15	0

(ICWA, December, 1995)

28. Colour-Tech Ltd. has just started to hire out a total of 160 colour TV sets. According to market survey, the distribution of length of hire of sets is,

Length of hire (yrs.)	:	1	2	3	4
Percentage of hiring	:	25	40	25	10

(a) Determine the number of new rentals needed each year for the next four years to maintain the company's total rental at 160. Also determine the average length of the hire period and the average number of new rentals each year.

(b) In an attempt to reduce their administrative overhead, the company is launching an advertising campaign aimed at their existing customers so as to encourage them to hire TV sets for longer periods. The company would regard their campaign successful if the following distribution of hiring was achieved in the long run:

Length of hire (yrs.)	:	1	2	3	4	5
Percentage of hiring	:	10	20	40	15	15

If the administrative cost of arranging a hire is Rs 40, determine the maximum amount the company could spend on their advertisement campaign every year.

29. There are 100 machines of a certain type which need repairing as they fail. Each machine repaired costs Rs 400 if it is serviced on individual basis. On the other hand, if all the machines are serviced simultaneously, it costs Rs 20,000, averaging Rs 200 per machine. From experience, the mortality rates have been observed as:

Month	:	5	6	7	8	9
Probability of failure by month ending	:	0.10	0.40	0.80	0.90	1.00

Presently, the machines are repaired as and when they fail and, also, all the machines are serviced simultaneously, irrespective of their condition, at the end of every seventh month.

Determine if the company's policy is optimal. If yes, how? If not, what would the optimal policy be?

30. A public sector bank requires 400 clerks, 250 officers and 50 managers for a certain state. The employces are recruited at the age of 21 years as clerks. The promotions take place only from within on the basis of the length of service. You are given the following life table.

Age (years):	21	22	23	24	25	26	27	28	29	30
No. of employees in service:	1,000	640	560	450	400	355	300	205	190	182
Age (years):	31	32	33	34	35	36	37	38	39	40
No. of employees in service:	175	170	162	155	150	148	145	142	140	132
Age (years):	41	42	43	44	45	46	47	48	49	50
No. of employees in service:	125	118	112	105	101	100	95	90	88	85
Age (years):	51	52	53	54	55	56	57	58	59	60
No. of employees in service:	80	72	70	63	55	40	32	25	20	0

Determine

(a) how many employees should be recruited each year.

(b) at what age should the promotions take place.

APPENDIX 11A

Suppose that the costs incurred in respect of some equipment in time periods (of equal duration) 1, 2, 3, ..., n of the life of the equipment are, respectively, $C_1, C_2, C_3, ..., C_n$. Further, suppose that these costs are increasing monotonically. In addition, the costs are assumed to be incurred in the beginning of the respective time periods. If C be the initial cost of the new equipment and the discount rate be r, the discounted value K_n of all future costs associated with the policy of replacing the equipment in question after every n periods would be given by the following equations:

$$K_n = \left(C + C_1 + \frac{C_2}{(1+r)} + \frac{C_3}{(1+r)^2} + \cdots + \frac{C_n}{(1+r)^{n-1}}\right) + \left(\frac{C+C_1}{(1+r)^n} + \frac{C_2}{(1+r)^{n+1}} + \frac{C_3}{(1+r)^{n+2}} + \cdots + \frac{C_n}{(1+r)^{2n+1}}\right) + \cdots$$

$$= \left(C + \sum_{t=1}^{n} \frac{C_t}{(1+r)^{t-1}}\right) + \left(\frac{1}{(1+r)^n}\right)\left(C + \sum_{t=1}^{n} \frac{C_t}{(1+r)^{t-1}}\right) + \left(\frac{1}{(1+r)^{2n}}\right)\left(C + \sum_{t=1}^{n} \frac{C_t}{(1+r)^{t-1}}\right) + \cdots$$

$$= \left[C + \sum_{t=1}^{n} \frac{C_t}{(1+r)^{t-1}}\right]\left[1 + \frac{1}{(1+r)^n} + \frac{1}{(1+r)^{2n}} + \cdots\right]$$

The expression in the right bracket of the RHS being a convergent geometric series, we can express K_n as

$$K_n = \frac{C + \sum_{t=1}^{n} (C_t/(1+r)^{t-1}}{1 - [1/(1+r)]^n}$$

K_n represents the amount of money required to pay currently for all future costs of acquiring and operating the equipment when it is renewed every n periods. Naturally, if K_n is less than K_{n+1}, then replacing the equipment every n years is preferable to replacing it every $n + 1$ years. Thus, if replacing the equipment every n periods is to be the optimal policy, then $K_{n+1} - K_n > 0$, and $K_{n-1} - K_n > 0$ must hold.

Of the two inequalities, it can be shown that the inequality $K_{n+1} - K_n > 0$ is equivalent to $C_{n+1}/(1 - (1/(1 + r))) > K_n$. To show this, we proceed as follows:

Let $X = 1/(1 + r)$, so that $K_n = \dfrac{C + \sum_{t=1}^{n} C_t X^{t-1}}{1 - X^n}$

By substituting $n + 1$ for n, we get

$$K_{n+1} = \frac{C + \sum_{t=0}^{n+1} C_t X^{t-1}}{1 - X^{n+1}} = \frac{C + \sum_{t=1}^{n} C_t X^{-1} + C_{n+1} X^n}{1 - X^{n+1}}$$

$$= \frac{(1-X^n)K_n + C_{n+1}X^n}{1-X^{n+1}}$$

$$= \frac{1-X^n}{1-X^{n+1}}K_n + \frac{C_{n+1}X^n}{1-X^{n+1}}$$

Therefore,

$$K_{n+1} - K_n = K_n\left(\frac{1-X^n}{1-X^{n+1}} - 1\right) + \frac{C_{n+1}X^n}{1-X^{n+1}}$$

$$= \frac{K_n\left(X^{n+1} - X^n\right) + C_{n+1}X^n}{1-X^{n+1}}$$

If $K_{n+1} - K_n > 0$, then

$$[K_n(X^{n+1} - X^n) + C_{n+1}X^n] > 0$$

Since $X < 1$, and therefore $(1 - X^{n+1}) > 0$, by dividing the above expression by X^n, we get

$$K_n(X-1) + C_{n+1} < 0$$

Hence, $C_{n+1} > (1-X)K_n$ or $\dfrac{C_{n+1}}{1-X} > K_n$.

Thus,

$$\frac{C_{n+1}}{1-[1/1+r]} < K_n \qquad \text{QED}$$

Similarly, it can be shown that

$$\frac{C_n}{1-[1/(1+r)]} > K_{n-1}$$

Both these conditions (inequalities) must hold for K_n to be a minimum. They are both necessary and sufficient conditions (because of the assumption of monotonically increasing costs, $C_n < C_{n+1}$, for all n). The inequality $C_n/(1-X) < K_{n-1}$ (X being equal to $1/(1+r)$, as defined earlier) can be expressed as $C_n < (1-X)K_{n-1}$. Substituting the expression for K_{n-1}, obtained by substituting $n-1$ for n in the expression for K_n, we get

$$C_n < (1-X)\,\frac{C + C_1 + C_2X + C_3X^2 + \cdots + C_{n-1}X^{n-2}}{1-X^{n-1}}$$

or

$$C_n < \frac{(C + C_1) + C_2X + C_3X^2 + \cdots + C_{n-1}X^{n-2}}{1 + X + X^2 + \cdots + X^{n-2}} \qquad \text{(i)}$$

The expression on the RHS of this inequality is the weighted average of all costs upto and including period $n - 1$. The weights, $1, X, X^2, ..., X^{n-2}$ are the discount factors applied to the costs in each period.

Similarly, for the other inequality,

$$C_{n+1} > K_n(1-X)$$

or $$C_{n+1} > \frac{1-X}{1-X^n}\,(C + C_1 + C_2X + C_3X^2 + \cdots + C_nX^{n-1})$$

or $$C_{n+1} > \frac{(C + C_1) + C_2X + C_3X^2 + \cdots + C_nX^{n-1}}{1 + X + X^2 + \cdots + X^{n-1}} \tag{ii}$$

From these pairs of inequalities (i), and (ii) the following rules for cost minimisation may be derived:

(1) Do not replace if the next period's cost is less than the weighted average of the previous costs, and

(2) Replace if the next period's cost is greater than the weighted average of the previous costs.

CHAPTER

12

PERT and CPM

CHAPTER OVERVIEW

While dealing with a project like construction of a building, introducing a new product, or buying and installing a new machine, what comes to mind includes the various activities needed to be performed—project completion deadlines, availability and requirement of resources for various activities, project costs, project delay penalties, etc. In fact, undertaking any project requires breaking it down into constituent activities—establishing precedence relationships between them (that is, which activity will precede/succeed which other activity or activities), time and resources needed for each activity, availability of the resources and the costs involved in various activities and project supervision along with any penalties for late completion and cash rewards for early completion of the project. *Programme Evaluation and Review Technique* and *Critical Path Method* are the techniques which are available to a manager for getting answers to questions like the following:

- Given the activities to be done in respect of a project along with time needed for doing each one of them, how long will the project take to complete?
- What are the earliest and latest times for performing various activities?
- By what date can the work on a particular stage of the project be expected to be completed?
- What activies of the project can be delayed and by how much, while still meeting the project completion deadline?
- What are the chances of completing the project by a certain due date when time durations for the activities involved are not known for sure?
- How best can the limited amounts of resources available be allocated to the different activities so that the project can be completed at the earliest?
- How can the activities of the project be scheduled so that variations in the demand for one or more key resources may be reduced or even eliminated?
- How can the activities be planned so that the commitment for employing financial resources can be postponed as much as possible?

All in all, PERT and CPM are important and potent techniques which have wide practical applications. This chapter requires knowledge of the concepts of expected value, variance and calculation of probabililites using normal distribution. Also, it involves drawing of graphs and charts. You should master the skills of drawing network diagrams especially those involving dummy activities.

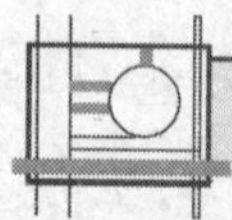

12.1 INTRODUCTION

Network analysis plays an important role in project management. By analysing a network, which is a graphic depiction of 'activities' and 'events', the planning, scheduling and control of a project becomes much easier. Programme Evaluation and Review Technique (PERT), and Critical Path Method (CPM) represent the two well-known network analysis techniques used to assist the managers in planning and controlling larger-scale construction projects, research and development, and so forth. Often, these projects are very large and complex involving performance of a host of tasks or jobs by several departments. Examples of such projects are construction of a residential complex; a shopping center; ships or aircraft repair projects; development of a new drug; implementation of a computer system from ordering and site preparation through installation and checkout, satellite mission development and countdown procedures; installation of a pipeline project; designing and laying of a new manufacturing plant, and the like. While working on such projects, large amounts of money, manpower and other equipment are involved. The techniques of PERT and CPM prove extremely valuable in assisting the managers in handling such projects and thus discharging their project management responsibilities.

The use of the PERT and CPM techniques is made in both planning and controlling of the projects. Planning in this context implies developing the overall layout of the project with estimates of the time and resources required and the detailed scheduling of the timing and sequence of various jobs to be performed. The control, on the other hand, takes place during the work on the project. Gradually as actual resource use and completion times are obtained, project management techniques can be used to reallocate, if necessary, the resources, according to the revised criticality rankings of the jobs remaining to be done. In particular, these techniques help the project managers to determine the expected project completion date; the scheduled start and completion time for the different activities comprising the project; the key activities of the project which must be completed at the scheduled time; the time period by which the non-key activities may be delayed without causing a delay in the completion of the whole project, etc.

Both the techniques use similar terminology and have the same general purpose, but they were developed independently of each other, during the late 1950s. PERT was developed and used in conjunction with the planning and designing of the Polaris Submarine System. It was credited with saving substantial time and cost on Polaris, which caused its use to be made mandatory on all the significant development projects undertaken for and on behalf of the Department of Defence of the United States. The CPM, on the other hand, was developed by the Du Pont Company and the Univac Division of Remington Rand Corporation as a device to control the maintenance of chemical plants.

While both PERT and CPM have similarities in terms of concepts and methodology, a basic difference exists between the two techniques. PERT is useful for analysing project scheduling problems in which the completion time of the different activities, and therefore the whole project, is not certain. It thus emphasises the uncertainties of the completion times of the activities. On the other hand, CPM is most appropriately used in projects in which the activity durations are known with certainty. Not only the amount of times needed to complete the various facets of the project but also the amounts of resources required for performing each of the activities are assumed to be known. This technique is basically concerned with obtaining the trade-offs between the project duration and cost. Thus, whereas variation in the project time is inherent in the projects where PERT is used, the time is systematically varied (using additional resources) where CPM is employed. In essence, then, while PERT is probabilistic in nature and as such is used more in research and development projects, the CPM is a deterministic technique and thus finds application mostly in the construction projects.

As seen earlier, widely diverse kind of projects can be analysed by the techniques of PERT and CPM. In fact, they are suitable for any situation where:

(a) The *project* consists of well-defined collection of *activities* or tasks.
(b) The activities can be started and terminated independently of each other, even if the resources employed on the various activities are not independent.
(c) The activities are ordered so that they can be performed in a technological sequence. Thus, precedence relationships exist which preclude the start of certain activities until others are completed. For instance, road levelling cannot start unless the road bed is laid.

We now proceed to discuss the techniques to provide answers to the types of questions stated earlier. The initial step in both of these is to portray the given project graphically by means of a network which provides the basic tool for analysis. We shall first consider the construction of networks and principles underlying them. Next we see how to determine which of the activities of a given project are *critical*, and, therefore, lie on the *critical path*, together with the scheduling times of various activities. It will be followed by the questions of time-cost trade-off and scheduling of activities when resources required for their performance are in limited supply. Finally, we discuss the PERT and PERT/cost.

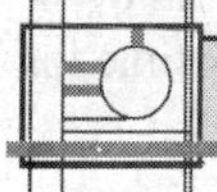

12.2 PERT/CPM NETWORKS

As mentioned earlier, a project can be viewed as a set of activities or jobs (also called *tasks* or *operations*) that are performed in a certain sequence determined logically or technologically. Therefore, the initial step in PERT/CPM project scheduling process is the determination of all specific activities that comprise the project and their interdependence relationships. Let us take a simple example. Suppose that a new machine is required by a department for which budget approval is needed. It is known that the use of a new machine necessitates employment of an operator who would be trained for operating this machine. The operator can be hired as soon as the proposal of buying the machine is cleared, and trained on a similar machine in the training division of the company. Once the machine is installed and the worker trained, the trial production can commence. In this project, the various activities required to be performed, along with the time needed for their execution, are given in Table 12.1.

TABLE 12.1 List of Activities and Precedence Relationships

Activity	*Description*	*Duration (weeks)*	*Immediate Predecessor(s)*
A	Obtain the budget approval	2	–
B	Obtain the machine	5	*A*
C	Hire the operator	1	*A*
D	Install the machine	1	*B*
E	Train the operator	6	*C*
F	Produce the goods	1	*D, E*

Note that this table contains additional information in the column headed *Immediate Predecessor(s)*. The immediate predecessors for a particular activity are those that must be completed immediately before this may start. For example, from the information contained in the table we can start on activity *A*—obtaining the budget approval—at any time because this is the first activity and it does not depend upon the completion of any prior activities. However, activities *B* and *C* cannot start before *A* is completed, *D* and *E* require, respectively, *B* and *C*, while for *F* to start, *D* and *E* must have been completed.

Once the activities comprising a project, as also the interdependency relationship among them, are clearly identified, they can be portrayed graphically by a *network* or an *arrow diagram*. Because the project planning function begins with a list of all the activities and their precedence relationships, the network can be constructed through the use of a series of *arrows* and *nodes*, thus conveniently expressing the sequential nature of the project.

Each of the activities that make up a project consumes time and resources and has a definable beginning and ending. The arrows, also called *arcs*, in a network represent the various activities of the project. Along with each arrow is given the description and the time estimate of the particular activity it is representing. For example, an activity *A* requiring 3 days is shown represented in Figure 12.1. The circles at the beginning and at the end of the arrow represent the *nodes*, or the *events*, of beginning and completion, respectively, of the activity in question. The events are points in time and can be considered as milestones of a project. The difference between activities and events should clearly be noted. Whereas an activity is a recognisable part of a project, involving mental or physical work and requiring time and resources for its completion; an event connotes an accomplishment occurring at an instantaneous point in time which neither requires any time for itself nor consumes any resources.

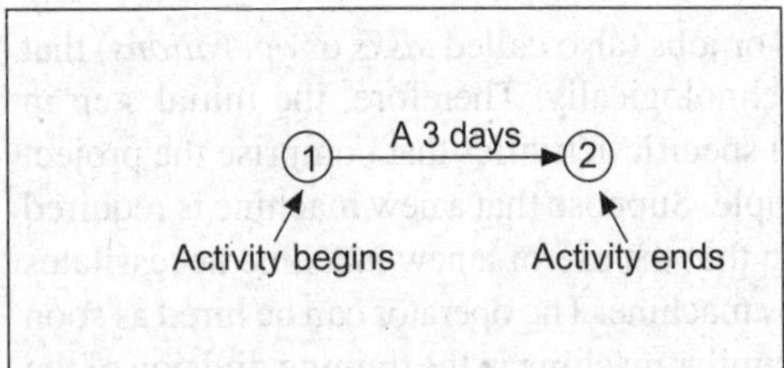

Figure 12.1 *Presentation of an Activity*

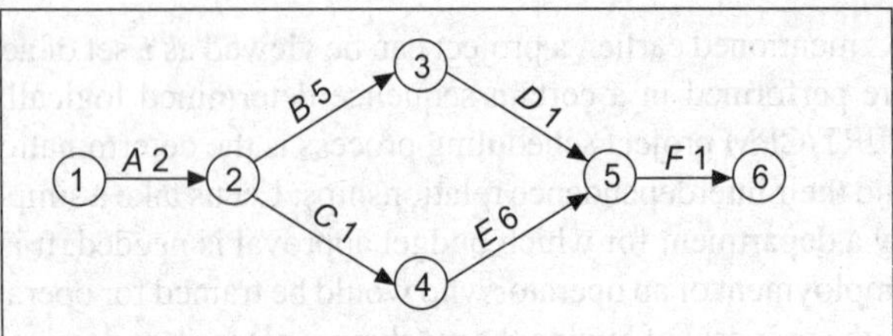

Figure 12.2 *Arrow Diagram*

Figure 12.2 depicts the network in respect of the illustrative example of machine installation. In this figure, the activities listed in Table 12.1 and their predecessor relationships are represented.

12.2.1 Rules of Network Construction

There are a number of concepts and ground rules in connection with the handling of events and activities of a project network that should be followed. This helps to develop a correct structure of the network. These are discussed here.

(a) Each defined activity is represented by one and only one arrow in the network. Therefore, no single activity can be represented more than once in the network.

(b) Before an activity can be undertaken, all activities preceding it must be completed. Thus, a network should be developed on the basis of logical or technical dependencies between various activities of the project. The discipline of networking requires that the project be considered in a thorough and analytic manner and the predecessor-successor relationships between the various activities clearly laid.

(c) The arrows depicting various activities are indicative of the logical precedence only. The length and bearing of the arrows are of no significance.

(d) The arrow direction indicates the general progression in time. The arrow head represents the point in time at which the 'activity completion' event takes place, while the arrow tail represents the point in time at which the event 'activity start' occurs. The events marking the start of activities are called *tail-events* while those marking the completion of activities are known as *head-events*.

(e) When a number of activities terminate at one event, it indicates that no activity emanating from that event may start unless all activities terminating there have been completed. For example, in the Figure 12.3 (a), activities *B*, *C*, and *D* must be completed before undertaking the activity *E*. It may be noted, however, that it only implies that they should be completed before *E* begins, and not that they must be completed simultaneously. Similarly, in part (b) of the figure, activity *A* should be completed before activities *B*, *C*, and *D* can be started. The events of the completion of the activity *A* marks the milestone for the start of the activities *B*, *C*, and *D*.

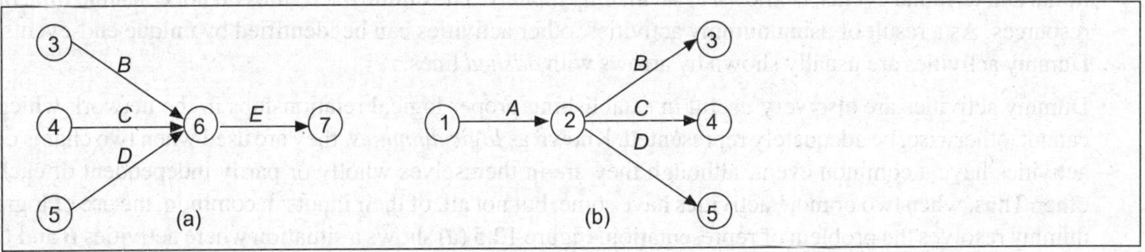

Figure 12.3 *Events and Activities*

(f) Events are identified by numbers. An effort should be made to have each event identified by a number higher than that allotted to the event immediately preceding. In assigning numbers to the events, care should be taken that there is no duplication of event numbers in a network.

(g) The activities are identified by the numbers of their starting and the ending events. They are expressed as *i-j*, where *i* represents the starting event, or the initial node, number, and *j* represents the ending event, or the terminal node, number. Notice that each of the activities shown in the network segments in Figure 12.3 can be identified by distinct node numbers. Thus, the activities *B, C, D* and *E* in part (a) of the figure can be expressed as 3–6, 4–6, 5–6, and 6–7, respectively. Event 3 represents the beginning of the activity *B* while event 6 represents the 'completion of activities *B, C* and *D*, and the beginning of the activity *E*.'

(h) A network should have only one initial and one terminal node.

(i) An event which represents the joint completion of more than one activity is known as a *merge event*, while an event which portrays the initiation of more than one activity is called the *burst event*. The event 6 in the Figure 12.3 (a) is an example of a *merge event* and event 2 in part (b) of the figure demonstrates a *burst event*.

(j) Parallel activities between two events, without intervening events, are prohibited. Thus two or more activities cannot be identified by the same beginning and ending events. By implication, any two events should not be directly connected with more than one arrow. When two or more parallel activities in a project have the same head and tail events, *dummy activities* are needed in constructing the network. To illustrate, in Figure 12.4 we have a situation in which both the activities *A* and *B* have the same end events. Of course, it is incorrect to represent the activities *A* and *B* as shown in part (i) because 1–2 can be used to represent either *A* or *B*. This problem is resolved by introducing a dummy activity, as shown in parts (ii) and (iii) of the figure. As a result of introducing the dummy activity, the activities *A* and *B* can be identified distinctly as 1–2 and 1–3 respectively (or as 1–3 and 1–2 respectively).

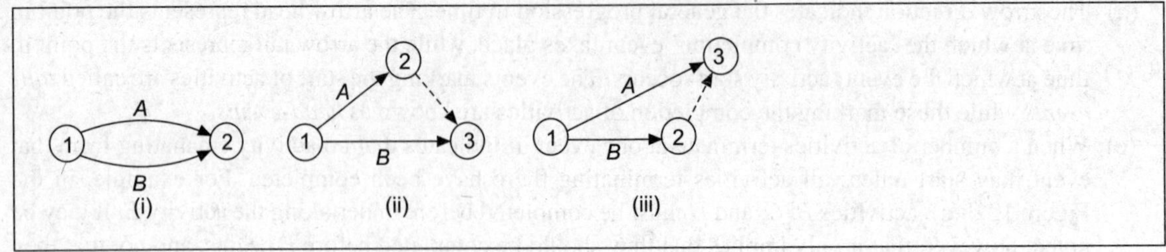

Figure 12.4 *Using a Dummy Activity*

A dummy activity of the type demonstrated just now, which is used when two or more activities have same initial and terminal events, is known as an *identity dummy*. The dummy activities do not consume time or resources. As a result of using dummy activities, other activities can be identified by unique end-events. Dummy activities are usually shown by arrows with *dashed* lines.

Dummy activities are also very useful in establishing proper logical relationships in the network which cannot, otherwise, be adequately represented. Known as *logic dummies,* they are used when two chains of activities have a common event, although they are in themselves wholly or partly independent of each other. Thus, when two or more activities have some, but not all, of their inputs in common, the use of logic dummy resolves the problem of representation. Figure 12.5 (a) shows a situation where activities *B* and *C* both must precede activity *D*, whereas activity *B* only need to precede activity *E*. Similarly, suppose activity *P* depends on activity *A*, activity *Q* depends on activities *A* and *B*, and activity *R* depends on activities *B* and *C*. We can represent this situation as shown in part (b) of the figure.

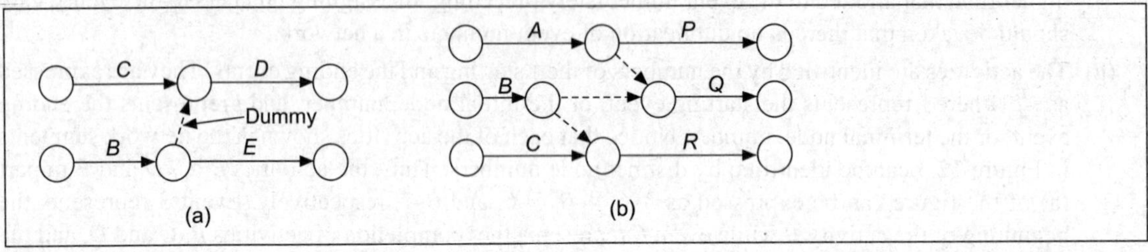

Figure 12.5 *Using Logic Dummies*

Thus, dummy activities, that are drawn only to adequately and correctly represent the given precedence relationships, are needed when (i) two or more activities in a project have identical immediate predecessor and successor activities, or (ii) two or more activities have some (and not all) of their immediate predecessor activities in common.

It may be mentioned that the use of dummies provides initially a convenient means of depicting the dependency relationships between the activities. However, only a minimum as may be needed to draw the network should be used. An efficient network contains a minimum number of dummy activities required to portray the correct precedence relationships between the activities of a given project. Thus, while constructing an arrow diagram, the dummy activities may be included liberally and then on completion of the diagram the redundant ones eliminated. In particular,

(i) If a dummy activity is the only activity emanating from a node, it can be eliminated. In the network shown in Figure 12.6 (a), the dummy activity DM_1 is redundant and can be eliminated, and the network redrawn.

(ii) If a dummy activity is the only activity converging to its final node, it can be eliminated. For example, the activity DM_1 is only a single activity going into its ending node, in Figure 12.6 (b). Thus, it can be eliminated. Here in fact we can eliminate either of the activities DM_1 and DM_2, but not both of them as it would violate the rule (j) in discussion.

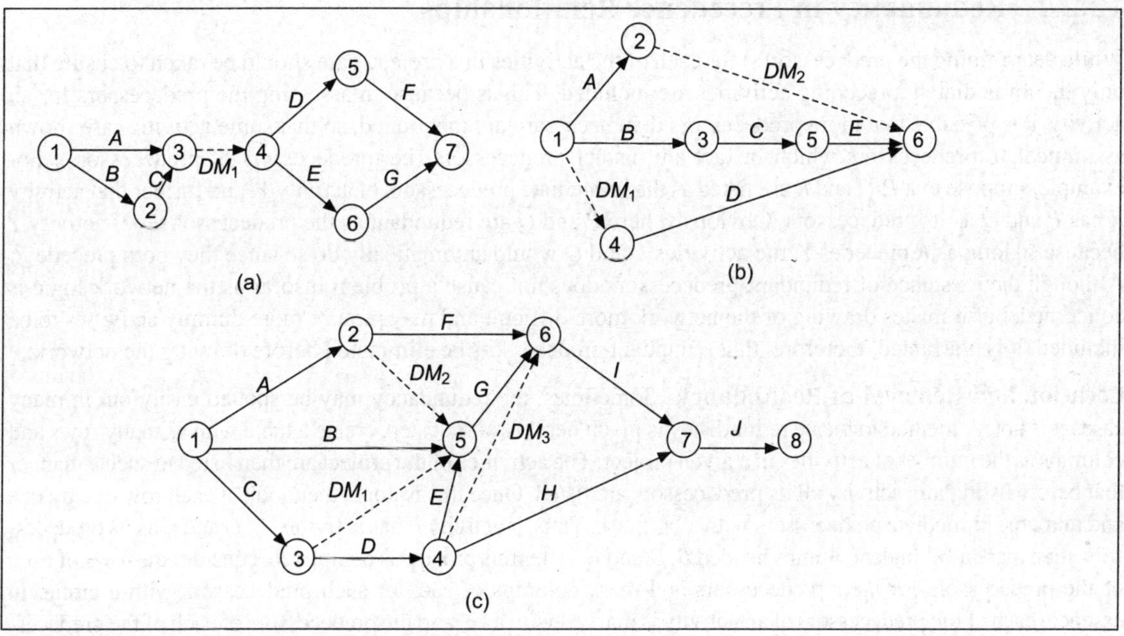

Figure 12.6 *Redundant Dummy Activities*

(iii) The dummy activities which portray predecessor relations that are already implied by other activities (regular or dummy) may be removed as being redundant (see more details on redundancy in the next section). Thus, activities DM_1 and DM_3 in the network shown in part (c) of the figure are redundant. For instance, activity DM_1 suggests that activity *G* requires performace of *C* and *E* for it. But since *E* has *D* as its immediate predecessor, preceding which is the activity *C*, therefore, to state that *E* precedes *G* implies that *C* precedes it.

(k) Looping is not permitted in a network. Thus, if *A* precedes *B*, and *B* precedes *C*, then *C* cannot precede *A*. In developing a network, it must be ensured that loops are not present. In Figure 12.7 activities *E*, *I* and *J* form a loop. Activity *E* requires that activities *C* and *J* should be completed before it begins. But *J* depends on *I* which in turn depends on *E*. Thus, activity *E* can never start because it is dependent on itself!

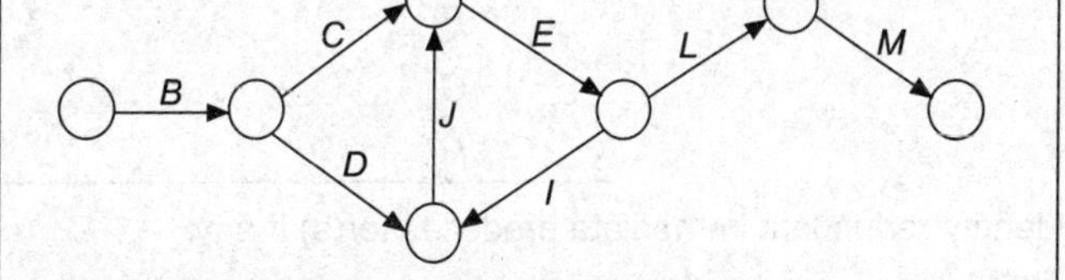

Figure 12.7 *Appearance of a Loop in the Network*

The appearance of a loop in a network suggests a fault in the logic, so that the definition of one or more of the dependency relationships is not valid. In case of a loop, precedence relationships, or dependencies, should be redefined to be related correctly.

12.2.2 Redundancy in Precedence Relationships

While determining the predecessor(s) for each of the activities in a project, care should be taken to ensure that only the immediately preceding activities are included. This is because in assessing the predecessors for an activity, it is possible that more predecessors than necessary are mentioned, so that some activities are shown as immediate predecessors which in fact are distant predecessors (i.e. predecessors of predecessors). For example, suppose that *P, Q* and *R* are listed as the immediate predecessors of activity *Y*, and further that activity *R* has *P* and *Q* as its predecessors. Obviously, here *P* and *Q* are redundant as the predecessor set of activity *Y* because so long as *R* precedes *Y*, the activities *P* and *Q* would automatically do so since they both precede *R*. Although the existence of redundant predecessors does not cause a problem insofar as the network logic is concerned, but it makes drawing of the network more difficult and may require more dummy activities to be included. It is warranted, therefore, that redundant predecessors be eliminated before drawing the network.

Location and Removal of Redundancy Sometimes, the redundancy may be spotted easily but in many cases it is not. A method to locate redundancy is given here. As a first step, create a table with as many rows and columns as the number of activities in a given project. The activities of the project are then listed in such a manner that before listing any activity all its predecessors are listed. Once this is done, then look at each row one by one and mark the immediate predecessors with a '×' mark. Thus, if activity *F* has activities *B, D* and *E* as its predecessors, then mark a '×' under columns headed *B, D* and *E*. After this process is completed, consider the rows of each of the predecessors for *their* predecessors and mark columns of each of such predecessors with a circle. In essence, each of the predecessors of an activity is marked with a '×' and the predecessors of each of the predecessors are marked with a circle. Finally, focus on the circled ×'s—they indicate redundancy.

Example 12.1 The following precedence relationships are given for a project consisting of eight activities:

Activity	*Immediate Predecessor(s)*
A	–
B	*A*
C	*A*
D	*A, C*
E	*B, D*
F	*B, D, E*
G	*B, D*
H	*E, F, G*

Identify redundant immediate predecessor(s) if any.

The activities are shown in Table 12.2, both vertically and horizontally. First of all, immediate predecessors, as given, are shown with ×'s opposite each of the activities. The first row, for activity *A*, is obviously blank since *A* has no predecessor. Then, in each row, all predecessors, marked with '×' are considered for their predecessors, both immediate and distant, and indicated in that row with circles.

TABLE 12.2 Redundancy in Precedence Relationships

Activity				*Predecessors*				
	A	*B*	*C*	*D*	*E*	*F*	*G*	*H*
A								
B	×							
C	×							
D	⊗		×					
E	○	×	○	×				
F	○	⊗	○	⊗	×			
G	○	×	○	×				
H	○	○	○	○	⊗	×	×	

To illustrate, in row *D, A* and *C* are shown as predecessors. Since *C* has *A* as the predecessor, a circle is marked in column headed *A*, which causes the '×' to be circled. Similarly, in row *F*, activities *B, D* and *E* as shown marked by '×', being the predecessors. Now, *B* has *A* as the predecessor; *D* has *A* and *C* as the predecessors; *C* has *A* as the predecessor and *E* has *B* and *D* as the predecessors. Thus, a circle appears in column headed *A, C, B* and *D*, and it is found that the ×'s under *B* and *D* are circled. Similarly, other rows are considered. Now, each of the activities where the '×' is circled are redundant, and can be safely removed.

Accordingly, the precedence relationship in the given project may be restated as follows:

Activity	*Immediate Predecessor(s)*
A	–
B	*A*
C	*A*
D	*C*
E	*B, D*
F	*E*
G	*B, D*
H	*F, G*

Example 12.2 Given the following information on a small project: *A* is the first activity of the project and precedes the activities *B* and *C*. The activity *D* succeeds both *B* and *C* whereas only *C* is required to start activity *E*. *D* precedes *F* while *G* succeeds *E*. *H* is the last activity of the project and succeeds *F* and *G*. Draw a network based on this information.

On the basis of the given information, the precedence relationships are provided here, and the network is shown in Figure 12.8.

Activity	*Immediate predecessor*
A	–
B	*A*
C	*A*
D	*B, C*
E	*C*
F	*D*
G	*E*
H	*F, G*

The dummy activity 3–4 has been introduced here because the activities *C* and *B* both precede *D* while only activity *C* precedes the activity *E*.

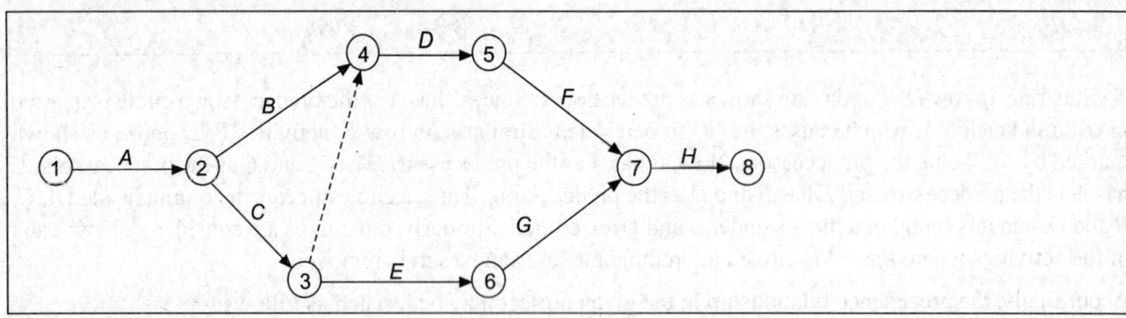

Figure 12.8 *Network*

Example 12.3 Draw an arrow diagram showing the following relationships:

Activity	:	*A*	*B*	*C*	*D*	*E*	*F*	*G*	*H*	*I*	*J*	*K*	*L*	*M*	*N*
Immediate prede.	:	–	–	–	*A, B*	*B, C*	*A, B*	*C*	*D, E, F*	*D*	*G*	*G*	*H, J*	*K*	*I, L*

The arrow diagram depicting these relationships is shown in Figure 12.9.

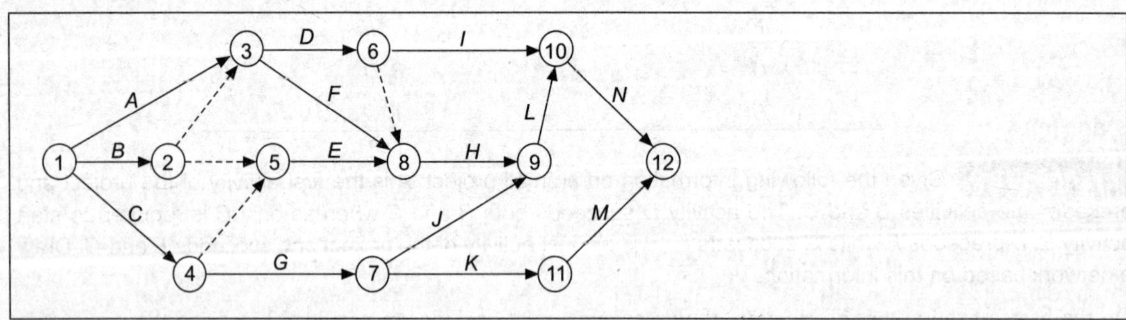

Figure 12.9 *Arrow Diagram*

The use of the dummies may carefully be noted here. In particular, the dummy activity 2–5 is necessary here because if it is eliminated, node 5 becomes the ending node of activity *B* and the initial node of activity *E*, implying that *D* and *F* require all *A*, *B*, and *C* to be completed before their start, which is not the case. Inclusion of this activity thus enables us to present the precedence relationships in a correct manner.

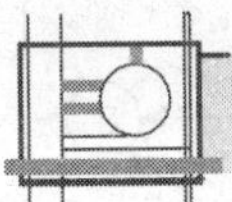

12.3 NETWORK ANALYSIS

A project network provides a means to derive a lot of information about the project involved. After the network plan is completed and activity times are known, we analyse it to obtain answers to questions like when the various activities can be scheduled to be performed, how long it will take the project work to be completed, and how much cushion is available for performing the activities in the sense that performance of which of the activities can be delayed and by how much, and which ones cannot be delayed to meet project deadlines. Such analysis is based on the implicit assumption that all resources needed for performing various activities are available in required amounts at the needed times. We consider this analysis first. Later, the question of resources is considered in detail.

12.3.1 Scheduling the Activities: Earliest and Latest Times

After the project network plan is completed and activity times are known, we consider the questions how long the project would take to complete and when the activities may be scheduled. The answers to these questions are provided by an arrow diagram and the time duration of the various activities. These computations involve a forward and a backward pass through the network. The *forward pass* calculations yield the *earliest start* and the *earliest finish* times for each activity, while the *backward pass* calculations render the *latest allowable start* and the *latest finish* times for each activity. We shall demonstrate the calculation of earliest start, earliest finish, latest start and latest finish times of various activities of a project with the help of the following example.

Example 12.4 Information on the activities required for a project is as follows:

Name	:	*A*	*B*	*C*	*D*	*E*	*F*	*G*	*H*	*I*	*J*	*K*
Activities Node	:	1–2	1–3	1–4	2–5	3–5	3–6	3–7	4–6	5–7	6–8	7–8
Duration (days)	:	2	7	8	3	6	10	4	6	2	5	6

Draw the network and calculate the earliest start (*ES*), earliest finish (*EF*), latest start (*LS*), and latest finish (*LF*) times of each of the activities.

The network corresponding to the given information is contained in Figure 12.10. Also given are the earliest and the latest start and completion times for the activities.

Determination of ES and EF Times: Forward Pass Calculations As mentioned earlier, the purpose of forward pass is to obtain the earliest start, *ES*, and earliest finish, *EF*, times. The *ES* time indicates the earliest that a given activity can be scheduled and *EF* indicates the time which the activity can be completed, at the earliest. For this purpose, to begin with, a value of 0 (zero) is assigned to the initial event of the project. Thus, each of the activities initiated from the starting node of the network is assumed to start at time 0 (the beginning of the day 1, say). The earliest finish time for each activity is obtained by adding the time duration of the activity to its earliest start time. In the example network, the activities *A*, *B*, and *C* are all the first activities of the project and, therefore, the *ES* time for each one of them is set equal to 0. Since the activity *A* has a duration of 2 days, its earliest finish time equals $0 + 2 = 2$. Similarly, *EF* times for *B* and *C* would be 7 and 8, respectively.

Now, all the subsequent activities are assumed to start as soon as possible—that is, as soon as all of their respective predecessor activities are completed. For a given activity, the *ES* would be taken as the maximum of the *EF*'s of activities preceding this activity. Thus, the *ES* time for an activity leaving a particular node equals the largest value of the *EF* times for all activities entering that node. In the example, the *ES* for activity *D*, 2-5, would equal 2, the *EF* of the preceding activity 1–2 because it is the only predecessor activity; whereas for the activity 5–7, the *ES* time would be taken to be 13, because of two activities terminating at event 5, being 2–5 whose *EF* time is 5, and 3–5 which has *EF* time of 13. The *ES* and *EF* times for various activities are written in brackets on the top of arrows, following the description and time duration of the activities. They are also tabulated and presented in Table 12.3. It is clear that the earliest that the events 8 may be reached is 22.

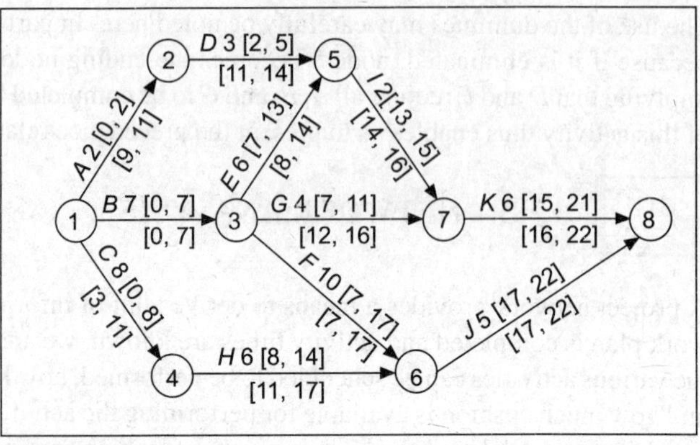

Figure 12.10 *Determination of ES, EF, LS and LF*

Determination of the LS and LF Times: Backward Pass Calculations The idea of the backward pass is to compute the latest allowable times of starting and finishing, *LS* and *LF*, of each of the activities of a project. The computations in the backward pass proceed in a similar manner as, and are the mirror image of, the forward pass calculations. The term 'latest allowable' means that the project's terminal event is desired to occur on or before some arbitrarily scheduled time. In view of this, the latest time is assigned to the ending event of the project and then the calculations for the latest start and finish times of various activities are made by rolling backward. These times would ensure that if the activities are scheduled by these, then the project would be completed by the scheduled time.

In case no specific time is given for the completion of the project, the terminal node of the project is assigned the latest of *EF* times of the activities merging into it. We shall use this concept in all of our problems in this chapter. In our example under consideration, the activities 7–8 and 6–8 are the terminal activities with *EF* times of 21 and 22 respectively. Thus, we assign a time 22 to the event 8. This implies that both these activities would be assigned their *LF* times equal to 22. The *LS* time for an activity being equal to its *LF* time minus its duration, for 7–8 the *LS* would be 22 – 6 = 16, while for 6–8 it would be 22 – 5 = 17.

In respect of others, the latest (allowable) finish time for an activity would be set as equal to the smallest, or earliest, of the latest allowable start times of its successor activities. Thus, the *LF* times of activities 3–6 and 4–6 would be taken as equal to 17, the *LS* of the only succeeding activity; whereas for the activity 1–3, the *LF* shall be taken to be 7, the lowest of the *LS* times of the successor activities: 3–5 : 8; 3–7 : 12; and 3–6 : 7. The *LS* and *LF* times for each of the activities are shown beneath the arrow representing it.

The earliest and latest start and finish times for various activities are also given in Table 12.3.

In addition to depicting the earliest and the latest timings for each of the activities, the earliest and latest times of various events may be determined as shown in Figure 12.11, in which the network of the example in consideration is reproduced. Notice that each circle representing an event is divided into three parts containing the event number, its earliest time and its latest time, as shown in the index. These times are obtained as discussed here.

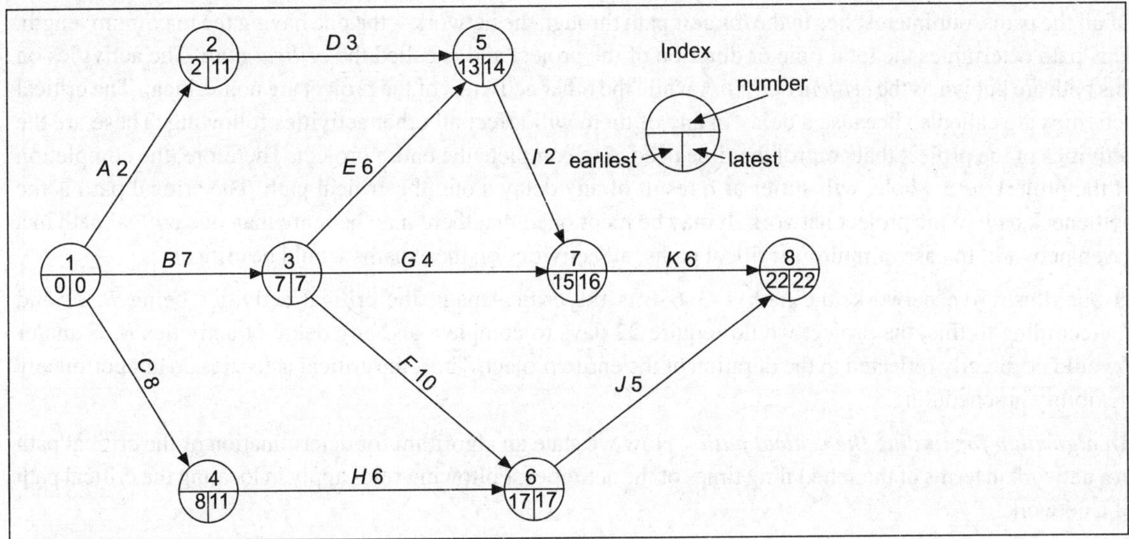

Figure 12.11 *Determination of Event Times*

For the activities emanating from a given event, the *ES* time would be given by the earliest time of it and the *LS* time by the latest time of it. In the forward pass, a 0 would be taken as the earliest time for the initial event of the project and then for each subsequent event, the earliest time would be taken as the latest of the *EF* time of the activities concluding on that event. Similarly, the terminal event of the project would be assigned the latest time equal to its earliest time (or other time if it is given and desired). Then, rolling back, the events are assigned the latest times. If only one activity starts from the node representing a given event, then the latest time for the event is taken to be the difference between the latest time of the head event of this activity and the activity duration time. In case, however, more than one activity starts from this node, then the minimum of such differences, as mentioned above, would be taken as the latest time for the event. For instance, for event 3, the differences are: 14 – 6 = 8; 16 – 4 = 12; and 17 – 10 = 7. Therefore, 7 is taken as the latest time for this event.

12.3.2 Determining the Critical Path

After having considered the determination of the earliest and the latest scheduled times for various activities, we now proceed to calculate the minimum time required for the completion of the whole project. In order to arrive at the duration of the project, we have to analyse the network and determine what is known as its *critical path*. A *path* is a set of continuous series of activities through the network that lead from the initial node of the network to its terminal node. For example, the connected activities defined by the nodes 1–2–5–7–8 in the figure form a path which consists of activities *A*, *D*, *I*, *K*. The 'length' of this path is 13 days—which is the aggregate of the time duration of the activities *A*, *D*, *I*, and *K*. Since all the paths are to be traversed in order to complete the project, we should find the lengths of all of them. For the network in reference, all the paths and their lengths are as follows:

Path	*Activities*	*Length*
1–2–5–7–8	*ADIK*	13
1–3–5–7–8	*BEIK*	21
1–3–7–8	*BGK*	17
1–3–6–8	*BFJ*	22
1–4–6–8	*CHJ*	19

Of all the paths, our interest lies in the *longest* path through the network—the one having the maximum length. This path determines the total time or duration of the project and is called the *critical path.* The activities on this path are known as the *critical activities* while the other activities of the project are non-critical. The critical activities are called so because a delay in any of them will affect all other activities following. These are the activities of the project that control the time needed to complete the entire project. Therefore, the completion of the project, as a whole, will suffer as a result of any delay along the critical path. The critical path is the bottleneck path in the project network. It may be mentioned that there may be more than one critical path in a given network. In case of multiple critical paths, all activities on these paths would be critical.

In our illustrative network, the path 1–3–6–8 is the critical path, the critical activities being *B*, *F*, and *J*. According to this, the project would require 22 days to complete and any delay of activities *B*, *F* and/or *J* would be directly reflected in the duration of the entire project. Thus, the critical activities do not permit any flexibility in scheduling.

An algorithm for finding the critical path Now we state an algorithm for determination of the critical path in a network in terms of the scheduling times of the activities. Following rules apply in locating the critical path of a network.

1. If, for an activity, the *ES* time equals the *LF* time at the head of the arrow and if the *ES* time equals the *LF* time at the tail of the arrow, the activity is possibly a critical activity, lying on the critical path.

   This is the first *necessary* condition for criticality.
2. If the first condition is met and the difference between the *ES* time at the head of the arrow and the *ES* time at the tail of the arrow is equal to the duration of the activity, then the activity is critical and lies on the critical path. This represents a *sufficient* condition for a critical activity.

To illustrate, in the network in question, for the activity 3–6, the *ES* at head of the arrow (for the activity 6–8) is 17 as also is the *LF* here, while the *ES* at the tail of the arrow is 7, which coincides with the *LF* at this event (for the preceding activity 1–3). Thus, the first of the above stated conditions is met. Also, the difference between these two times, 17 – 7 = 10, is the same as the duration of the activity (condition 2). This is, therefore, a *critical activity.* On the other hand, for the activity 3–5, though the *ES* time at the tail is the same as the *LF* time (= 7), the *ES* time at the head is 13 while the *LF* is 14. This, therefore, is *not* a critical activity.

Example 12.5 Draw a network corresponding to the following information. Obtain the early and late start and completion times, and determine the critical activities.

Activity	:	1–2	1–3	2–6	3–4	3–5	4–6	5–6	5–7	6–7
Duration	:	4	6	8	7	4	6	5	19	10

The network corresponding to the given information, along with the earliest start and finish times and the latest start and finish times for the various activities, is given in Figure 12.12. From the figure it is clear that there are two critical paths: 1–3–5–7, and 1–3–4–6–7, each with a length of 29. The critical activities are those represented by 1–3, 3–4, 3–5, 4–6, 5–7, and 6–7.

In particular, note the activity 5–6. For this activity, the *ES* and *LF* at the head are both equal to 19, and the *ES* and *LF* at the tail are both equal to 10, which means that the first of the conditions for criticality is met. However, the difference between these two times, 19 – 10 = 9, is not the same as the duration of this activity, which equals 5. Thus, the second of the conditions is not satisfied. Hence, it is not a critical activity.

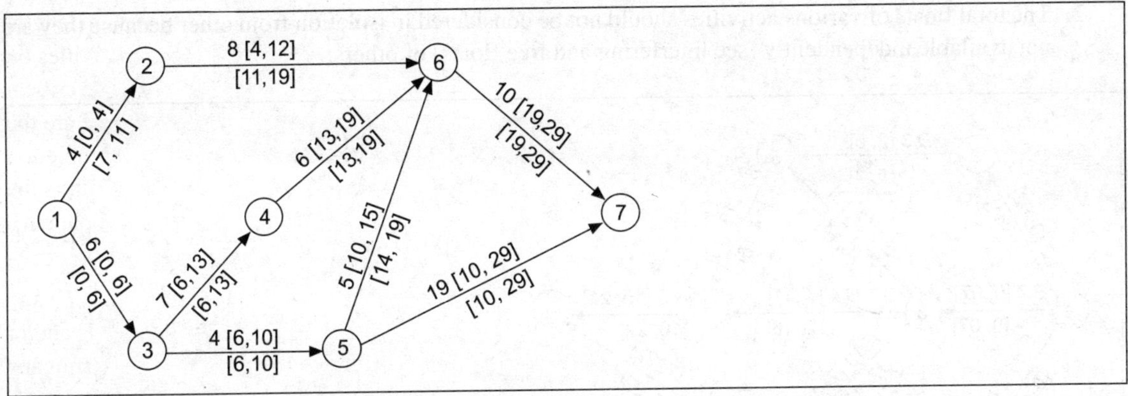

Figure 12.12 *Network, and ES, EF, LS and LF Times*

12.3.3 Calculation of Floats

It may be observed from these illustrations that for every critical activity in a network, the earliest start and the latest start times are the same. Since the critical activities cannot be scheduled later than their earliest schedule time, without delaying the project duration, they do not exhibit any flexibility in scheduling. However, while exercising control over time, resources, or cost it is necessary to know as to what flexibility exists for scheduling the non-critical activities of the project. The flexibility is seen in terms of the *float* or *slack* that any activity has. Needless to say, the critical activities of a project have no float. Work on non-critical activities, on the other hand, may be postponed for sometime without affecting the project duration.

Four types of floats can be identified. These are: total float, interfering float, free float, and independent float, which are discussed hereafter.

Total Float The total float of an activity represents the amount of time by which it can be delayed without delaying the project completion date. In other words, it refers to the amount of free time associated with an activity which can be used before, during or after the performance of this activity. It is equal to the difference between the total time *available* for the performance of an activity and the time *required* for its performance.

For activity 2–5 of the network of the Figure. 12.13 (reproduced from Figure 12.10), for example, the total available time (measured from the *ES* to the *LF* time) is 12 days while the time required to perform it is 3 days. Thus, the total float for this activity is 9 days. This is shown in Figure 12.14.

For any activity, *i–j*, the total float can be calculated as follows:

Total Float = Latest finish time – Earliest finish time

= Latest start time – Earliest start time

= Latest finish time – Earliest start time – Duration of the activity

The total float in respect of each of the activities of the network in question is contained in Table 12.3.

Two points may be noted here.

1. If an activity is delayed by more than the amount of total float available with it, it would delay completion of the project. For instance, if an activity in a project has 5 days of total float and the raw material needed for this activity is received with a delay of 7 days (from its earliest start), then other things remaining the same, the project would be delayed by 2 days.

2. The total floats of various activities should not be considered in isolation from other because they are not available independently (see interfering and free floats) of others.

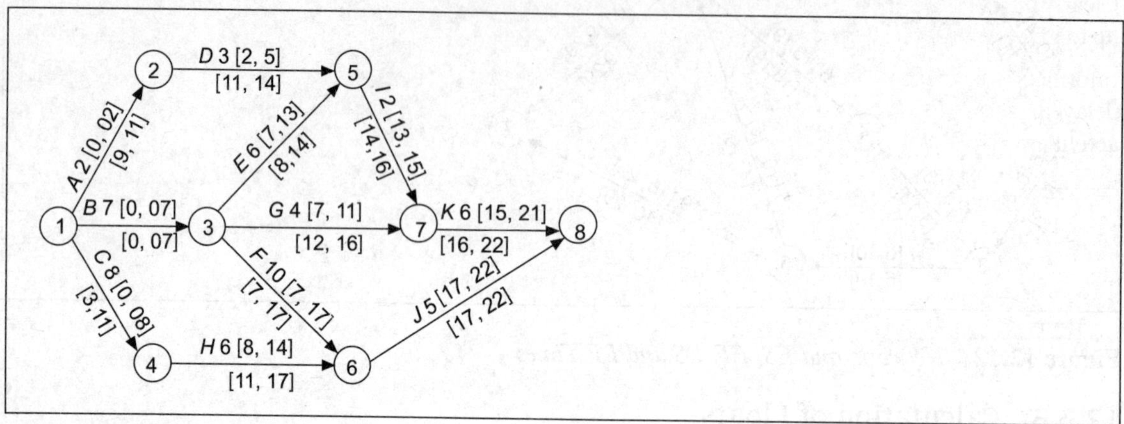

Figure 12.13 *Arrow Diagram*

Interfering Float Utilisation of the float of an activity may, and is likely to, affect the float times of the other activities in the network. The part of the total float which causes a reduction in the float of the successor activities is called *interfering float*. To illustrate, suppose the activity 1–4 in the Figure 12.13 is delayed by 2 days and it begins at $t = 2$ instead of at $t = 0$. It would finish then at $t = 10$. With $LS = 11$ for it, the float for 4–6 is reduced to 11 – 10 = 1 from 11 – 8 = 3. Thus, utilisation of float of 1–4 is seen to interfere into the float of its successor 4–6.

Formulated as the difference between the latest finish time of the activity in question and the earliest starting time of the following activity, or zero, whichever is larger; it indicates the portion of the activity float which cannot be consumed without affecting adversely the float of the subsequent activity or activities.

To continue, for the activity 2–5, the latest completion time is 14 while for the activity 5–7, its successor, the earliest starting time is 13. This gives the value of interfering float as equal to 14 – 13 = 1. For other activities the interfering float times are given in Table 12.3.

Free Float The free float is that part of the total float which can be used without affecting the float of the succeeding activities. Thus, it is that value of the float which is consumable when the succeeding activities (of the activity in question) can be started at their earliest starting times. To illustrate, consider activity 2–5 with a total float of 9 days. Suppose that it is delayed by 7 days so that it starts at $t = 9$ and completes at $t = 12$. The succeeding activity, however, can begin only at $t = 13$. Thus, the delay in 2–5 does not cause a delay in the start of its successor activity and hence does not interfere in its float. The free float is calculated as the earliest start time for the following activity minus the earliest completion time for this activity.

For activity 2–5, we have free float equal to 8 days because the earliest start of the succeeding activity 5–7 is 13 while the earliest finish time of this activity is 5.

Alternately, free float may be computed as follows. If the slack or float of an *event* is defined as the difference between the earliest and latest event times, we can calculate the slack of the head event and that of the tail event in respect of any activity. In that case,

$$\text{Free float} = \text{Total float} - \text{Head slack}$$

The slack of the event 5 in the network, which is the head event for activity 2–5, is equal to 14 – 13 = 1. Thus, free float for the activity 2–5 is 9 – 1 = 8.

Clearly then, the head event slack indicates the interfering slack, and the free float and the interfering float add up to give the total float. Figure 12.14 shows the free float in respect of the activity 2–5.

Information on the free float of an activity is useful in that it indicates how far the activity in question can be delayed beyond its earliest starting point without affecting the earliest start, and therefore, the total float of the activities following.

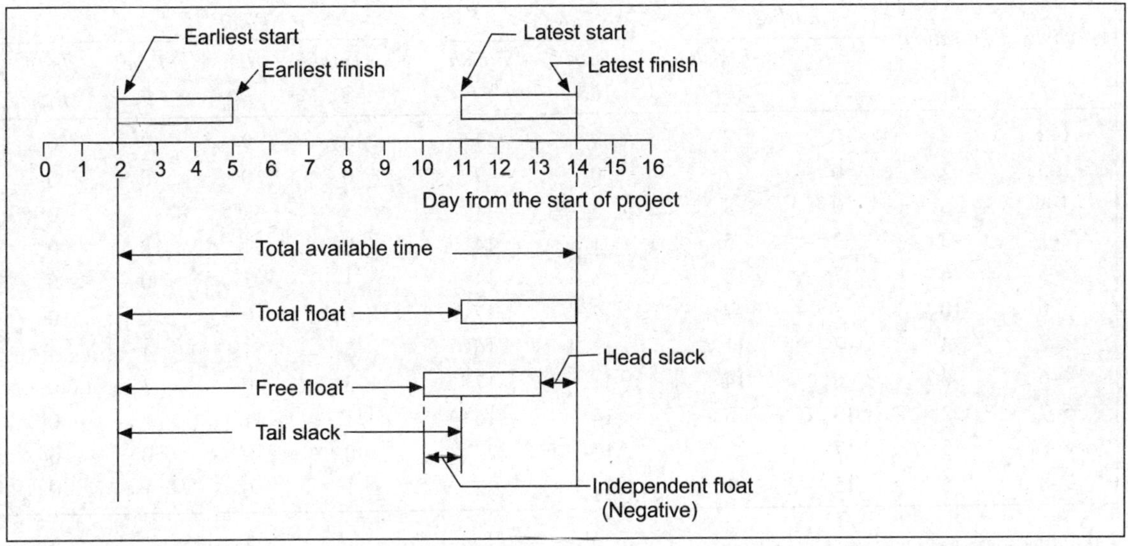

Figure 12.14 *Total, Free and Independent Floats*

Independent Float The independent float time of an activity is the amount of float time which can be used without affecting either the head or the tail events. It represents the amount of float time available for an activity when its preceding activities are completed at their latest and its succeeding activities begin at their earliest time—leaving the *minimum* time available for its performance. Any excess of this minimum time over the duration of the activity is termed as the independent float associated with it. The value of independent float is taken as follows:

Independent float = Earliest start time for the following activity – Latest finish time for the preceding activity – Duration of the present activity

Alternatively,

Independent float = Free float – Tail slack

Thus, the independent float is always either equal to or less than the free float of an activity. Besides, a negative value of the independent float may be obtained. If a negative value is obtained, the independent float is taken to be zero. This is obvious because where the activity duration equals or exceeds the minimum time available for its performance, there will be no independent float.

For the activity 2–5, we have

Earliest starting time for the following activity (5–7) = 13

Latest finishing time for the preceding activity (1–2) = 11

Duration of the activity = 3

Thus, Independent Float = 13 – 11 – 3 = – 1

Similarly, Free float = 8, Tail event slack = 11 – 2 = 9, and Independent float = 8 – 9 = – 1.

Being negative, the independent float is taken to be equal to zero.

TABLE 12.3 Determination of Earliest and Latest Times, and Floats

Activity	*Duration*	*Earliest*		*Latest*		*Float*			
		Start	*Finish*	*Start*	*Finish*	*Total*	*Int.*	*Free*	*Ind.*
i–j	t_{ij}	*ES*	*EF*	*LS*	*LF*	F_t	F_{int}	F_f	F_{ind}
1–2	2	0	2	9	11	9	9	0	0
1–3	7	0	7	0	7	0	0	0	0
1-4	8	0	8	3	11	3	3	0	0
2–5	3	2	5	11	14	9	1	8	0*
3–5	6	7	13	8	14	1	1	0	0
3–6	10	7	17	7	17	0	0	0	0
3–7	4	7	11	12	16	5	1	4	4
4–6	6	8	14	11	17	3	0	3	0
5–7	2	13	15	14	16	1	1	0	0
6–8	5	17	22	17	22	0	0	0	0
7–8	6	15	21	16	22	1	0	1	0

* It works out to be – 1. Hence, it is taken as equal to zero.

Symbolically, for an activity *i–j*, let

t_{ij} = duration of the activity,
ES_{ij} = earliest start of the activity,
EF_{ij} = earliest finish of the activity,
LS_{ij} = latest start of the activity, and
LF_{ij} = latest finish of the activity

We have,

$$\text{Total Float, } F_t = LS_{ij} - ES_{ij}$$

$$\text{Interfering Float, } F_{int} = \text{Max}\ \{0, LF_{ij} - ES_{jk}\}$$

$$\text{Free Float, } F_f = ES_{jk} - EF_{ij};\ \text{or } F_t - \text{Head Slack}$$

$$\text{Independent Float, } F_{ind} = \text{Max}\ \{0, ES_{jk} - LF_{hi} - t_{ij}\}$$

$$= \text{Max}\ \{0, F_f - \text{Tail Slack}\}$$

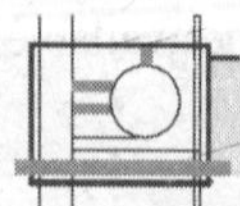

12.4 RESOURCE ANALYSIS AND ALLOCATION

In our discussion on the scheduling of activities in determining the scheduling timings, we have considered only the technological restrictions which lay that an activity in a project cannot start unless all its predecessors have been scheduled, and ignored the question of resources required for performing the various activities. Now we shall consider the question of resource requirements for the different activities, the availability of resources and their allocation to the activities, and some other problems. A feature common to all these problems is the recognition that the limitations on resources needed for carrying out the activities of a project can affect the scheduling of a project either because of their limited availability or by some other restrictions. Obviously, for our analysis, we assume that the resources required for each of the activities of a project and the resource restrictions are known.

The Question of Project Cost Linked with the time and resources required in completing a project is the question of costs. The execution of a project involves two types of costs:

Activity Direct Costs

The cost of materials, equipment and direct labour (payroll, overtime, hiring and firing costs) required in performing an activity is called its direct cost. If the activity in question is subcontracted and is being performed in its entirety by a contractor, then the activity direct cost is equal to the price of the subcontract. The sum total of the direct costs of all the activities of a project is the project direct cost.

Project Indirect Costs

The project indirect costs imply the overhead charges related to the project which include the supervision and other charges, late completion penalties and awards for early completion, and so on.

The project indirect costs are generally a function of the time the project takes to complete. Thus, the shorter the period the lesser the overhead charges. However, it may also be recognised that the direct cost of performing an activity would tend to increase if we desire to perform it in a time shorter than what it requires. Thus, if a project is set to be performed in a shorter time, then the direct costs would be larger and we can economise on the indirect costs. Similarly, a slower pace of work might, to some extent, mean lower direct costs, coupled with higher indirect costs.

Now we consider the following three problems relating to the resources:

(i) The time-cost trade-off;
(ii) The resource levelling; and
(iii) The resource allocation.

12.4.1 Time-cost Trade-off: Crashing

The critical path method is a very effective tool for obtaining useful information like the total completion time of a project, the critical activities, the non-critical activities, the total, free, and independent slack of various non-critical activities. A significant aspect of the CPM network analysis lies in its capacity to evaluate alternate ways to expedite some (or possibly all) of the activities indicated in a network and then analyse their cost implications. This analysis, as stated earlier, is referred to as the time-cost trade-off analysis.

As mentioned earlier, the time-cost trade-off problem is based on the conception that the duration of some of the activities of a project can be cut down, if some *additional* resources—men, material and/or equipment—are employed on them. For example, a particular activity might take 12 days to complete with 3 people assigned to it and only 8 days with 5 people assigned. Of course, as additional personnel are assigned to an activity, the cost of performing that task normally increases. In our example, the cost of completing the activity with 3 people assigned is 36 man-days while with 5 people it jumps to 40 man-days. Diseconomies of scale may be exhibited with too much labour on the job with inefficiency increasing very rapidly. In general, the more the time by which an activity time is required to be cut, the greater the amount of resources required to be employed on it. Thus, higher amounts of direct activity cost would be associated with smaller activity duration times, while longer duration time would involve comparatively lower direct cost. Such deliberate reduction of activity times by putting in extra effort is called *crashing*.

It is significant to note that for technical reasons, the duration of an activity cannot be reduced indefinitely. The *crash time* represents the fully expedited or the minimum activity time duration that is possible, and any attempts to further 'crash' would only raise the activity direct costs without reducing the time. The activity cost corresponding to the crash time is called the *crash cost* which equals the minimum direct cost required to achieve the crash performance time. These are in contrast to the *normal* activity *time* and the *normal* activity *cost*. The normal cost is equal to the absolute minimum of the direct costs required to perform an activity. The corresponding activity duration is known as the normal time—which may be stated more specifically as the shortest time required to perform an activity under the minimum direct cost constraint. An increase in the normal activity time would raise the direct costs.

The time-cost relationships can be visualised graphically in the form of a time versus cost curve which is, for a limited portion at least, sloping downward, as depicted in Figure 12.15. The widely accepted convex shape of the curve between the crash time-cost point and the normal time-cost point indicates that it is marginally costlier to induce the last percentages of reduction in activity time duration than the first percentages. Clearly, our interest lies in the central part of the curve contained between these two important points. Although the cost curve is convex in nature, for simplification in network scheduling it is commonly accepted that a linear approximation of the time-cost relationship be used. This is shown in the figure by two-point approximation from the normal time-cost point to the crash time-cost point, the points being shown joined by a dotted line. This linear approximation represents a linear incremental cost per unit of time saved under conditions of crashing. The incremental cost for an activity can be determined using the following equation:

$$\text{Incremental cost} = \frac{\text{crash cost} - \text{normal cost}}{\text{normal time} - \text{crash time}}$$

Thus, for an activity, if crash cost and time are Rs 700 and 8 days, and if the normal cost and time are Rs 500 and 12 days, we have

$$\text{Incremental cost per day} = \frac{700-500}{12-8} = \frac{200}{4} = \text{Rs } 50.$$

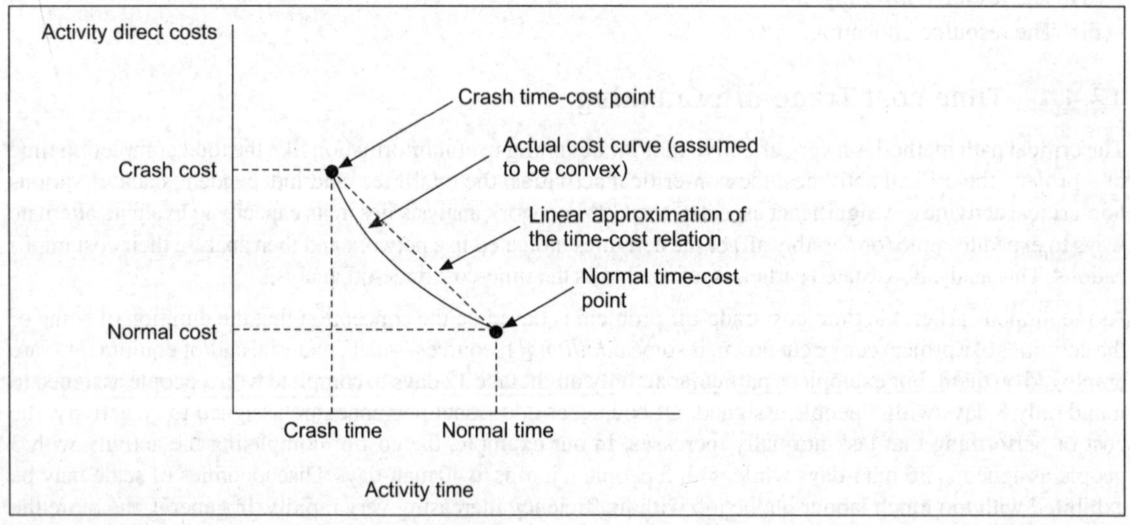

Figure 12.15 *Time-Cost Curve*

Quite naturally, if one had the option of crashing from two (or more) activities, the activity with a relatively flat time-cost curve would be more advantageous to choose. Thus, in case of a choice, the activity with the least incremental cost of crashing would be chosen to be crashed.

Once the crashing cost per unit time for each of the activities which can be crashed is obtained, we can proceed to determine the extent to which the entire project can be crashed and the extent to which the project should be crashed on economic considerations. The following points should be noted in this connection.

(a) In crashing a project, it is the activities on the critical path only that need to be considered. It is no use crashing non-critical activities because that would not reduce the project duration (we know that the project length is determined by the critical activities only).

(b) The critical path of the project may change while crashing.

(c) If at any stage of crashing, there are multiple critical paths in a project, then activities from each of the paths should be chosen for crashing.

Now we may examine the project crashing procedure. The time-cost trade-off analysis comprises the following steps.

Step 1 The first step is to identify and crash the critical activity that has the minimum incremental cost of crashing. In the event of multiple critical paths, an activity from each such path is chosen. Of the various combinations available, the one with the least cost is selected. In particular, it may be economical to consider joint critical activities—activities that are common to two or more critical paths.

In each case, the crashing is done for one time unit—by a day if the activity times are given in days.

Step 2 In the second step, the network is revised by adjusting the time and cost of the crashed activity. The critical path(s) is identified again, and we revert to the step 1. This process is continued till no more crashing of the project is possible.

Now the optimal duration of the project can be determined. It would be the time duration corresponding to which the total cost—direct cost plus indirect cost—is the minimum.

Let us consider the following example.

Example 12.6 The following table shows for each activity needed to complete the project the normal time, the shortest time in which the activity can be completed of a building contract and the cost per day for reducing the time of each activity. The contract includes a penalty clause of Rs 100 per day over 17 days. The overhead cost per day is Rs 160.

Activity	*Normal Time in Days*	*Shortest Time in Days*	*Cost of Reduction per Day (Rs.)*
1–2	6	4	80
1–3	8	4	90
1–4	5	3	30
2–4	3	3	–
2–5	5	3	40
3–6	12	8	200
4–6	8	5	50
5–6	6	6	–

The cost of completing the eight activities in normal time is Rs 6,500.

(a) Calculate the normal duration of the project, its cost and the critical path.

(b) Calculate and plot on a graph the cost/time function for the project and state:

(i) the lowest cost and associated time.

(ii) the shortest time and associated cost. *(ICWA, June, 1985)*

The network, according to the given information, is drawn and shown in Figure 12.16(a) in which the normal and the shortest (crash) times for various activities are given.

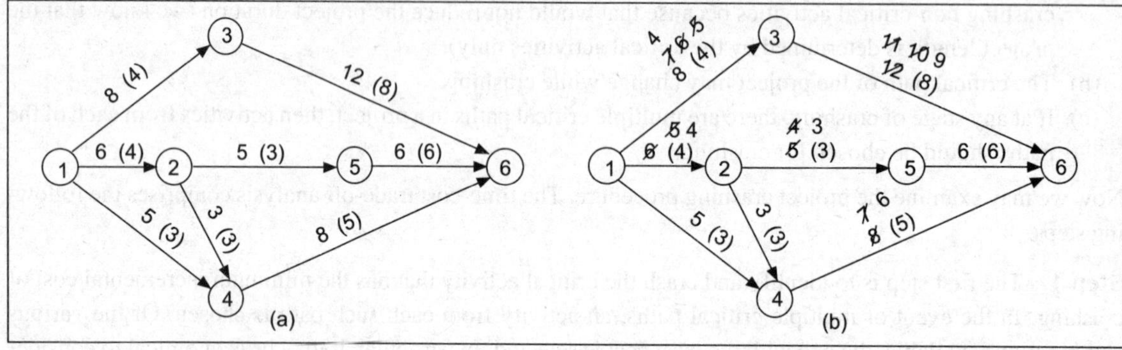

Figure 12.16 *(a) Arrow Diagram (b) Crashing the Activities*

From the network, we can determine the normal project duration by the length of the critical path using normal activity times. The minimum duration of the project is given by the length of the critical path based on the crash times as shown here:

Path	*Normal Length*	*Crash Length*
1–3–6	20 (critical)	12
1–4–6	13	8
1–2–4–6	17	12
1–2–5–6	17	13 (critical)

Accordingly, we have the normal and minimum duration of the project equal to 20 and 13 days respectively. Now we shall consider the time-cost relationship for this project when it is crashed successively by a period of one day, to know the total cost of the project for durations of 20 days through 13 days.

The First Crashing In this example, the critical activities are 1–3 and 3–6 for which the cost of reduction per day is Rs 90 and Rs 200. Obviously, we would decide to crash the activity 1–3. Crashing it by a day, the project length is reduced to 19 days and the total cost is equal to Rs 9,830. This is depicted in Table 12.4. For the normal duration of the project it would cost Rs 10,000, equal to the direct cost, the overhead and the penalty cost, which are, respectively, Rs 6,500; Rs 3,200 (= 160 × 20); and Rs 300 (= 100 × 3).

Now we change the duration of the activity 1-3 from 8 to 7 days, as shown in part (b) of the figure. At this stage also, the critical path remains 1–3–6.

The Second, and the Third Crashing For the second crashing, we are faced with the same activities to choose from as in the first crashing, viz. 1–3 and 3–6. The situation is the same in the third crashing. The total project cost equals Rs 9,660 and Rs. 9,490 after the second and the third crashings. Notice that the crashing cost at any given stage is equal to the cumulative cost of crashings till that point.

After the third crashing, the critical paths, each with a length of 17 days, are: 1–3–6; 1–2–4–6; and 1–2–5–6.

The Fourth Crashing To reduce the project length from 17 days to 16, an activity from each of these paths should be chosen. The various alternatives, along with their cost are as follows:

Alternative	*Activities*	*Total Crashing Cost*
1	1–3, 1–2	90 + 80 = 170
2	1–3, 4–6, 2–5	90 + 50 + 40 = 180
3	3–6, 1-2	200 + 80 = 280
4	3–6, 4–6, 2–5	200 + 50 + 40 = 290

Thus, we would crash activities 1–3 and 1–2 at a cost of Rs 170. The total cost of the project at this stage is Rs 9,500, and the critical paths, after adjusting the activity timings, are the same as above.

The Fifth Crashing For reducing the length of the project time to 15 days, we have the following alternatives. Notice that the activity 1–3 cannot be crashed anymore.

Alternative	*Activities*	*Total Crashing Cost*
1	3–6, 1–2	200 + 80 = 280
2	3–6, 4–6, 2–5	200 + 50 + 40 = 290

Now we decide to crash activities 3–6 and 1–2 by a day each, at an additional cost of Rs 280. The project cost now equals Rs 9,620, the critical paths still being 1–3–6; 1–2–4–6; and 1–2–5–6.

The Sixth and the Seventh Crashing At each of these crashings, the only choice open is to crash each of the following activities—one activity from every path at a cost of Rs 290: 3–6, 4–6, and 2–5.

The total cost of the project is Rs 9,750 and Rs 9,880, respectively, after these crashings.

From the table, it is clear that the lowest total cost is Rs 9,490 corresponding to the project duration equal to 17 days, whereas the shortest time to complete the project is 13 days at a total cost of Rs 9,880.

TABLE 12.4 Determination of Time-cost Relationship

Project Duration (days)	*Direct Cost*			*Indirect*			*Total Cost*
	Normal	*Crashing*	*Total*	*Overhead*	*Penalty*	*Total*	
20	6,500	–	6,500	3,200	300	3,500	10,000
19	6,500	90	6,590	3,040	200	3,240	9,830
18	6,500	180	6,680	2,880	100	2,980	9,660
17	6,500	270	6,770	2,720	–	2,720	9,490
16	6,500	440	6,940	2,560	–	2,560	9,500
15	6,500	720	7,220	2,400	–	2,400	9,620
14	6,500	1,010	7,510	2,240	–	2,240	9,750
13	6,500	1,300	7,800	2,080	–	2,080	9,880

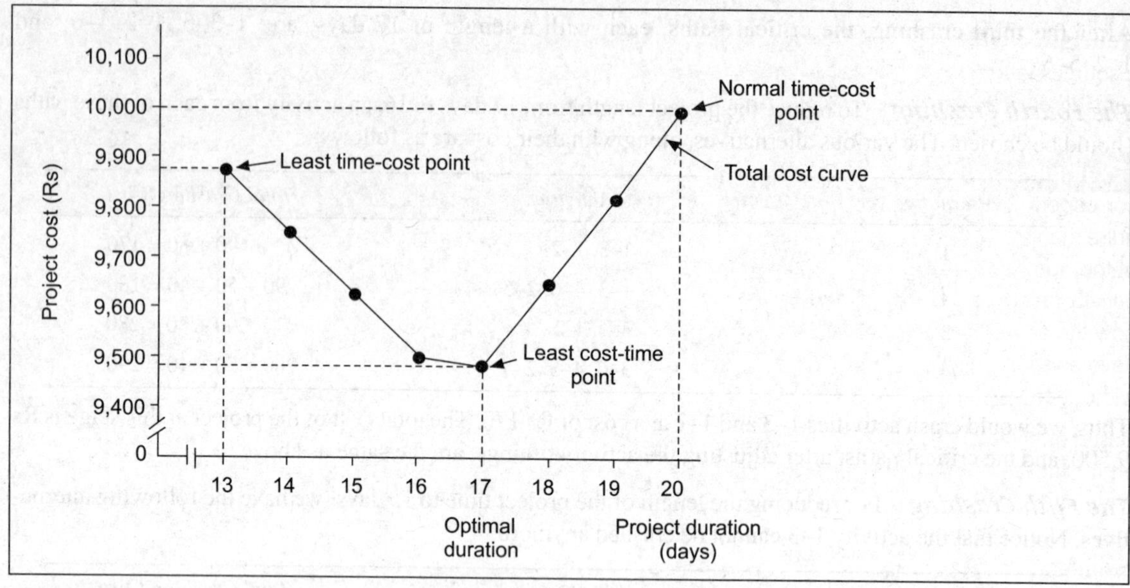

Figure 12.17 *Time-Cost Function for the Project*

The time cost function is shown graphically in Figure 12.17.

12.4.2 Resource Levelling

Networks are not simple sequences of events: activities can be performed only when resources, i.e. manpower, machines, materials and so forth, are allocated to them. Of course we cannot assume unlimited availability of the resources. Therefore, whatever resources are available, they must be utilised to the maximum extent. For a given project we can determine the earliest and the latest possible starting times for all the activities and all the slacks. The starting times for activities may be adjusted to achieve objectives other than just completing the project early. One such objective is the levelling of the resources allocated to the total project.

Often the managements want to keep the total amount of resources in use during the project performance as nearly constant as possible through time and avoid major shifts in manpower and other resources. And if constant amounts of resources are given, managerial planning decisions might aim at their maximum utilisation, in terms of the least idle time and, as far as possible, allowing for the smallest variation. The management may wish to level resources to maintain steady progress, expenditure patterns, and cash flows. This analysis, aiming at constancy of the resource usage is called *resource levelling*. In resource levelling, then, the aim is to reduce the peak resource requirements and smooth out period-to-period assignments within a constraint on the project duration. It may be recalled that like resource levelling, the crashing also involves reallocation of resources. But whereas in crashing, shortening of the project time is the objective, in resource levelling, smoothening of the resource usage rate without changing the project duration, is the objective.

It is possible to smooth either individual key resources or all the resources employed in the project on a daily basis. It may easily be visualised that such an analysis would be very complex when activities and resources involved are numerous, and it sometimes might require revising the entire project schedule. However, this might be worth investigating before key resources are committed.

Heuristic Procedure for Resource Levelling We now discuss a heuristic method for the levelling of resources in a given situation. To make the problem simple, suppose that personnel is the only resource to be considered. The method is explained with the following example.

Example 12.7 You are given in Figure 12.18 the network corresponding to a project. The activities of the project require manpower of similar type and the requirements, in terms of the number of men, are shown in parentheses next to the activity identification and timing (in days) for each. Determine how the activities of the project may be scheduled in order to have as smooth a requirement for the manpower as possible, without causing a delay in the project completion.

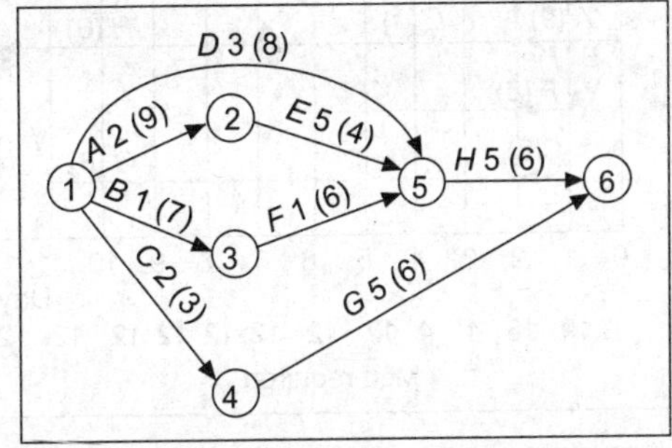

Figure 12.18 *Project Network Manpower Requirement*

Based on the given network, a schedule graph may be drawn on a time scale. It is helpful to exhibit activities on a time scale so that ranges of time allowable for various activities may be known. This is shown in Figure 12.19, drawn on the assumption that the various activities may be started as soon as possible, at their *ES* times. On this basis, the number of men required everyday are determined and are given in the lower part of the figure. These requirements are charted in the graph shown in part (b) of the figure.

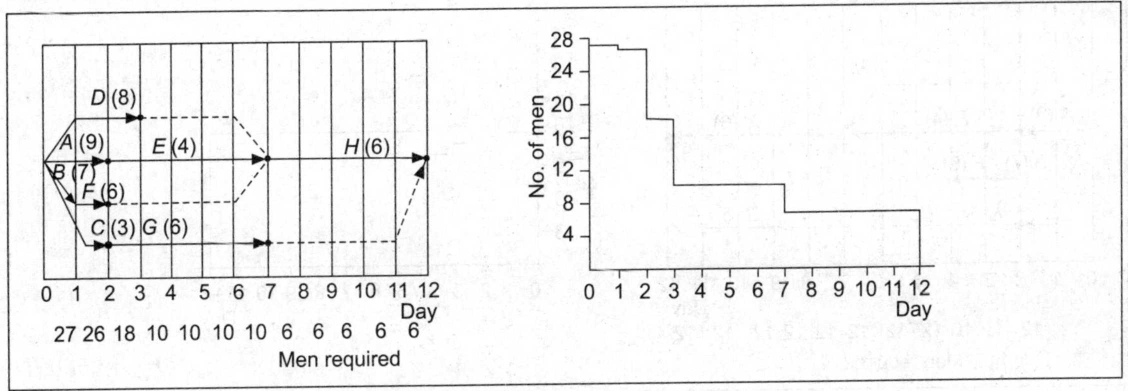

Figure 12.19 *Time Schedule Graph and Manpower Requirement*

A large variation is observed in manpower requirements of the different days that the project is worked upon. An effort would be made to shift the performance of some of the activities so that the requirements may be evened.

The continuous line in part (a) of the figure, represents the critical path of the project. Accordingly, the project length is 12 days and the activities *A*, *E*, and *H* are the critical activities. Naturally, if the project duration is not to be affected then none of these activities could be rescheduled. The remaining activities have slack and therefore can be shifted. Now, to lessen the manpower requirements in the earlier days and to raise it in the last days, we may postpone the start of the activity *D* to the beginning of the day 5 and that of activity *G* to the start of the 8th day. The result of this is depicted in Figure 12.20, wherein it may be observed that the requirement is less uneven in comparison with the previous situation.

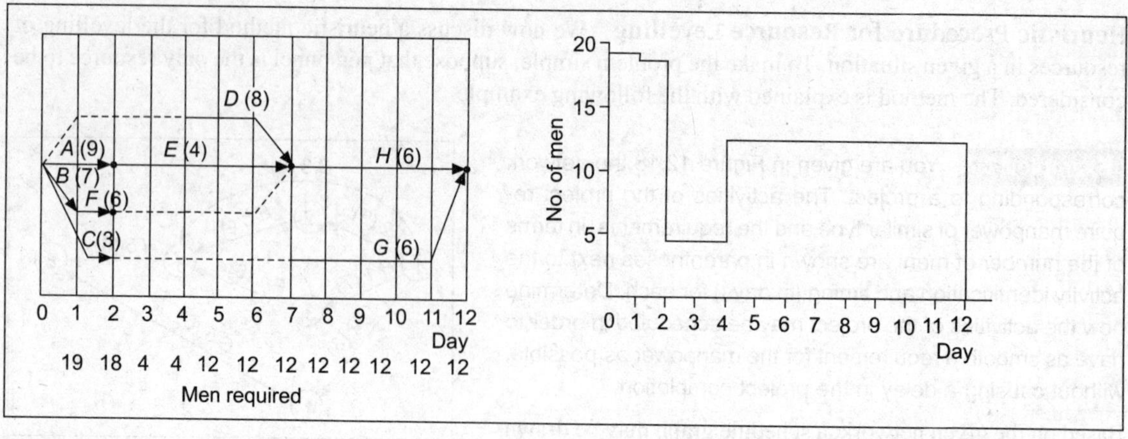

Figure 12.20 *Revised Time Schedule Graph and Manpower Requirement*

Further, by shifting activities *B* and *F* by 2 days each, the resource requirements may be made more balanced. This situation is depicted in Figure 12.21. Thus, the management would do well to employ 12 people for the project as all of them would be employed for the 12 days that the project would be worked upon, except on the third day and the fourth day when, respectively, one and two of the persons would remain idle insofar as this project is concerned.

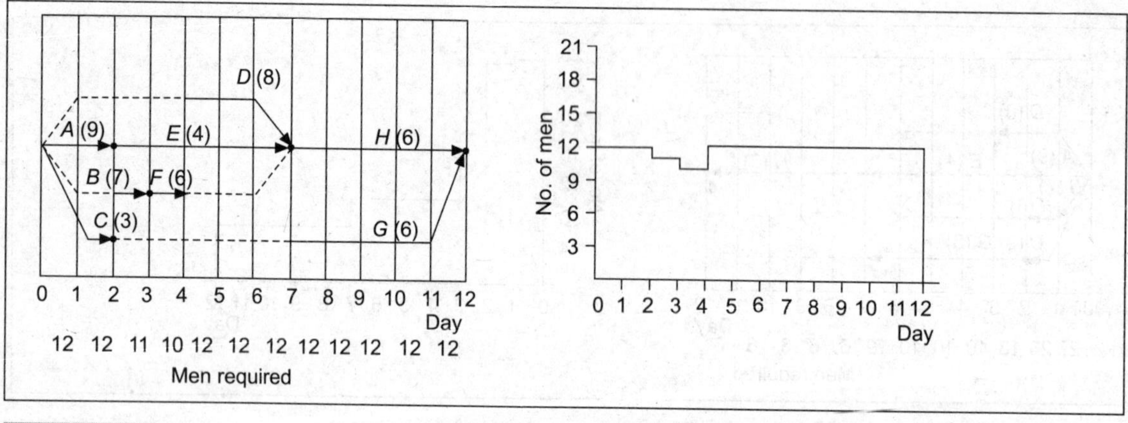

Figure 12.21 *Final Time Schedule Graph and Manpower Requirement*

Obviously, this heuristic procedure becomes quite complicated when many activities, with many combinations are required to be considered. In such situations it may be necessary and useful to employ a computer for the purpose.

12.4.3 Resource Allocation

After having considered the resource levelling problem, wherein stabilisation of rate of resource utilisation is aimed at within a constraint on the project duration, we now discuss the problem of resource allocation. In resource allocation programmes, it is aimed to allocate available resources to activities of a project in an attempt to determine the shortest project duration, consistent with the fixed resource limits.

It is easy to visualise that if the resources required for performing the various activities of a project are constrained in supply, then the project is likely to be delayed. Thus, in the problem of resource allocation, we explore whether the available resources are adequate for the project to be completed in time, or they are insufficient so that some of the activities of the project can be scheduled only after their latest allowable start times, resulting in a delay in its completion. Further, if at all the project would be delayed, then by how much time would it be so.

In simple cases, a check for resource loading can be made by visually examining the project network to find out the occurrence of demand for the resources in question. Assuming that all the activities start on their respective earliest start times, *ES*, we may observe the overlaps in the time periods when the resource is scheduled and note from there as to when, if at all, the demand for the resource is excessive. In case the resource overlaps are observed and the resouces are limited, the activities would have to be rescheduled. In such situations, alternate activity schedules may be prepared, in keeping with the limits dictated by the respective activity slack figures. For example, suppose that the network shown in the Figure 12.22 is given. It is known that each of the activities 5–7, 6–7, and 7–8, require the use of a large crane to be employed. The crane can be used on a single activity at a time and has to be employed from the start of the activity to its finish. Now, if we had two cranes, then they would be employed on these activities as soon as needed, and there would be no delay. However, if only one crane is available, then the question to be answered is whether, and by how much, the project would be delayed.

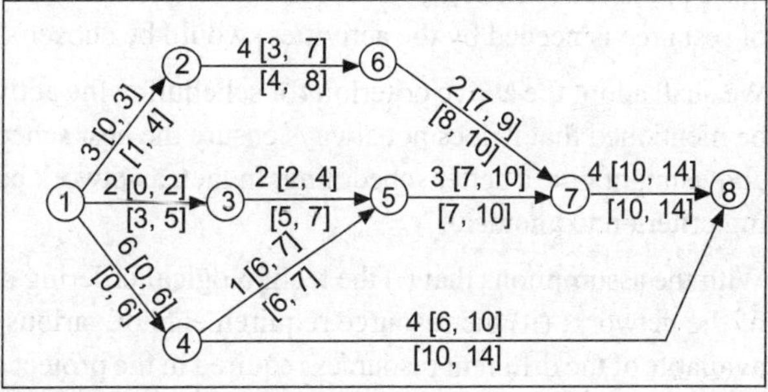

Figure 12.22 *Arrow Diagram*

From the network, we observe that technologically, activity 7–8 can be performed only when both, 5–7 and 6–7 are completed. However, activity 5–7 can be performed before or after the activity 6–7. Thus we have to decide which of the two activities 5–7 and 6–7 should be scheduled first. Since both of them have the same early start times, the choice would rest on the slack. We observe that the activity 5–7 has no slack while the activity 6–7 has a slack of one day. These activities could therefore be scheduled in the order 5–7, 6–7. The ordering of the activities would affect the length of the project. When the crane is used on activity 5–7 for 3 days, the activity 6–7 cannot start before the time 10 with the result that the activity 7–8 would start only at time 12 (the beginning of the day 13) and not at 10 as scheduled earlier. Thus, it would delay the project by 2 days. Notice that if the activities were scheduled as 6–7 followed by 5–7 instead, the project would have been delayed by 2 days then as well.

To extend the analysis, supposing different activities of a given project require varying amounts of a particular type of labour, then we can obtain the project length as a function of the number of workers employed. For a project with given labour requirements for its various activities, it is easy to visualise that the project length would be shorter when higher number of workers are available, and vice versa.

The question of allocation of resources and determination of the project duration, given resource availabilities and requirements, become intricate when multiple resources are involved. Allocation of resources in such cases requires systematic analysis which we shall discuss now. In particular, we shall consider how long a project would take when, for given resource requirements, fixed quantities of resources are given.

Allocation of limited resources to different activities of a project calls for laying priorities in clear cut terms. The most generally accepted criterion for scheduling the activities of a project is the slack that they have—so that at a given time if resources are to be allocated to some activities, then the activities shall be arranged in the order of the amount of float in them, beginning with the activity having the minimum float. Next, if there is a tie between the floats of two or more activities, then the activity with the smallest duration should be scheduled first. This criterion has the advantage that if one of the two (or more) activities can be scheduled at a given time, the waiting time is minimised by scheduling the shorter activity first. Further, if the duration of two or more of the activities is also tied, then the activity requiring a larger amount of resources (when the same type of resource is needed by the activities) would be chosen first for scheduling.

We shall adopt the above criterion for scheduling the activities of a project in the discussion to follow. It may be mentioned that it does not always ensure the best scheduling. In a given situation a criterion other than the above might give a better scheduling. In fact, a network can always be contrived that would favour one ordering criterion to another.

With the assumptions that (i) the technological ordering of the various activities is as indicated by the arrows of the network; (ii) the resource requirements of various activities are given and constant; (iii) the amounts available of the different resources required in the project are known; and (iv) the activities cannot be 'broken' in the sense that an activity, once started, has to be completed fully before the resources employed on it are diverted elsewhere, the allocation of resources proceeds as illustrated with the following example.

Example 12.8 For a project consisting of several activities, the duration and required resources of carrying out each of the activities and their availabilities are given below.

Activities	*Resources required*		*Duration*
	Equipment	*Operators*	*(Days)*
1–2	*X*	30	4
1–3	*Y*	20	3
1–4	*Z*	20	6
2–4	*X*	30	4
2–5	*Z*	20	8
3–4	*Y*	20	4
3–5	*Y*	20	4
4–5	*X*	30	6

Resources Availability:

No. of operators = 50

Equipment *X* = 1

Equipment *Y* = 1

Equipment *Z* = 1

(a) Draw the network, identify critical path and compute the total float for each of the activities.

(b) Find the project completion time under the given resource constraints. *(CA, November, 1985)*

Corresponding to the given information the network is shown in Figure 12.23. Also shown in the Table 12.5 are the earliest start and the latest allowable start times of the various activities as also their total float values.

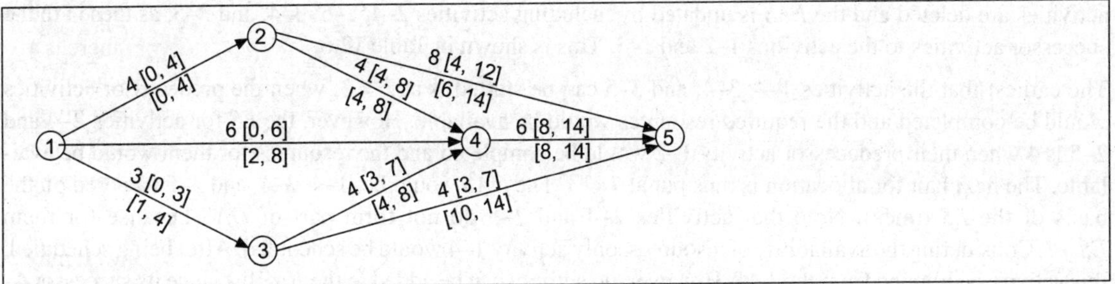

Figure 12.23 *Network*

Resource Allocation The first step in the resource allocation procedure is to determine the *ES*, and the *LS* times for each of the activities. In respect of the problem under consideration, these are given in the Table 12.5. Also mentioned are the resource requirements for all the activities.

TABLE 12.5 ES and LS Times and Resource Requirements of Activities

Activity	*Duration*	*ES*	*LS*	*Float*	*Operators*	*Equipment*
1–2	4	0	0	0	30	*X*
1–3	3	0	1	1	20	*Y*
1–4	6	0	2	2	20	*Z*
2–4	4	4	4	0	30	*X*
2–5	8	4	6	2	20	*Z*
3–4	4	3	4	1	20	*Y*
3–5	4	3	10	7	20	*Y*
4–5	6	8	8	0	30	*X*

The resources are allocated by stepping through time, scheduling the various activities as soon as their respective predecessors are scheduled and the resources required for them are available. For this purpose, two sets of activities are defined. The first of these comprises those activities which are eligible for assignment as their predecessors have been scheduled. This may be termed as the *Eligible Activity Set* (EAS). From the eligible activities, those activities are selected, which can start on the particular time when resources are to be allocated, and they are ordered in accordance with the criterion laid earlier. These activities constitute the *Ordered Activities Set* (OAS). Here, it is significant to note that the ordering of the activities on the basis of slack would be the same as obtained on the basis of their late start, *LS*, times. In fact, ordering on the basis of the *LS* values has the advantage that they do not change from one time period to another, whereas slack values would change as the activity that is ready to be scheduled but is not scheduled on a given day. The ordering of the activities in our discussion would be on the basis of their *LS* values.

In our example we start at $T = 0$ (the beginning of the day 1) when the *EAS* comprises activities 1–2, 1–3, and 1–4, since they all can start at the beginning of the project as they do not have any predecessors. These activities also constitute the OAS. Ordering these activities on the basis of their *LS* times gives the priorities as 1–2, 1–3, and 1–4 (their *LS* times being 0, 1, and 2, respectively). Since 50 men, and one unit of each of the equipment *X*, *Y* and *Z* are available, we shall allocate 30 men and *X* to the activity 1–2 for 4 days, and 20 men and *Y* to the activity 1–3 for 3 days. Being scheduled, as indicated by * mark on them, these activities are deleted and the *EAS* is updated by including activities 2–4, 2–5, 3–4, and 3–5, as these are the successor activities to the activities 1–2 and 1–3. This is shown in Table 12.6.

The earliest that the activities 1–4, 3–4, and 3–5 can be started is at $T = 3$, when the predecessor activities would be completed and the required resources would be available. However, the *ES* for activities 2–4 and 2–5 is 4 when their predecessor activity 1–2 would be completed and the resources for them would be available. The next halt for allocation is thus put at $T = 3$. The *OAS* would be: 1–4, 3–4, and 3–5 (ordered on the basis of the *LS* times). Note that activities 2–4 and 2–5 do not form part of *OAS* because for them $ES > T$. Considering the availability of resources, only activity 1–4 would be scheduled. After being scheduled, this activity is dropped from the *EAS*. However, no activity can be added to the *EAS* list since its successor 4–5 cannot be scheduled before 3–4 and 2–4 are scheduled.

Next, the *ES* times for the activities currently in the *EAS* are revised in accordance with the technological requirements are resource availabilities and we halt at $T = 4$, when activity 2–4 is scheduled. We continue in this manner until all the activities are scheduled. The details of all the resource assignments are given in Table 12.6 where the last halt is shown at $T = 17$. The *loading chart* given in Figure 12.24 portrays the assignment of the resources to the different activities. We can also observe from this as to which of the resources would be unutilised during various time periods.

TABLE 12.6 Scheduling of Activities and Allocation of Resources

Halt 1		$T = 0$						
EAS	:	1–2*	1–3*	1–4	2–4	2–5	3–4	3–5
ES	:	0	0	0				
LS	:	0	1	2				
OAS	:	1–2	1–3	1–4				
Schedule	:	1–2 for 4 days, Resources: 30 Persons, and *X*;						
		1–3 for 3 days, Resources: 20 Persons, and *Y*:						
Halt 2		$T = 3$						
EAS	:	1–4*	2–4	2–5	3–4	3–5		
ES	:	3	4	4	3	3		
LS	:	2	4	6	4	10		
OAS	:	1–4	3–4	3–5				
Schedule	:	1–4 for 6 days, Resources: 20 Persons, and *Z*.						
Halt 3		$T = 4$						
EAS	:	2–4*	2–5	3–4	3–5			
ES	:	4	9	4	4			
LS	:	4	6	4	10			
OAS	:	2–4	3–4	3–5				
Schedule	:	2–4 for 4 days, Resources: 30 Persons, and *X*.						
Halt 4		$T = 8$						
EAS	:	2–5	3–4*	3–5	4–5			

(Contd.)

TABLE 12.6 Contd

ES	:	9	8	8
LS	:	6	4	10
OAS	:	3–4	3–5	
Schedule	:	3–4 for 4 days, Resources: 20 Persons, and *Y*.		
Halt 5		$T = 9$		
EAS	:	2–5*	3–5	4–5
ES	:	9	12	12
LS	:	6	10	8
OAS	:	2–5		
Schedule	:	2–5 for 8 days, Resources: 20 Persons, and *Z*.		
Halt 6		$T = 12$		
EAS	:	3–5	4–5*	
ES	:	12	12	
LS	:	10	8	
OAS	:	4–5	3–5	
Schedule	:	4–5 for 6 days, Resources: 30 Persons, and *X*.		
Halt 7		$T = 17$		
EAS	:	3–5		
ES	:	17		
Schedule	:	3–5 for 4 days, Resources: 20 Persons, and *Y*.		

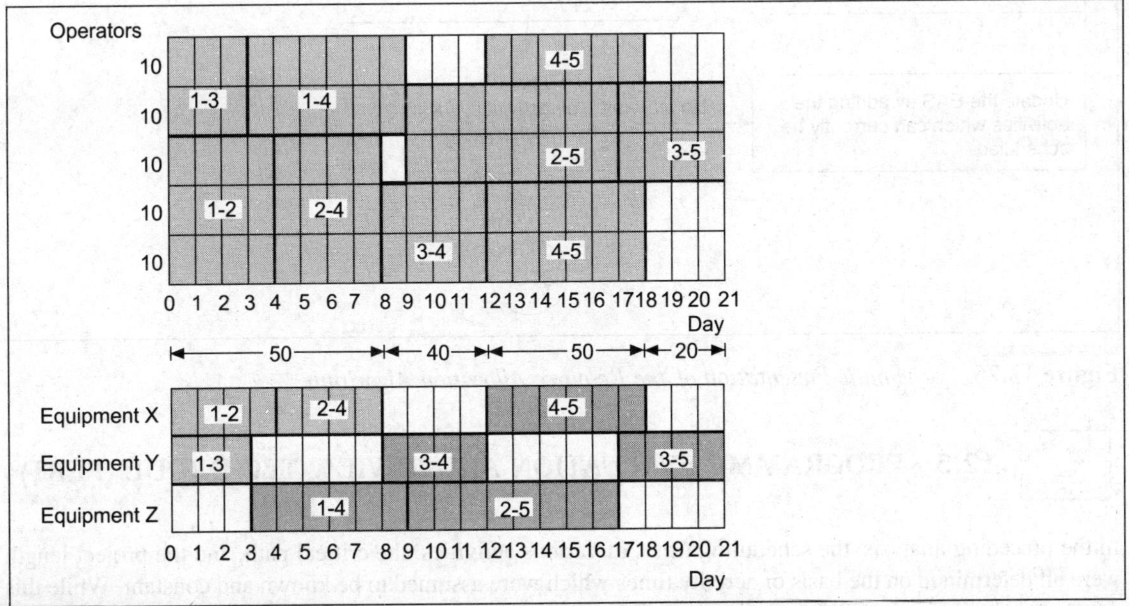

Figure 12.24 *Loading Chart*

We observe from the figure that the project would be completed in a total of 21 days. The number of men employed on each of the 21 days is also indicated at the bottom of the figure.

The resource allocation procedure discussed here is shown schematically in Figure 12.25.

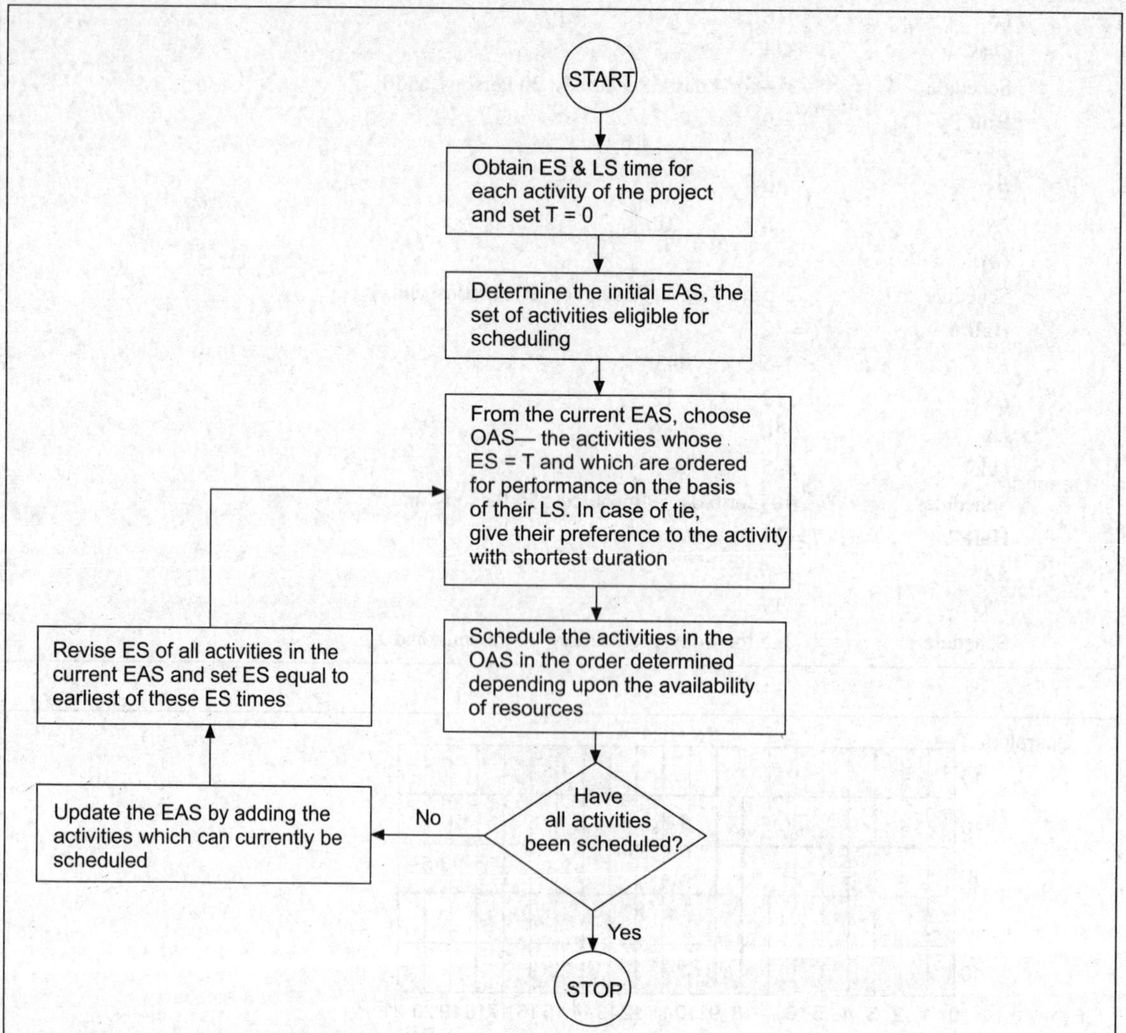

START

Obtain ES & LS time for each activity of the project and set T = 0

Determine the initial EAS, the set of activities eligible for scheduling

From the current EAS, choose OAS— the activities whose ES = T and which are ordered for performance on the basis of their LS. In case of tie, give their preference to the activity with shortest duration

Schedule the activities in the OAS in the order determined depending upon the availability of resources

Have all activities been scheduled?

No

Update the EAS by adding the activities which can currently be scheduled

Revise ES of all activities in the current EAS and set ES equal to earliest of these ES times

Yes

STOP

Figure 12.25 *Schematic Presentation of the Resource Allocation Algorithm*

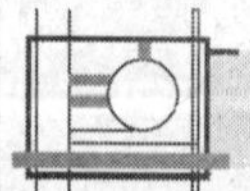

12.5 PROGRAMME EVALUATION AND REVIEW TECHNIQUE (PERT)

In the preceding analysis, the scheduling times of various activities, the critical path, and the project length were all determined on the basis of activity times which were assumed to be known and constant. While this assumption holds for the CPM analysis, according to the originators of the PERT, any given activity delineated in a network is unlikely to be completed on time. Sometimes all aspects of a job may be easier to complete than expected, while at other times, unexpected snags may occur, causing unplanned delays. This method uses three time estimates for an activity, rather than a single estimate. They are:

(a) *Optimistic time (a)* This is the shortest time the activity can take to complete. There is more than one chance in a hundred of completing the activity in this amount of time. It represents an ideal estimate.
(b) *Most likely time (m)* This refers to the time that would be expected to occur most often if the activity were frequently repeated under exactly the same conditions. It is the modal time.
(c) *Pessimistic time (b)* This is the longest time the activity could take to finish. It is the worst time estimate and represents the time the activity would take if bad luck was faced. It occurs with a probability of less than one per cent.

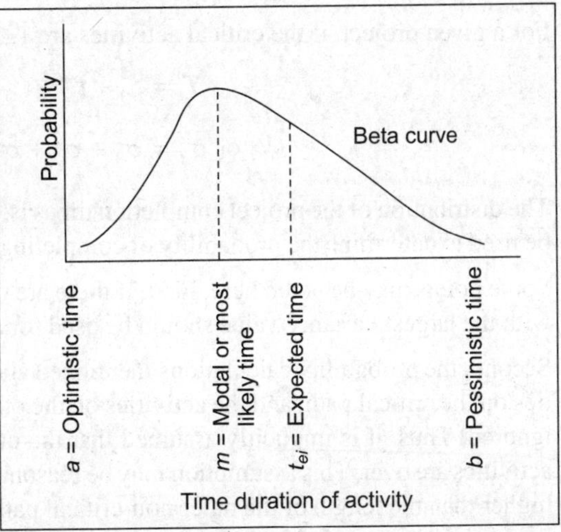

Figure 12.26 *Beta Distribution of Activity Times*

Depending on the values of a, m, and b, the resulting distribution of activity duration can take a variety of forms. Typically, completion of a given activity is assumed to follow beta distribution, as shown in Figure 12.26.

Using the values of a, b, and m, the expected times of various activities and their standard deviations are calculated as follows. The three time estimates are reduced into a single expected time (t_{ei}) with the weighted average formula:

$$t_{ei} = \frac{a+4m+b}{6}$$

where

t_{ei} = expected time of the ith activity
a = optimistic time
m = most likely, or modal time
b = pessimistic time

The standard deviation, σ_i, of the completion time of an activity is calculated as follows:

$$\sigma_i = \frac{b-a}{6}$$

From this, the variance

$$\sigma_i^2 = \left(\frac{b-a}{6}\right)^2$$

Once the expected times of the activities are obtained, the critical path of the project network is determined using these time estimates. Having found the critical path, the PERT methodology assumes that the aggregation of the mean times and the summation of the variances of critical jobs would yield the project duration expected and its variance. An assumption invoked is that of the *Central Limit Theorem* which states that the sum of several independent activity durations will tend to be normally distributed, with a mean equal to the sum of their individual job times and the variance equal to the sum of their individual activity variances. Accordingly, the probability distribution of times for completing a project can be approximated by using a normal distribution curve which becomes more exact as the number of activities increases. Estimates using the curve can be fairly exact if and when there are at least 30 activities along a given path.

For a given project, if the critical activities are 1, 2, . . ., k, we have

$$T_e = t_{e1} + t_{e2} + t_{e3} + \ldots + t_{ek}, \text{ and}$$

$$V_T \text{ or } \sigma_T^2 = \sigma_1^2 + \sigma_2^2 + \sigma_3^2 + \ldots + \sigma_k^2$$

The distribution of the project completion times is, then, normally distributed with $\mu = T_e$ and $\sigma = \sigma_T$. This can be used to determine the probability of completing the project by a given date, or during a given time interval.

Some points may be noted here. First, if there are two (or more) critical paths in a given network, then the one with the largest variance value should be used for determining T_e and V_T.

Second, the probability calculations mentioned above represent only the probability of completing the activities on the critical path, and the activities on the other critical path (if it exists) and the non-critical path(s) are ignored. Thus, it is implicitly assumed that the other activities would be completed by the time the critical activities are over. This assumption may be reasonable if the expected length of the critical path is significantly higher than the length of the other non-critical paths. However, if multiple paths exist and/or if the lengths of the critical and non-critical paths are nearly equal, then their separate probabilities of completion may be obtained, and their product calculated to get a more reasonable probability value of the project completion time. Strictly speaking, this way of multiplying the probabilities may only be undertaken if the paths are truly independent, i.e. they do not have common activities. Nevertheless, the method does give a reasonable approximation.

Third, the PERT is based on the assumption that the critical path is fixed and given once and for all, which is wrong in principle. This is because due to the variation in the time durations of the project activities if some activities take more or less time than their expected times, then this might render some of non-critical activities critical and vice versa.

An example on PERT methodology follows.

Example 12.9 The owner of a chain of fast-food restaurants is considering a new computer system for accounting and inventory control. A computer company sent the following information about the system installation:

Activity Identification	*Activity Description*	*Immediate Predecessor*	*Time*		
			Most Optimistic	*Most Likely*	*Most Pessimistic*
A	Select the computer model	–	4	6	8
B	Design input/output system	*A*	5	7	15
C	Design monitoring system	*A*	4	8	12
D	Assemble computer hardware	*B*	15	20	25
E	Develop the main programs	*B*	10	18	26
F	Develop input/output routines	*C*	8	9	16
G	Create database	*E*	4	8	12
H	Install the system	*D, F*	1	2	3
I	Test and implement	*G, H*	6	7	8

(a) Construct an arrow diagram for this problem, determine the critical path and state the expected project completion time.

(b) Determine the probability that the project will be completed in 55 days.

(c) If the company wants to be 90% sure that the system will be installed by a certain due date, how many days prior to that should it start the work?

(d) Suppose the company agrees to install the computer system in 50 days, failing which it would pay a penalty of Rs 500 per day. What is the probability that a penalty, but not exceeding Rs 2,000, will be paid?

(e) Obtain in earliest and the latest scheduling times of the various activities.

(a) The arrow diagram corresponding to the given information is depicted in Fig. 12.27. Also given are the expected times of the various activities as shown calculated in Table 12.7.

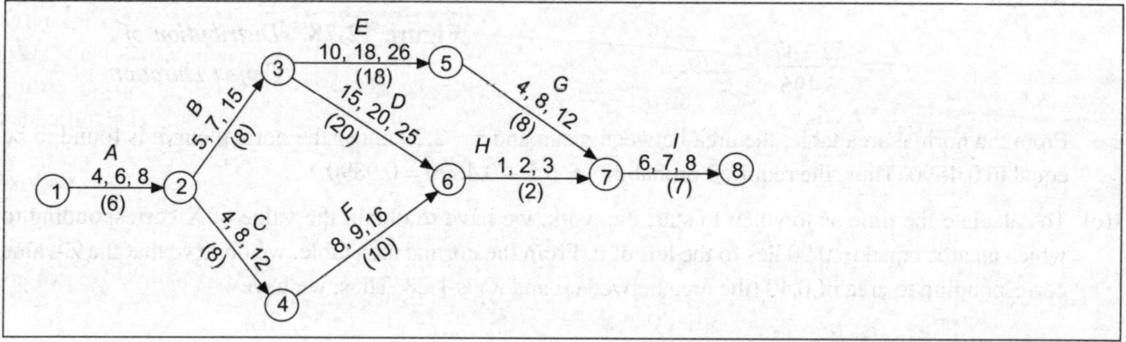

Figure 12.27 *Arrow Diagram Related to the Computer Project*

TABLE 12.7 Calculation of Expected Times and Variances

Activity		*Time* a	m	b	$t_{ei} = \dfrac{a+4m+b}{6}$	$\sigma_i = \dfrac{b-a}{6}$	σ_i^2
A	1–2	4	6	8	6	4/6	4/9
B	2–3	5	7	15	8	10/6	25/9
C	2–4	4	8	12	8	8/6	16/9
D	3–6	15	20	25	20	10/6	25/9
E	3–5	10	18	26	18	16/6	64/9
F	4–6	8	9	16	10	8/6	16/9
G	5–7	4	8	12	8	8/6	16/9
H	6–7	1	2	3	2	2/6	1/9
I	7–8	6	7	8	7	2/6	1/9

Using the expected times of activity duration, we obtain the critical path as 1–2–3–5–7–8. Thus, we have the expected project length,

$$T_e = 6 + 8 + 18 + 8 + 7 = 47 \text{ days},$$

and the variance of the project length,

$$V_T = \frac{4}{9} + \frac{25}{9} + \frac{64}{9} + \frac{16}{9} + \frac{1}{9} = \frac{110}{9}$$

(b) Now, the project duration being normally distributed with mean (T_e) = 47 days and standard deviation $\sigma = \sqrt{V_T} = \sqrt{110/9}$ = 3.496 days, we can determine the probability of the project being completed in 55 days. This would be equal to the area under the normal curve lying to the left of $X = 55$, as shown in Figure. 12.28.

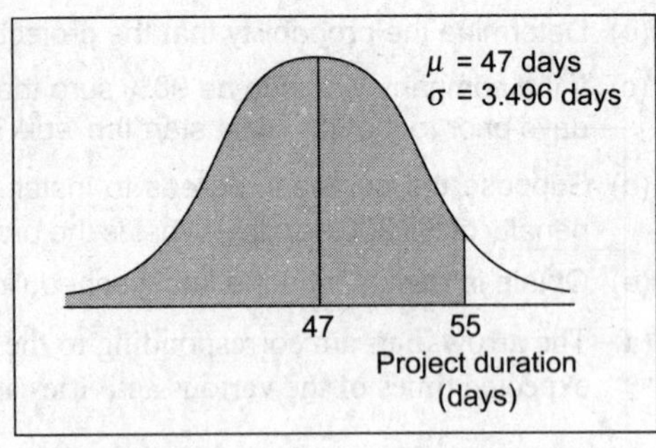

Figure 12.28 *Distribution of Project Duration*

We have,

$$z = \frac{55-47}{3.496} = 2.29$$

From the normal area table, the area between mean and $z = 2.29$ under the normal curve is found to be equal to 0.4890. Thus, the required probability = 0.5 + 0.4890 = 0.9890.

(c) To calculate the time as to when to start the work, we have to obtain the value of X corresponding to which an area equal to 0.90 lies to the left of it. From the normal area table, we observe that the Z-value corresponding to area of 0.40 (the area between μ and X) is 1.28. Thus, we have

$$1.28 = \frac{X - 47}{3.496}$$

$$X = 1.28 \times 3.496 + 47 = 51.47 \text{ or } 52 \text{ days app.}$$

(d) A penalty would be paid if the project is not completed in 50 days. For a penalty to be paid but not exceeding a sum of Rs 2,000, the project would complete between 50 and 54 days. The probability of this would be given by the area marked in Figure. 12.29. We have,

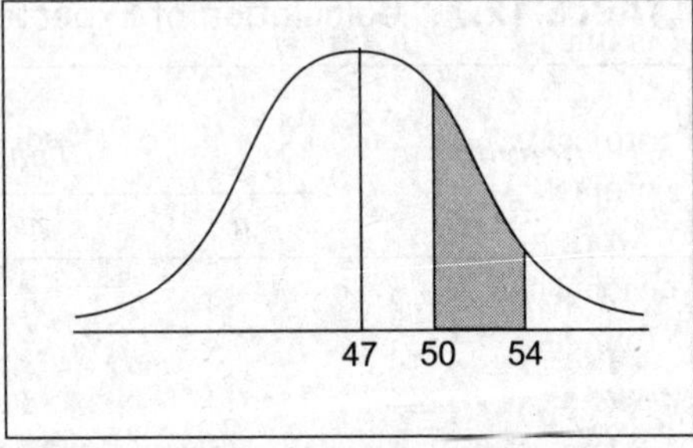

Figure 12.29 *Distribution of Project Duration*

	Area
$z_1 = \dfrac{54 - 47}{3.496} = 2.00$	0.4772
$z_2 = \dfrac{50 - 47}{3.496} = 0.86$	0.3051
Required area	0.1721

Thus, the probability that a penalty would be paid but not more than Rs 2,000 is equal to 0.1721.

(a) The earliest and the latest scheduling times for a PERT project are calculated using the expected times for various activities. For the given project, these are shown in Table 12.8.

TABLE 12.8 Earliest and Latest Scheduling Times

Activity	*Expected Time*	*ES*	*EF*	*LS*	*LF*
A	6	0	6	0	6
B	8	6	14	6	14
C	8	6	14	20	28
D	20	14	34	18	38
E	18	14	32	14	32
F	10	14	24	28	38
G	8	32	40	32	40
H	2	34	36	38	40
I	7	40	47	40	47

PERT/Cost PERT is an excellent method of planning, monitoring and controlling project duration. But this is also its limitation in the sense that it is only time-oriented and gives no consideration to the other important factor of the project cost. PERT/Cost is a modification of PERT which allows the planning, monitoring and controlling of project duration as well as *project cost*. We first see how project costs are planned and scheduled. Then we examine how they can be monitored and controlled.

Planning and Scheduling Project Costs: The Budgeting Process The basic idea underlying the budgeting process is to determine how much money is to be spent every week (or month) over the planned duration of a given project. For this, the manager has to first identify all costs associated with each of the activities of project. These costs are then added together to obtain one estimated cost or budget for each activity. In case the project is very large consisting of a large number of activities, several activities may be combined to form 'work packages' so that the project may be viewed as a combination of several work packages and cost for each work package is calculated. Actually, the measurement and control of costs by work packages is the very foundation of the PERT/Cost system. The work packages are so designed that the activities in a package represent parts of a project for which responsibility may be easily determined.

While dealing with relatively simpler and smaller projects, each of the activities may be regarded as a distinct package. Once the cost for each activity (or work package) is estimated, one usually makes a simplifying assumption that the expenditures for each activity are made at a constant rate during the planned duration of the activity.

To illustrate, if the budgeted cost for a given activity is Rs 12,000 and its expected time is 3 weeks, the budgeted cost per week is Rs 4,000 (= Rs 12,000/3 weeks).

Once the cost estimates for every activity on a weekly (monthly) basis are obtained, we can calculate how much money should be spent during each week or month during the time the project is done.

Budgeting Choices We have already seen from a technological viewpoint that the project manager can choose when to start each activity of a project. The manager may decide that every activity be started at its earliest start time or may as well decide that every activity be started at the latest start time. Scheduling the activities on their permissible latest times does not cause any delay in the completion of the project. The significance of this choice in scheduling the activities lies in the fact that funds are committed as and when the activities are performed. An early scheduling implies an early deployment of funds so that average funds committed at any given point would be higher. Similarly, a later scheduling will have the effect of commitment of relatively lower average of funds.

To illustrate, consider a project about which information is presented in Table 12.9, wherein precedence relationships and expected duration and funds required for each activity are given. Also given in the table are the earliest and latest start times for each activity, as obtained from the network reflecting the given precedence relationships and shown in Figure 12.30.

TABLE 12.9 Project Data

Activity	*Immediate Predecessor*	*Expected Time*	*ES*	*LS*	*Cost of Activity ('000 Rs)*	
					Total	*Per Week*
A	–	4	0	0	60	15
B	–	2	0	9	280	140
C	A	1	4	10	40	40
D	A	4	4	4	28	7
E	D	5	8	8	25	5
F	B, C	2	5	11	160	80
G	B, C	1	5	11	75	75
H	E, F	3	13	13	18	6
I	G	4	6	12	240	60

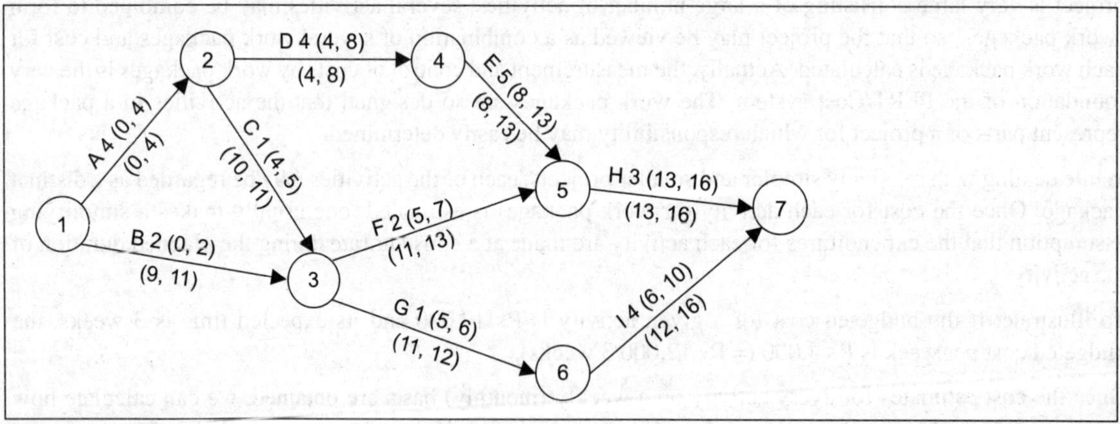

Figure 12.30 *Network Diagram*

For example, activity A has a duration of 4 weeks and it requires a total expenditure of Rs 60,000. Assuming a linear relationship between time and cost, it involves an expenditure of Rs 15,000 per week. Similarly, for the other activities the given information can be interpreted.

Now, based on the given information, two budgets are proposed. The first one is given in Table 12.10. This is drawn on the assumption that every activity starts at its earliest time. Based on this assumption and the one relating to uniform expenditure per week for each activity, the budget for every week of the total duration of the project is drawn up.

To illustrate, the ES time of activities A and B both is equal to 0. Hence, both of these can be started during the first week and a total of Rs 15 + Rs 140 = Rs 155 (thousand) should be spent. Because the two activities are still being done in the second week, a total of Rs 155 (thousand) should also be spent during the second week. Work on activity B only will continue in the third and fourth weeks, in each of which a sum of Rs 15 (thousand) will, therefore, be spent. Similar computations are done for all activities to determine the total budget for the entire project for each week. Finally, the weekly totals are accumulated gradually to obtain the total expenditure required till date. These are shown in the bottom row of the table. Also given in the table are the cumulative amounts of money which would be spent during the project.

In a similar manner, the budget, when all activities are scheduled at their latest times, is shown in Table 12.11. Here also, the amount of money spent in each of the 16 weeks of the project is shown, as also the cumulative money invested. It is evident from the two tables that there is a significant difference between the funds employed at various points in time.

Interestingly, beginning all the activities on a late start basis allows the project to continue on schedule and simultaneously reduces the average commitment of funds on the project by a considerable margin. To illustrate, at the end of 10 weeks, using the late start alternative would have needed a commitment of Rs 238 thousand as against Rs 893 thousand under the alternative of early start. The difference between the corresponding values for any given week indicates the amount of funds not committed to the project and hence available for use elsewhere. This assumes greater significance if borrowed funds are being used in the project.

In case of early scheduling of activities, the average funds invested in the project at any given point in time amount to Rs 6,69,688, while this average is Rs 3,19,063 in case of late scheduling of the activities.

TABLE 12.10 Budget with Early Starts (Amounts in'000 Rs)

Activity	*Week*															
	1	*2*	*3*	*4*	*5*	*6*	*7*	*8*	*9*	*10*	*11*	*12*	*13*	*14*	*15*	*16*
A	15	15	15	15												
B	140	140														
C					40											
D					7	7	7	7								
E									5	5	5	5	5			
F						80	80									
G						75										
H														6	6	6
I							60	60	60	60						
Total	155	155	15	15	47	162	147	67	65	65	5	5	5	6	6	6
Cumulative	155	310	325	340	387	549	696	763	828	893	898	903	908	914	920	926

TABLE 12.11 Budget with Latest Starts (Amounts in '000 Rs)

Activity	*Week*															
	1	*2*	*3*	*4*	*5*	*6*	*7*	*8*	*9*	*10*	*11*	*12*	*13*	*14*	*15*	*16*
A	15	15	15	15												
B										140	140					
C											40					
D					7	7	7	7								
E									5	5	5	5	5			
F												80	80			
G												75				
H														6	6	6
I													60	60	60	60
Total	15	15	15	15	7	7	7	7	5	145	185	160	145	66	66	66
Cumulative	15	30	45	60	67	74	81	88	93	238	423	583	728	794	860	926

The budgets given in Tables 12.10 and 12.11 form feasible budget ranges. Thus, once they are drawn, a manager can choose any budget that falls between them.

This idea is illustrated in Figure 12.31 where cumulative totals of ES and LS budgets are shown plotted. The figure shows the cumulative funds committed in the project under each of the two assumptions of scheduling of the activities of the project. The gap between the two provides the range of feasible budgets. The manager can use any budget between the two and still complete the project on time. Budgets like the ones shown in the figure are usually prepared before a project begins. Then as the work progresses, the funds expended are monitored and controlled.

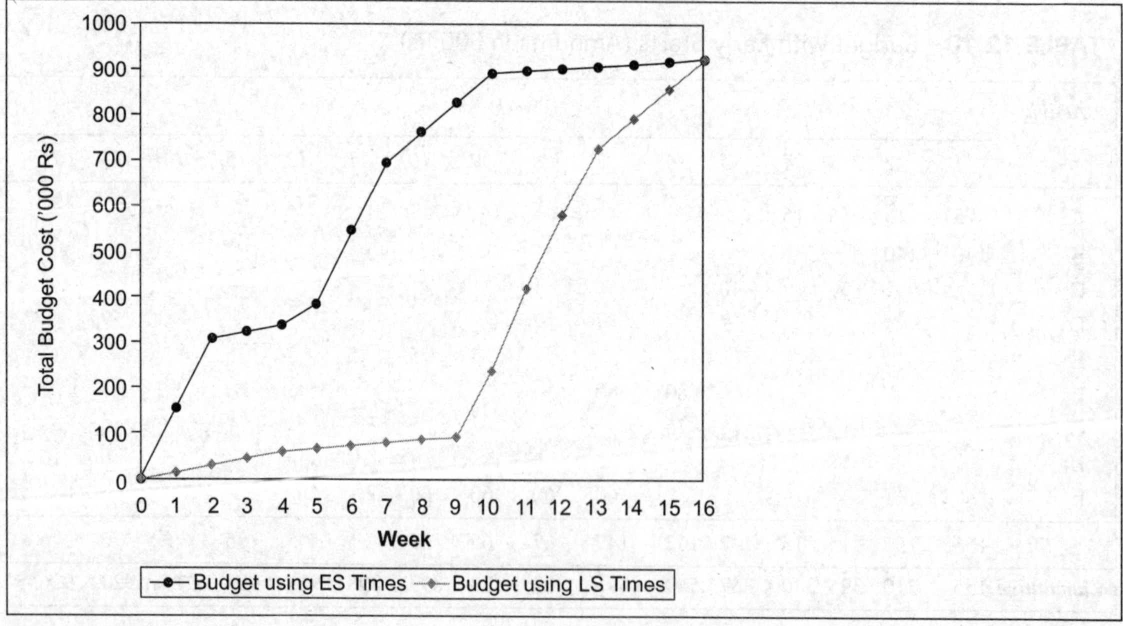

Figure 12.31 *Total Budget Amounts*

Controlling the Project Costs The PERT/Cost method of monitoring and controlling costs extends beyond the ordinary comparisons between actual costs and budgeted costs. Since time and costs both are considered by this method, actual work done can also be compared with projected work. With the underlying assumption that time and cost are directly proportional for each activity, we can usefully compare the work completed and the cost incurred. This serves as a useful input to cost control. Typically, if a given activity has incurred 80 per cent of its budgeted cost but is only 60 per cent complete, we have what is termed as *budget overrun*.

In addition, periodic assessments are made of the status of the whole project to establish whether it is on schedule; what is the value of work completed; are there any cost overruns; etc. In context of the example we are considering, suppose we are 11 weeks into the project and have chosen the budget as given in Table 12.11. The status of the work is as follows: Activities A, B and D have been completed. These activities incurred costs of Rs 56,000; Rs 2,92,000; and Rs 28,000 respectively. Activity C is 40% completed and so far the cost expended on it is Rs 24,000 while activity E is 70% complete and a sum of Rs 19,500 has been spent on this. Activities F, G, H and I have not started as yet. To see whether the project is on schedule and whether there are any cost overruns, we first calculate activity differences,

Activity Difference = Actual cost – Value of work done

where value of work done is equal to the product of the percentage of work completed and the total activity budget. A negative activity difference is termed as *cost under-run* while a positive difference is called *cost overrun.*

To illustrate, the activity C is 40% completed and its budget is Rs 40,000, while the actual cost incurred is Rs 24,000. Thus, for this activity,

Activity Difference = Rs 24,000 – 40% of Rs 40,000
= Rs 24,000 – Rs 16,000
= Rs 8,000

In Table 12.12, information about all activities is provided.

TABLE 12.12 Calculation of Activity Differences

Activity	*Budget*	*Percentage of Completion*	*Value of Work Done*	*Actual Cost*	*Activity Difference*
A	60,000	100	60,000	56,000	–4,000
B	280,000	100	280,000	292,000	12,000
C	40,000	40	16,000	24,000	8,000
D	28,000	100	28,000	28,000	–
E	25,000	70	17,500	19,500	2,000
F	160,000	0	0	0	0
G	75,000	0	0	0	0
H	18,000	0	0	0	0
I	240,000	0	0	0	0
Total			401,500	419,500	18,000

Here all activity differences are shown to be calculated and they are added together to obtain total cost overrun (or under-run) for the entire project. In this case, it is observed that the project has a cost overrun of Rs 18,000 because the value of the work done is only Rs 401,500 while the actual cost of the work done is Rs 419,000. Accordingly, the project is behind schedule and there are cost overruns. The manager needs to move faster on the project to finish on time and control the future costs in a careful manner so as to eliminate the current cost overrun of Rs 18,000. Calculations about value of work done and actual costs should be done at regular intervals so as to have an effective monitoring and cost control.

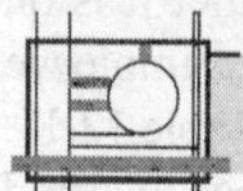

12.6 DIFFERENCE BETWEEN PERT AND CPM

As seen in the consideration of the two techniques PERT and CPM, although developed independently, have similarities and they share many concepts. To illustrate, they share the notion of critical path and are based on the network plan in a bid to know the critical activities to be controlled so as to meet the completion deadlines. However, there are some noticeable differences in the two, which are discussed here:

1. The PERT is probabilistic in nature. It acknowledges and considers the variability in completion times of activities and, in turn, the project. Accordingly, it is useful for analysing project scheduling problems with uncertain completion times of the activities involved. On the other hand, the CPM is deterministic in nature. It is most appropriately used in projects in which activity durations are known with certainty. Not only the amount of time needed to complete the various facets of the project but also the amounts of resources required for performing each of the activities are assumed to be known and certain.
2. The PERT is useful for projects that are new, non-repetitive, or which involve research and development, while CPM is of great value for projects that are repetitive and standardised, like those involving construction activities.
3. The PERT focuses primarily on time element and attaches lesser significance to the cost. On the contrary, the CPM puts strong emphasis on cost and specifically considers the time-cost relationship and trade-off. It considers systematic reduction of project duration along with cost implications.
4. From (1) and (3), it is evident that whereas variation in the project time is inherent in the projects where PERT is used, the time is systematically varied using additional resources where CPM is employed.
5. The PERT is event-oriented so that probabilities of reaching various events by certain dates may be calculated through event-variances. The CPM is activity-oriented.

REVIEW ILLUSTRATIONS

Example 12.10 A work project consists of twelve activities labelled *A* through *L*. Upon being asked to specify the order in which the jobs had to be done, the manager answered as follows:

A, *B*, and *C* are the first activities of the project and can start simultaneously and immediately; *A* and *B* precede *D* while *B* precedes *E*, *F*, and *H*. Activities *F* and *C* precede *G* while *E* and *H* precede *I* and *J*. *C*, *D*, *F* and *J* precede *K* which, in turn, precedes *L*. Further, *I*, *G*, and *L* are the terminal activities of the project.

The completion times of the various activities are listed as follows:

Activity :	*A*	*B*	*C*	*D*	*E*	*F*	*G*	*H*	*I*	*J*	*K*	*L*
Time (days) :	6	4	10	1	1	3	14	6	9	2	7	5

(a) Draw a network diagram corresponding to this project.
(b) Obtain the lengths of all the paths and determine critical path.

The precedence relationships implied from the given information are:

Activity	*Immediate Predecessor(s)*	*Activity*	*Immediate Predecessor(s)*
A	–	*G*	*C, F*
B	–	*H*	*B*
C	–	*I*	*E, H*
D	*A, B*	*J*	*E, H*
E	*B*	*K*	*C, D, F, J*
F	*B*	*L*	*K*

The network is shown in Figure 12.32.

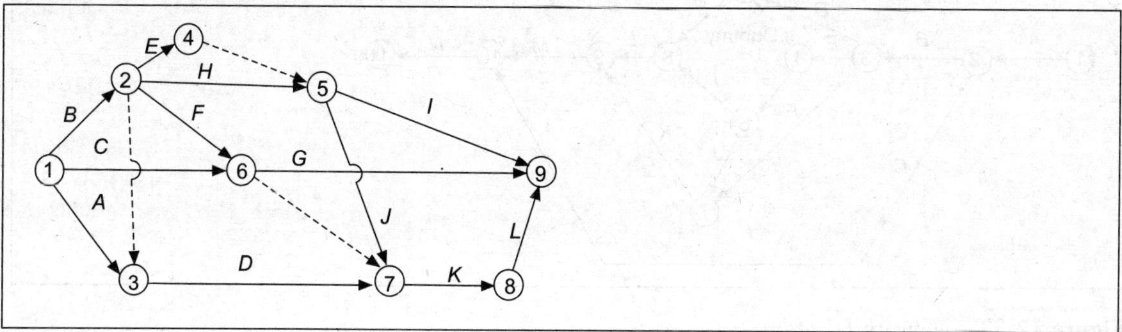

Figure 12.32 *Network Diagram*

The various paths and their lengths are as follows:

Path	*Length*
1–2–3–7–8–9	17
1–2–4–5–7–8–9	19
1–2–5–7–8–9	24* *Critical*
1–2–5–9	19
1–2–4–5–9	14
1–2–6–7–8–9	19
1–2–6–9	21
1–3–7–8–9	19
1–6–7–8–9	22
1–6–9	24* *Critical*

Example 12.11 A project has fourteen activities *A* through *M*. The relationships which obtain among these activities are given here. Construct the network and number them.

A is the first operation

B and *C* can be performed in parallel and are immediate successors to *A*
D, *E*, and *F* follow *B*
G follows *E*
H follows *D*, but it cannot start until *E* is complete
I and *J* succeed *G*
F and *J* precede *K*
H and *I* precede *L*
M succeeds *L* and *K*
The last operation *N* succeeds *M* and *C*. *(ICWA, December, 1985)*

The network corresponding to the given project is shown in Figure 12.33.

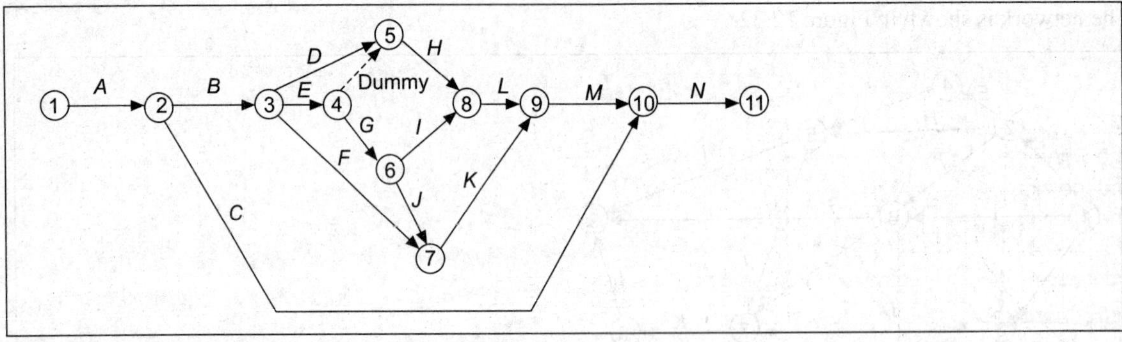

Figure 12.33 *Network Diagram*

Example 12.12 Point out the flaws in the network given in Figure 12.34.

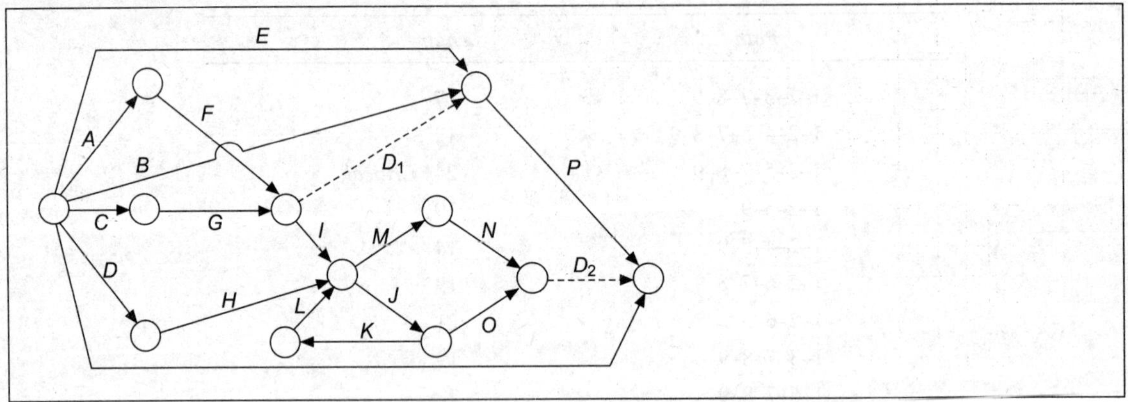

Figure 12.34 *Network Diagram*

On observation, we find that

(a) the activities *B* and *E* have the same initial and ending nodes which is not correct,

(b) a loop is formed by activities *J*, *K* and *L*, which is not permitted, and

(c) the dummy activity D_2 is unnecessary.

Example 12.13 A cabinet manufacturing company is planning to introduce a new model of cabinets which requires the following tasks:

Task	Description	Task Time (minutes)
A	Prepare the wheels	10
B	Mount the wheels	5
C	Assemble the sides	15
D	Attach the top	11
E	Attach the base	10
F	Insert the brackets	5
G	Insert the shelves	5
H	Attach the doors	10
I	Attach the back panel	10
J	Paint the unit	15

The wheels are mounted after they are prepared. The base cannot be attached until the sides are assembled and the wheels mounted. The top cannot be attached nor the brackets inserted until the sides are assembled. The shelves are inserted after the brackets are installed. The back panel is attached after the base and top are attached. The doors are attached after the shelves are inserted and the top and base are attached. The unit is painted after the back and doors are attached.

(i) Identify the immediate predecessors of each task and draw the network.

(ii) Find the critical path (s), and list the critical activities.

(iii) Obtain the earliest and latest start and completion times of all the activities and their total, free and independent floats. *(MIB, Delhi, 2010)*

Based on the given information, the immediate predecessors of each task are given in Table 12.13 and the network is shown in Figure 12.35. The table also gives the earliest and latest start and finish times together with floats. There are two critical paths: 1–3–6–8–9 and 1–3–6–7–8–9 and the critical activities are C, D, H, I and J.

TABLE 12.13 Precedence Relationships, Scheduling Times and Floats

Activity	*Imm. Pred.*	*Time*	*Earliest*		*Latest*		*Float*		
			Start	*Finish*	*Start*	*Finish*	*Total*	*Free*	*Ind.*
A	–	10	0	10	1	11	1	0	0
B	A	5	10	15	11	16	1	0	0
C	–	15	0	15	0	15	0	0	0
D	C	11	15	26	15	26	0	0	0
E	B,C	10	15	25	16	26	1	1	0
F	C	5	15	20	16	21	1	0	0
G	F	5	20	25	21	26	1	1	0
H	D, E, G	10	26	36	26	36	0	0	0
I	D, E	10	26	36	26	36	0	0	0
J	H, I	15	36	51	36	51	0	0	0

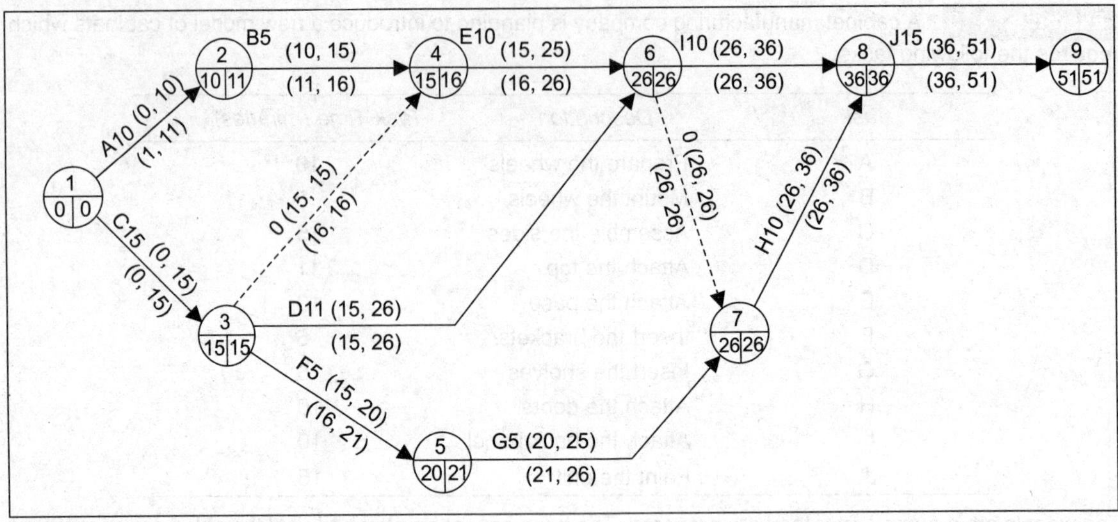

Figure 12.35 *Network*

Example 12.14 (a) Draw a network corresponding to the following information.

(b) Find the earliest and latest scheduling times of various activities.

(c) Obtain the total, interfering, free, and independent floats for each of the activities.

(d) Can this project be completed within 65 days?

(e) What would be the effect on the project length of reducing the resources to be used for activity 8–10 by such an amount as would increase the time for this activity by 5 days?

(f) The head of the department in which activity 6–8 is to be performed requests that he be allowed to work overtime so that the activity can be completed in 6 days. Should his request be considered in the interest of project completion at an earlier date? How about a similar request from the manager of the activity 2–7?

(g) It has come to be known that due to non-availability of resouces in time, activity 3–5 would be delayed by 9 days. Will it affect (i) the project completion time? (ii) the start of its successor activity? By how much?

Activity	*Time (Days)*	*Activity*	*Time (Days)*
1–2	8	6–8	10
1–3	2	7–10	12
1–4	6	8–9	3
1–5	12	8–10	6
2–4	5	9–12	8
2–7	9	10–12	18
3–5	3	10–14	9
3–6	7	11–12	7
4–10	4	11–14	4
5–11	10	12–13	11
6–7	2	13–14	4

(a) The network diagram is shown in the Figure 12.36.

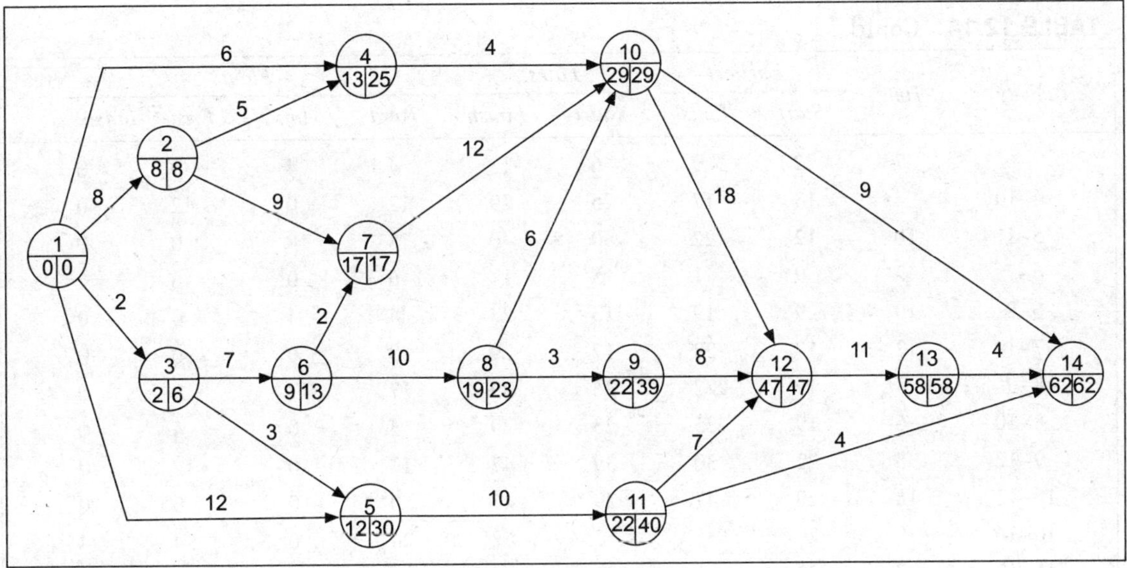

Figure 12.36 *Network Diagram*

(b) and (c) The earliest and the latest starting and finishing times of various activities are indicated on the figure and are given in Table 12.14. Also given in the table are the floats.

(d) Yes, the project can be completed in 65 days.

(e) Since the activity 8–10 has a total float equal to 4 days, a delay by 5 days in it would increase the project length by one day.

(f) The activity 6–8 is non-critical and has a total float equal to 6 days. Therefore, it is no use working overtime on it as a decrease in its time would not cause a change in the project duration. However, activity 2–7 is worthwhile working overtime on. If this activity is completed in 6 days, the project length would be reduced by 3 days that are saved on it. This is because activity 2–7 is a critical one.

(g) (i) This activity has a total float equal to 25 days. A delay of 9 days would not affect the project completion time. However, the delay of 9 days, would (ii) affect the start of its successor activity by 2 days because it has a free float equal to 7 days.

TABLE 12.14 Determination of Scheduling Times and Floats

Activity	*Time*	*Earliest*		*Latest*		*Float*			
		Start	*Finish*	*Start*	*Finish*	*Total*	*Int.*	*Free*	*Indep.*
1–2	8	0	8	0	8	0	0	0	0
1–3	2	0	2	4	6	4	4	0	0
1–4	6	0	6	19	25	19	12	7	7
1–5	12	0	12	18	30	18	18	0	0
2–4	5	8	13	20	25	12	12	0	0
2–7	9	8	17	8	17	0	0	0	0
3–5	3	2	5	27	30	25	18	7	3

(Contd.)

TABLE 12.14 Contd.

Activity	Time	Earliest Start	Earliest Finish	Latest Start	Latest Finish	Float Total	Float Int.	Float Free	Float Indep.
3–6	7	2	9	6	13	4	4	0	0
4–10	4	13	17	25	29	12	0	12	0
5–11	10	12	22	30	40	18	18	0	0
6–7	2	9	11	15	17	6	0	6	2
6–8	10	9	19	13	23	4	4	0	0
7–10	12	17	29	17	29	0	0	0	0
8–9	3	19	22	36	39	17	17	0	0
8–10	6	19	25	23	29	4	0	4	0
9–12	8	22	30	39	47	17	0	17	0
10–12	18	29	47	29	47	0	0	0	0
10–14	9	29	38	53	62	24	0	24	24
11–12	7	22	29	40	47	18	0	18	0
11–14	4	22	26	58	62	36	0	36	18
12–13	11	47	58	47	58	0	0	0	0
13–14	4	58	62	58	62	0	0	0	0

Example 12.15 A project has the following characteristics:

Activity	*Preceding Activity*	*Expected Completion Time (in Weeks)*
A	None	5
B	*A*	2
C	*A*	6
D	*B*	12
E	*D*	10
F	*D*	9
G	*D*	5
H	*B*	9
I	*C, E*	1
J	*G*	2
K	*F, I, J*	3
L	*K*	9
M	*H, G*	7
N	*M*	8

(a) Draw a PERT network for this project.

(b) Find the critical path and the project completion time.

(c) Prepare an activity schedule showing the *ES*, *EF*, *LS*, *LF*, and slack for each activity.

(d) Will the critical path change if activity *G* takes 10 weeks instead of 5 weeks? If so, what will be the new critical path?

(M Com, Delhi, 1987)

(a) The network for the given project is shown in Figure 12.37.

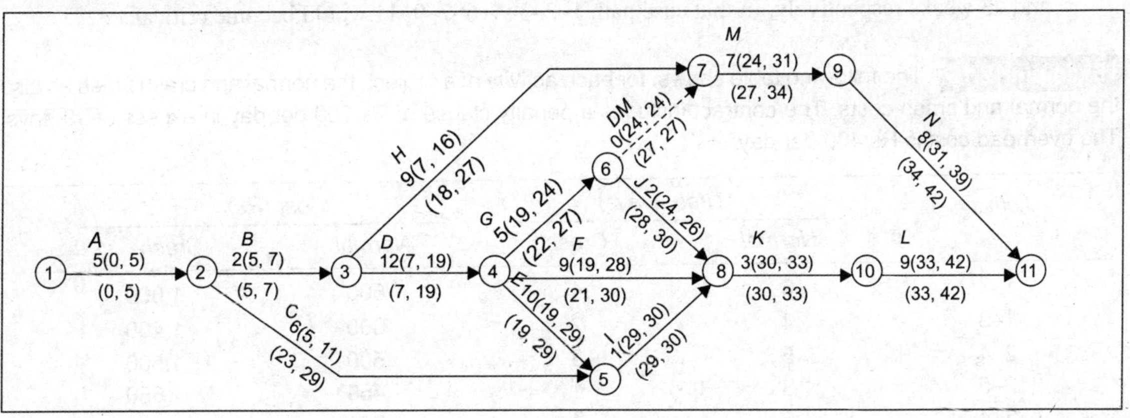

Figure 12.37 *Network Diagram*

(b) The various paths and their lengths are as follows:

Path	*Length (Weeks)*
(i) 1–2–5–8–10–11	24
(ii) 1–2–3–7–9–11	31
(iii) 1–2–3–4–6–7–9–11	39
(iv) 1–2–3–4–6–8–10–11	38
(v) 1–2–3–4–8–10–11	40
(vi) 1–2–3–4–5–8–10–11	42*

Thus, the critical path is 1–2–3–4–5–8–10–11 and the duration of the project is 42 weeks.

(c) The required information is given in Table 12.15.

TABLE 12.15 Determination of Earliest and Latest Times

Activity	*Duration*	*ES*	*EF*	*LS*	*LF*	*Slack*
A	5	0	5	0	5	0
B	2	5	7	5	7	0
C	6	5	11	23	29	18
D	12	7	19	7	19	0
E	10	19	29	19	29	0
F	9	19	28	21	30	2
G	5	19	24	22	27	3
H	9	7	16	18	27	11
I	1	29	30	29	30	0
J	2	24	26	28	30	4
K	3	30	33	30	33	0
L	9	33	42	33	42	0
M	7	24	31	27	34	3
N	8	31	39	34	42	3

(d) If activity G(4–6) takes 10 weeks instead of 5, the lengths of the paths (iii) and (iv) would become 44 and 43 weeks respectively. In that case path 1–2–3–4–6–7–9–11 would become critical.

Example 12.16 The following table shows, for each activity of a project, the normal and crash times as also the normal and crash costs. The contract includes a penalty clause of Rs 200 per day in excess of 19 days. The overhead cost is Rs 400 per day.

Activity	*Time (Days)*		*Cost (Rs)*	
	Normal	*Crash*	*Normal*	*Crash*
1–2	6	4	600	1,000
1–3	4	2	600	1,400
2–4	5	3	500	1,500
2–5	3	1	450	650
3–4	6	4	900	2,000
4–6	8	4	800	3,000
5–6	4	2	400	1,000
6–7	3	2	450	800

(a) Draw the project network and determine the critical path.

(b) Find the cost of completing the project in normal time.

(c) Crash the project activities and determine the cost of completing the project in the minimum time.

(d) What is the optimal duration of the project and what is the cost involved?

The network for the given project is shown in Figure 12.38.

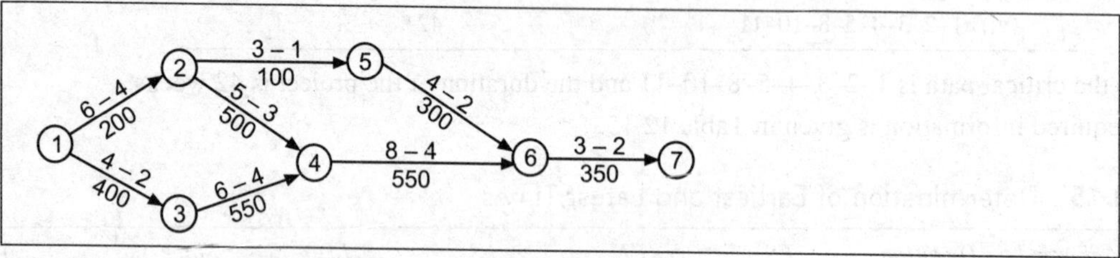

Figure 12.38 *Arrow Diagram*

From the diagram, we have:

Path	*Length*	
	Normal Time	*Crash Time*
1–2–5–6–7	16	9
1–2–4–6–7	22* (Critical)	13* (Critical)
1–3–4–6–7	21	12

For each path, the length is determined on the basis of (i) normal times and (ii) crash times. It is evident that the normal duration of the project is 22 days while the minimum duration is 13 days. The cost table is accordingly set up to show the cost of completing the project in 22, 21, 20,…13 days, as shown in Table 12.17. The total normal cost works out to be Rs 4,700, while the overhead is calculated at Rs 400 per day. These are shown in this table along with the given penalty values. The crashing cost per day for every activity is calculated and shown on lower side of the respective arrows in the network. Entries in the crashing cost column are made on the basis of the activity or activities crashed in a particular step. The activities chosen at every step are those which would reduce the length of the path(s) under consideration at minimum cost.

The step-wise crashing is shown in Table 12.16.

TABLE 12.16 Crashing the Project

Crashing	*Critical Path(s)*	*Decision*	*Cost*	*Duration after Crashing*
I	1–2–4–6–7	1–2	200	21
II	1–2–4–6–7	6–7	350	20
	1–3–4–6–7			
III	1–2–4–6–7	4–6	550	19
	1–3–4–6–7			
IV	1–2–4–6–7	4–6	550	18
	1–3–4–6–7			
V	1–2–4–6–7	4–6	550	17
	1–3–4–6–7			
VI	1–2–4–6–7	4–6	550	16
	1–3–4–6–7			
VII	1–2–4–6–7	1–2 and 1–3	600	15
	1–3–4–6–7			
VIII	1–2–4–6–7	1–3 and 2–4	900	14
	1–3–4–6–7			
IX	1–2–4–6–7	2–4 and 3–4	1,050	13
	1–3–4–6–7			

The length of various paths after crashing is given here:

Path	*Normal Length*	*Length after crashing*								
		I	*II*	*III*	*IV*	*V*	*VI*	*VII*	*VIII*	*IX*
1–2–5–6–7	16	15	14	14	14	14	14	13	13	13
1–2–4–6–7	22	21	20	19	18	17	16	15	14	13
1–3–4–6–7	21	21	20	19	18	17	16	15	14	13

TABLE 12.17 Determination of Total Cost

Project Duration (Days)	Direct Cost		Overhead	Penalty	Total Cost (Rs)
	Normal	Crashing			
22	4,700	0	8,800	600	14,100
21	4,700	200	8,400	400	13,700
20	4,700	550	8,000	200	13,450
19	4,700	1,100	7,600	–	13,400
18	4,700	1,650	7,200	–	13,550
17	4,700	2,200	6,800	–	13,700
16	4,700	2,750	6,400	–	13,850
15	4,700	3,350	6,000	–	14,050
14	4,700	4,250	5,600	–	14,550
13	4,700	5,300	5,200	–	15,200

From the table, it is evident that normal duration of the project is 22 days and the cost involved is Rs 14,100. The minimum time is 13 days at a total cost of Rs 15,200, while the optimal project duration is 19 days involving a cost of Rs 13,400.

Example 12.17 The data for a project are:

Activity	Preceding Activity	Time (in Weeks)		Cost (in Rs)	
		Normal	Crash	Normal	Crash
A	None	3	2	18,000	19,000
B	None	8	6	600	1,000
C	*B*	6	4	10,000	12,000
D	*B*	5	2	4,000	10,000
E	*A*	13	10	3,000	9,000
F	*A*	4	4	15,000	15,000
G	*F*	2	1	1,200	1,400
H	*C, E, G*	6	4	3,500	4,500
I	*F*	2	1	7,000	8,000

(a) Draw a project network diagram and find the critical path.

(b) If a deadline of 17 weeks is imposed for completion of the project, what activities will be crashed, what would be the additional costs, and what would be the critical activities of the network after crashing?

(M Com, Delhi, 1987)

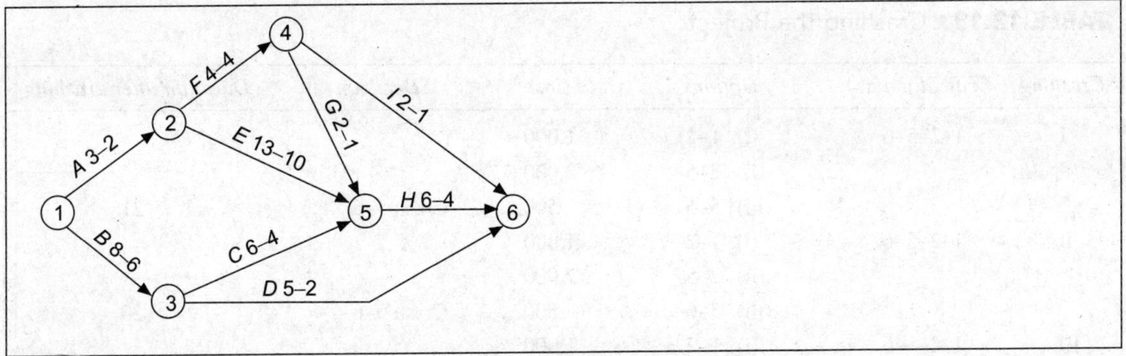

Figure 12.39 *Arrow Diagram*

The network diagram for the project is shown in Figure 12.39. The critical path of the project is 1–2–5-6, with a length of 22 weeks.

To complete the project in the stipulated 17 weeks, crashing would have to be done. In the first instance, we shall determine the weekly crashing cost for each of the activities. We have,

$$\text{Crashing cost per week} = \frac{\text{Crash Cost} - \text{Normal Cost}}{\text{Normal Time} - \text{Crash Time}}$$

Using this, we get the following information.

Activity		*Crashing Cost (Rs) per Week*
A	1–2	1,000/1 = 1,000
B	1–3	400/2 = 200
C	3–5	2,000/2 = 1,000
D	3–6	6,000/3 = 2,000
E	2–5	6,000/3 = 2,000
F	2–4	–
G	4–5	200/1 = 200
H	5–6	1,000/2 = 500
I	4–6	1,000/1 = 1,000

The activities to be crashed and the cost involved is shown in Table 12.18.

Thus, we have,

(a) Activities to be crashed: *A*–One week; *B*–One week; *E*–Two weeks; and *H*–Two weeks.

(b) Total Cost of the Project = Normal Cost + Crashing Cost

Normal Cost = 18,000 + 600 + 10,000 + 4,000 + 3,000 + 15,000 + 1,200 + 3,500 + 7,000 = 62,300

Crashing Cost = 500 + 500 + 1,000 + 2,000 + 2,200 = 6,200

∴ Total Cost = 62,300 + 6,200 = Rs 68,500

(c) Activities which are critical after crashing are: *A*, *B*, *C*, *E* and *H*.

TABLE 12.18 Crashing the Project

Crashing	Critical path(s)	Options	Cost	Decision	Duration after crashing
I	1–2–5–6	(i) 1–2	1,000		
		(ii) 2–5	2,000		
		(iii) 5–6	500	Crash 5–6	21
II	1–2–5–6	(i) 1–2	1,000		
		(ii) 2–5	2,000		
		(iii) 5–6	500	Crash 5–6	20
III	1–2–5–6	(i) 1–2	1,000		
		(ii) 2–5	2,000	Crash 1–2	19
IV	1–2–5–6	(i) 2–5	2,000	Crash 2–5	18
V	1–2–5–6	(i) 2–5			
	1–3–5–6	1–3	2,200		
		(ii) 2–5			
		3–5	3,000	Crash 2–5, 1–3	17

The length of various paths after each crashing is given here:

Path	Normal Length	Length after Crashing				
		I	II	III	IV	V
1–2–4–6	9	9	9	8	8	8
1–2–4–5–6	15	14	13	12	12	12
1–2–5–6	22	21	20	19	18	17
1–3–5–6	20	19	18	18	18	17
1–3–6	13	13	13	13	13	12

Example 12.18 Following are the manpower requirements for each activity in a project:

Activity	Normal time (day)	Manpower required per day
1–2	10	2
1–3	11	3
2–4	13	4
2–6	14	3
3–4	10	1
4–5	7	3
4–6	17	3
5–7	13	5
6–7	9	8
7–8	1	11

(a) Draw the network and find out total float and free float for each activity.

(b) The contractor stipulates that during the first 26 days only 4 to 5 men and during remaining days 8 to 11 men only can be made available. Rearrange the activities suitably for levelling the manpower resources, satisfying the above condition. *(ICWA, December, 1985)*

The network corresponding to the above project is shown in Figure 12.40. Also given are the *ES*, *EF*, *LS* and *LF* times of the various activities.

The earliest and the latest start and finishing times are given in Table 12.19, which also contains the information on the total and free float for each of the activities.

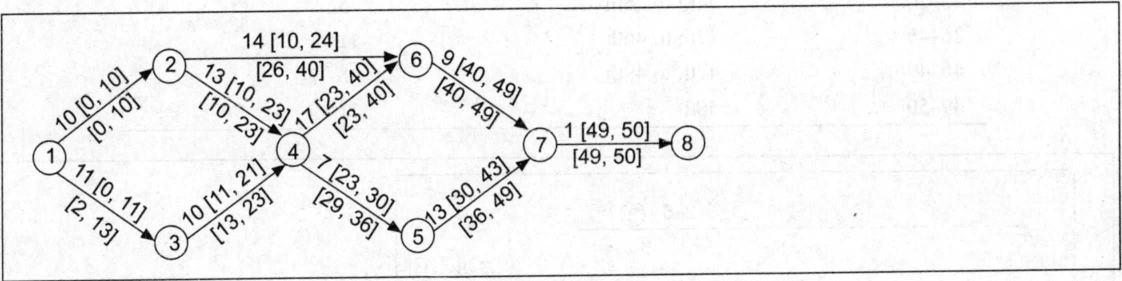

Figure 12.40 *Network Diagram*

TABLE 12.19 Activity Scheduling Times and Float

Activity	*Duration*	*ES*	*EF*	*LS*	*LF*	*Total Float*	*Free Float*
1–2	10	0	10	0	10	0	0
1–3	11	0	11	2	13	2	0
2–4	13	10	23	10	23	0	0
2–6	14	10	24	26	40	16	16
3–4	10	11	21	13	23	2	2
4–5	7	23	30	29	36	6	0
4–6	17	23	40	23	40	0	0
5–7	13	30	43	36	49	6	6
6–7	9	40	49	40	49	0	0
7–8	1	49	50	49	50	0	0

Levelling the Manpower It is sought here to rearrange the activities so that they can be performed with the given resource availabilities, in the stipulated time of 50 days (the project length). The activities 1–2, 2–4, 4–6, 6–7, and 7–8, being critical, would be scheduled first, as shown in Figure 12.41. Next, activity 1–3 is also scheduled at $t = 0$, followed by the activity 3–4. Although scheduling 1–3 at $t = 0$ would imply that the manpower requirement would rise to 7 on the 11th day of the project, which is in excess of the availability. However, we have no choice and have to schedule it because, otherwise, it would result in a delay of the project.

Since activities 2–6 and 4–5 can both be delayed to $t = 26$ and $t = 29$ respectively, they can both be scheduled at $t = 26$, when more men would be available. Finally, activity 5–7 is scheduled at $t = 33$.

The manpower requirements of various activities are shown in circles, in the Figure 12.41. Accordingly, the total daily requirements of the manpower are as follows:

t	Day(s)	Requirement
0–10	1st to 10th	5
10–11	11th	7 (overtime required)
11–21	12th to 21st	5
21–23	22nd to 23rd	4
23–26	24th to 26th	5
26–46	27th to 46th	11
46–49	47th to 49th	8
49–50	50th	11

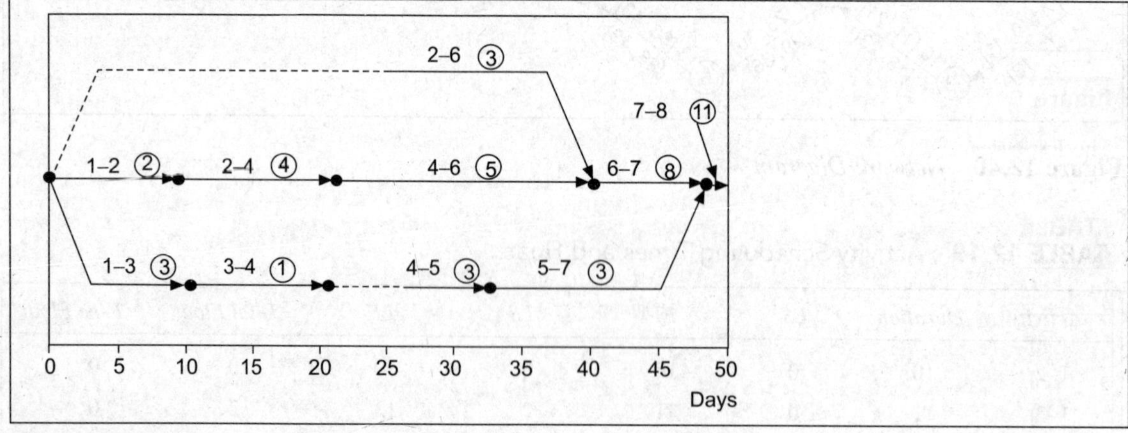

Figure 12.41 *Scheduling the Activities*

Example 12.19 A project consists of 10 activities, each of which requires either, or both, of the two types of resources R_1 and R_2 for its performance. The duration of the activities and their resource requirements are as follows:

Activity	Duration (Days)	Resource Requirement	
		R_1	R_2
1–2	3	3	2
1–3	2	6	
1–4	6	4	–
2–6	4	–	4
3–5	2	2	2
4–5	1	4	–
4–8	4	4	–
5–7	3	3	2
6–7	2	1	3
7–8	4	4	5

Resource availability: R_1 : 7 units, R_2 : 5 units

Determine the duration of the project under the given resource constraint. If the resources were not a problem, how long would the project take to complete in the normal course?

The network corresponding to the given project is shown in Figure 12.42.

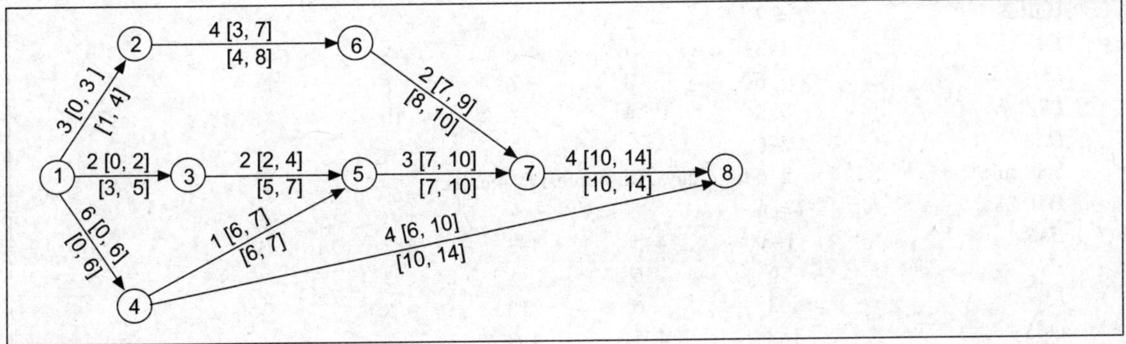

Figure 12.42 *Network for the Project*

The *ES*, *EF*, *LS* and *LF* times for each of the activities are given in the Table 12.20.

TABLE 12.20 Resource Requirements and Scheduling Times

Activity	*Duration*	R_1	R_2	*ES*	*EF*	*LS*	*LF*
1–2	3	3	2	0	3	1	4
1–3	2	6	–	0	2	3	5
1–4	6	4	–	0	6	0	6
2–6	4	–	4	3	7	4	8
3–5	2	2	2	2	4	5	7
4–5	1	4	–	6	7	6	7
4–8	4	4	–	6	10	10	14
5–7	3	3	2	7	10	7	10
6–7	2	1	3	7	9	8	10
7–8	4	4	5	10	14	10	14

Now we shall allocate the given resources in accordance with the rules given earlier. This is shown in Table 12.21.

TABLE 12.21 Scheduling of Activities and Allocation of Resources

Halt 1		$T = 0$					
EAS	:	1–2*	1–3	1–4*	2–6	4–5	4–8
ES	:	0	0	0			
LS	:	1	3	0			
OAS	:	1–4	1–2	1–3			

(Contd.)

TABLE 12.21 Contd.

Schedule	1–4 for 6 days, Resources: R_1 4 units;				
	1–2 for 3 days, Resources: R_2 3 units, R_2 2 units				
Halt 2	$T = 3$				
EAS	1–3	2–6*	4–5	4–8	6–7
ES	6	3	6	6	
LS	3	4	6	10	
OAS	2–6				
Schedule	2–6 for 4 days, Resources: R_2 4 units				
Halt 3	$T = 6$				
EAS	1–3*	4–5	4–8	6–7	3–5
ES	6	6	6	7	
LS	3	6	10	8	
OAS	1–3	4–5	4–8		
Schedule	1–3 for 2 days, Resources: R_1 6 units				
Halt 4	$T = 7$				
EAS	4–5	4–8	6–7*	3–5	
ES	8	8	7	8	
LS	6	10	8	5	
OAS	6–7				
Schedule	6–7 for 2 days, Resources: R_1 1 units, R_2 3 units				
Halt 5	$T = 8$				
EAS	4–5*	4–8	3–5*	5–7	
ES	8	8	8		
LS	6	10	5		
OAS	3–5	4–5	4–8		
Schedule	3–5 for 2 days, Resources: R_1 2 units, R_2 2 units				
	4–5 for 1 day, Resources: R_1 4 units				
Halt 6	$T = 9$				
EAS	4–8	5–7			
ES	9	10			
LS	10	7			
OAS	4–8				
Schedule	4–8 for 4 days, Resources: R_1 4 units				
Halt 7	$T = 10$				
EAS	5–7	7–8			
ES	10				
LS	7				
OAS	5–7				
Schedule	5–7 for 3 days, Resources: R_1 3 units, R_2 2 units				
Halt 8	$T = 13$				
EAS	7–8				
Schedule	7–8 for 4 days, Resources: R_1 4 units, R_2 5 units				

The loading chart is shown in Figure 12.43.

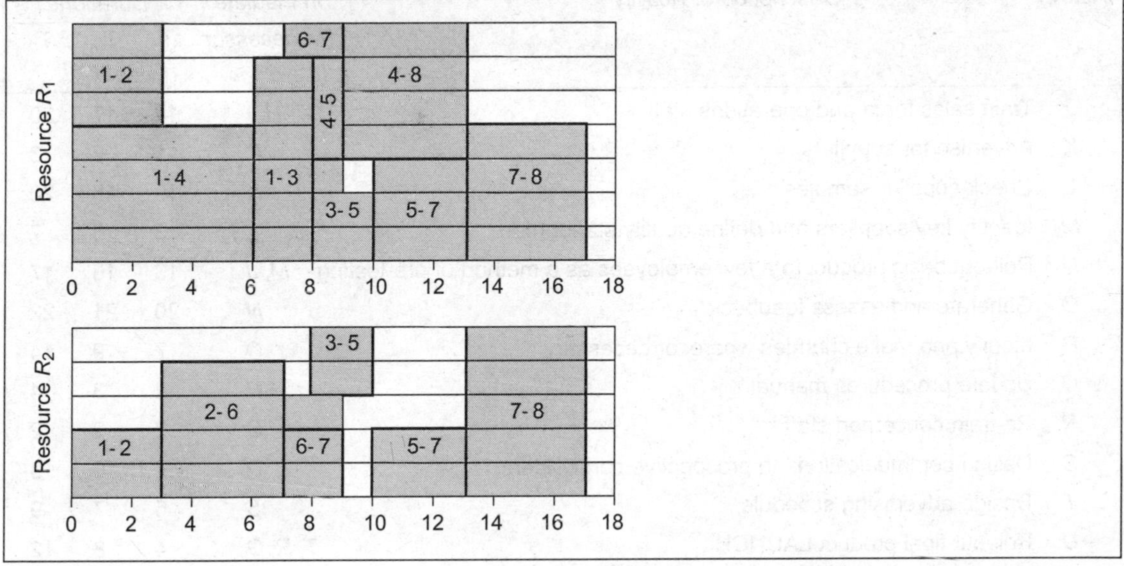

Figure 12.43 *Loading Chart*

With the given resource constraints, the project would be completed in 17 days. However, if there were no constraints on the resources, the project would be completed in 14 days, under normal conditions (implying when there is no crashing).

Example 12.20 Following table gives the list of various activities involved in the launch of a new CREDIT CARD service by a company, their immediate predecessors and their expected durations (in days):

Activity	*Description of Activity*	*Immediate Predecessor*	*Expected Durations* T_o	T_m *(in days)*	T_P
A	Conduct Market Research to determine size, potential and competition in the market	–	10	12	14
B	Define Marketing Strategy in terms of positioning and product features	*A*	14	15	17
C	Estimate expected volumes	*B*	2	3	4
D	Estimate additional manpower required	*C*	4	6	8
E	Identify modifications required to the present computer system based on expected volumes	*C*	10	12	14
F	Implement computer system modifications	*E*	20	25	27
G	Update physical facilities	*C*	10	17	20
H	Prepare instruction manuals for modified system	*F*	5	6	7
I	Hire additional Sales force Operations Staff	*D*	7	12	14

(Contd.)

Activity	Description of Activity	Immediate Predecessor	Expected Durations (in days) T_o	T_m	T_P
J	Train sales force and operations staff	H, I	14	17	20
K	Advertise for suppliers	C	1	2	3
L	Check supplier samples	K	10	15	20
M	Identify key suppliers and define quality standards	L	3	5	7
N	Roll-out basic product to a few employees as a method of pre-testing	M, J	13	15	17
O	Generate and assess feedback	N	20	21	22
P	Modify and make changes wherever necessary	O	7	9	14
Q	Update procedures manual	P	2	3	4
R	Re-train concerned staff	Q	2	2	2
S	Design communications to prospective cardholders	P	7	10	13
T	Design advertising schedule	S	5	7	9
U	Roll-out final product-LAUNCH	T, R, G	4	8	12

where T_o = most optimistic time, T_m = most likely time, and T_p = most pessimistic time

(a) Draw an arrow diagram for the project.

(b) Find the expected project completion time of the project.

(c) Determine the probability of completing the project in 165 days.

(MBA, Delhi, November, 1998)

As a first step, the expected duration, t_e, and variance, σ^2, for each of the activities are calculated and shown below. Here,

$$t_e = \frac{a + 4m + b}{6} \quad \text{and} \quad \sigma^2 = \left(\frac{b-a}{6}\right)^2$$

where $a = T_o$, $m = T_m$ and $b = T_p$

Activity	a	m	b	t_e	σ^2
A	10	12	14	12	4/9
B	14	15	17	91/6	1/4
C	2	3	4	3	1/9
D	4	6	8	6	4/9
E	10	12	14	12	4/9
F	20	25	27	49/2	49/36
G	10	17	20	49/3	25/9
H	5	6	7	6	1/9
I	7	12	14	23/2	49/36
J	14	17	20	17	1
K	1	2	3	2	1/9

(Contd.)

Activity	a	m	b	t_e	σ^2
L	10	15	20	15	25/9
M	3	5	7	5	4/9
N	13	15	17	15	4/9
O	20	21	22	21	1/9
P	7	9	14	19/2	49/36
Q	2	3	4	3	1/9
R	2	2	2	2	0
S	7	10	13	10	1
T	5	7	9	7	4/9
U	4	8	12	8	16/9

Based on the given precedence relationships, the network is shown in Figure 12.44. The expected durations for various activities are also given.

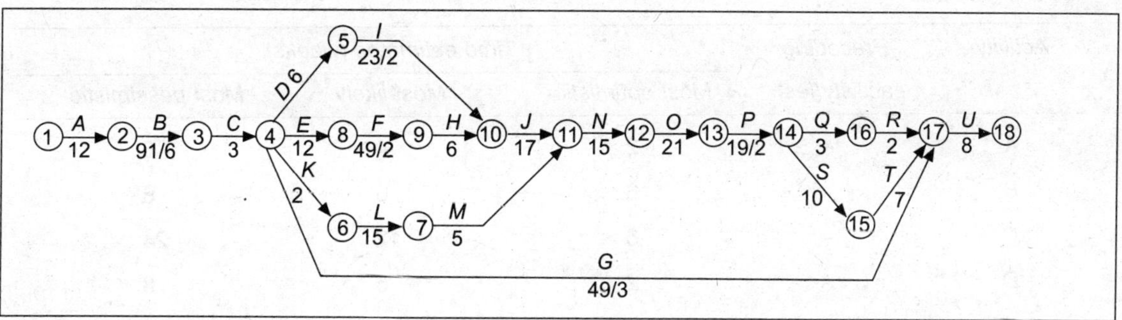

Figure 12.44 *Arrow Diagram*

The various paths and their lengths are as follows:

Path	Length
1–2–3–4–5–10–11–12–13–14–16–17–18	123⅙
1–2–3–4–5–10–11–12–13–14–15–17–18	135⅙
1–2–3–4–8–9–10–11–12–13–14–16–17–18	148⅙
1–2–3–4–8–9–10–11–12–13–14–15–17–18	160⅙
1–2–3–4–6–7–11–12–13–14–15–17–18	110⅔
1–2–3–4–6–7–11–12–13–14–16–17–18	122⅔
1–2–3–4–17–18	54½

The longest path has a duration of $160\frac{1}{6}$ or 160.17 days. The critical activities are: *A, B, C, E, F, H, J, N, O, P, S, T* and *U*. Adding up the variances of these activities, we get

$$\sigma^2 = \frac{4}{9}+\frac{1}{4}+\frac{1}{9}+\frac{4}{9}+\frac{49}{36}+\frac{1}{9}+1+\frac{4}{9}+\frac{1}{9}+\frac{49}{36}+1+\frac{4}{9}+\frac{16}{9} = 319/36$$

$$\therefore \qquad \sigma = \sqrt{319/36} = 2.98$$

Thus, this project has an expected duration of 160.17 days with a standard deviation of 2.98 days.

The probability of completing the project in 165 days is given by the area under the normal curve to the left of $X = 165$, which has $\mu = 160.17$ and $\sigma = 2.98$. This is shown in Figure 12.45. Now, $Z = \frac{X-\mu}{\sigma} = \frac{165-160.17}{2.98} = 1.62$. From the normal area table, it is observed that the area between μ and $Z = 1.62$ is 0.4474. Thus, total area to the left of $X = 165$ would equal $0.5 + 0.4474 = 0.9474$. Accordingly, the probability that the project would be completed within 165 days = 0.9474.

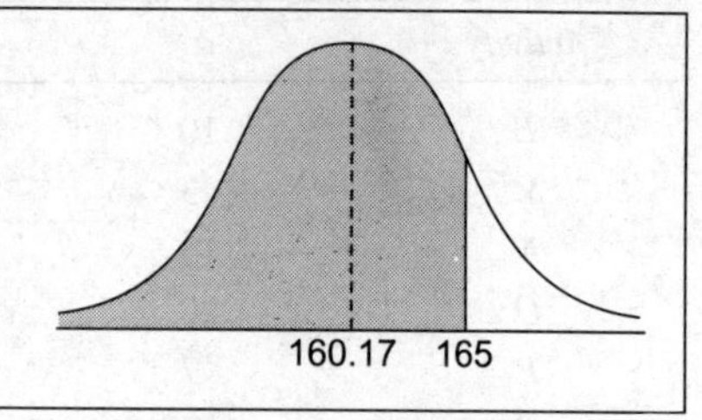

Figure 12.45 *Distribution of Project Duration*

Example 12.21 A project consists of nine activities whose time estimates (in weeks) and other characteristics are given below:

Activity	*Preceding activity(ies)*	*Time estimates (weeks)*		
		Most optimistic	*Most likely*	*Most pessimistic*
A	–	2	4	6
B	–	6	6	6
C	–	6	12	24
D	*A*	2	5	8
E	*A*	11	14	23
F	*B, D*	8	10	12
G	*B, D*	3	6	9
H	*C, F*	9	15	27
I	*E*	4	10	16

(i) Show the PERT network for the project.

(ii) Identify the critical activities.

(iii) What is the expected project completion time and its variance?

(iv) What is the probability of completing the project one week before the expected time?

(v) If the project is required to be completed by December 31 of a given year and the manager wants to be 95% sure of meeting the deadline, when should he start the project work?

(vi) A penalty of Rs 15,000 per week is to be imposed on the contractor if the project is not completed in 36 weeks. What is the probability that he has to pay a penalty? a penalty of Rs 45,000?

To use as input, we first obtain the expected duration and variance for each of the activities as follows:

Activity	a	m	b	t_e	σ^2
A	2	4	6	4	4/9
B	6	6	6	6	0
C	6	12	24	13	9
D	2	5	8	5	1
E	11	14	23	15	4
F	8	10	12	10	4/9
G	3	6	9	6	1
H	9	15	27	16	9
I	4	10	16	10	4

(i) The project network is shown in Figure 12.46.

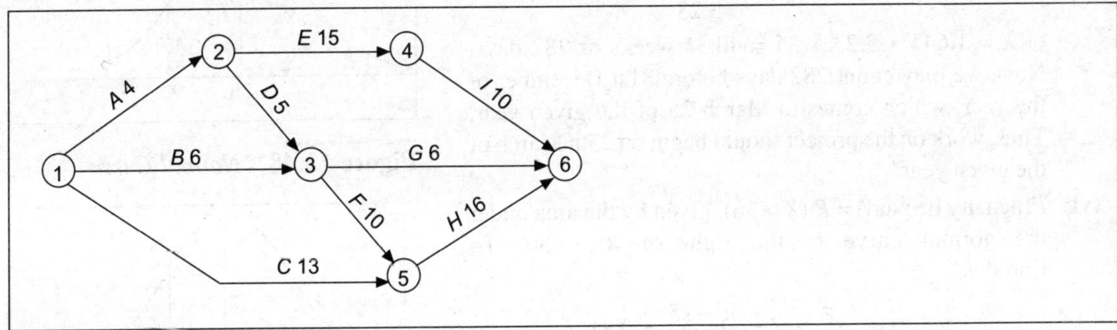

Figure 12.46 *Arrow Diagram*

(ii) and (iii) The various paths and their lengths are given below:

Path	*Length*
1–2–4–6	29
1–2–3–6	15
1–2–3–5–6	35*
1–3–6	12
1–3–5–6	32
1–5–6	29

Thus, the expected project completion time = 35 weeks. The critical path is 1–2–3–5–6 and critical activities are:

$$1\text{–}2\ (A),\ 2\text{–}3\ (D),\ 3\text{–}5\ (F) \text{ and } 5\text{–}6\ (H).$$

The variance of the project completion time = $\frac{4}{9} + 1 + \frac{4}{9} + 9 = \frac{98}{9}$ weeks^2 and the standard deviation = $\sqrt{98/9}$ = 3.23 weeks.

(iv) The required probability is given by the area to the left of $X = 34$, under normal curve with mean = 35 and standard deviation, $\sigma = 3.23$, as shown in Figure 12.47.

To find it,

$$Z = \frac{X-\mu}{\sigma} = \frac{34-35}{3.23} = -0.31$$

From the normal area table, area between μ and $Z = -0.31$ is 0.1217. Thus, $P(X < 34) = 0.5 - 0.1217 = 0.3783$.

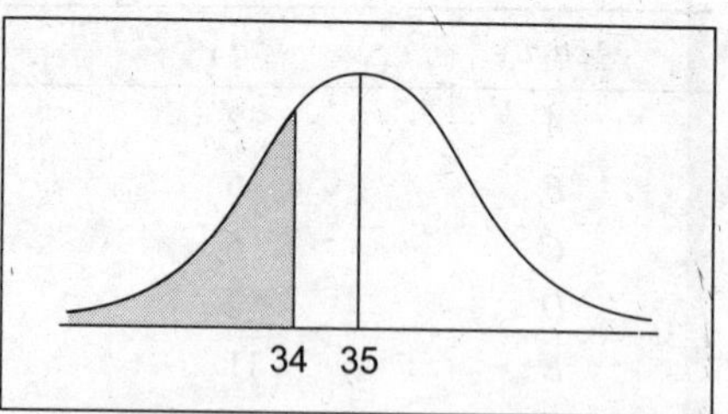

Figure 12.47 *Determination of Area to Left of X = 34*

(v) Let the project begin X weeks prior to the deadline. Here X is the value, to the left of which 0.95 area under the normal curve lies, as shown in Figure 12.48. Obviously then, the area between $\mu = 35$ and X equals $0.95 - 0.50 = 0.45$. From the normal area table, the Z-value corresponding to area = 0.45 works out to be 1.645 (the average of 1.64 and 1.65). Thus,

$$1.645 = \frac{X-35}{3.23}$$

or $X = 1.645 \times 3.23 + 35 = 40.31$ weeks or 282 days. Now, we may count 282 days before 31st December of the year, which comes to March 25 of the given year. Thus, work on the project should begin on 25th March of the given year.

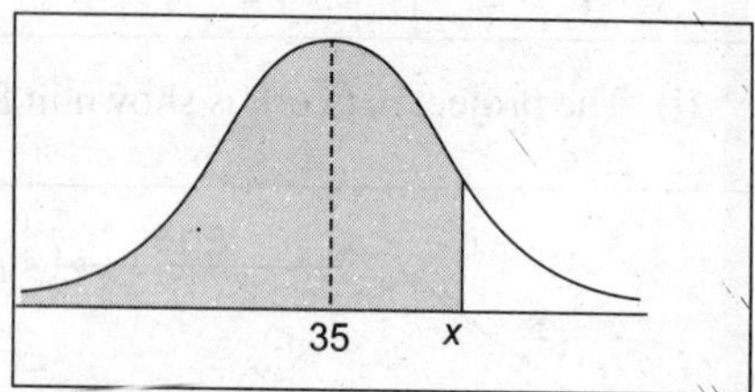

Figure 12.48 *Normal Curve*

(vi) P(penalty be paid) = $P(X > 36)$, given by the area under the normal curve to the right of $X = 36$. To find this,

$$Z = \frac{36-35}{3.23} = 0.31$$

From the normal area table, the area is found to be 0.1217 for $Z = 0.31$. Thus, the required area $= 0.5 - 0.1217 = 0.378$, which is the desired probability.

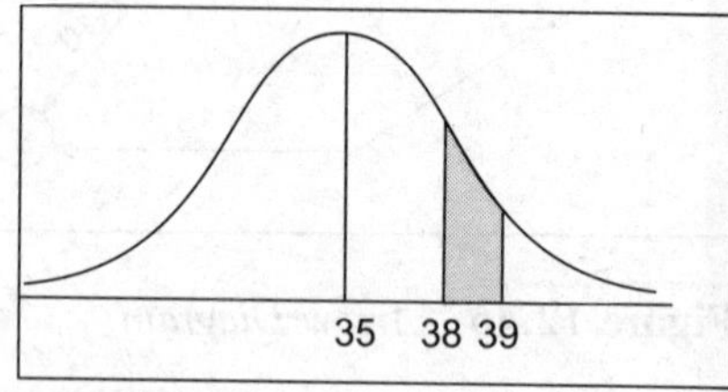

Figure 12.49 *Normal Curve*

Rs 45,000 Penalty A penalty of Rs 45,000 would be paid when the project would be completed in 39th week, or between $X = 38$ and $X = 39$, as shown in Figure 12.49. To find the probability, we have to determine the area between $X = 38$ and $X = 39$.

$$Z_1\ (X = 39) = \frac{39-35}{3.23} = 1.24 \quad \text{Area} = 0.3925$$

$$Z_2\ (X = 38) = \frac{38-35}{3.23} = 0.93 \quad \text{Area} = 0.3238$$

Thus, area between $X = 38$ and $X = 39$ is equal to $0.3925 - 0.3238 = 0.0687$. Accordingly, the probability that a penalty of Rs 45,000 be paid = 0.0687.

KEY POINTS TO REMEMBER

- PERT and CPM are two network analysis techniques used in planning and controlling projects. They both use similar concepts. While CPM is deterministic, the PERT is probabilistic in nature.

- A project can be viewed as a set of activities that are performed in a logically determined sequence. Thus, in dealing with a given project, first the activities are identified and precedence relationships are established.
- The set of activities is depicted digrammatically by means of a network/arrow diagram, wherein each activity is represented by an arrow, with a circle on either end (known as 'initial' and 'final' nodes). It may be necessary to use dummy activities for depicting correctly the given precedence relationships.
- When the times for all the activities are given, length of each of the paths is determined. The longest path is called the critical path and activities on it are called critical activities. Its length gives the project duration.
- Besides project duration, scheduling times of various activities are also determined: 'earliest' start and finish times are obtained by forward pass calculations and 'latest' start and finish times by backward pass calculations. The earliest and latest event times are also determined similarly.
- Critical activities have no float. For non-critical activities, total, free, interfering and independent floats are calculated as follows:

  Total float = latest start – earliest start of the activity

  Free float = earliest start of the next activity – earliest finish of the activity

  Interfering float = total float – free float of the activity

  Independent float = earliest start of the next activity – latest finish of the preceding activity – duration of the activity, or zero, whichever is higher
- Crashing of a project implies attempting to reduce the duration of a project by employing additional resources. For this, the cost of reduction per unit time for every activity is obtained first. Then the time reduction is attempted step-by-step (that is, day-by-day) by concentrating on the critical path, choosing the cheapest activity every time. In case of multiple critical paths, the cheapest combination of activities is taken.
- Resource levelling implies scheduling the performance of various activities of a given project in such a way that the requirement of resource/s may be levelled, to the extent possible, over the duration of the project. For this, the given project is redrawn on time scale, assuming that the activities would begin at their earliest and the resource requirement is determined. After this, the activities are re-scheduled in such a way that the requirement is balanced.
- Resource allocation deals with the allocation of limited amount of available resources in the best possible manner so that the project delay is kept at the minimum. The criteria for resource allocation are float, activity duration and quantum of resources needed, in that order.
- In PERT analysis, for every activity three time estimates are taken: a (optimistic), m (most likely) and b (pessimistic). With these, expected duration and variance for activity completion are obtained as

$$t_e = \frac{a + 4m + b}{6} \text{ and } \sigma^2 = \left(\frac{b - a}{6}\right)^2$$

- Using expected times, the critical path is obtained and critical activities are determined. The length of critical path indicates the expected duration of the project. Further, the summation of the critical activities' variances yields variance of the project completion time. Its square root gives the standard deviation.
- Probabilities of completion of the project are calculated by considering the project completion time to be normally distributed with parameters mean and standard deviation as obtained above.
- PERT/Cost is an extension of PERT. It deals with planning, monitoring and controlling project costs. Using PERT/Cost, it is possible to gain information about cost overrun or cost under-run and whether the project is moving on schedule.

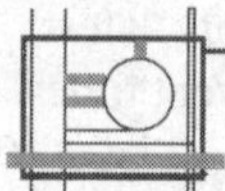

TEST YOUR UNDERSTANDING

Mark the following statements as T (*True*) or F (*False*).

1. A project network cannot have multiple initial and final nodes.
2. A dummy activity in a project network always has a zero duration.
3. It is desirable to have as many dummy activities in a project network as possible.
4. Two activities, P and Q, share common predecessors and successors. Therefore, they can have common initial and final nodes.
5. A dummy activity emanating from a given node is seen to extend *all* the activities which are terminating at that node.
6. Inclusion of more-than-desired number of dummy activities in a network can cause a delay in the completion of the project.
7. Events marking the start of activities are called *head-events*.
8. An activity cannot start unless *all* its immediate predecessors are completed.
9. Every activity which starts from an event has all the activities ending on that node as its immediate predecessors.
10. For every activity, whether real or dummy, the terminal node should bear a number higher than the one born by the initial node.
11. Occasionally, a project network may contain a sequence of activities that form a loop.
12. Critical path is the longest path in a project network.
13. It is not possible for a network to have more than one critical path.
14. The critical path of a project network represents the minimum time needed to complete the project.
15. A delay in the completion of critical activities need not cause a delay in the completion of the whole project.
16. Earliest start and completion times of the various activities of a project are obtained by forward pass calculations while the latest start and completion times are obtained by backward pass calculations.
17. The minimum time required to complete a project is given by the length of the shortest path of its network.
18. The difference between the earliest start of the successor activity and the latest completion of the given activity is termed as *free float*.

19. The latest finish (*LF*) time of an activity is equal to the lowest of the latest start (*LS*) times of the successor activities.
20. Every critical activity has identical earliest and latest start times, and earliest and latest finish times.
21. Interfering float is the excess of the total float over the free float.
22. Independent float, $IF_{i-j} = \text{Max}\ \{0, ES_{j-k} - LF_{h-i} - t_{i-j}\}$
23. Total float is the aggregate of the free, interfering and independent floats.
24. Every non-critical activity can be scheduled at any time between its earliest and latest start times.
25. For each critical activity, total, free and independent floats are equal to zero.
26. A non-critical activity must have a positive total float.
27. An activity with some free float cannot be a critical activity.
28. If a non-critical activity has no free float, it implies that the start of one or more of its immediately succeeding activities is dependent on the completion time of this activity.
29. Information about the float times for various activities is essential in effecting resource levelling.
30. It is possible to reduce the normal duration of a project by employing additional resources.
31. The total cost of doing a project is equal to the sum of the direct cost of completing the critical activities and the indirect cost, determined from the time taken to complete the project.
32. While the direct cost is concerned with individual activities of a project, the indirect cost, or overhead, is related to the whole project and is usually a function of time.
33. The *optimal duration* of a project is the minimum time in which it can be completed.
34. In case crashing is possible for a project, the minimum time in which a project can be completed would be given by the length of the critical path determined using crash times for various activities.
35. For an activity, incremental cost can be calculated as follows:

    Incremental cost per day = (normal cost – crash cost)/(normal time – crash time)
36. In crashing, we should always concentrate on critical path and choose between critical activities only.
37. An activity, may or may not be critical, with the minimum crashing cost should be crashed first.
38. When there are multiple critical paths, any of them may be reduced in duration when crashing.
39. It is possible to reduce an activity's duration below its crash time by allocating more resources and funds to it.
40. Resource levelling aims at smoothening of the resource usage rate without changing the project duration.
41. Ordering the activities, which are eligible to be allocated the resources at a given point of time, on the basis of slack is identical to ordering them on the basis of their latest start times.
42. The resource allocation programmes aim to allocate the variable resource supplies to activities of project with a view to complete the project in the minimum time.
43. PERT is probabilistic in nature while CPM is deterministic.
44. Completion time of each activity in a PERT network is assumed to follow beta distribution.
45. The expected completion time of an activity is the weighted average of the optimistic, pessimistic and most-likely times, wherein the "most-likely time" is assigned four times as much weightage as the weightage to the optimistic and the pessimistic times each.
46. The pessimistic time for every activity in a PERT project must always be greater than its optimistic time.
47. In PERT calculations, the critical path is determined by using the optimistic times for various activities.

48. Probability calculations in PERT represent only the probability of completing the activities on the critical path. It is implicitly assumed that the remaining activities would also be over by the time the critical path activities are completed.
49. The standard deviation of the project completion time is obtained by adding the standard deviations of the times of completing the critical activities.
50. In case of more than one critical path for a PERT network, the one with the least variance is selected for making probability calculations.
51. While activity times are variable in PERT, in CPM they are fixed and cannot be varied at all.
52. PERT/Cost aims at cost reduction.
53. PERT/Cost deals with planning, monitoring and controlling project costs.
54. If the actual time and actual cost are different from expected time and cost respectively, then it results in budget overrun.
55. Cost overrun refers to the excess of actual cost over the value of work done on a given activity.

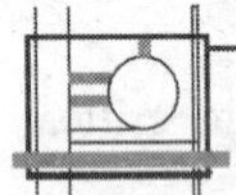

EXERCISES

1. What kinds of decision-making situations may be analysed using PERT and CPM techniques? State the major similarities between PERT and CPM. Under what circumstances is CPM a better technique of project management than PERT?
2. A construction company has received a contract to build an office complex. It has frequently engaged itself in constructing such buildings. Which of the two network techniques, PERT and CPM, should in your opinion, be employed by the company? Why?
3. What are the major comparative characteristics of the PERT model and the CPM model? What are their limitations, if any? Discuss.
4. State the rules of constructing a project network.
5. What is a dummy activity? Why do we need dummy activities? In this context, bring out clearly the difference between identity and logic dummies.
6. Illustrate (i) merge and burst events; (ii) appearance of a loop in a project network.
7. What are forward pass and backward pass? Explain the determination of the earliest and the latest start and completion times of the different activities of a project. Also explain how are the earliest and the latest event times obtained?
8. What is critical path? State the necessary and sufficient conditions of critical path. Can a project have multiple critical paths?
9. Why is the critical path of such importance in large project scheduling and control? Can a critical path change during the course of project? Why?
10. Explain the concept of float. Distinguish clearly between the free, interfering and independent floats.
11. What are the three time estimates used in the context of PERT? How are the expected duration of a project, and its standard deviation calculated?
12. Write a note on crashing.
13. (a) How do you distinguish between resource levelling and resource allocation problems? State and explain an algorithm for resource allocation.

(b) Explain the following as they are used in PERT/CPM

(i) Beta distribution, and (ii) Budget overrun.

14. "PERT network will act as only wall decorators in a business enterprise, since there are slippages of the project despite all the PERT chart." Do you agree with this view? Explain. *(CA, May, 1994)*

15. Consider each of the following statements and state whether it is true or false. In case it is false, write the correct position.

(i) Every path in a PERT network would be critical if all activities were to start at their earliest finish time.

(ii) PERT assumes that the distribution of completion times for a project is a normal distribution.

(iii) If an activity is completed after its earliest finish (EF) time, the project will necessarily be extended beyond its scheduled completion time.

(iv) Irrespective of the method of resource-levelling used to satisfy resource limitations, we must always preserve precedence relationships in the network. *(M Com, Delhi, 1997)*

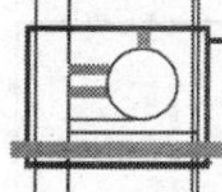

PRACTICAL PROBLEMS

1. Draw a network:

Activity	*Immediate predecessor(s)*	*Activity*	*Immediate predecessor(s)*
A	–	*H*	*F*
B	–	*I*	*C, D, G, H*
C	–	*J*	*I*
D	–	*K*	*I*
E	*A*	*L*	*J, K*
F	*B*	*M*	*J, K*
G	*E*	*N*	*M*

2. The Commerce Department of a university is considering to hold a faculty development programme. It has planned the following activities. Prepare a network diagram showing the interrelationships of the various activities:

Activity	*Description*	*Immediate predecessors*
A	Design conference meetings and theme	–
B	Design front cover of the conference proceedings	*A*
C	Prepare brochure and send request for papers	*A*
D	Compile list of distinguished speakers/guests	*A*
E	Finalise brochure and print it	*C, D*
F	Make travel arrangements for distinguished speakers/guests	*D*
G	Despatch brochures	*E*
H	Receive papers for conference	*G*
I	Edit papers and assemble proceedings	*F, H*
J	Print proceedings	*B, I*

3. The information about various activities of a project and the precedence relationships are given below. Draw a network using as few dummy activities as possible.

Activity	*Immediate predecessors*	*Activity*	*Immediate predecessors*
A	–	*I*	*F, G, H*
B	–	*J*	*O, E, N*
C	–	*K*	*B, C, D*
D	–	*L*	*K*
E	*B, C, D*	*M*	*B, C, D*
F	*A, B, C, D*	*N*	*B, C, D*
G	*A, B, C, D*	*O*	*A, B, C, D*
H	*A, B, C, D*	*P*	*I, J, M, L*

4. Bring out the dependency relationships from the network given in Figure 12.50.

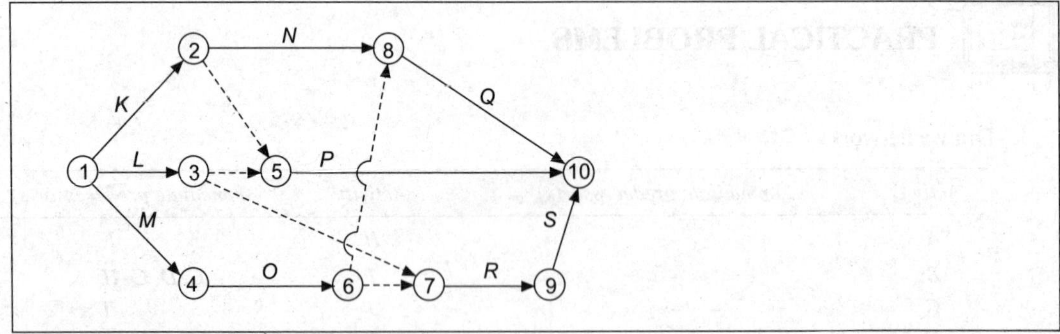

Figure 12.50 *Arrow Diagram*

5. For a small project of 12 activities, the details are given below. Draw the network and compute earliest occurrence time, latest occurrence time, critical activities and project completion time:

Activity	: A	B	C	D	E	F	G	H	I	J	K	L
Dependence	: –	–	–	*B, C*	*A*	*C*	*E*	*E*	*D, F, H*	*E*	*I, J*	*G*
Duration (Days)	: 9	4	7	8	7	5	10	8	6	9	10	2

6. Given that a project involves activities *A*, *B*, *C*, …, *I*; each requiring completion time, in days, as per the following table:

Activity:	*A*	*B*	*C*	*D*	*E*	*F*	*G*	*H*	*I*
Time:	23	8	20	16	24	18	19	4	10

Given that activity A precedes activities D and E; activities B and D precede activity F; activity C precedes activity G; activities B and G precede H, and activities F and G precede activity I, draw the network and calculate:

(i) Total float, (ii) Free float, (iii) Independent float, (iv) Critical path, and (v) Project completion time.

(M. Com., Delhi, 2007)

7. Activities involved in Generator Modernisation at BHEL Haridwar with their durations and imme diate predecessors are as under:

Job	Activity detail	Preceding activities	Duration (No. of days)
a	Preparation for hydraulic and cyclic test	–	2
b	Preparation for HV test	–	3
c	Generator dismantling	–	3
d	Rotor withdrawal	*c*	1
e	Insulation cover opening	*d*	6
f	HV test	*b*	1
g	De-wedging of slots	*c, d*	5
h	Debriding	*c, d*	8
i	Overhang bracket modification	*c, d*	5
j	Hydraulic and cyclic test	*a*	4
k	Bus bar modification	*d, i*	1
l	Epoxy injection in overhang	*d, k*	1
m	Ripple spring fixing	*c, d, j, e*	10
n	Slot wedging	*e, j, m*	8
o	Overhang braiding	*d, h*	15
p	Final hydraulic test	*k, l, n, o*	1
q	Box filling	*g, p*	4
r	Final HV test	*f, q*	3
s	Putty ring formation	*r*	2
t	Cleaning and painting	*s*	1

(i) Identify and remove redundant immediate predecessor(s) if any.

(ii) Develop an arrow diagram and find the project completion time.

(iii) Calculate total float for the non-critical activities. *(MBA, Delhi, April, 1994)*

8. (a) The following list of activities must be accomplished in order to complete a construction project:

Activity	Time (in weeks)	Predecessors
A	3	None
B	8	None
C	4	*A, B*
D	2	*B*
E	1	*A*
F	7	*C*
G	5	*E, F*
H	6	*D, F*
I	8	*G, H*
J	9	*I*

Construct a network for this project.

Find the critical path and the duration the project.

(b) Assume that you are the project manager of the project mentioned above. The project has progressed to the end of week 10 and the status is as follows:

Activities completed: *A*, *B*, *E*

Activities not started: *C*, *D*, *F*, *G*, *H*, *I*, *J*.

(i) If no managerial action is taken at all, when will the project get completed?

(ii) What action might you take to get the project back to a schedule that can be completed by the end of week 42? *(CA, May, 1993)*

9. Given the following information:

Activity	*Immediate Predecessors*	*Time (Days)*	*Activity*	*Immediate Predecessors*	*Time (Days)*
A	none	9	*H*	*D*, *G*	4
B	none	4	*I*	*E*, *G*	3
C	none	15	*J*	*H*, *I*	3
D	none	10	*K*	*D*	10
E	*A*	7	*L*	*J*, *K*	6
F	*B*	5	*M*	*L*	1
G	*C*, *F*	4			

Determine *ES*, *EF*, *LS*, *LF*, *IF*, and *FF* for each of the activities.

10. A small project consists of seven activities for which the relevant data are given below:

Activity	*Preceding activities*	*Activity duration (Days)*
A	–	4
B	–	7
C	–	6
D	*A*, *B*	5
E	*A*, *B*	7
F	*C*, *D*, *E*	6
G	*C*, *D*, *E*	5

(i) Draw the network and find the project completion time.

(ii) Calculate total float for each of the activities.

(iii) Draw the time scaled diagram. *(CA, November, 1996)*

11. Consider the network diagrams given in Figures 12.51(a) and (b). Bring out the effect of the change in the direction of dummy activity on the scheduling of the activities and the project duration.

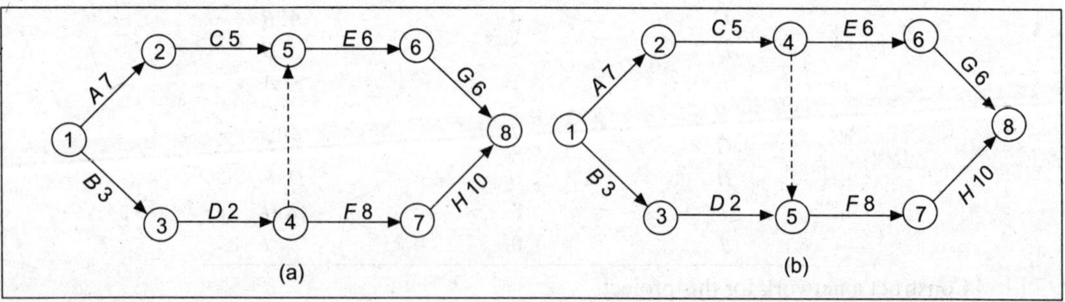

Figure 12.51 *Network Diagrams*

12. Following are the various activities involved in a project. The cost and time information for these activities is given below:

Activity	Immediately Preceding Activity	Normal Time (days)	Normal Cost (Rs)	Crash Time (days)	Crash Cost (Rs)
A	–	3	140	2	210
B	–	6	215	5	275
C	–	2	160	1	240
D	*A*	4	130	3	180
E	*A*	2	170	1	250
F	*A*	7	165	4	285
G	*B, D*	4	210	3	290
H	*C, E*	3	110	2	160

(a) Draw the PERT network.

(b) Find out the critical path and the expected project completion time.

(c) What is the minimum possible project completion time after crashing the activities involved in the project and the associated cost of completing the project? *(MIB, Delhi, 2005)*

13. The following table gives data on normal time and cost, and crash time and cost for a project.

Activity	Duration (Weeks) Normal	Duration (Weeks) Crash	Total Cost (Rs) Normal	Total Cost (Rs) Crash
1–2	3	2	300	450
2–3	3	3	75	75
2–4	5	3	200	300
2–5	4	4	120	120
3–4	4	1	100	190
4–6	3	2	90	130
5–6	3	1	60	110

(i) Draw the network and find out the critical path and the normal project duration.

(ii) Find out the total float associated with each activity.

(iii) If the indirect costs are Rs 100 per week, find out the optimum duration by crashing and the corresponding project costs.

(iv) With the crash duration indicated, what would be the minimum crash duration possible, ignoring indirect costs?

14. In planning a project to introduce a new product in the market, a company lists the various activities, their normal times and costs, and their crash times and costs. They are:

Activity	Immediate predecessor	Normal		Crash	
		Time (Weeks)	Cost (Rs)	Time (Weeks)	Cost (Rs)
A	–	5	10,000	4	12,000
B	–	2	6,000	2	6,000
C	A	4	8,000	3	10,000
D	A	4	10,000	3	15,000
E	A	3	11,000	1	16,000
F	C	1	7,000	1	7,000
G	D	4	8,000	2	12,000
H	B, E	5	9,000	3	12,000
I	H	2	8,500	2	8,500
J	F, G, I	3	7,500	2	10,000

If the indirect cost is Rs 4,000 per week, find the optimal project duration.

15. A small project is having seven activities. The relevant data about these activities is given below:

Activity	Dependence	Normal Duration (Days)	Crash Duration (Days)	Normal Cost (Rs)	Crash Cost (Rs)
A	–	7	5	500	900
B	A	4	2	400	600
C	A	5	5	500	500
D	A	6	4	800	1,000
E	B, C	7	4	700	1,000
F	C, D	5	2	800	1,400
G	E, F	6	4	800	1,600

(i) Find out the normal duration and the minimum duration.
(ii) What is the percentage increase in cost to complete the project in 21 days?

(CA, November, 1997)

16. The time and cost estimates of different activities of a project and their precedence relationships are given below:

Activity	Preceding Activity	Time (in Weeks)		Cost (Rs)	
		Normal	Crash	Normal	Crash
A	–	6	4	10,000	14,000
B	–	4	3	5,000	8,000
C	A	3	2	4,000	5,000
D	B	8	3	1,000	6,000
E	B	14	6	9,000	13,000
F	C, D	8	4	7,000	8,000

Overhead costs amount to Rs 1,000 per week.

It is stipulated that the contractor will have to pay a penalty of Rs 2,000 per week for completing the project beyond 12 weeks.

Crash the project to optimal extent.

What will be the critical activities after such crashing? *(M Com, Delhi, 1992)*

17. Given the following information about a project:

Activity	*Nodes*	*Normal Time (Days)*	*Shortest Time (Days)*	*Cost of Reduction per Day*
A	1–2	8	6	200
B	1–3	5	4	600
C	1–4	9	6	100
D	2–3	3	3	—
E	2–5	6	4	250
F	3–7	9	7	800
G	4–6	12	8	400
H	5–7	5	2	700
I	6–8	3	3	—
J	7–8	2	1	300

It is further given that the normal cost of doing all activities of the project in normal time is Rs 18,400 and the overhead cost is Rs 500 per day. Further, a penalty of Rs 2,000 per day in excess of 21 days would be levied. You are required to calculate:

(a) The normal duration of the project and the associated cost.

(b) The minimum duration of the project and the associated cost.

(c) The minimum cost at which the project can be done and the time corresponding to it.

(M Com, Delhi, 2008)

18. The following data relate to a project:

Activity	*Activity time (Days)*		*Activity cost (Rs)*	
	Regular	*Crash*	*Regular*	*Crash*
1–2	7	6	700	750
1–3	8	6	650	750
2–4	9	7	800	900
2–5	11	8	1,600	1,800
3–5	8	5	750	1,000
3–6	10	7	1,900	2,100
4–7	12	10	1,200	1,300
5–7	13	11	1,300	1,400
6–7	14	10	1,500	1,800

(a) Draw the network, find the critical path and the duration of the project (the minimum as well as the regular).

(b) If the project manager wants to reduce the completion time to 25 days, which activities would be required to be crashed and by how much? What shall be the cost involved?

(c) Can this project be completed in 22 days? If not, why? If yes, what is the cost involved?

19. The following are the various activities involved in a small project. The cost and time information for these activities is given here:

Activity	*Immediately Preceding Activity*	*Normal*		*Crash*	
		Time (Days)	*Cost (Rs)*	*Time (Days)*	*Cost (Rs)*
A	–	6	300	5	400
B	–	8	400	6	600
C	A	7	400	5	600
D	B	12	1,000	4	1,400
E	C	8	800	8	800
F	B	7	400	6	500
G	D, E	5	1,000	3	1,400
H	F	8	500	5	700

(a) Draw the project network.

(b) Find out the critical path and the expected project completion time.

(c) If the project is to be completed within 21 days with minimum crash cost, which activities should be crashed and by how many days? *(MIB, Delhi, 2008)*

20. The following table gives the activities in a small project and other relevant information:

Activity	*Duration (Days)*	*Immediate predecessor*	*Resources required*	
			Operators	*Mechanics*
A	3	None	2	–
B	2	*A*	2	2
C	4	*A*	4	4
D	6	*A*	5	5
E	3	*B*	2	2
F	2	*E*	2	–
G	6	*C*	–	2
H	4	*D*	2	2
I	4	*G*	4	2
J	2	*D*	2	–
K	2	*J*	2	2
L	4	*F, H, I*	4	4

(a) Draw the Network, compute Earliest Start Time (E), Latest Finish Time (F) for each of the activities and find out the project completion time and identify the Critical Path.

(b) Draw the Time Scaled Diagram along with resource accumulation table. Comment on demand for the operators and mechanics for the entire project duration and suggest the method for smoothening the resources. *(CA, May, 1988)*

21. The activities of a project and the labour requirements in respect thereof are:

Activity	*Time*	*Workers*	*Activity*	*Time*	*Workers*
1–2	6	6	3–6	3	3
1–3	6	18	4–7	5	8
1–4	2	7	5–7	7	16
2–5	3	10	6–7	4	9
3–5	7	15			

How should the activities be scheduled if it is desired to smoothen the labour employment on the project, as far as possible? What is the labour requirement?

22. The Madras Construction Company is bidding on a contract to install a line of microwave towers. It has identified the activities, along with their expected times, predecessor restrictions, and worker requirements:

Activity	*Duration (weeks)*	*Predecessors*	*Crew size (workers)*
A	4	None	4
B	7	None	2
C	3	*A*	2
D	3	*A*	4
E	2	*B*	6
F	2	*B*	3
G	2	*D, E*	3
H	3	*F, G*	4

The contract specifies that the project must be completed in 14 weeks. The company will assign a fixed number of workers to the project for its entire duration, and so it would like to ensure that the minimum number of workers is assigned and that the project will be completed in the 14 weeks. Find a schedule which will do this. *(CA, May, 1995)*

23. For a project, the following information is given:

Activity Description	*Node*	*Time*	*Activity Description*	*Node*	*Time*
A	1–2	4	*G*	3–6	10
B	1–3	8	*H*	4–6	8
C	1–4	12	*I*	5–6	6
D	1–5	6	*J*	6–7	12
E	2–6	6	*K*	6–8	14
F	3–7	4	*L*	7–8	8

The firm loses Rs 500 for every week beyond a total of 34 weeks. Consider each of the proposals given here individually and state whether it should be accepted by the firm.

(a) Activity *J* can be reduced to 9 weeks by taking resources from activity *K*, thereby raising its expected time to 17 weeks, at no extra cost.

(b) Activity *H* can be reduced to 5 weeks by using overtime, and extra labour at a cost of Rs 1,000.

(c) Activity *G* can be reduced to 7 weeks at a total cost of Rs 500.

(d) Activity *L* can be reduced to 5 weeks by taking resources from activity *K*, thereby increasing its expected time to 16 weeks at no additional cost.

24. Consider the following project. Assume that each activity requires manpower only and that the manpower requirements for the various activities are as follows:

Activity	*Duration (Weeks)*	*Manpower required*	*Activity*	*Duration (Weeks)*	*Manpower required*
1–2	2	10	5–7	3	6
2–3	8	18	5–8	5	10
2–4	6	10	6–7	4	8
3–5	12	15	6–9	3	9
3–6	5	6	7–9	4	12
4–5	5	8	8–9	6	8

Further, assume that idle workers are never kept on the payroll, that excess manpower is laid off immediately, and that inadequacies are accommodated by immediate hiring.

(a) Show graphically the level of resource employment through time, assuming that all activities are scheduled as soon as possible.

(b) Show graphically the level of resource employment through time, assuming that non-critical activities are scheduled as late as possible.

25. The following table lists the activities of a maintenance project.

Activity	*Duration (in months)*	*Activity*	*Duration (in months)*
1–2	2	4–7	3
1–3	2	5–8	1
1–4	1	6–8	4
2–5	4	7–9	5
3–6	5	8–9	3
3–7	8		

(i) Draw the project network.

(ii) Find the critical path and duration of the project.

(iii) Suppose we are required to employ a special piece of equipment on activities 1–3, 3–6, 2–5, 5–8 and 8–9, one at a time. Will it affect the duration of the project? Explain.

(CA, November, 1989)

26. The following activities are related to a given project.

Activity	*Time*	*Activity*	*Time*
1–2	8	3–6	2
1–3	6	4–6	6
1–4	4	5–7	5
2–8	4	6–7	9
3–5	7	7–8	5

A certain type of crane is needed in carrying out activities 2–8, 3–6, 4–6 and 5–7. Only one crane of this type is available.

Do you think the project completion would be delayed due to this? If yes, by how much?

27. The time durations and labour requirements of different activities of a project are:

Activity	*Duration*	*No. of Workers*
1–2	5	4
1–3	4	3
1–4	3	2
2–4	2	6
2–6	8	3
3–4	4	5
3–5	4	5
4–6	5	3
5–6	10	5

Additional information:

1. Labour rate is Rs 8 per day.
2. Overhead charge for the project is Rs 50 per day.
3. No activity can be started unless the needed labour force is available.
4. The firm has the option of employing 8, 9 or 10 workers.
5. Workers employed for the project would be paid for complete duration of the project.

Determine,

(a) the optimal number of workers to be employed, and

(b) the duration of the project, corresponding thereto.

28. A firm has 50 workers and one unit of each of machines M_1, M_2 and M_3 available for carrying out a project. The activities comprising the project, their durations and resource requirements are as follows:

Activity Description	*Node*	*Duration (No. of Days)*	*No. of Workers*	*Machine(s)*
A	1–2	5	30	M_1
B	1–3	2	20	M_2
C	1–4	5	30	M_1, M_3
D	2–6	6	20	M_2
E	2–5	8	20	M_1
F	3–5	3	20	M_3
G	4–7	4	40	M_2, M_3
H	5–6	2	20	M_2
I	6–7	3	20	M_2

Allocate the resources to the various activities and determine the duration of the project.

29. A small project consisting of eight activities has the following characteristics:

Activity	*Immediately Preceding Activity*	*Time Estimates (Weeks)*		
		Most Optimistic	*Most Likely*	*Most Pessimistic*
A	–	2	4	12
B	–	10	12	26
C	A	8	9	10
D	A	10	15	20
E	A	7	7.5	11
F	B, C	9	9	9
G	D	3	3.5	7
H	E, F, G	5	5	5

(a) Draw the PERT network.

(b) Find out the critical path and the expected project completion time.

(c) If a 30-week deadline is imposed, what is the probability that the project will be finished within the time limit?

(MIB, Delhi, 2005)

30. (a) For an activity in a PERT network, the following information is available:

Expected time = 9.5 days

Optimistic time = 5 days

Variance = 6.25 days^2

Calculate most likely and pessimistic times of this activity.

(b) The critical activities of a project, their expected durations and standard deviations (in weeks) are given here:

Activity	:	B	E	G	H	K	L
Expected duration	:	12	3	8	7	5	6
Standard deviation	:	2/3	1/3	2	5/3	4/3	0

(i) Find the probability that the project will be completed in 45 weeks.

(ii) If the manager wants the probability to be 90 per cent of completing the project in time, how many weeks prior to the deadline should she start the project?

31. The data for a PERT network is given in the following table:

Activity-nodes (i–j)	Time duration (days)		
	o	*m*	*p*
1–2	2	4	6
1–3	6	6	6
1–4	6	12	24
2–3	2	5	8
2–5	11	14	28
3–4	15	24	45
3–6	3	6	9
4–6	9	15	27
5–6	4	10	16

Note: *o*—optimistic; *m*—most likely; *p*—pessimistic

(i) Draw a network and estimate the earliest and latest event times for all nodes and hence derive critical path.
(ii) Estimate the expected duration of the project and corresponding variance.
(iii) What is the probability that the project duration will exceed 60 days?
(iv) What is the chance of completing the project between 45 and 54 days?
(v) What is the probability of completing the project within 30 days? *(MBA, Ruhailkhand, 2003)*

32. A new version of a popular spreadsheet software is due to be released in a year. However, the preliminary project team has not been able to come up with an expected completion date. But, as the table that follows shows, they have been able to estimate duration for each activity (in weeks) involved. Because disagreements arose, three time estimates are given: optimistic, most likely and pessimistic. Analyse this situation addressing the questions given below:
(a) Construct a network for the project.
(b) Compute the mean time of each activity.
(c) Compute the expected project completion time.
(d) Compute the standard deviation of the project completion time.
(e) Determine the likelihood that the project will be completed in 52 weeks.
(f) Determine the likelihood that the project will be completed in 65 weeks.
(g) Determine the probability that the project will not be ready within 70 weeks.

Activity	Predecessor(s)	Optimistic	Most Likely	Pessimistic
A Complaints review	—	2	4	8
B Suggestions review	—	1	2	3
C Initial specifications	AB	3	5	8
D Team completion	C	2	2	3
E Clarification requests	A, B	3	4	7
F Detailed specifications	D, E	4	6	9
G First coding phase	F	8	12	20
H Preliminary testing	G	2	3	5

(Contd.)

	Activity	*Predecessor(s)*	*Optimistic*	*Most Likely*	*Pessimistic*
I	First press release	G	2	2	3
J	External alpha testing	H	5	8	12
K	Second coding phase	H	6	9	14
L	Commercial response	I, J	3	4	7
M	Consolidation review	K, L	3	4	6
N	Secondary testing	K	2	3	5
O	External beta testing	N	4	6	9
P	Second press release	M	2	2	3
Q	Final revision coding	M, O	3	5	10
R	Final testing	Q	2	4	6
S	Package design	O, P	4	7	12
T	Production completion	R, S	2	3	4

(MBA, Delhi, 2007)

33. (a) Given that an activity takes generally 20 minutes to perform while the optimistic and pessimistic estimates of its performance are, respectively, one-half and thrice this time, answer the following:
    (i) What is the expected duration of this activity?
    (ii) What is its variance?
    (iii) In scheduling the project, of which this activity is a part, how much time should be provided for this activity?

(b) You are given the following information about a project:

Activity	*Duration (days)*	*Activity*	*Duration (days)*
1–2	4	2–6	18
1–3	7	3–5	10
1–4	10	3–6	16
2–3	3	4–5	9
2–4	8	5–6	6
2–5	11	5–7	11
		6–7	8

(i) Draw the network, obtain the scheduling times of various activities, and calculate their total and free slacks.
(ii) Obtain the critical path and find the duration of the project.
(iii) Activities 2– 6 and 4– 5 can each be speeded up and reduced by two days at no cost. Do you think it is worthwhile to reduce these times in terms of the completion time of the project?

(c) Assume that for the project given in (b), the activity durations given represent their expected durations. It is further given that variance along the critical path is 81 days^2. What is the probability of completing the project in 33 days? 44 days?

34. A project consisting of twelve distinct activities is to be analysed by using PERT. The following information is given (time estimates are in days):

Activity	*Predecessor activity*	*Optimistic time*	*Normal time*	*Pessimistic time*
A	–	2	2	2
B	–	1	3	7
C	*A*	4	7	8
D	*A*	3	5	7
E	*B*	2	6	9
F	*B*	5	9	11
G	*C, D*	3	6	8
H	*E*	2	6	9
I	*C, D*	3	5	8
J	*G, H*	1	3	4
K	*F*	4	8	11
L	*J, K*	2	5	7

Draw the PERT network. Indicate the expected total slack for each activity and hence indicate the average critical path. Within how many days would you expect the project to be completed with 99% chance?

35. (a) Using PERT, a manager was able to determine that the expected project completion time for the construction of a pleasure yatch is 21 months and the variance is 4 months^2?

   (i) What is the probability that the project will be completed in over 22 months?

   (ii) What is the probability that the project will be completed in the 24th month?

   (b) Given the following information:

Activity	*Nodes*	*Time Estimates (weeks)*		
		Optimistic	*Pessimistic*	*Expected*
A	1–2	3	15	8
B	1–3	4	8	6
C	1–4	6	20	12
D	2–4	6	6	6
E	2–6	5	9	7
F	3–4	6	15	9
G	3–5	4	12	6
H	4–7	10	12	11
I	5–7	4	8	6
J	5–8	10	10	10
K	6–7	10	18	14
L	7–8	2	6	3

   (i) Draw the project network. Obtain the standard deviation of completion time of each activity.

   (ii) Calculate expected duration of the project and its standard deviation.

(iii) Find the probability that the project will be completed in 38 weeks.

(iv) If the project manager wants to be 90 per cent sure that the project should be completed in time, how many weeks prior to the deadline should he start it? *(M Com, Delhi, 2004)*

36. A construction company is preparing a PERT network for laying the foundation of a new art museum. Given the following set of activities, their predecessor requirements and three time estimates of completion time:

Activity	*Predecessors*	*Time Estimates (Weeks)*		
		Optimistic	*Pessimistic*	*Most likely*
A	none	2	4	3
B	none	8	8	8
C	*A*	7	11	9
D	*B*	6	6	6
E	*C*	9	11	10
F	*C*	10	18	14
G	*C, D*	11	11	11
H	*F, G*	6	14	10
I	*E*	4	6	5
J	*I*	3	5	4
K	*H*	1	1	1

(i) Draw the PERT network.

(ii) Compute the slack for each activity and determine the critical path.

(iii) The contract specified a Rs 5,000 penalty for each week the completion of the project extends beyond 37 weeks. What is the probability that this company will have to pay a maximum penalty of Rs 15,000? *(CA, May, 1994)*

37. The data for a small PERT project is as given below, where *a* represents optimistic time, *m* most likely time and *b* pessimistic time estimates (in days) of the activities *A, B, ..., J, K*.

	A	*B*	*C*	*D*	*E*	*F*	*G*	*H*	*I*	*J*	*K*
a :	3	2	6	2	5	3	3	1	4	1	2
m :	6	5	12	5	11	6	9	4	19	2	4
b :	15	14	30	8	17	15	27	7	28	9	12

Precedence relationships: *A, B, C* can start simultaneously. $A < D, I$; $B < G, F$; $D < G, F$; $C < E$; $E < H, K$; $F < H, K$; $G, H < J$.

Draw the arrow network of the project. Calculate the earliest and latest expected times to reach each node and find critical path. What is the probability that the project will be completed 2 days later than expected? *(MBA, Chennai, 1996)*

38. A civil engineering firm has to bid for the construction of a dam. The activities and estimates are given as follow:

Activity	*Duration*		
	Optimistic	*Most likely*	*Pessimistic*
1–2	14	17	25
2–3	14	18	21
2–4	13	15	18
2–8	16	19	28
3–4 (dummy)			
3–5	15	18	27
4–6	13	17	21
5–7 (dummy)			
5–9	14	18	20
6–7 (dummy)			
6–8 (dummy)			
7–9	16	20	41
8–9	14	16	22

The policy of the firm with respect to submitting bids is to bid the minimum amount that will provide a 95% of probability of at best breaking even. The fixed costs for the project are eight lakhs and the variable costs are 9,000 everyday spent working on the project. The duration is in days and the costs are in terms of Rupees. What amount should the firm bid under this policy? (You may perform the calculations on duration and so on, up to two decimal places). *(CA, May, 1990)*

39. (a) PERT calculations yield the critical path length of a project to be 24 months with a variance of 9. What is the probability of its completion in 20 months? Within how many months would you expect the project to be completed with probability of 0.90?

(b) In a PERT network, the expected duration of the critical path is 36 weeks and the sum of the variances of the activities on the critical path is 36 weeks^2. Calculate the probability that the project may be completed not earlier than 30 weeks and not later than 42 weeks.

(c) If, in a PERT network, the expected duration of the critical path is 42 months and the sum of the variances of the activities on the critical path is 36 months^2, then the probability that the project may be completed not earlier than 36 months and not later than 48 months is approximately

(i) 50 per cent (ii) 68 per cent (iii) 95 per cent (iv) none of these

(d) The earliest time of completion of a project is the
   (i) time on the shortest path between start and end events of the project
   (ii) time on the longest path between start and end events of the project
   (iii) time on the longest path between any two events of the project
   (iv) none of the above

(e) An activity in a PERT project has expected duration of 19/3 days, with a standard deviation of 1 day. Obtain the optimistic and pessimistic durations of it, given that the activity has most likely duration of 6 days.

(f) One of the activities in a PERT project has an estimated duration of 17 weeks and a variance of 16 weeks^2. The most likely duration of this activity is shown to be 16 weeks. It can then be inferred that the estimates of the optimistic and pessimistic durations would respectively be:
   (i) 12 and 20 weeks
   (ii) 7 and 31 weeks
   (iii) 5 and 28 weeks
   (iv) Not determinable from the given data

40. An engineering firm is tendering for a contract to supply a steel fabrication with a target duration of 60 days. The tasks have been analysed as follows:

Task	*Immediate Predecessors*	*Duration (Days)*
A	–	10
B	–	12
C	*A*	10
D	*A*	9
E	*A*	13
F	*A, B*	17
G	*C*	12
H	*C, D*	14
I	*E*	13
J	*G, H*	12
K	*H*	10
L	*H, I*	14
M	*H, I, F*	13

(a) You are required to calculate the expected project duration and identify the critical path.

The firm is awarded the contract and starts work with all activities commencing on their earliest start times but after work on 23rd day there is a fire which destroys all the work done on task *C* and the work-in-progress on tasks *G* and *H*. Fortunately, no other completed tasks are affected but it is estimated that task *M* will now need 15 days. The project manager thinks that after the fire there will be variability in the task times and has made some uncertainty estimates which are shown as task standard deviations in days:

Task	*Standard deviation (in days)*
C	0.33
F	0.82
G	0.17
H	1.33
I	2.17
J	0.50
K	2.17
L	1.33
M	0.33

You are required

(b) to calculate the new expected duration and the critical path through the remaining activities after the fire;

(c) to calculate the probability of the project being completed on time after the fire.

(ICMA, November, 1990, Adapted)

Chapter 13

Decision Theory

CHAPTER OVERVIEW

Decision-making is the essence of management. In general, the process of making decisions calls for identifying the alternatives, gathering all the relevant information about them, and selecting the best alternative on the basis of some criterion. We discuss in this chapter the process of decision-making where outcomes are uncertain, in three sections. The first section deals with what is typically called newsboy's problem. A newsboy has to decide how many copies of a newspaper he should stock everyday. This is because, on a given day, the demand is uncertain and any overstocking will result in losses on account of unsold copies, while an under stocking will cause him to regret his decision for the profits he could have made.

In the second section, the decision analysis covers those situations where a sequence of decisions is needed to be taken, where the outcome of one decision has an influence on the next decision. Such kinds of problems are handled with the help of decision trees.

The final section is devoted to utility theory, which postulates that money means differently to different people. It says a given amount of money has different satisfaction levels for different people. Thus, in this approach to decision-making, the possible monetary outcomes of a proposal are translated in utility terms, and a decision is taken on the basis of expected returns in terms of level of utility rather than amounts of money.

This chapter enables a manager to respond to questions like the following:

- How can a pay-off matrix be developed using given information and how can a regret matrix be derived from there?
- What are the different principles on the basis of which decision may be reached in a given situation?
- How can any sample information be incorporated into the decision-making process?
- How can a decision tree be constructed and analysed when dealing with a sequential decision-making situation?
- How is a utility scale constructed in order to be used as a tool for reaching decisions?

This chapter requires knowledge of the concepts of probability, probability distribution and expected value. An understanding and usage of Bayes' Theorem is also essential. The decision problems, where decision trees are needed to be drawn, generally require a greater degree of skill for comprehension. Drawing a decision tree correctly may sometimes prove to be somewhat tricky.

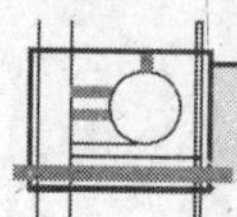

13.1 INTRODUCTION

So far, we have studied models for decision-making under conditions of certainty and uncertainty in the environment. In Chapters 2 to 8, we have considered models where different parameters were assumed to be known with certainty, while in others, uncertainty, wherever occurring, was incorporated into the analysis by specifying the probability distributions that the variables under consideration follow. We now turn to a specialised area, known as *decision theory*, which provides a formal analytic framework for decision-making under conditions of uncertainty.

The decision theory, also called the *decision analysis*, is used to determine optimal strategies where a decision-maker is faced with several decision alternatives and an uncertain, or risky, pattern of future events. To recapitulate, all decision-making situations are characterised by the fact that two or more alternate courses of action are available to the decision-maker to choose from. Further, a decision may be defined as the selection by the decision-maker of an act, considered to be best according to some predesignated standard, from among the available options. The decision-making process, thus, involves the following steps:

(a) Identification of the various possible outcomes, called *states of nature* or *events*, E_i's, for the decision problem. The events are beyond the control of the decision-maker.

(b) Identification of all the courses of action, A_j's, or the strategies that are available to the decision-maker. The decision-maker has control over choice of these.

(c) Determination of the pay-off function which describes the consequences resulting from the different combinations of the acts and events. The pay-offs may be designated as V_{ij}'s—the pay-off resulting from *i*th event and *j*th strategy.

(d) Choosing from among the various alternatives on the basis of some criterion, which may involve the information given in step (c) only or which may require and incorporate some additional information.

We shall detail the decision analysis in three parts in this chapter. The first part deals with single stage decision making problems, where decisions are taken by considering the (monetary) pays-offs resulting from various combinations of alternative courses of action and outcomes possible. The second part considers multi-stage decision problems wherein multiple decisions need to be taken one after another. The idea in such cases is to choose the optimal sequence (of decisions) from among the various alternatives. Finally, the third part uses utility, instead of monetary pay-offs, as the criterion for decision-making. We consider these in turn now.

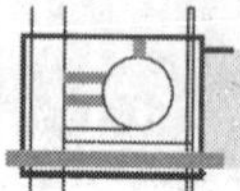

13.2 ONE-STAGE DECISION-MAKING PROBLEMS

As indicated, decision-making in case of single stage decision problems calls for (i) identification of the courses of action available to the decision-maker in the face of various possible events, (ii) developing a pay-off matrix, and (iii) choosing a particular course of action in accordance with some principle.

To understand the decision process under uncertain conditions, let us consider the following example.

Example 13.1 A bookstore sells a particular book of tax laws for Rs 100. It purchases the book for Rs 80 per copy. Since some of the tax laws change every year, the copies unsold at the end of a year become outdated and can be disposed of for Rs 30 each. According to past experience, the annual demand for this book is between 18 and 23 copies.

Assuming that the order for this book can be placed only once during the year, the problem before the store's manager is to decide how many copies of the book should be purchased for the next year.

For this problem, since the annual demand varies between 18 and 23 copies, there are six possible events:

E_1 :	18 copies are demanded,	E_4 :	21 copies are demanded,
E_2 :	19 copies are demanded,	E_5 :	22 copies are demanded,
E_3 :	20 copies are demanded,	E_6 :	23 copies are demanded.

Also, there are six possible strategies, or courses of action. They are:

A_1 :	buy 18 copies,	A_4 :	buy 21 copies,
A_2 :	buy 19 copies,	A_5 :	buy 22 copies,
A_3 :	buy 20 copies,	A_6 :	buy 23 copies.

Thus, in this problem there are 6 possible alternatives to choose from, and an equal number of states of nature, or events.

Having listed the possible acts and events, the next step is to construct the pay-off table.

13.2.1 Developing Pay-off and Regret Tables

A pay-off table depicts the economics of the given problem. A pay-off is a conditional value—a conditional profit, loss, or may be, a conditional cost. It is *conditional* in the sense that associated with each course of action is a certain profit/loss, given that certain event has occurred. Thus, the profit or loss resulting by the adoption of a certain strategy is dependent upon, and is therefore associated with, the particular event that may occur. A pay-off table thus represents the matrix of the conditional values associated with all the possible combinations of the acts and the events.

To consider how the pay-off table can be constructed for our example, let D denote the demand in units for the book, and let Q denote the quantity decided to be purchased (the course of action). The profit, P, for a total demand equal to, or greater than, the quantity purchased shall be equal to the difference of the total revenue from the sale of all the copies that were purchased and the total cost of procuring them. Thus, when $D \geq Q$, we have $P = 100Q - 80Q$ or $P = 20Q$. On the other hand, when the quantity demanded is less than the quantity purchased (i.e. $D < Q$), the profit shall be equal to the total revenue obtained from selling D copies *plus* the revenue from the unsold copies, equal to $30(Q - D)$, *minus* the total cost of buying Q copies. Thus, $P = 100D + 30(Q - D) - 80Q$ or $P = 70D - 50Q$. Alternatively, since each copy sold yields a profit of Rs 20 while every unsold copy involves a loss of Rs 50, the profit function, when $D < Q$, can be stated as: $P = 20D - 50(Q - D)$ which on simplification reduces to $P = 70D - 50Q$, the same as given above. Thus, we have

$$P = 20Q, \text{ when } \quad D \geq Q,$$

and

$$P = 70D - 50Q, \text{ when } \quad D < Q$$

The pay-offs for all combination A_j's and E_i's are given in Table 13.1.

TABLE 13.1 Pay-off Table

Event E_i	Act, A_j					
	A_1 : 18	A_2 : 19	A_3 : 20	A_4 : 21	A_5 : 22	A_6 : 23
E_1 : 18	360	310	260	210	160	110
E_2 : 19	360	380	330	280	230	180
E_3 : 20	360	380	400	350	300	250
E_4 : 21	360	380	400	420	370	320
E_5 : 22	360	380	400	420	440	390
E_6 : 23	360	380	400	420	440	460

Opportunity Loss or Regret Table The resultant outcomes of the various combinations of the acts and events (the states of nature) can alternatively be expressed in terms of the *opportunity loss*. Also called regret, the opportunity loss is defined as the amount of pay-off foregone by not adopting the optimal course of action—that which would give the highest pay-off, for each possible event. Thus, in the context of our example, the opportunity loss represents the amount of profit foregone by not stocking as many copies of the book as would yield the highest profit for each level of demand. For instance, if the demand is 18 copies, then the optimal act is to buy 18 copies for a profit of Rs 360. Note that in the first row, the highest profit is Rs 360 corresponding to the act A_1. If any other strategy is adopted, the profit earned would be less and, the greater the departure from the optimal strategy the lesser the profit earned and, consequently, the greater the opportunity loss (or regret). With the strategy of buying 20 copies, A_3, the profit is Rs 260 which is Rs 100 less than the profit that could be earned at this level of demand. Similarly, if the demand turns out for 20 copies, the optimal course would be to order 20 copies—the profit being Rs 400. Any order size other than this causes lesser profit than this. When, for example, 18 units, are stocked, the profit is Rs 360 which is Rs 40 less than the highest profit.

The pay-off matrix can be transformed into opportunity loss matrix by subtracting from the highest profit value in each row, all the other values in that row. The opportunity loss matrix is given in Table 13.2.

TABLE 13.2 Opportunity Loss or Regret Table

Event E_i	Act, A_j					
	A_1 : 18	A_2 : 19	A_3 : 20	A_4 : 21	A_5 : 22	A_6 : 23
E_1 : 18	0	50	100	150	200	250
E_2 : 19	20	0	50	100	150	200
E_3 : 20	40	20	0	50	100	150
E_4 : 21	60	40	20	0	50	100
E_5 : 22	80	60	40	20	0	50
E_6 : 23	100	80	60	40	20	0

Now, before proceeding to a discussion of the method of solution to this problem, a few observations follow.

Although the pay-off matrix (or the opportunity loss matrix) serves as an adequate representation of many a practical business decision situation, all types of decision situations cannot be represented by this approach. In a subsequent section of this chapter (see: The Decision Trees) those situations are discussed where a sequence of decisions is involved. Such situations cannot usually be handled by this model. Further, sometimes the events judged are under the control of a rational opponent. For example, the outcome of a certain strategy by a businessman depends upon the strategy adopted by his business competitor, who can influence the result by his own action. In such cases, although the decision can be represented by a pay-off matrix, the solution procedure is quite different. Such decision situations are discussed in the chapter titled Theory of Games.

While constructing a pay-off matrix, the alternative courses of action and the possible outcomes (events) must be clearly determined. In any given situation, the listing of the events and the actions must be distinct, mutually exclusive and collectively exhaustive. Alternatives a_1 and a_2, for instance, cannot occur together. If their joint occurrence was possible, then it would be defined as another alternative a_3.

Besides, the pay-off matrix in respect of the given example indicates the profit obtainable under different action-event combinations, the pay-offs in some cases are required to be expressed in terms of cost (see Example 13.2).

13.2.2 Decision Rules

After setting up the pay-off table (or the opportunity loss table) we proceed to take the decision. There are several rules, or criteria, on the basis of which decision may be taken. The selection of an appropriate criterion depends on factors like the nature of decision situation, attitude of the decision-maker, and so on. We shall first discuss the decision rules for taking decisions in conditions of uncertainty and then for the conditions of risk.

A. Decisions Under Uncertainty The decision situations where there is no way in which the decision-maker can assess the probabilities of the various states of nature are called decisions under uncertainty. In such situations, the decision-maker has no idea at all as to which of the possible states of nature would occur nor has he a reason to believe why a given state is more, or less, likely to occur as another. With probabilities of the various outcomes unknown, the actual decisions are based on specific criteria. The several principles which may be employed for taking decisions in such conditions are discussed below. Their discussion in the context of the bookstore problem is obviously based on the assumption that the bookstore has only this information that the demand of the book varies between 18 and 23 inclusive, and no more.

(a) Laplace Principle

The Laplace Principle is based on the simple philosophy that if we are uncertain about the various events then we may treat them as equally probable. Under this assumption, the expected (mean) value of pay-off for each strategy is determined and the strategy with highest mean value is adopted. Of course, if the pay-offs are in terms of costs, we choose the strategy with the lowest average cost.

For Example 13.1, expected pay-offs for different acts are as follows:

Act	*Mean (Expected) Pay-off*
A_1	(360 + 360 + 360 + 360 + 360 + 360)/6 = Rs 360.0
A_2	(310 + 380 + 380 + 380 + 380 + 380)/6 = Rs 368.3
A_3	(260 + 330 + 400 + 400 + 400 + 400)/6 = Rs 365.0

A_4 (210 + 280 + 350 + 420 + 420 + 420)/6 = Rs 350.0

A_5 (160 + 230 + 300 + 370 + 440 + 440)/6 = Rs 323.3

A_6 (110 + 180 + 250 + 320 + 390 + 460)/6 = Rs 285.0

Since the expected pay-off for A_2 is the maximum, it would be adopted. Thus, the bookstore manager would buy 19 copies of the book if he chooses to adopt the Laplace rule for taking decision.

(b) Maximin or Minimax Principle

This principle is adopted by pessimistic decision-makers who are conservative in their approach. Using this approach, the minimum pay-offs resulting from adoption of various strategies are considered and among these values the maximum one is selected. It involves, therefore, choosing the best (the maximum) profit from the set of worst (the minimum) profits.

When dealing with the costs, the maximum cost associated with each alternative is considered and the alternative which minimises this maximum cost is chosen. In this context, therefore, the principle used is *minimax*—the best (the minimum cost) of the worst (the maximum cost).

For our example, the minimum profit associated with various strategies is as follows:

A_1 :	Rs 360	A_3 :	Rs 260	A_5 :	Rs 160
A_2 :	Rs 310	A_4 :	Rs 210	A_6 :	Rs 110

Since the maximum of these is Rs 360, the strategy A_1 is selected corresponding to the maximin principle of choice.

(c) Maximax or Minimin Principle

The maximax principle is optimists' principle of choice. It suggests that for each strategy, the maximum profit should be considered and the strategy with which the highest of these values is associated should be chosen. The optimist obviously desires a chance for the maximum pay-off in the decision matrix.

For Example 13.1, the maximum pay-off associated with the different strategies is as follows:

A_1 :	Rs 360	A_3 :	Rs 400	A_5 :	Rs 440
A_2 :	Rs 380	A_4 :	Rs 420	A_6 :	Rs 460

The highest profit being Rs 460, strategy A_6 of ordering 23 copies of the book is the decision according to the maximax principle.

In decision problems dealing with costs, the minimum cost for each alternative is considered and then the alternative which minimises the minimum cost is selected. The principle in this case is obviously termed *minimin.*

(d) Hurwicz Principle

The Hurwicz principle of decision-making stipulates that a decision-maker's view may fall somewhere between the extreme pessimism of the maximin principle and the extreme optimism of the maximax principle. This principle provides a mechanism by which different levels of optimism and pessimism may be shown. For this, an index of optimism, α, is defined on scale ranging from 0 to 1. An $\alpha = 0$ indicates extreme pessimism while $\alpha = 1$ represents extreme optimism.

For taking a decision using this principle, first the decision-maker's degree of optimism is indicated on the scale. Assuming that the decision-maker is able to reflect a degree of optimism by assigning a particular value of α, we multiply the maximum profit for each strategy A_j by α, and the minimum profit for it by

$1 - \alpha$. The sum of these products, called the *Hurwicz Criterion*, is obtained for each strategy and we select the alternative which maximises this quantity. Obviously, when $\alpha = 0$, only the minimum of the profits for each strategy would be considered and the decision would in effect be according to the maximin criterion. Similarly, when $\alpha = 1$, the decision would be identical to that arrived through the maximax principle.

Suppose that for Example 13.1, the decision-maker's degree of optimism is reflected adequately by $\alpha = 0.6$. With this, we can obtain Hurwicz Criterion associated with various courses of action as given in Table 13.3.

TABLE 13.3 Hurwicz Criterion for Various Acts

Act	*Max*	*Min*	*Criterion Value* = α*(Max Value)* + *(1 –* α*)(Min Value)*
A_1	360	360	$0.6 \times 360 + 0.4 \times 360 = 360$
A_2	380	310	$0.6 \times 380 + 0.4 \times 310 = 352$
A_3	400	260	$0.6 \times 400 + 0.4 \times 260 = 344$
A_4	420	210	$0.6 \times 420 + 0.4 \times 210 = 336$
A_5	440	160	$0.6 \times 440 + 0.4 \times 160 = 328$
A_6	460	110	$0.6 \times 460 + 0.4 \times 110 = 320$

Since the value associated with the strategy A_1 is the maximum, the decision is to choose this strategy under the given principle.

In the case of costs, the principle works like this. The minimum of the costs for each course of action is multiplied by α (the indicator of the degree of optimism of the decision-maker), and the maximum of the costs for each alternative is multiplied by $1 - \alpha$. Then the sum of the products for each action strategy is obtained. The alternative for which the sum is the least is selected.

(e) Savage Principle

The Savage principle is based on the concept of regret and calls for selecting the course of action that minimises the maximum regret. It is alternatively known as the *principle of minimax regret*.

As a first step, the regret matrix is derived from the pay-off matrix, as explained earlier. Then the maximum regret value corresponding of each of the strategies is determined and the strategy which minimises the maximum regret is chosen. The principle of choice is also conservative in approach and is very close to the minimax principle applied to the original matrix containing pay-off values. It may, however, be noted that the results obtained by the two methods need not be the same.

From the regret matrix given in Table 13.2, we get the following maximum regret values associated with the various courses of action.

A_1:	Rs 100	A_3:	Rs 100	A_5:	Rs 200
A_2:	Rs 80	A_4:	Rs 150	A_6:	Rs 250

The maximum regret value for the strategy A_2 being the least, it represents the optimal choice.

B. Decision Under Risk The decision situations wherein the decision-maker chooses to consider several possible outcomes and the probabilities of their occurrence can be stated are called *decisions under risk*. The probabilities of various outcomes may be determined objectively from the past records. For Example 13.1, suppose that the bookstore observes from the past sales data that the proportion of times the number of copies sold is 18, 19, 20, 21, 22, and 23 are, respectively, 0.05, 0.10, 0.30, 0.40, 0.10 and 0.05. Of course, as the case here is, the sum of the probabilities must be 1.0—in keeping with the earlier observation that the set of possible outcomes in a decision problem should be mutually exclusive and collectively exhaustive.

However, past records may not be available in many cases to arrive at objective probabilities. In case the decision-maker may, on the basis of his experience and judgement, be able to assign subjective probabilities to the various outcomes, the problem can yet be solved as a decision problem under risk.

Under conditions of risk, there are generally two criteria to choose from. These are discussed below.

(a) Maximum Likelihood Principle

Under this principle, the decision-maker first considers the event that is most likely to occur. He then decides for the course of action which has the maximum conditional pay-off, corresponding to this event (of course, when the pay-off matrix is in terms of costs, then the action with the least conditional pay-off would be chosen). For our example, it is known that the probability (equal to 0.40) is the highest for a demand level of 21 copies. Therefore, we would consider the pay-offs resulting from adopting different strategies for this demand level, and observe that it is highest at Rs 420 when 21 copies are ordered for. Thus, the decision according to this criterion is to buy 21 copies of the book.

This principle may seem reasonable in many situations, specially where the probability of a particular event may be predominantly larger than the probabilities of the other possible events. However, it has the demerit that it ignores the available information, that is to say, the other possible events and their consequences are neglected.

(b) Expectation Principle

More generally, the decision-making in situations of risk is on the basis of the expectation principle. With the event probabilities assigned, objectively or subjectively as the case may be, the expected pay-off for each strategy is calculated by multiplying the pay-off values with their respective probabilities and then adding up these products. The strategy with the highest expected pay-off represents the optimal choice. It goes without saying that in problems involving pay-off matrix in terms of costs, optimal strategy is that corresponding to which the expected value is the least.

Symbolically, for a decision problem involving n events and m strategies, the expected pay-offs, EP, can be expressed as under:

$$EP_j = \sum_{i=1}^{n} p_i a_{ij} \qquad j = 1, 2, ..., m$$

wherein a_{ij} represents the pay-off resulting from the combination of ith event and jth act, while p_i represents the probability of ith event.

To illustrate the use of the principle, we consider Example 13.1 again, the pay-off matrix in respect of which is reproduced in Table 13.4, along with the probabilities of the occurrence of various events as mentioned previously.

TABLE 13.4 Calculation of Expected Pay-Offs

Events	*Probability*	*Act, A_j*					
E_i	p_i	$A_1: 18$	$A_2: 19$	$A_3: 20$	$A_4: 21$	$A_5: 22$	$A_6: 23$
$E_1: 18$	0.05	360	310	260	210	160	110
$E_2: 19$	0.10	360	380	330	280	230	180
$E_3: 20$	0.30	360	380	400	350	300	250
$E_4: 21$	0.40	360	380	400	420	370	320
$E_5: 22$	0.10	360	380	400	420	440	390
$E_6: 23$	0.05	360	380	400	420	440	460
Expected Pay-off		360	376.5	386	374.5	335	288.5

The calculation of the expected pay-off values, *EP*, for some of the acts is shown here:

For A_1, $EP_1 = 0.05 \times 360 + 0.10 \times 360 + 0.30 \times 360 + 0.40 \times 360 + 0.10 \times 360 + 0.05 \times 360 =$ Rs 360

For A_2, $EP_2 = 0.05 \times 310 + 0.10 \times 380 + 0.30 \times 380 + 0.40 \times 380 + 0.10 \times 380 + 0.05 \times 380 =$ Rs 376.5

For A_6, $EP_6 = 0.05 \times 110 + 0.10 \times 180 + 0.30 \times 250 + 0.40 \times 320 + 0.10 \times 390 + 0.05 \times 460 =$ Rs 288.5

Since the maximum expected pay-off is associated with strategy A_3, the best course of action is to buy 20 copies of the book.

Expected Opportunity Loss or Expected Regret The expected opportunity loss or expected regret criterion is another basis on which a decision may be taken. As we shall observe, this criterion leads to the same conclusion as the expected pay-off criterion.

For our example, the conditional opportunity loss, or regret, matrix along with the probability distribution of demand are reproduced in Table 13.5. As observed already, for any given event, the conditional regret of the optimal act is zero while the conditional regret in respect of any act other than this is positive and equals the difference of the pay-offs of the optimal act and the act adopted.

TABLE 13.5 Calculation of Expected Regret

Events	*Probability*	*Act, A_j*					
E_i	p_i	$A_1: 18$	$A_2: 19$	$A_3: 20$	$A_4: 21$	$A_5: 22$	$A_6: 23$
$E_1: 18$	0.05	0	50	100	150	200	250
$E_2: 19$	0.10	20	0	50	100	150	200
$E_3: 20$	0.30	40	20	0	50	100	150
$E_4: 21$	0.40	60	40	20	0	50	100
$E_5: 22$	0.10	80	60	40	20	0	50
$E_6: 23$	0.05	100	80	60	40	20	0
Expected Regret		51	34.5	25	36.5	76	122.5

The expected regret for any strategy is determined by summing up the products of the regret values and their respective probabilities. For example, for A_1, we have expected regret,

$$ER_1 = 0.05 \times 0 + 0.10 \times 20 + 0.30 \times 40 + 0.40 \times 60 + 0.10 \times 80 + 0.05 \times 100 = 51.$$

Obviously, under this criterion, the optimal strategy is the one which minimises the expected regret. Since the minimum value occurs at A_3 in the example under consideration, it represents the optimal decision, the same as under the expectation principle.

Expected pay-off of perfect information (EPPI) This is calculated as follows. When the bookstore manager knows that next year's demand is 18 copies, he adopts the strategy of ordering 18 copies of the book and obtains a profit of Rs 360. This occurs 5 per cent of the time since the probability of a demand of 18 copies is known as 0.05. Therefore, the expected profit is $0.05 \times 360 =$ Rs 18. When the manager knows that next year's demand is 19 copies, he orders for an equal number of copies for a profit of Rs 380, and this happens 10 per cent of the time. For this, the expected pay-off equals $0.10 \times 380 =$ Rs 38. Similarly, the expected pay-off for each level of demand can be obtained which, when aggregated, yields the EPPI, as shown here.

$$EPPI = 0.05 \times 360 + 0.10 \times 380 + 0.30 \times 400 + 0.40 \times 420 + 0.10 \times 440 + 0.05 \times 460 = \text{Rs } 411$$

Thus, if the manager can, in any way, know perfectly the demand for the book in advance, and order for as many copies, the store can average a net profit equal to Rs 411. This represents the highest profit that the store can make when perfect information is available. Now, suppose that an agency undertakes to supply this information. We may ask the question as to how much should the store be prepared to pay to the agency for it. In other words, what is the worth of this information? Since the bookstore can make an expected profit of Rs 386 when no such information is available, and an expected profit of Rs 411 when the information is available, the worth of this information is $EPPI - EP = 411 - 386 = 25$. The bookstore should, therefore, pay no more than Rs 25 (per year) for obtaining such information. This value is called the *expected value of perfect information*, (*EVPI*), and it equals the expected regret value of the optimal act.

It is interesting to observe that for each course of action, A_j, the expected profit plus expected regret value equals the expected pay-off of perfect information. That is to say, $EP_j + ER_j = EPPI$, for $j = 1, 2, \ldots$. For example, for A_1, $EP_1 = 360$, $ER_1 = 51$ and $360 + 51 = 411$ or $EPPI$, and for A_2, $EP_2 = 376.5$, $ER_2 = 34.5$ which add up to 411.

Example 13.2 Technico Ltd. has installed a machine costing Rs 4 lacs and is in the process of deciding on an appropriate number of a certain spare parts required for repairs. The spare parts cost Rs 4,000 each but are available only if they are ordered now. In case the machine fails and no spares are available, the cost to the company of mending the plant would be Rs 18,000. The plant has an estimated life of 8 years and the probability distribution of failures during this time, based on experience with similar machines, is as follows:

No. of failures during 8-yearly period :	0	1	2	3	4	5	6+
Probability :	0.1	0.2	0.3	0.2	0.1	0.1	0

Ignoring any discounting for time value of money, determine the following:

(a) The optimal number of units of the spare part on the basis of (i) minimax principle, (ii) minimin principle, (iii) Laplace principle, (iv) Hurwicz principle (taking $a = 0.7$), and (v) expected cost principle.

(b) The expected number of failures in the 8-year period.

(c) The regret table, and the optimal choice on the basis of least expected regret criterion.

(d) EVPI.

If we let F represent the number of failures, S the number of spares, and C the total cost, the cost function can be stated as follows:

$$\left.\begin{aligned} C &= 4{,}000S, && \text{when } F \leq S \\ &= 4{,}000S + 18{,}000\,(F - S) && \text{when } F > S \end{aligned}\right\}$$

On the basis of this information, the cost matrix has been constructed and shown in Table 13.6. Also shown in the table are probabilities of the various numbers of failures.

TABLE 13.6 Cost Matrix

(All cost figures in '000 Rs)

No. of Failures E_i	*Probability* p_i	*No. of Spares,* A_j					
		$A_1 : 0$	$A_2 : 1$	$A_3 : 2$	$A_4 : 3$	$A_5 : 4$	$A_6 : 5$
$E_1 : 0$	0.1	0	4	8	12	16	20
$E_2 : 1$	0.2	18	4	8	12	16	20
$E_3 : 2$	0.3	36	22	8	12	16	20
$E_4 : 3$	0.2	54	40	26	12	16	20
$E_5 : 4$	0.1	72	58	44	30	16	20
$E_6 : 5$	0.1	90	76	62	48	34	20
Column Minima		0	4	8	12	16	20
Column Maxima		90	76	62	48	34	20
Simple Average Cost		45	34	26	21	19	20
Expected Cost		41.4	29.2	20.6	17.4	17.8	20.0

(a) No. of units on the basis of different principles:

(i) *Minimax* The maximum values in each of the columns are indicated by the *column maxima* row. The minimum of these being 20, the decision on the basis of the minimax rule would be to buy 5 spare parts.

(ii) *Minimin* From the minimum values contained in the row entitled *column minima*, it may be observed that the least value is equal to zero. Thus, the optimal number of spare parts is nil.

(iii) *Laplace Principle* According to this principle, the different events, E_i's, are assumed to be equally probable. Thus, the decision is taken on the basis of the simple average cost values (determined without using the given probability values). Since the simple average cost is the minimum for A_5, it follows that the optimal number of spares, according to the Laplace rule, is 4.

(iv) *Hurwicz Principle* We shall first calculate the Hurwicz Criterion for each of the strategies. In the context of cost data, Hurwicz Criterion, $HC = \alpha$ (Min Value) + $(1 - \alpha)$ (Max Value). Its value for various strategies is as follows:

For A_1 : $0.7 \times 0 + 0.3 \times 90 = 27.0$
For A_2 : $0.7 \times 4 + 0.3 \times 76 = 25.6$
For A_3 : $0.7 \times 8 + 0.3 \times 62 = 24.2$
For A_4 : $0.7 \times 12 + 0.3 \times 48 = 22.8$
For A_5 : $0.7 \times 16 + 0.3 \times 34 = 21.4$
For A_6 : $0.7 \times 20 + 0.3 \times 20 = 20.0$

Being *lowest* for A_6, optimal strategy is to keep 5 spare parts.

(v) *Expected Cost Principle* From Table 13.6, we observe that the minimum value appears against the strategy A_4. Thus, according to the expectation principle, the optimal policy is to store 3 spare parts, the expected cost being Rs 17.4 thousand.

(b) Expected number of failures in the 8-year period,

$$E(F) = \sum_{i=1}^{6} p_i E_i$$

Thus, $E(F) = 0.1 \times 0 + 0.2 \times 1 + 0.3 \times 2 + 0.2 \times 3 + 0.1 \times 4 + 0.1 \times 5 = 2.3$

(c) The regret table can be derived from Table 13.6. For it, the *least* cost value in each row would be subtracted from other values. The differences represent the regret values. Table 13.7 depicts these.

TABLE 13.7 Regret Table

No. of Failures E_i	Probability p_i	No. of Spares, A_j $A_1 : 0$	$A_2 : 1$	$A_3 : 2$	$A_4 : 3$	$A_5 : 4$	$A_6 : 5$
$E_1 : 0$	0.1	0	4	8	12	16	20
$E_2 : 1$	0.2	14	0	4	8	12	16
$E_3 : 2$	0.3	28	14	0	4	8	12
$E_4 : 3$	0.2	42	28	14	0	4	8
$E_5 : 4$	0.1	56	42	28	14	0	4
$E_6 : 5$	0.1	70	56	42	28	14	0
Expected Regret		32.2	20	11.4	8.2	8.6	10.8

Since the expected regret for strategy A_4 is the least, the optimal policy according to this criterion, as in case of the expected cost principle, is to buy 3 spare parts.

(d) The expected value of perfect information, EVPI, when the pay-off matrix indicates cost, is defined as follows:

EVPI = Expected cost with optimal policy – Expected cost with perfect information

For our example, the expected cost with the optimal policy is 17.4 thousand rupees while the expected cost with perfect information is 9.2 thousand rupees, as follows:

Event	*Cost*	*Prob.*	*Prob. × Cost*
$E_1 : 0$	0	0.1	0.0
$E_2 : 1$	4	0.2	0.8
$E_3 : 2$	8	0.3	2.4
$E_4 : 3$	12	0.2	2.4
$E_5 : 4$	16	0.1	1.6
$E_6 : 5$	20	0.1	2.0
		Expected Cost	9.2

Thus, EVPI = 17.4 – 9.2 = 8.2 thousand rupees.

13.2.3 Bayesian Decision Rule: Posterior Analysis

In the preceding analysis of decision-making under risk, we have seen how probability information about the various states of nature is used for determining the expected pay-off values resulting from different courses of action. The Bayesian decision rule represents an extension of this. In this approach, the optimal strategy is chosen using the expected value criterion while the expected pay-offs are calculated by using *posterior probabilities*.

In using the Bayesian rule, the preliminary or prior information (from the past experience) of the decision-maker is revised on the basis of some additional information about the states of nature: the prior probabilities are converted into the posterior probabilities using the information. The use of these posterior probabilities for taking the decisions is likely to enable better decisions. In a given situation, the new information may be obtained through test research, raw material sample testing, etc.

We shall illustrate the use of this rule by means of the following example.

Example 13.3 Suppose Delhi Developers Limited (DDL), a construction company, has recently acquired a piece of land in a city on which it plans to construct a shopping complex. The company has now to decide the size of the complex. It is considering three options: a small-sized complex with 40 condominiums and a multiplex, say A_1; a medium-sized complex consisting of 60 condominiums and a multiplex, call it A_2; and a large-sized complex with 100 condominiums, say A_3. The company feels that the overall demand for the condominiums built would be either high or low. The returns from the project will obviously depend on what size of complex is developed and what the level of demand eventually turns out to be. The pay-offs (in thousands of rupees) expected under various event-action combinations, together with estimated probabilities of the likely demand, are given here:

		Act, A_j		
Event	*Probability*	*A_1: Small complex*	*A_2: Medium complex*	*A_3: Large complex*
High demand	0.4	1,800	2,200	4,200
Low demand	0.6	1,000	600	–1,200

Further, this company has the choice of engaging a marketing research firm to conduct a survey for it so that it can take a 'more informed' decision. Suppose the outcomes of the marketing research study may be indicated as I_1 and I_2 as follows:

I_1: A favourable report, indicating high demand for condominiums

I_2: An unfavourable report, indicating a low demand for condominiums

The company knows that, in all likelihood, the information provided by the marketing research firm would not always be cent per cent correct. Suppose that past record of the research firm on similar studies has led to the following estimates of the relevent probabilities:

Event	*Marketing Research Report*	
	Favourable (I_1)	*Unfavourable (I_2)*
E_1: High demand	0.9	0.1
E_2: Low demand	0.2	0.8

The marketing research firm has asked for Rs 300,000 as the fee for undertaking the study. How should the construction company proceed?

Here the company has two options to decide on the size of the shopping complex: one, on the basis of only its own estimates of the likely demand, and two, engaging and taking into consideration the report of the marketing research firm. The evaluation of the former is the same as the analysis that we have considered so far. It is also called *prior analysis* because it uses only prior probabilities. On the other hand, the latter involves incorporating the information obtained through marketing research into the analysis and take decision thereafter. This is termed as the *posterior analysis*. We consider them in turn now.

Prior analysis The pay-off table for the problem is reproduced in Table 13.8 and the expected values for the various acts are shown calculated.

TABLE 13.8 Calculation of Expected Pay-off

Event	*Probability*	*Act, A_j*		
		A_1: Small complex	*A_2: Medium complex*	*A_3: Large complex*
E_i	p_i			
E_1: High demand	0.4	1,800	2,200	4,200
E_2: Low demand	0.6	1,000	600	–1,200
Expected Pay-off		1,320	1,240	960

Here the expected pay-off for action A_1 is the maximum; it represents the optimal course of action. With optimal policy of A_1, EP_1 = Rs 1,320 thousand. Thus, according to the given probabilities of high and low demands, the company should construct a small-sized shopping complex.

It is evident from Table 13.8 that in the event of high demand for condominiums, the best strategy is a large-sized complex with a pay-off of Rs 4,200 thousand and, in case of low demand, a small-sized complex involving a pay-off of Rs 1,000 thousand is the best course. Using these values, we can determine the expected pay-off of perfect information, EPPI, for this decision problem as shown here:

Event	*Probability*	*Pay-off ('000 Rs)*	*Pay-off × Probability*
E_1: High demand	0.4	4,200	1,680
E_2: Low demand	0.6	1,000	600
		EPPI =	2,280

Thus, under conditions of complete and certain information, the expected profit would be Rs 2,280 thousand. From this, the expected value of perfect information, EVPI, can be obtained as follows:

$$EVPI = EPPI - EP = 2{,}280 - 1{,}320 = \text{Rs } 960 \text{ thousand}$$

Posterior analysis We now consider what decision the company would take if it engages the marketing research firm. The additional information provided by the firm may call for a revision of the decision of small-sized complex earlier or reinforce it. The approach involves revising the prior probabilitites of high and low demand to calculate posterior probabilities in the first place, and then using these posterior probabilities to reach a decision. Accordingly, this is called *posterior analysis*.

Let us consider how posterior probabilities may be calculated and used. In the first place, it may be noted that the probabilities given in the context of marketing firm's past records are in fact conditional probabilities. To illustrate, 0.90 is the answer to the following question: given that the state of nature eventually turns out to be a high demand, what is the probability that the marketing research study shall give a favourable report? This is, in other words, the conditional probability $P(I_1/E_1)$. Similarly, $P(I_1/E_2) = 0.20$, $P(I_2/E_1) = 0.10$ and $P(I_2/E_2) = 0.80$ can be appropriately interpreted. Obviously, on the basis of the probability estimates, we can place a high degree of confidence on the market research report. This is because when the true state of nature is high level of demand for the condominiums, E_1, the report shall be favourable 90 per cent of time, and unfavourable the 10 per cent. Similarly, when the true state is low demand, there is an 80 per cent likelihood of the report also being an unfavourable one, giving correct information.

Using the prior and conditional probabilities, we determine the total probability that the research report will be favourable, and the total probability that the research report will be unfavaourable. Since both, the favourable and unfavourable report may be associated with the events of high and low demands for condominiums, we can determine these probabilities as follows:

$$P(I_1) = P(E_1 \cap I_1) + P(E_2 \cap I_1)$$
$$= P(E_1) \times P(I_1/E_1) + P(E_2) \times P(I_1/E_2)$$

Substituting all values, we get

$$P(I_1) = 0.4 \times 0.9 + 0.6 \times 0.20$$
$$= 0.36 + 0.12 = 0.48$$

Similarly,

$$P(I_2) = P(E_1 \cap I_2) + P(E_2 \cap I_2)$$
$$= P(E_1) \times P(I_2/E_1) + P(E_2) \times P(I_2/E_2)$$
$$= 0.4 \times 0.1 + 0.6 \times 0.80 = 0.52$$

Now, we calculate the posterior probabilities for the events E_1 and E_2 under the conditions (a) that a favourable report is given; and (b) an unfavourable report is given by the marketing research firm. These probabilities would then be used to determine optimal course of action under each of the two situations represented by I_1 and I_2.

(a) When a Favourable Report is Given (Indicator I_1)

The posterior probabilities for the events E_1 and E_2, when a favourable report is given by the marketing research firm, are shown calculated in Table 13.9.

TABLE 13.9 Calculation of Posterior Probabilities

Event E_i	*Prior Probability,* $P(E_i)$	*Conditional Probability,* $P(I_1/E_i)$	*Joint Probability,* $P(I_1 \cap E_i)$	*Posterior Probability,* $P(E_i/I_1)$
E_1	0.4	0.90	0.36	0.36/0.48 = 0.75
E_2	0.6	0.20	0.12	0.12/0.48 = 0.25

The posterior probabilities, calculated by using Bayes' theorem, suggest that if marketing research report was in fact a favourable one, then the probability of the event of high demand, E_1 would be 0.75 and that of the event of low demand, E_2, would be 0.25. We shall use these probabilities instead of 0.4 and 0.6 respectively to determine the optimal course of action. The posterior probabilities of the event, the pay-off matrix and the expected values using these probabilities are contained in Table 13.10.

TABLE 13.10 Determination of Optimal Cource of Action

Event E_i	Probability p_i	Act, A_j — A_1: Small complex	A_2: Medium complex	A_3: Large complex
E_1 : High demand	0.75	1,800	2,200	4,200
E_2 : Low demand	0.25	1,000	600	–1,200
Expected Pay-off		1,600	1,800	2,850

Since the expected pay-off for the act A_3 is the highest, we conclude that given that a favaourable research report is obtained, the best course of action would be the construction of a large-sized shopping complex. This has expected pay-off of Rs 2,850 thousand.

(b) When an Unfavourable Report is Given (Indicator I_2)

Posterior probabilities in the event of indicator I_2–unfavourable research report–are calculated and shown in Table 13.11.

TABLE 13.11 Calculation of Posterior Probabilities

Event E_i	Prior Probability, $P(E_i)$	Conditional Probability, $P(I_2/E_i)$	Joint Probability, $P(I_2 \cap E_i)$	Posterior Probability, $P(E_i/I_2)$
E_1	0.4	0.10	0.04	0.04/0.52 = 0.0769
E_2	0.6	0.80	0.48	0.48/0.52 = 0.9231
		Total	0.52	

From this table, it is clear that if an unfavourable report is obtained, then the probabilities of high and low demand for condominiums would stand revised at 0.0769 and 0.9231 respectively. Using these probabilities, we can calculate the expected pay-offs of various courses of action and choose the appropriate one. The pay-off values and the expected pay-offs using the posterior probabilities are given in Table 13.12.

TABLE 13.12 Determination of Optimal Course of Action

Event E_i	Probability p_i	Act, A_j — A_1: Small complex	A_2: Medium complex	A_3: Large complex
E_1 : High demand	0.0769	1,800	2,200	4,200
E_2 : Low demand	0.9231	1,000	600	–1,200
Expected Pay-off		1,061.52	723.04	–784.74

Since the expected pay-off of the act A_1 is the largest, the company would decide to construct a small-sized shopping complex in case an unfavourable report indicating low demand is given by the marketing research firm.

Thus, when a favourable market research report is obtained, the optimal course of action would be A_3 with an expected pay-off of Rs 2,850 thousand, and in the event of an unfavourable report, the optimal act would be A_1, with an associated expected pay-off equal to Rs 1,061.52 thousand. But these decisions are conditional in the sense that either of them can be taken only when the nature of report is known. Since there is a likelihood of 0.48 for the report to be favourable, and 0.52 for it to be otherwise, we can get the expected pay-off value as shown in Table 13.13.

TABLE 13.13 Calculation of Expected Pay-off

Indicator	*Probability*	*Conditional Pay-off of the Best Act*	*Expected Value*
I_1 : Favourable Report	0.48	2,850	1,368
I_2 : Unfavourable Report	0.52	1,061.52	552
Expected Pay-off			1,920

To conclude, the expected pay-off of the optimal decision is Rs 1,920 thousand, if it is based on the market research information, as against Rs 1,320 thousand without such information.

Expected Value of Sample Information, EVSI As discussed earlier, the company in question would expectedly do best to build a large-sized shopping complex when a favourable research report is obtained, and a small-sized complex when the report is not favourable. In view of the fact that the company would have to incur extra cost by payment to the marketing research firm, it would like to know the value to be placed on the information provided by the research firm. In general terms, the information on the basis of which prior probabilities are revised is called the *sample information*. Thus, we have to determine here the expected value of sample information, EVSI. It is defined as follows:

EVSI = Expected pay-off *with* sample information *minus* Expected pay-off *without* sample information

As we have already calculated, the expected pay-off with sample information is Rs 1,920 thousand while the expected pay-off when no sample information is called for (i.e. without calling for market research report) is Rs 1,320 thousand. Therefore, EVSI = 1,920 – 1,320 = Rs 600 thousand. The company can, therefore, pay an amount up to Rs 600,000 to the marketing research firm for the research report. Since it is given here that the marketing research firm has asked a fee of Rs 300,000, it is worth engaging it.

Efficiency of EVSI

We have already determined that the expected value of perfect information, EVPI, for the given problem is Rs 960 thousand. In case the market research would always yield absolutely correct information, its value would match this amount, and thus it would be cent per cent efficient. However, as the report is not always likely to yield correct information, we can measure its efficiency by relating EVSI to EVPI. Thus, an efficiency index, *EI*, of smaple information can be obtained as follows:

$$EI = \frac{\text{EVSI}}{\text{EVPI}} \times 100$$

In our example,

$$EI = \frac{600}{960} \times 100 = 62.5\%$$

Thus, the sample information is about 63 per cent as efficient as perfect information.

Calculation of efficiency index enables the decision-maker to decide whether or not to seek the information and compare various sources through which the needed information may be obtained.

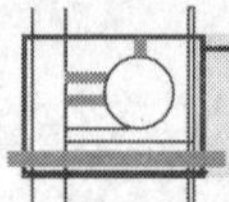

13.3 MULTI-STAGE DECISION-MAKING PROBLEMS: DECISION TREE

In the discussion of decision problems uptil now, our concern has been with the single stage problems wherein the decision-maker has to select the best course of action on the basis of whatever information is achievable at a point in time. We shall now consider the decision situations that involve multiple stages. Also called the *sequential decision* problems, they are characterised by a sequence of decisions in which following each decision, a chance event occurs which in turn influences the next decision.

In analysing multiple stage decision situations, we have to evaluate the decision proceeding in a backward manner by evaluating the best course of action at the later stages to decide the best action at the earlier stages. For this purpose, the *decision tree* or the *decision flow diagram* as it is sometimes called, is a very effective device.

A decision tree is a graphic representation of the sequences of action-event combinations available to the decision-maker. It depicts in a systematic manner all possible sequences of decisions and consequences. Each such sequence is shown by a distinct path through the tree. A decision tree enables the decision-maker to see the various elements of his problem in proper perspective and in a systematic manner. It may be mentioned that the criterion on the basis of which the decisions are made in the decision tree approach is generally the expectation principle. Thus, we may choose the alternative that maximises the expected profit, or the alternative that minimises the expected cost . . . and so on.

Let us consider the decision tree analysis with the help of the following example.

Example 13.4 An oil company has recently acquired rights in a certain area to conduct surveys and test drillings to lead to lifting oil if it is found in commercially exploitable quantities.

The area is considered to have good potential for finding oil in commercial quantities. At the outset, the company has the choice to conduct further geological tests or to carry out a drilling programme immediately. On the known conditions, the company estimates that there is a 70 : 30 chance of further tests showing a 'success'.

Whether the tests show the possibility of ultimate success or not or even if no tests are undertaken at all, the company could still pursue its drilling programme or alternatively consider selling its rights to drill in the area. Thereafter, however, if it carries out the drilling programme, the likelihood of final success or failure is considered dependent on the foregoing stages. Thus:

- if 'successful' tests have been carried out, the expectation of success in drilling is given as 80 : 20.
- if the tests indicate 'failure', then the expectation of success in drilling is given as 20 : 80.
- if no tests have been carried out at all, the expectation of success in drilling is given as 55 : 45.

Costs and revenues have been estimated for all possible outcomes and the net present value of each is as follows:

Outcome	*Net Present Value (Rs million)*
Success:	
With prior tests	100
Without prior tests	120
Failure:	
With prior tests	– 50
Without prior tests	– 40
Sale of exploitation rights:	
Prior tests show 'success'	65
Prior test show 'failure'	15
Without prior tests	45

(a) Draw up a decision (probability) tree diagram to represent the above information; and

(b) Evaluate the tree in order to advise the management of the company on its best course of action.

(MBA, Delhi, 2007)

The decision tree corresponding to the given problem is depicted in Figure 13.1.

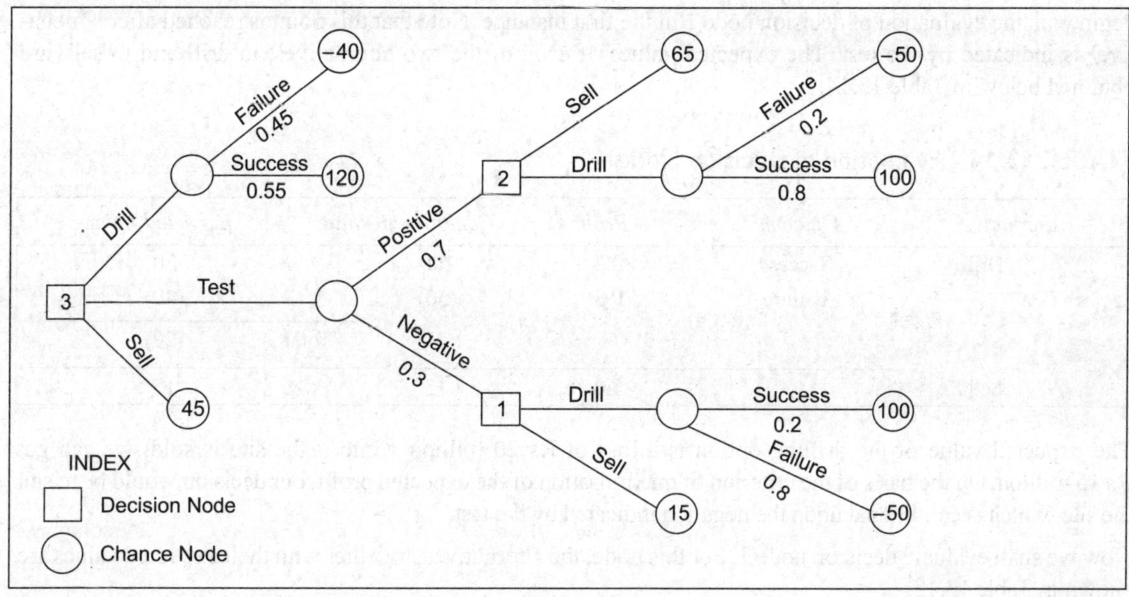

Figure 13.1 *Decision Tree*

Observe that the tree has several branches which originate from squares or from circles. A square represents a *decision node* or *decision fork* at which the decision-maker has to take a decision, while a circle represents a *chance node* or *chance fork* at which events (i.e. the states of nature) are branched out. At decision node marked 3, there are branches, representing three alternatives of drilling, testing for oil, and selling of rights, of which the decision-maker has to select one. Now, if the company decides to drill, there are two possible events—it may get oil or not, which are shown as branches emanating from circle, whose probabilities are 0.55 and 0.45, respectively. A profit of Rs 120 million would result in case oil is obtained and a loss of Rs 40 million if it is not.

The second alternative at the decision node 3 is to go for a test, which may give positive or negative results, with probabilities 0.7 and 0.3, respectively. In case a positive result is indicated, a choice has to be made as to whether to sell the rights for Rs 65 million or to drill, which is likely to succeed and fail with chances 80 : 20. Therefore, a decision node number 2 is shown by a square. Similarly, for a negative indication, there are two options—to sell for Rs 15 million or to drill, which has a 20 per cent chance of success. Thus, decision node 1 is indicated here.

The third alternative is to sell the rights for Rs 45 million.

The Method of Solution As mentioned earlier, the decisions have to be evaluated in a backward manner by evaluating the best course of action at the later stages so as to decide on the best course of action on the earlier stages. Thus, a solution to the problem is obtained by working backwards, from right to left, through the tree. This is called the *rollback* technique. In this technique, we assume that we have reached various decision nodes on the tree and then decide on the optimal act conditional on having reached the node being analysed. Thus, on reaching any decision node, we evaluate each of the alternative courses of action available there and select the most appropriate one. We begin with the rightmost decision node and, after having analysed it, we move backward and analyse the preceding decision node in a similar manner. The process is continued till the first, the leftmost, decision node is analysed. The decision at the first node represents the best initial decision.

Let us consider the solution to our problem for which the decision nodes have been marked as 1, 2, and 3. We begin with the evaluation of decision node 1 in the first instance. Note that this point is reached after a 'negative' is indicated by the test. The expected values of each of the two alternatives, to drill and to sell, are obtained below in Table 13.14.

TABLE 13.14 Evaluation of Decision Node 1

Alternative		Outcome	Prob.	Conditional Value	Expected Value
1.	Drill	Success	0.2	100	20
		Failure	0.8	(50)	(40)
				Total	(20)
2.	Sell		1.0	15	15

The expected value of the drilling option is a loss of Rs 20 million while if the site is sold, we can get Rs 15 million. On the basis of the criterion of maximisation of the expected profit, our decision would be to sell the site, which is conditional upon the negation indicated by the test.

Now we shall evaluate decision node 2. For this node, the alternatives, together with their expected values are shown in Table 13.15.

TABLE 13.15 Evaluation of Decision Node 2

Alternative		Outcome	Prob.	Conditional Value	Expected Value
1.	Sell		1.0	65	65
2.	Drill	Success	0.8	100	80
		Failure	0.2	(50)	(10)
				Total	70

The two alternative courses of action, namely selling and drilling, have expected values equal to Rs 65 million and Rs 70 million, respectively. Obviously, therefore, *provided* that a positive result is indicated by the test, the best course would be to go in for oil drilling.

Now, at this stage we have two conditional decisions—sell the site if a 'negative' is obtained on the test and drill in case the test indicates a 'positive'. At each node, the branches on which we have not to move, representing the options ruled out, have been shown cancelled. Next, we move to the decision node 1, where a decision has to be taken whether to drill at the outset, to undertake a test or to sell the rights outright. The alternatives, along with their expected values are shown in Table 13.16. Also, the expected values associated with the different chance nodes and the decision nodes are indicated in Figure 13.2 in which the decision tree given in Figure 13.1 is reproduced.

TABLE 13.16 Evaluation of Decision Node 3

Alternative		*Outcome*	*Prob.*	*Conditional Value*	*Expected Value*
1.	Drill	Success	0.55	120	66
		Failure	0.45	(40)	(18)
				Total	48
2.	Test	Positive	0.7	70	49.0
		Negative	0.3	15	4.5
				Total	53.5
3.	Sell		1.0	45	45

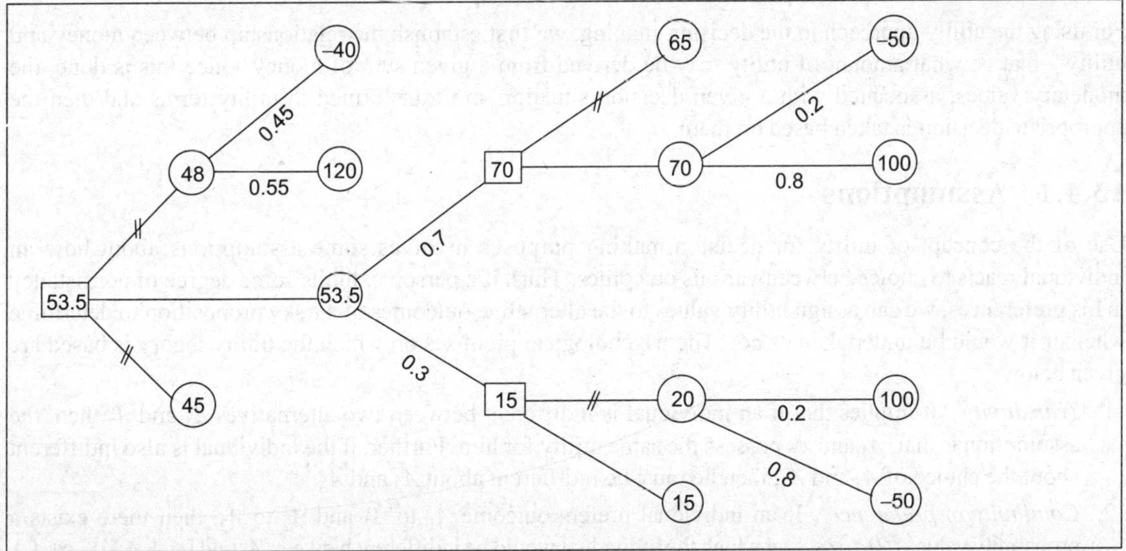

Figure 13.2 *Decision Tree*

As we may observe, the expected value of the alternative of carrying out a test is Rs 53.5 million, which is the highest of the three. Therefore, it is better to test before drilling. This is the initial decision. The overall decision can now be stated as: The test be carried out. If it proves negative, the rights should be sold to give a return of Rs 15 million. To proceed with drilling, if that happens, would expectedly lead to a loss. However, if the test proves positive, the drilling should be undertaken.

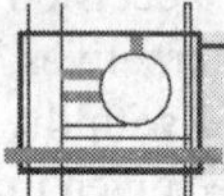

13.4 UTILITY THEORY: UTILITY AS BASIS FOR DECISION-MAKING

Suppose that a choice is offered between the following two alternatives: A_1—a certain gift of Rs 10,000, and A_2—a gift of Rs 25,000 if a coin, when tossed, falls head up, and nothing if the coin shows up tail. Clearly, most of us are likely to choose alternative A_1 with a sure amount of Rs 10,000 in preference to a risky A_2, although the expected (monetary) value for A_2, equal to $0.5 \times 25000 + 0.5 \times 0 =$ Rs 12,500, is larger than that of alternative A_1. It may therefore be reasonably argued that people do not always take decisions as will maximise their expected monetary value. Thus, the expectation principle, or the expected monetary value (EMV) criterion, which we have considered so far, does not always provide an adequate and satisfactory basis for decision-making. An alternative criterion that could be used for decision-making, which is consistent with the choice of alternatives like A_1 in our example is the one provided by Von Neumann and Morgenstern. According to them, the decisions are made so as to maximise *expected utility* rather than expected monetary value. In preferring A_1 over A_2, they contend, the decision-maker derives greater utility from alternative A_1 in comparison with the utility he would derive from alternative A_2. If he was indifferent between the two alternatives, the contention is that his expected utility is same for both the alternatives. It is possible to make generalisation about a person's utility function for a commodity, money in the present context, which are consistent logically and with observation of repeated decisions. Thus, it may be reasonably taken that decisions are made to maximise expected utility and not the expected monetary value.

For using the utility approach to the decision-making, we first establish the relationship between money and utility—that is, what amount of utility may be derived from a given sum of money. Once this is done, the monetary values, associated with a given decision situation, are transformed in utility terms and then the appropriate decision is taken based on them.

13.4.1 Assumptions

Use of the concept of utility for decision-making purposes involves some assumptions about how an individual reacts to choice between various outcomes. Thus, if a person exhibits some degree of consistency in his preferences, we can assign utility values to the alternative outcomes of a risky proposition to determine whether it would be undertaken or not. The psychological premises on which the utility theory is based are given below.

1. *Transitivity* It implies that if an individual is indifferent between two alternatives A_1 and A_2 then, the assumption is that, A_1 and A_2 possess the same utility for him. Further, if the individual is also indifferent about the choice of A_1 and A_3, then he must be indifferent about A_2 and A_3.
2. *Continuity of preference* If an individual prefers outcome A_1 to A_2 and A_2 to A_3, then there exists a probability value $\alpha, 0 < \alpha < 1$, at which the individual would be indifferent between A_2 and $[\alpha A_1 + (1 - \alpha) A_3]$.
3. *Independence* If an individual is indifferent between alternatives A_1 and A_2, and he is also indifferent between A_3 and A_4, then for any probability $\alpha\,(0 < \alpha < 1)$ he would be indifferent between $\alpha A_1 + (1 - \alpha)A_3$ and $\alpha A_2 + (1 - \alpha)A_4$.

4. *Desire for higher probability of success* According to this, if an individual prefers A_1 to A_2 and there are two probabilities α_1 and α_2 such that $\alpha_1 > \alpha_2$, then he would prefer $\alpha_1 A_1 + (1 - \alpha_1)A_2$ to $\alpha_2 A_1 + (1 - \alpha_2)A_2$. Also, for two alternatives with identical rewards, the one with the higher probability of success would be desired by the decision-maker.
5. *Compound probability* If A_1 and A_2 are two alternatives and α_1, α_2 and α_3 are three probabilities, an individual will be indifferent between $\alpha_1[\alpha_2 A_1 + (1 - \alpha_2)A_2] + (1 - \alpha_1)[\alpha_3 A_1 + (1 - \alpha_3)A_2]$ and $\alpha_1\alpha_2 + (1 - \alpha_1)\alpha_3](A_1) + [1 - \alpha_1\alpha_2 + (1 - \alpha_1)\alpha_3)](A_2)$.

13.4.2 Utility Measures and Utility Function

With the assumptions of utility theory, we now proceed to discuss how utility can be measured.

Von Neumann and Morgenstern have proposed an index for measurement of utility, which is a special type of cardinal measure. It measures utility in situations that involve risk for the decision-maker. This utility index is designed for predictive purposes, and allows to predict which of several bets a person would prefer and thus enables him to take decisions.

As stated earlier, the theory of utility postulates that a rational decision-maker will always decide to maximise utility or expected utility. The expected utility of a risky alternative is defined as the aggregate of the products of the utility values of all its possible outcomes and their respective probabilities. For an alternative A, if there are two possible outcomes x_1 and x_2 with respective probabilities of α and $1 - \alpha$, and that respective utilities are Ux_1 and Ux_2, we have,

Expected utility of A,

$$EU_A = \alpha Ux_1 + (1 - \alpha)Ux_2$$

The Utility Function Before we consider how the utility function, relating utility and money for a decision-maker, may actually be derived, it may be mentioned that the utility function should possess the property of *completeness*. A utility function is said to be complete if it measures the utility for all the possible alternatives that are available. Thus, whatever outcomes of a proposition or set of propositions are likely, it should be possible to obtain the utility associated with each one of them. If a utility function is complete, it allows for a comparison of the different outcomes.

For deriving a utility function, first we select two points of reference and assign arbitrary utility values to both of them. The largest and the smallest monetary values involved in a given situation may be taken to be the reference points. Then we may arbitrarily assign the values of 0 utils to the smallest value and 10 utils to the other one—the *utils* being the utility index or the units in which utility may be measured and expressed. For instance, if the lowest monetary value in a decision situation is – Rs 10,000 and the highest one is Rs 50,000, we shall assign a utility of zero utils to – Rs 10,000 and of 10 utils to Rs 50,000.

Thus,

$$U_{-10,000} = 0; \qquad \text{and} \qquad U_{50,000} = 10$$

This choice of 0 and 10, as mentioned, is purely arbitrary. The numbers chosen could as will be 20 and 752, for example. The utility scale in this sense may be compared with the scale of measuring temperature, where centigrade and fahrenheit, both the scales measure temperature but give different readings for the freezing and boiling points of water. The only point to remember in selecting utility values for the two points is that the greater monetary value should be assigned a larger number and the smaller value be assigned a smaller one. It is based on the assumption that money, like any other commodity, is desirable and people prefer more of it to less of it, and, therefore, higher amounts of money have greater utility than the smaller ones.

Continuing with the above situation where $U_{-10,000} = 0$ and $U_{50,000} = 10$, suppose that the decision-maker is asked the question that if he is given a choice between a cash-certain award of Rs 20,000 on the one hand, and bet involving a loss of Rs 10,000 or a gain of Rs 50,000, occurring with equal probabilities on the other, which one would he prefer? If he opts for the first alternative, it is interpreted that according to his preference, the utility of a sum of Rs 20,000 is greater than the expected utility of the bet which equals $0.5\ U_{-10,000} + 0.5\ U_{50,000} = 0.5 \times 0 + 0.5 \times 10 = 5$ units. For him, then $U_{20,000} > 5$ utils. Next, if he is given a choice between a sure amount of Rs 10,000 and the bet as before (that is, the amount – Rs 10,000 and Rs 50,000 with equal probabilities), and if he prefers the second alternative, then it would be similarly concluded that for him $U_{10,000}$ is less than the expected utility of the bet, 5 utils. In a similar way, if we keep on lowering the amount from Rs 20,000, or increasing from Rs 10,000, and asking for the preferences of the decision-maker as before, a value shall be obtained at which he would be indifferent between the two. For him, if this amount equals Rs 16,000, then we say that the utility of Rs 16,000 to that individual equals the expected utility of the given bet. Thus,

$$U_{16,000} = 0.5\ U_{-10,000} + 0.5\ U_{50,000} = 5 \text{ utils.}$$

Next we vary the probabilities in the bet. If the decision-maker is offered a bet involving a loss of Rs 10,000 with a probability, say 1/10 and a gain with a probability 9/10, or, alternatively, a gift of Rs 40,000, and if he prefers the sure award of Rs 40,000 then it is reckoned that the utility of this amount is greater than the expected utility of the bet. Then we keep on lowering down the amount of sure gift to arrive at a value for which he would be indifferent. If this amount is Rs 36,000, we have

$$\begin{aligned} U_{36,000} &= 0.1\ U_{-10,000} + 0.9\ U_{50,000} \\ &= 0.1 \times 0 + 0.9 \times 10 = 9 \text{ utils} \end{aligned}$$

In a similar manner, the different probability values for the reference points may be taken and the particular amounts at which the decision-maker would be indifferent to each of the values determined. The utility values for each of the amounts are then obtained in the same manner as discussed above.

Once the utility values for several amounts of money are derived, they are plotted on the graph as shown in Figure 13.3.

The amounts of money are shown on the X-axis while the utility measure is depicted vertically. The pairs of money amount and corresponding utility value are plotted on the graph and then they are joined by a *continuous curve*. This is the utility function.

Notice from this function that $U_{16000} = 5$ utils, and $(4/9)\ U_{-10000} + (5/9)U_{36000} = (4/9)0 + (5/9)9 = 5$ utils. As a check on the utility function derived, the decision-maker should be indifferent between a sure amount of Rs 16,000 and a bet involving a loss of Rs 10,000 with a probability 4/9 and a gain of Rs 36,000 with a probability equal to 5/9. If he is not, it implies that the individual in question is not exhibiting consistency in his preferences and the assumptions stated previously are not being met with, and so the utility function needs revision. In case the individual is consistent, the utility function can be used for decision-making.

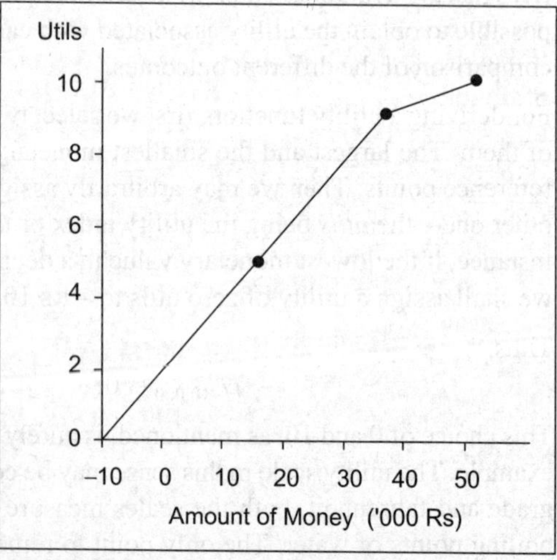

Figure 13.3 *Utility Function*

Before we discuss the use of utility function for decision-making purposes, some observations on the shape of utility functions follow. As stated already, money is considered as a desirable commodity and, therefore, more money is preferable to less money. Consequently, the utility function for money would always be upward sloping. However, as the utility function depicts the subjective attitude of a decision-maker to risk, the utility functions for different kinds of decision-makers would have different slopes. In this context, the decision-makers are classified in accordance with their psychological reactions to risk into three classes: risk-averse, risk-seekers, and risk-neutral.

A risk-averse is a person who is always willing to accept a small cash-certain amount than the expected values of a bet. Most of the people fall in this category. Our above analysis of derivation of utility function has been for a risk-averse individual. The utility function for such a person would be, like in Figure 13.3, the one shown in Figure 13.4.

For a risk-seeker, a person who demands an amount of cash-certain in excess of the expected monetary pay-off, the utility function would be rising at an increasing rate. Thus, a risk taker would gain more satisfaction from a bet where he can have a Rs 50,000 win (which has 0.5 chance of occurrence) or a loss of Rs 10,000 with an equal probability, than a cash-certain amount of Rs 20,000—the expected pay-off of the bet. He may demand, say, Rs 25,000 in cash to leave the bet.

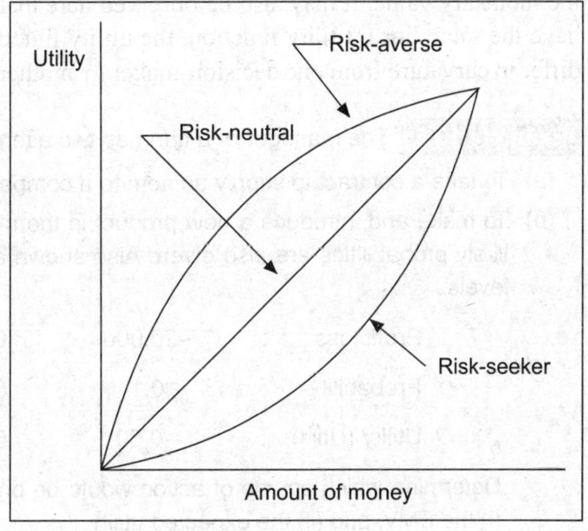

Figure 13.4 *Utility Functions*

Also shown in the figure in reference is the utility function of a risk-neutral, the one who is neutral between a cash-certain amount and a bet whose expected value is equal to that. For instance, if a person is neutral between a sure payment of a sum of Rs 20,000 and a bet with a gain of Rs 50,000 and a loss of Rs 10,000 with equal probabilities, he is said to be risk-neutral. A risk-neutral is neither averse to risk nor does he encourage it. For such decision-maker, the utility function is always linear.

The utility functions of risk-averse, risk-seeker and risk-neutral decision-makers are reproduced in Figure 13.5. It may be observed from part *a* of the figure that, if *M* be a certain sum of money the utility of an amount of double than that represented by *M*, equal to 2*M*, is less than twice the utility of the amount *M*. Thus, $U_{2M} < 2U_M$. Similarly, from the parts *b* and *c*, respectively, it is clear that for a risk-seeker $U_{2M} > 2U_M$ while $U_{2M} = 2U_M$ for a risk-neutral individual.

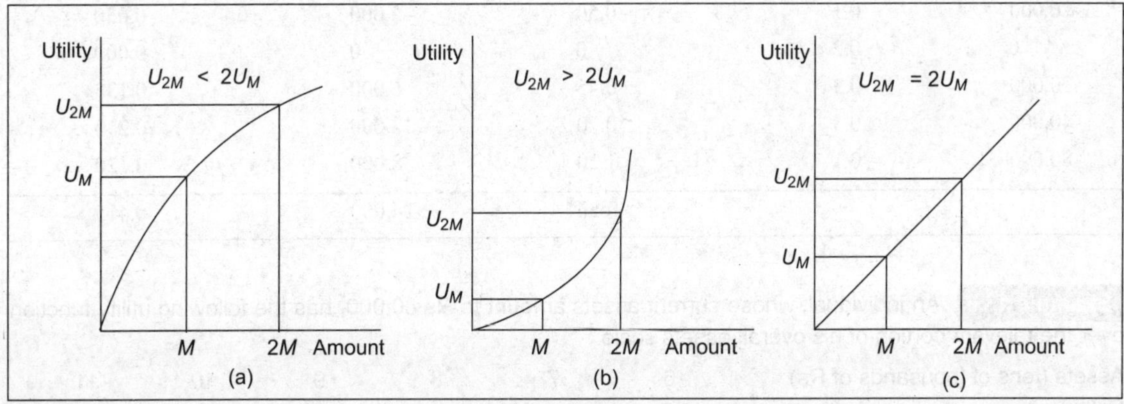

Figure 13.5 *Utility Functions*

The increasing slope of the utility function of a risk-seeker indicates that he has increasing marginal utility of money. Thus when his cash position improves, he places more value on each additional rupee that he would get. For a risk-averse, on the other hand, the marginal utility of money is decreasing. A linear utility function for a risk-neutral implies a constant marginal utility of money for him. His utility varies in direct proportion to the monetary value. It may also be observed here that while the risk-neutral decision makers would always have the same linear utility function, the utility functions for the risk-evaders or risk-preferers are likely to differ in curvature from one decision-maker to another.

Example 13.5 The manager of a firm has two alternatives to choose from, for the next quarter.

(a) To take a contract to supply an item to a company which would result in a sure profit of Rs 20,000.

(b) To make and introduce a new product in the market. The likely profit/loss possibilities along with the likely probabilities are also given. Also shown are the utility values associated with the various profit levels.

Profit/loss	:	–20,000	0	20,000	40,000	80,000
Probability	:	0.1	0.2	0.3	0.3	0.1
Utility (Utils)	:	–0.50	0	0.45	0.70	1.20

Determine which course of action would be preferred by the manager when he wanted to maximise (i) the EMV, and (ii) the expected utility.

(i) The expected profit associated with the alternative (a) is given to be Rs 20,000 while for the alternative (b) it equals Rs 24,000, as given in Table 13.17. Thus, the manager should decide for introducing the new product if he is seeking to maximise expected monetary value.

(ii) The expected utility, *EU*, for alternative (b) is 0.415 as shown in the table, while the utility associated with the profit of Rs 20,000 associated with alternative (a) is 0.45. Therefore, if the decision-maker seeks to maximise expected utility, he would decide in favour of the alternative (a).

TABLE 13.17 Calculation of EMV and EU

Profit/loss (i)	*Prob.* (ii)	*Utility* (iii)	*Expected Profit* (i) × (ii)	*Expected Utility* (ii) × (iii)
–20,000	0.1	–0.50	–2,000	–0.050
0	0.2	0	0	0.000
20,000	0.3	0.45	6,000	0.135
40,000	0.3	0.70	12,000	0.210
80,000	0.1	1.20	8,000	0.120
		Total	24,000	0.415

Example 13.6 An individual, whose current assets amount to Rs 80,000, has the following utility function over the relevant portion of his overall assets scale.

Assets (tens of thousands of Rs)	:	6	7	8	9	10	11
Utility	:	0.24	0.38	0.50	0.60	0.67	0.72

(a) He is offered a bet in which he has a sixty per cent chance of gaining Rs 20,000 and a forty per cent chance of losing Rs 20,000. Should he accept the offer?

(b) Alternatively, he is offered participation in two bets each involving a gain of Rs 10,000 with a probability of 0.6 and a loss of Rs 10,000 with a forty per cent chance. Should he decide differently than in case (a)?

(a) For a bet involving a gain of Rs 20,000 and a loss of Rs 20,000, we can calculate the expected utility as follows:

Outcome	*Assets*	*Utility*	*Probability*	*Expected Utility*
Gain	100,000	0.67	0.6	0.402
Loss	60,000	0.24	0.4	0.096
			Total	0.498

Presently, he has a utility of 0.50, while if he accepts the bet, his expected utility is 0.498. Thus, he would do well not to accept the bet.

(b) When he is invited to participate in two bets, the resulting expected utility can be obtained as follows:

Outcome	*Assets*	*Utility*	*Probability*	*Expected Utility*
Loses both	60,000	0.24	0.16 (= 0.4 × 0.4)	0.0384
Loses one	80,000	0.50	0.48 (= 2 × 0.6 × 0.4)	0.2400
Wins both	100,000	0.67	0.36 (= 0.6 × 0.6)	0.2412
			Total	0.5196

Since the expected utility is in excess of 0.50 (the utility corresponding to his present assets level), he should accept the offer.

Review Illustrations

Example 13.7 The relevant data for three alternatives to invest Rs one lakh are given below:

Investment		*Relevant Data*
A (Dealership)	:	The probability of success in exporting a product is 0.7 and in that case the return could be Rs 40,000 per year. Otherwise it could be sold in domestic market with a profit of Rs 20,000 per year. In any case, the expenditure per year is Rs 10,000.
B (Property)	:	If an expenditure of Rs 25,000 is incurred for renovation, a net profit of Rs 50,000 could be made. For this, the probability is 0.8. Otherwise it can be sold as it is, for a profit of Rs 30,000. The whole transaction would take one year from the date of investment.
C (Fixed deposit)	:	The company in which investment is made would pay 25% if it runs successfully for a year. Otherwise it would pay the principal amount only. The probability for failure of the company is 0.2.

Which investment should be chosen and what would be the rate of return? *(ICWA, December, 1993)*

For each of the investments, the expected percentage return is shown calculated below:

Investment	*Return*	*Probability*	*Expected Value*	*% Return*
A	Rs 40,000	0.7	28,000	
	Rs 20,000	0.3	6,000	
			34,000	
			– 10,000	
			24,000	24%
B	Rs 25,000	0.8	20,000	
	Rs 30,000	0.2	6,000	
			26,000	26%
C	Rs 25,000	0.8	20,000	
	Rs 0	0.2	0	
			20,000	20%

Notes:

1. For investment *A*, expected value of return is Rs 34,000. With a given expenditure of Rs 10,000, the net return works out to be Rs 24,000 or 24%.
2. The return of Rs 25,000 on investment *B* is calculated as the profit on sale *less* cost of innovation = Rs 50,000 – Rs 25,000.
3. A 25% return on Rs 100,000 in investment *C* would mean Rs 25,000. In case of company failure, no return is indicated. Thus, expected return is $0.8 \times 25{,}000 + 0.2 \times 0 =$ Rs 20,000.

Conclusion: It is evident from the above calculations that investment *B* is the best one.

Example 13.8 An investor is given the following investment alternatives and percentage rates of return.

	States of Nature (Market Conditions)		
	Low	*Medium*	*High*
Regular shares	2%	5%	8%
Risky shares	–5%	7%	15%
Property	–10%	10%	20%

Over the past 300 days, 150 days have been medium market conditions and 60 days have had high market increases.

On the basis of these data, state the optimal investment strategy for the investor.

According to the given information, the probabilities of low, medium, and high market conditions would be 90/300 or 0.30, 150/300 or 0.50, and 60/300 or 0.20, respectively. The expected pay-offs for each of the alternatives are calculated and shown in Table 13.18.

TABLE 13.18 Determination of Expected Return

Market Conditions	*Prob.*	*Strategy*		
		Regular Shares	*Risky Shares*	*Property*
Low	0.30	0.02	−0.05	−0.10
Medium	0.50	0.05	0.07	0.10
High	0.20	0.08	0.15	0.20
Expected Return		0.053	0.050	0.060

Since the expected return of 6 per cent is the highest for property, the investor should invest in this alternative.

Example 13.9 Informatics Corporation summarises international information reports (on a weekly basis), prints sophisticated data and forecasts, which are purchased weekly by mutual funds, banks and insurance companies. This information is very expensive and the demand for the reports is limited to a maximum of 30 units. The possible demands are 0, 10, 20 and 30 reports per week. The profit per report sold is Rs 30 and the loss per report unsold is Rs 20. No production of extra reports during a week is possible. Further, there is a penalty cost of Rs 250 for not meeting the demand. Unsold reports cannot be carried on to the next week. Using the pay-off table, find out the number of reports to be produced if:

(i) Maximin or pessimistic strategy is adopted.

(ii) Maximax or optimistic strategy is used. *(CA, May, 1991)*

From the given information,

Profit per report sold = Rs 30

Loss per report unsold = Rs 20

Penalty for not meeting the demand = Rs 250

Using these values, the pay-off matrix can be derived as shown in Table 13.19.

TABLE 13.19 Pay-off Matrix

Demand: No. of Reports per week	*Number of Reports Produced*			
	0	*10*	*20*	*30*
0	0	−200	−400	−600
10	−250	300	100	−100
20	−250	50	600	400
30	−250	50	350	900
Minimum Pay-off	−250	−200	−400	−600
Maximum Profit	0	300	600	900

(i) The maximum of the minimum values is –200. Hence, according to the *maximin* strategy, the decision is to produce 10 copies of the report with a loss of Rs 200.

(ii) The maximum among the maximum values is 900 when 30 reports are produced per week. Thus, according to the *maximax* strategy, the decision is to produce 30 reports.

Example 13.10 ABC Company needs to increase its production beyond its existing capacity. It has narrowed the alternatives to two approaches to increase the production capacity:

(1) expansion at a cost of Rs 8 million, or

(2) modernisation, at a cost of Rs 5 million.

Both approaches would require the same amount of time for implementation. Management believes that over the required payback period, demand will either be high or moderate. Since high demand is considered to be somewhat less likely than moderate demand, the probability of high demand has been set at 0.35. If the demand is high, expansion would gross an additional Rs 12 million but modernisation only an additional amount of Rs 6 million, due to lower maximum production capacity. On the other hand, if the demand is moderate, the comparable figures would be Rs 7 million for expansion and Rs 5 million for modernisation.

(i) Calculate the conditional profit in relation to various action-and-outcome combination and states of nature.

(ii) If the company wants to maximise its EMV, should it modernise or expand?

(iii) Calculate the EVPI and EOL. *(MBA, Delhi, April 1998)*

From the given information, the conditional profit matrix is obtained as shown in Table 13.20.

TABLE 13.20 Conditional Profit (in millions of Rs)

Demand	Probability	Course of Action	
		Expand	*Modernise*
High	0.35	4	1
Moderate	0.65	(1)	0
Expected Profit		0.75	0.35

From the expected values, it is evident that the company should decide to expand.

Calculation of EVPI

$$\text{EVPI} = \text{EPPI} - \text{Expected profit for optimal decision}$$

$$\text{EPPI} = 0.35 \times 4 + 0.65 \times 0 = \text{Rs } 1.4 \text{ (million)}$$

$$\therefore \quad \text{EVPI} = \text{Rs } 1.4 - \text{Rs } 0.75 = \text{Rs } 0.65 \text{ (million)}$$

Calculation of EOL

We can obtain opportunity loss by subtracting every value in a row from the larger value in the row. For the strategy of expansion, we get 0 and 1; while for the other strategy, we obtain 3 and 0. Accordingly,

EOL for Expansion : $0 \times 0.35 + 1 \times 0.65 = 0.65$

EOL for Modernisation : $3 \times 0.35 + 0 \times 0.65 = 1.05$

Example 13.11 A stockist of a particular commodity makes a profit of Rs 30 on each sale made within the same week of purchase, otherwise he incurs a loss of Rs 30 on each item. The data on the past sales are given here:

No. of items sold within the same week	:	5	6	7	8	9	10	11
Frequency	:	0	9	12	24	9	6	0

(a) Find out the optimum number of items the stockist should buy every week in order to maximise the profit.

(b) Calculate the expected value of perfect information. *(CA, May, 1996)*

(a) Since the number of items sold in a week are between 6 and 10, we may consider stocking 6 through 10 units as the courses of action. Also, the given frequencies may be converted into probabilities by dividing each of them by their sum total, that is 60. Now, using the given data, the conditional pay-off matrix is given in Table 13.21.

TABLE 13.21 Conditional Pay-off Matrix

No. of Units Sold	*Freq.*	*Prob.*	*No. of Units Stocked*				
			6	*7*	*8*	*9*	*10*
6	9	0.15	180	150	120	90	60
7	12	0.20	180	210	180	150	120
8	24	0.40	180	210	240	210	180
9	9	0.15	180	210	240	270	240
10	6	0.10	180	210	240	270	300
Expected Value			180	201	210	195	171

The expected pay-off for each of the actions is shown in the last row of the table. It is highest for the strategy of stocking 8 units. Thus, optimal strategy: stock 8 units.

(b) To obtain EVPI, we first calculate expected pay-off under perfect information (EPPI) as follows:

$$\text{EPPI} = 0.15 \times 180 + 0.20 + 210 + 0.40 \times 240 + 0.15 \times 270 + 0.10 \times 300 = \text{Rs } 235.50$$

$$\therefore \quad \text{EVPI} = \text{EPPI} - \text{Expected pay-off under optimal policy} = \text{Rs } 235.50 - \text{Rs } 210 = \text{Rs } 25.50$$

Example 13.12 The Rs 8,00,000 property of the Goodwill India Co has one-tenth of one per cent chance of catching fire that will cause damage to the property to the extent of Rs 1,00,000; and a one-twentieth of one per cent chance of catching fire that will completely destroy the property. The management of the company decides to insure the property and is reviewing two alternative insurance policies:

(a) A policy with Rs 50,000-deductible, that is, the insurance company covers all losses expect the initial Rs 50,000. The annual premium for such a policy is known to be one-tenth of one per cent of the value of the property.

(b) A no-deduction policy with full compensation having an annual premium of Rs 1,000.

If the company's objective is cost minimisation, which policy should it opt for? Sketch both, the pay-off table and the opportunity loss table, for the situation and solve both of them.

According to the given information, the management has two choices open:

A_1 : The Rs 50,000-deductible policy, and

A_2 : The full compensation policy

The various states of nature and their probabilities are:

State of Nature, E_i	*Prob., p_i*
E_1: A damage of Rs 1,00,000	0.0010
E_2: A damage of Rs 8,00,000	0.0005
E_3: No damage	0.9985

The pay-offs associated with different act-event combinations are contained in Table 13.22 and the expected pay-offs are calculated.

TABLE 13.22 Calculation of Expected Pay-offs

Event	*Probability*	*Act, A_j*	
		A_1	A_2
E_1	0.0010	49,200	99,000
E_2	0.0005	749,200	799,000
E_3	0.9985	(800)	(1,000)
Expected Pay-off		(375)	(500)

From the table we observe that the expected pay-offs associated with the acts are: A_1 (375) and A_2 (500). The negative pay-offs here imply the cost. Since the cost for A_1 is lower, the management should go in for the Rs 50,000-deductible policy. The opportunity loss values, derived from the pay-off values, are given in Table 13.23 and the expected values are also calculated.

TABLE 13.23 Calculation of Expected Opportunity Loss

Event E_i	*Probability*	*Act, A_j*	
		A_1	A_2
E_1	0.0010	49,800	0
E_2	0.0005	49,800	0
E_3	0.9985	0	200
	EOL	74.7	199.7

A_1 is the optimal act (as before), since the expected opportunity loss, EOL, is lower for it.

Example 13.13 Leisure Hotels Ltd. is planning to build another 700-room complex. It has been suggested that because the existing hotels average only 70% occupancy on an annual basis, the new complex should have only 500 rooms. It has been estimated that the cost per room per annum is Rs 21,000.

The following data based on demand at similar complexes has been obtained:

	Number of Days	Daily Demand	Average Price per Occupied Room per Day
Peak season	200	800	Rs 100
In between	80	600	Rs 80
Slack season	85	500	Rs 60

Now,

(a) prepare a pay-off table for a complex with 500, 600, 700 and 800 rooms;

(b) on the basis of this data, advise management as to the number of rooms it should include;

(c) state with reasons two further categories of data that would be useful before making the final decision.

(ICMA, May 1983, Adapted)

(a) For each of the different number of rooms, say R, the annual cost would be given by 21,000 R while the annual revenue could be broken down according to the particular season. Thus, if $R = 500$, all rooms would be let in each season and all revenues will be:

Peak season $200 \times 500 \times 100$ = Rs 100,00,000

In between $80 \times 500 \times 80$ = Rs 32,00,000

Slack season $85 \times 500 \times 60$ = Rs 25,50,000

The revenues in each season can be calculated for each of different sizes of the complex to get the pay-off table (Table 13.24).

TABLE 13.24 Pay-off Table (Rupees in lac)

Season	*No. of Rooms*			
	500	*600*	*700*	*800*
Peak	100.0	120.0	140.0	160.0
In between	32.0	38.4	38.4	38.4
Slack	25.5	25.5	25.5	25.5

(b) The annual profits for each room size are as follows:

$R = 500$: Profit = 100.0 + 32.0 + 25.5 – 105.0 = Rs 52.5 lac

$R = 600$: Profit = 120.0 + 38.4 + 25.5 – 126.0 = Rs 57.9 lac

$R = 700$: Profit = 140.0 + 38.4 + 25.5 – 147.0 = Rs 56.9 lac

$R = 800$: Profit = 160.0 + 38.4 + 25.5 – 168.0 = Rs 55.9 lac

The size of complex yielding the maximum expected annual profit is 600 rooms with an associated value of Rs 57,90,000 per annum.

(c) The data on demand, cost and revenue are seemingly based on current estimates only. In keeping with the nature of investment, it would be advisable to estimate the likely trends in these figures to ascertain the profitability in the years to come.

Besides, the revenues are presumably based on only the occupied rooms, having no regard to the unoccupied rooms. It may be possible, for instance, to stimulate extra demand by providing convention facilities and thereby attracting revenue from rooms which otherwise would be empty.

Example 13.14 The annual demand for a seasonal product follows the distribution shown here.

Demand (units) :	3,000	3,500	4,000	4,500	5,000
Probability :	0.10	0.20	0.30	0.30	0.10

The manufacturer of this item can produce it by one of the three methods:

(a) Using the existing equipment at a cost of Rs 8 per unit.

(b) Buy special equipment for Rs 22,000 whose salvage value at the end of the year would be Rs 2,000. The variable cost per unit, using this equipment is Rs 2.

(c) Buy special equipment for Rs 90,000, which would be depreciated on straight-line basis over a period of 4 years. The variable cost using this equipment is Rs 1.20 per unit.

Which method of production should the manufacturer follow in order to maximise profit, assuming that production must meet all the demand?

Since the cost data are given in this problem, we shall determine the conditional cost values and then calculate the expected cost in respect of each of the methods of production. The method with the least expected cost would be chosen. The calculations are given in Table 13.25.

TABLE 13.25 Calculation of Expected Cost

Event	*Prob.*	*Courses of Action, A_j*		
		A_1 : Existing	*A_2 : Special Equip. I*	*A_3 : Special Equip. II*
E_1 : 3,000	0.10	24,000	26,000	26,100
E_2 : 3,500	0.20	28,000	27,000	26,700
E_3 : 4,000	0.30	32,000	28,000	27,300
E_4 : 4,500	0.30	36,000	29,000	27,900
E_5 : 5,000	0.10	40,000	30,000	28,500
Expected cost		32,400	28,100	27,360

Result: Buy special equipment with a 4-year life.

Example 13.15 A company is considering the introduction of a new product to its existing product range. It has defined two levels of sales as 'high' and 'low' on which to base its decision and has estimated the chances that each market level will occur, together with their costs and consequent profits or losses. This information is summarised below:

States of Nature	*Probability*	*Course of Action*	
		Market Product (Rs '000)	*Do Not Market Product (Rs '000)*
High sales	0.3	150	0
Low sales	0.7	–40	0

The company's marketing manager suggests a market research survey be undertaken to provide further information on which to base the decision. On past experience with a certain market research organisation, the marketing manager assesses its ability to give good information in the light of subsequent actual sales achievements as follows:

Market Research Survey Outcome	*Actual Sales*	
	Market 'High'	*Market 'Low'*
'High' sales forecast	0.5	0.1
'Indecisive' survey report	0.3	0.4
'Low' sales forecast	0.2	0.5

Given that to undertake the market research survey will cost Rs 20,000, determine whether or not there is a case for employing the market research organisation. *(MBA, Delhi, October, 1996)*

If the survey is not undertaken, the expected profit is:

$$0.3 \times 150 + 0.70 \times (-40) = 17 \text{ that is, Rs } 17{,}000$$

If the survey is undertaken, the following outcomes are possible, with probabilities as shown:

Actual sales	*Survey outcome*	*Probability*
High	High	$0.3 \times 0.5 = 0.15$
High	Indecisive	$0.3 \times 0.3 = 0.09$
High	Low	$0.3 \times 0.2 = 0.06$
Low	High	$0.7 \times 0.1 = 0.07$
Low	Indecisive	$0.7 \times 0.4 = 0.28$
Low	Low	$0.7 \times 0.5 = 0.35$

For the three possible survey outcomes,

$$P(\text{High}) = 0.15 + 0.07 = 0.22$$

$$P(\text{Indecisive}) = 0.09 + 0.28 = 0.37$$

$$P(\text{Low}) = 0.06 + 0.35 = 0.41$$

If the survey were to give a *high sales forecast*, then the probabilities of sales actually being high and low are:

$P(\text{High Sales/High Forecast}) = 0.15/0.22 = 0.682$

$P(\text{Low Sales/High Forecast}) = 0.07/0.22 = 0.318$

and the expected profit is $0.682 \times 150 + 0.318 \times (-40) = 89.58$

For an indecisive forecast:

$P(\text{High Sales/Indecisive Forecast}) = 0.09/0.37 = 0.243$

$P(\text{Low Sales/Indecisive Forecast}) = 0.28/0.37 = 0.757$

and the expected profit is $0.243 \times 150 + 0.757 \times (-40) = 6.17$

For a low forecast:

P(High Sales/Low Forecast) = 0.06/0.41 = 0.146

P(Low Sales/Low Forecast) = 0.35/0.41 = 0.854

and the expected profit is $0.146 \times 150 + 0.854 \times (-40) = -12.26$

Since this last value is negative, it implies that the company would expect to make a loss if it launched the new product following a low sales forecast. It should not, therefore, go ahead and, consequently, in this case the expected profit would be taken to be zero. Now, taking all possible survey forecasts into account, the overall expected profit is

$$0.22 \times 89.58 + 0.37 \times 6.17 + 0.41 \times 0 = 22.0 \text{ or Rs } 22{,}000.$$

Therefore, as a result of taking the survey, expected profit has increased from Rs 17,000 to Rs 22,000. As the survey costs Rs 20,000 to undertake, it is evidently not worthwhile.

Example 13.16 An ice cream manufacturer sells soft scoop ice cream in special pressurised containers and is planning production for the summer, which is the peak period. The company wishes to ensure that it has the best quantity of containers on hand: too few and sales will be lost; too many and the surplus will have to be stored over the winter at a substantial cost. The containers can only be purchased in lots of 500. The following table shows the estimated lost contributions for various ordering patterns:

	Number of New Containers Bought			
	0	*500*	*1,000*	*1,500*
Poor summers—low sales	0	20	20	30
Fair summer—reasonable sales	15	0	15	20
Good summer—good sales	20	20	0	15
Very good summer—very high sales	30	25	15	0

Based on past data, the probabilities of the different types of weather are:

Poor : 0.3 Fair : 0.4 Good : 0.2 Very good : 0.1

The firm has obtained a copy of the long-range weather forecast for the summer which indicates that there will be good summer, but past experience states that the forecasts are not 100% accurate, as follows:

P(forecast *good* but weather *poor*) = 0.3

P(forecast *good* but weather *fair*) = 0.4

P(forecast *good* and weather *good*) = 0.7

P(forecast *good* but weather *very good*) = 0.2

You are required

(a) to calculate the number of containers that should be purchased based on past data only (i.e. ignore the forecast);

(b) to calculate whether the decision in (a) would need to be altered if the forecast is taken into account;

(c) to explain any changes made in the purchase decision as a result of comparing your answers to (a) and (b) above. *(ICMA, November, 1993, adapted)*

(a) *When the decision is on the basis of past data only: prior analysis*

The conditional loss contributions are given in Table 13.26. Also, the expected loss contributions are shown calculated. From the calculations, it is evident that 500 new pressurised containers should be ordered.

TABLE 13.26 Calculation of Expected Loss Contribution (Prior)

States of Nature	*Probability*	*Course of Action: No of Containers*			
		0	*500*	*1,000*	*1,500*
Poor Summer	0.3	0	20	20	30
Fair Summer	0.4	15	0	15	20
Good Summer	0.2	20	20	0	15
Very Good Summer	0.1	30	25	15	0
Expected Value		13	12.5	13.5	20

(b) *When decision is based on forecasts: posterior analysis*

We may first calculate the probability of poor, fair, good and very good summer, given that the forecast is good (E), as $P(A/E)$, $P(B/E)$, $P(C/E)$ and $P(D/E)$, respectively, using Bayes' theorem as follows:

Given $P(A) = 0.3$, $P(E/A) = 0.3$, $P(B) = 0.4$, $P(E/B) = 0.4$, $P(C) = 0.2$, $P(E/C) = 0.7$, $P(D) = 0.1$, and $P(E/D) = 0.2$. Accordingly,

$$\begin{aligned} P(E) &= P(A \cap E) + P(B \cap E) + P(C \cap E) + P(D \cap E) \\ &= P(A) \times P(E/A) + P(B) \times (E/B) + P(C) \times (E/C) + P(D) \times (E/D) \\ &= 0.3 \times 0.3 + 0.4 \times 0.4 + 0.2 \times 0.7 + 0.1 \times 0.2 = 0.41 \end{aligned}$$

Now,

$$P(A/E) = P(A \cap E)/P(E) = 0.3 \times 0.3/0.41 = 0.22$$

$$P(B/E) = P(B \cap E)/P(E) = 0.4 \times 0.4/0.41 = 0.39$$

$$P(C/E) = P(C \cap E)/P(E) = 0.2 \times 0.7/0.41 = 0.34$$

$$P(D/E) = P(D \cap E)/P(E) = 0.1 \times 0.2/0.41 = 0.05$$

We may now recalculate the expected values using these posterior probabilities. This is shown in Table 13.27.

TABLE 13.27 Calculation of Expected Loss Contribution (Posterior)

State of Nature	*Probability*	*Course of Action; No. of Containers*			
		0	*500*	*1,000*	*1,500*
Poor Summer	0.22	0	20	20	30
Fair Summer	0.39	15	0	15	20
Good Summer	0.34	20	20	0	15
Very Good Summer	0.05	30	25	15	0
Expected Value		14.15	12.45	11.0	19.50

It is clear from the table that since the expected lost contribution is the lowest with 1,000, the order should be in the lots of 1,000.

(c) Evidently, the purchasing decision is changed now in view of the increased ability of forecast good weather. It has resulted in deciding to purchase 1,000 pressurised containers in the expectation of a good summer associated with good sales.

Example 13.17 An engineering company's plant maintenance budget for the month of May is showing an adverse cost variance of Rs 1,200 on a total budget of Rs 20,000. The factory manager is uncertain whether it would be worthwhile investigating the cost variance and taking any corrective action since he estimates this will cost Rs 1,000. Further, he estimates that given the usual variation on his maintenance costs, he would only have a 50% chance in any case of being in the range of Rs 19,500 – Rs 20,500.

Higher than planned maintenance costs could be due to engineering failures more complex than expected. These would be rectified after investigation. The manager estimates that consequential losses of not investigating such a situation could amount to Rs 4,000.

(a) Calculate the standard deviation of the estimated maintenance costs;

(b) set out the pay-off table for action on the decision as to investigation of variance;

(c) state the decision rule and the probability that the process is out of control which would lead to a decision to investigate;

(d) advise the manager as to whether he should investigate the causes of the variances in May budget. (Assume that the variances in the maintenance budget follow a normal probability pattern).

(ICMA, May, 1982, Adapted)

(a) Given that variances in the maintenance budget follow normal distribution, the maintenance cost would be normally distributed with mean, μ = Rs 20,000, as shown in Figure 13.6. Further, since there is a 50 per cent chance that the cost would be between Rs 19,500 and Rs 20,500, there is a 0.25 probability that its value would lie between Rs 20,000 and Rs 20,500.

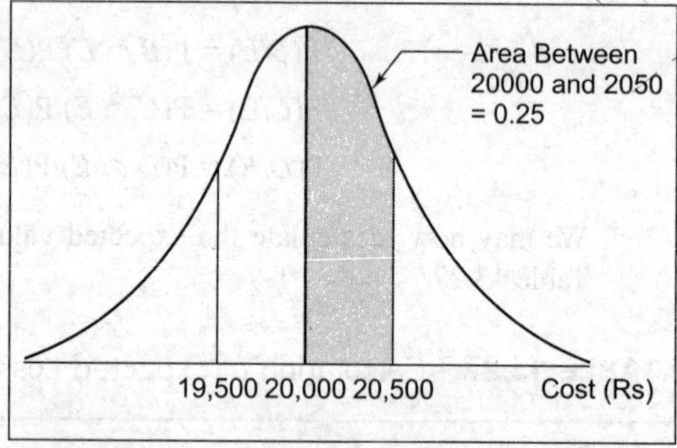

Figure 13.6 *Distribution of Cost*

We know, $z = \dfrac{X - \mu}{\sigma}$

From the normal real table, the z-value corresponding to area = 0.25 is 0.675. Thus, we have

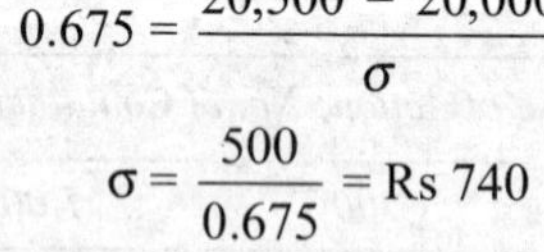

$$0.675 = \frac{20{,}500 - 20{,}000}{\sigma}$$

or $$\sigma = \frac{500}{0.675} = \text{Rs } 740$$

(b) There are two possible states of nature, or events:

E_1—the process is out of control (i.e. maintenance cost ≠ 20,000)

E_2—the process is in control (i.e. maintenance costs = 20,000)

Similarly, there are two courses of action, namely

A_1—investigate the variance,

A_2—do not investigate the variance.

According to the given information, the pay-off matrix is given here, the pay-offs being the cost values.

TABLE 13.28 Pay-off Table

Event	Act, A_j	
	A_1	A_2
E_1	1,000	4,000
E_2	1,000	0

(c) If the decision rule to be used is that of minimising the expected cost, the optimal course of action would depend upon the likelihood of each event. Let p be the probability of E_1, so that

$$P(E_1) = p \quad \text{and} \quad P(E_2) = 1 - p.$$

The expected values for each of the acts would be as follows:

$$E(A_1) = 1{,}000p + 1{,}000(1 - p) = 1{,}000$$

$$E(A_2) = 4{,}000p + 0(1 - p) = 4{,}000p.$$

For indifference,

$$1{,}000 = 4{,}000p \quad \text{or} \quad p = 1{,}000/4{,}000 = 0.25.$$

Thus, if the probability that the process is out of control exceeds 0.25, then the variance calls for investigation.

(d) When the process is under control, the variances in the cost will follow normal distribution (as given) with mean = 0 and a standard deviation = Rs 740. The probability of observing a variance of upto Rs 1,200 can be determined as follows:

$$z = \frac{1{,}200 - 0}{740} = 1.62$$

Area beyond $z = 1.62$ equals 0.053. Thus, the probability that a cost variance as great as Rs 1,200 (favourable or adverse) will occur $= 2 \times 0.053 = 0.11$ or 11%. It suggests that the variance should be investigated as the low probability of 11% is indicative of the state that the process might be out of control or at least that the probability of its being out of control is greater than 0.25.

Note here the probability value 0.11 should not be taken to mean that it represents the conditional probability of observing a variance in excess of Rs 1,200 *assuming* that the process is in control. If we denote the event process in control by C and 'process not in control' by C, we have,

$$P(\text{var.} = 1{,}200/C) = 0.11.$$

To obtain the probability that the process is in control having observed the variance of Rs 1,200, we need to use the Bayes' Theorem as follows:

$$P(C/\text{var.} = 1{,}200) = \frac{P\,(\text{var.} = 1{,}200/C) \times P\,(C)}{P\,(\text{var.} = 1{,}200/C) \times P\,(C) + P\,(\text{var.} = 1{,}200/\overline{C}) \times P(\overline{C})}$$

However, it is possible to calculate this probability only if we knew $P(C)$ and $P(\text{var.} = 1{,}200/\overline{C})$.

Example 13.18 The investment staff of TNC Bank is considering four investment proposals for a client: shares, bonds, real estate and savings certificate. These investments will be held for one year. The past data regarding the four proposals are given below:

Shares: There is 25 per cent chance that shares will decline by 10 per cent, a 30 per cent chance that they will remain stable and a 45 per cent chance that they will increase in value by 15 per cent. Also, the shares under consideration do not pay any dividends.

Bonds: These bonds stand a 40 per cent chance of increase in value by 5 per cent and 60 per cent chance of remaining stable and they yield 12 per cent.

Real Estate: This proposal has a 20 per cent chance of increasing 30 per cent in value, a 25 per cent chance of increasing in 20 per cent value, a 40 per cent chance of increasing in 10 per cent value, a 10 per cent chance of remaining stable and a 5 per cent chance of losing 5 per cent of its value.

Savings Certificates: These certificates yield 8.5 per cent with certainty.

Use a decision tree to structure the alternatives available to the investment staff, and using the expected value criterion, choose the alternative with the highest expected value. *(MBA, Delhi, Nov. 2006)*

The tree diagram is shown in Figure 13.7.

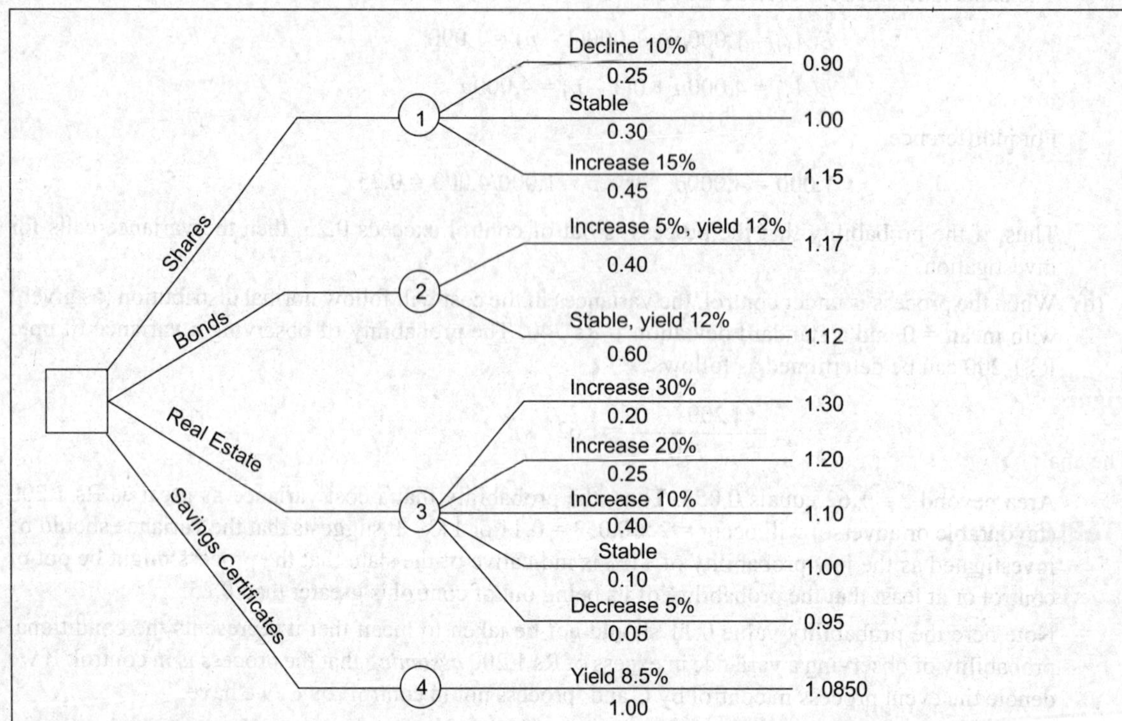

Figure 13.7 *Decision Tree: Investment Problem*

The end values give the amount a rupee 1 invested would become in each case.

The expected value at each chance node is shown calculated below:

Node 1 : $0.25 \times 0.90 + 0.30 \times 1.00 + 0.45 \times 1.15 = 1.0425$

Node 2 : $0.40 \times 1.17 + 0.6 \times 1.12 = 1.1400$

Node 3 : $0.20 \times 1.30 + 0.25 \times 1.20 + 0.40 \times 1.10 + 0.10 \times 1.00 + 0.05 \times 0.95 = 1.1475$

Node 4 : $1.00 \times 1.0850 = 1.0850$

It is evident that the maximum expected pay-off is at node 3. Hence, investment should be made in real estate.

Example 13.19 A Finance Manager is considering drilling a well. In the past, only 70% of wells drilled were successful at 20 metres depth in that area. Moreover, on finding no water at 20 metres, some persons in that area drilled it further up to 25 metres but only 20% struck water at that level. The prevailing cost of drilling is Rs 500 per metre. The Finance Manager estimated that in case he does not get water in his own well, he will have to pay Rs 15,000 to buy water from outside for the same period of getting water from the well. The following decisions are considered:

(i) Do not drill any well;
(ii) Drill up to 20 metres, and
(iii) If no water is found at 20 metres, drill further upto 25 metres.

Draw an appropriate decision tree and determine the Finance Manager's optimal strategy.

(CA, May, 1992)

Based on the given information, the decision tree is shown in Figure 13.8.

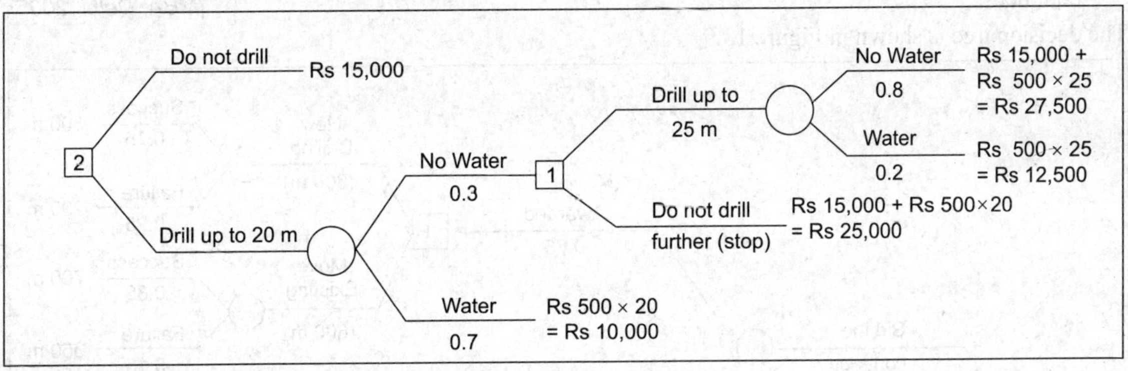

Figure 13.8 *Decision Tree: Drilling Problem*

The analysis of the tree is given in Table 13.29.

TABLE 13.29 Analysis Table: Decision Tree

Decision Node	*Options*	*Expected Cost*	*Decision*
1	Drill up to 25 metres	$0.8 \times 27{,}500 + 0.2 \times 12{,}500$ = Rs 24,500	Drill up to 25 metres
	Stop	Rs 25,000	
2	Do not drill	Rs 15,000	
	Drill up to 20 metres	$0.3 \times 24{,}500 + 0.7 \times 10{,}000$ = Rs 14,350	Drill up to 20 metres

From the analysis table, it may be observed that decision at node 2 implies that if it is decided to drill up to 20 metres and water is not found, then drilling up to 25 metres should be done. At node 1, the decision taken is to drill up to 20 metres as it involved lower expected cost. Thus, the optimal strategy is to drill up to 20 metres and if water is not struck then drill further to 25 metres.

Example 13.20 The Oil India Corporation is considering whether to go for an offshore drilling contract to be awarded in Bombay High. If they bid, value would be Rs 600 million with 65% chance of gaining the contract. The Corporation may set up a new drilling operation or move already existing operation, which has proved successful, to new site. The probability of success and expected returns are as follows:

Outcome	*New Drilling Operation*		*Existing Operation*	
	Probability	*Expected Revenue (Rs million)*	*Probability*	*Expected Revenue (Rs million)*
Success	0.75	800	0.85	700
Failure	0.25	200	0.15	350

If the Corporation does not bid or lose the contract, they can use Rs 600 million to modernise their operations. This would result in a return of either 5% or 8% on the sum invested with probabilities 0.45 and 0.55, respectively.

(a) Construct a decision tree for the problem showing clearly the courses of action.

(b) By applying an appropriate decision criterion recommend whether or not the Corporation should bid the contract.

(MBA, Delhi, 2005)

The decision tree is shown in Figure 13.9.

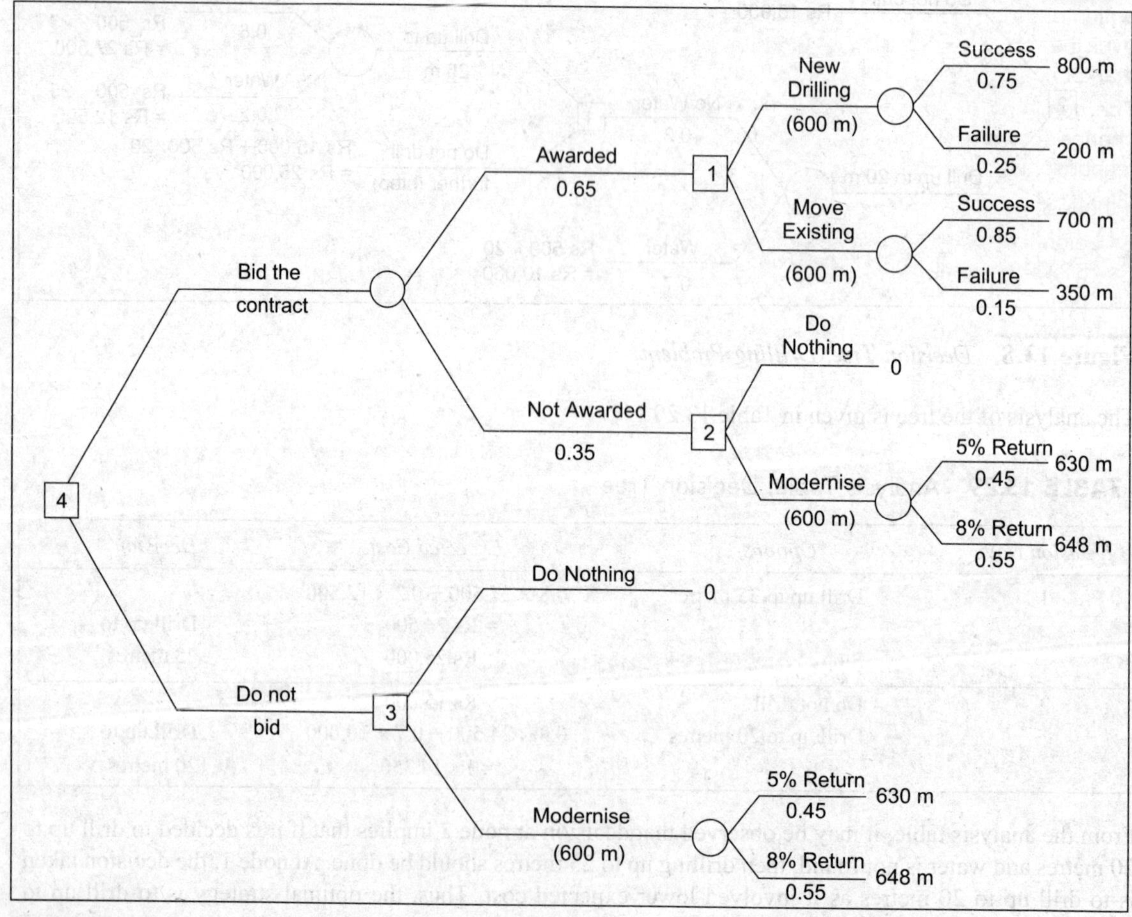

Figure 13.9 *Decision Tree: Bidding Problem*

The decision tree is analysed below:

Decision Node	*Options*	*EMV*	*Decision*
1	(a) New drilling	0.75 × 800 + 0.25 × 200 – 600 = 50 m	New drilling
	(b) Move existing operations	0.85 × 700 + 0.15 × 350 – 600 = 47.5 m	
2	(a) Do nothing	0	Modernise
	(b) Modernise	0.45 × 630 + 0.55 × 648 – 600 = 39.9 m	
3	(a) Do nothing	0	Modernise
	(b) Modernise	0.45 × 630 + 0.55 × 648 – 600 = 39.9 m	
4	(a) Bid the contract	0.65 × 50 + 0.35 × 39.9 = 46.465 m	Bid the contract
	(b) Do not bid the contract	39.9 m	

Decision: Bid for the contract. If the contract is awarded, then set up new drilling operation. If not, then modernise.

Example 13.21 A company has developed a new product in its R&D laboratory. The company has the option of setting up production facility to market this product straight away. If the product is successful, then over the three years expected product life, the returns will be Rs 120 lakh with a probability of 0.70. If the market does not respond favourable, then the returns will be only Rs 15 lakh with probability of 0.30.

The company is considering whether it should test market this product building a small pilot plant. The chance that the test market will yield favourable response is 0.80. If the test market gives favourable response, then the chance of successful total market improves to 0.85.

If the test market gives poor response then the chance of success in the total market is only 0.30.

As before, the returns from a successful market will be Rs 120 lakh and from an unsuccessful market only Rs 15 lakh. The installation cost to produce for the total market is Rs 40 lakh and the cost of the test marketing pilot plant is Rs 5 lakh. Using decision-tree analysis, draw a decision-tree diagram, carry out necessary analysis to determine the optimal decisions. *(MBA, Delhi, Nov. 2008)*

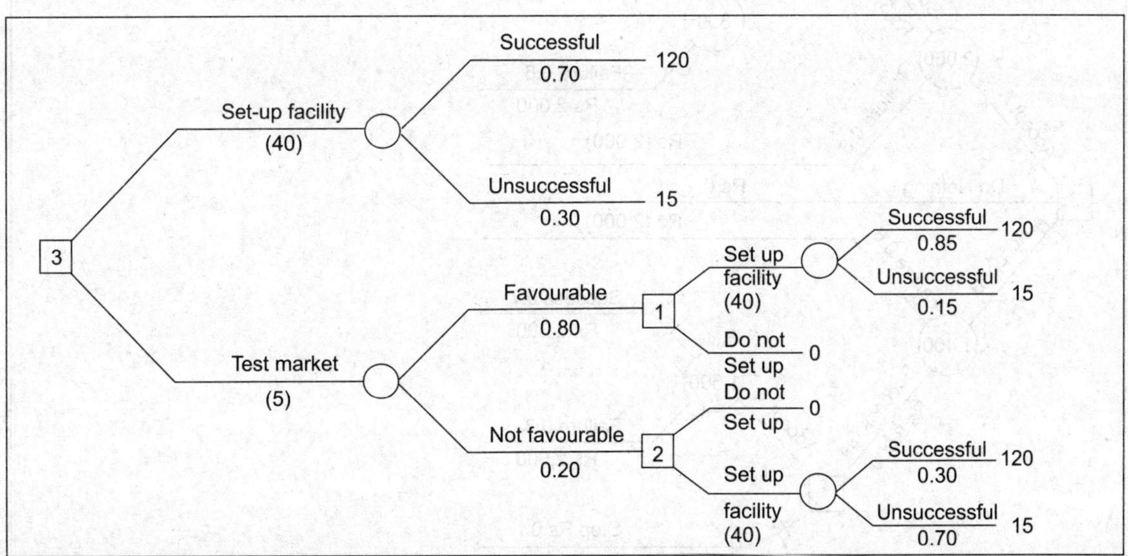

Figure 13.10 *Decision Tree: Setting up Facility*

Using the given information, the decision tree is shown in Figure 13.10. Note that, to begin with, the company has two options—to set up production facility or go for test market. In case it decides for test market, the company may set up the facility or quit, whether the result of test marketing is favourable or not.

The decision tree is analysed below.

Decision Node	*Options*	*EMV (in lakh of Rs)*	*Decision*
1	Set up facility Do not set up	$0.85 \times 120 + 0.15 \times 15 - 40 = 64.25$ 0	Set up facility
2	Set up facility Do not set up	$0.30 \times 120 + 0.70 \times 15 - 40 = 6.50$ 0	Set up facility
3	Set up facility Test market	$0.70 \times 120 + 0.30 \times 15 - 40 = 48.5$ $0.80 \times 64.25 + 0.2 \times 6.50 - 5 = 47.7$	Set up facility

Thus, the company should set up production facility straight way and not undertake test market.

Example 13.22 A businessman has two independent investments *A* and *B* available to him, but he lacks the capital to undertake both of them simultaneously. He can choose to take *A* first and then stop, or if *A* is successful then take *B*, or vice versa. The probability of success on *A* is 0.7, while for *B* it is 0.4. Both investments require an initial capital outlay of Rs 2,000, and both return nothing if the venture is unsuccessful. Successful completion of *A* will return Rs 3,000 (over cost), successful completion of *B* will return Rs 5,000 (over cost). Draw the decision tree and determine the best strategy. *(CA, May, 1985)*

The decision tree corresponding to the given information is depicted in Figure 13.11.

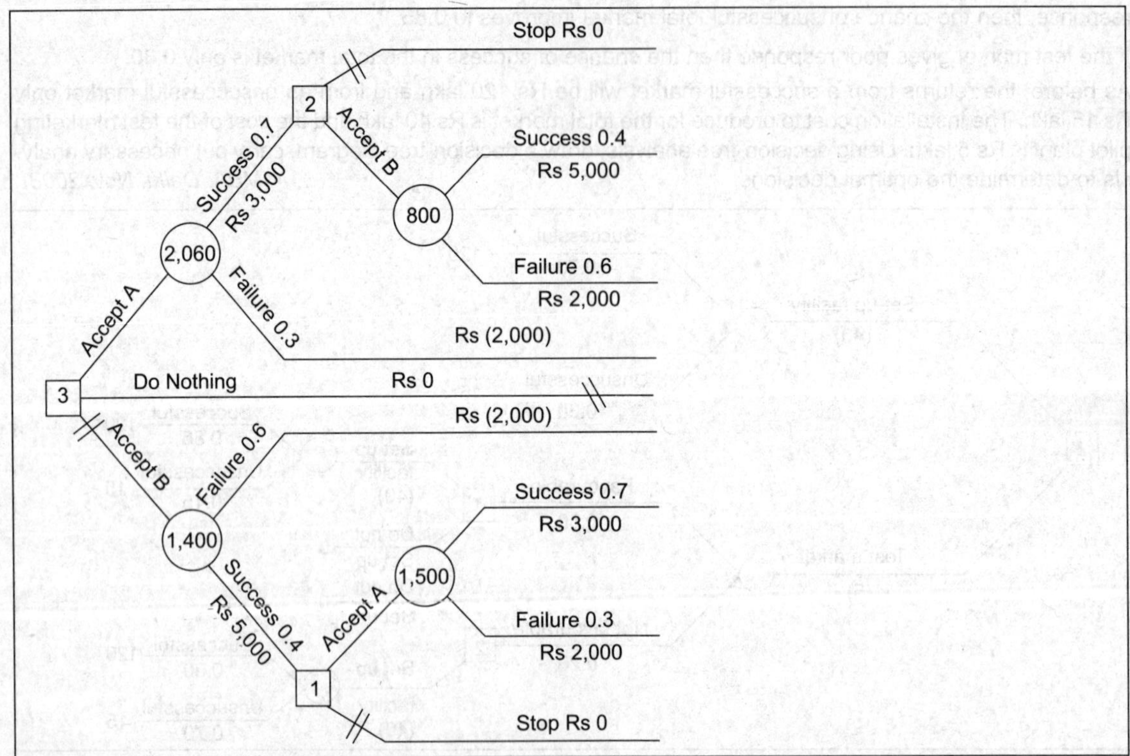

Figure 13.11 *Decision Tree: Investment Plan*

According to this, there are five strategies: (a) do nothing, (b) accept *A* and then stop, (c) accept *B* and then stop, (d) accept *A* and, if successful, then accept *B*, and (e) accept *B* and, if successful, accept *A*. The expected pay-off values are indicated in the circles representing chance nodes. There are three decision points marked 1, 2 and 3.

The evaluation of decision points by calculating expected values is contained in Table 13.30. The optimal decision is to accept *A* and, if successful, then accept *B*.

TABLE 13.30 Evaluation of Decision Points

Decision Point	*Outcome*	*Probability*	*Conditional Value*	*Expected Value*
1. Accept *A*	Success	0.7	Rs 3,000	2,100
	Failure	0.3	Rs (2,000)	(600)
				1,500
Stop				0
2. Accept *B*	Success	0.4	Rs 5,000	2,000
	Failure	0.6	Rs (2,000)	(1,200)
				800
Stop				0
3. Accept *A*	Success	0.7	Rs 3,000 + 800	2,660
	Failure	0.3	Rs (2,000)	(600)
				2,060
Accept *B*	Success	0.4	Rs 5,000 + 1,500	2,600
	Failure	0.6	Rs (2,000)	(1,200)
				1,400
Do nothing				0

Example 13.23 Re-solve Example 13.22 by preparing a pay-off table.

For the given problem, the various acts and states of nature, the events, are given here:

Courses of action:
- A_1 : do nothing
- A_2 : accept *A* and then stop
- A_3 : accept *B* and then stop
- A_4 : accept *A* and, if successful, then accept *B*
- A_5 : accept *B* and, if successful, then accept *A*

Events:
- E_1 : both *A* and *B* are successful
- E_2 : *A* will be successful but not *B*
- E_3 : *B* will be successful but not *A*
- E_4 : neither *A* nor *B* will be successful

The probabilities of various events are:

$$E_1 : 0.7 \times 0.4 = 0.28;\ E_2 : 0.7 \times 0.6 = 0.42;\ E_3 : 0.3 \times 0.4 = 0.12, \text{ and } E_4 : 0.3 \times 0.6 = 0.18.$$

The conditional pay-offs, resulting from different combinations of actions and events are given in Table 13.31. Since the expected value for the act A_4 is the largest, it represents the optimal choice.

TABLE 13.31 Calculation of Expected Pay-offs

Event, E_i	*Prob.*	*Act, A_j*				
		A_1	A_2	A_3	A_4	A_5
E_1	0.28	0	3,000	5,000	8,000	8,000
E_2	0.42	0	3,000	(2,000)	1,000	(2,000)
E_3	0.12	0	(2,000)	5,000	(2,000)	3,000
E_4	0.18	0	(2,000)	(2,000)	(2,000)	(2,000)
Expected Pay-off		0	1,500	800	2,060	1,400

Example 13.24 Mr X is considering whether to make an investment in a project with the following likely returns:

Amount (Rs)	*Probability*
2,00,000	0.6
–40,000	0.4

The utility function of Mr X is approximated as follows:

$$U = -0.0003M^2 \quad \text{when} \quad M < -5{,}000$$
$$= 1.05M \quad \text{when} \quad M \geq -5{,}000$$

Should the project be undertaken by him? Consider (a) Expected monetary value criterion, and (b) Expected utility criterion.

Calculation of expected monetary value (EMV) and expected utility (EU) is shown in Table 13.32.

TABLE 13.32 Calculation of EMV and EU

Conditional Monetary Value (i)	*Conditional Utility* (ii)*	*Probability (iii)*	*Expected Monetary Value (i) × (iii)*	*Expected Utility (ii) × (iii)*
200,000	210,000	0.6	120,000	126,000
– 40,000	– 480,000	0.4	–16,000	–192,000
Total			104,000	–66,000

* Obtained by substituting monetary values in the utility function.

Since EMV is positive, the project should be undertaken according to the EMV criterion, while it should not be accepted on the basis of the EU criterion since EU is negative.

KEY POINTS TO REMEMBER

- Decision theory is concerned with decision-making under conditions of uncertainty. The decision process involves the steps of identification of states of nature, courses of action and pay-off function depicting consequences resulting from their interaction, and then choosing the appropriate action on the basis of some given criterion. In addition to pay-offs, the outcome of event-action combinations can also be expressed in terms of regret.
- Among the rules for taking decisions are included the Laplace principle; maximin/minimax; maximax/minimin; Hurwicz principle; Savage principle; maximum likelihood and expectation principles. Expected regret criterion is employed when regret matrix is used.
- In addition to the optimal course of action, expected value of perfect information, EVPI, is also calculated. It equals the expected regret value of the optimal act and represents the maximum amount that a decision-maker would be willing to pay in case he is provided with perfect information.
- The decision analysis using expectation principle can be extended to cover situations where given probabilities of various states of nature are revised on the basis of some additional information about them (the states). The prior probabilities are converted into posterior probabilities using the conditional, joint and total probabilities. Such analysis is called *posterior analysis*.
- Besides determining optimal course of action using posterior probabilities, expected value of sample information, EVSI, can also be calculated as the difference of expected pay-off *with* and *without* sample information. Its efficiency is measured by expressing its value as a percentage of EVPI. The efficiency index enables the decision-maker to decide whether or not to seek the information.
- The decision-tree approach to decision-making is used in situations where multi-stage decisions are needed. The sequences of action-event combinations available to the decision-maker are presented graphically in the form of a decision tree. In analysing such situations, alternatives are evaluated by proceeding in a backward manner—by evaluating the best course at later stages to decide the best action at the earlier stages. The decision criterion is expected monetary value, EMV.
- As an alternative to the widely used EMV criterion, the expected utility criterion is also sometimes employed for taking decisions. In this approach, the monetary outcomes of various actions are replaced by *utils*, which are not real in nature and decision is taken by computing expected utility, EU, rather than expected monetary value, EMV.

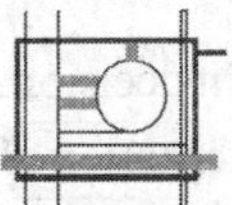

TEST YOUR UNDERSTANDING

Mark the following statements as T (*True*) or F (*False*).

1. Decision theory is concerned with determining optimal strategies where a decision-maker is faced with a number of alternatives and a risky pattern of future events.

2. Regret is the amount of money paid for not adopting the optimal course of action.
3. Different criteria for making decisions under risk would always lead to the same optimal choice.
4. *Laplace principle* is based on the premise of equal probability of occurrence of possible events.
5. *Minimax* is an optimist's choice while *Minimin* is a pessimist's criterion.
6. In *Hurwicz* criterion, the maximum pay-off is multiplied by α and the minimum by $1 - \alpha$, where α may be any number between zero and 100.
7. In situations of decisions under uncertainty, the *Laplace criterion* is the least conservative while the *minimax criterion* is the most conservative.
8. In case of pay-offs represented as profits, the *Savage criterion* for selecting optimal course of action will be based on the maximin principle.
9. *Maximum likelihood principle* is expected to yield good decisions when the probability of a particular event is predominantly larger in relation to probabilities of other events.
10. Expected pay-off and expected regret criteria would both lead to identical decisions in a given case.
11. All criteria for decisions under risk inherently assume that the particular (optimal) decision reached will be repeated a large number of times.
12. EPPI is the expected pay-off under certainty.
13. EVPI is the expected regret value of any strategy.
14. For each strategy in a given case, the following would hold:

    Expected pay-off + Expected regret = a constant (EPPI)
15. In Bayesian approach to decision making, the optimal strategy is determined using posterior probabilities.
16. Posterior probabilities are obtained by modifying prior probabilities taking into consideration the information from a sample.
17. Expected value of sample information is equal to expected cost with sample information minus expected cost without sample information.
18. Efficiency index is the ratio of expected value of sample information to expected value of perfect information.
19. The decision tree approach to decision-making is appropriate in those situations where a sequence of decisions are involved.
20. In the decision trees, no more than two alternative courses of action can emanate from a decision node.
21. In a given situation, *do nothing* can also be the best course of action.
22. In decision trees, the probabilities of all events at chance nodes and the monetary evaluations of different alternatives must all be known in advance.
23. The *rollback* technique of analysing decision trees involves working through it from right to left, evaluating the best course of action at later stages so as to decide on such action at the earlier stages.
24. The probabilities of various outcomes at each chance node should always add up to one.
25. A decision taken on the basis of expected monetary value would always prove to be the right decision.
26. The utility approach considers expected utility, rather than expected monetary value as the criterion of decision making.
27. A risk-averse individual is always willing to accept a smaller cash-certain amount than the expected value of a bet.
28. The expected utility and expected monetary value criteria would always yield same optimal solution in any given problem.

29. For any utility function, the utils, the units of measuring utility, should always range between zero and 100.
30. A given utility function holds for a given individual at a given point in time only.

EXERCISES

1. Describe the steps involved in the process of decision-making.
2. What are pay-off and regret functions? How can entries in a regret table be derived from a pay-off table?
3. Explain and illustrate the following principles of decision-making:
   (a) Laplace, (b) Maximin, (c) Maximax, (d) Hurwicz, (e) Savage, and (f) Expectation.
4. (a) How are maximum likelihood and expectation principles of choice differentiated?
   (b) Do the principles of minimisation of expected cost and minimisation of expected regret yield the same decision?
5. Define EVPI. How is it calculated?
6. Explain the posterior analysis of decision-making. Define the expected value of sample information. Give a measure of efficiency of EVSI.
7. What are decision trees? How and in what type of situations are they employed for decision-making?
8. What is the significance of utility as a basis of decision-making? State the assumptions underlying the theory of utility.

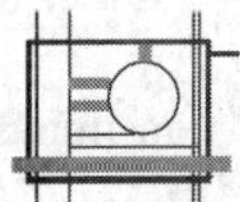

PRACTICAL PROBLEMS

1. The research department of Hindustan Lever has recommended the marketing department to launch a shampoo of three different types. The marketing manager has to decide one of the types of shampoo to be launched under the following estimated pay-offs (in millions of Rs) for various levels of sales:

Type of Shampoo	*Estimated level of sale (Units)*		
	15,000	*10,000*	*5,000*
Egg shampoo	30	10	10
Clinic shampoo	40	15	5
Deluxe shampoo	55	20	3

What will be the marketing manager's decision if (a) Maximin, (b) Maximax, (c) Laplace, and (d) Regret criterion is applied?

2. A toy company is bringing out a new type of toy. The company is attempting to decide whether to bring out a full, partial or minimal product line. The company has three levels of product acceptance. The management will make its decision on the basis of expected profit from the first year of production. The relevant data are shown in the following table:

Anticipated First Year Profit (in '000 Rs)

Product Acceptance	*Product Line*		
	Full	*Partial*	*Minimal*
Good	80	70	50
Fair	50	45	40
Poor	–25	–10	0

Take optimal decision under each of the following decision rules:

(a) Maximax (b) Maximin (c) Laplace (d) Minimax regret

(MBA, GGSIP Uni., 2002)

3. The oil company of India is interested in acquiring a piece of land which is considered likely to contain oil deposits. The company has the option of (a) buying the land outright, (b) obtaining an option to buy, drill for oil and if found exercise the option, and (c) not buying or obtaining option. There are three possibilities on such land: large oil reserves may be found; minor reserves may be found, or there may be no oil. The pay-offs (in lacs of Rs) resulting from various combinations of acts and events are tabulated below:

	Acts		
	Buy land	*Obtain option*	*No action*
Large Reserves	40	28	0
Minor Reserves	10	1	0
No Oil	–25	–2	0

What action should be taken by the company when the decision criterion is:

(i) Laplace, (ii) Maximin, (iii) Maximax, (iv) Minimax Regret, and (v) Expected pay-off (when the probabilities of obtaining large, minor, and no reserves are estimated to be 0.2, 0.5 and 0.3, respectively)?

4. A firm is considering the purchase of a complex equipment from either of the two suppliers S_1 and S_2. Supplier S_1 is capable of supplying the equipment on time to meet a certain desired deadline. The price chargeable by S_1 is, however, considerably higher than that of S_2. It is felt by the management of the firm that S_2 may deliver the equipment or may not be able to deliver on time. It is even suspected that Supplier S_2 may never be able to deliver the equipment to the specifications. However, the management believes that if it waits for some months, it may get better information on S_2's capabilities of supplying the equipment.

The management is considering three alternative courses of action.

A_1: Order from S_1. If later it is clear that S_2 can supply, order from S_1 can be cancelled. Of course, delay would be caused when order is given to S_2.

A_2: Order from supplier S_2. If it is known later on that S_2 cannot supply the equipment, the order may be switched to S_1.

A_3: Wait till the time information on S_2's capabilities is known. This would obviously cause delay.

The outcomes (profits) in the various possible situations are:

Event	Course of action		
	A_1	A_2	A_3
E_1	250	100	200
E_2	250	125	300
E_3	250	625	450

E_1 : S_2 fails to deliver
E_2 : S_2 delivers late
E_3 : S_2 delivers on time

What would be the management's decision according to each of the following criteria:

(a) Laplace, (b) Maximin, (c) Hurwicz (with $\alpha = 0.5$), and (d) Minimax Regret.

5. (a) From the following matrix, the elements of which indicate profits, obtain the decisions using the following principles of decision-making:

(i) Maximax (ii) Maximin (iii) Laplace

	a_1	a_2	a_3	a_4	a_5
S_1	26	22	13	22	18
S_2	26	22	34	30	20
S_3	18	22	18	18	20
S_4	22	22	18	18	18

(b) How would your answer in (a) be different if the elements of the matrix represented costs?

6. Find the best alternative in the following decision table using expected value criterion and the criterion of expected regret. The entries in the matrix indicate profits associated with different combinations of acts and states of nature.

State of nature (Q)	Prob.	Act a_1	Act a_2	Act a_3
Q_1	0.6	0	3	–3
Q_2	0.1	2	2	2
Q_3	0.2	5	–1	3
Q_4	0.1	–4	–3	1

7. A dealer in second-hand machinery is offered 5 machines by a company for Rs 5,000 only. He expects to sell each of the machines for Rs 2,200 at a fair but he also knows that any machines not sold would be a waste and not fetch anything later on. Obtain the pay-off matrix under each of the following assumptions:

(a) that the dealer has the option of buying the entire lot only, and

(b) that the dealer has the option of buying as many machines as he chooses, at the rate of Rs 1,000 per machine.

Using expectation principle of choice, determine the optimal strategy for the dealer in each of the two cases given.

8. Three types of souvenirs can be sold outside a stadium. From the following conditional pay-off table, construct the opportunity loss table. (Sales are dependent on the winning team.)

	Types of souvenir		
	I (Rs)	*II (Rs)*	*III (Rs)*
Team *A* wins	1,200	800	300
Team *B* wins	250	700	1,100

Point out which type of souvenir should be bought if probability of Team *A*'s winning is 0.6.

(M Com, Delhi, 1994)

9. Chemical Products Ltd. produces a compound which must be sold within the month it is produced, if the normal price of Rs 100 per drum is to be obtained. Anything unsold in that month is sold in a different market for Rs 20 per drum. The variable cost is Rs 55 per drum.

During the last three years, monthly demand was recorded and showed the following frequencies:

Monthly demand (No. of drums)	:	2,000	3,000	6,000
Frequency (No. of months)	:	8	16	12

(i) prepare an appropriate pay-off table;

(ii) advise the production management on the number of drums that should be produced next month. *(ICMA, May, 1983, Adapted)*

10. For the past 200 days, the sales of bread from the Fresh Food Bakery have been as follows:

Daily sales (loaves)	:	0	100	200	300	400
No. of days	:	10	60	60	50	20

(a) Determine the expected sales of bread.

(b) The bakery's production costs are Rs 2 per loaf, sale price is Rs 4 per loaf, and any bread unsold at the end of the day is contracted to a local farmer who pays Re 1 per loaf. Draw up a pay-off table for each sales/production combination.

(c) Compute the expected profit arising from each level of production and determine the optimal policy.

11. A physician purchases a particular vaccine on Monday of each week. The vaccine must be used within the week following, otherwise it becomes worthless. The vaccine costs Rs 20 per dose and the physician charges Rs 60 per dose. In the past 50 weeks, the physician has administered the vaccine in the following quantities:

Doses per week	:	20	25	40	60
No. of weeks	:	5	15	25	5

(a) Draw up a pay-off matrix,

(b) Obtain a regret matrix,

(c) Determine the optimum number of doses the physician should buy,

(d) The maximum amount the physician would be willing to pay per week for a perfect information about the number of doses expected to be demanded in a week.

12. A merchant buys a certain item for Rs 20 per case and sells it for Rs 50 per case. The high mark-up reflects the probability of the item and the great risk in stocking it. Any unsold items must be scrapped at the end of the day and have no salvage value. The merchant wants to know how many units of the item should be stocked at the beginning of the day. The past hundred days' sales are tabulated here:

Daily sales (No. of cases)	:	10	11	12	13
No. of days	:	15	20	40	25

(MBA, Delhi, April, 1995)

13. The manufacturer of designer jeans must decide whether to build a large factory or a small factory in a particular location. The profit per pair of jeans manufactured is estimated as Rs 10. A small factory will have an amortised annual cost of Rs 200,000 with a production capacity of 50,000 jeans per year. A large factory will have an amortised annualised cost of Rs 400,000 with a production capacity of 100,000 jeans per year. Four levels of manufacturing demand are considered equally likely, namely, 10,000; 20,000; 50,000 and 100,000 pairs of jeans per year.
    (i) Set up a pay-off table for building a small factory and a large factory for manufacturing designer jeans.
    (ii) Compute the expected monetary value (EMV) for building a small factory and building a large factory.
    (iii) Explain the meaning of the expected value of perfect information (EVPI).
    (iv) Which would you choose to build, a small factory or a large factory? Why?

(MMS, Mumbai, 2001)

14. A businessman has Rs 10,00,000 available to invest. He can either put the money in the bank at six per cent interest or he can start a new business venture. For the second act, he expects to make 12 per cent return on his investment if the business is a success, but if the business is a failure, he is expected to lose two per cent of his investment for the coming year. What probability must he assign to the occurrence of success (of the business venture) to make the expected monetary return of the two possible acts equally attractive? *(M Com, Delhi, 2000)*
15. A company has Rs 1,00,000 available for investment. It is considering two competing investment strategies: (a) invest in Reliable Company's bonds with a guaranteed 8% rate of interest per annum, and (b) purchase new equipment costing Rs 1,00,000 which, if a favourable condition of the economy occurs, will give a net return of Rs 12,000 in a year. However, if an unfavourable condition of the economy occurs, the proposal would yield a net loss of Rs 2,000 during the year. What probability can be assigned to the favourable condition of the economy in the next year so that the company is neutral between the two strategies? How would the probability value obtained help the company in making a decision between the two?
16. The Indian Yatch Company has developed a new cabin cruiser which they have earmarked for the medium-to-large boat market. A market analysis has put a 30% probability of annual sales being 5,000 boats, a 40% probability of 4,000 annual sales and a 30% probability of 3,000 annual sales. The company can go into limited production, where variable costs are Rs 10,000 per boat, and fixed costs are Rs 8,00,000 annually. Alternatively, they can go into full scale production, where variable costs are Rs 9,000 per boat, and fixed costs are Rs 50,00,000 annually. If the new boat is to be sold for Rs 11,000, should the company go into limited or full scale production when their objective is to maximise the expected profit? *(CA, November, 1993)*
17. The pay-off functions in respect of two acts, A_1 and A_2, are given here:

$$P_{A1} = 25\alpha - 50$$
$$P_{A2} = 35\alpha - 90$$

(a) For what value of α would the two be the same?
(b) Which of the acts is better for $\alpha = 10$?
(c) What is the regret of a poor strategy when $\alpha = 15$ and $\alpha = 4$?

18. A man has the choice of running either a hot snack stall or an ice cream stall at a sea-side resort during the summer season. If it is a fairly cool summer, he should make Rs 5,000 by running the hot snack stall, but if the summer is quite hot he can only expect to make Rs 1,000. On the other hand, if he operates the ice cream stall, his profit is estimated at Rs 6,500 if the summer is hot, but only Rs 1,000 if it is cool. There is 40% chance of the summer being hot. Should he opt for running the hot snack stall or the ice cream stall? Give a mathematical argument. *(ICWA, December, 1979)*

19. Company B has ordered a special purpose generator from BHEL at a cost of Rs 90 million. A critical component in the generator is subject to random failure as per probability pattern shown below. If a spare is ordered along with the order for the generator, the cost per component will be Rs 5 lacs. However, if the component is ordered at any later date the cost along with the loss in production during the period the generator is down will be Rs 40 lacs.

Number of breakdowns	:	0	1	2	3	4
Probability	:	0.80	0.08	0.06	0.04	0.02

Using EMV criterion, assess whether it would be beneficial to order extra components along with the initial order for the generator, and if so, the optimum number of spares to be ordered.

(MMS, Mumbai, 1999)

20. A television dealer finds that the cost of a TV in stock for a week is Rs 30 and the cost of a unit shortage is Rs 70. For one particular model of television the probability distribution of weekly sales is as follows:

Weekly sales	:	0	1	2	3	4	5	6
Probability	:	0.10	0.10	0.20	0.25	0.15	0.15	0.05

How many units per week should the dealer order? Also, find EVPI. *(CA, November, 1999)*

21. A company has a patent on a new medical device selling for Rs 28 per unit. However, the demand for the device is uncertain. The company can build its own manufacturing facilities and incur fixed cost of Rs 24 lacs per annum and a variable cost of Rs 8 per unit. On the other hand, the company could hire a subcontractor who would meet the demand for a cost of Rs 24 per unit.

At what demand level would management be indifferent between the two alternatives? (Ignore risk considerations.)

22. The XYZ company has total excess cash funds of Rs 60,000 to invest in various projects during this month and the next, according to the cash-flow statement prepared by the accounting department. The firm has been offered the following investment opportunities. It can participate immediately (at the start of this month) in a project by investing Rs 60,000 which is equally likely to result in a net profit of Rs 20,000 or a loss of Rs 10,000 within the month. In effect, the company will be able to reclaim its principal, with either a profit or loss, by the month's end. At the same time, the firm is informed that in one month from now, it will be given the opportunity of investing Rs 55,000 in another investment, which is equally likely to result in a net profit of Rs 15,000 or a net loss of Rs 5,000. Assuming that this company examines its cash position every two months to determine the feasibility of investing excess cash, advise this company as to whether it should invest in project 1 of this month or invest in project 2 of the next month or invest in projects 1 and 2 together if the objective is to maximise the expected profits over the next two months. *(CA, November, 1991)*

23. A mineral processing company wants to decide about the number of spare gear trains it has to order out at the time of placing order, for a high horsepower gearbox connected to a grinding unit. Although the life of a gear can be as high as 30 years and more, sudden failures cannot be ruled out. In case of a failure it would be expensive and time consuming to get a spare gear train. The cost would be Rs 200,000 including the loss of production due to down time of the equipment. If ordered-out with gear box, a gear train would cost only Rs 10,000 per unit. The following data is based on an analysis of past failures of 100 gearboxes:

No. of spare gear trains required	*No. of gearboxes requiring the spares*
0	93
1	4
2	1
3	1
4	1

You are expected to advise the optimal order size. *(MBA, Poona, 1982)*

24. A national boutique chain sells a wide range of high quality customised fashion goods. One particular outfit is bought in at Rs 80 and sold at Rs 130. Mean holding costs per season per outfit work out at Rs 5 and it costs Rs 800 to order and receive goods into stock. The manufacturers require orders in advance and once a batch has been made, it is not possible to be staggered over the fashion season. When a customer buys an outfit, she tries the outfit for size and design, alterations or adjustments, if any, are made, and she collects the outfits a day or so later. Generally, if the outfit is out of stock at one boutique it can be readily obtained from another branch, usually in a matter of hours. However, if the chain as a whole runs out of an item, then not only is the profit not earned, but the Rs 20 or so profit that comes from the extras that customers buy is also lost. Should the chain over-buy for a season then it is expected that the chain will be able to dispose of the surplus outfits at Rs 50 each.

The pattern of past sales of a comparable outfit shows the following probability distribution for the chain as a whole:

Outfits sold	*Probability*
1100	0.30
1200	0.40
1300	0.20
1400	0.10

The problem facing the management accountants of the chain is to decide how many outfits to order for the season ahead in order to maximise expected profit, bearing in mind the penalties for over and under ordering.

You are required to

(a) determine the number of outfits to order to maximise expected profits;

(b) compare and contrast the model that you have developed with the classical economic order model.

(ICMA, November, 1987, Adapted)

25. The manager of the boutique has to decide how many luxury gift packs of cosmetics to buy for the forthcoming Diwali season. The gift packs have to be bought from a manufacturer in cases of 50. The profit per pack is Rs 20. On the basis of trading conditions last Diwali, the current economic conditions and competitive market factors, the manager's estimate of the sales of this gift pack this season is as follows:

Sales (Cases)	:	5	10	15	20	25
Probability	:	0.20	0.20	0.30	0.20	0.10

Any unsatisfied demand does not affect the probability of future sales. Any unsold packs will be sold after Diwali at a loss of Rs 10 per pack.

(a) Assuming that the quantity of stock is bought in units of 5 cases and ignoring storage costs, how many cases should the manager buy to maximise her expected profit?

(b) If the manager was completely uncertain about the likelihood of sales between 5 and 25 cases (inclusive), how many cases should she stock?

26. Using the following pay-off function, develop a pay-off matrix and determine the optimal policy.

$$M(a_i, d_j) = \begin{cases} 400d_j - 300\,(a_i - d_j) & \text{if } a_i > d_j \\ 400a_i & \text{if } a_i \leq d_j \end{cases}$$

$$i = 3, 4, ..., 8; \quad j = 3, 4, ..., 6$$

$M(a_i, d_j)$ indicates the pay-off associated with *i*th act and *j*th state of nature.

State of Nature, j	:	3	4	5	6	7	8
Probability, p_j	:	0.05	0.10	0.30	0.40	0.10	0.05

27. Jackston, Inc. manufactures and distributes a line of Christmas toys. The company had neglected to keep its doll house line current. As a result, sales decreased to approximately 10,000 units per year from a previous high of 50,000 units. The doll house has been redesigned recently and is considered by company officials to be comparable to its competitors' models. The company plans to redesign the doll house each year in order to compete effectively. Joan Blocke, the sales manager, is not sure how many units can be sold next year, but she is willing to place probabilities on her estimates. Blocke's estimates of the number of units which can be sold during the next year and the related probabilities are as follows:

Estimated sales in units	*Probability*
20,000	0.10
30,000	0.40
40,000	0.30
50,000	0.20

The units would be sold for Rs 20 each.

The inability to estimate the sales more precisely is a problem for Jackston. The number of units of this product is small enough to schedule the entire year's sales in one production run. If the demand is greater than the number of units manufactured, then sales would be lost. If the demand is below supply, the extra units cannot be carried over to the next session and would be given away to various charitable organisations. The production and distribution cost estimates are listed here.

		Units manufactured			
		20,000	*30,000*	*40,000*	*50,000*
Variable costs	Rs	180,000	270,000	360,000	450,000
Fixed costs	Rs	140,000	140,000	160,000	160,000
Total cost		320,000	410,000	520,000	610,000

The company intends to analyse the data to facilitate making a decision as to the proper size of the production run. Prepare a pay-off table for the different sizes of production runs required to meet the four sales estimates prepared by Joan Blocke for Jackston, Inc. If Jackston, Inc. relied solely on the expected monetary value approach to make decision, what size of production run would be selected?

(CMA, Adapted)

28. **Z Ltd. is considering various product pricing and material purchasing options with regard to a new product it has developed. Estimates of demand and costs are as follows:**

If selling price per unit is		*Rs 15 per unit*	*Rs 20 per unit*
Forecasts	*Probability*	*Sales volume (000 units)*	*Sales volume (000 units)*
Optimistic	0.3	36	28
Most likely	0.5	28	23
Pessimistic	0.2	18	13
Variable manufacturing costs			
(excluding materials) per unit		Rs 3	Rs 3
Advertising and selling costs		Rs 25,000	Rs 96,000
General fixed cost		Rs 40,000	Rs 40,000

Each unit requires 3 kg of material and because of shortage problems any unused material must be sold at Re 1 per kg. The sole suppliers of material offer three purchase options, which must be decided at the outset, as follows:

(i) Any quantity at Rs 3 per kg, or

(ii) A price of Rs 2.75 per kg for a minimum quantity of 50,000 kg, or

(iii) A price of Rs 2.50 per kg for a minimum quantity of 70,000 kg.

You are required, assuming that the company is risk neutral, to

(a) prepare calculations to show what pricing and purchasing decisions the company should make, clearly indicating the recommended decisions;

(b) calculate the maximum price you would pay for perfect information as to whether the demand would be optimistic or most likely or pessimistic. *(ICMA, May, 1988, Adapted)*

29. **The Good-neighbours' Apartment has eight washing machines. The probability of machines failing in any given year is as follows:**

No. of machines failing	:	3	4	5	6
Probability	:	0.15	0.30	0.50	0.05

Once a machine has failed, its repair will cost either Rs 30, the minimum service charge; Rs 70 for a moderate repair, or Rs 120 for a major repair. The chance for a major repair is 36 per cent while for a moderate repair the chance is 44 per cent for any washer.

Keepfit offers a prepaid one-year maintenance policy that costs Rs 45 per machine. Should the owner of the apartment buy the maintenance insurance?

30. Mr Prime, the owner of a chain of facilities manufacturing detergents, is contemplating opening a new facility in a certain area. He usually builds 2,500 tonnes or 5,000 tonnes plants, depending on whether the anticipated demand is low or high. He has been able to construct the following net pay-off table from his experience:

	Net Pay-off (in Rs lacs)	
Demand	*2,500 Tonnes*	*5,000 Tonnes*
Low	30	– 20
High	40	55

Based on the information concerning the area at his disposal, Mr Prime feels that the probabilities of low and high demand are 0.3 and 0.7 respectively. He also feels that a survey must be conducted to get updated estimates of demand. A research firm agrees to conduct the survey for Rs 2 lacs and provide him with the information of either low or high demand, depending on the results of the survey. The research firm described the reliability of their estimates by means of the following table:

True	*Survey Estimate*	
Event	*Low Demand*	*High Demand*
Low Demand	0.80	0.20
High Demand	0.10	0.90

Should Mr Prime spend money on research or rely on his judgement? What capacity plant should he install? *(MMS Mumbai, 2001)*

31. A company receives shipments of certain items. It should decide whether to accept or reject the shipment, on the basis of inspection of a sample selected from the shipment. From the past experience, it is known that the percentage of defective items in a batch of shipment is either 1, 2 or 5, the probabilities for which are 0.5, 0.3 and 0.2, respectively. The company can accept only those batches which have 1 per cent defectives. The cost of rejecting a good batch, that is, batch with 1 per cent defectives is Rs 600. The cost of accepting a batch with 2 per cent defectives is Rs 400, and the cost of accepting a batch with 5% defectives is Rs 600.

A sample of 10 items has been selected from the shipment and two items are found to be defective. The conditional probabilities of getting 2 defectives in a sample of 10 items from a batch of 1%, 2% and 5% defective are calculated as 0.083, 0.185, and 0.265 respectively.

Determine whether the shipment should be accepted.

32. A company has just received a big lot of electronic components for inspection. Experience indicates that there are only two possible compositions of the lot: D_1 5% defective, or D_2 10% defective. Empirical evidence also tells that $P(D_1) = 0.7$ and $P(D_2) = 0.3$. The pay-off matrix for the decision problem is given here (in thousands of rupees of cost of error). These are the only relevant costs.

	D_1	D_2
Accept the lot	0	5
Reject the lot	3	0

(a) Should the lot be accepted if there is no additional evidence?

(b) Should the lot be accepted if a sample of two units is drawn and both are defective?

(c) What is the expected value of the information provided by a sample of two units?

(d) If testing costs Rs 40 per unit tested, should there be any testing or should lots just be accepted?

33. A company is contemplating the introduction of a new product and is considering the advisability of test marketing done before finally deciding on it. The options on introduction are: to introduce (call it A_1) and not to introduce (call it A_2). The management has estimated the following outcomes and the corresponding prior probabilities.

Outcome	*Profit (Loss)*	*Prob.*
d_1 : the product captures 20% of the market	Rs 10 m	0.70
d_2 : the product captures 7% of the market	Rs 1 m	0.10
d_3 : the product captures 2% of the market	(Rs 5 m)	0.20

In the event of test marketing, there are three outcomes to be considered. They are I_1-test sales of greater than 20% of the market, I_2-test sales of 10 – 20% of the market, and I_3-test sales of less than 10% of the market. The conditional probabilities of the test results are given as follows:

	I_1	I_2	I_3
$P(I/d_1)$	0.6	0.3	0.1
$P(I/d_2)$	0.3	0.6	0.1
$P(I/d_3)$	0.1	0.1	0.8

(a) What is the best option of the company before the test?

(b) How much would it be worth to gain perfect knowledge of the outcome?

(c) What action is the best for each possible result of the test marketing?

(d) Determine the value of the test marketing.

34. Mr Webbley, the Marketing Planning Manager of Computer and Allied Co. Ltd. is given the assignment of recommending whether his firm should produce a new type of pocket calculator. His boss feels that a chance of 60:40 exists that the calculator will be successful. The pay-off table in lacs of rupees is as follows:

Strategy	N_1	N_2
	(Success)	*(Failure)*
S_1: Produce	150	–100
S_2: Do not produce	0	0

Two market survey firms have submitted bids to do the field work of interviewing the target audience. The Alpha Company is known to conduct highly reliable surveys (but at a rather high cost), whereas Beta Company's reliability is somewhat lower but so is its cost.

Pertinent conditional probabilities of survey results Z_1 (indicating N_1) and Z_2 (indicating N_2) are shown below:

Firm	*Conditional Probabilities*	*Cost*
Alpha Company	$P(Z_1/N_1) = P(Z_2/N_2) = 0.9$	50,000
	$P(Z_1/N_2) = P(Z_2/N_1) = 0.1$	
Beta Company	$P(Z_1/N_1) = P(Z_2/N_2) = 0.7$	30,000
	$P(Z_1/N_2) = P(Z_2/N_1) = 0.3$	

Mr Webbley has the following options: conduct no survey, buy Alpha survey or buy Beta survey.

(i) Evaluate the expected payoff of each option, and suggest which option Mr Webbley should follow.

(ii) What is the expected value of perfect information?

(MMS, Mumbai, 1999)

35. You are given the following estimates concerning a Research and Development programme:

Decision number, j *j = 1: develop* *j = 2: do not develop*	*Probability of decision D_j given Research, R*	*Outcome number*	*Probability of outcome x_i given D_j, $P(x_i/D_j)$*	*Pay-off value of outcome x_i (Rs '000)*
		1	0.6	600
1	0.5	2	0.3	–100
		3	0.1	0
		1	0.0	600
2	0.5	2	0.0	–100
		3	1.0	0

Construct and evaluate the decision tree diagram for the data. Show your working for the evaluation.

(ICMA, November, 1983, Adapted)

36. The Unique Candles Company has developed and patented a unique process for manufacturing ornamental candles. The firm has three options: (1) sell the patent, (2) produce and market the product nationally at once, or (3) test market the product in one area and decide either to sell the patent or go in for a national launch. The following information is available:

(i)	Gross Margins (Rs thousands)	
	Good Market	Rs 400
	Fair Market	100
	Poor Market	10
(ii)	Marketing Costs (Rs thousands)	
	Test Marketing	15
	National Promotion	50
(iii)	Sale of Patent (Rs thousands)	
	Without Test Marketing	30
	Following Favourable Test	60
	Following Unfavourable Test	10
(iv)	Probability of Test Marketing Results	
	Favourable Test	0.60
	Unfavourable Test	0.40

(v) Probability of Market Reception

	Good	Fair	Poor
Without Test Marketing	0.30	0.40	0.30
Following Favourable Test	0.60	0.30	0.10
Following Unfavourable Test	0.10	0.30	0.60

What strategy should the firm follow? *(MMS, Mumbai, 2000)*

37. Portland p.l.c. is planning a new product launch and has to evaluate the following alternatives:

(1) To introduce the product in the South-East initially and, after a two-year period, to evaluate its performance and decide whether to go on a national distribution or remain on a regional distribution basis only.

(2) To introduce the product nationally from the outset. Consider the following sections in turn:

(a) Based on market research reports, the following tabulation of expected results over an estimated six-year product-life has been obtained:

	Expected demand level	*Present value of*	
		Initial capital outlays	*Net cash inflows*
		(Rs' 000)	(Rs' 000)
National distribution from the outset	H	500	750
	M	500	510.5
	L	500	450
Regional distribution only from the outset	H	150	275
	M	150	198
	L	150	122
Initial regional distribution followed by national distribution			
High regional followed by:			
high national demand	HH	600	794
medium national demand	HM	600	672
low national demand	HL	600	520
Medium regional followed by:			
high national demand	MH	620	732
medium national demand	MM	620	654
low national demand	ML	620	490
Low regional followed by:			
high national demand	LH	635	708
medium national demand	LM	635	638
low national demand	LL	635	460

Note: H = high demand, M = medium demand, L = low demand

Draw a decision-tree diagram to illustrate the relationships between the decision points and chance events.

(b) Estimates have also been made relating to the chances that each outcome in turn will occur. An extract of this information is given below:

Regional distribution		*Initial regional distribution*	
Market demand level	*Probability of occurrence*	*National demand given high regional demand*	*Conditional probability*
H	0.40	H	0.5
M	0.35	M	0.3
L	0.25	L	0.2

(i) Illustrate how you would employ the information given in this section (b) to evaluate the information contained in section (a).

(ii) Show how you would measure whether alternative 1 was riskier than alternative 2, indicating what additional data would be required. *(ICMA, May, 1983, Adapted)*

38. XYZ Company, dealing with a newly invented telephone device, is faced with the problem of selecting out of the following courses of action available:
    (i) manufacture the device itself; or
    (ii) be paid on a royalty basis by another manufacturer; or
    (iii) sell the rights for its invention for a lump sum.

    The profit (in Rs' 000s) which can be expected in each case and the probabilities associated with the level of sales are shown in the following table:

Outcome	*Probability*	*Manufacture itself*	*Royalties*	*Sell all rights*
High sales	0.1	75	35	15
Medium sales	0.3	25	20	15
Low sales	0.6	−10	10	15

Represent the company's problem in the form of a decision tree. Redraw further the decision tree by introducing the following additional information:

1. If it manufactures itself and sales are medium or high, then company has the opportunity of developing a new version of its telephone.
2. From the past experience, the company estimates that there is a 50 per cent chance of successful development.
3. The cost of development is 15 and returns after deduction of development cost are 30 and 10 for high and medium sales, respectively. *(MBA, Delhi, 2003)*

39. A company is currently working with a process, which, after paying for materials, labour and so on brings a profit of Rs 12,000. The company has the following alternatives:
    (i) The company can conduct research R_1 which is expected to cost Rs 10,000 and having 90% probability of success. If successful, the gross income will be Rs 26,000.
    (ii) The company can conduct research R_2, expected to cost Rs 6,000 and having a probability of 60% success. If successful, the gross income will be Rs 24,000.
    (iii) The company can pay Rs 5,000 as royalty of a new process which will bring a gross income of Rs 20,000.

    Because of limited resources, only one of the two types of research can be carried out at a time. Draw the decision tree and bring out the optimal strategy for the company. *(CA, May, 1994)*

40. B&L Inc. has been invited to bid on a government contract for a lens system for a satellite camera. B&L management estimates the cost of preparing a bid will be Rs 500,000. They also estimate that if they bid on this contract, there is a 60 per cent chance that they will win the contract. The contract calls for the delivery of a working lens system 10 months after the awarding of the bid. B&L researchers feel that three possible designs can be used for this system. Design *A* is a mechanical system that would cost Rs 600,000 and would have a 55 per cent probability of working. Design *B* is a mechanical system that would cost Rs 900,000 and would have a 70 per cent chance of working. Design *C* is an electronic system that would cost Rs 12,50,000 and since B&L has built such a system before, would have a 100 per cent chance of working. The goverment will pay the contractor Rs 25,00,000 upon delivering of a working system. The researchers feel that there is only enough time to build two systems; in other words, if mechanical system *A* or *B* is built and does not work, B&L will have to build the electronic system to ensure delivery.
    (a) Draw the decision tree for this problem.
    (b) Determine which decision(s) B&L should make and the expected monetary value of those decisions.
    (MBA, Delhi, 2004)
41. The Dragon Group owns a large store in Mumbai, which is rather old-fashioned and has declining profits. The management of the company is considering some options, and three possibilities worked out:
    (i) Shut down and sell the site for Rs 15 million.
    (ii) Continue as before with profits declining at 10 per cent per annum.
    (iii) Upgrade the store.

    The group has had similar problems in the past and experience suggests tha. wnen stores are upgraded, 60 per cent achieve good results and 40 per cent poor results. Unsure, the management is considering whether to engage a leading market research company to carry out consumer research in Mumbai for Rs 1 million. It is known from the work record of the research company that the actual outcome has been good, where the aptitude predicted by it is good, with a probability of 0.85. Also, when its prediction has been negative, the actual outcome has been recorded as poor in 90 per cent of the cases and good in 10 per cent of them.

    If the research indicates positive attitude, the management will consider deluxe upgrading that will generate more profit but will cost Rs 12 million, as compared with standard renovation costing Rs 6 million. On the other hand, if the research indicates negative attitude, then the management will consider standard renovation compared with shutting down and selling the site.

    The time horizon for the analysis is 10 years and the following estimates of returns (in present value terms) have been made:

    With deluxe upgrading:

Good results	Rs 40 million
Poor results	Rs 20 million

    With standard renovation:

Good results	Rs 25 million
Poor results	Rs 10 million

    If operations continue as before, returns next year will be Rs 2 million in present value terms, falling by 10 per cent per annum in the subsequent years.

    Draw a decision-tree diagram and determine the optimal course of action for the group.

42. Lots of a complicated machine part are obtained from a manufacturer. They may be tested only by actually using a sample of a few pieces. Suppose that it is known that the parts are produced by a process which turns out either a 10% defectives or 4% defectives. Suppose further that it is agreed with the manufacturer that a sample of two parts may be taken from every lot for test checking. The cost of testing is Rs 2,000, the cost of accepting a defective part Rs 20,000 while the cost of rejecting a non-defective part Rs 2,000. Each lot consists of 50 parts and it is assumed that the 10% or 4% defective lots have 3:2 chances of occurrence.
    (a) Draw an appropriate decision tree.
    (b) State the optimal course of action for the firm indicating whether it is worthwhile to take a sample or not.
    (c) Determine the cost of sampling that will make the firm in question neutral between taking and not taking a sample.
    (d) Calculate the expected value of perfect information.
43. The utility function of a manager is as follows:

Amount (Rs)	*Utility index*
–120,000	0.00
–100,000	0.20
–60,000	0.30
0	0.55
40,000	0.62
100,000	0.72
200,000	1.00

The manager is offered a contract which promises a net gain of Rs 1,00,000 or a net loss of Rs 60,000 with equal probabilities.
    (a) Should he accept the contract? Use (i) expected monetary value criterion, (ii) expected utility criterion.
    (b) Would his decision in (a) above be altered if he is offered to accept the contract twice?
44. Utility function for money of some decision-maker is as follows:

Money (Rs)	(–) 1,000	0	5,000	20,000	30,000
Utility	(–) 2	0	10	20	30

A preference test given to him shows that he is indifferent between an investment that will yield Rs 10,000 certainly and a risky venture with a 50% chance of Rs 30,000 profit and a 50% chance of a loss of Rs 1,000.

A new risky venture is proposed with possible pay-offs of either Rs 0 or Rs 20,000. The probabilities of gain cannot be determined. Find the probability combination of Rs 0 and Rs 20,000 which will yield Rs 10,000 with certainty. *(CA, November, 1994)*

45. A manager has following utility function depicting her attitude towards uncertainty:

Amount (x) Rs	*Utility U(x)*
0	0.00
1,000	0.25
2,000	0.40
3,000	0.52
4,000	0.68
5,000	0.80
6,000	0.90
7,000	1.00

(a) The manager is offered an opportunity with following likely returns along with their respective chances.

Amount (Rs) :	0	5,000	7,000
Probability :	0.54	0.30	0.16

What is the maximum amount she will be prepared to pay for the opportunity?

(b) How would she decide in the following case involving three mutually exclusive alternatives?

Alternative A_1

Rs 5,000 with a probability 0.50
Rs 2,000 with a probability 0.50

Alternative A_2

Rs 6,000 with a probability 0.50
Rs 1,000 with a probability 0.50

Alternative A_3

Rs 7,000 with a probability 0.50
Rs 0 with a probability 0.50

46. The utility function of an individual is as given here:

Amount ('000 of Rs) :	400	200	100	70	50	15	0
Utility Index :	1	0.90	.83	.80	.78	.73	.70
Amount ('000 of Rs) :	– 25	– 40	– 50	– 60	– 100	– 200	
Utility Index :	.64	.60	.58	.50	.40	.00	

(a) The individual in question is invited to participate in a contract which can result in a gain of Rs 2 lac or a loss of half this amount with equal probabilities. Show that he would not accept this contract.

Now, suppose that there are four individuals, each with utility functions like the above, who are deciding whether to enter this contract as a group, sharing profits or losses equally. Should they accept it?

(b) Reconsider the utility function. The individual involved is asked to choose one or none of the two contracts C_1 and C_2. Contract C_1 is likely to involve a loss of Rs 40 thousand with a probability of 0.40 and profit of Rs 70 thousand with a probability of 0.60. Contract C_2 is composed of two contracts each of which promises a profit of Rs 35 thousand with a probability of 0.60 and a loss of Rs 20,000 with a probability of 0.40. The outcomes of each of these two contracts making up C_2 are independent.

Should the individual accept the contract C_1? C_2? None?

Chapter 14

Markov Chains

CHAPTER OVERVIEW

Managers are always concerned about the market share of their products. Market shares change due to changes in consumer preferences caused by several factors. Over a period of time, some customers of a particular brand of a product stick to their brand, some others switch over to existing or new brands, while yet others get attracted to this brand and switch away from other brands. Markov chains deal essentially with such problems. If different brands of a product are termed as *states* (as they are called in Markovian analysis), and we watch the market shares of the competing brands at some regular intervals, we can ascertain how customers move between different states over time. In fact, the Markov chains analysis can be extended to cover all situations where the variable (or subjects like customers) changes states (or categories) over time. These may include price of a chare, employment status of the employees of a company (which may remain the same or change due to promotion, demotion, layoff, retirement, etc.) and accounts receivables, etc. (if we observe accounts receivables on a month-to-month basis, then a one-month old account would become, by the next month, either a two-month old account, an account paid-off or a bad debt).

Given the relevant data in each case, the Markov analysis helps a manager to obtain answers to questions like the following:

- What is the probability that a customer using my brand will also be using it three months from now?
- What are the chances that a customer using a particular brand will be using my brand two months hence?
- What are likely to be the market shares of various brands in the long run?
- What is the probability that a new accounts receivable shall ultimately be written off as bad debt?
- What are the chances that a two-month old account will eventually be collected?
- If we make credit sales of, Rs 60,000 every month, what amount shall be lost per annum by way of bad debts?

This chapter requires an understanding of the concepts of probability, transition matrix, matrix multiplication and solution of linear equations simultaneously using matrix algebra or otherwise.

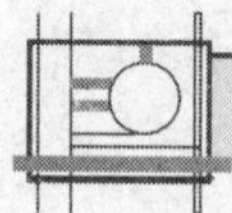

14.1 INTRODUCTION

Many a times, we are interested to know as to how a random variable changes over time. For example, we may like to know the way the market share of a detergent or soft drinks manufacturer changes year after year. Similarly, our interest may be in the behaviour of the price of a share during a certain period of time. The study of how a random variable evolves over a period of time includes *stochastic processes*. We shall discuss in this chapter the stochastic processes, focusing on a particular type of stochastic process, called *Markov chains*. Markov chains are used extensively with applications in marketing, finance, accounting, production, education, and so on.

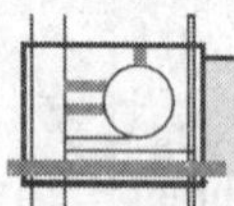

14.2 BRAND SWITCHING EXAMPLE

Let us begin with an example from marketing. Suppose that there are three brands of detergents D_1, D_2 and D_3 selling in a market. The market has been observed continuously month-after-month for changes in the brand loyalty—that is to say, whether and how customers change their brand of detergent over time. Let us say that the rate of brand switching has settled over time as follows: every month, customers are being drawn from brand D_1 to brand D_2 to the extent of 30 per cent, 10 per cent drawn to brand D_3 and the remaining 60 per cent staying with D_1 itself. However, 20 per cent of those using brand D_2 in a given month are drawn to brand D_1; one-half of them do not change the brand and use this same brand in the next month as well, while the remaining 30 per cent shift to brand D_3. Similarly, the behaviour with regard to D_3 is found to be like this—80 per cent of the customers stick to the brand; 15 per cent shift to brand D_1, while the remaining 5 per cent shift to brand D_2, each month. These results are shown in Figure 14.1.

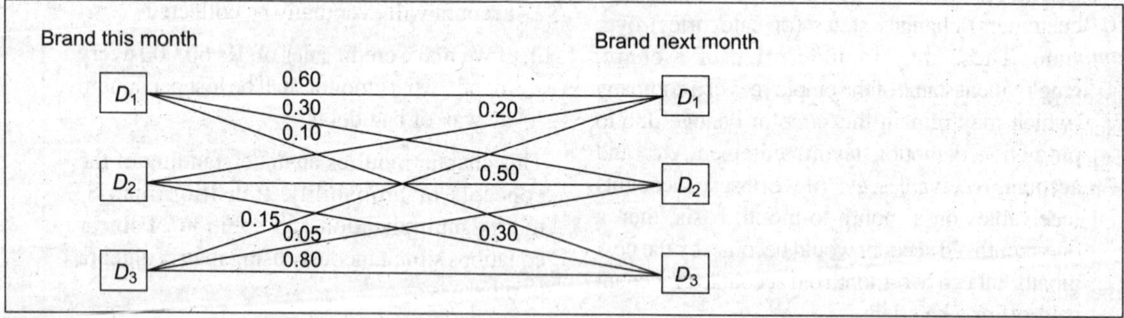

Figure 14.1 *Brand Switching*

Now, given these conditions about brand switching, assuming no further entry or exit, and given further that the market share for the three brands on a certain date, say March 1, is 30%, 45% and 25% for brands D_1, D_2 and D_3, respectively, the question to ask may be as to what would the ultimate market shares of the three brands be if and when the conditions settle down. This kind of problems (of determining market shares in a dynamic market, for example) are solved using the tool of Markov analysis. However, before we attempt to solve this problem, the following may be noted:

1. In the given problem, the aim is to determine the behaviour of the system. The objective is not to determine an appropriate marketing strategy for a detergent manufacturer. Rather, it is an attempt to know as to how the system would behave over time in terms of the actions of the households buying detergents. It follows from this that the Markovian analysis is *descriptive* in nature and not normative one, like linear programming or other tools where the objective is to determine the optimal course of action.
2. The situation is dynamic in the sense that it involves multiple time periods (different months here) requiring the customers to make a sequence of decisions in a chance environment, involving two or more possible outcomes at fixed intervals of time. In short, the given situation may be described as a *stochastic process*.

A stochastic process, in turn, may be described like this. Suppose we observe some characteristic of a system at discrete points in time, labelled as 0, 1, 2, 3, ..., and so on. Let X_t be the value of the system characteristic at time t, and it may be viewed as a random variable. A description of the relation between the random variables X_0, X_1, X_2 and so on is termed as a discrete-time stochastic process. Notice that in the given problem, we examine the respective market shares for the three brands at intervals of one month. The brand switching properties of the customers are examined for many months and noted when they stabilize. The probabilities of switching are termed as *transition probabilities*. For instance, the probabilities of transition from brand D_2 to D_1, D_2 and D_3 are, respectively, 0.20, 0.50, and 0.30.

As another example of discrete-time stochastic process, assume that a person has Rs 20 at time $t = 0$. At times 1, 2, ..., he plays a game wherein he can gain or lose Rs 10, with probabilities p and $1 - p$, respectively. He would quit if and when he loses his initial amount of Rs 20, or doubles this amount to Rs 40. Now, if we define X_i as his capital position after the time t game is played, then $X_0, X_1, X_2, ..., X_t$ may be viewed as a discrete-time stochastic process. While X_0 is known to be constant equal to Rs 20, we are not sure about X_1, X_2, X_3 and so on. They are random in nature. For example, X_1 shall be equal to Rs 30 (if he gains) with a probability p and Rs 10 with a probability of $1-p$. It may be noted that X_t = Rs 40, then X_{t+1} and all other later X_t's shall be equal to Rs 40, since he quits after the initial amount is doubled. Similarly, whenever he loses his initial capital so that $X_t = 0$, then X_{t+1} and all other later X_t's shall be equal to zero. This kind of situation is appropriately termed as a *gambler's ruin* problem.

Observing the closing prices of a share of a certain company for a number of days also is an example of discrete-time stochastic processes. Let X_0 be the price of the share at today's closing. If we let X_t be the price of the share at the closing of the ith trading day in the future, then our interest may be to know about the probability distribution of the X_{t+1}, with the values of $X_0, X_1, X_2, ..., X_t$ being known. In other words, what do the historical prices of a share point to its future price?

Now, in a stochastic process, if we can view the state of the system at *any* time, and not just at discrete point in time, such as observing prices of a share at any time during the trading day rather than only at the day-end, then such a process is termed as a *continuous-time stochastic process*. In fact, viewing the price of a share as a continuous-time stochastic process has led some important results in the theory of finance. The path-breaking Black and Scholes formula used for the valuation of stock options is one example of this. In this book, we shall discuss only discrete-time stochastic processes.

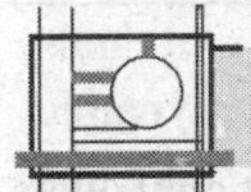

14.3 MARKOV PROCESSES

As indicated earlier, the detergent brand switching example is a stochastic process and since a customer's brand choice is dependent only on his choice of the immediately preceding month, and not on his choice two, three or four months ago, this stochastic process is termed as *Markov process*. Further, if the transition probabilities of the customers, switching from one brand (known as a *state*) to another remains constant from time to time, then the Markov process is termed as *stationary*, or *homogeneous*, Markov chain. In this chapter, we shall be dealing only with homogeneous Markov chains.

14.3.1 Assumptions Underlying Markov Analysis

The Markov chain analysis to follow is based on the following assumptions:

1. Finite States The given system has a finite number of states, none of which is "absorbing" in nature. In our detergent example, there are three states—three brands of detergent which the customers can switch between. They are all non-absorbing. To understand this, let us consider what an absorbing state is. In our example, if there was one brand of detergent from which a customer would never switch, then that brand would have represented an absorbing state. Similarly, in the gambler's ruin example discussed earlier, once the person involved reaches the state where he has Rs 40 (having doubled his capital) or Rs 0 (having vanished all his capital), he would not play more, and quit. Each of these two, therefore, represents an absorbing state. The analysis to follow assumes that all the states are non-absorbing. We shall later relax this assumption.

In regard to the classification of states, it may further be noted that

(i) For two states i and j, a sequence of transitions that begins in i and ends in j is called a *path* from i to j.

(ii) A state of j is known as *reachable* from state i if there is a path leading from i to j.

(iii) Two states i and j are said to *communicate* if each is reachable from the other: j is reachable from i, and i from j.

(iv) A state i is known as a *transient* state if there exists a state j which is reachable from i, but i is not reachable from state j. Thus, a state i is transient if there is a way to leave state i that never returns to state i.

(v) If a state is not transient, it is known as a *recurrent* state.

Figure 14.2 gives the transition from one state to another for the detergent problem in the graphical form. It is evident that each of the three states is "reachable" from all others, and hence communicative. None of these is absorbing and each one is recurrent.

Similarly, Figure 14.3 depicts the inter-state transition for the gambler's ruin problem. Here, states 0 and 4 are absorbing states, as mentioned earlier, and states 1, 2 and 3 are transient states. For instance, it is possible to go from state 2 to state 4 along the path 2-3-4, but there is no way to return to state 2 from state 4.

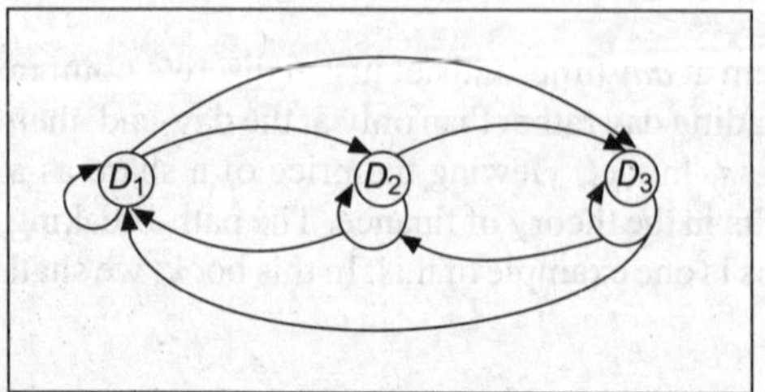

Figure 14.2 *Communicative States*

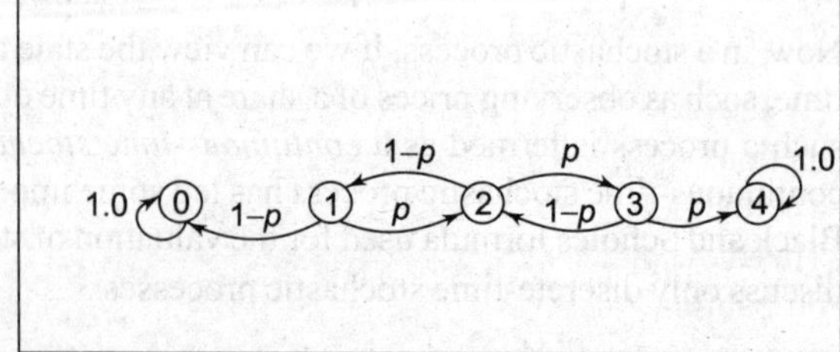

Figure 14.3 *Inter-state Transition*

2. First-order Process The condition (or state) of the system in any given period is dependent only on its condition prevailing in the previous period and the transition probabilities. In our brand switching case, it is assumed that the choice of a particular brand of detergent is dependent upon and influenced only by the choice in the previous month. Similarly, in the gambler's ruin problem, the amount of money that the person in question has after $t + 1$ plays of the game depends on the past history of the game *only* through the amount of money he has after t plays.

In this context, it may be mentioned that where the probability of the next event depends upon the outcome of the last event, like a customer's choice of detergent brand in a given month depends on the choice in the last month, the Markov process is termed as a *first-order* Markov process. A *second-order* Markov process assumes that the customers' choices next month may depend upon their choices during the immediately preceding two months. Similarly, a *third-order* process is based on the assumption that the customers' choices of detergent brands are a function of the choices made in the preceding three months. In this book, we confine to a discussion of first-order Markov processes only.

3. Stationarity The transition probabilities are constant over time. As already indicated, it is assumed that the system has settled down so that the switching among different brands takes place at the given rates in each time period. Needless to add, this makes the Markov chain here stationary. With the rules of the game unchanged over time for the gambler's ruin problem, the problem also is an example of stationary Markov chains.

4. Uniform Time Periods The changes from one state to another take place only once during each time period, and the time periods are equal in duration. Thus, in our example, it is assumed that the customers change their brands of detergent on a monthly basis and accordingly, the monitoring is also done on a month-to-month basis. If it is believed that the customers change the brands within a shorter period, then we may reduce the time periods to, say, weeks.

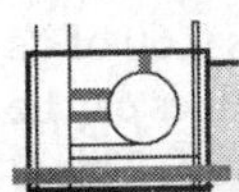

14.4 MARKOV ANALYSIS: INPUT AND OUTPUT

In the Markovian analysis, the analysis of a given system is based on the following two sets of input data—the transition matrix (containing the transition probabilities) and the initial condition in which the system is. Based on these inputs, the model provides for the following predictions:

(a) The probability of the system being in a given state at a given future time.

(b) The steady state (that is, long run or equilibrium) probabilities.

We shall consider first the inputs and then the analysis and the output.

14.4.1 Input

The two inputs, viz. transition probabilities and the initial conditions are discussed below.

(i) Transition Probabilities The transition probabilities are required for obtaining both the types of predictions mentioned above. It may be recalled that the Markov process describes movement of the system from a certain state in the current stage (may be current time period) to one of the n possible states in the next stage. This movement is in an uncertain environment but we are given the probability associated with any move. This probability is known as the transition probability, and expressed as p_{ij}, being the probability that the system moves from current state i to another state j in the next time period. The transition probabilities may be presented in the tabular or matrix form.

Example 14.1 Refer to the detergent example and construct the transition probabilities matrix.

The transition probabilities for the problem are given in Table 14.1.

TABLE 14.1 Transition Probabilities Matrix

From Brand (this month)	*To Brand (next month)*		
	D_1	D_2	D_3
D_1	0.60	0.30	0.10
D_2	0.20	0.50	0.30
D_3	0.15	0.05	0.80

Alternately,

$$P = \begin{pmatrix} 0.60 & 0.30 & 0.10 \\ 0.20 & 0.50 & 0.30 \\ 0.15 & 0.05 & 0.80 \end{pmatrix}$$

It may be noted that the rows of the matrix of transition probabilities indicate the retention and loss of market to other brands, while the columns contain the retention and gain of market. For example, row 1 indicates that detergent D_1 retains 0.60 (or sixty per cent) of its customers, loses 0.30 of its customers to D_2 and loses 0.10 to brand D_3. Row 2 indicates that detergent D_2 retains one-half of its customers and loses 0.20 and 0.30 of its customers to brand D_1 and D_3, respectively. Similarly, row 3 indicates a loss of, respectively, 0.15 and 0.05 of customers to D_1 and D_2, and a retention of 0.80 of the customers of D_3. Now, proceeding column-wise, it may be seen that the column 1 of the matrix indicates that detergent D_1 retains 0.60 of its customers while gains 0.20 of the customers of D_2 and 0.15 of the D_3's customers. Other column values can be similarly interpreted. Notice that for every row in the matrix, the sum of the probabilities is equal to 1. This is because, there are only three detergent brands—one may stick to the brand used or shift to either of the two other brands. They are collectively exhaustive and therefore, the sum of the probabilities would equal one.

Example 14.2 Obtain the transition matrix for the gambler's ruin problem discussed earlier.

From the given information, it can be seen that there are five states viz. Rs 0, Rs 10, Rs 20, Rs 30, and Rs 40. From all the states, except the states of Rs 0 and Rs 40, we know that the person would be in the next state (with Rs 10 more) with a probability of p, and in the previous state (with Rs 10 less) with a probability of $1 - p$. Of course, if the person reaches the state of Rs 0 or Rs 40, then he does not play the game anymore. In such a case, the state does not change. Hence, for each of these states, the transition probability would be equal to 1. Accordingly, $p_{00} = p_{44} = 1$.

The transition probabilities are contained in Table 14.2.

TABLE 14.2 Transition Probabilities

	State	*State*				
		Rs 0	*Rs 10*	*Rs 20*	*Rs 30*	*Rs 40*
	Rs 0	1	0	0	0	0
	Rs 10	$1-p$	0	p	0	0
$P =$	Rs 20	0	$1-p$	0	p	0
	Rs 30	0	0	$1-p$	0	p
	Rs 40	0	0	0	0	1

In general terms, the transition probabilities matrix, where there are n states in the system, can be represented as follows:

$$P = \begin{bmatrix} p_{11} & p_{12} & p_{13} & \cdots & p_{1j} & \cdots & p_{1n} \\ p_{21} & p_{22} & p_{23} & \cdots & p_{2j} & \cdots & p_{2n} \\ \cdot & & & & & & \\ \cdot & & & & & & \\ \cdot & & & & & & \\ p_{i1} & p_{i2} & p_{i3} & \cdots & p_{ij} & \cdots & p_{in} \\ \cdot & & & & & & \\ \cdot & & & & & & \\ \cdot & & & & & & \\ p_{n1} & p_{n2} & p_{n3} & \cdots & p_{nj} & \cdots & p_{nn} \end{bmatrix}$$

Here, the sum for each of the rows shall be equal to one. Thus,

$$p_{11} + p_{12} + p_{13} + \cdots + p_{1j} \cdots + p_{1n} = 1 \text{ and so on.}$$

Symbolically,

$$\sum_{j=1}^{n} p_{ij} = 1, \quad \text{for } i = 1, 2, ..., n$$

(ii) The Initial Conditions The initial conditions describe the situation the system presently is in. For instance, as indicated earlier, as the market is divided 30%, 45% and 25% between the brands D_1, D_2 and D_3 respectively on March 1, the current date, it describes the initial condition. It may be expressed in terms of a row vector [0.30 0.45 0.25]. In case the initial condition is described as [0 0 1] for the *market*, it implies that the brand D_3 holds the entire market, while for a *customer*, it would imply that the given customer has currently bought brand D_3. Further, for the gambler's ruin problem, the initial condition is given by [0 0 1 0 0], which implies that he currently is in the state where his capital is Rs 20.

14.4.2 Output

As stated earlier, there are two predictions which a Markov analysis provides. The first of these is the probability of the system being in a particular state at a future time, while the other is the steady state probabilities.

1. Specific-state Probabilities For calculating the probabilities for the system in specific states, we let $q_i(k)$ to represent the probability (q) of the system being in a certain state (i) in a certain period (k), called the state probability. Since the system would occupy one and only one state at a given point in time, it is obvious that the sum of all q_i values would be equal to 1. In general terms, with a total of n states,

$$q_1(k) + q_2(k) + q_3(k) + \cdots + q_n(k) = 1, \text{ for every } k$$

in which k is the number of transitions (0, 1, 2, ...).

Let us consider the calculation of the $q_{i(k)}$ probabilities for the detergent example. With states of the system designated as D_1, D_2 and D_3, q_{D1} (0) represents the probability of a customer choosing brand D_1 this month (at $t = 0$) and q_{D1} (1) represents the probability of choosing this brand after one transition, that is, the next month.

Similarly, $q_{D1}(2)$ is the probability of choosing this brand after two transitions (in the next to next month) and so on. Using these symbols, the probability distribution of the customer choosing any given brand (D_1, D_2, D_3) in any given month (k) may be expressed as a row vector as follows:

$$Q(k) = [q_{D1}(k)\ q_{D2}(k)\ q_{D3}(k)]$$

In general, for n states

$$Q(k) = [q_1(k)\ q_2(k)\ q_3(k)\ \cdots\ q_n(k)]$$

The initial condition is obviously expressed as $Q(0)$.

For the detergent example, since the market share for the three brands D_1, D_2, and D_3 initially (on March 1) is given to be 30%, 45% and 25%, respectively, we can write initial state probabilities as

$$Q(0) = [q_{D1}(0)\ q_{D2}(0)\ q_{D3}(0)] = [0.30\ 0.45\ 0.25]$$

Now, the managers of the three brands of detergents would benefit from knowing the market shares that would occur at a given future time (k). This information would be given by $q(k)$ where $k = 1, 2, 3, \ldots$ and so on. To be specific, let us calculate the share of the market likely to be held by each of the brands on April 1 (since the time period considered by us is one month, as mentioned earlier). This would be represented by $Q(1)$, since $k = 1$ for the next month. For this purpose, we use the matrix of transition probabilities, P. The row vector $Q(0)$ would be post-multiplied by the matrix P to get $Q(1)$. Thus,

$$Q(1) = (0.30\ 0.45\ 0.25)\begin{pmatrix} 0.60 & 0.30 & 0.10 \\ 0.20 & 0.50 & 0.30 \\ 0.15 & 0.05 & 0.80 \end{pmatrix} = (0.3075 \quad 0.3275 \quad 0.3650)$$

To understand matrix multiplication, we may consider the first element 0.3075, which has been obtained as follows:

(i) D_1's share of the market × D_1's propensity to retain its customers $= 0.30 \times 0.60 = 0.1800$
(ii) D_2's share of the market × D_1's propensity to attract D_2's customers $= 0.45 \times 0.20 = 0.0900$
(iii) D_3's share of the market × D_1's propensity to attract D_3's customers $= 0.25 \times 0.15 = 0.0375$

Total probability (i) + (ii) + (iii) $= 0.1800 + 0.0900 + 0.0375 = 0.3075$. This is $q_{D1}(1)$.

In a similar way, the market share of 0.3275 for D_2 is obtained as $0.30 \times 0.30 + 0.45 \times 0.50 + 0.25 \times 0.05$; while for D_3, it is $0.30 \times 0.10 + 0.45 \times 0.30 + 0.25 \times 0.80 = 0.3650$. These are, respectively, the values of $q_{D2}(1)$ and $q_{D3}(1)$.

In general terms, then,

$$Q(1) = [q_1(1)\ q_2(1)\ q_3(1)\ \cdots\ q_n(1)] = Q(0)P$$

By similar reasoning, it may be shown that

$$Q(2) = Q(1)P, \text{ which may be expressed as}$$

$$Q(2) = Q(0)PP = Q(0)P^2$$

In the same way, we can obtain $Q(3)$, $Q(4)$ and so on. Thus,

$$Q(k) = Q(k-1)P = Q(k-2)P^2 = \cdots = Q(0)P^k$$

Accordingly, the market shares of the three detergents two months hence, on May 1, may be obtained as follows:

$$P^2 = P \times P = \begin{pmatrix} 0.60 & 0.30 & 0.10 \\ 0.20 & 0.50 & 0.30 \\ 0.15 & 0.05 & 0.80 \end{pmatrix} \begin{pmatrix} 0.60 & 0.30 & 0.10 \\ 0.20 & 0.50 & 0.30 \\ 0.15 & 0.05 & 0.80 \end{pmatrix}$$

$$= \begin{pmatrix} 0.435 & 0.335 & 0.230 \\ 0.265 & 0.325 & 0.410 \\ 0.220 & 0.110 & 0.670 \end{pmatrix}$$

Now,
$$Q(2) = Q(0)\,P^2 = [0.30\ 0.45\ 0.25] \begin{pmatrix} 0.435 & 0.335 & 0.230 \\ 0.265 & 0.325 & 0.410 \\ 0.220 & 0.110 & 0.670 \end{pmatrix}$$

$$= [0.30475 \quad 0.27425 \quad 0.42100]$$

Thus, market shares of the three brands of detergents D_1, D_2 and D_3 are expected to be 30.5%, 27.4% and 42.1%, respectively, two months from now.

Besides calculating the market shares of the three brands at different points in time as represented by the state probabilities, information may also be obtained of the chances of a particular buyer buying a specific brand at a specific time, given that he previously purchased a particular brand. This situation is reflected when we desire to calculate, say, the probability that a customer, who has last purchased brand D_2, would purchase brand D_3 in two months from now. Such probabilities are termed as *conditional state probabilities*.

To calculate the probability of a customer to buy D_3 two months hence, given that his latest purchase has been D_2, we may recognise that the desired event of the buyer buying D_3 may result in any of the following ways:

(i) Customer switches from brand D_2 to D_1 in month 1, and to brand D_3 in month 2

(ii) Customer stays on brand D_2 in month 1, and switches to brand D_3 in month 2

(iii) Customer switches from brand D_2 to D_3 in month 1 and then stays on to D_3 in month 2

The various probabilities are obtained from the transition probability matrix. For example, the probability of switching from D_2 to D_1 in a month is 0.20; D_1 to D_3 is 0.10; D_2 to D_3 is 0.30, and staying at D_2 and D_3 is 0.50 and 0.80, respectively. Since the transition probabilities are assumed to be constant over time, the probabilities for each month are obtained from the same matrix.

The information is shown by means of a tree diagram given in Figure 14.4.

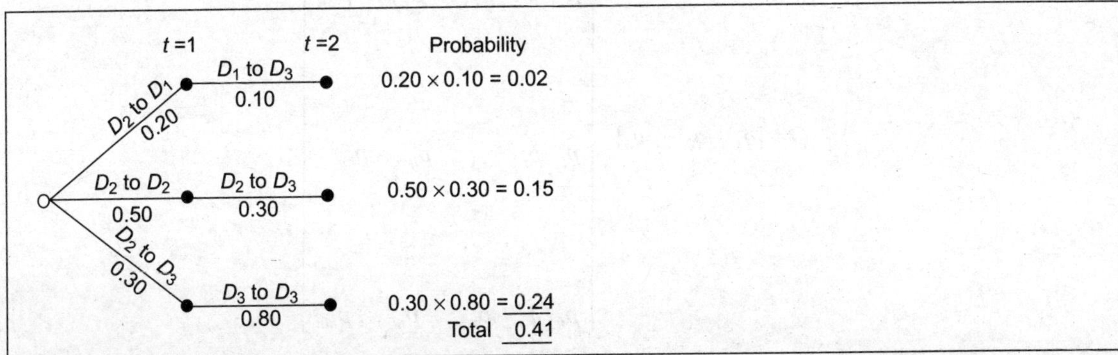

Figure 14.4 *Tree-diagram: Calculation of Probability*

The required probability is obtained by the summation of all the three probabilities calculated from the three forks. It is interesting to observe that this probability can also be obtained from the matrix P_2. In this matrix, the element 2, 3 is equal to 0.410 which is the same as probability calculated above.

In the same way, we can obtain other similar probabilities. For example, if a customer's last purchase has been detergent D_1, the probability two months from now he would buy brand D_3 would be given by the element 1, 3 of the matrix P_2, which is equal to 0.230. In general terms, the conditional state probabilities can be expressed as $q_{ij}(k)$; which means the probability of being in state j, given initial condition i, after k transitions. To illustrate, if a customer has bought brand D_2 in the last period, the chances that he would buy the brand D_3 after 3 periods, would be represented as $q_{23}(3)$ and shall be the element 2, 3 in the matrix P^3 (obtained as $P \times P \times P$).

2. Steady State Probabilities A significant property of the Markov chains is that the process tends to stabilise in the long run. A stabilised system is said to be in a *steady state* or in *equilibrium*, so that the system's operating characteristics can become independent of time. It is important to note that for a Markov chain to reach steady state condition, the chain must be *ergodic*. In an ergodic chain, it is possible to go from one state to any other state in a finite number of steps, irrespective of the present state. A transition probability matrix describing a Markov chain may be examined to see whether it is ergodic by checking if it is possible to move from every starting/present state to all other states. In the brand switching example considered earlier, it is easy to see that one can reach to every other state starting from any state.

The phenomenon of equilibrium can be expressed symbolically as $Q(k) = Q(k-1)$, so that the state probabilities in period k are the same as in the previous period. In the context of our detergent example, it means the following. We have seen in the preceding paragraphs as to how, given the current shares, the shares of various brands one month hence, two months hence, three months hence and so on can be calculated. We can extend this process to any number of periods. However, it may be observed that as calculations are made for larger and larger values of k, then each of the probability values, q_{ij}, tends to converge and stabilise. After this, successive calculations of q_{ij} for higher values of k do not show evidence of change. These are the *equilibrium*, or *steady state* probabilities.

We may now see as to how the equilibrium probabilities can be obtained.

To begin with, the formulation for steady state, as mentioned above, may be expressed as $Q(k) = Q(k)P$ [since $Q(k) = Q(k-1)$]. Deleting index k would yield the equilibrium probabilities (because they are independent of time). Thus, $Q = QP$. In matrix notation,

$$Q = [q_1 \quad q_2 \cdots q_n] \begin{bmatrix} p_{11}\ p_{12}\ p_{13} \cdots p_{1j} \cdots p_{1n} \\ p_{21}\ p_{22}\ p_{23} \cdots p_{2j} \cdots p_{2n} \\ \cdot \\ \cdot \\ \cdot \\ p_{i1}\ p_{i2}\ p_{i3} \cdots p_{ij} \cdots p_{in} \\ \cdot \\ \cdot \\ \cdot \\ p_{n1}\ p_{n2}\ p_{n3} \cdots p_{nj} \cdots p_{nn} \end{bmatrix}$$

The matrix multiplication shown here results in a system of n simultaneous linear equations as follows:

$$q_1 = p_{11}q_1 + p_{21}q_2 + \cdots + p_{n1}q_n$$
$$q_2 = p_{12}q_1 + p_{22}q_2 + \cdots + p_{n2}q_n$$
$$q_3 = p_{13}q_1 + p_{23}q_2 + \cdots + p_{n3}q_n$$
$$\vdots$$
$$q_n = p_{1n}q_1 + p_{2n}q_2 + \cdots + p_{nn}q_n$$

In fact, it so happens that one of the equations in this set of equations is redundant. Therefore, this system of equations cannot be solved for a unique solution. To circumvent the problem, we make use of the fact that the sum of the probabilities q_1, q_2, $\cdots$ and q_n must be unity. Thus, the equation $q_1 + q_2 + q_3 + \cdots + q_n = 1$ is added to the system of equations above. Next, they are solved simultaneously for the values of q_1, q_2, $\cdots$ and q_n, which depict the equilibrium probabilities.

It is interesting to note that the equilibrium probabilities will be the same (provided the transition probabilities remain fixed) regardless of the initial condition. In the context of the problems of market shares of different firms, it implies that if the given transition probabilities remain unchanged, the relative market shares of the firms in the long run would be the same no matter what the original shares were. This may be visualized from the fact that the initial condition is not considered in determining equilibrium probabilities. Of course, the initial condition is only relevant to the extent that the nearer the initial market shares happen to be to the final equilibrium shares, the faster would the equilibrium be approached.

Calculation of steady state probabilities For our example of detergents, we can state the system of linear equations as follows:

$$q_1 = 0.60q_1 + 0.20q_2 + 0.15q_3$$
$$q_2 = 0.30q_1 + 0.50q_2 + 0.05q_3$$
$$q_3 = 0.10q_1 + 0.30q_2 + 0.80q_3$$

Further, $q_1 + q_2 + q_3 = 1$

To obtain the values of q_1, q_2 and q_3 we may consider any two of the first three equations, and the fourth one. For example, considering the first two of the three equations, we have

$$q_1 = 0.60q_1 + 0.20q_2 + 0.15q_3$$
$$q_2 = 0.30q_1 + 0.50q_2 + 0.05q_3$$
$$q_1 + q_2 + q_3 = 1$$

Restating the equations, we get

$$0.40q_1 - 0.20q_2 - 0.15q_3 = 0$$
$$-0.30q_1 + 0.50q_2 - 0.05q_3 = 0$$
$$q_1 + q_2 + q_3 = 1$$

In matrix notation,

$$\begin{pmatrix} 0.40 & -0.20 & -0.15 \\ -0.30 & 0.50 & -0.05 \\ 1 & 1 & 1 \end{pmatrix} \begin{pmatrix} q_1 \\ q_2 \\ q_3 \end{pmatrix} \begin{matrix} = \\ = \\ = \end{matrix} \begin{pmatrix} 0 \\ 0 \\ 1 \end{pmatrix}$$

Using the method of determinants, we can solve the equations to get,

$$\Delta = 0.290$$
$$\Delta_1 = 0.085$$
$$\Delta_2 = 0.065$$
$$\Delta_3 = 0.140$$

Accordingly,
$$q_1 = \Delta_1/\Delta = 0.085/0.290 = 0.293$$
$$q_2 = \Delta_2/\Delta = 0.065/0.290 = 0.224$$
$$q_3 = \Delta_3/\Delta = 0.140/0.290 = 0.483$$

Thus, if the given transition probabilities remain constant over time, the market shares of the brands shall in the long run be 29.3%, 22.4% and 48.3% respectively. These values are independent of the present market shares of the three brands.

14.4.3 Absorbing Chains

In our discussion so far, we have analysed situations where none of the states was absorbing in nature. Now let us consider the Markov Chains where some of the states would be absorbing in nature. The *accounts receivable* situation of a firm may be modelled as an absorbing Markov chain.

Example 14.3 Consider the accounts receivable problem of a firm. Suppose that, for this firm, an accounts receivable turns into bad debt if the account is more than three months overdue. Now, at the beginning of each month, each of the accounts may be classified into one of the following states:

State 1 : New account
State 2 : Account is one month overdue for payment
State 3 : Account is two months overdue for payment
State 4 : Account is three months overdue for payment
State 5 : Account has been paid
State 6 : Account is written off as bad debt.

Suppose further that on the basis of the past data, the following transition matrix, indicating how the status of an account changes from one month to the next, has been derived:

State	*State*					
	1	*2*	*3*	*4*	*5*	*6*
1	0	0.7	0	0	0.3	0
2	0	0	0.6	0	0.4	0
3	0	0	0	0.4	0.6	0
4	0	0	0	0	0.8	0.2
5	0	0	0	0	1	0
6	0	0	0	0	0	1

The various entries in the matrix can be interpreted in the usual way. For example, the probability of a new account to be paid by the beginning of the next month is 30% while the probability of it being not paid and, therefore, become a one-month overdue is 70%; and if an account is two months overdue at the beginning of a month, there is a 40% chance that at the beginning of the next month the account shall not be paid up and hence be three months overdue, and a 60% chance that the account would be paid up. It is assumed for simplicity that after the three-months period, an account is either paid up or is written off as bad debt. Thus, if and when an account is paid up or written off as bad debt, the account gets closed and no more transitions can occur. States 5 and 6, account paid up and account written off, respectively, are therefore, the absorbing states—every new account shall eventually be absorbed as either a paid up debt or a bad debt. The other states, 1 through 4, are transient states.

The questions that we seek to answer include the following:

(i) What is the probability that a new account shall eventually be collected?

(ii) What is the probability that a new account shall be written off as bad debt?

(iii) What is the probability that an account which is one-month overdue shall eventually be a bad debt?

(iv) What is the probability that a two-month overdue account shall finally be collected?

(v) If the firm makes a credit sales of Rs 400,000 per month, on the average, what amount can it expect to lose per annum by way of bad debts?

These questions are different versions of a general question, namely, if the chain begins in a given transient state, what is the probability of ending up in each absorbing state? Besides, we may also like to know that if a chain begins in a particular transient state, and before reaching a particular state, what is the expected number of times that each state will be entered, and how many periods do we expect to spend in a transient state before absorption takes place. To answer such questions, we first re-write the transition probability matrix with the states listed as transient states and absorbing states, in that order. To be specific, if there are n states, out of which m are absorbing states, and $n - m$ transient states, then we may express the matrix as follows:

$$P = \begin{array}{c c} & \begin{array}{cc} n-m \text{ columns} & m \text{ columns} \end{array} \\ \begin{array}{c} n-m \text{ rows} \\ m \text{ rows} \end{array} & \left[\begin{array}{c|c} Q & R \\ \hline O & I \end{array}\right] \end{array}$$

In this, rows and columns of P correspond to states $1, 2, \cdots, n-m$; Q represents the square matrix, of the order $(n-m)$, of transient probabilities, representing transitions between transient states; R is an $(n-m)$ by m matrix representing transitions from transient to absorbing states; I is an identity matrix of the order m, depicting the fact that it is not possible to go from one absorbing state to another; and O is a null matrix consisting of all zeros.

Now, if we let the transient states be represented as t_i's and absorbing states by a_i's, the transition probability matrix for the problem can be written in the designated order, as follows:

State	*State*					
	t_1	t_2	t_3	t_4	a_1	a_2
t_1	0	0.7	0	0	0.3	0
t_2	0	0	0.6	0	0.4	0
t_3	0	0	0	0.4	0.6	0
t_4	0	0	0	0	0.8	0.2
a_1	0	0	0	0	1	0
a_2	0	0	0	0	0	1

where t_1 : New account

t_2 : Account is one month overdue for payment

t_3 : Account is two months overdue for payment

t_4 : Account is three months overdue for payment

a_1 : Account has been paid

a_2 : Account is written off as bad debt.

With $n = 6$ and $m = 2$, we have,

$$Q = \begin{pmatrix} 0 & 0.7 & 0 & 0 \\ 0 & 0 & 0.6 & 0 \\ 0 & 0 & 0 & 0.4 \\ 0 & 0 & 0 & 0 \end{pmatrix} \quad R = \begin{pmatrix} 0.3 & 0 \\ 0.4 & 0 \\ 0.6 & 0 \\ 0.8 & 0.2 \end{pmatrix}$$

After this, we obtain the inverse of the matrix $(I-Q)$, that is, $(I-Q)^{-1}$ (with identity matrix I being of the same order as the matrix Q) and a matrix $(I-Q)^{-1}R$. The matrix $(I-Q)^{-1}$ is known as the *fundamental matrix* of the Markov chains. The *ij*th element of this matrix gives the expected number of periods that will be spent in transient state t_i before reaching an absorbing state.

Similarly, if the chain begins in a given transient state, t_i, the probability that it shall be eventually absorbed in absorbing state a_j would be given by the *ij*th element of the matrix $(I-Q)^{-1}R$.

For our example, we have,

$$I - Q = \begin{pmatrix} 1 & 0 & 0 & 0 \\ 0 & 1 & 0 & 0 \\ 0 & 0 & 1 & 0 \\ 0 & 0 & 0 & 1 \end{pmatrix} - \begin{pmatrix} 0 & 0.7 & 0 & 0 \\ 0 & 0 & 0.6 & 0 \\ 0 & 0 & 0 & 0.4 \\ 0 & 0 & 0 & 0 \end{pmatrix} = \begin{pmatrix} 1 & -0.7 & 0 & 0 \\ 0 & 1 & -0.6 & 0 \\ 0 & 0 & 1 & -0.4 \\ 0 & 0 & 0 & 1 \end{pmatrix}$$

Further,

$$(I-Q)^{-1} = \begin{array}{c} \\ t_1 \\ t_2 \\ t_3 \\ t_4 \end{array} \begin{array}{c} \begin{array}{cccc} t_1 & t_2 & t_3 & t_4 \end{array} \\ \begin{bmatrix} 1 & 0.7 & 0.42 & 0.168 \\ 0 & 1 & 0.6 & 0.24 \\ 0 & 0 & 1 & 0.4 \\ 0 & 0 & 0 & 1 \end{bmatrix} \end{array}$$

$$(I-Q)^{-1}R = \begin{bmatrix} 1 & 0.7 & 0.42 & 0.168 \\ 0 & 1 & 0.6 & 0.24 \\ 0 & 0 & 1 & 0.4 \\ 0 & 0 & 0 & 1 \end{bmatrix} \begin{bmatrix} 0.3 & 0 \\ 0.4 & 0 \\ 0.6 & 0 \\ 0.8 & 0.2 \end{bmatrix} = \begin{array}{c} \\ t_1 \\ t_2 \\ t_3 \\ t_4 \end{array} \begin{array}{c} \begin{array}{cc} a_1 & a_2 \end{array} \\ \begin{bmatrix} 0.9664 & 0.0336 \\ 0.9520 & 0.0480 \\ 0.9200 & 0.0800 \\ 0.8000 & 0.2000 \end{bmatrix} \end{array}$$

The answers to the questions posed earlier may be obtained with reference to above matrix as follows:

(i) Probability that a new account (t_1) shall eventually be collected $(a_1) = 0.9664$ (element 1, 1 of the matrix)

(ii) Probability that a new account (t_1) shall be written off as bad debt $(a_2) = 0.0336$ (element 1, 2 of the matrix)

(iii) Probability that an account which is one-month overdue (t_2) shall eventually be written off as bad debt $(a_2) = 0.048$ (element 2, 2 of the matrix)

(iv) Probability that a two-month overdue account (t_3) shall finally be collected (a_1) = 0.92 (element 3, 1 of the matrix)

(v) For a credit sales of Rs 400,000 per month, on the average, the annual sales shall be Rs 48,00,000. The amount the firm can expect to lose by way of bad debts = 48,00,000 × 0.0336 = Rs 1,61,280.

REVIEW ILLUSTRATIONS

Example 14.4 A market survey is made on two brands of breakfast foods *A* and *B*. Every time a customer purchases, he may buy the same brand or switch to another brand. The transition matrix is given below:

From	*To*	
	A	*B*
A	0.8	0.2
B	0.6	0.4

At present, it is estimated that 60 per cent of the people buy brand *A* and 40 per cent buy brand *B*. Determine the market shares of brand *A* and brand *B* in the steady state. *(MIB, Delhi, 1999)*

Let q_1 and q_2 be the market shares (also called steady state probabilities) for brands *A* and *B*, respectively, in the steady state. With given transition matrix, we have

$$(q_1 \;\; q_2) = (q_1 \;\; q_2)\begin{pmatrix} 0.8 & 0.2 \\ 0.6 & 0.4 \end{pmatrix}$$

On simplification, we have

$$q_1 = 0.8q_1 + 0.6q_2 \qquad \text{(i)}$$

$$q_2 = 0.2q_1 + 0.4q_2 \qquad \text{(ii)}$$

Further, we have $q_1 + q_2 = 1$. Replacing Eq. (ii) by this equation, we get

$$q_1 = 0.8q_1 + 0.6q_2$$

$$1 = q_1 + q_2$$

Solving these equations for q_1 and q_2, we get $q_1 = 0.75$ and $q_2 = 0.25$. Accordingly, in the long run, the share of brands *A* and *B* would be 75% and 25%, respectively. It may be mentioned again that the steady state probabilities (market shares), q_1 and q_2, are not affected by the current market shares of the two brands.

Example 14.5 In a certain market, only two brands of lipsticks, *A* and *B*, are sold. Given that a lady last purchased lipstick *A*, there is 80% chance that she would buy the same brand in the next purchase, while if a lady purchased brand *B*, there is 90% chance that her next purchase would be brand *B*. Using this information, develop the transition probability matrix. Now, calculate

(i) the probability that if a customer is currently a brand *A* purchaser, she will purchase brand *B* two purchases from now;

(ii) the probability that if a customer is a brand *B* purchaser, she will purchase brand *A* three periods from now;

(iii) the probability that three periods from now, a customer shall buy brand *B*, given that the market share of the two brands is as follows: Brand *A* – 70%, Brand *B* – 30%;

(iv) the steady state probabilities.

From the given information, the transition probability matrix can be obtained as follows:

$$P = \begin{matrix} & A & B \\ A & 0.8 & 0.2 \\ B & 0.1 & 0.9 \end{matrix}$$

(i) The probability of a customer, who is currently a brand A purchaser, to purchase brand B two purchases from now can be obtained from the matrix P^2. The element 1, 2 of this matrix shall be the required value. Now.

$$P^2 = \begin{pmatrix} 0.8 & 0.2 \\ 0.1 & 0.9 \end{pmatrix} \begin{pmatrix} 0.8 & 0.2 \\ 0.1 & 0.9 \end{pmatrix} = \begin{pmatrix} 0.66 & 0.34 \\ 0.17 & 0.83 \end{pmatrix}$$

Thus, the required probability is 0.34.

(ii) Since the probability required is for purchases three periods from now, we shall first obtain P^3. Thus,

$$P^3 = PP^2 = \begin{pmatrix} 0.8 & 0.2 \\ 0.1 & 0.9 \end{pmatrix} \begin{pmatrix} 0.66 & 0.34 \\ 0.17 & 0.83 \end{pmatrix}$$

$$= \begin{pmatrix} 0.562 & 0.438 \\ 0.219 & 0.781 \end{pmatrix}$$

The element 2, 1 of this matrix has a value equal to 0.219. This is the required probability.

(iii) With present shares of the two brands being 70% and 30%, we have $q = [0.70 \quad 0.30]$. Now, to find the probability that three periods from now, a customer shall buy brand B, we shall post-multiply the vector q by the column 2 of the P^3 matrix to get

$$(0.7 \quad 0.3) \begin{pmatrix} 0.438 \\ 0.781 \end{pmatrix} = 0.5409$$

Hence, three periods from now, 54% of all purchasers will be purchasing brand B lipsticks.

(iv) If we let q_1 and q_2 be the steady state probabilities for brands A and B, respectively, we have

$$(q_1 \quad q_2) = (q_1 \quad q_2) \begin{pmatrix} 0.80 & 0.20 \\ 0.10 & 0.90 \end{pmatrix}$$

On simplification,

$$q_1 = 0.80q_1 + 0.10q_2$$

$$q_2 = 0.20q_1 + 0.90q_2$$

Replacing the second equation by the condition $q_1 + q_2 = 1$, we get

$$q_1 = 0.80q_1 + 0.10q_2$$

$$1 = q_1 + q_2$$

Solving for q_1 and q_2, we get $q_1 = 1/3$ and $q_2 = 2/3$. Hence, after a long time, there is a 1/3 probability that a given customer shall buy lipstick A, and 2/3 that a customer shall buy lipstick B.

Example 14.6 Consider the following transition probability matrix:

$$P = \begin{pmatrix} 0.3 & 0.7 & 0 & 0 & 0 \\ 0.5 & 0.5 & 0 & 0 & 0 \\ 0 & 0 & 0.8 & 0.2 & 0 \\ 0 & 0 & 0.4 & 0.3 & 0.3 \\ 0 & 0 & 0 & 0.1 & 0.9 \end{pmatrix}$$

You are required to

(i) provide an example of path in this;

(ii) give an example of states which communicate;

(iii) provide an example of recurrent state;

(iv) provide an example of transient state; and

(v) identify if there exists an absorbing state.

The transition probability matrix is reproduced in Figure 14.5 graphically.

(i) For two states i and j, a path from i to j is a sequence of transitions that starts in i and ends in j, such that each transition has a positive probability of occurring. An example here is 3–4–5, a path between 3 and 5.

(ii) States i and j are said to communicate if j is reachable from i, and i is reachable from j. For example, here states 4 and 5 communicate.

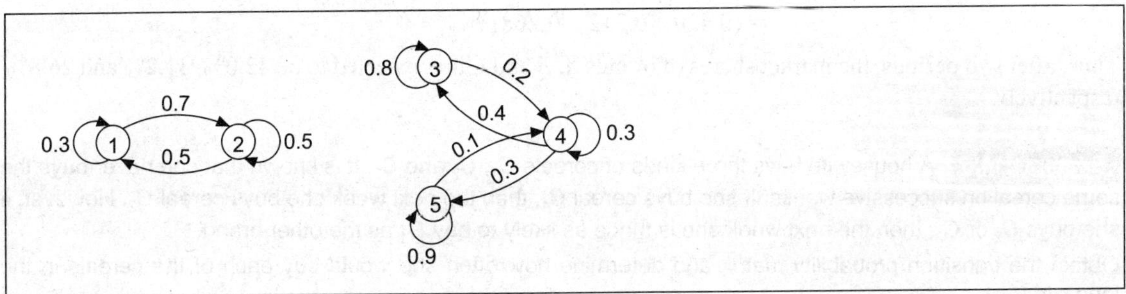

Figure 14.5 *Graphic Presentation of Transition Probabilities*

(iii) Each state in this matrix is recurrent, since for each one (i) there exists a state (j) that is reachable from it, and the state (i) is reachable from that (j).

(iv) None of the states here is transient since all of them are recurrent.

(v) For an absorbing state i, $p_{ii} = 1$. Since none of the p_{ii} values is equal to unity, none of the states is absorbing.

Example 14.7 Recently, a market research team has conducted a survey of consumer buying habits with respect to three brands of talcum powder in an area. It estimates that, at present, 20 per cent of the customers buy brand A, 50 per cent of the customers buy brand B and 30 per cent of them buy brand C. In addition, the firm has analysed its survey data and has determined the following brand switching matrix:

Brand Just Bought	Brand Next Bought		
	A	B	C
A	0.6	0.3	0.1
B	0.4	0.5	0.1
C	0.2	0.1	0.7

What will be the expected distribution of customers two time periods later?

The initial condition $Q(0)$, indicating the present market shares, is: $Q(0) = (0.20 \quad 0.50 \quad 0.30)$. Further, the transition probability matrix, P, is given as:

$$P = \begin{pmatrix} 0.6 & 0.3 & 0.1 \\ 0.4 & 0.5 & 0.1 \\ 0.2 & 0.1 & 0.7 \end{pmatrix}$$

The expected distribution of customers two periods hence, $Q(2)$, would be given by $Q(0) \times P^2$. Accordingly,

$$Q(2) = (0.20 \quad 0.50 \quad 0.30) \begin{pmatrix} 0.6 & 0.3 & 0.1 \\ 0.4 & 0.5 & 0.1 \\ 0.2 & 0.1 & 0.7 \end{pmatrix}^2$$

$$= (0.420 \quad 0.312 \quad 0.268)$$

Thus, after two periods, the market shares of brands A, B and C are expected to be 42.0%, 31.2% and 26.8%, respectively.

Example 14.8 A housewife buys three kinds of cereals C_1, C_2 and C_3. It is known that she never buys the same cereal on successive weeks. If she buys cereal C_1, then the next week she buys cereal C_2. However, if she buys C_2 or C_3, then the next week she is thrice as likely to buy C_1 as the other brand.

Obtain the transition probability matrix and determine how often she would buy each of the cereals in the long run.

The given information is presented in Table 14.3 in the form of a transition probability matrix.

TABLE 14.3 Transition Probabilities Matrix

State	*State*		
	C_1	C_2	C_3
C_1	0	1	0
C_2	0.75	0	0.25
C_3	0.75	0.25	0

If q_1, q_2 and q_3, denote the steady state probabilities of states C_1, C_2 and C_3, respectively, we have

$$q_1 = 0.00q_1 + 0.75q_2 + 0.75q_3$$
$$q_2 = 1.00q_1 + 0.00q_2 + 0.25q_3$$
$$q_3 = 0.00q_1 + 0.25q_2 + 0.00q_3 \text{ and}$$
$$q_1 + q_2 + q_3 = 1$$

Considering the last two of the above equations and the fourth one, and re-arranging, we get

$$q_1 - q_2 + 0.25q_3 = 0$$
$$0q_1 + 0.25q_2 - q_3 = 0$$
$$q_1 + q_2 + q_3 = 1$$

In matrix notation,

$$\begin{pmatrix} 1 & -1 & 0.25 \\ 0 & 0.25 & -1 \\ 1 & 1 & 1 \end{pmatrix} \begin{pmatrix} q_1 \\ q_2 \\ q_3 \end{pmatrix} = \begin{pmatrix} 0 \\ 0 \\ 1 \end{pmatrix}$$

Using the method of determinants, we get

$$\Delta = 2.1875$$
$$\Delta_1 = 0.9375$$
$$\Delta_2 = 1.0000$$
$$\Delta_3 = 0.2500$$

Accordingly,

$$q_1 = \Delta_1/\Delta = 0.9375/2.1875 = 0.4286$$
$$q_2 = \Delta_2/\Delta = 1.0000/2.1875 = 0.4571$$
$$q_3 = \Delta_3/\Delta = 0.2500/2.1875 = 0.1143$$

Thus, in the long run, the housewife would buy the three cereals C_1, C_2 and C_3 with a frequency of 42.86%, 45.71% and 11.43%, respectively.

Example 14.9 To investigate brand switching between different brands of detergent powder, a research company surveyed shoppers in an area to discover their behaviour. The survey on a number of shoppers in relation to three leading brands Superwash, Clean and Shine has revealed the following information.

Current Buy	*Previous Buy*	*Number*
Superwash	Superwash	200
Clean	Clean	150
Shine	Shine	100
Clean	Superwash	50
Shine	Superwash	25
Superwash	Clean	80
Shine	Clean	45
Superwash	Shine	130
Clean	Shine	20

(a) Using the information, develop transition probability matrix.

(b) Calculate the market share for each brand after two purchases.

(c) Calculate the market share for each brand in the long run.

(a) The given information is shown tabulated here.

From Brand	*To Brand*			
	Superwash	*Clean*	*Shine*	*Total*
Superwash	200	50	25	275
Clean	80	150	45	275
Shine	130	20	100	250
Total	410	220	170	800

From these data, the transition probability matrix, P, is produced as given below. The values in each row are divided by the respective row totals to get the probabilities.

$$P = \begin{bmatrix} 8/11 & 2/11 & 1/11 \\ 16/55 & 30/55 & 9/55 \\ 13/25 & 2/25 & 10/25 \end{bmatrix}$$

Further, the current market shares of the three brands may be obtained by dividing respective column totals by the grand total as 410/800, 220/800 and 170/800. Accordingly, $Q(0) = (41/80\ 22/80\ 17/80)$.

(b) Market share for each brand after two purchases, $Q(2)$, may be obtained as follows:

$$Q(2) = Q(0) \times P^2 = (41/80 \quad 22/80 \quad 17/80) \begin{bmatrix} 8/11 & 2/11 & 1/11 \\ 16/55 & 30/55 & 9/55 \\ 13/25 & 2/25 & 10/25 \end{bmatrix}^2$$

$$= (520/901 \qquad 130/503 \qquad 73/144)$$

$$= (0.5771 \qquad 0.2584 \qquad 0.1644)$$

Thus, expected shares after two months are 57.71%, 25.84% and 16.44% respectively for the three brands.

(c) If q_1, q_2 and q_3 be the long-run market shares of Superwash, Clean and Shine respectively, we have

$$q_1 = \frac{8}{11} q_1 + \frac{16}{55} q_2 + \frac{13}{25} q_3 \qquad \text{(i)}$$

$$q_2 = \frac{2}{11} q_1 + \frac{30}{55} q_2 + \frac{2}{25} q_3 \qquad \text{(ii)}$$

$$q_3 = \frac{1}{11} q_1 + \frac{9}{55} q_2 + \frac{10}{25} q_3 \qquad \text{(iii)}$$

Also, $q_1 + q_2 + q_3 = 1$ (iv)

Considering equations (i), (ii) and (iv) and re-arranging these, we get

$$75q_1 - 80q_2 - 143q_3 = 0 \qquad \text{(v)}$$

$$-50q_1 + 125q_2 - 22q_3 = 0 \qquad \text{(vi)}$$

$$q_1 + q_2 + q_3 = 1 \qquad \text{(vii)}$$

In matrix notation, we have

$$\begin{bmatrix} 75 & -80 & -143 \\ -50 & 125 & -22 \\ 1 & 1 & 1 \end{bmatrix} \begin{bmatrix} q_1 \\ q_2 \\ q_3 \end{bmatrix} = \begin{bmatrix} 0 \\ 0 \\ 1 \end{bmatrix}$$

Using the method of determinants,

$\Delta = 33{,}810$, $\Delta_1 = 19{,}635$, $\Delta_2 = 8{,}800$ and $\Delta_3 = 5{,}375$.

Accordingly,

$$q_1 = 19{,}635/33{,}810 = 0.5807 \text{ or } 58.07\%$$
$$q_2 = 8{,}800/33{,}810 = 0.2603 \text{ or } 26.03\%$$
$$q_3 = 5{,}375/33{,}810 = 0.1590 \text{ or } 15.90\%$$

Thus, long-term market shares of three brands are expected to be 58.07%, 26.03% and 15.90%, respectively.

Example 14.10 Suppose that new razor blades were introduced in the market by three companies at the same time. When they were introduced, each company had an equal share of the market, but during the first year the following changes took place:

(i) Company *A* retained 90 per cent of its customers, lost 3 per cent to *B* and 7 per cent to *C*.
(ii) Company *B* retained 70 per cent of its customers, lost 10 per cent to *A* and 20 per cent to *C*.
(iii) Company *C* retained 80 per cent of its customers, lost 10 per cent to *A* and 10 per cent to *B*.

Assuming that no changes in the buying habits of the consumer occur,

(1) what are the market shares of the three companies at the end of the first year and the second year?
(2) what are the long-run market shares of the three companies? *(MBA, Delhi, April, 1997)*

From the given data, we have

Initial condition, $Q(0) = (0.333 \quad 0.333 \quad 0.333)$, and

Transition probabilities matrix as

$$P = \begin{pmatrix} 0.90 & 0.03 & 0.07 \\ 0.10 & 0.70 & 0.20 \\ 0.10 & 0.10 & 0.80 \end{pmatrix}$$

Now, market shares of companies *A*, *B* and *C*, respectively, at the end of one year, $Q(1)$, and at the end of two years are calculated below:

$$Q(1) = Q(0) \times P$$
$$= (0.333 \quad 0.333 \quad 0.333) \begin{pmatrix} 0.90 & 0.03 & 0.07 \\ 0.10 & 0.70 & 0.20 \\ 0.10 & 0.10 & 0.80 \end{pmatrix}$$
$$= (0.367 \quad 0.277 \quad 0.356)$$

$$Q(2) = Q(0) \times P^2$$
$$= (0.333 \quad 0.333 \quad 0.333) \begin{pmatrix} 0.90 & 0.03 & 0.07 \\ 0.10 & 0.70 & 0.20 \\ 0.10 & 0.10 & 0.80 \end{pmatrix} \begin{pmatrix} 0.90 & 0.03 & 0.07 \\ 0.10 & 0.70 & 0.20 \\ 0.10 & 0.10 & 0.80 \end{pmatrix}$$
$$= (0.3933 \quad 0.2403 \quad 0.3664)$$

Accordingly, market shares would be:

	Company A	*Company B*	*Company C*
End of One Year	36.7%	27.7%	35.6%
End of Two Years	39.3%	24.0%	36.6%

To calculate long-term market shares q_1, q_2 and q_3, for companies *A*, *B* and *C*, respectively, we set the linear equations as follows:

$$q_1 = 0.90q_1 + 0.10q_2 + 0.10q_3$$

$$q_2 = 0.03q_1 + 0.70q_2 + 0.10q_3$$

$$q_3 = 0.07q_1 + 0.20q_2 + 0.80q_3 \text{ and}$$

$$q_1 + q_2 + q_3 = 1$$

Considering the first two of the above equations and the fourth one, and re-arranging, we get

$$0.10q_1 - 0.10q_2 - 0.10q_3 = 0$$

$$-0.03q_1 + 0.30q_2 - 0.10q_3 = 0$$

$$q_1 + q_2 + q_3 = 1$$

In matrix notation,

$$\begin{pmatrix} 0.10 & -0.10 & -0.10 \\ -0.03 & 0.30 & -0.10 \\ 1.00 & 1.00 & 1.00 \end{pmatrix} \begin{pmatrix} q_1 \\ q_2 \\ q_3 \end{pmatrix} = \begin{pmatrix} 0 \\ 0 \\ 1 \end{pmatrix}$$

To solve for q_1, q_2 and q_3, we have

$$\Delta = 0.080$$

$$\Delta_1 = 0.040$$

$$\Delta_2 = 0.013$$

$$\Delta_3 = 0.027$$

Accordingly,

$$q_1 = \Delta_1/\Delta = 0.040/0.080 = 0.50$$

$$q_2 = \Delta_2/\Delta = 0.013/0.080 = 0.16$$

$$q_3 = \Delta_3/\Delta = 0.027/0.080 = 0.34$$

Thus, at equilibrium, market shares would be: *A* 50%, *B* 16% and *C* 34%.

Example 14.11 Care-for-all Hospital operates on a charity basis. All expenses are paid for by the government. Recently, the board of governors of the hospital has been complaining about the size of the budget and insisting that the hospital cut expenses. The major area of concern has been the cost of keeping patients in the intensive care. This cost has averaged Rs 1,000 per week per person, compared to only Rs 500 per week per person for keeping patients in the wards. Past history shows that of those patients in ICU at the beginning of the week, 50 per cent will be there at the end of the week, and 50 per cent will be moved to a ward. Of the patients in the wards at the beginning of the week, 50 per cent will be there at the end of the week, 10 per cent will get worse and be transferred to the ICU and 40 per cent will become outpatients. Of those persons who are outpatients at the beginning of the week, 85 per cent will remain outpatients at the end of the week, 10 per cent will be admitted to a ward, and 5 per cent will be admitted to ICU.

The board of governors believes the criteria for keeping patients in ICU are too strict and has instructed the ICU staff to relax the creteria so that only 40 per cent of the ICU patients remain in ICU each week and 60 per cent are transferred to wards.

The staff insists that if this is done, 20 per cent of the ward patients will be going into ICU each week and only 30 per cent will be transferred to outpatient status. The percentage of ward patients remaining will still be 50 per cent. There will be no change in the outpatient status.

Will the policy advocated by the board of governors actually save money? *(MIB, Delhi, 2010)*

Using the given information, we can derive the transition probability matrices as follows:

Present conditions:

Beginning of week	*End of week*		
	ICU	*Wards*	*Outpatients*
ICU	0.50	0.50	0.00
Wards	0.10	0.50	0.40
Outpatients	0.05	0.10	0.85

Proposed conditions:

Beginning of week	*End of week*		
	ICU	*Wards*	*Outpatients*
ICU	0.40	0.60	0.00
Wards	0.20	0.50	0.30
Outpatients	0.05	0.10	0.85

With these, we can determine proportion of total patients in each of the three states, in equilibrium, under each of the two conditions. To calculate q_1, q_2, and q_3 as the long-run shares respectively of ICU, wards, and outpatients, we set the linear equations as follows:

Present conditions:

$$q_1 = 0.50q_1 + 0.10q_2 + 0.05q_3$$

$$q_2 = 0.50q_1 + 0.50q_2 + 0.10q_3$$

$$q_3 = 0.00q_1 + 0.40q_2 + 0.85q_3$$

$$q_1 + q_2 + q_3 = 1$$

Considering the first two and the fourth of the above equations, and re-arranging them, we get

$$0.50q_1 - 0.10q_2 - 0.05q_3 = 0$$

$$-0.50q_1 + 0.50q_2 - 0.10q_3 = 0$$

$$q_1 + q_2 + q_3 = 1$$

In matrix notation, $AQ = B$,

$$\begin{pmatrix} 0.50 & -0.10 & -0.05 \\ -0.50 & 0.50 & -0.10 \\ 1 & 1 & 1 \end{pmatrix} \begin{pmatrix} q_1 \\ q_2 \\ q_3 \end{pmatrix} = \begin{pmatrix} 0 \\ 0 \\ 1 \end{pmatrix}$$

We have, $Q = A^{-1} B$

$$|A| = 0.31 \text{ and } A^{-1} = \frac{1}{0.31}\begin{bmatrix} 0.60 & 0.05 & 0.04 \\ 0.40 & 0.55 & 0.08 \\ -1.00 & -0.60 & 0.20 \end{bmatrix}$$

Finally,

$$Q = \frac{1}{0.31}\begin{bmatrix} 0.60 & 0.05 & 0.04 \\ 0.40 & 0.55 & 0.08 \\ -1.00 & -0.60 & 0.20 \end{bmatrix}\begin{bmatrix} 0 \\ 0 \\ 1 \end{bmatrix} = \begin{bmatrix} 0.1129 \\ 0.2419 \\ 0.6452 \end{bmatrix}$$

Thus, $q_1 = 0.1129$, $q_2 = 0.2419$ and $q_3 = 0.6452$.

Proposed conditions:

Here,

$$q_1 = 0.40q_1 + 0.20q_2 + 0.05q_3$$
$$q_2 = 0.60q_1 + 0.50q_2 + 0.10q_3$$
$$q_3 = 0.00q_1 + 0.30q_2 + 0.85q_3$$
$$q_1 + q_2 + q_3 = 1$$

Using the equations above except the third one, we can determine the values of q_1, q_2, q_3 as follows:

$$Q = A^{-1}B = \frac{1}{0.315}\begin{bmatrix} 0.60 & 0.15 & 0.05 \\ 0.50 & 0.65 & 0.09 \\ -0.10 & -0.80 & 0.18 \end{bmatrix}\begin{bmatrix} 0 \\ 0 \\ 1 \end{bmatrix}$$

$$= \begin{bmatrix} 0.1429 \\ 0.2857 \\ 0.5714 \end{bmatrix}$$

Thus, $q_1 = 0.1429$, $q_2 = 0.2857$ and $q_3 = 0.5714$.

From the calculations above, we can obtain the expected cost per patient per week for each of the two conditions as follows:

$$\text{Expected cost} = 1{,}000\ q_1 + 500q_2$$

For present conditions,

$$\text{Expected cost} = 1{,}000 \times 0.1129 + 500 \times 0.2419 = \text{Rs } 233.85$$

For proposed conditions,

$$\text{Expected cost} = 1{,}000 \times 0.1429 + 500 \times 0.2857 = \text{Rs } 285.75$$

Thus, the proposal of the board of governors is not likely to save money.

Example 14.12 A manufacturing company has two production departments, designated below as P_1 and P_2, and three service departments, designated below as S_1, S_2 and S_3, and wishes to allocate all factory overhead to production departments. A primary distribution of overhead to all departments has been made and is indicated below. The company makes the secondary distribution of overhead from service departments to production departments on a reciprocal basis, recognising the fact that services of one service department are utilised by another. Data regarding costs and allocation percentages are as follows:

Department	Direct cost	Percentage allocation of total cost of the department		
		S_1	S_2	S_3
S_1	Rs 25,000	0	30	10
S_2	Rs 124,000	20	0	10
S_3	Rs 77,000	10	10	0
P_1	Rs 120,000	40	30	30
P_2	Rs 290,000	30	30	50
		Total 100	100	100

For example, S_1 spends 20% of its time and efforts on S_2, 10% on S_3, 40% on P_1 and 30% on P_2. Similarly, the values in respect of S_2 and S_3 may be interpreted. Now, once the expenses of S_1 are apportioned, part of them would go to S_2 and S_3 while it would also receive a part of the expenses of each of them. The process of apportionment needs to be repeated over and over again until all the expenses get absorbed by the production departments.

Draft this problem as absorbing chains to figure out how to allocate the costs of service departments to production departments and calculate the total costs for each of the two departments.

To consider this problem as absorbing chains, the transition probability matrix may be obtained by considering the service department as transient states and the production departments as absorbing states. This is because once any expenses of a service department are allocated to a production department, they are absorbed and there is no question of their being allocated back. The proportion of expenses of service departments to be allocated to the various departments serve as the transition probability.

Accordingly, the transition probability matrix may be presented as follows:

States

$$P = \begin{array}{c} \\ S_1 \\ S_2 \\ S_3 \\ P_1 \\ P_2 \end{array} \begin{pmatrix} S_1 & S_2 & S_3 & P_1 & P_2 \\ 0 & 0.2 & 0.1 & 0.4 & 0.3 \\ 0.3 & 0 & 0.1 & 0.3 & 0.3 \\ 0.1 & 0.1 & 0 & 0.3 & 0.5 \\ 0 & 0 & 0 & 1 & 0 \\ 0 & 0 & 0 & 0 & 1 \end{pmatrix}$$

Here,

$$Q = \begin{pmatrix} 0 & 0.2 & 0.1 \\ 0.3 & 0 & 0.1 \\ 0.1 & 0.1 & 0 \end{pmatrix} \quad R = \begin{pmatrix} 0.4 & 0.3 \\ 0.3 & 0.3 \\ 0.3 & 0.5 \end{pmatrix}$$

Now,

$$I - Q = \begin{pmatrix} 1 & 0 & 0 \\ 0 & 1 & 0 \\ 0 & 0 & 1 \end{pmatrix} - \begin{pmatrix} 0 & 0.2 & 0.1 \\ 0.3 & 0 & 0.1 \\ 0.1 & 0.1 & 0 \end{pmatrix}$$

$$= \begin{pmatrix} 1 & -0.2 & -0.1 \\ -0.3 & 1 & -0.1 \\ -0.1 & -0.1 & 1 \end{pmatrix}$$

Taking the inverse of $(I - Q)$, we get

$$(I-Q)^{-1} = \frac{1000}{915}\begin{pmatrix} 0.99 & 0.21 & 0.12 \\ 0.31 & 0.99 & 0.13 \\ 0.13 & 0.12 & 0.94 \end{pmatrix}$$

With this,

$$(I-Q)^{-1}R = \frac{1}{915}\begin{pmatrix} 990 & 210 & 120 \\ 310 & 990 & 130 \\ 130 & 120 & 940 \end{pmatrix}\begin{pmatrix} 0.4 & 0.3 \\ 0.3 & 0.3 \\ 0.3 & 0.5 \end{pmatrix}$$

The vector of the expenses of the three service departments E would be post-multiplied by $(I-Q)^{-1}R$ to get the allocation of expenses to the two production departments. Thus,

$$E = D(I-Q)^{-1}R = (25{,}000\ \ 1{,}24{,}000\ \ 77{,}000)\begin{pmatrix} 495/915 & 420/915 \\ 460/915 & 455/915 \\ 370/915 & 545/915 \end{pmatrix}$$

$$= (1{,}07{,}000\ \ 1{,}19{,}000)$$

Accordingly, the departments P_1 and P_2 would get Rs 1,07,000 and Rs 1,19,000, respectively, as the share of overhead charges of the service departments S_1, S_2 and S_3.

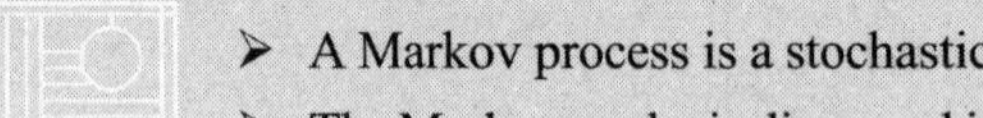

KEY POINTS TO REMEMBER

- A Markov process is a stochastic process.
- The Markov analysis discussed in this chapter is based on the assumptions that (i) there are a finite number of states; (ii) it is a first-order process where the state of the system in any given period is dependent only on its state prevailing in the previous period; (iii) the transition probabilities are stationary; and (iv) the successive time periods are uniform.
- Given the transition probabilities of moving from one state to another and the initial condition, the Markovian analysis gives the probability of the system being in a particular state at the particular time and the steady state, or equilibrium, probabilities which are independent of time. The steady state probabilities are obtained by (i) expressing the steady formulation as $Q = QP$ and obtaining the system of n simultaneous linear cqua-tions from it; (ii) solving for the steady state probabilities by solving the equations simultaneously together with the condition that sum of the probabilities is equal to one.
- Markov analysis can be extended to cover absorbing states as well. Analysis of absorbing Markov chains helps to answer questions like what would the probability of the system ending up in each absorbing state be, when the chain begins in a given transient state.

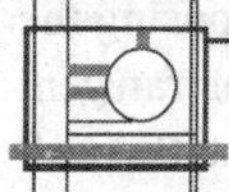

TEST YOUR UNDERSTANDING

Mark the following statements as T (*True*) or F (*False*).

1. Markov chains exemplify stochastic processes.
2. Like other operations research tools, such as linear programming, the Markovian analysis is used to determine the optimal course of action in a given problem.
3. Markovian analysis deals with those situations which involve multiple time periods.
4. In a discrete stochastic process, the state of the system can be viewed at *any* point in time.
5. Black and Scholes' formula for valuation of options is based on the assumption of viewing the share prices as a continuous-time stochastic process.
6. A transition probability matrix indicates the probabilities of inter-state moving.
7. Homogeneous Markov chains are based on the assumption of constancy of transition probabilities.
8. A sequence of transitions beginning in state i and ending in state j is called a transition from i to j.
9. If state j is reachable from state i, and i is reachable from j, then i and j are said to be communicative.
10. A state i is said to be transient if there is a way to leave it but it never returns to it.
11. An absorbing state is onc which once reached, does not allow transition to another state.
12. Two states i and j are not communicative if j is an absorbing state.
13. In a system, recurrent states are those which are not transient.
14. In a transition probability matrix, the sum of probabilities for each row and column is equal to 1.
15. A transition probability equal to 1 in a transition probability matrix indicates an absorbing state.
16. A market where only three brands of soft drinks are sold and the choice of a particular brand of soft drinks is assumed to depend on the choice in the immediately preceding time period, can be modelled as a first-order Markov process.
17. The inputs required for Markovian analysis include the transition probabilities and the initial condition in which the given system is.
18. In a market with four brands of bread, the initial condition described by the vector [0.25 0.25 0.50 0] implies that, currently, the first two brands enjoy an equal share of the market, the third brand has a share equal to the total share of the first two brands, and the fourth brand enjoys none.
19. The probability distribution of a system being in different states 1, 2, ..., n, at any given time k gives the steady state probabilities, and is expressed as $Q(k) = [q_1(k) \quad q_2(k) \quad \ldots \quad q_n(k)]$.
20. Steady state probabilities are independent of time.
21. In a Markov chain with three states, 1, 2 and 3, with given transition probability matrix, P, the probability that the system, presently in state 3, shall be in the state 1, three periods from now, would be given by element 3,1 of the matrix P^3.
22. For a Markov process with two states, 1 and 2, transition probabilities as p_{11}, p_{12}, p_{21} and p_{22}, the equilibrium probabilities q_1 and q_2 can be obtained by solving the following pair of equations simultaneously: $q_1 = p_{11}q_1 + p_{21}q_2$ and $q_2 = p_{12}q_1 + p_{22}q_2$.
23. It is true to say that in equilibrium, the state probabilities in period k, $[Q(k)]$, are the same as the state probabilities in period $k-1$, $[Q(k-1)]$.

24. The analysis of Markov chains with n states, out of which m are transient, requires decomposing the transition probability matrix into four sub-matrices: Q (of the order $n - m$ by $n - m$) matrix of transient probabilities; R (of the order $n - m$ by m) matrix of transitions from transient to absorbing states; null matrix O (of the order m by $n - m$) and an identity matrix I (of the order m by m).
25. If a chain begins in a given transient state, t_i, the probability that it shall eventually be absorbed in absorbing state a_j would be given by the ijth element of the "fundamental matrix" of the Markov chains.

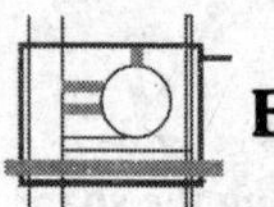

EXERCISES

1. What do you understand by Markov processes? In what areas of management can they be applied successfully?
2. What do you understand by transition probabilities? Is the assumption of stationary transition probabilities realistic, in your opinion? Why or why not?
3. Distinguish between a recurrent and an absorbing state. How can you tell in a transition matrix as to which states are recurrent and which are not?
4. Discuss the following assumptions in relation to the Markov chains:
   (i) Finite states
   (ii) First-order process
   (iii) Stationarity of transition probabilities
   (iv) Uniform time periods
5. How are steady state probabilities calculated? Do you agree that they are independent of the initial condition? Explain.
6. Explain how the probability tree helps to understand the problem of Markov processes.
7. Explain the method of calculation of ending up in each absorbing state when a chain begins in a particular transient state.
8. What is *fundamental matrix* of Markov chains? What does it calculate?
9. "The Markov analysis may be understood as the way of analysing the past and present movement of some variable in an effort to forecast the future movement of that variable." Explain this statement by taking examples from marketing and personnel. *(MBA, Delhi, April, 1998)*
10. How can Markov analysis be used by a company to
    (i) predict its manpower needs;
    (ii) allocate service department overhead to production departments; and
    (iii) analyse its accounts receivables.
11. "The Markov chain method analyses the current behaviour of a process and relates the existing characters to the future." Elucidate this statement by taking examples from functional areas of management.

(MBA, Delhi, March, 2004)

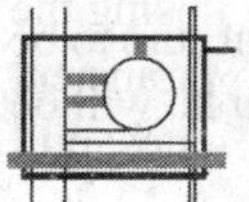

PRACTICAL PROBLEMS

1. A person plays a game in which he gains Rs 20 with a probability of 0.4 or loses Rs 10 with a probability of 0.6. He has an amount of Rs 20 with him and plays the game repeatedly until he loses all the amount he has or adds Rs 30 or Rs 40 to the initial amount. Draw up a transition probability matrix of him.
2. A small city with 1,700 customers has three shoe stores S_1, S_2 and S_3. Suppose that due to its location, there are no new customers added and the old ones also do not leave.
   A survey of the number of customers visiting each store in the past year reveals the following:

Store	*Customers as on Jan 1, Year 1*	*Change during the year*		*Customers as on Jan 1, Year 2*
		Gains	*Losses*	
S_1	800	250	350	700
S_2	400	275	200	475
S_3	500	150	125	525

   Using the following information about gain/loss summary, obtain the transition probability matrix that defines the market structure for the three stores.

Customers as on Jan 1, Year 1	*Gains from*			*Losses to*		
	S_1	S_2	S_3	S_1	S_2	S_3
S_1	0	150	100	0	250	100
S_2	250	0	25	150	0	50
S_3	100	50	0	100	25	0

3. A researcher is analysing switching between two competing products. The shares for the two products are found to be as given here:

Period	*Percentage Market Share*	
	Product A	*Product B*
1	60.0	40.0
2	64.0	36.0
3	65.6	34.4

   The researcher believes that an accurate representation of the market share in any period can be obtained by using Markov processes. Assuming his belief is correct:
   (a) Estimate the transition probability matrix.
   (b) Obtain the market shares in period 4 using the estimated transition matrix.
   (c) If the actual market shares for period 4 were 66% and 34%, would you revise your estimate of the transition matrix? Give reason.

4. The arrival behaviour of a bank employee reveals that if he is late on a day, he is 90 per cent sure to be in-time the next day. Similarly, if he is on time on a day, there is a 30 per cent chance that he will be late the next day. How often, in the long run, is he expected to be late for the office?
5. (a) The purchase patterns of two brands of toothpaste can be expressed as a Markov process with the following probabilities:

	Brand X	*Brand Y*
Brand X	0.90	0.10
Brand Y	0.05	0.95

(i) Which brand appears to have most loyal customers? Explain.
(ii) What are the projected market shares for the two brands?

(b) Suppose that in part (a) a new toothpaste brand enters the market such that the following transition probabilities exist:

	Brand X	*Brand Y*	*Brand Z*
Brand X	0.80	0.10	0.10
Brand Y	0.05	0.75	0.20
Brand Z	0.40	0.30	0.30

What are the new long run market shares? Which brand will suffer from the introduction of the new brand of toothpaste? *(MBA, Delhi, March, 2007)*

6. Consider the following transition probability matrix:

	A	*B*	*C*
A	0.80	0.12	0.08
B	0.15	0.70	0.15
C	0	0	1

Without making calculations, can you say what would the market share of each of the firms, *A*, *B* and *C* be in equilibrium? Now, verify it by calculations. Here, *C* is an absorbing state. True?

7. In a city, only two brands of cola are sold: *AA* and *BB*. If a buyer bought cola *AA* last time, there is 0.75 chance that he would buy the same cola in the next purchase. Similarly, it is known that if a buyer bought brand *BB* last time, the probability for him to buy the brand *AA* next time is 0.4.
   (a) Using this information, develop the transition probability matrix.
   (b) Calculate the following:
   (i) The probability that a customer, who is currently a cola *AA* purchaser, shall take cola *BB* in the next-to-next purchase.
   (ii) The probability that three periods from now a customer shall buy cola *AA*, given that the current market shares of the two are as follows—cola *AA* : 60%, cola *BB* : 40%.
   (iii) The respective market shares the two colas would have in the long run, assuming that the market and their preferences do not change over time.

8. A researcher is analysing brand switching between different airlines, operating on the Delhi-Mumbai route by frequent fliers. On the basis of the data collected by her, the researcher has developed the following transition probability matrix:

		To Airline		
		AA	*BB*	*CC*
	AA	0.90	0.03	0.07
From	BB	0.15	0.80	0.05
Airline	CC	0.20	0.30	0.50

It is found that currently the airlines AA, BB and CC have 20%, 50% and 30% of the market respectively.

(a) Obtain the market share for each airline in two months' time, and

(b) Calculate the long-run market share for each airline, assuming

(i) Frequent fliers make one flight in a month.

(ii) Frequent fliers make two flights in a month.

9. A salesman makes all sales in three cities X, Y and Z only. It is known that he visits each city on a weekly basis and never visits the same city in successive weeks. If he visits city X in a given week, then he visits city Z in the next week. However, if he visits city Y or Z, he is twice as likely to visit city X than the other city. Obtain the transition probability matrix. Also determine the proportionate visits by him to each of the cities in the long run.

10. Three business magazines, *Business Today, Business Line and Business Life*, compete for advertising shares in a city. At the end of last year, *Business Today* had 40 per cent of the market while its competitors *Business Line* and *Business Life* each had 30 per cent of the market. During the last year, total revenue for the three magazines was Rs 5 million and *Business Today* had a net profit of 15 per cent of its revenue. Both these figures are expected to remain the same for next year. The manager of *Business Today* has suggested a policy whereby it would retain 88 per cent of its customers while gaining 12 per cent of customers of *Business Line* and 8 per cent of customers of *Business Life*. *Business Line* is expected to retain 85 per cent of its customers while gaining 7 per cent customers of *Business Today*, and 10 per cent of *Business Life*. Finally, *Business Life* would retain 82 per cent of its customers and gain 5 and 3 per cent of *Business Today* and *Business Line*, respectively. If the above policy would cost Rs 50,000, do you think it is advisable for *Business Today* to adopt it?

11. On January 1 (this year), Bakery A had 30% of its local market share while the other two bakeries B and C had 40% and 30%, respectively, of the market share. Based upon a study by a marketing research firm, the following facts were compiled. Bakery A retains 85% of its customers while gaining 5% of competitor B's customers and 15% of C's customers. Bakery B retains 90% of its customers while gaining 8% of A's customers and 7% of C's customers. Bakery C retains 80% of its customers and gains 7% of A's customers and 5% of B's customers. What will each firm's share be on January 1, next year; January 1, next to next year; and what will each firm's market share be at equilibrium?

12. There are three firms ABC, PQR and XYZ sharing a market as 40%, 40% and 20% respectively on January 1, 2009. Over a year, the following developments take place:

ABC retains 80 per cent of its customers, loses 16 per cent to PQR and 4 per cent to XYZ.

PQR retains 84 per cent of its customers, loses 12 per cent to ABC and 4 per cent to XYZ.

XYZ retains 76 per cent of its customers, loses 18 per cent to ABC and 6 per cent to PQR.

Assuming that the market does not change,

(a) what share of the market shall be held by each firm on January 1, 2011?

(b) what would the long-run shares of the firms at equilibrium be if the buying habits do not change?

13. A company is using Markov theory to analyse brand switching between three different brands of DVDs. Data has been obtained and used to estimate the following transition matrix for the probability of moving between brands each month:

		To Brand		
		X	*Y*	*Z*
From	X	0.80	0.10	0.10
Brand	Y	0.03	0.95	0.02
	Z	0.20	0.05	0.75

The current (i.e. month 1) market shares are 45%, 25% and 30% for brands X, Y and Z respectively.

(a) Obtain the expected market shares after two months have elapsed.

(b) What market shares can the three brands expect in the long run?

(c) Would you expect the actual market shares to approach the expected long-run prediction for the market or not? Why?

14. ABC Co. had 45% of the local market for its cosmetics, while the two other manufacturers XYZ Co. and PQR Co. have 30% and 25% shares, respectively, in the local market, as on 1st January of this year. A study by a market research firm has disclosed the following: every year, ABC retains 80% of its customers, while it gains 3% and 6% of the customers from its two competitors, XYZ and PQR, respectively. XYZ retains 90% of its customers, and gains 14% and 9% of customers respectively from ABC and PQR. PQR retains 85% of its customers and gains 7% and 6% of customers from XYZ and ABC respectively.

What will be the share of each firm on January 1 next year? What will be the market share of each firm at equilibrium?

15. A transport company with a large fleet of buses periodically inspects the bearings on its buses and categorises them in one of the four states: (1) Good Bearings, (2) Lightly-worn Bearings, (3) Moderately-worn Bearings and (4) Worn-out Bearings. Replacement of bearings during inspection costs the company Rs 250. In case, however, a bus develops bad bearings during operations, the company estimates the cost to be Rs 1,200. On the basis of past experience, it has been found that a set of bearings which is good (i.e. new) at a given inspection has a 0.85 probability of being in state 2 and a 0.15 probability of being in state 3 by the next inspection. The bearings which are classified as state 2 in an inspection are likely to be remaining in the same state with a probability of 0.5, and the likelihood of them going to states 3 and 4 by the next inspection is 0.3 and 0.2, respectively. A set of bearings classified as state 3 has a probability of 0.6 of remaining in the state 3 and a 0.4 probability of wearing out by the next inspection.

The company is considering the adoption of one of the following policies:

(i) Replace all bearings in state 2 or worse at each inspection. (ii) Replace all bearings in state 3 or worse at each inspection. Which of the policies is more economical?

16. Formulate alternative Markov chain models for two policies for repairing a piece of equipment that must be finely adjusted. The equipment can be in any of three conditions: running (or just repaired), running fairly well, and running poorly. If running well, it will continue in that condition with prob-

ability 0.7, will be running fairly well next week with probability 0.2 and will be running poorly with probability 0.1 If running fairly well, it will continue in the same condition next week with probability 0.6 and will deteriorate with probability 0.4. Under policy 1, the equipment is repaired whenever it starts a week running poorly. It takes a whole week for repairs. Under policy 2, the equipment is repaired whenever it starts a week running fairly well or poor and takes a whole week of repair.

Also determine what percentage of time does the machine spend 'down' under each policy.

17. A manufacturing company has a certain piece of equipment that is inspected at the end of each day and classified as just overhauled, good, fair or inoperative. If the piece is inoperative, it is overhauled, a procedure that takes one day. Assume that the working condition of the equipment follows a Markov process with the following transition matrix:

	From state	*To state* 1	2	3	4
	1	0	3/4	1/4	0
	2	0	1/2	1/2	0
$P =$	3	0	0	1/2	1/2
	4	1	0	0	0

It costs Rs 125 to overhaul a machine (including lost time) on the average and Rs 75 in production is lost if a machine is found inoperative. Using the steady state probabilities, compute the expected per day cost of maintenance. *(MBA, Delhi, March, 2003)*

18. ABC company has three service departments, S_1, S_2 and S_3, and two production departments P_1 and P_2. Overhead is allocated to the production departments for inclusion in the stock valuation. The analysis of benefits received by each department during the last quarter and the overhead expense incurred by each department were as follows:

Service departments	*Percentages to be allocated to departments* S_1	S_2	S_3	P_1	P_2
S_1	0	25	5	40	30
S_2	10	0	25	35	30
S_3	30	15	0	15	40
Direct overhead expense (Rs '000)	60	25.5	60.5	26	84

You are required to formulate this problem as absorbing chains and determine the total overhead to be allocated from each of S_1, S_2 and S_3 to the production departments.

19. A company has two production departments, P_1 and P_2 and three service departments, S_1, S_2 and S_3. The direct cost allocated to each of the departments and percentage of total cost of each service department apportioned to various departments are given here:

Department	Direct cost (Rs)	Percentage allocation of total cost		
		S_1	S_2	S_3
P_1	60,000	40	35	25
P_2	74,000	20	40	20
S_1	6,000	0	20	35
S_2	8,000	15	0	20
S_3	68,500	25	5	0

Draft this problem as absorbing chains and determine the total cost (allocated and apportioned) for each production department.

20. **A holding company H has two subsidiaries S_1 and S_2. The subsidiaries also own a portion of the share capital of H. The percentage ownership of the group is as given below:**

Owners of shares	Percentage of share capital held in company		
	H	S_1	S_2
H	0	60	80
S_1	3	0	10
S_2	6	10	0
Outside shareholders	91	30	10
Total	100	100	100

If the separately earned net profits of H, S_1 and S_2 are Rs 30,000, Rs 17,500 and Rs 5,000, respectively, find the total net profit of each company (separately earned profit *plus* share in the profits of other companies) using Markov chains analysis. Also, verify that the sum of the profits allocated to the outside shareholders equals the sum of the separately earned profits of all the companies.

21. **A credit card company is attempting to determine a more effective set of credit control policies. It has traditionally classified all its accounts receivables into one of the five categories listed below:**

Accounts receivable category	Status of accounts receivable
1	0–30 days late
2	31–90 days late
3	91 + days late
4	paid in full
5	defaulted

Based on the historical data, the company has derived the following transition matrix describing the behaviour of various categories on a weekly basis.

From category	To category				
	1	2	3	4	5
1	0.4	0.2	0.1	0.2	0.1
2	0.3	0.4	0.1	0.1	0.1
3	0.2	0.4	0.1	0.1	0.2
4	0.0	0.0	0.0	1.0	0.0
5	0.0	0.0	0.0	0.0	1.0

Currently, the company has the following accounts receivables in various categories:

Category	*Amount*
1	Rs 2 million
2	Rs 4 million
3	Rs 3 million

You are required to calculate the expected amounts which may eventually be (i) collected, and (ii) defaulted.

CHAPTER 15

Theory of Games

CHAPTER OVERVIEW

There are several situations in which the outcome of a particular course of action is dependent upon the action taken by another party. Examples include a chess game being played by two players, a war between two countries, competing firms operating in a market, and the management and workers' union negotiations, etc. In such cases, the parties involved are the players who have conflicting interests so that a gain to one party is a loss to the other, with each of them having a certain number of strategies to choose from, and the outcome depends upon the particular pair of strategies chosen by the parties. All these ingredients make it a game situation, hence the name *Theory of Games*. The theory helps to determine the best decision for each of the parties involved.

An understanding of the theory helps the manager to answer questions like:

- What are the consequences of the interplay of each combination of strategies of both the players?
- Is there one (or more) strategy for each of the players that are clearly the best one to play in the given situation? In the language of the theory, does the game have a *saddle point* so that players play pure strategies?
- If there is no clear cut strategy for each player, then what combination of strategies should be played by each one? Thus, when no saddle point exists, what is the optimal mix of the strategies for the players? And what is the expected pay-off resulting from the game?
- Is a particular strategy or a combination of strategies for a player superior to another strategy, so that the dominated strategy can be eliminated? By this process, if the game reduces to a small size, then how to solve it analytically?
- If the game size cannot be reduced to a small size, then how to formulate the game as a linear programming problem and solve accordingly? How to use the primal-dual relationship to obtain the answers?

For this chapter, you should know the concept and calculation of expected value and an application of simplex method to solve linear programming problems, together with the primal-dual connection. The ideas and concepts to focus on in this chapter include developing a pay-off matrix using the given information, obtaining a saddle point if it exists, the reduction of game size by the principle of dominance, and formulation of a game as an LPP (especially when some negative pay-offs are involved) and its solution by simplex algorithm.

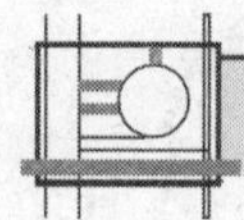

15.1 INTRODUCTION

Game theory is a body of knowledge which is concerned with the study of decision-making in situations where two or more rational opponents are involved under conditions of competition and conflicting interests. It deals with human processes in which an individual decision making unit who can be an individual, a group, a formal or informal organisation, or a society, is not in complete control of the other decision-making units, the opponent(s), and is addressed to problems involving conflict, co-operation or both at various levels.

The main objective in the theory of games is to determine the rules of rational behaviour in the game situations, in which the outcomes are dependent on the actions of the interdependent players. A game refers to a situation in which two or more players are competing. It involves the players (the decision-makers) who have different goals or objectives and whose fates are intertwined. They are in a situation in which there may be a number of possible outcomes with different values to them. Although they might have some control that would influence the outcome, they do not have complete control over others. Labour unions striking against the company management, players in a chess game, army generals engaged in fighting the enemy, firm striving for larger share of market in duopolistic market conditions and so on, all illustrate situations which can be viewed as games.

In a game situation, each of the players has a set of *strategies* available. A strategy refers to the action to be taken by a player in various contingencies in playing a game. There is a set of outcomes each of which is the result of the particular choices of strategies made by the players on a given play of the game, and *pay-offs* are accorded to each player in each of the possible outcomes. Each of the players being assumed to be rational, his preference ordering of the different outcomes is determined by the order of magnitudes of the associated pay-offs, and since, in general, the orders of magnitudes of the pay-offs accruing to the players in different outcomes do not coincide, a game models a situation in which there are conflicts of interests.

The players in the game strive for *optimal strategies*. An optimal strategy is such as provides the best situation in the game in the sense that it involves maximal pay-off to the players.

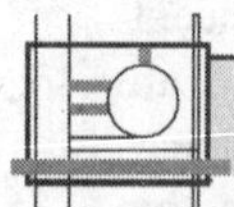

15.2 GAME MODELS

There are several game theory models which can be classified on the basis of factors like the number of players involved, the sum of gains or losses, and the number of strategies employed in the game.

In a game situation, interests of two or more than two participants may be in conflict. The former of these is called a *two-person game* and the latter one is known as a *n-person game*, where it does not necessarily imply that in its play exactly *n* people would be involved but rather that the participants can be classified into *n* mutually exclusive categories and members of each of the categories have identical interests. Participants may form coalitions and in such cases one of the objectives is to determine the reasonable coalitions.

In a game, if the sum of gains and losses is equal to zero, it is called *zero-sum* or *constant-sum* game. For example, if two chess players agree that at the end of the game the loser would pay Rs 50 to the winner, then it would mean a zero-sum: a gain of one exactly matches the loss of the other. If the sum of gains and loses is not equal to zero, it would obviously be called a *non-zero-sum* game.

The games can also be classified on the basis of the number of strategies. A game is said to be *finite* if each player has the option of choosing from only a finite number of strategies otherwise it is called *infinite*.

In our analysis, we shall deal with two-person zero-sum games involving finite strategies open to the players.

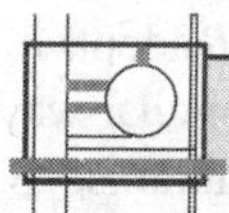

15.3 TWO-PERSON ZERO-SUM GAMES AND THEIR SOLUTION

As stated earlier, a two-person zero-sum game is the one which involves two persons (players) and where the gain made by one equals the loss incurred by the other. To illustrate, suppose that there are two firms *A* and *B* in an area which, for a long period in the past, have been selling a competing product and are now engaged in struggle for a larger share of the market. Now with the total market of a given size, any share of the market gained by one firm must be lost by the other, and, therefore, the sum of the gains and losses equals zero.

Let us assume that both these firms are considering the same three strategies in a bid to gain the share in the market; low advertising, high advertising, and quality improvement. We assume also that currently they are sharing the market equally and further that each of the firms can employ only one of the strategies at a time.

Under the conditions postulated, there are a total of $3 \times 3 = 9$ combinations of the moves possible—thus, low advertising by firm *A* may be accompanied by low advertising, high advertising or quality improvement by firm *B* and so on for other strategies. Each pair of the moves shall affect the sharing of the market in a particular way. For example, low advertising both by firm *A* and by firm *B* shall lead to 12 points (implying 12 per cent of the total market) in favour of firm *A*, while low advertising by *A* and high advertising by *B* would lead to a shift of 8 points in favour of *B*. Similarly, there are pay-offs corresponding to other pairs of moves. These are shown in the pay-off matrix that follows. The strategies of low advertising, high advertising and quality improvement have been marked as a_1, a_2, and a_3, respectively for the firm *A*, and b_1, b_2 and b_3, respectively for the firm *B*.

		B's Strategy		
		b_1	b_2	b_3
	a_1	12	–8	–2
A's Strategy	a_2	6	7	3
	a_3	–10	–6	2

This pay-off matrix is drawn from *A*'s point of view—a positive pay-off indicates that the firm *A* has gained the market share at the expense of firm *B* while negative pay-offs imply *B*'s gain at *A*'s expense.

Now the problem before us is to determine the best strategy for *A* and *B*, assuming that they both are acquainted with the information contained in the pay-off matrix and that each one is not aware of the move the other is likely to take. The conservative approach to the selection of the best strategy would call for a cautious attitude of assuming the worst and acting accordingly. In reference to the pay-off matrix, if firm *A* employs strategy a_1, it would expect the firm *B* to employ strategy b_2, thereby reducing *A*'s pay-off from the strategy a_1 to its minimum value, equal to –8, representing a loss to firm *A*. If firm *A* employs strategy a_2, it would assume the competing firm would employ strategy b_3 that would give *A* a 3 per cent increase in the share of market. Similarly, corresponding to its strategy a_3, firm *A* would expect the firm *B* to use b_1, which puts the firm *A* at a loss of 10 points.

The firm *A* would like to make the best use of the situation by choosing the maximum of these minimal pay-offs. Obviously, among the three alternate strategies open to it, firm *A* would choose the one which would ensure it the highest of the minimum rewards associated with these. This decision rule of selection is called the *maximin strategy*.* Since the minimal pay-offs corresponding to a_1, a_2, and a_3 are, respectively, – 8, 3, and – 10, firm *A* would select a_2 as its strategy.

* See also Chapter 13.

Similarly, the firm *B* would also employ a cautious approach in choosing its strategy. When firm *B* adopts b_1, it can expect firm *A* to employ a_1 which would ensure it (firm *A*) the maximum advantage. In a similar way, adoption of b_2 and b_3 would, in firm *B*'s expectation, be associated with employment of a_2. To minimise the advantage accruing to *A*, the firm would select a strategy that would yield the least advantage to its competitor—the strategy b_3. The decision of *B* is said to be in accordance with the *minimax strategy*.

It may be noted that corresponding to maximin rule of firm *A* and the minimax rule of firm *B*, the pay-off is 3. This amount is the worth, or the *value*, of the game, *V*, and represents the final pay-off to the winning player by the losing player. Since the pay-off matrix is drawn from *A*'s point of view, if the value of the game is positive, as is in the present case, it is favourable to *A* while if the game value is negative, the game is favourable to *B*. The game is said to be *fair* or equitable if its value is equal to zero, for then it favours none.

15.3.1 Saddle Point

From the preceding discussion on the choice of strategies by the players, it is clear that maximin strategy a_2 of *A* and the minimax strategy, b_3 of *B*, both lead to the same pay-off. Thus, the pay-off resulting from the adoption of maximin strategy by *A* is exactly what *B* would expect *A* to receive if it (*B*) adopted the minimax strategy. We may argue why *A* should not employ strategy a_1 which can give it a 12 per cent increase in the market share? The reason is simple: in that event *B* shall counter by adopting the strategy b_2 which would put *A* to a loss of 8 points or the strategy b_3 involving a loss of 2 points to *A*. In a similar way, if we argue that *B* should adopt the strategy b_1 that could give it a gain in the market share. This would, however, happen when *A* adopts the strategy a_3, and not otherwise. If *A* responds by choosing a_1 or a_2, it would cause a shift of, respectively, 12 or 6 points in its favour. We can similarly evaluate the other strategies available to each of them only to establish that the strategies a_2 and b_3 adopted by *A* and *B* are the best for each one. There is no reason for them to deviate. By adopting the maximin strategy, *A* can stop *B* from lowering its (*A*'s) gain in the share of the market below 3 per cent. Similarly, by adopting the minimax strategy b_3, *B* can prevent *A* from taking any more share than 3 per cent. The situation, therefore, is one of equilibrium. The maximin value is equal to the minimax value and the game is solved. The point of equilibrium is known as the *saddle point*.

Operationally, to obtain the saddle point, if it exists, we determine the minimum pay-off value for each row and also the maximum pay-off value for each of the columns. Now, in case the largest of the row minima is equal to the smallest of the column maxima, then it represents the saddle point. For our illustration, the row minima and column maxima are shown in the following table.

		B's Strategy			
		b_1	b_2	b_3	Row Minima
	a_1	12	−8	−2	−8
A's Strategy	a_2	6	7	3	3 *
	a_3	−10	−6	2	−10
Column Maxima		12	7	3*	

Here 3 represents the saddle point.

It is also possible that there are more than one saddle points for a given problem. Consider, for instance, the pay-off matrix given in the following table:

		B's Strategy				
		b_1	b_2	b_3	b_4	Row Minima
	a_1	4	–16	14	–15	–16
A's Strategy	a_2	–6	7	–4	–6	–6*
	a_3	6	–2	0	–6	–6*
Column Maxima		6	7	14	–6 *	

In relation to *B*'s minimax strategy, b_4 in the present case, *A* could employ either a_2 or a_3, each one of which represents a maximin strategy for him. Since the pay-off corresponding to *B*'s minimax strategy and *A*'s either maximin strategies, is identical, there are two saddle points, represented by a_2b_4 and a_3b_4. The value of game in the present case is –6, a net loss of 6 points to *A* and an equivalent profit to *B*.

Example 15.1 Two leading firms, Nirmala Textiles Ltd. and Swati Rayons Ltd., for years have been selling shirting, which is but a small part of both firms' total sales. The Marketing Director of Nirmala Textiles raised the question, "What should the firm's strategies be in terms of advertising for the product in question?" The system group of Nirmala Textiles developed the following data for varying degrees of advertising:

(a) No advertising, medium advertising and heavy advertising for both firms will result in equal market share.
(b) Nirmala Textiles with no advertising: 40 per cent of the market with medium advertising by Swati Rayons and 28 percent of the market with heavy advertising by Swati Rayons.
(c) Nirmala Textiles using medium advertising: 70 per cent of the market with no advertising by Swati Rayons and 45 per cent of the market with heavy advertising by Swati Rayons.
(d) Nirmala Textiles using heavy advertising: 75 per cent of the market with no advertising by Swati Rayons and 52.5 per cent of the market with medium advertising by Swati Rayons.

Based upon the above information, answer the marketing director's question. *(MBA, Delhi, 1985)*

The pay-off matrix from the viewpoint of Nirmala Textiles Ltd., showing its market share under several combinations of the strategies, is given below. Also, row minima and column maxima have been obtained to see if saddle point exists.

			Swati Rayons Ltd.'s Strategy			
			No Advt.	*Med. Advt.*	*Hvy. Advt*	
			b_1	b_2	b_3	Row Minima
Nirmala Textiles Ltd's Strategy	No. Advt.	a_1	50	40	28	28
	Med. Advt.	a_2	70	50	45	45
	Hvy. Advt.	a_3	75	52.5	50	50*
	Column Maxima		75	52.5	50*	

We observe that saddle point exists at the intersection of a_3 and b_3. Thus, the optimal strategy for each one is to engage in heavy advertising and it will result in an even distribution of the market between the firms. Nirmala Textiles Ltd's marketing director should, therefore, resort to heavy advertising.

15.3.2 When No Saddle Point Exists

We have considered game situations which have a saddle point(s) and observed that in such cases we are provided with an adequate theory of how best to play the game. It is possible that there is no saddle point of a game and so it is not possible to find its solution in terms of the *pure strategies*—the maximin and the minimax. Games without saddle points are not strictly determined. The solution to such problems calls for employing *mixed strategies*.* A mixed strategy represents a combination of two or more strategies that are selected one at a time, according to pre-determined probabilities. Thus, in employing a mixed strategy, a player decides to mix his choices among several alternatives in a certain ratio. Now we shall discuss the determination of mixed strategy—first for the 2 × 2 games and then for problems of higher order. An example follows.

Example 15.2 The following is the pay-off matrix of a game being played by *A* and *B*. Determine the optimal strategies for the players and the value of the game.

		B's Strategy	
		b_1	b_2
A's Strategy	a_1	8	–7
	a_2	–6	4

This given problem does not have a saddle point. Therefore, the method discussed before does not suffice to enable us to determine optimal ways for *A* and *B* to play. If *A* plays a_1, then *B* would play b_2 while for a_2 played by *A*, *B* would choose to play b_1. So if *B* knows what choice *A* will make then *B* can ensure that he gains by choosing a strategy opposite to the one desired by *A*. Thus, it is of utmost importance for *A* to make it difficult for *B* to guess as to what choice he is going to make. Similarly, *B* would like to make it very difficult for *A* to assess the strategy he is likely to adopt. It would pay each one of them to play either of the respective strategies open to each with certain probability.

Now suppose that *A* plays strategy a_1 with probability x and plays strategy a_2 with probability $1 - x$. If *B* plays strategy b_1, then *A*'s expected pay-off can be determined in reference to the figures given in the first column of the payoff matrix as follows:

$$\text{Expected pay-off (given that } B \text{ plays } b_1) = 8x - 6(1 - x)$$

Similarly, if *B* plays strategy b_2, the expected payoff of *A* can be determined as follows:

$$\text{Expected pay-off (given that } B \text{ plays } b_2) = -7x + 4(1 - x)$$

Now we shall determine a value x so that the expected pay-off for *A* is the same irrespective of the strategy adopted by *B*. This value can be obtained by equating these two equations.

Thus, $\quad 8x - 6(1 - x) = -7x + 4(1 - x)$

or $\quad 8x - 6 + 6x = -7x + 4 - 4x$

or $\quad x = 10/25 = 2/5.$

A would do best to adopt the strategies a_1 and a_2, choosing in a random manner, in the proportion 2 : 3 (i.e. 2/5 and 3/5). The expected pay-off for *A* using this mixed strategy equals

* It may therefore be noted that all games of any size and irrespective of whether they have saddle points or not are solvable.

$$8 \times (2/5) - 6(3/5) = -2/5$$

or
$$-7(2/5) + 4(3/5) = -2/5$$

Thus, he shall net a loss of 2/5 per play in the long run.

We can determine mixed strategy for B in a similar manner as for A. Thus, if he plays strategy b_1 with probability y and strategy b_2 with probability $1 - y$, we have

$$\text{Expected pay-off (given that } A \text{ plays } a_1) = 8y - 7(1 - y),$$

and
$$\text{Expected pay-off (given that } A \text{ plays } a_2) = -6y + 4(1 - y)$$

We can determine the value of y as will ensure equal pay-off irrespective of the strategy of A, as follows:

$$8y - 7(1 - y) = -6y + 4(1 - y)$$

or
$$8y - 7 + 7y = -6y + 4 - 4y$$

or
$$y = 11/25$$

Thus, B should play strategies b_1 and b_2 in the ratio of 11 : 14 in a random manner.

B's expected pay-off (loss) per play shall be:

$$8(11/25) - 7(14/25) = -10/25 = -2/5$$

or
$$-6(11/25) + 4(14/25) = -10/25 = -2/5$$

It implies that B shall gain 2/5 per play in the long run.

Thus, we conclude that A and B should both use mixed strategies as given here and the value of game equals –2/5.

	Strategy	*Probability*
For A,	a_1	2/5
	a_2	3/5
For B,	b_1	11/25
	b_2	14/25

In general, for a zero-sum two-persons game in which each of the players, say A and B, have strategies a_1 & a_2 and b_1 & b_2 respectively, and the pay-offs as given below then, if x is the probability with which A chooses strategy a_1 and if y is the probability that B plays strategy b_1, we have,

		B's Strategy	
		b_1	b_2
	a_1	a_{11}	a_{12}
A's Strategy	a_2	a_{21}	a_{22}

$$x = \frac{a_{22} - a_{21}}{(a_{11} + a_{22}) - (a_{12} + a_{21})}; \quad y = \frac{a_{22} - a_{12}}{(a_{11} + a_{22}) - (a_{12} + a_{21})};$$

and
$$V = \frac{a_{11}\,a_{22} - a_{12}\,a_{21}}{(a_{11} + a_{22}) - (a_{12} + a_{21})}.$$

For the Example 15.2, $a_{11} = 8$, $a_{12} = -7$, $a_{21} = -6$, and $a_{22} = 4$. Substituting these values in the formulae, we get

$$x = \frac{4-(-6)}{(8+4)-(-7-6)} = \frac{10}{25} = \frac{2}{5};$$

$$y = \frac{4-(-7)}{(8+4)-(-7-6)} = \frac{11}{25};$$

$$V = \frac{8 \times 4 - (-7)(-6)}{(8+4) - (-7-6)} = \frac{-10}{25} = \frac{-2}{5}$$

These values match with the ones already obtained.

15.3.3 Dominance Rule

In a game, sometimes a strategy available to a player might be found to be preferable to some other strategy/strategies. Such a strategy is said to *dominate* the other one(s). This concept of domination is very usefully employed in simplifying the games and thus helps in finding solutions to the games. Let us consider the pay-off matrix in respect of our first illustration of two firms *A* and *B*, which is reproduced hereafter.

		B's Strategy		
		b_1	b_2	b_3
	a_1	12	–8	–2
A's Strategy	a_2	6	7	3
	a_3	–10	–6	2

Consider the strategies open to firm *A*. We notice that every element of the second row exceeds the corresponding element of the third row. Clearly, *A* shall never prefer to play a_3, because, in comparison to this strategy, it shall be better off by adopting a_2, regardless of what strategy is adopted by *B*. Thus, a_3 is dominated by a_2 and hence, it can be deleted.

		B's Strategy		
		b_1	b_2	b_3
A's Strategy	a_1	12	–8	–2
	a_2	6	7	3

From the reduced matrix, we observe that the first column values are larger than their counterparts in the third column. Since *B* would like to minimise the pay-offs to *A*, it would always prefer to choose b_3 to b_1. Thus, strategy b_1 is dominated by strategy b_3. Now, deleting b_1, we have,

		B's Strategy	
		b_2	b_3
A's Strategy	a_1	–8	–2
	a_2	7	3

The second row values in this matrix are larger than the corresponding values of the first one and, therefore, a_2 dominates a_1. Further, deletion of a_1 yields only a_2 open to firm A. Since firm B loses more by choosing b_2, it would adopt b_3. Thus, the strategies are a_2 and b_3, and a value 3 as the solution to this reduced 1×1 game. This is also the solution to the original game, obtained earlier.

Thus, by using the dominance rule on one player's strategies we can get a reduced game. Then we can use domination on the strategies of the other player to further reduce the game, if possible. In this way, moving forward and backward we reduce the game as much as possible. If saddle point of the game exists, then naturally the game shall be reduced to a 1×1 game—giving the same solution, as given by the saddle point. When the saddle point does not exist then, if after the application of dominance rule, a game becomes of the order 2×2, then the mixed strategies can be determined by using the analytical method discussed in the preceding section. And, when after the domination rule is applied, the game is reduced to a $2 \times n$ or $m \times 2$ size, it can be handled graphically, as discussed later, and when the reduced game has more than two strategies available to each of the players, then it can be formulated as a linear programming problem. It may be mentioned here that all the two-person zero-sum games can be directly transformed as LPP and solved using simplex algorithm. The only advantage in applying domination rule is that the game could be reduced in size and hence solved in an easier way. In any event, whenever a game is reduced in size by applying the dominance rule then the solution of the reduced game is also the solution of the original game.

The schematic presentation of the solution procedure for two-person zero-sum game is given in Figure 15.1.

Example 15.3 Find the optimal strategies for A and B in the following game. Also obtain the value of the game.

		B's Strategy		
		b_1	b_2	b_3
A's Strategy	a_1	9	8	–7
	a_2	3	–6	4
	a_3	6	7	–7

The pay-offs in relation to a_1 are equal to or greater than the corresponding pay-offs from a_3. Thus, a_1 dominates a_3. Deleting a_3, we get,

		B's Strategy		
		b_1	b_2	b_3
A's Strategy	a_1	9	8	–7
	a_2	3	–6	4

Now, we observe that strategy b_1 is dominated by b_2. Thus, we delete b_1. This gives the following matrix:

		B's Strategy	
		b_2	b_3
A's Strategy	a_1	8	–7
	a_2	–6	4

This problem is now exactly the same as that in Example 15.2. It has, therefore, the following solution:

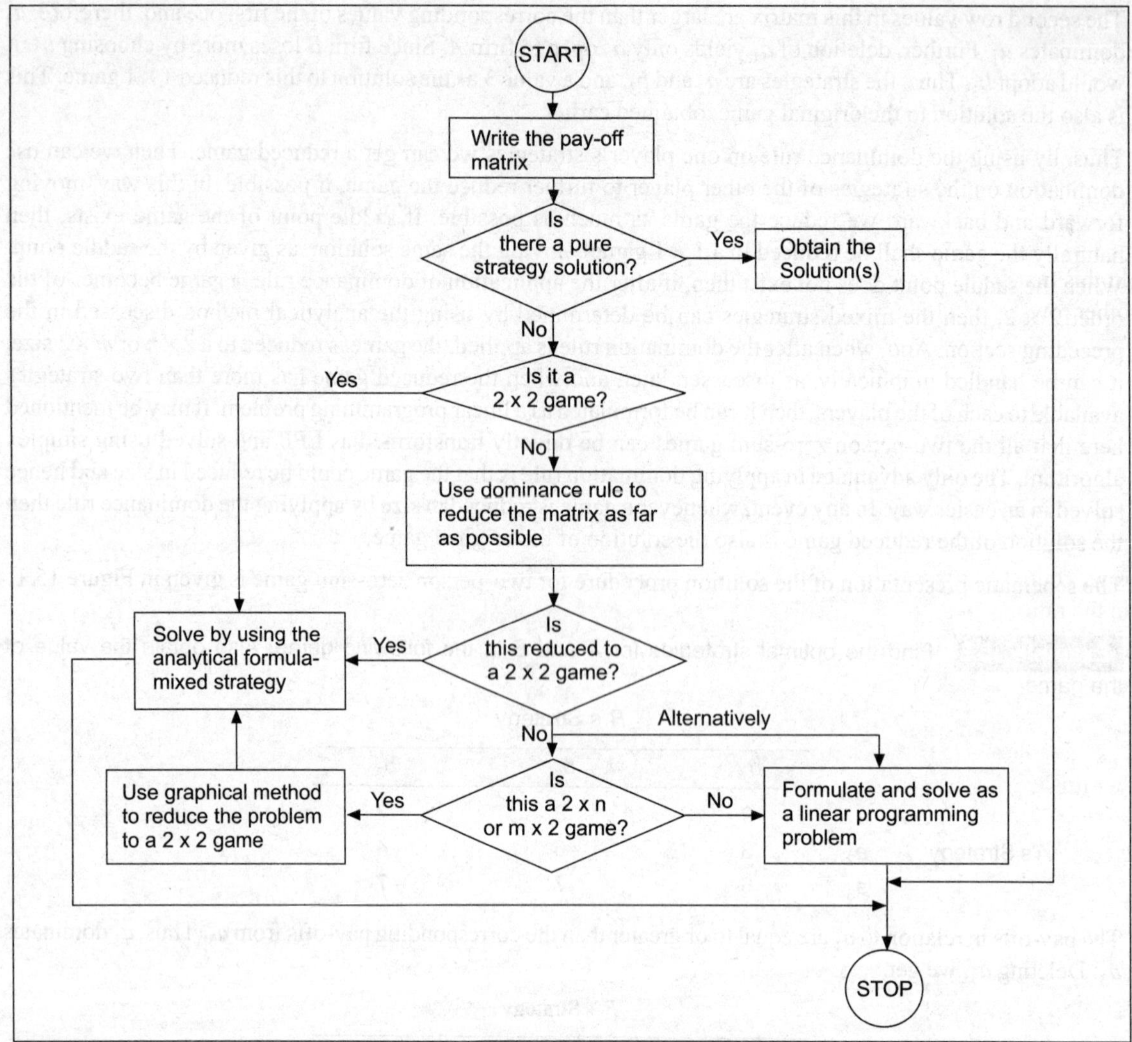

Figure 15.1 *Schematic Presentation of the Method of Solution of Two-Person Zero-Sum Games*

For *A*:	a_1	2/5	For *B*:	b_1	0	Value of the game, $V = -2/5$
	a_2	3/5		b_2	11/25	
	a_3	0		b_3	14/25	

Before concluding our discussion on dominance, it may be mentioned that we have only looked at domination of a pure strategy by a pure strategy. Also, if a strategy is dominated by a convex linear combination of two strategies, then it can be ruled out.* Consider the following example.

---

*To illustrate, if R_1 and R_2 be the two rows, their convex linear combination implies the following: $\alpha_1 R_1 + \alpha_2 R_2 = R$ such that $\alpha_1 + \alpha_2 = 1$. Here the coefficients α_1 and α_2 are non-negative fractional values, and R is the row resulting from the combination of R_1 and R_2.
In general terms, if $V_1, V_2, \ldots V_n$ are n vectors, their convex linear combination means that $\alpha_1 V_1 + \alpha_2 V_2 + \ldots + \alpha_n V_n$ equals a vector V, subject to the condition that $\alpha_1 + \alpha_2 + \ldots + \alpha_n$ is equal to 1, and further that $0 \le \alpha_i \le 1$.

Example 15.4 Solve the following game:

		B's Strategy	
		B_1	B_2
	A_1	28	0
A's Strategy	A_2	2	12
	A_3	4	7

In this matrix, the elements in the third row are not smaller than the corresponding elements of either of the other rows. However, we notice that

$$4 < \frac{1}{4} \times 28 + \frac{3}{4} \times 2, \text{ and}$$

$$7 < \frac{1}{4} \times 0 + \frac{3}{4} \times 12$$

Therefore, it seems that A would never play A_3 because he would always do better by dividing between the first two rows, representing A_1 and A_2 (in the ratio 1:3) any probability that he might consider assigning to the third row. Thus, the third row is dominated by the combination of the first and second rows (assigning a weightage in the ratio 1 to 3 respectively to them). The problem is reduced to the following:

	B_1	B_2
A_1	28	0
A_2	2	12

For this game,

$$x = \frac{a_{22} - a_{21}}{(a_{11} + a_{22}) - (a_{12} + a_{21})} = \frac{12 - 2}{(28 + 12) - (0 + 2)} = \frac{10}{38} = \frac{5}{19};$$

$$y = \frac{a_{22} - a_{12}}{(a_{11} + a_{22}) - (a_{12} + a_{21})} = \frac{12 - 0}{(28 + 12) - (0 + 2)} = \frac{12}{38} = \frac{6}{19}; \text{ and}$$

$$V = \frac{a_{11} a_{22} - a_{12} a_{21}}{(a_{11} + a_{22}) - (a_{12} + a_{21})} = \frac{28 \times 12 - 0 \times 2}{(28 + 12) - (0 + 2)} = \frac{336}{38} = \frac{168}{19}$$

Thus, optimal strategy for A is (5/19, 14/19, 0); for B it is (6/19, 13/19) and the game value is 168/19.

15.4 SOLUTION OF 2 × *n* AND *m* × 2 GAMES

Now we shall consider the solution of games where either of the players has only two strategies available. When the player A, for example, has only 2 strategies to choose from and the player B has n, the game shall be of the order $2 \times n$, whereas in case B has only two strategies available to him and A has m strategies, the game shall be a $m \times 2$ game.

The problem may originally be a $2 \times n$ or a $m \times 2$ game or a problem might have been reduced to such size after applying the dominance rule. In either case, we can use graphical method to solve the problem. By using the graphical approach, it is aimed to reduce a game to the order of 2×2 by identifying and eliminating the dominated strategies, and then solve it by the analytical method used for solving such games. The resultant solution is also the solution to the original problem. Although the game value and the optimal strategy can be read off from the graph, we generally adopt the analytical method (for 2×2 games) to get the answer.

We shall illustrate the solution of the $2 \times n$ and $m \times 2$ games in turn with the help of the following examples.

Example 15.5 Solve the following game using graphical approach:

		B's Strategy			
		b_1	b_2	b_3	b_4
A's Strategy	a_1	8	5	−7	9
	a_2	−6	6	4	−2

Here A has two strategies a_1 and a_2 which, suppose, he plays with probabilities x and $1 - x$ respectively. When B chooses to play b_1, the expected payoff for A shall be $8x + (-6)(1 - x)$ or $14x - 6$. Similarly, the expected pay-off functions in respect of b_2, b_3 and b_4 can be derived as being $6 - x$; $4 - 11x$; and $11x - 2$, respectively. We can represent these by graphically plotting each pay-off as a function of x. This is shown in Figure 15.2.

The lines are marked b_1, b_2, b_3 and b_4 and they represent the respective strategies. For each value of x, the height of the lines at that point denotes the pay-offs of each of B's strategies against $(x, 1 - x)$ for A. A is concerned with his least pay-off when he plays a particular strategy, which is represented by the lowest of the four lines at that point, and wishes to choose x so as to maximise this minimum pay-off. This is at

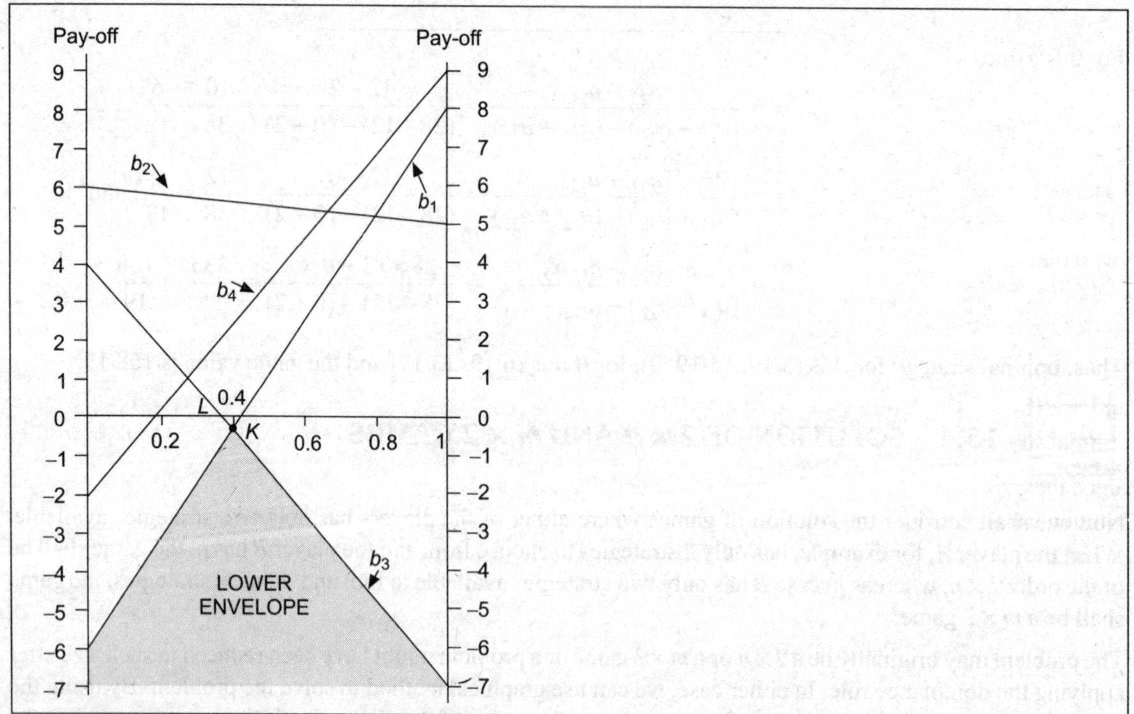

Figure 15.2 *Graphic Solution to the Game*

K in the figure where the lower envelope (represented by the shaded region), the lowest of the lines at point, is the highest. This point lies at the intersection of the lines representing strategies b_1 and b_3. The distance $KL = -0.4$ (or $-2/5$) in the figure represents the game value, V, and $x = OL(= 0.4$ or $2/5)$ is the optimal strategy for A.

Alternatively, the game can be written as a 2×2 game as follows, with strategies a_1 and a_2 for A, and b_1 and b_3 for B.

	b_1	b_3
a_1	8	−7
a_2	−6	4

This problem is the same as Example 15.2, and has

$$x = \frac{4-(-6)}{(8+4)-(-7-6)} = \frac{2}{5}, \quad y = \frac{4-(-7)}{(8+4)-(-7-6)} = \frac{11}{25}, \text{ and}$$

$$V = \frac{8 \times 4 - (-7)(-6)}{(8+4)-(-7-6)} = \frac{-2}{5}$$

Thus, optimal strategy for A is (2/5, 3/5) and for B it is (11/25, 0, 14/25, 0).

Example 15.6 Solve the following game using the graphical method:

		B's Strategy	
		b_1	b_2
	a_1	−7	6
A's Strategy	a_2	7	−4
	a_3	−4	−2
	a_4	8	−6

Let B play the strategies b_1 and b_2 with respective probabilities y and $1-y$, the expected pay-off for which, when A chooses to play a_1, shall be $-7y + 6(1-y)$ or $-13y + 6$. Similarly, pay-offs in respect of other strategy plays can be determined. These are presented graphically in Figure 15.3, where each pay-off is plotted as function of y.

Here we shall be concerned with the upper envelope which is formed by the lines representing strategies a_1, a_2 and a_4. The value of the game is determined by the lowest point, P, on this convex set. We can obtain the optimal strategies and the game value algebraically as follows.

Since the point P is determined by the lines representing strategies a_1 and a_2, these strategies would enter the solution with non-zero probabilities. These are, thus, worthwhile in this situation. Now, considering strategies a_1 and a_2 for A, and b_1 and b_2 for B, we have the pay-off matrix as follows:

	b_1	b_2
a_1	−7	6
a_2	7	−4

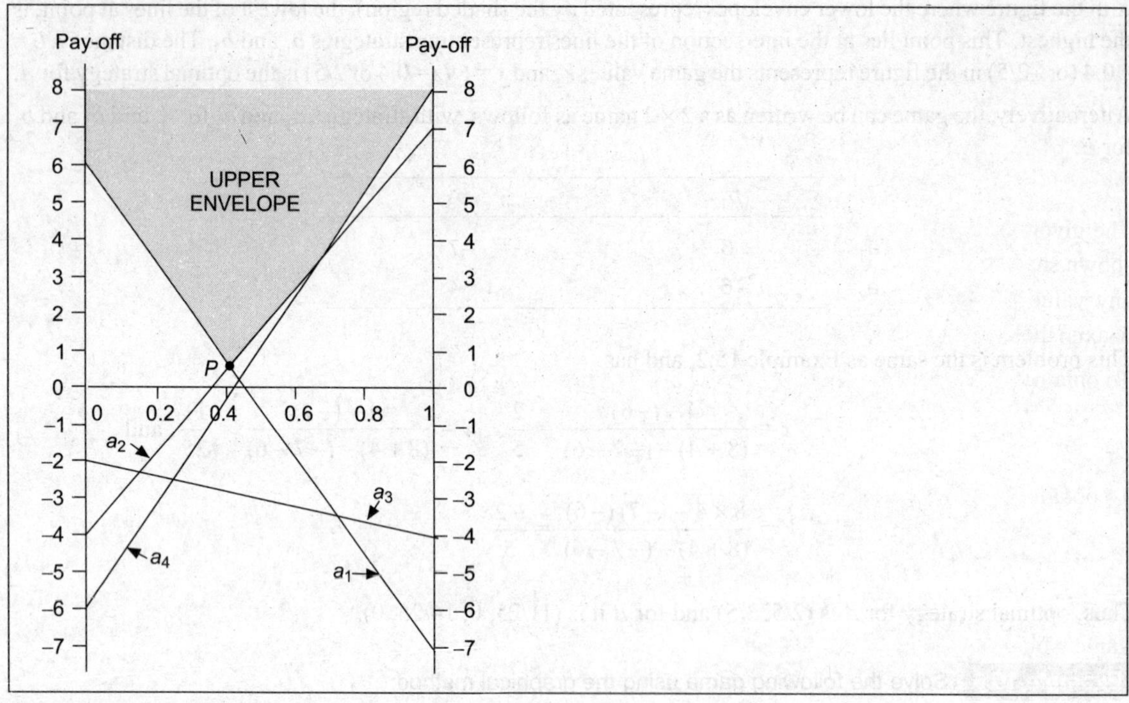

Figure 15.3 *Graphic Solution to the Game*

Further,

$$x = \frac{a_{22} - a_{21}}{(a_{11} + a_{22}) - (a_{12} + a_{21})} = \frac{-4-7}{(-7-4)-(6+7)} = \frac{11}{24},$$

$$y = \frac{a_{22} - a_{12}}{(a_{11} + a_{22}) - (a_{12} + a_{21})} = \frac{-4-6}{(-7-4)-(6+7)} = \frac{5}{12},$$

and $$V = \frac{a_{11}a_{22} - a_{12}a_{21}}{(a_{11} + a_{22}) - (a_{12} + a_{21})} = \frac{(-7)(-4)-(7)(6)}{(-7-4)-(6+7)} = \frac{7}{12}$$

Thus, optimal strategy for *A* is (11/24, 13/24, 0, 0), and for *B* it is (5/12, 7/12), whereas the game value is 7/12.

We observe that the game value is determined by the maximum ordinate of the convex set* bounded above by various lines, and by the minimum ordinate of the convex set which is bounded below (lower envelope) by various lines. It is also possible that a player might have many optimal strategies. This would be the case when there is no single point to represent the highest/lowest ordinate in an envelope. Consider the following example.

Example 15.7 Solve the game with the following pay-off matrix:

* Refer Chapter 2.

		B's Strategy		
		b_1	b_2	b_3
A's Strategy	a_1	6	4	3
	a_2	2	4	8

The given information is displayed in Figure 15.4. The lower envelope representing the feasible region is shown shaded. There is no single point representing the highest point. The value of the game is clearly 4, and any value of x, between the points represented by K and L shall be optimal for A. We can read (on the X-axis) the values of K and L from the graph or obtain them by considering b_1 and b_2, and b_2 and b_3 respectively.

To obtain K,

$$x = \frac{a_{22} - a_{21}}{(a_{11} + a_{22}) - (a_{12} + a_{21})} = \frac{4 - 2}{(6 + 4) - (4 + 2)} = \frac{2}{4} = 0.5$$

To obtain L,

$$x = \frac{8 - 4}{(4 + 8) - (3 + 4)} = \frac{4}{5} = 0.8$$

Thus, an optimal strategy for A is any pair of $(x, 1 - x)$ where $0.5 \le x \le 0.8$. An optimal strategy for B in this game would be (0, 1, 0).

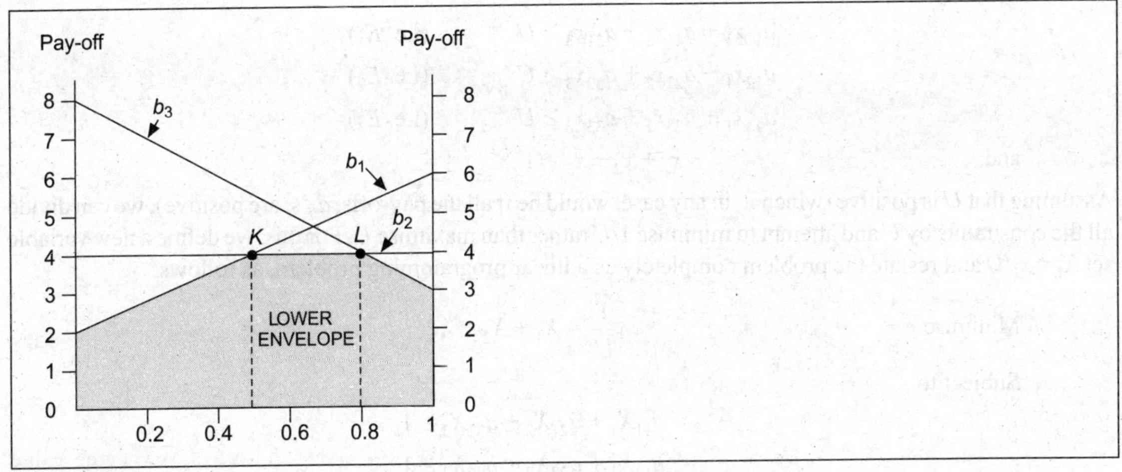

Figure 15.4 *Graphic Solution to the Game*

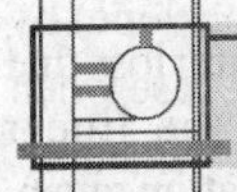

15.5 SOLUTION OF $m \times n$ GAMES—FORMULATION AND SOLUTION AS AN LPP

We have already stated that a game can also be formulated and solved as a linear programming problem. We shall illustrate this formulation by considering a simple 3×3 pay-off matrix. Here each of the players A and B has three strategies and pay-offs corresponding to various combinations of the strategies of the players represented by a_{11}, a_{12}, etc.

		B's Strategy		
		B_1	B_2	B_3
	A_1	a_{11}	a_{12}	a_{13}
A's Strategy	A_2	a_{21}	a_{22}	a_{23}
	A_3	a_{31}	a_{32}	a_{33}

Let us consider the problem first from A's point of view. A is interested in maximising the minimum gain that can be obtained by playing this game. The objective, then, is to maximise the minimum value of the game, U, subject to the constraints that regardless of the alternative that B chooses, the expected pay-off shall be at least equal to U. Now, let x_1, x_2 and x_3 be the probabilities with which A plays the strategies A_1, A_2, and A_3; E_1 be the expected pay-off of A when he plays the three strategies with these probabilities and B decides B_1. Similarly, let E_2 and E_3 be the expected pay-offs associated with the strategies B_2 and B_3 respectively, adopted by B.

$$E_1 = a_{11}\,x_1 + a_{21}\,x_2 + a_{31}\,x_3$$
$$E_2 = a_{12}\,x_1 + a_{22}\,x_2 + a_{32}\,x_3$$
$$E_3 = a_{13}\,x_1 + a_{23}\,x_2 + a_{33}\,x_3$$

We can now express the problem as
Maximise U (i.e. maximise the minimum value U)

Subject to

$$a_{11}x_1 + a_{21}x_2 + a_{31}x_3 \geq U \qquad \text{(i.e. } E_1\text{)}$$
$$a_{12}x_1 + a_{22}x_2 + a_{32}x_3 \geq U \qquad \text{(i.e. } E_2\text{)}$$
$$a_{13}x_1 + a_{23}x_2 + a_{33}x_3 \geq U \qquad \text{(i.e. } E_3\text{)}$$

and

$$x_1 + x_2 + x_3 = 1$$

Assuming that U is positive (which it, in any case, would be if all the pay-offs, a_{ij}'s, are positive), we can divide all the constraints by U and attempt to minimise $1/U$ rather than maximise U. For this, we define a new variable set $X_i = x_i/U$ and restate the problem completely as a linear programming problem, as follows:

Minimise
$$\frac{1}{U} = X_1 + X_2 + X_3$$

Subject to

$$a_{11}X_1 + a_{21}X_2 + a_{31}X_3 \geq 1$$
$$a_{12}X_1 + a_{22}X_2 + a_{32}X_3 \geq 1$$
$$a_{13}X_1 + a_{23}X_2 + a_{33}X_3 \geq 1$$
$$X_1, X_2, X_3 \geq 0$$

Looking from B's point of view, it is easy to visualize that his problem is to minimise, V, the maximum loss that he is to incur by playing this game. B's problem can be interpreted as the dual to the linear programming model obtained above in respect of A. Also, we can derive it directly from the information given. It follows.

Let y_1, y_2, and y_3 be the probabilities with which B plays strategies B_1, B_2 and B_3, respectively and E_1', E_2', and E_3' be the expected pay-offs associated with the events that A chooses, respectively, A_1, A_2, and A_3 while B plays the strategies B_1, B_2 and B_3 with the given probabilities. Accordingly,

$$E'_1 = a_{11}y_1 + a_{12}y_2 + a_{13}y_3,$$

$$E'_2 = a_{21}y_1 + a_{22}y_2 + a_{23}y_3,$$

and
$$E'_3 = a_{31}y_1 + a_{32}y_2 + a_{33}y_3$$

Now, the problem is to minimise V, subject to

$$a_{11}y_1 + a_{12}y_2 + a_{13}y_3 \leq V$$

$$a_{21}y_1 + a_{22}y_2 + a_{23}y_3 \leq V$$

$$a_{31}y_1 + a_{32}y_2 + a_{33}y_3 \leq V$$

and
$$y_1 + y_2 + y_3 = 1$$

If we define (assuming as before that V is positive) $Y_i = y_i/V$, and divide all constraints by V, we can express the LPP as follows:

Maximise
$$\frac{1}{V} = Y_1 + Y_2 + Y_3$$

Subject to

$$a_{11}Y_1 + a_{12}Y_2 + a_{13}Y_3 \leq 1$$

$$a_{21}Y_1 + a_{22}Y_2 + a_{23}Y_3 \leq 1$$

$$a_{31}Y_1 + a_{32}Y_2 + a_{33}Y_3 \leq 1$$

$$Y_1, Y_2, Y_3 \geq 0$$

This, clearly, is the dual of the LPP given earlier.

We have stated that U and V should be positive. In cases where some a_{ij}'s are negative, U and V may not be positive. In such cases, a constant is added to all the entries in the pay-off matrix as will ensure that each of them would become positive. After this, the problem is solved in usual manner and then the objective function is adjusted by subtracting that constant from it, which is the required answer.

Example 15.8 Formulate the following game as an LPP and obtain its solution:

		B's Strategy		
		b_1	b_2	b_3
	a_1	8	9	3
A's Strategy	a_2	2	5	6
	a_3	4	1	7

The given problem can be formulated as an LPP from A's and B's point of view as follows:

Let x_1, x_2, and x_3 be the probabilities with which A chooses respectively the strategies a_1, a_2 and a_3, and y_1, y_2, and y_3 be the probabilities in respect of B choosing b_1, b_2, and b_3.

From A's point of view, we have to

Minimise
$$\frac{1}{U} = X_1 + X_2 + X_3$$

Subject to

$$a_{11}X_1 + a_{21}X_2 + a_{31}X_3 \geq 1$$
$$a_{12}X_1 + a_{22}X_2 + a_{32}X_3 \geq 1$$
$$a_{13}X_1 + a_{23}X_2 + a_{33}X_3 \geq 1$$
$$X_1, X_2, X_3 \geq 0$$

where U represents the value of the game, and $X_i = x_i/U$. We have to determine the values of U, x_1, x_2, and x_3. Thus, the problem is

Minimise $$\frac{1}{U} = X_1 + X_2 + X_3$$

Subject to

$$8X_1 + 2X_2 + 4X_3 \geq 1$$
$$9X_1 + 5X_2 + X_3 \geq 1$$
$$3X_1 + 6X_2 + 7X_3 \geq 1$$
$$X_1, X_2, X_3 \geq 0$$

From B's point of view, we have to

Maximise $$\frac{1}{V} = Y_1 + Y_2 + Y_3$$

Subject to

$$a_{11}Y_1 + a_{12}Y_2 + a_{13}Y_3 \leq 1$$
$$a_{21}Y_1 + a_{22}Y_2 + a_{23}Y_3 \leq 1$$
$$a_{31}Y_1 + a_{32}Y_2 + a_{33}Y_3 \leq 1$$
$$Y_1, Y_2, Y_3 \geq 0$$

where V is the game value and $Y_i = y_i/V$.

Thus, the LPP is

Maximise $$\frac{1}{V} = Y_1 + Y_2 + Y_3$$

Subject to

$$8Y_1 + 9Y_2 + 3Y_3 \leq 1$$
$$2Y_1 + 5Y_2 + 6Y_3 \leq 1$$
$$4Y_1 + Y_2 + 7Y_3 \leq 1$$
$$Y_1, Y_2, Y_3 \geq 0$$

To obtain the required values we can solve either of these LPPs, and read the solution to the other from it as each one is the dual of the other. We shall solve the game from B's point of view.

Introducing the slack variables, we have

Maximise $$\frac{1}{V} = Y_1 + Y_2 + Y_3 + 0S_1 + 0S_2 + 0S_3$$

Subject to

$$8Y_1 + 9Y_2 + 3Y_3 + S_1 = 1$$
$$2Y_1 + 5Y_2 + 6Y_3 + S_2 = 1$$
$$4Y_1 + \ Y_2 + 7Y_3 + S_3 = 1$$
$$Y_i \geq 0, S_i \geq 0, i = 1, 2, 3$$

The solution is contained in Tables 15.1 through 15.4.

TABLE 15.1 Simplex Tableau 1: Non-optimal Solution

Basis		Y_1	Y_2	Y_3	S_1	S_2	S_3	b_i	b_i/a_{ij}
S_1	0	8*	9	3	1	0	0	1	1/8 ← Outgoing variable
S_2	0	2	5	6	0	1	0	1	1/2
S_3	0	4	1	7	0	0	1	1	1/4
c_j		1	1	1	0	0	0		
Solution		0	0	0	1	1	1		
Δ_j		1	1	1	0	0	0		
		↑ Incoming variable							

TABLE 15.2 Simplex Tableau 2: Non-optimal Solution

Basis		Y_1	Y_2	Y_3	S_1	S_2	S_3	b_i	b_i/a_{ij}
Y_1	1	1	9/8	3/8	1/8	0	0	1/8	1/3
S_2	0	0	11/4	21/4	–1/4	1	0	3/4	1/7
S_3	0	0	–7/2	11/2*	–1/2	0	1	1/2	1/11 ← Outgoing variable
c_j		1	1	1	0	0	0		
Solution		1/8	0	0	0	3/4	1/2		
Δ_j		0	–1/8	5/8	–1/8	0	0		
				↑ Incoming variable					

TABLE 15.3 Simplex Tableau 3: Non-optimal Solution

Basis		Y_1	Y_2	Y_3	S_1	S_2	S_3	b_i	b_i/a_{ij}
Y_1	1	1	15/11	0	7/44	0	–3/44	1/11	1/15
S_2	0	0	67/11*	0	5/22	1	–21/22	3/11	3/67 ← Outgoing variable
Y_3	1	0	–7/11	1	–1/11	0	2/11	1/11	–1/67
c_j		1	1	1	0	0	0		
Solution		1/11	0	1/11	0	3/11	0		
Δ_j		0	3/11	0	–3/44	0	–5/44		
			↑ Incoming variable						

TABLE 15.4 Simplex Tableau 4: Optimal Solution

Basis		Y_1	Y_2	Y_3	S_1	S_2	S_3	b_i
Y_1	1	1	0	0	29/268	–15/67	39/268	2/67
Y_2	1	0	1	0	5/134	11/67	–21/134	3/67
Y_3	1	0	0	1	–9/134	7/67	11/134	8/67
c_j		1	1	1	0	0	0	
Solution		2/67	3/67	8/67	0	0	0	
Δ_j		0	0	0	–21/268	–12/268	–19/268	

Substituting the values of Y_1, Y_2 and Y_3 in the objective function, we have,

$$\frac{1}{V} = \frac{2}{67} + \frac{3}{37} + \frac{8}{67} = \frac{13}{67}$$

Therefore V, the game value = 67/13.

Since $Y_i = y_i/V$, we have $y_i = Y_i \times V$. Thus,

$$y_1 = \frac{2}{67} \times \frac{67}{13} = \frac{2}{13}; \quad y_2 = \frac{3}{67} \times \frac{67}{13} = \frac{3}{13}; \quad \text{and} \quad y_3 = \frac{8}{67} \times \frac{67}{13} = \frac{8}{13}.$$

We can read the values of the dual variables X_1, X_2 and X_3 from the Δ_j row of the Table 15.4. These are, respectively, 21/268, 12/268, and 19/268. From these,

$$\frac{1}{U} = \frac{21}{268} + \frac{12}{268} + \frac{19}{268} = \frac{52}{268} = \frac{13}{67}$$

Thus, U = 67/13 (same as shown earlier).

Now, we have

$$x_1 = \frac{21}{268} \times \frac{67}{13} = \frac{21}{52}; \quad x_2 = \frac{12}{268} \times \frac{67}{13} = \frac{12}{52}; \quad \text{and} \quad x_3 = \frac{19}{268} \times \frac{67}{13} = \frac{19}{52}.$$

Thus, the optimal strategy for A is (21/52, 12/52, 19/52); for B it is (2/13, 3/13, 8/13) while the game value is 67/13.

Example 15.9 Solve the following game:

B's Strategy

		B_1	B_2	B_3
	A_1	3	4	−2
A's Strategy	A_2	−3	0	1
	A_3	−1	−4	2

Since some of the entries in the pay-off matrix are negative, we shall add a constant to each of the values to convert them all into positive values. Adding 5 to each of them, we get the following revised pay-off matrix:

B's Strategy

		B_1	B_2	B_3
	A_1	8	9	3
A's Strategy	A_2	2	5	6
	A_3	4	1	7

This problem now is the same as Example 15.8, for which we have the value of the game equal to 67/13. Since a constant 5 was added to each of the pay-offs, we shall subtract this value from 67/13 to obtain 2/13 which is the value of the game under consideration. Thus, the solution to the given game is:

Optimal Strategy of A: 21/52, 12/52, 19/52;

Optimal Strategy of B: 2/13, 3/13, 8/13;

and Game Value, $V = 2/13$.

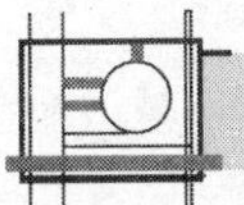

15.6 LIMITATIONS OF THE GAME THEORY

Our discussion of game theory has been restricted to the two-person zero-sum games. There are practically no applications of game theory to the real world situations. This is because of the assumptions underlying the theory. As mentioned earlier, the game model that we have considered is based on the assumption that each of the individuals involved not only acts rationally so that his preference ordering of the outcomes is determined by the order of magnitudes of their associated pay-offs, but also he assumes that the opponent is also rational in that sense. Besides, it is also assumed that both the players are having complete and equal knowledge about the strategies available to them and the resulting pay-offs. It can easily be visualised that in real world situations, it is only but rarely that each of the persons would have complete knowledge about all the strategies available to his competitor, as also of the exact pay-off values associated with various combinations of strategies.

Further, the solution to the games is based on the application of conservative principles of maximin and minimax strategies. This also rests on the assumption that complete information about strategies and the associated pay-offs is available and is certain. In real conditions, complete information may not generally be available and so these principles may not work. Even otherwise, the business managers do not always adopt conservative approach to their decisions and prefer to take risks.

Besides, in a given situation, there are likely to be more than two participants and the sum of gains and losses of the opponents may not be equal to zero. A great deal of mathematical sophistication is involved in games beyond the two-person zero-sum games, like the non-zero games, n-person games, games with uncertain pay-offs, games involving an infinite number of pure strategies and so on.

REVIEW ILLUSTRATIONS

Example 15.10 Shruti Ltd. has developed a sales forecasting function for its products and the products of its competitors, Purnima Ltd. There are four strategies S_1, S_2, S_3 and S_4 available to Shruti Ltd. and three strategies P_1, P_2 and P_3 to Purnima Ltd. The pay-offs corresponding to all the twelve combinations of the strategies are given below. From the table we can see that, for example, if strategy S_1 is employed by Shruti Ltd. and strategy P_1 by Purnima Ltd., then there shall be a gain of Rs 30,000 in quarterly sales to the former. Other entries can be similarly interpreted.

Considering this information, state what would be the optimal strategy for Shruti Ltd.? Purnima Ltd.? What is the value of the game? Is the game fair?

		Purnima Ltd.'s Strategy		
		P_1	P_2	P_3
	S_1	30,000	–21,000	1,000
Shruti's Strategy	S_2	18,000	14,000	12,000
	S_3	–6,000	28,000	4,000
	S_4	18,000	6,000	2,000

For determining the optimal strategies for the players, we shall examine if saddle point exists for this problem.

		Purnima's Strategy			
		P_1	P_2	P_3	Row Minima
	S_1	30,000	–21,000	1,000	–21,000
Shruti's Strategy	S_2	18,000	14,000	12,000	12,000*
	S_3	– 6,000	28,000	4,000	–6,000
	S_4	18,000	6,000	2,000	2,000
Column Maxima		30,000	28,000	12,000*	

Here saddle point exists at the intersection of S_2 and P_3. These represent optimal policies, respectively, of Shruti Ltd. and Purnima Ltd. Correspondingly, the value of game, $V = 12,000$. Since $V \neq 0$, the game is not a fair one.

Example 15.11 There is a famous Italian game called *Two-finger Morra*. This game is played by two people each of whom shows one or two fingers and simultaneously guesses the number of fingers his opponent will show. If just one player guesses correctly, he wins an amount equal to the sum of the fingers shown by himself and by his opponent, otherwise the game is considered a draw.
Obtain the pay-off matrix corresponding to it.

If (1, 1) represents a situation that a player holds one finger and guesses that his opponent will also show one finger; (1, 2) indicates that a player shows one finger and his opponent will show two fingers, ... and so on, the pay-off matrix can be written as follows:

		Player *II*			
		(1, 1)	*(1, 2)*	*(2, 1)*	*(2, 2)*
	(1, 1)	0	2	–3	0
Player *I*	(1, 2)	–2	0	0	3
	(2, 1)	3	0	0	– 4
	(2, 2)	0	–3	4	0

Example 15.12 Can there be a 'two-person zero-sum' game which has a value of zero? If so, write the pay-off matrix of one such game and find out the optimal strategies of the players.

(M Com, Delhi, 1999)

Yes, a 'two-person zero-sum' game can have a value of zero. One such game is given below, where player *A* has 3 and player *B* has 4 strategies available. Along with the pay-off matrix, the row minima and column maxima are also provided.

		Player *B*				
		B_1	B_2	B_3	B_4	*Row Minima*
	A_1	42	18	–5	–20	–20
Player *A*	A_2	20	20	0	18	0
	A_3	24	–15	–12	30	–15
Column Maxima		42	20	0	30	

From the above matrix, it may be observed that the maximum value in the row minima is zero, while the minimum value is the column maxima is also zero. Hence, the given game has a saddle point. The optimal strategies are:

Player $A : A_2$ (maximin).
Player $B : B_3$ (minimax), and
Value of game = 0.

Example 15.13 In a small town, there are only two stores that handle sundry goods—*ABC* and *XYZ*. The total number of customers is equally divided between the two, because price and quality of goods sold are equal. Both stores have good reputation in the community, and they render equally good customer service. Assume that a gain of customer by *ABC* is a loss to *XYZ* and vice versa. Both stores plan to run annual pre-Diwali sales during the first week of November. Sales are advertised through a local newspaper, radio and television media. With the aid of an advertising firm, store *ABC* constructed the game matrix given below. (Figures in the matrix represent a gain or loss of customers.)

Strategy of ABC	*Strategy of XYZ*		
	Newspaper	*Radio*	*Television*
Newspaper	30	40	– 80
Radio	0	15	– 20
Television	90	20	50

Determine optimal strategies and worth of such strategies for both *ABC* and *XYZ*.

(MBA, Delhi, November, 1999)

The game matrix, with row minima and column maxima, is reproduced below:

Strategy of ABC	*Strategy of XYZ*			*Row Minima*
	Newspaper	*Radio*	*Television*	
Newspaper	30	40	–80	–80
Radio	0	15	–20	–20
Television	90	20	50	20
Column Maxima	90	40	50	

Evidently, saddle point does not exist. Now, it may be observed that row 3 dominates row 2, and column 3 dominates column 1. Deleting R_2 and C_1, the game reduces to the order 2×2, as shown below:

	Radio	*Television*
Newspaper	40	–80
Television	20	50

With usual meaning of the symbols used, we have

$$x = \frac{a_{22} - a_{21}}{(a_{11} + a_{22}) - (a_{12} + a_{21})} = \frac{50 - 20}{(40 + 50) - (-80 + 20)} = \frac{30}{150} = 1/5$$

$$y = \frac{a_{22} - a_{12}}{(a_{11} + a_{22}) - (a_{12} + a_{21})} = \frac{50 - (-80)}{(40 + 50) - (-80 + 20)} = \frac{130}{150} = 13/15$$

$$V = \frac{(a_{11} \times a_{22}) - (a_{12} \times a_{21})}{(a_{11} + a_{22}) - (a_{12} + a_{21})} = \frac{(40 \times 50) + (80 \times 20)}{(40 + 50) - (-80 + 20)} = \frac{3{,}600}{150} = 24$$

Thus, optimal strategies are:

	Newspaper	*Radio*	*Television*
For *ABC* :	1/5	0	4/5
For *XYZ* :	0	13/15	2/15

Value of the game = 24.

Example 15.14 (a) Given the following pay-off matrix for a 2-person zero-sum game, the second row second column entry will be a saddle point only if:

(i) $p \leq 6$ and $q \leq 6$

(ii) $p \geq 6$ and $q \leq 6$

(iii) $p \leq 6$ and $q \geq 6$

(iv) None of the choices given above

2	q	3
p	6	10
7	3	5

(b) The pay-off matrix for a 2-person zero-sum game is given below:

	Player *B*		
	1	–1	–2
Player *A*	1	0	–1
	2	–1	2

From the analysis of the structure of the matrix it can be concluded that the optimal strategies for this play would be:

(i) with saddle point.

(ii) player *A* should assign probabilities (0, 0.75, 0.25) to his strategies and *B* should assign (0, 0.75, 0.25) to his.

(iii) player *A* should assign probabilities (0.40, 0, 0.60) to his strategies and *B* should assign (0.80, 0, 0.20) to his.

(iv) *A* and *B* should play all the given strategies with equal probabilities.

(v) none of the above. *(DMS, Bombay, 1993)*

(a) For the given data, the minimum value in the third row is 3 and maximum value in the third column is 10. Since the second-row-second-column entry, equal to 6, is intended to be a saddle point, it is implied that the maximum value in row minima and the minimum value in the column maxima be 6. It is possible when p is at least 6 and q does not exceed 6. Thus, statement (ii) is correct.

(b) Here, the row minima are –2, –1 and –1, respectively for the three rows, and column maxima are 2, 0 and 2 respectively for the columns. Clearly, there is no saddle point, Thus, statement (i) is ruled out.

Next, it may be observed that column 2 dominates column 1. The reduced matrix is:

	C_2	C_3
R_1	–1	–2
R_2	0	–1
R_3	–1	2

Now, R_2 is seen to dominate R_1. Deleting this row, we get

	C_2	C_3
R_2	0	–1
R_3	–1	2

To solve this 2 × 2 problem analytically, we have

$$x = \frac{a_{22} - a_{21}}{(a_{11} + a_{22}) - (a_{21} + a_{12})} = \frac{2-(-1)}{(0+2)-(-1-1)} = \frac{3}{4} = 0.75$$

$$y = \frac{a_{22} - a_{12}}{(a_{11} + a_{22}) - (a_{21} + a_{12})} = \frac{2-(-1)}{(0+2)-(-1-1)} = \frac{3}{4} = 0.75$$

Thus, optimal strategy for player A is (0, 0.75, 0.25) and for player B, it is (0, 0.75, 0.25). Hence, statement (ii) is correct. It rules out, thus, the remaining options.

Example 15.15 Reduce the following two-person zero-sum game to 2 × 2 order, and obtain the optimal strategies for each player and the value of the game:

		Player B			
		B_1	B_2	B_3	B_4
	A_1	3	2	4	0
	A_2	3	4	2	4
Player A	A_3	4	2	4	0
	A_4	0	4	0	8

(M Com, Delhi, 1984)

We observe that all entries in the third row of the given matrix are greater than, or equal to, the corresponding entries in the first row. Thus the first row is dominated by the third row and as such can be deleted. The deletion of the first row leads to the following matrix:

	B_1	B_2	B_3	B_4
A_2	3	4	2	4
A_3	4	2	4	0
A_4	0	4	0	8

Next, all elements of the first column are at least equal to their counterparts in the third column. We can, thus, delete the first column to get the following matrix:

	B_2	B_3	B_4
A_2	4	2	4
A_3	2	4	0
A_4	4	0	8

We notice now that the first column is dominated by a convex linear combination of the second and the third columns because

$$4 > \frac{1}{2}(2) + \frac{1}{2}(4);\ 2 = \frac{1}{2}(4) + \frac{1}{2}(0);\ \text{and } 4 = \frac{1}{2}(0) + \frac{1}{2}(8).$$

Its deletion results in the following:

	B_3	B_4
A_2	2	4
A_3	4	0
A_4	0	8

Similarly, the first row is equated to the convex linear combination of the other two rows as shown.

$$2 = \frac{1}{2}(4) + \frac{1}{2}(0); \text{ and } 4 = \frac{1}{2}(0) + \frac{1}{2}(8)$$

Hence, we delete the first row to get the following 2 × 2 matrix which can be solved analytically.

	B_3	B_4
A_3	4	0
A_4	0	8

For this game,

$$x = \frac{8-0}{(4+8)-(0+0)} = \frac{8}{12} = \frac{2}{3}; \quad y = \frac{8-0}{(4+8)-(0+0)} = \frac{8}{12} = \frac{2}{3}; \text{ and}$$

$$V = \frac{4\times 8 - 0\times 0}{(4+8)-(0+0)} = \frac{32}{12} = \frac{8}{3}.$$

Thus, the optimal strategy for A is (0, 0, 2/3, 1/3), for B it is (0, 0, 2/3, 1/3) and the game value V = 8/3.

Example 15.16 A company is currently involved in negotiations with its union on the upcoming wage contract. With the aid of an outside mediator, the table below was constructed by the management group. The pluses are to be interpreted as proposed wage increases while a minus figure indicates that a wage reduction is proposed. The mediator informs the management group that he has been in touch with the union and that they have constructed a table that is comparable to the table developed by the management. Both the company and the union must decide on an overall strategy before negotiations begin. The management group understands the relationship of company strategies to union strategies in the following table but lacks specific knowledge of game theory to select the best strategy (or strategies) for the firm. Assist the management on this problem. What game value and strategies are available to the opposing groups?

Conditional costs to company

(in Lac Rs)

Company strategies	*Union strategies*			
	U_1	U_2	U_3	U_4
C_1	+ 0.25	+ 0.27	+ 0.35	– 0.02
C_2	+ 0.20	+ 0.16	+ 0.08	+ 0.08
C_3	+ 0.14	+ 0.12	+ 0.15	+ 0.13
C_4	+ 0.30	+ 0.14	+ 0.19	0.00

(MBA, Delhi, November, 2008)

Since the company represents the 'minimising player' and the union the 'maximising player,', we shall recast the pay-off matrix (by taking transpose of the given matrix) as follows:

Union strategies	*Company strategies*			
	C_1	C_2	C_3	C_4
U_1	0.25	0.20	0.14	0.30
U_2	0.27	0.16	0.12	0.14
U_3	0.35	0.08	0.15	0.19
U_4	– 0.02	0.08	0.13	0.00

This game has no saddle point.

We observe that all entries in the third row of this matrix are greater than, or equal to, the corresponding entries in the fourth row. Thus, fourth row is dominated by the third row and hence can be deleted. Deleting it we get,

	C_1	C_2	C_3	C_4
U_1	0.25	0.20	0.14	0.30
U_2	0.27	0.16	0.12	0.14
U_3	0.35	0.08	0.15	0.19

In this matrix, the first column is dominated both by the second and the third columns; and the fourth column is dominated by the third column. Deleting the dominated columns, the matrix is reduced to the following:

	C_2	C_3
U_1	0.20	0.14
U_2	0.16	0.12
U_3	0.08	0.15

Here, the second row is dominated by the first. Deleting this row we get the following matrix of the order 2×2, and obtain the solution to the game analytically.

	C_2	C_3
U_1	0.20	0.14
U_3	0.08	0.15

If x be the probability with which the union adopts policy U_1 and y be the probability of adoption of C_2 by the company, we have

$$x = \frac{0.15 - 0.08}{(0.20 + 0.15) - (0.08 + 0.14)} = \frac{7}{13};$$

$$y = \frac{0.15 - 0.14}{(0.20 + 0.15) - (0.08 + 0.14)} = \frac{1}{13};$$

and
$$V = \frac{0.20 \times 0.15 - 0.08 \times 0.14}{(0.20 + 0.15) - (0.08 + 0.14)} = \frac{0.0188}{0.1300} = \frac{47}{325}$$

Thus, optimal strategy for the company is (0, 1/13, 12/13, 0); for the union it is (7/13, 0, 6/13, 0) and the game value is 47/325 (representing increased wages).

Example 15.17 For the following pay-off matrix, find the value of the game and the strategies of players *A* and *B* using linear programming:

		Player *B*		
		1	2	3
Player *A*	1	3	–1	4
	2	6	7	–2

(MBA, Delhi, 1986)

Since two of the entries in the pay-off matrix are negative, we shall add a constant, say 3, to each of the values, by which each one of them would become a positive value. The pay-off matrix then becomes as shown here:

		Player *B*		
		1	2	3
Player *A*	1	6	2	7
	2	9	10	1

Now, let x_1 and x_2 represent the probabilities with which *A* chooses strategies 1 and 2 respectively, while y_1, y_2 and y_3 be the probabilities in respect of *B* choosing strategies 1, 2 and 3.

From *A*'s point of view, the problem is,

Minimise $$\frac{1}{U} = X_1 + X_2$$

Subject to

$$6X_1 + 9X_2 \geq 1$$
$$2X_1 + 10X_2 \geq 1$$
$$7X_1 + X_2 \geq 1$$
$$X_1, X_2 \geq 0$$

where $X_1 = x_1/U$, and $X_2 = x_2/U$.

Similarly, from *B*'s view point, the problem is,

Maximise $$\frac{1}{V} = Y_1 + Y_2 + Y_3$$

Subject to

$$6Y_1 + 2Y_2 + 7Y_3 \leq 1$$
$$9Y_1 + 10Y_2 + Y_3 \leq 1$$
$$Y_1, Y_2, Y_3 \geq 0$$

where $Y_i = y_i/V$.

Now we shall solve the problem from *B*'s point of view. The problem can be augmented by introducing slack variables S_1 and S_2 as follows:

Maximise $$\frac{1}{V} = Y_1 + Y_2 + Y_3 + 0S_1 + 0S_2$$

Subject to

$$6Y_1 + 2Y_2 + 7Y_3 + S_1 = 1$$
$$9Y_1 + 10Y_2 + Y_3 + S_2 = 1$$
$$Y_1, Y_2, Y_3, S_1, S_2 \geq 0$$

The solution is contained in Table 15.5 through 15.8.

TABLE 15.5 Simplex Tableau 1: Non-optimal Solution

Basis		Y_1	Y_2	Y_3	S_1	S_2	b_i	b_i/a_{ij}
S_1	0	6	2	7	1	0	1	1/6
S_2	0	9*	10	1	0	1	1	1/9 ← Outgoing variable
c_j		1	1	1	0	0		
Solution		0	0	0	1	1		
Δ_j		1	1	1	0	0		
		↑ Incoming variable						

TABLE 15.6 Simplex Tableau 2: Non-optimal Solution

Basis		Y_1	Y_2	Y_3	S_1	S_2	b_i	b_i/a_{ij}
S_1	0	0	–14/3	19/3*	1	–2/3	1/3	1/19 ← Outgoing variable
Y_1	1	1	10/9	1/9	0	1/9	1/9	1
c_j		1	1	1	0	0		
Solution		1/9	0	0	1/3	0		
Δ_j		0	–1/9	8/9	0	–1/9		
				↑ Incoming variable				

TABLE 15.7 Simplex Tableau 3: Non-optimal Solution

Basis		Y_1	Y_2	Y_3	S_1	S_2	b_i	b_i/a_{ij}
Y_3	1	0	–14/19	1	3/19	–2/19	1/19	–
Y_1	1	1	68/57*	0	–1/57	7/57	2/19	3/34 ← Outgoing variable
c_j		1	1	1	0	0		
Solution		2/19	0	1/19	0	0		
Δ_j		0	31/57	0	–8/57	–1/57		
			↑ Incoming variable					

TABLE 15.8 Simplex Tableau 4: Optimal Solution

Basis		Y_1	Y_2	Y_3	S_1	S_2	b_i
Y_3	1	42/68	0	1	5/34	–1/34	2/17
Y_2	1	57/68	1	0	–1/68	7/68	3/34
c_j		1	1	1	0	0	
Solution		0	3/34	2/17	0	0	
Δ_j		–31/68	0	0	–9/68	–5/68	

The optimal values for Y_1, Y_2, and Y_3 are 0, 3/34 and 2/17, respectively. From these, we have

$$\frac{1}{V} = 0 + \frac{3}{34} + \frac{2}{17} = \frac{7}{34}$$

Therefore, $V = 34/7$. Since a value 3 was added to the original pay-off values, the game value is equal to $V - 3$ or $34/7 - 3 = 13/7$. Further, since $y_i = Y_iV$, we have $y_1 = 0 \times 34/7 = 0$, $y_2 = (3/34)(34/7) = 3/7$, and $y_3 = (2/17)(34/7) = 4/7$.

The values of the dual variables X_1 and X_2 can be read from the Δ_j row of the Simplex Tableau 4. From this, $X_1 = 9/68$ and $X_2 = 5/68$. From these, $1/U = 9/68 + 5/68 = 7/34$. Therefore, $U = V = 34/7$. Thus, $x_1 = (9/68)(34/7) = 9/14$ and $x_2 = (5/68)(34/7) = 5/14$.

The solution to the problem, therefore, is:

Player A		*Player B*		*Value of game*
Strategy	*Probability*	*Strategy*	*Probability*	
1	9/14	1	0	13/7
2	5/14	2	3/7	
		3	4/7	

Example 15.18 Consider the problem given in Example 15.17 and re-solve by algebraic method.

The pay-off matrix is reproduced here:

		Player B		
		1	2	3
Player *A*	1	3	–1	4
	2	6	7	–2

The first column of the given matrix is dominated by a linear combination of the second and the third columns, as

$$\frac{1}{2}(-1) + \frac{1}{2}(4) = \frac{3}{2} < 3$$

$$\frac{1}{2}(7) + \frac{1}{2}(-2) = \frac{5}{2} < 6$$

Deleting the first column values, we get the reduced 2 × 2 matrix which would be used for finding solution to the given problem. Using the usual notations, we have

$$x = \frac{a_{22} - a_{21}}{(a_{11} + a_{22}) - (a_{12} + a_{21})} = \frac{-2-7}{(-1-2)-(4+7)} = \frac{9}{14}$$

$$y = \frac{a_{22} - a_{12}}{(a_{11} + a_{22}) - (a_{12} + a_{21})} = \frac{-2-4}{(-1-2)-(4+7)} = \frac{6}{14} = \frac{3}{7}$$

$$V = \frac{a_{11}a_{22} - a_{12}a_{21}}{(a_{11} + a_{22}) - (a_{12} + a_{21})} = \frac{(-1)(-2)-(4 \times 7)}{(-1-2)-(4+7)} = \frac{26}{14} = \frac{13}{7}$$

Accordingly, the optimal policy for A is (9/14, 5/14), for B it is (0, 3/7, 4/7), and the game value is 13/7, the results being the same as obtained earlier.

KEY POINTS TO REMEMBER

- Theory of games is concerned with decision-making in situations where two or more rational players, each with a set of strategies, are involved in conditions of competition and conflicting interests.
- With the pay-offs resulting from the play of various combinations of the strategies by different players being given and known to all of them, the solution to a game calls for determining the optimal strategies for the players.
- For a game involving two-persons and zero-sum (where the gain for one is equal to the loss of the other) with a given set of a pay-offs expressed from the viewpoint of the 'maximising' player, obtaining the solution calls for determining, in the first place, whether the saddle point exists.
- For the 'maximising' players, the maximin strategy is determined. For this, the minimum pay-offs of all the strategies are obtained (as row minima) and then the strategy with the maximum among those minimum values is determined. Similarly, the maximum pay-offs of all the strategies open to the 'minimising' player (column maxima) are obtained and the least value in them is considered. The strategy is called 'minimax'. If the pay-off of the maximin strategy matches with the pay-off of the minimax strategy, then the saddle point exists. In such a case, the strategies so determined are the respective optimal strategies for the players and the pay-off involved is the *value* of the game. The strategies involved are termed *pure*.
- If the saddle point does not exist, then the players have to employ *mixed* strategies. For each player, the optimal mix is obtained in such a manner that the pay-off is the same no matter what strategy the opponent chooses to play.
- For a 2 × 2 game, involving two strategies for each of the players, the optimal mix can be obtained by using analytical method.
- Games of the order 2 × n or m × 2 may be reduced to 2 × 2 games by using graphical method and then solved as such.
- The games involving more than two strategies available to each of the players may be attempted to be reduced to the order of 2 × 2, by using the rule of dominance. If a game can be so reduced, it is solved by using the analytical method.

- If a strategy is found to be better than, or at least as good as another strategy in terms of the comparable pay-offs, then the former is said to "dominate" the latter. Also, a weighted average of two (or more) strategies may be seen to dominate a given strategy. In any event, a dominated strategy can be deleted.
- If a game cannot be reduced to a 2×2 game, it can be formulated as a linear programming problem and solved accordingly. In general, any game can be formulated and solved as an LPP.

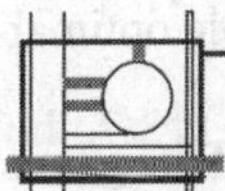

TEST YOUR UNDERSTANDING

Mark the following statements as T (*True*) or F (*False*).

1. In a two-person game, both the players must have an equal number of strategies.
2. The *zero-sum* game implies that any gain of one player is exactly matched by a loss to the other so that their sum is equal to zero.
3. The solution to a game implies determining optimal strategies for both the players and value of the game.
4. The solution to a two-person game is based on the assumption that player A will always play his strategy first and the other player, B, would play his strategy thereafter.
5. In the two-person games, both the players are assumed to be in full knowledge of the strategies available to each one and the pay-offs resulting from each combination of strategies.
6. For a two-person game with, respectively, 6 and 5 strategies available to the two players, a total of 30 conditional pay-offs would be involved.
7. Player A's strategy is determined on the basis of maximin criterion whereas player B's is on minimax criterion.
8. Saddle point is the point of equilibrium.
9. It is possible that multiple minimax/maximin strategies might exist in a two-player game.
10. If the largest element in the pay-off matrix is negative, the game should favour player B irrespective of what the optimal strategies of the players are.
11. If the value of the game is a negative value, it implies that the game is favouring player B.
12. A game is said to be fair when both the players know about the optimal strategy of each other and the losing player pays off the amount involved to the gaining player.
13. Mixed strategy for a player can involve no more than two strategies.
14. Mixed strategies for each of the players are determined such that same pay-off would be expected irrespective of the strategy adopted by the opponent.
15. Mixed strategies are adopted by the players when they are doubtful about the optimal strategies for them.
16. Strategies involved in mixed strategies of the players are played in a random manner in the ratio determined in the optimal mix.
17. An optimal strategy (0, 2/9, 7/9) implies that in every nine plays, the first strategy should never be played, the second strategy twice and the third strategy be adopted in the remaining seven times.
18. The graphic approach to the solution of games can be applied when one of the players has two strategies available.

19. Every game is solvable irrespective of whether the players adopt pure or mixed strategies.
20. In case of m by 2 games, the value of the game is given by the highest point in the lower envelope.
21. In a two-person zero-sum game, if the optimal solution requires one player to use a pure strategy, the other player must also do the same.
22. The dominance principle suggests that strategies of one player are dominating the strategies of the other.
23. When each value in a column is less than or equal to the corresponding value in another column, the strategy of the first column is said to dominate that of the second column.
24. It is possible that a convex linear combination of the two strategies dominates a third strategy.
25. In a 2 by n or m by 2 game, each of the players can mix at most two strategies if the multiple optimal solutions do not exist.
26. If a constant is added to all the elements of a pay-off matrix, it would affect only the value of the game and not the optimal strategies of the players.
27. Every two-person zero-sum game can be represented by a pair of linear programming problems with primal-dual relationship.
28. If some elements are negative in a pay-off matrix of a two-person, zero-sum game, a constant should be added to each of the elements of the matrix such that all of them become positive.
29. The constant value, if added to each element of the pay-off matrix to formulate the given problem as an LPP, should be subtracted from the value of the game, determined from the solution to the LPP.
30. The theory of games is very restrictive in real world applications because of the nature of assumptions made.

EXERCISES

1. What is game theory? State the assumptions underlying it. Discuss its importance to business decisions.
2. What do you understand by 'zero-sum' in the context of game theory? Can there be a non-zero-sum game also?
3. 'Game Theory provides a systematic quantitative approach for analysing competitive situations in which the competitors make use of logical processes and techniques in order to determine an optimal strategy for winning.' Comment.
4. Explain the following:

   (a) saddle point (b) pure strategy (c) mixed strategy.

   Give the linear programming equivalent to a Game Theory problem.
5. What is a 'game' in game theory? What are the properties of a game? Explain the "best strategy" on the basis of minimax criterion of optimality. *(MBA, Delhi, November, 1999)*
6. Describe the maximin and minimax principles of game theory.
7. Explain with examples the concept of dominance in games theory.
8. "The primary contribution of the game theory has been its concepts rather than its formal application to solving real problems." Explain. *(M Com, Delhi, 1984)*
9. "The two-person, zero-sum game is unrealistic." Elucidate the statement bringing out the limitations of game theory, if any.

10. What are the major limitations of game theory?
11. How can a 'Two-Person, Zero-Sum Game' problem be converted into a linear programming problem? Illustrate with an example.

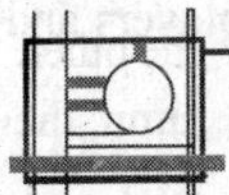

PRACTICAL PROBLEMS

1. Find the saddle point, if it exists, for each of the following games:

(a)

	b_1	b_2
a_1	3	7
a_2	–5	5

(b)

	b_1	b_2	b_3	b_4
a_1	5	8	2	4
a_2	2	6	1	3

2. Given the following pay-off matrix of a zero-sum game, determine the optimal strategies for the players and the value of the game:

A's strategy	*B's strategy*			
	b_1	b_2	b_3	b_4
a_1	5	– 4	5	8
a_2	6	2	0	– 5
a_3	7	12	8	7
a_4	2	8	–6	5

3. Is the following game strictly determinable? Is it fair?

	b_1	b_2	b_3	b_4	b_5	b_6
a_1	18	8	18	8	18	8
a_2	15	6	15	6	15	6
a_3	18	8	18	8	18	8
a_4	–15	–6	–15	–6	–15	–6

4. You are given the pay-off matrix in respect of a two-person, zero-sum game, as follows:

A's strategy	*B's strategy*				
	B_1	B_2	B_3	B_4	B_5
A_1	8	10	–3	–8	–12
A_2	3	6	0	6	12
A_3	7	5	–2	–8	17
A_4	–11	12	–10	10	20
A_5	–7	0	0	6	2

(a) Write the maximin and minimax strategies.
(b) Is it a strictly determinable game?
(c) What is the value of the game?
(d) Is this game a fair one?

5. For the following 'two-person, zero-sum' game, find the optimal strategies for the two players and value of the game:

		Player *B*		
		B_1	B_2	B_3
	A_1	5	9	3
Player *A*	A_2	6	–12	–11
	A_3	8	16	10

If the saddle point exists, determine it using the principle of dominance.

6. The two major scooter companies of India, *ABC* and *XYZ*, are competing for a fixed market. If both the manufacturers make major model changes in a year, then their shares of the market do not change. Also, if they both do not make major model changes, their shares of the market remain constant. If *ABC* makes a major model change and *XYZ* does not, then *ABC* is able to take away *a*% of the market away from *XYZ*, and if *XYZ* makes a major model change and *ABC* does not, *XYZ* is able to take away *b*% of the market away from *ABC*.

   Express it as a two-by-two game and solve for the optimal strategy for each of the producers.

7. The squadron of Sqdn. Leader Kumar is equipped with six Big-27 fighters. The enemy is using *X*-16 fighters. In combat the two types of planes are matching and they fight to draw if they are in equal numbers. However, if either side uses a larger number of planes, the other side will be destroyed. In the early hours of a day, Kumar is informed by the radar operator that five enemy planes are approaching the side. From past experience and the report of the intelligence agency, he knows that the planes would split up at the last minute and attack two installations. He, of course, wants to destroy the maximum number of enemy planes at the least cost to his own.

   Assuming that Kumar does not want any of the installations remain undefended, what strategies are available to him? Also list the strategies of enemy and construct the pay-off matrix taking the pay-offs to be the number of enemy planes minus the number of planes Kumar loses. Does the matrix have a saddle point?

8. M/S Godrej & Boyce and Hindustan Lever Ltd. have been selling competing products Cinthol and Liril respectively. The brand manager of Cinthol raised the following question:

   What should be the firm's strategy in terms of advertising for Cinthol? The market Research group of Godrej & Boyce developed data for varying degrees of advertising by Liril and Cinthol:

   1. No advertising, medium advertising and high advertising by both the firms will result in equal marketing share for Liril and Cinthol.
   2. No advertising by Cinthol will result in a market share of 40 per cent and 28 per cent if medium level of advertising and heavy level of advertising respectively is adopted by Liril.
   3. If Cinthol adopts medium level of advertising it will achieve a market share of 70 per cent with no advertising by Liril and 45 per cent market share with heavy advertising by Liril.
   4. If Cinthol resorts to heavy advertising it would achieve a market share of 75 per cent with no advertising by Liril and would get a 47.5 per cent market share with medium advertising by Liril.

Based upon the foregoing information, what strategy should the brand manager of Cinthol adopt? *(MMS, Mumbai, 1999)*

9. A partnership firm with two partners has a net worth of W. Each of the partners owns half of the business and, in addition, has a personal bank account of amount K. Of late, the partners are not on talking terms. Finally, they have agreed through a mediator that one of them would buy other's share of business according to the following plan. Each partner submits a sealed bid B ($0 \le B \le K$). The partner making the higher bid would pay the amount bid to his associate in return of full ownership of the business. In case the bids are equal, the mediator would assign business to any one of them through a flip of a coin and without any payment. Taking $K = 4$, $W = 4$ and $B = 0, 1, 2, 3, 4$, draft the above as a two-person, zero-sum game. Find the optimal strategy(ies) of the partners and the value of the game. Is the game a fair one?

10. Two competing firms must open their next branch at one of the three cities: A, B and C, whose distance profile is given here. If both companies open their branches in the same city, they will split the business evenly. If, however, they build the branches in different cities, the company that is closer to a given city will get all that city's business. If all the three cities have the same amount of business, where should the firm open the branches?

Distance profile	
From A to B	25 km
From A to C	18 km
From B to C	22 km

11. A and B play a game in which each has three coins: a 5 paise, 10 paise and 20 paise coin. Each player selects a coin without the knowledge of the other's choice. If the sum of the coins is an odd amount, A wins B's coin, and if the sum is even, B wins A's coin. Find the best strategy for each player and the value of the game. *(CS, December, 1993)*

12. Assume that two firms are competing for market share for a particular product. Each firm is considering what promotional strategy to employ for the coming period. Assume that the following pay-off matrix describes the increase in market share for firm A and decrease in market share for firm B.

Firm A's strategy	*Firm B's strategy*		
	No promotion	*Moderate promotion*	*High promotion*
No promotion	0	2	–15
Moderate promotion	12	8	–4
Price cut	20	15	6

Now,

(a) determine the optimal strategies for both the firms.

(b) determine the value of this game and state whether it is a strictly determinable and fair game.

(c) state whether as a manager of firm B would you adopt the policy prescribed in (a). Why or why not?

(d) state whether the solution strategies obtained in (a) above would necessarily maximize profit for either of them.

13. Two competitors are competing for the market share of the similar product. The pay-off matrix in terms of their advertising plan is as follows:

Competitor A's Strategies	*Competitor B's Strategies*		
	No Advertising	*Medium Advertising*	*Heavy Advertising*
No Advertising	10	5	–2
Medium Advertising	13	12	13
Heavy Advertising	16	14	10

Suggest the optimal strategies for the two competitors and the net outcome thereof.

(MIB, Delhi, 2008)

14. Two companies, *A* and *P*, competing for the same product, are sharing a market. Each of these attempts to raise its share of the market. An advertising campaign by *A* can increase its share of the market by 7 per cent provided that *P* does nothing. Company *P* contemplates invoking a price cut if it is worthwhile. The price cut shall cause a 5 per cent gain to the company *P*, but the gain is subject to the condition that *A* takes no action. A price cut by *P*, accompanied by the advertising campaign of *A* shall lead to a 2 per cent rise in *P*'s share of market. No action on part of both the companies shall leave the market share for them undisturbed at 40 : 60.

    Write the information in form of a pay-off table and suggest the best strategy for each of the companies to follow.

15. Solve the rectangular game whose pay-off matrix for player *A* is:

	B_1	B_2	B_3
A_1	8	–4	–2
A_2	9	17	16
A_3	10	20	–5

16. Two competitors are competing for the similar product. The pay-off matrix in terms of their advertising plan is shown below:

		Competitor B		
		Large advt.	*Med. advt.*	*Small advt.*
	Large advt.	70	80	50
Competitor A	Med. advt.	90	60	95
	Small advt.	105	90	65

Find the optimal strategies and the value of the game. *(MBA, Delhi, November, 1997)*

17. The management of a corporation is in the process of deciding whether to agree to negotiate with the striking union now or delay. The decision is difficult because the management does not know the union leadership's position. The union leaders may be adamant and insist on their original demands, they may be ready to compromise or they may be ready to yield and accept the original management offer. The matrix of pay-offs to management, as management sees it, is (in Rs 1 million units).

	Union position		
	B_1 Adamant	B_2 Compromise	B_3 Yield
A_1 Negotiate	–2	1	2
A_2 Delay	5	–2	–3

(a) Solve the management's problem.
(b) What should be the union's strategy?
(c) Discuss the implications of a conclusion to adopt a random strategy.

(MBA, Delhi, April, 1978)

18. Two firms, F_1 and F_2, make colour, and black and white television sets. F_1 can make either 300 per colour sets in a month or an equal number of black and white sets, and make a profit of Rs 200 per colour set and Rs 150 per black and white set. F_2 can, on the other hand, make either 600 colour sets, or 300 black and white and 300 colour sets, or 600 black and white sets per month. It also has the same profit margin on the two sets as F_1. Each month there is a market of 300 colour sets and 600 black and white sets and the manufacturers would share market in proportion in which they manufacture a particular type of set.
Write the pay-off matrix of F_1 per month. Obtain F_1 and F_2's optimal strategies and value of the game. Attempt graphically.

19. Determine the optimal strategies and the value of the game from the following 4×4 pay-off matrix game:

X	Y			
	Y_1	Y_2	Y_3	Y_4
X_1	–5	16	13	15
X_2	20	–5	60	–70
X_3	–5	9	12	10
X_4	–20	–2	50	–80

(CS, December, 1995)

20. The pay-off table for a zero-sum game between A and B is shown below, where A is playing to maximise his pay-off. Solve the game for the optimal strategies of the two players and for the value of the game.

	B_1	B_2	B_3	B_4
A_1	150	–18	78	90
A_2	6	102	54	70
A_3	130	–30	78	80

(DMS, Mumbai, 1992)

21. The following matrix represents player A's pay-off in a two-person, zero-sum game:

	Player B		
	30	40	–80
Player A	0	15	–20
	90	20	50

Find the optimal strategies for the two players and value of the game. *(MCom, Delhi, 1985)*

22. From the following two-person, zero-sum game, find the optimal strategy for each player and the value of the game:

		Player B			
		B_1	B_2	B_3	B_4
	A_1	0	–1	3	5
Player A	A_2	–5	2	4	5
	A_3	–2	–3	–4	–2

23. Solve the following game and determine the value of the game also. A's pay-off matrix is as:

		B		
		b_1	b_2	b_3
A	a_1	2	5	6
	a_2	7	3	4

24. The following pay-off matrix describes increase in sales of Firm A from each of its strategies A_1, A_2, A_3, and A_4 against each of the strategies B_1, B_2, B_3 and B_4 of the rival Firm B.

		Firm B			
		B_1	B_2	B_3	B_4
	A_1	43	43	33	13
Firm A	A_2	38	28	23	8
	A_3	48	58	8	18
	A_4	63	68	18	23

Assuming a zero-sum game, show that no optimal strategy-mix exists in pure strategies irrespective of whether we use the principle of dominance or that of maximin and minimax. The game, however, has a solution in mixed strategies but not without the aid of the principle of dominance. Solve the game and find its value. *(M Com, Delhi, 2007)*

25. The effectivity of advertisement in different media by two competitive firms A and B is given by the pay-off matrix (in Rs '00) as follows:

A	*B*		
	Radio	*Television*	*Press*
Radio	3	–2	4
Television	–1	4	2
Press	2	2	6

(i) Test whether there is a saddle point and a pure strategy for optimum returns, with reasons.

(ii) Use dominance property to evaluate the mixed strategies of A and B and the value of the game. Test the results from A's as well as B's point of view. *(ICWA, December, 1995)*

26. There are two companies A and B in a certain city. Both companies have similar reputation and the total number of customers is equally divided between the two companies. Both the companies want to attract a greater number of customers by using different media of advertisement. By seeing the market trend, the company A constructed the following pay-off matrix where numbers in the matrix indicate a gain or a loss of customers to it.

Company A	*Company B*		
	Newspaper	*Radio*	*TV*
Newspaper	40	50	–17
Radio	10	25	–10
TV	100	30	60

Find optimal strategies for both the companies and also the value of the game.

27. Solve the following zero-sum game where player B pays to player A:

Player A	*Player B*				
	B_1	B_2	B_3	B_4	B_5
A_1	2	–4	6	–3	5
A_2	–3	4	–4	1	0

(CA, May, 1985)

28. Solve graphically the following game:

Player A	Player B	
	B_1	B_2
A_1	3	4
A_2	–3	12
A_3	6	–2
A_4	–4	–9
A_5	5	–3

29. Consider the following 'two-person, zero-sum' game:

		Player B's strategies		
		B_1	B_2	B_3
Player A's strategies	A_1	6	11	8
	A_2	9	5	9
	A_3	7	8	8

Suggest optimal strategies for the two players and the value of the game. *(M Com, Delhi, 1995)*

30. Firm X is fighting for its life against the determination of firm Y to drive it out of the industry. Firm X has the choice of increasing price, leaving it unchanged, or lowering it. Firm Y has the same three options. Firm X's gross sales in the event of each of the pairs of choices are shown below:

		Firm Y's pricing strategy		
		Increase price	*Do not change*	*Reduce price*
	Increase price	90	80	110
Firm X's pricing strategies	Do not change	110	100	90
	Reduce price	120	70	80

Assuming firm X as the maximising one, formulate the problem as a linear programming problem. Find the value of the game by using any method. *(MBA, Delhi, December, 1986)*

31. Convert the following game problem into a linear programming problem.

	Player B's Strategies		
Player A's Strategies	5	7	2
	10	4	9
	6	2	0

(M Com, Delhi, 2000)

32. Convert the following game problem, involving 'two-person, zero-sum game' into a linear programming problem:

	Player B			
Player A	8	20	−3	1
	6	25	4	2
	0	−8	12	9
	16	9	21	0

Don't solve.

33. In respect of a two-person, zero-sum game, the following pay-off matrix is given:

Player X	*Player Y*		
	Y_1	Y_2	Y_3
X_1	5	−2	−2
X_2	−2	−2	2
X_3	−2	1	−2

(a) Show that the above game has no saddle point.
(b) Show that the game cannot be reduced by applying dominance rule.
(c) Formulate the game as a linear programming problem for player Y and obtain his optimal strategy.
(d) Obtain the value of the game.

34. Formulate the following two-person, zero-sum game as an LPP and solve for the optimal strategies for both players and the value of the game.

Player A	*Player B*		
	B_1	B_2	B_3
A_1	5	4	7
A_2	5	8	4
A_3	8	5	6

35. The matrix below shows the pay-off in a 3×3 game between two players X and Y. X is maximising player and Y the minimising player.

(a) Formulate a suitable Linear Programme model of the game with respect to the minimising player Y and derive the optimal strategy for Y.

(b) Write down the dual of the above Linear Programme and derive the solution to the dual from the optimal tableau of the primal and interpret the meaning of this solution.

		Y	
	1	–1	–1
X	–1	–1	3
	–1	4	–1

(DMS, Mumbai, 1984)

36. A car manufacturer, Hindustan Motor Company, in a competitive market, has a plant which can produce 5 models of cars in any desired ratio. The company's profit depends upon the models of cars produced by its competitor, Indian Motor Company, and are given in the following table (the figures are in lacs of rupees).

		India Motor Co's model			
		K_1	K_2	K_3	K_4
	J_1	3	4	2	8
	J_2	5	2	7	4
Hindustan Motor Co's model	J_3	2	1	4	5
	J_4	6	4	2	3
	J_5	2	1	9	4

Although the management of the Hindustan Motor Co. knows the models India Motor Company can produce (along with the resulting pay-off), it does not know what model(s) would actually be produced and sold by that company. The management of Hindustan wisheds to maximise the profits. What should its strategy be? If it were known that India Motor Co. would produce model K_4 only, what should Hindustan Motor's decision be?

CHAPTER 16

Dynamic Programming

CHAPTER OVERVIEW

There are certain situations in which a decision is reached by considering the problem in a backward manner, beginning from the end. Consider this. One day, a customer asked for six ounces of oil from a shop. The shop assistant was in a fix as he had only two cups available at the time—a nine ounce and a four ounce cup. He gave it a thought but could not make out how to fill the order. He asked the manager for help. The manager thought for a while and then proceeded to solve the problem. By starting from near the end, he realised that the problem could be easily solved if he could somehow get one ounce of oil filled in the four ounce cup. Because then he could fill the nine ounce cup and empty three ounces from the nine ounce cup into partially filled four ounce cup. At that point, he would be left with six ounces of oil. With this in mind, he evolved the solution as shown below, wherein the initial solution is writen in the last and the final one is written first.

9–oz Cup	4–oz Cup
6	0
6	4
9	1
0	1
1	0
1	4
5	0
5	4
9	0
0	0

Actually, the manager's thinking followed the general thought process underlying dynamic programming, an important tool of quantitative analysis. Since he could not solve the problem directly in one shot, he, in fact, created a second problem namely, if there was one ounce of oil in the four-ounce cup, and then in order to get six ounces oil all that would be done was to add three ounces from the nine-ounce cup. The manager, in turn, created a third problem namely, how to get one ounce oil in the four-ounce cup. Overall, the problem was segmented into smaller problems and their solution led to the solution of the original problem.

Thus, dynamic programming requires a manager to understand as to:

- ❑ How a given problem can be broken into sub-problems when it cannot be solved directly in a single shot.
- ❑ Beginning with the last of the sub-problems, how each of the successive sub-problems may be solved and the result woven into the immediately preceding sub-problem.

While it may sound simple to break a given problem into sub-problems for solution, its mathematical presentation and solution may be found to be relatively tricky.

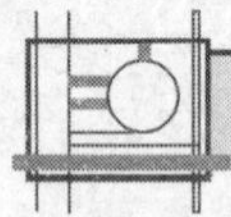

16.1 INTRODUCTION

Dynamic Programming is a useful quantitative analysis technique that can be used to solve many optimisation problems. It deals with relatively large and complex problems and involves making a sequence of interrelated decisions. The technique provides a systematic procedure for determining optimal combination of decisions.

A given problem may be complex if it is not possible to solve it in one go. It may be solved by breaking it into a series of smaller and more tractable sub-problems. This is exactly what dynamic programming does. In most applications, it obtains solutions by working backward from the end of the problem toward the beginning. For this, it divides the given problem into a number of decision stages; the outcome of a decision at a given stage affects the decision at each of the following stages. Solving problems with dynamic programming involves the following four steps:

1. Divide the original problem into sub-problems, called stages.
2. Solve the last stage of the problem for all possible conditions or states.
3. Moving backward from the last stage, solve each intermediate stage successively. This is done by determining optimal policies from that stage to the end of the problem.
4. Obtain the optimal solution to the original problem by solving all stages sequentially.

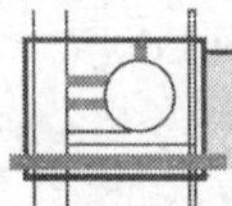

16.2 DYNAMIC PROGRAMMING VERSUS LINEAR PROGRAMMING

Dynamic programming is different from linear programming on two counts. First, there does not exist a standard mathematical formulation of dynamic programming problem. Accordingly, there is no algorithm, like the Simplex method, that may be programmed to solve all problems. Dynamic programming is, instead, a technique that permits us to dissect difficult problems into a sequence of sub-problems, which are then evaluated by stages. This provides a generalised approach to problem solving. Secondly, while linear programming is a method which gives single stage, that is, one time-period solutions, dynamic programming has the power to determine the optimal solution over, say, a one-year time horizon by breaking the problem into twelve smaller one-month time horizon problems and to solve each of these optimally. Thus, it uses a multistage approach to problem solving.

There is a wide variety of problems which can be handled using dynamic programming. Some examples include shortest (or longest) route problems, knapsack problems, distribution of resources, like allocating capital funds, and assigning sales personnel to various marketing areas, smoothing production employment and evaluating investment opportunities.

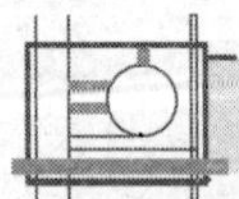

16.3 DYNAMIC PROGRAMMING: A NETWORK EXAMPLE

Many applications of dynamic programming involve locating the shortest (or longest) path joining two points in a given network. We discuss here an application of dynamic programming by considering a prototype example involving the shortest route on a network.

Example 16.1 Let us suppose that a traveller wants to go from city *A* to city *B*. Suppose, further, that the two cities are at a long distance from each other and also that there is no direct link between them. The traveller is, in fact, required to break his journey thrice during the course of his travel. At each break, there are options available, as shown in Figure 16.1. Each circle in the figure represents a *node* or *exchange point*. At an exchange point, the traveller has to leave the old means of transport and shift to a new one. For convenience of reference, the exchange points are numbered from 1 to 11, and the travel time, in hours, is indicated along with every route.

To illustrate, the traveller would begin from city *A* (node 1) and go to either node 2 or node 3. If he moves to node 2, he has to take a new means of transport and decide whether to move to node 4 or node 5. Similarly, if he chooses to go to node 3, he must decide whether to go to node 5 or node 6. In the same manner, he would move from 4, 5 or 6, whichever he lands in, and go to one of the nodes 7, 8, 9, or 10. Finally, he would reach node 11, City *B*, his destination. The problem is to determine the route from City *A* to City *B* (node 1 to node 11) that involves the minimum travelling time.

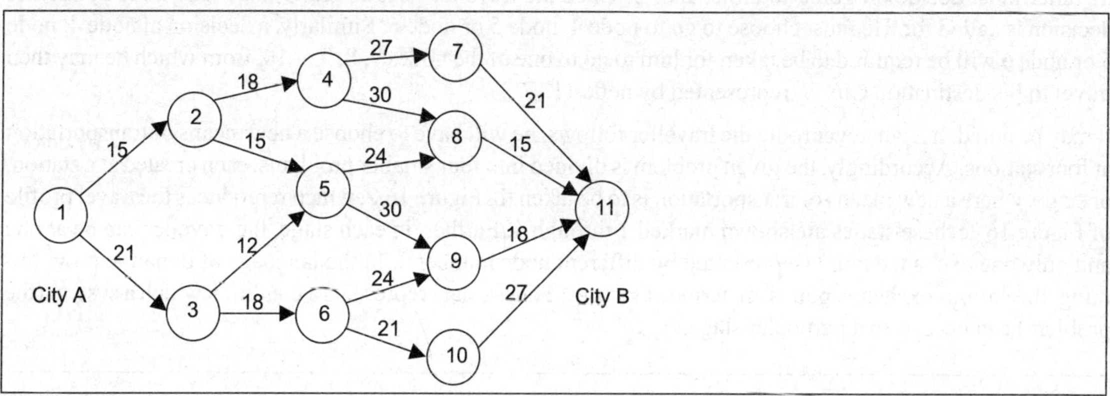

Figure 16.1 *Travel Profile*

Solution by Complete Enumeration

One way to solve this problem is to identify all the routes connecting the two cities and choose the one that involves the least time. In this case, there is a total of eight routes available, as indicated in Table 16.1. Given along with each route is the time needed to cover the total distance.

TABLE 16.1 Route Profile and Time Taken

Route	*Travel time (in hours)*
1–2–4–7–11	81
1–2–4–8–11	78
1–2–5–8–11	69
1–2–5–9–11	78
1–3–5–8–11	72
1–3–5–9–11	81
1–3–6–9–11	81
1–3–6–10–11	87

It is evident from the table that the route 1–2–5–8–11 involves the minimum time, equal to 69 hours. Hence, of all the routes, this is the best one.

It may be observed that in this case the answer has been obtained by complete enumeration and examination of all the routes possible. The enumeration method does work well in relatively small and simple problems like this. In large travel networks, however, a complete listing may become cumbersome and even impractical. Besides, when constraints on travel, like fares, safety and the chances of delays are required to be considered, this approach may not work. Dynamic programming is the answer in such cases.

Solution by Dynamic Programming

As explained earlier, the first step in the dynamic programming solution to a problem calls for decomposition of the given problem into smaller problems, called *stages*. In our problem, a stage is a decision point where the traveller must decide what means of transportation to take next. This is illustrated here. In City *A* (node 1), the traveller must decide on going to either 2 or 3. Once the traveller goes to either node 2 or node 3, another decision is called for. He must choose to go to node 4, node 5 or node 6. Similarly, a decision at node 4, node 5 or node 6 will be required to be taken for him to go to one of the nodes 7, 8, 9 or 10, from which he may then travel to his destination City *B*, represented by node 11.

It may be noted that whatever route the traveller follows, he will have to choose a new means of transportation at four stations. Accordingly, the given problem is divided into four smaller problems, each created at a station, or stage, where a new means of transportation is to be taken. In Figure 16.2, which reproduces the travel profile of Figure 16.1, these stages are shown marked 1 through 4. Further, in each stage, the traveller can be at one and only one exchange point (represented by different node numbers). In the language of dynamic programming, the various exchange points are termed as *states*. Thus, a state represents a condition which a system (the problem) can occupy in a particular stage.

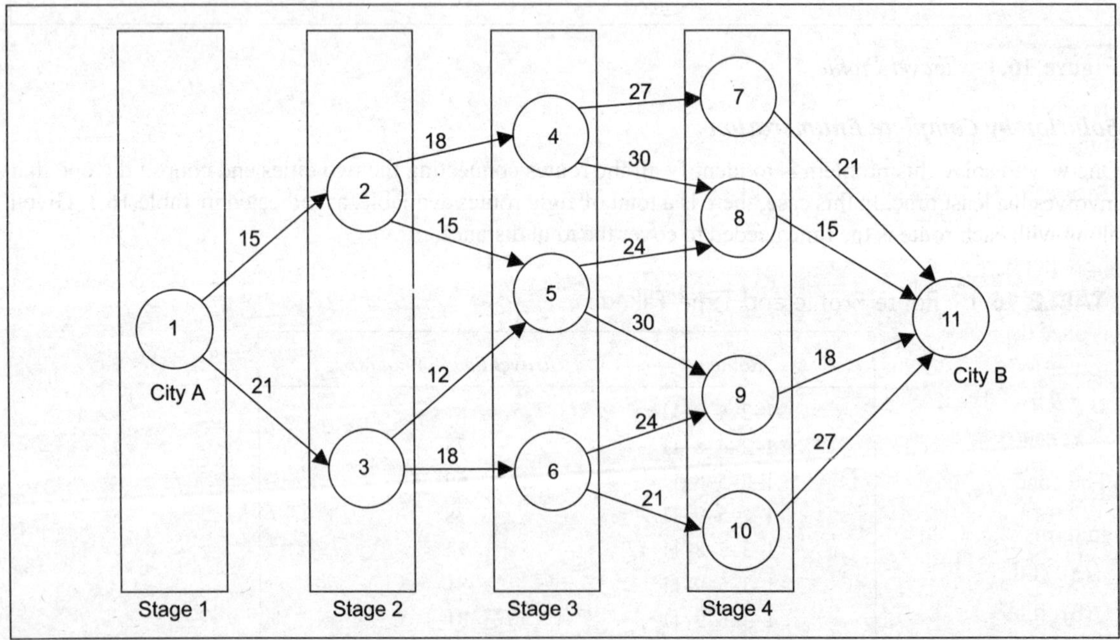

Figure 16.2 *Travel Profile: Stages and States*

Let us now consider the solution to the problem. In dynamic programming, the solution procedure uses roll-back approach. This is to say that the end of the problem is analysed first and then decisions are taken moving backwards until we reach the initial node. To illustrate, the traveller first assumes that he is in City *B*. He then considers how to proceed from any given exchange point to this city so as to involve least travel time. The process starts with the last stage, which is labelled as stage 4. Once this decision is taken, the next decision to be taken is how best to reach in each of the exchange points of stage 4 from those in stage 3. The process is then repeated until City *A* is reached. The detailed analysis of decision-making is given here.

Stage 4 When the traveller has reached City *B*, he has attained his goal. However, immediately before that, he must be in one of the exchange points 7, 8, 9 or 10 of stage 4. It may be seen that there is only one possible way of going from each of these points to the destination, City B. Accordingly, if the traveller is at exchange point 7, then the best (and only) way for him is to travel 7–11 in 21 hours. Thus, no matter how he reaches, once if the traveller reaches at exchange point 7, then he must travel by the route 7–11 for reaching City *B*. *A* similar analysis can be done for all decision points 8, 9 and 10 in stage 4, as there is only one way to reach City *B* from each of them. The results are shown in Table 16.2. it is drawn on *if-then* basis. To illustrate, if the traveller is in exchange point 7, then it is best to continue travelling 7–11, for 21 hours.

TABLE 16.2 Payoffs at Stage 4

State	*Alternative Routes*	*Travel Time (hrs)*	*Best Route Time (hrs)*
7	7–11	21	21
8	8–11	15	15
9	9–11	18	18
10	10–11	27	27

It may be mentioned that the criterion of effectiveness in this case in the travel time, so that the smaller the time, the better it is. Thus, hours of travel is the payoff which is sought to be minimised.

Stage 3 The exchange points 4, 5 and 6 are the states of stage 3. The traveller has to be at one of these points in order to reach any of the exchange points in stage 4: 7, 8, 9 or 10. From each of the states, we examine the best possibility of getting to City *B*. But, instead of considering all routes to this final destination, we examine only the states of stage 4. This is because we have already established as to what the best route to take is once the traveller is at a given exchange point in stage 4. Beginning with the state 4, there are two alternative routes to reach stage 4, namely 4-7 and 4-8. The first of these requires 27 hours plus the optimal time to reach city *B* from state 7 in stage 4, equal to 21 hours, making a total of 48 hours. Similarly, the second of the routes, 4–8, needs 30 hours plus 15 hours from state 8 in stage 4 to the destination city, equalling 45 hours.

The conditional payoffs in respect of each of the states in stage 3 are indicated in Table 16.3.

In terms of Table 16.3, the following are the conditional optimal decisions:

(a) if the traveller is at exchange point 4, he would travel 4-8-11,

(b) if he is at exchange point 5, he would go 5-8-11, and

(c) in case the traveller is at point 6, the best route to follow is 6-9-11.

TABLE 16.3 Payoffs at Stage 3

State	*Alternative Routes*	*Time (hrs)*			*Best Route*
		To stage 4	*From stage 4 to city B*	*Total*	
4	4–7	27	21	48	
	4–8	30	15	45	√
5	5–8	24	15	39	√
	5–9	30	18	48	
6	6–9	24	18	42	√
	6–10	21	27	48	

Stage 2 In order to get to either exchange point 4, 5 or 6 in stage 3, the traveller has to be at exchange point 2 or 3–the states of stage 2. Here also, we seek to find the optimal route from each of these points to City *B*. But, instead of enumerating all routes to the destination city, only the routes to the states of stage 3 are examined and the results obtained in stage 3 serve as input for computation in stage 2. The computations are shown in Table 16.4.

TABLE 16.4 Payoffs at Stage 2

State	*Alternative Routes*	*Time (hrs)*			*Best Route*
		To stage 3	*From stage 3 to city B*	*Total*	
2	2–4	18	45	63	
	2–5	15	39	54	√
3	3–5	12	39	51	√
	3–6	18	42	60	

Stage 1 Now the traveller is at the initial point, City *A*, where he has two alternatives: either go to exchange point 2 or exchange point 3. There is only one state to examine here. The results are given in Table 16.5.

TABLE 16.5 Payoffs at Stage 1

State	*Alternative Routes*	*Time (hrs)*			*Best Route*
		To stage 2	*From stage 2 to city B*	*Total*	
1	1–2	15	54	69	√
	1–3	21	51	72	

Optimal solution: Now we are in a position to obtain the optimal solution to the whole problem. This is done by a forward pass, i.e., moving forward from stage 1 through stage 4. To illustrate, in stage 1 we decide to go 1–2. In the stage 2, we know that if one is at state 3, the best course is to go to state 5 (Table 16.4). Further, from Table 16.3, we know that once one is at state 5, the best route to follow is 5–8. Finally, from state 8, the route to be followed is 8–11 (Table 16.2). Accordingly, the optimal route for the traveller is 1–2–5–8–11, involving a total distance of 69 hours, which is identical to the one obtained earlier.

It may be observed that in working for the solution to the problem, at every state of a given stage, the total time has been calculated as equal to the time to previous (i.e. previously analysed) stage *plus* the best route time from previous stage to City *B*, wherein the *best* route time refers to the *smallest* route time. Now since time is the variable of interest here, we say that the *payoff* (or reward) is in terms of time, which is sought to be minimised. Accordingly, we can formally express the above calculation as Total payoff (or reward) at stage *n* = Immediate payoff + Optimal payoff in the previous stage (stage $n + 1$, since analysis follows in a backward manner). Further, optimal payoff is the smallest payoff of all, keeping in view nature of the variable of interest.

Further, it may also be observed that the use of dynamic programming obviates the necessity to list and evaluate all possible routes as is donc in complete enumeration. Although computational work in the traveller's example is seen to be much larger in using dynamic programming than that required for complete enumeration approach, in larger problems it usually is the other way.

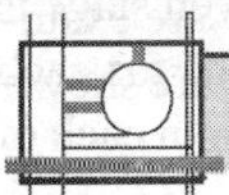

16.4 TERMINOLOGY

We can now formally define the basic concepts and terms used in dynamic programming. These are discussed here.

1. Stages A fundamental idea employed by dynamic programming is to divide the given problem into sub-problems, which are called *stages*. Actually, a stage represents a decision point. It means that at every stage, a policy decision is required to be made. As illustrated, the traveller's example was solved by breaking the problem into four sub-problems, represented as stages 1 through 4. At every stage, a decision was to be taken as to which exchange point to select for the next means of transportation.

2. States Each stage has a number of *states* associated with it. In general terms, states are the various possible conditions in which the system might be in that stage of the problem. To illustrate, in the traveller's case, at stage 2, there are two states, with nodes numbered 2 and 3, while at stage 3, the three states are numbered 4, 5 and 6. At stage 2, the traveller can be only at node 2 or at node 3. Similarly, at stage 3, the traveller can be at one of the three possible states—nodes 4, 5 or 6. While travelling along the route, the traveller moves from one state in one stage to another state in the next stage. To make the correct decision at any stage, the traveller does not need to know how he got to that location. For example, if the traveller is in state 5 at stage 3, then his remaining decision does not depend on how he goes to state 5, rather his future decisions depend on just the fact that now he is at the place represented as state 5.

3. Payoff The payoff, also known as reward, is the variable of interest. To illustrate, in the traveller's example, our concern was travel time which was sought to be minimised. In each stage, the decision taken is such as would minimise the total travel time from each of the exchange points (states) to the City *B*, the destination. It goes without saying that depending on the nature of the underlying variable, the solution may call for maximisation or minimisation of payoff.

4. Decision Process The solution to a problem by dynamic programming may be visualised as a decision process wherein we move from one stage to another, making a decision at each stage. The effect of a policy decision at each stage is to transform the current state to a state associated with the beginning of the next stage. The solution procedure is designed to finding an *optimal policy* for the entire problem, which consists of determining the optimal policy decision at each stage for each of the possible states. To illustrate, in the traveller's example, the solution procedure involved constructing a table for each stage that prescribed the optimal decision for each possible state.

The determination of optimal policy at a given state requires calculation of *immediate* and *total* payoffs. It may be mentioned that the immediate payoff is the payoff associated with a move between two adjacent stages. Similarly, the total payoff for each state in every stage is the sum of immediate payoff and the *optimal payoff* obtained in the previous stage. To illustrate, in Table 16.4, the column headed "Time to stage 3" represents immediate payoff for each alternative route (state to state) between two adjacent stages 2 and 3. Similarly, the column titled "Total time" which contains sum total of the "Time to stage 3" and "Time from stage 3 to City B" indicates total payoff. The later of these, viz. "Time from stage 3 to City B" is in fact the optimal payoff at the previous stage, stage 3. (Remember that the solution process involves reverse movement–stage 4, stage 3, stage 2, etc.) The optimal payoff at the previous stage is obtained with reference to the preceding table (Table 16.3).

It may further be noted that the best of all the total payoffs in respect of the various states in a given stage is called the optimal total payoff. In the context of our traveller's example, the optimal total payoff for each exchange point is the shortest route (in terms of the time taken) from that exchange point to City B. In each of Table 16.2 through 16.5, the optimal total payoffs are shown marked in the last column.

5. Principle of Optimality Given the current state, the optimal policy for the remaining stages does not depend on the policy decisions adopted in the previous stages. Accordingly, the optimal immediate decision depends only on the current state and not how you got there. This is known as the *principle of optimality.* To illustrate, in the traveller's example, if the traveller is in state 5 at stage 3, then his remaining decision does not depend on how he goes to state 5, rather his future decisions just depend on the fact that now he is at exchange point represented as state 5.

6. Recursive Relationship Formulation of a dynamic programming problem requires the creation of recursive relationship. It needs us to define an optimal payoff function for every stage and state. The recursive relationship establishes the relationship between the optimal payoff functions at successive stages. In dynamic programming, it is possible to write a recursive relationship for every problem. Once it is written, it becomes much easier to solve the given problem.

In other words, the recursive relationship expresses the notion that optimal payoff at any given stage, for any given state, is given as the value of the best alternative, when each alternative includes the total of the immediate payoff and the optimal payoff obtained in the previous stage. Accordingly, a recursive relationship identifies the optimal policy for stage n, given the optimal policy for stage $n + 1$ is available.

In general terms, the recursive relationship may be formulated and stated as follows. Let

N = number of stages

n = index for current stage ($n = 1, 2, \ldots N$)

$n + 1$ = previous stage (remember that successive stages are numbered in descending order)

S_n = state of the system in the current stage for which the recursive relationship holds

S_{n+1} = state in the previous stage

$f_n(S_n)$ = total payoff for each alternative, starting from state S_n, in stage n, to the end of the process

$f_n^*(S_n)$ = optimal total payoff (i.e. the best $f_n(S_n)$) from stage S_n in stage n.

$f_{n+1}^*(S_{n+1})$ = optimal total payoff obtained in the previous stage

$C_n(S_n, x_n)$ = immediate payoff in stage n when decision x_n is made for a specific value S_n of the state variable

x_n = decision among alternatives made at stage n in the state under consideration

The recursive relationship (for minimisation) for state S_n at stage n may be stated as under:

$$f_n^*(S_n) = \min_{x_n} \{C_n(S_n, x_n) + f_{n+1}^*(S_{n+1})\}$$

where $C_n(S_n, x_n)$ is obtained by the values of c_{ij} when i is the current state and j is the current destination. For problems in which the variable of interest (payoff) is sought to be maximised, the recurssive relationship would be:

$$f_n^*(S_n) = \max_{x_n} \{C_n(S_n, x_n) + f_{n+1}^*(S_{n+1})\}$$

We now consider to fit the recursive relationship to the traveller's problem and solve the problem in terms of this relationship by rollback, or *backward recursion*, that is from the right to the left.

Beginning with stage 4, the possible states are 7, 8, 9 and 10. Since this is the only the first stage to be evaluated, we have $f_4^*(7) = C_{7,11} = 21$, $f_4^*(8) = C_{8,11} = 15$, $f_4^*(9) = C_{9,11} = 18$, and $f_4^*(10) = C_{10,11} = 27$, as shown in Table 16.2.

Moving to stage 3, the states are 4, 5 and 6. From state 4, the traveller can go to state 7 or state 8; from 5 he can move to states 8 or 9, while from state 6, the traveller can move either to state 9 or to state 10. The travel times are obtained for each of these and then optimal times at each of the states in the relevant previous stage are added. To illustrate, total payoffs are shown calculated here:

for 4 to 7 : 27 + 21 = 48, and

(immediate payoff) ($f_4^*(7)$)

for 4 to 8 : 30 + 15 = 45

(immediate payoff) ($f_4^*(8)$)

Since the lower of the two payoffs is 45, the decision at state 4 (i.e. node 4) in stage 3, (x_n), is 4–8. Accordingly.

$$f_3^*(4) = C_3(4, 8) + f_4^*(8) = 30 + 15 = 45$$

Similarly, for state 5, the optimal payoff would be lower of the two payoffs:

(i) 5 to 8: $[C_{5,8} + f_4^*(8)] = (24 + 15) = 39$

(ii) 5 to 9: $[C_{5,9} + f_4^*(9)] = (30 + 18) = 48$

Accordingly,

$$f_3^*(5) = C_3(5, 8) + f_4^*(8) = 39$$

In the same manner, for state 6, we have min (24 + 18, 21 + 27) = 42. Thus,

$$f_3^*(6) = C_3(6, 9) + f_4^*(9) = 42$$

Next, proceeding from stage 2, for state 2 we have

$$f_2^*(2) = \min [C_{2,4} + f_3^*(4), C_{2,5} + f_3^*(5)]$$
$$= \min [18 + 45, 15 + 39)] = 54,$$

while for state 3,

$$f_2^*(3) = \min [C_{3,5} + f_3^*(5), C_{3,6} + f_3^*(6)]$$
$$= \min [12 + 39, 18 + 42] = 51.$$

Thus,

$$f_2^*(2) = C_2(2, 5) + f_3^*(5) = 54, \text{ and}$$
$$f_2^*(3) = C_2(3, 5) + f_3^*(5) = 51$$

Finally, in stage 1, the only state is state 1. Thus,

$$f_1^*(1) = \min [C_{1,2} + f_2^*(2); C_{1,3} + f_2^*(3)]$$
$$= \min [15 + 54, 21 + 51] = 69$$

Accordingly,

$$f_1^*(1) = C_1(1, 2) + f_2^*(2) = 69$$

Now, we have the optimal decision which requires the traveller to move from node 1 to node 2, from node 2 to node 5, from node 5 to node 8 and, finally, from there to node 11, to minimise the total travel time. The minimum time is 69 hours.

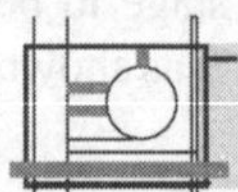

16.5 DETERMINISTIC VERSUS PROBABILISTIC DYNAMIC PROGRAMMING

Dynamic programming is said to be deterministic when it is applied to the solution of deterministic problems, while it is termed probabilistic when applied in relation to probabilistic problems. We explain these in turn.

16.5.1 Deterministic Dynamic Programming

The traveller's example discussed earlier illustrates application of dynamic programming approach to the solution of deterministic problems. A deterministic problem is one in which the state at the next stage is completely determined by the state and the policy decision taken at the current stage. To illustrate, let us reconsider Table 16.4 in traveller's example. It is evident that if the traveller is at exchange point 2 (i.e. state 2 in stage 2) then he should go to state 5 (and not 4) in the next stage. Similarly, if he is in state 3, he should move to state 5 and not state 6.

In general terms, the deterministic dynamic programming may be described as shown in Fig. 16.3.

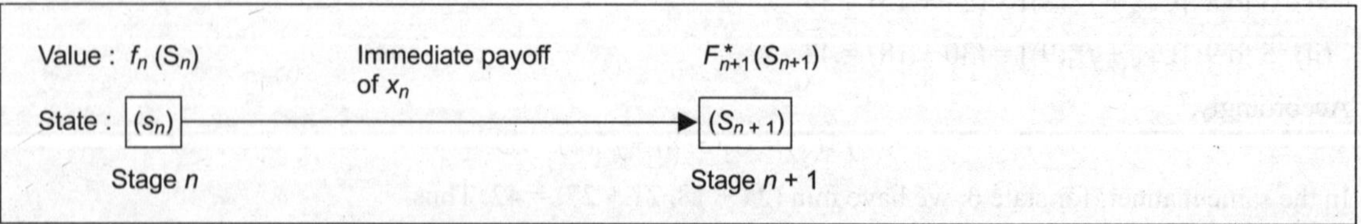

Figure 16.3 *Basic Structure of Deterministic Dynamic Programming*

Accordingly, at stage n, the system is in some state S_n. A policy decision then moves the system to state $n+1$ at stage $n+1$, where the payoff under optimal policy from that point until the end $f^*_{n+1}(S_{n+1})$ is available. To this payoff is added the immediate payoff resulting from the policy decision x_n. The combination of these two payoffs yields the total payoff $f_n(S_n, x_n)$. Finally, optimising with respect to x_n, we get $f^*_n(S_n)x^*_n$. After x^*_n and $f^*_n(S_n)$ are obtained for each possible value of S_n, the solution procedure can move back one stage.

We now consider another example of deterministic dynamic programming.

Example 16.2 A company has Rs 6,00,000 to invest, and a total of three investments are available to it. If x_j rupees (in lakhs) are invested in investment j, then a net present value (in lakhs of rupees) of $v_j(x_j)$ is obtained, where $v_j(x_j)$'s are given below:

$$v_1(x_1) = 7x_1 + 2$$

$$v_2(x_2) = 3x_2 + 7$$

$$v_3(x_3) = 4x_3 + 5$$

$$(x_1, x_2, x_3 > 0) \text{ and } x_1(0) = x_2(0) = x_3(0) = 0$$

The amount placed in each investment has to be in multiples of Rs 1 lakh only. How should the company invest Rs 6,00,000 in order to maximise the net present value obtained from the investments?

To solve this problem, we first define the parameters as given here.

Stages: Each investment may be considered as a stage. Accordingly, let the three investments be stages 1, 2 and 3 respectively. Thus, $N = 3$.

States: The amount of funds available for investing in a particular investment. For each of the stages 2 and 3, the states are 0, 1, 2, 3, 4, 5 and 6 (amount in lakhs of rupees). For stage 1, there is only one state, state 6, the total amount of funds available.

Decision: The amount of funds to be invested in each of the investment options.

Recursive Relationship:

$$f^*_n(S_n) = \max_{x_n} \{v_n x_n + f^*_{n+1}(S_n - x_n)\}$$

wherein

S_n is the amount of funds available for investment

$v_n x_n$ is the immediate payoff—the payoff from current investment

$f^*_n(S_n)$ is the optimal payoff at stage n, for stage S_n

$f^*_{n+1}(S_n - x_n)$ is the optimal payoff at the previous stage.

The solution by rollback follows.

Stage 3 We begin by stage 3, which involves investment in the third option. The solution at this stage is trivial and we have the (net present value) payoff function for various states as defined in Table 16.6.

Notice that if no investment is made, the payoff would be zero. For other investment amounts, the payoff are obtained by substituting the amount x_n in the equation $4x_n + 5$.

TABLE 16.6 Determination of Optimal Payoff at Stage 3

Funds available for allocation in the third investment (Rs lakhs) State S_3	$f_3(S_3) = v_n x_n = 4x_n + 5$								
	$x_n = 0$	$x_n = 1$	$x_n = 2$	$x_n = 3$	$x_n = 4$	$x_n = 5$	$x_n = 6$	$f^*_3(S_3)$	x^*_n
0	0							0	0
1	0	9						9	1
2	0	9	13					13	2
3	0	9	13	17				17	3
4	0	9	13	17	21			21	4
5	0	9	13	17	21	25		25	5
6	0	9	13	17	21	25	29	29	6

Stage 2 The payoffs at this stage are as shown in Table 16.7 and the optimal payoffs for each option arc indicated.

TABLE 16.7 Determination of Optimal Payoff at Stage 2

Funds available for allocation in second and the third investments (in lakhs of Rs) State S_2	$f_2(S_2) = (3x_2 + 7) + f^*_3(S_2 - x_2)$								
	$x_2 = 0$	$x_2 = 1$	$x_2 = 2$	$x_2 = 3$	$x_2 = 4$	$x_2 = 5$	$x_2 = 6$	$f^*_2(s_2)$	x^*_n
0	0							0	0
1	9	10						10	1
2	13	19	13					19	1
3	17	23	22	16				23	1
4	21	27	26	25	19			27	1
5	25	31	30	29	28	22		31	1
6	29	35	34	33	32	31	25	35	1

To illustrate, suppose we consider $S_2 = 5$, that is, Rs 5 lakh are available for investments second and third. Now, if $x_2 = 1$, that is to say, Rs 1 lakh are put in second investment, then return from this would be $3 \times 1 + 7 =$ Rs 10 lakh, while $S_2 - x_2$ or $5 - 1 =$ Rs 4 lakh would be available for the third investment. From Table 16.6, it is evident that the optimal payoff for this state is 21. Thus, total payoff $= 10 + 21 = 31$.

Stage 1 Finally, we obtain the payoffs at stage 1 and determine the optimal policy. This is shown in Table 16.8.

TABLE 16.8 Determination of Optimal Payoff at Stage 1

State S_1 *Funds available for three investments*	$f_1(S_1) = (7x_1 + 2) + f_2^*(6 - x_1)$								
	$x_1 = 0$	$x_1 = 1$	$x_1 = 2$	$x_1 = 3$	$x_1 = 4$	$x_1 = 5$	$x_1 = 6$	$f_1^*(s_1)$	x_n^*
6	35	40	43	46	49	47	44	49	4

Optimal Solution We observe from Table 16.8 that the optimal policy calls for placing Rs 4 lakh in the first proposal since $x_1^* = 4$. Further, it leaves Rs 2 lakh for the second and third proposals. From Table 16.7, we observe that in such a condition, the optimal policy calls for investing Rs 1 lakh in the second investment proposal. With this, the amount left for the third investment is Rs 1 lakh. Thus, the optimal policy calls for investing Rs 4 lakh, Rs 1 lakh and Rs 1 lakh respectively in the three investments respectively. The maximum NPV obtainable is Rs 49 lakh.

Some valuable observations can be made from the solution to this example.

1. For every state in each stage, the optimal return in terms of net present value (NPV) is computed.
2. A sensitivity analysis may be performed. To illustrate, if the company chooses to consider only two proposals, then the optimal solution can be found in the intermediate computations. Thus, if only the second and third investment proposals are considered, the best policy, as may be read from Table 16.7, is to invest Rs 1 lakh and Rs 5 lakh in the second and the third proposals respectively.
3. From the dynamic programming solution procedure, we can also identify the second best alternative. For example, from Table 16.8, the second highest payoff 47 (or NPV = Rs 47 lakh) is seen to be when Rs 5 lakh are invested in the first investment option and the Rs 1 lakh remaining are invested in the second investment option (Table 16.7).
4. The addition of a new investment proposal merely adds an additional stage while the availability of more funds adds more states to the calculations.

Dynamic Programming Solution to LPPs Dynamic programming can be used for solving linear programming problems as well, although only relatively small-sized problems can be handled this way. Even non-linear programming problems (where the assumption of linearity is not satisfied) can also be handled by this method. We discuss the solution to an LPP here. Consider Example 3.1 again, which is reproduced below.

Example 16.3

Maximise $Z = 40x_1 + 35x_2$

subject to

$$2x_1 + 3x_2 \leq 60$$
$$4x_1 + 3x_2 \leq 96$$
$$x_1, x_2 \geq 0$$

Solve this LPP using dynamic programming.

In this problem, there are two interrelated decisions to be made, the number of units of products A and B, respectively, to be produced. In other words, the problem is to determine level of activity $1(x_1)$ and activity 2 (x_2). In the context of dynamic programming, the two can be interpreted as the two stages. Thus, we let n = activity n (where $n = 1, 2$). Accordingly, x_n represents the decision variable at stage n.

The states here are the total amount of resources available for allocation to the activities remaining at any stage. Accordingly,

State S_n = amount of respective resources available for allocation to remaining activities. Thus,
$S_n = (R_1, R_2)$

where R_i is the amount of resource i remaining to be allocated ($i = 1, 2$). We have, therefore

$$S_1 = (60, 96)$$
$$S_2 = (60 - 2x_1, 96 - 4x_1)$$

However, when we begin to solve at stage 2, we do not know the value of x_1. Hence, we begin the solution by taking $S_2 = (R_1, R_2)$ at that stage.

It may be mentioned that both the state variables are continuous and not discrete as in the previous examples considered. When the variables are discrete, only a limited number of specific values are needed to be considered for function maximisation/minimisation. But since continuous variables may take *any* value in a given range, we have to find the *point* at which the relevant payoff function is optimised. Now, we may define the payoff functions as follows:

$f_2(R_1, R_2; x_2)$ = contribution of activity 2 to Z (profit) if the system begins at stage 2 in state (R_1, R_2) and the decision is x_2.
$= 35x_2$

$f_1(60, 96; x_1)$ = contribution of the two activities 1 and 2 to Z when system begins in state (60, 96) at stage 1, with immediate decision as x_1 and optimal decision at stage 2.

$$= 40x_1 + \max_{\substack{3x_2 \le 60 - 2x_1 \\ 3x_2 \le 96 - 4x_1 \\ x_2 \ge 0}} (35x_2)$$

The optimal payoff at stage $n (n = 1, 2)$,

$$f_n^*(R_1, R_2) = \max_{x_n} f_n(R_1, R_2; x_n)$$

Accordingly,

$$f_2^*(R_1, R_2) = \max_{\substack{3x_2 \le R_1 \\ 3x_2 \le R_2 \\ x_2 \ge 0}} \{35x_2\} \quad \text{(i)}$$

$$f_1(60, 96; x_1) = 40x_1 + f_2^*(60 - 2x_1, 96 - 4x_1), \quad \text{(ii)}$$

and

$$f_1^*(60, 96) = \max_{\substack{2x_1 \le 60 \\ 4x_1 \le 96 \\ x_1 \ge 0}} \{40x_1 + f_2^*(60 - 2x_1, 96 - 4x_1)\} \quad \text{(iii)}$$

In this, equation (i) will be used to solve the problem at stage 2, while equation (iii) shows the recursive relationship. The solution follows.

Stage 2 Equation (i) reveals that x_2^* would be the largest value of x_2, which simultaneously satisfies the conditions $3x_2 \le R_1$, $3x_2 \le R_2$, and $x_2 \ge 0$. For feasible solutions to exist, $R_1 \ge 0$ and $R_2 \ge 0$ should also hold. Here the largest such value of x_2 would be equal to the smaller of the values $R_1/3$ and $R_2/3$. Accordingly, we have the results shown in Table 16.9.

TABLE 16.9 Determination of Optimal Payoff at Stage 2

State (R_1, R_2)	$f_2^*(R_1, R_2)$	x_2^*
$R_1 \geq 0\ R_2 \geq 0$	$35 \min \left\{\frac{R_1}{3}, \frac{R_2}{3}\right\}$	$\text{Min}\left\{\frac{R_1}{3}, \frac{R_2}{3}\right\}$

Stage 1 To solve the two stage problem (for $n = 1$), we substitute the solution just obtained for $f_2^*(R_1, R_2)$ in equation (iii). We know that for stage 2,

$$(R_1, R_2) = (60 - 2x_1, 96 - 4x_1) \text{ so that}$$

$$f_2^*(60 - 2x_1, 96 - 4x_1) = 35 \min \left\{\frac{R_1}{3}, \frac{R_2}{3}\right\}$$

$$= 35 \min \left\{\frac{60 - 2x_1}{3}, \frac{96 - 4x_1}{3}\right\}$$

Now,

$$f_1^*(60, 96) = \max_{x_1 \geq 0}\left[40x_1 + 35 \min\left\{\frac{60 - 2x_1}{3}, \frac{96 - 4x_1}{3}\right\}\right]$$

To solve it, we first set $x_1 = 0$. This gives $f_1(60, 96) = 700$. Next, we observe that the maximum value that x_1 can assume (such that none of the terms in the expression, namely, $(60 - 2x_1)/3$ and $(96 - 4x_1)/3$ becomes negative) is 24. When $x_1 = 24$, $f_1(60, 96) = 960$. Now, we determine if any value of x_1 between 0 and 24 yields a higher value of the function. For this, if we set $(60 - 2x_1)/3 = (96 - 4x_1)/3$, we get $x_1 = 18$. When $x_1 = 18$, we have $f_1(60, 96) = 1000$, which is higher than the values obtained earlier. Further, when $x_1 = 18, x_2^* = 8$. Accordingly, we have $x_1^* = 18$, $x_2^* = 8$ and $f_1^*(60, 96) = 1000$. This is the optimal solution and it matches with the solution obtained in Chapter 3.

16.5.2 Probabilistic Dynamic Programming

Probabilistic dynamic programming is different from deterministic dynamic programming in the following way. While in the deterministic dynamic programming, the state in the next stage is completely determined by the state and policy decision at the current stage, in probabilistic dynamic programming there is a probability distribution for what the next state would be. This idea is presented in Figure 16.4.

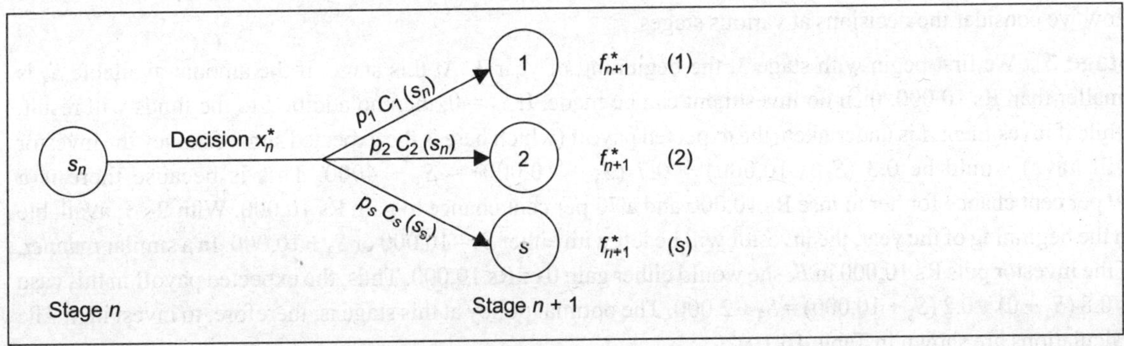

Figure 16.4 *Probabilistic Dynamic Programming*

As indicated in the figure, let the number of states at stage $n + 1$ be S, indicated as 1, 2, ..., S, and the system goes to a state i with a probability p_i, given state n and decision x_n at stage n. Further, if the system goes to state i, the immediate payoff of stage n to the function is c_i. For such a situation, we obtain the expected payoff at stage n as $f_n(S_n) = \Sigma p_i(c_i + f^*_{n+1}(S_{n+1}))$. Consider the following example.

Example 16.4 An investor has Rs 10,000 to invest and she has an opportunity to invest the amount in either of two investments *A* or *B*, at the beginning of each of the following three years. The investment *A* results either in a total loss of the amount invested with probability 0.3, or a gain of Rs 10,000 over and above the amount invested, with probability 0.7. On the other hand, the investment *B* results in a zero gain with a probability 0.8 or a gain of Rs 10,000 with probability 0.2. The conditions allow to make at the most one investment each year and she can invest only Rs 10,000 each time and any additional funds accumulated are left idle.

Using dynamic programming, find the investment policy that maximises the expected amount of money she will have after three years.

This problem involves a sequence of decisions. In each of the years, a decision has to be made whether or not the investment can be made and, if the investment is to be made then whether it is *A* or *B*. Accordingly, we define the following parameters:

Stages : Each year is considered a stage. Thus, there are three stages, $n = 1, 2, 3$.

States : The states are the funds available for investment. Thus, S_n is the amount at hand at the beginning of a stage, n.

Payoff : If x_n be the investment made in the year n_j (so that $x_n = O, A, B$), the payoff function is $f_n(S_n, x_n)$ which is the expected amount of money at the end of the third year, given S_n and x_n in the year. From the given information, we have

$$\text{For} \quad S_n \geq 10{,}000 \quad f_n(S_n, x_n) = \begin{cases} f^*_{n+1}(S_n) & x_n = 0 \\ 0.3 f^*_{n+1}(S_n - 10{,}000) + 0.7 f^*_{n+1}(S_n + 10{,}000) & x_n = A \\ 0.8 f^*_{n+1}(S_n) + 0.2 f^*_{n+1}(S_n + 10{,}000) & x_n = B \end{cases}$$

For $\quad 0 \leq S_n < 10{,}000 \quad f_n(S_n, x_n) = f^*_{n+1}(S_n)$ and $\quad x_n = 0$

(since she cannot invest less than Rs 10,000)

Now we consider the decisions at various stages.

Stage 3 We first begin with stage 3, the beginning of year 3. At this stage, if the amount available S_3 is smaller than Rs 10,000, then no investment can be made. If $x_3 = 0$, then no addition to the funds will result, while if investment *A* is undertaken, the expected payoff (which here is the expected sum of money the investor will have) would be $0.3\,(S_3 - 10{,}000) + 0.7\,(S_3 + 10{,}000) = S_3 + 4000$. This is because there is a 30 per cent chance for her to lose Rs 10,000 and a 70 per cent chance to gain Rs 10,000. With Rs S_3 available in the beginning of the year, the investor will be left with either $S_3 - 10{,}000$ or $S_3 + 10{,}000$. In a similar manner, if the investor puts Rs 10,000 in *B*, she would either gain 0 or Rs 10,000. Thus, the expected payoff in this case is $0.8\,(S_3 + 0) + 0.2\,(S_3 + 10{,}000) = S_3 + 2{,}000$. The optimal policy at this stage is, therefore, to invest in *A*. The calculations are shown in Table 16.10.

TABLE 16.10 Determination of Optimal Policy at Stage 3

State $S_{(3)}$ *Amount available for investment in year 3*	$f_3(S_3\ x_3)$			$f_3^*(S_3)$	x_n^*
	$x_3 = 0$	$x_3 = A$	$x_3 = B$		
$0 \le S_3 < 10{,}000$	S_3	–	–	S_3	0
$S_3 \ge 10{,}000$	S_3	$S_3 + 4{,}000$	$S_3 + 2{,}000$	$S_3 + 4{,}000$	A

Stage 2 At this stage, the options available are again 0, A and B, while the amount available for investment is indicated in the first column of Table 16.11. The calculation of payoffs indicated in this table is shown here. If the amount available, S_2, is less than Rs 10,000, then no investment can be made and it would remain the same S_2. If the amount available is Rs 10,000 or more but less than Rs 20,000 and

(i) *If no investment is made* ($x_2 = 0$): This would imply that $S_3 = S_2$.

(ii) *If investment is made in* $A(x_2 = A)$*:* This would mean that the total payoff $f_2(S_2, A) = 0.3\, f_3^*(S_2 - 10{,}000) + 0.7 f_3^*(S_2 + 10{,}000)$.

Now, since $S_3 = (S_2 - 10{,}000) \le 10{,}000, f_3^*(S_2 - 10{,}000) = S_2 - 10{,}000$.

Also, since $S_3 = (S_2 + 10{,}000) \ge 10{,}000, f_3^*(S_2 + 10{,}000) = (S_2 + 10{,}000) + 4{,}000 = S_2 + 14{,}000$. Accordingly,

$$f_2^*(S_2, A) = 0.3\,(S_2 - 10{,}000) + 0.7\,(S_2 + 14{,}000) = S_2 + 6{,}800$$

(iii) *If investment is made in* B ($x_2 = B$)*:* It would result in total payoff $f_2(S_2, B) = 0.8\, f_3^*(S_2) + 0.2\, f_3^*(S_2 + 10{,}000)$. Since $S_3 = S_2, f_3^*(S_2) = S_2 + 4000$ (because $S_2 \ge 10{,}000$). Similarly, since $S_3 = S_2 + 10{,}000$, $f_3^*(S_2 + 10{,}000) = (S_2 + 10{,}000) + 4{,}000$. Accordingly, $f_2(S_2, B) = 0.8\,(S_2 + 4{,}000) + 0.2\,(S_2 + 14{,}000) = S_2 + 6{,}000$.

Further, if at least Rs 20,000 is available for investment, and

(i) *If no investment is made* ($x_2 = 0$)*:* It implies $S_3 = S_2$. The total payoff $f_2(S_2, 0) = S_2 + 4{,}000$.

(ii) *If investment is made in* $A(x_2 = A)$*:* In either outcome of the investment, S_3 would be at least Rs 10,000. Thus, $f_2^*(S_2, A) = 0.3\, f_3^*(S_2 - 10{,}000) + 0.7\, f_3^*(S_2 + 10{,}000) = 0.3\,[S_2 - 10{,}000 + 4{,}000] + 0.7\,[S_2 + 10{,}000 + 4{,}000] = S_2 + 8{,}000$.

(iii) *If investment is made in* $B(x_2 = B)$*:* For either outcome of investment B, the amount available for investment at stage 3, S_3, would be at least Rs 20,000. Thus, $f_2(S_2, B) = 0.8\, f_3^*(S_2 - 10{,}000) + 0.2\, f_3^*(S_2 + 10{,}000) = 0.8\,(S_2 + 4{,}000) + 0.2\,(S_2 + 10{,}000 + 4{,}000) = S_2 + 6{,}000$.

TABLE 16.11 Determination of Optimal Policy at Stage 2

State S_2 *Amount available for investment in year 2*	$f_2(S_2, x_2)$			$f_2^*(S_2)$	x_n^*
	$x_2 = 0$	$x_2 = A$	$x_2 = B$		
$0 \le S_2 < 10{,}000$	S_2	–	–	S_2	0
$10{,}000 \le S_2 < 20{,}000$	$S_2 + 4{,}000$	$S_2 + 6{,}800$	$S_2 + 6{,}000$	$S_2 + 6{,}800$	A
$S_2 \ge 20{,}000$	$S_2 + 4{,}000$	$S_2 + 8{,}000$	$S_2 + 6{,}000$	$S_2 + 8{,}000$	A

Stage 1 At this stage, there is only one state since Rs 10,000 are available for investment. Thus, $S_1 = 10{,}000$. If $x_1 = 0$, so that no investment is made, then $S_2 = 10{,}000$. Accordingly, $f_1^*(S_1, 0) = 10{,}000 + 6800 = 16{,}800$. Now, if $x_1 = A$, then $S_2 = 10{,}000 - 10{,}000 = 0$ or $S_2 = 10{,}000 + 10{,}000 = 20{,}000$. Accordingly, $f_2^*(S_1 - 10{,}000) = 0$ and $f_2^*(S_1 + 10{,}000) = S_2 + 8{,}000 = 20{,}000 + 8{,}000 = 28{,}000$. Thus, $f_1^*(S_1, A) = 0.3 f_2^*(S_1 - 10{,}000) + 0.7 f_2^*(S_1 + 10{,}000) = 0.3 \times 0 + 0.7 \times 28{,}000 =$ Rs 19,600. Similarly, if $x_1 = B$, $f_1^*(S_1, B) = 0.8 f_2^*(S_1) + 0.2 f_2^*(S_1 + 10{,}000) = 0.8\ (10{,}000 + 6{,}800) + 0.2\ (20{,}000 + 8{,}000) =$ Rs 19,040. The results are presented in Table 16.12.

TABLE 16.12 Determination of Optimal Policy at Stage 1

State S_1 – Amount available for investment	$f_1(S_1, x_1)$			$f_1^*(S_1)$	x_n^*
	$x_1 = 0$	$x_1 = A$	$x_1 = B$		
$S_1 = 10{,}000$	16,800	19,600	19,040	19,600	A

Thus, the optimal policy, considering Tables 16.12, 16.11 and 16.10 is to always invest in A. This yields highest expected amount of money equal to Rs 19,600.

REVIEW ILLUSTRATIONS

Example 16.5 Four types of items are under consideration for loading on a plane, with an unlimited supply of each type. The weights and values of various types of items are given below:

Item	*Weight (tons)*	*Value*
A	3	22
B	6	44
C	5	35
D	4	24

Determine as to which items should be loaded on the plane, and in what quantities, if the maximum capacity of the plane is 13 tons and the objective is to maximise the value of the shipment. Give a mathematical representation of the problem and solve it by dynamic programming.

Here the stages may be represented by the items under consideration. Thus, if

n = the stage (item) under consideration
x_n = the number of units of item of type n to load
w_n = the weight of one unit of item of type n
v_n = the value of one unit of item of type n
k = maximum capacity available
S_n = state at stage n, indicating the weight available

The problem is:

Maximise $\quad Z = \Sigma v_n x_n,\ n = 1, 2, 3, 4$

Subject to $\quad \Sigma\, w_n\, v_n \le k$

and $\quad x_n \ge 0$ and integer

Accordingly, we have

$$\text{Maximise} \quad Z = 22x_1 + 44x_2 + 35x_3 + 24x_4$$
$$\text{Subject to} \quad 3x_1 + 6x_2 + 5x_3 + 4x_4 \le 13$$
$$x_1, x_2, x_3, x_4 \ge 0 \text{ and integers}$$

This, however, is integer programming problem formulation, as discussed in Chapter 7. The dynamic programming problem formulation for it is shown now.

Stage : Each type of item is a stage.

States : The capacity (in tons) available for allocation in a given stage.

Decision : The quantity of each item to include in the optimal mix.

Recursive relationship :

$$f_n^*(S_n) = \max_{x_n} \{v_n x_n + f_{n+1}^*(S_n - w_n x_n)\}$$

where

S_n is the amount of weight available for allocation

$v_n x_n$ is the immediate payoff

$f_n^*(S_n)$ is the optimal total payoff at stage n for state n

$f_{n+1}^*(S_n - w_n x_n)$ is the optimal total payoff at the previous stage analysed.

The solution by backward recursion follows.

Stage 4 We first consider item *D*. Table 16.13 shows the states in the first column, while on the top are indicated the possible number of units of *D* that can be loaded. With a maximum capacity to load 13 tons and a load of 4 tons per unit, the maximum number of units of *D* which can be loaded is 3. Accordingly, x_4 can take values 0, 1, 2 or 3. The payoff function is given by $v_4 x_4 = 24x_4$, where x_4 is the number of units of *D*.

TABLE 16.13 Determination of Optimal Policy at Stage 4

State S_4 (tons available for allocation to item D)	$f_4(S_4) = v_4 x_4 = 24x_4$ *(No. of units of items D to load, x_4)*					
	$x_4 = 0$	$x_4 = 1$	$x_4 = 2$	$x_4 = 3$	$f_4^*(S_4)$	x_n^*
0	0				0	0
1	0				0	0
2	0				0	0
3	0				0	0
4	0	24			24	1
5	0	24			24	1
6	0	24			24	1
7	0	24			24	1
8	0	24	48		48	2
9	0	24	48		48	2
10	0	24	48		48	2
11	0	24	48		48	2
12	0	24	48	72	72	3
13	0	24	48	72	72	3

Stage 3 In this stage, an allocation is made to item *C*, with the remaining weight utilised according to optimal policy recommended in stage 4. The payoffs corresponding to various states are shown in Table 16.14. With 5 tons per unit, a maximum of 2 units of *C* can be loaded. The total payoff function is $f_3(S_3) = 35x_3 + f_4^*(S_3 - 5x_3)$. Note that if x_3 units of *C* are loaded, $S_3 - 5x_3$ tons would be left. For example, in state 11, when 11 tons are available, then the following alternatives are there:

(a) Zero to *C*, then 11 goes to *D*. From Table 16.13, we see that the optimal allocation of11 to *D* yields 48.

(b) One unit of *C* results in 6 tons remaining. The optimal allocation of 6 (from stage 4) yields 24 plus 35 obtained from the allocation of one unit of *C*, resulting in a total payoff of 59.

(c) Two units to *C* requires 10 tons, leaving one ton for *D*. However, no units of *D* can be loaded as it weighs 4 tons. Accordingly, the total payoff would be 70 + 0 = 70.

TABLE 16.14 Determination of Optimal Policy at Stage 3

State, S_3 (tons available for C and D)	$f_3(S_3) = 35x_3 + f_4^*(S_3 - 5x_3)$				
	$x_3 = 0$	$x_3 = 1$	$x_3 = 2$	$f_3^*(S_3)$	x_n^*
0	0			0	0
1	0			0	0
2	0			0	0
3	0			0	0
4	24			24	0
5	24	35		35	1
6	24	35		35	1
7	24	35		35	1
8	48	35		48	0
9	48	59		59	1
10	48	59	70	70	2
11	48	59	70	70	2
12	72	59	70	72	0
13	72	83	70	83	1

Stage 2 Since a unit of product *B* involves 6 tons, a maximum of 2 units of it can be loaded. At this stage, the total payoff function is $f_2(S_2) = 44x_2 + f_3^*(S_2 - 6x_2)$, wherein x_2 is the number of units of item *B* loaded. The calculations are shown in Table 16.15.

TABLE 16.15 Determination of Optimal Policy at Stage 2

State S_2 *(tons available for B, C and D)*	$f_2(S_2) = 44x_2 + f_3^*(S_2 - 6x_2)$				
	$x_2 = 0$	$x_2 = 1$	$x_2 = 2$	$f_2^*(S_2)$	x_n^*
0	0			0	0
1	0			0	0
2	0			0	0
3	0			0	0
4	24			24	0
5	35			35	0
6	35	44		44	1
7	35	44		44	1
8	48	44		48	0
9	59	44		59	0
10	70	68		70	0
11	70	79		79	1
12	72	79	88	88	2
13	83	79	88	88	2

Stage 1 Finally, we consider stage 1, where 13 tons are available to be loaded. If x_1 be the number of units of item *A*, then x_1 can take values 0, 1, 2, 3 or 4 since a maximum of 4 units of item *A*, each of 3 tons, can be loaded. The payoff function at this stage if $f_1(S_1) = 22x_1 + f_2^*(S_1 - 3x_1)$. The results are shown in Table 16.16.

TABLE 16.16 Determination of Optimal Policy at Stage 1

State S_1 *(tons available for A, B, C and D)*	$f_1(S_1) = 22x_1 + f_2^*(S_1 - 3x_1)$						
	$x_1 = 0$	$x_1 = 1$	$x_1 = 2$	$x_1 = 3$	$x_1 = 4$	$f_1^*(S_1)$	x_n^*
13	88	92	88	90	88	92	1

Optimal solution From the preceding analysis and results, we can read the optimal solution to the problem as follows. From Table 16.16 we get $x_1 = 1$, that is to say, one unit of item *A* (weighing 3 tons). The remaining 13 – 3 = 10 tons are allocated in an optimal manner according to stage 2 (Table 16.15). For this quantity, we have $x_2 = 0$ so that no units of item *B* are included. Proceeding to stage 3 (Table 16.14), the optimal allocation is x_3 = 2, meaning that 2 units of item *C* weighing 2 × 5 = 10 tons. With no capacity left, $x_4 = 0$. Thus, overall solution to the problem is to load 1 unit of *A* and 2 units of *C*, and none of items *B* and *D*. The maximum value is 1 × 22 + 2 × 35 = 92 units.

Example 16.6 A company has trained five salespersons in the past three months, whom it wants to assign to three sales regions: North, West and South. It has been estimated that increases in sales (in lakhs of rupees) over the next year in each region, if each were allocated different number of salesmen, are as indicated below.

Salespersons	*Region*		
	North	*West*	*South*
0	0	0	0
1	61	36	66
2	86	61	86
3	106	91	96
4	121	126	116
5	136	166	146

Use dynamic programming to determine which allocation would maximise the increase in sales.

To solve the problem using dynamic programming, we first define stages, states, etc.

Stages : The problem requires making three interrelated decisions of determining how many salespersons to allocate to each of three regions. While there is no fixed sequence of regions, we may consider the scquence of regions in which they are given–North, West and South. The three regions may be taken as three stages, n ($n = 1, 2, 3$). It is required to determine the number of salespersons to be allocated at a given stage, x_n.

States : At each stage, the 'states' represents the number of sales persons still available for allocation to the regions remaining. To illustrate, at stage 1, when all the regions remain to be considered, $S_1 = 5$. At stage 2 or 3 (region West or South), S_n would be equal to 5 minus the number of salespersons allocated at the preceding stage. Accordingly, the sequence of states at the three stages is: $S_1 = 5$, $S_2 = 5 - x_1$, and $S_3 = 5 - x_2$.

Decision : The number of salesperson to assign to each of the three regions, North, West and South, as would maximise the total increase in sales.

Recursive Relationship:

$$f_n^*(S_n) = \max_{x_n} \{C_n(S_n, x_n) + f_{n+1}^*(S_n + 1)\}$$

wherein

S_n is the number of salespersons available for allocation at stage n,

$C_n(S_n, x_n)$ is the increment in sales resulting from current allocation of salespersons—the immediate payoff,

$f_n^*(S_n)$ is the optimal payoff at stage n, and

$f_{n+1}^*(S_{n+1})$ is the optimal payoff at previous stage

The solution follows.

Stage 3 At this stage, the solution is trivial because allocation of salespersons to the South region only is to be analysed. For various number of salespersons available for allocation to South, the payoff function is defined in Table 16.17. From the table, it is clearly evident that as many people will be assigned as are available for employment to this region.

TABLE 16.17 Determination of Optimal Policy at Stage 3

State, S_3 *No. of persons available for allocation*	$f_3(S_3) = C_3(S_3, x_3)$						$f_3^*(S_3)$	x_3^*
	$x_3 = 0$	$x_3 = 1$	$x_3 = 2$	$x_3 = 3$	$x_3 = 4$	$x_3 = 5$		
0	0						0	0
1	0	66					66	1
2	0	66	86				86	2
3	0	66	86	96			96	3
4	0	66	86	96	116		116	4
5	0	66	86	96	116	146	146	5

Stage 2 At this stage, obtaining x_2^* requires calculating and comparing $C_2(S_2, x_2)$ for the alternative values of x_2, namely, $x_2 = 0, 1, \ldots 5$. Thus, we consider various possible number of salespersons available at this stage, obtain the current payoff for each one and add it to the optimal payoff at stage 3. To illustrate, we consider $S_2 = 2$. If $x_2 = 0$, then resulting state at stage 3 will be $S_2 - x_2 = 2 - 0 = 2$, whereas $x_2 = 1$ leads to state 1 and $x_2 = 2$ amounts to state 0 at stage 3. The corresponding values of $C_2(S_2, x_2)$ from the column under West in the given problem, along with values of $f_3^*(S_3)$ [or $f_3^*(S_2 - x_2)$] from Table 16.18 are considered and added together. For $S_2 = 2$, the calculations are summarised here:

$$f_2(2) = C_2(2, x_2) + f_3^*(2 - x_2)$$

for $x_2 = 0, f_2(2, 0) = C_2(2, 0) + f_3^*(2) = 0 + 86 = 86$

for $x_2 = 1, f_2(2, 1) = C_2(2, 1) + f_3^*(1) = 36 + 66 = 102$

for $x_2 = 2, f_2(2, 2) = C_2(2, 2) + f_3^*(0) = 61 + 0 = 61$

Since the objective is to maximise, $x_2^* = 1$.

TABLE 16.18 Determination of Optimal Policy at Stage 2

State S_2 *No. of salespersons available for allocating to West and South*	$f_2(S_2) = C_2(S_2, x_2) + f_3^*(S_2 - x_2)$						$f_2^*(S_2)$	x_2^*
	$x_2 = 0$	$x_2 = 1$	$x_2 = 2$	$x_2 = 3$	$x_2 = 4$	$x_2 = 5$		
0	0						0	0
1	66	36					66	0
2	86	102	61				102	1
3	96	122	127	91			127	2
4	116	132	147	157	126		157	3
5	146	152	157	177	192	166	192	4

Stage 1 We now move backward to stage 1. At this stage, we have the beginning state, $S_1 = 5$. The calculations of payoffs in respect of different values of x_1 are shown in Table 16.19. The payoff function is $f_1(5) = C_1(1, x_n) + f_2^*(5 - x_1)$. To illustrate, for $x_1 = 2$, that is, when two salespersons are assigned to North, the current payoff $C_1(1, 2)$ is 86 (from the given data table) while $f_2^*(5 - 2) = f_2^*(3)$ equals 127 (from Table 16.18), the total payoff would be $86 + 127 = 213$.

TABLE 16.19 Determination of Optimal Policy at Stage 1

State S_1	$f_1(S_1) = C_1(1, x_n) + f_2^*(5 - x_1)$							
No. of salesmen	$x_1 = 0$	$x_1 = 1$	$x_1 = 2$	$x_1 = 3$	$x_1 = 4$	$x_1 = 5$	$f_1^*(S_1)$	x_1^*
5	192	218	213	208	187	136	218	1

Accordingly, the optimal solution has $x_1^* = 1$, which makes $S_2 = 5 - 1 = 4$, so that $x_2^* = 3$ and $S_3 = 4 - 3 = 1$. In simple terms, the optimal solution calls for employing 1 salesman in North, 3 in West and 1 in South. Further, since $f_1^*(5) = 218$, the allocation of 5 salesmen to the three regions will lead to a total of 218 lakhs of rupees. The next best alternative is when $x_1^* = 2$, yielding an increment of Rs 213 lakh. When $x_1^* = 2$, we have $S_2 = 5 - 2 = 3$, for which $x_2^* = 2$, and in turn $S_3 = 1$ and $x_3^* = 1$. Thus, employment of 2 salespersons each in the North and West, and 1 in the South is second best alternative, which would yield a total of Rs 213 lakh increase in sales.

Example 16.7 You are given the following network wherein the numbers indicate the distance between various pairs of nodes. Use dynamic programming to determine the shortest path between the origin and destination. What is the second shortest path and the what is the distance involved?

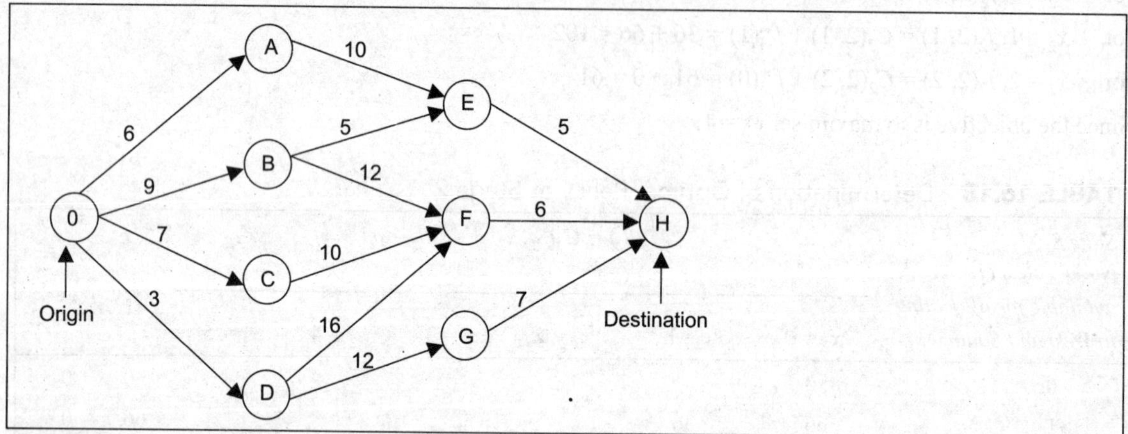

Figure 16.5 *Network Diagram*

We first define input parameters and recursive relationship for this problem.

Stages : In this case, we may take stages as given by vertical columns. There are three stages: stage 1 at origin, stage 2 represented by nodes *A*, *B*, *C* and *D*, and stage 3 comprising the nodes *E*, *F* and *G*.

States : The nodes at various stages constitute the states. Accordingly, $S_1 = (0)$, $S_2 = (A, B, C, D)$ and $S_3 = (E, F, G)$.

Recursive Relationship:

$$f_n^*(S_n) = \min_{x_n} \{C_n(S_n, x_n)\} + f_{n+1}^*(S_{n+1})\}$$

wherein

$C_n(S_n, x_n)$ = current payoff (distance from a current node to a node on the next stage).

$f_n^*(S_n)$ = optimal payoff at stage n, i.e. total minimum distance from current state (S_n) to the destination

$f_{n+1}^*(S_n + 1)$ = optimal payoff at previous stage, i.e. total minimum distance from previous stage ($n + 1$) to the destination

The solution follows.

Stage 3 The solution at this stage is trivial because from each of the nodes in this stage, there is one and only one link with the destination. There being no choice, each of them indicates optimal policy, as indicated in Table 16.20.

TABLE 16.20 Determination of Optimal Policy at Stage 3

State, S_3 *Node*	$f_3^*(S_3)$	x_3^*
E	5	E
F	6	F
G	7	G

Stage 2 At this stage, we analyse various options from each of the nodes and determine the optimal ones. This is shown in Table 16.21. The solution provides that if one is at node A, one should move to E, if one is at node B, then one should go to node F and so on, as indicated in the last column of the table. The minimum distance from each of the nodes A, B, C and D to the destination is given in the fourth column.

TABLE 16.21 Determination of Optimal Policy at Stage 2

State, S_2 *Node*	$f_2(S_2) = C_2(S_2, x_2) + f_3^*(S_3)$		$f_2^*(S_2)$	x_2^*
A	$x_2 = E$	10 + 5 = 15	15	E
B	$x_2 = E$	5 + 5 = 10	10	E
	$x_2 = F$	12 + 5 = 17		
C	$x_2 = F$	10 + 6 = 16	16	F
D	$x_2 = F$	16 + 6 = 22		
	$x_2 = G$	12 + 7 = 19	19	G

Stage 1 In this stage, there are four values that x_1 may take, namely, *A*, *B*, *C* and *D*. For all options, the current payoffs are added to the optimal solution values at the previous stage. As is evident from Table 16.22, which provides these calculations, the least of the distance values equals 19, which is the optimal value $f_1^*(S_1)$.

TABLE 16.22 Determination of Optimal Policy at Stage 1

State, S_1	$f_1(S_1) = C_1(S_1, x_1) + f_2^*(S_2)$					
	$x_1 = A$	$x_1 = B$	$x_1 = C$	$x_1 = D$	$f_1^*(S_1)$	x_1^*
0	6 + 15 = 21	9 + 10 = 19	7 + 16 = 23	3 + 19 = 22	19	B

Now, we may derive the optimal solution as follows. From origin, the movement should be to node *B* (Table 16.22), from node *B* to node *E* (Table 16.21) and from node E to the final destination (Table 16.20). This route O-B-E-H involves the minimum distance equal to 19.

From Table 16.22 it is evident that the second-best route has a total distance of 21. The movement in this route is along the path O-A-E-H, as indicated by Table 16.22 through 16.20.

Example 16.8 Solve the following LPP using dynamic programming:

Maximise $Z = 8x_1 + 16x_2$

Subject to

$$x_1 + x_2 \leq 200$$
$$x_2 \leq 125$$
$$3x_1 + 6x_2 \leq 900$$
$$x_1, x_2 \geq 0$$

To solve the problem, we first define various parameters.

Stages : The solution calls for two interrelated decisions of determining the level of activities 1 and 2 (x_1 and x_2 respectively). These are the two stages involved here, $n = 1, 2$. Here x_n is defined as the decision variable at stage *n*.

States : The states S_n indicate the amounts of resources available for allocation. If R_i is the amount of resource *i* remaining to be allocated ($i = 1, 2, 3$), we have $S_n = (R_1, R_2, R_3)$. Thus, we have

$$S_1 = (200, 125, 900)$$
$$S_2 = (200 - x_1, 125, 900 - 3x_1)$$

However, since the values of x_1 are not known at stage 2, we take $S_2 = (R_1, R_2, R_3)$.

Recursive Relationship:

Since optimal payoff at stage *n* ($n = 1, 2$) is given by

$$f^*(R_1, R_2, R_3) = \max_{x_n} (R_1, R_2, R_3; x_n),$$

we have

$$f_2^*(R_1, R_2, R_3) = \max_{\substack{x_2 \leq R_1 \\ x_2 \leq R_2 \\ 6x_2 \geq R_3 \\ x_2 \geq 0}} (16x_2) \quad \text{(i)}$$

Similarly, $f_1(200, 125, 900) = 8x_1 + f_2^*(200 - x_1, 125, 900 - 3x_1)$ (ii)

and $f_1^*(200, 125, 900) = \max_{\substack{x_1 \le 200 \\ 3x_1 \le 900 \\ x_1 \ge 0}} \{8x_1 + f_2^*(200 - x_1, 125, 900 - 3x_1)\}$ (iii)

Now, the solution follows.

Stage 2 From the equation (i), it is evident that x_2^* is the largest value which satisfies the condition $x_2 \le R_1$, $x_2 \le R_2$ (i.e. 125), $6x_2 \le R_3$ and $x_2 \ge 0$. For the feasible solution to exist, we also should have $R_1, R_2, R_3 \ge 0$. In this case, the largest value of x_2 is going to be the least of the values R_1, 125 and $R_3/6$. Thus, at this stage, the solution is as given in Table 16.23.

TABLE 16.23 Determination of Optimal Policy at Stage 2

State (R_1, R_2, R_3)	$f_2^*(R_1, R_2, R_3)$	x_2^*
$R_1 \ge 0$ $R_3 \ge 0$	16 min $\{R_1, 125, R_3/6\}$	min $\{R_1, 125, R_3/6\}$

Stage 1 To solve the two-stage problem ($n = 1$), we substitute the solution obtained above for $f^*_2(R_1, R_2, R_3)$ into equation (iii) to get

$$f_2^*(200 - x_1, 125, 900 - 3x_1) = 16 \min \{R_1, 125, R_3/6\}$$

$$= 16 \min \left\{200 - x_1, 125, \frac{900 - 3x_1}{6}\right\}$$

$$= 16 \min \{200 - x_1, 125, 150 - 0.5x_1\}$$

Further,

$$f_1^*(200, 125, 900) = \max_{x_1 \ge 0} [8x_1 + 16 \min \{200 - x_1, 125, 150 - 0.5x_1\}]$$

To obtain the solution, we first set $x_1 = 0$. For this value of x_1, $f_1(200, 125, 900) = 8 \times 0 + 16 \times 125 = 2000$. Similarly, the maximum value that x_1 can assume is 200. For this value of x_1, the function equals 1600. Further, let us see if any values of x_1 between zero and 200 give a higher value of the function. For this, we set different pairs of R_i values equal and determine x_1. To illustrate, for $200 - x_1 = 125$, $x_1 = 75$; while for $150 - 0.5x_1 = 125$, $x_1 = 50$; and for $200 - x_1 = 150 - 0.5x_1$, $x_1 = 100$. Let us evaluate the payoff function for these values of x_1.

When $x_1 = 75$, $f_1(200, 125, 900) = 8 \times 75 + 16 \times 112.5 = 2400$,

When $x_1 = 50$, $f_1(200, 125, 900) = 8 \times 50 + 16 \times 125 = 2400$, and

When $x_1 = 100$, $f_1(200, 125, 900) = 8 \times 100 + 16 \times 100 = 2400$.

The values are all equal and highest. It implies that the problem has multiple optimal solutions. From Table 16.23 when $x_1^* = 75$, $x_2^* = 112.5$; when $x_1^* = 50$, $x_2^* = 125$; and when $x_1^* = 100$, $x_2^* = 100$. Further, $f_1^*(200, 125, 900) - 2400$.

KEY POINTS TO REMEMBER

- Dynamic programming solves problems by breaking them into smaller sub-problems. The sub-problems are also called stages.
- The outcome of a decision at a given stage affects the decision at each of the following stages.
- Dynamic programming is different from linear programming in that there is no standard mathematical formulation of a dynamic programming problem unlike as for an LPP.
- Dynamic programming requires that we identify and specify stages, states, decision variables, decision criterion and recursive relationship.
- While stages are logical sub-problems, states are the possible input values or the beginning conditions. Decision variables are the alternatives we have at every stage. The decision criterion states the objective of the problem optimisation of a payoff function. The recursive relationship is an algebraic statement that shows relationship between stages. It identifies the optimal policy for stage n, given the optimal policy for stage $n + 1$.
- Dynamic programming may be deterministic or probabilistic.
- A deterministic dynamic programming problem is one in which the state at the next stage is determined by the state and policy decision of the current stage.
- In a probabilistic dynamic programming problem, the state in the next stage is not sure and there is a probability distribution of what the next state would be.
- Dynamic programming can be used for solving a wide variety of problems related to networks and resource allocation, knapsack problems and other such problems.

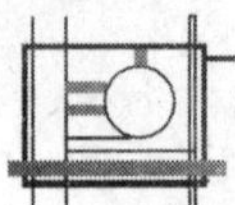

TEST YOUR UNDERSTANDING

Mark the following statements as T (*True*) or F (*False*).

1. Dynamic programming is a quantitative analysis technique which is applied to large, complex problems.
2. A dynamic programming problem consists of multiple smaller problems, known as states, with interconnected decisions.
3. In dynamic programming, the given problem is solved in stages beginning at the last stage and working backward to initial stage.
4. Dynamic programming problems consist of a series of independent sub-problems.
5. The optimal solution to a dynamic programming problem is obtained by solving all stages of the problem sequentially.
6. By dynamic programming it is possible to obtain the best and next-best solution to a problem.
7. In dynamic programming, there is a standard algorithm which can be programmed to solve all problems.
8. Dynamic programming is similar to linear programming in its approach to solving problems.
9. In a shortest route problem, the optimal solution indicates the length of most direct route from origin to destination that involves least number of nodes on the route.

10. Probabilistic dynamic programming is used where it is not possible to break the given problem into clear cut sub-problems.
11. Each sub-problem in a dynamic programming problem is called a stage.
12. Every stage in a dynamic programming problem calls for making a policy decision.
13. The states of a given stage refer to the various possible conditions in which the system can be in that stage.
14. All dynamic programming problems call for maximising the given payoff function.
15. The *payoff* is the variable of interest in a dynamic programming problem.
16. In dynamic programming, the decision process is designed to obtain an optimal policy for the entire problem, which consists of determining optimal policy decision at each stage for each of its possible states.
17. To solve a dynamic programming problem, it is necessary to define an optimal payoff function for every stage and state.
18. The recursive relationship identifies the optimal policy for stage n, irrespective of whether or not the optimal policy for stage $n + 1$ is available.
19. In dynamic programming, the optimal solution depends on the current state and how you reach there.
20. Working forward from the initial stage to the final stage is the last step of a dynamic programming problem solution.
21. A deterministic dynamic programming problem is one in which the state at the next stage is completely determined by state and policy decision taken at the current stage.
22. In probabilistic dynamic programming, there is a probability distribution for what the next state would be.
23. A state also means the information that is needed at any stage to make an optimal decision.
24. The recursive relationship essentially formalizes the working backward or rollback procedure.
25. An LPP with minimisation type of objective function cannot be solved using dynamic programming.

EXERCISES

1. Explain dynamic programming. How is it different from linear programming?
2. What is a stage in dynamic programming?
3. Explain the steps involved in solution to dynamic programming problems.
4. Explain the following in the context of dynamic programming:
   (a) Stages
   (b) States
   (c) Payoff function
   (d) Recursive relationship
5. Give some examples where dynamic programming may be used.
6. Distinguish between deterministic and probabilistic dynamic programming.
7. "Solution to dynamic programming problems is obtained by backward recursion." Explain and illustrate backward recursion in this statement.

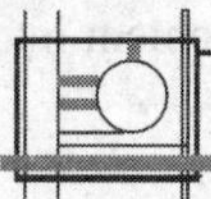

PRACTICAL PROBLEMS

1. You are given the following network for a project, for which it is desired to obtain the longest path from start to finish. Required:
   (i) Identify the stages and states for the dynamic programming formulation of this problem.
   (ii) Use dynamic programming to solve this problem.

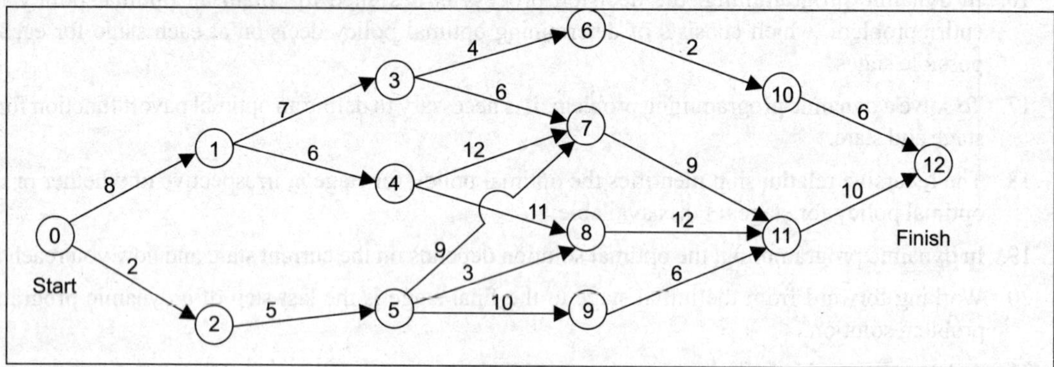

Figure 16.6 *Network*

2. Obtain the shortest route (i) from A to E, and (ii) A to any point on D, for the following network, using dynamic programming.

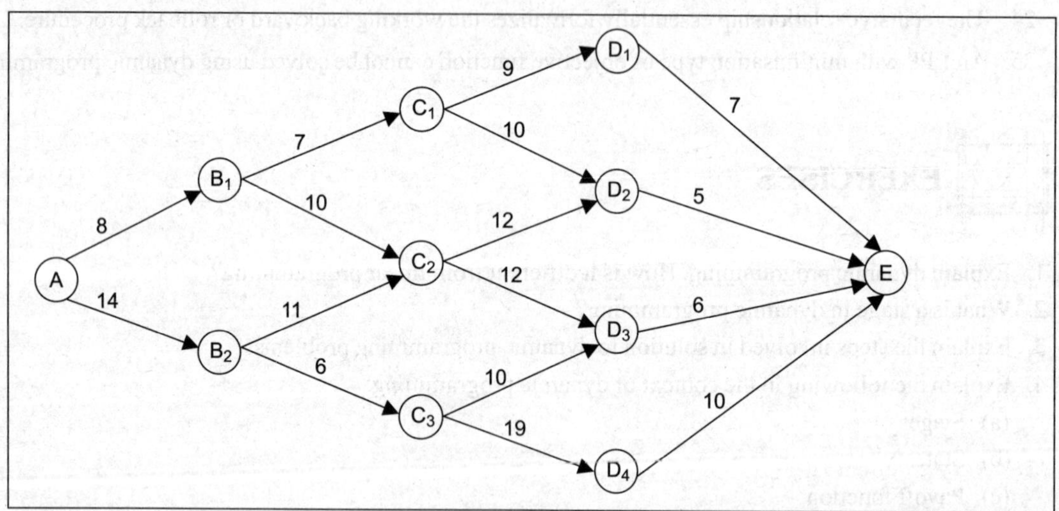

Figure 16.7 *Network*

3. (a) Solve the following network problem for shortest route between the origin and destination. The numbers on the arrows connecting different pairs of nodes represent distance in kms. Use dynamic programming.

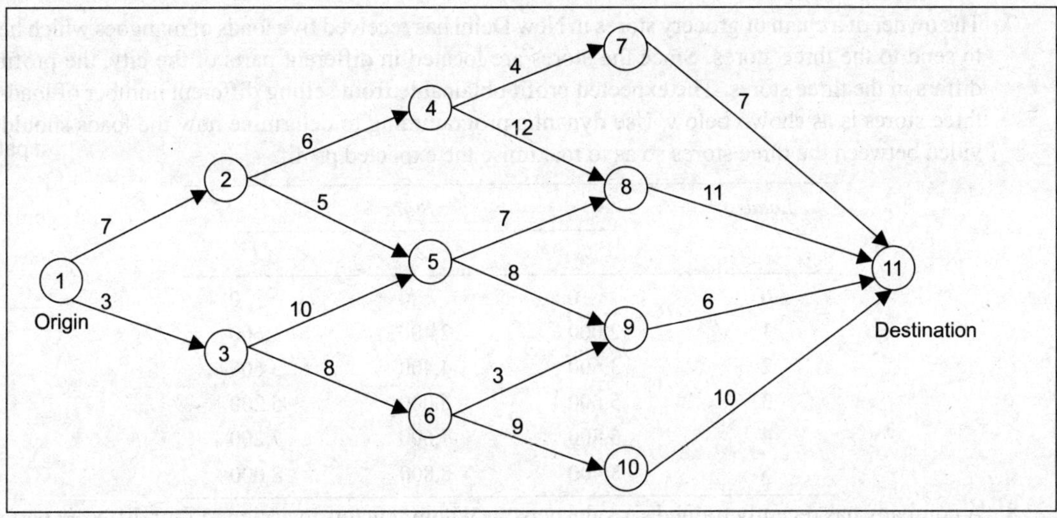

Figure 16.8 *Network*

(b) Reconsider the network given in (a). If the road between nodes 6 and 9 is blown up and is therefore not available, how will the shortest route change, if at all?

4. Solve the following LPP using dynamic programming.

Minimise $Z = 40x_1 + 24x_2$

Subject to

$$20x_1 + 50x_2 \geq 4800$$
$$80x_1 + 50x_2 \geq 7200$$
$$x_1, x_2 \geq 0$$

5. Solve the following LPP using dynamic programming.

Maximise $Z = 30x_1 + 20x_2$

Subject to

$$2x_1 + 4x_2 \leq 12$$
$$15x_1 + 5x_2 \leq 40$$
$$x_1, x_2 \geq 0$$

6. Maximise $Z = 12x_1 + 20x_2$

Subject to

$$2x_1 \leq 8$$
$$6x_2 \leq 36$$
$$6x_1 + 4x_2 \leq 36$$
$$x_1, x_2 \geq 0$$

Solve this linear programming problem using dynamic programming.

7. The owner of a chain of grocery stores in New Delhi has received five loads of mangoes which he wants to send to the three stores. Since the stores are located in different parts of the city, the profitability differs in the three stores. The expected profit obtainable from selling different number of loads in the three stores is as shown below. Use dynamic programming to determine how the loads should be divided between the three stores so as to maximise the expected profit.

Loads	*Store*		
	1	*2*	*3*
0	0	0	0
1	2,000	2,400	1,600
2	3,600	4,400	3,600
3	5,600	6,000	5,200
4	6,800	7,600	7,200
5	8,400	8,800	8,000

8. A company has recently trained six salespersons whom it wants to assign to four different parts of the country. The following table gives the estimated increase in sales (in lakhs of rupees) over the next year in each region if it were allocated different numbers of salespersons.

	Region			
Salespersons	*1*	*2*	*3*	*4*
0	0	0	0	0
1	8	14	10	12
2	18	22	20	22
3	30	32	30	28
4	36	36	36	32
5	44	40	42	34
6	48	42	44	36

Use dynamic programming to determine the number of salespersons to be assigned to the respective regions in order to maximise increase in sales.

9. A political campaign for election to the parliament is entering its final stage and pre-poll surveys are indicating a very close contest in a certain constituency. One of the candidates in the constituency has sufficient funds to give five full-page advertisements in four different areas. Based on the polling information, an estimate has been made of the approximate number (in thousands) of additional votes that can be polled in different areas. This is shown below.

No. of	*Area*			
Commercial Ads	*A*	*B*	*C*	*D*
0	0	0	0	0
1	9	13	11	7
2	15	17	19	15
3	19	21	23	25
4	25	23	21	29
5	31	25	27	33

Using dynamic programming, determine how the five commercial ads be distributed between the four areas so as to maximise the estimated number of votes.

10. The chairman of certain political party is making plans for the parliament. He has the services of six volunteers (party workers) and wishes to assign them to four districts in such a way as to maximise the effectiveness. He feels that it would be ineffective to assign a worker to more than one district but he is not willing to assign a worker to any district if he or she accomplishes more in another district. The following table gives the estimated increase in votes in his favour in each district if it were allocated to various workers. Using dynamic programming technique determine how many of the six workers should be assigned to each of the four districts in order to maximise the total estimated increase in the number of votes.

Workers	*District*			
	1	*2*	*3*	*4*
0	0	0	0	0
1	25	20	33	13
2	42	38	43	24
3	55	54	47	32
4	63	65	50	39
5	69	73	52	45
6	72	80	53	50

(MFC, Delhi, 2001)

11. A transport company delivers four types of containers between New Delhi and Mumbai. The company wishes to load a minimum of 15 tons on each truck. The weight and handling cost paid to porters for each type of container is given below:

Container	*Weight (tons)*	*Handling cost (Rs)*
A	5	780
B	3	600
C	4	720
D	6	840

It is required to determine how many containers should be loaded on each truck in order that the total handling cost may be minimised. Solve by dynamic programming and obtain the total minimum cost.

12. A company is planning to spend Rs 7 lakh on advertising next year to promote four of its products. It plans to spend to multiples of Rs 1 lakh only on the products and also to spend at least Rs 1 lakh on each of them. Given in the following table are the estimates of increase in the sales volume (in lakhs of Rs) as a result of the advertisement expenses on various products.

Amount (in lakh of Rs)	*Product*			
	A	*B*	*C*	*D*
1	9	15	6	9
2	15	17	12	13
3	18	20	18	19
4	21	27	24	25

Use dynamic programming to determine the optimal allocation of the advertising budget. What maximum increment in sales can be expected? What is the next best solution? What minimum increase in sales can the company expect to attain from this advertising campaign, when the entire amount of Rs 7 lakh in spent?

13. The management of a company is considering allocation of Rs 80 lakh among its three plants. It has been decided that the allocation to a plant would be either Rs 0, Rs 20, Rs 40, Rs 60 or Rs 80 lakh. A forecast of the yearly returns corresponding to different levels of funds employed is made and presented in the following table.

Amount allocated (in lakhs of Rs)	*Return (in lakhs of Rs)*		
	Plant X	*Plant Y*	*Plant Z*
0	0	0	0
20	4	6	8
40	10	12	18
60	30	24	22
80	28	30	32

Determine the optimal allocation of funds to each plant so as to maximise the estimated overall allocation. Also indicate the second best allocation.

14. Three contractors have submitted bids to build six sub-stations for power supply to NDMC as indicated in the following table:

No. of sub-stations	*Contractor A*	*Contractor B*	*Contractor C*
1	70	60	65
2	140	120	120
3	200	170	200
4	260	250	270
5	330	320	340
6	400	410	400

Use dynamic programming to determine the best allocation of six sub-stations among the contractors in a manner that minimises the total cost.

15. Four types of items, I_1, I_2, I_3 and I_4 are available in unlimited supply. These are to be loaded on planes and a plane can take a maximum load of 14 tons. The weight and value (in appropriate units) of each unit of these items are given below:

Item	*Weight (tons)*	*Value (V)*
I_1	3	40
I_2	4	50
I_3	5	60
I_4	2	36

The requirement is to find which of the items should be loaded on a plane and in what quantities so that the total value may be maximised. You are required to

(a) formulate the problem as an integer programming problem.

(b) formulate and solve the problem by dynamic programming.

16. An investor has 5,000 shares of a particular company which she wants to sell on or before the fifth day from this day, i.e., Monday. Assume that price of the share fluctuates between Rs 40 and Rs 42. On a given day, there is a 25 per cent chance of selling the shares at Rs 40; 30 per cent chance of selling them at Rs 41; and 45 per cent of selling them at Rs 42. Required:

    (a) Suggest an optimal policy. Assume that all the shares are sold in a single trade.

    (b) How much money can the investor expect to receive if she has to pay a brokerage of 20 paisa per share?

17. The purchase manager of Sigma Steel Company has to buy 1,000 tons of scrap for the next month's production. The purchases can be done only on Mondays, the day when ships arrive at the port. The weekly prices are independent and each week there is a 20 per cent chance that the scrap will cost Rs 1,000 per ton, 50 per cent chance that it will cost Rs 1,100 and 30 per cent that it will cost Rs 1,200. The manager fears that if he waits too long, then price increases may force him to buy at a premium, while if he buys early then he may miss an opportunity to economise on the purchase. Use dynamic programming to help the manager to decide as to when to purchase the scarp.

18. Use dynamic programming to solve the following travelling salesman problem:

Inter-city Distance Profile

City	*A*	*B*	*C*	*D*
A	–	1297	1522	772
B	1297	–	1306	1360
C	1522	1306	–	884
D	772	1360	884	–

CHAPTER

17

Simulation

CHAPTER OVERVIEW

In quantitative approach to decision-making, a manager can format a problem mathematically and then apply an appropriate tool to find the optimal solution to it. But a mathematical formulation of a decision problem always requires some assumptions to be made. Further, a solution to a mathematically formulated problem is obtained by application of some tool or algorithm and does not involve any experimentation. There is another approach to the solution of a variety of problems where simulation is taken recourse to. Simulation is different in the sense that it requires experimentation. Unlike mathematical approach, however, it does not attempt to provide the optimal solution to the problem. Rather, it aims at describing a system and providing insight into the likely output resulting from a set of given inputs. This can be applied to numerous areas like inventory control, queuing systems, project management, capital budgeting and so forth. Accordingly, simulation is a very potent tool available to manager, which can be employed where mathematical methods may not be applied as also where they can be employed for decision-making.

After reading this chapter, a manager may be able to get answers to questions like the following:

- What is Monte Carlo simulation?
- What are random numbers, what are their properties and how can they be obtained?
- What is the use of random numbers in simulation?
- How can demand for an item be simulated?
- How can an inventory system be simulated in an attempt to determine the appropriate re-order level and re-order quantity, given the probability distributions of demand and lead time along with necessary cost data?
- How can a queuing system be simulated given distributions of the arrival time and service time in an attempt to obtain the operating characteristics of the system?
- How can simulation be used to estimate the completion of a project, given the probability distributions of the completion times of its constituent activities?

A key thing to remember in simulation is that it does not provide the optimal solution to a problem. It instead provides sufficient basis to reach a decision.

The chapter is relatively easy to grasp but one needs to be familiar with the ideas of probability distributions and random numbers.

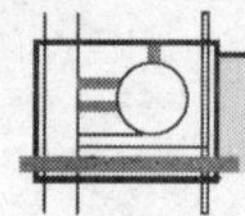

17.1 INTRODUCTION

In the earlier chapters of this book, we have discussed mathematical models to describe and analyse the characteristics of a given system. Such models could be solved analytically to determine the optimal solutions or to describe the system. For example, we considered in Chapters 2 and 3 how a given decision-making problem could be modelled as a linear programming problem and then solved using simplex technique to obtain an optimal solution. Similarly, in Chapter 10 we saw how a queuing system, that meets a certain set of assumptions, can be described in terms of the queue length, waiting time and so on. However, despite the fact that mathematical modelling can, and does help to analyse a wide variety of problems, there are many situations which are too complex to be handled this way. For instance, a given queuing situation may not meet all the assumptions of any of the queuing models and, therefore, not lend itself to modelling. In these circumstances, it is often possible to simulate the given system and study its behaviour. Not only this, simulation may also be employed to estimate the impact of system changes when experimentation with the real system may not be feasible.

To simulate is to imitate. In general terms, simulation involves developing a model of some real phenomenon and then performing experiments on the model evolved. It is a descriptive, and not optimising, technique. In simulation, a given system is copied and the variables and constants associated with it are manipulated in that artificial environment to examine the behaviour of the system. To illustrate, for aerodynamic testing, scaled down models of aeroplanes are built and placed in the wind tunnels. Using a wind tunnel, air is blown to examine the aerodynamic properties of the model.

Using simulation, an analyst can introduce the constants and variables related to the problem, set up the possible courses of action and establish criteria which act as measures of effectiveness. The benefit of simulation from the viewpoint of the analyst stems from the fact that the results of taking a particular course of action can be estimated prior to its implementation in the real world. Instead of using hunches and intuition to determine what may happen, the analyst using simulation can test and evaluate various alternatives and select the one that gives the best results.

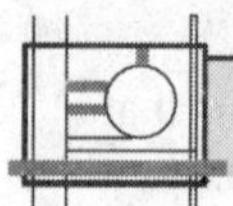

17.2 PROCESS OF SIMULATION

Broadly, there are four phases of the simulation process. They are:

(a) definition of the problem and statement of objectives,

(b) construction of an appropriate model,

(c) experimentation with the model constructed; and

(d) evaluation of the results of simulation.

Each of the phases calls for the performance of a number of preliminary tasks. Of these, the two major tasks are collection of data and selection of means by which thc simulation activity would replicate the random behaviour of the real world.

The first step in problem solving of any situation is to identify and clearly define the problem and list the objective(s) that the solution is intended to achieve. This is true of simulation as well. A clear statement not only facilitates the development of an appropriate model but also provides a basis for evaluation of the simulation results. In general, simulation aims to determine how the system under consideration would behave under certain conditions. Naturally, the more specific the analyst is about what he is looking for, the greater the chances that the simulation model will be designed to accomplish that. Thus, the scope and the level of detail of the simulation should be decided upon carefully.

The next step in simultaion is the development of a suitable model. During the course of a simulation, the model mimics the important elements of what is being simulated. A simulation model may be a physical or mathematical model, a mental conception, or a combination. Many simulations involve physical models. Examples include a scaled down model of an aeroplane or ship constructed out of wood or other material. Since physical models are relatively expensive to build, mathematical models are often preferred. In such a model, mathematical symbols or equations are used to represent the relationships in the system.

Collection of data is a significant aspect of model development, and the quantum and type of data needed are directly governed by the scope and extent of the detail of the simulation. The data are needed both for model development and evaluation. Obviously, the model for simulation must be so designed that it would enable evaluation of the key decision alternatives. An ancillary step here is of designing experiments. The experiments help answer the 'what if...' types of questions in simulation studies. By going through this process, the analyst is able to learn about the system behaviour.

Once the simulation model is developed, the next step is to run it. If the model is deterministic, with all its parameters known and constant, then only a single run would suffice. On the other hand, if the simulation is stochastic in nature, with the parameters subject to random variation, then a number of runs would be needed to get a clear picture of the model performance. The probabilistic simulation is akin to the random sampling where each run represents one observation. Thus, statistical theory can be used to determine the optimal sample sizes. Evidently, the greater the variability inherent in the simulation results, the larger would be the simulation runs needed to obtain a reasonable degree of confidence that the results are truly indicative of the system behaviour.

The last step in the process of simulation is to analyse and interpret the results of the runs. The interpretation of results is, in a large measure, dependent on the extent to which the simulation model portrays the reality. Obviously, closer the approximation of the real system by the simulation model, lesser will be the need for adjusting the results and also lesser will be the risk inherent in applying the results.

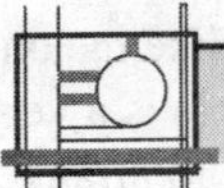

17.3 MONTE CARLO SIMULATION

Although simulation can be of many types, our discussion will focus on the probabilistic simulation using the *Monte Carlo* method. Also called *computer simulation*, it can be described as a numerical technique that involves modelling a stochastic system with the objective of predicting the system's behaviour. The chance element is a very significant feature of Monte Carlo simulation and this approach can be used when the given process has a *random*, or *chance*, component.

In using the Monte Carlo method, a given problem is solved by simulating the original data with random number generators. Basically, its use requires two things. First, as mentioned earlier, we must have a model, that represents and image of the reality of the situation. Here the model refers to the probability distribution of the variable in question. What is significant here is that the variable may not be known to explicitly follow any of the theoretical distributions like Poisson, Normal, and so on. The distribution may be obtained by direct observation or from past records. To illustrate, suppose that a bakery keeps a record of the sale of the number of cakes of a certain type. Information relating to 200 days' sales is,

Demand (No. of cakes)	:	5	6	7	8	9	10	11	12	*Total*
(No. of days)	:	4	10	16	50	62	38	12	8	200

Assuming that this is an adequate representation of the distribution of demand for the cake, we can derive the probability distribution of demand by expressing each of the frequencies in terms of proportions. This is done by dividing each one of the values by 200—the total frequency. The resultant distribution follows:

Demand (No. of cakes)	:	5	6	7	8	9	10	11	12
Probability	:	0.02	0.05	0.08	0.25	0.31	0.19	0.06	0.04

Thus, there is 0.02 or 2% chance that 5 cakes would be demanded on a day, a 0.05 or 5% chance that the demand would be for 6 cakes . . . and so on. This distribution would serve as the model of the situation under consideration.

The second thing required for simulation is a mechanism to simulate the model—something to capture the random nature of the given system. Thus, we should have available a procedure that would help us to select, at random, values for the variables which can be used to approximate the state of the system. Such a mechanism can be any *random number generator* consisting of a device or a procedure by which random numbers can be determined and/or selected.

There are various ways in which random numbers (or apparently random, but not truly so) may be generated. These could be: result of some device like coin or die; published tables of random numbers, midsquare method, or some other sophisticated method. It may be mentioned here that the 'random' numbers generated by some methods may not be really random in nature. In fact, such numbers are called *pseudo-random numbers*. There are some tests with which numbers may be tested for their randomness but we shall not consider them here and consider only briefly how the numbers may be obtained and used.

One way to generate random numbers is to fix up a spinning arrow on a common clock. When the arrow is spun, the number on which it stops would be taken to be random number for that trial. Naturally, any number of spinnings of the arrow would result in an equal number of random numbers. In a similar way, random numbers can be generated using spinning of a roulette wheel, tossing dice . . . and like that. Although simple, these are very slow methods and cannot meet the practical requirements where a large number of random numbers may be needed.

A more fast and convenient method is to make use of the published tables of random numbers, like the one published by the Rand Corporation (of USA): *A Million Random Digits*. A random number table is a very efficient way to generate random data in most situations. The numbers in this table are in random arrangements. The underlying theory is that each number has an equal opportunity of being selected.

Still on a more sophisticated level, computers are used for generating the random numbers. With computers, it is typically easier to generate random numbers by an arithmetic process as needed rather than to read the numbers from stored table. An early, probably the earliest, method proposed for use on digital computers to generate random numbers is the *midsquare*. To illustrate this method, suppose that we wish to generate four-digit integers and the last number generated was 8937. To obtain the next number, in the sequence, we square the last one and use the middle four digits of the product. In this case the product is 79869969 so that the next pseudo-number is 8699. The next few numbers in the sequence are 6726, 2390, 7121 and so on. Thus, using this method, having drawn up a suitable computer program, a four-digit number may be fed into the computer and a list of pseudo-random numbers obtained.

Of all the random number generators, we shall make use of random number tables for demonstrating the simulation process. In particular, we shall use the random number table (Table B7) given in Appendix B of the book. To consider how the table can be used for generating data relating to our bakery problem, we proceed as follows:

Step 1 An assignment has to be worked out so that the intervals of random numbers will correspond to the probability distribution. Here, since the probabilities have been calculated to two decimal places, which add up to 1.00, we need 100 numbers of two digits to represent each point of probability. Thus, we take random numbers 00 through 99 to represent them. Now, as the probability of 5 cakes is equal to 0.02, we assign two random numbers 00–01 to this demand level; the probability of 6 cakes being equal to 0.05, the next five numbers, 02–06 would be assigned to this level. In a similar manner, each of the demand levels would be assigned appropriate intervals as given here. It may be mentioned that cumulative probabilities shown are calculated to ease the determination of the random number intervals. The cumulative probabilities column allows the assigned numbers to correspond to the same probability range for each event.

Demand (No. of cakes)	*Probability*	*Cumulative Probability*	*Random Number Interval*
5	0.02	0.02	00 01
6	0.05	0.07	02–06
7	0.08	0.15	07–14
8	0.25	0.40	15–39
9	0.31	0.71	40–70
10	0.19	0.90	71–89
11	0.06	0.96	90–95
12	0.04	1.00	96–99

Instead, if probabilities are calculated to three decimal places, then 3-digit random numbers would be required ... and so on.

Step 2 Once the random number intervals are determined, we select a tracking pattern for drawing random numbers from the random number table. We may start with any column and row of the table and read the values in any set manner—horizontally, vertically, or diagonally. Using the pattern, we draw the random numbers and match them with the assigned events. We may decide, for example, to read every third value horizontally, starting with the fifth column and fourth row of the table of random numbers. The random numbers, according to this pattern are 61, 74, 24, 03, 59, 16, 84, 92, 52, 07 and so on. We draw as many random numbers as the number of day's demand is required to be simulated.

The first of the list of the numbers, 61, lies in the interval 40–70 corresponding to the demand level of 9 units. Thus, the simulated demand for the first day is 9 cakes. In a similar manner, we can obtain the demand for each of the days. For the 10-day period, we have the following demand:

Day	:	1	2	3	4	5	6	7	8	9	10
Random Number	:	61	74	24	03	59	16	84	92	52	07
Demand (cakes)	:	9	10	8	6	9	8	10	11	9	7

Having considered the method of generating data with random numbers, we now proceed to illustrate the simulation of an inventory and a queuing system. They follow in turn.

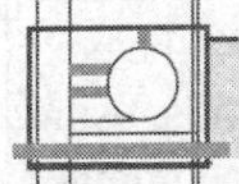

17.4 SIMULATION OF AN INVENTORY SYSTEM

In our discussion of inventory management in Chapter 9, it was pointed out that when variation is observed in both, the demand and the lead time, then the analytical approach gives way to the technique of simulation. We now consider such a situation and see how simulation can be employed to model an inventory system.

Example 17.1 Consider the case of a dealer of a certain product for which the probability distribution of daily demand and the probability distribution of the lead time, both developed empirically by observations made over a long span of period, are as follows:

Probability distribution of daily demand

Units demanded :	3	4	5	6	7	8	9	10	11	12
Probability :	0.02	0.08	0.11	0.16	0.19	0.13	0.10	0.08	0.07	0.06

Probability distribution of lead time

Lead time (days) :	2	3	4	5
Probability :	0.20	0.30	0.35	0.15

The ordering cost is known to be Rs 80 per order, the holding cost per unit per day is estimated at Rs 2, while the unit shortage cost, representing the loss in profits is Rs 20 per unit per day.

The dealer is anxious to known, for specific re-order levels and re-order quantities, what would be the total inventory costs (made up of the ordering, holding, and shortage costs), and thereby selecting an appropriate combination of the two.

For our illustration, we shall evaluate a simulation plan which calls for a re-order quantity of 40 units and a re-order level of 20 units, with a beginning inventory balance of 30 units.

As a first step, we shall assign a coding system that relates the identified values of the variable(s) to the random numbers. As discussed before, since the probabilities in both the given distributions are calculated and given to two decimal places, the coding system to be used would be based on these probabilities and consist of two-digit random numbers: 00 through 99. The random number coding for each of the distributions shall be as given in Tables 17.1 and 17.2.

TABLE 17.1 Random Number Coding: Demand Distribution

Daily Demand	*Probability*	*Cumulative Probability*	*Random Number Interval*
3	0.02	0.02	00–01
4	0.08	0.10	02–09
5	0.11	0.21	10–20
6	0.16	0.37	21–36
7	0.19	0.56	37–55
8	0.13	0.69	56–68
9	0.10	0.79	69–78
10	0.08	0.87	79–86
11	0.07	0.94	87–93
12	0.06	1.00	94–99

TABLE 17.2 Random Number Coding: Lead Time Distribution

Lead Time (days)	*Probability*	*Cumulative Probability*	*Random Number Interval*
2	0.20	0.20	00–19
3	0.30	0.50	20–49
4	0.35	0.85	50–84
5	0.15	1.00	85–99

Having determined the random number intervals, we select an appropriate random number table to draw numbers from. Also, we decide the tracking pattern for identifying the random numbers. For this illustration, again we shall use the random number table (Table B7) and simulate the data on demand using random number read column-wise from the top of the sixth column. The simulation of the lead time would be similarly done by reading numbers column-wise from the last column of the table. The results of the 40-day simulation are given in Table 17.3.

The second column of the table contains the random numbers drawn in respect of the daily demand while the corresponding demand for each day, determined in accordance with the random number coding given in Table 17.1. For instance, the random number 68, for the day 1, lies in the interval 56–68 indicating a demand of 8 units. With an initial balance of 30 units as indicated in column 7 of the table, and a demand of 8 units on the first day, the number of units in the stock would be 22 for this day, involving a holding cost of Rs 44 given in column 9. The balance would reduce to 17 units when the 5 units of demand simulated for the day 2 are sold. On this day, an order for 40 units be given for an ordering cost of Rs 80 (column 8). The

TABLE 17.3 Simulation Worksheet

Day (1)	*R. No. (2)*	*Demand (3)*	*R. No. (4)*	*L. Time (5)*	*Receipts (6)*	*Balance (7)*	*O (8)*	*H (9)*	*S (10)*
0						30			
1	68	8				22		44	
2	13	5	47	3		17	80	34	
3	09	4				13		26	
4	20	5				8		16	
5	73	9			40	39		78	
6	07	4				35		70	
7	92	11				24		48	
8	99	12	74	4		12	80	24	
9	93	11				1		2	
10	18	5				0		0	80
11	24	6				0		0	120
12	22	6			40	34		68	

Contd

TABLE 17.3 Contd.

Day (1)	*R. No.* (2)	*Demand* (3)	*R. No.* (4)	*L. Time* (5)	*Receipts* (6)	*Balance* (7)	*O* (8)	*H* (9)	*S* (10)
13	07	4				30		60	
14	29	6				24		48	
15	57	8	25	3		16	80	32	
16	33	6				10		20	
17	49	7				3		6	
18	65	8			40	35		70	
19	92	11				24		48	
20	98	12	21	3		12	80	24	
21	00	3				9		18	
22	57	8				1		2	
23	12	5			40	36		72	
24	31	6				30		60	
25	96	12	47	3		18	80	36	
26	85	10				8		16	
27	92	11				0		0	60
28	91	11			40	29		58	
29	77	9	69	4		20	80	40	
30	37	7				13		26	
31	34	6				7		14	
32	11	5				2		4	
33	27	6			40	36		72	
34	10	5				31		62	
35	59	8				23		46	
36	33	6	09	2		17	80	34	
37	87	11				6		12	
38	72	9			40	37		74	
39	73	9				28		56	
40	79	10	80	4		18	80	36	
					Total		640	1,456	260

O : Ordering cost H : Holding cost S : Shortage cost

random number chosen for simulating the lead time being 47, the lead time would be 3 days, and the 40 units ordered would, therefore, be received on day 5, as entered in column 6. We proceed in this manner to discover that on day 9 there is one unit in the stock. The demand of 5 units on day 10 indicates that it would be an out-of-stock situation. The shortage of 4 units would result in a stock out cost of Rs 80, given in the last column of the table. The condition of being out-of-stock would again prevail on the next day.

Completing a 40-day period, we find that the total ordering cost = Rs 640, the holding cost = Rs 1,456 and the out-of-stock cost = Rs 260, the three adding upto Rs 2,356.

Performing the same feat with re-order level of 15 units and the same re-order quantity of 40 units (the random numbers being the same) we obtain a total cost equal to Rs 2,358 comprising: ordering cost Rs 560, holding cost Rs 1,198, and out-of-stock cost Rs 600. In a similar manner, the total cost obtained from having a policy of re-order level equal to 20 units and the re-order quantity of 30 units, equals Rs 2,730 (ordering cost: Rs 720, holding cost: Rs 1,090 and stockout cost: Rs 920).

In a problem like this, which requires the determination of two parameters (order quantity and re-order level), the approach is to set up a two-way table of costs with various combinations of the two parameter values. We would simulate each combination a large number of times, compute the total cost corresponding to each one and enter it into the table. This procedure will enable us to locate the neighborhood of the optimum and select the plan with approximately the lowest cost.

It may be observed here that the cost estimates for a given combination of order quantity, re-order level and starting inventory level depend upon the tracking pattern adopted and the starting point chosen. It is desirable, therefore, that a number of cost estimates be made by varying the starting point and the tracking pattern, and then and average value may be determined. Alternatively, a large number of days' data may be simulated in order to ensure more dependable cost estimates.

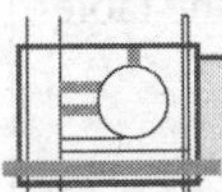

17.5 SIMULATION OF QUEUING SYSTEM

In Chapter 10 we considered the queuing situations. The characteristics of the situation described were based on the assumptions, among others, that the arrivals are Poisson distributed while the service times follow an exponential distribution. Although such assumptions are often appropriate for the real life queuing systems, they are not invariably so. Thus, where such assumptions fail to apply, formal mathematical analysis can be extremely difficult and even impossible. Simulation provides the answer in such cases.

We shall illustrate the use of simulation in the study of queues through the following example.

Example 17.2 In a large workshop undertaking servicing jobs, the mechanics obtain their tools and requirements from a central store. The manager, perturbed about the waiting time of mechanics, is in the process of determining whether more attendants be hired for the store for raising the level of service. The idle time cost for the mechanics and the wages required to be paid to the attendants being known, he wishes to ascertain how many attendants may be employed to minimise the total cost involved. For helping the manager solve his problem using simulation, we proceed as follows. The following data on the times between successive arrivals and the service times for the mechanics have been obtained from the past 200 observations made on the system presently in operation.

Distribution of inter-arrival time:

Time (Minutes)	*Frequency*	*Probability*
0	12	0.06
3	18	0.09
6	50	0.25
9	74	0.37
12	32	0.16
15	14	0.07
Total	200	

Distribution of service time:

Time (Minutes)	Frequency	Probability
4	8	0.04
6	20	0.10
8	36	0.18
10	88	0.44
12	48	0.24
Total	200	

As usual, in the first step we assign random numbers to each observed arrival time interval based on the likelihood of the occurrence of each time interval as shown in Table 17.4. For instance, the probability of arrival within 0 minutes of the last arrival (implying simultaneous arrivals) is 0.06. Therefore, we assign the two-digit random numbers 00 through 05 to the event of 0 minutes between the arrivals. To facilitate this process, as before, cumulative probabilities have been determined as given in the third column of the table.

TABLE 17.4 Random Number Coding: Inter-arrival Times

Time (minutes)	*Probability*	*Cumulative Probability*	*Random Number Interval*
0	0.06	0.06	00–05
3	0.09	0.15	06–14
6	0.25	0.40	15–39
9	0.37	0.77	40–76
12	0.16	0.93	77–92
15	0.07	1.00	93–99

The same procedure is followed for the service time distribution. Table 17.5 depicts the cumulative probabilities and the random number coding for each of the observed service times. The table shows, for instance, that the probability of service time of 4 minutes is 0.04. Thus, the random numbers 00–03 are assigned to the event that the service time is 4 minutes. If a random number in the range of 00–03 is drawn from the random number table, the event that 'the service time is 4 minutes' is assumed to have occurred.

TABLE 17.5 Random Number Coding: Service Times

Time (minutes)	*Probability*	*Cumulative Probability*	*Random Number Interval*
4	0.04	0.04	00–03
6	0.10	0.14	04–13
8	0.18	0.32	14–31
10	0.44	0.76	32–75
12	0.24	1.00	76–99

Now we are ready to simulate the operation. To determine the arrival times of the successive mechanics, we start with the fifth major column of the table of random numbers, and for the service times, we begin with the fourth major column, and read vertically down in each case. Supposing that a total of 25 observations are decided to be made, we draw 25 random numbers for each, the inter-arrival times and the service times. These numbers are given in columns 2 and 4, respectively of Table 17.6. The times corresponding to the random numbers, as determined from the Tables 17.4 and 17.5, are contained in columns 3 and 5, respectively.

Now, we start with the first mechanic and observe that the random number for the arrival is 58. This lies in the interval 40–76 and therefore corresponds to the time period of 9 minutes. Assuming that the workshop starts at 8.00 am, the first arrival at the store would be at 8.09 am. Further, the random number for service being 87, which lies on the interval 76–99 in the Table 17.5, a service time of 12 minutes is suggested. Thus, the service to him begins at 8.09 am and ends at 8.21 am. Obviously, there would be no waiting time for the mechanic. The second of the arrivals' random number is 47 which also indicates an inter-arrival time of 9 minutes. Thus, he would come to the store at 8.18 am—9 minutes after the first arrival. Since the store would be free at 8.21 am from the first mechanic, the service to the second would start at that time. Thus, the waiting time for him would be 3 minutes. Corresponding to the random number 39 for service time for this mechanic, the time needed is 10 minutes. This means that the service to this mechanic would go from 8.21 to 8.31 am.

Notice that when the first mechanic is being serviced between 8.09 and 8.21 am, then the second mechanic would have to wait. Thus, the queue length over this interval would be taken to bc 1.

In the same way, we proceed further and determine the arrival and waiting times for the mechanics, as contained in the columns 4 and 7 of the Table 17.6. Queue length for each interval of the service times is obtained by considering the number of mechanics reaching upto the upper limit of that interval after the one being currently serviced.

From the column showing the pattern of the mechanics' waiting time, and the last column containing the queue length, we can obtain the average waiting time and the average queue length by dividing the total of each one by the number of observations, 25. We get, the average waiting time = 623/25 = 24.92 minutes, and the average queue length = 73/25 = 2.92 mechanics.

Using this information, the manager can determine the total cost in the manner as we considered in Chapter 10. For examining other alternatives, he can do so by estimating the new service time patterns based on additional attendants. He would have to simulate the new alternative to determine the outcome based on the new service patterns and choose the best one.

TABLE 17.6 Simulation Worksheet

S. No.	*Arrivals*			*Service*		*Waiting*	*Service*		*Queue*
	R. No.	*Time*	*O'clock*	*R. No.*	*Time*	*Time (mts)*	*Begins*	*Ends*	*Length*
(1)	*(2)*	*(3)*	*(4)*	*(5)*	*(6)*	*(7)*	*(8)*	*(9)*	*(10)*
1	58	9	8.09 am	87	12	00	8.09 am	8.21 am	1
2	47	9	8.18	39	10	03	8.21	8.31	1
3	23	6	8.24	28	8	07	8.31	8.39	2
4	69	9	8.33	97	12	06	8.39	8.51	2
5	35	6	8.39	69	10	12	8.51	9.01	2

Contd.

TABLE 17.6 Contd.

S. No	Arrivals			Service		Waiting	Service		Queue
	R. No.	Time	O'clock	R. No.	Time	Time (mts)	Begins	Ends	Length
(1)	(2)	(3)	(4)	(5)	(6)	(7)	(8)	(9)	(10)
6	55	9	8.48	87	12	11	9.01	9.13	2
7	69	9	8.57	52	10	16	9.13	9.23	2
8	90	12	9.09	52	10	14	9.23	9.33	2
9	86	12	9.21	15	8	12	9.33	9.41	2
10	74	9	9.30	85	12	11	9.41	9.53	2
11	39	6	9.36	41	10	17	9.53	10.03	2
12	15	6	9.42	82	12	21	10.03	10.15	3
13	90	12	9.54	98	12	21	10.15	10.27	4
14	98	15	10.09	99	12	18	10.27	10.39	4
15	39	6	10.15	23	8	24	10.39	10.47	4
16	16	3	10.18	77	12	29	10.47	10.59	5
17	52	9	10.27	42	10	32	10.59	11.09	6
18	56	9	10.36	60	10	33	11.09	11.19	6
19	21	6	10.42	22	8	37	11.19	11.27	6
20	23	6	10.48	91	12	39	11.27	11.39	5
21	00	0	10.48	68	10	51	11.39	11.49	4
22	87	12	11.00	36	10	49	11.49	11.59	3
23	20	6	11.06	22	8	53	11.59	12.07 pm	2
24	40	9	11.15	92	12	52	12.07 pm	12.19	1
25	73	9	11.24	34	10	55	12.19	12.29	0
				Total		623 mts			73

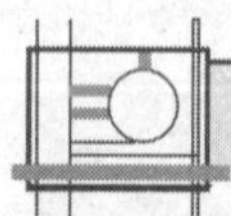

17.6 ADVANTAGES AND DISADVANTAGES OF SIMULATION

The chief merit of the simulation technique is its capacity to lend itself to problems that are cumbersome, or impossible to handle mathematically using analytical methods. Not only this, the technique allows the analyst to experiment with the system behaviour without subjecting it to the risks that would be inherent in experimenting with the real system, it also compresses time to enable the manager visualise the long-term effects in a quick manner. Besides, simulation is often used to test proposed analytical solutions as well.

However, it must be emphasised that the simulation technique does not represent a methodology for derivation of optimal solutions to the given problems. This approach is designed merely to provide a characterisation of the behaviour of the system in question for a given set of inputs. Further, the simulation approach is not precise in the sense that it yields only estimates which are subject to sampling error. Of course, the sampling error can be reduced by increasing the sample size (the number of runs) or by lengthening the computer run time.

Another drawback, associated with the technique, is that it may not prove economical where large-scale simulation, requiring considerable effort to develop a suitable model, and computer time, may be required to be done, as is the case in most of the practical situations. Also, Monte Carlo simulation is restricted to situations that contain elements which can be described by random variables.

Probably the most serious limitation of simulation stems from the fact that it is a tool of solution evaluation and thus does not generate problem solution. Therefore, the analyst has to develop the proposed solution; then simulation can be used to test the relative desirability of those solutions.

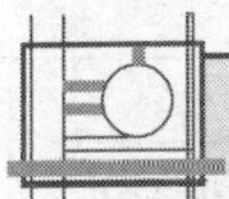

17.7 APPLICATIONS OF SIMULATION

In spite of its limitations, simulation is a very potent, flexible, and, therefore, a commonly employed quantitative tool for solving decision problems. It has been applied successfully to a broad spectrum of problems. To count a few, its use extends to areas like police dispatching and beat design; location of emergency vehicles like ambulances; making inventory policy decisions; evaluation of operating alternatives at airports; in financial planning—both portfolio selection and capital budgeting, scheduling the production processes; large-scale military battles as well as individual weapons systems for aiding in the designing of both the weapon systems and of the strategic and tactical operations. Simulation is of paramount importance where the experimentation with the real situation is risky. The use of computer simulation for studying the likely behaviour of the nuclear reactors during accidents is a case in point. Obviously, testing actual reactors or even scaled down reactor models under emergency conditions would involve excessive risks.

Thus, its application in situations which are too complex to permit mathematical solution, its simplicity and use in varied fields, its power in enabling the decision-maker to conduct experiments that help in understanding the system's behaviour while avoiding the risks associated with conducting tests on the model's real-life counterpart, and also availability of extensive computer software packages, are the main factors which account for the popularity of simulation technique.

REVIEW ILLUSTRATIONS

Example 17.3 The Tit-Fit Scientific Laboratories is engaged in producing different types of high class equipment for use in science laboratories. The company has two different assembly lines to produce its most popular product 'Pressurex'. The processing time for each of the assembly lines is regarded as a random variable and is described by the following distributions.

Process Time (minutes)	*Assembly A_1*	*Assembly A_2*
10	0.10	0.20
11	0.15	0.40
12	0.40	0.20
13	0.25	0.15
14	0.10	0.05

Using the following random numbers, generate data on the process times for 15 units of the item and compute the expected process time for the product. For the purpose, read the numbers vertically taking the first two digits for the processing time on assembly A_1 and the last two digits for processing time on assembly A_2.

4134	8343	3602	7505	7428
7476	1183	9445	0089	3424
4943	1915	5415	0880	9309

In the first stage, we assign random number intervals to the processing times on each of the assemblies. This is shown in Table 17.7.

TABLE 17.7 Random Number Coding for Process Times

Time (mts)	Assembly A_1			Assembly A_2		
	Prob.	Cum. Prob.	RN Interval	Prob.	Cum. Prob.	RN Interval
10	0.10	0.10	00–09	0.20	0.20	00–19
11	0.15	0.25	10–24	0.40	0.60	20–59
12	0.40	0.65	25–64	0.20	0.80	60–79
13	0.25	0.90	65–89	0.15	0.95	80–94
14	0.10	1.00	90–99	0.05	1.00	95–99

The simulation worksheet given in Table 17.8 shows the simulated data on the process times. Thus, the random numbers for the first unit are 41 and 34, respectively, for the assemblies A_1 and A_2. From the Table 17.7 we observe that the times corresponding to these are 12 and 11 minutes respectively. Thus, the total time required for the unit is 23 minutes. In the same way, the times for the other 14 units are determined and shown in the last column of the table. The expected completion time for a unit works out to be 23.27 minutes.

TABLE 17.8 Simulation Worksheet

Unit	Assembly A_1		Assembly A_2		Total Time
	R. Number	Time	R. Number	Time	(mts.)
1	41	12	34	11	23
2	74	13	76	12	25
3	49	12	43	11	23
4	83	13	43	11	24
5	11	11	83	13	24
6	19	11	15	10	21
7	36	12	02	10	22
8	94	14	45	11	25
9	54	12	15	10	22
10	75	13	05	10	23
11	00	10	89	13	23
12	08	10	80	13	23
13	74	13	28	11	24
14	34	12	24	11	23
15	93	14	09	10	24
					Expected time = 349/15 = 23.27

Example 17.4 Haggins Plumbing and Heating maintains a stock of 30–gallon hot water heaters that it sells to home-owners and installs for them. The owner likes the idea of having large supply on hand so as to meet all customer demand, but he also recognizes that it is expensive to do so. He examines hot water heater sales over the past 50 weeks and notes the following:

Hot water heater sales per week	*Number of weeks this number was sold*
4	6
5	5
6	9
7	12
8	8
9	7
10	3
Total	50

Using the random numbers given below, simulate demand for 20 weeks and answer the following questions:

(i) If Higgins maintains a constant supply of 8 hot water heaters in any given week, how many times will he be out of stock during the 20-week simulation period?

(ii) What is the average number of heaters demanded per week over the 20-week interval?

Random Numbers:

10, 24, 03, 32, 23, 59, 95, 34, 34, 51

08, 48, 66, 97, 03, 96, 46, 74, 77, 44

(M Com, Delhi, 2009)

As a first step, we allocate random numbers 00–99 in proportion to the probabilities associated with demand of the heaters per day, as shown in Table 17.9.

TABLE 17.9 Allocation of Random Numbers

Demand per Week	*No. of Weeks*	*Probability*	*Cumulative Probability*	*Random Number Interval*
4	6	0.12	0.12	00–11
5	5	0.10	0.22	12–21
6	9	0.18	0.40	22–39
7	12	0.24	0.64	40–63
8	8	0.16	0.80	64–79
9	7	0.14	0.94	80–93
10	3	0.06	1.00	94–99
Total	50	1.00		

Based on the given random numbers, we may simulate the demand for the heaters for 20 weeks as shown in Table 17.10.

TABLE 17.10 Simulation Worksheet

Day	*Random Number*	*Demand*	*Day*	*Random Number*	*Demand*
1	10	4	11	08	4
2	24	6	12	48	7
3	03	4	13	66	8
4	32	6	14	97	10
5	23	6	15	03	4
6	59	7	16	96	10
7	95	10	17	46	7
8	34	6	18	74	8
9	34	6	19	77	8
10	51	7	20	44	7
				Total	135

From the table, it is evident that

(i) Number of weeks he is expected to be out-of-stock = 3

(ii) Average demand for heaters per week $= \dfrac{\text{Total demand}}{\text{Number of weeks}}$

$$= \frac{135}{20} = 6.75$$

Example 17.5 A company manufactures 30 units per day. The sale of these items depends upon demand which has the following distribution:

Sales (Units)	*Probability*
27	0.10
28	0.15
29	0.20
30	0.35
31	0.15
32	0.05

The production cost and sale price of each unit are Rs 40 and Rs 50, respectively. Any unsold product is to be disposed off at a loss of Rs 15 per unit. There is a penalty of Rs 5 per unit if the demand is not met. Using the following random numbers, estimate the total profit/loss for the company for the next ten days:

10, 99, 65, 99, 95, 01, 79, 11, 16, 20

If the company decides to produce 29 units per day, what is the advantage or disadvantage of the company?

(CA, November, 1997)

As a first step, random numbers 00–99 are allocated to various possible sale values in proportion to the probabilities associated with them. This is shown in Table 17.11.

TABLE 17.11 Assignment of Random Numbers

Sales (Units)	*Probability*	*Cumulative Probability*	*Random Number Interval*
27	0.10	0.10	00–09
28	0.15	0.25	10–24
29	0.20	0.45	25–44
30	0.35	0.80	45–79
31	0.15	0.95	80–94
32	0.05	1.00	95–99

Now, we simulate the demand for the next ten days using the given random numbers, as shown in Table 17.12. From the given information, we have,

$$\text{profit per unit sold} = \text{Rs } 50 - \text{Rs } 40 = \text{Rs } 10$$

$$\text{loss per unit unsold} = \text{Rs } 15$$

$$\text{penalty for refusing demand} = \text{Rs } 5 \text{ per unit}$$

Using these inputs, the profit/loss for the ten days is calculated as given in the table, first when production is 30 units per day and then when it is 29 units.

It is evident that the total profit for ten days is Rs 2,695 when 30 units are produced. Also, if the company decides to produce 29 units per day, the total profit works out to be the same, as shown calculated in the last column of the table.

TABLE 17.12 Simulation Worksheet: Calculation of Profit/Loss

Day	*Random Number*	*Estimated Sales (units)*	*Profit/Loss per Day with Production*			
			30 units		*29 units*	
1	10	28	28 × 10 – 2 × 15	= Rs 250	28 × 10 – 1 × 15	= Rs 265
2	99	32	30 × 10 – 2 × 5	= Rs 290	29 × 10 – 3 × 5	= Rs 275
3	65	30	30 × 10	= Rs 300	29 × 10 – 1 × 5	= Rs 285
4	99	32	30 × 10 – 2 × 5	= Rs 290	29 × 10 – 3 × 5	= Rs 275
5	95	32	30 × 10 – 2 × 5	= Rs 290	29 × 10 – 3 × 5	= Rs 275
6	01	27	27 × 10 – 3 × 15	= Rs 225	27 × 10 – 2 × 15	= Rs 240
7	79	30	30 × 10	= Rs 300	29 × 10 – 1 × 5	= Rs 285
8	11	28	28 × 10 – 2 × 15	= Rs 250	28 × 10 – 1 × 15	= Rs 265
9	16	28	28 × 10 – 2 × 15	= Rs 250	28 × 10 – 1 × 15	= Rs 265
10	20	28	28 × 10 – 2 × 15	= Rs 250	28 × 10 – 1 × 15	= Rs 265
			Total Profit	Rs 2,695		Rs 2,695

Example 17.6 Dr Strong is a dentist who schedules all her patients for 30-minutes appointments. Some of the patients take more or less than 30 minutes depending on the type of dental work to be done. The following summary shows the various categories of work, their probabilities and the time actually needed to complete the work.

Category	*Time Required*	*Probability of Category*
Filling	45 Minutes	0.40
Crown	60 Minutes	0.15
Cleaning	15 Minutes	0.15
Extraction	45 Minutes	0.10
Check-up	15 Minutes	0.20

Simulate the dentist's clinic for four hours and determine average waiting time for the patients as well as the idleness of the doctor. Assume that all the patients show up at the clinic at exactly their scheduled arrival time starting at 8 a.m. Use the following random numbers for handling the above problem: 40, 82, 11, 34, 25, 66, 17, 79.

(CA, November, 1990)

As a first step, we determine random number intervals. This is done below:

Category	*Time Required*	*Prob.*	*Cum. Prob.*	*Random Number Interval*
Filling	45 m	0.40	0.40	00–39
Crown	60 m	0.15	0.55	40–54
Cleaning	15 m	0.15	0.70	55–69
Extraction	45 m	0.10	0.80	70–79
Check-up	15 m	0.20	1.00	80–99

The simulation is shown in Table 17.13. Each of the patients is assumed to require service in accordance with the random number associated with him/her. For example, the first patient with random number 40 is assumed to need 'crown', as the number falls in the group 40–54.

TABLE 17.13 Simulation Worksheet: Dentist's Clinic

Patient	*Arrival Time*	*Random Number*	*Service Time Needed*	*Service*		*Waiting Time (Minutes)*	*Idle Time (Minutes)*
				Begins	*Ends*		
1	8.00 AM	40	60	8.00 AM	9.00 AM	0	0
2	8.30 AM	82	15	9.00 AM	9.15 AM	30	0
3	9.00 AM	11	45	9.15 AM	10.00 AM	15	0
4	9.30 AM	34	45	10.00 AM	10.45 AM	30	0
5	10.00 AM	25	45	10.45 AM	11.30 AM	45	0
6	10.30 AM	66	15	11.30 AM	11.45 AM	60	0
7	11.00 AM	17	45	11.45 AM	12.30 PM	45	0
8	11.30 AM	79	45	12.30 PM	1.15 PM	60	0
				Total		285	0

From the table, we get

$$\text{Doctor's idle time} = \text{nil}$$

$$\text{Average waiting time of patients} = \frac{285}{8} = 35.6 \text{ minutes}$$

Example 17.7 The materials manager of a firm wishes to determine the expected mean demand for a particular item in stock during the reorder lead time. This information is needed to determine how far in advance to reorder, before the stock level is reduced to zero. However, both the lead time, in days, and the demand per day for the item are random variables, described by the probability distribution that follows:

Lead Time (in days)	*Probability*	*Demand per Day (units)*	*Probability*
1	0.45	1	0.15
2	0.30	2	0.25
3	0.25	3	0.40
		4	0.20

Manually simulate the problem for 30 reorders, to estimate the demand during lead time.

(MBA, Delhi, April, 1987)

As a first step, random number intervals for each of the given probability distributions would be determined. This is shown in Table 17.14.

TABLE 17.14 Determination of Random Number Intervals

Lead Time (days)	*Probability*	*R. No. Interval*	*Demand per Day (units)*	*Probability*	*R. No. Interval*
1	0.45	00–44	1	0.15	00–14
2	0.30	45–74	2	0.25	15–39
3	0.25	75–99	3	0.40	40–79
			4	0.20	80–99

Since we are required to simulate the problem for 30 reorders, we shall draw as many random numbers to estimate the lead time for each of the reorders. Depending on the lead time simulated, we shall draw random numbers in respect of the demand. Thus, if we draw a number which indicates a lead time of 3 days, we shall pick three numbers—each of which shall be an estimator of the demand for a particular day.

The simulation worksheet given in Table 17.15 contains the estimation of demand during lead time. The random numbers for lead time, picked up from the fifth column of the table of random numbers, and those for demand, read from the tenth column of the table, are given in the parentheses and the corresponding lead time/demand values are mentioned along with each of them. The demand during each of the lead times is contained in the last column of the table. The average demand is found to be 4.6 units.

TABLE 17.15 Simulation Worksheet

S. No.	Lead Time		Daily Demand During Lead Time			Demand During Lead Time
	R. No.	Days				
1	(84)	3	(35)2	(55)3	(21)2	7
2	(46)	2	(64)3	(05)1		4
3	(77)	3	(35)2	(92)4	(28)2	8
4	(61)	2	(65)3	(27)2		5
5	(08)	1	(09)1			1
6	(39)	1	(52)3			3
7	(74)	2	(66)3	(51)3		6
8	(00)	1	(07)1			1
9	(99)	3	(47)3	(70)3	(83)4	10
10	(24)	1	(76)3			3
11	(72)	2	(07)1	(79)3		4
12	(48)	2	(37)2	(02)1		3
13	(03)	1	(46)3			3
14	(26)	1	(80)4			4
15	(31)	1	(08)1			1
16	(59)	2	(90)4	(53)3		7
17	(29)	1	(47)3			3
18	(31)	1	(19)2			2
19	(16)	1	(35)2			2
20	(98)	3	(18)2	(71)3	(59)3	8
21	(72)	2	(32)2	(87)4		6
22	(84)	3	(96)4	(40)3	(52)3	10
23	(05)	1	(10)1			1
24	(53)	2	(86)4	(73)3		7
25	(92)	3	(52)3	(31)2	(83)4	9
26	(41)	1	(44)3			3
27	(82)	3	(16)2	(15)2	(21)2	6
28	(52)	2	(74)3	(23)2		5
29	(09)	1	(18)2			2
30	(66)	2	(48)3	(08)1		4
				Expected demand during lead time = 138/30 = 4.6		

Example 17.8 The occurrence of rain in a city on a day is dependent upon whether it rained on the previous day. If it rained on the previous day, the rain distribution is given by:

Event	*Probability*
No rain	0.50
1 cm rain	0.25
2 cm rain	0.15
3 cm rain	0.05
4 cm rain	0.03
5 cm rain	0.02

it did not rain the previous day, the rain distribution is given by:

Event	*Probability*
No rain	0.75
1 cm rain	0.15
2 cm rain	0.06
3 cm rain	0.04

Simulate the city's weather for 10 days and determine by simulation the total days without rain as well as the total rainfall during the period. Use the following random numbers:

67 63 39 55 29 78 06 78 76

for simulation. Assume that for the first day of the simulation it had not rained the day before.

(CA, November, 1993)

The random number intervals are determined in the first place is as shown in Tables 17.16 and 17.17.

TABLE 17.16 Random Number Intervals (Rained Previously)

Event	*Prob.*	*Cum. Prob.*	*Random Number Interval*
No rain	0.50	0.50	00–49
1 cm rain	0.25	0.75	50–74
2 cm rain	0.15	0.90	75–89
3 cm rain	0.05	0.95	90–94
4 cm rain	0.03	0.98	95–97
5 cm rain	0.02	1.00	98–99

TABLE 17.17 Random Number Intervals (No Rain Previously)

Event	*Prob.*	*Cum. Prob.*	*Random Number Interval*
No rain	0.75	0.75	00–74
1 cm rain	0.15	0.90	75–89
2 cm rain	0.06	0.96	90–95
3 cm rain	0.04	1.00	96–99

We may now predict the rainfall on various days in accordance with the random numbers for them by having reference to the appropriate table. For example, it is given that for the first day it may be assumed that it did not rain the previous day, we locate the random number for day 1 in Table 17.17. Similarly, each day's random number may be located in Table 17.16 or Table 17.17, depending on previous day's weather. The simulation of weather is given in Table 17.18, where rainfall is predicted with reference to the table indicated.

TABLE 17.18 Simulation Worksheet: Weather

Day	*Random Number*	*Rainfall*	*Table Reference*
1	67	No rain	Table 17.17
2	63	No rain	Table 17.17
3	39	No rain	Table 17.17
4	55	No rain	Table 17.17
5	29	No rain	Table 17.17
6	78	1 cm	Table 17.17
7	70	1 cm	Table 17.16
8	06	No rain	Table 17.16
9	78	1 cm	Table 17.17
10	76	2 cm	Table 17.16
	Total	5 cm	

Accordingly, total days without rain = 6, and total rainfall during the period = 5 cm.

Example 17.9 A project consists of 7 activities. The time for performance of each of the activities is a random variable with the respective probability distribution given here.

Activity	*Immediate Predecessor*	*Time (in days)* / *Probability*
A	–	3, 4, 5 / 0.20, 0.60, 0.20
B	–	4, 5, 6, 7, 8 / 0.10, 0.30, 0.30, 0.20, 0.10
C	*A*	1, 3, 5 / 0.15, 0.75, 0.10
D	*B, C*	4, 5 / 0.80, 0.20
E	*D*	3, 4, 5, 6 / 0.10, 0.30, 0.30, 0.30
F	*D*	5, 7 / 0.20, 0.80
G	*E, F*	2, 3 / 0.50, 0.50

Now,

(a) draw the network diagram and identify the critical path using the expected activity times.

(b) simulate the project using random numbers to determine the activity times. Find the critical path and the project duration.

(c) repeat the simulation four more times. Is the same path critical in all the simulations?

(a) The network corresponding to the given information is shown in Figure 17.1. The time for each of the activities is the expected time obtained by the summation of the products of the activity time and the corresponding probability. For example, for activity A, the expected time $= 3 \times 0.20 + 4 \times 0.60 + 5 \times 0.20 = 4$ days.

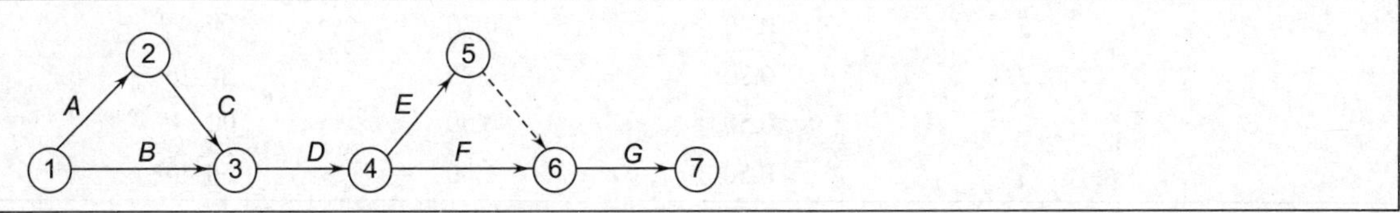

Figure 17.1 *Network Diagram*

From the figure, we observe that the critical path of the project is 1–2–3–4-6–7, with an expected duration of 20.2 days.

(b) and (c) To use simulation for obtaining the project length we determine, in the first stage, the random number intervals in respect of each of the activities. This is shown in Table 17.19. The five runs of simulation made by taking random numbers, read from the sixth column of the random number table, B7, are then shown in Table 17.20. For each run, the critical path and the project durations are obtained. For each run, the project time is obtained as follows:

Total time = larger of times for activities A & C and B + time for activity D + larger of times for activities E and F + time for activity G

TABLE 17.19 Random Number Coding for Activity Times

Activity	*Time*	*Prob.*	*Cum. Prob.*	*Random Number Interval*
A	3	0.20	0.20	00–19
	4	0.60	0.80	20–79
	5	0.20	1.00	80–99
B	4	0.10	0.10	00–09
	5	0.30	0.40	10–39
	6	0.30	0.70	40–69
	7	0.20	0.90	70–89
	8	0.10	1.00	90–99
C	1	0.15	0.15	00–14
	3	0.75	0.90	15–89
	5	0.10	1.00	90–99

Contd.

TABLE 17.19 Contd.

Activity	*Time*	*Prob.*	*Cum. Prob.*	*Random Number Interval*
D	4	0.80	0.80	00–79
	5	0.20	1.00	80–99
E	3	0.10	0.10	00–09
	4	0.30	0.40	10–39
	5	0.30	0.70	40–69
	6	0.30	1.00	70–99
F	5	0.20	0.20	00–19
	7	0.80	1.00	20–99
G	2	0.50	0.50	00–49
	3	0.50	1.00	50–99

TABLE 17.20 Simulation Worksheet

Run	*Activity*													
	A		*B*		*C*		*D*		*E*		*F*		*G*	
	R. No.	*Time*	*R. No.*	*Time*	*R.No.*	*Time*	*R. No.*	*Time*	*R. No.*	*Time*	*R. No.*	*Time*	*R. No.*	*Time*
1	68	4	13	5	09	1	20	4	73	6	07	5	92	3
2	99	5	93	8	18	3	24	4	22	4	07	5	29	2
3	57	4	33	5	49	3	65	4	92	6	98	7	00	2
4	57	4	12	5	31	3	96	5	85	6	92	7	91	3
5	77	4	37	5	34	3	11	4	27	4	10	5	59	3

From the data given in Table 17.20 we have the following information:

Simulation Run	*Activity Times*	*Project Duration*	*Critical Path(s)*
1	5 + 4 + 6 + 3	18 days	1–2–3–4–5–6–7
			1–3–4–5–6–7
2	8 + 4 + 5 + 2	19 days	1–2–3–4–6–7
			1–3–4–6–7
3	7 + 4 + 7 + 2	20 days	1–2–3–4–6–7
4	7 + 5 + 7 + 3	22 days	1–2–3–4–6–7
5	7 + 4 + 5 + 3	19 days	1–2–3–4–6–7

As may be observed, same path is not critical in each of the runs.

Example 17.10 A plant has a large number of similar machines. The machines breakdown randomly and the breakdowns are independent of each other. Once a machine breaksdown, it has to be taken out of production till the time it is repaired. On the basis of the past data, the following distributions have been constructed.

No. of Breakdowns per Hour	*Probability*	*No. of Hours Required for Repair per Breakdown*	*Probability*
0	0.900	1	0.100
1	0.085	2	0.240
2	0.012	3	0.450
3	0.003	4	0.165
		5	0.040
		6	0.005

Each hour that a machine remains idle due to being, or waiting to be repaired, it costs the plant Rs 80 per hour by way of lost production. If a repairman is paid at Rs 8 per hour, how many repairmen should be hired by the company to service the machine breakdowns? For the purpose, simulate the system for a 50-hour period and use the following random numbers, reading row-wise starting with the NW corner.

For breakdowns

100	375	084	990	128	660	310	852	635	737
985	118	834	886	995	654	801	743	699	098
914	803	441	125	636	611	154	945	424	235
044	005	359	598	460	321	692	195	451	948
980	331	809	797	186	740	541	116	483	690

For repair times

765	648	196	093	801	340	455	020	053	035
672	121	099	195	981	783	389	421	125	623

The random number coding for the hourly breakdowns and the repair times is shown in Table 17.21.

TABLE 17.21 Random Number Coding: Breakdowns and Repairs

Hourly Breakdowns	*Prob.*	*R. No. Interval*	*Repair Time (hrs)*	*Probability*	*R. No. Interval*
0	0.900	000–899	1	0.100	000–099
1	0.085	900–984	2	0.240	100–339
2	0.012	985–996	3	0.450	340–789
3	0.003	997–999	4	0.165	790–954
			5	0.040	955–994
			6	0.005	995–999

The simulation worksheet in Table 17.22 displays the data on breakdown and the repair times needed. Assuming that all breakdowns occur at the beginning of an hour, we observe that two machines would break at the beginning of the fourth hour, two at the stroke of the 15th hour, and so on. In all, 10 machines are estimated to breakdown during the period in review. The repair times simulated in respect of the breakdowns are shown in part (b) of the worksheet.

Waiting time with one repairman When only one repairman is hired, the waiting time for the machines and the total cost involved would be obtained as follows. The first breakdown is estimated to occur in the beginning of the fourth hour. M_1, the breakdown, would take 3 hours for repairs. The other breakdown, M_2, would have to wait for 3 hours before its repair starts (which would also take 3 hours to complete). Both these machines would be ready, however, before the machines M_3 and M_4 breakdown at the beginning of the 11th hour. Here, machine M_4 would have to wait for 2 hours during which M_3 would be repaired. The machines M_5 and M_6 are likely to fail at the beginning of the hour 15 and are scheduled to need 4 and 3 hours respectively. Of these, M_6 would wait for 4 hours before repair work on it can start. The machine M_7, failing at the start of the 21st hour would have to wait for one hour before it is serviced. None of the other machines has to wait. Thus, the waiting time of the machines in the queue adds upto 10 hours as shown here.

TABLE 17.22 Simulation Worksheet

(a) Breakdowns

Hour	*Random No.*	*Breakdowns*	*Hour*	*Random No.*	*Breakdowns*
1	100	0	26	611	0
2	375	0	27	154	0
3	084	0	28	945	1 M_8
4	990	2 M_1, M_2	29	424	0
5	128	0	30	235	0
6	660	0	31	044	0
7	310	0	32	005	0
8	852	0	33	359	0
9	635	0	34	598	0
10	737	0	35	460	0
11	985	2 M_3, M_4	36	321	0
12	118	0	37	692	0
13	834	0	38	195	0
14	886	0	39	451	0
15	995	2 M_5, M_6	40	948	1 M_9
16	654	0	41	980	1 M_{10}
17	801	0	42	331	0
18	743	0	43	809	0
19	699	0	44	797	0
20	098	0	45	186	0
21	914	1 M_7	46	740	0
22	803	0	47	541	0
23	441	0	48	116	0
24	125	0	49	483	0
25	636	0	50	690	0

(b) **Repair Times**

Machine	*Random Number*	*Repair Time (hrs)*
M_1	765	3
M_2	648	3
M_3	196	2
M_4	093	1
M_5	801	4
M_6	340	3
M_7	455	3
M_8	020	1
M_9	053	1
M_{10}	035	1

Machine	*Waiting Time*
M_2	3 hours
M_4	2 hours
M_6	4 hours
M_7	1 hour
Total	10 hours

Now,

Total cost = Idle time cost + Repairma(e)n's wages

= Idle time × Hourly rate + Total hours × Hourly rate

Idle time = Repairing time + Waiting time

In the present case, idle time = 22 + 10 = 32 hours

∴ Total cost = 32 × 80 + 50 × 8 = 2,560 + 400 = Rs 2,960.

For 2 repairmen In this case, none of the machines would have to wait because not more than two machines have been found to breakdown simultaneously in the above data. Also, each of the machines would be serviced before another breaks down. Thus, the total idle time would be only the repairing time of 22 hours.

∴ Total cost = 22 × 80 + 50 × 8 × 2 = 1,760 + 800 = Rs 2,560.

Hiring more than 2 repairmen would only add to cost. The cost when two repairmen are engaged being smaller, as many repairmen should be hired.

Example 17.11 A company receives consignments each of 25 components. From the past experience of the company it is determined that of all the consignments, 40% contain no defectives; 35% contain one defective; 12% contain two defectives; 8% contain three defectives; and 5% contain four defectives. Due to the expensive nature of the testing process, the company policy is to test two of the components, and refuse to accept delivery of the consignment if either of the tested components is found to be defective. However, if the first tested component is found to be defective, the other need not be examined.

It is required to simulate this system. Also making 15 runs, it is required to estimate the proportion of shipments for which delivery is likely to be refused.

The random number coding for the distribution of defectives in the consignments is given in Table 17.23. Also given in the table are the random number intervals for the probability values of the nature of the items examined. It may be observed that the probability that an item examined would be defective is dependent upon the number of defectives the consignment contains. For instance, when the consignment contains one defective item, the probability of an item to be defective is 1/25, which equals 2/25 when two defectives are contained. Similarly, the probabilities for the second item are conditional in the sense that they are based on the assumption that the first item examined was non-defective.

TABLE 17.23 Random Number Coding

No. of Defectives	*Prob.*	*R. No. Interval*	*Item 1*			*Item 2*		
			Nature	*Prob.*	*R. No. Interval*	*Nature*	*Prob.*	*R. No. Interval*
0	0.40	00–39	*D*	0/25	–	*D*	0/24	–
			N–D	25/25	00–24	*N–D*	24/24	00–23
1	0.35	40–74	*D*	1/25	00	*D*	1/24	00
			N–D	24/25	01–24	*N–D*	23/24	01–23
2	0.12	75–86	*D*	2/25	00–01	*D*	2/24	00–01
			N–D	23/25	02–24	*N–D*	22/24	02–23
3	0.08	87–94	*D*	3/25	00–02	*D*	3/24	00–02
			N–D	22/25	03–24	*N–D*	21/24	03–23
4	0.05	95–99	*D*	4/25	00–03	*D*	4/24	00–03
			N–D	21/25	04–24	*N–D*	20/24	04–23

We shall consider here two-digit random numbers upto 24 in respect of the first item and upto 23 in case of the second item examination. While drawing random numbers, we shall consider only those which do not exceed these values and ignore those above them.

The simulation worksheet in Table 17.24 contains results of the 15 runs.

The random numbers used in the second, fourth and sixth columns are drawn respectively from the first, tenth and twentieth column of the random number table. Of the 15 runs of simulation, we observe that the decision to reject is expected to occur once.

TABLE 17.24 Simulation Worksheet

Consignment	*R. No.*	*No. of Defectives*	*Item 1*		*Item 2*		*Accept/ Reject*
			R. No.	*Nature*	*R. No.*	*Nature*	
1	22	0	21	*N–D*	09	*N–D*	Accept
2	19	0	05	*N–D*	05	*N–D*	Accept
3	16	0	09	*N–D*	20	*N–D*	Accept
4	78	2	07	*N–D*	18	*N–D*	Accept

Contd.

TABLE 17.24 Contd.

Consignment	R. No.	No. of Defectives	Item 1		Item 2		Accept/ Reject
			R. No.	*Nature*	*R. No.*	*Nature*	
5	03	0	07	*N–D*	20	*N–D*	Accept
6	93	3	02	*D*			Reject
7	78	2	08	*N–D*	03	*N–D*	Accept
8	23	0	19	*N–D*	13	*N–D*	Accept
9	15	0	18	*N–D*	00	*N–D*	Accept
10	58	1	10	*N–D*	19	*N–D*	Accept
11	57	1	16	*N–D*	20	*N–D*	Accept
12	48	1	15	*N–D*	17	*N–D*	Accept
13	61	1	21	*N–D*	06	*N–D*	Accept
14	36	0	23	*N–D*	19	*N–D*	Accept
15	18	0	18	*N–D*	17	*N–D*	Accept

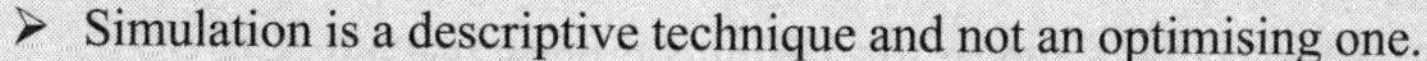

KEY POINTS TO REMEMBER

- Simulation is a descriptive technique and not an optimising one.
- To simulate is to copy a system. It involves developing a model of some real phenomenon and then performing experiment on the model, with a view to take appropriate decisions.
- Its benefit stems from the fact that results of taking a particular course of action can be estimated prior to its implementing in a real world situation.
- The four phases of simulation are: problem definition and objective setting, model formulation, experimentation with model and result evaluation.
- Monte Carlo simulation method uses the random numbers, which comprises numbers in which digits are distributed randomly.
- Simulation calls for determining random number intervals on the basis of given probabilities and then, with the help of random numbers, obtained from the table or as given, place the given inputs into relevant groups or categories.
- It is an extensively used tool and finds applications in areas like inventory control, capital budgeting, queuing, project scheduling, and so on.

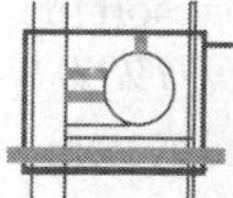

TEST YOUR UNDERSTANDING

Mark the following statements as T (*True*) or F (*False*).

1. A number of decision-making situations do not lend themselves to application of mathematical methods.

2. Simulation is a useful tool to employ to study the impact of system changes when experimentation with the real system is not feasible.
3. Simulation is an optimising technique.
4. In simulation, an attempt is made to copy a real system and the parameters are manipulated in an artificial environment.
5. Simulation always involves developing a physical model of some real phenomenon and then performing experiments on the model evolved.
6. In general terms, simulation aims to determine how the system under consideration would behave in certain conditions.
7. Construction of an appropriate model and experimentation with the model constructed are the core activities in simulation.
8. A simulation may be physical or mathematical, a mental conception or a combination.
9. The output of a simulation model is independent of the size of the simulation run.
10. Any simulation model is akin to a statistical experiment whose output is subject to statistical error.
11. *Monte Carlo* simulation involves modelling a stochastic system with the objective of predicting the system's behaviour.
12. Averaging the observations yields a better estimate of the output of simulation model.
13. Randomness is a key requirement of the *Monte Carlo* simulation.
14. A *random number generator* is a device or procedure by which random numbers are obtained.
15. In random numbers, various digits appear in a definite, orderly fashion.
16. The *mid-square* method can be used without fail for generating random numbers series of any length.
17. The random numbers generated by mid-square method are actually *pseudo-random numbers*.
18. If random numbers are available in the range of (0, 1), they can be converted into random numbers in the range of (0, 100) by multiplication by a constant 100.
19. Simulation cannot be used in those cases where mathematical methods can be used.
20. Solutions to decision problems by simulation would be identical to the solutions obtained by applying mathematical techniques.

EXERCISES

1. What is simulation? Describe the simulation process. State the major two reasons for using simulation to solve a problem. What are the advantages and limitations of simulation?
2. "When it becomes difficult to use an optimisation technique for solving a problem, one has to resort to simulation". Discuss. *(M.Com., Rajasthan, 1987)*
3. "Simulation is typically the process of carrying out sampling experiments on the models of the system rather than the system itself." Elucidate this statement by taking some examples. *(MBA, Delhi, April, 1994)*
4. Define a simulation model. Distinguish between deterministic and stochastic simulation models.
5. Discuss Monte Carlo Simulation. Illustrate how would you use it in situations of (i) queuing, and (ii) inventory control.

6. What are pseudo-random numbers? Are they not random in nature?

PRACTICAL PROBLEMS

1. Customers arrive at a milk booth for the required service. Assume that inter-arrival and service times are constants and given by 1.8 and 4 time units respectively. Simulate the system by hand computation for 14 time units. What is the average waiting time per customer? What is the percentage idle time of the facility? (Assume that the system starts at $t = 0$). *(Andhra Productivity Council, 1978)*
2. A bakery keeps stock of a popular brand of cakes. Previous experience shows the daily demand pattern for the item with associated probabilities, as given:

Daily demand (Nos.):	0	10	20	30	40	50
Probability :	0.01	0.20	0.15	0.50	0.12	0.02

Use the following sequence of random numbers to simulate the demand for next 10 days. Also find out the average demand per day.

Random numbers: 25, 39, 65, 76, 12, 05, 73, 89, 19, 49,

(ICWA, December, 1986)

3. A confectioner sells confectionery items. Past data of demand per week in hundred kilograms with frequency is given below:

Demand/week :	0	5	10	15	20	25
Frequency :	2	11	8	21	5	3

Using the following sequence of random numbers, generate the demand for next 15 weeks. Also find out the average demand per week.

Random numbers: 35 52 90 13 23 73 34 57

35 83 94 56 67 66 60

(Grad. Prog. in OR, ORSI, May, 1983)

4. The manager of a book has to decide the number of copies of a particular tax law book to order. A book costs Rs 60 and is sold for Rs 80. Since some of the tax laws change year after year, any copies unsold while the edition is current must be sold for Rs 30. From past records, the distribution of demand for this book has obtained as follows:

Demand (No. of copies) :	15	16	17	18	19	20	21	22
Proportion :	0.05	0.08	0.20	0.45	0.10	0.07	0.03	0.02

Using the following sequence of random numbers, generate data on demand for 20 time periods (years). Calculate the average profit obtainable under each of the course of action open to the manager. What is the optimal policy?

14, 02, 93, 99, 18, 71, 37, 30, 12, 10,
88, 13, 00, 57, 69, 32, 18, 08, 92, 73

5. In the morning of each working day, the manager of the Small Departmental Store has to decide whether he needs additional help to restock the shelves. It is known from experience that if extra help is needed one day, the chances are 60% that extra help would be needed the next day, while if extra help is not needed on a day, then the chances are only thirty per cent that it would be needed the next day. You are required to simulate the system, run the model for 20 days and estimate the proportion of days on which extra help is needed. Assume that extra help was not needed the day before the simulation run begins.
6. The Good Grain Cooperative Society is attempting to evaluate the risk associated with growing wheat in a new piece of land it has recently acquired. The data on estimated price per quintal, the yield per acre, and the cost per acre are given as follows. The probabilities of the three distributions are assumed to be independent.

Price per quintal		*Yield per acre*		*Cost per acre*	
Price (Rs)	*Prob.*	*Yield (Quin)*	*Prob.*	*Cost (Rs)*	*Prob.*
240	0.1	70	0.2	12,000	0.3
250	0.2	75	0.4	14,000	0.3
260	0.5	80	0.3	16,000	0.2
270	0.1	85	0.1	18,000	0.2
280	0.1				

Simulate 20 runs and estimate the expected profit per acre.

7. A small retailer has studied the weekly receipts and payments over the past 200 weeks and has developed the following set of information:

Weekly receipts (Rs)	*Probability*	*Weekly payment (Rs)*	*Probability*
3,000	0.20	4,000	0.30
5,000	0.30	6,000	0.40
7,000	0.40	8,000	0.20
12,000	0.10	10,000	0.10

Using the following sequence of random numbers, simulate the weekly pattern of receipts and payments for the 12 weeks of the next quarter, assuming further that the beginning bank balance is Rs 8,000. What is the estimated balance at the end of the 12 weekly period? What is the highest weekly balance during the quarter? What is the average weekly balance for the quarter?

Random numbers

For Receipts : 03, 91, 38, 55, 17, 46, 32, 43, 69, 72, 24, 22

For Payments : 61, 96, 30, 32, 03, 88, 48, 28, 88, 18, 71, 99

8. Mr. X is invited to play a game in which he is to toss a coin until the difference in the number of heads and tails tossed is equal to 5. He is required to pay rupee one for each toss of the coin and would get Rs 20 on conclusion of the game. Using simulation, find the expected gain to him in 5 such games.

9. In a simple queuing situation, the following formulae may be used to some of the system parameters:

$$\text{average queuing time} = \lambda/\mu(\mu - \lambda)$$

$$\text{average system process time (or time in the system)} = 1/(\mu - \lambda)$$

$$\text{average number in the queue (including zeros)} = \lambda/\mu(\mu - \lambda)$$

$$\text{average number in the system} = \lambda/(\mu - \lambda)$$

where λ = mean (average) rate of arrivals

μ = mean (average) rate of service

You have observed a queuing system for several days and one sequence of observations had the results as shown:

Arrival number	*Time between start of observation, first and subsequent arrivals*	*Service time*
1	41	70
2	46	12
3	38	23
4	57	36
5	87	136
6	221	55
7	20	65
8	99	4
9	3	17
10	10	17

Note: All times are expressed in units of 20 seconds.

Now,

(a) Calculate the four system parameters quoted above by a simulation of the system using the given data;

(b) Calculate the four system parameters using the formulae; and comment on the value obtained in (a) and (b) and the use of the simulation technique. *(ICMA, May, 1982)*

10. Inter-arrival and service durations studied over past few years for a single channel queuing system revealed the following patterns:

Inter-arrival time		*Service time*	
Minutes	*Probability*	*Minutes*	*Probability*
2	0.19	1	0.15
4	0.22	3	0.28
6	0.32	5	0.30
8	0.17	7	0.17
10	0.10	9	0.10

Using the random number table given below, simulate the queue behaviour for a period of 60 minutes.

1908 3227 5974 8196 2748 4507 2665 5278 7792
4649 8512 8363 3410 4206 4397 0753 1748 2375

Use the above simulation to estimate:

(i) the probability of the server being idle.

(ii) the mean time spent by the customer waiting for service. *(DMS, Mumbai, 1996)*

11. The output of a production line is checked by an inspector for one or more of the three different types of defects called defects *A, B* and *C*. If defect *A* occurs, the item is scrapped. If defect *B* or *C* occurs, the item must be reworked. The time required to rework a *B* defect is 15 minutes and the time required to rework a *C* defect is 30 minutes. The probabilities of an *A*, a *B* and a *C* defects are 0.15, 0.20 and 0.10 respectively. For 10 items coming of the assembly line, determine the number of items without any defects, the number scrapped, and the total minutes of rework time. Use the following random numbers:

Random numbers for defect A

48 55 91 40 93 01 83 63 47 52

Random numbers for defect B

47 36 57 04 79 55 10 13 57 09

Random numbers for defect C

82 95 18 96 20 84 56 11 52 03

(CA, May 1994)

12. (a) Patients arriving at a village dispensary are treated by a doctor on a first-come first-served basis. The inter-arrival time of the patients is known to be uniformly distributed between 0 and 80 minutes, while their service time is known to be uniformly distributed between 15 and 40 minutes. It is desired to simulate the system and determine the average time a patient has to be in the queue for getting service and the proportion of time the doctor would be idle.

Carry out the simulation using the following sequences of random numbers. The numbers have been selected between 00 and 80 to estimate inter-arrival times and between 15 and 40 to estimate the service times required by the patients.

Series 1 : 07, 21, 12, 80, 08, 03, 32, 65, 43, 74,
Series 2 : 23, 37, 16, 28, 30, 18, 25, 34, 19, 21,

(b) How would your answer in (a) change if there were two doctors instead, to attend the patients?

13. A retailer deals in a perishable commodity. The daily demand and supply are variable. The data for the past 500 days show the following demand and supply:

Supply		*Demand*	
Availability (kgs)	*No. of days*	*Demand (kgs)*	*No. of days*
10	40	10	50
20	50	20	110
30	190	30	200
40	150	40	100
50	70	50	40

The retailer buys the commodity at Rs 20 per kg and sells it at Rs 30 per kg. Any commodity remaining at the end of the day has no saleable value. Moreover, the loss (unearned profit) on any unsatisfied demand is Rs 8 per kg. Given the following pairs of random numbers, simulate six days' sales, demand, and profit.

(31, 18); (63, 84); (15, 79); (07, 32); (43, 75); (81, 27)

The first random number in the pair is for supply and the second random number is for demand viz. in the first pair (31, 18), use 31 to simulate supply and 18 to simulate demand. *(CA, Nov. 2000)*

14. For a project comprising activities *A*, *B*, ..., *H*, the following information is available:

Precedence Relationships:

A and *B* are the first activities of the project. *C* succeeds *A* while *B* precede *D*. Both *C* and *D* precede *E* and *F*. Activity *G* follows activity *F* while *H* is the last activity of the project and succeeds *E* and *G*.

Time Estimates and Probabilities:

Activity	*Time (days)*					
	1	*2*	*3*	*4*	*5*	*6*
A		0.2	0.4	0.4		
B				0.3		0.7
C			0.3	0.3	0.4	
D		0.2	0.6		0.2	
E				0.6	0.4	
F		0.8	0.2			
G	0.3		0.5	0.2		
H		0.1	0.2	0.3	0.4	

Simulate the project and determine, using random numbers, the activity times. Find the critical path and the project duration. Repeat five times. Does the critical path change? State the estimated duration of the project in each of the trials.

15. An investment company wants to study the investment projects based on market demand, profit and the investment required, which are independent of each other. Following probability distributions have been estimated for each of the three factors:

Annual demand							
(Units in thousands) :	25	30	35	40	45	50	55
Probability :	0.05	0.10	0.20	0.30	0.20	0.10	0.05
Profit per unit (Rs) :	3.00	5.00	7.00	9.00	10.00		
Probability :	0.10	0.20	0.40	0.20	0.10		
Investment required							
(in thousands of Rs) :	2,750		3,000		3,500		
Probability :	0.25		0.50		0.25		

Using simulation process, repeat the trial 10 times, compute the return on investment on each trial taking these factors into consideration. What is the most likely return?

(30, 12, 16); (59, 09, 69); (63, 94, 26); (27, 08, 74); (64, 60, 61);

(28, 28, 72); (31, 23, 57); (54, 85, 20); (64, 68, 18); (32, 31, 87).

In each bracket above, the first random number is for annual demand, the second one is for profit, and the last one is for the investment required.

(CA, May 2001)

16. For a washing powder manufacturing factory, frequency distribution of contribution (= sales price – variable cost) per unit, annual demand, and requirement of investment were found as follows:

Contribution per unit (Rs)	*Relative frequency*	*Average demand (in '000 units)*	*Relative frequency*
3	0.1	20	0.05
5	0.2	25	0.10
7	0.4	30	0.20
9	0.2	35	0.30
10	0.1	40	0.20
		45	0.10
		50	0.05

Required investment (Rs '000)	*Relative frequency*
1,750	0.25
2,000	0.50
2,500	0.25

Consider the random numbers 93, 03, 51, 59, 77, 61, 71, 62, 99, 15 for Monte Carlo simulation for 10 runs, to estimate the percentage of return on investment (ROI%) defined as

$$\text{ROI\%} = \frac{\text{Cash inflow}}{\text{Investment}} \times 100, \text{ for each run.}$$

Recommend an optimum investment strategy based on the modal value of ROI%.

(CA, Nov. 1995)

17. A company has planned to introduce a new product in the market, for which it is not certain about the selling price, the first year sales off-take, the variable cost, and the fixed cost on new facilities. Probabilistic estimates of these parameters have been made as detailed below. Using the random numbers given below, simulate the possibilities for 12 iterations and compute the first year profit volume. From the simulated results, find the probability that

    (i) the company will not break-even on this product in the first year.

    (ii) the profit volume of the first year will be more than Rs 3 lakh.

    (iii) the profit volume will not be more than 4 lakh.

Selling price per unit	Probability	Sales off-take Units	Probability
Rs 25	0.55	45,000	0.20
		50,000	0.35
		55,000	0.45
Rs 30	0.45	40,000	0.35
		45,000	0.40
		50,000	0.25

Variable cost per unit	Probability	Fixed cost	Probability
Rs 10	0.25	Rs 3 lakh	0.35
Rs 12	0.35	Rs 4 lakh	0.45
Rs 14	0.40	Rs 5 lakh	0.20

Use the following random number digits sequentially starting from first digit on first row. (Do not omit any digit).

120933 658779 154328 467211 989205 132567

549042 387645 980633 286410 749271 275304

010667 964852 374580 331267 545395 238952

(DMS, Mumbai, 1992)

18. A book store wishes to carry Systems Analysis and Design in stock. Demand is probabilistic and replenishment of stock takes 2 days (i.e. if an order is placed on March 1, it will be delivered at the end of the day on March 3). The probabilities of demand are given below:

Demand (Daily)	:	0	1	2	3	4
Probability	:	0.05	0.10	0.30	0.45	0.10

Each time an order is placed, the store incurs an ordering cost of Rs 10 per order. The store also incurs a carrying cost of Re 0.50 per book per day. The inventory carrying cost is calculated on the basis of stock at the end of each day.

The manager of the book store wishes to compare two options for his inventory decision.

(A) Order 5 books when the inventory at the beginning of the day plus orders outstanding is less than 8 books.

(B) Order 8 books when the inventory at the beginning of the day plus orders outstanding is less than 8.

Currently (beginning 1st day) the store has a stock of 8 books plus 6 books ordered two days ago and expected to arrive next day.

Using Monte-Carlo Simulation for 10 cycles, recommend which option the manager should choose. The two-digit random numbers are given below:

89, 34, 70, 63, 61, 81, 39, 16, 13, 73

(CA, May, 2000)

19. A company has been having problems with stockouts for one of its components and is contemplating making alterations either to the reorder quantity or to the reorder level, or to both.

Before making any decision, the company wishes to explore whether any guidance can be obtained by simulating the operation of the system.

The pattern of weekly demand over the past few years has been as follows:

Weekly demand (Units)	*Frequency*
500	10
525	15
550	30
575	50
600	55
625	60
650	40
675	20
700	10
725	5

Ordering costs are Rs 20 per order and the carrying cost is Rs 5 per unit. The estimated loss when an order cannot be met is Rs 12 per unit. When stock reaches the pre-set order point, a replenishment order is issued.

(a) You are required to describe, using a flow chart or other means, how a simulation model for this problem might work.

Using a re-order point of 2,500 and an order quantity of 2,000, a simulation of 20 weeks' operations has been run on a computer and the following summary produced:

Simulation Summary

Number of Periods	20 weeks	Carrying Cost	Rs 1,984.13
Average Inventory	1,031.8 units	Ordering Cost	Rs 120.00
Orders	6	Stockout Cost	Rs 20,580.00
Average Demand	606.3 units	Total Cost	Rs 22,684.13
Lead Time	4 weeks		

(b) You are required to interpret the above summary and suggest what could be done next to make the simulation more realistic. *(ICMA, May, 1992, adapted)*

20. (a) A businessman is considering taking over a certain new business. Based on past information and his own knowledge of the business, he works out the probability distributions of the monthly costs and sales revenues, as given here:

Cost (in Rs)	*Probability*	*Sales revenue (Rs)*	*Probability*
17,000	0.10	19,000	0.10
18,000	0.10	20,000	0.10
19,000	0.40	21,000	0.20
20,000	0.20	22,000	0.40
21,000	0.20	23,000	0.15
		24,000	0.05

Use the following sequences of random numbers to be used for estimating costs and revenues. Obtain the probability distribution of the monthly net revenue.

Sequence 1 : 82, 84, 28, 82, 36, 92, 73, 91, 63, 29,
27, 26, 92, 63, 83, 03, 10, 39, 10, 10,

Sequence 2 : 39, 72, 38, 29, 71, 83, 19, 72, 92, 59,
49, 39, 72, 94, 01, 92, 72, 18, 09, 00.

(b) Repeat the analysis in (a) using the following random number streams:

Sequence 1 : 20, 63, 46, 16, 45, 41, 44, 66, 87, 26,
78, 40, 29, 92, 21, 36, 57, 03, 28, 08,

Sequence 2 : 23, 57, 99, 84, 51, 29, 41, 11, 66, 30,
41, 80, 62, 74, 64, 26, 41, 40, 97, 15,

21. The management of *ABC* company is considering the question of marketing a new product. The fixed cost required in the project is Rs 4,000. Three factors are uncertain, viz. the selling price, variable cost and the annual sales volume. The product has a life of only one year. The management has the data on these three factors as under:

Selling price (Rs)	*Probability*	*Variable cost (Rs)*	*Probability*	*Sales volume (Units)*	*Probability*
3	0.2	1	0.3	2,000	0.3
4	0.5	2	0.6	3,000	0.3
5	0.3	3	0.1	4,000	0.4

Consider the following sequence of thirty random numbers:

81, 32, 60, 04, 46, 31, 67, 25, 24, 10, 40, 02, 39, 68, 08, 59, 66, 90, 12, 64, 79, 31, 86, 68, 82, 89, 25, 11, 98, 16.

Using the sequence (first three numbers for the first trial and so on), simulate the average profit for the above project on the basis of 10 trials. *(CA, November, 1994)*

22. M.C. Company is evaluating an investment proposal that has uncertainty associated with the three important aspects: the original cost, the useful life and the annual net cash flows. The three probability distributions for these variables are as shown below:

Original cost		*Useful life*		*Annual cash flows*	
Value	*Probability*	*Period*	*Probability*	*Value*	*Probability*
Rs 60,000	0.3	5 years	0.4	Rs 10,000	0.1
Rs 70,000	0.6	6 years	0.4	Rs 15,000	0.3
Rs 90,000	0.1	7 years	0.2	Rs 20,000	0.4
				Rs 25,000	0.2

The firm wants to perform five simulation runs of this project's life. The firm's cost of capital is 15 per cent and the risk-free rate is 6 per cent. For simplicity, it is assumed that these two values are known for certain, and will remain constant over the life of the project.

To simulate the probability distributions of the original cost, the useful life and the annual net cash flows, use the following sets of random numbers : 09, 84, 41, 92, 65; 24, 38, 73, 07, 04; 07, 48, 57, 64, 72 respectively. Determine the NPV and the payback period for each of the five simulation runs. *(CA, Nov. 1992)*

CHAPTER

18

Investment Analysis and Break-Even Analysis

CHAPTER OVERVIEW

The major focus in this chapter is how capital budgeting decisions are taken. The common sense solution to an investment problem is that if the returns, or inflows,would be more than the money invested, or outflows, then it makes sense to go ahead with the proposal, otherwise not. However, since ordinarily funds are needed to be invested immediately while the returns come at later dates, we need to match the inflows with the outflows by considering the time value of money because, after all, a rupee today is worth more than a rupee a year later. This chapter provides an account of the various methods that are used for taking capital budgeting decisions, covering situations where the cash flows are certain and where they are not. The other part of the chapter discusses the break-even analysis, also called the cost-volume-profit analysis.

The present chapter helps a manager in arriving at answers to the following questions:

- ❑ What is the time value of money? How is the concept used in obtaining the compounded and discounted values of various sums of money?
- ❑ What are the methods of capital budgeting that are used to determine whether a project is acceptable or not, and which of the alternative projects is better than the others?
- ❑ How do various methods of capital budgeting compare with each other, and which of them is most preferable?
- ❑ How are capital budgeting decisions taken when the given project has returns that are not certain? How is the uncertainty measured?
- ❑ How can simulation be used in making capital budgeting decisions for projects involving uncertainty?
- ❑ What is break-even point and how is it determined?
- ❑ What level of sales is needed to attain a given amount of profit and how much profit or loss will result from a given level of sales?
- ❑ How is the break-even point determined for a multi-product firm?

The chapter requires a variety of concepts and tools to be understood and to apply. They include compounding, discounting, calculation of net present value, internal rate of return, expected value, standard deviation and variance, normal distribution—obtaining areas and normal loss integral, decision trees, random numbers and their application in simulation.

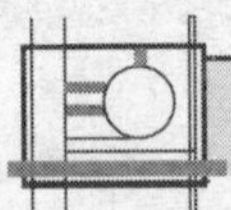

18.1 INTRODUCTION

In today's business environment, financial management is, in most cases, the single most important factor in the success or otherwise of a firm. The financial managers realise that when a firm makes a major decision, the effect of the action will be felt across the organisation. For example, an increase in the plant and equipment expenditures would affect the firm's profitability, its cash position, its borrowing ability, its dividend payments, and so forth. Capital expenditures have such widespread ramifications because the activities of a firm are systematically interrelated.

The present chapter is addressed to some aspects of the financial management. It is broadly divided into two parts. In the first part, we begin with a consideration of the time value of money. This is followed by the *investment analysis*—the questions of making capital budgeting decisions. The second part deals with the interrelations between cost, revenue, volume, and the profit planning. It is labelled *break-even analysis*.

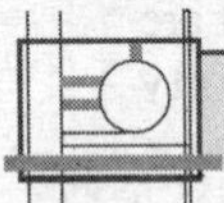

18.2 TIME VALUE OF MONEY

The notion that money has *time value* is one of the basic concepts of finance. In a world in which the expected cash inflows and outflows associated with a business decision would all take place at the same moment in time, and with the same degree of certainty, no problem will be faced in analysing the decision.

Any decision with a positive net cash flow would be acceptable, and in case of two or more alternative decisions, the one with the largest net inflow would be picked. However, in the real world when the cash flows relating to a project may occur at different points in time, we have to consider the time value of money.

The idea of time value of money is like this. We can easily visualise that a rupee received today is more valuable than a rupee to be received a year later. This is because as soon as funds are received they can be invested to earn interest thereon, with the result that a larger sum would be available later on. In a way, it is the price one would expect for foregoing purchasing power currently. A rational decision-maker would not value the opportunity to receive some amount of money now equally with the opportunity to have the same amount of money at some later date. This phenomenon is known as the decision-maker's *time preference of money*.

It may be observed here that in the real world which is full of uncertainties, the appropriate rate of exchange between having a certain amount of money now and the expectation of having a larger sum of money at a later date will include not just the time value of money but an additional amount to cover the uncertainty involved. In the following discussion, we shall first consider the question of time value of money only and analyse how decisions regarding investment may be made assuming that there is no uncertainty about the cash flow related to the proposals. This shall be followed by bringing the question of uncertainty in the analysis.

18.2.1 Time Preference or Interest Rate

The time preference for money may be expressed by means of an interest rate. Thus, the time preference for an individual (or for a firm for that matter) may be given to be 8% per annum. In that case he would be indifferent to Rs 100 now or Rs 108 a year later.

The information on the time preference of an individual is helpful in making decisions in many situations. For example, if an individual with an 8% per annum interest rate is offered either Rs 10,000 now or Rs 11,000 after one year's time, he would decide for the second option because, according to his time preference, he would be satisfied with 10,000 × 108/100 = Rs 10,800 received one year from now.

18.2.2 Compound Value

We have seen how the equivalent amounts at the beginning and at the end of a certain time period (also called *conversion period*) may be obtained, when the interest rate for the period is given. However, usually the investments involve more than one period. In such a case, compounding needs to be done. To illustrate, suppose an investor accepts an interest rate equal to 10% p.a. If he invests Rs 100 now, he would receive Rs 110 (= 100 + 10% of 100) at the end of the first year. In case he invests for two years, he would reinvest the amount of Rs 110 for one year (i.e. the second year) to receive 121 = (110 + 10% of 110) at the end of the second year. The same reasoning may be extended to determine the *compound value* at the end of any year—third, fourth, *n*th. Thus, in compounding, the interest for each period is added (compounded or converted) to the principal before the interest is calculated for the next period, and, finally, the compound value is obtained at the end of the period. In general, if P = the original principal, i = interest rate per unit per conversion period, n = the number of conversion periods and S = compound amount at the end of the nth period, we have,

$$S = P(1 + i)^n$$

The factor $(1 + i)^n$ is also sometimes called the *accumulation factor*.

Example 18.1 (a) Find the compound value of Rs 5,000 at the end of 4 years when the interest rate is 6% per annum.

(b) What is the compound amount at the end of 9 months if Rs 20,000 is borrowed at 8% per annum compounded quarterly?

(c) How would your answer in (b) be different if compounding were done bi-annually instead?

(a) $P = 5{,}000$, $i = 6\% = 0.06$, $n = 4$, we get

$S = 5{,}000\ (1 + 0.06)^4$ = Rs 6,312.38

(b) Here $P = 20{,}000$, $i = 0.08/4 = 0.02$ per quarter, $n = 3$ quarters, $S = 20{,}000\ (1 + 0.02)^3$ = Rs 21,224.16

(c) Now, $P = 20{,}000$, $i = 0.04$, and $n = 3/2 = 1.5$

Thus, $S = 20{,}000\ (1 + 0.04)^{1.5}$ = Rs 21,211.92

The compound values of rupee one at a given interest rate i per conversion period for different n conversion periods are tabulated in Table B5, appendix B, (at the end of the book). When this table is used, we have S = Principal (P) × An appropriate entry from the table. To illustrate, for $i = 6\%$ and $n = 4$, we get from the table the value equal to 1.2625. Thus, for (a), $S = 5{,}000 \times 1.2625$ = Rs 6,312.50.

18.2.3 Present Value

On many occasions, we are required to discount an amount which is due on a future date. To discount a given amount due in future means to find its present value on the date of discount. This can be achieved by using the compound interest relationship given earlier to determine the compound amount. Thus, since $S = P(1 + i)^n$, we have

$$P = \frac{S}{(1+i)^n} \quad \text{or} \quad S(1+i)^{-n}$$

The factor $(1 + i)^{-n}$ is referred to as the *present value factor* in discounting a given compound amount. The Table B3 in the appendix B gives the present value of one rupee receivable (or payable) at the end of n conversion periods, when the rate of interest is i per cent.

Example 18.2 Find the present value of Rs 8,000 payable at the end of 3 years when the interest rate is 10% per annum.

Here S = Rs 8,000, $i = 0.10$ and $n = 3$. Thus,

$$P = 8{,}000\ (1 + 0.10)^{-3} = \text{Rs } 6{,}010.40.$$

Alternatively, for $i = 10\%$ and $n = 3$, the present value factor (PV factor) for one rupee is 0.7513. Thus,

$$P = 8{,}000 \times 0.7513 = \text{Rs } 6{,}010.40.$$

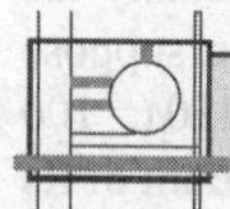

18.3 ANNUITIES

An annuity refers to a series of payments, usually in equal amount. The question of periodical payments is of interest in many situations, examples of which include installment payments on TV, refrigerator or other household appliances, life insurance premia, lease payments, interest payments on debentures, etc. For annuities, the payments are computed by the compound interest method and they are made at equal intervals of time, such as annually, quarterly, semi-annually, bi-annually, and so on. The time between two successive dates of payment is called the *payment interval*, and the time between the start of first payment and the end of the last payment is called the *term* of the annuity. On the basis of their terms, annuities may be classified as follows:

(a) Annuity Certain

When the term of an annuity begins and ends at certain fixed dates, it is known as annuity certain. For example, a 5-year annuity may begin on April 1, 2009 and end by April 1, 2015.

(b) Perpetuity

In case an annuity begins at a particular date and never terminates, it is known as a perpetuity. In a case like this, the principal is never touched and only interest is drawn, so that the length of term of the annuity is infinite.

(c) Contingent Annuity

For a contingent annuity, the term is not fixed in advance. It begins at a particular time and its end is contingent upon the happening of a certain event in future. The payment of life insurance premium, for example, stops as soon as the insured person dies.

The annuities can also be classified on the basis of the time of payments as follows:

(a) Ordinary Annuity

In an ordinary annuity, payments are made at the end of every payment interval. For example, for an annuity with a 5-year term beginning on January 1st, 2009, the payment would be made on 1st January of the years 2010 through 2014.

(b) Annuity Due

When payments are made at the beginning of the payment interval, the annuity is called *annuity due*. Premia on life insurance policies illustrate the point.

(c) Deferred Annuity

When the payments are made at the end of each payment interval but the term of annuity does not begin until after a specified period of time, it is referred to as a *deferred annuity*. For example, when the house building loan is granted to a person, the repayment of the loan begins after the expiry of, may be, a year or two from the date of grant of loan.

In the discussion to follow, we shall presume the annuity to be an ordinary one, unless otherwise mentioned.

18.3.1 Amount of Annuity

The amount of annuity refers to the terminal value of the series of payments, including principal and compound interest, at the end of the term of the annuity. To obtain the amount, we add up the compounded amounts of the individual periodic payments. For an annuity, let

A = the periodic payment,

n = the interval, or the term of the annuity, or the number of payments,

i = the rate of interest.

The first payment, being made at the end of the first time period, shall carry accumulated interest for $n - 1$ time periods, the second one would similarly carry interest for $n - 2$ time periods, and so on. The last payment would not carry any interest because it is made at the end of the term. If we let $(1 + i) = R$, we have the amount of the annuity M, as follows:

$$M = AR^{n-1} + AR^{n-2} + ... + AR^2 + AR + A \quad \text{(i)}$$

To determine this, we first multiply both sides by R to get

$$MR = AR^n + AR^{n-1} + ... + AR^2 + AR \quad \text{(ii)}$$

Subtracting Eq. (i) from Eq. (ii), we get

$$MR - M = AR^n - A$$

or

$$M(R - 1) = A(R^n - 1)$$

or

$$M = \frac{A\left(R^n - 1\right)}{R - 1}$$

Example 18.3 Find the amount of an annuity of Rs 1,000 in 10 years allowing a compound interest of 5% per annum.

With $A = 1,000$, $n = 10$, $i = 0.05$, we have $R = 1 + i = 1 + 0.05 = 1.05$. Further,

$$M = \frac{1,000\left(1.05^{10} - 1\right)}{(1.05 - 1)} = \text{Rs } 12,577.9$$

For convenience, the values of $(R^n - 1)/(R - 1)$ for various interest rates i and the number of payments n are provided in Table B6 in the Appendix B. Each entry in this table gives the amount of the annuity in rupees when each payment is Re 1. Thus, when the annuity has a periodic payment of A, the amount of annuity = $A \times$ Amount Factor. For $n = 10$ and $i = 5$ the amount is given to be 12.5779. For this example, then, $M = 1,000 \times 12.5779$ = Rs 12,577.9.

Example 18.4 A company wants to establish a sinking fund, beginning at the end of the year. Annual deposits would be made at the end of this year and the following 9 years. If deposits earn interest at the rate of 8% compounded annually, how much money must be deposited each year in order to have an amount of 8 lakhs at the end of the tenth deposit?

Here

$$M = 8,00,000,\ i = 0.08 \text{ and } n = 10.$$

$$M = A \times \text{Amount Factor}$$

The amount factor for $n = 10$ and $i = 0.08$ is found to be 14.4866. Thus, $A = 8{,}00{,}000/14.4866 =$ Rs 55,223.4, the amount to be deposited every year.

18.3.2 Present Value of an Annuity

For an annuity, its present value equals the sum total of the present values of all the periodic payments. Using the same symbols, the present value of the first payment would be AR^{-1}, that of the second payment would be AR^{-2}, and similarly, for the last payment, it would equal AR^{-n}.

Thus, the present value, V, would be:

$$V = AR^{-1} + AR^{-2} + \ldots + AR^{-n} \tag{iii}$$

Multiplying both sides by R, we get

$$VR = A + AR^{-1} + \ldots + Ar^{-n+1} \tag{iv}$$

Subtracting Eq. (iii) from Eq. (iv), we get

$$VR - V = A - AR^{-n}$$

or

$$V(R-1) = A(1 - R^{-n})$$

or

$$V = \frac{A\left(1 - R^{-n}\right)}{R - 1}$$

Example 18.5 Determine the present value of an annuity if the amount of each payment is Rs 2,000, payable at the end of each year for 4 years and the interest rate is 6% compounded annually.

With $A = 2{,}000$, $n = 4$, $R = 1 + i = 1.06$, we have

$$V = \frac{2{,}000\left(1 - 1.06^{-4}\right)}{(1.06 - 1)} = \text{Rs } 6{,}930.2.$$

The values of $(1 - R^{-n})/(R - 1)$ for various interest rates i and the number of payments M have been tabulated in the Appendix B (Table B4) for convenience. In general, we have $V = A \times$ Present Value Factor for Annuity. For $n = 4$ and $i = 6$, we find the present value factor to be 3.4651.Thus,

$$V = 2{,}000 \times 3.4651 = \text{Rs } 6{,}930.2.$$

Example 18.6 X purchased a colour TV from the OWNATV shop for Rs 12,800. He made a down payment of Rs 4,000 and agreed to pay the balance in 24 equal monthly payments. If interest charged was 12% compounded monthly, how much should X pay every month?

Here present value of annuity, $V =$ Price of the TV – Down Payment = 12,800 – 4,000 = 8,800. Further, $i = 12\%/12 = 1\%$ (per month) and $n = 24$. For $i = 1$ and $n = 24$, we have the PV factor equal to 21.2434. Therefore, $A = V/PV$ factor for annuity = 8,800/21.2434 = Rs 414.25.

Thus, X should pay Rs 414.25 per month for the 24-month period.

18.3.3 Present Value of a Perpetuity

The present value of a perpetuity equals A/i. This is shown below. For a perpetuity, there is no end to the periodic payments. Under the condition, therefore, n approaches infinity. As $n \to \infty$, $R^{-n} \to 0$. Thus, the expression $V = A(1 - R^{-n})/(R - 1)$ reduces to $V = A/(R - 1)$ or A/i, (since $R = 1 + i$).

Example 18.7 A freehold estate is bought for Rs 4,00,000. At what rent should it be let so that the owner may receive 12% on the purchase money?

Here we are required to determine the periodic payment, A, given that V = Rs 4,00,000 and $i = 0.12$. Thus,

$$A = V \times i = 4{,}00{,}000 \times 0.12 = \text{Rs } 48{,}000 \text{ p.a.}$$

For a perpetuity, if the present value V is, say, Rs 10,000 and the annuity payment is Rs 2,000, then the value can be said to be equal to 10,000/2,000 = 5 years' purchase. In general, the value of a perpetuity is said to be *k years' purchase* where $V = RA$.

Example 18.8 How many years' purchase should be given for a freehold estate if the rate of interest is 8% p.a.?

For perpetuity, $V = kA = A/i$. Thus, $k = 1/i = 1/0.08 = 12.5$. Therefore, 12.5 years' purchase should be given for the estate.

Example 18.9 If a perpetual annuity is worth 25 years' purchase, find the annuity to continue for 3 years which can be purchased for Rs 5,000.

With $k = 25$ years, $i = 1/k = 1/25 = 0.04$. Now,

$$V = \frac{A\left(1 - R^{-n}\right)}{R - 1}$$

where $V = 5{,}000$, $R = 1 + 0.04 = 1.04$, and $n = 3$. For $i = 4\%$ and $n = 3$, the present value factor for annuity is found to be 2.7751. From this, $5{,}000 = 2.7751 \times A$. Thus, $A = 5{,}000/2.7751$ = Rs 1,801.75.

18.3.4 Present Value of a Deferred Annuity

For an annuity in respect of which the first payment is scheduled to begin after the lapse of a certain time period, the present value can be obtained in a similar way as for the ordinary annuity. To illustrate, if the annuity begins in p years from now (so that the first payment would, as usual, be made at the end of the first year at that time), the present values of the successive payments would be $AR^{-(p+1)}$, $AR^{-(p+2)}$, and so on. Thus,

$$V = AR^{-(p+1)} + AR^{-(p+2)} + \ldots + AR^{-(p+n)} \tag{v}$$

To obtain the value of V, we first multiply both sides of the Eq. (v) by R to get

$$VR = AR^{-p} + AR^{-(p+1)} + \ldots + AR^{-(p+n)+1} \tag{vi}$$

Subtracting Eq. (v) from Eq. (vi), we get

$$VR - V = AR^{-p} - AR^{-(p+n)}$$

or

$$V(R - 1) = AR^{-p} - AR^{-(p+n)}$$

or

$$V = \frac{AR^{-p}}{R-1} - \frac{AR^{-(p+n)}}{R-1}$$

Example 18.10 If an annuity of Rs 3,000 for 5 years at 8% is to begin in 3 years' time from now, what would be its present value?

With $A = 3{,}000$, $n = 5$, $R = 1 + i = 1.08$, and $p = 3$, we have

$$V = \frac{3,000\,(1.08)^{-3}}{1.08-1} - \frac{3,000\,(1.08)^{-(3+5)}}{1.08-1} = \text{Rs } 9,508.62$$

If a perpetuity is to begin after a certain period from now, its present value can be obtained as $V = AR^{-p}/R - 1$. This is because as $n \to \infty$, AR^{-p-n} tends to equal zero.

Example 18.11 Determine the present value of a perpetual annuity of Rs 1,000 starting in 4 years from now, and carrying an interest of 5%.

With $A = 1,000$, $p = 4$, and $R = 1 + 0.05 = 1.05$, we have,

$$V = \frac{AR^{-p}}{R-1} = \frac{1,000 \times 1.05^{-4}}{1.05 - 1} = \text{Rs } 16,454.04$$

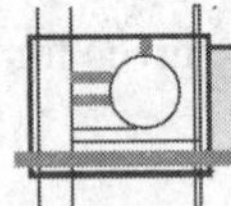

18.4 INVESTMENT ANALYSIS: CAPITAL BUDGETING

Companies invest in a variety of real assets. They include tangible assets such as plant and machinery, and intangible assets such as patents. The object of investment decisions is to identify the real assets which are worth more than what they cost. In order to make correct capital expenditure decision, a manager needs at least three sets of information, which are,

(a) Estimates of the net capital outlay required for a project and the future cash flows promised by it.
(b) Estimates of the availability and the cost of capital to the company.
(c) A correct set of standards by which to select projects for execution so that the long-term economic benefits to the owners are maximised.

Given the necessary information, a method has to be found for evaluating the profitability of new investments in assets. In selecting a suitable criterion, the following two fundamental principles should be kept in view.

1. *The bigger the better principle:* Other things being equal, bigger benefits are preferable to smaller ones.
2. *The bird in hand principle:* Other things being equal, early benefits are preferable to later benefits.

Of course, these principles by themselves can hardly be used as criteria for judging the acceptability of investment proposals, and means have to be found for taking both of these into account in a single yardstick.

There are four basic methods for the measurement of investment worth. They are: (a) Payback, (b) Average rate of return, (c) Net present value, and (d) Internal rate of return method.

18.4.1 Payback Method

The payback period of any investment project indicates the time period required to recover the original cash investment. Thus, payback occurs when the cumulative cash inflows from a project minus the initial investment is equal to zero. If C_0 is the initial investment (cash outflow), and C_t is cash flow in the tth year, then the payback period would be the value of k such that

$$\sum_{t=1}^{k} C_t - C_0 = 0$$

When the payback period calculated is less than some maximum acceptable period, then a proposal is accepted, otherwise not. For two or more mutually exclusive projects, the ranking would be done on the basis of their payback periods—the project with the least payback period would be preferred the most.

Example 18.12 An investment proposal requires an investment of Rs 40,000. The cash flows expected to be generated in different years are:

Year	:	1	2	3	4	5
Cash flow	:	18,000	20,000	10,000	10,000	8,000

Should the proposal be accepted if the maximum acceptable payback period be 3 years?

If the cash flows for a year are expected to be generated uniformly through the period, then the payback of Rs 40,000 investment (= 18,000 + 20,000 + 2,000 out of 10,000) can be estimated as follows:

$$2 + 2{,}000/10{,}000 = 2.2 \text{ years.}$$

Since it is less than the 3 years maximum, the proposal in an acceptable one.

18.4.2 Average Rate of Return Method

The average rate of return is an accounting method. The rate of return is calculated as the ratio of the average annual profits, after taxes, to the *average* investment in the project. For instance, if the average annual book profits are Rs 4,000 for five years, and the initial investment in the project is Rs 32,000, so that the average investment (assuming straight line depreciation) is Rs 32,000/2 = Rs 16,000, the rate of return on the project would be 4,000/16,000 = 25%. After computing the average rate of return, it is compared with a required rate of return to determine whether a particular project should be accepted or rejected.

Criticism of the Payback and Rate of Return Methods The payback method ignores the 'the bigger the better' principle completely. It does not take into account the cash inflows after the recovery of investment. It gives some attention to timing but does not satisfy the 'bird in hand' principle as it gives higher weights to all the receipts before the recovery of the investment and zero weights to all the subsequent receipts. In fact, it ignores all the dimensions of profitability and over-emphasises the liquidity.

The rate of return method also suffers from the limitation that it is unable to take account of the timing of cash proceeds from the investment.

The methods of net present value and internal rate of return take account of the timing of cash flows and are titled *discounted cash flow* (DCF) measures.

18.4.3 Net Present Value Method (NPV)

The net present value of an investment proposal is determined by discounting the cash inflows back to the present at the required rate of return, i, and then subtracting the initial investment, so that

Net Present Value, $$\text{NPV} = \sum_{t=1}^{n} \frac{C_t}{(1+i)^t} - C_0$$

Alternatively, $$\text{NPV} = \sum_{t=0}^{n} \frac{C_t}{(1+i)^t}$$

If the sum of these discounted cash flows in greater than zero, the proposal is accepted, if less than zero, rejected. And, naturally, the higher the NPV of a project, the better it is.

Example 18.13 A company is trying to decide which of the two machines to purchase. Each will involve an investment of Rs 10,000. The expected net incremental cash flows are given here.

Year	Machine A	Machine B
1	5,000	2,000
2	4,000	3,000
3	2,000	5,000
4	2,000	4,000

The company's cost of capital is 10%. Calculate the NPV of each machine. If this is used as the basis of selection, which machine should be purchased?

The calculation of NPV is shown in Table 18.1. It is clear that machine *B* should be preferred to *A* because of its higher NPV.

TABLE 18.1 Calculation of NPV

Year t	PV Factor @ 10%	Machine A		Machine B	
		Cash Flow C_t	Present Value	Cash Flow C_t	Present Value
0	1.0000	−10,000	−10,000.0	−10,000	−10,000.0
1	0.9091	5,000	4,545.5	2,000	1,818.2
2	0.8264	4,000	3,305.6	3,000	2,479.2
3	0.7513	2,000	1,502.6	5,000	3,756.5
4	0.6830	2,000	1,366.0	4,000	2,732.0
			NPV = 719.7		NPV = 785.9

In addition to solving for a project's net present value, the capital budgeting decision may be shown graphically. Using different discount rates, to discount the cash flows of a project, the NPV values may be obtained and plotted on a graph paper to obtain the *present value profile*. The illustrate, for project *A* in Example 18.13, the NPVs using selected discount rates are,

Discount Rate	NPV (Rs)
0	3,000.0
5	1,763.0
10	719.7
15	−169.0
20	−933.9
25	−1,596.8

This is shown graphically in Figure 18.1. The present value profile is helpful in that it depicts the sensitivity of NPV to the discount rate employed—the steeper the slope of the present value profile, the more sensitive the NPV to the rate employed.

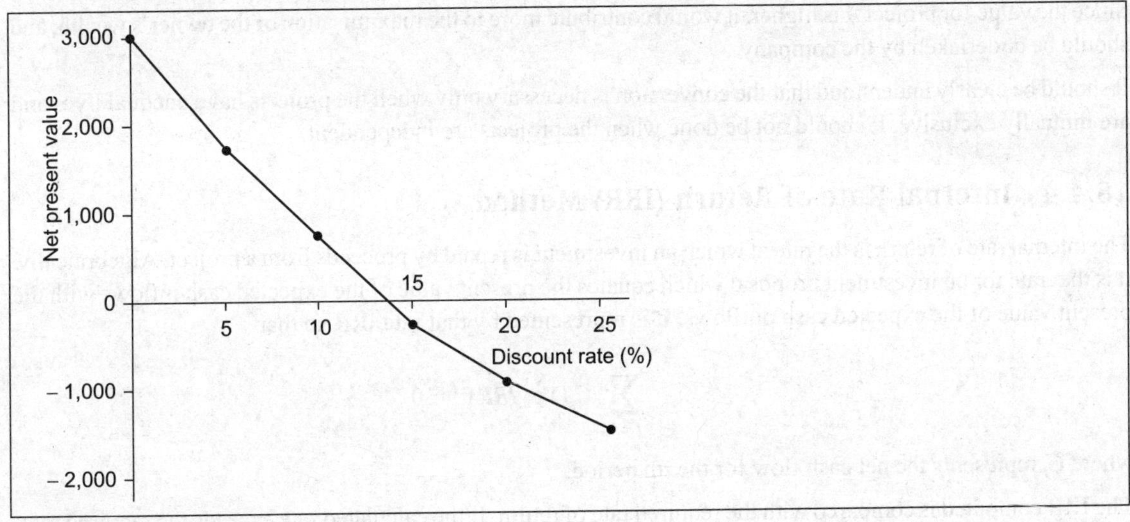

Figure 18.1 *Present Value Profile*

One thing that should be noted in connection with using the NPV approach to decide the choice of projects is that when a choice is required to be made between two (or more) mutually exclusive projects, due attention should be paid to their lives. When the lives are unequal, then adjustment is called for, for taking an appropriate decision because NPV is a function of the life of a project. To understand this, let us consider the following example.

Example 18.14 A company is deciding to choose between two mutually exclusive projects *A* and *B*. Project *A* requires an initial investment of Rs 3,00,000 and is expected to generate cash flow of Rs 1,50,000 per annum for the 3 years of its life. Project *B*, on the other hand, requires Rs 3,40,000, has a life of 6 years and would generate Rs 1,00,000 every year. Which proposal should be accepted? Assume a 10% rate of interest.

Since both the projects involve annuities of certain fixed amounts, the NPV's can be calculated using the present value factors of annuities.

Project *A* : For $n = 3$, $i = 0.10$, we have, *PV* factor = 2.487.

$\therefore$ $NPV_A = 1{,}50{,}000 \times 2.487 - 3{,}00{,}000 = \text{Rs } 73{,}050$

Project *B* : For $n = 6$, $i = 0.10$, we have, *PV* factor = 4.355.

$\therefore$ $NPV_B = 1{,}00{,}000 \times 4.355 - 3{,}40{,}000 = \text{Rs } 95{,}500$

Other things being equal, project *B* should be chosen because it has a higher NPV. However, to make a valid comparison, it is necessary to equalise the lives of the two projects. A way to accomplish this is the equivalent annual annuity method. Under this approach, the NPV is restated to be a yearly net present value. The equivalent annual annuity = NPV_n/PV factor for $\text{annuity}_{i,n}$ where NPV_n is the project's NPV over its original life, and PV factor for $\text{annuity}_{i,n}$ is the factor based on the required rate of return and the original life of the project. For the two projects in question, we have

Equivalent Annual Annuity for project *A* = 73,050/2.487 = Rs 29,372.73

Equivalent Annual Annuity for project *B* = 95,500/4.355 = Rs 21,928.82

Since the value for project *A* is higher, it would contribute more to the maximisation of the owner's wealth, and should be undertaken by the company.

It should be clearly understood that the conversion is necessary only when the projects have unequal lives and are mutually exclusive. It should not be done when the projects are independent.

18.4.4 Internal Rate of Return (IRR) Method

The internal rate of return is the rate at which an investment is repaid by proceeds from a project. Algebraically, it is that rate for an investment proposal which equates the present value of the expected cash inflows with the present value of the expected cash outflows. It is represented by that rate IRR so that

$$\sum_{t=0}^{n} C_t(1 + IRR)^{-t} = 0$$

where C_t represents the net cash flow for the *t*th period.

The IRR computed is compared with the required rate of return. If the calculated rate exceeds the required rate, the proposal is accepted. The required rate of return is the minimum return that the investors expect the firm to earn on projects. Therefore, accepting a project with an IRR in excess of the required rate of return should result in an increase in the wealth of the owners of the business.

Example 18.15 Calculate IRR of the project *A* in Example 18.13.

Here we are required to calculate the IRR of given series so that

$$\frac{5{,}000}{(1+IRR)^1} + \frac{4{,}000}{(1+IRR)^2} + \frac{2{,}000}{(1+IRR)^3} + \frac{2{,}000}{(1+IRR)^4} = 10{,}000$$

The IRR value can be determined by trial and error method. A convenient way is as follows:

(a) Find the average annual cash flow. Here $C_t = (5{,}000 + 4{,}000 + 2{,}000 + 2{,}000)/4 = 13{,}000/4 = 3{,}250$.

(b) Divide the initial investment outlay by this average annual annuity amount: $10{,}000/3{,}250 = 3.0769$.

(c) Locate the nearest PV factor in Table B4 from the row appropriate to the number of years. This is an approximate IRR which provides a starting point for further analysis. For the project, for $n = 4$, we find that the closest 4-year value is 3.102 at 11%, so that the starting IRR may be taken to be 11%.

(d) Using the discount rate obtained as in (c), we determine the present value of cash inflows. If the discounted present value is greater than cash outflow, we increase the discount rate, and if it is less than that, we lower the rate.

Using an 11% discount rate, the present value of the cash inflows is calculated to be Rs 10,530.7 as shown here. This is in excess of the cash outflow of Rs 10,000.

Cash Flow	*PV Factor @ 11%*	*Present Value*
5,000	0.9009	4,504.5
4,000	0.8116	3,246.4
2,000	0.7312	1,462.4
2,000	0.6587	1,317.4
		10,530.7

Since the present value of the cash inflows is found to be greater than the outflow, we increase the discount rate. Using a 12% discount rate, the present value of inflows may be shown to be equal to Rs 10,347.9, using 13% it equals Rs 10,170.2 and, finally, raising it to 14% it works out to Rs 9,998.2, almost equal to Rs 10,000.

Cash Flow	*PV Factor @ 14%*	*Present Value*
5,000	0.8772	4,386.0
4,000	0.7695	3,078.0
2,000	0.6750	1,350.0
2,000	0.5921	1,184.2
		9,998.2

Thus, we conclude that this project has an IRR (almost) equal to 14%. This is also evident from the Figure 18.1. The point where the present value profile intersects the horizontal axis is the internal rate of return on the project. This is obvious because IRR is defined to be the rate at which the NPV reduces to zero.

NPV Versus IRR Both the methods, the NPV and IRR are consistent with the objective of maximising owner's claim on the company. The two are, however, not in complete correspondence with one another. Of course, if the problem is that of dichotomy, that is, simply the selection of likely cases from among a number of independent proposals, both the methods give the same investment decision. Thus, if a proposal shows NPV at *i*% to be greater than zero, and is acceptable by NPV approach, it is also acceptable by IRR. If the projects are mutually exclusive, however, thc *ranking* of investment proposals has to be done to determine which of the projects will actually be undertaken. While the NPV and IRR methods may lead to similar accept-reject decisions, in some cases the two will provide different ranking. Thus, a choice has to be made as to which method gives a more accurate ranking of the projects that are mutually exclusive.

Projects have different rankings if their costs differ or if the timings of cash flows among them differ. For instance, consider two mutually exclusive projects *A* and *B* with cash flows as given here:

		Project A	*Project B*
Initial Investment		Rs 700	Rs 800
Cash Inflow,	Year 1	Rs 100	Rs 600
	Year 2	Rs 400	Rs 400
	Year 3	Rs 600	Rs 100

To portray the effect of differing cash flow patterns, the present value profiles of these projects are shown in Figure 18.2. The NPV of each project will change as the rate of return used to discount the periodic cash flows varies—moving along the horizontal axis to higher discount rates, the NPV of each project would diminish. However, the NPV for the project which has largest cash flows in future will be affected more severely by the increasing discount rate. Thus, the NPV for the project *A* decreases more rapidly than for the project *B*. As is clear from the figure, the ranking derived from the NPV differs from that of IRR if the discount rate used is lower than the rate determined by point *P*—the crossover rate which equates the two projects' NPVs. Thus, the conflict between the NPV and IRR rankings depends on whether the discount rate used by the firm is less or more than the crossover rate.

In a situation where such a conflict arises, the NPV based ranking should be used. The NPV measures

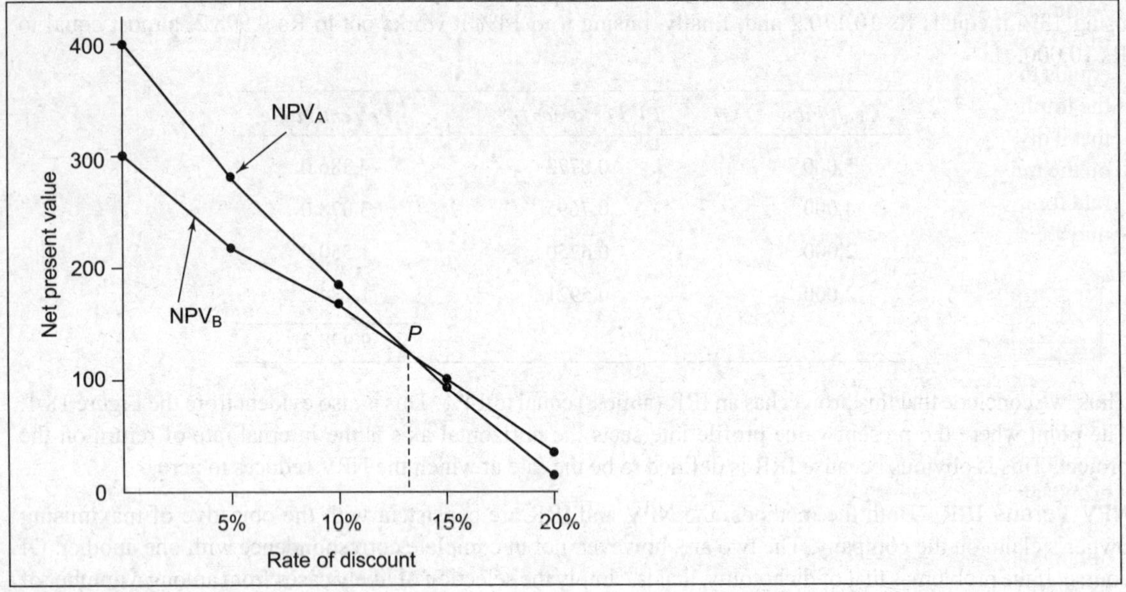

Figure 18.2 *NPV and IRR Conflict*

the added value accruing to the wealth of the shareholders of the company, when the project is undertaken. Thus the project which has the highest NPV (the one which maximises this value) should be the project to be undertaken. The IRR method is inferior to the NPV method because of certain theoretical disadvantages associated with it. Two of them are discussed here:

(a) Multiple Rates-of-Return

The IRR method is plagued by the fact that sometimes it gives multiple rates-of-return on a project, when cash flows with changing signs are observed, which may be necessitated, for example, by the requirement to inject additional capital into the project at some future periods. Not only this, sometimes imaginary values of IRR (of the type $\sqrt{-1}$, for instance) may be found which cannot be interpreted properly. To illustrate multiple rates-of-return, suppose there is a project which involves cash flows of –100, 800 and –900 at times 0, 1, and 2, respectively. Its IRR can be obtained as follows:

$$-100 + \frac{800}{(1+IRR)^1} - \frac{900}{(1+IRR)^2} = 0$$

or $$-100(1+IRR)^2 + 800(1+IRR) - 900 = 0$$

or $$100\,IRR^2 - 600\,IRR + 200 = 0.$$

This being a quadratic equation, we get the *IRR* values as,

$$IRR = \frac{600 \pm \sqrt{(-600)^2 - 4 \times 100 \times 200}}{200}$$

$$= 5.65 \text{ and } 0.3542$$

Thus, here two values of *IRR* are observed.

(b) Reinvestment Rate Issue

Another difficulty associated with the IRR is the underlying assumption of varying reinvestment rates. The use of NPV method implicitly assumes that the cash flows generated by a project can be reinvested at a rate which equals the 'cost of capital' to the firm (the rate used to discount the cash flows). With the use of IRR method, the implication is that the firm has the opportunity to reinvest at a rate equal to IRR. Thus, IRR has the demerit that it divorces the discounting process from the cost of capital and implies that reinvestment rates are contingent on the individual projects. To the extent that the rate of return may be regarded as a measure of the opportunity rate for investment, the NPV method is better than the IRR. Also, the discount rate used for NPV computation may be conservative, but it has the virtue of being applied consistently to all investment proposals.

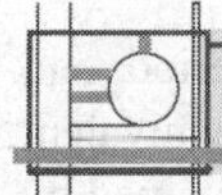

18.5 METHODS OF INCORPORATING RISK INTO CAPITAL BUDGETING

The preceding discussion has assumed that the financial manager makes investment decisions under conditions of complete certainty, and hence they are risk-free. This assumption implies that the NPV of an investment proposal is considered to be a fixed quantity and not a random variable, capable of assuming values other than the one specified. It is for this reason that once a positive value of the NPV of an investment proposal is obtained, it can be unequivocally stated that it is an acceptable proposal. Reality, however, is far from this, for the world is one of change and uncertainty. Thus, when we calculate that an investment would yield a particular rate of return per annum, we are aware that unforeseen events, like new and better technology, changes in the raw materials and so on may invalidate our estimates. Thus, some risk would usually be associated with a project so that variations in the cash flows may be observed, and that the degree of risk would vary with the different projects.

There are many ways in which risk can be taken into account while making investment decision. Basically, there are two approaches to risk adjustment. First, there is the *certainty equivalent method* which involves adjusting the numerator of the equation of the present value. In this method, we reduce the value of the expected cash inflows to adjust for the risk—the riskier the cash flow, the greater the reduction and, consequently, the lower the present value of the asset. Alternatively, the risk could be accounted for by adjusting the denominator of the present value equation—the greater the riskiness of the cash flows, the higher the discount rate and, therefore, the lower the present value of the asset.

Besides, there are three interrelated methods of analysing the investment proposals involving risk. They are: statistical distribution method (also called *mean-standard deviation approach*), decision-tree method, and simulation technique. These methods are different from both, the certainty equivalent and risk-adjusted discount rate methods, because they allow the statistical distributions of the net present value to be explicitly estimated. Using these techniques, an interval rather than a point estimate of the expected NPV is presented and, thus, they are more general and objective. In addition to these methods, sensitivity analysis is yet another method of analysing risky proposals.

We shall now discuss the various techniques in turn.

18.5.1 Certainty Equivalent Approach

The certainty equivalent approach to incorporating risk represents a direct attempt to incorporate the management's utility function into the analysis. The idea of certainty equivalent is as follows. When a cash flow C is risky, we ask, 'what is the *certain* return for which the manager could exchange the risky flow C?'

The amount that the manager feels is equivalent to the expected but risky cash flow offered by the investment is called the *certainty equivalent* of *C*. To illustrate, suppose that the forecasted value of a risky cash flow is Rs 1,000, but that one would be willing to trade it for a safe cash flow of Rs 850, then Rs 850 is the certainty equivalent of the risky cash flow of Rs 1,000. The ratio 850/1,000 = 0.85 is called the *certainty equivalent coefficient* for the period, and is represented by α. In general terms,

$$\alpha_t = \frac{\text{Certain Cash Flow}_t}{\text{Risky Cash Flow}_t}$$

The α's can very between 0 and one—0 indicating the extreme risk and 1 the absence of risk. The variations in α are thus inversely related to the degree of risk involved.

For a given problem, the certainty equivalent coefficients, α's, are determined for each of the time periods and then the given risky cash flows are multiplied by their respective coefficient values to obtain an equivalent certain cash flow stream. Once the risk is taken out of the cash flows of the project, the NPV is obtained by using the risk-free rate of discount, to take an appropriate decision regarding its acceptance. Symbolically,

$$\text{NPV} = \sum_{t=0}^{n} \frac{\alpha_t C_t}{(1+i)^t}$$

Example 18.16 A company uses certainty equivalent approach in its evaluation of risky investments. The company has now to make a choice between two proposals *X* and *Y*. The necessary information is as follows:

Year, t	*Proposal X*		*Proposal Y*	
	Cash Flow$_t$	α_t	*Cash Flow$_t$*	α_t
0	(40,000)	1.00	(30,000)	1.00
1	20,000	0.90	15,000	0.95
2	18,000	0.80	12,000	0.80
3	12,000	0.80	10,000	0.75
4	10,000	0.60	5,000	0.70

If the after-tax risk-free rate is 8%, which of the two alternatives should be selected?

The NPV for each of the projects is shown calculated as,

For Proposal X:

Year, t	*Cash Flow, C_t*	α_t	$\alpha_t C_t$	*PV Factor @ 8%*	*Present Value*
0	(40,000)	1.00	(40,000)	1.0000	(40,000)
1	20,000	0.90	18,000	0.9260	16,668.00
2	18,000	0.80	14,400	0.8573	12,345.12
3	12,000	0.80	9,600	0.7938	7,620.48
4	10,000	0.60	6,000	0.7350	4,410.00
					NPV = 1,043.60

For Proposal Y:

Year, t	Cash Flow, C_t	α_t	$\alpha_t C_t$	PV Factor @ 8%	Present Value
0	(30,000)	1.00	(30,000)	1.0000	(30,000)
1	15,000	0.95	14,250	0.9260	13,195.50
2	12,000	0.80	9,600	0.8573	8,230.08
3	10,000	0.75	7,500	0.7938	5,953.50
4	8,000	0.70	5,600	0.7350	4,116.00
					NPV = 1,495.08

NPV being higher for proposal *Y*, this should be preferred and accepted.

18.5.2 Risk-adjusted Discount Rates

The use of the risk-adjusted discount rates is based on the notion that the investors expect higher returns for more risky projects. In this method of incorporating risk, the risk-free rate of return, i, is adjusted upward by adding a suitable risk premium, Φ, representing a compensation, the risk-averse investors in the market would require before they will consent to the risk of the investment.

Thus, if k is the required rate of return, we have

$$k = i + \Phi$$

The relationship between risk and return is shown in Figure 18.3.

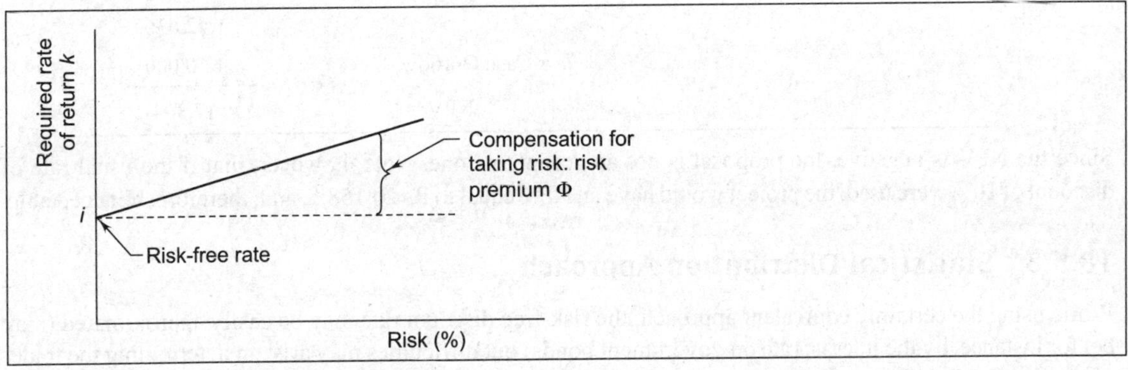

Figure 18.3 *Risk-Return Relationship*

The risk-free rate compensates the investors for deferring the consumption of goods and services to make investments. In effect, it is a reward only for waiting and applies only to those investments on which there is no chance that the realised rate of return will be different from the rate expected. If, however, risk is also to be borne, the risk premium adds the necessary compensation for bearing that risk.

In a given situation, the manager determines the required rate of return by adjusting the risk-free rate of return, by adding the necessary risk premium in keeping the risk that the proposal carries. Once the appropriate required rate of return for a project with a given level of risk is determined, the cash flows are discounted to present value terms using this risk-adjusted rate is obtained as:

$$\text{NPV} = \sum_{t=0}^{n} \frac{C_t}{(1+k)^t}$$

Example 18.17 The management of the Play Toy Factory, a manufacturer of toys in New Delhi, is considering the introduction of a new type of a toy-remote control motorbike. In the past, the management has been quite conservative in making investments in new products and considers this project quite a risky one. The management feels that the normally used required rate of return of 10% is not proper in this case and, instead, a return of 16% is expected on this project.

The project, requiring an outlay of Rs 1,50,000 has the following expected returns over its estimated life of 6 years.

Year	:	1	2	3	4	5	6
Cash flow ('000 Rs)	:	30	30	50	60	40	25

Should the project be undertaken?

The net present value, using the 16% discount rate, is shown calculated below.

Year	*Cash flow*	*PV factor @ 16%*	*Present value*
1	30,000	0.8621	28,563
2	30,000	0.7432	22,296
3	50,000	0.6407	32,035
4	60,000	0.5523	33,138
5	40,000	0.4761	19,044
6	25,000	0.4104	10,260
			1,42,636
		less Cash Outflow	1,50,000
		NPV =	(7,364)

Since the NPV is negative, the proposal is not an acceptable one. Note, however, that if the usual rate of discount of 10% were used, the project would have an NPV equal to Rs 20,158.5, and, therefore, be acceptable.

18.5.3 Statistical Distribution Approach

While using the certainty equivalent approach, the risk-free discount rate may be easily approximated (may be, for instance, by the interest rate on government bonds) but difficulties may arise in determining the trade-off between risk and return for the purposes of converting a particular distribution of NPV into its certainty equivalent. In a similar manner, in using the risk-adjusted discount rate method, the determination of the risk-premium to be added to the risk-free rate of return would pose difficulty. In using either of these approaches, it is important that we should be able to measure the degree of risk associated with the project(s) in question. In the statistical distribution approach, the degree of risk associated with a project is sought to be measured in terms of the variance (or standard deviation) of the NPV distribution, and the investment decisions are taken considering the expected (mean) value, and its standard deviation, of the net present value distribution. This information about the project risk may also be usefully employed for calculating certainty-equivalent for the uncertain returns from the investment proposal, as also it is a major factor in calculating the size of the risk-adjusted discount rate to use. The derivation of the probabilistic information about investment proposals owes its origin to the work of Frederick Hillier.

In this method of considering risky investment proposals, the net cash flow from an investment in each period is viewed as a random variable which can assume any one of the possible values. The method requires that probability distribution of cash flows for each of the years be obtained and considered. Using the cash flow distributions, the expected value of the NPV distribution and its variance are calculated in the first instance. These are calculated as discussed here.

(a) Expected Value of NPV

Suppose that there is an investment proposal with cash flows whose probability distributions are given for each of the n years of the project life. The cash flows have means equal to $\overline{C}_1, \overline{C}_2, ..., \overline{C}_n$, with standard deviations equal to $\sigma_1, \sigma_2, ..., \sigma_n$ respectively. For calculating the expected value of the NPV distribution, we shall add the discounted mean value of the cash flows for each of the n time periods. This gives,

$$E(\text{NPV}) = \overline{C}_0 + \frac{\overline{C}_1}{(1+i)^1} + \frac{\overline{C}_2}{(1+i)^2} + \cdots + \frac{\overline{C}_n}{(1+i)^n}$$

or

$$\mu = \sum_{t=0}^{n} \frac{\overline{C}_t}{(1+i)^t}$$

E(NPV) is also known as the *expected monetary value* (EMV) of the project. As before, if this NPV exceeds zero, the proposal becomes acceptable while if it is lower than zero, the proposal becomes rejectable.

Example 18.18 The New Electronic Company is considering to make investment in a proposal which requires an outlay of Rs 1,20,000. The project has a life of three years over which the following cash inflows are likely to be generated:

Year 1		Year 2		Year 3	
Cash Flow	*Probability*	*Cash Flow*	*Probability*	*Cash Flow*	*Probability*
30,000	0.2	30,000	0.1	40,000	0.3
40,000	0.4	50,000	0.4	60,000	0.3
50,000	0.3	80,000	0.4	80,000	0.2
60,000	0.1	90,000	0.1	1,00,000	0.2

The management feels that the expected cash flows in the various periods may be considered to base its decision about acceptance or rejection of the project. If the discount rate is 10%, should the proposal be accepted?

We shall first obtain the expected cash inflow for each of the years. This is calculated in Table 18.2.

TABLE 18.2 Calculation of Expected Cash Flows

Year t	*Cash flow* C_{ti}	*Probability* p_{ti}	*Expected Value*	*Expected Cash Flow* $\overline{C}_t = \Sigma C_{ti} p_{ti}$
1	30,000	0.2	6,000	
	40,000	0.4	16,000	
	50,000	0.3	15,000	
	60,000	0.1	6,000	43,000

TABLE 18.2 Contd.

Year t	Cash flow C_{ti}	Probability p_{ti}	Expected Value	Expected Cash Flow $\overline{C}_t = \Sigma C_{ti}p_{ti}$
2	30,000	0.1	3,000	
	50,000	0.4	20,000	
	80,000	0.4	32,000	
	90,000	0.1	9,000	64,000
3	40,000	0.3	12,000	
	60,000	0.3	18,000	
	80,000	0.2	16,000	
	1,00,000	0.2	20,000	66,000

Now the expected NPV can be calculated as:

Year	Expected cash flow	PV factor @ 10%	Present value
0	–1,20,000	1.0000	–1,20,000.0
1	43,000	0.9091	39,091.3
2	64,000	0.8264	52,889.6
3	66,000	0.7513	49,585.8
		Expected NPV =	21,566.8

Since the expected NPV of the proposal is greater than zero, it is an acceptable one.

(b) Variance of NPV

In the discussion on the variance of the NPV distribution, Hillier has given an analysis of three cases. In the first case, the cash flows between different periods are assumed to be independent of one another. This is to say, the cash flows of one period are not related to the cash flows of another period. In the second case, the cash flows between different periods are assumed to be perfectly correlated. The third case deals with the mixed situation in which part of the cash flows are perfectly correlated and part are independent. Obviously, when we consider more than two periods of time, the cash flows cannot all be perfectly negatively correlated with each other. This explains why Hillier's analysis is restricted to the case of positive correlation.

Now we consider the three cases in turn:

(i) Independent Net Cash Flows When net cash flows for the various years are independent of each other, then the calculation of the variance of the distribution of cash flows becomes a difficult task. This is because independence substantially increases the number of possible outcomes. To illustrate, suppose a project has a life of 3 years and in each of the years, there are four cash flow values possible with some given probabilities. Under the assumption of independence, a total of $4 \times 4 \times 4 = 64$ combinations are possible. The probability of occurrence of each of the combinations is given by the product of the probabilities of the particular cash flow values of different years entering into that combination. For this probability distribution, we can find the present values of each of the possible cash flow streams (64 in our example) and determine the expected value and the variance of present values in the usual way.

However, the complications in calculations can be avoided and instead the variance can be obtained directly using the following formulation:

$$V(\text{NPV}) = \sigma^2 = \sigma_0^2 + \frac{\sigma_1^2}{(1+i)^2} + \frac{\sigma_2^2}{(1+i)^4} + \cdots + \frac{\sigma_n^2}{(1+i)^{2n}}$$

$$= \sum_{t=0}^{n} \frac{\sigma_t^2}{(1+i)^{2t}}$$

Unless otherwise stated, the cash flows of a given project may be taken to be independent.

(ii) Perfectly Correlated Cash Flows The assumption of perfect correlation between the cash flows of the successive years implies, technically, that if random factors cause a cash flow C_j (for the jth year) to deviate from its mean value by a standard deviations, the same factors will cause the cash flow C_k (for the kth year) to deviate from its own mean in the same direction by a standard deviations. In the situation in which the periodic net cash flows are given to be perfectly correlated, the variance of NPV distribution can be computed as follows:

$$V(\text{NPV}) = \left[\sigma_0 + \frac{\sigma_1}{(1+i)} + \frac{\sigma_2}{(1+i)^2} + \cdots + \frac{\sigma_n}{(1+i)^n}\right]^2$$

or

$$\sigma^2 = \left[\sum_{t=0}^{n} \frac{\sigma_t}{(1+i)^t}\right]^2$$

It may be noted that the standard deviation value, when the cash flows are perfectly correlated, would be higher than when they are independent.

(iii) Mixed Case There are very few investments for which the net cash flows are either completely independent or perfectly correlated. Closer to the reality is the 'mixed' case is which cash flows are partially dependent and partially independent. To understand this situation, consider a very simple case where a company is contemplating to introduce a new product, whose life is expected to be only three years. The market acceptance of the product may be unsatisfactory, satisfactory, or excellent. How the product will be accepted in the first year will determine how it will be accepted in the second and third years. In this respect, it is a case of dependence. Also, the product sales are influenced by general economic conditions which may be poor, good, or excellent. However, economic conditions in one year do not affect the economic conditions in the next year. From this standpoint, it is a case of independence.

For the mixed case as well, all possible combinations of the cash flows are obtained and their joint probabilities calculated. The present value of each of the cash flow streams is then calculated and we can get the mean and the standard deviation as usual. In the simple case, however, where the cash flows C_j''s can be broken down into two components: C_j' and C_j^*, where C_j' is the part of C_j which varies independently and C_j^* is that part of C_j that is perfectly correlated with C^* in any other period, the variance can be obtained using the formulae given above jointly to get the following:

$$V(\text{NPV}) = \sigma^2 = V(C_0') + \frac{V(C_1')}{(1+i)^2} + \frac{V(C_2')}{(1+i)^4} + \cdots + \frac{V(C_n')}{(1+i)^{2n}}$$

$$+ \left[\sqrt{V(C_0^*)} + \frac{\sqrt{V(C_1^*)}}{(1+i)} + \frac{\sqrt{V(C_2^*)}}{(1+i)^2} + \cdots + \frac{\sqrt{V(C_n^*)}}{(1+i)^n}\right]^2$$

Alternatively,

$$\sigma^2 = \left(\sum_{t=0}^{n} \frac{V(C_t')}{(1+i)^{2t}} \right) + \left(\sum_{t=0}^{n} \frac{\sqrt{V(C_t^*)}}{(1+i)^t} \right)^2$$

Once the expected net present value and its standard deviation (from variance) value are obtained, the riskiness of the project can be measured. The standard deviation is a measure of absolute amount of risk associated with a given project. While it is a useful measure for the purpose of risk evaluation, it is not suitable when comparative riskiness of the projects is to be considered. In order to compare the various projects, we should compute their respective coefficients of variation. We have,

$$\text{Coefficient of variation} = \frac{\text{Standard deviation}}{\text{Expected (mean) value of NPV}} \times 100$$

A higher coefficient value points to a higher risk associated with a project.

Normally Distributed NPV

If the cash flows for each of the years be assumed to be normally distributed, then the NPV of the project would also be normally distributed, with a mean equal to the expected net present value and standard deviation equal to the square root of the variance of the NPV distribution described above. The assumption of normal distribution of the cash flows, and therefore of the NPV of the project, is a reasonable one, and enables us to make precise probabilistic statements regarding the various possible values which NPV may assume.

EVPI of a Project

In Chapter 13 we saw that the expected value of perfect information, EVPI, is the sum of the conditional opportunity losses for the optimal action multiplied by the respective prior probabilities. The cases we considered were discrete in nature so that only a finite number of outcomes (of the state of nature) could be possible. The concept of EVPI for the continuous case can be visualised intuitively in a similar manner. The continuous distribution can be conceived as being approximated by a histogram with a large number of bars with equal width and heights equal to the ordinate of the normal curve at the midpoint of each bar. The area of each bar, width times height, approximates the probability associated with the interval on the horizontal axis. If each probability is multiplied by the corresponding value of the opportunity loss at the midpoint of the bar and the interval width on the X-axis becomes small to approach zero, the limit of the sum of these products is EVPI for the continuous case.

The expression for EVPI can be written as follows:

$$\text{EVPI} = \sigma L_N(D)$$

where σ = the standard deviation of the distribution, and

$L_N(D)$ = the unit normal loss function. It is also called the *normal loss integral*.

The values of $L_N(D)$ representing the expected opportunity loss (or EVPI) for a (linear) loss function with unit slope and normal distribution are given in Table B2 in the Appendix B. The value of D represents the relative position of the break-even point. When using $L_N(D)$ for a given normal distribution, the value of D represents the deviation of the break-even point from the mean, expressed in standard deviation units.

In the context of the NPV distribution, which is assumed to follow normal distribution, the EVPI can be calculated as follows. First, the value of D is obtained by finding the difference between zero (the break-even point) and the expected NPV, μ, and dividing by the standard deviation σ. Then the value of $L_N(D)$ is read from the table corresponding to the D value. This value is multiplied by the standard deviation σ to get EVPI.

Example 18.19 Measure the risk associated with the project, considered in Example 18.18 by standard deviation. Assume that the cash flows of the three years are independent. Also determine EVPI of the investment proposal.

The pertinent information is reproduced here:

Year 1		*Year 2*		*Year 3*	
Cash Flow	*Probability*	*Cash Flow*	*Probability*	*Cash Flow*	*Probability*
30,000	0.2	30,000	0.1	40,000	0.3
40,000	0.4	50,000	0.4	60,000	0.3
50,000	0.3	80,000	0.4	80,000	0.2
60,000	0.1	90,000	0.1	1,00,000	0.2

Expected Value:

$$\overline{C}_1 = 43{,}000, \quad \overline{C}_2 = 64{,}000, \quad \overline{C}_3 = 66{,}000$$

To compute the standard deviation of the project, we shall first obtain the variance σ_t^2 for the cash flows for each year. We have,

$$\sigma_t^2 = \Sigma(C_{ti} - \overline{C}_1)^2 p_{ti}$$

where $t = 1, 2, 3$ and $i = 1, 2, ...,$ for the ith cash flow.

The calculations are given in Table 18.3.

TABLE 18.3 Calculation of Variance

Year t	*Cash Flow* C_{ti}	*Probability* p_{ti}	$(C_{ti} - \overline{C}_t)^2$	$(C_{ti} - \overline{C}_t)^2 p_{ti}$
1	30,000	0.2	169×10^6	33.8×10^6
	40,000	0.4	9×10^6	3.6×10^6
	50,000	0.3	49×10^6	14.7×10^6
	60,000	0.1	289×10^6	28.9×10^6
				$\sigma_1^2 = 81 \times 10^6$
2	30,000	0.1	$1{,}156 \times 10^6$	115.6×10^6
	50,000	0.4	196×10^6	78.4×10^6
	80,000	0.4	256×10^6	102.4×10^6
	90,000	0.1	676×10^6	67.6×10^6
				$\sigma_2^2 = 364 \times 10^6$
3	40,000	0.3	676×10^6	202.8×10^6
	60,000	0.3	36×10^6	10.8×10^6
	80,000	0.2	196×10^6	39.2×10^6
	1,00,000	0.2	$1{,}156 \times 10^6$	231.2×10^6
				$\sigma_3^2 = 484 \times 10^6$

Using the variance values obtained, we can determine the variance, and the standard deviation, for the project. We know,

$$\sigma^2 = \frac{\sigma_1^2}{(1+i)^2} + \frac{\sigma_2^2}{(1+i)^4} + \frac{\sigma_3^2}{(1+i)^6}$$

Year, t	*Variance,* σ_t^2	*PV Factor,* $(1+i)^{-2t}$	*Present Value*
1	81×10^6	0.8264	66.9384×10^6
2	364×10^6	0.6830	248.6120×10^6
3	484×10^6	0.5645	273.2180×10^6
		$\sigma^2 =$	588.7684×10^6

Now, standard deviation, $\sigma = \sqrt{588.7684} \times 10^3 =$ Rs 24,264.6.

Calculation of EVPI With $\mu = 21{,}566.7$, and $\sigma = 24{,}264.6$, we have

$$D = \frac{0-\mu}{\sigma} = \frac{0 - 21{,}566.7}{24{,}264.6} = -0.89$$

Corresponding to $D = 0.89$, $L_N(D) = 0.1023$.

Thus, EVPI $= 0.1023 \times 24{,}264.6 =$ Rs 2,482.27.

Example 18.20 A company is trying to choose between two investment proposals *A* and *B*. Project *A* has a standard deviation of Rs 6,500 while project *B* has a standard deviation of Rs 7,200. The firm's financial manager wishes to know which investment to choose, given each of the following combinations of the expected values.

(a) Project *A* and project *B* both have expected net present value of Rs 15,000.
(b) Project *A* has expected NPV of Rs 18,000 while for the project *B* it is Rs 22,000.

In case (a), the manager should choose the project *A* because the expected NPV of both the projects is the same while the standard deviation, and therefore the risk, is smaller for *A* than for *B*.

To determine the relative acceptability of the two projects in case (b), we shall compute the coefficient of variations for each one.

$$\text{Coefficient of variation} = \frac{\text{Standard Deviation}}{\text{Mean}} \times 100$$

For project *A*,

$$CV_A = \frac{6{,}500}{18{,}000} \times 100 = 36.11\%$$

For project *B*,

$$CV_B = \frac{7{,}200}{22{,}000} \times 100 = 32.73\%$$

Since $CV_B < CV_A$, the project *B* has greater acceptability than project *A*.

Example 18.21 The XYZ company is considering whether or not to invest Rs 1,20,000 in an investment proposal to add a new product to its existing range. The management forecasts that the new plant will generate incremental net cash flows, each of which is assumed to be normally distributed random variable with following parameters.

Year	Expected Value (Rs)	Standard Deviation (Rs)
1	44,000	5,800
2	40,000	5,600
3	40,000	6,000
4	35,000	5,000
5	30,000	4,000

Using the discount rate of 10%, determine:

(a) the mean of the present value distribution and its standard deviation,

(b) the probability that the investment shall have a negative NPV,

(c) whether the project should be undertaken if the management has laid down that no project be undertaken unless it has at least 70% chance of giving an NPV at least equal to Rs 25,000.

(a) The expected NPV is calculated as,

Year, t	*Cash Flow,* C_t	*PV Factor @ 10%*	*Present Value*
0	(1,20,000)	1.0000	(1,20,000)
1	44,000	0.9091	40,000
2	40,000	0.8264	33,056
3	40,000	0.7513	30,052
4	35,000	0.6830	23,905
5	30,000	0.6209	18,627
		Expected NPV =	25,640

Thus, the project has the expected NPV of Rs 25,640. Its standard deviation equals Rs 9,347.27, the value of which is derived from variance calculated as shown:

Year	*Standard Deviation* σ_t	*Variance* σ_t^2	*PV factor @ 10%* $(1+i)^{-2t}$	*Present value*
1	5,800	$3,364 \times 10^4$	0.8264	$2,780.00 \times 10^4$
2	5,600	$3,136 \times 10^4$	0.6830	$2,141.89 \times 10^4$
3	6,000	$3,600 \times 10^4$	0.5645	$2,032.20 \times 10^4$
4	5,000	$2,500 \times 10^4$	0.4665	$1,166.25 \times 10^4$
5	4,000	$1,600 \times 10^4$	0.3855	616.80×10^4
			$\sigma^2 =$	$8,737.14 \times 10^4$

Standard deviation = $\sqrt{8,737.14 \times 10^4}$ = Rs 9,347.27

(b) To calculate the required probability, we shall find the area under the normal curve, with $\mu = 25,640$ and $\sigma = 9,347.27$, to the left of $X = 0$. Thus,

$$z = \frac{0 - 25,640}{9,347.27} = -2.74$$

From the normal area table, the area to the left of $z = -2.74$ is 0.0031 (= 0.5 – 0.4969). Thus, there is hardly any chance for the NPV of the project to assume a negative value.

(c) For the given project, area to the right of $X = 25,000$ equals 0.5 plus are between 25,000 and 25,640. Thus,

$$z = \frac{25,000 - 25,640}{9,347.27} = -0.07$$

For $z = -0.07$, area = 0.0279. Thus, $P(\text{NPV} \geq 25,000) = 0.5 + 0.0279 = 0.53$ (app.) Since it is less than the required 70%, the project may not be acceptable to the management.

Example 18.22 Reconsider Example 18.21. Assume that the cash flows are the same as given but they are perfectly correlated. Work out the expected NPV and its standard deviation. Also calculate the probability that the project would have a net present value exceeding Rs 20,000.

The expected value of NPV, as before, is equal to Rs 25,640. The cash flows being perfectly correlated, standard deviation would be calculated as follows:

Year, t	*Standard Deviation*	*PV Factor* $(1 + i)^{-t}$	*Present Value*
1	5,800	0.9091	5,272.78
2	5,600	0.8264	4,627.84
3	6,000	0.7513	4,507.80
4	5,000	0.6830	3,415.00
5	4,000	0.6209	2,483.60
		σ =	20,307.02

Thus, $E(\text{NPV}) = \mu = \text{Rs } 25,640$ and $\sigma = \text{Rs } 20,307.02$.

To find $P(\text{NPV} > 20,000)$, we shall determine the area between $X = 20,000$ and mean = 25,640, and add it to 0.5 (the area beyond 25,640).

$$z = \frac{20,000 - 25,640}{20,307.02} = -0.28$$

From the normal area table, area corresponding to $z = 0.28$ is 0.1103. Thus, $P(\text{NPV} > 20,000) = 0.1103 + 0.5 = 0.6103$.

Example 18.23 The management of *ABC* wants to invest in a project requiring an initial outlay of Rs 1,20,000, whose cash flows are assumed to be made up of two components—one of which varies independently over time, C_t', and the other that is perfectly correlated, C_t^*. Each of them is normally distributed with the following parameters:

Year	Component C′		Component C*	
	Mean	Standard Deviation	Mean	Standard Deviation
1	24,000	3,000	20,000	2,800
2	20,000	2,800	20,000	2,800
3	20,000	3,000	20,000	3,000
4	20,000	3,000	15,000	2,000
5	18,000	2,000	12,000	2,000

Compute the expected NPV and its standard deviation. What probability shall be assigned to the NPV of the project exceeding Rs 20,000? (Use a discount rate of 10%.)

The expected NPV is calculated as follows:

Year, t	*Cash Flow, $C_t = C_t' + C_t^*$*	*PV Factor @ 10%*	*Present Value*
1	44,000	0.9091	40,000
2	40,000	0.8264	33,056
3	40,000	0.7513	30,052
4	35,000	0.6830	23,905
5	30,000	0.6209	18,627
			1,45,640
		less Cash Outflow	1,20,000
		NPV =	25,640

We know that the variance of the NPV distribution is given by the expression:

$$V(\text{NPV}) = \left(\sum_{t=0}^{n} \frac{V(C_t')}{(1+i)^{2t}}\right) + \left(\sum_{t=0}^{n} \frac{\sqrt{V(C_t^*)}}{(1+i)^t}\right)^2$$

The values of the two components would be calculated separately and added together.

For independent flows:

Year, t	*Standard Deviation, σ_t*	*Variance, σ_t^2*	*PV Factor*	*Present Value*
1	3,000	900×10^4	0.8264	743.760×10^4
2	2,800	784×10^4	0.6830	535.472×10^4
3	3,000	900×10^4	0.5645	508.050×10^4
4	3,000	900×10^4	0.4665	419.850×10^4
5	2,000	400×10^4	0.3855	154.200×10^4
			$\sigma^2 =$	$2{,}361.332 \times 10^4$

For perfectly correlated flows:

Year, t	*Standard Deviation, σ_t*	*PV Factor*	*Present Value*
1	2,800	0.9091	2,545.48
2	2,800	0.8264	2,313.92
3	3,000	0.7513	2,253.90
4	2,000	0.6830	1,366.00
5	2,000	0.6209	1,241.80
		σ =	9,721.10

$V(\text{NPV}) = (2{,}361.332 \times 10^4) + (9{,}721.10)^2 = 118{,}113{,}105$

∴ Standard deviation = Rs 10,868.

18.5.4 Decision Trees

The idea of decision trees discussed in Chapter 13 can be extended to risk situations in investment theory. The decision tree approach provides a convenient means to use the probability distribution information to assess risk inherent in a given project. It is of immense importance in cases when the cash flows cannot be easily classified as either independent or perfectly correlated, and the correlation is of moderate order.

To illustrate, suppose that a firm has a two-year project that requires an initial investment of Rs 1,00,000. The cash flows expected in each of the years along with their probabilities are given in Figure 18.4. It may be noted that in this example both the cash flows and the probabilities are conditional (a case where the cash flows are not independent) on what happens in the first year.

It is evident from this figure that the decision tree shows nine different combinations of outcomes as possible. One possibility is that the first year will have an inflow of Rs 40,000 which shall be followed by a Rs 60,000 cash flow in the second year. As shown in the figure, the net present value associated with this set of cash flows, discounted using a rate of 8%, is $40{,}000 \times (1.08)^{-1} + 60{,}000 \times (1.08)^{-2} - 1{,}00{,}000 = -11{,}526$. In a similar manner, the net present value of each of the other eight combinations are given. The (joint) probability of each combination is obtained by multiplying the probabilities of occurrences of each of the parts. For example, for the first combination, the probability is 0.08 (= 0.2 × 0.4).

By multiplying the joint probability for each of the nine combinations times their associated NPVs and summing, we obtain the project's expected net present value, *E*(NPV). The standard deviation of the project equals Rs 18,648 app., obtained as follows:

$$\sigma = \sqrt{\sum_{j=1}^{n} \left[\text{NPV}_j - E(\text{NPV})\right]^2 p_j}$$

NPV_j = NPV of the *j*th combination,

p_j = probability of the *j*th combination

$\sigma = [(-11{,}526 - 20{,}177)^2 \times 0.08 + (-2{,}953 - 20{,}177)^2 \times 0.08 + \ldots]^{1/2}$ = Rs 18,684.

Although useful for setting out all possible combinations of a proposed project, the decision-tree approach suffers from a shortcoming that in situations involving a large number of possible outcomes, it may be too complex to handle.

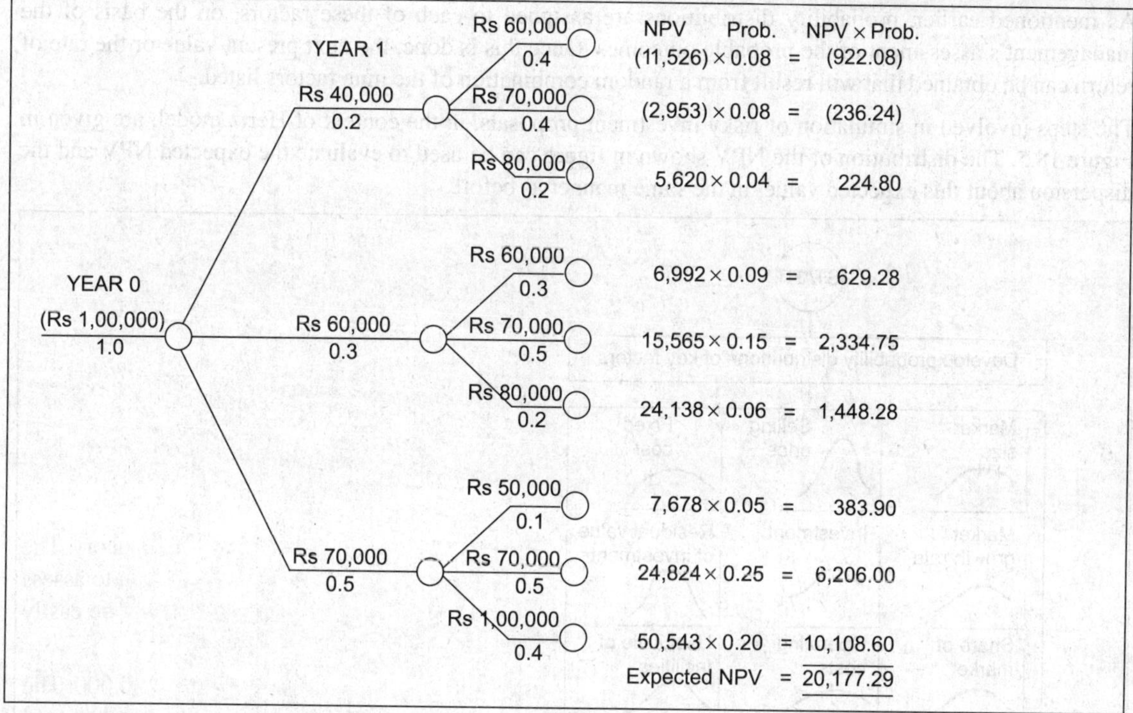

Figure 18.4 *Decision Tree*

18.5.5 Simulation Approach to Risky Investments

Another technique by which risk in investment may be considered is the simulation technique that was discussed in Chapter 17. Simulation can be used to approximate the NPV or the expected return and its dispersion, about the expected value. For its application, in the first stage, the various factors influencing the cash flows relating to a project are identified. Then the probability distributions of each of them are obtained, either using the historical data, if available, or the decision-maker's judgment and expert knowledge. In the next stage, values for each of the variables are simulated in accordance with the procedure laid in the Chapter 17, and the NPV, or the expected return, resulting from a combination thereof is computed. This procedure is repeated a large number of times and the probability distribution of the NPV (or the rate of return) is obtained. This distribution is then used as a basis for taking appropriate decisions relating to the proposal.

A simulation model proposed by Hertz* considers the following factors in analysing risky investment projects:

Market Analysis:

1. Market size
2. Selling price
3. Market growth rate
4. Share of market (which results in physical sales volume)

Investment Cost Analysis:

1. Investment required
2. Residual value of investment

Operating and Fixed Cost:

1. Operating costs
2. Fixed costs
3. Useful life of facilities.

* David B. Hertz, "Risk Analysis in Capital Investment" *Harvard Business Review*, 42 (Jan-Feb 1964) pp. 95-106.

As mentioned earlier, probability distributions are assigned to each of these factors, on the basis of the management's assessment of the probable outcomes. Once this is done, the next present value or the rate of return can be obtained that will result from a random combination of the nine factors listed.

The steps involved in simulation of risky investment proposals, in the context of Hertz model, are given in Figure 18.5. The distribution of the NPV shown in figure can be used to evaluate the expected NPV and the dispersion about this expected value, in the same manner as before.

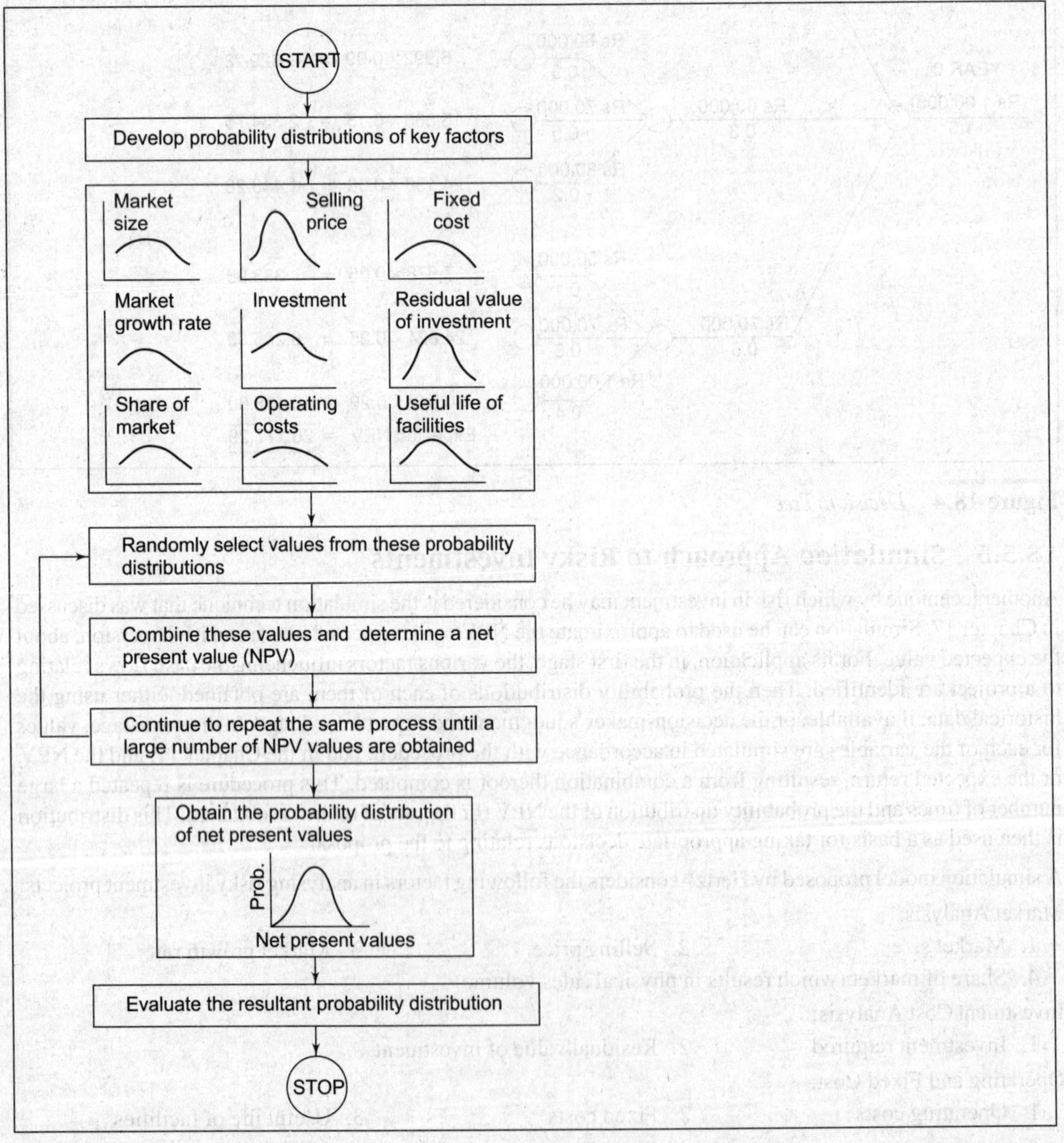

Figure 18.5 *Simulation*

There are two problems that have impeded the widespread use of simulation. First, the identification of the variables that will affect the cash flows, and their probabilities of occurrence is a difficult task particularly in situations where little prior information is available. Also, the variables affecting the cash flows may not be independent, as assumed in the above analysis, and be correlated instead. Of course the interdependence can be incorporated into the simulation itself with some effort. But, what is more difficult is the estimates of the nature and degree of dependence. The second problem in using simulation is that it does not decide about the acceptability or otherwise of the project—it merely tells the manager as to what the entire NPV (or the rate of return) distribution looks like, leaving for him to take the decision.

18.5.6 Sensitivity Analysis

Sensitivity analysis represents yet another technique used in capital budgeting. It does not formally attempt to quantify risk. Rather, it focuses on how sensitive is NPV (or IRR) to changes in any of the input variables. In its simplest form, sensitivity analysis consists of the following steps.

(a) In the first step, the data relating to cash flows are broken down into a number of major variables. In other words, using some estimated values of the variables like initial outlay required for the project, production hours. sales quantities, selling prices, wage rates, material prices, and so on, a usable model of the project is drawn up.

(b) Each variable is considered in turn and its value is modified by a fairly large arbitrary percentage, both upward and downward from the original estimate.

(c) The NPV (or IRR) values are recalculated by taking each new estimate in turn.

For example, the IRR of a project using estimated values of five major variables were originally calculated as 16%. If each of the variable values was modified one at a time and the IRR recalculated, we might end up with a table like this:

Rate of return when the variable value is altered by 10%:

Variable	*Adversely*	*Favourably*
Initial outlay	13%	18%
Sales volume	8%	22%
Selling price	12%	19%
Raw material prices	14%	19%
Production efficiency	9%	21%

Now it is evident that the rate of return achievable from the given project is particularly sensitive to fluctuations in sales volume and the production efficiency. Further, the degree of uncertainty of each of these variables may be evaluated and the resulting information used to assess the project's risk, or as a guide to where additional efforts in refining estimates may be made.

In effect, then, sensitivity analysis is a systematic way to determine which of the various factors affecting cash flows are most important and calls for recalculating the IRR or the NPV using assumptions that differ from those used to produce the original net cash flow estimates. We can use this method to assess the degree of risk in the project and indicate areas where additional effort to produce better (that is less uncertain) estimates might be justified.

Sensitivity analysis is widely used in practice. The great advantage of the method is that in adoption of this technique, the management have to investigate thoroughly the estimates of a few variables only before making a final choice of acceptance or otherwise of an investment proposal. Although a flexible method providing useful information, sensitivity analysis fails to provide an answer to what decision should be made in a given situation. Its conclusions tend to suffer from a lack of conciseness, precision and comprehensiveness.

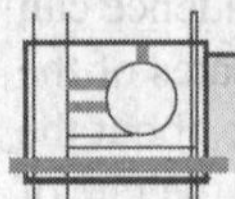

18.6 BREAK-EVEN ANALYSIS

We have so far considered the questions of time value of money and its use in the capital budgeting decisions for determining whether or not an investment project be undertaken. Closely related to the question of assessing the cash flows related to a project is the systematic analysis of costs and revenues and the resultant profit/loss to the firm. In this section we discuss the relationships between costs, revenues and profits which must be thoroughly understood by a manager so that he can determine the potential effects of his decisions on the total revenues, costs, profits and production or sales volume necessary to achieve or maintain various levels of financial success. This relationship is called the *cost-volume-profit* (*CVP*) *relationship*. The analysis is also labelled as *break-even analysis*.

It is obvious that profits are a function of prices, costs and the volume of activity (i.e. the number of units of output), and the three are related to each other as well—volume is a function of price and cost is a function of volume. There are several decisions for which these interrelationships are significant. To illustrate, a manager's problem could be such that it is useful to know the break-even volume—the level of activity at which the firm has just enough revenues to recover all the costs. At this level of activity, the firm will not make any money, however it would not lose money either. It is a no-profit no-loss activity level. This level acts as a reference point for the decision maker who knows then the level of output that must be exceeded if some profit is to be made. Similarly, a manager may want to know the volume necessary to achieve a target profit. For instance, the target return on investment may be 20%. Knowing the investment, this could be translated into a target rupee profit that must be achieved. Then the CVP analysis can be employed to make necessary estimates. Another application of the CVP relationship is the sensitivity analysis wherein the manager may be able to estimate the impact of changes in any or all the key variables—the selling prices, costs, activity levels, target profit. Thus, CVP analysis can provide data to help answer the questions like—if the price is lowered by 5%, by how much must volume increase in order for the profit to be 8% higher than last year's?

The central thing in CVP analysis is to predict how costs will behave in response to activity level changes. This requires that costs should be classified as either *fixed* or *variable*. We assume that all costs can be segregated into these two categories. Fixed costs are taken to be those for which the total amount remains constant (within a specified range) regardless of the variations in the activity levels. Depreciation on plant and machinery, salaries of executives and managers, property taxes, etc. are examples of fixed costs. Variable costs are those for which the total amount varies in direct proportion to the changes in the levels of activity. Thus the variable costs are uniformly variable so that they are incurred at a constant rate per unit of output. Raw materials and labour cost of the workers engaged in the production of a product are typical examples of it.

The primary source of revenue for a firm is the sale of its product or servicc. In a similar manner, the primary cost of doing business is a direct effort of the firm to produce some product(s) or offer services. Thus, both the costs and revenue, are related to the volume or level of activity which shall be measured in terms of the number of units of product/service. For our analysis, it is assumed that the price of the product per unit would remain unchanged, irrespective of the number of units sold.

Another concept which is of significance in the CVP analysis is that of *contribution margin*. It is defined as the excess of selling price over the variable cost. The selling price and the variable cost being constant per unit of output, the contribution margin per unit is also a constant amount. The total contribution margin is used to meet the fixed costs, and the excess, if any, represents the profit. Thus, we have

Total Sales – Total Variable Cost = Contribution Margin,

and Contribution Margin – Fixed Cost = Profit.

Also, Contribution Margin = Fixed Cost + Profit.

When the contribution margin is related to the sales, either on a unit basis or on the aggregate basis, we get the *contribution margin ratio* or the *P/V ratio* (Profit/Volume ratio). The P/V ratio is an indicator of the profitability of the product. We have,

$$P/V \text{ Ratio} = \frac{\text{Contribution margin per unit}}{\text{Selling price}} \times 100$$

$$= \frac{S-V}{S} \times 100 \text{ or } 1 - \frac{V}{S}$$

where S = selling price per unit, and V = variable cost per unit.

Alternatively,

$$P/V \text{ Ratio} = \frac{\text{Contribution margin}}{\text{Sales}} \times 100$$

The Basic CVP Model Suppose a firm, which is engaged in producing and selling a single product, is attempting to coordinate its sales price, production volume, and costs in order to achieve its profit goals.

The two criteria used in the CVP analysis are the break-even point and the target profit it seeks to earn. The *break-even point* refers to the level of activity at which total sales revenue matches the total costs incurred in the production of output sold (or produced). The target profit is the amount of profit that the firm has set to achieve.

18.6.1 Calculation of BEP

The break-even condition exists when the total revenue equals total costs and therefore, there is no profit or no loss. Under the conditions postulated, let

F = Fixed cost per period,

V = Variable cost per unit, and

S = Selling price per unit

If X units of the product are produced and sold, the profit function can be obtained as follows:

$$\text{Total revenue} = XS, \text{ Total cost} = F + VX,$$

$$\therefore \text{ Profit, } P = XS - (F + VX) = X(S - V) - F$$

At break-even point, $P = 0$. Setting $P = 0$ and solving for X, we get

$$X(S - V) - F = 0$$

or

$$X = \frac{F}{S-V} = \frac{F}{CM}$$

Thus, break-even point (in units) = F/CM

The break-even point can be expressed in monetary terms as well. Called break-even sales, it can be obtained by multiplying the BEP by the selling price so that

$$BE \text{ Sales} = \frac{F}{CM} \times S \quad \text{or} \quad \frac{F}{1-\dfrac{V}{S}}$$

Alternatively,

$$BES = \frac{F}{P/V \text{ Ratio}}$$

Example 18.24 The Hi-Hi Electronics is engaged in producing a certain component which is sold at a uniform price of Rs 8 each. The variable cost of producing the component amounts to Rs 4.80 per unit while the fixed costs amount to Rs 24,000. How many units of the component must be produced and sold so that the company breaks even? How much sales would be made at this level of activity?

Here $F = 24{,}000$, $S = 8$ and $V = 4.80$. Thus, $CM = 8 - 4.80 = 3.20$ (per unit).

$$P/V \text{ ratio} = \frac{CM}{S} \times 100 = \frac{3.20}{8} \times 100 = 40\%$$

Accordingly,

$$BEP = \frac{F}{CM} = \frac{24{,}000}{3.20} = 7{,}500 \text{ units},$$

and $$BES = \frac{F}{P/V \text{ Ratio}} = \frac{24{,}000}{40\%} = \text{Rs } 60{,}000.$$

Target Profit Continuing with the analysis, suppose that a firm desires to earn a certain profit π. We know,

$$\text{Profit } P = X(S - V) - F$$

Setting $P = \pi$, we have

$$\pi = X(S - V) - F$$

Solving for X $$X = \frac{F + \pi}{S - V} = \frac{F + \pi}{CM}$$

Here X would give the number of units required to be sold in order to earn a desired profit π. In a similar way,

$$\text{Desired rupee sales} = \frac{F + \pi}{P/V \text{ Ratio}}$$

In Example 18.24, suppose the firm desires a profit of Rs 16,000. We can determine the number of units required to be sold to achieve this profit, as follows:

$$\text{Required sales, } X = \frac{F + \pi}{CM} = \frac{24{,}000 + 16{,}000}{3.2} = 12{,}500 \text{ units}$$

Similarly, we can determine the rupee sales required for this profit level.

$$\text{Required rupee sales} = \frac{F + \pi}{P/V \text{ Ratio}} = \frac{24{,}000 + 16{,}000}{40\%} = \text{Rs } 1{,}00{,}000$$

This relationship can be used to obtain the profit resulting from a particular sales level. To illustrate, at a sales volume of Rs 80,000, the profit π would be:

$$80{,}000 = \frac{24{,}000 + \pi}{40\%}$$

or $$\pi = 80{,}000 \times 40\% - 24{,}000 = \text{Rs } 8{,}000$$

18.6.2 Margin of Safety

The margin of safety is the excess of the actual (or budgeted) sales over the break-even sales. This can be expressed in units or in rupee terms.Thus,

$$\text{Margin of Safety} = \text{Actual Sales} - \text{Break-Even Sales}$$

It can also be calculated and expressed in relative terms, known as the *margin of safety ratio*. This refers to the margin of safety expressed as a percentage of the actual sales.

$$\text{Margin of Safety Ratio} = \frac{\text{Actual Sales} - \text{Break-Even Sales}}{\text{Actual Sales}} \times 100$$

A manager may desire to know the margin of safety in actual or planned activity because it indicates the amount by which sales could decrease before losses occur.

In Example 18.24, suppose that the manager, on the basis of general economic and market conditions, anticipates a sales of 9,000 units. The margin of safety for the company would in that case be 9,000 – 7,500 = 1,500 units. Also margin of safety = 9,000 × 8 – 60,000 = Rs 12,000.

The margin of safety ratio would be $16\frac{2}{3}\%$ obtained as follows:

$$\text{Margin of Safety Ratio} = \frac{9{,}000 - 7{,}500}{9{,}000} \times 100 = 16\frac{2}{3}\%$$

Graphic Analysis The determination of the break-even point can be depicted graphically also. For this, the level of activity (in terms of the number of units, capacity level, sales volume in rupees, etc.) is shown on the X-axis, and the Y-axis is taken to show costs and revenue. The fixed cost is shown, first of all, by means of a straight line parallel to the horizontal axis at a height given by the amount of fixed cost. The variable cost component is then added to it, to get the total cost line. Then the sales revenue function is plotted on the graph. The point where the two intersect, gives the break-even point. For the earlier example considered, the data arc plotted in the Figure 18.6. From the graph, we can read the BEP to be 7,500 units, as before. Also depicted on this graph is the margin of safety when the budgeted sales are 9,000 units.

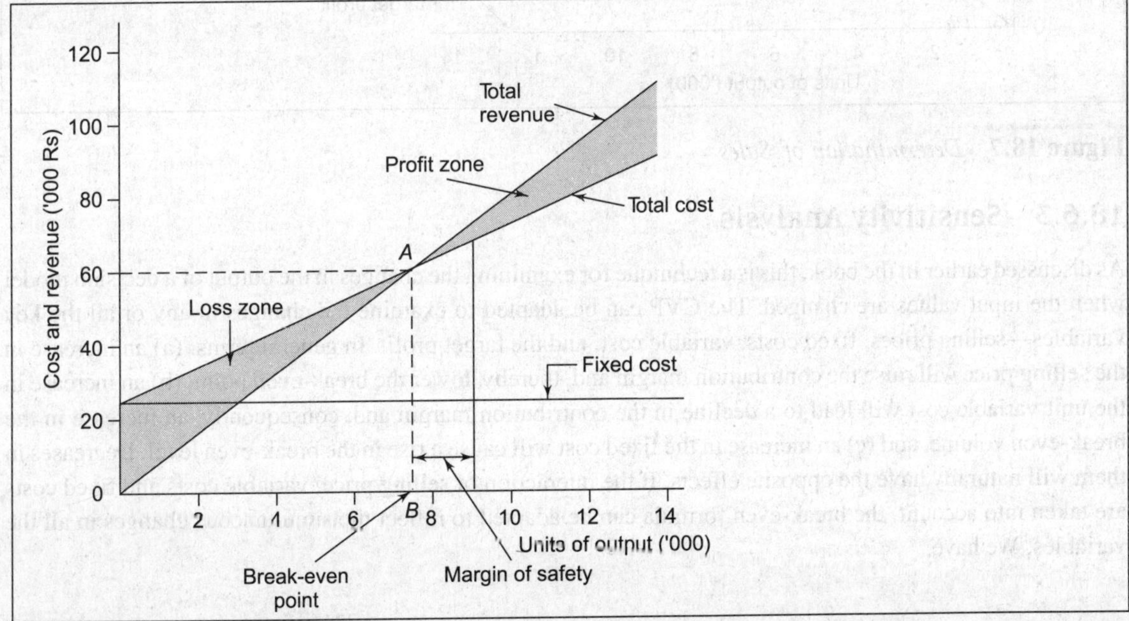

Figure 18.6 *Break-Even Chart*

Called the *break-even chart* of the *cost-volume-profit* graph, it gives us more information than the break-even point and the margin of safety. In Figure 18.6, we notice the shaded profit and loss zones. Below *B*, the total cost function lies above the total revenue function indicating that all activity levels in the range *OB* are unprofitable. On the other hand, any activity level, larger than given by *B* would yield profit since in that region the total revenue function lies above the total cost line. In any case, the larger the deviation from the *BEP*, in either case, the greater the loss or profit (as the case may be) involved.

The cost-volume-profit graph can be adapted to represent a target profit instead of a break-even condition. In respect of the example under consideration, the target profit of Rs 16,000 is added to the total cost function and shown in Figure 18.7. It intersects the total revenue function at point *k*. The sales corresponding to this point is observed to be 12,500 units (or Rs 1,00,000). This is the required sales volume.

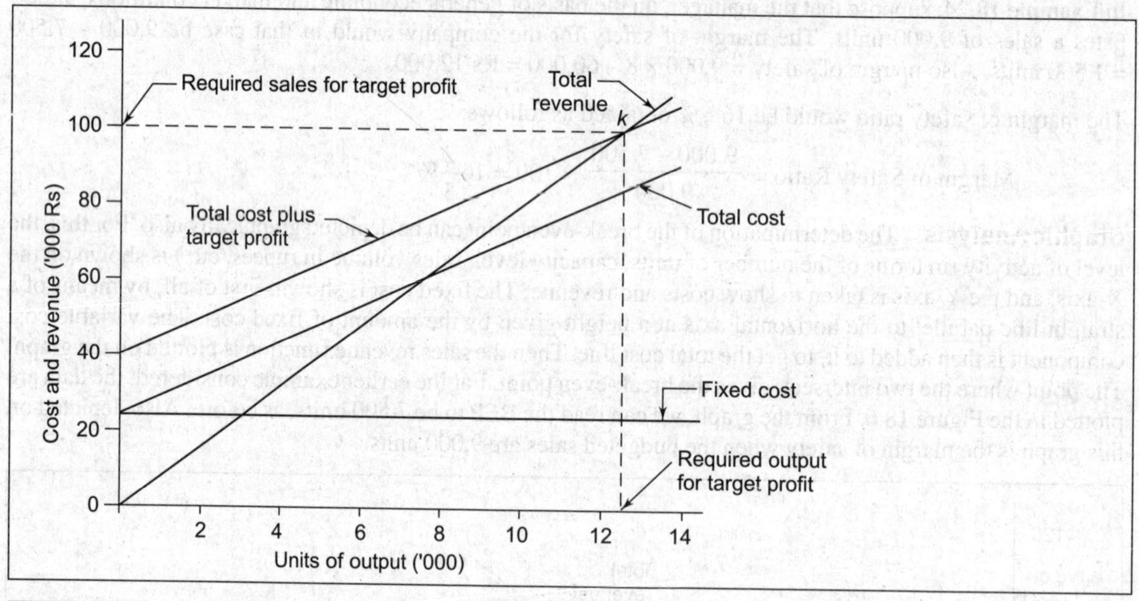

Figure 18.7 *Determination of Sales*

18.6.3 Sensitivity Analysis

As discussed earlier in the book, this is a technique for examining the changes in the output of a decision model when the input values are changed. The CVP can be adapted to examine the changes in any or all the key variables—selling prices, fixed costs, variable cost, and the target profit. In general terms: (a) an increase in the selling price will raise the contribution margin and, thereby, lower the break-even point, (b) an increase in the unit variable cost will lead to a decline in the contribution margin and, consequently, an increase in the break-even volume, and (c) an increase in the fixed cost will cause a rise in the break-even level. Decreases in them will naturally have the opposite effects. If the interaction of selling price, variable costs and fixed costs are taken into account, the break-even formula can be adapted to reflect the simultaneous changes in all the variables. We have,

$$\text{Break-Even Sales (New)} = \frac{F \pm \Delta F}{1 - \dfrac{V \pm \Delta V}{S \pm \Delta S}}$$

where Δ stands for changes.

Example 18.25 Considering the data given in Example 18.24, recalculate the break-even sales in each of the following situations:

(a) when the selling price increases by 25%,

(b) when the variable cost decreases to Rs 4 per unit,

(c) when the fixed cost increases by Rs 6,000, and

(d) when all the three (a)–(c) occur.

Given $S = 8$, $V = 4.80$, $F = 24{,}000$, $\Delta S = +\,25\%$ of $8 = 2$, $\Delta V = -\,0.80$, $\Delta F = +\,6{,}000$.

$$\text{(a) } BES = \frac{F}{CM\text{ (New)}} \times S\text{ (New)} = \frac{24{,}000}{10 - 4.80} \times 10 = \text{Rs } 46{,}154$$

$$\text{(b) } BES = \frac{F}{CM\text{ (New)}} \times S = \frac{24{,}000}{8 - 4} \times 8 = \text{Rs } 48{,}000$$

$$\text{(c) } BES = \frac{\text{(New) } F}{CM} \times S = \frac{30{,}000}{8 - 4.80} \times 8 = \text{Rs } 75{,}000$$

$$\text{(d) } BES = \frac{24{,}000 + 6{,}000}{1 - \dfrac{4.80 - 0.80}{(8 + 2)}} = \frac{30{,}000}{6} \times 10 = \text{Rs } 50{,}000.$$

18.6.4 Volume Profit Graph

We have considered earlier how the required sales for a certain targeted profit may be obtained. This information can be had much more easily using a volume-profit graph, also called *profit graph* sometimes. For this, profit/loss is shown vertically on a graph paper and the level of activity (volume of output) is shown horizontally. Then, using the given information on prices, variable cost and fixed cost, the profit at any two levels of output may be calculated. These profit (or loss) values are plotted on the graph corresponding to the output levels chosen. The points so obtained are joined by a straight line which is extended on both sides. This gives the profit line. The amount of profit or loss corresponding to any output level can be read directly from this graph.

To illustrate, from the data given in Example 18.24, we have $F = 24{,}000$, $V =$ Rs 4.80 per unit and $S =$ Rs 8 per unit. Also, the profit function is: $P = X(S - V) - F$.

Taking $X = 1{,}000$ and $10{,}000$, (say), we get

$$P(1{,}000) = 1{,}000(8 - 4.80) - 24{,}000 = -20{,}800$$

$$P(10{,}000) = 10{,}000(8 - 4.80) - 24{,}000 = 8{,}000$$

These points are plotted on the graph shown in Figure 18.8, and joined together and extended to get the profit line. The point where the profit line intersects the horizontal line, gives the break-even point.

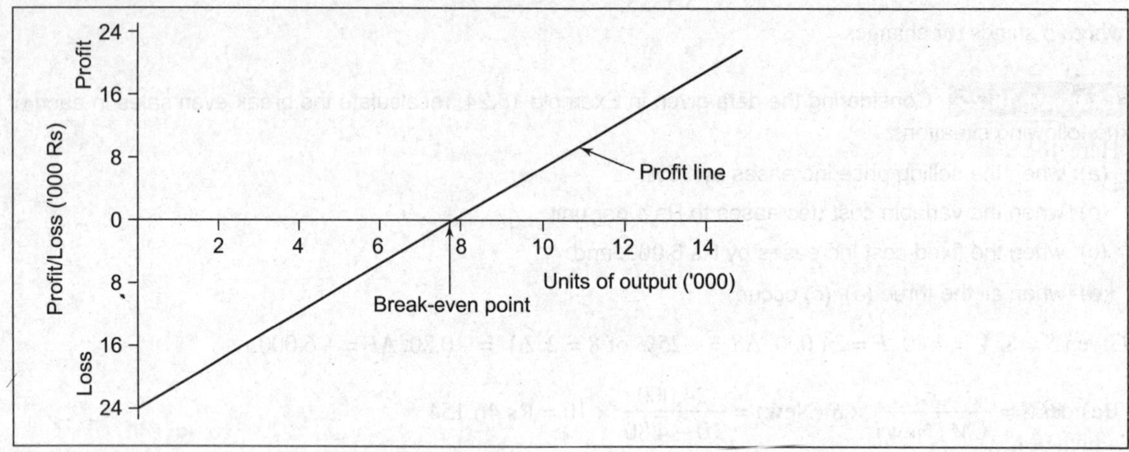

Figure 18.8 *Volume-Profit Graph*

18.6.5 Break-even Analysis for the Multiproduct Case

In our discussions so far, we have implicitly assumed that the firm in question deals only in a single product. We now consider how the break-even sales can be determined for a firm that is engaged in multiple products, each of which has its own contribution margin and any changes in the contribution margin of a product shall be reflected in the total contribution margin. In order to assess the impact of the product mix on the overall profit rates, an idea of the break-even sales is desirable.

For a given product mix, the contribution margin ratio for a multi-product firm equals the weighted summation of the individual contribution margin ratios. The overall contribution margin ratio is given by the following:

$$CM \text{ Ratio} = \frac{\sum_{i=1}^{n}[q_i(S_i - V_i)]}{\sum_{i=1}^{n} S_i q_i}$$

where n = number of different products,

q_i = estimated sales volume for the ith product,

S_i = unit selling price for the ith product,

V_i = unit variable cost for the ith product.

From this,

$$\text{Break-Even Sales} = \frac{\text{Fixed Cost}}{CM \text{ Ratio}}$$

Example 18.26 Determine the break-even sales in the following case:

		Product		
		A	B	C
Sales (units)		5,000	6,000	4,000
Unit selling price	(Rs)	10	15	18
Unit variable cost	(Rs)	6	4	13
Fixed cost	(Rs) 4,000			

Here, the overall contribution margin ratio is 50%, as calculated here.

$$CM\text{ Ratio} = \frac{5{,}000(10-6)+6{,}000(15-4)+4{,}000(18-13)}{(5{,}000\times 10)+(6{,}000\times 15)+(4{,}000\times 18)} = \frac{1{,}06{,}000}{2{,}12{,}000} = 50\%$$

Thus,

$$BES = \frac{\text{Fixed cost}}{CM\text{ Ratio}} = \frac{40{,}000}{50\%} = \text{Rs } 80{,}000$$

Besides obtaining the break-even point in the multi-product situations, calculations can be made to show which of the alternate product-mixes would be most profitable.

Example 18.27 A manufacturer has three products with the following data:

Product	*Price*	*Variable Cost/Unit*	*Per cent of Sales*
Book cases	Rs 60	Rs 40	30
Tables	Rs 100	Rs 60	20
Chairs	Rs 200	Rs 120	50
Total fixed costs	= Rs 75,000		
Sales	= Rs 2,50,000		

The manufacturer also has the following alternative with the following data:

Product	*Price*	*Variable Cost/Unit*	*Per cent of Sales*
Book cases	Rs 60	Rs 40	44
Beds	Rs 216	Rs 180	16
Chairs	Rs 200	Rs 110	40
Total fixed costs	= Rs 75,000		
Sales	= Rs 2,70,000		

Which of the two alternatives would you recommend and Why? (*CA, November, 1981*)

First alternative: Total sales = Rs 2,50,000

Product	*Sales (Rs)*	*Quantity sold*
Book cases	30% of 2,50,000 = 75,000	75,000 ÷ 60 = 1,250
Tables	20% of 2,50,000 = 50,000	50,000 ÷ 100 = 500
Chairs	50% of 2,50,000 = 1,25,000	1,25,000 ÷ 200 = 625

The profit is determined as follows:

Total Sales						Rs 2,50,000
Total Variable Costs:						
Book cases	1,250	×	40	=	30,000	
Tables	500	×	60	=	30,000	
Chairs	625	×	120	=	75,000	1,55,000
					Contribution Margin	95,000
					less Fixed Cost	75,000
					Profit =	20,000

Second alternative: Total Sales = 2,70,000

Product	*Sales (Rs)*					*Quantity Sold*		
Book cases	44% of 2,70,000	=	1,18,800	1,18,800	÷	60	=	1,980
Beds	16% of 2,70,000	=	43,200	43,200	÷	216	=	200
Chairs	40% of 2,70,000	=	1,08,000	1,08,000	÷	200	=	540

The profit is determined as follows:

Total Sales						Rs 2,70,000
Total Variable Costs:						
Book cases	1,980	×	40	=	79,200	
Beds	200	×	180	=	36,000	
Chairs	540	×	110	=	59,400	1,74,600
					Contribution Margin	95,400
					less Fixed Cost	75,000
					Profit =	20,400

The second alternative is preferable as it entails greater amount of profit.

REVIEW ILLUSTRATIONS

Example 18.28 (a) A freehold estate is bought for Rs 2,750. At what rent should it be let so that the owner may receive 4% pa, compounded annually, on his purchase money?

(b) A man borrows Rs 500 at 5% rate of interest compounded annually. If the principal and the interest are to be repaid in 10 equal instalments, what should be the amount of each instalment?

(c) An investor has a capital of Rs 20,000 on which he earns interest @ 5% pa. If he spends Rs 1,800 per year, show that he will be ruined of his capital before the end of 17th year.

(d) Determine the present value of a perpetuity of Rs 100 payable at the end of the 1st year, Rs 200 at the end of the 2nd year, and Rs 300 at the end of the 3rd year, and so on, increasing by Rs 100 payable at the end of each subsequent year. Assume a time preference rate of 5% pa compounded annually.

(e) A pipeline is due for repairs. It will cost Rs 10,000 and last for 3 years. Alternatively, a new pipeline can be laid at a cost of Rs 30,000 and lasts for 10 years. Assuming cost of capital to be 10% and ignoring salvage value, which alternative should be chosen? *(CA, May, 1981)*

(a) For a freehold estate, the rent is the perpetual annuity. We know that, for a perpetuity of Rs A per year, with a rate of interest i, we have the present value, $V = A/i$. Thus, $A = Vi$.

For the given problem, $V =$ Rs 2,750 and $i = 0.04$

$\therefore$ Rent, $A = 2{,}750 \times 0.04 =$ Rs 110.

(b) Here we are given a problem of annuity for 10 years, carrying an interest rate of 5% and has a present value of Rs 5,000. To determine A, we proceed as follows.

We have, $$V = \frac{A(1 - R^{-n})}{R - 1}$$

From Table B4, the present value factor for annuity when $i = 5\%$ and $n = 10$ is 7.7217. Thus, $5{,}000 = A \times 7.7217$.

$\therefore$ $A = 5{,}000/7.7217 =$ Rs 647.53.

(c) The given problem can be viewed as an annuity of Rs 1,800 per annum @ 5% with a period of 17 years. If the present value of this annuity can be shown to be greater than Rs 20,000, the statement that he will be ruined of his capital before the end of the 17th year shall be proved.

From the Table B4, Appendix B, the PV factor for annuity for $n = 17$ and $i = 5\%$ is obtained as 11.2741. We have,

$$V = A \times PV \text{ factor for annuity} = 1{,}800 \times 11.2741 = \text{Rs } 20{,}293.38$$

It may be noted that for $n = 16$, this factor is 10.8378, so that

$$V = 1{,}800 \times 10.8378 = \text{Rs } 19{,}508.04$$

Since for $n = 16$, $V < 20{,}000$ while for $n = 17$, $V > 20{,}000$, it shows that the person shall be ruined in the 17th year, and, hence, the statement is correct.

(d) The information can be presented as:

$$V = \frac{100}{(1+0.05)^1} + \frac{200}{(1+0.05)^2} + \frac{300}{(1+0.05)^3} + \frac{400}{(1+0.05)^4} + \ldots$$

The series on the RHS is an arithmetico-geometric series. We can rewrite it as follows:

$$\frac{V}{100} = \frac{1}{(1.05)^1} + \frac{2}{(1.05)^2} + \frac{3}{(1.05)^3} + \frac{4}{(1.05)^4} + \ldots$$

If we let $(1.05)^{-1} = x$, and denote the sum of the series by S, we get,

$$S = x + 2x^2 + 3x^3 + 4x^4 + \ldots,$$

and $$Sx = x^2 + 2x^3 + 3x^4 \ldots$$

By subtraction of the second equation from the first,

$$S - Sx = x + x^2 + x^3 + x^4 + \ldots$$

or $$S(1 - x) = x(1 + x + x^2 + x^3 + \ldots)$$

or $$S(1 - x) = x/1 - x$$

$\therefore$ $$S = \frac{x}{(1 - x)^2}$$

Since we have put $S = V/100$, we have

$$V = \frac{100(x)}{(1-x)^2}$$

Substituting the value of x, we get

$$V = \frac{100(1.05)^{-1}}{[1-(1.05)^{-1}]^2} = \frac{100 \times 1.05}{0.05^2} = \text{Rs } 42{,}000$$

(e) The choice between the two alternatives can be made by comparing the present value of the future costs associated with each one. We have,

$$V = A + \frac{A}{(1+i)^1} + \frac{A}{(1+i)^2} + \ldots = \frac{A}{1-(1+i)^{-n}}$$

Here $i = 0.10$, and $n = 3$ for the first alternative while we have $n = 10$ for the second one. Thus, for the alternative of existing pipeline,

$$V = \frac{10{,}000}{1-(1+0.10)^{-3}} = \frac{10{,}000}{1-0.7513} = \text{Rs } 40{,}209$$

For the alternative of new pipeline,

$$V = \frac{30{,}000}{1-(1+0.10)^{-10}} = \frac{30{,}000}{1-0.3855} = \text{Rs } 48{,}820$$

Since the cost for the first alternative is lower, the existing pipeline is preferable.

Example 18.29 Companies are required to pay income-tax on book profits which are affected by the depreciation policy adopted by them. Suppose that a company has bought a machine for Rs 1,20,000, to be depreciated over a period of 5 years. The tax authorities allow either of the following three methods of providing for depreciation: (a) straight line, (b) sum of years' digits, and (c) double declining balance. Since the balance of the asset never becomes zero under the method (c), a company following it is allowed to switch to the method (a) in any year before the project life ends.

Determine the amount of depreciation in different years, under each of the methods given above and state which method would yield the best results, assuming that the management considers an 8% rate as a reasonable return on this type of investment.

It may be recalled that the calculation of depreciation under different methods proceeds as follows:

If an amount A is spent on a machine having a life of n years, then the depreciation in the jth year, D_j would be:

Straight Line Method: $D_j = A/n$,

Sum of Years' Digits Method: $D_j = \frac{n-j+1}{n(n+1)} 2A$

Double Declining Balance Method: $D_j = \frac{2A}{n}\left[1-\frac{2}{n}\right]^{j-1}$

The depreciation, according to each of the methods, for different years is given in Table 18.4. For the third method, three sets of calculations are shown. They indicate the depreciation amounts to be charged when the change to the straight line method is made in third, fourth, and fifth year respectively.

TABLE 18.4 Determination of Depreciation

Year	*Method a*	*Method b*	*Method c*		
			k = 3	*k = 4*	*k = 5*
1	24,000	40,000	48,000	48,000	48,000
2	24,000	32,000	28,800	28,800	28,800
3	24,000	24,000	14,400	17,280	17,280
4	24,000	16,000	14,400	12,960	10,368
5	24,000	8,000	14,400	12,960	15,552
Total	1,20,000	1,20,000	1,20,000	1,20,000	1,20,000

Since the rate of interest is given to be 8%, we shall determine the present value of each of the streams of depreciation. The stream with the highest present value would be the best as the idea is to adopt such a depreciation policy as will give the maximum tax advantage.

The present value of each of the series would be obtained as follows:

Method a: The PV factor for annuity with $i = 8\%$ and $n = 5$ is 3.9927.

Thus, present value of depreciation = 24,000 × 3.9927 = Rs 95,825.

Method b: For this we have,

$$\frac{40{,}000}{(1.08)^1} + \frac{32{,}000}{(1.08)^2} + \frac{24{,}000}{(1.08)^3} + \frac{16{,}000}{(1.08)^4} + \frac{8{,}000}{(1.08)^5} = \text{Rs } 1{,}00{,}726$$

Similarly, computations for method c yield the following results:

Present Value for Method c, $(k = 3)$ = Rs 1,00,949

Present Value for Method c, $(k = 4)$ = Rs 1,01,196

Present Value for Method c, $(k = 5)$ = Rs 101,056

The present value of depreciation under method $c(k = 4)$ being the largest, the optimal depreciation policy is the double declining balance method with changeover to the straight line method in the fourth year.

Example 18.30 An instrument transformer manufacturer in an industrial town is considering the purchase of an ultrasonic welding machine WELDAL, to replace an existing manually operated machine. The existing machine cost Rs 12,000 two years ago and has been depreciated down to book value of Rs 10,000, using a 12-year life and no salvage. However, the machine can be sold in the market at present for Rs 4,000 only. The ultrasonic welder would improve the product quality enough to boost revenue from a present Rs 80,000 to Rs 1,00,000 per year. It would cost Rs 44,000 and have a ten-year life. Any salvage value on it would be consumed in the removal expenses. Another advantage of the ultrasonic machine is that by reducing annual labour costs, it would cut operating expenses from Rs 8,000 to Rs 3,000 annually. The manufacturer is in a 50% tax bracket and estimates the firm's cost of capital at 12%. Use present value analysis to determine whether the manufacturer should buy the new machine.

To decide as to whether the existing machine should be replaced by the new one, we shall compute the NPV of the cash flows. For this, we shall first calculate the net cash outflow at the time when the new machine is purchased (i.e. $t = 0$). This will be followed by the calculation of the net inflows of the life of the new machine (i.e. $t = 1$ through 10).

	Amount before tax	Effect (Net of taxes)	Time of Occurrence	PV factor @ 12%	Present Value
Cost of new machine	44,000	44,000	0	1.0000	44,000
Salvage value of old machine	(4,000)	(4,000)	0	1.0000	(4,000)
Tax effect of the sale of old machine	(6,000)	(3,000)	0	1.0000	(3,000)
Loss × t = [10,000 – 4,000]0.5					
			Total Present Value of Outflows =		37,000
Decrease in operating costs	5,000	2,500	1–10	5.6502	14,125.50
[Cost reduction = Decrease × (1 – t)]					
Increase in Revenue	20,000	10,000	1–10	5.6502	56,502.00
Depreciation on new machine	4,400				
Depreciation on old machine	1,000				
Net change in tax savings from depreciation	3,400	1,700	1–10	5.6502	9,605.34
			Total Present Value of Inflows =		80,232.84

Net Present Value = Inflows – Outflows = 80,232.84 – 37,000 = Rs 43,232.84.

Since the NPV is greater than zero, the new machine should be purchased to replace the old one.

Example 18.31 A project requires an initial cash outlay of Rs 60,000 and offers an annual expected cash inflow of Rs 30,000 for three years, and has no salvage value. The risk coefficients for the three years are estimated to be 0.90, 0.80 and 0.75 respectively. The risk-free rate of interest is estimated to be 10%. Compute the NPV of the project.

The calculation of NPV is shown in Table 18.5.

TABLE 18.5 Determination of Net Present Value

Year, t	Cash Flow C_t	Risk Coefficient α	Certainty Equivalent	PV Factor @ 10%	Present Value
1	30,000	0.90	27,000	0.9091	24,545.70
2	30,000	0.80	24,000	0.8264	19,833.60
3	30,000	0.75	22,500	0.7513	16,904.25
					61,283.55
			less Cash Outflow		60,000.00
			Net Present Value =		1,283.55

Example 18.32 For a project the relevant data on investment and projected cash flows are presented here:

Year	*Mean Cash Flow*	*Standard Deviation of Cash Flow Distribution (Rs)*
0	–40,000	7,000
1	30,000	4,000
2	30,000	3,000

Assume that the cash flows are independently and normally distributed and the discount rate is 0.20. Compute:

(a) the probability of the investment having a net present value of zero or less.

(b) the expected value of loss.

Area under normal curve

z	:	0.68	0.69	0.70	0.71	0.72	0.73	0.74	0.75
Area	:	0.2518	0.2549	0.2580	0.2611	0.2642	0.2673	0.2704	0.2734

Unit normal loss integral

z	:	0.68	0.69	0.70	0.71	0.72	0.73	0.74	0.75
Loss Integral	:	0.1478	0.1453	0.1429	0.1405	0.1381	0.1358	0.1334	0.1312

(CA, November, 1986)

We shall first determine the expected value of the NPV distribution and its standard deviation.

Mean:

Year	*Cash Flow (Rs)*	*PV Factor @ 10%*	*Present Value (Rs)*
0	– 40,000	1.0000	– 40,000
1	30,000	0.8333	24,999
2	30,000	0.6944	20,832
		Expected NPV =	5,831

Standard Deviation:

Year t	*Standard Deviation* σ_t	*Variance* σ^2_t	*PV factor,* $(1 + i)^{-2t}$ $(1.20)^{-2t}$	*Present Value*
0	7,000	49×10^6	1.0000	49.0000×10^6
1	4,000	16×10^6	0.6944	11.1104×10^6
2	3,000	9×10^6	0.4823	4.3407×10^6
			$\sigma^2 =$	64.4511×10^6

$$\text{Standard Deviation} = \sqrt{\text{Variance}} = \sqrt{64.4511 \times 10^6} = 8{,}028.14$$

To determine the probability of the investment having an NPV of zero or less, we shall determine the area to the left of $X = 0$, under the normal curve with $\mu = 5{,}831$ and $\sigma = 8{,}028.14$. We have,

$$z = \frac{X - \mu}{\sigma} = \frac{0 - 5,831}{8,028.14} = -0.73$$

From the given information, area between mean and $z = 0.73$ is 0.2673. Thus, $P(X \leq 0) = 0.5000 - 0.2673 = 0.2327$.

Also,
$$D = \frac{0 - \mu}{\sigma} = \frac{0 - 5,831}{8,028.14} = -0.73$$

Since $L_N(0.73) = 0.1358$ (given), we have,

$$\text{EVPI} = \sigma L_N(D) = 8,028.14 \times 0.1358 = \text{Rs } 1,090.22.$$

Example 18.33 You are given the information on cash flows about a project to introduce a new product:

Year	Source	Expected Value of Net Cash Flow (In thousand Rs)	Standard Deviation (In thousand Rs)
0	Initial Investment	(700)	–
1	Production Cash Outflow	(200)	20
2	Production Cash Outflow	(200)	10
3	Production Cash Outflow	(250)	10
4	Production Cash Outflow	(250)	10
5	Production Cash Outflow	(100)	10
1	Marketing	400	50
2	Marketing	700	100
3	Marketing	600	100
4	Marketing	600	100
5	Marketing	500	100

Assume all the outflows to be independent and the marketing flows to be perfectly correlated. Taking 10% as the risk-free rate, determine the expected value of the NPV for the proposal and its standard deviation.

The *NPV* of the distribution can be obtained as follows:

Year, t	*Net Cash Flow, C_t*	*PV Factor @ 10%*	*Present Value*
1	400 – 200 = 200	0.9091	181.820
2	700 – 200 = 500	0.8264	413.200
3	600 – 250 = 350	0.7513	262.955
4	600 – 250 = 350	0.6830	239.050
5	500 – 100 = 400	0.6209	248.360
			1,345.385
		less Initial Investment	700.000
		NPV	645.385

For this project in which some the cash flows are independent while others are perfectly correlated, the variance can be obtained using the following rule:

$$V(\text{NPV}) = \left(\sum_{t=0}^{n} \frac{V(C_t')}{(1+i)^{2t}}\right) + \left(\sum_{t=0}^{n} \frac{V(C_t^*)}{(1+i)^{t}}\right)^2$$

Accordingly,

$$V(\text{NPV}) = \frac{20^2}{(1.10)^2} + \frac{10^2}{(1.10)^4} + \frac{10^2}{(1.10)^6} + \frac{10^2}{(1.10)^8} + \frac{10^2}{(1.10)^{10}}$$

$$+ \left[\frac{50}{(1.10)} + \frac{100}{(1.10)^2} + \frac{100}{(1.10)^3} + \frac{100}{(1.10)^4} + \frac{100}{(1.10)^5}\right]^2$$

$$= 1{,}11{,}839.47$$

From this, standard deviation, $\sigma = \sqrt{1{,}11{,}839.47}$ = Rs 334.42 (thousand)

Example 18.34 Two mutually exclusive investment proposals are being considered. The following information is available:

	Project A	*Project B*
Cost	Rs 20,000	Rs 20,000
Life	2 years	2 years
Cash flow each year	Rs 12,000	Rs 12,000
Salvage value	0	0

Upon further analysis it was found that the probability of cash inflow each year, for the projects is as follows:

Project A		*Project B*	
Possible inflow	*Probability*	*Possible inflow*	*Probability*
10,000	0.2	11,000	0.2
12,000	0.6	12,000	0.6
14,000	0.2	13,000	0.2

Advise for selecting the proposal, assuming cost of capital 10%. *(CA, May, 1981)*

We shall first calculate the mean and variance of cash flows for each of the projects.

	Inflow, C_{ti}	*Prob.* p_{ti}	$C_{ti}p_{ti}$	$(C_{ti} - \overline{C}_t)$	$(C_{ti} - \overline{C})^2 p_{ti}$
Project *A*:	10,000	0.2	2,000	2,000	0.8×10^6
	12,000	0.6	7,200	0	0
	14,000	0.2	2,800	2,000	0.8×10^6
		$\overline{C}$ =	12,000	σ_1^2 =	1.6×10^6

	Inflow, C_{ti}	*Prob.* p_{ti}	$C_{ti}p_{ti}$	$(C_{ti} - \overline{C}_t)$	$(C_{ti} - \overline{C})^2 p_{ti}$
Project *B*:	11,000	0.2	2,200	1,000	0.2×10^6
	12,000	0.6	7,200	0	0
	13,000	0.2	2,600	1,000	0.2×10^6
		$\overline{C}$ =	12,000	$\sigma_1^2 =$	0.4×10^6

Calculation of NPV for each of the projects is the same because of identical cash flows. This is equal to Rs 826.

Year	*Cash flow*	*PV factor @ 10%*	*Present value*
1	12,000	0.9091	10,909.2
2	12,000	0.8264	9,916.8
			20,826.0
		less Cost	20,000.0
		NPV =	826.0

The values of variances for each project's cash flows are calculated as follows:

Year	*Present Value Factor* $(1 + i)^{-2t}$	*Project A* Variance	*Project A* Present Value	*Project B* Variance	*Project B* Present Value
1	0.8264	1.6×10^6	1.32224×10^6	0.4×10^6	0.33056×10^6
2	0.6830	1.6×10^6	1.09280×10^6	0.4×10^6	0.27320×10^6
		$\sigma_A^2 =$	2.41504×10^6	$\sigma_B^2 =$	0.60376×10^6

$$\sigma_A = \sqrt{2.4150 \times 10^6} = \text{Rs } 1{,}554 \text{ app.}$$

$$\sigma_B = \sqrt{0.60376 \times 10^6} = \text{Rs } 777 \text{ app.}$$

To determine the relative risk degree of the two projects, we shall compare their coefficients of variation. We have,

$$\text{Coefficient of Variation} = \frac{\text{Standard Deviation}}{\text{Expected Value}} \times 100$$

Thus, $$CV_A = \frac{1{,}554}{826} \times 100 = 188.1\%, \text{ and } CV_B = \frac{777}{826} \times 100 = 94.1\%.$$

The project *B*, is, therefore, preferable.

Example 18.35 A company has invested Rs 300 in a new machinery with expected cash flow over two years as follows:

Year 1		Year 2	
Cash Flow	*Probability*	*Cash Flow*	*Probability*
Rs 200	0.4	Rs 100	0.7
		Rs 200	0.3
Rs 300	0.6	Rs 200	0.4
		Rs 300	0.6

Find expected NPV and its SD, assuming cost of capital as 20%. *(CA, May, 1987)*

The given information is depicted by means of a decision tree in Figure 18.9.

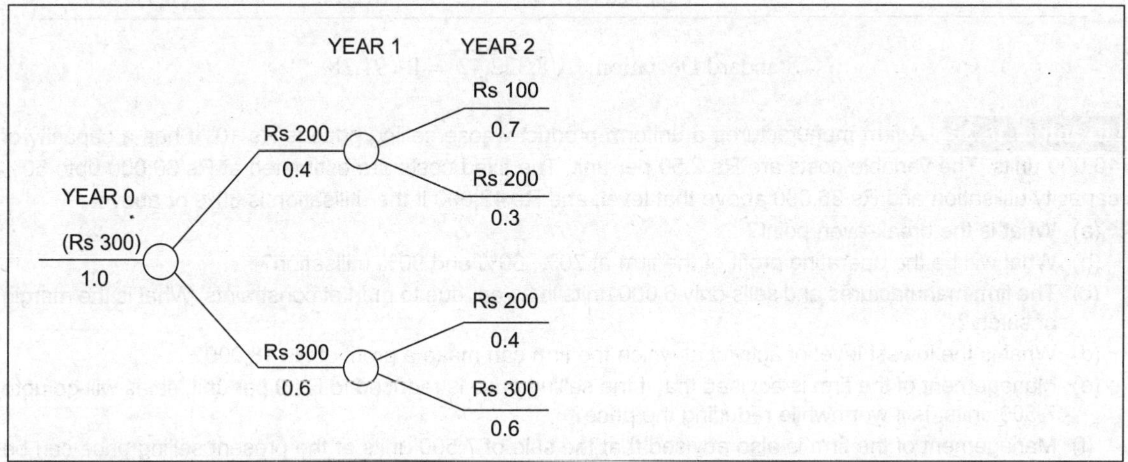

Figure 18.9 *Decision Tree*

It may be noticed from the figure that four streams of cash inflows are possible: Rs 200 in the first year followed by Rs 100 in the second year; Rs 200 in the first year followed by an equal amount in the second year ... and so on. The NPV of the first stream is a negative Rs 63.9, as calculated here:

Year	*Cash flow*	*PV factor @ 20%*	*Present value (Rs)*
1	200	0.8333	166.66
2	100	0.6944	69.44
			236.10
		less Cash outflow	300.00
		NPV =	– 63.90

In a similar way, the net present value of each of the other three streams is calculated and presented here. From these, we observe that the expected NPV of the project is Rs 61.09, obtained by summing up the products of the NPVs of the different streams and the corresponding joint probabilities (equal to products of probabilities on each branch). The standard deviation of the distribution works out to be Rs 91.28, also shown below.

Stream	*Net Present Value, NPV_i*	*Joint Probability p_i*	*Expected Value $NPV_i p_i$*	$(NPV_i - \overline{NPV})^2 p_i$
1	–63.90	0.28	–17.8920	4,374.52
2	5.54	0.12	0.6648	370.34
3	88.87	0.24	21.3288	185.17
4	158.31	0.36	56.9916	3,402.39
		Expected NPV =	61.0932	8,332.42

$$\text{Standard Deviation} = \sqrt{8,332.42} = \text{Rs } 91.28.$$

Example 18.36 A firm manufactures a uniform product whose selling price is Rs 10. It has a capacity of 10,000 units. The variable costs are Rs 2.50 per unit. The fixed costs are estimated at Rs 30,000 upto 50% capacity utilisation and Rs 36,000 above that level, and Rs 42,000 if the utilisation is 80% or above.

(a) What is the break-even point?
(b) What will be the operating profit of the firm at 70%, 80% and 90% utilisation?
(c) The firm manufactures and sells only 6,000 units in a year due to market constraints. What is the margin of safety?
(d) What is the lowest level of activity at which the firm can make a profit of Rs 18,000?
(e) Management of the firm is advised that if the selling price is reduced to Rs 9 per unit, sales will go upto 7,500 units. Is it worthwhile reducing the price?
(f) Management of the firm is also advised that the sale of 7,500 units at the present selling price can be achieved if a sum of Rs 5,000 is spent on advertisement.

Which alternative is more profitable—reducing the selling price or spending on advertisement?

(SAS, August, 1980)

(a) With S = Rs 10, V = Rs 2.50, and F = Rs 30,000, we have CM = 10 – 2.50 = 7.50 per unit, and

$$BEP = \frac{F}{CM} = \frac{30,000}{7.50} = 4,000 \text{ units, i.e. } 40\% \text{ capacity utilisation.}$$

(b) The operating profit at different activity levels is given in Table 18.6.

TABLE 18.6 Determination of Operating Profit

Particulars	*Capacity Utilisation*		
	70%	*80%*	*90%*
Production and Sales in units	7,000	8,000	9,000
Sales @ Rs 10 per unit	Rs 70,000	Rs 80,000	Rs 90,000
Variable Cost @ Rs 2.50 per unit	17,500	20,000	22,500
Contribution Margin	52,500	60,000	67,500
less Fixed Cost	36,000	42,000	42,000
Operating Profit	16,500	18,000	25,500

(c) Margin of Safety = Actual Sales – Break-even Sales
$$= 6{,}000 \times 10 - 4{,}000 \times 10 = \text{Rs } 20{,}000$$

(d) For a profit of Rs 18,000, the sales required are calculated first using the minimum fixed cost amount:

$$\text{Sales} = \frac{F + P}{CM} = \frac{30{,}000 + 18{,}000}{7.5} = 6{,}400 \text{ units,}$$

This implies that the capacity level is 64%. Since the capacity level is greater than 50%, the fixed cost appropriate to this level is Rs 36,000.

Thus, $$\text{Sales} = \frac{36{,}000 + 18{,}000}{7.50} = 7{,}200 \text{ units.}$$

Corresponding to this the capacity utilisation level is 72%. This is the lowest level of activity at which the firm can make a profit of Rs 18,000.

(e) When selling price is reduced to Rs 9, we have

$$CM \text{ per unit} = 9 - 2.50 = \text{Rs } 6.50$$
$$\text{Fixed Cost} = \text{Rs } 36{,}000 \text{ (for 7,500 units)}$$

Now we can determine the profit as follows:

$$7{,}500 = \frac{36{,}000 + \text{Profit}}{6.50}$$

or $$\text{Profit} = 7{,}500 \times 6.50 - 36{,}000 = \text{Rs } 12{,}750.$$

The profit corresponding to the current output of 6,000 is Rs 9,000.

$$6{,}000 = \frac{36{,}000 + \text{Profit}}{7.5}$$

or $$\text{Profit} = 6{,}000 \times 7.5 - 36{,}000 = \text{Rs } 9{,}000.$$

It is thus worthwhile to reduce the price to Rs 9.

(f) The profit when Rs 5,000 would be spent on advertisement, is calculated here:

$$7{,}500 = \frac{36{,}000 + 5{,}000 + \text{Profit}}{7.5}$$
$$\text{Profit} = 7{,}500 \times 7.5 - 36{,}000 - 5{,}000 = \text{Rs } 15{,}250.$$

Of the two alternatives—price reduction and advertisement spending—the latter one is preferable because the profit associated with it is higher.

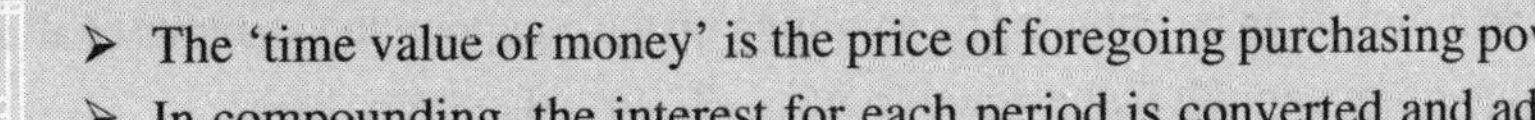

KEY POINTS TO REMEMBER

- The 'time value of money' is the price of foregoing purchasing power currently.
- In compounding, the interest for each period is converted and added to the principal before interest is calculated for the next period.
- The accumulation factor is meant to find the compounded values while the present value factor is used for discounting purposes.
- Annuities are the equally-spaced periodical payments—*ordinary*, if payments are made at the period ends; *due*, if payments are made at the beginnings; and *deferred* if payments begin after some time. A perpetuity is an annuity forever.

- Amount of annuity is $A(R^n - 1)/R - 1$ and present value is $A(1 - R^{-n})/R - 1$. For a perpetuity, the present value is equal to $1/r$.
- Methods of measuring investment worth include (a) payback period, (b) average rate of return (ARR), (c) net present value (NPV), and (d) internal rate of return (IRR).
- Payback period is the time period required to recover original investment.

  ARR is the average annual profits expressed as a percentage of the average investment.

  NPV is the excess (shortage) of the present value of inflows over present value of outflows. If NPV > 0, the project is acceptable.

  IRR is the rate of return which equates the outflows with the inflows. If in excess of the required rate, the project is accepted.
- Risk may be incorporated into capital budgeting decisions mainly by certainty equivalent approach, risk-adjusted discount rates or through statistical distributions approach, which considers mean and standard deviation of the NPV distribution. Decision-trees are also an important aid in studying the risk. An alternative method of considering risky investments is simulation.
- Break-even analysis, also known as cost-volume-profit analysis, deals with inter-relationships between total revenue, cost and profit.
- Break-even sales, which involves no-profit no-loss, is obtained by the amount of fixed cost divided into P/V ratio.
- In a multi-product case, the break-even point is calculated by dividing the amount of fixed cost into weighted contribution margin.

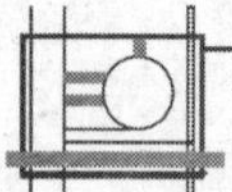

TEST YOUR UNDERSTANDING

Mark the following statements as T (*True*) or F (*False*).

1. Time value of money is the bridge connecting the present and future amounts of money.
2. Compounding can be with any periodicity—annual, semi-annual, quarterly, monthly and so forth.
3. Annuities are defined as payments made only at the end of every year.
4. *Annuity certain* refers to an annuity without default.
5. Perpetuity is a series of periodical payments which never terminates.
6. Deferred annuity is an annuity contingent upon the happening of some event.
7. The amount of an annuity can be obtained by summing up the compounded amounts of all the payments.
8. The present value of an annuity is the equal to $A(1 - R^{-n})/r$, where $R = 1 - r$.
9. The (final) amount of a perpetuity is infinity.
10. If $r = 0.05$, then a perpetuity would be equal to 20 years' purchase.
11. The criteria for judging the acceptability of the investment proposals include (i) 'the bigger the better' principle, and (ii) 'the bird in hand' principle.
12. The payback period of a proposal is the time period required to recover the original investment made.
13. The payback method ignores 'the bigger the better' principle.

14. The NPV of a project is the aggregate of present value of all the inflows resulting from it.
15. When using the NPV approach, choice between mutually exclusive projects should be made after due consideration of their lives.
16. The internal rate of return is the rate at which an investment is repaid by proceeds from a project.
17. Both the NPV and IRR approaches yield identical accept–reject decisions and rankings of the given set of investment projects.
18. IRR is the rate of return which makes NPV = 0.
19. It is possible to obtain multiple rates of return for a series of cash flows relating to a project.
20. The certainty-equivalent approach calls for adjusting the numerator and denominator of the equation of the present value.
21. The risk-adjusted discount rate is obtained by adjusting the risk-free rate of return upward by a suitable risk premium.
22. In statistical distribution approach of considering risky investment proposals, the net cash flow from an investment in each period is viewed as a random variable.
23. When cash flows are independent, the variance of the NPV distribution is obtained as the summation of the discounted variances of various periods.
24. For a normally distributed NPV, the EVPI is obtained by finding the product of standard deviation and the normal loss integral.
25. Decision-tree approach to analysis of risky proposals is suitable when a large number of possible outcomes are involved.
26. The technique of simulation calls for a reasonable assessment of the probability distributions of the various factors determining the net present value of a project.
27. Sensitivity analysis deals with determining how are NPV and IRR likely to be affected if the inputs are changed by a certain measure.
28. An essential requirement of the BEP analysis is that all costs should be classified as being either fixed or variable.
29. The P/V ratio is defined as the ratio of sales to contribution margin.
30. Margin of safety is the excess of break-even sales over actual sales.
31. Margin of safety ratio is the margin of safety expressed as a percentage of actual sales.
32. It is possible to determine the required sales for earning a target level of profit from the break-even chart.
33. Other things being equal, an increase in the selling price would raise the break-even point.
34. The formula for calculating the break-even point can be adapted to reflect simultaneous changes in the selling price, variable cost and fixed cost.
35. Division of fixed cost *plus* desired profit by P/V ratio gives the required volume of sales necessary to earn such amount of profit.

EXERCISES

1. 'In financial decision-making, time is of a paramount importance'. Discuss, giving a detailed account of the concept of time value of money.

2. What is an annuity? Distinguish between various types of annuities on the basis of (a) their terms, and (b) the payment times.
3. Derive expressions for obtaining (a) the amount of an annuity, (b) the present value of an annuity, and (c) the present value of a deferred annuity.
4. What is a perpetuity? How can we determine the present value of a perpetuity at a given interest rate?
5. What is investment analysis? Why is it of great significance to a firm? Write a note on *OR* and investment decision. *(CA, May, 1978)*
6. Discuss the various methods of evaluating investment proposals.
7. Discuss the discounted cash flow method and its role in investment decision.
8. How do you differentiate between *NPV* and *IRR* methods of evaluating investment proposals? Do they always lead to same conclusion with respect to (a) accept/reject decision for different proposals, and (b) ranking of various proposals that are mutually exclusive in nature?
9. Critically examine the Net Present Value and Internal Rate of Return methods used in appraising capital projects. Also discuss a few simple approaches for incorporating risk factor in such decision-making. *(M Com, Delhi, 1984)*
10. What do you mean by risk in investment proposals? How can it be measured?
11. Discuss the certainty-equivalent and risk-adjusted discount rate methods of incorporating risk into capital budgeting process. Which of the two is superior? Why?
12. Explain the statistical distribution approach to evaluate risky proposals. How would the variance be obtained when the cash flows of the successive time periods are (a) perfectly correlated, and (b) independent in nature?
13. How can the *EVPI* of an investment proposal be obtained?
14. Discuss the decision-tree and simulation approaches to the consideration of risky investment proposals. Illustrate each of these.
15. What do you understand by decision-tree analysis? Give its role for investment decisions.
16. What is sensitivity analysis in the context of investment decisions? Explain.
17. What is break-even analysis? How is it useful? How can the break-even point for the multi-product firm be calculated?
18. What is *P*/*V* ratio? What does it measure and how is it useful?
19. Explain the concept of 'margin of safety'. How can the margin of safety be measured?

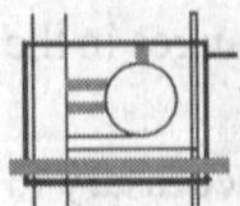

PRACTICAL PROBLEMS

1. A company has three offers for its existing equipment in one of the divisions. The first buyer is willing to pay Rs 50,000 at the end of 8 years' period. The second buyer offers Rs 39,000—consisting of an immediate payment of Rs 14,000 and Rs 25,000 after 6 years. The third buyer agrees to buy the equipment for Rs 29,000 payable right away. Which is the best offer for the company if it can earn an interest @ 8% per annum on the money received?
2. Determine the amount of an annuity of Rs 120 payable at the end of each year for 20 years from now, if money is worth 5% per annum.

3. Mr K has just retired and received a sum of Rs 3 lakhs as retirement benefits. He wants to invest so that the amount may be repaid to him in 10 equal annual instalments. An insurance company offers an annuity at 8% interest rate. Determine the amount of annuity receivable by K if he invests with the insurance company.
4. A housewife is looking at ways of producing domestic hot water and considers two possibilities—an electrical immersion heater having an installation cost of Rs 160 and estimated annual electrical charges of Rs 200, and a gas boiler with an installation cost of Rs 760 with annual fuel bills of Rs 80. Assuming yourself as a consultant to this cost-conscious housewife, advise her suitably by comparing the two systems, on the basis of (i) total expenditure, and (ii) present value, over a 5 years period. Take interest at 9%.

   What will be your recommendation if you consider both the equipments for a 8-year period?

   (CA, May, 1993)
5. Mr Green buys a colour TV set for which he has to make a down payment of Rs 2,000 and Rs 500 as monthly instalment for a total of 30 months. What is the cash down price of the TV set if the money is 12% compounded monthly?
6. A building owner gets two offers from a prospective buyer:
   (a) Rs 5 lakhs payable now.
   (b) Rs 6,20,000 in five equal instalments, payable at the year-ends, the first of the payments to be made 3 years from now.

   If the time value of money is reckoned to be 8%, which of the two offers is more attractive?
7. An employee of a factory takes a loan of an amount of Rs 75,000 from Our-Own-Society. It is to be repaid in yearly instalments of Rs 3,750. If the rate of interest charged is 3.5%, how many payments does he have to make?
8. A company wishes to establish a sinking fund to replace a machine by a new one at the end of 8 years. The payments to the sinking refund shall be made at the year-ends, the first one being at the end of this year. How much payment should the company make if it wishes to have an amount of Rs 12,00,000 by the end of the 8-year period and the time value of money is 8% pa?
9. A person is invited to quote the annual rent for a freehold property he has just bought for Rs 30 lakhs. What should the quotation be if the money's worth is 10% pa?
10. (a) If a perpetuity is worth 20 years' purchase, find the annuity of Rs 1,000 to continue for 2 years.
    (b) If 20 year' purchase must be paid for an annuity to continue k years and 25 years' purchase for an annuity to continue for $2k$ years, find the interest rate.
    (c) Determine the present worth of perpetual annuity of Rs 100 payable at the end of the first year, Rs 200 at the end of the second, Rs 300 at the end of the third year and so on (increasing by Rs 100 each year), when the rate of interest is given to be 5% pa.
11. Anil Corporation provides a multitude of benefits to its employees, including a company funded pension plan. Biren is due to retire in five years and will receive annual pension cheques starting one year from the date of his retirement. Each year's pension cheque will equal one-half of his present annual salary of Rs 12,000. The company will make annual deposits to a Pension Fund Trust during the next 5 years. Assume that the pension fund's assets are expected to earn 10% annually and that Biren is expected to collect payments for 10 years after retirement. What should be the annual contribution of the corporation to the pension fund? *(CA, November, 1977)*

12. The SOFTOUCH Company is engaged in manufacturing software equipment for computers. The company can either build its own factory building for Rs 20 lakhs or rent an equivalent building with a 20-year lease for an annual rental of Rs 2 lakhs, payable at year-ends, together with an option to buy the building for Rs 5 lakhs at the expiry of the lease. The owner of that building wishes to retain a small portion of the building for running his small jewellery shop for which he agrees to grant a concession of the Rs 2,000 per month during the lease time, and he agrees to surrender the shop after the lease period. Should the company construct its own building or rent the facility? For your calculations, take an 8% rate of discount.

13. A company can acquire a new machine at a cost of Rs 72,000 to replace a manual process. The annual net cash flow saving would be Rs 18,000 for a period of 7 years at the end of which there would be no salvage value of the machine.
    (a) Calculate the net present value using a discount rate of 15%.
    (b) Determine the IRR of the project.
    (c) Should the company accept the proposal? Why?

14. The Bluebird Corporation is considering a capital investment of Rs 10,00,000 in a project that is expected to yield the following cash flows:

Years	:	1	2	3	4 through 7	8
Cash inflow ('000 Rs)	:	160	190	170	150	200

If the required rate of return is 14%, determine (a) NPV of the project, and (b) IRR of the project. Is the proposal an acceptable one?

15. The Omega Company is considering two alternative capital investment proposals (a) renovating the existing machine at an initial cash outlay of Rs 25,000, or (b) replacing the existing machine with a new machine costing Rs 70,000. The renovation of the existing machine would result in a saving of Rs 8,000 per annum in each of the next 6 years while the new machine, being more efficient, would save Rs 22,000 per annum in the six-yearly period. If the required rate of return be 16%, what is the net present value of each project?
Which proposal would be acceptable to the company?

16. Agfa p.l.c. is considering purchasing for Rs 50 thousand a machine that will reduce cash outlays by Rs 10,500 each year for the next seven years. At the end of the 7-year period, the machine will be worthless. What rate of discount will equate the purchase price of the machine and the cost savings for the 7 years? If the required rate of return is 12%, should the company buy the machine?

17. A company is considering the following two mutually exclusive projects:
Project X : Initial cost Rs 2,00,000; Rs 30,000 per year after-tax inflows for 10 years
Project Y : Initial cost Rs 2,00,000; Rs 20,000 per year after-tax inflows for 20 ycars
Compute the net present value for each project assuming the required rate of return equal 5%. Which of the two is better?

18. A company is considering two mutually exclusive projects. The initial investment required for both the projects is Rs 10,00,000 each. The projects have a life of 5 years and straight line method of depreciation is used. The required rate of return is 15% and the company pays 50% tax. The cash inflows before depreciation and taxation are:

Year	*Project A*	*Project B*
1	Rs 7,00,000	Rs 8,00,000
2	8,00,000	6,00,000
3	8,00,000	10,00,000
4	9,00,000	7,00,000
5	6,00,000	6,00,000

Which project is acceptable under NPV and IRR methods?

19. The Risk-Less Company uses the certainty-equivalent approach to adjust for the project risk. It is considering a project which requires an initial outlay of Rs 64,000 with a useful life of 5 years, and no salvage value thereafter. Thc expected cash flows and their certainty equivalent coefficients are as follows:

Year	:	1	2	3	4	5
Cash inflow ('000 Rs)	:	18	20	21	22	12
C.E. Coefficient	:	0.95	0.90	0.85	0.85	0.70

Using a discount rate of 12%, compute the risk-adjusted net present value of the investment project.

20. A firm is considering the introduction of a new product which will have a life of 5 years. Two alternative methods of promoting the product have been identified:

Alternative 1: This will involve employing a large number of agents. An immediate expenditure of Rs 5,00,000 will be required to advertise the product. This will produce net annual cash inflows of Rs 3,00,000 at the end of each of the subsequent five years. However, the agents will have to be paid Rs 50,000 each year. On termination of the contract, the agents will have to be paid a lump sum of Rs 1,00,000 at the end of the fifth year.

Alternative 2: Under this alternative, the firm will not employ agents but will sell directly to the consumers. The initial expenditure on advertising will be Rs 2,50,000. This will bring in cash inflows at the end of each year of Rs 1,50,000. However, this alternative will involve out-of-pocket costs for sales administration to the extent of Rs 50,000. The firm also proposes to allocate fixed costs worth Rs 20,000 per year to this product if this alternative is pursued.

(a) Advise the management as to the method of promotion to be adopted.

(b) Calculate the internal rate of return for alternative 2. You may assume that the firm's cost of capital is 20% pa. *(CA, May, 1986)*

21. A company is presently considering an investment proposal involving Rs 25,000. It is found that probabilities for various net cash inflows expected to be generated by the proposal for three-year period are as follows:

Year 1		*Year 2*		*Year 3*	
Cash flow	*Prob.*	*Cash flow*	*Prob.*	*Cash flow*	*Prob.*
11,000	0.3	11,000	0.4	11,000	0.2
12,000	0.1	12,000	0.2	12,000	0.3
13,000	0.2	13,000	0.3	13,000	0.4
14,000	0.4	14,000	0.1	14,000	0.1

Compute expected monetary value EMV for each. Determine also (a) the expected net value of the investment proposal, assuming a discount rate of 10%; (b) risk associated with the project as measured by the standard deviation.

22. The All Good Company is considering two mutually exclusive capital budgeting projects. Each of these projects has a life of two years, and costs Rs 80,000. The cash flow data for the two projects are as follows:

Project P_1

Year 1:	Cash flow (Rs)	Prob.	Year 2:	Cash flow (Rs)	Prob.
	45,000	0.35		50,000	0.25
	50,000	0.40		60,000	0.50
	60,000	0.25		70,000	0.25

Project P_2

Year 1:	Cash flow (Rs)	Prob.	Year 2:	Cash flow (Rs)	Prob.
	(2,000)	0.20		10,000	0.10
	40,000	0.30		45,000	0.30
	50,000	0.30		65,000	0.35
	72,000	0.20		80,000	0.25

(a) Calculate the expected value of the annual cash flow for each of the projects.

(b) Using variability of the cash flows as the criterion, calculate the risk-adjusted NPV of each project. Use a 14% discount rate for the more risky project and a 12% discount rate for the less risky project.

(c) Determine the internal rate of return for each of the projects.

23. Obtain expected DCF and the associated risk as measured by standard deviation for the project *A*, if the cost of capital equals 10%, given probability distribution of cash flows as:

Year 1		*Year 2*		*Year 3*	
Cash flow	*Prob.*	*Cash flow*	*Prob.*	*Cash flow*	*Prob.*
1,000	0.10	1,900	0.20	1,500	0.10
1,500	0.20	2,500	0.30	2,250	0.70
2,000	0.40	2,750	0.20	2,500	0.10
2,500	0.20	3,150	0.30	3,000	0.10
3,000	0.10				

24. The possible outcomes (in thousand rupees) for three alternative investment proposals and their probabilities of occurrence are given here:

Alternative A_1		*Alternative A_2*		*Alternative A_3*	
Outcome	*Prob.*	*Outcome*	*Prob.*	*Outcome*	*Prob.*
125	0.2	225	0.3	200	0.4
200	0.4	400	0.5	500	0.5
300	0.4	500	0.2	1,000	0.1

Rank the three alternatives in terms of (i) expected value of outcome, and (ii) risk.

25. Consider the alternative investment proposals 1 and 2. The cash flow estimates for these projects are:

Project 1		*Project 2*	
Annual cash flow (Rs)	*Prob.*	*Annual cash flow (Rs)*	*Prob.*
20,000	0.10	30,000	0.20
30,000	0.20	50,000	0.30
60,000	0.40	70,000	0.40
90,000	0.20	90,000	0.10
1,00,000	0.10		

The initial cost of each of the proposals is Rs 2,00,000, while the cash flow estimates are valid for a period of six years in respect of each of them. Determine expected NPV for each of the projects and its standard deviation. Use a discount rate of 10 per cent.

26. An investment proposal involves an outlay of Rs 2,20,000. The proposal is expected to have a life of 3 years only, and is likely to generate the following cash inflows at the year-ends.

Year 1		*Year 2*		*Year 3*	
Cash inflow (Rs)	*Prob.*	*Cash inflow (Rs)*	*Prob.*	*Cash inflow (Rs)*	*Prob.*
50,000	0.2	70,000	0.2	1,40,000	0.3
80,000	0.7	1,00,000	0.5	1,50,000	0.3
1,20,000	0.1	1,30,000	0.3	1,62,500	0.4

(a) Determine the expected cash flow for each of the 3 years and its standard deviation.

(b) Using a 12% discount rate, calculate the expected net present value of the project.

(c) Determine the risk associated with the project as measured by the standard deviation, assuming that the cash flows of the various years are independent of each other.

(d) Assuming the NPV is distributed normally, determine the probability that the project shall have a negative NPV.

(e) Obtain the EVPI of the project.

27. An investment proposal had the following discrete probability distributions of expected cash flows:

State (j)	Year 1		Year 2		Year 3	
	Return	Prob.	Return	Prob.	Return	Prob.
1	50	0.10	20	0.10	– 40	0.10
2	60	0.20	40	0.25	30	0.30
3	70	0.40	60	0.30	50	0.30
4	80	0.20	80	0.25	80	0.20
5	90	0.10	100	0.10	140	0.10

Give (a) the expected return each year and net present worth of the total expected return, (b) standard deviation of the expected returns in each year; and also standard deviation of present value of expected returns. (Discount rate = 10%). *(CA, May, 1978)*

28. **A company has developed the following discrete probability distribution of cash flows for a particular capital budgeting proposal:**

Year 1		Year 2		Year 3	
Cash flow	Prob.	Cash flow	Prob.	Cash flow	Prob.
100	0.10	300	0.20	500	0.10
200	0.20	400	0.40	600	0.30
300	0.30	500	0.30	700	0.30
400	0.40	600	0.10	1,000	0.30

Assume that the probability distributions of the flows for different years are perfectly correlated over time; the initial cost of the project is 1,000; and the appropriate discount rate is 10%. Now determine,

(a) the mean and standard deviation of the cash flows for each year.

(b) the expected NPV of the project.

(c) the probability that the project shall have an NPV greater than zero, assuming that the NPV distribution follows 'normal' pattern.

(d) the probability that the project would succeed, assuming that instead of being perfectly correlated, the successive flows are independent of each other.

(e) the EVPI for the project under the assumption of (i) perfectly correlated, and (ii) independent cash flows.

29. **Assuming a discount rate of 0.10, and the following projected cash flows, determine (a) the expected NPV of the investment proposal, (b) the standard deviation of the present value distribution, and (c) the probability of the project to have a net present value lying between zero and Rs 10,000.**

Year	Expected cash flow	Standard deviation
0	– 20,000	
1	12,000	5,000
2	12,000	4,000
3	10,000	5,000

Consider the flows to be independent.

30. The Fair Fax Company is attempting to decide whether or not to invest in a project that requires an initial outlay of Rs 3.4 lakhs. The cash flows of the project are known to be made up of two parts, one of which varies independently over time and the other one which displays perfect positive correlation. The cash flows for the six-year life of the project are:

Year	'Perfectly correlated' component		'Independent' component	
	Mean	Standard deviation	Mean	Standard deviation
1	Rs 40,000	Rs 4,400	Rs 42,000	Rs 4,000
2	50,000	4,500	50,000	4,400
3	48,000	3,000	50,000	4,800
4	48,000	3,200	50,000	4,000
5	55,000	4,000	52,000	4,000
6	60,000	4,000	52,000	3,600

Find out the expected value of the NPV and its standard deviation, using a discount rate of 15%. Also find the probability that the project will be successful. What is the value of its EVPI?

31. Determine expected value of perfect information (EVPI) in respect of a project regarding which the following information is available:

Initial outlay : Rs 1,00,000
Life : 4 years
Nature of cash flows: Perfectly correlated
Cash flow amounts :

Year	Expected cash flow	Standard deviation
1	Rs 36,000	Rs 3,200
2	42,000	3,600
3	50,000	3,600
4	45,000	3,300

Discount rate : 0.10

32. The management of Superfast Transport Company is considering whether to buy or lease an additional small truck it needs in order to extend its service. An analysis of costs and potential demand yields the following estimates of net cash flows (NCF) for each of the first two years.

Year	Lease		Buy	
	Net cash flow (Rs)	Prob.	Net cash flow (Rs)	Prob.
1	– 5,000	0.25	– 10,000	0.20
	5,000	0.40	0	0.50
	15,000	0.35	10,000	0.30
2	5,000	0.30	10,000	0.30
	10,000	0.50	15,000	0.50
	15,000	0.20	20,000	0.20

The two-year time horizon is considered by the management because it is decided that the truck would be replaced in 2 years whether it is leased or purchased. Lease, once taken, cannot be cancelled before 2 years. Also, if purchased, the truck would have a salvage value of Rs 5,000, which is included in the above figures of net cash flows. Assume that appropriate rate of discounting be 15% and there are no taxes.

(a) Using decision-tree analysis, find the expected NPV of each alternative.

(b) Which of the two proposals be accepted by the management?

33. The manager of Good-Morning Stores is considering opening a new store in the rapidly expanding NOIDA. Two sizes of the store are under consideration: Regular and Large. The initial costs, expected demand situation and profits along with their probabilities are:

Store Size	Initial costs ('000 Rs)	Demand situation	Year 1 Profit ('000)	Year 1 Prob.	Year 2 Profit ('000)	Year 2 Prob.
(a) Regular	160	Low	60	0.2	100	0.2
		Medium	100	0.5	160	0.4
		High	150	0.3	200	0.4
(b) Large	200	Low	50	0.6	120	0.4
		Medium	120	0.3	180	0.3
		High	200	0.1	220	0.3

Assuming a discount rate of 10%, you are required to calculate expected value of NPV for each proposal and its standard deviation. Use decision-tree approach.

34. Calculate the break-even point and the profit made in the following case:
The fixed costs for the year are Rs 96,000; variable cost per unit of the single product being made is Rs 4 while each unit sells for Rs 20. Estimated sales for the year are valued at Rs 2,00,000. The number of units involved coincides with the expected volume of output.

35. Two businesses, *AB* Ltd. and *CD* Ltd. sell the same type of product in the same market. Their budgeted profit and loss accounts for the year ending 30th June 20XX are as follows:

		AB Ltd.		CD Ltd.	
Sale			Rs 1,50,000		Rs 1,50,000
Less Cost					
	Variable	1,20,000		1,00,000	
	Fixed	15,000	1,35,000	35,000	1,35,000
Net budgeted profit			15,000		15,000

Now, (a) calculate the break-even point of each business, and

(b) state which business is likely to earn greater profits in conditions of (i) heavy demand for the product, (ii) low demand for the product.

36. Caravan Company sold 1,00,000 units of its product at Rs 20 per unit. Variable costs are Rs 14 per unit (manufacturing costs of Rs 11 and selling costs of Rs 3). Fixed costs are incurred uniformly throughout the year and amount to Rs 7,92,000 (manufacturing costs of Rs 5,00,000 and selling costs of Rs 2,92,000). There are no inventories at the beginning or at the end of the accounting period.
With this information, you are required to determine (i) break-even level for this product, (ii) the number of units that must be sold to earn an income of Rs 60,000 (before income-tax), (iii) the number of units that must be sold to earn an after-tax profit of Rs 90,000 (assuming a tax-rate of 50 per cent), and (iv) the break-even point for this product after a 10 per cent increase in the wages and salaries (assuming wages and salaries are 50 per cent of variable costs and 20 per cent of fixed costs).

(M Com., Delhi, 1986)

37. A firm is engaged in manufacturing four products: P_1, P_2, P_3, and P_4. The data on the product-mix, selling prices and variable costs are:

	Product			
	P_1	P_2	P_3	P_4
Sales (units)	3,800	4,800	6,000	2,000
Selling price per unit (Rs)	20	25	15	40
Variable cost as a percentage of selling price	50	40	80	60

Compute (a) the contribution margin ratio, and (b) the break-even sales. What is the margin of safety at the given sales level?

For making computations, take the fixed cost to be equal to Rs 80,000.

38. The P/V ratio for an item is 40% and the margin of safety is 50%. Find out the net profit and the break-even point if the sales volume is Rs 8,00,000.

39. Aggarwal and Company sold in two successive years 7,000 and 9,000 units and incurred a loss of Rs 10,000 and earned Rs 10,000 as profit, respectively. The selling price per unit is Rs 100. Calculate:
    (a) the amount of fixed cost,
    (b) the number of units to break-even, and
    (c) the number of units to earn a profit of Rs 50,000.

40. (a) Kolkata Company Ltd. manufactures and sells four types of products under the brand names ACE, UTILITY, LUXURY, and SUPREME. The sales mix in value comprises:

Brand	Percentage
ACE	33.33%
UTILITY	41.67%
LUXURY	16.67%
SUPREME	8.33%

The total budgeted sales (100%) are Rs 6,00,000 per month. The operating costs are:

ACE	:	60% of selling price;
UTILITY	:	68% of selling price;
LUXURY	:	80% of selling price; and
SUPREME	:	40% of selling price.

The fixed costs are Rs 1,59,000 per month. Calculate the break-even point for the products on an overall basis.

(b) It has been proposed to change sales mix as follows, the total sales per month remaining Rs 6,00,000.

Brand	Percentage
ACE	25%
UTILITY	40%
LUXURY	30%
SUPREME	5%

Assuming that this proposal is implemented, calculate the new break-even point.

(CA, November, 1986)

CHAPTER

19

Forecasting

CHAPTER OVERVIEW

The only thing certain about the future is that it is uncertain. Innumerable situations call for making forecasts for the future. In fact, forecasting is an integral part of the jobs of many managers. There are several methods by which forecasting may be done. Some of these are fairly simple while others are relatively more complex. This chapter gives a brief account of the forecasting methods, which are categorised as being qualitative and quantitative. The quantitative models are statistical in nature, which entails either an analysis of the time-based data or the cross-sectional data. In case of time-based data, the variable of interest is forecasted in the future on the basis of the past behaviour, while in the cross-sectional data, the forecasted variable is thought to be dependent on one or more other variables.

The contents of this chapter will help a manager to find answers to questions like the following.

- What are the qualitative models of forecasting?
- What are the components of a time series?
- How can moving average and exponential smoothing be used for the purposes of forecasting? How is the appropriate value of smoothing constant chosen?
- How can exponential smoothing be adjusted to account for trend?
- How is trend equation obtained and used for making forecasts? Further, how are seasonal variations taken into consideration for the purpose of making forecasts?
- What are causal models of forecasting? How can simple and multiple regression analyses be differentiated?
- How can sales be predicted when it is known to be dependent on number of salesmen employed and the amount spent on advertisement, and we are given the number of sales people employed and a certain amount spent as advertisement?
- How to calculate the beta of a given company's shares by relating movements in its prices to movements in a stock price index?

Most of the material included in this chapter is statistical in nature, using ideas of time series and regression analysis. Knowledge of these topics would facilitate the understanding of the contents. Besides, some algebraic manipulations are included.

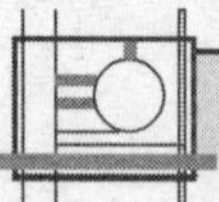

19.1 INTRODUCTION

Forecasting is a basic tool to help managerial decision-making. Managerial decisions are seldom made in the absence of some form of forecasting. Everyday, managers have to take decisions in the face of uncertainty, without knowing what would happen in the future. For example, keeping inventory without knowing what the sales will be, buying equipment and machinery having no knowledge about the demand for the products, and making investments in shares by a fund manager when no one knows what profits are going to result. Managers strive to reduce this uncertainty and make better estimates of what is likely to happen in the future. This is what forecasting aims to accomplish. Thus, whether it is inventory control, marketing strategy formulation, financial planning, production planning or any other such area of operations, managers have to employ the tool of forecasting. While forecasting can be in relation to anything, our focus here is on demand forecasting.

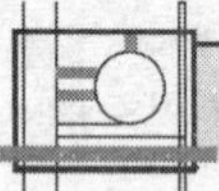

19.2 FORECASTING MODELS

Forecasts can be obtained using a variety of techniques/models. The forecasting models discussed here fall into one of the two categories: qualitative models and quantitative models, as shown in Figure 19.1.

The qualitative models use personal judgement and involve qualities like intuition and experience as the bases of forecasts, and are subjective by their very nature. On the other hand, quantitative models are objective in nature and they employ numerical information as the basis of making forecasts. Quantitative models include time series models and causal models. The *time series models* attempt to predict the future values using the historical data. In such models, the demand forecast is done on the basis of the past demand values. They hold

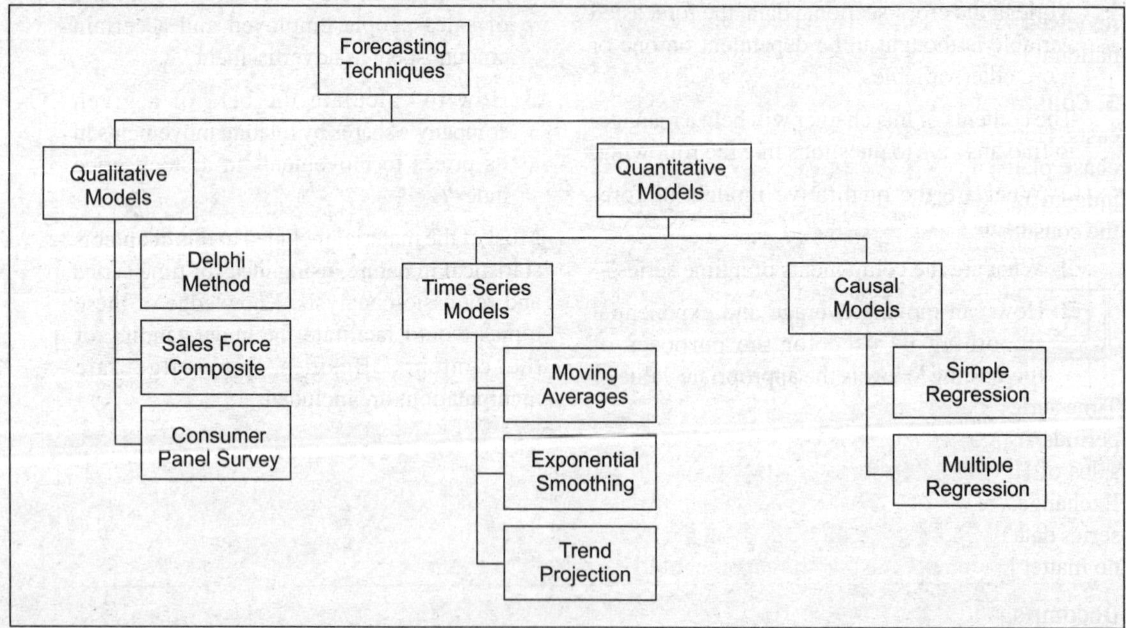

Figure 19.1 *Forecasting Techniques*

the future to be a continuum of the past and are based on the premise that what would happen in the future is a function of what has happened in the past. They, therefore, involve predicting for future in keeping with the past behaviour. On the other hand, the *casual models* are used where one variable is related to and, therefore, dependent upon, the values of some other variable(s). The demand for a product may, for example, be seen to be dependent upon the price charged, amount spent on advertisement, incomes of the people, and so on. The analysis is done in terms of what is called as regression analysis.

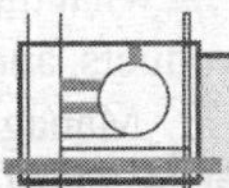

19.3 QUALITATIVE MODELS OF FORECASTING

Some of the qualitative methods of forecasting are discussed below:

1. Delphi Method Perhaps the most important method of forecasting involving judgements is called the *Delphi method*. The method is an iterative group process and it employs a group of experts, not an oracle(!), to obtain forecasts. These experts are usually not known to each other and their interaction takes place through a coordinator. The other participants in the Delphi process are the staff and respondents. The staff personnel assist the experts by preparing, distributing, collecting, and summarising and tabulating the data, while the respondents are the subjects whose judgements are being sought and are valued. In this method, the coordinator obtains the forecasts from the experts and then assimilates each of their forecasts. Each of the experts whose estimate is not included in the middle 50 per cent of the forecasts is asked to explain the analysis underlying the forecast. The coordinator disseminates to each expert a summary of all the forecasts and of the explanations of forecasts that lie outside the middle 50 per cent. Based on this additional information, each of the experts may revise his/her original forecast. After repeating this process appropriately, the experts tend to reach a consensus forecast.

2. Sales Force Composite In this method of demand forecasting, all of the members comprising sales force of a company, are asked to estimate the likely sales in their respective areas. The estimates are then reviewed to ensure that they are realistic. After this, the estimates are combined at the district, regional and national level to obtain the overall forecast.

3. Consumer Panel Survey Some marketing research methods employ consumer panels for making forecasts. Here, a consumer panel is maintained and consumers on such a panel are questioned about their purchase plans. The goal here is to forecast demand for products and services on the basis of the subjective judgements of the subjects involved, about possible purchases. This method works on a strong assumption that the consumers on the panel are truly representative of the ultimate purchasers.

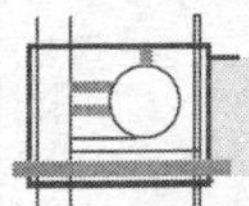

19.4 TIME SERIES MODELS OF FORECASTING

Time series models use time-based data. A time series is a collection of readings belonging to different time periods, usually equally spaced, which may be months, weeks or years, of some economic variable. It may be sales of HP personal computers, production of steel, per capita income, price of a share on National Stock Exchange, and so on. Thus, it is a series in which the data are put in a chronological order. Forecasting time series data implies that predictions about the future values are made only from the past data, and other factors, no matter how important they might potentially be, are not considered.

Decomposition of a Time Series A time series is the result of a number of movements which are caused by numerous economic, political, natural and other factors. The analysis of the time series means decomposing the past data into components and then projecting them forward. A time series typically has four components, though on occasions only one or two of these may eclipse the others.

(i) Secular Trend

Over a long period, a time series will have an overall tendency either to move upwards or downwards or sometimes a blend thereof, though the actual movement will not be regular. This tendency of the time series is known as the secular trend. For example, a glance at the sales of a popular soft drink manufacturer is likely to reveal an increasing trend.

(ii) Seasonal Variations

The fluctuations that occur periodically—the movements recurring within a definite period, may be every week or month or quarter—with reasonably high degree of predictability are called seasonal variations. For a soft drink manufacturer, while the yearly sales may be on the increase, the sales are likely to be relatively high every summer and low every winter.

(iii) Cyclical Movements

They are caused by business cycles or trade cycles. These are oscillatory movements—a series of repeated sequences—superimposed on the original data. These movements are of longer duration than a year. The sales of a company, for example, may be high because the level of economic activity may be high. Similarly, the sales may be low due to overall subdued economic activity.

(iv) Random Movements

Random movements are residual, or erratic, movements that do not have any set pattern and are usually caused by some unpredictable reason, like wild-cat strike, earthquake, fire, and so on.

The time series models deal first with projections of values without decomposition and then relating to trend, also making adjustment for the seasonal variations. The three time series models discussed in this chapter include moving averages, exponential smoothing and trend projections.

19.4.1 Moving Averages

As a tool of forecasting, the method of moving averages attempts to forecast values on the basis of the average of the values of past few periods. Successive values are calculated by considering the new value and dropping the old one. To illustrate, a four-month moving average is obtained by adding the demand during the past four months and dividing the sum by 4. With each passing month, the most recent month's data are added to the previous three month's values, thereby dropping the value of the earliest month.

Example 19.1 The demand for an item is observed for 15 months and recorded below:

Month	Demand	Month	Demand
1	280	9	309
2	288	10	315
3	266	11	320
4	295	12	332
5	302	13	310
6	310	14	308
7	303	15	320
8	328		

Calculate (i) 3-monthly and (ii) 4-monthly moving averages. What is the forecast for month 16 for each one?

The given data are reproduced in Table 19.1. The three-monthly moving averages and four-monthly moving averages are shown in the third and the fourth columns of the table. From the third column, the forecast for month 16 is about 313 units, while a four-monthly moving average, in the fourth column, yields a forecast of about 318 units for this month.

Weighted Moving Averages A careful analysis of the moving average method reveals that the moving average with a base of n periods is in fact an equal-weighted average with a weightage of $1/n$ to each of the preceding n values and a zero weightage to all the previous values. Thus, in a three-monthly forecast, the immediately last three months' values are given a weightage of 1/3 each and the remaining values a weightage of zero. However, one may like to forecast values by giving differential weights to the values entering into moving average calculation. For example, a three-monthly moving average may be calculated by assigning weightage of 3, 2 and 1 respectively for the three months' values. To illustrate the weighted average using demand for months 1, 2 and 3, given in Table 19.1, would be 275.7 as shown below:

$$\text{(Weighted) Moving average} = \frac{266 \times 3 + 288 \times 2 + 280 \times 1}{3 + 2 + 1} = 275.7$$

Evidently, the numerator is the summation of the products of the input values and the respective weights, and the denominator is the summation of the weights used. In effect, weights used for the three months are, respectively, 1/2, 1/3 and 1/6. For a four-month average, the weights would be 4/10, 3/10, 2/10, and 1/10. The weighted three-monthly moving averages, using weights for three months as 3, 2 and 1 are given in the last column of the Table 19.1.

TABLE 19.1 Forecasting Using Moving Averages

Month	*Demand (Y)*	*3-Monthly Moving Average*	*4-Monthly Moving Average*	*3-Monthly Weighted Moving Average*
1	280			
2	288			
3	266			
4	295	278.0		275.7
5	302	283.0	282.3	284.2
6	310	287.7	287.8	293.7
7	303	302.3	293.3	304.8
8	328	305.0	302.5	305.2
9	309	313.7	310.8	316.7
10	315	313.3	312.5	314.3
11	320	317.3	313.8	315.2
12	332	314.7	318.0	316.5
13	310	322.3	319.0	325.2
14	308	320.7	319.3	319.0
15	320	316.7	317.5	312.7
16		312.7	317.5	314.3

19.4.2 Exponential Smoothing

This is another time series forecasting technique where the forecast for the next period is calculated as weighted average of all the previous values. It is based on the premise that the most recent value is the most important for predicting the future value. Also, it presumes that values prior to the current value are also relevant but in declining importance as we go back in time. The weights decline exponentially as we consider the older values.

Symbolically,

$$F_{t+1} = \alpha Y_t + \alpha(1-\alpha)Y_{t-1} + \alpha(1-\alpha)^2 Y_{t-2} + \alpha(1-\alpha)^3 Y_{t-3} + \alpha(1-\alpha)^4 Y_{t-4} + \cdots$$

where F_{t+1} is the forecast for the next period, Y_t is the actual value for the current period, and Y_{t-1}, Y_{t-2}, and so on are the values of the successive preceding periods, and α is a fractional value between zero and one, inclusive. Thus, $0 \le \alpha \le 1$. It is called the *smoothing constant.*

From this expression, two points may be noted:

1. *Weights are declining* Since $\alpha < 1$, successive weights (of preceding time period values) are declining since each weight is obtained by multiplying previous weight by $(1-\alpha)$, which is also a fraction.

2. *Summation of weights is equal to unity* It can be shown that, like in case of moving average method, the summation of weights given to various values in the method of exponential smoothing also equals 1. This is shown below:

$$\begin{aligned}\Sigma w &= \alpha + \alpha(1-\alpha) + \alpha(1-\alpha)^2 + \alpha(1-\alpha)^3 + \cdots \\ &= \alpha[1 + (1-\alpha) + (1-\alpha)^2 + (1-\alpha)^3 + \cdots] \\ &= \alpha\left[\frac{1}{1-1+\alpha}\right] \\ &= 1\end{aligned}$$

[Since the sum of a geometric progression, $1 + r + r^2 + ...$, to infinity is equal to $1/(1-r)$]

Making Forecasts While the calculation of the forecast seems a difficult proposition in terms of the equation given earlier, it can be simplified as shown below:

$$F_{t+1} = \alpha Y_t + \alpha(1-\alpha)Y_{t-1} + \alpha(1-\alpha)^2 Y_{t-2} + \alpha(1-\alpha)^3 Y_{t-3} + \alpha(1-\alpha)^4 Y_{t-4} + \cdots \quad \text{(i)}$$

$$F_t = \alpha Y_{t-1} + \alpha(1-\alpha)Y_{t-2} + \alpha(1-\alpha)^2 Y_{t-3} + \alpha(1-\alpha)^3 Y_{t-4} + \alpha(1-\alpha)^4 Y_{t-5} + \cdots \quad \text{(ii)}$$

Multiplying both sides of the Eq. (ii) by $(1-\alpha)$, we get

$$(1-\alpha)F_t = \alpha(1-\alpha)Y_{t-1} + \alpha(1-\alpha)^2 Y_{t-2} + \alpha(1-\alpha)^3 Y_{t-3} + \alpha(1-\alpha)^4 Y_{t-4} + \alpha(1-\alpha)^5 Y_{t-5} + \cdots \quad \text{(iii)}$$

Subtracting Eq. (iii) from Eq. (i), we have

$$F_{t+1} - (1-\alpha)F_t = \alpha Y_t$$

or

$$\begin{aligned}F_{t+1} &= \alpha Y_t + (1-\alpha)F_t \\ &= F_t + \alpha(Y_t - F_t)\end{aligned}$$

Y_t being the actual value of period t and F_t being the forecasted value for the same period, the difference $(Y_t - F_t)$ represents the error in the previous forecast.

It is evident from the above that the current forecast may be obtained by adjusting the previous forecast and it is equal to the previous forecast plus α times the error margin. Thus, in words,

Current forecast = Last period's forecast + α(Last period's actual value – Last period's forecasted value)

To begin with the process of forecasting, an initial forecast value, if given, is used as the forecast for period 1. If no such value is given, then the first forecast obtained is for period 2, which is taken as the actual value for the period 1. Subsequent forecasts are made on the basis of this forecast, as indicated above.

Exponential smoothing is better than the moving averages method, both conceptually and operationally. On the conceptual front, unlike in case of the moving averages method where forecast value is obtained on the basis of only a few periods' data and ignoring all the previous values, the method of exponential smoothing considers all the past data, as indicated by the expression used for obtaining forecast value. But, while all data are considered for making the forecast, it is possible alternatively to make forecast in current period easily once the previous forecast is available. The existence of this recursive relationship makes the method of exponential smoothing of making forecasts an easily usable one.

The smoothing constant, α, may be altered to change relative weightage of different values entering into the forecast. A high value of the constant would give a high weightage to the recent value while its low value would give a relatively large weightage to the past data. For example, if $\alpha = 0.8$, then the two most recent values would get a weightage of 0.8 and $0.8\ (1 - 0.8) = 0.16$. The new forecast is based practically on the values of the last two periods. Similarly, if $\alpha = 0.5$, then the forecast is based almost entirely on the previous three values. And, in case $\alpha = 0.1$, the forecast places little weightage on the recent value and takes many periods (about 19) of historic values into consideration.

The Choice of Smoothing Constant, α The exponential smoothing approach to forecasting is easy to use but choosing an appropriate value of the smoothing constant, α, is an important matter. This is because the choice of this value can make the difference between an accurate forecast and an inaccurate forecast. In selecting an appropriate value of the smoothing constant, the objective is to obtain the most accurate forecast. The overall accuracy of a forecasting model is determined by comparing the actual values with the forecasted values. The difference between an actual value and a forecasted value is called *forecast error*. It is an error of over-estimation if the forecasted value exceeds the actual value. If the former is lower than the latter, it represents an error of under-estimation.

A measure of overall error of the forecasts made is the *mean absolute deviation*, MAD. It is calculated by dividing the summation of the absolute values of the forecast errors by the number of time periods. Thus,

$$\text{MAD} = \frac{\Sigma\,|\text{forecast error}|}{n}$$

Alternatively,

$$\text{MAD} = \frac{\Sigma\,|Y_t - F_t|}{n}$$

It is significant to note that in calculating MAD, absolute values of the errors are considered. This is because if the signs are not ignored then positive and negative forecasting errors would tend to cancel out each other, thereby giving the erroneous impression of the forecast being good. Since over-estimation and under-estimation are equally unattractive, it is required that they are all averaged to find the mean error. This is achieved by considering the absolute error values. In a given situation, we choose between selected values of the smoothing constant by calculating the MAD of each one of them, selecting the one with the least MAD value.

Example 19.2 The demand for a particular item during the ten months of a year is as given below. The manager is considering how well the exponential smoothing serves as an appropriate technique in forecasting the demand of this item. She is testing three values of the smoothing constant: $\alpha = 0.2$, $\alpha = 0.5$ and $\alpha = 0.8$. You are required to (a) calculate forecasted values using each of the given α values, assuming the initial forecast as 208, and (b) calculate MAD for each of the these series of estimates and suggest which of them is most appropriate.

Month	:	1	2	3	4	5	6	7	8	9	10
Demand	:	213	201	198	207	220	232	210	217	212	225

The forecasts using each of these values are shown calculated in Table 19.2. To illustrate, the forecast for $t = 1$ is taken to be equal to the initial value 208.0. Now, with $Y_1 = 213$ and $\alpha = 0.2$, the forecast for the period 2, $F_2 = 208 + 0.2(213 - 208) = 209$.

For each case, the absolute difference between actual and forecasted values are obtained and averaged to get MAD. Of the three cases, MAD is found to be minimum for $\alpha = 0.2$ and, therefore, this smoothing constant is most preferable of these in the present case.

TABLE 19.2 Calculation of Forecasted Values and Forecast Errors

Month	*Demand*	$\alpha = 0.2$		$\alpha = 0.5$		$\alpha = 0.8$	
t	Y_t	F_t	$\lvert Y_t - F_t \rvert$	F_t	$\lvert Y_t - F_t \rvert$	F_t	$\lvert Y_t - F_t \rvert$
1	213	208.0	5.0	208.0	5.0	208.0	5.0
2	201	209.0	8.0	210.5	9.5	212.0	11.0
3	198	207.4	9.4	205.8	7.8	203.2	5.2
4	207	205.5	1.5	201.9	5.1	199.0	8.0
5	220	205.8	14.2	204.4	15.6	205.4	14.6
6	232	208.7	23.3	212.2	19.8	217.1	14.9
7	210	213.3	3.3	222.1	12.1	229.0	19.0
8	217	212.7	4.3	216.1	0.9	213.8	3.2
9	212	213.5	1.5	216.5	4.5	216.4	4.4
10	225	213.2	11.8	214.3	10.7	212.9	12.1
11		215.6		219.6			
	Total		82.3		91.0		97.4
	MAD		8.23		9.10		9.74

Other Measures In addition to MAD, there are two other measures for judging the accuracy of forccasts: *mean squared error* (MSE), and *mean absolute percentage error* (MAPE). The MSE is the mean of the squared differences between the actual and forecasted values, while MAPE is the mean of the absolute differences between the actual and forecasted values, expressed as a percentage of the actual values.

Example 19.3 Calculate mean squared error (MSE), and mean absolute percentage error (MAPE) for the forecasts obtained in Example 19.2 using $\alpha = 0.5$.

These measures are shown calculated in Table 19.3.

TABLE 19.3 Calculation of MSE and MAPE

Month t	Demand Y_t	Forecasted Value $F_t(\alpha = 0.5)$	$\lvert Y_t - F_t\rvert$	$(\lvert Y_t - F_t\rvert)^2$	$\frac{\lvert Y_t - F_t\rvert}{Y_t} \times 100$
1	213	208.0	5.0	25.00	2.35
2	201	210.5	9.5	90.25	4.73
3	198	205.8	7.8	60.84	3.94
4	207	201.9	5.1	26.01	2.46
5	220	204.4	15.6	243.36	7.09
6	232	212.2	19.8	392.04	8.53
7	210	222.1	12.1	146.41	5.76
8	217	216.1	0.9	0.81	0.41
9	212	216.5	4.5	20.25	2.12
10	225	214.3	10.7	114.49	4.76
		Total		1,119.46	42.15
		Mean		111.946	4.215

The squared errors are shown in the fifth column of the table. These add up to 1,119.46. Thus, MSE = 1,119.46/10 = 111.946. Similarly, the absolute errors, as percentage of observed values are shown in the last column. From these, the mean absolute percentage error (MAPE) works out to be 42.15/10 = 4.215.

Trend-adjusted Exponential Smoothing The simple exponential smoothing and moving averages methods fail to respond to the presence of trends, if any, in the series since they are 'constant-level' models. A constant-level model assumes that the variable, say demand, shows no trend. Obviously, such a method is likely to lag behind if the underlying process has a continuing trend. The method of exponential smoothing, however, can be modified so that it makes adjustment for trend. It involves computing a simple exponential smoothing forecast and then adjusting for positive or negative trend.

In a constant-level model, the forecast made for period $t + 1$, on the basis of the past t periods' data also provides the best current forecast for time periods $t + 1 + m$, for $m = 1, 2, 3, \ldots$ (that is to say, for periods beyond $t + 1$). However, in case of presence of trend component, this does not hold and we can get estimates for the future time periods considering the trend component. The method of forecasting, when linear trend is assumed, is given below.

If Y_t is the observed (actual) value of the time series at time t, then first obtain the smoothed level S_t, at time t as follows:

$$S_t = \alpha Y_t + (1 - \alpha)(S_{t-1} + B_{t-1})$$

and

$$B_t = \beta(S_t - S_{t-1}) + (1 - \beta)B_{t-1}$$

In the above equations,

S_t = smoothed level at time t (which is calculated the same way as the forecasted values obtained earlier, with addition of the trend/slope to indicate the passage of a unit of time.

α = smoothing constant as defined earlier, with a value between 0 and 1.

B_t = smoothed value of the slope at time t.

β = smoothing constant, like α, for obtaining smoothed value of the slope, with values ranging between 0 and 1.

Having obtained S_t and B_t, we can calculate the forecasted values for m periods ahead as:

$$F_{t+m} = S_t + mB_t$$

Evidently, the value of trend smoothing constant, β, resembles the constant α so that a high β is more responsive to recent changes in the trend, while low β gives less weightage to the current trend changes.

Example 19.4 Calculate trend-adjusted forecasts using the following data:

Quarter :	1	2	3	4	5	6	7	8	9	10
Demand :	213	201	198	207	220	232	210	217	212	225

Further, given initial estimate = 208, initial trend = 0, α = 0.2 and β = 0.1

We have,

$$S_t = \alpha Y_t + (1 - \alpha)(S_{t-1} + B_{t-1}),$$

$$B_t = \beta(S_t - S_{t-1}) + (1 - \beta)B_{t-1}, \text{ and}$$

$$F_{t+m} = S_t + mB_t$$

At $t = 1$,

$$S_1 = \alpha Y_1 + (1 - \alpha)(S_0 + B_0)$$

With $\alpha = 0.2$, $Y_1 = 213$, $S_0 = 208$ and $B_0 = 0$, we get

$$S_1 = 0.2 \times 213 + (1 - 0.2)(208 + 0) = 209.0$$

$$B_1 = 0.1(209 - 208) + (1 - 0.1) \times 0 = 0.1$$

$$F_{1+1} = 209 + 0.1 = 209.1$$

$$S_2 = 0.2 \times 201 + (1 - 0.2)(209 + 0.1) = 207.5$$

$$B_2 = 0.1(207.5 - 209) + (1 - 0.1) \times 0.1 = -0.06$$

$$F_{2+1} = 207.5 + (-0.6) = 207.4$$

In a similar manner, further calculations are made and shown in Table 19.4.

TABLE 19.4 Calculation of Trend-adjusted Exponential Smoothing Forecasts

Quarter t	*Demand* Y_t	*Smoothed Value* S_t	*Trend* B_t	*Forecasted Value* F_t
1	213	209.0	0.100	208.0
2	201	207.5	– 0.060	209.1
3	198	205.5	– 0.250	207.4
4	207	205.6	– 0.220	205.3
5	220	208.3	0.076	205.4
6	232	213.1	0.548	208.4
7	210	212.9	0.474	213.7
8	217	214.1	0.546	213.4

TABLE 19.4 Contd.

Quarter t	*Demand* Y_t	*Smoothed Value* S_t	*Trend* B_t	*Forecasted Value* F_t
9	212	214.1	0.492	214.7
10	225	216.7	0.700	214.6
11				217.4
12				218.1
13				218.8
14				219.5
15				220.2

It may be noted that after $t = 10$ (where no historical values are given) the successive forecasted values are obtained by adding the trend value 0.7. Thus, $F_{10+1} = S_{10} + B_{10} = 216.7 + 0.7 = 217.4$; $F_{10+2} = S_{10} + 2B_{10} = 216.7 + 2 \times 0.7 = 218.1$ and so forth. The forecasted values are shown in the table upto $t = 15$ and they may be obtained easily for any periods desired.

The simple exponential smoothing is called the *first order* smoothing, while the trend-adjusted exponential smoothing is termed as the *second order* smoothing. The model can be extended to incorporate seasonal variations as well. However, it is beyond the scope of the book.

19.4.3 Trend Projections

We now consider the last of time-series forecasting techniques, in the form of trend projections. For this method, a *trend line* is fitted to the given time series data and then projections are made into future using this line. The trend line may be linear (straight line) or curvilinear in nature. There is a wide variety of curvilinear trend lines possible to draw, but our focus in this chapter is on linear trend only. For obtaining the trend line, the given historical data are first plotted on the graph, representing time scale on the X-axis. Then a line is drawn through these points in such a way that (i) the sum of deviations above the line is equal to the sum of the deviations below the line so that the sum of deviations is equal to zero, and (ii) the sum of the squares of these (vertical) deviations is the minimum. In essence, the trend line is drawn on the basis of the *principle of least squares*. Such a line, like any other straight line, is represented by the equation:

$$Y_t = a + bX$$

where Y_t = the trend value (which is to be predicted)

a = the Y-axis intercept

b = slope of the trend line

X = the independent variable, the time

Parameters a and b of the trend line drawn on the principle of least squares are obtained using the following pair of normal equations:

$$\Sigma Y = na + b\Sigma X$$

$$\Sigma XY = a\Sigma X + b\Sigma X^2$$

Alternatively,

$$b = \frac{\Sigma XY - n\overline{X}\overline{Y}}{\Sigma X^2 - n\overline{X}^2}$$

and

$$a = \overline{Y} - b\overline{X}$$

Here, ΣY = summation of the values of the dependent variable (the variable whose values are to be forecasted)

ΣX = summation of the values of the independent variable (time in this case)

ΣXY = summation of the products of the X and corresponding values

ΣX^2 = summation of the squares of values of the independent variable

n = number of data points

The value of b in the equation indicates the change in the variable under consideration per unit of time.

A note on the X variable In a given time series of demand of an item over past ten years, for example, where the demand represents the Y-variable, the X-variable needs to be transformed. This is because the years 2003, 2004, 2005, 2006 and so forth are by themselves not the X-variable values. They may be changed to a time scale as 0, 1, 2, 3 ... and so on. These values are then used as the X-variable values. This is demonstrated in Example 19.5.

Example 19.5 The sales of a company, in millions of rupees, are given below:

Year :	2001	2002	2003	2004	2005	2006	2007	2008
Sale :	82	80	90	92	83	94	99	92

Using the principle of least squares, fit a straight line trend equation to the above data. Show that the sum of deviations is equal to zero. Also determine the sum of squares of deviations. Forecast the sales for the years 2009 and 2010. Show the data and the trend line graphically.

The given data are reproduced in Table 19.5. Also shown in the table is the calculation about the inputs for obtaining the values of a and b. Notice that the years have been replaced by time scale with values 0, 1, 2, 3, and so on.

TABLE 19.5 Calculation of Trend Equation Parameters, and Deviations

Year	X	*Sales, Y*	XY	X^2	Y_t	$Y - Y_t$	$(Y - Y_t)^2$
2001	0	82	0	0	82	0	0
2002	1	80	80	1	84	– 4	16
2003	2	90	180	4	86	4	16
2004	3	92	276	9	88	4	16
2005	4	83	332	16	90	– 7	49
2006	5	94	470	25	92	2	4
2007	6	99	594	36	94	5	25
2008	7	92	644	49	96	– 4	16
Total	28	712	2,576	140		0	142

We have, $\overline{X} = \frac{\Sigma X}{n} = \frac{28}{8} = 3.5,\ \overline{Y} = \frac{\Sigma Y}{n} = \frac{712}{8} = 89$

$$b = \frac{\Sigma XY - n\overline{X}\overline{Y}}{\Sigma X^2 - n\overline{X}^2} = \frac{2{,}576 - 8 \times 3.5 \times 89}{140 - 8\,(3.5)^2} = 2$$

$$a = \overline{Y} - b\overline{X} = 89 - 2 \times 3.5 = 82$$

Accordingly, the trend equation is: $Y_t = 82 + 2X$, with origin in 2001. Using this equation, trend values for various years are obtained (by substituting the corresponding X-values in the equation) and shown in the sixth column of the table. The next column gives the deviations of Y-values from the corresponding Y_t-values. Their sum is found to be equal to zero. The last column of the table contains the square of deviations which sum upto 142.

Forecasting

$$Y_{2009} = 82 + 2 \times 8 = \text{Rs 98 million (for year 2009, } X = 2009 - 2001 = 8)$$

$$Y_{2010} = 82 + 2 \times 9 = \text{Rs 100 million}$$

Graphic Presentation The given data and the trend line are shown in Figure 19.2. The projections for the years 2009 and 2010 using the trend line are also shown. These projections have the same values as obtained earlier.

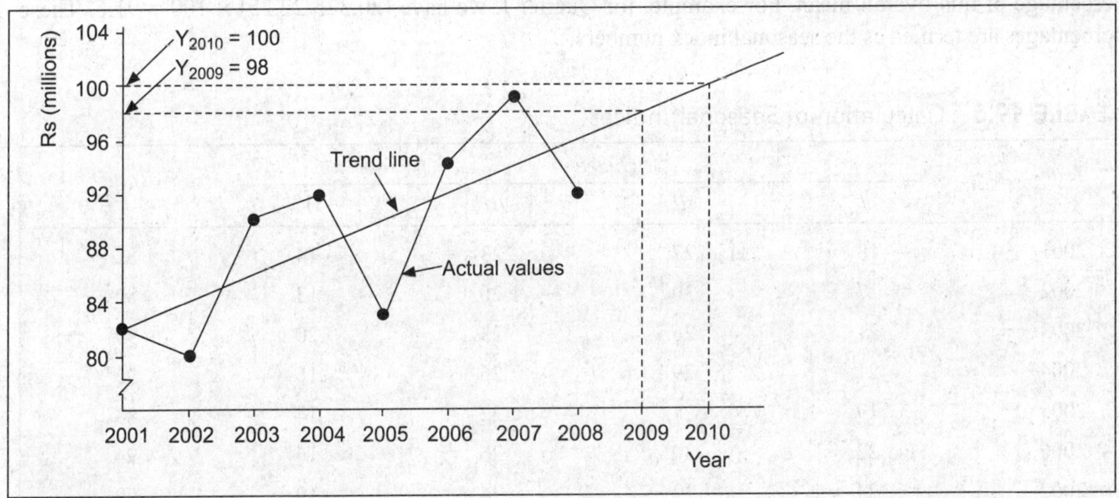

Figure 19.2 *Trend Line and Trend Projections*

Seasonal Variations The trend projections of time series data discussed earlier involves focusing at the trend (long term tendency) of the data over a series of time periods. Thus, if yearly sales data are given, we can obtain forecasts for likely yearly sales using the trend line. However, if projections are to be made on, say, monthly or quarterly basis, then recurring variations at certain seasons of the year make a seasonal adjustment in the trend forecast necessary. Sales of soft drinks, woollen garments, cooking gas, and so on are certainly not expected to be the same in all months of the year. A consideration of monthly or quarterly data makes it easy to study seasonal patterns. Seasonal indices, which are used to study seasonal variations, can be obtained by applying a number of methods. Here, we demonstrate the method of *simple averages*.

Example 19.6 The sales data given in Example 19.5 are reproduced below, with the break-up of the yearly sales on a quarterly basis. You are required to (a) obtain seasonal indices, and (b) forecast the sales for the years 2009 and 2010 on a quarterly basis.

Year	*Quarter*				*Total*
	I	*II*	*III*	*IV*	
2001	18	27	23	14	82
2002	17	30	20	13	80
2003	21	26	23	20	90
2004	21	29	26	16	92
2005	19	28	22	14	83
2006	22	30	26	16	94
2007	24	32	24	19	99
2008	21	29	29	13	92

The given data are reproduced in Table 19.6. To obtain seasonal indices, we first calculate average sales for each of the four quarters. The overall mean of these quarterly averages is calculated, which here works out to be $(20.375 + 28.875 + 24.125 + 15.625)/4 = 89/4 = 22.25$. Each of the quarterly values is then expressed as a percentage of this overall mean. For example, for Quarter I, we have $(20.375/22.25) \times 100 = 91.6$. These percentages are termed as the seasonal index numbers.

TABLE 19.6 Calculation of Seasonal Indices

Year	*Quarter*				*Total*
	I	*II*	*III*	*IV*	
2001	18	27	23	14	82
2002	17	30	20	13	80
2003	21	26	23	20	90
2004	21	29	26	16	92
2005	19	28	22	14	83
2006	22	30	26	16	94
2007	24	32	24	19	99
2008	21	29	29	13	92
Total	163	231	193	125	
Average	20.375	28.875	24.125	15.625	89/4 = 22.25
Seasonal index	91.6	129.8	108.4	70.2	

Estimation of Quarterly Sales The sales for the years 2009 and 2010 are estimated to be Rs 98 million and Rs 100 million respectively, which yield a corresponding average sales of Rs 24.5 million and Rs 25 million. Based on the seasonal indices, the quarterly sales are estimated as follows:

Quarter	Year 2009	Year 2010
I	24.5 × 0.916 = Rs 22.44 m	25 × 0.916 = Rs 22.90 m
II	24.5 × 1.298 = Rs 31.80 m	25 × 1.298 = Rs 32.45 m
III	24.5 × 1.084 = Rs 26.56 m	25 × 1.084 = Rs 27.10 m
IV	24.5 × 0.702 = Rs 17.20 m	25 × 0.702 = Rs 17.55 m

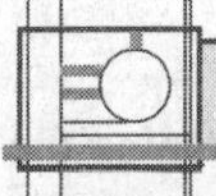

19.5 CAUSAL MODELS OF FORECASTING

As indicated earlier, the causal forecasting models consider situations where the variable to be forecasted, called the *dependent variable*, is related to some variable(s), known as *independent variable(s)*. For example, the sales of a company depends on, and is related to, price charged, prices of the substitute goods, competitors' prices, level of economic activity, and so on. Similarly, agricultural production depends upon the amount of rainfall, grade and amount of seeds used, use of fertilisers and the like. The forecasting of the dependent variable is done by developing a statistical relationship between it and the independent variable(s). This is done in terms of regression analysis.

Here we consider simple regression analysis and then have some idea about multiple regression analysis.

19.5.1 Simple Regression Analysis

The simple regression analysis is employed where there is one independent or explanatory variable. Here, an estimate of the dependent variable is made corresponding to a given value of the independent variable, by placing the relationship between the two variables in the form of a *regression line*. The regression line is obtained on the basis of the given data, involving paired observations of the X and Y variables. (It is, of course, assumed here that the relationship between the variables is linear in nature. The non-linear regression is not discussed in this book.) For this purpose, the given paired data are plotted on a graph by means of points. The line is obtained on the basis of the principle of least squares, discussed earlier in this chapter. The regression line is represented by the following equation, which is obviously called the regression equation:

$$Y_c = a + bX$$

where a = y-intercept of the regression line

b = slope of the regression line

X = the independent variable

Y_c = the predicted value of the dependent variable

As in case of a trend line, the values of a and b are obtained as follows:

$$b = \frac{\Sigma XY - n\overline{X}\overline{Y}}{\Sigma X^2 - n\overline{X}^2}; \quad \text{and} \quad a = \overline{Y} - b\overline{X}$$

Once parameters a and b of the line are obtained, predictions of the dependent variable can be made for the given values of the independent variable, X, in the equation. Parameter b is called the regression coefficient. It indicates the expected change in Y resulting from a unit change in the independent variable.

Example 19.7 A car manufacturer has recently held 3-day road side exhibits on the introduction of a new model of its deluxe cars. The number of sales personnel employed at each of a sample of 10 exhibitions and the number of cars booked at each one are given as follows:

No. of Salesmen	:	5	8	6	8	9	3	5	4	6	6
No. of Cars Booked	:	132	160	148	156	168	102	142	98	152	142

Using these data, regress the number of cars booked on the number of salesmen, and obtain the regression equation. Show the original data and the regression line on graph. Estimate the number of cars booked if 10 salesmen are employed on an exhibition.

The given data are reproduced in Table 19.7. The table also shows other inputs required for obtaining the regression equation.

TABLE 19.7 Obtaining Regression Equation

X	Y	XY	X^2	Y^2
5	132	660	25	17,424
8	160	1,280	64	25,600
6	148	888	36	21,904
8	156	1,248	64	24,336
9	168	1,512	81	28,224
3	102	306	9	10,404
5	142	392	16	20,164
4	98	392	16	9,604
6	152	912	36	23,104
6	142	852	36	20,164
Total 60	1,400	8,760	392	200,298

We have, $\overline{X} = \dfrac{\Sigma X}{n} = \dfrac{60}{10} = 6,\quad \overline{Y} = \dfrac{\Sigma Y}{n} = \dfrac{1{,}400}{10} = 140$

$$b = \frac{\Sigma XY - n\overline{X}\overline{Y}}{\Sigma X^2 - n\overline{X}^2} = \frac{8{,}760 - 10 \times 6 \times 140}{392 - 10 \times 6^2} = \frac{360}{32} = 11.25$$

$$a = \overline{Y} - b\overline{X} = 140 - 11.25 \times 6 = 72.5$$

Thus, the regression equation of Y on X is: $Y_c = 72.5 + 11.25X$.

Forecast: For $X = 10$, we have, $Y = 72.5 + 11.25 \times 10 = 185$. Thus, when 10 sales persons are employed, a total of 185 cars can be expected to be booked.

Graphic presentation: The given data and the regression line are shown graphically in Figure 19.3.

Goodness of Fit

To determine how well the regression line obtained fits the given data points, we consider three components of variation. The sum of squares of the deviations of the different observations from their arithmetic means is called the *total variation*. It shows the extent to which the arithmetic mean fails to describe the data. This is made up of two parts: (a) thc variation that is 'explained' by the line of relationship, called *explained variation*; and (b) the variation which is not explained by the said line, known as *unexplained variation*. The former is calculated as the sum of the squares of the deviations of different points on the line from the arithmetic mean whereas the latter is found by adding the squares of the deviations of different observations from corresponding points on the line. Thus,

Total variation $= \Sigma(Y - \bar{Y})^2$, called Sum of Squares Total (SST)

Explained variation $= \Sigma(Y_c - \bar{Y})^2$, called Sum of Squares Regression (SSR)

Unexplained variation $= \Sigma(Y - Y_c)^2$, called Sum of Squares Error (SSE)

Accordingly, total variation = explained variation + unexplained variation

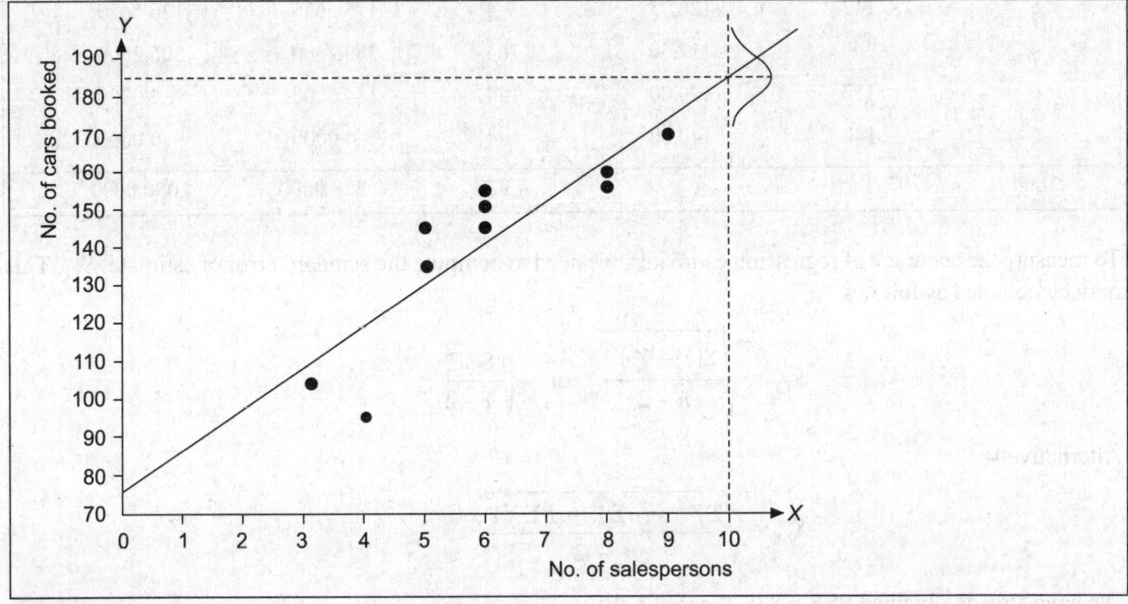

Figure 19.3 *Regression Line and Forecasting*

A fit is good if SSE is small and, obviously, SSR is large. The ratio of explained variation to total variation is termed as *coefficient of determination*. Thus, R^2 = SSR/SST. On the other hand, the ratio of unexplained variation to total variation is called *coefficient of non-determination*, $1 - R^2$. It tells what proportion of the total variation in Y which is not explained by X. Thus, $1 - R^2$ = SSE/SST. For the data given in Example 19.7, the calculation of SST, SSR and SSE is shown in Table 19.8.

Here, SST = 4,928, SSR = 4,050 and SSE = 878. Further, we obtain R^2 = SSR/SST = 4,050/4,928 = 0.8218. This implies that the number of salespersons employed explains about 82% of the variation in the number of cars booked. All other factors combined can explain at most 20% of the variations. Thus, we may say that the linear relationship between the two variables, X and Y, is strong.

Forecasting accuracy Continuing with the example, the forecast of a booking of 185 cars when 10 sales persons are employed is known as a *point estimate* of the variable Y. Such a point estimate is actually the expected value (or mean) of the distribution of the possible values of Y, as illustrated in Figure 19.3.

TABLE 19.8 Calculation of Total, Explained and Unexplained Variation

X	Y	Y_c	$(Y-\bar{Y})^2$	$(Y-Y_c)^2$	$(Y_c-\bar{Y})^2$
5	132	128.75	64	10.5625	126.5625
8	160	162.50	400	6.2500	506.2500
6	148	140.75	64	64.0000	0.0000
8	156	162.50	256	42.2500	506.2500
9	168	173.75	784	33.0625	1,139.0625
3	102	106.25	1,444	18.0625	1,139.0625
5	142	128.75	4	175.5625	126.5625
4	98	117.50	1,764	360.2500	506.2500
6	152	140.00	144	144.0000	0.0000
6	142	140.00	4	4.0000	0.0000
Total			4,928	878.0000	4,050.0000

To measure the accuracy of regression estimates, we need to compute the standard error of estimate, S_{YX}. This may be obtained as follows:

$$S_{YX} = \sqrt{\frac{\Sigma(Y-Y_c)^2}{n-2}} \quad \text{or} \quad \sqrt{\frac{\text{SSE}}{n-2}}$$

Alternatively,

$$S_{YX} = \sqrt{\frac{\Sigma Y^2 - a\Sigma Y - b\Sigma XY}{n-2}}$$

We have already obtained SSE = 878. Accordingly,

$$S_{YX} = \sqrt{\frac{\text{SSE}}{n-2}} = \sqrt{\frac{878}{10-2}} = 10.48$$

Using the alternative formulation,

$$S_{YX} = \sqrt{\frac{2{,}00{,}988 - 72.5 \times 1{,}400 - 11.25 \times 8{,}760}{10-2}} = 10.48$$

It is usually true that about 68% of the values of Y would lies within S_{YX} of the predicted values of Y_c, and about 95% of the values of Y would lie within $2S_{YX}$ of the predicted values of Y_c. In context of the present example, we expected that 68% of data points related to cars booked would lie within 10.48 of the expected number of cars booked and 95% will be within 20.96. It may be observed that 70% of the actual values of Y lie within S_{YX} of the estimates (the three values not lying are samples 7, 8 and 9), while 100% of the actual values lie within $2S_{YX}$ limits.

19.5.2 Multiple Regression Analysis

The multiple regression is an extension of the simple regression analysis and it allows to build a model with more than one independent variable. The issues involved in multiple regression analysis are many and mathematics involved is rigorous, specially when more than two variables are involved. We state here the case which has two independent variables. A two-variable regression equation is:

$$Y_c = a + b_1X_1 + b_2X_2$$

The parameters a, b_1, and b_2 are obtained by solving the following three equations simultaneously:

$$\Sigma Y = na + b_1\Sigma X_1 + b_2\Sigma X_2$$

$$\Sigma X_1Y = a\Sigma X_1 + b_1\Sigma X_1^2 + b_2\Sigma X_1X_2$$

$$\Sigma X_2Y = a\Sigma X_2 + b_1\Sigma X_1X_2 + b_2\Sigma X_2^2$$

Example 19.8 Given the following data:

Performance Evaluation	(Y) :	28	33	21	40	38	46
Aptitude Test Score	(X_1) :	74	87	69	69	81	97
Prior Experience	(X_2) :	5	11	4	9	7	10

(i) Develop the estimating equation best describing these data.

(ii) If an employee scored 83 on the aptitude test and had a prior experience of 7, what performance evaluation would be expected? *(M.Com., Delhi, 1999)*

The data and the calculations for regression equation are presented in Table 19.9.

TABLE 19.9 Obtaining Multiple Regression Equation

Y	X_1	X_2	X_1Y	X_2Y	X_1X_2	X_1^2	X_2^2
28	74	5	2,072	140	370	5,476	25
33	87	11	2,871	363	957	7,569	121
21	69	4	1,449	84	276	4,761	16
40	69	9	2,760	360	621	4,761	81
38	81	7	3,078	266	567	6,561	49
46	97	10	4,462	460	970	9,409	100
206	477	46	16,692	1,673	3,761	38,537	392

Accordingly, the three equations are:

$$206 = 6a + 477b_1 + 46b_2$$

$$16{,}692 = 477a + 38{,}537b_1 + 3{,}761b_2$$

$$1{,}673 = 46a + 3{,}761b_1 + 392b_2$$

Solving the three equations simultaneously, we obtain $a = 4.364$, $b_1 = 0.1978$ and $b_2 = 1.858$. The regression equation, therefore, is

$$Y_c = 4.364 + 0.1978X_1 + 1.858X_2$$

In this equation, b_1 and b_2 are partial regression coefficients. The value $b_1 = 0.1978$ implies that a unit increase in aptitude test score will lead to 0.1978 units increase of performance, while a unit of more experience (one year, if the units of experience are years) would mean 1.858 units of increase in performance, on the average.

Estimation of Y With $X_1 = 83$ and $X_2 = 7$, we have

$$Y_c = 4.364 + 0.1978 \times 83 + 1.858 \times 7 = 33.79$$

Goodness of Fit For multiple regression analysis, the SST, SSR and SSE are defined the same way as shown earlier in case of simple regression analysis. Also, $R^2 = \text{SSR/SST}$, gives the proportion of variation in Y which is explained by the k independent variables and $1 - R^2 = \text{SSE/SST}$ as the proportion of variation in Y which is not explained by the k independent variables. Similarly, the standard error of estimate is defined as follows:

$$S_{Y.X_1, X_2, \ldots X_k} = \sqrt{\frac{\text{SSE}}{n-k-1}}$$

For the data given in Example 19.8, calculation about variation yields the following results:

$$\text{SST} = 383.00, \text{SSE} = 164.963 \text{ and SSR} = 218.037.$$

Accordingly, $R^2 = 218.037/383 = 56.93\%,$

and $S_{Y.X_1, X_2} = \sqrt{\frac{164.963}{6-2-1}} = \sqrt{54.9877} = 7.415$

Further, as stated earlier, we expect about 68% of the Y values to be within $S_{Y.X_1, X_2}$ and approximately 95% of the values to be within $2S_{Y.X_1, X_2}$.

Review Illustrations

Example 19.9 Using the following data on the demand of a product, calculate (i) 3-monthly moving averages, (ii) exponentially smoothed averages, assuming initial forecast = 250 and $\alpha = 0.3$, and (iii) trend-adjusted exponentially smoothed averages, assuming initial forecast = 250, initial trend = 0.7, $\alpha = 0.3$ and $\beta = 0.1$.

Month :	1	2	3	4	5	6	7	8	9	10
Demand :	252	259	258	263	262	260	258	263	270	272

The given data are presented in Table 19.10. The third column contains the 3-monthly moving averages, while in the fourth column are given the exponentially smoothed averages. Calculation of trend and trend-adjusted forecasts are given in the last three columns of the table.

TABLE 19.10 Calculation of Forecasts Using Moving Averages and Exponential Smoothing

Month	*Demand*	*3-Monthly Moving Average*	$F_{t+1} = F_t + \alpha(Y_t - F_t)$	*Trend-adjusted Forecasts*		
t	Y			S_t	B_t	$F_{t+1} = S_t + B_t$
1	252		250.0	250.60	0.060	250.00
2	259		250.6	253.16	0.310	250.66
3	258		253.1	254.83	0.446	253.47
4	263	256.3	254.6	257.59	0.678	255.28

Contd.

TABLE 19.10 Contd

Month	*Demand*	*3-Monthly*	$F_{t+1} = F_t +$	*Trend-adjusted Forecasts*		
t	Y	*Moving Average*	$\alpha(Y_t - F_t)$	S_t	B_t	$F_{t+1} = S_t + B_t$
5	262	260.0	257.1	259.39	0.790	258.27
6	260	261.0	258.6	260.13	0.784	260.18
7	258	261.7	259.0	260.04	0.697	260.91
8	263	260.0	258.7	261.41	0.765	260.73
9	270	260.3	260.0	264.53	1.000	262.18
10	272	263.7	263.0	267.47	1.194	265.52
11		268.3	265.7			268.66

Example 19.10 An analysis of the past data reveals the following information:

Trend equation: $Y_t = 1{,}008 + 15.8X$

Origin : January, 2009
X Unit : One Month
Y Unit : Monthly Demand

Seasonal Indices:

January	103.4	July	105.2
February	112.2	August	104.0
March	101.7	September	97.4
April	95.6	October	100.2
May	96.8	November	94.6
June	99.8	December	88.7

On the basis of the given data, forecast the demand for the months of January, 2011 through June, 2011.

The forecast of demand for the given time period can be obtained as follows:

(i) Find trend values using the trend equation. For this purpose, we substitute $X = 24$ for January, 2011 in the equation to get $Y_t = 1{,}008 + 15.8 \times 24 = 138.72$. Similarly, for the following months, $X = 25$, 26, and so on.

(ii) Adjust the trend values by seasonal indices. For example, for January, 2011, the trend forecast would be adjusted by multiplying by 1.034 (= 103.4/100), to get $1{,}387.2 \times 1.034 = 1{,}434.4$.

The calculations are shown in Table 19.11.

TABLE 19.11 Calculation of Forecast Values

Month	*X*	*Trend Value* $Y_t = 1008 + 15.8X$	*Seasonal Index*	*Forecast* F_t
January	24	1,387.2	103.4	1,434.4
February	25	1,403.0	112.2	1,574.2
March	26	1,418.8	101.7	1,442.9
April	27	1,434.6	95.6	1,371.5
May	28	1,450.4	96.8	1,404.0
June	29	1,466.2	99.8	1,463.3

Example 19.11 In the following table are recorded data showing the experience of machine operators and their performance rating as given by the number of good parts turned out per 100 pieces:

Operator	:	1	2	3	4	5	6	7	8
Experience (years)	:	16	12	18	4	3	10	5	12
Performance Rating	:	87	88	89	68	78	80	75	82

Obtain the regression equation of performance rating on experience. Use this equation to estimate the probable performance if an operator has 7 years of experience.

Let the performance rating be denoted by Y and the experience by X. We are to fit the equation $Y_c = a + bX$. The calculations are shown in Table 19.12.

TABLE 19.12 Obtaining Regression Equation

X	Y	XY	X^2	Y^2
16	87	1,392	256	7,569
12	88	1,056	144	7,744
18	89	1,602	324	7,921
4	68	272	16	4,624
3	78	234	9	6,084
10	80	800	100	6,400
5	75	375	25	5,625
12	82	984	144	6,724
Total 80	647	6,715	1,018	5,2691

We have, $\overline{X} = \frac{\Sigma X}{n} = \frac{80}{8} = 10, \quad \overline{Y} = \frac{\Sigma Y}{n} = \frac{647}{8} = 80.875$

$$b = \frac{\Sigma XY - n\overline{X}\overline{Y}}{\Sigma X^2 - n\overline{X}^2} = \frac{6{,}715 - 8 \times 10 \times 80.875}{1{,}018 - 8 \times 10^2} = \frac{245}{218} = 1.124$$

$$a = \overline{Y} - b\overline{X} = 80.875 - 1.124 \times 10 = 69.635$$

Thus, the regression equation of Y on X is: $Y_c = 69.635 + 1.124X$.

Forecast: For $X = 7$, we have, $Y = 69.635 + 1.124 \times 7 = 77.5$. Thus, an operator with a 7 years' experience is likely to have a performance rating of about 77.5.

Example 19.12 A company keeps track of its expenditure on advertising (X_1) and its spending on in-store displays (X_2) over a period of 10 weeks. The resulting sales for each week are recorded as the dependent variable Y. Use the data given below to estimate the least squares equation relating these variables.

Y	72	76	78	70	68	80	82	65	62	90
X_1	12	11	15	10	11	16	14	8	8	18
X_2	5	8	6	5	3	9	12	4	3	10

(M Com, Delhi, 2008)

The data and the calculations for regression equation are given in Table 19.13.

TABLE 19.13 Obtaining Multiple Regression Equation

Y	X_1	X_2	X_1Y	X_2Y	X_1X_2	X_1^2	X_2^2
72	12	5	864	360	60	144	25
76	11	8	836	608	88	121	64
78	15	6	1,170	468	90	225	36
70	10	5	700	350	50	100	25
68	11	3	748	204	33	121	9
80	16	9	1,280	720	144	256	81
82	14	12	1,148	984	168	196	144
65	8	4	520	260	32	64	16
62	8	3	496	186	24	64	9
90	18	10	1,620	900	180	324	100
743	123	65	9,382	5,040	869	1,615	509

The parameters a, b_1 and b_2 for the regression equation $Y_c = a + b_1X_1 + b_2X_2$ are obtainable from the following equations:

$$\Sigma Y = na + b_1\Sigma X_1 + b_2\Sigma X_2$$

$$\Sigma X_1Y = a\Sigma X_1 + b_1\Sigma X_1^2 + b_2\Sigma X_1X_2$$

$$\Sigma X_2Y = a\Sigma X_2 + b_1\Sigma X_1X_2 + b_2\Sigma X_2^2$$

Substituting the calculated values, we get

$$743 = 10a + 123b_1 + 65b_2$$

$$9{,}382 = 123a + 1{,}615b_1 + 869b_2$$

$$5{,}040 = 65a + 869b_1 + 509b_2$$

Solving these equations simultaneously, we get a = 47.165, b_1 = 1.599 and b_2 = 1.149. Accordingly, the required regression equation is

$$Y_c = 47.165 + 1.599X_1 + 1.149X_2$$

KEY POINTS TO REMEMBER

- Forecasting is a vital aid to managerial decision-making.
- Broadly, forecasts can be made by using qualitative methods, using judgement, or by quantitative methods using numerical information.
- Quantitative techniques comprise time series models, where the historical data are used for making forecasts, and causal models in which forecasts of a variable are made on the basis of the values of the other variable or variables to which it is related.
- Time series models include: (a) moving averages, where each period's forecast is obtained as the simple, or weighted, mean of the actual values for the immediately preceding n periods; (b) exponential smoothing, where a forecast is obtained as the weighted

average of the values of all the preceding periods; and (c) trend projections, in which the past trend is projected into future to make the forecasts.

- Regression analysis represents a causal technique to make forecasts. In a one-independent variable case, a regression line is fitted to the given pairs of the data points, which is drawn on the principle of least squares. Once the slope and the *Y*-intercept of the line are obtained, forecasts can be made by substituting known values of the independent variable. In multiple regression also, the *Y*-intercept and the partial regression coefficients are obtained similarly as in case of the simple regression analysis. The standard error of estimate represents the overall accuracy of the forecasting in each case.

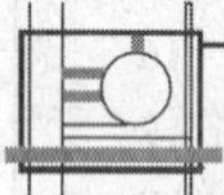

TEST YOUR UNDERSTANDING

Mark the following statements as T (*True*) or F (*False*).

1. Qualitative methods of forecasting are employed where the variable under consideration cannot be quantified.
2. Opinions and judgements play no role in quantitative forecasting.
3. Delphi method is the most important method of forecasting involving judgements.
4. Exponential smoothing is a causal method of forecasting.
5. Decomposition of a time series involves breaking down the historical data into explained and unexplained variation.
6. In moving averages method, the broader the base (for calculating moving average), the better the results, since forecasts are then made on the basis of more data.
7. In exponential smoothing, the weight applied to actual value of a period k is $(1 - \alpha)$ times the weight applied to the value of the period $k - 1$.
8. In the exponential smoothing method of forecasting, the previous forecast is adjusted by a fraction of the error in that forecast to obtain the next forecast.
9. Exponential smoothing method of forecasting is the appropriate one under all conditions.
10. A high value of α would always lead to better forecasts since it allows making quick adjustements to the forecasts.
11. For demand forecasting purposes, a company should always use the same value of α for reasons of stability.
12. MAPE represents the mean of absolute forecast errors expressed as a percentage of the forecasted values.
13. The choice of α determined by MAD, MSE and MAPE is always identical and therefore, only one of these may be considered.
14. In a regression-based forecast, the standard error of estimate gives the maximum error of forecast.
15. The simple regression analysis is based on the assumption that the changes in the dependent variable are caused by changes in the independent variable only and none else.
16. The coefficient of determination is equal to the ratio SSR/SST.
17. Multiple regression analysis allows to have multiple dependent and independent variables.

18. In multiple regression analysis, constants b_1, b_2, b_3 and so on are termed as partial regression coefficients.
19. The coefficient of determination measures the percentage variation in the dependent variable which is explained by the independent variable/s.
20. For a given set of data, about 95% of the actual data points are likely to fall within twice the standard error of estimate.

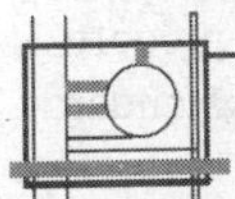

EXERCISES

1. What is the difference between qualitative and quantitative techniques of forecasting? When is a qualitative model appropriate?
2. Briefly discuss the Delphi method of making forecasts.
3. What is a time-series forecasting model? Explain the principle underlying the trend line.
4. Discuss the moving average model of forecasting and mention its drawbacks.
5. Do you think the exponential smoothing model of forecasting is in any way better than the moving average model?
6. How would you make a choice between selected values of smoothing constants for making forecasts?
7. Discuss the method of obtaining trend-adjusted exponential smoothed forecasts.
8. What is MAD? Discuss its importance in selection and use of forecasting models.
9. What is coefficient of determination in the regression model of forecasting? How is it measured?
10. What is standard error of estimate? What does it indicate?

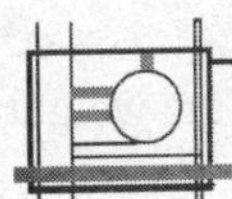

PRACTICAL PROBLEMS

1. You are given the following information about demand of an item:

Month	:	1	2	3	4	5	6	7	8	9	10	11
Demand	:	220	228	217	219	258	241	239	244	256	260	265

Calculate forecasted values using (i) 3-monthly moving averages, (ii) 5-monthly moving averages, and (iii) 4-monthly weighted moving averages with weights as 4 : 3 : 2 : 1, the largest weight being for the most recent value.

2. Obtain the profit forecasts using (i) 4-yearly moving averages, and (ii) 5-yearly moving averages, from the following data relating to profits (in '000 Rs):

Year	*Profit*	*Year*	*Profit*	*Year*	*Profit*	*Year*	*Profit*
1994	48	1998	58	2002	61	2006	70
1995	53	1999	63	2003	68	2007	76
1996	55	2000	68	2004	58	2008	83
1997	56	2001	60	2005	63	2009	88

3. Given the following data about the demand of an item, use exponential smoothing to forecast demand taking $\alpha = 0.1$ and $\alpha = 0.3$. Also, plot the original data and the forecasts on graph.

Year	*Quarter*			
	I	*II*	*III*	*IV*
2005	70	160	110	200
2006	90	120	60	110
2007	100	190	150	300
2008	270	350	320	

Further, calculate MAD for each of the series and suggest which value of α is more appropriate. Does the method of exponential smoothing appear to be satisfactory one in this case?

4. Using the following data, compute the exponential forecasts using $\alpha = 0.3$ but use MAD as a tracking signal. Whenever the MAD exceeds 50, switch to $\alpha = 0.8$, and when the MAD is less than or equal to 50, return to $\alpha = 0.3$.

Year	*Quarter*			
	I	*II*	*III*	*IV*
2005	70	160	110	200
2006	90	120	60	110
2007	100	190	150	300
2008	270	350	320	

5. Demand (in '000 metric tonnes) for sugar of Sweet India is given below:

Year :	2002	2003	2004	2005	2006	2007	2008
Demand :	77	88	94	85	91	98	90

(i) Fit a straight line trend by method of least squares.

(ii) Calculate trend values and plot observed and trend values on a graph.

(iii) Obtain the forecast of demand for the year 2010.

6. Fit a straight line trend to the following data on demand of steel ingots (in millions) and project the demand for the year 2009.

Year :	2002	2003	2004	2005	2006	2007	2008
Demand :	80	84	90	93	98	100	104

7. Fit a straight line trend to the following data on average monthly domestic demand (in millions of barrels) for motor fuel:

Year :	1998	1999	2000	2001	2002	2003	2004	2005	2006	2007	2008
Demand :	61	66	72	76	82	90	96	100	103	110	114

What is the forecast of demand for the year 2009?

8. The data on monthly sales (in thousands of Rs) of a company are given as follows:

Year	Jan	Feb	Mar	Apr	May	Jun	Jul	Aug	Sep	Oct	Nov	Dec
2006	46	45	44	46	45	47	46	43	40	40	41	45
2007	45	44	43	46	46	45	47	42	43	42	43	44
2008	42	41	40	44	45	45	46	43	41	40	42	45

Assuming there is no trend, compute the monthly seasonal indices. Using these indices, prepare the monthly sales schedule for the year 2009 if the total sales for the year are projected to be Rs 5,60,000.

9. Obtain the least squares regression equation of Y on X from the following data:

X :	89	86	74	65	64	63	66	67	72	79
Y :	92	91	84	75	73	72	71	75	78	84

Use the regression equation to forecast values of Y when (i) $X = 70$, and (ii) $X = 85$.

10. In security analysis, the returns on a share are related to the returns on the market portfolio to calculate the *beta* of that security. For a share the return is usually calculated as $(P_t - P_{t-1})/P_{t-1}$, where P_t is the current day's closing price while P_{t-1} is the closing price of the previous day. The return on market portfolio is taken as the return on share price index. The return on index is calculated in the same way as return on a share. This is equal to $(I - I_{t-1})/I_{t-1}$. Each of these may be expressed as percentages as well. For a given share, the prices are observed for 13 days and are recorded below along with the index values on those days.
You are required to regress the returns on the share on the returns on the index. What does the beta, the regression coefficient indicate? Now, estimate the return on this share if the market were to increase by 12%.

Day	*Share price index*	*Price of share*	*Day*	*Share price index*	*Price of share*
1	1376.15	818.35	8	1447.55	792.30
2	1388.75	811.75	9	1439.70	778.30
3	1408.85	819.85	10	1427.65	740.95
4	1418.00	836.05	11	1398.25	718.35
5	1442.85	815.65	12	1401.40	737.50
6	1445.15	804.30	13	1419.70	735.55
7	1438.65	801.30			

11. Ashish, owner of a business unit is concerned about the sales behaviour of his product. He realises that there are many factors that might help explain sales, but believes that advertising and prices are major determinants. He has collected the following data:

Sales (units sold)	:	33	61	70	82	17	24
Advertising (no. of ads)	:	3	6	10	13	9	6
Price (Rs)	:	125	115	140	130	145	140

(i) Calculate the regression equation to predict sales from advertising and price.

(ii) If advertising is 7 and price is 132, what sales would you predict? *(M Com., Delhi, 1996)*

12. A company recently embarked on an in-store promotional compaign, with displays of its antacid featured prominently in the supermarkets. The company also ran its usual radio and television commer-

cials. Over a period of 10 weeks, the company kept track of its expenditure on radio and television advertising, variable X_1 as well as its expenditure on in-store displays, variable X_2. The resulting sales for each week in the area studied were recorded as the dependent variable Y. The company analyst hypothesised a linear regression linking sales volume with the two independent variables. Using the following data (all values in thousands of Rs), fit the desired regression equation.

Y	:	72	76	78	70	68	80	82	65	62	90
X_1	:	12	11	15	10	11	16	14	8	8	18
X_2	:	5	8	6	5	3	9	12	4	3	10

Now, estimate the sales if X_1 = Rs 13,000 and X_2 = Rs 7,000. *(M Com., Delhi, 2000)*

APPENDIX A

Key to 'Test Your Understanding'

CHAPTER 1

1. F	2. F	3. T	4. T	5. F	6. T	7. T
8. T	9. F	10. F	11. T	12. T	13. F	14. T
15. F						

CHAPTER 2

1. T	2. F	3. T	4. T	5. T	6. T	7. F
8. F	9. F	10. F	11. F	12. F	13. T	14. T
15. F	16. F	17. T	18. F	19. T	20. F	21. F
22. F	23. T	24. F	25. T	26. T	27. F	28. F
29. F	30. T					

CHAPTER 3

1. T	2. T	3. T	4. T	5. T	6. F	7. T
8. F	9. F	10. F	11. T	12. F	13. F	14. T
15. F	16. F	17. T	18. T	19. T	20. T	21. T
22. T	23. T	24. T	25. F	26. F	27. F	28. F
29. F	30. F					

CHAPTER 4

1. T	2. T	3. T	4. F	5. T	6. T	7. T
8. F	9. T	10. T	11. T	12. T	13. T	14. T
15. F	16. F	17. F	18. T	19. T	20. T	

CHAPTER 5

1. T	2. T	3. F	4. T	5. F	6. F	7. T
8. F	9. F	10. T	11. F	12. T	13. T	14. T
15. T	16. T	17. T	18. T	19. F	20. T	21. T
22. T	23. F	24. T	25. T	26. T	27. T	28. F
29. T	30. F	31. F	32. F	33. T	34. T	35. F

CHAPTER 6

1. T	2. T	3. T	4. T	5. F	6. T	7. T
8. F	9. T	10. F	11. T	12. F	13. T	14. F
15. T	16. F	17. T	18. F	19. T	20. F	

CHAPTER 7

1. F	2. F	3. F	4. T	5. T	6. F	7. F
8. T	9. T	10. F	11. F	12. T	13. T	14. F
15. F	16. T	17. T	18. F	19. F	20. T	21. T
22. F	23. T	24. F	25. T	26. T	27. T	28. T
29. F	30. F					

CHAPTER 8

1. T	2. F	3. T	4. T	5. F	6. T	7. T
8. F	9. T	10. F	11. F	12. T	13. F	14. T
15. F	16. T	17. T	18. F	19. T	20. T	

CHAPTER 9

1. T	2. F	3. T	4. T	5. T	6. F	7. T
8. T	9. T	10. T	11. T	12. T	13. F	14. T
15. T	16. T	17. T	18. F	19. F	20. T	21. T
22. T	23. T	24. F	25. F	26. T	27. T	28. T
29. T	30. T	31. T	32. F	33. T	34. T	35. T
36. F	37. T	38. F	39. F	40. T		

CHAPTER 10

1. T	2. T	3. F	4. F	5. T	6. T	7. F
8. T	9. T	10. T	11. T	12. F	13. F	14. T
15. T	16. T	17. T	18. T	19. F	20. T	21. F
22. F	23. T	24. T	25. F	26. F	27. F	28. T
29. T	30. T					

CHAPTER 11

1. F	2. T	3. T	4. F	5. F	6. T	7. F
8. T	9. T	10. T	11. T	12. T	13. T	14. T
15. F	16. T	17. F	18. T	19. F	20. F	

CHAPTER 12

1. T	2. T	3. F	4. F	5. T	6. F	7. F
8. T	9. T	10. T	11. F	12. T	13. F	14. T
15. F	16. T	17. F	18. F	19. T	20. T	21. T
22. T	23. F	24. F	25. T	26. T	27. T	28. T
29. T	30. T	31. F	32. T	33. F	34. T	35. F
36. T	37. F	38. F	39. F	40. T	41. T	42. T
43. T	44. T	45. T	46. F	47. F	48. T	49. F
50. F	51. F	52. F	53. T	54. F	55. T	

CHAPTER 13

1. T	2. F	3. F	4. T	5. F	6. F	7. T
8. F	9. T	10. T	11. T	12. T	13. F	14. T
15. T	16. T	17. F	18. T	19. T	20. F	21. T
22. T	23. T	24. T	25. F	26. T	27. T	28. F
29. F	30. T					

CHAPTER 14

1. T	2. F	3. T	4. F	5. T	6. T	7. T
8. F	9. T	10. T	11. T	12. T	13. T	14. F
15. T	16. T	17. T	18. T	19. F	20. T	21. T
22. F	23. T	24. T	25. F			

CHAPTER 15

1. F	2. T	3. T	4. F	5. T	6. T	7. T
8. T	9. T	10. T	11. T	12. F	13. F	14. T
15. F	16. T	17. F	18. T	19. T	20. F	21. F
22. F	23. T	24. T	25. T	26. T	27. T	28. T
29. T	30. T					

CHAPTER 16

1. T	2. F	3. T	4. F	5. T	6. T	7. F
8. F	9. F	10. F	11. T	12. T	13. T	14. F
15. T	16. T	17. T	18. F	19. F	20. T	21. T
22. T	23. T	24. T	25. F			

CHAPTER 17

1. T	2. T	3. F	4. T	5. F	6. T	7. T
8. T	9. F	10. T	11. T	12. T	13. T	14. T
15. F	16. F	17. T	18. T	19. F	20. F	

CHAPTER 18

1. T	2. T	3. F	4. F	5. T	6. F	7. T
8. F	9. T	10. T	11. F	12. T	13. T	14. F
15. T	16. T	17. F	18. T	19. T	20. F	21. T
22. T	23. T	24. T	25. F	26. T	27. T	28. T
29. F	30. F	31. T	32. T	33. F	34. T	35. T

CHAPTER 19

1. F	2. T	3. T	4. F	5. F	6. F	7. T
8. T	9. F	10. F	11. F	12. F	13. F	14. F
15. T	16. T	17. F	18. T	19. T	20. T	

APPENDIX B

Statistical and Other Tables

TABLE B1 Areas under the Standard Normal Curve from 0 to z

z	0	1	2	3	4	5	6	7	8	9
0.0	.0000	.0040	.0080	.0120	.0160	.0199	.0239	.0279	.0319	.0359
0.1	.0398	.0438	.0478	.0517	.0557	.0596	.0636	.0675	.0714	.0753
0.2	.0793	.0832	.0871	.0910	.0948	.0987	.1026	.1064	.1103	.1141
0.3	.1179	.1217	.1255	.1293	.1331	.1368	.1406	.1443	.1480	.1517
0.4	.1554	.1591	.1628	.1664	.1700	.1736	.1772	.1808	.1844	.1879
0.5	.1915	.1950	.1985	.2019	.2054	.2088	.2123	.2157	.2190	.2224
0.6	.2258	.2291	.2324	.2357	.2389	.2422	.2454	.2486	.2518	.2549
0.7	.2580	.2612	.2642	.2673	.2704	.2734	.2764	.2794	.2823	.2852
0.8	.2881	.2910	.2939	.2967	.2996	.3023	.3051	.3078	.3106	.3133
0.9	.3159	.3186	.3212	.3238	.3264	.3289	.3315	.3340	.3365	.3389
1.0	.3413	.3438	.3461	.3485	.3508	.3531	.3554	.3577	.3599	.3621
1.1	.3643	.3665	.3686	.3708	.3729	.3749	.3770	.3790	.3810	.3830
1.2	.3849	.3869	.3888	.3907	.3925	.3944	.3962	.3980	.3997	.4015
1.3	.4032	.4049	.4066	.4082	.4099	.4115	.4131	.4147	.4162	.4177
1.4	.4192	.4207	.4222	.4236	.4251	.4265	.4279	.4292	.4306	.4319
1.5	.4332	.4345	.4357	.4370	.4382	.4394	.4406	.4418	.4429	.4441
1.6	.4452	.4463	.4474	.4484	.4495	.4505	.4515	.4525	.4535	.4545
1.7	.4554	.4564	.4573	.4582	.4591	.4599	.4608	.4616	.4625	.4633
1.8	.4641	.4649	.4656	.4664	.4671	.4678	.4686	.4693	.4699	.4706
1.9	.4713	.4719	.4726	.4732	.4738	.4744	.4750	.4756	.4761	.4767
2.0	.4772	.4778	.4783	.4788	.4793	.4798	.4803	.4808	.4812	.4817
2.1	.4821	.4826	.4830	.4834	.4838	.4842	.4846	.4850	.4854	.4857
2.2	.4861	.4864	.4868	.4871	.4875	.4878	.4881	.4884	.4887	.4890
2.3	.4893	.4896	.4898	.4901	.4904	.4906	.4909	.4911	.4913	.4916
2.4	.4918	.4920	.4922	.4925	.4927	.4929	.4931	.4932	.4934	.4936
2.5	.4938	.4940	.4941	.4943	.4945	.4946	.4948	.4949	.4951	.4952
2.6	.4953	.4955	.4956	.4957	.4959	.4960	.4961	.4962	.4963	.4964
2.7	.4965	.4966	.4967	.4968	.4969	.4970	.4971	.4972	.4973	.4974
2.8	.4974	.4975	.4976	.4977	.4977	.4978	.4979	.4979	.4980	.4981
2.9	.4981	.4982	.4982	.4983	.4984	.4984	.4985	.4985	.4986	.4986
3.0	.4987	.4987	.4987	.4988	.4988	.4989	.4989	.4989	.4990	.4990
3.1	.4990	.4991	.4991	.4991	.4992	.4992	.4992	.4992	.4993	.4993
3.2	.4993	.4993	.4994	.4994	.4994	.4994	.4994	.4995	.4995	.4995
3.3	.4995	.4995	.4995	.4996	.4996	.4996	.4996	.4996	.4996	.4997
3.4	.4997	.4997	.4997	.4997	.4997	.4997	.4997	.4997	.4997	.4998
3.5	.4998	.4998	.4998	.4998	.4998	.4998	.4998	.4998	.4998	.4998
3.6	.4998	.4998	.4999	.4999	.4999	.4999	.4999	.4999	.4999	.4999
3.7	.4999	.4999	.4999	.4999	.4999	.4999	.4999	.4999	.4999	.4999
3.8	.4999	.4999	.4999	.4999	.4999	.4999	.4999	.4999	.4999	.4999
3.9	.5000	.5000	.5000	.5000	.5000	.5000	.5000	.5000	.5000	.5000

TABLE B2 Unit Normal Loss Function

D	*.00*	*.01*	*.02*	*.03*	*.04*	*.05*	*.06*	*.07*	*.08*	*.09*
0.0	.3989	.3940	.3890	.3841	.3793	.3744	.3697	.3649	.3602	.3556
0.1	.3509	.3464	.3418	.3373	.3328	.3284	.3240	.3197	.3154	.3111
0.2	.3069	.3027	.2986	.2944	.2904	.2863	.2824	.2784	.2745	.2706
0.3	.2668	.2630	.2602	.2555	.2518	.2481	.2445	.2409	.2374	.2339
0.4	.2304	.2270	.2236	.2203	.2169	.2137	.2104	.2072	.2040	.2009
0.5	.1978	.1947	.1917	.1887	.1857	.1828	.1799	.1771	.1742	.1714
0.6	.1687	.1659	.1633	.1606	.1580	.1554	.1528	.1503	.1478	.1453
0.7	.1429	.1405	.1381	.1358	.1334	.1312	.1289	.1267	.1245	.1223
0.8	.1202	.1181	.1160	.1140	.1120	.1100	.1080	.1061	.1042	.1023
0.9	.1004	.09860	.09680	.09503	.09328	.09156	.08986	.08819	.08654	.08491
1.0	.08332	.08174	.08019	.07866	.07716	.07568	.07422	.07279	.07138	.06999
1.1	.06862	.06727	.06595	.06465	.06336	.06210	.06086	.05964	.05844	.05726
1.2	.05610	.05496	.05384	.05274	.05165	.05059	.04954	.04851	.04750	.04650
1.3	.04553	.04457	.04363	.04270	.04179	.04090	.04002	.03916	.03831	.03748
1.4	.03667	.03587	.03508	.03431	.03356	.03281	.03208	.03137	.03067	.02996
1.5	.02931	.02865	.02800	.02736	.02674	.02612	.02552	.02494	.02436	.02380
1.6	.02324	.02270	.02217	.02165	.02114	.02064	.02015	.01967	.01920	.01874
1.7	.01820	.01785	.01742	.01699	.01656	.01617	.01578	.01539	.01501	.01464
1.8	.01428	.01392	.01357	.01323	.01290	.01257	.01226	.01195	.01164	.01134
1.9	.01105	.01077	.01049	.01022	$.0^{2}9957$	$.0^{2}9698$	$.0^{2}9445$	$.0^{2}9198$	$.0^{2}8957$	$.0^{2}8721$
2.0	$.0^{2}8491$	$.0^{2}8266$	$.0^{2}8046$	$.0^{2}7832$	$.0^{2}7623$	$.0^{2}7416$	$.0^{2}7219$	$.0^{2}7024$	$.0^{2}6835$	$.0^{2}6649$
2.1	$.0^{2}6468$	$.0^{2}6292$	$.0^{2}6120$	$.0^{2}5952$	$.0^{2}5788$	$.0^{2}5628$	$.0^{2}5472$	$.0^{2}5320$	$.0^{2}5172$	$.0^{2}5028$
2.2	$.0^{2}4887$	$.0^{2}4750$	$.0^{2}4616$	$.0^{2}4486$	$.0^{2}4358$	$.0^{2}4235$	$.0^{2}4114$	$.0^{2}3996$	$.0^{2}3882$	$.0^{2}3770$
2.3	$.0^{2}3662$	$.0^{2}3556$	$.0^{2}3453$	$.0^{2}3352$	$.0^{2}3255$	$.0^{2}3159$	$.0^{2}3067$	$.0^{2}2977$	$.0^{2}2889$	$.0^{2}2804$
2.4	$.0^{2}2720$	$.0^{2}2640$	$.0^{2}2561$	$.0^{2}2484$	$.0^{2}2410$	$.0^{2}2337$	$.0^{2}2267$	$.0^{2}2199$	$.0^{2}2132$	$.0^{2}2067$
2.5	$.0^{2}2005$	$.0^{2}1943$	$.0^{2}1883$	$.0^{2}1826$	$.0^{2}1769$	$.0^{2}1715$	$.0^{2}1662$	$.0^{2}1610$	$.0^{2}1560$	$.0^{2}1511$
3.0	$.0^{3}3822$	$.0^{3}3689$	$.0^{3}3560$	$.0^{3}3436$	$.0^{3}3316$	$.0^{3}3199$	$.0^{3}3087$	$.0^{3}2978$	$.0^{3}2873$	$.0^{3}2771$
3.5	$.0^{4}5848$	$.0^{4}5620$	$.0^{4}5400$	$.0^{4}5188$	$.0^{4}4984$	$.0^{4}4788$	$.0^{4}4599$	$.0^{4}4417$	$.0^{4}4242$	$.0^{4}4073$
4.0	$.0^{5}7145$	$.0^{5}6835$	$.0^{5}6538$	$.0^{5}6253$	$.0^{5}5980$	$.0^{5}5718$	$.0^{5}5468$	$.0^{5}5227$	$.0^{5}4997$	$.0^{5}4777$

TABLE B3 Present Value of Re 1 Due at the End of n Periods

Periods	*1%*	*2%*	*3%*	*4%*	*5%*	*6%*	*7%*	*8%*	*9%*	*10%*
1	.9901	.9804	.9709	.9615	.9524	.9434	.9346	.9259	.9174	.9091
2	.9803	.9612	.9426	.9246	.9070	.8900	.8734	.8573	.8417	.8264
3	.9706	.9423	.9151	.8890	.8638	.8396	.8163	.7938	.7722	.7513
4	.9610	.9238	.8885	.8548	.8227	.7921	.7629	.7350	.7084	.6830
5	.9515	.9057	.8626	.8219	.7835	.7473	.7130	.6806	.6499	.6209
6	.9420	.8880	.8375	.7903	.7462	.7050	.6663	.6302	.5963	.5645
7	.9327	.8706	.8131	.7599	.7107	.6651	.6227	.5835	.5470	.5132
8	.9235	.8535	.7894	.7307	.6768	.6274	.5820	.5403	.5019	.4665
9	.9143	.8368	.7664	.7026	.6446	.5919	.5439	.5002	.4604	.4241
10	.9053	.8203	.7441	.6756	.6139	.5584	.5083	.4632	.4224	.3855
11	.8963	.8043	.7224	.6496	.5847	.5268	.4751	.4289	.3875	.3505
12	.8874	.7885	.7014	.6246	.5568	.4970	.4440	.3971	.3555	.3186
13	.8787	.7730	.6810	.6006	.5303	.4688	.4150	.3677	.3262	.2897
14	.8700	.7579	.6611	.5775	.5051	.4423	.3878	.3405	.2992	.2633
15	.8613	.7430	.6419	.5553	.4810	.4173	.3624	.3152	.2745	.2394
16	.8528	.7284	.6232	.5339	.4581	.3936	.3387	.2919	.2519	.2176
17	.8444	.7142	.6050	.5134	.4363	.3714	.3166	.2703	.2311	.1978
18	.8360	.7002	.5874	.4936	.4155	.3503	.2959	.2502	.2120	.1799
19	.8277	.6864	.5703	.4746	.3957	.3305	.2765	.2317	.1945	.1635
20	.8195	.6730	.5537	.4564	.3769	.3118	.2584	.2145	.1784	.1486
21	.8114	.6598	.5375	.4388	.3589	.2942	.2415	.1987	.1637	.1351
22	.8034	.6468	.5219	.4220	.3418	.2775	.2257	.1839	.1502	.1228
23	.7954	.6342	.5067	.4057	.3256	.2618	.2109	.1703	.1378	.1117
24	.7876	.6217	.4919	.3901	.3101	.2470	.1971	.1577	.1264	.1015
25	.7798	.6095	.4776	.3751	.2953	.2330	.1842	.1460	.1160	.0923
26	.7720	.5976	.4637	.3607	.2812	.2198	.1722	.1352	.1064	.0839
27	.7644	.5859	.4502	.3468	.2678	.2074	.1609	.1252	.0976	.0763
28	.7568	.5744	.4371	.3335	.2551	.1956	.1504	.1159	.0895	.0693
29	.7493	.5631	.4243	.3207	.2429	.1846	.1406	.1073	.0822	.0630
30	.7419	.5521	.4120	.3083	.2314	.1741	.1314	.0994	.0754	.0573
35	.7059	.5000	.3554	.2534	.1813	.1301	.0937	.0676	.0490	.0356
40	.6717	.4529	.3066	.2083	.1420	.0972	.0668	.0460	.0318	.0221
45	.6391	.4102	.2644	.1712	.1113	.0727	.0476	.0313	.0207	.0137
50	.6080	.3715	.2281	.1407	.0872	.0543	.0339	.0213	.0134	.0085
55	.5785	.3365	.1968	.1157	.0683	.0406	.0242	.0145	.0087	.0053

(*Contd.*)

TABLE B3 (Contd.)

Periods	12%	14%	15%	16%	18%	20%	24%	28%	32%	36%
1	.8929	.8772	.8696	.8621	.8475	.8333	.8065	.7813	.7576	.7353
2	.7972	.7695	.7561	.7432	.7182	.6944	.6504	.6104	.5739	.5407
3	.7118	.6750	.6575	.6407	.6086	.5787	.5245	.4768	.4348	.3975
4	.6355	.5921	.5718	.5523	.5158	.4823	.4230	.3725	.3294	.2923
5	.5674	.5194	.4972	.4761	.4371	.4019	.3411	.2910	.2495	.2149
6.	.5066	.4556	.4323	.4104	.3704	.3349	.2751	.2274	.1890	.1580
7	.4523	.3996	.3759	.3538	.3139	.2791	.2218	.1776	.1432	.1162
8	.4039	.3506	.3269	.3050	.2660	.2326	.1789	.1388	.1085	.0854
9	.3606	.3075	.2843	.2630	.2255	.1938	.1443	.1084	.0822	.0628
10	.3220	.2697	.2472	.2267	.1911	.1615	.1164	.0847	.0623	.0462
11	.2875	.2366	.2149	.1954	.1619	.1346	.0938	.0662	.0472	.0340
12	.2567	.2076	.1869	.1685	.1372	.1122	.0757	.0517	.0357	.0250
13	.2292	.1821	.1625	.1452	.1163	.0935	.0610	.0404	.0271	.0184
14	.2046	.1597	.1413	.1252	.0985	.0779	.0492	.0316	.0205	.0135
15	.1827	.1401	.1229	.1079	.0835	.0649	.0397	.0247	.0155	.0099
16	.1631	.1229	.1069	.0930	.0708	.0541	.0320	.0193	.0118	.0073
17	.1456	.1078	.0929	.0802	.0600	.0451	.0258	.0150	.0089	.0054
18	.1300	.0946	.0808	.0691	.0508	.0376	.0208	.0118	.0068	.0039
19	.1161	.0829	.0703	.0596	.0431	.0313	.0168	.0092	.0051	.0029
20	.1037	.0728	.0611	.0514	.0365	.0261	.0135	.0072	.0039	.0021
21	.0926	.0638	.0531	.0443	.0309	.0217	.0109	.0056	.0029	.0016
22	.0826	.0560	.0462	.0382	.0262	.0181	.0088	.0044	.0022	.0012
23	.0738	.0491	.0402	.0329	.0222	.0151	.0071	.0034	.0017	.0008
24	.0659	.0431	.0349	.0284	.0188	.0126	.0057	.0027	.0013	.0006
25	.0588	.0378	.0304	.0245	.0160	.0105	.0046	.0021	.0010	.0005
26	.0525	.0331	.0264	.0211	.0135	.0087	.0037	.0016	.0007	.0003
27	.0469	.0291	.0230	.0182	.0115	.0073	.0030	.0013	.0006	.0002
28	.0419	.0255	.0200	.0157	.0097	.0061	.0024	.0010	.0004	.0002
29	.0374	.0224	.0174	.0135	.0082	.0051	.0020	.0008	.0003	.0001
30	.0334	.0196	.0151	.0116	.0070	.0042	.0016	.0006	.0002	.0001
35	.0189	.0102	.0075	.0055	.0030	.0017	.0005	.0002	.0001	*
40	.0107	.0053	.0037	.0026	.0013	.0007	.0002	.0001	*	*
45	.0061	.0027	.0019	.0013	.0006	.0003	.0001	*	*	*
50	.0035	.0014	.0009	.0006	.0003	.0001	*	*	*	*
55	.0020	.0007	.0005	.0003	.0001	*	*	*	*	*

*The factor is zero to four decimal places.

TABLE B4 Present Value of an Annuity of Re 1 Per Period for n Periods

Periods	*1%*	*2%*	*3%*	*4%*	*5%*	*6%*	*7%*	*8%*	*9%*
1	0.9901	0.9804	0.9709	0.9615	0.9524	0.9434	0.9346	0.9259	0.9174
2	1.9704	1.9416	1.9135	1.8861	1.8594	1.8334	1.8080	1.7833	1.7591
3	2.9410	2.8839	2.8286	2.7751	2.7232	2.6730	2.6243	2.5771	2.5313
4	3.9020	3.8077	3.7171	3.6299	3.5460	3.4651	3.3872	3.3121	3.2397
5	4.8534	4.7135	4.5797	4.4518	4.3295	4.2124	4.1002	3.9927	3.8897
6	5.7955	5.6041	5.4172	5.2421	5.0757	4.9173	4.7665	4.6229	4.4859
7	6.7282	6.4720	6.2303	6.0021	5.7864	5.5824	5.3893	5.2064	5.0330
8	7.6517	7.3255	7.0197	6.7327	6.4632	6.2098	5.9713	5.7466	5.5348
9	8.5660	8.1622	7.7861	7.4353	7.1078	6.8017	6.5152	6.2469	5.9952
10	9.4713	8.9826	8.5302	8.1109	7.7217	7.3601	7.0236	6.7101	6.4177
11	10.3676	9.7868	9.2526	8.7605	8.3064	7.8869	7.4987	7.1390	6.8052
12	11.2551	10.5753	9.9540	9.3851	8.8633	8.3838	7.9427	7.5361	7.1607
13	12.1337	11.3484	10.6350	9.9856	9.3936	8.8527	8.3577	7.9038	7.4869
14	13.0037	12.1062	11.2961	10.5631	9.8986	9.2950	8.7455	8.2442	7.7862
15	13.8651	12.8493	11.9379	11.1184	10.3797	9.7122	9.1079	8.5595	8.0607
16	14.7179	13.5777	12.5611	11.6523	10.8378	10.1059	9.4466	8.8514	8.3126
17	15.5623	14.2919	13.1661	12.1657	11.2741	10.4773	9.7632	9.1216	8.5436
18	16.3983	14.9920	13.7535	12.6593	11.6896	10.8276	10.0591	9.3719	8.7556
19	17.2260	15.6785	14.3238	13.1339	12.0853	11.1581	10.3356	9.6036	8.9501
20	18.0456	16.3514	14.8775	13.5903	12.4622	11.4699	10.5940	9.8181	9.1285
21	18.8570	17.0112	15.4150	14.0292	12.8212	11.7641	10.8355	10.0168	9.2922
22	19.6604	17.6580	15.9369	14.4511	13.1630	12.0416	11.0612	10.2007	9.4424
23	20.4558	18.2922	16.4436	14.8568	13.4886	12.3034	11.2722	10.3711	9.5802
24	21.2434	18.9139	16.9355	15.2470	13.7986	12.5504	11.4693	10.5288	9.7066
25	22.0232	19.5235	17.4131	15.6221	14.0939	12.7834	11.6536	10.6748	9.8226
26	22.7952	20.1210	17.8768	15.9828	14.3752	13.0032	11.8258	10.8100	9.9290
27	23.5596	20.7069	18.3270	16.3296	14.6430	13.2105	11.9867	10.9352	10.0266
28	24.3164	21.2813	18.7641	16.6631	14.8981	13.4062	12.1371	11.0511	10.1161
29	25.0658	21.8444	19.1885	16.9837	15.1411	13.5907	12.2777	11.1584	10.1983
30	25.8077	22.3965	19.6004	17.2920	15.3725	13.7648	12.4090	11.2578	10.2737
35	29.4086	24.9986	21.4872	18.6646	16.3742	14.4982	12.9477	11.6546	10.5668
40	32.8347	27.3555	23.1148	19.7928	17.1591	15.0463	13.3317	11.9246	10.7574
45	36.0945	29.4902	24.5187	20.7200	17.7741	15.4558	13.6055	12.1084	10.8812
50	39.1961	31.4236	25.7298	21.4822	18.2559	15.7619	13.8007	12.2335	10.9617
55	42.1472	33.1748	26.7744	22.1086	18.6335	15.9905	13.9399	12.3186	11.0140

(Contd.)

TABLE B4 (Contd.)

Periods	*10%*	*12%*	*4%*	*15%*	*16%*	*18%*	*20%*	*24%*	*28%*	*32%*
1	0.9091	0.8929	0.8772	0.8696	0.8621	0.8475	0.8333	0.8065	0.7813	0.7576
2	1.7355	1.6901	1.6467	1.6257	1.6052	1.5656	1.5278	1.4568	1.3916	1.3315
3	2.4869	2.4018	2.3216	2.2832	2.2459	2.1743	2.1065	1.9813	1.8684	1.7663
4	3.1699	3.0373	2.9137	2.8550	2.7982	2.6901	2.5887	2.4043	2.2410	2.0957
5	3.7908	3.6048	3.4331	3.3522	3.2743	3.1272	2.9906	2.7454	2.5320	2.3452
6	4.3553	4.1114	3.8887	3.7845	3.6847	3.4976	3.3255	3.0205	2.7594	2.5342
7	4.8684	4.5638	4.2883	4.1604	4.0386	3.8115	3.6046	3.2423	2.9370	2.6775
8	5.3349	4.9676	4.6389	4.4873	4.3436	4.0776	3.8372	3.4212	3.0758	2.7860
9	5.7590	5.3282	4.9464	4.7716	4.6065	4.3030	4.0310	3.5655	3.1842	2.8681
10	6.1446	5.6502	5.2161	5.0188	4.8332	4.4941	4.1925	3.6819	3.2689	2.9304
11	6.4951	5.9377	5.4527	5.2337	5.0286	4.6560	4.3271	3.7757	3.3351	2.9776
12	6.8137	6.1944	5.6603	5.4206	5.1971	4.7932	4.4392	3.8514	3.3868	3.0133
13	7.1034	6.4235	5.8424	5.5831	5.3423	4.9095	4.5327	3.9124	3.4272	3.0404
14	7.3667	6.6282	6.0021	5.7245	5.4675	5.0081	4.6106	3.9616	3.4587	3.0609
15	7.6061	6.8109	6.1422	5.8474	5.5755	5.0916	4.6755	4.0013	3.4834	3.0764
16	7.8237	6.9740	6.2651	5.9542	5.6685	5.1624	4.7296	4.0333	3.5026	3.0882
17	8.0216	7.1196	6.3729	6.0472	5.7487	5.2223	4.7746	4.0591	3.5177	3.0971
18	8.2014	7.2497	6.4674	6.1280	5.8178	5.2732	4.8122	4.0799	3.5294	3.1039
19	8.3649	7.3658	6.5504	6.1982	5.8775	5.3162	4.8435	4.0967	3.5386	3.1090
20	8.5136	7.4694	6.6231	6.2593	5.9288	5.3527	4.8696	4.1103	3.5458	3.1129
21	8.6487	7.5620	6.6870	6.3125	5.9731	5.3837	4.8913	4.1212	3.5514	3.1158
22	8.7715	7.6446	6.7429	6.3587	6.0113	5.4099	4.9094	4.1300	3.5558	3.1180
23	8.8832	7.7184	6.7921	6.3988	6.0442	5.4321	4.9245	4.1371	3.5592	3.1197
24	8.9847	7.7843	6.8351	6.4338	6.0726	5.4510	4.9371	4.1428	3.5619	3.1210
25	9.0770	7.8431	6.8729	6.4642	6.0971	5.4669	4.9476	4.1474	3.5640	3.1220
26	9.1609	7.8957	6.9061	6.4906	6.1182	5.4804	4.9563	4.1511	3.5656	3.1227
27	9.2372	7.9426	6.9352	6.5135	6.1364	5.4919	4.9636	4.1542	3.5669	3.1233
28	9.3066	7.9844	6.9607	6.5335	6.1520	5.5016	4.9697	4.1566	3.5679	3.1237
29	9.3696	8.0218	6.9830	6.5509	6.1656	5.5098	4.9747	4.1585	3.5687	3.1240
30	9.4269	8.0552	7.0027	6.5660	6.1772	5.5168	4.9789	4.1601	3.5693	3.1242
35	9.6442	8.1755	7.0700	6.6166	6.2153	5.5386	4.9915	4.1644	3.5708	3.1248
40	9.7791	8.2438	7.1050	6.6418	6.2335	5.5482	4.9966	4.1659	3.5712	3.1250
45	9.8628	8.2825	7.1232	6.6543	6.2421	5.5523	4.9986	4.1664	3.5714	3.1250
50	9.9148	8.3045	7.1327	6.6605	6.2463	5.5541	4.9995	4.1666	3.5714	3.1250
55	9.9471	8.3170	7.1376	6.6636	6.2482	5.5549	4.9998	4.1666	3.5714	3.1250

TABLE B5 Compound Value of Re 1 at the End of n Periods

Periods	1%	2%	3%	4%	5%	6%	7%	8%	9%	10%
1	1.0100	1.0200	1.0300	1.0400	1.0500	1.0600	1.0700	1.0800	1.0900	1.1000
2	1.0201	1.0404	1.0609	1.0816	1.1025	1.1236	1.1449	1.1664	1.1881	1.2100
3	1.0303	1.0612	1.0927	1.1249	1.1576	1.1910	1.2250	1.2597	1.2950	1.3310
4	1.0406	1.0824	1.1255	1.1699	1.2155	1.2625	1.3108	1.3605	1.4116	1.4641
5	1.0510	1.1041	1.1593	1.2167	1.2763	1.3382	1.4026	1.4693	1.5386	1.6105
6	1.0615	1.1262	1.1941	1.2653	1.3401	1.4185	1.5007	1.5869	1.6771	1.7716
7	1.0721	1.1487	1.2299	1.3159	1.4071	1.5036	1.6058	1.7138	1.8280	1.9487
8	1.0829	1.1717	1.2668	1.3686	1.4775	1.5938	1.7182	1.8509	1.9926	2.1436
9	1.0937	1.1951	1.3048	1.4233	1.5513	1.6895	1.8385	1.9990	2.1719	2.3579
10	1.1046	1.2190	1.3439	1.4802	1.6289	1.7908	1.9672	2.1589	2.3674	2.5937
11	1.1157	1.2434	1.3842	1.5395	1.7103	1.8983	2.1049	2.3316	2.5804	2.8531
12	1.1268	1.2682	1.4258	1.6010	1.7959	2.0122	2.2522	2.5182	2.8127	3.1384
13	1.1381	1.2936	1.4685	1.6651	1.8856	2.1329	2.4098	2.7196	3.0658	3.4523
14	1.1495	1.3195	1.5126	1.7317	1.9799	2.2609	2.5785	2.9372	3.3417	3.7975
15	1.1610	1.3459	1.5580	1.8009	2.0789	2.3966	2.7590	3.1722	3.6425	4.1772
16	1.1726	1.3728	1.6047	1.8730	2.1829	2.5404	2.9522	3.4259	3.9703	4.5950
17	1.1843	1.4002	1.6528	1.9479	2.2920	2.6928	3.1588	3.7000	4.3276	5.0545
18	1.1961	1.4282	1.7024	2.0258	2.4066	2.8543	3.3799	3.9960	4.7171	5.5599
19	1.2081	1.4568	1.7535	2.1068	2.5270	3.0256	3.6165	4.3157	5.1417	6.1159
20	1.2202	1.4859	1.8061	2.1911	2.6533	3.2071	3.8697	4.6610	5.6044	6.7275
21	1.2324	1.5157	1.8603	2.2788	2.7860	3.3996	4.1406	5.0338	6.1088	7.4002
22	1.2447	1.5460	1.9161	2.3699	2.9253	3.6035	4.4304	5.4365	6.6586	8.1403
23	1.2572	1.5769	1.9736	2.4647	3.0715	3.8197	4.7405	5.8715	7.2579	8.9543
24	1.2697	1.6084	2.0328	2.5633	3.2251	4.0489	5.0724	6.3412	7.9111	9.8497
25	1.2824	1.6406	2.0938	2.6658	3.3864	4.2919	5.4274	6.8485	8.6231	10.834
26	1.2953	1.6734	2.1566	2.7725	3.5557	4.5494	5.8074	7.3964	9.3992	11.918
27	1.3082	1.7069	2.2213	2.8834	3.7335	4.8223	6.2139	7.9881	10.245	13.110
28	1.3213	1.7410	2.2879	2.9987	3.9201	5.1117	6.6488	8.6271	11.167	14.421
29	1.3345	1.7758	2.3566	3.1187	4.1161	5.4184	7.1143	9.3173	12.172	15.863
30	1.3478	1.8114	2.4273	3.2434	4.3219	5.7435	7.6123	10.062	13.267	17.449
40	1.4889	2.2080	3.2620	4.8010	7.0400	10.285	14.974	21.724	31.409	45.259
50	1.6446	2.6916	4.3839	7.1067	11.467	18.420	29.457	46.901	74.357	117.39
60	1.8167	3.2810	5.8916	10.519	18.679	32.987	57.946	101.25	176.03	304.48

(*Contd.*)

TABLE B5 (Contd.)

Periods	*12%*	*14%*	*15%*	*16%*	*18%*	*20%*	*24%*	*28%*	*32%*	*36%*
1	1.1200	1.1400	1.1500	1.1600	1.1800	1.2000	1.2400	1.2800	1.3200	1.3600
2	1.2544	1.2996	1.3225	1.3456	1.3924	1.4400	1.5376	1.6384	1.7424	1.8496
3	1.4049	1.4815	1.5209	1.5609	1.6430	1.7280	1.9066	2.0972	2.3000	2.5155
4	1.5735	1.6890	1.7490	1.8106	1.9388	2.0736	2.3642	2.6844	3.0360	3.4210
5	1.7623	1.9254	2.0114	2.1003	2.2878	2.4883	2.9316	3.4360	4.0075	4.6526
6	1.9738	2.1950	2.3131	2.4364	2.6996	2.9860	3.6352	4.3980	5.2899	6.3275
7	2.2107	2.5023	2.6600	2.8262	3.1855	3.5832	4.5077	5.6295	6.9826	8.6054
8	2.4760	2.8526	3.0590	3.2784	3.7589	4.2998	5.5895	7.2058	9.2170	11.703
9	2.7731	3.2519	3.5179	3.8030	4.4355	5.1598	6.9310	9.2234	12.166	15.916
10	3.1058	3.7072	4.0456	4.4114	5.2338	6.1917	8.5944	11.805	16.059	21.646
11	3.4785	4.2262	4.6524	5.1173	6.1759	7.4301	10.657	15.111	21.198	29.439
12	3.8960	4.8179	5.3502	5.9360	7.2876	8.9161	13.214	19.342	27.982	40.037
13	4.3635	5.4924	6.1528	6.8858	8.5994	10.699	16.386	24.758	36.937	54.451
14	4.8871	6.2613	7.0757	7.9875	10.147	12.839	20.319	31.691	48.756	74.053
15	5.4736	7.1379	8.1371	9.2655	11.973	15.407	25.195	40.564	64.358	100.71
16	6.1304	8.1372	9.3576	10.748	14.129	18.488	31.242	51.923	84.953	136.96
17	6.8660	9.2765	10.761	12.467	16.672	22.186	38.740	66.461	112.13	186.27
18	7.6900	10.575	12.375	14.462	19.673	26.623	48.038	85.070	148.02	253.33
19	8.6128	12.055	14.231	16.776	23.214	31.948	59.567	108.89	195.39	344.53
20	9.6463	13.743	16.366	19.460	27.393	38.337	73.864	139.37	257.91	468.57
21	10.803	15.667	18.821	22.574	32.323	46.005	91.591	178.40	340.44	637.26
22	12.100	17.861	21.644	26.186	38.142	55.206	113.57	228.35	449.39	866.67
23	13.552	20.361	24.891	30.376	45.007	66.247	140.83	292.30	593.19	1178.6
24	15.178	23.212	28.625	35.236	53.108	79.496	174.63	374.14	783.02	1602.9
25	17.000	26.461	32.918	40.874	62.668	95.396	216.54	478.90	1033.5	2180.0
26	19.040	30.166	37.856	47.414	73.948	114.47	268.51	612.99	1364.3	2964.9
27	21.324	34.389	43.535	55.000	87.259	137.37	332.95	784.63	1800.9	4032.2
28	23.883	39.204	50.065	63.800	102.96	164.84	412.86	1004.3	2377.2	5483.8
29	26.749	44.693	57.575	74.008	121.50	197.81	511.95	1285.5	3137.9	7458.0
30	29.959	50.950	66.211	85.849	143.37	237.37	634.81	1645.5	4142.0	10143.0
40	93.050	188.88	267.86	378.72	750.37	1469.7	5455.9	19426	66520	*
50	289.00	700.23	1083.6	1670.7	3927.3	9100.4	46890	*	*	*
60	897.59	2595.9	4383.9	7370.1	20555	56347	*	*	*	*

* Greater than 99999

TABLE B6 Compound Value of an Annuity of Re 1 Per Period at the End of n Periods

Periods	*1%*	*2%*	*3%*	*4%*	*5%*	*6%*	*7%*	*8%*	*9%*	*10%*
1	1.0000	1.0000	1.0000	1.0000	1.0000	1.0000	1.0000	1.0000	1.0000	1.0000
2	2.0100	2.0200	2.0300	2.0400	2.0500	2.0600	2.0700	2.0800	2.0900	2.1000
3	3.0301	3.0604	3.0909	3.1216	3.1525	3.1836	3.2149	3.2464	3.2781	3.3100
4	4.0604	4.1216	4.1836	4.2465	4.3101	4.3746	4.4399	4.5061	4.5731	4.6410
5	5.1010	5.2040	5.3091	5.4163	5.5256	5.6371	5.7507	5.8666	5.9847	6.1051
6	6.1520	6.3081	6.4684	6.6330	6.8019	6.9753	7.1533	7.3359	7.5233	7.7156
7	7.2135	7.4343	7.6625	7.8983	8.1420	8.3938	8.6540	8.9228	9.2004	9.4872
8	8.2857	8.5830	8.8923	9.2142	9.5491	9.8975	10.259	10.636	11.028	11.435
9	9.3685	9.7546	10.159	10.582	11.026	11.491	11.978	12.487	13.021	13.579
10	10.462	10.949	11.463	12.006	12.577	13.180	13.816	14.486	15.192	15.937
11	11.566	12.168	12.807	13.486	14.206	14.971	15.783	16.645	17.560	18.531
12	12.682	13.412	14.192	15.025	15.917	16.869	17.888	18.977	20.140	21.384
13	13.809	14.680	15.617	16.626	17.713	18.882	20.140	21.495	22.953	24.522
14	14.947	15.973	17.086	18.291	19.598	21.015	22.550	24.214	26.019	27.975
15	16.096	17.293	18.598	20.023	21.578	23.276	25.129	27.152	29.360	31.772
16	17.257	18.639	20.156	21.824	23.657	25.672	27.888	30.324	33.003	35.949
17	18.430	20.012	21.761	23.697	25.840	28.212	30.840	33.750	36.973	40.544
18	19.614	21.412	23.414	25.645	28.132	30.905	33.999	37.450	41.301	45.599
19	20.810	22.840	25.116	27.671	30.539	33.760	37.379	41.446	46.018	51.159
20	22.019	24.297	26.870	29.778	33.066	36.785	40.995	45.762	51.160	57.275
21	23.239	25.783	28.676	31.969	35.719	39.992	44.865	50.422	56.764	64.002
22	24.471	27.299	30.536	34.248	38.505	43.392	49.005	55.456	62.873	71.402
23	25.716	28.845	32.452	36.617	41.430	46.995	53.436	60.893	69.531	79.543
24	26.973	30.421	34.426	39.082	44.502	50.185	58.176	66.764	76.789	88.497
25	28.243	32.030	36.459	41.645	47.727	54.864	63.249	73.105	84.700	98.347
26	29.525	33.670	38.553	44.311	51.113	59.156	68.676	79.954	93.323	109.18
27	30.820	35.344	40.709	47.084	54.669	63.705	74.483	87.350	102.72	121.09
28	32.129	37.051	42.930	49.967	58.402	68.528	80.697	95.338	112.96	134.20
29	33.450	38.792	45.218	52.966	62.322	73.639	87.346	103.96	124.13	148.63
30	34.784	40.568	47.575	56.084	66.438	79.058	94.460	113.28	136.30	164.49
40	48.886	60.402	75.401	95.025	120.79	154.76	199.63	259.05	337.88	442.59
50	64.463	84.579	112.79	152.66	209.34	290.33	406.52	573.76	815.08	1163.9
60	81.669	114.05	163.05	237.99	353.58	533.12	813.52	1253.2	1944.7	3034.8

(*Contd.*)

TABLE B6 (Contd.)

Periods	*12%*	*14%*	*15%*	*16%*	*18%*	*20%*	*24%*	*28%*	*32%*	*36%*
1	1.0000	1.0000	1.0000	1.0000	1.0000	1.0000	1.0000	1.0000	1.0000	1.0000
2	2.1200	2.1400	2.1500	2.1600	2.1800	2.2000	2.2400	2.2800	2.3200	2.3600
3	3.3744	3.4396	3.4725	3.5056	3.5724	3.6400	3.7776	3.9184	4.0624	4.2096
4	4.7793	4.9211	4.9934	5.0665	5.2154	5.3680	5.6842	6.0156	6.3624	6.7251
5	6.3528	6.6101	6.7424	6.8771	7.1542	7.4416	8.0484	8.6999	9.3983	10.146
6	8.1152	8.5355	8.7537	8.9775	9.4420	9.9299	10.980	12.135	13.405	14.798
7	10.089	10.730	11.066	11.413	12.141	12.915	14.615	16.533	18.695	21.126
8	12.299	13.232	13.726	14.240	15.327	16.499	19.122	22.163	25.678	29.731
9	14.775	16.085	16.785	17.518	19.085	20.798	24.712	29.369	34.895	41.435
10	17.548	19.337	20.303	21.321	23.521	25.958	31.643	38.592	47.061	57.351
11	20.654	23.044	24.349	25.732	28.755	32.150	40.237	50.398	63.121	78.998
12	24.133	27.270	29.001	30.850	34.931	39.580	50.894	65.510	84.320	108.43
13	28.029	32.088	34.351	36.786	42.218	48.496	64.109	84.852	112.30	148.47
14	32.392	37.581	40.504	43.672	50.818	59.195	80.496	109.61	149.23	202.92
15	37.279	43.842	47.580	51.659	60.965	72.035	100.81	141.30	197.99	276.97
16	42.753	50.980	55.717	60.925	72.939	87.442	126.01	181.86	262.35	377.69
17	48.883	59.117	65.075	71.673	87.068	105.93	157.25	233.79	347.30	514.66
18	55.749	68.394	75.836	84.140	103.74	128.11	195.99	300.25	459.44	700.93
19	63.439	78.969	88.211	98.603	123.41	154.74	244.03	385.32	607.47	954.27
20	72.052	91.024	102.44	115.37	146.62	186.68	303.60	494.21	802.86	1298.8
21	81.698	104.76	118.81	134.84	174.02	225.02	377.46	633.59	1060.7	1767.3
22	92.502	120.43	137.63	157.41	206.34	271.03	469.05	811.99	1401.2	2404.6
23	104.60	138.29	159.27	183.60	244.48	326.23	582.62	1040.3	1850.6	3271.3
24	118.15	158.65	184.16	213.97	289.49	392.48	723.46	1332.6	2443.8	4449.9
25	133.33	181.87	212.79	249.21	342.60	471.98	898.09	1706.8	3226.8	6052.9
26	150.33	208.33	245.71	290.08	405.27	567.37	1114.6	2185.7	4260.4	8233.0
27	169.37	238.49	283.56	337.50	479.22	681.85	1383.1	2798.7	5624.7	11197.9
28	190.69	272.88	327.10	392.50	566.48	819.22	1716.0	3583.3	7425.6	15230.2
29	214.58	312.09	377.16	456.30	669.44	984.06	2128.9	4587.6	9802.9	20714.1
30	241.33	356.78	434.74	530.31	790.94	1181.8	2640.9	5873.2	12940	28172.2
40	767.09	1342.0	1779.0	2360.7	4163.2	7343.8	22728	69377	*	*
50	2400.0	4994.5	7217.7	10435	21813	45497	*	*	*	*
60	7471.6	18535	29219	46057	*	*	*	*	*	*

* Greater than 99999

TABLE B7 Random Numbers

22 17 68 65 84	68 95 23 92 35	61 09 43 95 06	87 02 22 57 51	58 24 82 03 47
19 36 27 59 46	13 79 93 37 55	85 52 05 30 62	39 77 32 77 09	47 83 51 62 74
16 77 23 02 77	09 61 87 25 21	16 71 13 59 78	28 06 24 25 93	23 05 47 47 25
78 43 76 71 61	20 44 90 32 64	46 38 03 93 22	97 67 63 99 61	69 81 21 99 21
03 28 28 26 08	73 37 32 04 05	88 69 58 28 99	69 30 16 09 05	35 07 44 75 47
93 22 53 64 39	07 10 63 76 35	08 13 13 85 51	87 03 04 79 88	55 34 57 72 69
78 76 58 54 74	92 38 70 96 92	82 63 18 27 44	52 06 79 79 45	69 66 92 19 09
23 68 35 26 00	99 53 93 61 28	56 65 05 61 86	52 70 05 48 34	90 92 10 70 80
15 39 25 70 99	93 86 52 77 65	22 87 26 07 47	15 33 59 05 28	86 96 98 29 06
58 71 96 30 24	18 46 23 34 27	49 18 09 79 49	85 13 99 24 44	74 16 32 23 02
57 35 27 33 72	24 53 63 94 09	44 04 95 49 66	41 10 76 47 91	39 60 04 59 81
48 50 86 54 48	22 06 34 72 52	33 29 94 71 11	82 21 15 65 20	15 91 29 12 02
61 96 48 95 03	07 16 39 33 66	77 21 30 27 12	98 56 10 56 79	90 49 22 23 62
36 93 89 41 26	29 70 83 63 51	87 09 41 15 09	99 74 20 52 36	98 60 16 03 03
18 87 00 42 31	57 90 12 02 07	54 08 01 88 63	23 47 37 17 31	39 41 88 92 10
88 56 53 27 59	33 35 72 67 47	08 18 27 38 90	77 34 55 45 70	16 95 86 70 75
09 72 95 84 29	49 41 31 06 70	64 84 73 31 65	42 38 06 45 18	52 53 37 97 15
12 96 88 17 31	65 19 69 02 83	24 64 19 35 51	60 75 86 90 68	56 61 87 39 12
85 94 57 24 16	92 09 84 38 76	29 81 94 78 70	22 00 27 69 85	21 94 47 90 12
38 64 43 59 98	98 77 87 68 07	40 98 05 93 78	91 51 67 62 44	23 32 65 41 18
53 44 09 42 72	00 41 86 79 79	35 55 31 51 51	68 47 22 00 20	00 83 63 22 55
40 76 66 26 84	57 99 99 90 37	37 40 13 68 97	36 63 32 08 58	87 64 81 07 83
02 17 79 18 05	12 59 52 57 02	28 14 11 30 79	22 07 90 47 03	20 69 22 40 98
95 17 82 06 53	31 51 10 96 46	56 11 50 81 69	92 06 88 07 77	40 23 72 51 39
35 76 22 42 92	96 11 83 44 80	33 42 40 90 60	34 68 35 48 77	73 96 53 97 86
26 29 13 56 41	85 47 04 66 08	82 43 80 46 15	34 72 57 59 13	38 26 61 70 04
77 80 20 75 82	72 82 32 99 90	89 73 44 99 05	63 95 73 76 63	48 67 26 43 18
46 40 66 44 52	91 36 74 43 53	78 45 63 98 35	30 82 13 54 00	55 03 36 67 68
37 56 08 18 09	77 53 84 46 47	24 16 74 11 53	31 91 18 95 58	44 10 13 85 57
61 65 61 68 66	37 27 47 39 19	53 21 40 06 71	84 83 70 07 48	95 06 79 88 54
93 43 69 64 07	34 18 04 52 35	61 85 53 83 45	56 27 09 24 86	19 90 70 99 00
21 96 60 12 99	11 20 99 45 18	18 37 79 49 90	48 13 93 55 34	65 97 38 20 46
95 20 47 97 97	27 37 83 28 71	45 89 09 39 84	00 06 41 41 74	51 67 11 52 49
97 86 21 78 73	10 65 81 92 59	04 76 62 16 17	58 76 17 14 97	17 95 70 45 80
69 92 06 34 13	59 71 74 17 32	23 71 82 13 74	27 55 10 24 19	63 52 52 01 41

TABLE B7 Random Numbers

04 31 17 21 56	33 73 99 19 87	53 77 57 68 93	26 72 39 27 67	60 61 97 22 61
61 06 98 03 91	87 14 77 43 96	45 60 33 01 07	43 00 65 98 50	98 99 46 50 47
85 93 85 86 88	72 87 08 62 40	23 21 34 74 97	16 06 10 89 20	76 38 03 29 63
21 74 32 47 45	73 96 07 94 52	25 76 16 19 33	09 65 90 77 47	53 05 70 53 30
15 69 53 82 80	79 96 23 53 10	45 33 02 43 70	65 39 07 16 29	02 87 40 41 45
02 89 08 04 49	20 21 14 68 86	11 29 01 95 80	87 63 93 95 17	35 14 97 35 33
87 18 15 89 79	85 43 01 72 73	89 74 39 82 15	08 61 74 51 69	94 51 33 41 67
98 83 71 94 22	59 97 50 99 52	87 80 61 65 31	08 52 85 08 40	91 51 80 32 44
10 08 58 21 66	72 68 49 29 31	59 73 19 85 23	89 85 84 46 06	65 09 29 75 25
47 90 56 10 08	88 02 84 27 83	66 56 45 65 79	42 29 72 23 19	20 71 53 20 25
22 85 61 68 90	49 64 92 85 44	50 14 49 81 06	16 40 12 89 88	01 82 77 45 12
67 80 43 79 33	12 83 11 41 16	77 02 54 00 52	25 58 19 68 70	53 43 37 15 26
27 62 50 96 72	79 44 61 40 15	27 31 58 50 28	14 53 40 65 39	11 39 03 34 25
33 78 80 87 15	38 30 06 38 21	54 96 87 53 32	14 47 47 07 26	40 36 40 96 76
13 13 92 66 99	47 24 49 57 74	10 97 11 69 84	32 25 43 62 17	99 63 22 32 98
10 27 53 96 23	71 50 54 36 23	04 14 12 15 09	54 31 04 82 98	26 78 25 47 47
28 41 50 61 88	64 85 27 20 18	39 71 65 09 62	83 36 36 05 56	94 76 62 11 89
34 21 42 57 02	59 19 18 97 48	05 24 67 70 07	80 30 03 30 98	84 97 50 87 46
61 81 77 23 23	82 82 11 54 08	44 07 39 55 43	53 28 70 58 96	42 34 43 39 28
61 15 18 13 54	16 86 20 26 88	14 53 90 51 17	90 74 80 55 09	52 01 63 01 59

APPENDIX C

Answers to Practical Problems

CHAPTER 2

1. Max $Z = 40x_1 + 30x_2$ s.t. $400x_1 + 350x_2 \leq 250{,}000$; $85x_1 + 50x_2 \leq 26{,}100$; $55x_1 + 30x_2 \leq 43{,}500$; $20x_1 \leq 17{,}400$; $x_1 \leq x_2$; $x_1, x_2 \geq 0$
2. Max $Z = 500x_1 + 600x_1 + 1{,}200x_3$ s.t. $2x_1 + 4x_2 + 6x_3 \leq 160$; $3x_1 + 2x_2 + 4x_3 \leq 120$; $x_1, x_2, x_3 \geq 0$
3. Max $Z = 16x_1 + 25x_2 + 12x_3 + 12x_4$ s.t. $45x_1 + 60x_2 + 45x_3 + 30x_4 \leq 1{,}920$; $x_4 \leq 30$; $x_1 + x_2 + x_3 \leq 70$; $x_3 + x_4 \leq 60$; $x_4 \geq 6$; $x_2 \geq 10$; $x_1, x_2, x_3, x_4 \geq 0$
4. Max $Z = 20x_1 + 30x_2$ s.t. $10x_1 + 40x_2 \leq 4{,}000$; $3x_1 + 2x_2 \leq 2{,}000$; $2x_1 + 3x_2 \leq 1{,}400$; $x_1, x_2 \geq 0$
5. Max $Z = 1.72x_{14} + 1.45x_{23} + 1.02x_{32} + 1.005x_{41}$ s.t. $x_{11} + x_{12} + x_{13} + x_{14} + 36{,}000 = 30{,}000 + 28{,}000$; $x_{21} + x_{22} + x_{23} + 31{,}000 = 1.005x_{11} + 52{,}000$; $x_{31} + x_{32} + 40{,}000 = 1.02x_{12} + 1.005x_{21} + 3{,}400$; $x_{41} + 20{,}000 = 1.45x_{13} + 1.02x_{22} + 1.005x_{31} + 22{,}000$; $x_{ij} \geq 0$
6. Max $Z = 22.50x_1 + 40x_2 + 15x_3$ s.t. $0.01x_1 + 0.03x_2 + 0.02x_3 \leq 120$; $0.20x_1 + 0.50x_2 + 0.30x_3 \leq 3{,}000$; $x_1 + x_2 + x_3 \geq 5{,}000$; $x_1 \leq 3{,}000$; $x_2, \leq 2{,}000$; $x_3 \leq 6{,}000$; $x_1; x_2, x_3 \geq 0$
7. Max $Z = 0.14x_1 + 0.19x_2 + 0.23x_3 + 0.12x_4$ s.t. $x_1 + x_2 + x_3 + x_4 \leq 20$; $(12x_1 + 24x_2 + 48x_3 + 64x_4)/(x_1 + x_2 + x_3 + x_4) \leq 42$; $x_4 \geq 2$; $x_3 \leq 0.20\ (x_1 + x_2 + x_3 + x_4)$; $x_i \geq 0$
8. Max $Z = 0.25x_1 + x_2 + 0.95x_3$ s.t. $x_1/25 + x_2/40 + x_3/25 \leq 1$; $x_1/25 + x_2/20 + x_3/20 \leq 1$; $x_1/40 + x_2/30 + x_3/40 \leq 1$; $x_1, x_2, x_3 \geq 0$
9. Min $Z = x_1 + x_2 + x_3 + x_4 + x_5 + x_6 + x_7 + x_8 + x_9 + x_{10} + x_{11} + x_{12}$ s.t. $x_1 + x_2 + x_3 + x_4 \geq 50$; $x_2 + x_3 + x_4 + x_5 \geq 60$; $x_3 + x_4 + x_5 + x_6 \geq 80$; $x_4 + x_5 + x_6 + x_7 \geq 80$; $x_5 + x_6 + x_7 + x_8 \geq 70$; $x_6 + x_7 + x_8 + x_9 \geq 70$; $x_7 + x_8 + x_9 + x_{10} \geq 60$; $x_8 + x_9 + x_{10} + x_{11} \geq 50$; $x_9 + x_{10} + x_{11} + x_{12} \geq 50$; $x_{10} + x_{11} + x_{12} + x_1 \geq 30$; $x_{11} + x_{12} + x_1 + x_2 \geq 20$; $x_{12} + x_1 + x_2 + x_3 \geq 40$; $x_i \geq 0\ i = 1, 2 \ldots 12$
10. Min $Z = x_1 + x_2 + x_3 + x_4 + x_5 + x_6 + x_7$ s.t. $x_1 + x_4 + x_5 + x_6 + x_7 \geq 35$; $x_1 + x_2 + x_5 + x_6 + x_7 \geq 55$; $x_1 + x_2 + x_3 + x_6 + x_7 \geq 60$; $x_1 + x_2 + x_3 + x_4 + x_7 \geq 50$; $x_1 + x_2 + x_3 + x_4 + x_5 \geq 60$; $x_2 + x_3 + x_4 + x_5 + x_6 \geq 50$; $x_3 + x_4 + x_5 + x_6 + x_7 \geq 45$; $0 \leq x_i \leq 40$
11. Min $Z = 600x_1 + 500x_2 + 400x_3$ s.t. $150x_1 + 200x_2 + 160x_3 \geq 2{,}000$; $100x_1 + 100x_2 + 80x_3 \geq 3{,}000$; $500x_1 + 760x_2 + 890x_3 \geq 3{,}000$; $x_1 \leq 20$; $x_2 \leq 20$; $x_3 \leq 18$; $x_i \geq 0$
12. Min $Z = 25x_{11} + 30x_{12} + 40x_{13} + 45x_{14} + 51x_{21} + 41x_{22} + 36x_{23} + 31x_{24}$ s.t. $0.10x_{11} + 0.10x_{12} + 0.10x_{13} + 0.10x_{14} \leq 120$; $0.20x_{11} + 0.20x_{12} + 0.20x_{13} + 0.20x_{14} \leq 260$; $0.11x_{21} + 0.11x_{22} + 0.11x_{23} + 0.11x_{24} \leq 140$; $0.22x_{21} + 0.22x_{22} + 0.22x_{23} + 0.22x_{24} \leq 250$; $x_{11} + x_{21} = 500$; $x_{12} + x_{22} = 300$; $x_{13} + x_{23} = 1{,}000$; $x_{14} + x_{24} = 200$; $x_{ij} \geq 0$
13. Max $Z = 6{,}300x_1 + 10{,}800x_2 + 11{,}340x_3$ s.t. $80{,}000x_1 + 130{,}000x_2 + 150{,}000x_3 \leq 40{,}00{,}000$; $x_1 + x_2 + x_3 \leq 30$; $3x_1 + 6x_2 + 6x_3 \leq 150$; $x_i \geq 0$
14. Max $Z = -1.5x_{11} + 3.5x_{21} + 2.9x_{31} - 3.0x_{21} + 2.0x_{22} + 1.4x_{32} - 4.0x_{13} + x_{23} + 0.4x_{33}$ s.t. $x_{11} + x_{12} + x_{13} \leq 500{,}000$; $x_{21} + x_{22} + x_{23} \leq 500{,}000$; $x_{31} + x_{32} + x_{33} \leq 360{,}000$; $x_{11} \geq 0.50\ (x_{11} + x_{21} + x_{31})$; $x_{21} \leq 0.25\ (x_{11} + x_{21} + x_{31})$; $x_{12} \geq 0.25\ (x_{12} + x_{22} + x_{32})$; $x_{22} \leq 0.50\ (x_{12} + x_{22} + x_{32})$; $x_{ij} \geq 0$
15. Min $Z = 16x_{111} + 20x_{121} + 18x_{112} + 22x_{122} + 20x_{113} + 24x_{123} + 22x_{114} + 26x_{124} + 16x_{212} + 20x_{222} + 18x_{213} + 22x_{223} + 20x_{214} + 24x_{224} + 16x_{313} + 20x_{323} + 18x_{314} + 22x_{324} + 16x_{414} + 18x_{424}$ s.t. $x_{111} + x_{112} + x_{113} + x_{114} \leq 80$; $x_{212} + x_{213} + x_{214} \leq 90$; $x_{313} + x_{314} \leq 95$; $x_{414} \leq 70$; $x_{121} + x_{122} + x_{123} + x_{124} \leq 10$; $x_{222} + x_{223} + x_{224} \leq 10$; $x_{323} + x_{324} \leq 20$; $x_{424} \leq 10$; $x_{111} + x_{121} = 65$; $x_{112} + x_{122} + x_{212} + x_{222} = 80$; $x_{113} + x_{123} + x_{213} + x_{223} + x_{313} + x_{323} = 135$; $x_{114} + x_{124} + x_{214} + x_{224} + x_{314} + x_{324} + x_{414} + x_{424} = 75$; $x_{ijk} \geq 0$
16. Max $Z = 74{,}400x_{11} + 49{,}600x_{12} + 46{,}800x_{21} + 31{,}200x_{22} + 48{,}300x_{31} + 32{,}200x_{32}$ s.t. $18{,}000x_{11} + 12{,}000x_{12} + 16{,}000x_{21} + 10{,}000x_{22} + 19{,}000x_{31} + 15{,}000x_{32} \leq 500{,}000$; $x_{11} + x_{12} \leq 12$; $x_{21} + x_{22} \leq 24$; $x_{31} + x_{32} \leq 12$; $x_{11} + x_{12} \geq 5$; $x_{21} + x_{22} \geq 4$; $x_{31} + x_{32} \geq 5$; $x_{ij} \geq 0$

19. Max $Z = 1{,}600x_1 + 3{,}000x_2 + 5{,}600x_3$ s.t. $15x_1 + 12x_2 + 14x_3 \le 3{,}000$; $4x_1 + 3x_2 + 5x_3 \le 1{,}200$; $2{,}500x_1 + 4{,}500x_2 + 9{,}000x_3 \le 136{,}800$; $x_1 \ge 2$; $x_3 \ge 8$; $x_i \ge 0$
20. $x_1 = 7, x_2 = 1; Z = 10$
21. $x_1 = 6, x_2 = 8; Z = 124$
22. $x_1 = 3, x_2 = 6; Z = 960$
23. (a) Max $Z = 5x_1 + 8x_2$ s.t. $0.250x_1 + 0.500x_2 \le 750$; $0.750x_1 + 0.500x_2 \le 1{,}200$; $x_1, x_2 \ge 0$
    (b) $x_1 = 900$, $x_2 = 1{,}050$ and $Z = 12{,}900$; $x_1 = 0$ and $x_2 = 1{,}500$, or $x_1 = 900$ and $x_2 = 1{,}050$; with $Z = 15{,}000$
24. $x_1 = 150$, $x_2 = 700$, and $Z = 2{,}850$
25. Min $Z = 68.80x_1 + 68.00x_2$ s.t. $320x_1 + 240x_2 \ge 2{,}000$; $x_1 \le 9$; $x_2 \le 11$; $x_1, x_2 \ge 0$; $x_1 = 6.25$, $x_2 = 0$ with $Z = 430$.
26. $x_1 = 2, x_2 = 8$; and $Z = 22$
27. $x_1 = 4, x_2 = 1$; and $Z = 22$
28. $x_1 = 0, x_2 = 5$; and $Z = 15$
29. Max $Z = 400x_1 + 200x_2$ s.t. $18x_1 + 3x_2 \le 800$; $9x_1 + 4x_2 \le 600$; $x_2 \le 150$; $x_1, x_2 \ge 0$. Optimal solution: $x_1 = 0, x_2 = 150$; $Z = 30{,}000$.
30. Min $Z = 20x_1 + 40x_2$ s.t. $36x_1 + 6x_2 \ge 108$; $3x_1 + 12x_2 \ge 36$; $20x_1 + 10x_2 \ge 100$; $x_1, x_2 \ge 0$; $x_1 = 4, x_2 = 2$ with $Z = 160$
31. $x_1 = 4, x_2 = 2$; and $Z = 160$
32. (a) No
    (b) $x_1 = 3, x_2 = 4$; and $Z = 10$
33. Max $Z = 2.80x_1 + 2.20x_2$ s.t. $x_1 \le 20{,}000$; $x_2 \le 40{,}000$; $0.003x_1 + 0.001x_2 \le 66$; $x_1 + x_2 \le 45{,}000$; $x_1, x_2 \ge 0$; $x_1 = 10{,}500$, $x_2 = 34{,}500$; with $Z = 105{,}300$
34. Plant I: 4 days, Plant II: 12 days; TC = Rs 72,000
35. (a) Max $Z = 100x_1 + 115x_2$ s.t. $8x_1 + 10x_2 \le 40{,}000$; $10x_1 + 6x_2 \le 30{,}000$; $4x_1 + 12x_2 \le 36{,}000$; $x_1, x_2 \ge 0$ P_1: 1500 units; P_2: 2500 units, Profit = Rs 152,500
    (b) Labour hours in Department *A*
36. Unbounded solution
37. (a) $x_1 = 2, x_2 = 6$
    (b) $x_1 = 6, x_2 = 2$
    (c) $x_1 = 2, x_2 = 6$
    (d) $x_1 = 2, x_2 = 2/3$
38. $x_1 = 0$ and $x_2 = 8$, or $x_1 = 1$ and $x_2 = 4$; with $Z = 24$
39. No feasible solution
40. Maximisation: Unbounded solution
    Minimisation: $x_1 = 56/17$ and $x_2 = 33/17$
41. Max $Z = x_1 + 6x_2$ s.t. $800x_1 + 4{,}000x_2 \le 16{,}000$; $x_1 \ge 5$; $x_2 \le 4$; $x_1, x_2 \ge 0$; Optimal solution: $x_1 = 5$, $x_2 = 3$ for $Z = 23$. TV spots is redundant.
42. $x_1 = 25, x_2 = 15, Z = 14{,}375$
43. (a) Pixie: 1,000 units, King; 1,000 units; Contribution = Rs 139,000
    (b) Extra profit for every 1,000 hours overtime = Rs 1,125
44. Infeasible solution

CHAPTER 3

1. Max $Z = 8x_1 - 6x_2 + 7x_3 + 0S_1 + 0S_2 + 0S_3 - MA_1 - MA_2$ s.t. $4x_1 + 3x_2 + 6x_3 + x_4 + S_1 = 40; -x_1 + 2x_2 + 3x_3 + x_4 + S_2 = 5; 9x_1 - 5x_2 + 7x_3 - x_4 - S_3 + A_1 = 60; 6x_2 + 2x_3 + 4x_4 + A_2 = 47$; all variables ≥ 0
2. $x_1 = 10, x_2 = 0; Z = 70$
3. $x_1 = 3, x_2 = 0; Z = 63$
4. $x_1 = 0, x_2 = 28/5, x_3 = 116/5; Z = 284$
5. $x_1 = 175/17, x_2 = -72/17; Z = 1688/17$
6. Drug A: 2,000 units, Drug B: 2,000 units; Total profit = Rs 10,000
7. Belt A: 200, Belt B: 600; Profit = Rs 13,000
8. $x_1 = 3/2, x_2 = 2; Z = 8.5$
9. $x_1 = 0, x_2 = 2; Z = 4$; Unique optimal solution
10. Product *A*: 800/7 Units, Product *B*; nil, Product *C* = 480/7 Units. Total Profit = Rs 5280/7
11. Doll *A*: 89/41, Doll *B*: 50/41, Doll *C*: 62/41 Total profit = Rs 765/41. None of the machines would remain idle.
12. Pistons = 100/3, Rings = 200/3, Valves = 0, Profit = Rs 733.33
13. (a) Max $Z = 12x_1 + 3x_2 + x_3$ s.t. $10x_1 + 2x_2 + x_3 \leq 100; 7x_1 + 3x_2 + 2x_3 \leq 77; 2x_1 + 4x_2 + x_3 \leq 80; x_1, x_2, x_3 \geq 0$
    (b) *A*: 73/8 units, *B*: 35/8 units, *C*: nil. Total profit = Rs 981/8
    (c) $S_1 = S_2 = 0, S_3 = 177/4$
14. $x_1 = 30, x_2 = 1185, x_3 = 0; Z = 20,625$
15. (a) Max $Z = 4,000x_1 + 2,000x_2 + 5,000x_3$ s.t. $12x_1 + 7x_2 + 9x_3 \leq 1,260; 22x_1 + 18x_2 + 16x_3 \leq 19,008; 2x_1 + 4x_2 + 3x_3 \leq 396; x_i \geq 0$
    (b) $x_1 = 12, x_2 = 0, x_3 = 124$
    (c) Row = 12, Kayaks = 124, Canoes = 0, Revenue = Rs 668,000
    (d) Wood is spare to the extent of 16,760 board feet.
    (e) 2248 board feet
16. Type *A*: 10, Type *B*: nil, Type *C*: 20; Capacity: 289, 800 tkm
17. $x_1 = 0, x_2 = 70, x_3 = 0, x_4 = 235/2$; and $Z = 610$
18. Desk I: 0, Desk II: 4,000/7, Desk III: 0, Desk IV: 600/7; Total profit = 1,04,000/7
19. $x_1 = 8, x_2 = 0$; with $Z = 48$
20. $x_1 = 21/13, x_2 = 10/13; Z = 31/13$
21. $x_1 = 1/5, x_2 = 6/5, x_3 = 0; Z = 210$
22. $x_1 = 15, x_2 = 20$ for $Z = 2,500$
23. (a) Max $Z = 20x_1 + 30x_2$ s.t. $10x_1 + 20x_2 \leq 30,000; 15x_1 + 5x_2 \leq 30,000; 10x_1 + 8x_2 \leq 40,000; x_1, x_2 \geq 0$
    (b) Ash trays: 1,800, Tea trays: 600; Daily profit = Rs 190
    (c) Ash trays: 1,000, Tea trays: 1,000; Profit = Rs 150/day
24. $x_1 = 15, x_2 = 20$; and $Z = 1,250$
25. (a) Max $Z = 20x_1 + 40x_2$ s.t. $2x_1 + 4x_2 \leq 100; -8x_1 + 24x_2 \leq 0; x_1, x_2 \geq 0$
    (b) $x_1 = 30, x_2 = 10$, with $Z = 1,000$
    (c) $x_1 = 50, x_2 = 0$, with $Z = 1,000$

26. X: 3 tonnes Y: 6 tonnes, Total profit = Rs 960
27. Optimal mix: $I_1 = 15$ and $I_2 = 25$ units. Increase in profit = Rs 300. Idle time on $M_2 = 15$ hours
28. $x_1 = 3, x_2 = 3$, Z = 24,00,000; Type A: 15,90,000, Type B: 8,10,000
29. Max $Z = 1{,}000x_1 + 900x_2 + 280x_3$ s.t. $10{,}000x_1 + 5{,}000x_2 + 6{,}000x_3 \leq 100{,}000$; $x_1 \geq 2$; $0 \leq x_2 \leq 5$; $x_3 \geq$ 2 A: 6.30, B: 5, C = 2 Expected exposure = 11,360 (thousand)
30. $x_1 = 15, x_2 = 20$, with $Z = 5{,}000$ Unbounded solution in case of maximisation
31. Max $Z = 10x_1 + 12x_2$ s.t. $2x_1 + 3x_2 \leq 1{,}500$; $3x_1 + 2x_2 \leq 1{,}500$; $x_1 + x_2 \leq 1{,}000$; $x_1, x_2 \geq 0$ Optimal solution: $x_1 = 300, x_2 = 300$. Hours unused on $M_3 = 400$; Total profit = Rs 7,200
32. C_1: 60 units, C_2: nil, C_3: nil; Profit = 360. No change in answer by the given statement.
33. (a) $x_1 = 0, x_2 = 8/3, x_3 = 56/3$; Profit = 488/3
    (b) Yes. There is no artificial variable in the basis.
    (c) No. It is unique.
    (d) Not degenerate.
    (e) 2 units of x_2 and 5 units of x_3
34. $x_1 = 0, x_2 = 28$ for $Z = 560$
35. $X = 0, Y = 50, Z = 50, Z = 50; \pi = 2{,}750$
36. Unbounded solution
37. Factor A = 2 kg, Factor B = 3 kg; Profit = Rs 28
38. $x_1 = 5/2$ and $x_2 = 45/4$; or $x_1 = 11/3$ and $x_2 = 533/12$; with $Z = 185$
39. Not feasible
40. $x_1 = 0, x_2 = 0, x_3 = 20$; $Z = 2{,}400$; Degenerate solution

CHAPTER 4

1. Min $G = 10y_1 + 2y_2 + 6y_3$ s.t. $y_1 + 2y_2 + 2y_3 \geq 1$; $y_1 - 2y_3 \geq -1$; $y_1 - y_2 + y_3 \geq 3$; $y_i \geq 0$
2. (a) Min $G = 10x_1 - 12x_2$ s.t. $3x_1 + 2x_2 \geq 10$; $x_1 - 3x_2 \geq 8$; $2x_1 - x_2 \leq 6$; $x_1, x_2 \geq 0$
   (b) Max $G = -4y_1 + 13y_2 + 4y_3$ s.t. $y_1 + 2y_2 + y_3 \geq -4$; $-2y_1 + 3y_2 - y_3 \geq 3$; $y_1, y_2, y_3 \geq 0$
3. Min $G = 4a + 26b + 35c$ s.t. $a + 3b + 10c \geq 5$; $a + 8b + 7c \geq 7$; $a, b, c \geq 0$
4. Min $G = 4y_1 + 6y_2 + 5y_3 - y_4$ s.t. $y_1 + y_3 \geq 3$; $y_2 + y_3 - y_4 \geq -2$; $y_i \geq 0$
5. Max $G = 5y_1 + 4y_2$ s.t. $y_1 + 3y_2 \leq 2$; $4y_1 + y_2 \leq 9$; $2y_1 + 2y_2 = 3$; $y_1, y_2 \geq 0$
6. Min $G = 10y_1 - 15y_2 + 7y_3$ s.t. $y_1 - 4y_2 - y_3 \geq 3$; $y_1 + y_2 + y_3 \geq 4$; $y_1 + y_2 + y_3 = 7$; $y_1, y_2 \geq 0$ y_3: unrest. in sign
7. Max $G = 2y_1^* + 6y_2 - 3y_3$ s.t. $3y_1^* + 4y_2 - 4y_3 \leq 4$; $y_1^* + 3y_2 - 2y_3 \leq 1$; $y_2, y_3 \geq 0$; y_1^*: unrest. in sign
8. (a) Min $G = 30y_1 + 10y_2$ s.t. $2y_1 + y_2 + y_3 \geq 2$; $3y_1 + 2y_2 - y_3 \geq 3$; $y_i \geq 0$
   (b) Solution 2 is optimal
   (c) Min $G = 10y + 16y_3$ s.t. $y + 2y_3 - 3y_4 \geq 7$; $y - y_3 - y_4 \geq 5$; $3y + 3y_3 + 2y_4 = 2$; $y_3, y_4 \geq 0$, y: unrest. in sign
9. (i) Models A and B: none, Model C: 15 cars; Max profit = Rs 450,000
   (ii) Process I: Rs 150/worker day 0 – 10,000/3
   Process II: nil 5,400 – ∞
   Process III: nil 2,400 – ∞

(iii) Min $G = 3{,}000y_1 + 6{,}000y_2 + 3{,}000y_3$ s.t. $60y_1 + 100y_2 + 80y_3 \geq 7{,}500$; $100y_1 + 240y_2 + 100y_3 \geq 15{,}000$; $200y_1 + 360y_2 + 160y_3 \geq 30{,}000$; $y_i \geq 0$ optimal values; $y_1 = 150$, $y_2 = y_3 = 0$

(iv) $x_1 = 0, x_2 = 10, x_3 = 10$

10. (a) Max $Z = 30x_1 + 20x_2 + 12x_3$ s.t. $8x_1 + 4x_2 + 3x_3 \leq 640$; $4x_1 + 6x_2 + 2x_3 \leq 540$; $x_1 + x_2 + x_3 \leq 100$; $x_i \geq 0$
 (b) Tables = 60, Chairs = 40, Book cases = 0, Profit = Rs 2,600
 (c) Timber: Rs 2.50/cf Assembly: nil Finishing: Rs 10/mh
 (d) For tables (x_1): 20–40; For chairs (x_2): 15.60–30
 For resources: Timber = 520 – 800, Assembly = 480 – ∞, Finishing = 80 – 107.50
 (e) Book cases not being produced. Optimal solution unique.
11. $x_1 = 10, x_2 = 0, Z = 8{,}000$
12. (a) $x = 60, y = 50$, Increase = Rs 1,720
 (b) Deptt. I 75 hours
 (c) Rs 5.60/unit
13. Model A: nil, Model B: 400, Model C: nil; Assembly: nil, Wiring: Rs 633.33/hour, Packaging: nil
 Min $G = 80y_1 + 120y_2 + 100y_3$ s.t. $0.1y_1 + 0.2y_2 + 0.1y_3 \geq 120$; $0.2y_1 + 0.3y_2 + 0.1y_3 \geq 190$; $0.3y_1 + 0.4y_2 + 0.1y_3 \geq 210$; $y_i \geq 0$; $y_1 = 0, y_2 = 633.33, y_3 = 0$
14. Primal: $x_1 = 0, x_2 = 95/4, x_3 = 0, x_4 = 0$; Dual $y_1 = 0, y_2 = 0, y_3 = 50$; $Z = G = 9{,}500$
15. (a) Max $Z = 21X_1 + 25X_2 + 16X_3$ s.t. $0.1X_1 + 0.1X_2 + 0.2X_3 \leq 1{,}200$; $0.1X_1 + 0.2X_2 + 0.1X_3 \leq 2{,}000$; $0.2X_1 + 0.1X_2 + 0.1X_3 \leq 2{,}200$; $X_i \geq 0$
 (c) Entering variable X_2 (with largest Δ_j), leaving variable X_5
 (d) Product mix: $X_1 = 4{,}000$, $X_2 = 8{,}000$, X_3: nil; contribution = 284,000 p.m.
 (e) (i) $X_1 = 6{,}000$, $X_2 = 7{,}000$, Contribution = 301,000
 (ii) $X_1 = 3{,}400$, $X_2 = 8{,}200$, $X_3 = 200$, Contribution = Rs 279,600
16.

Variable	y_1	y_2	y_3	y_4	y_5	y_6	OF value = 76/3
Solution	0	4/15	1/15	0	0	0	
Δ_j	–80/3	0	0	–160/3	–8/3	–20/3	

17. (a) Min $C = 2x_1 + 4x_2$ s.t. $x_1 + x_2 \geq 200$; $0.25x_1 + 0.75x_2 \geq 100$; $0.01x_1 + 0.20x_2 \leq 35$; $x_1, x_2 \geq 0$
 (b) Max $Z = 200y_1 + 100y_2 - 35y_3$ s.t. $y_1 + 0.25y_2 - 0.10y_3 \leq 2$; $y_1 + 0.75y_2 - 0.20y_3 \leq 4$, $y_i \geq 0$; Primal: $x_1 = 100, x_2 = 100, C = 600$
18. (a) Max $Z = 400x_1 + 200x_2 + 100x_3$ s.t. $2x_1 + 3x_2 + 2.5x_3 \leq 1{,}920$; $3x_1 + 2x_2 + 2x_3 \leq 2{,}200$; $x_1 \leq 200$; $x_i \geq 0$
 (b) x_4: Unused hours PI, x_5: Unused hours PII, x_6: Unused sales potential
 (c) $x_1 = 200, x_2 = 506.7, x_3 =$ nil
 (i) increase in $x_2 = 6.6$ units, contribution increase = Rs 1,333.4
 (ii) increase in $x_1 = 10$ units, $x_2 = 6.7$ units, contribution = Rs 2,667
 (iii) decrease in $x_2 = 8.3$ units, contribution reduction = Rs 666.7
19. Primal: $x_1 = 4, x_2 = 1$; Dual: $y_1 = 5/26, y_2 = 0, y_3 = 1/13, Z = 14$
20. Primal: $x_1 = 15, x_2 = 10, x_3 = 0$. The dual is: Min $G = 25y_1 + 40y_2 + 25y_3 + 60y_4$ s.t. $y_1 + 2y_2 + y_3 \geq 70$; $y_1 + y_2 + y_3 \geq 50$; $y_1 + y_2 + y_4 \geq 50$; $y_i \geq 0$. $y_1 = 30, y_2 = 20, y_3 =$ nil, $y_4 =$ nil
21. $x_1 = 100/3, x_2 = 200/3, x_3 = 0$; $Z = 7{,}333.33$ Dual: Engg. time = Rs 100/3 ph, Labour time = Rs 20/3 ph, Administrative time = nil

22. (a) $x_1 = 0, x_2 = 40/3, x_3 = 100/3$, Z = Rs 920/3
    (b) Centre 1: nil, centre 2: Rs 5/3 per hour, Centre 3: Rs 4/3 per hour; Centre 2
    (c) 320/3 to infinity; 80 to 200; 30 to 120
    (d) –infinity to 23/3; 3 to 12; 4 to 16
    (e) $y_1 = 0, y_2 = 5/3, y_3 = 4/3$
23. (i) Yes (ii) No (iii)No (iv) Yes (v) Yes (v) Re 1 (vii) $x_1 = 0, x_2 = 6$, Rs 24 (viii) Rs 8 (ix) A: Rs 4/hour B: nil (x) A: Rs 4/hour B: nil
24. $x_1 = 0, x_2 = 50, x_3 = 50; Z = 700$
    For solution 1: Milling: nil, Lathe = Rs 2/hour, Grinder = Rs 8/hour
    Ranges: Milling 250 – ∞; Lathe 0 – 150; Grinder 0 – 50
    For solution 2: Milling: Rs 8/9 ph, Lathe: Rs 2/9 ph, Grinder: nil; Ranges: Milling 100 – 250, Lathe 150 – 375, Grinder 50 – ∞
25. Primal: Infeasibility, Dual: Infeasibility
26. (a) If demand of A increases, contribution increases by 2 units. B has a zero shadow price. Product of C would reduce profit @ 3.
27. (a) Yes, all $\Delta_j \le 0$
    (b) Yes, no artificial variable in basis
    (c) No
    (d) Rs 5,700
    (e) R_1: Rs 2.9/unit, R_2: Re 0.9/unit, R_3: nil
    (f) Rs 2.9
    (g) $x_1 = 312, x_2 = 292$; Rs 5,718
28. (a) $x_1 = 0, x_2 = 25, x_3 = 125, x_4 = 0, Z = 525$
    (b) No
    (c) $y_1 = 3/10, y_2 = 0, y_3 = 9/5, G = 525$
    (d) Re 3/10 per hour; nil; and Rs 9/5 per hour; Machine III
    (e) Five paise
    (f) Yes, since price increase is more than Δ_j value of 5 paise
29. (a) $x_1 = 120, x_2 = 300, S_1 = S_2 = 0, S_3 = 240$; Z = Rs 39,600
    (b) Min $G = 720y_1 + 1{,}800y_2 + 900y_3$ s.t. $y_1 + 5y_2 + 3y_3 \ge 80$, $2y_1 + 4y_2 + y_3 \ge 100$; $y_i \ge 0$
    (c) $y_1 = 30, y_2 = 10, y_3 = 0$; G = Rs 39,600
    (d) Machining Rs 30/per hour, Fabrication Rs 10 per hour, Assembly nil
    (e) Machining, upto 180 hours
    (f) 288 hours
    (g) Valid for 50–125. No change
    (h) 40–160
    (i) Rs 90
30. (i) Lawn movers 100, snow blowers 200; Rs 19,000
    (ii) Labour Rs 15/hour, Steel nil, Snowblower Rs 20/engine
    (iii) Labour hours 800–1,067, Steel 1,000 – ∞, Engines 180 – 250

(iv) x_1: 0 – 40 x_2: 60 – ∞

(v) Min $G = 1{,}000y_1 + 1{,}200y_2 + 200y_3$ s.t. $2y_1 + 6y_2 \geq 30$; $4y_1 + 2y_2 + y_3 \geq 80$; $y_i \geq 0$. Solution: $y_1 = 15$, $y_2 = 0$, $y_3 = 20$.

CHAPTER 5

1. $x_{13} = 25, x_{14} = 30, x_{21} = 10, x_{22} = 20, x_{23} = 15, x_{31} = 30, x_{43} = 10, x_{45} = 40$. TC = 695. Alternate solution: $x_{12} = 20, x_{13} = 5, x_{14} = 30, x_{21} = 10, x_{23} = 35, x_{31} = 30, x_{43} = 10, x_{45} = 40$
2. Non-optimal, TC = 1,335. Optimal: $x_{13} = 25, x_{14} = 30, x_{21} = 10, x_{22} = 20, x_{23} = 15, x_{31} = 30, x_{43} = 10, x_{45} = 40$, TC = 1,295
3. $x_{12} = 4, x_{14} = 4, x_{24} = 4, x_{25} = 8, x_{31} = 4, x_{33} = 6$ TC = 80
4. $x_{12} = 7, x_{21} = 3, x_{24} = 4, x_{33} = 8, x_{34} = 2, x_{44} = 6, x_{52} = 3, x_{54} = 2, x_{61} = 6$; TC = 336
5. $x_{12} = 18, x_{13} = 2, x_{23} = 10, x_{24} = 6, x_{31} = 15, x_{34} = 8$
6. $x_{12} = 10, x_{13} = 20, x_{23} = 5, x_{24} = 35, x_{31} = 20, x_{32} = 10$; AS: $x_{11} = 10, x_{13} = 20, x_{23} = 5, x_{24} = 35, x_{31} = 10, x_{32} = 20$
7. $x_{12} = 10, x_{21} = 20, x_{22} = 10, x_{23} = 50, x_{31} = 15$; Rs 830 + 160 (Penalty)
8. $x_{11} = 5, x_{14} = 2, x_{22} = 3, x_{23} = 7, x_{32} = 5, x_{34} = 13$; Max Savings = Rs 296
9. $x_{13} = 10, x_{21} = 8, x_{24} = 4, x_{32} = 8$; TC = Rs 67,000,000; When CX is not allowed: $x_{11} = 8, x_{13} = 2, x_{21} = 1, x_{22} = 7, x_{24} = 4, x_{33} = 8$; TC = Rs 106,000,000
10. (a) $x_{12} = 60, x_{22} = 50, x_{24} = 90, x_{31} = 30, x_{33} = 120, x_{42} = 50$; TC = Rs 12,300
    (b) No cost reduction
11. (ii) $x_{12} = 56, x_{22} = 41, x_{23} = 41, x_{31} = 72, x_{32} = 5$; TC = 2,744
    (iii) Yes. $x_{12} = 56, x_{22} = 41, x_{23} = 41, x_{31} = 72, x_{32} = 5$
    (iv) They provide operational flexibility.
    (v) $x_{12} = 36, x_{13} = 20, x_{22} = 61, x_{23} = 21, x_{31} = 72, x_{32} = 5$; Cost up Rs 20
    (vi) No change
12. (i) No
    (ii) Yes, unique since all $\Delta_{ij} \leq 0$
    (iii) Rs 15
    (iv) Increase Rs 8
13. (a) $x_{12} = 15, x_{13} = 85, x_{22} = 20, x_{24} = 105, x_{31} = 60, x_{32} = 45, x_{35} = 70$; TC = Rs 7,605
    (b) $x_{12} = 15, x_{13} = 85, x_{21} = 20, x_{24} = 105, x_{31} = 40, x_{32} = 65$ and $x_{35} = 70$
    (c) Max $Z = 100u_1 + 125u_2 + 175u_3 + 60v_1 + 80v_2 + 85v_3 + 105v_4 + 70v_5$ s.t. $u_1 + v_1 \leq 20$; $u_1 + v_2 \leq 18$; $u_1 + v_3 \leq 18$; $u_1 + v_4 \leq 21$; $u_1 + v_5 \leq 19$; $u_2 + v_1 \leq 21$; $u_2 + v_2 \leq 22$; $u_2 + v_3 \leq 23$; $u_2 + v_4 \leq 20$; $u_2 + v_5 \leq 24$; $u_3 + v_1 \leq 18$; $u_3 + v_2 \leq 19$; $u_3 + v_3 \leq 21$; $u_3 + v_4 \leq 18$; $u_3 + v_5 \leq 19$; u_i, v_j unrestricted in sign; Optimal values $u_1 = 18, u_2 = 22, u_3 = 19, v_1 = -1, v_2 = 0, v_3 = 0, v_4 = -2$ and $v_5 = 0$. With these, Z = 7,605.
14. $x_{12} = 55{,}000, x_{13} = 220{,}000, x_{22} = 165{,}000, x_{31} = 110{,}000, x_{33} = 110{,}000, x_{34} = 440{,}000$; TC = Rs 51,70,000
15. (a) $x_{12} = 20, x_{13} = 20, x_{22} = 30, x_{31} = 20, x_{33} = 10$; TC = 700
    (b) Max $Z = 40u_1 + 30u_2 + 30u_3 + 20v_1 + 50v_2 + 30v_3$ s.t. $u_1 + v_1 \leq 12$; $u_1 + v_2 \leq 8$; $u_1 + v_3 \leq 2$; $u_2 + v_1 \leq 9$; $u_2 + v_2 \leq 10$; $u_2 + v_3 \leq 9$; $u_3 + v_1 \leq 7$; $u_3 + v_2 \leq 15$; $u_3 + v_3 \leq 6$; u_i, v_j; unrestricted in sign; $Z = 700$

16. (a) Not optimal
    (b) $x_{12} = 12, x_{13} = 2, x_{14} = 8, x_{23} = 15, x_{31} = 7, x_{34} = 1$, TC = 149
    (c) At least Rs 2
17. $x_{11} = 10, x_{14} = 90, x_{15} = 50, x_{25} = 30, x_{34} = 120, x_{41} = 70, x_{42} = 40, x_{43} = 20$; TC = Rs 2,340
18. Use Rs 18 *l* of AR Oct to pay AP of Oct; Rs 27 *l* of AR Nov to pay Rs 25 *l* of Nov and Rs 2 *l* AP of Oct; Rs 35 *l* of AR Dec to pay Rs 28 *l* of AP of Dec and Rs 7 *l* of AP of Nov; Rs 15 *l* of bank loan Dec to pay AP of Dec
19. (a) Yes (b) No (c) Yes (d) Rs 15 (e) $u_1 = 10, u_2 = 5, u_3 = 7, v_1 = 1, v_2 = -2, v_3 = 10, v_4 = 0$ (f) u_i: Value of product at origin, v_j: delivered value (g) cost increases by Rs 8 (h) Not advisable (i) cost reduction of Rs 10 (j) D_2
20. $x_{11} = 200, x_{14} = 200, x_{24} = 300, x_{31} = 100, x_{32} = 400, x_{33} = 300$; Total profit = Rs 15,800
21. Year-Investment type-Amount (000 Rs): 1–E–500; 2–B–600; 3–D–750; 4–A–250; 4–C–500; 4–D–50
22. Plant at C: Total profit = Rs 62,100; Plant at D: Total profit = Rs 58,100; Setup in city C
23. $x_{12} = 100, x_{21} = 200, x_{22} = 50, x_{23} = 150, x_{33} = 50, x_{34} = 125, x_{35} = 75; x_{12} = 25, x_{15} = 75, x_{21} = 200, x_{22} = 125, x_{23} = 75, x_{33} = 125, x_{34} = 125$; Total return = Rs 116,225
24. Year 1: 40 *l* in *P*, 30 *l* in *Q*; Year 2: 20 *l* in *Q*, 20 *l* in *R*; Year 3: 40 *l* in *R*, 50 *l* in *S*; Year 4: 10 *l* in *S*; Return Rs 130,00,000
25. $X_{IC} = 11, X_{ID} = 40, X_{IE} = 3, X_{IIE} = 2, X_{IIIE} = 28, X_{IVA} = 25, X_{IVB} = 29, X_{IVC} = 20$
26. Prod.-Supply-Qty: J-J-200, J-F-25, F-F-225, M-M-150, A-A-80, M-M-120; TC = Rs 22,880
27. Prod.-Supply-Qty: 1-1-20, 1-2-20, 2-2-10, 2-3-20, 3-3-30, 4-4-40; TC = Rs 2,210
28. JN-JS-200, JN-JD-100, JOT-JS-50, FN-FS-120, FN-FD-180, FOT-FS-180, MN-MS-125, MN-MD-175, MOT-MS-195; TC = Rs 643,000
29. (a) $A_1B_3 = 60, A_1B_4 = 340, A_2B_1 = 280, A_2B_2 = 180, A_2B_4 = 340, A_3B_3 = 400, A_4B_2 = 180, A_4B_5 = 220$; TC = Rs 88,440
    (b) $A_1B_4 = 680, A_2A_1 = 280, A_2B_1 = 280, A_2B_2 = 240, A_3B_3 = 400, A_4B_2 = 120, A_4B_5 = 280, B_5B_3 = 60$; TC = Rs 85,260
30. (a) $x_{12} = 40, x_{13} = 200, x_{21} = 80, x_{22} = 80; x_{12} = 120; x_{13} = 120, x_{21} = 80, x_{23} = 80$. Total cost = Rs 36,400
    (b) Same as in (a)
31. *A* to *F* : 50; *B* to *D* : 40; *C* to *F* : 60; *D* to *E* : 20; and *F* to *E* : 75. Total cost = Rs 405

CHAPTER 6

1. (a)

Worker	*Job*			*Supply*
	1	*2*	*3*	
1	3	4.5	7	1
2	6	4	6	1
3	5	M	4	1
Demand	1	1	1	3

   (b) Min $Z = 3x_{11} + 4.5x_{12} + 7x_{13} + 6x_{21} + 4x_{22} + 6x_{23} + 5x_{31} + Mx_{32} + 4x_{33}$ s.t. $x_{11} + x_{12} + x_{13} = 1; x_{21} + x_{22} + x_{23} = 1; x_{31} + x_{32} + x_{33} = 1; x_{11} + x_{21} + x_{31} = 1; x_{12} + x_{22} + x_{32} = 1; x_{13} + x_{23} + x_{33} = 1; x_{ij} \geq 0$

2. $A-2, B-3, C-1; T = 14$
3. $A-2, B-3, C-1; T = 9$
4. $B_1S_1, B_2S_5, B_3S_4, B_4S_3, B_5S_2$; T = 27
5. (a) 1–C, 2–B, 3–A, 4–D, 5–E; T = 11
   (b) 1–D, 2–B, 3–A, 4–C, 5–E; TC = 155
6. $J_1 - M_3, J_2 - M_2, J_3 - M_1, J_4 - M_4, J_5 - M_5; J_1 - M_3, J_2 - M_5, J_3 - M_1, J_4 - M_4, J_5 - M_2$; TC = 365
7. (a) Multiple solutions. Units 1, 3, 4 and 5 respond
   (b) 3.5 minutes
8. $S_1 - C_1, S_2 - C_3, S_3 - C_2, S_4 - C_1$, TC = 325.4 lakh
9. $T_1 -, T_2 -, T_3 - 1, T_4 - 3, T_5 - 5, T_6 -, T_7 - 4, T_8 - 2$; T = 380
10. 1–A, 2–E, 3–C, 4–D; TC = Rs 98 (thousand)
11. I–C, II–B, III–A, IV–D; I–B, II–D, III–A, IV–C; T = 18 hours
12. $M_1 - A, M_2 - B, M_3 - E, M_4 - D$; T = 32
13. Anand-Breast stroke, Bhaskar-Freestyle, Chandru-Backstroke, Easwar-Butterfly.
14. (i) 3–A, 4–D, 2–E
    (ii) 1–D, 3–B, 4–C, 5–A, 6–E
    (iii) Rs 42 thousand
    (iv) 1–B, 2–E, 3–A, 4–D, 5–C; TC = Rs 31 thousand
15. (a) Add I_2 same as I_1
    (b) 1 – 3, 2 – 1, 3 – 5, 4 – 6, I_1 – 2, I_2 – 4; TC = 302(0,000 Rs)
16. 1 – 4, 2 – 1, 3 – 6, 4 – 5, 5 – 2, 6 – 3
17. P_1-Education, P_2-Housing, P_3-Health; Score 108
18. A – I, B – IV, C – III; Total sales = 247
19. 1–D, 2–B, 3–C, 4–E, Decline A; Profit = 376
20. A–Y, B–W, C–Z, D–X; Rs 595 lakh
21. A–V, B–I, C–II, D–III; Rs 359 lakh; A–V, B–I, C–II, D–IV; Rs 356 lakh
22. $P_1 - C_1, P_2 - C_2, P_3 - C_4$, TA – C_3; Multiple solutions
23. (i) P : V, Q : I, R : IV, S : III, T : II; Total runs = 232
    (ii) P : V, Q : I, R : IV, T : III, U : II; Total runs = 263
24. $G_1 - C_1, G_2 - C_3, G_3 - C_4, G_4 - C_2$; Total sales = Rs 242,000
25. A – P_2, B – P_1, C – P_5; Profit = Rs 55,200
26. $C_1 - S_4, C_2 - S_3, C_3 - S_2$; Profit = Rs 1,156
27. Sat Mor: City 2, Sun Mor: City 1, Sat Eve: City 3, Sun Eve: City 4; Expected response = 560
28.

Crew	:	1	2	3	4	5
Residence	:	C	B	B	C	B
Route	:	1–d	2–e	3–a	4–b	5–c

Waiting time = 33.5 hours

CHAPTER 7

1. Max $Z = 18,00,000x_1 + 2,00,000x_2 + 7,20,000x_3 + 8,00,000x_4$ s.t. $3,00,000x_1 + 1,20,000x_2 + 3,00,000x_3 + 2,00,000x_4 \le 6,50,000$; $4,00,000x_1 + 80,000x_2 + 2,00,000x_3 + 4,00,000x_4 \le 8,00,000$; $4,00,000x_1 + 2,00,000x_3 + 4,00,000x_4 \le 8,00,000$; $3,00,000x_1 + 40,000x_2 + 2,00,000x_3 + 1,00,000x_4 \le 5,00,000$; $x_i = 0$ or 1
2. Min $Z = 9,000d_1 + 6,000d_2 + 4,500d_3 + 11x_1 + 10x_2 + 16x_3$ s.t. $x_1 + x_2 + x_3 \ge 5,000$; $x_1 \le 4,000d_1$; $x_2 \le 3,000d_2$; $x_3 \le 1,000d_3$; $x_1, x_2, x_3 \ge 0$; $d_1, d_2, d_3 = 0, 1$
3. Max $Z = 8x_1 + 3x_2$ s.t. $6x_1 + 4x_2 \le 25$; $120x_1 + 82x_2 \le 240$; $x_1, x_2 \ge 0$, integer
4. (a) $x_3 + x_8 = 1$
   (b) $x_4 - x_7 \ge 0$
   (c) $x_1 + x_3 + x_5 + x_8 \le 2$
   (d) $x_3 - x_6 = 0$
   (e) $2 \le x_1 + x_2 + x_4 + x_5 + x_6 + x_8 \le 4$
   (f) $x_1 + x_2 + x_3 + x_4 + x_5 + x_6 + x_7 + x_8 = 5$
5. Min $Z = 1397 - 93.5x_1 - 41.8x_2 - 27.5x_3 - 38.5x_4$ s.t. $0.04x_1 + 0.02x_3 + 0.06x_4 \le 40$; $0.02x_1 + 0.01x_2 + 0.06x_3 + 0.04x_4 \le 40$; $0.02x_1 + 0.05x_2 + 0.15x_4 \le 40$; $0.15x_2 + 0.06x_3 \le 40$; $0.03x_1 + 0.09x_2 + 0.20x_3 \le 40$; $0.06x_1 + 0.06x_2 + 0.02x_3 + 0.05x_4 \le 40$; $x_i \ge 0$
6. Max $Z = 1,250x_1 + 1,320x_2 + 620x_3 + 740x_4$ s.t. $720x_1 + 880x_2 + 300x_3 + 350x_4 \le 1,400$; $560x_1 + 550x_2 + 170x_3 + 210x_4 \le 800$; $400x_1 + 360x_2 + 110x_3 \le 440$; $x_1 + x_2 \le 1$; $x_3 - x_4 \ge 0$; $x_i = 0$ or 1
7. $x_1 = 1, x_2 = 1; Z = 50$
8. $x_1 = 13, x_2 = 10; Z = 181$
9. $x_1 = 0, x_2 = 22; Z = 396$
10. $x_1 = 0, x_2 = 7, x_3 = 16, Z = 76$
11. $x_1 = 5, x_2 = 0; Z = 125$
12. $x_1 = 16, x_2 = 0; Z = 480$
13. $x_1 = 2, x_2 = 3; Z = 10$
14. Tour: 1 – 4 – 3 – 5 – 2 – 1; Cost = 84
15. Tour: A – C – D – B – E – A; Distance = 80 (hundred) km
16. Tour: A – D – B – C – A; Cost = 19
17. Tour: $C_1 - C_2 - C_3 - C_5 - C_4 - C_1$; Distance = 54 km
18. Tour: 1 – 5 – 2 – 4 – 3 – 1; Distance = 15 km
19. Min $Z = d_1^+ + d_2^- + d_2^+ + d_3^- + d_4^- + d_5^- + d_6^- + d_7^-$ s.t. $4x_1 + 5x_2 + 5x_3 + d_1^- - d_1^+ = 600$; $3x_1 + 4x_2 + 3x_3 + 3x_4 + d_2^- - d_2^+ = 360$; $x_1 + 2x_2 + 2x_3 + x_4 + d_3^- - d_3^+ = 120$; $x_2 + x_3 + x_4 + d_4^- - d_4^+ = 120$; $x_1 + x_3 + x_4 + d_5^- - d_5^+ = 120$; $x_1 + x_2 + x_3 + x_4 + d_6^- - d_6^+ = 120$; $x_1 + x_2 + x_4 + d_7^- - d_7^+ = 120$; all variables ≥ 0
20. Min $Z = P_1d_1^- + P_2d_2^+ + P_3d_3^- + P_4d_4^-$ s.t. $x_2 + d_1^- - d_1^+ = 30$; $40x_1 + 20x_2 + 150x_3 + d_2^- - d_2^+ = 1,200$; $x_1 + x_3 + d_3^- - d_3^+ = 20$; $x_3 + d_4^- - d_4^+ = 6$; all variables ≥ 0

21. Min Z = $P_1d_1^- + P_2d_2^+ + P_3d_3^- + P_4d_1^+ + P_5d_4^- + 3P_6d_5^- + 4P_6d_6^- + 2P_6d_7^-$ s.t. $7x_1 + 10x_2 + 8x_3 + d_1^- - d_1^+ = 6{,}000$; $10{,}000x_1 + 12{,}000x_2 + 7{,}000x_3 + d_2^- - d_2^+ = 80{,}00{,}000$; $3{,}000x_1 + 4{,}000x_2 + 2{,}000x_3 + d_3^- - d_3^+ = 1{,}05{,}00{,}000$; $x_1 + x_2 + x_3 + d_4^- = 1{,}000$; $x_1 + d_5^- - d_5^+ = 200$; $x_2 + d_6^- - d_6^+ = 500$; $x_3 + d_7^- - d_7^+ = 300$; all variables ≥ 0
22. Min Z = $0.32d_1^- + 0.16d_2^+ + 0.08d_3^- + 0.02d_4^-$ s.t. $x_1 + x_2 + x_3 + x_4 \leq 20{,}00{,}000$; $x_4 \leq 0.15\,(x_1 + x_2 + x_3 + x_4)$; $0.08x_1 + 0.12x_2 + 0.16x_3 + 0.14x_4 + d_1^- - d_1^+ = 2{,}50{,}000$; $x_3 + d_2^- - d_2^+ = 4{,}00{,}000$; $x_1 + d_3^- - d_3^+ = 3{,}00{,}000$; $0.05x_2 + 0.12x_3 + 0.14x_4 + d_4^- - d_4^+ + = 60{,}000$; all variables ≥ 0
23. Min Z = $P_1d_1^- + P_2d_2^- + P_3d_1^+ + P_4d_3^+$ s.t. $2.5x_1 + 1.0x_2 + 2.0x_3 + d_1^- - d_1^+ = 120$; $50x_1 + 80x_2 + 35x_3 + d_2^- - d_2^+ = 8{,}000$; $6x_1 + 9x_2 + 4.5x_3 + d_3^- - d_3^+ = 600$; all variables ≥ 0
24. Min Z = $P_1d_1^+ + P_1d_2^+ + P_1d_3^+ + P_1d_4^+ + P_2d_5^- + P_2d_6^- + P_2d_7^- + P_2d_8^- + P_3d_9^- + P_4d_{10}^+$ s.t. $0.06x_1 + 0.17x_2 + 0.10x_3 + 0.14x_4 + d_1^- - d_1^+ = 600$; $0.18x_1 + 0.20x_2 + 0.14x_4 + d_2^- - d_2^+ = 500$; $0.07x_1 + 0.20x_2 + 0.08x_3 + 0.12x_4 + d_3^- - d_3^+ = 550$; $0.09x_1 + 0.12x_2 + 0.07x_3 + 0.15x_4 + d_4^- - d_4^+ = 450$; $x_1 + d_5^- = 2{,}600$; $x_2 + d_6^- = 1{,}800$; $x_3 + d_7^- = 4{,}100$; $x_4 + d_8^- = 1{,}200$; $9{,}000x_1 + 10{,}000x_2 + 8{,}000x_3 + 12{,}000x_4 + d_9^- - d_9^+ = 700{,}00{,}000$; $2.6x_1 + 1.4x_2 + 2.5x_3 + 3.2x_4 + d_{10}^- - d_{10}^+ = 150{,}000$; all variables ≥ 0

    $x_1 = 40/3$, $x_2 = 40/3$; first and second priority goals achieved; under-achievement of third priority goal with $d_5^- = 800/3$
26. $x_1 = 4$, $x_2 = 4/3$; goal 3 under-achieved by 5/3
27. Line 1 = 8 hours, line 2 = 12 hours; output = 224 units

CHAPTER 8

1. $J_2 - J_1$, 17 hours
2. J_3 - J_1 - J_2, T = 20 hours
3. 2 – 8 – 7 – 6 – 1 – 5 – 4 – 3, T = 51
4. A – C – I – B – H – F – D – E – G, T = 61
5. 7 – 6 – 2 – 4 – 5 – 1 – 3, T = 101 hours, Idle time on B = 21 hours
6. C – F – A – G – D – B or C – F – G – A – D – B, T = 51 hours
7. 2 – 4 – 5 – 3 – 6 – 1; T = 85
8. Present schedule: Total cost = Rs 64,630; Optimal schedule: Total cost = Rs 60,310; Downtime = 71
9. 4 – 1 – 3 – 2 – 5 – 6, T = 420 hours
10. $J_1 - J_5 - J_7 - J_3 - J_4 - J_2 - J_6$; T = 68 minutes; Idle time (knurling) = 18 min.
11. $J_3 - J_5 - J_2 - J_4 - J_1 - J_6$; T = 62
12. 1 – 4 – 5 – 3 – 2; T = 42 hours
13. 5 – 1 – 2 – 3 – 4; T = 51 hours
14. Optimal sequence = 6 – 1 – 3 – 2 – 4 – 5; T = 67
15. (a) 3 – 4 – 5 – 7 – 2 – 6 – 1 (b) 1 – 4 – 3 – 6 – 2 – 5 – 7, T = 86
16. $J_4 - J_2 - J_5 - J_1 - J_3 - J_6$; T = 83
17. D – C – B – A; T = 82

CHAPTER 9

1. (a) 600 units (b) 22 days app.
2. (i) 15 days (ii) Rs 3,600
3. (a) (i) 300 units (ii) 346 units app. (iii) 346 units app. (iv) 400 units (v) 25%
   (b) 1,162 units
   (c) 223.6 units
4. (a) Rs 10,000 or 400 units (b) 800 units
5. (a) 4,660 units app.
   (b) (i) 11.81% increase (ii) 13.43% decrease
6. Machine A: Rs 1,47,394; Machine B: Rs 1,47,207; ELS = 665 units.
7. EOQ = 1,000 units; ROL = 1,400 units; ROL(7) = 700 units, ROL(21) = 2,100 units
8. A: EOQ = 2,000, ROL = 320; B: D = 9,000, EOQ = 2,000; C: A = Rs 180; LT = 7
9. (i) Outside supplier: EOQ = 548, TC = 121,095; Subsidiary: EOQ = 500, TC = 131,000.
   (ii) Buy from outside supplier
10. (a) 900 gallons (b) 94,500
11. (a) (i) $T(C) = (D/Q)C_o + (Q/2)C_h$ (ii) Rs 1,58,11,388; 6.32 times (iii) Rs 6,32,455 (iv) Rs 19,60,784
    (b) Rs 1,29,09,945; 7.75; Rs 7,74,596
12. TC (566) = Rs 32,11,314; TC(2,000) = Rs 30,60,000
13. Cost saving with Q = 500 is Rs 42,400
14. Optimal order: 20,000 units, TC = Rs 14,04,240
15. 894 units; TC = 22,878
16. EOQ = 100, every two months; Yes, order 200 units, TC = Rs 3,564 p.a.
17. Order 5,000 units, TC = Rs 19,25,150
18. 1.6 percent
19. Current policy: TC = Rs 16,991; EOQ policy: TC = Rs 17,324
20. (a) (i) 1,000 units (ii) 5 (iii) Rs 2,000
    (b) No cost saving. Order 1,000 or 2,000 units TC = Rs 52,000
21. Less than or equal to 5.85%
22. ELS = 31,623 units, 0.158 year
23. ELS = 1,000 units, Cost saving = Rs 300
24. (a) 3,600 units (b) 0.15 year (c) Rs 2,92,320
25. Rs 20/unit/year; 262 units
26. (a) 24,000 packs (b) 8 (c) 30 working days (d) Rs 17,280
27. Rs 8.56
28. (a) 2,000 units
    (b) Rs 1,200
    (c) 10
    (d) 10 days
    (e) 1,200 units
29. Outside: TC = Rs 80,480; Internally: TC = Rs 75,900; Produce internally

30. (i) EOQ = 137; 18 app (ii) EPLS = 808 units; 0.168 years (iii) Manufacture internally; TC = Rs 61,548
31. 67 days; 44 units app
32. (a) 500 units; TC = Rs 122,650
    (b) (i) 2,000 units (ii) Rs 150,000 (iii) 5 (iv) 0.1 year (v) 1,000 units
    (c) EOQ larger in case (ii) since P > D
    (d) (i) 1,225 units (ii) 968
    (e) (i) 67 units (ii) 24 units (iii) 99.38%
33. (i) 82 chairs (ii) 66 (iii) Rs 1,314.5
34. (a) (i) 400 (ii) Rs 2,000
    (b) (i) 490 (ii) 327 (iii) 163 (iv) Rs 1,633
35. 1,897 units; 1,000 units
36. (i) 577 units, (ii) 433 units (iii) Rs 4,330
37. (i) 1155 units (ii) 866 units (iii) 289 units (iv) Rs 8,660; EOQ = 1,000, Max stock 1,000, Rs 10,000
38. (a) 300 units (b) 300 units (c) 150 units
39. (i) 3,000 units (ii) 12 (iii) 4,500 units (iv) 3,000 units
40. 56 units, 400 units
41. SS = 100 units, ROL = 400 units
42. 76.42%; 439 units
43. ROQ = 1,570 units, ROP = 469 units
44. (a) 200 units (b) 5 (c) 40 units (d) 50 or 60 units
45. (a) 48 (b) 30 units, Rs 25 (c) 28 units (d) 35 units
46. (a) 150 units (b) SS = 41 units (c) 3 app
47. (a) 727 units (b) 1,527 units (c) Rs 378
48. 330 units
49. (a) 180 units (b) 50 per cent (c) 204 units
50. (i) 88 units app (ii) 72 units app
51. Buy no spares

CHAPTER 10

1. (a) $\lambda = 12$ customers/hour, $\mu = 30$ customers/hour
   (b) 0.40
   (c) 0.01536
   (d) 0.01024
   (e) 4/15 customer
   (f) 2/3 customer
   (g) 4/3 minutes
   (h) 10/3 minutes
   (i) 20 customers/hour
2. (a) 0.4232; 0.9502 (b) 6

3. (a) 0.00248; 0.01487; 0.04462; 0.08924; 0.13385; 0.16062
   (b) 40%
   (c) 3.2 hours
   (d) 0.4; 0.24; 0.144; 0.0864; 0.05184; 0.031104
4. (a) 4 minutes
   (b) 0.75 hour or 45 minutes
   (c) False
   (d) 3 minutes
   (e) 2.08 minutes
5. 3 hours; 1.67
6. (i) 0.6 (ii) 1.5 (iii) 0.075 hour (iv) 0.125 hour
7. (a) 0.67 (b) 2 (c) 60 min (d) 40 min (e) 0.1975
8. (i) 0.6 (ii) 7.5 minutes (iii) 10.8 hours
9. Not workable since $\lambda > 12$
10. (i) 0.784 (ii) 0.216 (iii) 9 minutes
11. (a) 0.75
    (b) 0.578125
    (c) 3 customers
    (d) 2.25 customers
    (e) 0.75 hour
    (f) 1 hour
    (g) 1 hour
    (h) 0.6348
    (i) 0.8465
12. Peak: 13; Normal: 8; Low: 5
13. (a) 0.60
    (b) 0.40
    (c) 2/3 customers
    (d) 1/6 hour
14. (a) 0.3
    (b) 0.09
    (c) 1.29 minutes (present W_q); 11.43 customers/hour
15. Present: Rs 4,900/day; Proposed: Rs 1,800/day; install.
16. Present: TC = Rs 600/day; Proposed: TC = Rs 367/day
17. (i) 4/3 and 1/2 (ii) 2 and 1 (iii) 2/3 and 1/4 (iv) 1 and 1/2; TC (I) = Rs 54 per hour, TC (II) = Rs 35 per hour
18. Mechanic B; TC = Rs 1,050/day
19. F_1: TC = Rs 154,000 p.a.; F_2 TC = 108,000 p.a.
20. Mechanic B; TC = Rs 544/day; for A, TC = Rs 1,120/day
21. (a) 0.47
    (b) 0.31; 0.16; 0.05; 0.01
    (c) 0.29

(d) 0.82
(e) 1.09 months
(f) 3.09 months

22. (a) 0217; 0.244; 0.229; 0.171; 0.096; 0.036; 0.007
(b) 10.50 hours
(c) 1.822 machines
(d) 1.039 machines

23. (i) 0.4807 (ii) 3.21 minutes

24. (a) 0.80
(b) 0.0130
(c) 0.1386
(d) 2.219
(e) 6.219
(f) 0.0567
(g) 3.33 minutes
(h) 9.33 minutes

25. Yes. W_q reduces from 7.5 minutes to 3.75 minutes

26. (a) 45 minutes
(b) 19.29 minutes

27. (a) 22 hours
(b) 0.74 hour

28. (a) 0.3747
(b) 0.20414 hour
(c) 0.03747 hour
(d) 1.333

29. Present: W_q (1) = 8 minutes; W_q (2) = 16 minutes
Proposed: W_q = 4.6 minutes

30. (a) For $K = 2$: P(0) = 0.0638; $L_s = 7.22$; $L_q = 5.47$; $W_s = 1.03$ m; $W_q = 0.78$ m. For $K = 3$: P(0) = 0.1576; $L_s = 2.22$; $L_q = 0.47$; $W_s = 0.32$ m; $W_q = 0.07$ m
(b) For $K = 2$: TC = 0.50 rupee/min; For $K = 3$; TC = 0.26 rupee/min

31. Present: TC = Rs 41.03/hr; Proposed: TC = Rs 48.67/hr

32. (a) 0.2608; 0.2608; 0.1739; 0.0869; 0.0435
(b) 2.1739
(c) 0.1739
(d) 2
(e) 6.52 minutes
(f) 0.52 minutes

33. (a) (i) 0.125 (ii) 0.1429 (iii) 0.0179 (iv) 0.2857 m (v) 0.0357 m (vi) 0.875 (vii) 0.125
(b) (i) 0.0625; 0.0625
(ii) 0.0667; 0.1255
(iii) 0.0042; 0.0005
(iv) 0.1333 mts; 0.2510 mts

(v) 0.0083 mts; 0.0010 mts
(vi) 0.9375; 0.8824
(vii) 0.0625; 0.0074

34. (a) 0.1429
(b) 0.2143
(c) 0.0678
(d) 1.9286
(e) 3.4286
(f) 1.9286 min
(g) 3.4286 min
35. One channel: 10 min; Three channels: 12.5 min
36. Three servers. TC = Rs 93.91/hour

CHAPTER 11

1. 5 years; AC = Rs 3,520
2. 6 years; AC = Rs 1,862.50
3. 4 years; AC = Rs 3,125
4. 6 years; AC = Rs 3,167
5. 6 years; AC = Rs 1,583
6. 1 year; AC = Rs 810
7. 5 or 6 years; AC = Rs 106,000
8. Yes, after 1 year
9. (i) 5 years (ii) Yes; after 2 years
10. Buy now, replace after 4 years; AC = Rs 762 p.a.
11. 3 years; AC = 143.3
12. 5 years; AC = Rs 27,000; Replace old by new after 2 years
13. 5 years; AC = Rs 5,400
14. (i) 5 or 6 years; AC = Rs 2,900 (ii) 6 years; AC = Rs 3,077
15. (i) M_1: 9 years; M_2: 8 years (ii) M_2
16. Rs 1,481.5
17. (i) 3 months; AC = Rs 2,623 (ii) GR policy is better.
18. 3 months; AC = Rs 4,605.33
19. (a) GR policy better. 3 months; AC = Rs 123.92
(b) GR policy still better. 6 months; AC = Rs 133.20
20. (a) 3.72 months
(b) IR policy better than GR policy.
21. 'IR only' policy: Rs 5,000/month
GR policy: Rs 4,571/month (3-months interval)
22. (a) Rs 8,510
(b) Rs 7,450; 4-months GR

23. IR: Rs 5,000; Do not switch
24. (a) Rs 160,000/quarter
    (b) GR every two quarters, AC = Rs 125,600
25. IR policy : Rs 3,521.20/week; GR policy : Rs 2,958.7/week for optimal policy of GR every 3 weeks
26. IR policy is better. AC = Rs 516,000/month
27. IR policy: Rs 111,000/week; GR = Rs 89,000/week (2-weeks GR)
28. (a) 73 app.
    (b) Rs 808
29. GS policy every 5 months. AC = Rs 4,800
30. (a) 96
    (b) 31 years; 52 years

CHAPTER 12

4.

Activity	:	*K*	*L*	*M*	*N*	*O*	*P*	*Q*	*R*	*S*
Imm. Pred.	:	–	–	–	K	M	K,L	N,O	L,O	R

5.

Activity	:	*A*	*B*	*C*	*D*	*E*	*F*	*G*	*H*	*I*	*J*	*K*	*L*
ES	:	0	0	0	7	9	7	16	16	24	16	30	26
LS	:	0	12	9	16	9	19	28	16	24	21	30	38

Critical activities: *A*, *E*, *H*, *I*, *K*; Duration = 40 days

6. (TF, FF, IF): *A*(0, 0, 0), *B*(31, 0, 0), *C*(18, 0, 0), *D*(0, 0, 0), *E*(20, 0, 0), *F*(0, 0, 0), *G*(18, 0, 0), *H*(24, 24, 0), *I*(0, 0, 0); *C* Path: *A*–*D*–*F*–*I*; Duration = 67

7. (i)

Activity	:	*g*	*h*	*i*	*k*	*l*	*m*	*n*	*o*	*p*
Red. IP	:	*c*	*c*	*c*	*d*	*d*	*c,d*	*e,j*	*d*	*k,l*

(ii) 39 days (iii)*a* : 4, *b* : 29, *f* : 29, *g* : 20, *h* : 1, *i* : 17, *j* : 4, *k* : 17, *l* : 17; *o* : 1

8. (a) B–C–F–H–I–J; 42 weeks
   (b) (i) 44 weeks (ii) reduce path C–F–H–I–J by 2 weeks.

9.

Activity	:	*A*	*B*	*C*	*D*	*E*	*F*	*G*	*H*	*I*	*J*	*K*	*L*	*M*
ES	:	0	0	0	0	9	4	15	19	19	23	10	26	32
LS	:	4	6	0	6	13	10	15	19	20	23	16	26	32
TF	:	4	6	0	6	4	6	0	0	1	0	6	0	0
FF	:	0	0	0	0	3	6	0	0	1	0	6	0	0
IF	:	0	0	0	0	0	0	0	0	0	0	0	0	0

10. (i) 20 days (ii) (TF) A:3, B:0, C:8, D:2, E:0, F:0, G:1
11. Duration increases from 24 to 30
12. (b) A–D–G, 11 days (c) 8 days, cost = Rs 1,600
13. (i) 1–2–3–4–6; 13 weeks (ii) 2–4 : 2, 2–5 : 3, 5–6 : 3, others : zero (iii) 10 weeks; Rs 2,045 (iv) 8 weeks
14. 14 weeks, Rs 148,500
15. (i) 25 days, 18 days (ii) 15.5%
16. Optimal duration = 12 weeks; all

17. (a) 24 days, Rs 36,400
    (b) 17 days, Rs 31,350
    (c) 21 days, Rs 29,400
18. (a) 1–3–6–7; 31 days, 25 days (b) 1–2:1 day, 1–3:2 days, 2–4:2 days, 2–5:3 days, 3–5: none, 3–6:3 days, 4–7: none, 5–7:2 days, 6–7: 2 days; Rs 11,302 (c) No
19. A – C– E–G; A: 1 day; B: 2 days; C: 2 days; G: 2 days; Cost = Rs 900
20. (ES, LF) *A*:(0, 3), *B*:(3, 12), *C*:(3, 7), *D*:(3, 13), *E*:(5, 15), *F*:(8, 17), *G*:(7, 13), *H*:(9, 17), *I*:(13, 17), *J*:(9, 19), *K*:(11, 21), *L*:(17,21); A–C–G–I–L, 21 days
21.

Days :	1–2	3–8	9–11	12–13	14–15	16–20
Workers :	25	24	25	24	25	24

22. Schedule C 7 weeks after ES, F 2 weeks after ES; 6 workers
23. (a) Yes, Rs 1,000 saving (b) No saving (c) No (d) Yes, Rs 1,000 saving
24.

E Days :	1–2	3–8	9–10	11–13	14–15	16–18	19	20–22	23–25	26–27	28–29	30–33
No. :	10	28	26	29	21	32	23	15	16	22	20	8
L Days :	1–2	3–10	11	12–17	18–20	21–22	23–25	26	27	28–29	30	31–33
No. :	10	18	15	25	23	29	16	18	24	22	20	29

25. (ii) 1–3–7–9; 15 months (iii) No delay
26. Delay of 2 days
27. (a) 10 workers, TC = Rs 3,250 (b) 25 days
28. 22 days
29. Critical path: A–D–G–H; 30 weeks; 0.6591
30. (a) $m = 14$ and $b = 24.5$
    (b) (i) 0.9075 (ii) 44 weeks 6 days
31. (i) 1–2–3–4–6 (ii) 51 days, 35.44 days^2 (iii) 0.0655 (iv) 0.5353 (v) 0.0002
32. (c) 64.83 weeks
    (d) 3.3624 weeks
    (e) 0.0001
    (f) 0.5199
    (g) 0.9382
33. (a) (i) 25 minutes (ii) 69.44 min^2 (iii) 25 minutes
    (b) (i) (ES, LS, TF, FF) 1-2: (0, 0, 0, 0), 1–3: (0, 4, 4, 0), 1–4: (0, 2, 2, 2), 2–3: (4, 8, 4, 0), 2–4: (4, 4, 0, 0), 2–5: (4, 10, 6, 6), 2–6: (4, 9, 5, 5), 3–5: (7, 11, 4, 4), 3–6: (7, 11, 4, 4), 4–5: (12, 12, 0, 0), 5–6: (21, 21, 0, 0), 5–7: (21, 24, 3, 3), 6–7: (27, 27, 0, 0)
    (ii) 1–2–4–5–6–7; 35 days (iii) 2-6: no effect, 4-5: duration less 2 days
    (c) 0.4129; 0.8413
34. B–F–K–L; 30 days app
35. (a) (i) 0.3085 (ii) 0.0919
    (b) (ii) A–E–K–L, 32 weeks, 2.582 weeks (iii) 0.9898 (iv) 35.3 weeks
36. (ii) 0, 1, 0, 1, 6, 0, 1, 0, 6, 6, 0; A–C–F–H–K (iii) 0.9306
37. C–E–H–J; 0.6628
38. Rs 15,74,000

39. (a) 0.0918; 2y 3m 25 d (b) 0.6826 (c) 68% (d) statement (ii) (e) 4,10 (f) 7, 31
40. (a) A–E–I–L; 50 days (c) 0.0778

CHAPTER 13

1. (a) Egg shampoo
   (b) Deluxe shampoo
   (c) Deluxe shampoo
   (d) Deluxe shampoo
2. (a) Full
   (b) Minimal
   (c) Full or Partial
   (d) Partial
3. (i) Obtain option (ii) No action (iii) By land (iv) Obtain option (v) Buy land or obtain option
4. (a) A_3
   (b) A_1
   (c) A_2
   (d) A_2 or A_3
5. (a) (i) a_3 (ii) a_2 (iii) a_1
   (b) (i) a_3 (ii) a_5 (iii) a_5
6. a_2 under both criteria. Exp. payoff = 1.5, Expected regret = 1.6
7. (a) Buy EMV = Rs 500
   (b) Buy 3, EMV = Rs 1,400
8. Type I. Expected regret = Rs 340
9. Produce 3,000 units
10. (a) 205
    (c) Expected payoff: for 0 : Rs 0; for 100 : Rs 475; for 200 : Rs 680; for 300 : Rs 660, for 400 : Rs 440. Produce 200.
11. (c) 40, EMV = Rs 1,210
    (d) EVP1 = Rs 210
12. Stock 12 cases. Exp. payoff = Rs 335
13. (ii) EMV (Small) = Rs 150,000; EMV (Large) = Rs 100,000
    (iii) Maximum price payable by a decision-maker for a perfect forecast
    (iv) Small factory (since its EMV is higher)
14. 4/7
15. 5/7
16. Limited production. EMV = Rs 32,00,000
17. (a) 4
    (b) P_{A2}
    (c) Nil
18. Hot snack stall. EMV = Rs 3,400

19. Yes; 1
20. 3; EVP1 = Rs 45
21. 150,000
22. Invest in both. EMV = Rs 7,500
23. 1
24. (a) 1,200 units
25. (a) 15 cases, Expected payoff: Rs 10,500
    (b) 20 cases, Expected payoff: Rs 11,000
26. Optimal policy: $a_i = 6$, Exp. value = 1,945
27. 30,000 units. EMV = Rs 170,000
28. (a) Selling price Rs 20, purchase option 3, cost Rs 2.75/kg for a minimum 70,000 units
    (b) EVP1 = Rs 12,300
29. Don't buy policy
30. Advisable to spend. For a 'low' indication : build 2,500-tonnes plant; for a 'high' indication : build 5,000-tonnes plant.
31. Reject the lot
32. (a) Accept
    (b) Reject
    (c) EVSI = Rs 396
    (d) Testing be done
33. (a) A_1
    (b) Rs 1 m
    (c) For $I_1 : A_1$; For $I_2 : A_1$; For $I_3 : A_2$
    (d) Rs 90,000
34. Alpha : Exp. payoff= Rs 76.35 lacs
    Beta : Exp. payoff= Rs 51 lacs
    No survey : Rs 50 lacs
    Best option : Survey by Alpha Company
    EVP1 = Rs 40 lacs
35. Expected payoff = Rs 165,000
36. Go for test marketing and then launch nationally, whether it is favourable or unfavourable. EMV = Rs 128,000
37.

	Regional Distribution	*National Distribution*
EMV	59.8	93.575
Co-efficient of variation	101.4% (Less risky)	143.9% (More risky)

38. Optimal policy: Royalty basis in both cases, EMV = Rs 15,000
39. Pay Rs 5,000 as royalty. Expected payoff = Rs 15,000
40. Bid and design A. If it fails, then design C. EMV = Rs 302,500
41. Choose survey. If positive, choose de-luxe upgrade, and if negative, close down.

42. (b) EC (i) Do not take sample : Rs 76,000 (ii) Take sample : Rs 79,997. Do not take sample
    (c) Rs 1.50
    (d) Rs 6,000
43. (a) Using EMV : Accept; Using EU : Do not accept
    (b) Using EMV : Accept; Using EU : Accept
44. 0.30 and 0.70 for Rs 0 and Rs 20,000 respectively
45. A_1
46. (a) Single individual : do not accept; four individuals as a group : accept
    (b) Accept C_2

CHAPTER 14

1.

	Rs 0	Rs 10	Rs 20	Rs 30	Rs 40	Rs 50	Rs 60
Rs 0	1	0	0	0	0	0	0
Rs 10	0.6	0	0	0.4	0	0	0
Rs 20	0	0.6	0	0	0.4	0	0
Rs 30	0	0	0.6	0	0	0.4	0
Rs 40	0	0	0	0.6	0	0	0.4
Rs 50	0	0	0	0	0	1	0
Rs 60	0	0	0	0	0	0	1

2.
$$\begin{array}{c} \\ S_1 \\ S_2 \\ S_3 \end{array} \begin{array}{c} \begin{array}{ccc} S_1 & S_2 & S_3 \end{array} \\ \begin{bmatrix} 0.5625 & 0.3125 & 0.1250 \\ 0.3750 & 0.5000 & 0.1250 \\ 0.2000 & 0.0500 & 0.7500 \end{bmatrix} \end{array}$$

3. (a) $P = \begin{pmatrix} 0.8 & 0.2 \\ 0.4 & 0.6 \end{pmatrix}$
   (b) 66.24% and 33.76%
   (c) No (because values are close)
4. 0.25
5. (a) (i) Y (ii) X : 1/3, Y : 2/3
   (b) 44%, 39%, 17%, Y
6. 0, 0, 1; Yes
7. (a) $\begin{pmatrix} 0.75 & 0.25 \\ 0.40 & 0.60 \end{pmatrix}$
   (b) (i) 0.3375 (ii) 0.615 (iii) AA : 0.615 BB : 0.385
8. (a) (i) 39.57%, 46.29%, 14.14%; (ii) 49.65%, 38.97%, 11.38%
   (b) (i) and (ii) 62.27%, 26.37%, 11.36%
9. 40%, 15%, 45% in X, Y and Z respectively
10. Not advisable

11. A: 35.58%, B: 43.27%, C: 21.15%
12. (a) ABC: 40.6%, PQR = 42.2%, XYZ = 17.2%
    (b) ABC: 40.18%, PQR = 45.54%, XYZ = 14.28%
13. (a) X: 40.59%, Y: 33.91%, Z = 25.50%
    (b) X: 23.71%, Y: 61.86% Z = 14.43%
    (c) Generally not, due to changing conditions.
14. ABC: 38.40%, XYZ: 35.55%, PQR: 26.05%; 17.09%, 51.87% and 31.04% respectively
15. Policy 1 : Cost = Rs 250/bus/inspection
    Policy 2 : Cost = Rs 212.22/bus/inspection (Better policy)
16. Policy 1 : 16.7%, Policy 2 : 23.1% (Downtime)
17. Rs 36.36
18. P_1: Rs 69,000 P_2: Rs 77,000
19. P_1: Rs 106,500 P_2: 110,000
20. (45,500 6,000 1,000)
    Here 45,500 + 6,000 + 1,000 = 52,500 = 30,000 + 17,500 + 5,000
21. Rs 48,34,862 and Rs 41,65,138

CHAPTER 15

1. (a) $a_1\ b_1$ (b) $a_1\ b_3$
2. A: a_3, B: b_1, $v = 7$
3. Yes, $v = 8$, Not fair
4. (a) Maximin: A_2, Minimax: B_3 (b) Yes, a saddle point exists (c) $v = 0$ (d) yes, since $v = 0$
5. A_3B_1
6. Major change by each, $v = 0$
7. No saddle point
8. (0, 1/3, 2/3), Market share 48.33
9. Four saddle points: 1–1, 1–2, 2–1, 2–2; game value = 0, fair
10. City C for both
11. A: (1/2, 1/2, 0), B:(2/3, 1/3, 0), $v = 0$
12. (a) A: Price cut, B: High promotion, (b) $v = 6$, strictly determinable, not fair (c) Yes (d) No more profits under given conditions
13. A: (0, 0.8, 0.2), B: (0, 0.6, 0.4), $v = 12.4$
14. A: advertise, P: price out
15. A: (0, 15/22, 7/22), B:(21/22, 0, 1/22), $v = 205/22$
16. A : (0, 5/12, 7/12), B : (0, 1/2, 1/2), $v = 77\frac{1}{2}$
17. Management: (7/10, 3/10), Union: (3/10 7/10, 0), $v = 1/10$
18. F_1 : (3/11, 8/11), F_2 : (6/11, 0, 5/11), $v = 38,182$
19. X : (9/11, 2/11, 0, 0), Y : (17/22, 0, 0, 5/22), $v = -5/11$
20. A : (4/11, 7/11, 0), B:(5/11, 6/11, 0, 0), $v = 642/11$

21. $A:(1/5, 0, 4/5), B:(0, 13/15, 2/15), v = 24$
22. $A:(7/8, 1/8, 0), B:(3/8, 5/8, 0, 0), v = -5/8$
23. $A:(4/7, 3/7), B:(2/7, 5/7, 0), v = 29/7$
24. $A:(0.2, 0, 0, 0.8), B:(0, 0, 0.4, 0.6), v = 21$
25. (i) No saddle point
26. $A:(30/97, 0, 67/97), B:(0, 77/97, 20/97), v = 3{,}510/97$
27. $A:(4/9, 5/9), B:(4/9, 0, 5/9, 0), v = -7/9$
28. $A:(8/9, 0, 1/9, 0, 0), B:(2/3, 1/3), v = 10/3$
29. $A:(4/9, 5/9, 0), B:(2/3, 1/3, 0), v = 23/3$
30. $X:(1/4, 3/4, 0), Y:(0, 1/2, 1/2), v = 95$
31. For A : Min $1/U = X_1 + X_2 + X_3$ s.t. $5X_1 + 10X_2 + 6X_3 \geq 1$; $7X_1 + 4X_2 + 2X_3 \geq 1$; $2X_1 + 9X_2 \geq 1$; $X_1, X_2, X_3 \geq 0$ For B : Max $1/V = Y_1 + Y_2 + Y_3$ s.t. $5Y_1 + 7Y_2 + 2Y_3 \leq 1$; $10Y_1 + 4Y_2 + 9Y_3 \leq 1$; $6Y_1 + 2Y_2 \leq 1$; $Y_1, Y_2, Y_3 \geq 0$
32. Add constant (say 10) for entries to be non-negative. For A : Min $1/U = X_1 + X_2 + X_3 + X_4$ s.t. $18X_1 + 16X_2 + 10X_3 + 26X_4 \geq 1$; $30X_1 + 35X_2 + 2X_3 + 19X_4 \geq 1$; $7X_1 + 14X_2 + 22X_3 + 31X_4 \geq 1$; $11X_1 + 12X_2 + 19X_3 + 10X_4 \geq 1$; $X_1, X_2, X_3, X_4 \geq 0$. Game value = $U - 10$.
33. $y_1 = 12/61, y_2 = 28/61, y_3 = 21/61; v = -38/61$
34. For A : (9/22, 8/22, 5/22), For B : (1/22, 9/22, 12/22), $v = 125/22$
35. (a) $y_1 = 10/19, y_2 = 4/19, y_3 = 5/19; v = 1/19$ (b) $x_1 = 10/19, x_2 = 5/19, x_3 = 4/19$
36. HMC: 10/42, 23/42, 0, 9/42, 0; IMC: 25/42, 0, 2/42, 15/42; $v = 199/42$
    HMC to produce J_1 if K_4 is produced by IMC.

CHAPTER 16

1. 0–1–4–8–11–12, Length = 47
2. (i) A–B_1–C_1–D_2–E; 30 (ii) A–B_1–C_1–D_1; 24
3. (a) 1–3–6–9–11; 20 (b) 1–2–4–7–11; 23
4. $x_1 = 0, x_2 = 144; Z = 3{,}456$
5. $x_1 = 2, x_2 = 2; Z = 100$
6. $x_1 = 2, x_2 = 6; Z = 144$
7. Store (1, 2, 3) : (1, 2, 2) or (3, 2, 0); Exp. profit = Rs 10,000
8. Region (1, 2, 3, 4) : (3, 1, 0, 2); (3, 1, 1, 1) or (0, 1, 3, 2); Exp. increase = Rs 66 lakh
9. A : none, B : 1, C : 1, D : 3, No. of votes = 49 thousand
10. District 1 = 2, 2 = 3, 3 = 1, 4 = 1; Exp. increase = 129
11. Container $A = 0, B = 1, C = 0, D = 2$; Cost = Rs 2,280
12. Product $A = 1, B = 1, C = 4, D = 1$; or $A = 2, B = 1, C = 3, D = 1$; Rs 57 lakh
13. Plant $X = 60, Y = 0, Z = 20$; Rs 38 lakh; Second best: $X = 0, Y = 0, Z = 80$; or $X = 0, Y = 60, Z = 20$; Rs 32 lakh
14. Contractor $A = 1, B = 3, C = 2$; Total cost = 366
15. (b) $I_1 = 0, I_2 = 0, I_3 = 0, I_4 = 7$; Total value = 252
16. (a) Monday through Thursday : Sell if the share price is Rs 22, else wait. Friday : Sell at prevailing price. (b) Rs 2,08,050

17. First and second week : Buy if the price = Rs 1,000, wait if price = Rs 1,100 or Rs 1,200; Third week : Buy if the price = Rs 1,000 or Rs 1,100, wait if price = Rs 1,200; Fourth week : Buy at prevailing price.
18. A–B–C–D; Total distance = 4,259

CHAPTER 17

1. $W = 7.7$ min/customer, Idle time = nil
2. Average demand = 24 units/day
3. Average demand = 12.6 units/week
4. Optimal policy: 17 copies, exp. profit = Rs 6,400
5. Using random numbers from second row of B7, required proportion = 9/20
6. Random numbers from first three columns, exp. profit = Rs 5,537.50
7. End balance = (Rs 3,000), Highest bal. = Rs 7,000, Average weekly bal. = Rs 3,750
8. Use 0 for head and 1 for tail. Select random numbers 0 and 1 and proceed.
9. (a) $W_q = 142$ sec, $W_s = 1{,}012$ sec, $L_q = 0.10$, $L_s = 0.77$ (b) $W_q = 2{,}024$, sec, $W_s = 2{,}894$ sec, $L_q = 1.63$, $L_s = 2.33$
10. (i) 9/64 (ii) 1.3 minutes
11. 5 out of 10 : no defects; one scrapped, 90 min of re-work time
12. (a) $W_q = 12.8$ min; Percentage idle time = 31 (b) $W_q = 1.7$ min; Percentage idle time = 32 (Doc. 1), 62 (Doc. 2)
13. P/L 0, 400, 40, (60), 220, (200)
14. Random numbers from N–E corner of Table B7. C.Paths: B–D–F–G–H (16); B–D–F–G–H (19); B–D–F–G–H (18); B–D–F–G–H (17); A–C–F–G–H/B–D–F–G–H (16 and 18)
15. 5.83 percent
16. Optimal policy: invest Rs 20,00,000, Modal ROI = 12.25%
17. (i) 1/12 (ii) 5/12 (iii) 2/3
18. Policy 1: TC = Rs 59.50; Policy 2: TC = Rs 42.50; Adopt policy 2.
20. (a)

NR ('000 Rs)	:	(2)	(1)	0	1	2	3	4	5	6
Prob	:	0.05	0.00	0.15	0.30	0.15	0.25	0.05	0.00	0.05

(b)

NR ('000 Rs)	:	0	1	2	3	4	5
Prob	:	0.05	0.10	0.30	0.30	0.05	0.20

21. Average profit per trial = Rs 2,100
22. NPV: (17,880); 14,240; 28,340; (5,760); 14,240

CHAPTER 18

1. Second buyer, Rs 29,755
2. Rs 3,967.90
3. Rs 44,708.85
4. (i) either (ii) electric heater; For 8-year period : gas boiler
5. Rs 14,904

6. Offer (a)
7. 35
8. Rs 1,12,817.74 p.a.
9. Rs 3,00,000 p.a.
10. (a) Rs 2,050 (b) 3.75% (c) Rs 42,000
11. Rs 6,039
12. Rent the facility (PV = Rs 18,35,236)
13. (a) NPV = Rs 2,887, (b) IRR = 16.2% (c) Yes, if required return $\leq$ 16%
14. (a) – Rs 233.59 thousand (b) 6.6%
15. a: NPV = Rs 4,478; b: NPV = Rs 11,063
16. IRR = 10.7%, do not buy
17. X: NPV = Rs 31,651, Y: NPV = Rs 49,244; Prefer X
18. A: NPV = Rs 6,11,490, IRR = 38.6%; B: NPV = Rs 5,87,930, IRR = 38.3%
19. NPV = (Rs 5,026)
20. (a) Alternative 1 (b) IRR = 28.68%
21. Exp. NPV = Rs 5, 862, SD = Rs 1,597
22. (a) P_1: 50,750 and 60,000 P_2: 41,000 and 57,250 (b) P_1: NPV (12%) = Rs 13,147, P_2: NPV (14%) = Rs 19 (c) P_1: 23.9%, P_2: 14.0%
23. DCF = Rs 5,697, SD = Rs 670
24. CV for A_1 = 29.8%, for A_2 = 41.8%, for A_3 = 55.1%; (i) A_3, A_2, A_1 (ii) A_1, A_2, A_3
25. Project 1: Rs 61,318, Rs 46,971; Project 2: Rs 52,607, Rs 33,017
26. Exp. NPV = Rs 39,951, SD = Rs 24,351, $P(X < 0) = 0.0505$, EVP1 = Rs 514.8
27. EMV = 150.8, SD = 39.2
28. (a) Yr 1 : 300, 100; Yr 2 : 430, 90; Yr 3 : 740, 180 (b) Exp NPV = 184 (c) 0.7291 (d) 0.8485 (e) (i) 49.86 (ii) 14.09
29. EXP NPV = Rs 8,340, SD = Rs 6,760; Prob = 0.4894
30. EXP NPV = Rs 29,007, SD = Rs 16,195, $P(X > 0) = 0.9633$, EVP1 = Rs 237.1
31. Exp NPV = Rs 5,004, SD = Rs 10,843, EVP1 = Rs 2,281.4
32. Buy: CV = 56.1%, Lease: CV = 57.9%; Prefer Lease (EMV = 12,401)
33. Regular: EV = 74, SD = 41.8, Large: EV = 17, SD = 56.7
34. BES = Rs 120,000; Profit = Rs 64,000
35. (a) BES for AB = Rs 75,000, for CD = Rs 105,000 (b) CD Ltd in high demand, AB Ltd in low demand
36. (i) BES = Rs 26,40,000, (ii) 1,42,000 units (iii) 1,62,000 units (iv) 1,47,156 units
37. BES = Rs 1,83,000; MS = 50%
38. BES = Rs 400,000; Profit = Rs 160,000
39. (a) Rs 80,000 (b) 8,000 (c) 13,000
40. (a) Rs 4,54,286 (b) Rs 5,00,000

CHAPTER 19

1. (i) 221.67; 221.33; 231.33; 239.33; 246; 241.33; 246.33; 253.33; 260.33
   (ii) 228.40; 232.60; 234.80; 240.20; 247.60; 248.00; 252.80
   (iii) 157.20; 199.20; 220.30; 235.10; 239.30; 241.40; 243.30; 247.50
2. (i) 53.00; 55.50; 58.00; 61.25; 62.25; 63.00; 64.25; 61.75; 62.50; 64.75; 66.75; 73.00; 79.25
   (ii) 54; 57; 60; 61; 62; 64; 63; 62; 64; 67; 70; 76
3. For $\alpha = 0.1$: 70; 79; 82.10; 93.89; 93.50; 96.15; 92.54; 94.28; 94.85; 104.37; 108.93; 128.04; 142.23; 163.01; 178.71

   For $\alpha = 0.3$: 70; 97; 100.90; 130.63; 118.44; 118.91; 101.24; 103.87; 102.71; 128.89; 135.23; 184.66; 210.26; 252.18; 272.53

   For $\alpha = 0.1$: MAD = 83.37; For $\alpha = 0.3$: MAD = 62.99
4. 70.00; 142.00; 116.40; 183.28; 108.66; 117.73; 71.55; 83.09; 88.16; 118.71; 128.10; 265.62; 269.12; 333.82; 329.67
5. (i) $Y_t = 83 + 2X$ (origin : 2002, X unit : 1 year, Y unit : Annual demand in '000 mt)
   (ii) $Y_t = 99$ ('000 mt)
6. $Y_t = 92.7 + 4X$ (2005 = 0, X unit = 1 year, Y unit = Annual demand in millions); 108.7 million
7. $Y_t = 1{,}058.18 + 65.02\,X$ (2003 = 0; X unit = 1 year, Y unit = Annual demand); 1,513.32 mb
8. 47,379; 46,310; 45,242; 48,448; 48,804; 49,517; 45,598; 44,173; 43,461; 44,886; 47,735
9. $Y = 20.9405 + 0.8077X$

   For $X = 70$, $Y = 77.48$; For $X = 85$, $Y = 89.60$
10. $Y = -1.0907 + 0.8519X$; For $X = 12$, $Y = 9.1321$ (percent)
11. (i) $Y = 219.23 + 6.3815X_1 - 1.6708X_2$
    (ii) 43.25
12. $Y = 47.1649 + 1.599X_1 + 1.1487X_2$; Rs 76,000

Bibliography

Ackoff, R.L. and Sasieni, *Fundamentals of Operations Research*, John Wiley and Sons Inc., New York, 1968.

Agee, M.H., Taylor, R.E. and Torgersen, P.E., *Quantitative Analysis for Management Decisions*, Prentice-Hall Inc., Englewood Cliffs, NJ, 1976.

Anderson, D.R., Sweeny, D.J., and Williams, T.A., *An introduction to Management Science:Quantitative Approaches to Decision-Making*, South-Western Education Publishing, 2005.

Brigham, E.F., *Fundamentals of Financial Management*, The Dryden Press, Chicago, 1986.

Bronson, Richard and Govindasami Naadimuthu, *Schaum's Outline of Operations Research*, McGraw-Hill Book Co., New York, 2004.

Brown R.G., *Decision Rules for Inventory Management*, Holt, Rinehart and Winston Inc., New York, 1967.

Brown, K.S., and ReVelle, J.B., *Quantitative Methods for Managerial Decisions*, Addison Wesley Publishing Co., 1978.

Buchan, J., and Koenigsberg, E., *Scientific Inventory Management*, Prentice-Hall of India, New Delhi, 1976.

Budnick, F.S., Mogena, R., and Vollmann, T.E., *Principles of Operations Research for Management*, Richard D. Irwin Inc. Homewood, Illinois, 2003.

Buffa, E.S. and Dyer, J.S., *Essentials of Management Science/Operations Research*, John Wiley and Sons Inc., New York, 1978.

Churchman, C.W., Ackoff, R.L., and Arnoff, E.L., *Introduction to Operations Research*, John Wiley and Sons Inc., New York, 1957.

Cook, W.P., *Quantitative Methods for Management Decisions*, McGraw-Hill Book Co., New York, 1985.

Eck, R.D., *An Introduction to Quantitative Methods for Business Application*, Wadsworth Publishing Co. Inc., California, 1979.

Fabrycky, W.J., Ghare, P.M., and Torgersen, P.E., *Applied Operations Research and Management Science*, Prentice-Hall of India, New Delhi, 1987.

Gass, S.I., *An Illustrated Guide to Linear Programming*, McGraw-Hill Book Co., New York, 1970.

Gass, Saul I. and Carl M. Harris, *Encyclopedia of Operations Research and Management Science*, Kluwer Academic Publishers, Massachusetts, 2004.

Glicksman, A.M., *An Introduction to Linear Programming and the Theory of Games*, John Wiley and Sons Inc., New York, 1963.

Godfrey, A., *Quantitative Methods for Managers*, Edward Arnold, London, 1977.

Gupta, S.K. and Cozzolino, J.M., *Fundamentals of Operations Research for Management*, Holden Day Inc., San Francisco, 1974.

Harrison, P., *Operational Research*, Mitchell Beazley Publishers, London, 1983.

Hillier, Frederick et al, *Introduction to Management Science*, McGraw-Hill Book Co. /Irwin, 2003.

Hillier, Fredcrick S., and Gerald J. Lieberman, *Introduction to Operations Research*, McGraw-Hill Book Co., 2005

Isaacson D. and R. Madsen, *Markov Chains, Theory and Applications*, John Wiley and Sons Inc., New York, 1976.

Kim, C., *Introduction to Linear Programming*, Holt, Rinehart and Winston Inc., New York, 1971.

Knotts, U.S., Jr. and Swift, E.W., *Management Science for Management Decisions*, Allyn and Becon Inc., Massachusetts, 1978.

Krajewski, L.J. and Thompson, H.E., *Management Science: Quantitative Methods in Context*, John Wiley and Sons Inc., New York, 1981.

Lapin, L.L., *Quantitative Methods for Business Decisions*, Harcourt Brace, Jovanovich, San Diego, 1988.

Law, Averill M., and Kelton, W. David, *Simulation Modeling and Analysis*, McGraw-Hill Book Co., New York, 1982.

Levin, R.I., Rubin D.S., and Stinson J., *Quantitative Approaches to Management*, McGraw-Hill Book Co., New York, 1989.

Loomba, N.P., *Management - A Quantitative Perspective*, Macmillan Publishing Company Inc., 1978.

Madura, J. and Veit, E.T., *Introduction to Financial Management*, West Publishing Co. St. Paul, 1988.

Markland, Robert E., *Topics in Management Science*, John Wiley and Sons Inc., New York, 1989.

Mize, J.H. and Cox J.G., *Essentials of Simulation*, Prentice-Hall Inc., Englewood Cliffs, NJ, 1968.

Moder J.J., and Phillips C.R., *Project Management with CPM and PERT*, Van Nostrand Reinhold Company, New York, 1970.

Moskowitz, H. and Wright, G.P., *Operations Research: Techniques for Management*, Prentice-Hall Inc., Englewood Cliffs, NJ, 1979.

Murty, Katta G., *Operations Research: Deterministic Optimization Models*, Prentice-Hall Inc., Englewood Cliffs, NJ, 1995.

Paik, C.M., *Quantitative Methods for Managerial Decisions*, McGraw-Hill Book Co., New York, 1973.

Panico, J.A., *Queuing Theory: A Study of Waiting Lines for Business, Economics and Science*, Prentice-Hall Inc., Englewood Cliffs, NJ, 1969.

Payne, T.A., *Quantitative Techniques for Management: A Practical Approach*, Reston Publishing Co. Inc., Virginia, 1982.

Plossl, G.W. and Wight, O.W., *Production and Inventory Control*, Prentice-Hall Inc., Englewood Cliffs, NJ, 1967.

Raiffa, H., *Decision Analysis Reading Mass*, Addison Wesley Publishing Company, Philippines, 1968.

Render B. and Ralph M. Stair, *Quantitative Analysis for Management*, 9th Ed., PHI, 2007.

Richmond S.B., *Operations Research for Management Decisions*, The Ronald Press Company, New York, 1968.

Riggs, J.L. and Inoue, M.S., *Introduction to Operations Research and Management Science*, McGraw-Hill Book Co., New York, 1975.

Saaty, T., *Elements of Queuing Theory with Applications*, Dover, New York, 1983.

Sargeaunt M.J., *Operational Research for Management*, Heinemann, London, 1965.

Starr M.K. and Miller D.W., *Inventory Control: Theory and Practice*, Prentice-Hall Inc., Englewood Cliffs, NJ, 1962.

Strum J.E., *Introduction to Linear Programming*, Holden Day Inc., San Francisco, 1972.

Taha, Hamdy A., *Operations Research: An Introduction*, 8th Ed, Prentice-Hall of India, New Delhi, 2008.

Taylor, Bernard W., *Principles of Management Science*, Prentice-Hall Inc., 2006.

Thierauf, R.J. and Klekamp, R.C., *Decision Making Through Operations Research*, John Wiley and Sons Inc., New York, 1975.

Thomopoloulos N.T. *Applied Forecasting Methods*, Prentice-Hall Inc., Englewood Cliffs, NJ, 1980.

Thompson G.E., *Management Science*, McGraw-Hill Book Co., New York, 1976.

Trueman, Richard E., *An Introduction to Quantitative Methods for Decision Making*, Holt, Rinehart and Winston, New York, 1981.

Turban E. and Meredith, J.R., *Fundamentals of Management Science*, Business Publications Inc., Plano, Texas, 1985.

Wagner, Harvey M., *Principles of Management Science: With Applications to Executive Decisions*, Prentice-Hall Inc., Englewood Cliffs, NJ, 1975.

Waters, D., *Inventory Control and Management*, John Wiley and Sons Inc., New York, 2003.

Weist, J.D. and Levy, F.K., *A Management Guide to PERT/CPM*, Prentice-Hall Inc., Englewood Cliffs, NJ, 1977.

Wilkes, F.M., *Elements of Operational Research*, McGraw-Hill Book Co., (UK) Ltd., London, 1980.

Winston W.L., *Operations Research - Applications and Algorithms*, Thomson, Indian Ed., 2007.

Zipkin, Paul H., *Foundations of Inventory Management*, McGraw-Hill Book Co., New York, 2000.

Zuwaylif, F.H., Baum P. and Smith, G.D., *Management Science: An Introduction*, John Wiley and Sons Inc., Santa Barbara, 1979.

Index